Wetting Theory

Wetting: Theory and Experiments
Two-Volume Set

Eli Ruckenstein
Gersh Berim

Volumes in the Set:

Wetting Theory (ISBN: 9781138393301)

Wetting Experiments (ISBN: 9781138393332)

Wetting Theory

Eli Ruckenstein
Gersh Berim

CRC Press
Taylor & Francis Group
Boca Raton London New York

CRC Press is an imprint of the
Taylor & Francis Group, an **informa** business

CRC Press
Taylor & Francis Group
6000 Broken Sound Parkway NW, Suite 300
Boca Raton, FL 33487-2742

Printed on acid-free paper

International Standard Book Number-13: 978-1-1383-9330-1 (Hardback)

Library of Congress Cataloging-in-Publication Data

Names: Ruckenstein, Eli, 1925- author. | Berim, G. O. (Gersh Osievich), author.
Title: Wetting theory / Eli Ruckenstein and Gersh Berim.
Description: Boca Raton : Taylor & Francis, a CRC title, part of the Taylor & Francis imprint, a member of the Taylor & Francis Group, the academic division of T&F Informa, plc, 2018. | Includes bibliographical references and index.
Identifiers: LCCN 2018034120| ISBN 9781138393301 (hardback : alk. paper) | ISBN 9780429401848 (ebook)
Subjects: LCSH: Wetting. | Surface chemistry.
Classification: LCC QD506 .R83 2018 | DDC 541/.33--dc23
LC record available at https://lccn.loc.gov/2018034120

Visit the Taylor & Francis Web site at
http://www.taylorandfrancis.com

and the CRC Press Web site at
http://www.crcpress.com

Contents

Preface

This book contains papers published by Professor Ruckenstein and his coworkers on the theoretical and experimental investigation of wetting of solid surfaces. It is one of two standalone books that comprise *Wetting: Theory and Experiments, Two-Volume Set*, which contains six chapters, each of which is preceded by a short introduction. Each volume making up the set is available to be read and understood on its own. Reading both volumes together provides the reader with a comprehensive view of the subject. The papers of the chapters are selected according to the specific features being considered and they are arranged in logical rather than chronological order. The main attention is given to the wetting on the nanoscale (nanodrops on solid surfaces, liquid in the nanoslit) considered on the basis of microscopic density functional theory, and to dynamics of fluid on the solid surface considered on the basis of hydrodynamic equations. Along with this, experimental studies of wetting related to various applications are presented. A description of the contents of each volume within the Set follows.

***Wetting Theory* (ISBN: 9781138393301):**

In Chapter 1, various microscopic processes (static and dynamic) in a liquid in contact with a solid are considered. They are the flow of liquid along horizontal and inclined surfaces, slipping of the contact line of a liquid on a solid, etc.

Chapter 2 is about the symmetry breaking of fluid density distribution in nanoslit between parallel solid walls. One component classical and quantum fluids and binary mixtures are considered and conditions for symmetry breaking to occur are examined.

In Chapter 3, the microscopic approach is applied to the treatment of macroscopic drops on smooth or rough, planar or curved, solid surfaces. It is based on fluid–fluid and fluid–solid interaction potentials and considers the drop equilibrium state as that having the minimum of total potential energy. The concept of microscopic contact angle is introduced and both macroscopic (classical) contact and microscopic angles are calculated.

In the next chapter, Chapter 4, a liquid drop on a smooth or rough planar solid surface is examined on the basis of a microscopic nonlocal density functional theory in canonical ensemble. The variety of characteristic features are examined including nonuniform fluid density distribution inside the drop, drop profile, microscopic contact angle, sticking force, etc. The results are compared with predictions of classical theories for macroscopic drops and similarities and differences between them are analyzed.

In the last chapter, Chapter 5, the theory of the rupture of liquid and solid films is developed, first in the linear approximation, and then extended to the case of perturbations of finite amplitude. The theory provides, in particular, the conditions for the instability, the dominant wavelength of the disturbances, and the time of rupture of the films. Along with this, the rupture of liquid films supported on a solid surface is examined on the basis of a thermodynamic approach which considers the change of the free energy of the film after the formation of a hole in it. The theory is applied to the practically important problem of tear film stability and rupture.

***Wetting Experiments* (ISBN: 9781138393332):**

This volume focuses on experimental studies of wetting that are related to biological problems, polymers, and catalysts. The biology-related studies are devoted to the problem of selecting synthetic materials for use in biological media. The polymers are examined to estimate experimentally various surface characteristics such as the ability of polymeric solids to alter their surface structures between different environments in order to minimize their interfacial free energy. The investigation of catalysts concentrates on their physical and chemical changes, formed of small crystallites of Pt, Pd, Ni, Co, Fe, or Ag supported on alumina.

Authors

Eli Ruckenstein, National Academy of Engineering and National Academy of Art and Science member, National Medal of Science winner, SUNY Distinguished Professor of CBE Eli Ruckenstein's copious and pioneering contributions to chemical engineering have been rewarded with numerous distinctions, including the 2004 National Academy of Engineering Founders Award, the 2002 American Institute of Chemical Engineers (AIChE) Founder's Award, AIChE's 1988 Walker Award, the 1977 AIChE's Alpha Chi Sigma Award, the 1986 American Chemical Society's (ACS) Kendall Award, the 1994 ACS Langmuir Lecture Award, the 1996 ACS E.V. Murphree Award, the 1985 the Alexander von Humboldt Foundation's Senior Humboldt Award, and the 1985 NSF Creativity Award. Ruckenstein joined the School's faculty in 1973, and was the first full-time SUNY system professor elected to the NAE. A leading influential chemical engineer, he has made numerous contributions to modernizing research and development in key areas of chemical engineering. He is a fellow of AIChE which with the occasion of its 100th anniversary designated him as one of 50 eminent chemical engineers of the Foundation age. Dr. Ruckenstein published about 900 papers in various areas of Chemical and Biological Engineering.

Gersh Berim earned a Ph.D. degree in Physics in 1978 from Kazan State University, Russia. Till 2001, his research was focused on nonequilibrium properties of low dimensional spin system. In 2001 he joined the group of Professor Eli Ruckenstein at SUNY at Buffalo where have been studying various topics of Chemical Physics especially related to nanosystem. He has authored or co-authored more than 70 papers.

1 Wetting of Solid by Liquid

Dynamics and Statics

Eli Ruckenstein and Gersh Berim

INTRODUCTION TO CHAPTER 1

Chapter 1 consists of two parts where various microscopic processes in the liquid (droplet) which is in contact with a solid are considered. In the first part (Sec. 1.1–1.4), attention is given to the flow of liquid along horizontal and inclined surfaces. Combining thermodynamic and dynamic approaches, the role of intermolecular forces and gravity (Sec. 1.1–1.3) in the spreading kinetics were examined. In Sec. 1.4, a new expression for the driving force for wetting when the droplet is not in equilibrium is provided. In the second part (Sec. 1.5–1.8), a slip velocity at the contact line of a liquid on a solid surface is examined in the presence of the gradient of the chemical potential in the liquid along the solid-liquid interface (Secs. 1.5, 1.7). It is also argued (Sec. 1.6) that slip of the liquid over solid does not occur directly over the solid surface but over a gap between solid and liquid which is generated when the liquid and the solid have different natures. The summary of the obtained results is provided in Sec. 1.8. In addition to the above topics, fluid adsorption and absorption by a solid, the density of which near the interface depends on the amount adsorbed fluid are examined on the basis of the density functional theory (Sec. 1.9).

1.1 The Origin of Flow during Wetting of Solids*

Clarence A. Miller[a] *and Eli Ruckenstein*[b]

[a] Department of Chemical Engineering, Carnegie-Mellon University, Pittsburgh, Pennsylvania 15213

[b] Department of Chemical Engineering, State University of New York at Buffalo, Buffalo, New York 14214

Received April 4, 1973; accepted December 26, 1973

Wetting processes involve motion of a fluid interface along a solid surface which it contacts. This motion can occur only if flow arises near the contact line. We propose here a physical explanation for the origin of such flow and present a simple theoretical analysis which, for a given system and contact angle, predicts the direction of flow along the fluid interface near the contact line.

Figure 1 illustrates a situation where liquid either displaces gas or is displaced by gas as the contact line moves along the solid surface. We focus on a point Q at the liquid-gas interface. If the radii of curvature of the fluid interface near the contact line are much greater than the effective range of intermolecular forces, the interface can be considered flat for purposes of the present discussion. We consider interaction of liquid molecules at Q with other liquid molecules and with molecules of the solid but neglect interaction with gas molecules on account of the low density of the gas phase.

It is evident from symmetry that the net force F_S on a molecule at Q due to its interaction with molecules of the solid is perpendicular to the solid-liquid interface. It has components $F_{ST} = F_S \sin \alpha$ along the fluid interface directed toward the contact line and $F_{SN} = F_S \cos \alpha$ normal to the fluid interface, as shown in Fig. 1. The magnitude of F_S decreases as Q becomes more distant from the contact line, becoming negligible when Q is farther from the nearest molecules of the solid than the effective range of intermolecular forces.

The net force F_L on a molecule at Q due to its interaction with other liquid molecules varies in both magnitude and direction with the position of Q. It is clear, nevertheless, that its component F_{LT} along the fluid interface is directed away from the contact line and that its normal component F_{LN} is directed toward the liquid as shown. The tangential component F_{LT} becomes vanishingly small for points Q sufficiently far from the contact line.

We wish to determine whether tangential flow arises along the fluid interface for a given value of the contact angle. If it does, we also wish to know whether it is directed toward or away from the contact line. Because pressure in the gas is uniform and because, as noted previously, the interface can be considered flat near the contact line, pressure in the liquid phase is the same for all points Q along the interface. As a result, there is no tangential pressure gradient to cause flow along the interface. Were it not for the tangential forces F_{ST} and F_{LT} due to interaction among molecules, no flow would arise. From the directions of these forces as shown in Fig. 1, we see that if F_{ST} exceeds F_{LT} in magnitude, flow along the interface is toward the contact line. If, on the other hand, F_{LT} is greater than F_{ST}, flow is away from the contact line.

The normal components F_{SN} and F_{LN} are also, of course, driving forces for flow. But they are not the only driving forces in the direction normal to the fluid interface. Indeed, they are counteracted by normal pressure gradients in the fluid just as the familiar gravitational body force is counteracted by hydrostatic pressure effects. In general, there will be additional normal pressure gradients due to

* *Journal of Colloid and Interface Science*, Vol. 48, No. 3, p. 368, September 1974. Republished with permission.

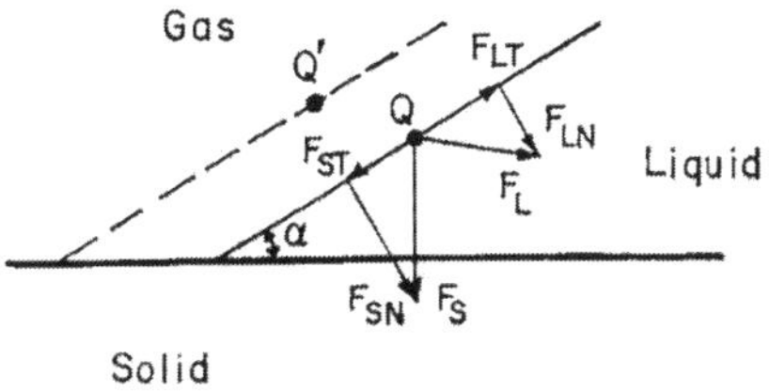

FIG. 1 Forces at a point Q near a contact line due to interaction among molecules. F_S and F_L are net forces due to molecules of the solid and molecules of the liquid, respectively.

dynamic effects so that the overall normal driving force depends on the solution of the differential equations governing flow in the liquid. What happens physically is that motion in the normal direction at Q is simply that due to movement of the interface itself along the solid. Thus, if liquid displaces gas in contact with the solid, a point Q near the contact line moves to some position Q′ as the fluid interface moves to the position of the dashed line in Fig. 1. In this case the normal component of fluid motion at Q is directed toward the gas phase.

We could determine the forces F_S and F_L by summing vectorially the interaction forces a molecule at Q has with all individual molecules of the solid and liquid phases. It proves more convenient, however, to calculate the potential energy of interaction Φ between Q and these molecules. The tangential driving force for flow, $F_{ST} - F_{LT}$, can then be found by evaluating $\partial\Phi/\partial x_T$, where x_T is the distance between Q and the contact line (see Fig. 2).

We shall calculate the difference between the potential energy, Φ, per unit volume at a point Q on the fluid interface near a contact line as shown in Fig. 2, and the potential energy, Φ_F, that would exist were Q simply a point on a flat interface POQ of infinite extent with no solid present. It is clear from Fig. 2 that the former situation differs from the latter in that: (a) Solid is present instead of liquid in region A; and (b) solid is present instead of gas in region B.

For most situations of interest the interaction energy between two molecules of materials i and j, separated by a distance, $\boldsymbol{r}$, has the form

$$\phi_{ij} = -\beta_{ij}/\boldsymbol{r}^m. \qquad [1]$$

If the total potential energy at Q can be found by simply summing contributions of individual molecules with which it interacts, the last statement of the preceding paragraph can be used to write the following equation:

$$\Phi - \Phi_F = \int_A \frac{n_L\left(n_L\beta_{LL} - n_S\beta_{SL}\right)}{\left|\mathbf{r} - \mathbf{r}_Q\right|^m} dV - \int_B \frac{n_L n_S \beta_{SL} dV}{\left|\mathbf{r} - \mathbf{r}_Q\right|^m}. \qquad [2]$$

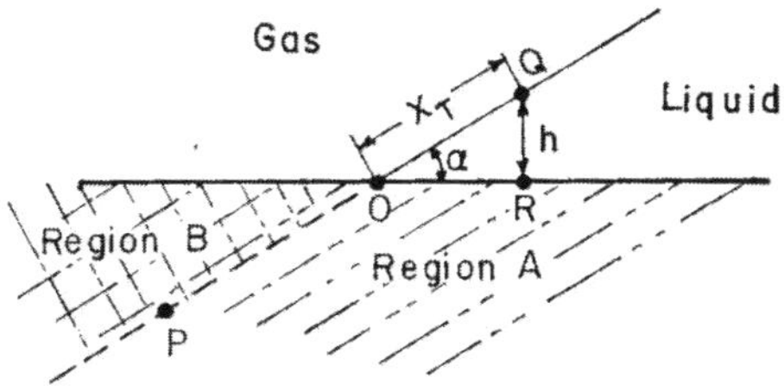

FIG. 2 Relation between intermolecular interactions near a contact line and in a large pool of liquid. The former configuration may be formed from the latter by replacing liquid with solid in region A and gas with solid in region B.

Here n_i is the molecular density in phase i, $\mathbf{r}$ is the position vector of a general point in the region over which integration is being performed, and $\mathbf{r}_Q$ is the position vector of Q.

The integrals in [2] can be evaluated using a cylindrical coordinate system with its origin at point O of Fig. 2 and with the z-axis along the contact line. The distance $|\mathbf{r} - \mathbf{r}_Q|$ can be calculated in terms of the contact angle α and the distance x_T of Q from the contact line. For this initial calculation we choose the case of London-van der Waals forces and neglect retardation effects, so that $m = 6$. Under these conditions, [2] becomes

$$\Phi - \Phi_F = n_L\left(n_L\beta_{LL} - n_S\beta_{SL}\right)\int_{\pi+\alpha}^{2\pi} d\theta \int_0^{\infty} r dr$$

$$\times \int_{-\infty}^{\infty} \frac{dz}{\left[z^2 + x_T^2 + r^2 - 2x_T r\cos(\theta - \alpha)\right]^3}$$

$$-n_L n_S \beta_{SL}\int_{\pi}^{\pi+\alpha} d\theta \int_0^{\infty} r dr$$

$$\times \int_{-\infty}^{\infty} \frac{dz}{\left[z^2 + x_T^2 + r^2 - 2x_T r\cos(\theta - \alpha)\right]^3}. \qquad [3]$$

The integrals over z are identical and are readily evaluated. The result is

$$\int_{-\infty}^{\infty} \frac{dz}{\left[z^2 + x_T^2 + \boldsymbol{r}^2 - 2x_T r\cos(\theta - \alpha)\right]^3}$$

$$= \frac{3\pi}{8\left[\boldsymbol{r}^2 + x_T^2 - 2x_T r\cos(\theta - \alpha)\right]^{\frac{5}{2}}}. \qquad [4]$$

When [4] is substituted into [3], the integration over $\boldsymbol{r}$ can be carried out in a straightforward manner. The first integral in [3] takes, for example, the following form:

$$\frac{3\pi n_L}{8}\left(n_L\beta_{LL} - n_S\beta_{SL}\right)\int_{\pi+\alpha}^{2\pi} d\theta$$

$$\times \int_0^{\infty} \frac{r dr}{\left[r^2 + x_T^2 - 2r x_T\cos(\theta - \alpha)\right]^{\frac{5}{2}}}$$

$$= \frac{\pi n_L}{8x_T^3}\left(n_L\beta_{LL} - n_S\beta_{SL}\right)$$

$$\times \int_{\pi}^{2\pi-\alpha}\left[1 + \frac{2\cos\omega}{\sin^4\omega} + \cot^2\omega + 2\frac{\cos^2\omega}{\sin^4\omega}\right] d\omega. \qquad [5]$$

Here the angular variable has been changed, for convenience, from θ to ω, which is $(\theta - \alpha)$. Finally, the integration over ω is performed, the result being

$$\left[\int 1+\frac{2\cos\omega}{\sin^4\omega}+\cot^2\omega+2\frac{\cos^2\omega}{\sin^4\omega}\right]d\omega$$

$$=\omega-\frac{2}{3}\csc^3\omega-\left(\cot\omega+\omega\right)-\frac{2}{3}\cot^3\omega. \qquad [6]$$

Substituting [6] into [5] and the similar expression obtained from the second integral in [3] and evaluating at the appropriate limits of integration, we obtain the following equation:

$$\Phi-\Phi_{\mathrm{F}}=\frac{\pi}{12x_{\mathrm{T}}^3}\left(n_{\mathrm{L}}^2\beta_{\mathrm{LL}}-n_{\mathrm{L}}n_{\mathrm{S}}\beta_{\mathrm{SL}}\right)G(\alpha)$$

$$-\frac{\pi}{12x_{\mathrm{T}}^3}n_{\mathrm{L}}n_{\mathrm{S}}\beta_{\mathrm{SL}}G(\pi-\alpha) \qquad [7]$$

$$G(\alpha)=\csc^3\alpha+\cot^3\alpha+\frac{3}{2}\cot\alpha. \qquad [8]$$

As indicated previously, the tangential driving force F_{T} for flow at Q can be found by differentiating [7] with respect to x_{T}. The potential energy Φ_F of a point at the surface of a large, flat fluid interface is, of course, independent of x_{T}. The result is

$$F_{\mathrm{T}}=\frac{d\Phi}{dx_{\mathrm{T}}}=\frac{-\pi}{4x_{\mathrm{T}}^4}\left(n_{\mathrm{L}}^2\beta_{\mathrm{LL}}-n_{\mathrm{L}}n_{\mathrm{S}}\beta_{\mathrm{SL}}\right)G(\alpha)$$

$$+\frac{\pi}{4x_{\mathrm{T}}^4}n_{\mathrm{L}}n_{\mathrm{S}}\beta_{\mathrm{SL}}G(\pi-\alpha). \qquad [9]$$

A positive value of F_{T} corresponds to a force along the fluid interface directed toward the contact line.

A graph of the function $G(\alpha)$ is shown in Fig. 3. Clearly $G(\alpha)$ is positive and decreases monotonically with increasing α while $G(\pi - \alpha)$ is positive and increases monotonically with α. Let us consider situations in which $(n_{\mathrm{L}}n_{\mathrm{S}}\beta_{\mathrm{SL}} > n_{\mathrm{L}}^2\beta_{\mathrm{LL}})$, i.e., where a liquid molecule interacts more strongly with a given volume of solid than with the same volume of liquid at the same location. According to [9] F_{T} is positive, and a driving force for flow toward the contact line exists for all values of the contact angle α. From the point of view of whether spreading occurs, this result amounts to the usual criterion that spontaneous spreading takes place when the work of adhesion between liquid and solid exceeds the work of cohesion of the liquid. The novelty here is the insight obtained about the dynamics of spreading; viz., the demonstration of how inter-molecular forces generate flow in the liquid near the contact line and the quantitative expression [9] for the tangential driving force for flow. In a subsequent paper we will show how [9] can be combined with the governing equations of hydrodynamics to predict rates of spreading in simple cases.

When $(n_{\mathrm{L}}^2\beta_{\mathrm{LL}} > n_{\mathrm{L}}n_{\mathrm{S}}\,\beta_{\mathrm{SL}})$ the two terms in [9] have opposite signs. In view of the properties of the function G mentioned above and illustrated in Fig. 3, we see that there is precisely one value of α between 0 and π for which F_{T} vanishes. This angle is α_e, the equilibrium contact angle. The predicted existence of an equilibrium contact angle α_e in cases in which the work of cohesion exceeds the work of adhesion and the predicted increase in magnitude of α_e with increasing excess of cohesion over adhesion are again in accordance with well-known results of surface chemistry. As before, the novelty is in predictions about the dynamics of spreading. For [9] predicts that $\alpha < \alpha_e$ whenever flow along the fluid interface is away from the contact line, i.e., when liquid is moving away from the contact line and gas is replacing liquid in contact with the solid.

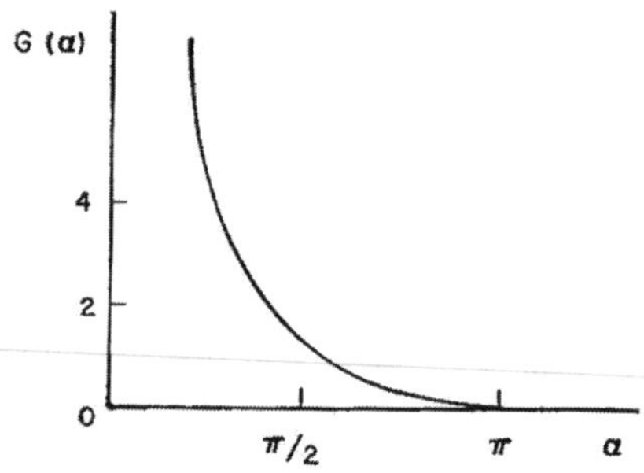

FIG. 3 The function $G(\alpha) = \csc^3\alpha + \cot^3\alpha + (3/2)\cot\alpha$. This function appears in [7] and [9] which describe how the contact angle influences the direction of flow near a contact line.

That $\alpha < \alpha_e$ for such "receding" contact lines is in accordance with experimental observations (1). Similarly, [9] predicts that $\alpha > \alpha_e$ when flow along the fluid interface is toward the contact line and liquid displaces gas in contact with the solid. This prediction about "advancing" contact lines also agrees with experimental results (1).

As mentioned above, there is no flow and the system is in equilibrium when the tangential force F_T vanishes. In this case [9] simplifies to the following expression involving the equilibrium contact angle α_e:

$$\frac{G(\pi - \alpha_e)}{G(\alpha_e)} + 1 = \frac{n_L{}^2\beta_{LL}}{n_L n_S \beta_{SL}}. \qquad [10]$$

One way of using [10] is to estimate β_{LL} and β_{SL} with some model of intermolecular forces. Another is to assume that $n_L{}^2\beta_{LL}/n_L n_S\beta_{SL}$ is approximately equal to the ratio of the liquid work of cohesion $2\gamma_L$ to the work of adhesion between liquid and solid.[1] Following Fowkes (2) we further assume that, if liquid and solid interact by dispersion forces alone, the work of adhesion may be expressed in terms of liquid and solid surface tensions as $2\sqrt{\gamma_L{}^d\gamma_S{}^d}$. The superscript d denotes that only the portions of γ_L and γ_S due to dispersion forces are involved in the interaction between phases. With this information [10] can be rewritten as

$$\frac{G(\pi - \alpha_e)}{G(\alpha_e)} = \frac{\gamma_L - \sqrt{\gamma_L{}^d\gamma_S{}^d}}{\sqrt{\gamma_L{}^d\gamma_S{}^d}}$$

$$= -1 + \frac{\gamma_L}{\sqrt{\gamma_L{}^d}} \cdot \frac{1}{\sqrt{\gamma_S{}^d}}. \qquad [11]$$

Thus, a plot of $[G(\pi - \alpha_e)/G(\alpha_e)]$ as a function of $\left(\gamma_L/\sqrt{\gamma_L{}^d}\right)$ for various liquids on the same solid should yield a straight line passing through the point (−1, 0).

Figure 4 shows such a plot of Fox and Zisman's data (3) for various liquids on solid $n\text{−}C_{36}H_{74}$. Only data for organic liquids where $\gamma_L \cong \gamma_L{}^d$ have been included. We note that the data do fall near a straight line and that $\gamma_S{}^d$, as determined from the value of $\left(\gamma_L/\sqrt{\gamma_L{}^d}\right)$ at the point corresponding to $\alpha_e = G(\pi - \alpha_e)/G(\alpha_e) = 0$, is about 23 dyn/cm. Fowkes (2) obtained a value for $\gamma_S{}^d$ of about 21 dyn/cm using the same data with his somewhat different expression for α_e.

[1] Young's equation may be used to express the ratio of the work of cohesion to the work of adhesion as $2/(1 + \cos\alpha_e)$. With the present assumption, this ratio is also approximately equal to the left-hand side of [10]. These two expressions differ by a maximum of about 35% for α_e in the range 0°–120°, indicating that the assumption is reasonable. The difference is somewhat larger if retardation effects are included in the expression for intermolecular interactions used in the present analysis, i.e., if $m = 7$ instead of $m = 6$ is used in Eq. [2].

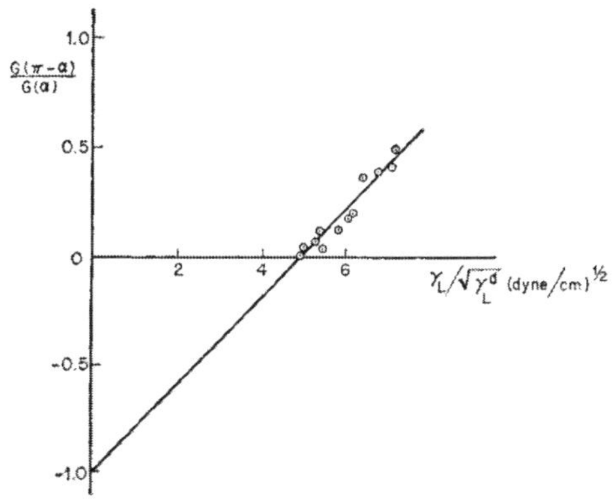

FIG. 4 Comparison of data of Fox and Zisman (3) with prediction of Eq. [11].

Fowkes' work was based on combining Young's equation with the following expression for the solid-liquid interfacial tension γ_{SL}:

$$\gamma_{SL} = \gamma_S + \gamma_L - 2\sqrt{\gamma_L{}^d \gamma_S{}^d} \qquad [12]$$

If we combine Eq. [10] above with Young's equation, we find a slightly different expression for γ_{SL} based on the present theory. Actually, a cubic equation is obtained, but only one of the three roots yields physically meaningful results for the special cases of no interaction between liquid and solid and interaction between liquid and solid having the same composition and density. This root has the form

$$\gamma_{SL} = \gamma_S + \gamma_L - \gamma_L\left[1 + \sqrt{3}\sin(\eta/3) - \cos(\eta/3)\right]; \qquad [13]$$

$$\cos\eta = 1 - 2\frac{n_S \beta_{SL}}{n_L \beta_{LL}}. \qquad [14]$$

Figure 4 indicates that the present theory gives results comparable to those of Fowkes in correlating data on equilibrium contact angles. We emphasize, however, that the main advantage of our approach is that values of the driving force F_T for tangential flow can be calculated for dynamic situations where the contact angle differs from its equilibrium value. We note that the general expression [9] for F_T simplifies to the following result when the contact angle deviates by only a small amount, δ, from its equilibrium value:

$$F_T = \frac{3\pi}{16 x_T{}^4} n_L{}^2 \beta_{LL} \delta. \qquad [15]$$

The equilibrium result [10] was used in deriving [15] to eliminate terms in β_{SL}.

Another situation of interest is when the contact angle, α, is small. In this case $G(\alpha)$ and $G(\pi - \alpha)$ approach $(2/\alpha^3)$ and zero, respectively, and [9] may be written

$$F_T = \frac{d\Phi}{dx_T} \cong \frac{d}{dx_T}\left[\frac{\pi\left(n_L{}^2\beta_{LL} - n_L n_S \beta_{SL}\right)}{6(\alpha x_T)^3}\right]. \qquad [16]$$

Under these conditions the liquid layer has a local thickness h approximately equal to (αx_T), and h varies slowly with distance from the contact line (see Fig. 2). It is not surprising, therefore, that the expression in brackets in [16] is the same as that obtained for the "disjoining pressure" of a liquid film of uniform thickness h on a solid surface (4). The disjoining pressure is usually defined in terms of the thermodynamics of thin films and reflects the difference in behavior between a film and a bulk liquid phase (4). Equation [16] states that for small contact angles the tangential driving force. F_T for flow along the fluid interface can be expressed in terms of the derivative of the disjoining pressure, a result which agrees with previous discussions of stability and wave motion for thin liquid films of nearly uniform thickness (5, 6). For larger contact angles, F_T bears no simple relation to the disjoining pressure because the liquid layer near the contact line no longer closely resembles a thin film of uniform thickness.

It is the case of small contact angles where [16] applies that is easiest to tackle in terms of solving the hydrodynamic equations. For since the liquid layer is of nearly uniform thickness, flow is nearly parallel to the fluid interface throughout the layer. Moreover, according to "lubrication theory" which is often used for solving such problems, the driving force F_T for tangential flow along the interface at Q is equal to the driving force for flow at all points along the line QR of Fig. 2 (7). Solutions predicting spreading rates of liquid drops on solids and the variation of drop shape with time will be published elsewhere.

To keep the above analysis simple, we have used a simple model for intermolecular forces and for the configuration of the various phases near the contact line. These simplifications could be removed, but the resulting equations would be considerably more complex. Since we have neglected surface roughness and local variations in composition within the solid, our equations cannot predict irregularities in motion along the contact line although such irregularities have frequently been observed. Yet the present analysis does illustrate the basic mechanism by which intermolecular forces generate flow near a contact line and provides the framework for including intermolecular forces in the equations describing fluid motion. While we have emphasized spreading on solids, the same basic ideas also apply to spreading on liquids.

ACKNOWLEDGMENT

Acknowledgment is made to the donors of the Petroleum Research Fund, administered by the American Chemical Society, for partial support of this research. Jaime Lopez assisted with some of the calculations.

REFERENCES

1 Schwartz, A. M., Rader, C. A., and Huey, E., *Advan. Chem. Ser.* **43**, 250, American Chemical Society, Washington, D. C., 1964.
2 Fowkes, F. M., *Advan. Chem. Ser.* **43**, 99, American Chemical Society, Washington, D. C., 1964.
3 Fox, H. W., and Zisman, W. A., *J. Colloid Sei.* **7**, 428 (1952).
4 Sheludko, A., *Adv. Colloid Interface Sei.* **1**, 391 (1967).
5 Vrij, A., *Disc. Faraday Soc.* **42**, 23 (1966).
6 Vrij, A., Hesselink, F., Lucassen, J., and van den Tempel, M., *Proc. Kon Ned. Akad. Wetensch., Ser. B.* **73**, 124 (1970).
7 Batchelor, G. K., "An Introduction to Fluid Dynamics," p. 219. Cambridge University Press, 1967.

1.2 The Appearance of Dry Patches on a Wetted Wall*

Bryon E. Anshus[a] *and Eli Ruckenstein*[b†]

[a] Department of Chemical Engineering, University of Delaware, Newark, Delaware

[b] Faculty of Engineering and Applied Sciences, State University of New York at Buffalo

Corresponding Author

[†] Now at Chevron Research Co., 576 Standard Ave., Richmond, California 94802.

Received December 3, 1973; accepted November 19, 1974

NOMENCLATURE

Roman

a	Growth parameter; see Eq. (5). Dimensionless
a_1	a/Γ Reduced growth parameter. Dimensionless
b	$a^2 - \alpha^2$
b_1^2	$a_1^2 - \alpha^2/\Gamma^2$
D	d/dy
D_1	d/dY
g	Acceleration of gravity; L/τ^2
h	Mean film thickness; L
k	Thermal conductivity of the liquid
N	Dimensionless thermocapillary parameter; see Eq. (21)
p	Isotropic pressure; F/L^2
P	Pressure in undisturbed flow; F/L^2
Re	$\bar{u}_a h/\nu = h^3 g \sin\beta/3\nu^2$; Reynolds number
Re_{mw}	Minimum wetting Reynolds number. Defines stability of dry patches [Eq. (1)]
Re_c	Critical Reynolds number
T	Temperature
U	Undisturbed flow velocity; L/τ
$\bar{u}_a$	Average undisturbed flow velocity; L/τ
$\bar{u}$	$U/\bar{u}_a$
u_x, u_y, u_z	Velocity components in the three spatial directions; L/τ
u, v, w	Dimensionless velocity components, $u_x/\bar{u}_a, u_y/\bar{u}_a, u_z/\bar{u}_a$

* *Journel of Colloid and Interface Science*, Vol. 51, No. 1, p. 12, April 1975. Republished with permission.

$\hat{u}, \hat{v}, \hat{w}$	y dependence of velocity components [Eq. (5)]; Dimensionless
x_1, y_1, z_1	Spatial coordinates; L
x, y, z	Dimensionless spatial coordinates; $x_1/h, y_1/h, z_1/h$
Y	$y\Gamma$, scaled spatial coordinate
Greek	
α	$2\pi h/\lambda$, cross stream wavenumber
β	Angle of inclination of the plate
Γ	$\zeta \mathrm{Re}^{-\frac{2}{3}}$, assumed to be large for these calculations
η	Dimensionless surface deflection; see Fig. 2
θ	Contact angle; see Fig. 1
λ	z-wavelength; L; see Fig. 2
μ	Fluid viscosity; $F\tau/L^2$
ν	Fluid kinematic viscosity; L^2/τ
ζ	$\sigma/\rho\left(\nu^4 g \sin\beta/3\right)^{\frac{1}{3}}$; surface tension parameter
ρ	Fluid density; M/L^3
σ	Surface tension of fluid; F/L
Superscript	
′(prime)	Denotes infinitesimal perturbation quantity

INTRODUCTION

Experiment shows that, when a thin liquid film flows down a wall, dry patches often form if the flow-rate is below a critical value and that heat and mass transfer have an important effect on the formation of these dry patches. Because of its importance in boiling heat transfer, in nuclear reactor burnout, and in the efficiency of packed towers, the formation and stability of dry patches on a wetted wall has received attention in recent years (1-7). Norman and McIntyre (2) and Simon and Hsu (5) have reported the results of experiments to determine the minimum wetting rate (MWR) required to ensure a completely water-wet surface on walls of several metals during heat transfer. They found that the MWR depends on the metal of the surface and its roughness as well as on the properties of the fluid and rate of heat transfer. Hartley and Murgatroyd (3) performed a semitheoretical analysis to determine the MWR. They assumed that a dry patch of the type shown in Fig. 1 will be stable if the surface forces at the stagnation point are greater than the kinetic head associated with the upstream stagnation streamline. The result, expressed in terms of a minimum wetting Reynolds number, Re_{mw}, is

$$\mathrm{Re}_{mw} = 0.875\left[\zeta\left(1-\cos\theta\right)\right]^{3/5}, \qquad [1]$$

where Re is defined to be the volumetric flow rate per unit width of plate divided by the kinematic viscosity ν,

$$\zeta = \frac{\sigma}{\rho\left(\nu^4 g \sin\beta/3\right)^{\frac{1}{3}}}$$

is the surface tension parameter and θ is the equilibrium contact angle of the liquid on the solid.

The Hartley and Murgatroyd analysis has to do with the stability of dry patches, once these are formed. The criterion indicates that if the contact angle, θ, is zero, dry patches are never stable, whereas for any nonzero contact angle there will be a finite range of small Reynolds numbers for which dry patches are stable. The Hartley and Murgatroyd analysis has nothing to say, however, concerning the mechanism of appearance of dry patches on a wetted surface.

Norman and McIntyre (2) describe in the following manner the events leading to the film breakdown during heat transfer. "In the first stage, which will be referred to as roping, the film became thicker in several vertical bands spaced around the periphery of the heated section. Intermittent dry patches then appeared between the bands but these were quickly covered again and the film became very disturbed. A further reduction in the liquid rate resulted in film breakdown with permanent dry patches appearing on the tube. The minimum wetting rate was taken as the liquid rate at which permanent dry patches first appeared." The occurrence of patches involves thinning of the film in some regions and a corresponding thickening in the neighboring regions. Thus one may consider that they are a consequence of the growth in the downstream direction of a "roping" perturbation applied to the steady laminar flow of the film. The main features suggested by experiment are the irregularity of the disturbance in the cross-stream (z) direction and its growing or decaying in the downstream direction. For the sake of simplicity, the perturbation in the present analysis is assumed to be steady, that is, it is assumed to grow or decay in the streamwise direction but to be independent of time. It will be shown in what follows that this form of disturbance leads to an instability below a critical Reynolds number and that the theoretical result is in satisfactory agreement with experiment. This proposed instability is different from the well known instability which leads to wave motion [see Ref. (9)]. In that case the laminar steady motion is subjected to a perturbation periodic in the streamwise direction, and is found to grow in time.[1] The roping perturbation leads to an instability which becomes stronger as the Reynolds number decreases whereas the time dependent instability becomes stronger as Reynolds number increases.

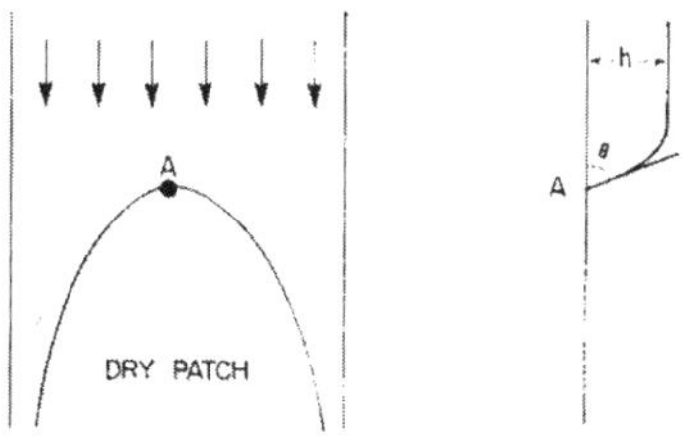

FIG. 1. Configuration of a dry patch on a wall.

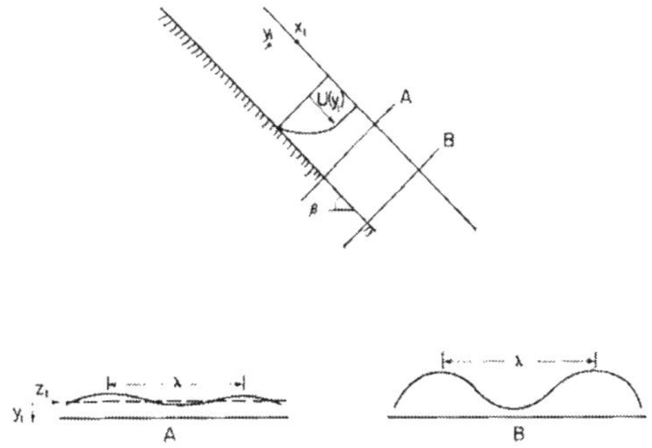

FIG. 2. Definition sketch and form of the perturbation.

---

[1] Krantz and Owens (12) have recently suggested considering wave instabilities in shear flows to grow spatially rather than in time.

FORMULATION OF THE PROBLEM

We consider a film of liquid flowing steadily under gravity on an inclined wall (Fig. 2).

It is assumed that the undisturbed flow is parallel to the wall; the undisturbed velocity is given by the Nusselt distribution

$$U\left(y_1\right)=\left(g\sin\beta/2\nu\right)\left(h^2-y_1^{\,2}\right), \tag{2}$$

or, in nondimensional form

$$\bar{u}=\frac{3}{2}\left(1-y^2\right).$$

The pressure field is given by

$$P\left(y\right)=\rho y_1 g\cos\beta.$$

The development of the disturbance equations of motion leading to the Orr-Sommerfeld equations is available elsewhere in the literature (8, 9), but because of the novel disturbance form used here, some details of the development are presented.

The velocity and pressure fields are expressed as

$$u_x=U+u_x',\ u_y=u_y',\ u_z=u_z',$$

and

$$p=P+p',$$

where the primed velocities are assumed to be small compared to U, and p' is small compared to P. The linearized momentum equations describing the perturbations are then

$$\frac{\partial u_x'}{\partial t_1}+U\frac{\partial u_x'}{\partial x_1}+u_y'\frac{dU}{dy_1}=-\frac{1}{\rho}\frac{\partial p'}{\partial x_1}+\nu\nabla^2 u_x', \tag{3a}$$

$$\frac{\partial u_y'}{\partial t_1}+U\frac{\partial u_y'}{\partial x_1}=-\frac{1}{\rho}\frac{\partial p'}{\partial y_1}+\nu\nabla^2 u_y'. \tag{3b}$$

$$\frac{\partial u_z'}{\partial t_1}+\frac{\partial u_z'}{\partial x_1}=-\frac{1}{\rho}\frac{\partial p'}{\partial z_1}+\nu\nabla^2 u_z', \tag{3c}$$

$$\frac{\partial u_x'}{\partial x_1}+\frac{\partial u_y'}{\partial y_1}+\frac{\partial u_z'}{\partial z_1}=0. \tag{3d}$$

These equations are to be solved subject to the condition of zero velocity at the solid surface and stress balance boundary conditions at the free surface. At the wall, $y_1=h$, the velocity must identically vanish:

$$u_x'=u_y'=u_z'=0 \text{ at } y_1=h. \tag{4a}$$

At the free surface, $y_1 = h\eta(x, z)$, which is assumed exposed to a nonviscous environment, the tangential stress must vanish (thermocapillary effects are considered later). Thus the x-y component and z-y component of the stress tensor must vanish at the surface. After linearization these become

$$\mu\left(\frac{\partial u'_x}{\partial y_1} + h\eta\frac{d^2U}{dy_1^{\,2}} + \frac{\partial u'_y}{\partial x_1}\right) = 0 \text{ at } y_1 = 0, \qquad [4b]$$

$$\mu\left(\frac{\partial u'_z}{\partial y_1} + \frac{\partial u'_y}{\partial z_1}\right) = 0 \text{ at } y_1 = 0. \qquad [4c]$$

Also at the free surface the force normal to the surface exerted by the fluid must be balanced by the surface tension force. This, upon linearization, becomes,

$$-p' - \eta\frac{dP}{dy_1} + 2\mu\frac{\partial u'_y}{\partial y_1}$$

$$= -h\sigma\left(\frac{\partial^2\eta}{\partial x_1^{\,2}} + \frac{\partial^2\eta}{\partial z_1^{\,2}}\right)\text{at } y_1 = 0. \qquad [4d]$$

The surface description is completed with a kinematic condition connecting the velocity of the surface to fluid velocity:

$$u_y(y_1 = 0) = \frac{Dh\eta}{Dt_1} = h\left(\frac{\partial\eta}{\partial t_1} + U(0)\frac{\partial\eta}{\partial x_1}\right). \qquad [4e]$$

As mentioned in the introduction, the form of disturbance which we now impose is not standard in flow stability problems, but is of a type consistent with experimental observations of this particular kind of instability. We assume a disturbance which, rather than being periodic in the streamwise direction and growing or decaying in time, is allowed to grow or decay in the downstream (x) direction. As in the usual small amplitude stability analysis, the solutions can be Fourier decomposed. It is sufficient to examine one term of the decomposition:

$$\begin{bmatrix} u \\ \upsilon \\ w \\ p \\ \eta \end{bmatrix} = \begin{bmatrix} \hat{u}(y) \\ \hat{\upsilon}(y) \\ \hat{w}(y) \\ \hat{p}(y) \\ \hat{\eta} \end{bmatrix} e^{i\alpha z} e^{ax}, \qquad [5]$$

where the velocities u, υ, w are nondimensional forms of u'_x, u'_y, u'_z respectively; x, y, z are nondimensional spatial coordinates, α is the wave number, $(2\pi h/\lambda)$, in the z-direction and "a" may be termed the "growth parameter." When the disturbance form [5] is inserted into equations [3] and their boundary conditions [4], the resulting system, after nondimensionalization and some algebraic manipulations, is

$$(D^2 + b^2)\hat{\upsilon}$$

$$= a\,Re\left[\bar{u}(D^2 + b^2)\hat{\upsilon} - (D^2\bar{u})\hat{\upsilon}\right], \qquad [6]$$

$$\hat{\upsilon} = D\hat{\upsilon} = 0 \text{ at } y = 1, \tag{7a,b}$$

$$\left(D^2 - b^2\right)\hat{\upsilon} - \frac{\left(D^2\bar{u}\right)}{\bar{u}}\hat{\upsilon} = 0, \text{ at } y = 0, \tag{7c}$$

$$\left(D^2 + 3b^2\right)D\hat{\upsilon} - a\mathrm{Re}\bar{u}D\hat{\upsilon}$$

$$+\frac{b^2}{a\bar{u}}\left(3\cot\beta + b^2\zeta\,\mathrm{Re}^{-\frac{2}{3}}\right)\hat{\upsilon} = 0,$$

$$\text{at } y = 0, \tag{7d}$$

where

$$D \equiv d/dy, \bar{u} = \frac{3}{2}\left(1 - y^2\right),\ D^2\bar{u} = -3,$$

and

$$b^2 = a^2 - \alpha^2.$$

The objective now, of course, is to determine the growth parameter eigenvalue, a, as a function of the surface tension parameter, ζ, the Reynolds number or nondimensional film thickness, Re, the angle of inclination, β, and the wave number, α. The calculation should lead to a critical film thickness or critical Reynolds number.

ASYMPTOTIC ANALYTICAL RESULTS Re → 0

For simplicity, the limiting case of small Reynolds number will be treated first. In the limit Re → 0, Eqs. [6] and boundary conditions [7] become

$$\left(D^2 + b^2\right)^2\hat{\upsilon} = 0, \tag{8}$$

$$\hat{\upsilon} = D\hat{\upsilon} = 0, \text{ at } y = 1, \tag{9a,b}$$

$$\left(D^2 - b^2\right)\hat{\upsilon} + 2\hat{\upsilon} = 0, \text{ at } y = 0, \tag{9c}$$

$$\left(D^2 + 3b^2\right)D\hat{\upsilon}$$

$$+\frac{2}{3}\left(\frac{b^2}{a}\right)\left(-3\cot\beta + b^2\zeta\mathrm{Re}^{-\frac{2}{3}}\right)\hat{\upsilon} = 0,$$

$$\text{at } y = 0. \tag{9d}$$

Equations [8] and [9a–c] are obvious limiting forms of [6] and [7a–c] but the form of Eq. [9d] is not so clear. In the limit of small Reynolds number (Re→ 0), or of large surface tension (ζ→∞), as in the following section, it would appear that Eq. [7d] should reduce to simply $\hat{\upsilon}(0) = 0$. If $\hat{\upsilon}(0)$ is taken to be zero, however, then Eqs. [8], [9] have only the trivial solution, i.e., $\hat{\upsilon} = 0$ everywhere. It seems correct, in fact, that if surface tension is very large then the surface is not very deformable so that the y-velocity should be nearly zero at the surface. However, if the Reynolds number is small but not identically zero, Eqs. [8–9d] lead to a solution from which it can be demonstrated that the normal velocity of the surface $\left[\hat{\upsilon}(0)\right]$ is indeed small compared to the tangential surface velocity $[D\hat{\upsilon}(0)]$ but it is not identically zero. Rather, $(b^4/a)\zeta\,\mathrm{Re}^{-\frac{2}{3}}\hat{\upsilon}(0)$ and $b^2D\hat{\upsilon}(0)$ in Eq. [9d] have the same magnitude.

The form of the solution to [8] is,

$$\hat{\upsilon}(y) = \cos by + A\sin by + Bby\cos by$$

$$+ Cby\sin by \qquad [10]$$

The boundary conditions [9] are now applied to Eq. [10]. Conditions [9c] and [9d] yield, respectively,

$$C = \left(b^2 - 1\right)/b^2, \qquad [10a]$$

$$A = +\frac{\cot\beta}{ab} - \frac{1}{3}\frac{b}{a}\zeta\,\mathrm{Re}^{-\frac{2}{3}}. \qquad [10b]$$

Conditions [9a] and [9b] may be used to eliminate the constant B:

$$(A + Cb)(b - \sin b\cos b) - \cos^2 b$$

$$+ Cb\sin b\cos b = 0. \qquad [10c]$$

We look now for the growth coefficient, a' from Eqs. [10a–c], Assuming that cot β is finite and that the wave number α is small compared to a (indeed a is found to be large), one finds that as the Reynolds number approaches zero, the growth parameter is given by,

$$a = \frac{1}{3}\zeta\,\mathrm{Re}^{-\frac{2}{3}} = \frac{\sigma}{\rho g h^2 \sin\beta}. \qquad [11]$$

The neglected terms in going from [10a–c] to [11] are of order $\mathrm{Re}^{\frac{2}{3}}/\zeta$ compared to 1, so that the range in Reynolds number in which Eq. [11] is applicable is dependent on the magnitude of ζ. If ζ is small, Eq. [11] will only be good for very small Re, whereas if ζ is large it may be good for Re of order 1. This matter is clarified somewhat in the following; what is clear at this point is that the stability analysis predicts a very strong instability effective at small Reynolds numbers. It should also be noted that the growth parameter, a, is independent of α, the cross-stream wave number, as long as α is not extremely large. One concludes, then, that there should be no preferred spacing of the roping instability.

The result, Eq. [11] is an exact solution to the equations of motion in the limit Re→0, ζ finite. A somewhat broader, but approximate result is obtained in the following section.

$\zeta \to \infty$

A simple result can be obtained for the physically interesting class of problems in which the surface tension parameter, ζ, is very large but in which the Reynolds number need not be extremely small. The analysis is carried out in the spirit of other asymptotic analyses (13). The equations are subjected to a scaling transformation which (in the asymptotic limit) is intended to bring out the most important terms and eliminate the unimportant ones. The transformed equations are then solved to obtain the essential result. The working premise in the present problem is that if such a simple transformation is to be of use it must preserve the highest order derivative in the differential equation and must also preserve dependence on surface tension. A candidate transformation is suggested by the results of the Re→0 asymptote, namely, that the growth parameter, "*a*," becomes large proportional to $\mathrm{T} \equiv \zeta\,\mathrm{Re}^{-\frac{2}{3}}$ either as Re→0 or as $\zeta \to \infty$. The following transformation is then made

$$\text{let } Y \equiv \Gamma y,$$

$$a_1 = a/\Gamma,$$

and

$$b_1^2 = a_1^2 - \alpha^2/\Gamma^2. \qquad [12]$$

Equations [6–7d] then become

$$\left[D_1^2 + a_1^2 - \frac{3}{2\Gamma}\left(1 - \frac{Y^2}{\Gamma^2}\right)a_1\mathrm{Re}\right] \times \left(D_1^2 + a_1^2\right)\hat{\upsilon} = \mathrm{O}\left(\Gamma^{-2}\right), \qquad [13]$$

$$\hat{\upsilon} = D_1\hat{\upsilon} = 0, \text{ at } Y = \Gamma, \qquad [14a,b]$$

$$\left(D_1^2 - a_1^2\right)\hat{\upsilon} = \mathrm{O}\left(\Gamma^{-2}\right), \text{ at } Y = 0, \qquad [14c]$$

$$\left(D_1^2 + 3a_1^2\right)D_1\hat{\upsilon} - \frac{3}{2\Gamma}a_1\mathrm{Re}D_1\hat{\upsilon} + \frac{2}{3}a_1^3\Gamma\hat{\upsilon} = \mathrm{O}\left(\Gamma^{-2}\right), \text{ at } Y = 0, \qquad [14d]$$

where $D_1 \equiv d/dY$.

In Eqs. [13–14d], all terms which are of higher order (proportional to $1/\Gamma^2$ or smaller) after transformation according to Eqs. [12] have been lumped into the order symbol "O" to indicate that if the transformation is valid, these are negligible.

A comparison of Eqs. [8] and [9] with Eqs. [13] and [14] reveals that in the latter equations the next order term of a series in the quantity $(\mathrm{Re}/a_1\Gamma)$ is being retained. In this section, therefore, it is assumed that $a_1\Gamma \gg \mathrm{Re}$ and that terms proportional to $(\mathrm{Re}/a_1\Gamma)^2$ may be neglected compared to those of order $(\mathrm{Re}/a_1\Gamma)$.

Here, as in the Re → 0 case, and for the same reason, terms which appear to be of quite different order are retained in the normal stress boundary condition [14d] or [9d]. These terms are, however, found, as before, to be of the same magnitude. It is seen again that the cross-stream wave number, *α,* appears only in higher order terms. This implies that there should be no preferred spacing for this instability.

The result implied by the transformation, Eqs. [12–14] is that the instability leading to dry patches becomes more intense as the surface tension increases. This is opposite to what is found for the instability leading to travelling waves on a film. In that case (see, e.g., *Yih* (9)) the spatial growth coefficient is found to be inversely proportional to the surface tension. The difference is in the nature of the instability. Sinusoidal travelling waves are stabilized by surface tension since the crests and troughs correspond, respectively, to positions of high and low pressure in the fluid film—the resulting flow tends to level the surface. For the nonsinusoidal surface deformation described by Eq. [5], however, capillary forces produce along the film a monotonic pressure gradient which is destabilizing. This pressure gradient increases with the surface tension. In other words, whereas surface travelling waves arise from a gravity-driven instability which is decreased by surface tension, dry patches result from a surface tension driven instability. One can now also understand why there is little dependence on crossstream wave number. A periodicity in the transverse direction ($\alpha \neq 0$), must act to stabilize the film; however, if the growth parameter, a, is large compared to the wave number, α, as is the case here since a is found to be extremely large, then the stabilization is minimal. Only if the wave number is very large (very short wave cross-stream structure) can there be an appreciable effect.

The eigenvalue problem, Eqs. [13–14d] can, at least in principle, be solved for the eigenvalue a for any given value of Reynolds number, Re. The calculation is somewhat tedious, however, and what is believed to be the essential result can be obtained by an approximate analytical technique. This approximate method, previously shown to be successful for the solution to the travelling wave stability problem (10), involves solving the *constant coefficient* equation obtained by simply dropping the term (Y^2/Γ^2) within the parentheses in Eq. [13]

$$\left(D_1^{\,2}+a_1^{\,2}-\frac{3a_1\mathrm{Re}}{2\Gamma}\right)\left(D_1^{\,2}+a_1^{\,2}\right)\hat{\upsilon}=0. \qquad [15]$$

This approximation cannot be rigorously justified. However, the neglected quantity is, in fact, only a part of a small correction term. That is, the small variable quantity

$$\left(3a_1\mathrm{Re}/2\Gamma\right)\left(1-\left(Y^2/\Gamma^2\right)\right),$$

an additive correction to the term a_1^2 in Eq. [13] is being approximated by the small constant quantity $3a_1\,\mathrm{Re}/\Gamma$ in Eq. [15]. Moreover, the approximation is exact in the neighborhood of the free surface where the instability is presumed to arise.

The solution to Eq. [15] may now be obtained analytically and the boundary conditions [14] may be employed to determine the eigenvalue, a_1. The procedure is briefly outlined. The form of solution to Eq. [15] is

$$\hat{\upsilon}=\cos a_1 y + A'\sin a_1 y$$
$$+B'\cos a_2 y + C'\sin a_2 y \qquad [16]$$

where,

$$a_2^{\,2}=a_1^{\,2}-\left(3a_1\mathrm{Re}/2\Gamma\right)$$

or, again neglecting terms of order Γ^{-2},

$$a_2 \cong a_1-\frac{3}{4}\left(\mathrm{Re}/\Gamma\right).$$

To lowest order in Γ^{-1}, the boundary conditions [14] yield,

$$\cos a_1\Gamma + A'\sin a_1\Gamma + B'\cos\left(a_1\Gamma-\frac{3}{4}\mathrm{Re}\right)$$
$$+\,C'\sin\left(a_1\Gamma-\frac{3}{4}\mathrm{Re}\right)=0, \qquad [16a]$$

$$-\sin a_1\Gamma + A'\cos a_1\Gamma - B'\sin\left(a_1\Gamma - \frac{3}{4}\text{Re}\right)$$

$$+ C'\cos\left(a_1\Gamma - \frac{3}{4}\text{Re}\right) = 0, \qquad [16b]$$

$$B' + 1 = -\frac{3}{4}(\text{Re}/a_1\Gamma), \qquad [16c]$$

$$A' + C' + \frac{1}{3}(1 + B')\Gamma = 0. \qquad [16d]$$

Equations [16a–d] then yield, again to lowest order in Γ^{-1},

$$a_1 = \frac{\text{Re}}{8}\frac{\sin\frac{3}{4}\text{Re}}{1-\cos\frac{3}{4}\text{Re}}. \qquad [17]$$

The reduced growth parameter, a_1 according to Eq. [17], vanishes at $\text{Re} = \frac{4}{3}\pi$. The calculations are, however, valid only when the product $a_1\Gamma$ is large compared to Reynolds number and consequently not valid for a strictly zero. Nevertheless, because Γ is large, even small values of a_1 satisfy the inequality $a_1\Gamma \gg \text{Re}$. The asymptotic analysis results in the conclusion that

$$a = \zeta\text{Re}^{-\frac{2}{3}} \cdot \frac{\text{Re}}{8} \cdot \frac{\sin\frac{3}{4}\text{Re}}{1-\cos\frac{3}{4}\text{Re}} \qquad [18]$$

so that

$$a \to \frac{1}{3}\zeta\,\text{Re}^{-\frac{2}{3}} \text{ as Re} \to 0,$$

and

$$a \to 0 \text{ as Re} \to \frac{4}{3}\pi.$$

Consequently the critical Reynolds number is $\frac{4}{3}\pi \cong 4.18$. Expressed in terms of a critical film thickness, the result is

$$h_c = 2.33\left(v^2/g\sin\beta\right)^{\frac{1}{3}}.$$

NUMERICAL RESULTS

To confirm the analytic result of Eq. [18], the eigenvalue problem for the constant coefficient form of Eq. [6] $\left(\bar{u} = \frac{3}{2}, D^2\bar{u} = -3\right)$ has been solved numerically to obtain the growth parameter, a, as a function of Reynolds number for the several cases in which Γ is large. Briefly, the procedure is to express the eigenfunction, $\hat{\upsilon}$, as a linear combination of four independent solutions to the fourth order constant-coefficient differential equation and then use the homogeneous boundary conditions to calculate a determinant of coefficients whose root is the eigenvalue, a. A search was made in the two-dimensional a-Re space for various large values of ζ to find the locus of eigenvalues. Figure 3 shows plots of the growth parameter, a, for several values of the surface tension parameter, ζ. It appears from this figure that, for sufficiently large values of ζ, there exists a critical value of the Reynolds number. Although, for reasons discussed in the previous section, the calculation becomes unreliable for small values of "a," there seems to be convincing evidence in Fig. 3 that the growth of a disturbance is very much stronger when the Reynolds number is below a critical value. Moreover, as indicated in Eq. [18], this critical Reynolds number is independent of the surface tension parameter, ζ, if ζ is sufficiently large. Figure 4 shows the results of a calculation of the reduced growth parameter, a_1, for an assumed ζ of 10^6. This value, although beyond the range of physical significance, allows for better resolution at small values of a_1. It is thus concluded that in agreement with the asymptotic results, the apparent critical Reynolds number is 4.18 for very large values of ζ, and that disturbances of the assumed type will grow very strongly for Reynolds numbers below this value. This particular type of instability is important for Reynolds numbers less than 4.18. At this Reynolds number the unsteady perturbation, also unstable for a vertical plate, has a very small growth coefficient, and for an inclined plate is even smaller.

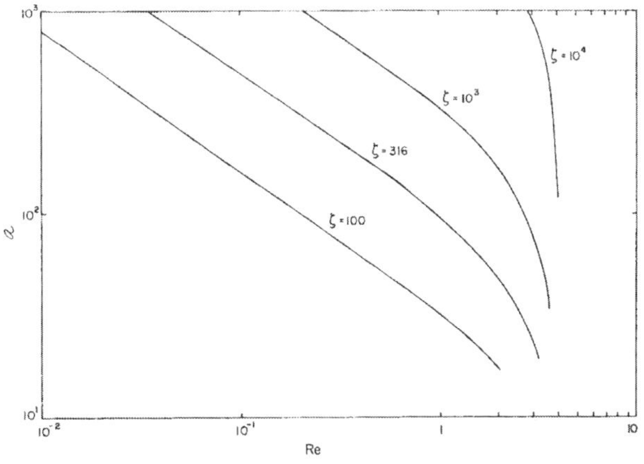

FIG. 3. The growth parameter dependence on Reynolds number (Re) and surface tension parameter calculated from the constant coefficient stability equation.

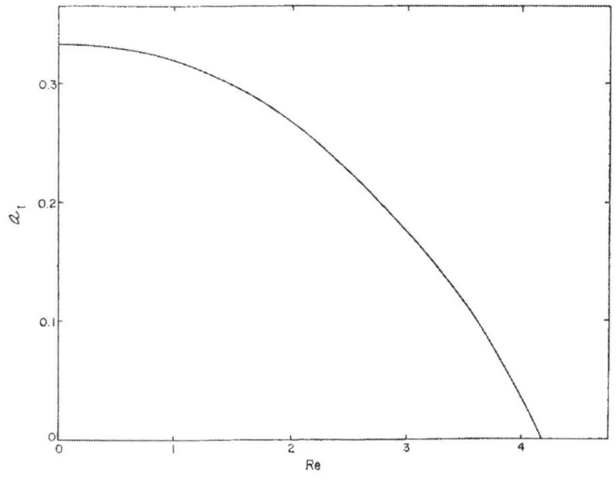

FIG. 4. The scaled growth parameter $a_1 = a\mathrm{Re}^{\frac{2}{3}}\zeta^{-1}$ for the limiting case of large ζ.

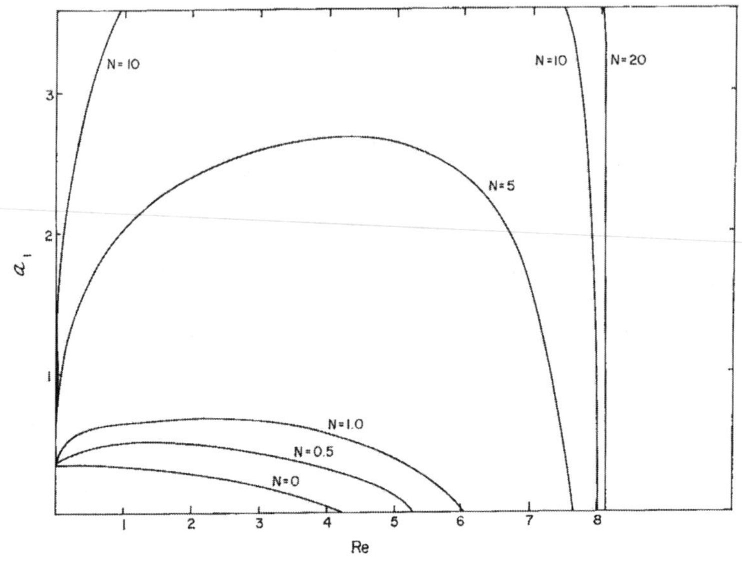

FIG. 5. Dependence of the scaled growth parameter, a_1, on Reynolds number and the thermocapillary parameter, N.

THERMOCAPILLARY EFFECT

It is well known (2, 5) that the presence of a nonzero heat flux into and through the liquid film from the solid plate has the effect of increasing the minimum wetting rate. Indeed, this is the central reason for studying the mechanism of breakdown. Simon and Hsu (5) have concluded, via a semiquantitative argument, that the principal effect of the finite heat flux on stability is to locally vary the surface tension through variations in surface temperature and thereby to impose capillary flow from thin regions of the film where the surface tension is smaller to thick regions where the surface tension is larger.

The complete analysis of temperature fields within a heated film is outside the scope of this paper. We consider only a well-developed thermal field in which the heat flux, q, and the wall temperature are constant and the surface temperature, T_i, varies because of the purturbation in the film thickness. We do not consider the effect of vaporization from the free surface, although it is recognized that this may have a significant destabilizing effect. Because of the surface tension gradient, the tangential stress conditions become (11):

$$\tau_{yx} = \partial\sigma/\partial x,\ \tau_{yz} = \partial\sigma/\partial z, \text{ at } y = \eta h,$$

and consequently,

$$\mu\left(\frac{\partial u'_x}{\partial y_1} + h\eta\frac{d^2U}{dy_1^{\ 2}} + \frac{\partial u'_y}{\partial x_1}\right) = -\left(-\frac{\partial\sigma}{\partial T}\right)\frac{\partial T_i}{\partial x_1}$$

$$= \left(-\frac{\partial\sigma}{\partial T}\right)\frac{qh}{k}\frac{\partial\eta}{\partial x_1}, \text{ at } y_1 = 0, \qquad [19a]$$

$$\mu\left(\frac{\partial u'_z}{\partial y_1} + \frac{\partial u'_y}{\partial z_1}\right) = \left(-\frac{\partial\sigma}{\partial T}\right)\frac{qh}{k}\frac{\partial\eta}{\partial z_1},$$

$$\text{at } y_1 = 0. \qquad [19b]$$

The only effect then on the eigenvalue problem is in the boundary condition numered [7c]; this is to be replaced by

$$\left(D^2-b^2\right)\hat{\upsilon}-\left(\frac{D^2\bar{u}}{\bar{u}}\right)\hat{\upsilon}+\frac{N\zeta\,\mathrm{Re}^{-\frac{1}{3}}}{\bar{u}}\frac{b^2}{a}\hat{\upsilon}=0,$$

$$\text{at } y=0, \qquad [20]$$

where

$$N\equiv\left(\frac{3\nu^2}{g}\right)^{\frac{1}{3}}\frac{q}{k}\left(-\frac{\partial\ln\sigma}{\partial T}\right). \qquad [21]$$

Because $d\sigma/dT<0$ this parameter possesses the algebraic sign of the heat flux, q. For any moderate heat flux, N is quite small for ordinary fluids. In the limit of small Reynolds number, the growth parameter to lowest order in N has the form

$$a=\frac{1}{3}\zeta\,\mathrm{Re}^{-\frac{2}{3}}\left(1+N\,\mathrm{Re}^{\frac{1}{3}}\right). \qquad [22]$$

Hence, if the heat flux is positive, the growth parameter is increased if the coefficient of surface tension with temperature is negative. The numerical calculations (in the limit of large values of Γ) go as before and the results are presented in Figs. 5 and 6. Figure 5 indicates that large N may increase

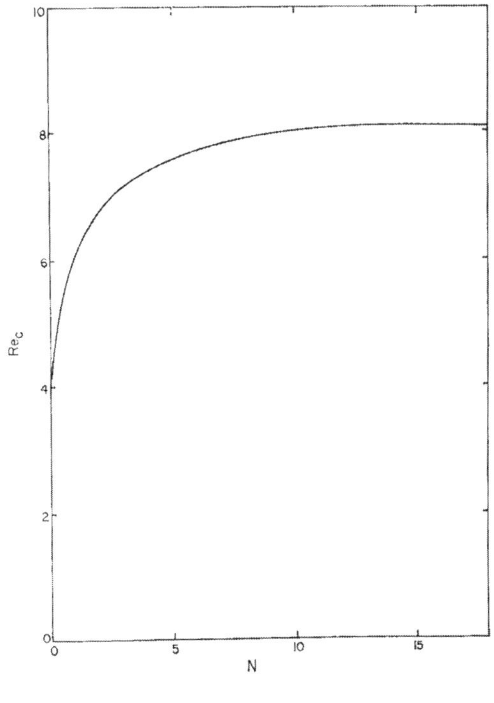

FIG. 6. Dependence of the critical Reynolds number on the thermocapillary number.

TABLE 1
Comparison of Theoretical Reynolds Number to Available Isothermal Data; Films on Metal Tubes

	Water at 30°C (2,5)	45% Glycerol in Water (5)
ζ	5×10^3	1×10^3
Critical Reynolds number, Re_c, calculated here	4.18	4.18
Critical Reynolds number, Re_{mw}, calculated from Eq. [1] assuming $\theta = 30°$	90	31
Experimental critical Reynolds number	13	5.5

the growth parameter by orders of magnitude but that its effect on the critical Reynolds number is quite minor. Apparently, from Fig. 6, the critical Reynolds number can be at most doubled when the thermocapillary effect is accounted for in this manner. The result is in the right direction, but is quantitatively in adequate since it is known from experiment (2, 5) that the effect of heat flux on the minimum wetting rate is quite large. It may be that evaporation at the interface is sufficiently destabilizing to account for the difference.

COMPARISON WITH AVAILABLE EXPERIMENTAL DATA

Norman and McIntyre (2) and Simon and Hsu (5) have conducted careful experiments to determine the flow rate of water below which it is impossible to maintain a continuous water film on a metal surface. The two sets of experiments seem to be in good quantitative agreement for the case of water at 30°C flowing isothermally over (in the two cases) copper pipe and nickel pipe. The observed critical Reynolds number is 13. Table 1 shows the comparison between the experimental and the calculated critical Reynolds number for the instability examined here. Table 1 also contains a comparison between the experimental and the calculated critical Reynolds number for the flow of a 45% glycerin in water solution on a nickel plate at 30°C.

CONCLUSION

It has been demonstrated that at sufficiently low Reynolds number a falling film has a "roping" instability which rapidly grows in the streamwise direction. If the surface tension parameter, ζ, is very large, the constant coefficient form of the linear stability equation shows a critical Reynolds number of 4.18, below which the flow is unstable. This result is in satisfactory agreement with available experiment on the appearance of dry patches on wetted walls.

REFERENCES

1. Norman, W. S., and Binns, D. T., *Trans. Inst. Chem. Eng.* London **38**, 294 (1960).
2. Norman, W. S., and McIntyre, V., *Trans. Inst. Chem. Eng.* London **38**, 301 (1960).
3. Hartley, D. E., and Murgatroyd, W., *Int. J. Heat Mass Transfer* **7**, 1003 (1964).
4. Zuber, N., and Staub, F. W., *Int. J. Heat Mass Transfer* **9**, 897 (1966).
5. Simon, F. F., and Hsu, Y.-Y., Thermocapillary induced breakdown of a falling liquid film, NASA, TN D-5624, 1970.
6. Ruckenstein, E., *Int. J. Heat Mass Transfer* **14**, 165 (1971).
7. Bankoff, S. G., *Int. J. Heat Mass Transfer* **14**, 2143 (1971).

8. Lin, C. C., "The Theory of Hydrodynamic Stability," Cambridge University Press, London/New York, 1953.
9. Yih, C.-S., *Phys. Fluids* **6**, 321 (1963).
10. Anshus, B. E., and Goren, S. L., *AIChE J.* **12**, 1004 (1966).
11. Levich, V. G., "Physicochemical Hydrodynamics," p. 379 ff. Prentice–Hall, Englewood Cliffs, NJ, (1962).
12. Krantz, W. B., and Owens, W. B., *AIChE J.* **19**, 1163 (1973).
13. Anshus, B. E., *Ind. Eng. Chem. Fundam.* **11**, 502 (1972).

1.3 Spreading Kinetics of Liquid Drops on Solids*

Jaime Lopez[a], *Clarence A. Miller*[b] *and Eli Ruckenstein*[c]

[a] Gulf Research & Development Co., Pittsburgh, Pa. 15230

[b] Department of Chemical Engineering, Carnegie-Mellon University, Pittsburgh, Pennsylvania 15213

[c] Faculty of Engineering and Applied Science, SUNY, Buffalo, Buffalo, New York 14214

Received August 12, 1975; accepted December 30, 1975

INTRODUCTION

Wetting and spreading phenomena occur during a wide variety of important processes. Removal of dirt by detergent solutions, drop-wise condensation of liquids, and displacement of one fluid by another in soils, underground oil reservoirs, packed beds, and other porous media are some examples.

Thermodynamic aspects of wetting and contact angles are generally well understood, but kinetic aspects are not. We present here a simple kinetic analysis of spreading of a thin liquid drop on a horizontal solid surface (Fig. 1). Our approach is to seek similarity solutions applicable when (a) the drop has spread over an area much greater than it initially covered and (b) the spreading rate is controlled by the balance between viscous resistance to flow and a single dominant driving factor, either gravity or surface forces near the drop's periphery. Unlike previous analyses of spreading of drops on solids (1) and of flow in the immediate vicinity of a moving contact line (2), our similarity solutions, when they apply, provide a complete velocity distribution throughout the entire drop. We note that a related mathematical technique has been used previously to describe spreading of an oil slick on the sea (3, 4).

With regard to spreading produced by surface forces, we do not simply invoke a line force at the drop's edge as in most previous work (1, 3). Instead we evaluate intermolecular forces throughout the thin liquid layer near the edge using a method developed previously (5). For simplicity we consider only cases where the liquid spreads "spontaneously." Under these conditions there is no equilibrium contact angle, and, as we shall see, our analysis yields better agreement with experiment than does the line force approach.

GENERAL EQUATIONS

We shall not attempt to analyze the early stages of spreading where inertial effects are important and drop curvature is appreciable. Instead we consider the drop after it has reached a condition where its thickness varies slowly with position. For then the flow is nearly one-dimensional and inertial effects can be neglected, the approximations of "lubrication theory" (6). Others (7, 8) have employed lubrication theory for spreading problems, and Huh and Scriven (2) even suggested

* *Journal of Colloid and Interface Science*, Vol. 56, No. 3, p. 460, September 1976. Republished with permission.

including the effect of intermolecular forces. The present analysis is the first, however, to use the lubrication approximation for an entire spreading drop. With this approximation the equations of motion for the drop shown in Fig. 1 become

$$0 = -\frac{\partial P}{\partial x} + \mu \frac{\partial^2 v_x}{\partial z^2} \qquad [1]$$

$$0 = -\partial P / \partial z \qquad [2]$$

$$P = p + \Phi_G + \Phi_M. \qquad [3]$$

Here Φ_G is the local gravitational potential energy per unit volume, while Φ_M is the difference between the potential energy density due to interaction with nearby solid and liquid molecules and that in a bulk liquid phase. This term is important only near the drop periphery where local thickness of the liquid layer is small. We note that the present formulation assumes that bulk liquid values for properties such as density, viscosity, and surface tension may be used in thin layers. No attempt is made, for example, to account for changes in local density which might result from the proximity of solid and gas phases. This approach seems a reasonable first approximation for describing flow in thin layers.

Equation [1] can be integrated twice with respect to z using boundary conditions of zero velocity at the solid surface and zero shear stress at the fluid surface. By averaging the resulting velocity profile over the film thickness, we obtain the mean velocity $\bar{v}$ in the liquid:

$$\bar{v} = \frac{1}{h}\int_0^h v_x dz = -\frac{h^2}{3\mu}\frac{\partial P}{\partial x}. \qquad [4]$$

To use this equation, we must evaluate $\partial P/\partial x$. As P is independent of z, according to Eq. [2], it suffices to evaluate P at $z = h$. Clearly, the gravitational potential energy there is $\rho g h$, while the pressure p differs from atmospheric pressure by the product of surface tension γ and local curvature. This difference is small for $\left(\gamma/\rho g x_0^2\right) \ll 1$, where x_0 is the drop radius. For the thin, flat drops of interest here this condition is satisfied, and, accordingly, we neglect curvature effects on the flow.

The final quantity needed to evaluate P from Eq. [3] is the potential energy density Φ_M due to interaction between molecules. Now the difference between Φ_{MF} at $z = h$ and the potential energy density Φ_{MF} at the flat surface of a deep pool of liquid can be found by direct integration of the interactions between a molecule at the fluid surface and all nearby molecules of the solid and liquid phases (5). The general result for a point on the fluid surface near a contact line is

$$\Phi_M - \Phi_{MF} = \frac{\pi}{12d^3}\left(n_L^2\beta_{LL} - n_L n_S \beta_{SL}\right)G(\alpha) - \frac{\pi}{12d^3} n_L n_S \beta_{SL} G(\pi - \alpha) \qquad [5]$$

$$G(\alpha) = \csc^3\alpha + \cot^3\alpha + \frac{3}{2}\cot\alpha. \qquad [6]$$

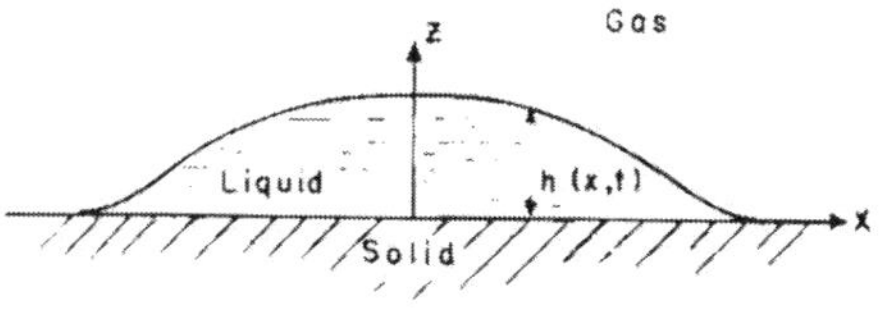

FIG. 1. Spreading of a liquid drop on a horizontal solid surface.

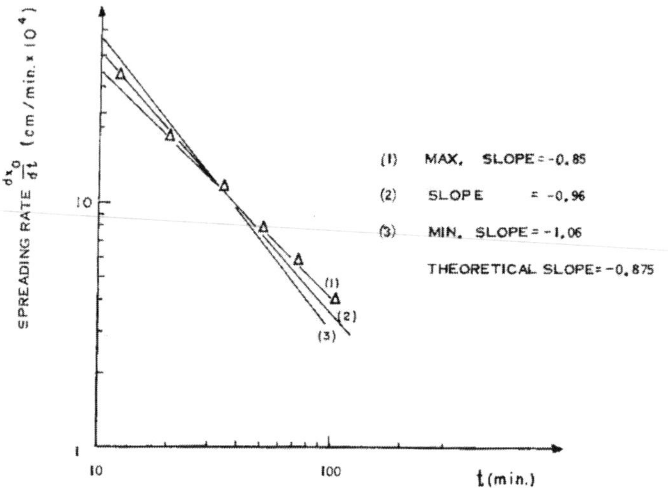

FIG. 2. Spreading of squalane on steel in the gravity-viscous regime. Data from Bascom *et al.* (15). The maximum and minimum values of the slope are the limits of the 98% confidence range.

Here d is the distance along the fluid surface from the point of interest to the contact line, α is the contact angle measured in the liquid phase, and n_L and n_S are the molecular densities of the liquid and solid phases, respectively. Also β_{LL} and β_{SL} are parameters in the following equations describing London-van der Waals interactions between two molecules of the appropriate types:

$$\phi_{LL} = -\beta_{LL} / r^6; \quad \phi_{SL} = -\beta_{SL} / r^6. \qquad [7]$$

For the present situation where the contact angle $\alpha \to 0$, the quantities $G(\alpha)$ and $G(\pi - \alpha)$ approach $(2/\alpha^3)$ and zero, respectively, and Eq. [5] simplifies to

$$\Phi_M - \Phi_{MF} = \frac{\pi\left(n_L^2\beta_{LL} - n_L n_S \beta_{SL}\right)}{6h^3} = -\frac{K}{h^3}. \qquad [8]$$

The right side of Eq. [8] is the same as the expression found for the "disjoining pressure" of a thin liquid film of uniform thickness h on a solid surface (9). The disjoining pressure is usually defined in terms of the thermodynamics of thin films and reflects the difference in behavior between a film and a bulk liquid phase (9). Equations [3], [4], and [8] imply that flow arises near the periphery of a spreading drop due to variations in local disjoining pressure. That such variations can cause flow is in agreement with previous work on stability and wave motion of thin liquid films of nearly uniform thickness (10–12).

From a physical point of view, we emphasize that flow caused by this mechanism is important only for liquid layers thinner than the effective range of intermolecular forces (about 1000 Å). For the present situation where liquid-solid interactions are stronger than liquid-liquid interactions, liquid flows outward from thicker to thinner regions near the drop's periphery because such flow brings more liquid molecules near the solid, reducing the system's free energy.

When the appropriate expressions for and Φ_G and Φ_M are substituted into Eq. [4], we find

$$\bar{v} = -\frac{h^2}{3\mu}\left(\rho g \frac{\partial h}{\partial x} + \frac{3K}{h^4}\frac{\partial h}{\partial x}\right). \qquad [9]$$

The equation specifying continuity of mass for the liquid layer at position x is also needed to solve for the velocity distribution in the drop:

$$\frac{\partial h}{\partial t} = -\frac{1}{x^{q-1}} \frac{\partial}{\partial x}\left(x^{q-1}\bar{\upsilon} h\right) \quad [10]$$

This equation applies for both cylindrical drops (shape invariant with respect to y-coordinate; $q = 1$) and axisymmetric drops ($q = 2$).

Substituting Eq. [9] into Eq. [10] yields a nonlinear, second-order partial-differential equation in $h(x, t)$:

$$\frac{\partial h}{\partial t} = \frac{1}{3\mu x^{q-1}} \frac{\partial}{\partial x}\left(\rho g h^3 x^{q-1} \frac{\partial h}{\partial x} + \frac{3K}{h} x^{q-1} \frac{\partial h}{\partial x}\right) \quad [11]$$

In principle, it can be solved if an initial drop shape is specified and if the following two boundary conditions are imposed:

Drop symmetry:

$$\partial h / \partial x = 0 \text{ at } x = 0 \quad [12]$$

Known drop volume:

$$V = \left(2\pi^{q-1}\right) \int_0^\infty x^{q-1} h dx \quad [13]$$

For cylindrical drops ($q = 1$), V is the volume per unit length.

Had we retained the effect of curvature on fluid pressure near the interface, Eq. [9] would have had an additional term involving $\partial^3 h / \partial x^3$, and two additional boundary conditions would have been needed. One of these would require $\partial^3 h / \partial x^3$ to vanish at $x = 0$ in order that $\bar{v}$ be zero there. The other would apply at the drop's edge and include effects such as diffusion in the region where the liquid layer ranges from a single to a few molecules thick. As this region is beyond the domain of continuum theory, the form the condition should take is, at present, unknown. By neglecting it here, we implicitly assume that surface diffusion in this adsorption layer of molecular dimensions does not appreciably influence the rate at which the bulk of the drop spreads.

SPREADING DUE TO GRAVITY

If gravity is the chief cause of spreading, we may simplify Eq. [11] by neglecting its last term. Then we seek a similarity solution of the form:

$$h(x,t) = F_1(t) F_2\left(x / x_0(t)\right) \quad [14]$$

The function $x_0(t)$ is a measure of how far the drop has spread. For an initial condition we replace the requirement for a complete initial drop profile with the simple condition $x_0(0) = 0$. Accordingly, our similarity solution does not apply during the early stages of spreading but only after the drop has spread over an area much greater than it initially occupied. This limitation of the similarity approach is not a serious one in the framework of our analysis because, as noted above, the whole lubrication approximation is questionable during the early stages of spreading.

A solution of the form [14] which satisfies the boundary conditions [12] and [13] and the initial condition $x_0(0) = 0$ is the following:

$$h(x,t)=\frac{V}{2w(q)x_0^q(t)}\left[1-\left(\frac{x}{x_0}\right)^2\right]^{\frac{1}{3}} \quad [15]$$

$$x_0(t)=\left[\frac{(3q+2)}{36[w(q)]^3}\cdot\frac{V^3\rho gt}{\mu}\right]^{[1/(3q+2)]} \quad [16]$$

$$w(q)=\frac{\pi^{q/2}}{3q+2}\cdot\frac{\Gamma\left(\frac{1}{3}\right)}{\Gamma\left\{\left(\frac{1}{2}\right)+\left[(-1)^{q-1}/3q\right]\right\}}3^{q-1} \quad [17]$$

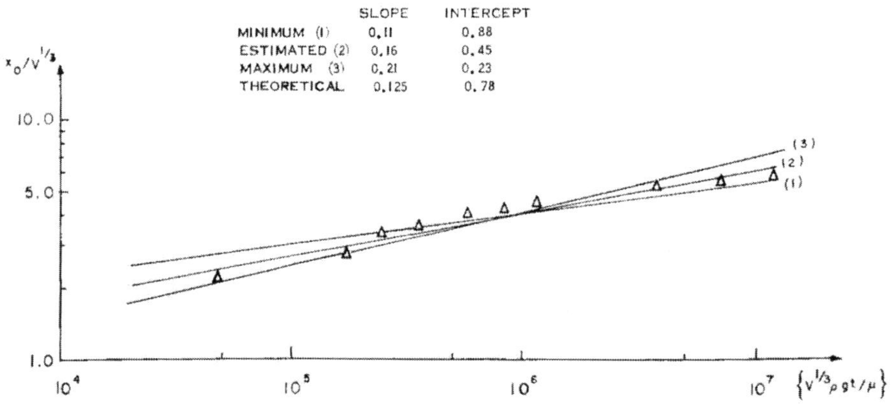

FIG. 3. Spreading of crude oil on ice in the gravity-viscous regime. Data from Chen *et al.* (13). The maximum and minimum values of the slope are the limits of the 99% confidence range. The intercepts are values obtained by extrapolating to $V^{\frac{1}{3}}\rho gt/\mu=1$.

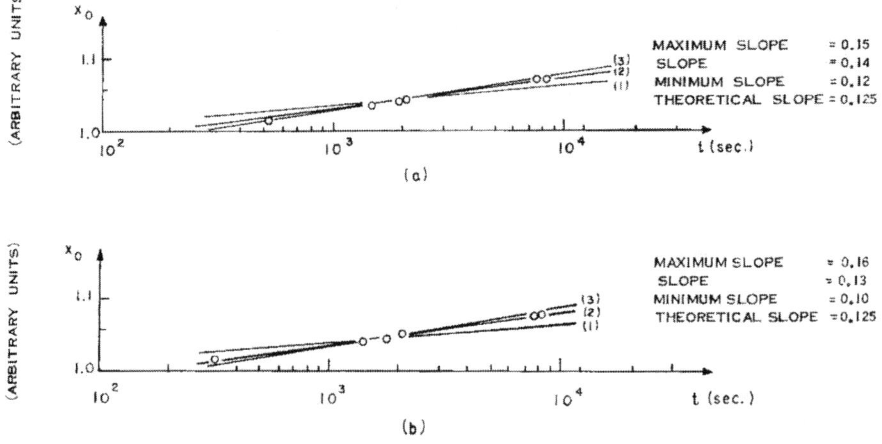

FIG. 4. Spreading of polydimethylsiloxane on mica in the gravity-viscous regime. Data from Ogarev *et al.* (16). (a) $V = 32 \times 10^{-4}$ cm^3, (b) $V = 17 \times 10^{-4}$ cm^3. The maximum and minimum values of the slope are the limits of the 99% confidence range.

In the last equation Γ denotes the usual gamma function defined by

$$\Gamma(x) = \int_0^\infty u^{x-1} e^{-u} du \qquad [18]$$

We note from Eq. [15] that $h(x_0, t) = 0$ so that $x_0(t)$ describes the advance of the drop's edge. The dependence of the drop's radius on its volume and physical properties and on the spreading time is thus given by Eq. [16]. It is somewhat different from the dependence predicted by the order-of-magnitude estimates of Chen *et al.* (13) because of an error in their analysis. A correct order-of-magnitude analysis similar to that described below for spreading produced by intermolecular forces (14) gives results in agreement with Eq. [16].

Equation [16] predicts that for axisymmetric drops ($q = 2$) a plot of drop radius x_0 as a function of time on log-log paper should yield a straight line with a slope of $\left(\frac{1}{8}\right)$. A similar plot for the spreading rate (dx_0/dt) should give a straight line having a slope of $\left(-\frac{7}{8}\right)$. Figures 2–4 show that these predictions are in good agreement with experimental data obtained by three different groups using three different liquid-solid systems (13, 15, 16). Figure 3, which shows the data of Chen *et al.* (13) for spreading of crude oil on ice, is particularly interesting because not only the slope, but the intercept as well, agrees with the predictions of Eq. [16]. No comparison of intercepts could be made for Figs. 2 and 4 because insufficient data were available.

SPREADING DUE TO INTER-MOLECULAR FORCES

We expect gravity to be the main factor causing spreading until the drop becomes rather thin. At this time intermolecular forces near the drop periphery come into play, and, as spreading continues, they ultimately become the dominant driving force. For this last situation the first or gravity term on the right side of Eq. [11] may be dropped. A similarity solution is sought as before, the result for a cylindrical drop being:

$$h(x,t) = \frac{V}{\pi x_1(t)}\left[1+\left(\frac{x}{x_1}\right)^2\right]^{-1} \qquad [19]$$

$$x_1(t) = 2\pi Kt/\mu V$$

Unlike $x_0(t)$ for the previous case of gravitational spreading, $x_1(t)$ does not represent the advance of the drop's edge. For $h(x_1, t)$ is easily seen to be $(V/2\pi x_1)$. Although the drop profile of Eq. [19] has no clearly defined edge in that h gradually approaches zero for large x, a measure of the spreading rate can be obtained following the point $x^*(t)$ on the drop corresponding to some small but fixed thickness h^*. From Eq. [19], we find that for $x^* \gg x_1$ (both distances being measured from the drop centerline), the advance of this point is given by

$$x^*(t) = \left(2Kt/\mu h^*\right)^{\frac{1}{2}}. \qquad [21]$$

Drop radius and spreading rate at time t are thus independent of drop volume.

Although no similarity solution was found for the axisymmetric case, the dependence of drop radius on time and other parameters can be obtained using order-of-magnitude arguments such as Fay (17) and Hoult (3) applied to spreading oil slicks. We imagine a drop whose outer portion is a long, thin "tail" in which intermolecular forces promote spreading and viscous forces resist it. Such tails are predicted by Eq. [19] and have, in fact, been observed for spreading oil drops (15).

They are sometimes, but not always, seen when polymers spread (18, 19). But factors not considered here may well be at work during polymer spreading.

If tail thickness is of order h^* the total viscous force on the spreading drop is of order

$$F_\upsilon \sim \mu x^{*q+1} / th^*. \qquad [22]$$

In deriving this expression, we have taken velocity in the drop as (x^*/t), to an order of magnitude. The net force F_M causing spreading may be viewed as the product of the disjoining pressure at the edge of the film and the area over which it acts. To an order of magnitude, we find

$$F_M \sim Kx^{*q-1} / h^{*2}. \qquad [23]$$

Equating [22] and [23], we obtain the following expression for both cylindrical and axisymmetric drops:

$$x^* \sim (Kt/\mu h^*)^{\frac{1}{2}}. \qquad [24]$$

As expected, this result agrees with Eq. [21].

The same approach can be used to estimate how drop radius varies with time when spreading is presumed due to a line force. The latter acts at the drop's edge and is equal to the "spreading coefficient" $S(=\gamma_{SG} - \gamma_{SL} - \gamma_{LG})$. This model for spreading has been used in some previous work (1, 3). The drop is now imagined to have a definite radius $\bar{x}$ and a thickness of order h. Hence, its volume V is, in order of magnitude, equal to $\bar{x}^q h$. This time viscous and spreading forces are given by

$$F_\upsilon \sim \mu \bar{x}^{q+1}/th \qquad [25]$$

$$F_S \sim S\bar{x}^{q-1} \qquad [26]$$

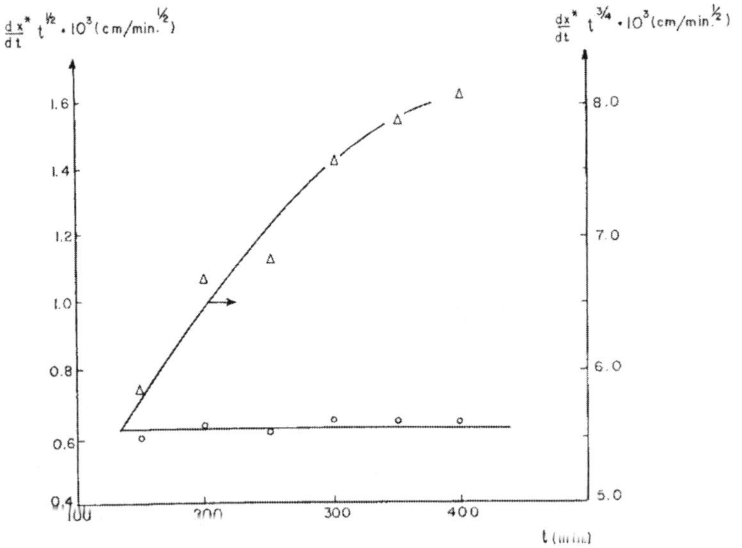

FIG. 5. Spreading of squalane on steel in the regime where intermolecular forces cause spreading. The nearly constant value of $\left[(dx^*/dt) \cdot t^{\frac{1}{2}}\right]$ is in agreement with Eq. (24) for times greater than about 100 min.

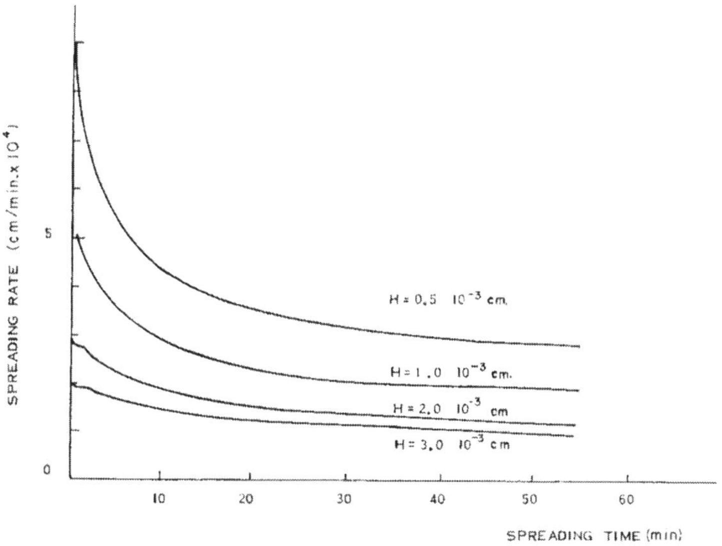

FIG. 6. Spreading rates of thin liquid drops as given by the numerical solution. H is the initial thickness at the drop centerline.

Equating these expressions and replacing h by $(V/\overline{x}^q)$, we find

$$\overline{x} \sim (SVt/\mu)^{1/q+2}. \qquad [27]$$

So the predicted time dependence of drop radius and spreading rate is different from that found above for intermolecular forces acting throughout a long tail.

Few workers have continued their spreading experiments into the regime where intermolecular forces dominate gravity as a driving force, at least for nonpolymeric liquids. We have found but one run from the work of Bascom *et al.* (15) with which to compare our predictions. As shown in Fig. 5, the data are consistent with a spreading rate proportional to $t^{-\frac{1}{2}}$ as predicted by our model but not with a spreading rate proportional to $t^{-\frac{3}{4}}$ as predicted by the line force model. Moreover, the drop did develop a tail, as noted previously. Further experiments are needed to confirm other predictions of Eqs. [21] and [24] such as the lack of dependence of the drop's radius $x^*(t)$ on its volume.

NUMERICAL SOLUTION

To supplement the similarity solutions, we solved Eq. [11] numerically for a cylindrical drop, keeping both the gravity and the intermolecular force terms. We used a finite difference scheme with spatial derivatives for a given time step evaluated at the average of initial and final times by a Crank-Nicolson technique (20). Further details of the numerical procedure are given elsewhere (14).

Initial profiles of the thin, wide drops were taken either as portions of circles or as given by the following cosine function:

$$h(x,0) = h^* + H\cos\left[\pi x / 2x^*(0)\right] \qquad [28]$$

In both cases a small finite value, usually 50 Å, was chosen for drop thickness h^* at its leading edge since mathematical difficulties precluded taking $h^* = 0$. Initial drop radius $x^*(0)$ was taken as 1 cm while initial thickness H at the drop's centerline ranged from 0.5 to 3.0×10^{-3} cm. A value of 3.5×10^{-11} cm^3/sec was chosen for (K/μ) based on reasonable values of the effective Hamaker constant K (3.5×10^{-13} erg) and fluid viscosity μ(1 cP).

The numerical results show a high initial spreading rate (dx^*/dt), defined as the rate of advance of the leading edge where $h = h^*$. The spreading rate decreases rapidly to an almost constant value (Fig. 6). The initial rapid transient is associated with formation of a thin film or tail along the solid surface ahead of the main portion of the drop. Evidently intermolecular forces in the tail control the spreading process, for once the tail forms, the spreading rate remains practically constant. Gravity has no appreciable influence on the spreading rate although it does affect somewhat the profile near the center of the drop.

As Fig. 6 shows, thinner drops spread faster, perhaps because intermolecular forces act over a longer tail. This interpretation is reinforced by comparison with results of the similarity solution. From Eq. [19] we can find the time required for a drop having a given volume V to reach a configuration with centerline thickness H. With this time and Eq. [21], we can calculate the drop's instantaneous spreading rate (dx^*/dt). Such calculations show that, for given V and H, the similarity solution predicts longer tails and spreading rates about 20% faster than are found in the constant rate regime of the numerical solution.

DISCUSSION

The similarity solutions presented above have their limitations, the main one being that shear stress increases without bound in the region near the drop's edge where its local thickness approaches zero. Such singular behavior is the rule in existing analyses of flow near moving contact lines (2). No satisfactory resolution of this problem has yet appeared.

On the other hand, we should recognize that the singular behavior exists only in a tiny region near the drop's edge for gravity-driven spreading and only in the extremely thin leading edge of the tail for spreading driven by intermolecular forces. The agreement of our predicted spreading rates with experimental data strongly suggests that the analysis is essentially correct over most of the drop in spite of difficulties in these small regions. Nevertheless, the similarity solutions are but a first approximation, and further research on flow during spreading is needed—especially for systems where a finite contact angle exists.

The picture of spreading on solids which emerges here is basically that found by Hoult (3) for spreading on deep liquid substrates. After an initial stage in which inertial effects are important, we find a stage where the balance between gravitational and viscous effects controls the spreading rate and finally a stage where intermolecular forces replace gravity as the chief factor causing spreading. This view is confirmed by the observation that a given drop passed successively through the last two stages (Figs. 2 and 5).

ACKNOWLEDGMENT

Acknowledgment is made to the donors of the Petroleum Research Fund, administered by the American Chemical Society, for support of this research.

REFERENCES

1. Yin, T. P., *J. Phys. Chem.* **73**, 2413 (1969).
2. Huh, C. and Scriven, L. E., *J. Colloid Interface Sci.* **35**, 85 (1971).
3. Hoult, D. P., *Ann. Rev. Fluid Mech.* **4**, 341 (1972).
4. Buckmaster, J., *J. Fluid Mech.* **59**, 481 (1973).
5. Miller, C. A. and Ruckenstein, E., *J. Colloid Interface Sci.* **48**, 368 (1974).
6. Batchelor, G. K., "An Introduction to Fluid Dynamics," p. 219. Cambridge University Press, Cambridge, England, 1967.
7. Friz, G., *Z. Angew Phys.* **19**, 374 (1965).
8. Ludviksson, V. and Lightfoot, E. N., *AIChE J.* **14**, 674 (1968).
9. Shelud'ko, A., *Adv. Colloid Interface Sci.* **1**, 391 (1967).

10. Vrij, A., *Disc. Faraday Soc.* **42**, 23 (1966).
11. Vrij, A., Hesselink, F., Lucassen, J., and van den Tempel, M., *Proc. Kon. Ned. Akad. Wetensch. B.* **73**, 124 (1970).
12. Ruckenstein, E. and Jain, R. K., *J. Chem. Soc. Faraday Trans. II* **70**, 132 (1974) (Section 5.1 of this volume).
13. Chen, E. C., Overall, J. C. K., and Phillips, C. R., *Canad. J. Chem. Eng.* **52**, 71 (1974).
14. Lopez, J., "Dynamics of Wetting Processes," Ph.D. thesis, Carnegie-Mellon University, 1974.
15. Bascom, W. D., Cottington, R. L., and Singleterry, C. R., *in* "Advances in Chemistry Series," Vol. 43, pp. 355 ff. American Chemical Society, Washington, D. C., 1964.
16. Ogarev, V. A., Ivanova, T. N., Arslanov, V. V., and Trapeznikov, A. A., *Izv. Akad. Nauk. SSSR Ser. Khim.* **7**, 1467 (1973).
17. Fay, J. A., *in* "Oil on the Sea," (D. P. Hoult, Ed.), pp. 5 ff. Plenum, New York, 1969.
18. Dettre, R. H. and Johnson, R. E., Jr., *J. Adhesion* **2**, 61 (1970).
19. Schonhorn, H., Frisch, H. L., and Kwei, T. K., *J. Appl. Phys.* **37**, 4967 (1966).
20. von Rosenberg, D. U., "Methods for the Numerical Solution of Differential Equations," p. 63. Elsevier, New York, 1969.

1.4 Concerning the Driving Force for Wetting*

Eli Ruckenstein and P. S. Lee

Faculty of Engineering and Applied Sciences, State University of New York at Buffalo, Buffalo, New York 14214, U.S.A.

Received 21 January 1975

In previous papers [1,2] the tangential force acting per molecule in the liquid—gas interface was used to describe the dynamics of wetting. An explicit expression for this driving force was computed taking into account the interaction potential between the molecules of the liquid and between those of the liquid and the solid. Whether flow is toward or away from the contact line depends both on the relative magnitude of liquid—liquid and liquid—solid intermolecular forces and on the size of the contact angle. The coupling of this tangential force with the hydrodynamic equations of motion permitted the time evolution of the shape of the droplet to be described in some cases [2]. The tangential force was computed directly from the interaction energy between the molecules however. The entropy term in the expression of the free energy as well as any temperature dependence was neglected. In what follows it will be shown that on the basis of thermodynamic considerations an exact expression for the tangential force acting per unit area of the liquid—gas interface can be obtained. For non-equilibrium conditions the calculated tangential force correctly shows when the flow of the liquid is toward or away from the contact line. For equilibrium conditions the tangential force is necessarily zero and the expression reduces to Young's equation.

The equilibrium of a macroscopic drop on a smooth, homogeneous and non-deformable solid surface was treated from a thermodynamic point of view by Gibbs [3] and more recently by Johnson [4]. The equilibrium condition leads to Laplace's and to Young's equations. Since Laplace's equation is equivalent to the vanishing of the normal force acting on a molecule in the liquid—gas interface, it is natural to expect that Young's equation is related to the vanishing of the tangential force acting on a molecule in the liquid—gas interface. The considerations which follow introduce some modifications in the previous treatments and prove the equivalence between Young's equation and the vanishing of the tangential force acting on the liquid—gas interface. The main interest in the new formulation however lies in the fact that it provides an expression for the driving force for wetting when the droplet is not in equilibrium.

The variation of the Helmholtz free energy F (hereupon referred to as free energy) for a system composed of a drop on a solid surface can be written as†:

$$F = F_V + F_\Omega + F_{\text{ext}}, \tag{1}$$

where F_V is the free energy of the volume, F_Ω the free energy of the surface and F_{ext} the free energy caused by an external field (e.g., the gravitational field). The solid surface is assumed plane, smooth, homogeneous and non-deformable.

* *Surface Science* 50 (1975) 597–604. Republished with permission.

† Cf. Notation at the end of this note.

Consider a reversible isothermal process during which the surface of the liquid is deformed so that the interfacial areas Ω^{ij} separating the phases i and j ($i, j = l$, g, s refer to the liquid, gas and solid phases respectively) change to $\Omega^{ij} + \delta\Omega^{ij}$ as shown in fig. 1. If the effect of the external fields is negligible, the variation of F can be written as

$$\delta F = -\int_V P\delta(\mathrm{d}V) + \int_\Omega \gamma\delta(\mathrm{d}\Omega) + \sum_i \int_V \mu_i \delta(\mathrm{d}N_i^V) + \sum_i \int_\Omega \mu_i \delta(\mathrm{d}N_i^\Omega), \tag{2}$$

where P is the pressure, γ the surface tension, μ_i the chemical potential of the ith component. The superscripts V and Ω refer to the volume and surface respectively, whereas V and Ω attached to the integrals show that the integrations have to be carried out over the volume V and the surface Ω respectively. We are interested here in transformations taking place under conditions of chemical equilibrium. Since under these conditions

$$\sum_i \int_V \mu_i \delta(\mathrm{d}N_i^V) + \sum_i \int_\Omega \mu_i \delta(\mathrm{d}N_i^\Omega) = 0, \tag{3}$$

eq. (2) becomes:

$$\delta F = -\int_V P\delta(\mathrm{d}V) + \int_\Omega \gamma\delta(\mathrm{d}\Omega). \tag{4}$$

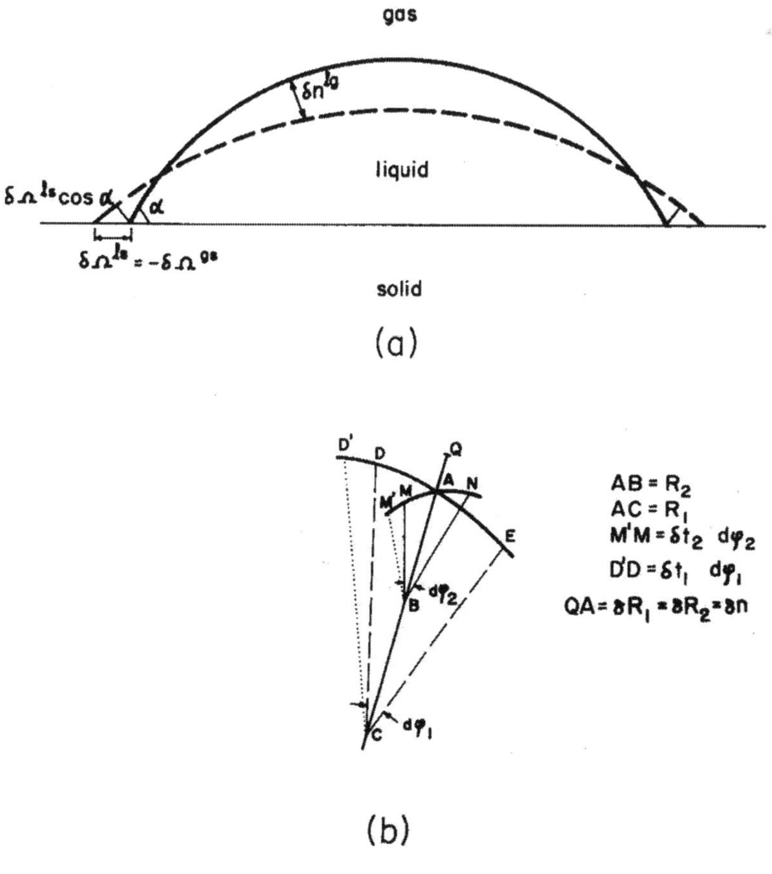

FIG. 1 Vertical deformation of the surface of the drop.

Following Johnson [4], the first term on the right of eq. (4) can be rewritten as

$$\int_V P\delta(\mathrm{d}V) = \delta\int_V P\mathrm{d}V - \int_V \delta P\mathrm{d}V$$

$$= \int_{\Omega^{lg}} \left(P^l - P^g\right)\delta n^{lg}\mathrm{d}\Omega^{lg} - \int_V \delta P\mathrm{d}V, \tag{5}$$

where P^l and P^g are the pressure in the liquid and the gas respectively.

Before rewriting the second term on the right of eq. (4), $\delta\Omega^{lg}$ is calculated,

$$\delta\Omega^{lg} = \delta\int_{\Omega^{lg}} R_1R_2\mathrm{d}\varphi_1\mathrm{d}\varphi_2 = \int_{\Omega^{lg}} \left\{(R_1+\delta n)(R_2+\delta n)\times(\mathrm{d}\varphi_1+\delta\mathrm{d}\varphi_1)(\mathrm{d}\varphi_2+\delta\mathrm{d}\varphi_2) - R_1R_2\mathrm{d}\varphi_1\mathrm{d}\varphi_2\right\}$$

$$= \int_{\Omega^{lg}} \left\{(R_1+R_2)\delta n\ \mathrm{d}\varphi_1\mathrm{d}\varphi_2 + R_1R_2\mathrm{d}\varphi_1\mathrm{d}\varphi_2\left(\frac{\delta\mathrm{d}\varphi_1}{\mathrm{d}\varphi_1}+\frac{\delta\mathrm{d}\varphi_2}{\mathrm{d}\varphi_2}\right)\right\}$$

$$= \int_{\Omega^{lg}} \left(\frac{1}{R_1}+\frac{1}{R_2}\right)\delta n + \left(\frac{\delta t_1}{R_1}+\frac{\delta t_2}{R_2}\right)\Bigg\} R_1R_2\mathrm{d}\varphi_1\mathrm{d}\varphi_2, \tag{6}$$

where R_1 and R_2 are the two principal radii of curvature,

$$\delta t_1 \equiv R_1\frac{\delta\mathrm{d}\varphi_1}{\mathrm{d}\varphi_1} \text{ and } \delta t_2 \equiv R_2\frac{\delta\mathrm{d}\varphi_2}{\mathrm{d}\varphi_2} \tag{7}$$

are the changes in length per unit sustended angle in the tangential direction along the two principal curvatures, δn is the variation along the normal to the surface, and $\mathrm{d}\varphi_k$ ($k = 1,2$) are the differentials of the angles corresponding to kth principal curvature. For convenience, we have omitted the superscripts lg in R_j, δn, $\delta\varphi_j$ and $\mathrm{d}\varphi_j$ in (6).

As first recognized by Gauss [6] (see also Hwang [5]) using differential geometry considerations, the variation $\delta\Omega^{lg}$ is composed of two parts (fig. 1): a change from R_j to $R_j + \delta n$ which is the change from the solid line to the dashed line, and a change due to spreading representing the contribution due to the change of the liquid—solid interface. The first part is given by the term containing δn in (6), whereas the second part has to be equal to $\delta\Omega^{ls}\cos\alpha$. Consequently

$$\int_{\Omega^{lg}} \left(\frac{\delta t_1}{R_1}+\frac{\delta t_2}{R_2}\right)^{lg} \mathrm{d}\Omega^{lg} = \delta\Omega^{ls}\cos\alpha = -\delta\Omega^{sg}\cos\alpha. \tag{8}$$

Since we assume that the solid retains its infinite radius of curvature, the variation of the surface area for liquid—solid and gas—solid interfaces are given by the expressions

$$\delta\Omega^{js} = \int_{\Omega^{js}} \left(\frac{\delta t_1}{R_1^{js}}+\frac{\delta t_2}{R_2^{js}}\right)^{js} \mathrm{d}\Omega^{js},\ j = l, g, \tag{9}$$

which when substituted into (8) lead to

$$\int_{\Omega^{ls}} \left(\frac{\delta t_1}{R_1}+\frac{\delta t_2}{R_2}\right)^{ls} \cos\alpha\,\mathrm{d}\Omega^{ls} = -\int_{\Omega^{gs}} \left(\frac{\delta t_1}{R_1}+\frac{\delta t_2}{R_2}\right)^{gs} \cos\alpha\,\mathrm{d}\Omega^{gs} = \int_{\Omega^{lg}} \left(\frac{\delta t_1}{R_1}+\frac{\delta t_2}{R_2}\right)^{lg} \mathrm{d}\Omega^{lg} \tag{10}$$

Taking the differentials on both sides of (10), one obtains

$$\left(\frac{\delta t_1}{R_1}+\frac{\delta t_2}{R_2}\right)^{ls}\cos\alpha\,\mathrm{d}\Omega^{ls}--\left(\frac{\delta t_1}{R_1}+\frac{\delta t_2}{R_2}\right)^{gs}\cos\alpha\,\mathrm{d}\Omega^{gs}=\left(\frac{\delta t_1}{R_1}+\frac{\delta t_2}{R_2}\right)^{lg}\mathrm{d}\Omega^{lg}. \tag{11}$$

Now consider the term $\int \gamma^{lg}\,\delta\,(\mathrm{d}\Omega^{lg})$. Because of the tangential expansion of $\mathrm{d}\Omega^{lg}$, the surface element, after the variation, will shift to a new position. Since the position of a surface element can be described by two independent coordinates, say $w_j = w_j\,(x, y, z)$, $j = 1, 2$, all surface quantities can be written as functions of these coordinates. Consequently

$$\int\limits_{\Omega^{lg}}\gamma^{lg}\delta\left(\mathrm{d}\Omega^{lg}\right)=\int\limits_{\Omega^{lg}}\gamma_{\mathrm{a}}\left(w_1,w_2\right)\left[\mathrm{d}\Omega_{\mathrm{b}}\left(w_1+\delta w_1,w_2+\delta w_2\right)-\mathrm{d}\Omega_{\mathrm{a}}\left(w_1,w_2\right)\right]$$

$$=\int\gamma_{\mathrm{a}}\left(w_1+\delta w_1-\delta w_1,w_2+\delta w_2-\delta w_2\right)\mathrm{d}\Omega_{\mathrm{b}}\left(w_1+\delta w_1,w_2+\delta w_2\right)-\int\gamma_{\mathrm{a}}\left(w_1,w_2\right)\mathrm{d}\Omega_{\mathrm{a}}\left(w_1,w_2\right)$$

$$=\int\left\{\gamma_{\mathrm{a}}\left(w_1',w_2'\right)-\frac{\partial\gamma_{\mathrm{a}}}{\partial w_1'}\delta w_1-\frac{\partial\gamma_{\mathrm{a}}}{\partial w_2'}\delta w_2\right\}\mathrm{d}\Omega_{\mathrm{b}}\left(w_1',w_2'\right)-\int\gamma_{\mathrm{a}}\left(w_1,w_2\right)\mathrm{d}\Omega_{\mathrm{a}}\left(w_1,w_2\right) \tag{12}$$

$$=\int\gamma_{\mathrm{a}}\left[\mathrm{d}\Omega_{\mathrm{b}}\left(w_1,w_2\right)-\mathrm{d}\Omega_{\mathrm{a}}\left(w_1,w_2\right)\right]-\int\left[\frac{\partial\gamma_{\mathrm{a}}}{\partial w_1}\delta w_1+\frac{\partial\gamma_{\mathrm{a}}}{\partial w_2}\delta w_2\right]\mathrm{d}\Omega_{\mathrm{b}}\left(w_1,w_2\right), \tag{13}$$

where $w_1' = w_1+\delta w_1$, $w_2' = w_2+\delta w_2$, and the subscript a refers to the state before the variation and b to the state after the variation.

The first term in (13) is just

$$\int\limits_{\Omega^{lg}}\gamma^{lg}\left[\mathrm{d}\Omega_{\mathrm{b}}^{lg}-\mathrm{d}\Omega_{\mathrm{a}}^{lg}\right]=\int\limits_{\Omega^{lg}}\gamma^{lg}\left\{\left(\frac{1}{R_1^{lg}}+\frac{1}{R_2^{lg}}\right)\delta n^{lg}+\left(\frac{\delta t_1^{lg}}{R_1^{lg}}+\frac{\delta t_2^{lg}}{R_2^{lg}}\right)\right\}\mathrm{d}\Omega^{lg}, \tag{14}$$

and the second term can be written as $-\int \delta\gamma\,\mathrm{d}\Omega$. In other words,

$$\int\limits_{\Omega^{lg}}\gamma^{lg}\delta\left(\mathrm{d}\Omega^{lg}\right)=\int\limits_{\Omega^{lg}}\gamma^{lg}\left\{\left(\frac{1}{R_1^{lg}}+\frac{1}{R_2^{lg}}\right)\delta n^{lg}+\left(\frac{\delta t_1^{lg}}{R_1^{lg}}+\frac{\delta t_2^{lg}}{R_2^{lg}}\right)\right\}\mathrm{d}\Omega^{lg}-\int\delta\gamma^{lg}\mathrm{d}\Omega^{lg}. \tag{15}$$

Similarly,

$$\int\limits_{\Omega^{sj}}\gamma^{sj}\delta\left(\mathrm{d}\Omega^{sj}\right)=\int\limits_{\Omega^{sj}}\gamma^{sj}\left(\frac{\delta t_1^{sj}}{R_1^{sj}}+\frac{\delta t_2^{sj}}{R_2^{sj}}\right)\mathrm{d}\Omega^{sj}-\int\delta\gamma^{sj}\mathrm{d}\Omega^{sj},\ j=l,\mathrm{g}. \tag{16}$$

Substituting (5), (12), (15), (16) into (4) gives

$$\delta F=\int\limits_{\Omega^{lg}}\left\{\left[\left[P^{g}-P^{l}+\gamma^{lg}\left(\frac{1}{R_1^{lg}}+\frac{1}{R_2^{lg}}\right)\right]\right]\delta n^{lg}+(\gamma^{lg}\cos\alpha-\gamma^{gs}+\gamma^{ls})\left(\frac{\delta t_1^{lg}}{R_1^{lg}}+\frac{\delta t_2^{lg}}{R_2^{lg}}\right)\sec\alpha\right\}\mathrm{d}\Omega^{lg}$$
$$+\int\limits_{V}\delta P\,\mathrm{d}V-\int\limits_{\Omega}\delta\gamma\,\mathrm{d}\Omega \tag{17}$$

It is noted once again in conjunction with eq. (17) for the variation of Helmholtz free energy that the effects of gravity on the drop shape have been ignored, that a chemically equilibrated multicomponent system is being considered, and that the solid surface retains its infinite radius of curvature. Because the mechanical force $\boldsymbol{f}$ is related to the free energy $\boldsymbol{F}$ by $\boldsymbol{f} = -\nabla \boldsymbol{F}$, the normal force component per unit surface area of the liquid—gas interface is

$$f_{\mathrm{n}} = -\frac{\delta F}{d\Omega^{lg}\delta n^{lg}} = P^{l} - P^{g} - \gamma^{lg}\left(\frac{1}{R_1^{lg}} + \frac{1}{R_2^{lg}}\right), \tag{18}$$

and the tangential force component per unit surface area of the liquid—gas interface along the kth principal curvature is

$$f_{\mathrm{t}k} = -\frac{\delta F}{d\Omega^{lg}\delta t_k^{lg}} = -\frac{1}{R_k^{lg}\cos\alpha}\left(\gamma^{lg}\cos\alpha - \gamma^{gs} + \gamma^{ls}\right). \tag{19}$$

For thermodynamic equilibrium, this force has to be zero. In this case, $f_{\mathrm{n}} = 0$ gives Laplace's equation and $f_{tk} = 0$ for both $k = 1$ and $k = 2$ give the same equation, namely Young's equation.

$$\gamma^{lg}\cos\alpha_0 - \gamma^{gs} + \gamma^{ls} = 0, \tag{20}$$

where α_0 is the equilibrium angle. This proves the equivalence between Young's equation and the vanishing of the tangential force acting upon a molecule in the liquid—gas interface.

For non-equilibrium conditions

$$f_{\mathrm{t}k} = \frac{\gamma^{lg}}{R_k^{lg}\cos\alpha}\left(\cos\alpha_0 - \cos\alpha\right). \tag{21}$$

For $0 < \alpha < \pi/2$, we have $\cos\alpha > 0$. Because R_k^{lg} are positive in this case, $f_{tk} > 0$ when $\alpha > \alpha_0$ and the motion of the liquid is toward the contact line. However, $f_{tk} < 0$ when $\alpha < \alpha_0$ and the liquid is moving away from the contact line. A similar discussion can be made if $\pi/2 < \alpha < \pi$.

Compared to the previous treatment, the present treatment has the advantage of providing an expression for the tangential force at the liquid—gas interface. When the droplet is in equilibrium, the tangential force has to be equated to zero and leads to Young's equation.

NOTATION

F	Helmholtz free energy (free energy).
F_{ext}	free energy caused by an external field.
F_V	free energy of the volume.
F_Ω	free energy of the surface.
∇F	gradient of the free energy F.
f	$f = -\nabla F$: the mechanical force.
f_{n}	normal force component per unit surface area of the liquid-gas interface.
f_{tk} $k = 1, 2$	the tangential force component per unit surface area of the liquid-gas interface along the kth principal curvature.
N_i^{V}	number of molecules of the ith component in the volume V.
N_i^{Ω}	number of molecules of the ith component in the surface Ω.
δn^{lg}	the variation in the direction normal to the liquid—gas interface.
P	the pressure.
P^l	the pressure of the liquid phase.
P^{g}	the pressure of the gas phase.

R_k^{ij}	$i, j = l$, g, s; $k = 1,2$: the radius of curvature of the kth principal curvature of the interface Ω^{ij}.
δt_k^{ij}	$i, j = l$, g, s; $k = 1,2$: $\delta t_k^{ij} \equiv R_k^{ij}\left(\delta \mathrm{d}\varphi_k^{ij}/\mathrm{d}\varphi_k^{ij}\right)$, is the changes in length per unit sustended angle in tangential direction along the kth principal curvature of the interface Ω^{ij}.
V	the volume.
w_j	$\equiv w_j(x, y, z)$ where $j = 1,2$ are the parameters defining the positions of a point (x, y, z) on the surface Ω^{lg}.
w'_j	$\equiv \mathrm{w}_\mathrm{j} + \delta \mathrm{w}_\mathrm{j}$.
α	the contact angle.
α_0	the contact angle at thermodynamic equilibrium.
γ	surface tension.
$\delta\gamma$	the variation of the surface tension.
γ^{ij}	$i, j = l$, g, s: the surface tension of the ij interface.
γ_a	the surface tension before the variation in shape of the droplet.
γ_b	the surface tension after the variation in shape of the droplet.
μ_i^V	the chemical potential of the ith component in the volume.
μ_i^Ω	the chemical potential of the ith component on the surface.
φ_k	$k = 1,2$; the angle corresponding to kth principal curvature.
Ω	the surface.
Ω^{ij}	$i, j = l$, g, s: total surface area of the interface separating phases i to j.
Ω_a	total liquid—gas interfacial surface area before the variation.
Ω_b	total surface area of the liquid—gas interface after variation.

The superscripts l, g and s refer to the liquid, the gas, and the solid phases respectively; two superscripts refer to the interface between the two phases.

This work was supported by NSF.

REFERENCES

[1] C.A. Miller and E. Ruckenstein, J. Colloid Interface Sci. 48 (1974) 368 (Section 1.1 of this volume).
[2] J. Lopez, C.A. Miller and E. Ruckenstein, Preprints of 48th National Colloid Symposium, Austin (1974) p. 65.
[3] J.W. Gibbs, The Collected Works of J. Willard Gibbs, Vol. 1, Thermodynamics (Yale University Press, New Haven, 1928).
[4] R.E. Johnson, Jr., J. Phys. Chem. 63 (1959) 1655.
[5] S.K. Hwang, Preprints of 48th National Colloid Symposium, Austin (1974), page 54.
[6] C.F. Gauss, Theorie der Gestalt von Flussigkeiten (Verlag von Wilhelm Engelmann, Leipzig, 1903) p. 46.

1.5 Slip Velocity during Wetting of Solids*

Eli Ruckenstein and C. S. Dunn
Faculty of Engineering and Applied Sciences, State University of New York at Buffalo, Buffalo, New York 14214

Received February 12, 1976; accepted September 10, 1976

The main difficulty in understanding the kinetics of wetting has been in describing the motion of the contact line over a solid surface. It is obvious that in the region of the contact line the liquid moves over the solid surface displacing a gas, or another immiscible liquid, from the solid surface and that the no-slip boundary condition employed in fluid mechanics is violated. It is therefore not surprising that hydrodynamic models of the motion predict unbounded stresses and viscous dissipation at the contact line when the no-slip condition is used (1). Dussan (2) has introduced a slip velocity near the contact line decaying to zero at larger distances. Choosing various reasonable, but arbitrary, forms for the dependence of the slip velocity on the distance to the contact line, she has developed a hydrodynamic analysis of wetting. However, no one has answered the basic question, which is to explain why slip occurs near the contact line. In the present paper a physical explanation is proposed for the origin of the slip velocity and a simple theoretical analysis is presented for its computation. In essence it is shown that near the contact line a net translational velocity of liquid molecules parallel to the solid surface originates because of the force generated by a gradient of the chemical potential in the liquid along the solid–liquid interface.

Let us consider the region near the contact line of a spreading droplet (Fig. 1). The shape of the liquid–fluid interface near the leading edge depends upon position in a complicated way as was shown before for a droplet at equilibrium (3). Although we have established equations for the slip velocity which take into account the real shape near the leading edge, the effective computations are very intricate. To illustrate the essential qualitative features of the problem it will be therefore assumed, for the sake of simplicity, that near the contact line the film thickness is linear with distance from the leading edge. We will focus now on a point located in the liquid on the solid–liquid interface. Consider first that the range of interaction forces between this molecule and other molecules of the liquid is smaller than the distance to any of the points on the liquid–fluid interface. In this case the chemical potential at the chosen point will be equal to that of a corresponding point situated at the solid–liquid interface of a semi-infinite liquid. This chemical potential is denoted by μ_0. However, if the range of interaction forces is larger than the distance to the liquid–fluid interface, then the molecule at the chosen point is additionally subject to a field ϕ_0 originating because the interaction potential due to the neighboring molecules differs from that of a semi-infinite liquid. This field, equal to the difference between the true interaction potential and that corresponding to a semi-infinite liquid, decays to zero for large enough distances from the leading edge; it becomes increasingly important when the leading edge is approached because liquid molecules are replaced by gas molecules within the effective range of the interaction forces. Thermodynamics allows one to express the chemical potential as

$$\mu = \mu_0 + \phi_0, \tag{1}$$

* *Journal of Colloid alld Lnterface Science*. Vol. 59, No. 1, p. 135, March 15, 1977. Republished with permission.

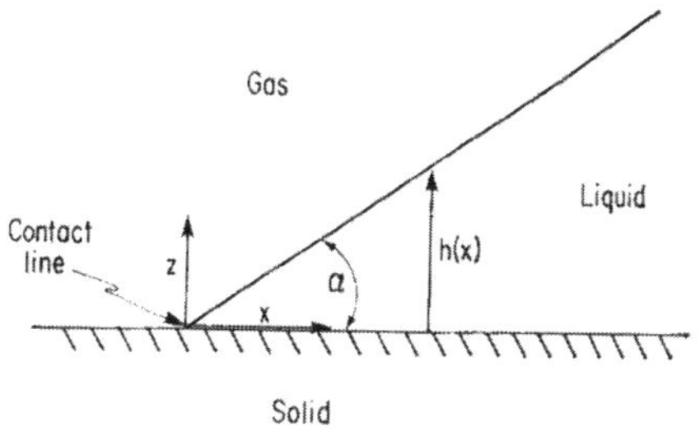

FIG. 1. Schematic representation of the region near the contact line of a spreading liquid drop.

Where

$$\phi_0 = n_0\left[\int_V nfudV - \int_\infty nfudV\right] \qquad [2]$$

The first integral is extended over the finite volume V of the spreading liquid, the second over a semi-infinite system composed of the same liquid. In Eq. [2], n is the molecular density, n_0 the molecular density at the chosen point, f the radial distribution function, and u the interaction potential between two molecules. The liquid is assumed incompressible and, therefore, n is a constant n_L. For convenience μ, μ_0, and ϕ_0 are expressed per unit volume.

A gradient of the chemical potential arises both because μ_0 is pressure dependent and because ϕ_0 depends upon the distance x from the leading edge. The gradient of the chemical potential along the interface induces a force per molecule $F = -(1/n_L)(\partial\mu/\partial x)$ which causes the slip.

The mechanism by which slip occurs can be understood using an approach similar to that developed in the treatment of electrical conductivity of ionic crystals (4). On this basis a relation is derived between the slip velocity v_S and the force $\boldsymbol{F}$. Consider a liquid molecule which moves in the x direction on a solid surface by jumping from one potential well to the next, over barriers of height E a distance δ apart. For a symmetrical potential barrier, the probability of the molecule jumping from a given well to one of its two neighbors in a unit time is given by (4)

$$C = v_0 e^{-E/kT},$$

where v_0 is the vibration frequency of a molecule in the potential well, k is Boltzman's constant, and T is the temperature. If an external force F acts upon the molecule, the potential barrier in the direction of the force is decreased by $\frac{1}{2}F\delta$, while in the opposite direction it is increased by the same amount. In this case, the probability for the molecule jumping per unit time is

$$C_+ = \frac{1}{2}v_0 e^{-\left(E-\frac{1}{2}F\delta\right)/kT}$$

in the positive x direction, and

$$C_- = \frac{1}{2}v_0 e^{-\left(E+\frac{1}{2}F\delta\right)/kT}$$

in the negative x direction. The mean velocity of the molecule is, therefore

$$\upsilon_S = \delta\left(C_+ - C_-\right) = \delta C \sinh\left(F\delta/2kT\right). \qquad [3]$$

When $F\delta \ll 2kT$, $\sinh(F\delta/2kT)$ can be expanded in series. Retaining the main term one obtains Einstein's equation

$$\upsilon_S = \left(\delta^2 C / 2kT\right) F \equiv qF, \tag{4a}$$

where q is the mobility of the molecule. Since $\delta^2 C/2 = D$, where D is the surface diffusion coefficient, [4a] becomes

$$\upsilon_S = -\left(D / n_L kT\right)\left(\partial\mu / \partial x\right) \tag{4b}$$

In equilibrium conditions μ is the same everywhere and υ_S must be zero. In non-equilibrium conditions, on the other hand,

$$\frac{\partial\mu}{\partial x} = \frac{\partial\mu_0}{\partial p}\frac{\partial p}{\partial x} + \frac{\partial\phi_0}{\partial x} = \frac{\partial\left(p+\phi_0\right)}{\partial x}, \quad z=0. \tag{5}$$

where p is the pressure, x is the distance along the solid surface and z is the distance from the solid surface. Consequently

$$\upsilon_S = -\frac{D}{kTn_L}\frac{\partial\left(p+\phi_0\right)}{\partial x}, \quad z=0. \tag{6}$$

We recognize the possibility of a nonuniform density distribution near the leading edge which may contribute to the gradient of the chemical potential and thus to the force F. It is difficult, for the time being, to evaluate its importance.

We now use the equations of motion to help us to obtain an equation for $p+\phi_0$ at $z=0$, The forces which act upon an element of liquid in a thin layer differ from those in a bulk fluid when the range of intermolecular forces is larger than the thickness of the film. The difference in the forces acting in the thin layer and in the bulk liquid is accounted for in the equations of motion by a body force. If ϕ is the potential energy function per unit volume accounting for the difference in the behavior of a thin liquid film and a bulk liquid, then the body force in the equations of motion has the components $-\partial\phi/\partial x$ and $-\partial\phi/\partial z$. Very near the leading edge the flow is nearly one-dimensional and inertial effects can be neglected. With some reasonable approximations the equations of motion become (7)

$$0 = -\frac{\partial P}{\partial x} + \eta\frac{\partial^2 \upsilon_x}{\partial z^2}, \tag{7}$$

$$0 = -\frac{\partial p}{\partial z} - \frac{\partial\phi}{\partial z} \equiv -\frac{\partial P}{\partial z}, \tag{8}$$

$$P = p+\phi, \tag{9}$$

where η is the viscosity of the liquid and υ_x is the velocity component along the x direction.

Equation [8] shows that P is independent of the distance z. This means that the value of $p+\phi$ for $z=0$ is equal to that of $p+\phi$ at $z=h$, where h is the thickness of the film. In terms of London forces the value of ϕ at the free interface has been found to be (5)

$$\phi_h = \frac{A_{SL}\cos^3\alpha}{12\pi x^3}\left[\frac{A_{LL}-A_{SL}}{A_{SL}}G(\alpha) - G(\pi-\alpha)\right], \tag{10}$$

$$G(\alpha) = \csc^3\alpha + \cot^3\alpha + \frac{3}{2}\cot\alpha \quad [11]$$

A_{LL} and A_{SL} are Hamaker's constants for the liquid–liquid and liquid–solid interactions related to the molecular densities in the liquid n_L and in the solid n_S and to the coefficients β_{LL} and β_{SL} in the London interaction potential between two molecules via $A_{LL} = \pi^2 n_L^{\,2} \beta_{LL}$ and $A_{SL} = \pi^2 n_S n_L \beta_{SL}$. Because the profile of the drop near the contact line was assumed linear, no curvature effect has to be considered and hence, the pressure for $z = h$ is the gas pressure p_0. Therefore

$$(\phi + p)_{z=0} = \phi_h + p_0. \quad [12]$$

Substituting Eqs. [10] and [12] in Eq. [6] leads to the following expression for the slip velocity:

$$v_s = \frac{D}{n_L kT}\frac{A_{SL}\cos^3\alpha}{4\pi x^4}\left[\frac{A_{LL} - A_{SL}}{A_{SL}} G(\alpha) - G(\pi - \alpha)\right]. \quad [13]$$

Of course the smallest value of x, denoted by x_0, has to correspond to a film thickness equal to the diameter a_{LL} of a molecule of liquid, hence

$$x_0 = a_{LL}/\tan\alpha \quad [14]$$

Strictly speaking the extrapolation of Eq. [13] to distances of molecular dimensions is not entirely valid because it was established on the basis of a quasi-continuum approach. This extrapolation is carried out in the same spirit in which the Stokes equation established for large bodies is used to relate the velocity of an ion in solution to the viscosity of the solution.

It is convenient to displace the origin and to rewrite Eq. [13] as

$$v_s = \frac{D}{n_L kT}\frac{A_{SL}\cos^3\alpha}{4\pi (x' + x_0)^4}\left[\frac{A_{LL} - A_{SL}}{A_{SL}} G(\alpha) - G(\pi - \alpha)\right] \quad [15]$$

This time the leading edge corresponds to $x' = 0$. At the equilibrium contact angle α_e the slip velocity must be zero because (5)

$$\left[(A_{LL} - A_{SL})/A_{SL}\right] G(\alpha_e) - G(\pi - \alpha_e) = 0 \quad [16]$$

In Fig. 2 we plot the slip velocity as a function of distance from the leading edge and contact angle for one set of values of the physical constants. The equilibrium angle was calculated from Eq. [16]. When $\alpha < \alpha_e$ the slip velocity is positive, the liquid moves away from the contact line and gas replaces liquid in contact with the solid. However, when $\alpha > \alpha_e$, the slip velocity is negative and liquid moves towards the contact line displacing gas in contact with the solid. Figure 2 shows that the slip velocities for receding or advancing contact lines are very different, a fact in agreement with previous observation (6).

The main conclusion is that the slip velocity decays as $(x' + x_0)^{-4}$ and that the rate of decay depends upon the wetting angle. Solutions predicting the rate of spreading based upon the present treatment of the slip velocity and of equations of motion (accounting also for the true shape of the drop near the leading edge) will be published elsewhere. While the present treatment is based on an idealized representation of the interaction forces and of the solid surface (ignoring for instance the surface roughness), it reveals, however, the basic mechanism by which intermolecular forces generate the movement of the contact line.

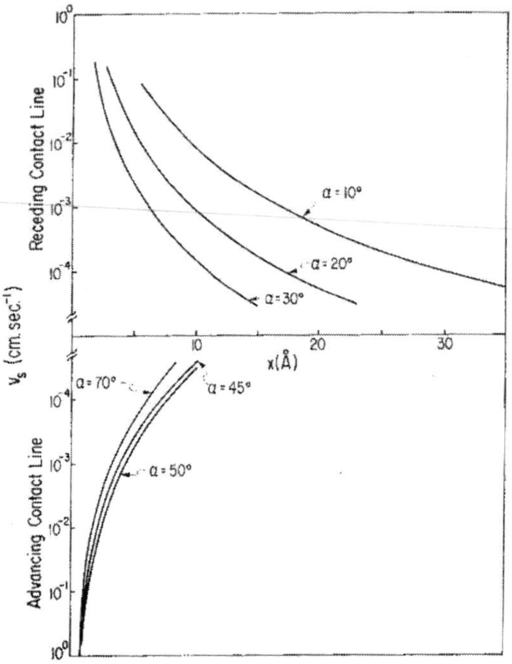

FIG. 2. Slip velocity as a function of distance from the leading edge. The values of the physical parameters are: $D = 10^{-8}\,cm^2\,sec^{-1}$, $n_L = 3.34 \times 10^{22}$ molecules cm^{-3}, $T = 300°K$, $a_{LL} = 10^{-8}cm$, $A_{SL} = 10^{-13}$ erg, $A_{LL} = 1.039 \times 10^{-13}$ erg, $\alpha_e = 40°$. (The slip velocity is negative for advancing contact lines.)

REFERENCES

1. DUSSAN, E. B., AND DAVIS, S. H., *J. Fluid Mech.* **65**, 71 (1974).
2. DUSSAN, E. B., Personal information, 1975.
3. RUCKENSTEIN, E., AND LEE, P. S., *Surface Sei.* **52**, 298 (1975) (Section 3.1 of this volume).
4. FRENKEL, J., "Kinetic Theory of Liquids." Oxford Univ. Press, London/New York, 1946.
5. MILLER, C. A., AND RUCKENSTEIN, E., *J. Colloid Interface Sei.* **48**, 368 (1974) (Section 1.1 of this volume).
6. SCHWARTZ, A. M., RAIDER, C. A., AND HUEY, E., *Advan. Chetn. Ser.* **43**, 250 (1964).
7. LOPEZ, J., MILLER, C. A., AND RUCKENSTEIN, E., *J. Colloid Interface Sei.* **56**, 460 (1976) (Section 1.3 of this volume).

1.6 On the No-Slip Boundary Condition of Hydrodynamics*

Eli Ruckenstein and P. Rajora

Department of Chemical Engineering, State University of New York at Buffalo, Amherst, New York 14260

Received March 9, 1983; accepted June 21, 1983

INTRODUCTION

A slip velocity was observed experimentally in flow through capillaries of small diameters, with walls made repellent to the liquid. More than a century ago, Helmholtz (1) found experimental evidence of slip for a liquid flowing over a solid surface. Later, Schnell (2) found experimental evidence of slip of water in a glass capillary pretreated with the vapor of dimethyldichlorosilane to make it water repellent. Tolstoi (3) and Somov (4) have also measured a slip velocity for the flow of mercury in glass capillary. The question which arises now is that whether the molecules of liquid slip directly on the solid wall or there is a gap between the two interfaces which is the cause of slip. An answer to this question is sought in this paper.

It is argued that the gradient of the chemical potential caused by the pressure drop gives rise to a velocity along the surface of the solid. The magnitude of this velocity will depend on the interaction between the solid and liquid and the applied pressure gradient. For example, a hydrophilic surface provides sites for strong adsorption of water molecule which inhibits their surface movement. On the other hand, a hydrophobic surface repels water molecules, thus enhancing the movement along the surface. It is clear that the slip is larger in the latter than in the former case. To establish whether the experimentally observed slip occurs directly on the solid surface, a model based on the slip over a solid surface is developed and compared with experiment. Agreement between the experimentally observed slip velocity and that predicted by the model would make plausible that the slip occurs directly over the solid surface. However, a disagreement would mean that the mechanism assumed in the model is not valid and an alternative mechanism has to be found to explain the experiment.

A slip velocity is also found to occur in the flow of polymer solutions. In this case, the polymer species dissolved in the fluid diffuses away from the solid surface where the stress is higher, to the center, where the stress is lower, leading to the creation of a thin layer of solvent over which the bulk appears to slide. The analysis for such systems is beyond the scope of the present work and some suitable references are (5–7).

SLIP DIRECTLY ON THE SOLID SURFACE

The slip velocity is caused by the gradient of the chemical potential. Consider a liquid flowing in the x-direction in a circular tube of radius R under a constant pressure gradient $\Delta p/L$. As an idealization the surface of the tube is assumed to be perfectly homogenous and smooth. The gradient of the

* *Journal of Colloid and Interface Science*. Vol. 96, No. 2, p. 488, December 1983. Republished with permission.

chemical potential μ at the solid surface, which can be taken equal to that in the bulk and which is defined per molecule, then gives rise along the interface to a force:

$$F = -\frac{d\mu}{dx}. \quad [1]$$

This force is the cause of slip. Note that the temperature gradient generates an additional force which is neglected here (8). In general, the gradient of the chemical potential is given by

$$\frac{d\mu}{dx} = v\frac{dp}{dx} - s\frac{dT}{dx}, \quad [2]$$

where v is the volume of a molecule, dp/dx is the pressure gradient, s is the entropy per molecule, and dT/dx is the temperature gradient. However, since all the experimental observations on slip have been carried out under isothermal conditions, Eq. [2] can be simplified to

$$\frac{d\mu}{dx} = v\frac{dp}{dx}, \quad [3]$$

which, when combined with Eq. [1], further reduces, for a constant pressure gradient, to

$$F = \frac{1}{n_L}\frac{\Delta p}{L}, \quad [4]$$

where n_L is the number of molecules per unit volume. The mechanism by which liquid molecules move on the surface of the solid can be developed along the lines similar to those used by Ruckenstein and Dunn (9) for calculating the slip velocity during the spreading of drops on solids. This treatment is similar to that used in the theory of electrical conductivity in ionic crystals. Here, the liquid molecules are modeled as moving parallel to the x-direction, by jumping from one potential well to another, over a potential barrier of height E. For barriers of uniform height E, the probability of a molecule jumping in unit time from a given well into another of the two neighboring ones is given by:

$$\alpha = v_0 e^{-E/kT}, \quad [5]$$

where v_0 is the vibrational frequency of the molecule in the well, k is the Boltzmann constant, and T is the absolute temperature. Now, if an external force of constant magnitude F acts upon the molecule in the positive direction of the x-axis, there results a decrease of the potential barrier by ½ $F\delta$ in the positive direction, and an increase by the same amount in the negative direction, δ being the distance between two neighboring wells. Hence, the probability α_+ that the particle will jump from the initial well to the next on the positive side is given by:

$$\alpha_+ = \tfrac{1}{2} v_0 e^{-(E-(F\delta/2))/kT}. \quad [6]$$

Similarly, the probability that the particle will jump from the initial well to the previous one on the negative side results from:

$$\alpha_- = \tfrac{1}{2} v_0 e^{-(E+(F\delta/2))/kT} \quad [7]$$

The slip velocity is given by the difference:

$$V_s = \delta(\alpha_+ - \alpha_-). \quad [8]$$

Combining Eqs. [6] to [8], one obtains

$$V_s = \delta\alpha \sinh\left(\frac{F\delta}{2kT}\right). \quad [9]$$

When $1/2F\delta \gg kT$

$$V_s = \frac{\delta\alpha}{2} e^{(F\delta/2kT)} \quad [10a]$$

and if $1/2F\delta \ll kT$, one obtains Einstein's equation:

$$V_s = \frac{\delta^2\alpha}{2kT} F = qF, \quad [10b]$$

where q is the mobility of the molecule. In our case $1/2F\delta$ is less than 10^{-23} erg/molecule and $kT \approx 10^{-14}$ erg/molecule. Consequently, the inequality $1/2F\delta \ll kT$ is satisfied and Eq. [10b] provides the relation between the slip velocity and the force. Denoting $D_s = \delta^2\alpha/2$, where D_s is the surface diffusion coefficient, we get

$$q = \frac{D_S}{kT}. \quad [11]$$

Combining Eqs. [4], [10b], and [11], one obtains

$$V_s = \frac{D_s}{kTn_L}\frac{\Delta p}{L}. \quad [12]$$

Equation [12] provides an expression for the slip velocity in terms of the surface diffusion coefficient and the pressure gradient.

For the flow in a cylindrical capillary, one easily obtains for the average velocity the expression:

$$V_m = \frac{R^2}{8\eta}\frac{\Delta p}{L} + \frac{D_s}{kTn_L}\frac{\Delta p}{L}, \quad [13]$$

where R is the radius of the tube and η is the dynamic viscosity of the fluid. In terms of the volumetric flow rate Q, one can write

$$Q = Q_0\left[1 + \frac{8\eta D_s}{kTR^2 n_L}\right], \quad [14]$$

where $Q_0 = \frac{\pi R^4}{8\eta}\frac{p}{L}$ is the volumetric flow rate for zero slip. A similar expression for the slip when there is also a temperature gradient is

$$Q = Q_0\left[1+\frac{8\eta D_s}{R^2kT}\left(\frac{1}{n_L}+\frac{sdT/dx}{p/L}\right)\right]. \qquad [15]$$

It is interesting to note that Eq. [15] shows that for water at an average temperature of 27°C a temperature gradient of 4.45°C/cm can produce as much slip as would be produced by a pressure drop of 1 atm/cm. The value of 6.73×10^{-18} erg/molecule °K was used for the entropy of water at an average temperature of 27°C. While the slip produced by the pressure gradient is always in the direction of the flow, that produced by the temperature gradient can be in the same or in the opposite direction of flow, depending on its sign.

DISCUSSION

The interactions of the solid surface with the flowing liquid determines the value of the surface diffusion coefficient D_s. When these interactions are strong, as when a hydrophobic (hydrophilic) surface interacts with a hydrophobic (hydrophilic) liquid, D_s is very small. On the other hand, when the nature of the solid surface and liquid are different, D_s is larger. In order to get an estimate of the surface diffusivity, the experimental data of Tolstoi (3) and Somov (4), who have studied the slip of mercury in a glass capillary, are used. Tolstoi's (3) data for a capillary radius of 13.3 μm gives $D_s = 1580$ cm²/sec, whereas for a radius of 3.48 μm. $D_s = 450$ cm²/sec. Somov's (4) data also lead to a value of D_s of the order of 100 cm²/sec for a capillary radius of about 1 μm. These values of D_s do not agree with the observed values of the surface diffusion coefficients which are much less than 10^{-5} cm²/sec for liquids and much less than 1 cm²/sec for gases. The large values of D_s suggest that there exists a kind of free motion of the liquid molecules on the solid-liquid interface. Such a free motion is indeed possible if there exists a gap between the liquid and solid surfaces. In addition, the dependence of D_s on the radius of the capillary contradicts the fact that the surface diffusion coefficient is a characteristic of the liquid-solid interface. This means that the slip velocity inside the capillary is governed by a process which is dependent on the size of the capillary, such as the hydrodynamic process itself. This is possible if there exists a gap between the liquid and solid surfaces. Such a gap can be generated when the liquid and solid have different natures (one of them hydrophobic and the other hydrophilic). No gap would, however, form when the liquid and solid are alike in nature. The motion of the liquid molecules over this gap does not occur by hopping from one site to another as was assumed in Eq. [14], We also note that the gas which is used to create the pressure drop across the capillaries in these experiments, can come to the gap, either because it is entrained by the flow of the liquid and/or because of the desorption of some of the soluble gas. This may lead to an increase in the thickness of the gap.

In contrast, however, one may note, that surface diffusion can explain the motion of a contact line on a smooth surface. Indeed, a value of $D_s = 10^{-8}$ cm²/sec was sufficient to evaluate the right order of magnitude of the velocity of the contact line (9).

CONCLUSION

The origin of the slip velocity measured experimentally in flow through capillaries of sufficiently small diameter, with walls made repellent to the flowing liquid is examined. For a slip occurring directly on the solid surface, the slip velocity is given by the product of the gradient of the chemical potential and the mobility coefficient. However, comparison with the experiment gives very high values for the surface diffusion coefficient and this suggests that slip occurs over a gap rather than directly on the solid surface. This gap forms when a hydrophobic liquid flows over a hydrophilic surface and vice versa. The thickness of the gap may be increased by the entrained and soluble gases used to create the pressure drop across the capillary.

REFERENCES

1. Helmholtz, H., and Pictowsky, G., *Sitzber Wien (2a)* **40**, 607 (1868).
2. Schnell, E., *J. Appl. Phys.* **27**, 1149 (1956).
3. Tolstoi, D. M., *Dokl. Akad. Nauk SSSR* **85**, 1329 (1952).
4. Somov, A. N., *Kolloidn. Zh.* **44**, 160 (1982).
5. Tirrel, M., and Malone, M. F., *J. Polym. Sci.-Polym. Phys.* **15**, 1569 (1977).
6. Cohen, Y., and Metzner, A. B., "Interfacial Phenom ena in Enhanced Oil Recovery," p. 77. AIChE Symposium Series, 1982.
7. Mashelkar, R., and Dutta, A., *Chem. Eng. Sei.* **37**, 969 (1982).
8. Landau, L. D., and Lifshitz, E. M., "Fluid Mechanics," Pergamon Press, London, 1959, p. 224.
9. Ruckenstein, E., and Dunn, C. S., *J. Colloid Interface Sei.* **59**, 135 (1977) (Section 1.5 of this volume).

1.7 Dynamics of Partial Wetting*

Eli Ruckenstein
Department of Chemical Engineering, State University of New York at Buffalo, Buffalo, New York 14260-4200

Received June 29, 1992

In a recent paper,[1] Brochard-Wyart and de Gennes divide in two classes the models that have been proposed to describe the motion of a contact line when the wetting angle θ deviates from its equilibrium value θ_e: (i) the Eyring approach, employed by Blake and Haynes,[2] which involves the microscopic jump of a single molecule at the tip, and (ii) de Gennes' hydrodynamic approach[3] that emphasizes the viscous losses inside an assumed liquid wedge of angle θ. In 1977, Ruckenstein and Dunn[4] derived an equation for the slip velocity during wetting of solids, and the goal of the present paper is to compare their approach to those in refs 2 and 3, and particularly to show that their equation can be reduced for small values of θ to that derived by de Gennes. In essence, in ref 4 it was shown that near the contact line a net translational velocity parallel to the solid is generated by a force caused by the gradient of the chemical potential in the liquid along the solid–liquid interface. In order to understand the origin of this gradient of the chemical potential, let us focus on a molecule located in the liquid on the solid–liquid interface and consider first that the range of interaction forces between this molecule and other molecules of the liquid is smaller than the distance to any of the points on the liquid–gas interface. In this case, the molecule does not feel the presence of the liquid–gas interface and its chemical potential can be equated to that in a semi-infinite liquid. If, however, a molecule on the solid–liquid interface is selected for which the range of interaction forces is larger, the molecule at the chosen point is additionally subject to a field that is generated because the number of molecules with which it interacts at its left differs from those with which it interacts at its right. This field, equal to the difference between the real interaction potential and that for a semi-infinite liquid, decays to zero for large enough distances from the leading edge, but becomes increasingly important when the leading edge is approached because liquid molecules are replaced by gas molecules within the range of the interaction forces. Consequently, near the leading edge along the solid surface there is a nonuniform region in which the interaction potential of a molecule located on the solid surface with all the other molecules varies with the distance to the leading edge.

The chemical potential per unit volume can be written in the form

$$\mu = \mu_0 + \phi_0$$

where μ_0 is the chemical potential in a point of the solid surface corresponding to a semifinite layer of liquid, defined per unit volume, and ϕ_0 is the additional interaction potential, also defined per unit volume. The gradient of the chemical potential along the interface generates a force F per molecule

$$F = -\frac{1}{n_L}\frac{\partial \mu}{\partial x} \tag{1}$$

* *Langmuir* 1992, 8, 3038–3039. Republished with permission.

where n_L is the concentration in the liquid, in molecules per unit volume, and x the distance along the liquid–solid interface measured from the leading edge. For the slip velocity along the solid–liquid interface, one can use Einstein's equation

$$\upsilon_s = qF$$

where $q = D/kT$ is the mobility of the molecule, D is the surface diffusion coefficient, k is the Boltzmann constant, and T is the temperature (K). Consequently

$$\upsilon_s = -\frac{D}{kTn_L}\frac{\partial \mu}{\partial x} \tag{2}$$

A more complex equation can also be derived,[4] namely

$$\upsilon_s = \delta C \sinh\left(F\delta/2kT\right) \tag{3}$$

where δ is the distance between two potential wells located on the solid surface over which a molecule jumps and $C = \upsilon_0 e^{-E/kT}$, ν_0 being the vibration frequency of a molecule in the potential well and E an activation energy. For $F\delta/2kT \ll 1$, eq 3 reduces to eq 2.

The main difference between our treatment and that of Blake and Haynes consists in the fact that eq 2 or 3 provides a velocity which is a function of x and not only the velocity caused by the microscopic jump of a single molecule at the tip. Compared to the treatment of de Gennes, eq 2 provides an explanation for the origin of the slip velocity.

In equilibrium conditions μ is independent of x and $\upsilon_s = 0$. In nonequilibrium conditions

$$\frac{\partial \mu}{\partial x} = \frac{\partial \mu_0}{\partial p}\frac{\partial p}{\partial x} + \frac{\partial \phi_0}{\partial x} = \frac{\partial\left(p+\phi_0\right)}{\partial x}$$

$$z = 0 \tag{4}$$

where p is the pressure and z is the distance to the plate. Consequently

$$\upsilon_s = -\frac{D}{kTn_L}\frac{\partial\left(p+\phi_0\right)}{\partial x}$$

$$z = 0 \tag{5}$$

In conventional fluid mechanics, the nonslip condition at the solid surface is employed as a boundary condition. It is clear that such a condition is violated by the spreading of a droplet on a solid surface. In such a case the nonslip boundary condition must be replaced, for reasons explained in the previous sections, by eq 5. A complete formulation of the problem of wetting must involve the equations of motion which, together with the boundary condition (eq 5), can provide the complete velocity field. In what follows a very simplified version of the equations of motion will be used, namely, the equations for a wedge flow of angle θ. The liquid layer in this wedge is considered thin. The forces which act upon an element of liquid in a thin layer differ from those in a bulk liquid because the range of intermolecular forces is larger than the thickness of the film. The difference in the forces acting in the thin film and in the bulk can be accounted for in the equations of motion by a body force. If ϕ is the potential energy per unit volume that accounts for the difference in the behavior of a thin liquid film and a bulk liquid, then the body force has the components $-\partial\phi/\partial x$ and $-\partial\phi/\partial z$. Near the leading edge, the flow is nearly one dimensional and inertial effects can be neglected. Consequently, one can write

$$0 = -\frac{\partial p}{\partial x} - \frac{\partial \phi}{\partial x} + \eta \frac{\partial^2 \upsilon_x}{\partial z^2} \tag{6}$$

and

$$0 = -\frac{\partial p}{\partial z} - \frac{\partial \phi}{\partial z} \tag{7}$$

where η is the viscosity of the liquid and υ_x is the velocity component along the x direction. Equation 7 shows that $p + \phi = P$ is constant, and hence that the value of $p + \phi$ for $z = 0$ is equal to the value of $p + \phi$ at $z = h$, where h is the thickness of the film. Assuming London interactions, an expression for ϕ for a wedge was derived[5] and the following result obtained:

$$\phi(h) = \frac{A_{\mathrm{SL}} \cos^3\theta}{12\pi x^3}\left[\frac{A_{\mathrm{LL}} - A_{\mathrm{SL}}}{A_{\mathrm{SL}}} G(\theta) - G(\pi - \theta)\right] \tag{8}$$

where

$$G(\theta) = \csc^3\theta + \cot^3\theta + (3/2)\cot\theta \tag{9}$$

A_{LL} and A_{SL} are the Hamaker constants for the liquid–liquid and liquid–solid interactions. Since the profile of the droplet near the contact line was approximated by a wedge, the pressure p for $z = h$ can be replaced by the gas pressure p_0. Consequently

$$\upsilon_{\mathrm{s}} = -\frac{D}{kTn_{\mathrm{L}}} \frac{\partial \phi(h)}{\partial x} \equiv \frac{D}{kTn_{\mathrm{L}}} \frac{\cos^3\theta}{4\pi x^4}\left[(A_{\mathrm{LL}} - A_{\mathrm{SL}})G(\theta) - A_{\mathrm{SL}}G(\pi - \theta)\right] \tag{10}$$

Obviously the smallest value of x, denoted x_0, should correspond to a film thickness equal to the diameter a of a molecule of liquid. Consequently, $x_0 = a/\tan\theta$, and the velocity at the tip is given by the expression

$$\upsilon_{\mathrm{s}} = \frac{D}{n_{\mathrm{L}}kT} \frac{(\cos^3\theta) tg^4\theta}{4\pi a^4}\left[(A_{\mathrm{LL}} - A_{\mathrm{SL}})G(\theta) - A_{\mathrm{SL}}G(\pi - \theta)\right] \tag{11}$$

At equilibrium, $\upsilon_{\mathrm{s}} = 0$ and

$$\frac{(A_{\mathrm{LL}} - A_{\mathrm{SL}})}{A_{\mathrm{SL}}} = \frac{G(\pi - \theta_e)}{G(\theta_e)} \tag{12}$$

where θ_{e} is the equilibrium angle.

Because eq 12 was derived for a wedge and not for a droplet which has a curvature, it is clear that eq 12 is an approximate one. A more rigorous thermodynamic equation valid in this case is

$$2A_{\mathrm{SL}}/A_{\mathrm{LL}} = 1 + \cos\theta_{\mathrm{e}} \tag{13}$$

On the basis of eq 13 one can improve eq 11 by replacing $(A_{\mathrm{LL}} - A_{\mathrm{SL}})/A_{\mathrm{SL}}$ by $2/(1+\cos\theta_{\mathrm{e}})-1$ and $G(\pi-\theta)/G(\theta)$ by $2/(1+\cos\theta)-1$. One thus obtains

$$\upsilon_{\mathrm{s}} = \frac{D}{n_{\mathrm{L}}kT} \frac{(\cos^3\theta) tg^4\theta}{4\pi a^4} A_{\mathrm{SL}} G(\theta)\left(\frac{2}{1+\cos\theta_{\mathrm{e}}} - \frac{2}{1+\cos\theta}\right) \tag{14}$$

For small values of θ

$$G(\theta) = 2\theta^{-3}$$

$$\sin\theta = \theta$$

$$\cos\theta = 1 - \theta^2/2$$

and eq 14 becomes

$$\upsilon_s = \frac{D}{n_L kT}\frac{A_{SL}}{8\pi a^4}\theta\left(\theta_e^{\,2} - \theta^2\right) \tag{15}$$

In this equation, the slip velocity is negative for an advancing contact line and positive for a receding contact line. Equation 15 coincides with that derived by de Gennes with the difference that the proportionality constant has a more clear physical meaning and the mechanism proposed explains the origin of the slip velocity. Because θ_e is small, $A_{SL} \cong A_{LL} \propto \gamma$, where γ is the surface tension of the liquid. On the other hand, D is expected to be inversely proportional to the viscosity of the liquid. Consequently

$$\upsilon_s \propto (\gamma/\eta)\theta\left(\theta_e^{\,2} - \theta^2\right) \tag{16}$$

and the slip velocity is directly proportional to the surface tension γ and inversely proportional to the liquid viscosity. Equation 16 involves the slip velocity caused by the chemical potential gradient and achieved via surface diffusion as well as the velocity field near the leading edge. (The fact that one of the equations of motion, eq 6, was not explicitly used is due to the wedge flow approximation employed.)

The two approaches lead to the same result because the velocity at the tip of the leading edge should be equal to the average velocity in the fluid near the leading edge. As noted in ref 4, the slip velocity at the solid–liquid interface decays to zero after a very short distance from the leading edge, about 10 Å or so. Consequently, the average velocity near the leading edge can be calculated neglecting the slip. It is surprising that the equations of Blake and Haynes, and of de Gennes, do not lead to the same result, because the velocity at the tip should be equal to the average velocity in the hydrodynamic layer in its neighborhood. A possible explanation for this discrepancy may be related to the selection of the driving force. It is plausible to consider the spreading (expressed in dynes per centimeter) as the integral of the shear stress, and this is the approach used by de Gennes, but it is less plausible to multiply the spreading with the square of the distance δ between two potential wells to obtain an energy to be included in the activation energy. In the latter case, the force should be related to the gradient of the chemical potential.

REFERENCES

(1) Brochard-Wyart, F.; de Gennes, P. G. *Adv. Colloid Interface Sci.* 1992, *39*, 1.
(2) Blake, T. D.; Haynes, J. M. *J. Colloid Interface Sci.* 1969, *30*, 421.
(3) de Gennes, P. G. *Kolloid Polym. Sci.* 1986, *264*, 463.
(4) Ruckenstein, E.; Dunn, C. S. *J. Colloid Interface Sci.* 1977, *59*, 135 (Section 1.5 of this volume).
(5) Miller, C. A.; Ruckenstein, E. *J. Colloid Interface Sci.* 1974, *48*, 368 (Section 1.1 of this volume).

1.8 Slip Velocity during the Flow of a Liquid over a Solid Surface*

Eli Ruckenstein[a]
Department of Chemical and Biological Engineering, State University of New York at Buffalo, Buffalo, NY 14260-4200, USA
Corresponding Author
[a]E-mail: feaeliru@buffalo.edu

Received 02 June 2011/Received in final form 15 June 2011 Published online 30 August 2011

1 INTRODUCTION

The goal of this review paper is to summarize the results obtained by Ruckenstein and coworkers regarding the slip velocity during wetting of solids [1–3]. It is obvious that the contact line between a nonequilibrated drop and a solid surface moves over the solid surface and that the no slip boundary condition usually employed in fluid mechanics is violated. It is therefore not surprising that the models of the motion of the contact line involving the no slip boundary condition predict unbounded stresses and viscous dissipation at the contact line [4]. In reality, near the contact line, there is a translational motion of liquid molecules induced by a chemical potential gradient along the solid liquid interface. This gradient generates a force which combined with Einstein's equation for the mobility of molecules provides the slip velocity.

A slip velocity also was observed during flow through capillaries of small diameters. In 1868, Helmholtz and Pictowsky [5] found experimental evidence of slip in such a case. Later, Schnell [6] provided evidence of slip of water in a glass capillary pretreated with dimethyldichlorosilane vapor to make it water repellent, and Tolstoi [7] and Somov [8] have measured the slip velocity for the flow of mercury in glass capillary. A possible explanation is that the gradient of the chemical potential due to the pressure drop gives rise to the slip velocity. The magnitude of this velocity is expected to depend on the interaction between the solid and liquid as well as on the applied pressure gradient. A hydrophilic surface provides sites for strong adsorption of water molecule which delay their surface movement. On the other hand, a hydrophobic surface repels the water molecules, thus enhancing the movement along the surface. To establish whether the observed slip occurs directly on the solid surface or is an apparent one, a model based on slip over a solid surface is developed and compared with experiment.

Two issues are examined in this review: – 1 the slip velocity observed experimentally during flow through capillaries; – 2 the slip velocity during partial wetting of solids by nonequilibrated drops.

* *Eur. Phys. J. Special Topics*. 197, 203–209 (2011). Republished with permission.

1.1 Slip Directly on the Solid Surface

Consider a liquid flowing in the x-direction in a circular tube of radius R under a constant pressure gradient $\Delta p/L$. The gradient of the chemical potential μ at the solid surface, which is defined per unit volume, gives rise along the interface to a force per molecule given by

$$F = -\frac{1}{n_L}\frac{d\mu}{dx}, \tag{1}$$

where n_L is the number of molecules per unit volume. This force is the cause of slip. In general, the gradient of the chemical potential is given by

$$\frac{d\mu}{dx} = v\frac{dp}{dx} - s\frac{dT}{dx}, \tag{2}$$

where v is the volume of n_L molecules (which is unity), $\frac{dp}{dx}$ is the pressure gradient, s is the entropy of the n_L molecules, and $\frac{dT}{dx}$ is the temperature gradient. However, since all the experimental observations on slip have been carried out under isothermal conditions, Eq. (2) becomes

$$\frac{d\mu}{dx} = \frac{dp}{dx}, \tag{3}$$

which, when conbined with Eq. (1) leads to

$$F = \frac{1}{n_L}\frac{\Delta p}{L}. \tag{4}$$

The liquid molecules are considered to move parallel to the x-direction, by jumping from one potential well to another, over potential barriers of height E. For barriers of uniform height E, the probability of a molecule jumping in unit time from a given well into another of the two neighboring ones in given by:

$$\alpha = v_0 e^{-E/kT}, \tag{5}$$

where v_0 is the vibrational frequency of the molecule in the well, k is the Boltzmann constant, and T is the absolute temperature. If an external force of constant magnitude F acts upon the molecule in the positive direction of the x-axis, there results a decrease of the potential barrier by $F\delta/2$ in the positive direction, and an increase by the same amount in the negative direction, δ being the distance between two neighboring wells. Hence, the probability α_+ that the particle will jump from the initial well to the next on the positive side is given by:

$$\alpha_+ = \frac{1}{2}v_0 e^{-[E-(F\delta/2)]/kT}. \tag{6}$$

Similarly, the probability that the particle will jump from the initial well to the neighboring one on the negative side results from:

$$\alpha_- = \frac{1}{2}v_0 e^{-[E+(F\delta/2)]/kT}. \tag{7}$$

The slip velocity is given by the difference:

$$V_s = \delta(\alpha_+ - \alpha_-). \tag{8}$$

Combining Eqs. (6) to (8), one obtains

$$V_s = \delta\alpha \sinh\left(\frac{F\delta}{2kT}\right), \tag{9}$$

and if $F\delta/2 \ll kT$, one obtains Einstein's equation:

$$V_s = \frac{\delta^2\alpha}{2kT}F = qF, \tag{10}$$

where q is the mobility of the molecule. In our case $F\delta/2$ is less than 10^{-23} erg/molecule and $kT \approx 10^{-14}$ erg/molecule. Consequently, the inequality $F\delta/2 \ll kT$ is satisfied and Eq. (10) provides the relation between the slip velocity and the force. Denoting $D_s = \delta^2\alpha/2$, where D_s is the surface diffusion coefficient, we get

$$q = \frac{D_s}{kT}. \tag{11}$$

Combining Eqs. (4), (10), and (11), one obtains

$$V_s = \frac{D_s}{kTn_L}\frac{\Delta p}{L}. \tag{12}$$

Equation (12) provides an expression for the slip velocity in terms of the surface diffusion coefficient and the pressure gradient.

For the flow in a cylindrical capillary, one easily obtains for the average velocity the expression:

$$V_m = \frac{R^2}{8\eta}\frac{\Delta p}{L} + \frac{D_s}{kTn_L}\frac{\Delta p}{L}, \tag{13}$$

where η is the dynamic viscosity of the fluid. In terms of the volumetric flow rate Q, one can write

$$Q = Q_0\left(1 + \frac{8\eta D_s}{kTR^2 n_L}\right), \tag{14}$$

where $Q_0 = \frac{\pi R^4}{8\eta}\frac{\Delta p}{L}$ is the volumetric flow rate for zero slip. A similar expression for the slip when there is also a temperature gradient is

$$Q = Q_0\left[1 + \frac{8\eta D_s}{kTR^2}\left(\frac{1}{n_L} + \frac{s dT/dx}{\Delta p/L}\right)\right]. \tag{15}$$

It is of interest to note that Eq. (15) shows that for water at an average temperature of 27 °C a temperature gradient of 4.45 °C/cm can produce as much slip as would be produced by a pressure drop of 1 atm/cm. The value of 6.73 × 10^{-18} erg/molecule °K was used for the entropy of water at an average temperature of 27 °C.

The interactions of the solid surface with the flowing liquid determines the value of the surface diffusion coefficient D_s. When these interactions are strong, as when a hydrophobic (hydrophilic) surface interacts with a hydrophobic (hydrophilic) liquid, D_s is very small. On the other hand, when the natures of the solid surface and liquid are different, D_s is expected to be larger. In order to estimate the surface diffusivity, the experimental data of Tolstoi [3] and Somov [4], who have studied the slip of mercury in a glass capillary, were used. Tolstoi's [3] data for a capillary radius of 13.3 μm provides $D_s = 1580$ cm^2/sec, whereas for a radius of 3.48 μm, $D_s = 450$ cm^2/sec. Somov's [4] data also lead to a value of D_s of the order of 100 cm^2/sec for a capillary radius of about 1 μm. These values of D_s do not agree with the observed values of the surface diffusion coefficient which are much smaller than 10^{-5} cm^2/sec for liquids and much smaller than 1 cm^2/sec for gases. In addition, the dependence of D_s on the radius of the capillary contradicts the fact that the surface diffusion coefficient is a characteristic of the liquid-solid interface. The large values of D_s suggest that there is a kind of free motion of the liquid molecules over the solid interface. Such a free motion is possible if there is a gap between the liquid and solid surfaces. This means that the "slip velocity" inside the capillary is governed by a process which is dependent on the size of the capillary, such as the hydrodynamic process itself. The motion of the liquid over this gap does not occur by hopping from one site to another as assumed in Eq. (13). The gas which is used to create the pressure drop across the capillaries in these experiments, can create the gap, either because it is entrained by the flow of the liquid and/or because of the desorption of some dissolved gas.

1.2 Partial Wetting of a Solid by a Droplet

In 1977, Ruckenstein and Dunn [1] derived an equation for the slip velocity during wetting of solids. In essence, it was shown that near the contact line a net translational velocity parallel to the solid is generated by a force caused by the gradient of the chemical potential in the liquid along the solid-liquid interface. In order to understand the origin of this gradient of the chemical potential, let us focus on a molecule located in the liquid on the solid-liquid interface and consider first that the range of interaction forces between this molecule and other molecules of the liquid is smaller than the distance to any of the points on the liquid-gas interface. In this case, the molecule does not feel the presence of the liquid-gas interface and its chemical potential can be equated to that in a semi-infinite liquid. If, however, the molecule on the solid-liquid interface has a larger range of interaction, it becomes additionally subject to a field that is generated because the number of molecules with which it interacts at its left differs from those with which it interacts at its right. This field, equal to the difference between the real interaction potential and that for a semi-infinite liquid, decays to zero for large enough distances from the leading edge, but becomes increasingly important when the leading edge is approached because liquid molecules are replaced by gas molecules within the range of the interaction forces. Consequently, near the leading edge along the solid surface there is a nonuniform region in which the interaction potential of a molecule located on the solid surface with all the other molecules varies with the distance x to the leading edge along the surface.

The chemical potential per unit volume can be written in the form

$$\mu = \mu_0 + \phi, \tag{16}$$

where μ_0 is the chemical potential in a point of the solid surface corresponding to a semifinite layer of liquid, defined per unit volume, and ϕ is the additional interaction potential, also defined per unit volume.

In equilibrium conditions μ is independent of x and $V_s = 0$ In nonequilibrium conditions

$$\frac{\partial \mu}{\partial x} = \frac{\partial \mu_0}{\partial p}\frac{\partial p}{\partial x} + \frac{\partial \phi}{\partial x} = \frac{\partial (p+\phi)}{\partial x} \qquad z = 0, \tag{17}$$

where p is the pressure and z is the distance to the plate.

Consequently, combining Eqs. (1), (10), (11) and (17), one obtains

$$V_s = -\frac{D_s}{kTn_L}\frac{\partial(p+\phi)}{\partial x} \qquad z = 0. \tag{18}$$

A complete formulation of the problem of wetting must involve the equations of motion which, together with Eq. (18), can provide the complete velocity field. In what follows a very simplified version of the equations of motion will be used, namely, the equations for a wedge flow of angle θ. The liquid layer in this wedge is considered thin. The difference in the forces acting in the thin film and in the bulk can be accounted for in the equations of motion by a body force. If ϕ is the potential energy per unit volume that accounts for the difference in the behavior of a thin liquid film and a bulk liquid, then the body force has the components $\partial\phi/\partial x$ and $\partial\phi/\partial z$. Near the leading edge, the flow is nearly one dimensional and inertial effects can be neglected. Consequently, one can write

$$0 = \frac{\partial p}{\partial x} - \frac{\partial \phi}{\partial x} + \eta\frac{\partial^2 v_x}{\partial z^2}, \tag{19}$$

and

$$0 = -\frac{\partial p}{\partial z} - \frac{\partial \phi}{\partial z}, \tag{20}$$

where v_x is the velocity component along the x direction. Equation (20) shows that $p+\phi = P$ is constant, and hence that the value of $p+\phi$ for $z = 0$ is equal to the value of $p+\phi$, at $z = h$, where h is the thickness of the film. Assuming London interactions, an expression for ϕ for a wedge was derived [9] and the following result obtained:

$$\phi(h) = \frac{A_{SL}\cos^3\theta}{12\pi x^3}\left[\frac{A_{LL-}A_{SL}}{A_{SL}}G(\theta) - G(\pi-\theta)\right], \tag{21}$$

where

$$G(\theta) = \csc^3\theta + \cot^3\theta + (3/2)\cot\theta, \tag{22}$$

And A_{LL} and A_{SL} are the Hamaker constants for the liquid-liquid and liquid-solid interactions. Since the profile of the droplet near the contact line was approximated by a wedge, the pressure p for $z = h$ can be replaced by the gas pressure p_0. Consequently

$$V_s = -\frac{D_s}{kTn_L}\frac{\partial\phi(h)}{\partial x} \equiv \frac{D_s}{kTn_L}\frac{\cos^3\theta}{4\pi x^4}\left[(A_{LL} - A_{SL})G(\theta) - A_{SL}G(\pi-\theta)\right]. \tag{23}$$

Obviously the smallest value of x, denoted x_0, should correspond to a film thickness equal to the diameter a of a molecule of liquid. Consequently, $x_0 = a/\text{tg}\ \theta$, and the velocity at the tip is given by the expression

$$V_s = -\frac{D_s}{kTn_L}\frac{(\cos^3\theta)tg^4\theta}{4\pi a^4}\left[(A_{LL} - A_{SL})G(\theta) - A_{SL}G(\pi-\theta)\right]. \tag{24}$$

At equilibrium, $V_s = 0$ and

$$\frac{A_{LL} - A_{SL}}{A_{SL}} = \frac{G(\pi - \theta_e)}{G(\theta_e)}, \tag{25}$$

where θ_e is the equilibrium angle.

Because Eq. (25) was derived for a wedge and not for a droplet which has a curvature, it is clear that Eq. (25) is an approximate one. A rigorous thermodynamic equation valid in this case is

$$\frac{2A_{SL}}{A_{LL}} = 1 + \cos\theta_e. \tag{26}$$

On the basis of Eq. (26), one can improve Eq. (24) by replacing $\frac{A_{LL}-A_{SL}}{A_{SL}}$ by $2/(1+\cos\theta_e)-1$ and $\frac{G(\pi-\theta)}{G(\theta)}$ by $2/(1+\cos\theta)-1$. One thus obtains

$$V_s = -\frac{D_s}{kTn_L}\frac{(\cos^3\theta)tg^4\theta}{4\pi a^4}A_{SL}G(\theta)\left(\frac{2}{1+\cos\theta_e} - \frac{2}{1+\cos\theta}\right). \tag{27}$$

For small values of θ

$$G(\theta) = 2\theta^{-3},\ \sin(\theta) = \theta,\ \cos(\theta) = 1 - \theta^2/2,$$

and Eq. (27) becomes

$$V_s = \frac{D_s}{kTn_L}\frac{A_{SL}}{8\pi a^4}\theta(\theta_e^2 - \theta^2). \tag{28}$$

Because θ_e is small, $A_{SL} \cong A_{LL} \propto \gamma$, where γ is the surface tension of the liquid. On the other hand, D_s is expected to be inversely proportional to the viscosity of the liquid. Consequently

$$V_s \propto (\gamma/\eta)\theta(\theta_e^2 - \theta^2), \tag{29}$$

and the slip velocity is directly proportional to the surface tension γ and inversely proportional to the liquid viscosity. Eq. (29) contains the slip velocity caused by the chemical potential gradient and achieved via surface diffusion as well as via the velocity field near the leading edge.

2 CONCLUSIONS

The main results obtained are: (i) an explanation of the apparent slip velocity observed in capillaries; (ii) a suggestion for the mechanism of generation of the force responsible for the slip velocity during spreading of nonequilibrated drops; (iii) the derivation of an equation for the latter velocity which shows that it decays as x^{-4}; (iv) the derivation of an equation for the latter velocity for small values of the wetting angle, which reveals that it depends on the ratio γ/η, that was found to be significant in the droplet spreading experiments of Schonhorn et al. [10]. Three papers should be mentioned additionally: That of Cherry and Holmes [11] in which the Eyring theory is applied to wetting; the theory of Blake and Haynes [12] which is concerned with the microscopic jump of a single molecule at the tip of the drop and uses the Eyring theory in the treatment of the above jump;

and finally, the de Gennes hydrodynamic approach [13] that accounts for the viscous dissipation inside a liquid wedge of angle θ and provides an expression for the slip velocity similar to Eq. (29).

REFERENCES

1. E. Ruckenstein, C.S. Dunn, J. Colloid Interface Sci. **59**, 135 (1977) (Section 1.5 of this volume).
2. E. Ruckenstein, P. Rajora, J. Colloid Interface Sci. **96**, 488 (1983) (Section 1.6 of this volume).
3. E. Ruckenstein, Langmuir. **8**, 3038 (1992) (Section 1.7 of this volume).
4. E.B. Dusan, S.H. Davis, J. Fluid Mech. **65**, 71 (1974)
5. H. Helmholtz, G. Pictowsky, Sitzber Wien (2a). **40**, 607 (1868)
6. Schnell, J. Appl. Phys. **27**, 1149 (1956)
7. D.M. Tolstoi, Dokl. Akad. Nauk SSSR. **85**, 1329 (1952)
8. A.N. Somov, Kolloidn. Zh. **44**, 160 (1980)
9. C.A. Miller, E. Ruckenstein, J. Colloid Interface Sci. **48**, 368 (1974) (Section 1.1 of this volume).
10. H. Schonhorn, H.L. Frisch, T.K. Kwei, J. Appl. Phys. **37**, 4967 (1966)
11. B.W. Cherry, C.M. Holmes, J. Colloid Interface Sci. **29**, 174 (1969)
12. T.D. Blake, J.M. Haynes, J. Colloid Interface Sci. **30**, 4211 (1969)
13. P.G. de Gennes, Colloid Poly. Sci. **264**, 463 (1986)

1.9 Sorption on Deformable Solids

*Density Functional Theory Approach**

Gersh Berim and Eli Ruckenstein[†]
Department of Chemical and Biological Engineering,
State University of New York at Buffalo, Buffalo,
New York 14260, United States

Corresponding Author

[†] E-mail: feaeliru@buffalo.edu.
Phone: (716)645-1179. Fax: (716)645–3822.

1. INTRODUCTION

Most existing microscopic theories describing a fluid in contact with a solid (such as, for instance, the density functional theory) treat the solid as a continuous uniform body. This implies that there is a well-defined surface of the solid at which its density drops down to zero in a discontinuous manner. The interactions between the solid and fluid molecules are usually described by a Lennard-Jones potential coupled with a hard core repulsion. In such cases, the density functional theory (DFT) predicts the formation of several molecular layers near the solid surface where oscillations occur in the fluid density distribution (FDD).[1–4] The consecutive formation of these layers during adsorption generates multiple steps in the adsorption isotherm, which could be predicted by the conventional DFT[3,5,6] and identified experimentally.[7] Experiment also reveals that the steps in the adsorption isotherms are often damped[6] by microscopic heterogeneities present on real surfaces. Several approaches were developed to account for such a heterogeneity (see refs 6 and 8, and references therein). One of the most successful approaches was developed by Neimark et al.,[8] the so-called quenched solid DFT (QSDFT) approach. This approach involves two density distributions: one for the adsorbed fluid and another one for the adsorbent. The latter distribution was assumed to be linear and its two parameters were selected such as to achieve optimum fit with the experimental adsorption isotherms. The density distribution of the fluid component was obtained by minimizing the grand canonical potential of the system with respect to the fluid density. Although QSDFT provides a good description of the experimental results (see ref 8), it involves a phenomenological modeling of the solid density distribution (SDD). Both an intrinsic (in the absence of adsorbate) solid density distribution and an additional deformation caused by the adsorbate occur.

They can affect the wetting characteristics of the solid, which are sensitive even to microscopic inhomogeneities of the order of a few angstroms. As examples, one can mention the results obtained in refs 9 and 10, where it was demonstrated theoretically[9] and via molecular dynamics simulation[10]

* *J. Phys. Chem. B* 2011, 115, 13271–13274. Republished with permission.

that the presence of asperities with dimensions of only a few angstroms on an ideal hydrophilic surface can considerably increase the contact angle of a nanodrop on such a surface.

In this paper, we outline a procedure which can be used to predict both (the solid and fluid) density distributions, SDD and FDD, on the basis of interaction potentials only. In the present approach, the traditional multicomponent DFT used to describe fluids confined in a pore in contact with an external reservoir of fluid molecules is combined with another one involving a closed system containing a fixed number of molecules that generate the solid. The approach implies that the solid is a continuous medium in which the crystal structure is not taken into account. Such an assumption is frequently used in the literature.[2,11]

2. THE MODEL

The considered system is treated as a mixture of spherical molecules of component A (CA) and component B (CB) confined in a slitlike pore presented schematically in Figure 1. The walls of the slit do not provide external potentials acting on the molecules present inside the slit but serve only as boundaries. One of the components (CB) is connected to a reservoir which contains only molecules of the same component at the chemical potential μ_B. Because the interactions between the molecules of CA are assumed to be very strong, it is reasonable to assume that their concentration in the above reservoir is negligible and that the slit contains a fixed number of molecules of CA interacting among themselves and with the molecules of CB. As explained in detail below, the molecules of CA form a solid body (shaded area in Figure 1) with a continuous nonuniform density distribution near the fluid—solid interface.

The interactions between molecules inside the slit are characterized by short-range hard-core repulsions and long-range van der Waals attractions. The latter is calculated using the Weeks—Chandler—Anderson approach[12]

$$U_{ij}(r) = \begin{cases} \varepsilon_{ij}, & r \le 2^{1/6}\sigma_{ij} \\ 4\varepsilon_{ij}\left[\left(\dfrac{\sigma_{ij}}{r}\right)^{12} - \left(\dfrac{\sigma_{ij}}{r}\right)^{6}\right], & r > 2^{1/6}\sigma_{ij} \end{cases} \tag{1}$$

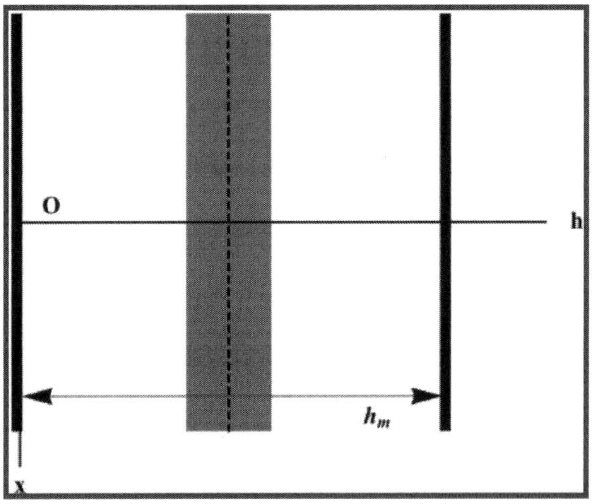

FIGURE 1 Schematic view of the considered system, consisting of a slit between two neutral, infinite, parallel boundaries separated by the distance h_m and filled with a binary mixture. The shaded rectangle in the middle of the slit represents the solid formed from one of the components. (The density distribution of the solid molecules will be found by solving the appropriate Euler—Lagrange equation considered below.) The vertical dashed line represents the plane of symmetry of the slit. The y-axis is normal to the plane of the figure.

where the subscripts i and j can be A and B corresponding to components A and B, r is the distance between the centers of a pair of interacting molecules, and the constants σ_{ij} and ε_{ij} are hard core diameters and energy parameters, respectively. Because the main purpose of this paper is to provide a procedure able to predict the density distributions of both the solid and fluid, rather than a specific case, the values of the parameters that appear in eq 1 were selected arbitrarily as follows: $\varepsilon_{BB}/k_B = 119.76$ K, $\sigma_{BB} = 3.405$ Å, $\varepsilon_{AA}/k_B = 295.0$ K, $\sigma_{AA} = 3.800$ Å, $\varepsilon_{AB}/k_B = 153.0$ K, $\sigma_{AB} = 3.530$ Å, where k_B is the Boltzmann constant. Note that the parameters for components B and A were selected close to those for argon and solid carbon dioxide, respectively.

The density distributions ρ_i(r) (i = A, B) are calculated using the Rosenfeld et al.[13] version of DFT. The Euler—Lagrange equations for the density distributions $\rho_{i;}$(r) (i = A, B) have been obtained by minimizing the thermodynamic potential

$$\Omega[\rho_A(r),\rho_B(r)] = F_{id}[\rho_A(r),\rho_B(r)] + \Phi[\rho_A(r),\rho_B(r)] + F_{attr}[\rho_A(r),\rho_B(r)] - \mu_B\int_V \rho_B(r)dr \tag{2}$$

with respect to the densities ρ_A (r) and ρ_B (r) under the constraint of fixed average density of component A in the slit.

In eq 2, $F_{id}[\rho_A(r),\rho_B(r)]$ is the Helmholtz free energy of a mixture of ideal gases, $\Phi[\rho_A(r),\rho_B(r)]$ and $F_{attr}[\rho_A(r),\rho_B(r)]$ are contributions to the excess free energy due to a system of hard spheres and to the attractive part of intermolecular interactions, respectively. Expressions for these contributions are provided in the Appendix.

The Euler—Lagrange equations for densities ρ_A (r) and ρ_B (r) have the form

$$\log[\Lambda_i^3\rho_i(r)] - Q_i(r) = \lambda_i/k_BT, \quad (i = A, B) \tag{3}$$

where the functions Q_i (r) are provided in the Appendix, $\Lambda_i = h_P/(2\pi m_i k_B T)^{1/2}$ is the thermal de Broglie wavelength of the molecules of component i, h_P is the Planck constant, T is the absolute temperature, and m_i is the molecular mass of component i, which were selected as for argon (i = B) and CO_2 (i = A). For component B, $\lambda_B = \mu_B$, where μ_B is the chemical potential in the reservoir. For component A, λ_A is a Lagrange multiplier arising because of the constraint of fixed number of molecules of CA in the slit and determined using this constraint. Equation 3 can be solved using a standard iteration procedure. Due to the uniformity of the system in x and y directions, the densities ρ_A(r) and ρ_B(r) depend only on the distance h from one of the slit walls (see Figure 1).

3. RESULTS AND CONCLUSION

The calculations were carried out at T = 85 K for a slit of width h_m = 100 σ_{BB} and for an average density of CA in the slit ρ_A^0, given by $\rho_A^0\sigma_{BB}^3 = 0.2246$. Because the system is symmetric, $\rho_B(h)$ and $\rho_A(h)$ are symmetric with respect to the middle of the slit, $\rho_A(h)$ being located almost completely in a 30 σ_{BB} range around the middle of the slit.

In Figure 2a, the density distributions of CB (dashed lines) and CA (solid lines) are plotted for $\mu_B/k_BT = -16.809$, -14.1454, and -12.706 corresponding to the bulk pressures P of CB equal to P/P_0 = 0.0058, 0.0835, and 0.288, respectively, where P_0 is the bulk pressure at liquid—vapor coexistence. The density $\rho_A(h)$ of component A increases rapidly over a distance of a few molecular diameters from almost zero to a constant value $\rho_{A,s} = 0.702/\sigma_{BB}^3$, where $\rho_{A,s}$ is the bulk density of the solid. Several oscillations are present in the h-dependence of $\rho_A(h)$.

One can define the location h_s of the planar solid surface as the position corresponding to an equimolar dividing surface given by the expression

$$\int_0^{h_m/2} \rho_A(h)\,dh = \rho_{A,s}(h_m/2 - h_s) \tag{4}$$

as illustrated in Figure 2b. (Note that, in addition to the surface located at $h = h_s$, there is also a surface located at $h = h_m - h_s$.) In Figure 2a, the locations h_s of the corresponding surfaces are indicated by dots on each of the solid density profiles. For the profile corresponding to $P/P_0 = 0.288$ the solid surface is represented by a vertical line.

Figure 2a indicates that the solid surface located via eq 4 is displaced, with increasing bulk pressure, toward the center of the slit, the displacement being of about $0.2\sigma_{BB}$ when P/P_0 varies from 0.0058 to 0.288. Our calculations have shown that the density profile $\rho_A(h)$ remains practically the same for P/P_0 greater than 0.288 and P/P_0 smaller than 0.0058. In the latter case, the solid density profile can be considered as the intrinsic one, because it is calculated for conditions close to the absence of fluid inside the system. The fluid density profile near a non- homogeneous solid differs from that near a uniform solid with an ideal planar surface. In Figure 3, density profiles are presented for CB near an ideal uniform planar solid (solid line) and a non-uniform solid (dashed line). The interaction potential between the molecules of the uniform

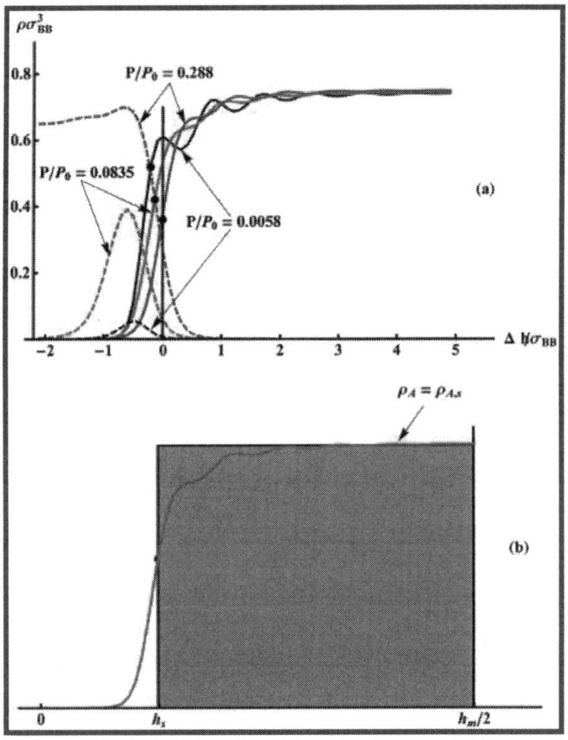

FIGURE 2 (a) Density distributions of component A (solid lines) and component B (dashed lines) at various bulk pressures P of component B. P_0 is the bulk pressure at the liquid—vapor coexistence. Only a part of the slit on the left-hand side of the plane of symmetry of the slit (see Figure 1) is presented. All densities are represented as functions of the distance $\Delta h = h - h_s^0$, where h_s^0, provides the position of a solid surface calculated using eq 4 for an arbitrarily selected $P/P_0 = 0.288$. (b) Illustration of the location of the planar solid surface at $h = h_s$. h_s is obtained by equating the area of the rectangle to the area under the curve of the actual fluid density distribution (solid line).

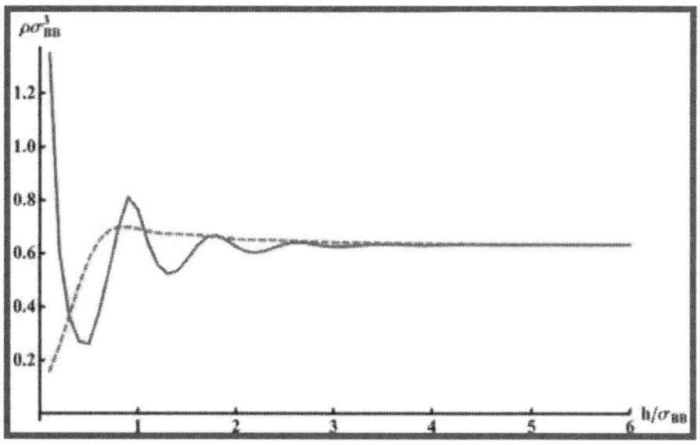

FIGURE 3 Density profiles of component B near a hard uniform planar wall made of molecules of component A (solid line) as well as near the solid surface considered in the present paper (dashed line).

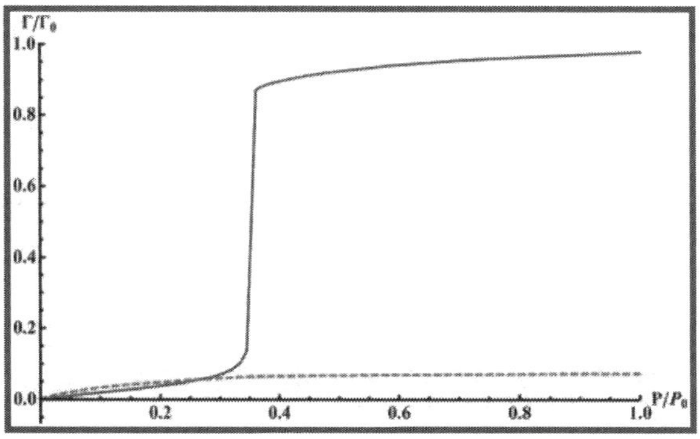

FIGURE 4 Adsorption isotherm (solid line) in the volume between a slit wall at $h = 0$ and the surface of the nonuniform solid located at $h = h_s$. The dashed line represents the amount of fluid absorbed by the solid multiplied by 50. Γ_0 is the adsorption at the bulk pressure P_0.

solid and those of CB inside the slit is assumed to be of the Lennard-Jones type coupled with a hard core repulsion. The energy and length parameters of the fluiduniform solid interaction were taken ε_{AB} and σ_{AB}, respectively, and the uniform solid density was assumed to be $\rho_{A,s}$. In this case, the interaction of a fluid molecule with the hard uniform solid is provided by the potential $\psi(h) = (2\pi/3)\varepsilon_{AB}\rho_s\sigma^3_{AB}\left[(2/15) - (\sigma_{AB}/(\sigma_{AB}+h))^9 - (\sigma_{AB}/(\sigma_{AB}+h))^3\right]$.

The main difference between the two fluid density profiles is the absence of oscillations in the vicinity of the nonuniform solid. Such a damping of density oscillations in the fluid density profile also was noted in ref 8, where argon in contact with solid carbon was considered. Such a difference in the behavior of the fluid density profile is expected. Indeed, as well-known, the oscillations of fluid density near the surface of a uniform solid is a consequence of close packing, which leads to the formation of molecular layers near the surface. If the solid is nonhomogeneous, such a layering does not occur because some molecules of CB are absorbed by the solid.

Figure 4 presents an adsorption isotherm of CB (solid curve) calculated for the volume between a slit boundary at $h = 0$ and the surface of the nonuniform solid located at $h = h_s$. The amount of fluid adsorbed per unit area of the solid is provided by equation $\Gamma = \int_0^{h_s} \rho_B(h)dh - \rho_{B,bulk}h_s$, where $\rho_{B,bulk}$ is the fluid density in the reservoir. The amount of fluid absorbed by the solid near that surface is presented by the dashed line. As expected, the adsorption isotherm is smooth, free of the multiple steps which were observed near the surface of a uniform solid. The amount of absorbed fluid is small, its fraction compared to that of adsorbed fluid decreasing from 6% at small fluid densities in the reservoir to 0.1% at large densities.

In conclusion, the suggested version of the DFT provides a unified approach for the description of a fluid in contact with a deformable solid surface. Both the fluid and the solid density distributions were obtained from the solution of the appropriate Euler—Lagrange equations. The fluid density profile near a deformable solid does not exhibit the strong oscillations present when the solid is uniform. Another consequence, which was observed experimentally is the absence of steps in the adsorption isotherms. The theory allows to calculate the amount of fluid adsorbed and the amount of fluid absorbed by the solid.

APPENDIX. DERIVATION OF THE EULER—LAGRANGE EQUATION

The density distributions $\rho_i(\mathbf{r})$ (i = A, B) of components A and B are calculated using a modified density functional approach. The total Helmholtz free energy $F_{tot}\left[\rho_{\mathrm{A}}(\mathrm{r}),\rho_{\mathrm{B}}(\mathrm{r})\right]$ is expressed as the sum of an ideal gas free energy $F_{\mathrm{id}}\left[\rho_{\mathrm{A}}(\mathrm{r}),\rho_{\mathrm{B}}(\mathrm{r})\right]$, and an excess free energy $F_{\mathrm{ex}}\left[\rho_{\mathrm{A}}(\mathrm{r}),\rho_{\mathrm{B}}(\mathrm{r})\right]$. The ideal gas free energy contribution has the form

$$F_{\mathrm{id}}\left[\rho_{\mathrm{A}}(\mathbf{r}),\rho_{\mathrm{B}}(\mathbf{r})\right]=k_{\mathrm{B}}T\sum_{i=\mathrm{A,B}}\int \mathrm{d}r\rho_i(\mathbf{r})\left\{\log\left[\Lambda_i^3\rho_i(\mathbf{r})\right]-1\right\} \tag{A.1}$$

where $\Lambda_i=h_{\mathrm{p}}/(2\pi m_i k_{\mathrm{B}}T)^{1/2}$ is the thermal de Broglie wavelength of the molecules of component i, k_{B} and h_P are the Boltzmann and Planck constants, respectively, T is the absolute temperature, and m_i is the molecular mass of component i. The excess free energy contains a contribution from a system of hard spheres $\Phi\left[\rho_{\mathrm{A}}(\mathrm{r}),\rho_{\mathrm{B}}(\mathrm{r})\right]$ and a contribution $F_{\mathrm{attr}}\left[\rho_{\mathrm{A}}(\mathrm{r}),\rho_{\mathrm{B}}(\mathrm{r})\right]$ due to the attractive interactions between molecules. The former contribution has, in Rosenfeld's approximation,[13] the form

$$\Phi\left[\rho_{\mathrm{A}}(\mathbf{r}),\rho_{\mathrm{B}}(\mathbf{r})\right]/k_{\mathrm{B}}T=-n_0\log(1-n_3)+\frac{n_1n_2-n_{1v}n_{2v}}{1-n_3}+\frac{n_2{}^3-3n_2\xi_0{}^2+2\xi_0^3}{24\pi(1-n_3)^2} \tag{A.2}$$

where $n_\alpha\ (\alpha=0,1,2,3)$ and $n_{\alpha v}\ (\alpha=1,2)$ are defined as follows

$$n_\alpha(\mathbf{r})=\sum_{i=\mathrm{A,B}}\int \mathrm{d}\mathbf{r}'\rho_i(\mathbf{r}')\omega_i^{(\alpha)}(\mathbf{r}-\mathbf{r}'),\ \alpha=0,1,2,3 \tag{A.3}$$

$$n_{\alpha v}(\mathbf{r})=\sum_{i=\mathrm{A,B}}\int \mathrm{d}\mathbf{r}'\rho_i(\mathbf{r}')\bar{\omega}_i^{(\alpha)}(\mathbf{r}-\mathbf{r}'),\ \alpha=1,2 \tag{A.4}$$

and $\xi_0=\left(n_{2v,x}{}^2+n_{2v,y}{}^2+n_{2v,h}{}^2\right)^{1/2}$. The scalar, $\omega_i^{(\alpha)}(\mathbf{r}-\mathbf{r}')$, and vector, $\bar{\omega}_i^{(\alpha)}(\mathbf{r}-\mathbf{r}')$, weight functions can be written in the form[14]

$$\omega_i^{(3)}(\mathbf{r})=\Theta\left(|\mathbf{r}|-\mathrm{R}_i\right),\quad \omega_i^{(2)}(\mathbf{r})=\delta\left(|\mathbf{r}|-\mathrm{R}_i\right) \tag{A.5}$$

$$\omega_i^{(1)}(\mathbf{r})=\omega_i^{(2)}(\mathbf{r})/4\pi R_i,\ \omega_i^{(0)}(\mathbf{r})=\omega_i^{(2)}(\mathbf{r})/4\pi R_i^2,\ \bar{\omega}_i^{(\alpha)}(\mathbf{r})=\left(\mathbf{r}/|\mathbf{r}|\right)\omega_i^{(\alpha)}(\mathbf{r})\ (\alpha=1,2),$$

where R_i is the molecular radius of component i, $\delta(x)$ and $\Theta(x)$ are the δ - Dirac and Heaviside functions, respectively.

The contribution to the excess free energy due to the attraction between the molecules of the mixture, calculated in the mean-field approximation, is given by

$$F_{\text{attr}}\left[\rho_{\text{A}}(\mathbf{r}),\rho_{\text{B}}(\mathbf{r})\right]=\frac{1}{2}\sum_{i,j=\text{A,B}}\iint d\mathbf{r}\, d\mathbf{r}'\rho_i(\mathbf{r})\rho_j(\mathbf{r}')U_{ij}\left(|\mathbf{r}-\mathbf{r}'|\right) \quad \text{(A.6)}$$

where the interaction potential $U_{ij}(|\mathbf{r}-\mathbf{r}'|)$ is provided by eq 1 in the text of the paper.

The Euler—Lagrange equations for the fluid density distributions $\rho(\text{r})$ (i = A, B) can be obtained by minimizing the thermodynamic potential eq 2 with respect to the densities $\rho_{\text{A}}(\text{r})$ and $\rho_{\text{B}}(\text{r})$ of components A and B, respectively, under the constraint of fixed average density of component A in the slit

$$\rho_{\text{A,av}}=\frac{1}{\text{V}}\int_{\text{V}} d\mathbf{r}\rho_{\text{A}}(\mathbf{r}) \quad \text{(A.7)}$$

where V is the volume of the slit.

These equations have the form of eq 3 where

$$k_{\text{B}}TQ_i(\mathbf{r})=-\sum_{\alpha=0}^{3}\int d\mathbf{r}'\frac{\partial\Phi}{\partial n_\alpha(\mathbf{r}')}\omega_i^{(\alpha)}(\mathbf{r}-\mathbf{r}')$$

$$-\sum_{\alpha=1}^{2}\int d\mathbf{r}'\frac{\partial\Phi}{\partial n_\alpha(\mathbf{r}')}\bar{\omega}_i^{(\alpha)}(\mathbf{r}-\mathbf{r}')$$

$$-\int d\mathbf{r}'\left[\rho_i(\mathbf{r}')U_{ii}\left(|\mathbf{r}-\mathbf{r}'|\right)+\rho_j(\mathbf{r}')U_{ij}\left(|\mathbf{r}-\mathbf{r}'|\right)\right],\ i,j=A,B\ (i\neq j) \quad \text{(A.8)}$$

For component B, $\lambda_{\text{B}}=\mu_{\text{B}}$ where μ_{B} is the chemical potential in the reservoir. For component A, λ_{A} is a Lagrange multiplier arising because of the constraint provided by eq A.7.

Using eqs A.7 and 3, the Lagrange multiplier can be rewritten in the form

$$\lambda_{\text{A}}=-k_{\text{B}}T\log\left[\frac{1}{\rho_{\text{A,av}}V\Lambda_{\text{A}}^{3}}\int_{V} d\text{r}\, e^{Q_{\text{A}}(\mathbf{r})}\right] \quad \text{(A.9)}$$

Eliminating λ_{A} between eqs 3 and A.9 and performing the integration in eq A.8 with respect to the y-direction, one obtains two integral equations for the density distributions $\rho_{\text{A}}(\text{r})$ and $\rho_{\text{B}}(\text{r})$ which can be solved by iterations. Due to the symmetry of the system, the densities $\rho_{\text{A}}(\text{r})$ and $\rho_{\text{B}}(\text{r})$ are functions of the distance h from a slit wall only (see Figure 1). The uniform distributions $\rho_{\text{A}}(h)=\rho_{\text{A}}^{0}$ and $\rho_{\text{B}}(h)=\rho_{\text{B}}^{0}$ were selected as initial guesses in the iteration procedure.

REFERENCES

(1) Tarazona, P.; Marconi, U. M. B.; Evans, R. *Mol. Phys.* **1987**, 60, 573–595.
(2) Evans, R. *J. Phys.: Condens. Matter* **1990**, 2, 8989–9007.
(3) Lastoskie, C.; Gubbins, K. E.; Quirke, N. *Langmuir* **1993**, 9, 2693–2702.
(4) Neimark, A. V.; Ravikovitch, P. I.; Vishnyakov, A. *Phys Condens. Matter* **2003**, 15, 347–365.
(5) Ravikovitch, P. I.; Vishnyakov, A.; Russo, R.; Neimark, A. V. *Langmuir* **2000**, 16, 2311–2320.
(6) Ravikovitch, P. I.; Vishnyakov, A.; Neimark, A. V. *Phvs. Rev. E* **2001**, 64, 011602.
(7) Larher, Y.; Angerand, F.; Maurice, Y. *Chem. Soc., Faradav Trans. 1* **1987**, 83, 3355–3366.
(8) Neimark, A. V.; Lin, Y. Z.; Ravikovitch, P. I.; Thommes, M. *Carbon* **2009**, 47, 1617–1628.

(9) Berim, G. O.; Ruckenstein, E. *Colloid Interface Sci.* **2011**, 359, 304–310 (Section 4.5 of this volume).
(10) Daub, C. D.; Wang, J. H.; Kudesia, S.; Bratko, D.; Luzar, A. *Faradav Discuss.* **2010**, 146, 67–77.
(11) Ebner, C.; Saam, W. F. *Phvs. Rev. Lett.* **1977**, 38, 1486–1489. Szybisz, L.; Sartarelli, S. A. J. *Chem. Phys.* **2008**, 128, 124702.
(12) Weeks, J. D.; Chandler, D.; Andersen, H. C. *Chem. Phvs.* **1971**, 54, 5237–5247.
(13) Rosenfeld, Y.; Schmidt, M.; Lowen, H.; Tarazona, P. *Phys. Rev. E* **1997**, 55, 4245–4263.
(14) Roth, R.; Dietrich, S. *Phvs. Rev. E* **2000**, 62, 6926–6936.

2 Fluid in a Nanoslit
Symmetry Breaking

Eli Ruckenstein and Gersh Berim

2.1 INTRODUCTION

In this chapter, the various aspects of symmetry breaking (the existence of a stable state of a system, in which the symmetry is lower than that of the system itself) for classical and quantum fluids is examined. The emphasis is on the conditions which cause symmetry breaking in the density distribution for one component fluids and binary mixtures confined in a closed nanoslit between identical solid walls. It was shown that two kinds of symmetry breaking can occur in such systems. First, a one-dimensional symmetry breaking which occurs only in the direction normal to the walls as a fluid density profile asymmetric with respect of the middle of the slit and uniform in any direction parallel to the walls (Secs. 2.1–2.3). Second, a two-dimensional symmetry breaking which occurs in the fluid density distribution that is nonuniform in one of the directions parallel to the walls and asymmetrical in the direction normal to the walls (Secs. 2.4–2.5). It manifests itself as liquid bumps and bridges in the fluid density distribution. For one component fluids, conditions of existence of symmetry breaking are provided in terms of the average fluid density, strength of fluid–solid interactions, distance at which the solid wall generates a hard core repulsion, and temperature (Secs. 2.1–2.4). In the case of binary mixtures, the occurrence of symmetry breaking also depends on the composition of the confined mixtures (Sec. 2.5). For quantum fluids, it was shown that the symmetry breaking has properties similar to those for classical fluids, and that symmetry breaking does not affect the superfluidity of the considered quantum fluid (Sec. 2.6). The results are summarized in Sec. 2.7

2.1 Fluid Density Profile Transitions and Symmetry Breaking in a Closed Nanoslit*

Gersh Berim and Eli Ruckenstein†
Department of Chemical and Biological Engineering,
State University of New York at Buffalo, Buffalo, New York 14260

Corresponding author

† E-mail: feaeliru@buffalo.edu.
Phone: (716)645-1179, Fax: (716)645-3822.

Received: August 11, 2006; In Final Form: January 16, 2007

1. INTRODUCTION

It is well-known that a semiinfinite fluid in contact with a solid substrate can be subjected to several kinds of transformations, such as, for example, the wetting and prewetting transitions that involve a change in the density profile of the fluid near the walls.[1–12] A wetting transition occurs for temperatures higher than the wetting temperature T_w at which a liquid film of infinite thickness is formed on the solid surface. The prewetting transition occurs in the temperature range $T_w < T < T_{sc}$, where T_{sc} is the surface critical temperature, as a discontinuous transition from a thin liquid film to a thick liquid film of finite thickness. In all the above transitions, the presence of an infinite reservoir of molecules of bulk fluid plays an important role because in all of them a change in fluid density occurs near the solid surface and the number of molecules in that region changes via an exchange with the reservoir. Inside an open slit, the chemical potential is equal to its value in the reservoir and the number of molecules per unit area of the slit is not constant. Therefore, the grand canonical ensemble was usually employed and two versions (local and nonlocal) of density functional theory (DFT) have been used[7,9,10,13] to describe the state of the fluid in the slit.

If the number of molecules is constant, as happens in a closed system, the behavior of the fluid is expected to be very different from that in an open system. First, in the absence of an external reservoir of molecules, neither wetting nor prewetting can occur in their conventional way. Only redistributions of molecules inside the closed volume, which nevertheless can bear some features of the phase transformations that are present in open systems, are expected to occur.

The behavior of a fluid confined in a close volume should be treated in a canonical ensemble. Few papers have examined such systems.[14–17] They were concerned with small spherical cavities containing a small fixed number of molecules, and suitable DFT approaches have been developed that accounted for the large fluctuations of the local density arising due to the small number of molecules. The density profile in the canonical ensemble was expressed in terms of that corresponding to a grand canonical ensemble with an additional term involving the reciprocal of the number of molecules. In a previous paper,[18] we examined some aspects of the problem by considering a

* *J. Phys. Chem. B* 2007, 111, 2514–2522. Republished with permission.

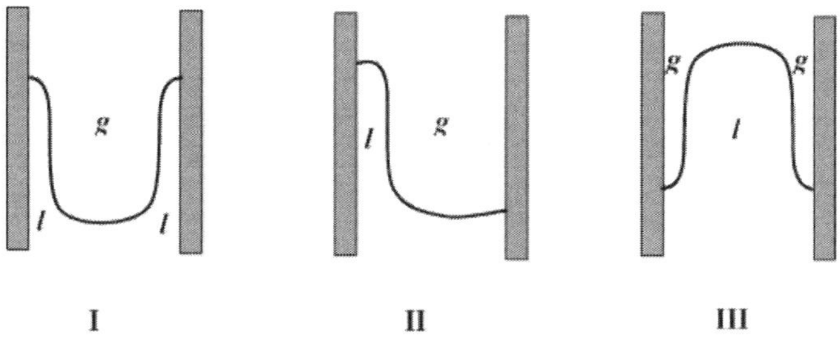

FIGURE 1. Schematic presentation of the three possible density profiles in the slit. Profile I corresponds to the state in which both walls are covered by thin films of fluid of liquid-like density, the remaining of the slit being filled by a gas-like fluid. In contrast, in the state with profile III, layers of gas-like fluid are present near the walls, the remaining of the slit being filled by a liquid-like fluid. Profile II corresponds to the state with asymmetric distribution of molecules (a mirror asymmetric profile is also possible).

methanol-like fluid in a long closed slit containing a large number of molecules. For that system, the effect of the small number of molecules, which was important in the cases considered in refs 14–17, was not relevant. The DFT version employed involved a canonical ensemble and the constraint of a constant number of molecules. The fluid—walls and fluid—fluid interactions were selected to be of the van der Waals type, and the cases of identical and nonidentical walls were examined. The main attention was paid to the calculation of the pressure tensor.

In the present paper, we continue the investigation of closed slits by focusing on the profile changes that occur when the number of molecules inside the slit and the temperature are changed. In particular, the possibility of symmetry breaking of the density distribution in the case of identical walls is examined. Simple intuitive arguments based on a macroscopic theory of wetting suggested such a possibility. Let us consider the three possible density profiles, shown in Figure 1. In the first profile, both walls of the slit are covered by a liquid-like density fluid, and a gas-like density fluid is present between them. In the second, a film of liquid-like density covers one of the walls and the remaining of the slit is occupied by a gas-like fluid. Finally, in the third profile, the walls are covered by a gas-like fluid and the remaining of the slit is filled with a liquid-like fluid. Denoted by γ_{lw}, γ_{gw}, and γ_{lg}, respectively, the liquid-wall, gas-wall, and liquid-gas surface tensions that satisfy the Young equation $\gamma_{gw} - \gamma_{lw} = \gamma_{lg} \cos \theta$, where θ is the contact angle, one can find the differences $\Delta F_{2,1}$ and $\Delta F_{2,3}$ between the free energy of the asymmetric profile II and the free energies of the symmetric profiles I and III, respectively

$$\Delta F_{2,1} = \gamma_{lg}\left(\cos\theta - 1\right) \leq 0 \tag{1a}$$

$$\Delta F_{2,1} = -\gamma_{lg}\left(\cos\theta + 1\right) \leq 0 \tag{1b}$$

Equations 1a, b indicate that the free energy for the asymmetric profile is smaller than those for the symmetric ones, with the exception of the cases $\theta = 0°$ or $180°$ when they are equal. This suggested that even for identical walls the asymmetric profile might be the stable one. Rigorously speaking, the above macroscopic considerations are not valid for nanoslits. However, they point out the possibility for an asymmetric profile to be the stable one. As another argument, supporting the possibility of symmetry breaking in a slit with identical walls, one can mention the results obtained by Merkel and Löwen[19] who found such a symmetry breaking in a fluid confined in two parallel identical planes (with no fluid between them).

In the present paper, the van der Waals (vdW) and Lennard-Jones (LJ) potentials were selected for the fluid—fluid and the fluid—wall interactions with the values of the parameters corresponding to argon between two parallel walls made of solid carbon dioxide. Open systems of this type were previously examined both theoretically[6,8,10,11] and experimentally.[20] Our calculations revealed a symmetry breaking of the density distribution in a closed slit with identical walls, which occurs in a temperature-dependent average density range specific for each potential. In addition, a new kind of transition between symmetric density profiles, which was named "cup-hill transition", was found.

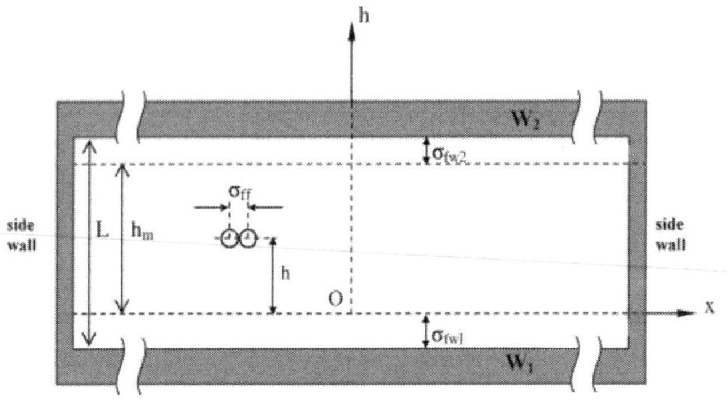

FIGURE 2. Closed slit of width L between two, generally nonidentical, solid walls W_1 and W_2. σ_{ff}, σ_{fw1}, and σ_{fw2} are the hard core diameters of the fluid—fluid, fluid—wall W_1, and fluid—wall W_2 interactions, respectively. The y axis is perpendicular to the plane of the figure.

2. MODEL

The system under consideration consists of a one-component fluid confined in a closed nanoslit between two parallel structureless identical solid walls W_1 and W_2 separated by a distance L. However, for reasons that will be understood later, we consider for the time being the more general case of nonidentical walls (see Figure 2). The distance between the side walls that close the slit is considered to be very large compared to L. As a result, the end effects can be considered negligible, and the density distribution can be assumed to be a function of a single coordinate h.

The interactions between the fluid molecules are described as the sum between a potential energy of a reference system and an attractive perturbation. As usual, the reference system is that of hard spheres, and the repulsive forces between the hard spheres provide the following main contribution to the Helmholtz free energy

$$F_{HS} = \int_V f_{\mathrm{H}}\left[\rho(\mathbf{r}),T\right]\mathrm{d}V \tag{2}$$

where the function $f_{\mathrm{H}}[\rho(\mathbf{r}),T]$ is provided by the Carnahan—Starling expression[21] in the form presented in ref 22

$$f_{\mathrm{H}}\left[\rho(\mathbf{r}),T\right] = k_{\mathrm{B}}T\rho(\mathbf{r})\left[\log(\eta_\rho) - 1 + \eta_\rho\frac{4-3\eta_\rho}{(1-\eta_\rho)^2}\right] \tag{3}$$

In eqs 2 and 3, V is the volume, k_{B} is the Boltzmann constant, T is the absolute temperature, $\rho(\mathbf{r})$ is the fluid density at point $\mathbf{r}$, which depends only on the distance of this point from the wall W_1, $\eta_\rho = (1/6)\pi\rho(\mathbf{r})\sigma_{\mathrm{ff}}^3$ is the packing fraction of the fluid molecules, and σ_{ff} is their hard core diameter.

The attractive part of the fluid—fluid interactions, which is considered as a perturbation, is described by a potential, $\phi_{\mathrm{ff}}(|\mathbf{r}-\mathbf{r}'|) = 4\epsilon_{\mathrm{ff}}\,[k_{\mathrm{LJ}}(\sigma_{\mathrm{ff}}/r)^{12} - (\sigma_{\mathrm{ff}}/r)^6]$ for $r \geq \sigma_{\mathrm{ff}}, r \equiv |\mathbf{r}-\mathbf{r}'|$ and $\phi_{\mathrm{ff}}(|\mathbf{r}-\mathbf{r}'|) = 0$ for $r < \sigma_{\mathrm{ff}}$, where $k_{\mathrm{LJ}} = 1$ (LJ potential) or $k_{\mathrm{LJ}} = 0$ (vdW potential). The interaction parameters ϵ_{ff} and σ_{ff} were selected as those of argon:[10] $\epsilon_{\mathrm{ff}}/k_{\mathrm{B}} = 119.76$ K, $\sigma_{\mathrm{ff}} = 3.405$ Å. Similarly, the interaction between the molecules of the fluid and the molecules of the walls are described by the potential $\phi_{\mathrm{fw}\alpha}(|\mathbf{r}-\mathbf{r}'|) = 4\epsilon_{\mathrm{fw}\alpha}\,[k_{\mathrm{LJ}}(\sigma_{\mathrm{fw}\alpha}/r)^{12} - (\sigma_{\mathrm{fw}\alpha}/r)^6]$ for $r \geq \sigma_{\mathrm{fw}\alpha}$ and $\phi_{\mathrm{fw}\alpha}(|\mathbf{r}-\mathbf{r}'|) = \infty$ for $r < \sigma_{\mathrm{fw}\alpha}$, where $\alpha = 1, 2$ refers to the walls W_1 and W_2, respectively, with the corresponding hard core diameters $\sigma_{fw\alpha}$ and interaction energy parameters $\epsilon_{\mathrm{fw}\alpha}$. The total fluid—walls interaction is provided by potential, $\Phi_{\mathrm{fw}}(h)$, obtained by integrating $\phi_{\mathrm{fw}\alpha}$ over the volume of both walls, assuming constant densities of the walls.[11] The explicit forms of this potential will be specified below.

The density distribution of the fluid in the slit will be calculated using the density functional theory in the local density approximation (LDA).[9,10,22] The LDA does not account for the short-ranged correlations arising from excluded volume effects and cannot predict, for this reason, the density oscillations that occur near solid walls. After LDA, several more sophisticated approximations were developed, such as the weighted density approximation[13,23] (WDA) and fundamental measure theory.[24–26] However, the LDA provides a reasonable (at least qualitative) description of a constrained fluid as was demonstrated, in particular, in refs 13 and 23, where LDA and WDA were compared for a fluid in an open slit.

The density functional $\Omega[\rho(\mathbf{r})]$ appropriate for the description of a fluid in a closed slit consists of a Helmholtz free energy of the system (all quantities are defined per unit area of one wall)

$$F(N,V,T) = F_{\text{HS}} + U_{\text{ff}} + U_{\text{fw}} \tag{4}$$

and a Lagrange term $-\lambda N$ that accounts for the constraint of a constant number, N, of molecules of fluid in the slit, λ being a Lagrange multiplier

$$N = \int_V \rho(\mathbf{r})\, d\mathbf{r} \tag{5}$$

Consequently

$$\Omega[\rho(\mathbf{r})] = F(N,V,T) - \lambda N \tag{6}$$

In the above equations, V is the volume of the slit per unit area of one wall, and U_{ff} and U_{fw} are the contributions to the total potential energy of the fluid—fluid and fluid—wall interactions, respectively.

In the mean-field approximation

$$U_{\text{ff}} = \frac{1}{2}\int_V\int_V d\mathbf{r}\, d\mathbf{r}'\, \rho(\mathbf{r})\rho(\mathbf{r}')\phi_{\text{ff}}(|\mathbf{r}-\mathbf{r}'|) \tag{7}$$

$$U_{\text{fs}} = \int_V d\mathbf{r}\rho(\mathbf{r})\left[\int_{V_{\text{w1}}} d\mathbf{r}'\rho_{\text{s1}}\phi_{\text{fs1}}(|\mathbf{r}-\mathbf{r}'|) + \int_{V_{\text{w2}}} d\mathbf{r}'\rho_{\text{s2}}\phi_{\text{fs2}}(|\mathbf{r}-\mathbf{r}'|)\right] \tag{8}$$

where V_{w1} and V_{w2} are the total volumes of the walls, which will be considered as semiinfinite uniform bodies of densities ρ_{s1} and ρ_{s2}, respectively. The Euler—Lagrange equation for the density profile $\rho(h)$, which can be derived from the minimization of the functional eq 6, has the following dimensionless form

$$1 + \log \eta_z + \frac{\partial}{\partial\bar{\rho}(z)}\left\{\bar{\rho}(z) f_{\text{H}}^{*}\left[\bar{\rho}(z)\right]\right\} = \frac{2\pi\epsilon_{\text{ff}}}{k_{\text{B}}T}\int_0^{z_{\text{m}}} dz'\bar{\rho}(z')K(z-z') - \frac{\epsilon_{\text{ff}}}{k_B T}\Phi_{\text{fw}}(z) + \bar{\lambda} \tag{9}$$

where $z_{\text{m}} = h_{\text{m}}/\sigma_{\text{ff}}$, $h_{\text{m}} = L - \sigma_{\text{fw1}} - \sigma_{\text{fw2}}$ (see Figure 2), $z = h/\sigma_{\text{ff}}$, $\bar{\rho}(z) = \rho(\text{h})\sigma_{\text{ff}}^{3}$, $\bar{\lambda} = \lambda/k_{\text{B}}T$, $\eta_z = (\pi/6)\bar{\rho}(z)$

$$f_{\mathrm{H}}^{*}\left[\bar{\rho}(z)\right]=-1+\eta_z\frac{4-3\eta_z}{\left(1-\eta_z\right)^2} \tag{10}$$

$$K\left(z-z'\right)=\begin{cases}1-\dfrac{2}{5}k_{\mathrm{LJ}} & \left|z-z'\right|<1\\ \dfrac{1}{\left(z-z'\right)^4}\left[1-\dfrac{2}{5}\dfrac{k_{\mathrm{LJ}}}{\left(z-z'\right)^6}\right] & \left|z-z'\right|\geq 1\end{cases} \tag{11}$$

$$\Phi_{\mathrm{fw}}\left(z\right)=\psi_1\left(z\right)+\psi_2\left(z\right) \tag{12}$$

$$\psi_1\left(z\right)=\frac{2\pi}{3}\varepsilon_{\mathrm{fw1}}\bar{\rho}_{\mathrm{fw1}}s_{\mathrm{fw1}}^3\left[\frac{2}{15}\frac{k_{\mathrm{LJ}}}{\left(1+z/s_{\mathrm{fw1}}\right)^9}-\frac{1}{\left(1+z/s_{\mathrm{fw1}}\right)^3}\right] \tag{13}$$

$$\psi_2\left(z\right)=\frac{2\pi}{3}\varepsilon_{\mathrm{fw2}}\bar{\rho}_{\mathrm{fw2}}s_{\mathrm{fw2}}^3\left[\frac{2}{15}\frac{k_{\mathrm{LJ}}}{\left[1+\left(z_{\mathrm{m}}-z\right)/s_{\mathrm{fw2}}\right]^9}-\frac{1}{\left[1+\left(z_{\mathrm{m}}-z\right)/s_{\mathrm{fw2}}\right]^3}\right] \tag{14}$$

In eqs 13 and 14, $\psi_\alpha(z)$ (α = 1,2) is the interaction potential between the wall W_α and the fluid, and $\varepsilon_{\mathrm{fw}\alpha}=\epsilon_{\mathrm{fw}\alpha}/\epsilon_{\mathrm{ff}}$, $\bar{\rho}_{w\alpha}=\rho_{w\alpha}\sigma_{\mathrm{ff}}^3$, $s_{\mathrm{fw}\alpha}=\sigma_{w\alpha}/\sigma_{\mathrm{ff}}$ are the dimensionless fluid—wall interaction energy parameters, densities of the walls, and hard core diameters, respectively.

Formally, eq 9 coincides with that for an open system without side walls. However, the Lagrange multiplier λ in closed systems is not known in advance, in contrast to the open systems, where λ is the chemical potential. The relation between λ and the number N of molecules is provided by the constraint eq 5, which has the following dimensionless form

$$\int_0^{z_{\mathrm{m}}}\bar{\rho}\left(z\right)\mathrm{d}z=N\sigma_{\mathrm{ff}}^2 \tag{15}$$

By solving eq 9 with respect to η_z and substituting the solution for $\bar{\rho}(z)$ into eq 15, one obtains for $\bar{\lambda}=\lambda/k_BT$ the expression

$$\bar{\lambda}=-\log\left[\frac{6}{\pi N\sigma_{\mathrm{ff}}^2}\int_0^{z_{\mathrm{m}}}\mathrm{d}z e^{Q(z)}\right] \tag{16}$$

where

$$Q\left(z\right)=-1-\frac{\partial}{\partial\bar{\rho}\left(z\right)}\left\{\bar{\rho}\left(z\right)f_{\mathrm{H}}^{*}\left[\bar{\rho}\left(z\right)\right]\right\}+\frac{2\pi\epsilon_{\mathrm{ff}}}{k_{\mathrm{B}}T}\times$$

$$\int_0^{z_{\mathrm{m}}}\mathrm{d}z'\bar{\rho}\left(z'\right)K\left(z-z'\right)-\frac{\epsilon_{\mathrm{ff}}}{k_BT}\Phi_{\mathrm{fw}}\left(z\right) \tag{17}$$

The nonlinear integral equation, eq 9, was solved by numerical iterations. The initial trial profile was selected as the homogeneous distribution of the fluid density for the selected number of molecules in the system. The calculations were carried out until the quantities $\delta_1=\left|\max\left\{\bar{\rho}_{i+1}\right\}-\min\left\{\bar{\rho}_i\right\}\right|/\max\left\{\bar{\rho}_i\right\}$ (the difference between two consecutive profiles) and $\delta_2=1-(1/N)\int_0^{\mathrm{hm}}\rho\left(h\right)\mathrm{d}h$ (the relative deviation of the real number of molecules from the selected one) became both smaller than 10^{-7}. In the expressions of δ_1 and δ_2, $\bar{\rho}_i$ represents the density profile after the *i-th* iteration and max $\{\bar{\rho}_{i+1}\}$ (min$\{\bar{\rho}_i\}$) is the largest (smallest) density value at the grid points of the corresponding profiles. The number of grid points was selected to be 10 per molecular diameter σ_{ff}. To control the consistency of calculations, this number was increased in some cases up to 20 per molecular diameter.

3. RESULTS

In the calculations, the specific values $\epsilon_{fw\alpha}/k_B = 153$ K, $\sigma_{fw\alpha} = 3.727$ Å, and $\rho_{w\alpha} = 2.19 \times 10^{28}$ m^{-3} ($\alpha = 1, 2$), which constitute the parameters for the solid carbon dioxide interacting with argon,[10] were selected. The width of the slit was taken to be $h_m = 15\sigma_{ff}$.

First, the case of vdW potential ($K_{LJ} = 0$) was examined. The fluid density profile obtained as solution of eq 9 with $\epsilon_{fw1} = \epsilon_{fw2}$ is symmetrical about the center of the slit, and its shape depends on the number N of molecules per unit area of one of the walls (or average density $\rho_{av} = N/h_m$ of the fluid) and temperature. (Note that the solution was symmetrical even when the initial profile guess

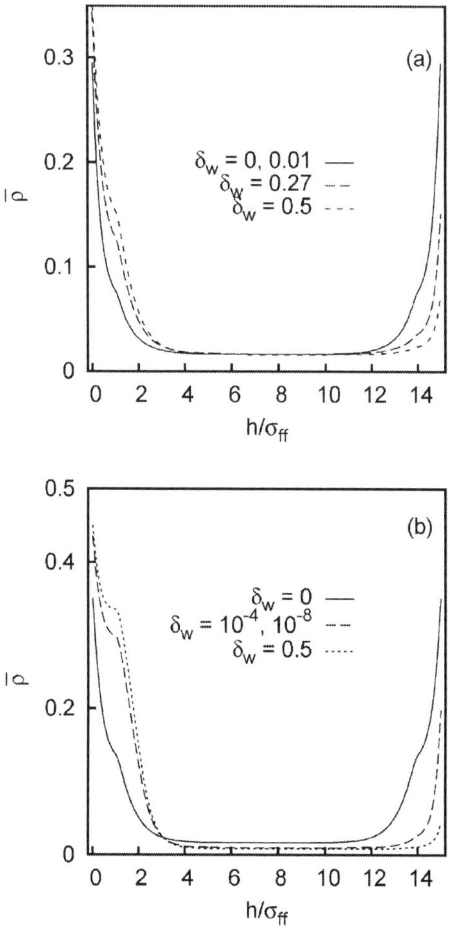

FIGURE 3. Two qualitatively different sequences of profiles for the van der Waals potential and $\delta_w \to 0$ at $T = 87$ K. (a) The sequence that has the symmetric profile ($\delta_w = 0$) as the asymptotic one. The profiles with $\delta_w \leq 10^{-2}$ are nondistinguishable; $\bar{\rho}_{av} = 0.03865$. (b) The sequence that has the asymmetric profile (long-dashed line) as the asymptotic one. All profiles with $\delta_w \leq 10^{-4}$ are represented by the same, long-dashed line. The symmetric profile for $\delta_w = 0$ is represented by the solid line; $\bar{\rho}_{av} = 0.05411$.

in the iterations was selected asymmetrical.) Typical profiles for small and large ρ_{av} are sketched in Figure 1 (profiles I and III). The first of them, profile I, has a cup-like shape and the second, profile III, a hill-like shape. (As shown below, depending on the value of ρ_{av}, the simple cup- and hill-like profiles may become more complicated). The hill-like profile is characterized by a relatively high density of fluid in the middle of the slit, which is of the order of the density of a liquid phase, and by a low-density fluid in the vicinity of the walls. This profile resembles the dried state of an infinite fluid in contact with a wall. In contrast, the density of the fluid in the middle of a cup-like profile is very small, comparable to the density of a gas, whereas in the vicinity of the walls it can have a liquid-like value. This profile resembles the case of a thin liquid film on a solid surface.

Because of the arguments presented in the Introduction, we checked the possibility of symmetry breaking, i.e., the existence of an asymmetric profile (such as profile II in Figure 1). The following procedure for the search of an asymmetric state of the system was employed: Let us consider two nonidentical walls ($\epsilon_{fw1} \neq \epsilon_{fw2}$) and introduce the quantity

$$\delta_{w} = \left| 1 - \frac{\epsilon_{fw2}}{\epsilon_{fw1}} \right| \tag{18}$$

which characterizes the difference between the walls. (For identical walls, $\delta_w = 0$). Only the case $\epsilon_{fw2} < \epsilon_{fw1}$ was considered. Then, a sequence of density profiles was calculated using the Euler-Lagrange equation for fixed values of the temperature T, average density ρ_{av}, and decreasing values of δ_w ($\delta_w \to 0$, $\delta_w > 0$). In the calculations, values of δ_w in the range $0.5 \geq \delta_w \geq 10^{-8}$ were used. For relatively large δ_w ($\delta_w \geq 0.1$), the profiles in all cases were asymmetrical due to the difference between walls. However, depending on the values of T and ρ_{av}, the sequence of profiles had two different tendencies as δ_w decreased to the order of 10^{-4}, 10^{-8}. In the first, the profiles became almost symmetrical and their difference from the profile obtained for $\delta_w = 0$ negligible. In the second, the profiles remained asymmetrical and there were no visible differences between those calculated for $10^{-8} \leq \delta_w \leq 10^{-4}$.

TABLE 1
Values (Per Unit Area of One Wall) of the Helmholtz Free Energies F_{sym} and F_{asym} of the Metastable (Symmetric Density Profile) and Stable (Asymmetric Density Profile) States of the Fluid in the Slit with $h_m = 15\sigma_{ff}$, Respectively, for Various Average Densities[a]

Van der Waals Potential

$\bar{\rho}_{av}$	F_{sym}	F_{asym}
0.2319	−15.10	−16.62
0.3092	−22.32	−22.77
0.3865	−28.46	−28.96
0.4638	−34.63	−35.17

Lennard-Jones Potential

$\bar{\rho}_{av}$	F_{sym}	F_{asym}
0.2319	−11.66	−11.90
0.3092	−15.53	−15.87
0.3865	−19.58	−19.85
0.4638	−23.80	−23.87

[a] $T = 87$ K. The free energies are expressed in $k_B T/\sigma_{ff}^2$ units.

In both cases, the limiting profile (either symmetrical or asymmetrical) was considered as the asymptotic solution of the Euler—Lagrange equation for $\delta_w \to 0$. If it is symmetrical, then the Euler—Lagrange equation has a unique stable solution. The asymmetric profile provides an additional solution to the Euler—Lagrange equation, which is the stable one if its free energy is smaller than that of the symmetric profile obtained for $\delta_w = 0$.

For illustration, in Figure 3a, b, two qualitatively different sequences of profiles calculated for different values of δ_w are presented for a slit that has one wall of solid carbon dioxide and another one of a hypothetical material for which the energy parameter ϵ_{fw2} can take any value smaller than ϵ_{fw1}. The average fluid densities are $\bar{\rho}_{av} = 0.03865$ for Figure 3a and $\bar{\rho}_{av} = 0.05411$ for Figure 3b, and $T = 87$ K. For the smaller average density (Figure 3a), the density profiles have asymmetric shapes for comparatively large δ_w and almost symmetrical shapes for small values of δ_w. The density profiles obtained for $\delta_w \leq 0.01$ practically coincide with that calculated for identical walls ($\delta_w = 0$).

For the higher average density (Figure 3b), the density profiles have highly asymmetric shapes for all values of δ_w, including the smallest $\delta_w = 10^{-8}$. This behavior is qualitatively different from that presented in Figure 3a for the smaller density. Analysis revealed that the change in the character of the density profile has occurred at an average density $\bar{\rho}_{sb} = 0.04799$, which represents the critical average density above which symmetry breaking occurs. It was also found that for $\bar{\rho}_{av} > \bar{\rho}_{sb}$ the asymmetric density profile provides a smaller Helmholtz free energy than the symmetric density profile, and hence, the former corresponds to the stable state. Several values of the free energy of the stable and metastable states are listed in Table 1.

In Figure 4a, the symmetric profiles obtained as solutions of eq 9 for various $\bar{\rho}_{av}$ and $\delta_w = 0$ are presented for $T = 87$ K. The solid curves represent stable and the dashed ones metastable profiles. The stable asymmetric profiles, corresponding to the same average densities as the metastable ones of Figure 4a, are presented in Figure 4b. As a quantitative characteristic of the symmetry breaking transition, one can consider the change of the fluid densities $\bar{\rho}(0)$ and $\bar{\rho}(h_m)$ at the walls of the slit for the stable state. In Figure 4c, the dependencies of $\bar{\rho}$ (0) and $\bar{\rho}(h_m)$ on the average density $\bar{\rho}_{av}$ are plotted. One can see that both densities change suddenly at $\bar{\rho}_{av} = \bar{\rho}_{sb} = 0.04799$, where symmetry breaking starts to occur.

The temperature dependence of $\bar{\rho}_{sb}$ is presented in Figure 4d. For the points $(T, \bar{\rho}_{av})$ under (above) the curve, the equilibrium density profile is symmetric (asymmetric).

For a fixed average density, symmetry breaking occurs at temperatures smaller than a critical value T_{sb}, which can be estimated from Figure 4d. Thus, for $\bar{\rho}_{av} = 0.4799$, $T_{sb} = 87$ K. In Figure 5a, the profiles calculated for $T \leq T_{sb}$ (dashed curves) and for $T > T_{sb}$ (solid curve) are presented for $\rho_{av} = 0.4799$. The first two, plotted for $T = 84$ and 87 K are asymmetric, and the third for $T = 89$ K is symmetric. The temperature dependence of the fluid densities at the walls is presented in Figure 5b. The rate of change of those densities with changing temperature has the largest value for $T = T_{sb}$.

It should be noted that the free energy difference between the metastable and stable states does not exceed a few percents (see Table 1), and therefore, it is not unlikely for the system to be in a metastable state. For this reason, the symmetric density profiles obtained from the solution of the Euler—Lagrange equation (eq 9) at $\delta_w = 0$ for $\bar{\rho}_{av} > \bar{\rho}_{sb}$, i.e., in the metastable region, will be considered below in some details. Those profiles undergo a change in shape with increasing $\bar{\rho}_{av}$, which can be associated with the transition between profiles I and III in Figure 1. Because of the characteristic shapes of those profiles, this transition will be named the cup-hill transition. It occurs within a very narrow range of values of ρ_{av} and has therefore an almost discontinuous character. As an example, in Figure 6a, the density profiles of argon are presented for various $\bar{\rho}_{av}$ at $T = 87$ K. In this case, the fluid—fluid interaction parameter is smaller than the fluid—walls interaction parameters ($\epsilon_{ff} < \epsilon_{fw\alpha}$). The transformation of the density profile from a cup- to a hill-like shape occurs when the average density changes from $\bar{\rho}_{av,1} = 0.25089$ (dashed curve) to $\bar{\rho}_{av,2} = 0.25120$ (dotted curve). Considering the arithmetic mean value of $\bar{\rho}_{av,1}$ and $\bar{\rho}_{av,2}$ as an appropriate estimate of the dimensionless average density $\bar{\rho}_{c-h}$ at which the cup-hill transition occurs, one obtains $\bar{\rho}_{c-h} = 0.25105$.

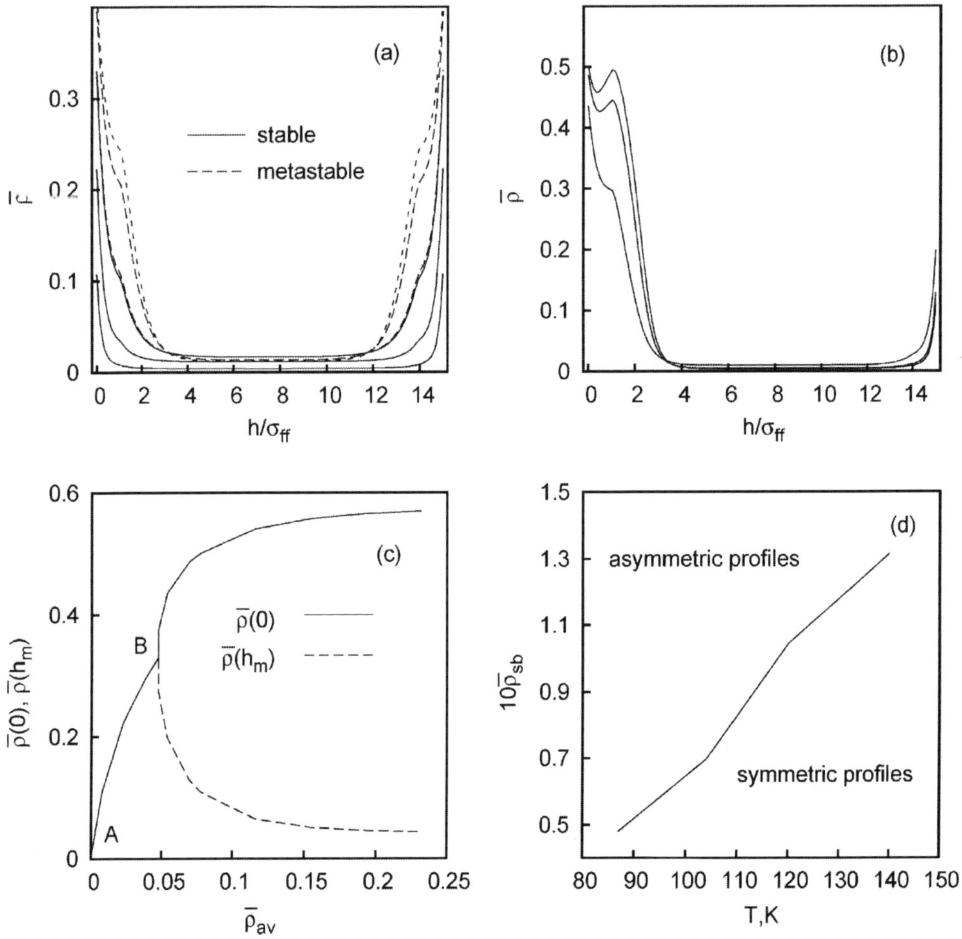

FIGURE 4. Slit with the van der Waals potential. (a) Stable (solid lines) and metastable (dashed lines) symmetric profiles at $T = 87$ K plotted for $\bar{\rho}_{av} = 0.00772, 0.02319, 0.04638, 0.04799, 0.06956$, and 0.07129, respectively. The fluid density near the walls increases with increasing average density. (b) Stable profiles corresponding to the same average densities as the metastable ones in panel a. The density near wall W_1 increases with increasing $\bar{\rho}_{av}$; $T = 87$ K. (c) The densities at the walls W_1 (solid line) and W_2 (dashed line) as functions of the average density in the slit; $T = 87$ K. On the segment AB, both curves coincide. (d) Temperature dependence of the critical average density $\bar{\rho}_{sb}$.

As in the case of symmetry breaking, one can consider the fluid density $\bar{\rho}(h_m/2)$ in the middle of the slit as well as that at the walls $\left(\bar{\rho}(0) = \bar{\rho}(h_m)\right)$ as quantitative characteristics of the cup-hill transition. In Figure 6b, the dependencies of $\bar{\rho}(h_m/2)$ and $\bar{\rho}(0)$ on $\bar{\rho}_{av}$ are presented. Both densities have a very rapid change near $\bar{\rho}_{c-h}$. The former becomes approximately 400 times larger, and the latter becomes 12 times smaller when $\bar{\rho}_{av}$ passes through $\bar{\rho}_{c-h}$.

The value of $\bar{\rho}_{c-h}$ is almost independent of temperature. As shown in Figure 6c, in the range 84 K $< T <$ 150 K it slightly decreases from $\bar{\rho}_{c-h} = 0.2516$ at $T = 84$ K to $\bar{\rho}_{c-h} = 0.2492$ at $T = 150$ K. Figure 6c can be considered as a kind of phase diagram in the $\bar{\rho}_{c-h} - T$ plane. In the entire range of temperatures, the density profile has a cup-like shape for all points below the curve AB and a hill-like shape for all points above the curve AB. If the actual average density $\bar{\rho}_{av}$ has a value in the range $\bar{\rho}_{av,min} < \bar{\rho}_{av} < \bar{\rho}_{av,max}$ ($\bar{\rho}_{av,max} = 0.2492$, $\bar{\rho}_{av,max} = 0.2516$, see Figure 6c), then the cup-hill transition

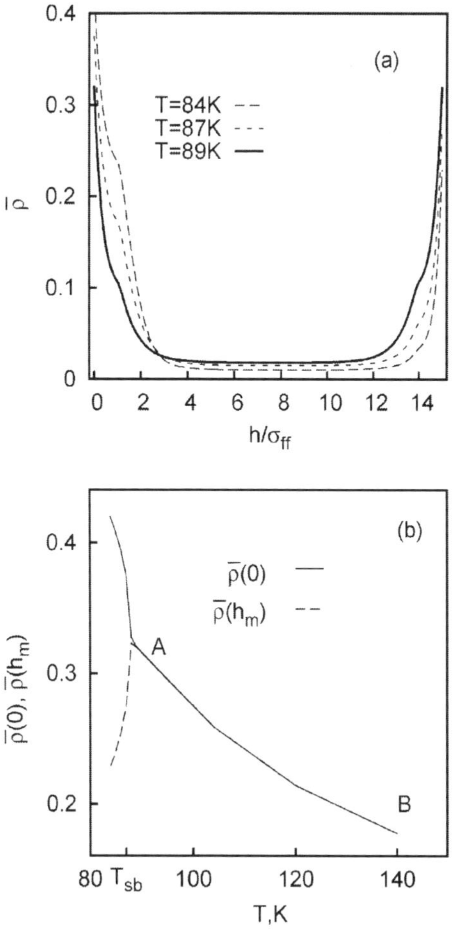

FIGURE 5. Slit with the van der Waals potential. (a) Density profiles for $\bar{\rho}_{av} = 0.04799$ and temperatures $T = 84$, 87 (dashed lines), and 89 K (solid line). The first two are asymmetric, whereas the latter, calculated for $T > T_{sb} \simeq 87$ K, is symmetric. (b) Densities at the walls W_1 (solid line) and W_2 (dashed line) as functions of temperature; $\bar{\rho}_{av} = 0.04799$. On the segment AB, both curves coincide.

occurs at a temperature $T_{c-h} > 84$ K. In Figure 7a, the density profiles are plotted for various temperatures and $\bar{\rho}_{av} = 0.24964$. The cup-hill transition occurs at $T_{c-h} \simeq 101.57$ K.

Figure 7b presents the temperature dependencies of the densities $\bar{\rho}(0)$ and $\bar{\rho}(h_m/2)$. Both densities change their behavior when the temperature passes through the temperature T_{c-h} of the cup-hill transition. The density $\bar{\rho}(0)$ of the fluid near the walls decreases almost discontinuously at the transition point. Such a change is similar to the drying transition (or wetting by the gas) that occurs in the system solid wall—bulk fluid.[27] For $T > T_{c-h}$ the density $\bar{\rho}(0)$ slightly increases with increasing temperature. In contrast, the density $\bar{\rho}(h_m/2)$ in the middle of the slit suddenly increases at $T = T_{c-h}$ but decreases with increasing temperature when $T > T_{c-h}$.

The temperature dependence of the density profile for an average density $\bar{\rho}_{av}$ outside the range between $\bar{\rho}_{av,min}$ and $\bar{\rho}_{av,max}$ differs from that observed above. The main difference consists in the absence of a transition from a cup- to hill-like profile at all temperatures. In Figure 8a, b, the density profiles and the temperature dependencies of the densities $\bar{\rho}(0)$ and $\bar{\rho}(h_m/2)$ are presented for $\bar{\rho}_{av} = 0.19323$, which is smaller than $\bar{\rho}_{av,min}$. At all temperatures, the density profile has a cup-like shape, and the temperature behaviors of $\bar{\rho}(0)$ and $\bar{\rho}(h_m/2)$ are monotonic with the density in the

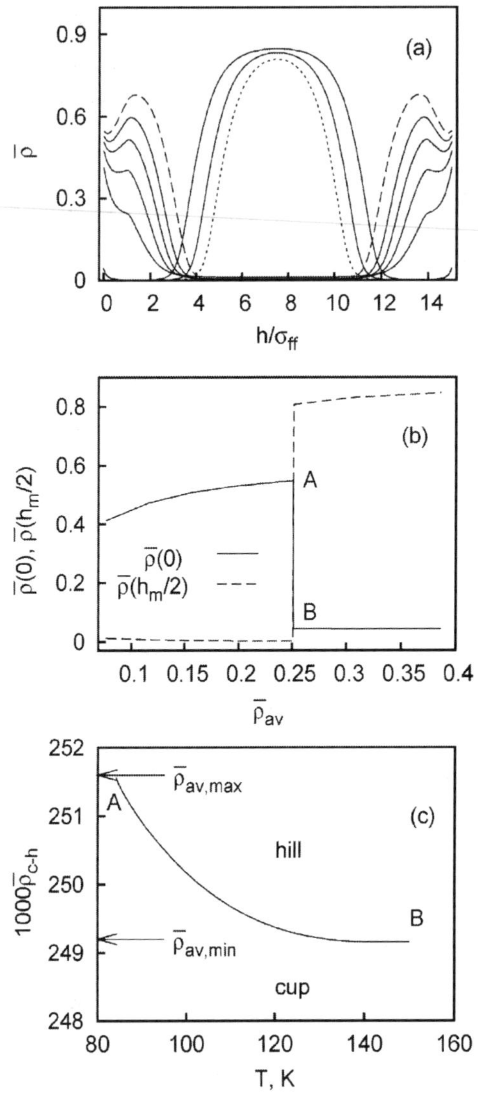

FIGURE 6. Slit with the van der Waals potential. (a) Density profiles for $\bar{\rho}_{av}$ equal to 0.077294, 0.11594, 0.15459, 0.19323 (solid curves), 0.25090 (dashed curve), corresponding to the cup-like profiles, equal to 0.25120 (dotted curve), 0.30917, 0.38647 (solid curves), and corresponding to the hill-like profiles; $T = 87$ K. For the cup-like profiles, the fluid density at the walls increases with increasing $\bar{\rho}_{av}$; for the hill-like profiles, the density in the middle of the slit increases with increasing $\bar{\rho}_{av}$. The cup-hill transition occurs at $\bar{\rho}_{c-h} = 0.25105$. (b) Dependence of the fluid density at the walls (solid curve) and in the middle of the slit (dashed curve) on the average density of the fluid for $T = 87$ K. On the AB segment, both curves coincide. (c) Temperature dependence of the critical average density $\bar{\rho}_{c-h}$ for a cup-hill transition; $\bar{\rho}_{av,min} = 0.2492$ and $\bar{\rho}_{av,max} = 0.2516$.

middle of the slit increasing and the density at the walls decreasing with increasing temperature. For $\bar{\rho}_{av} > \bar{\rho}_{av,mix}$ $\left(\bar{\rho}_{av} = 0.3865\right)$,the density profile has always a hill-like shape (Figure 9a). The temperature behavior of the densities in the middle of the slit and at the walls is in this case opposite to that for $\bar{\rho}_{av} > \bar{\rho}_{av,mix}$ the density $\bar{\rho}(0)$ increasing and the density $\bar{\rho}(h_m/2)$ decreasing with increasing temperature (see Figure 9b).

One should note that for the van der Waals potential the cup-hill transition always occurs between metastable states.

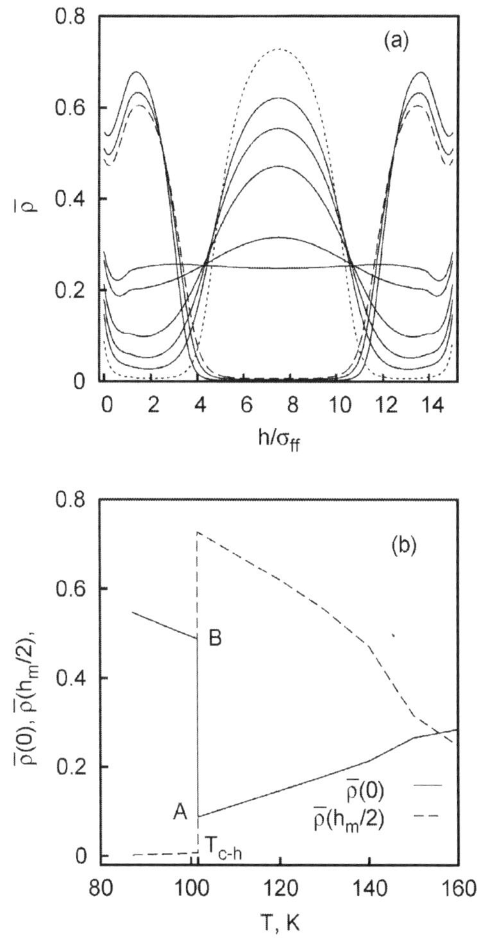

FIGURE 7. Slit with the van der Waals potential. (a) Density profiles for $\bar{\rho}_{av} = 0.24964$ and various temperatures. The cup-like profiles correspond to the temperatures 87, 96 (solid curves), and 101.55 K (dashed curve). The density at the wall for those profiles decreases with increasing temperature. The hill-like profiles correspond to the temperatures 101.6 (dotted curve), 120, 130, 140, 150, and 160 K (solid curves). The density in the middle of the slit for those profiles decreases with increasing temperature. A cup-hill transition occurs at $T_{c-h} \simeq 101.57$ K. (b) Temperature dependence of the fluid density at the walls (solid line) and in the middle of the slit (dashed line). On the AB segment, both curves coincide.

The use of the LJ potential $(k_{LJ} = 1)$ instead of the vdW one leads to the following changes in the fluid behavior. First, instead of a single critical density ρ_{sb} for the symmetry breaking transition, there are two critical densities, ρ_{sb1} and $\rho_{sb2} > \rho_{sb1}$, both being larger than ρ_{sb} at the same temperature. The stable profile is asymmetric for $\rho_{sb1} \leq \rho_{av} \leq \rho_{sb2}$ and symmetric for $\rho_{av} > \rho_{sb2}$ and $\rho_{av} < \rho_{sb1}$. The temperature dependence of ρ_{sb1} and ρ_{sb2} is shown in Figure 10. Symmetry breaking occurs only for those (ρ_{av}, T) pairs that are located between the two curves of Figure 10. One can see that the range of ρ_{av} where symmetry breaking occurs decreases with increasing temperature. For $T > T_{sb,0} \simeq 112$ K, no symmetry breaking occurs and the profiles are symmetrical about the middle of the slit. It should be noted that for the vdW potential the temperature $T_{sb,0}$ either does not exist or is located outside the considered temperature range (see Figure 4d). Table 1 compares the free energies of the symmetric and asymmetric profiles.

The second difference between the systems with vdW and LJ potentials concerns the cup-hill transition. At low temperature, e.g., $T = 87$ K, the cup-hill transition for the LJ potential remains

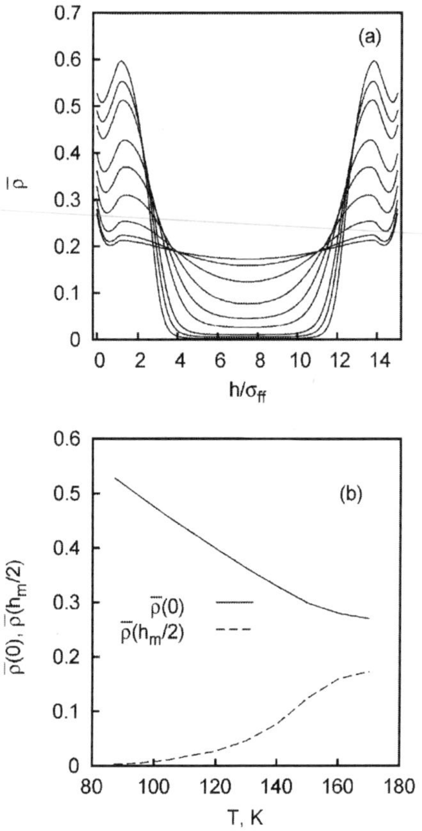

FIGURE 8. Slit with the van der Waals potential. (a) Density profiles for $\bar{\rho}_{av} = 0.19323$ and temperatures 87, 96, 104, 120, 130, 140, 150, 160, and 170 K. At all temperatures, the profiles have a cup-like shape. The density in the middle of the slit increases with increasing temperature. (b) Temperature dependence of the fluid density at the walls (solid curve) and in the middle of the slit (dashed curve).

similar to that for the vdW potential. However, at higher temperatures, e.g., $T = 104$ K, the change of the profile shape during the cup-hill transition becomes smoother and the variation of the fluid densities at the walls and in the middle of the slit is not as abrupt (see Figure 11). For $T \leq T_{sb,0}$ the cup-hill transition occurs between metastable states, whereas for $T > T_{sb,0}$ there is no symmetry breaking and the cup-hill transition occurs between stable states.

4. CONCLUSION

The main results of this paper can be summarized as follows:

For selected values of the parameters that characterize the fluid—fluid and the fluid—solid interactions, the density profile in a closed slit with identical walls depends on the average density of molecules inside the slit and temperature. The profiles can exhibit one of the three main characteristic shapes shown in Figure 1, namely the symmetric cup- and hill-like shapes (profiles I and III, respectively) and the asymmetric shape (profile II). The cup-like profiles occur mainly in slits with low average densities and the hill-like profiles in those with large average densities of the fluid molecules. The asymmetric profiles appear in systems described by symmetric equations as a consequence of symmetry breaking. At a given temperature, they occur in a certain range of the average density and at a given average density at temperatures smaller than a temperature T_{sb}, which depends on ρ_{av}.

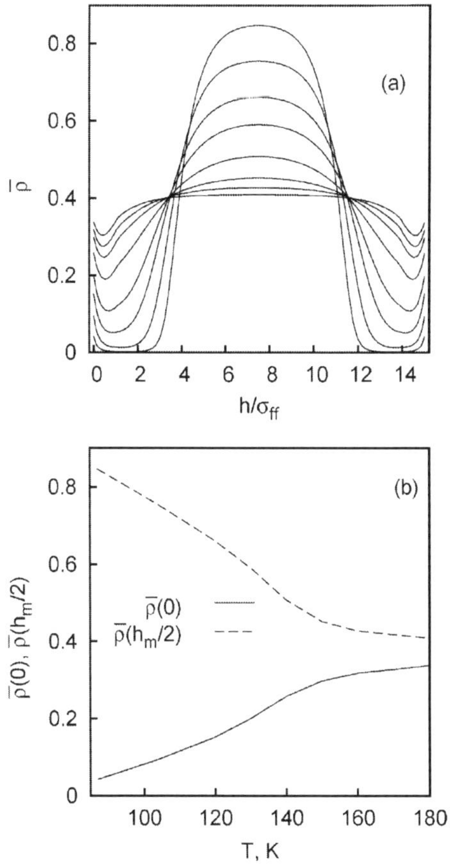

FIGURE 9. Slit with van der Waals potential. (a) Density profiles for $\bar{\rho}_{av} = 0.3865$ and temperatures 87, 104, 120, 130, 140, 150, 160, and 180 K. At all temperatures, the profiles have a hill-like shape. The density in the middle of the slit decreases with increasing temperature. (b) Temperature dependence of the fluid density at the walls (solid curve) and in the middle of the slit (dashed curve).

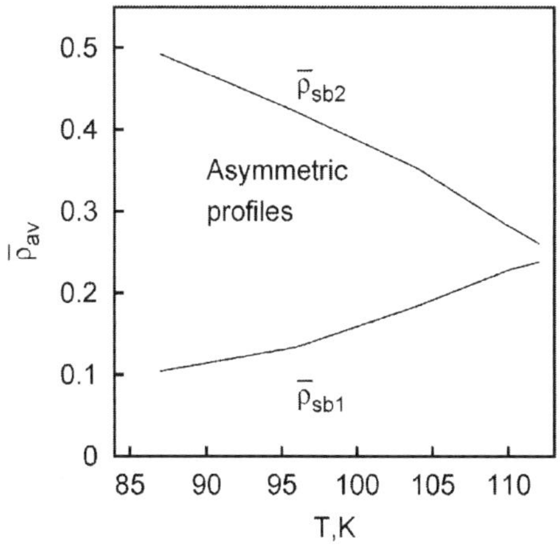

FIGURE 10. Temperature dependence of the critical values $\bar{\rho}_{sb1}$ and $\bar{\rho}_{sb2}$ on the average density in a closed slit with Lennard-Jones potential and $h_m = 15\sigma_{ff}$ for $T \geq 87$ K. Symmetry breaking occurs when $\bar{\rho}_{sb1} \leq \bar{\rho}_{av} \leq \bar{\rho}_{sb2}$. For temperatures higher than $T_{sb,0} = 112$ K, no symmetry breaking occurs.

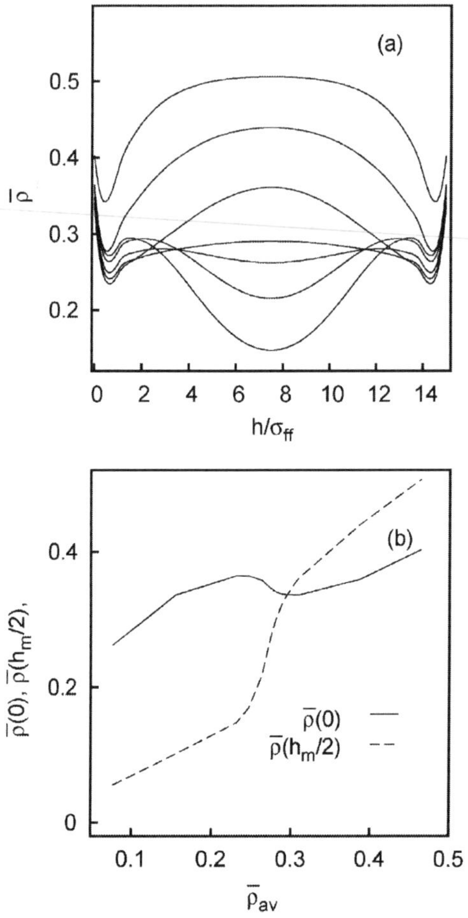

FIGURE 11. (a) Density profiles for Lennard-Jones potential and $\bar{\rho}_{av}$ equal to 0.2319, 0.2628, 0.2721, 0.2783, 0.3092 (cup-like profiles), and 0.4638 (hill-like profile); T = 104 K. The fluid density in the middle of the slit increases with increasing $\bar{\rho}_{av}$ (b) Dependence of the fluid density at the walls (solid curve) and in the middle of the slit (dashed curve) on the average density of the fluid for T = 104 K.

The possible density profiles of the system as function of ρ_{av} at a fixed temperature are presented in Figure 12 for the van der Waals potential. For $\rho_{av} < \rho_{sb}$, the stable profile is symmetric and both walls are covered by thin liquid-like films, a gas-like fluid filling the remaining of the slit. No asymmetric profiles are present in this range of ρ_{av}. When the average density increases and reaches the critical value ρ_{sb}, a spontaneous symmetry breaking of the density distribution inside the slit occurs. The asymmetric profile, obtained as the asymptotic solution of the Euler—Lagrange equation for $\delta_w \to 0$, provides a lower Helmholtz free energy than the symmetric profile obtained as the solution of the Euler—Lagrange equation for $\delta_w = 0$. As a result, the former profile provides a stable state, whereas the second provides a metastable one. The free energy of the metastable state differs little from the free energy of the stable state. If the energy barrier between the stable and metastable states is large enough (the estimation of the height of this barrier is a separate problem that was not considered in this paper), the system can have a long lifetime in the latter state. In this state, when ρ_{av} increases further, a transition from a cup- to a hill-like profile occurs at a second critical density $\rho_{c\text{-}h} > \rho_{sb}$ of the average fluid density.

The profiles as function of ρ_{av} for the Lennard-Jones potential are presented in Figure 13 for $T \leq T_{sb,0}$. One should note the presence of a second critical density ρ_{sb2} and the absence of a metastable state in the ranges $\rho_{av} < \rho_{sb1}$ and $\rho_{av} > \rho_{sb2}$. For $T > T_{sb,0}$ there is no symmetry breaking.

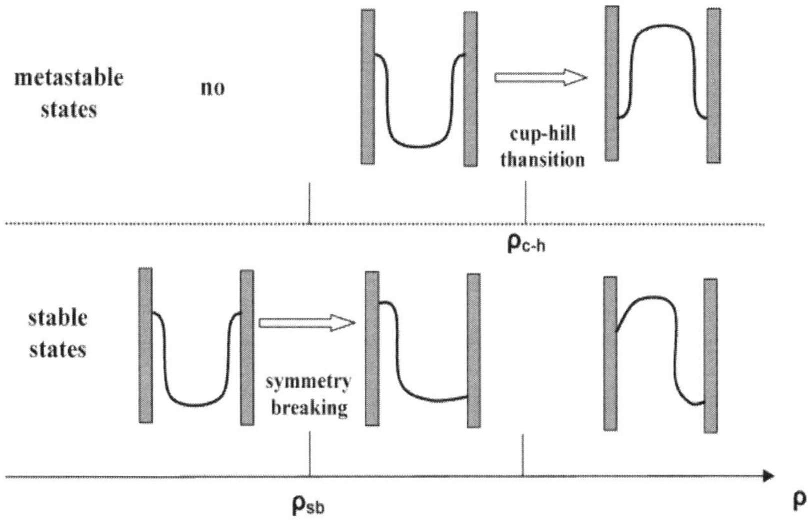

FIGURE 12. Possible states and the corresponding density profiles in the slit for the van der Waals potential at fixed temperature as function of the average fluid density. Note that the cup-hill transition at $\rho_{av} = \rho_{c\text{-}h}$ always occurs between metastable states of the system.

Several issues that were not addressed in this paper should be mentioned. The first one is the dependence of the symmetry breaking on the parameters of fluid—fluid and fluid—solid interactions. In the present paper, symmetry breaking was found for a fluid—fluid energy parameter ϵ_{ff} smaller than the fluid—solid one. However, in ref 18, where the opposite inequality was satisfied, symmetry breaking was absent. It is likely that there is a critical value of the fluid—solid energy parameter (for a given fluid—fluid one) above which symmetry breaking occurs and it is interesting to find that value. A second issue is related to the dependence of symmetry breaking and cup-hill transition on the size of the slit and, finally, a third one to the existence of symmetry breaking in open slits.

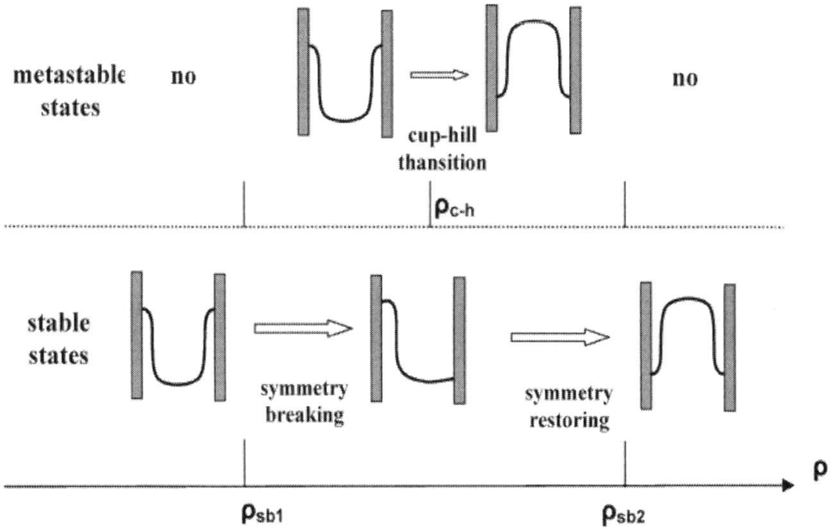

FIGURE 13. Possible states and the corresponding density profiles in the slit for Lennard-Jones potential at fixed temperature as function of the average fluid density for $T < T_{sb,0}$.

REFERENCES AND NOTES

(1) Muller, M.; MacDowell, L. G. *J. Phys.: Condens. Matter* **2003**, *15*, R609.
(2) Bonn, D.; Ross, D.; Bertrand, E.; Ragil, K.; Shahidzadeh, N.; Broseta, D.; Meunier, J. *Physica A* **2002**, *306*, 279.
(3) Yochelis, A.; Pismen, L. M. *Colloids Surf., A* **2006**, *274*, 170.
(4) Sweatman, M. B. *Phys. Rev. E* **2001**, *65*, 011102.
(5) Gelb, L. D.; Gubbins, K. E.; Radhakrishnan, R.; Sliwinska-Bartkowiak, M. *Rep. Prog. Phys.* **1999**, *62*, 1573.
(6) Ancilotto, F.; Toigo, F. *Phys. Rev. B* **1999**, *60*, 9019.
(7) *Evans, R.* J. Phys.: Condens. Matter **1990**, *2,* 8989.
(8) Dhawan, S.; Reimel, M. E.; Scriven, L. E.; Davis, H. T. *J. Chem. Phys.* **1991**, *94*, 4479.
(9) Evans, R.; Marconi, U. M. B.; Tarazona, P. *J. Chem. Phys.* **1986**, *84*, 2376.
(10) Evans, R.; Tarazona, P. *Phys. Rev. A* **1983**, *28*, 1864.
(11) Ebner, C.; Saam, W. F. *Phys. Rev. Lett.* **1977**, 38, 1486.
(12) Cahn, J. W. *J. Chem. Phys.* **1977**, 66, 3667.
(13) Tarazona, P.; Marconi, U. M. B.; Evans, R. *Mol. Phys.* **1987**, *60,* 573. Tarazona, P. *Phys. Rev. A* **1985**, *31*, 2672.
(14) Gonzalez, A.; White, J. A.; Roman, F. L.; Velasco, S. *Phys. Rev. Lett.* **1997**, *79*, 2466.
(15) Gonzalez, A.; White, J. A.; Roman, F. L.; Evans, R. *J. Chem. Phys.* **1998**, *109*, 3637.
(16) White, J. A.; Velasco, S. *Phys. Rev.* **2000**, 62, 4427.
(17) White, J. A.; Gonzalez, A.; Roman, F. L.; Velasco, S. *Phys. Rev. Lett.* **2000**, *84*, 1220.
(18) Berim, G. O.; Ruckenstein, E. *J. Chem. Phys.* **2006**, *125*, 164717.
(19) Merkel, M.; Löwen, H. *Phys. Rev. E* **1996**, *54*, 6623.
(20) Mistura, G.; Ancilotto, F.; Bruschi, L.; Toigo, F. *Phys. Rev. Lett.* **1999**, *82*, 795.
(21) Carnahan, N. F.; Starling, K. E. *J. Chem. Phys.* **1969**, *51*, 635.
(22) Bieker, T.; Dietrich, S. *Physica A* **1998**, *252*, 85.
(23) Maciolek, A.; Evans, R.; Wilding, N. B. *J. Chem. Phys.* **2003**, *119*, 8663.
(24) Rosenfeld, Y.; Schmidt, M.; Lowen, H.; Tarazona, P. *J. Phys.: Condens. Matter* **1996**, *8*, L577.
(25) Tarazona, P. *Phys. Rev. E* **1997**, 55, R4873.
(26) Tarazona, P. *Phys. Rev. Lett.* **2000**, *84*, 694.
(27) Tarazona, P.; Evans, R. *Mol. Phys.* **1984**, *52*, 847

2.2 Symmetry Breaking of the Fluid Density Profiles in Closed Nanoslits*

Gersh Berim and Eli Ruckenstein†

Department of Chemical and Biological Engineering,
State University of New York at Buffalo, Buffalo, New York 14260

Corresponding Author

† E-mail: feaeliru@buffalo.edu.
Phone: (716)645-1179; Fax: (716)645-3822.

(Received 5 January 2007; accepted 19 February 2007; published online 29 March 2007)

I. INTRODUCTION

The phenomenon of symmetry breaking is well known in physics, chemistry, biology, etc. It occurs when the symmetry of the stable state of the system is lower than the symmetry of the equations describing this system. A classical example from the physics of fluids is the freezing transition[1] in which the continuous translational symmetry of a fluid (in the absence of an external potential) is broken. The discrete symmetry, e.g., reflectional or the left-right one, can also be broken. A well known example is the occurence of spontaneous magnetization in ferromagnetic materials in the absence of an external magnetic field at temperatures smaller than the Curie temperature T_C. Recently, the possibility of a reflectional symmetry breaking was demonstrated in Ref. 2 where it was shown that in an open system consisting of two parallel identical layers of a two-dimensional liquid coupled by an interlayer potential, a symmetry breaking can occur for some strengths of the interlayer interaction. The fluid in one layer has a liquidlike density, while in the other it has a gaslike density.

Let us now consider a fluid confined in a narrow slit with identical solid walls in contact with a reservoir. In such a system which was intensively studied for several decades (see, e.g., Refs. 3–9), the fluid is exposed to an external potential due to fluid-solid interactions which possess a reflectional symmetry about the middle of the slit. The same symmetry is present in the basic Euler-Lagrange equation[5] describing the density profile across the slit. In all the above mentioned studies the density profile was symmetrical. However, the following qualitative arguments based on the macroscopic theory of wetting suggest that a symmetry breaking of the density profile may occur.

Let us consider the three possible density profiles, shown in Fig. 1. In the first profile, both walls of the slit are covered by a liquidlike density fluid, and a gaslike density fluid is present between them. In the second, a film of liquidlike density covers one of the walls (left or right one) and the remaining of the slit is occupied by a gaslike fluid. Finally, in the third profile the walls are covered by a gaslike fluid and the remaining of the slit is filled with a liquidlike fluid. Denoting by γ_{lw}, γ_{gw}, and γ_{lg}, respectively, the liquid-wall, gas-wall, and liquid-gas surface tensions which satisfy the

* *J. Chem. Phys.* 126, 124503 (2007). Republished with permission.

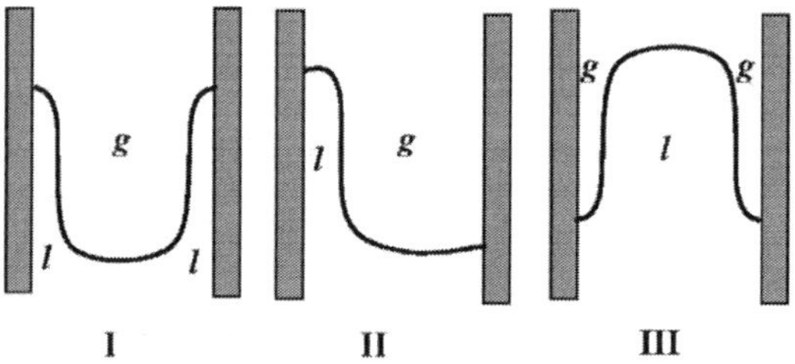

FIG. 1. Schematic presentation of the three possible density profiles in the slit. Profile I corresponds to the state where both walls are covered by thin films of fluid of liquidlike density, the remaining of the slit being filled by a gaslike fluid. In contrast, in the state with profile III layers of gaslike fluid are present near the walls, the remaining of the slit being filled by a liquidlike fluid. Profile II corresponds to the state with asymmetric distribution of molecules (a mirror asymmetric profile is also possible).

Young equation $\gamma_{\mathrm{gw}} - \gamma_{\mathrm{lw}} = \gamma_{\mathrm{lg}} \cos\theta$, where θ is the contact angle, one can calculate the differences $\Delta F_{2,1}$ and $\Delta F_{2,3}$ between the free energy of the asymmetric profile II and the free energies of the symmetric profiles I and III, respectively,

$$\Delta F_{2,1} = \gamma_{\mathrm{lg}}\left(\cos\theta - 1\right) \leqslant 0, \tag{1a}$$

$$\Delta F_{2,3} = \gamma_{\mathrm{lg}}\left(\cos\theta + 1\right) \leqslant 0. \tag{1b}$$

Equations (1) indicate that the free energy for the asymmetric profile is smaller than those for the symmetric ones, with the exception of the cases $\theta = 0°$ or $\theta = 180°$ when they are equal. While the above macroscopic considerations are not strictly valid for a nanoslit, they suggest that even for identical walls an asymmetric profile might be the stable one. In addition, because the density profile is provided by a nonlinear integral equation,[5,10] multiple solutions for the density profile can exist. Indeed, it was found in Ref. 11 that for open slits the equation for the density profile can have two solutions among which the one providing the smallest value of the grand canonical potential provides the stable state of the system. All those solutions posess reflectional symmetry and no symmetry breaking was detected. Nevertherless, it is not yet obvious that symmetry breaking does not occur in the slits at all.

In all the above mentioned papers only open slits were considered. Inside such slits, the chemical potentials are equal to those in the reservoirs with which they are in contact, and the number of molecules per unit area of the slit is not constant. Therefore, the grand canonical ensemble was usually employed and two versions (local and nonlocal) of the density functional theory (DFT) have been used[4–6,12] to describe the state of the fluid in the slit.

If the number of molecules is constant, as happens in a closed slit, the behavior of the fluid is expected to be very different from that in an open system because in the absence of an external reservoir of molecules only their redistribution inside the closed volume can occur. The behavior of a fluid confined in a closed volume has to be treated in a canonical ensemble. Only few papers examined such systems (see, e.g., Refs. 13–16). They were concerned with small spherical cavities containing small fixed numbers of molecules, and suitable DFT approaches have been developed which accounted for the large fluctuations of local density arising due to the small number of molecules. The density profile in a canonical ensemble was expressed in terms of that corresponding to a grand canonical ensemble and of an additional term involving the reciprocal of the number of molecules. In Ref. 17 we examined some features of the problem by considering a methanol-like fluid in a long closed slit containing a large number of molecules. In that case the effect of the small number of molecules, which was important in the cases considered in Refs. 13–16 was not relevant. The local DFT version employed involved a canonical ensemble and the constraint of a constant number of molecules. The fluid-wall and fluid-fluid interactions were selected to be of the van der Waals type, and the cases of identical and nonidentical walls were examined. The main attention was paid to the calculation of the pressure tensor. No symmetry breaking was found in that system.

In a previous paper[18] we examined among others the existence of symmetry breaking of the density profile of argon in a closed slit with identical solid walls composed of solid carbon dioxide in the framework of the local density approximation of the DFT.

Both van der Waals and Lennard-Jones potentials were selected for the fluid-fluid and fluid-solid interactions. It was found that, indeed, a stable asymmetric profile was provided by the completely symmetric integral equation as a consequence of symmetry breaking. For the van der Waals potential and at a given temperature symmetry breaking occurred at an average density $\rho_{av} > \rho_{sb}$, and at a given average density it occurred at a temperature smaller than T_{sb}. Here ρ_{av} is the average density of the fluid in the slit, ρ_{sb} and T_{sb} being critical values of the average density and temperature, respectively, which depend on the width of the slit. For the Lennard-Jones potential, two critical average densities ρ_{sb1} and ρ_{sb2} were found. Symmetry breaking occurred for $\rho_{sb1} \leqslant \rho_{av} \leqslant \rho_{sb2}$.

However, the local density approximation provides only a qualitative description of the fluid in contact with solid walls and, furthermore, is not suitable at large fluid densities. To acquire a more accurate picture of the symmetry breaking and of the density profile the application of a more sophisticated theory is required.

The purpose of the present paper is to examine for a closed slit the occurence of symmetry breaking on the basis of a nonlocal density approximation (NLDA) of the density functional theory. This approximation is valid for high fluid densities[5] and can predict such fine details as the strong oscillations of the fluid density at distances of several molecular diameters near the solid wall.

Lennard-Jones potentials were selected for the fluid-fluid and the fluid-wall interactions with the values of the parameters corresponding to argon between two identical parallel walls made of solid carbon dioxide. The open systems of this kind were previously examined both theoretically[12,19–21] and experimentally.[22] Our calculations confirmed the existence of symmetry breaking of the density distribution in a closed slit with identical walls and provided new details about this behavior.

II. MODEL

The system under consideration consists of a one component fluid confined in a closed nanoslit between two parallel structureless identical solid walls W_1 and W_2 separated by a distance L. However, for a reason which will be clarified later, we consider for the time being the more general case of nonidentical walls (see Fig. 2). The distance between the sidewalls which close the slit is considered to be very large compared to L. As a result, the end effects can be considered negligible, and consequently the density distribution can be assumed to be a function of a single coordinate h. (The closed slit with planar walls can be imagined also as a thin layer of thickness L between a solid sphere of large radius $R \to \infty$ embedded into a spherical cavity of radius $R + L$ in the same solid.)

The interaction between the fluid molecules is considered as the sum of a potential energy of a reference system and an attractive perturbation. As usual,[5,6,10] the reference system is that of hard spheres, and the repulsive forces between hard spheres provide the main contribution to the Helmholtz free energy $F(N, V, T)$.

The attractive part of the fluid-fluid interactions, which is treated as a perturbation in a mean field approximation, is described by the Lennard-Jones 6–12 potential, $\phi_{ff}(|\mathbf{r}-\mathbf{r}'|) = 4\epsilon_{ff}\left[(\sigma_{ff}/r)^{12} - (\sigma_{ff}/r)^{6}\right]$ for $r \geqslant \sigma_{ff}, (r = |\mathbf{r}-\mathbf{r}'|)$ and $\phi_{ff}(|\mathbf{r}-r'|) = 0$ for $r < \sigma_{ff}$. The interaction between the molecules of the fluid and those of the walls are described by the potential $\phi_{fw\alpha}(|\mathbf{r}-\mathbf{r}'|) = 4\epsilon_{fw\alpha}\left[(\sigma_{fw\alpha}/r)^{12} - (\sigma_{fw\alpha}/r)^{6}\right]$ for $r \geqslant \sigma_{fw\alpha}$ and $\phi_{fw\alpha}(|\mathbf{r}-\mathbf{r}'|) = \infty$ for $r < \sigma_{fw\alpha}$, where $\alpha = 1,2$ refers to the walls W_1 and W_2, respectively, with the corresponding hard core diameters $\sigma_{fw\alpha}$ and interaction energy parameters $\epsilon_{fw\alpha}$. The total fluid-wall interaction is therefore provided by the

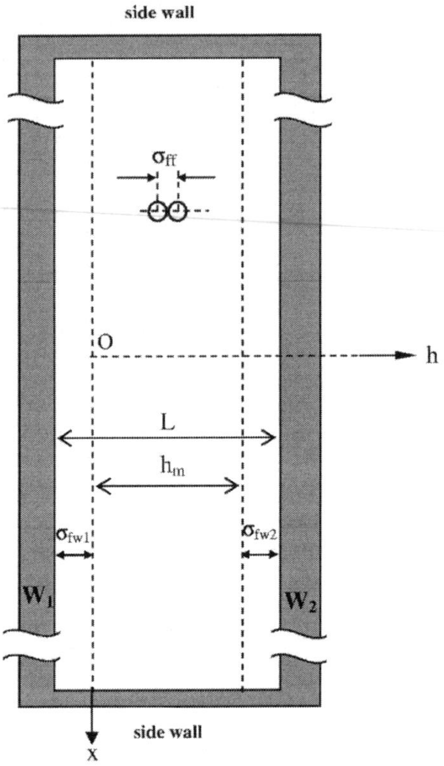

FIG. 2. Closed slit of width L between two, generally nonidentical, solid walls W_1 and W_2. σ_{ff}, σ_{fw1}, and σ_{fw2} are the hard core diameters of the fluid-fluid, fluid-wall W_1, and fluid-wall W_2 interactions, respectively. The y axis is perpendicular to the plane of the figure.

Lennard-Jones 3–9 potential, obtained by integrating $\phi_{fw\alpha}$ over the volume of the corresponding wall, by assuming constant densities of the walls.[21]

The density distribution of the fluid in the slit will be calculated using the nonlocal density functional theory in the form presented in Ref. 5. Such a theory was used to describe a fluid in an open slit,[5,8,9,11] both for small and large fluid densities.

The density functional $\Omega[\rho(\mathbf{r})]$ appropriate for the description of a fluid in a closed slit consists of a Helmholtz free energy of the system (all quantities are defined per unit area of one wall),

$$F(N,V,T) = F_{\mathrm{HS}} + U_{ff} + U_{fw} \tag{2}$$

and a Lagrange term $-\lambda N$ which accounts for the constraint of constant number N of molecules of fluid in the slit, λ being a Lagrange multiplier,

$$N = \int_V \rho(\mathbf{r})\, d\mathbf{r} \tag{3}$$

Consequently

$$\Omega[\rho(\mathbf{r})] = F(N,V,T) - \lambda N. \tag{4}$$

In the above equations, V is the volume of the slit, F_{HS} is the Helmholtz free energy for the hard spheres, and U_{ff} and U_{fw} are the contributions to the total potential energy of the fluid-fluid and fluid-wall interactions, respectively.

The Helmholtz free energy F_{HS} in the NLDA is given by

$$F_{HS} = \int_V f_{id}(\rho)dV + \int_V \Delta\psi_{hs}(\bar{\rho})dV, \tag{5}$$

where $\rho \equiv \rho(\mathbf{r})$ is the fluid density at point **r**, $\bar{\rho} \equiv \bar{\rho}(\mathbf{r})$ is a smoothed density, obtained by averaging the fluid density over a sphere with a diameter equal to the hard core diameter σ_{ff},

$$f_{id}(\rho) = k_B T \rho\left[\log\left(\Lambda^3 \rho\right) - 1\right] \tag{6}$$

is the free energy density of an ideal gas, $\Lambda = h_P/(2\pi m k_B T)^{1/2}$ is the thermal de Broglie wavelength, m is the molecular mass of argon, h_P and k_B are the Planck and Boltzmann constants, respectively, and T is the absolute temperature. The excess fluid repulsion, $\Delta\psi_{hs}(\bar{\rho})$, in the nonlocal DFT is provided by the expression,[5,10] obtained using the Carnahan-Starling equation of state,[23]

$$\Delta\psi_{hs}(\bar{\rho}) = k_B T \eta_\rho \frac{4 - 3\eta_\rho}{\left(1 - \eta_\rho\right)^2}, \tag{7}$$

where $\eta_\rho = (1/6)\pi\bar{\rho}(\mathbf{r})\sigma_{ff}^3$ is the packing fraction of the fluid molecules.

The smoothed density $\bar{\rho}(\mathrm{r})$ is provided by the following expression suggested by Tarazona:[5,10]

$$\bar{\rho}(\mathbf{r}) = \int_V d\mathbf{r}'\rho(\mathbf{r}')w\left[|\mathbf{r} - \mathbf{r}'|; \bar{\rho}(\mathbf{r})\right], \tag{8}$$

where the weighting function w is given by

$$w\left[|\mathbf{r} - \mathbf{r}'|; \bar{\rho}(\mathbf{r})\right] = w_0\left(|\mathbf{r} - \mathbf{r}'|\right) + w_1\left(|\mathbf{r} - \mathbf{r}'|\right)\bar{\rho}(\mathbf{r})$$
$$+ w_2\left(|\mathbf{r} - \mathbf{r}'|\right)\bar{\rho}(\mathbf{r})^2. \tag{9}$$

The final result for $\bar{\rho}(\mathbf{r})$, obtained in Ref. 5, has the form

$$\bar{\rho}(\mathbf{r}) = \frac{2\bar{\rho}_0(\mathbf{r})}{1 - \bar{\rho}_1(\mathbf{r}) + \left[\left|1 - \bar{\rho}_1(\mathbf{r})\right|^2 - 4\bar{\rho}_0(\mathbf{r})\bar{\rho}_2(\mathbf{r})\right]^{1/2}}, \tag{10}$$

where

$$\bar{\rho}_i(\mathbf{r}) = \int_V d\mathbf{r}'\rho(\mathbf{r}')w_i\left(|\mathbf{r} - \mathbf{r}'|\right), \ (i = 0,1,2). \tag{11}$$

Explicit expressions of the functions w_0, w_1, and w_2 and of the coefficients $\bar{\rho}_0(\mathbf{r}), \bar{\rho}_1(\mathbf{r})$, and $\bar{\rho}_2(\mathbf{r})$ are available in Ref. 5.

In the mean-field approximation

$$U_{ff} = \frac{1}{2}\int_V d\mathbf{r}\rho(\mathbf{r})U_f(\mathbf{r}), \tag{12}$$

$$U_{fw} = \int_V d\mathbf{r}\rho(\mathbf{r})U_s(\mathbf{r}), \tag{13}$$

where

$$U_f(\mathbf{r}) = \int_V d\mathbf{r}'\rho(\mathbf{r}')\phi_{ff}(|\mathbf{r}-\mathbf{r}'|) \tag{14}$$

and

$$U_s(\mathbf{r}) = \int_{V_{w1}} d\mathbf{r}'\rho_{w1}\phi_{fw1}(|\mathbf{r}-\mathbf{r}'|)$$

$$+\int_{V_{w2}} d\mathbf{r}'\rho_{w2}\phi_{fw2}(|\mathbf{r}-\mathbf{r}'|) \tag{15}$$

are the total potential energies of a molecule located at point **r** due to its interaction with other fluid molecules and the solid walls, respectively, V_{wl} and V_{w2} are the volumes of the walls which will be considered as semi-infinite uniform bodies with densities ρ_{wl} and ρ_{w2}, respectively.

The Euler-Lagrange equation for the density profile $\rho(\mathbf{r})$, derived through the minimization of the functional [Eq. (4)] has the following form:

$$k_BT\log(\Lambda^3\rho) + \Delta\psi_{hs} + \overline{\Delta\psi'_{hs}} + U_f + U_s = \lambda, \tag{16}$$

where $U_f \equiv U_f(\mathbf{r}), U_s \equiv U_s(\mathbf{r})$, and

$$\overline{\Delta\psi'_{hs}} \equiv \overline{\Delta\psi'_{hs}}(\mathbf{r}) = \sum_{n=0}^{2}\int_V d\mathbf{r}'\frac{\rho(\mathbf{r}')w_n(|\mathbf{r}-\mathbf{r}'|)[\bar{\rho}(\mathbf{r}')]^n}{1-\bar{\rho}_1(\mathbf{r}')-2\bar{\rho}(\mathbf{r}')\bar{\rho}_2(\mathbf{r}')}$$

$$\times\frac{\partial\Delta\psi_{hs}[\bar{\rho}(\mathbf{r}')]}{\partial\bar{\rho}(\mathbf{r}')}. \tag{17}$$

Explicit expressions for U_f, U_s and $\partial\Delta\psi_{hs}[\bar{\rho}(\mathbf{r}')]/\partial\bar{\rho}(\mathbf{r}')$ are provided in the Appendix. Equation (16) can be rearranged in the form

$$\rho(\mathbf{r}) = \rho_0\exp\left\{-\frac{1}{k_BT}\left[\Delta\psi_{hs}(\bar{\rho}) + \overline{\Delta\psi'_{hs}}(\mathbf{r}) + U_f(\mathbf{r}) + U_s(\mathbf{r})\right]\right\}, \tag{18}$$

where $\rho_0 = \exp(\lambda/k_BT)/\Lambda^3$.

Formally, Eq. (16) coincides with that for an open system without sidewalls. However, the Lagrange multiplier λ in closed systems is not known in advance, in contrast to the open systems, where λ is the chemical potential. The relation between λ and the number N of molecules is provided by the constraint Eq. (3). Substituting Eq. (18) into Eq. (3) one obtains for λ the expression

$$\lambda = -k_BT\log\left[\frac{1}{N\Lambda^3}\int_V d\mathbf{r}e^{Q(\mathbf{r})}\right], \tag{19}$$

where

$$Q(\mathbf{r}) = -\frac{1}{k_B T}\left[\Delta\psi_{hs}(\bar{\rho}) + \overline{\Delta\psi'_{hs}(\mathbf{r})} + U_f(\mathbf{r}) + U_s(\mathbf{r})\right]. \tag{20}$$

Equation (16) was solved by numerical iterations. The calculations were carried out until the quantities $\delta_1 = \sigma_{ff}^5 \int_V d\mathbf{r}[\rho_{i+1}(\mathbf{r}) - \rho_i(\mathbf{r})]^2$ (the measure of the difference between two consecutive profiles) and $\delta_2 = 1-(1/N)\int_V \rho(\mathbf{r})d\mathbf{r}$ (the relative deviation of the actual number of molecules from the selected one) became both smaller than $\varepsilon = 10^{-7}$ (in some cases $\varepsilon = 10^{-14}$). In the expressions of δ_1 and δ_2, $\rho_i(\mathbf{r})$ represents the density profile after the ith iteration. The number of grid points was taken equal to 10 per molecular diameter σ_{ff}.

First, an initial trial profile was selected as the homogeneous distribution of the fluid density for the selected number of molecules in the system with identical walls ($\epsilon_{fw1} = \epsilon_{fw2}$, $\sigma_{fw1} = \sigma_{fw2}$, $\rho_{w1} = \rho_{w2}$). In such a case, the iterations always converged to a symmetric profile which constitutes one of the solutions of the Euler-Lagrange equation.

To determine if another solution is also present two computational approaches were used. In the first, an asymmetric profile was selected as the initial guess for the system with identical walls. For some values of the parameters (temperature and average density of the fluid in the slit) and $\varepsilon = 10^{-7}$ the iterations converged to a symmetric or asymmetric profile. [Specifically, we used density profiles of the type $\rho(h) = \text{const}/h^n$, where n had the values from 4 to 40 and the const was selected to provide the selected number of molecules (or equivalently, the selected average density of the fluid) in the slit.] Then calculations were repeated with increasing precision. If the obtained profile did not change with increasing precision from $\varepsilon = 10^{-7}$ to $\varepsilon = 10^{-14}$, then it was accepted as a solution of the basic integral equation. When the profile was symmetrical it coincided with that obtained for a homogeneous initial guess. If the profile was asymmetrical, it constituted a second solution of the Euler-Lagrange equation.

A second computational approach was also employed by considering slightly nonidentical walls. The difference between them was introduced by changing the energy parameter ϵ_{fw2} of the fluid-solid interactions to a new value $\epsilon_{fw2} = \epsilon_{fw1} + \Delta\epsilon_{fw}$. The parameter

$$\delta_w = \left|\frac{\Delta\epsilon_{fw}}{\epsilon_{fw1}}\right| \tag{21}$$

was introduced as a characteristic of nonidentity of the walls. The initial guess was taken as a spatially homogeneous profile and iterations were carried out for several decreasing values of δ_w (e.g., $\delta_w = 10^{-2}, 10^{-4}, 10^{-6}, 10^{-8}$). The obtained profiles had a clear asymptotic behavior, providing either a symmetric or an asymmetric profile, which can be considered as a solution of the Euler-Lagrange equation for a system with identical walls. The results of the last two computational approaches were identical.

III. RESULTS

In the calculations, the interaction parameters ϵ_{ff} and σ_{ff} were selected for argon:[12] $\epsilon_{ff}/k_B = 119.76$ K, $\sigma_{ff} = 3.405$ Å; and the values $\epsilon_{fw\alpha}/k_B = 153$K, $\sigma_{fw\alpha} = 3.727$ Å, and $\rho_{w\alpha} = 1.91 \times 10^{28}$ m^{-3} ($\alpha = 1,2$) were those for the solid carbon dioxide interacting with argon.[12] The width of the slit was taken $h_m = 15\sigma_{ff}$. The density of the fluid $\rho(\mathbf{r})$ depends only on the distance h from the wall, $\rho(\mathbf{r}) \equiv \rho(h)$. The dimensionless quantities

$$\rho^* = \rho\sigma_{ff}^3,\ h^* = h/\sigma_{ff} \tag{22}$$

for the fluid density and distance from the wall, respectively, were employed in the calculations.

The fluid density profile obtained as solution of Eq. (16) for identical walls and an initial guess taken as a uniform density profile is symmetrical about the middle of the slit and its shape depends on the number N of molecules per unit area of one of the walls (or average density $\rho^*_{av} = N\sigma^2_{ff}/h^*_m$ of the fluid) and temperature. Typical profiles for small and large values of ρ^*_{av} are presented in Figs. 3(a) and 3(b). The profiles in Fig. 3(a) are plotted for various ρ^*_{av} in a slit with $h_m = 15\sigma_{ff}$ at $T = 87$ K. For small average densities, the oscillations of the density profiles are significant only near the walls. The density of the fluid in the very vicinity of the walls has liquidlike values ($\rho^* \sim 1$) and becomes gaslike ($\rho^* \sim 10^{-2}$) at distances from the wall of the order of several molecular diameters. (Note, that such profiles are present in open systems far from condensation conditions[11]). When ρ^*_{av} increases, the oscillations become larger and the region with gaslike fluid density decreases. For relatively high densities ($\rho^*_{av} \sim 0.5411$) the whole slit is filled with a liquidlike fluid, and the oscillations of the fluid density become significant everywhere in the slit. The fluid density at the walls increases with increasing ρ^*_{av}.

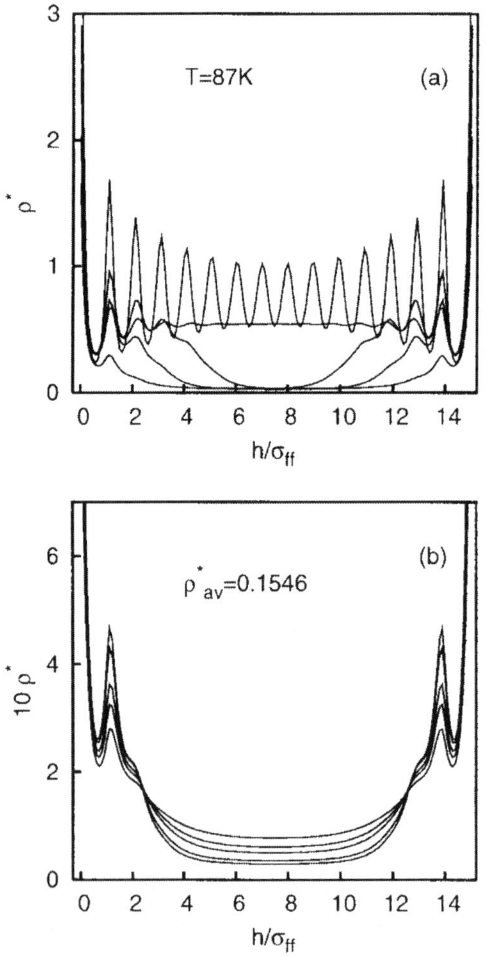

FIG. 3. Density profiles in the closed slit for $h^*_m = 15$. (a) The symmetric profiles at $T = 87$ K plotted for $\rho^*_{av} = 0.1159, 0.2319, 0.3865, 0.5411$, and 0.7729. The fluid density at any point inside the slit increases with increasing average density. (b) The symmetric profiles corresponding to the average density $\rho^*_{av} = 0.1546$ for temperatures $T = 87, 90, 96, 100$, and 106 K. The density in the middle of the slit increases with increasing temperature.

In Fig. 3(b) profiles corresponding to $\rho^*_{av} = 0.1546$ are presented for various temperatures. The amplitude of the oscillations of the density near the walls decreases and the density in the central part of the slit increases with increasing temperature. This is an obvious consequence of the penetration of the molecules from the walls into the internal part of the slit.

Asymmetric density profiles were obtained using the two procedures emphasized in a previous section. Several asymmetric profiles are shown in Fig. 4(a) for $T = 87$ K and various values of ρ^*_{av}. When ρ^*_{av} increases the number as well as the amplitude of oscillations near the left wall increases and the region where the fluid has a gaslike density decreases and moves toward the right wall. Note that the asymmetric solutions are twofold degenerate. The profiles obtained from those shown in Fig. 4(a) by reflection about the middle of the slit are also solutions of the Euler-Lagrange equation.

In Fig. 4(b) two profiles, asymmetric and symmetric ones, obtained for $\rho^*_{av} = 0.1546$ and $T = 87$ K are shown as an example of multiple solutions of Euler-Lagrange equation. It was found that the asymmetric density profile always provides a slightly smaller Helmholtz free energy than the symmetrical one (see Table 1) and, hence, the former corresponds to a stable state. This means that, indeed, symmetry breaking occurs in closed systems.

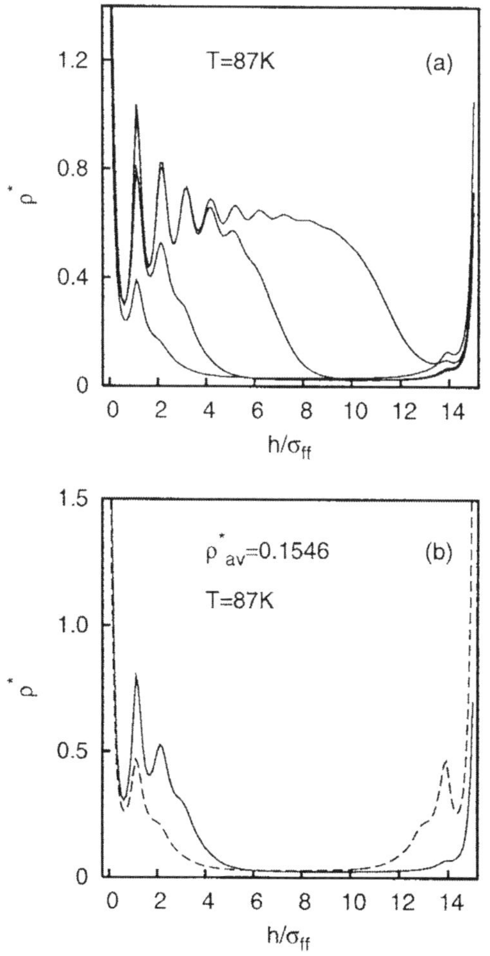

FIG. 4. Density profiles in the closed slit for $h^*_m = 15$. (a) The stable asymmetric profiles at $T = 87$ K plotted for $\rho^*_{av} = 0.1043, 0.1546, 0.3092$, and 0.4978. The fluid density near the left wall increases with increasing average density. (b) The stable (asymmetric profile) and metastable (symmetric profile) solutions of the Euler-Lagrange equation in closed slit for average density $\rho^*_{av} = 0.1546$ and temperatures $T = 87$ K.

TABLE 1
The values (per unit area of one wall) of the Helmholtz free energies F_{sym} and F_{asym} of the metastable (symmetric density profile) and stable (asymmetric density profile) states of the fluid in the slit with $h_m^* = 15$, respectively, for various average densities $\rho_{av1}^* < \rho_{av}^* < \rho_{av2}^*$. The temperature $T = 87$ K. Free energies are given in k_BT/σ_{ff}^2 units.

ρ_{av}^*	F_{sym}	F_{asym}
0.1546	–26.59	–26.67
0.2319	–39.66	–39.86
0.3092	–52.81	–53.06
0.3865	–66.01	–66.26
0.4638	–79.22	–79.48

To characterize quantitatively the asymmetry of the density profiles, the parameter

$$\Delta_N = \frac{1}{2N}\int_0^{h_m} dh\left|\rho(h) - \rho(h_m - h)\right| \quad (23)$$

was used. This parameter is equal to zero if the profile is symmetric about the middle of the slit and $0 < \Delta_N \leqslant 1$ in the opposite case. In the extremly asymmetric case where all molecules are located on one side of the plane of symmetry, $\Delta_N = 1$. The dependence of Δ_N on the average density ρ_{av}^* is presented in Fig. 5 for $T = 87$, 96, and 104 K. Two features should be mentioned. First, for each of the considered temperatures there is a range of average densities $\rho_{sb1}^* < \rho_{av}^* < \rho_{sb2}^*$, where Δ_N has significant values and becomes almost zero outside this range. Near ρ_{sb1}^* and ρ_{sb2}^* the parameter Δ_N changes very rapidly with ρ_{av}^* and, therefore, those values of ρ_{av}^* should be considered as "critical" values. Second, the critical value ρ_{sb1}^* increases, and ρ_{sb2}^* decreases with increasing temperature.

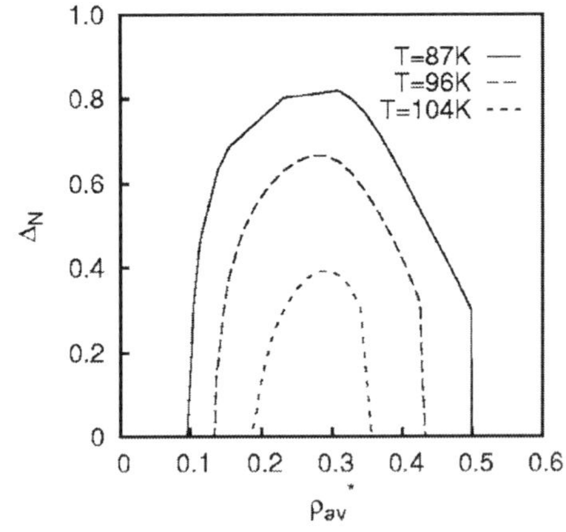

FIG. 5. The dependence of the parameter Δ_N, characterizing the asymmetical density distribution across a closed slit with $h_m^* = 15$, on the average density ρ_{av}^* of the fluid inside the slit for three different temperatures. For $T \geqslant 106$ K $\Delta_N = 0$ for all ρ_{av}^* and all density profiles are symmetrical about the middle of the slit.

The temperature dependencies of ρ^*_{sb1} and ρ^*_{sb2} are presented in Fig. 6. Symmetry breaking occurs only for those $(\rho^*_{av} - T)$ pairs which are located between the two curves on that figure. The difference $\rho^*_{sb1} - \rho^*_{sb2}$ which provides the range of existence of the asymmetric solution and hence, of the symmetry breaking, decreases with increasing temperature. For $T > T_{sb} \simeq 106$ K no symmetry breaking occurs and the profile is symmetrical about the middle of the slit.

At a selected ρ^*_{av}, Δ_N also has a critical-like behavior as function of temperature. As an example, the T dependence of Δ_N in a slit with $\rho^*_{av} = 0.1546$ is presented in Fig. 7 which shows that there is a very rapid change of Δ_N near $T = T_{sb1} \simeq 99.2$ K.

Some remarks should be made regarding the connection between the solutions obtained for a closed slit and those for an open slit as well as on the existence of symmetry breaking in open slits. As already mentioned, the Euler-Lagrange equation [Eq. (16)] for the density profile has the same form for both the closed and open slits. In the former case, λ is a Lagrange multiplier which is not known in advance. Its value $\lambda = \lambda_0$ can be calculated from Eq. (19) by selecting a number of molecules $N = N_0$ in the slit and using the density profile $\rho(\mathbf{r})$ obtained from Eq. (16). In this way, a pair (N_0, λ_0) can be related to each density profile. Obviously, that profile is also a solution of

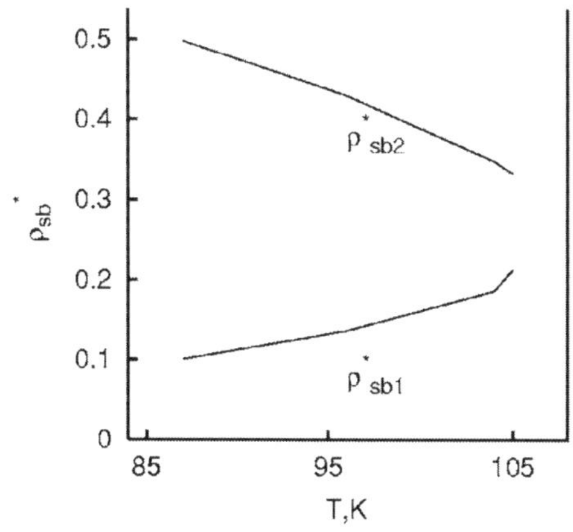

FIG. 6. Temperature dependence of the critical values ρ^*_{sb1}, and ρ^*_{sb2} of the average density in the closed slit with $h^*_m = 15$ for T$\geqslant$87 K. Symmetry breaking occurs when $\rho^*_{sb1} < \rho^*_{av} < \rho^*_{sb2}$. For temperatures higher than $T = 106$ K no symmetry breaking occurs.

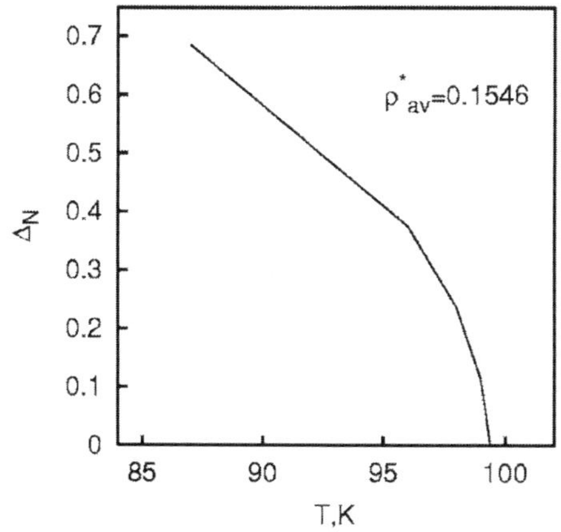

FIG. 7. Temperature dependence of the parameter Δ_N in the closed slit with $h^*_m = 15$ for T > 87 K and $\rho^*_{av} = 0.1546$. Symmetry breaking occurs (i.e., $\Delta_N > 0$) when $T \leqslant 99.4$ K.

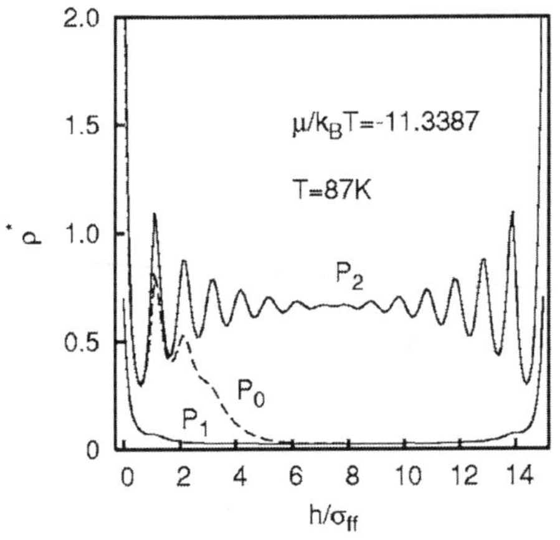

FIG. 8. Stable (P_2) and metastable (P_0 and P_1) density profiles for an open slit for $\mu = -11.3387k_BT$ and $T = 87$ K. The asymmetric density profile has the largest grand canonical free energy.

Eq. (16) for an open slit, corresponding to the chemical potential $\mu = \lambda_0$. In this case, the number of molecules in the slit can be obtained from Eq. (3). Vice versa, any solution of Eq. (16) for an open slit with a chemical potential $\mu = \mu_1$ provides a density profile with a certain number of molecules, say, $N = N_1$, and this solution is also the solution of Eq. (16) for a closed system with $N = N_1$ and $\lambda = \mu_1$. As a consequence, the asymmetric profile obtained from the solution of the Euler-Lagrange equation for a closed slit provides also a solution for an open slit. (Such solutions for open slits with identical walls were not yet reported). Let us consider as an example an asymmetric density profile P_0 obtained for a closed slit for $\rho^*_{av} = 0.1546\left(N = 2\times10^{19}\right)$ and $T = 87$K. This profile is presented in Fig. 8 by a dashed curve. In this case the value of the Lagrange multiplier λ is $\lambda_0 = -11.3387k_BT$. The usual iteration procedure applied to the Euler-Lagrange equation for an open slit with $\mu = \lambda_0$ and started with an arbitrary initial guess provided two other density profiles P_1 and P_2, both symmetrical, which are presented in Fig. 8 by the solid lines. Profile P_1 describes the state with $\rho^*_{av} = 0.04843$, and profile P_2 describes the state with $\rho^*_{av} = 0.6748$. They were obtained from initial guesses corresponding to the average densities $\rho^*_{av1} = 0.07729$ and $\rho^*_{av2} = 0.7729$, respectively. It should be noted that in both cases the final profile was the same whether the initial guess was symmetrical or asymmetrical about the middle of the slit. Profile P_0 could be obtained through such iterations only when the initial guess was near the final result. Because it is almost impossible to make such a guess, this explains why asymmetric solutions were not found for open slits. We could obtain the asymmetric profile for the open slit on the basis of the correspondence between the Euler-Lagrange equations for open and closed slits. Calculating for each of the obtained profiles the grand canonical free energy per unit area of one wall, $\Omega_{\mu,i}$ ($i = 0,1,2$) using the potential

$$\Omega_\mu(\rho) = F(N,V,T) - \mu\int_V d\mathbf{r}\rho(\mathbf{r}), \tag{24}$$

we obtained $\Omega_{\mu,0} = -0.3847$, $\Omega_{\mu,1} = -0.5539$, and $\Omega_{\mu,2} = -1.5749$ for the profiles P_0, P_1, and P_2, respectively. (The free energy is expressed in k_BT/σ_{ff}^2 units). The state with the smallest free energy is described by the profile P_2. The asymmetric profile has the highest free energy and therefore represents a metastable state. Hence, in the considered example, symmetry breaking does not provide an equilibrium state.

One should note that along with the asymmetrical, the symmetrical solution of the Euler-Lagrange equation for a closed slit is also a solution for an open slit for another value of the chemical potential, hence λ. In the considered example, the value of the Lagrange multiplier for the symmetrical profile is $\lambda_s = -11.2114k_BT \neq \lambda_0$. To determine the number of solutions of Eq. (16) for an open slit corresponding to the same chemical potential let us represent the Lagrange multipliers λ for symmetrical and asymmetrical profiles in closed slits as functions of the average density in the range of existence of asymmetrical solution (see Fig. 9). This figure shows that for any λ in the range $-11.49 \leqslant \lambda/k_BT \leqslant -11.15$ there are two solutions, one symmetrical and another asymmetrical, of the Euler-Lagrange equation for a closed slit, and hence, the same solutions are valid for an open slit with $\mu = \lambda$. In general, they do not coincide with the solutions obtained by iterations for an open slit. Consequently, at least in some cases, the Euler-Lagrange equation for an open slit can have four solutions. The existence and stability of those solutions for specific values of the parameters (chemical potential, temperature, and size of the slit) will be examined in a separate paper.

Asymmetrical profiles appear in a natural way in an open slit with nonidentical walls. As an example, two solutions for the slit with $\epsilon_{fw2} = \epsilon_{fw1}/2$ and $\mu/k_BT = -11.5$ are presented in Fig. 10. The stable solution is that with the larger average density.

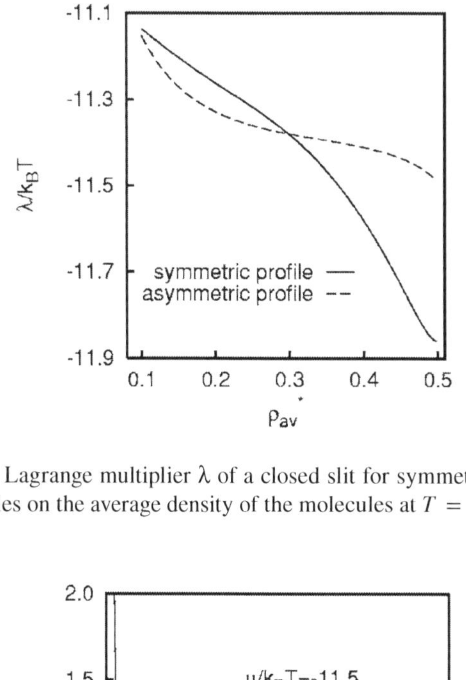

FIG. 9. Dependence of the Lagrange multiplier λ of a closed slit for symmetrical (solid curve) and asymmetrical (dashed curve) profiles on the average density of the molecules at $T = 87$ K.

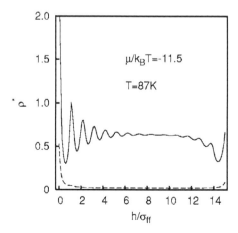

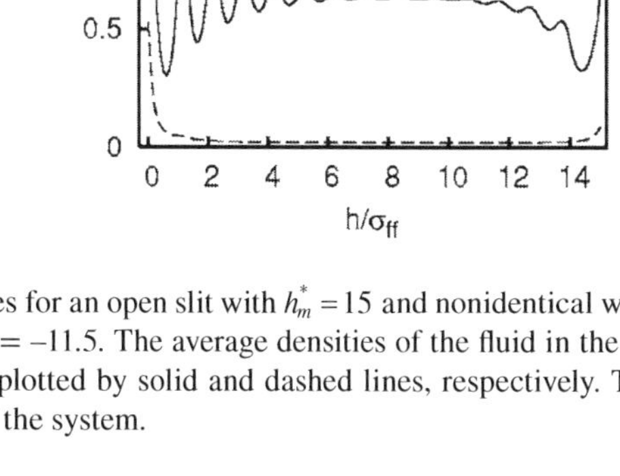

FIG. 10. Two density profiles for an open slit with $h_m^* = 15$ and nonidentical walls ($\epsilon_{fw2} = 0.5\epsilon_{fw1}$) at $T = 87$ K and chemical potential $\mu/k_BT = -11.5$. The average densities of the fluid in the slit are equal to $\rho_{av1}^* = 0.6114$ and $\rho_{av2}^* = 0.030$ for profiles plotted by solid and dashed lines, respectively. The profile with $\rho_{av}^* = \rho_{av1}^*$ corresponds to the stable state of the system.

IV. CONCLUSION

The main result of this paper is that in the considered system (argon in a closed slit with identical walls of solid carbon dioxide and $h_m = 15\sigma_{ff}$) there is a range of parameters (average density ρ_{av} and temperature T) in which the stable state of the system has a density profile asymmetric about the middle of the slit. Simultaneously, there is a metastable state characterized by a symmetric density profile. Symmetry breaking occurs only for temperatures $T < T_{sb}$ where $T_{sb} = 106$ K for the considered slit.

At any given temperature $T < T_{sb}$ the possible profiles of the system as function of ρ^*_{av} are presented schematically in Fig. 11 where for the sake of simplicity the density oscillations are neglected. For $\rho^*_{av} \leqslant \rho^*_{sb1}$ and $\rho^*_{av} \geqslant \rho^*_{sb2}$ the stable profile is symmetric and no asymmetric or metastable symmetric solutions exist. In the range $\rho^*_{sb1} \leqslant \rho^*_{av} \leqslant \rho^*_{sb2}$, a spontaneous symmetry breaking of the density distribution inside the slit occurs and the stable profile is asymmetric, being twofold degenerate with respect to the reflection about the middle of the slit. In the latter range, there is also a metastable symmetric profile the free energy of which differs very little from that of the stable state. If the energy barrier between the stable and metastable states is large enough (the estimation of this barrier is a separate problem which was not considered in this paper), the system can have a long lifetime in the latter state before achieving equilibrium.

The results obtained for closed slits could be extended to open slits in contact with reservoirs. In this manner, asymmetric solutions could be obtained for open slits. In the example presented this solution was metastable.

Several issues which were not adressed in this paper should be mentioned.

The first general issue is to relate the symmetry breaking transition to those investigated in the large number of studies of wetting and drying, e.g., wetting and prevetting transitions.

The second one is the dependence of the symmetry breaking on the parameters of the fluid-fluid and fluid-solid interactions. In the present paper, symmetry breaking was proven for the case where the fluid-fluid energy parameter ϵ_{ff} was smaller than the fluid-solid one. However, in Ref. 17, where the opposite inequality was valid, the symmetry breaking was absent. It is likely that there is a critical value of the fluid-solid energy parameter (for given fluid-fluid one) above which symmetry breaking occurs and it is interesting to find its value.

A third issue is related to the dependence of the symmetry breaking transition on the size of the slit, and finally, a forth one to the more complete examination of the existence of symmetry breaking in open slits.

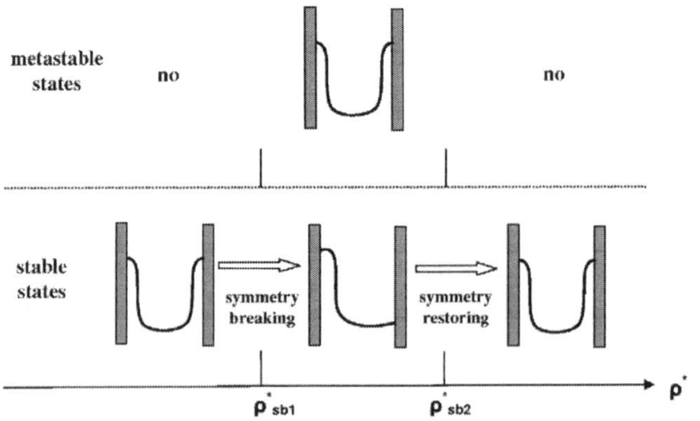

FIG. 11. The possible states and corresponding density profiles in the slit at fixed temperature $T < T_{sb}$ as function of the average fluid density.

APPENDIX: INTERACTION POTENTIALS

Taking into account that the density $\rho(\mathbf{r})$ depends only on the distance h of the point $\mathbf{r}$ from the left wall, the expression Eq. (14) for $U_f(\mathbf{r}) \equiv U_f(h)$ can be rewritten after integration over the lateral coordinates in the form

$$U_f(h) = \int_0^{h_m} dh' \rho(h') K(|h-h'|), \tag{A1}$$

where

$$K(h-h') = 2\pi\epsilon_{ff}\sigma_{ff}^2 \begin{cases} (2/5)\sigma_{ff}^{10}/(h-h')^{10} - \sigma_{ff}^4/(h-h')^4, & |h-h'| > \sigma_{ff}, \quad \text{(A2)} \\ -3/5, & |h-h'| > \sigma_{ff} \quad \text{(A3)} \end{cases}$$

The potential $U_s \equiv U_s(h)$ after integration over the volumes of the walls is given by the expression

$$U_s(h) = \psi_1(h) + \psi_2(h_m - h), \tag{A4}$$

where

$$\psi_i(h) = \frac{2\pi}{3}\epsilon_{fwi}\rho_{wi}\left[\frac{2}{15}\left(\frac{\sigma_{fwi}}{\sigma_{fwi}+h}\right)^9 - \left(\frac{\sigma_{fwi}}{\sigma_{fwi}+h}\right)^3\right], \tag{A5}$$

($i = 1, 2$).

The derivative $\partial\Delta\psi_{hs}[\bar{\rho}(\mathbf{r}')]/\partial\bar{\rho}(\mathbf{r}')$ is given by

$$\frac{\partial\Delta\psi_{hs}[\bar{\rho}(\mathbf{r}')]}{\partial\bar{\rho}(\mathbf{r}')} = k_B T \frac{\pi}{6}\sigma_{ff}^3 \frac{4-2\eta_\rho}{(1-\eta_\rho)^2}, \eta_\rho = \frac{\pi}{6}\bar{\rho}(\mathbf{r}')\sigma_{ff}^3. \tag{A6}$$

REFERENCES

1. H. Löwen, Phys. Rep. **237**, 249 (1994).
2. M. Merkel and H. Löwen, Phys. Rev. E **54**, 6623 (1996).
3. J. J. Magda, M. Tirrel, and H. T. Davis, J. Chem. Phys. **83**, 1888 (1985).
4. R. Evans, U. M. B. Marconi, and P. Tarazona, J. Chem. Phys. **84**, 2376 (1986).
5. P. Tarazona, U. M. B. Marconi, and R. Evans, Mol. Phys. **60**, 573 (1987).
6. R. Evans, J. Phys.: Condens. Matter **2**, 8989 (1990).
7. S. Sarman, J. Chem. Phys. **92**, 4447 (1990).
8. C. Lastoskie, K. E. Gubbins, and N. Quirke, Langmuir **9**, 2693 (1993).
9. A.Maciolek, R. Evans, and N. B. Wilding, J. Chem. Phys. **119**, 8663 (2003).
10. P. Tarazona, Phys. Rev. A **31**, 2672 (1985).
11. R. H. Nilson and S. K. Griffiths, J. Chem. Phys. **111**, 4281 (1999).
12. R. Evans and P. Tarazona, Phys. Rev. A **28**, 1864 (1983).
13. A. González, J. A. White, F. L. Román, and S. Velasco, Phys. Rev. Lett. **79**, 2466 (1997).
14. J. A. González, F. L. White, R. Román, and J. Evans, Chem. Phys. **109**, 3637 (1998).
15. J. A. White and S. Velasco, Phys. Rev. E **62**, 4427 (2000).
16. J. A. White, A. Gonzalez, F. L. Roman, and S. Velasco, Phys. Rev. Lett. **84**, 1220 (2000).
17. G. O. Berim and E. Ruckenstein, J. Chem. Phys. **125**, 164717 (2006).
18. G. O. Berim and E. Ruckenstein, J. Phys. Chem. B **111**, 2514 (2007) (Section 2.1 of this volume).

19. F. Ancilotto and F. Toigo, Phys. Rev. B **60**, 9019 (1999).
20. S. Dhawan, M. E. Reimel, L. E. Scriven, and H. T. Davis, J. Chem. Phys. **94**, 4479 (1991).
21. C. Ebner and W. F. Saam, Phys. Rev. Lett. **38**, 1486 (1977).
22. G. Mistura, F. Ancilotto, L. Bruschi, and F. Toigo, Phys. Rev. Lett. **82**, 795 (1999).
23. N. F. Carnahan and K. E. Starling, J. Chem. Phys. **51**, 635 (1969).

2.3 Effect of Fluid-Solid Interactions on Symmetry Breaking in Closed Nanoslits*

Gersh Berim and Eli Ruckenstein[†]

Department of Chemical and Biological Engineering,
State University of New York at Buffalo, Buffalo, New York 14260

Corresponding Author

[†] E-mail: feaeliru@buffalo.edu. Phone: (716) 645-1179,
Fax: (716) 645-3822.

Received: February 28, 2007; In Final Form: August 1, 2007

1. INTRODUCTION

A fluid in a slit with identical walls is subjected to an external potential due to the fluid—solid interactions which possess a reflectional symmetry about the middle of the slit. The same symmetry is present in the basic Euler—Lagrange equation,[1] which describes the equilibrium density profile across the slit and provides a symmetrical solution for the profile.[1–7] However, it is known that in some cases the symmetry of the stable state of a system can be lower than the symmetry of the equations describing the system. This symmetry breaking (SB) is well-known in physics, chemistry, biology, etc. A classical example from the physics of fluids is the freezing transition[8] in which the continuous translational symmetry of a fluid (in the absence of an external potential) is broken. The discrete symmetry, e.g., reflectional, or the left—right one, can also be broken. A well-known example is the occurrence of spontaneous magnetization in ferromagnetic materials in the absence of an external magnetic field at temperatures lower than the Curie temperature. The possibility of a reflectional SB was demonstrated in ref 9 for an open system consisting of two parallel identical layers of a two-dimensional liquid coupled by an interlayer potential.

In ref 10, the existence of SB for fluid density profiles across a long closed nanoslit with identical walls was predicted by employing the local density functional theory (DFT) and was confirmed later in ref 11 with a more accurate nonlocal DFT. Argon was the fluid, and the walls were considered composed of carbon dioxide. Lennard-Jones potentials were used for both the fluid—fluid and fluid—walls interactions. It was shown that, for selected values of the energy parameters of the fluid—fluid (ϵ_{ff}) and fluid—walls (ϵ_{fw}) interactions, the Euler—Lagrange equation for the fluid density profile has in some cases two solutions. One of them is symmetrical about the middle of the slit and the other one asymmetrical. The free energy of the state described by the asymmetrical profile was always lower than for the symmetrical profile. Therefore, the stable state of the system is that with the asymmetrical density distribution across the slit, and the state possessing a symmetrical density distribution is

* *J. Phys. Chem.* B 2007, 111, 12823–12828. Republished with permission.

metastable. Such a symmetry breaking occurred only in some domain of the $\rho_{av} - T$ plane, where ρ_{av} and T are the average fluid density in the slit and the absolute temperature, respectively. In the present paper, the considerations of ref 11 are extended to other wall materials which are characterized by the fluid—wall energy parameter ϵ_{fw}. It is shown that there are two temperature-dependent critical values of ϵ_{fw}, ϵ_{fw1} and ϵ_{fw2}, and that SB occurs when ϵ_{fw} is between those two critical values. In addition, in the last section, we compare some of our results with Monte Carlo simulations.

2. BACKGROUND

The system under consideration was described in detail in ref 11. It consists of one component fluid (argon) confined in a long closed nanoslit between two parallel structureless identical solid walls W_1 and W_2 separated by a distance L (see Figure 1). The distance between the side walls which close the slit is considered to be very large compared to L. As a result, the end effects can be considered negligible. It is also assumed that the density distribution is a function of only a single coordinate h (see Figure 1) that represents the distance of the fluid molecules from the left wall. The closed slit with planar walls can be also imagined as a thin layer of thickness L between a solid cylinder of large radius R $\left(R \to \infty\right)$ embedded into a cylindrical cavity of radius $R + L$.

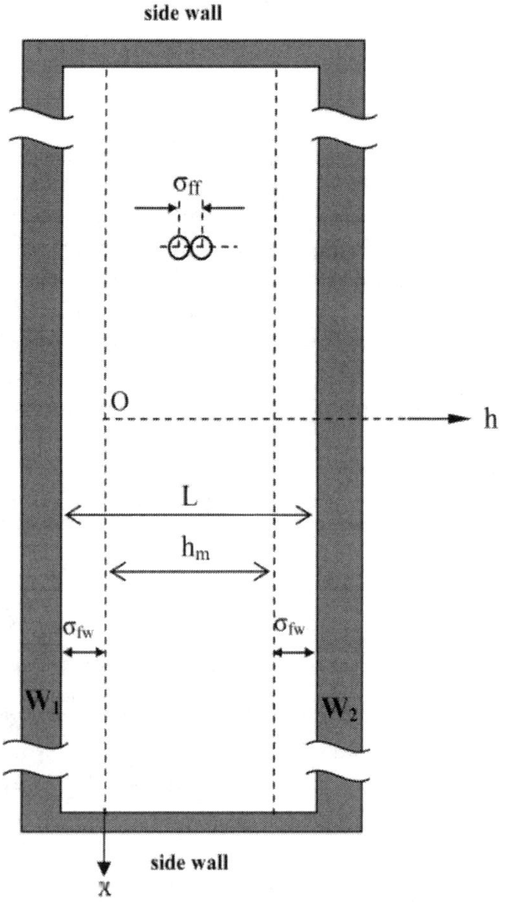

FIGURE 1. Closed slit of width L between two identical solid walls W_1 and W_2. σ_{ff} and σ_{fw} are the hard core diameters of the fluid—fluid and fluid—wall interactions, respectively. The y axis is perpendicular to the plane of the figure.

The interaction between fluid molecules is considered as the sum of a potential energy of a reference system (hard spheres) and an attractive perturbation which is treated in the mean-field approximation.

The attractive part of the fluid—fluid interactions is described by the Lennard-Jones 6–12 potential, $\phi_{ff}(|\mathbf{r}-\mathbf{r}'|)=4\epsilon_{ff}\left[(\sigma_{ff}/r)^{12}-(\sigma_{ff}/r)^{6}\right]$ for $r \geq \sigma_{ff}$ $(r=|\mathbf{r}-\mathbf{r}'|)$ and $\phi_{ff}(|\mathbf{r}-\mathbf{r}'|)=0$ for $r<\sigma_{ff}$. The interaction between the molecules of the fluid and those of the walls are described by the potential $\phi_{fw}(|\mathbf{r}-\mathbf{r}'|)=4\epsilon_{fw}\left[(\sigma_{fw}/r)^{12}-(\sigma_{fw}/r)^{6}\right]$ for $r \geq \sigma_{fw}$ and $\phi_{fw}(|\mathbf{r}-\mathbf{r}'|)=\infty$ for $r<\sigma_{fw}$. In the above potentials, σ_{ff} and σ_{fw} are hard core diameters and ϵ_{ff} and ϵ_{fw} are interaction energy parameters for fluid—fluid and fluid—walls interactions, respectively. The vectors $\mathbf{r}$ and $\mathbf{r}'$ provide the positions of the interacting molecules.

The density distribution of the fluid in the slit will be calculated using a nonlocal density functional theory of the form presented in ref 1. Such a theory was successfully used to describe a fluid in an open slit,[1,6,7,12] both for small and large fluid densities.

The Helmholtz free energy F of a system is given by

$$F=\int_0^{h_m} dh\left[k_B T\rho\left[\log\left(\Lambda^3\rho\right)-1\right]+\rho\Delta\Psi_{hs}(h)+\frac{1}{2}\rho U_f(h)+\rho U_S(h)\right] \tag{1}$$

where $\rho \equiv \rho(h)$ is the density profile, $\Lambda=h_P/(2\pi m k_B T)^{1/2}$ is the thermal de Broglie wavelength, h_P and k_B are the Planck and Boltzmann constants, respectively, m is the mass of a fluid molecule, T is the absolute temperature, $h_m=L-2\sigma_{fw}$ (see Figure 1). The function $\rho\Delta\Psi_{hs}(h)$ represents the free energy density of a hard core fluid and has the form

$$\rho\Delta\Psi_{hs}(h)=k_B T\rho\eta_{\bar{\rho}}\frac{4-3\eta_{\bar{\rho}}}{\left(1-\eta_{\bar{\rho}}\right)^2} \tag{2}$$

where $\eta_{\bar{\rho}}=(1/6)\pi\bar{\rho}(h)\sigma_{ff}^3$ is the packing fraction of the fluid molecules and $\bar{\rho}(h)$ is the smoothed density. The functions $U_f(h)$ and $U_s(h)$ account for the interactions between the fluid molecules and the fluid molecules with the solid walls, respectively. Their explicit expressions are provided in Appendix A. Here and below all extensive quantities (free energy, number of molecules N, volume V) are expressed per unit area of one wall.

The Euler—Lagrange equation for the density profile ρ in a closed slit, obtained by minimizing F, has the following form:[11]

$$k_B T\log\left(\Lambda^3\rho\right)+\Delta\Psi_{hs}(h)+\overline{\Delta\Psi'}_{hs}(h)+U_f(h)+U_s(h)=\lambda \tag{3}$$

where λ is the Lagrange multiplier arising because of the constraint

$$N=\int_0^{h_m}\rho(h)dh \tag{4}$$

of constant number of molecules in the closed slit and

$$\overline{\Delta\Psi'}_{hs}(h)=$$

$$\sum_{n=0}^{2}\int_V d\mathbf{r}'\frac{\rho(h')w_n(|\mathbf{r}-\mathbf{r}'|)\left[\bar{\rho}(h')\right]^n}{1-\bar{\rho}_1(h')-2\bar{\rho}(h')\bar{\rho}_2(h')}\frac{\partial\Delta\Psi_{hs}\left[\bar{\rho}(h')\right]}{\partial\bar{\rho}(h')} \tag{5}$$

In eqs 2 and 5

$$\bar{\rho}(h)=\int_V \mathrm{d}\mathbf{r}'\rho(h')w\left[|\mathbf{r}-\mathbf{r}'|;\bar{\rho}(h)\right] \tag{6}$$

the weighting functions $w\left[|\mathbf{r}-\mathbf{r}'|;\bar{\rho}(h)\right]$ and $w_n\left(|\mathbf{r}-\mathbf{r}'|\right)$ the coefficients $\bar{\rho}_n(h')\,(n=0,1,2)$ have been taken from refs 1 and 13, and an expression for $\left(\partial\Delta\Psi_{\mathrm{hs}}\left[\bar{\rho}(h')\right]/\partial\bar{\rho}(h')\right)$ is provided in Appendix A.

Formally, eq 3 coincides with that for an open system without side walls. However, the Lagrange multiplier λ in closed systems is not known in advance, in contrast to the open systems where λ is the chemical potential. The relation between λ and the number N of molecules is provided by the constraint eq 4 and has the form[11]

$$\lambda=-k_{\mathrm{B}}T\log\left[\frac{1}{N\Lambda^3}\int_0^{h_{\mathrm{m}}}\mathrm{d}h\,e^{Q(h)}\right] \tag{7}$$

where

$$Q(h)=\frac{1}{k_{\mathrm{B}}T}\left[\Delta\Psi_{\mathrm{hs}}(h)+\overline{\Delta\Psi'}_{\mathrm{hs}}(h)+U_{\mathrm{f}}(h)+U_{\mathrm{S}}(h)\right] \tag{8}$$

By eliminating λ between eqs 3 and 7, one obtains an integral equation for the density profile ρ(h) that can be solved by numerical iterations of the kind used in ref 13. The input density profile $\rho_i^{\mathrm{in}}(h)$ for the (i + 1)th iteration $\rho_{i+1}(h)$ generated by the Euler—Lagrange equation is given by the expression

$$\rho_{\mathrm{i}}^{\mathrm{in}}(h)=(1-x)\rho_{i-1}^{\mathrm{in}}(h)+x\rho_i(h) \tag{9}$$

where $\rho_i(\mathrm{h})$ is the ith iteration and the constant x was taken 0.01. The value of λ, if needed, can be found from eq 7 using the obtained equilibrium density profile. Owing to the mentioned identity of the Euler—Lagrange equations for $\rho(h)$ in open and closed systems, one can consider the obtained value of λ as the chemical potential of a fictitious open system which has the same density profile as the closed one. (Note, that the state of the open system with such a profile was metastable in all cases considered by us.[11])

For any selected set of parameters (temperature, average density of the fluid in the slit, and strength of the fluid—wall interaction), two initial guesses for the density profile were employed. A homogeneous density equal to a selected average density was employed as the first initial guess. In this case, the resulting profile was always symmetrical. As a second choice of the initial guess, an asymmetrical profile describing the state with most of the molecules located in the left (right) half of the slit was used. Starting with such a guess, either a symmetrical density profile that coincided with that obtained with the symmetrical initial guess, or an asymmetrical profile representing an additional solution to the Euler—Lagrange equation was obtained. When an asymmetrical density profile was obtained, it always provided a lower free energy than the symmetrical profile.

Let us make some remarks regarding the number of iterations in the numerical computations. As a measure of the precision of the iteration procedure, the dimensionless quantity $\delta=\sigma_{\mathrm{ff}}^{5}\int_0^{h_{\mathrm{m}}}\mathrm{d}h\left[\rho_{i+1}(h)-\rho_i^{\mathrm{in}}(h)\right]^2$ is introduced which characterizes the difference between (i + 1) th iteration $\rho_{i+1}(h)$ and the input profile $\rho_i^{\mathrm{in}}(h)$ from which that iteration was obtained. When the initial guess in computations was selected to be homogeneous, the iterations were carried out until δ became smaller than $\epsilon = 10^{-7}$. The increase in precision did not lead to appreciable changes in the density profile. However, during the computation which started with an asymmetrical initial guess, the precision $\epsilon = 10^{-7}$ was not sufficient, because the profile $\rho_i(h)$ which was asymmetrical for $\epsilon = 10^{-7}$ sometimes became symmetrical with increased precision. To increase the confidence that the density profile obtained is symmetrical or asymmetrical, the precision of calculations was increased up to $\epsilon = 10^{-12}$ and the dependence of the parameter

$$\Delta_N = \frac{1}{2N}\int_0^{h_m} dh\left|\rho(h)-\rho(h_m-h)\right| \tag{10}$$

on ϵ was examined by varying ϵ from 10^{-7} to 10^{-12}. The parameter Δ_N should be equal to zero if the profile is symmetrical about the middle of the slit and $0 < \Delta_N \leq 1$ in the asymmetrical case. The dependence of Δ_N on ϵ had one of the following two asymptotic behaviors. In the first case, Δ_N continuously decreased approaching zero. The resulting profile in this case was considered symmetrical even though the profile obtained after achiving the precision $\epsilon = 10^{-12}$ had a small asymmetry. In the second case, Δ_N reached a nonzero asymptotic value beginning with some $\epsilon > 10^{-12}$ (say, $\epsilon = 10^{-9}$), which remained unchanged with increasing precision. Only in such a case the resulting asymmetric profile was considered to be a solution of the Euler—Lagrange equation.

All obtained density profiles were tested with respect to their thermodynamic stability by examining their response to arbitrarily small perturbations of the density distribution. To perform such a test we used the second variation δ^2F of the free energy F with respect to the fluid density profile, which has the following general form

$$\delta^2F = \int_0^{h_m} dh\,\delta\rho(h)\int_0^{h_m} dh' X(h,h')\delta\rho(h') \tag{11}$$

where $\delta\rho(h)$ is a small perturbation of the density profile and $X(h, h')$ is a function of $\rho(h)$ and $\rho(h')$. The explicit form of $X(h, h')$ is provided in Appendix B. The state of the system described by a given density profile is stable if δ^2F is positive for any $\delta\rho(h)$. In this case, the free energy of the system has a local minimum for that profile. Because the interactions depend on the densities in two different points, it is difficult to provide a general proof of the stability condition. Instead, we checked the sign of δ^2F for a variety of small perturbations. For example, perturbations of the form

$$\delta\rho(h) = A(h) + B(h)\left[\sin\left(\frac{2\pi h}{h_m}n_1\right)\right]^{n_2} + C(h)\left[\cos\left(\frac{2\pi h}{h_m}n_3\right)\right]^{n_4} \tag{12}$$

were considered, where $A(h)$, $B(h)$, and $C(h)$ have the form of power or exponential functions of h and $n_1, \ldots, n_4$ are constants (not neccessarly integers). The coefficients have been adjusted to maintain the number of molecules in the slit constant. In all considered cases, the second variation was positive, confirming the thermodynamic stability of the obtained density profiles.

3. RESULTS

In the calculations, the parameters ϵ_{ff} and σ_{ff} were selected as for argon:[14] $\epsilon_{ff}/k_B = 119.76$ K, $\sigma_{ff} = 3.405$ Å. The chemical nature of the walls was taken into account only through the energy parameter ϵ_{fw}, whereas the hard core diameter σ_{fw} and the walls density ρ_w were kept constant, $\sigma_{fw} = 3.727$ Å, $\rho_w = 1.91 \times 10^{28}$ m^{-3}. The width of the slit was taken $h_m = 15\sigma_{ff}$. The dimensionless quantities $\rho^* = \rho\sigma_{ff}^3, h^* = h/\sigma_{ff}$ for the fluid density and distance from the wall, respectively, were employed in the calculations. In what follows, ϵ_{fw} will be expressed in units of the energy parameter ϵ_{ff} through the dimensionless quantity $\varepsilon_{fw} = \epsilon_{fw}/\epsilon_{ff}$.

In ref 11, it was shown that at a selected temperature $T < T_{sb}$, where T_{sb} is a critical temperature above which no symmetry breaking occurs, SB was observed only for an average fluid density $\rho^*_{av} = N\sigma^2_{ff}/h^*_m$ in the slit which satisfies the inequalities $\rho^*_{sb1} < \rho^*_{av} < \rho^*_{sb2}$, where the critical densities ρ^*_{sb1} and ρ^*_{sb2} depend on temperature. For instance, for $\varepsilon_{fw} = 1.277$ (walls of solid carbon dioxide) and $T = 104$ K, $\rho^*_{sb1} = 0.1971$ and $\rho^*_{sb2} = 0.3517$.[11] As expected, the present calculations show that at a selected temperature the critical values ρ^*_{sb1} and ρ^*_{sb2} depend also on ε_{fw}. In Figure 2a, the dependencies of ρ^*_{sb1} and ρ^*_{sb2} on also on ε_{fw} are plotted for $T = 104$ K. Symmetry breaking occurs only for those $\rho^*_{av} - \varepsilon_{fw}$ pairs which are located inside the area *ABCD*. For $\varepsilon_{fw} < \varepsilon_{fw1}$ or $\varepsilon_{fw} > \varepsilon_{fw2}$, where $\varepsilon_{fw1} = 0.34$ and $\varepsilon_{fw2} = 1.38$ are critical values, no SB occurs and the density profile is symmetric.

The dependence on ε_{fw} of the difference $\Delta\rho^*_{av} = \rho^*_{sb2} - \rho^*_{sb1}$ between the average critical densities in which SB occurs is presented in Figure 2b which shows that $\Delta\rho^*_{av}$ passses through a maximum $\Delta\rho^*_{av} = 0.986$. In the latter case, $\rho^*_{sb1} = 0.1662$, $\rho^*_{sb2} = 0.3517$.

As shown in Figure 3, the critical values ε_{fw1} and ε_{fw2} depend on temperature. The former slightly increases and the latter decreases with increasing temperature. As a result, the interval $\Delta\varepsilon_{fw} = \varepsilon_{fw2} - \varepsilon_{fw1}$ in which SB occurs decreases with increasing temperature.

By increasing the average fluid density ρ^*_{av} at a fixed value of ε_{fw}, the density profile of the stable state of the system first looses its symmetry for ρ^*_{av} larger than ρ^*_{sb1} and then restores it for ρ^*_{av} larger than ρ^*_{sb2}. In Figure 4a, the stable profiles corresponding to $\rho^*_{av} = 0.1005$, 0.1546, and 0.3092 are presented for $T = 87$ K and $\varepsilon_{fw} = 0.639$. (At that temperature, $\rho^*_{sb1} = 0.1051$, and $\rho^*_{sb2} = 0.5241$). For $\rho^*_{av} = 0.1005$ the profile is symmetrical and for $\rho^*_{av} = 0.1546$ and $\rho^*_{av} = 0.3092$ they are asymmetrical. The restoring of the profile symmetry near ρ^*_{sb2} is illustrated in Figure 4b where the profiles for $\rho^*_{av} = 0.5194$, 0.5225 (asymmetrical), and 0.5411 (symmetrical) are presented. Note that, in addition

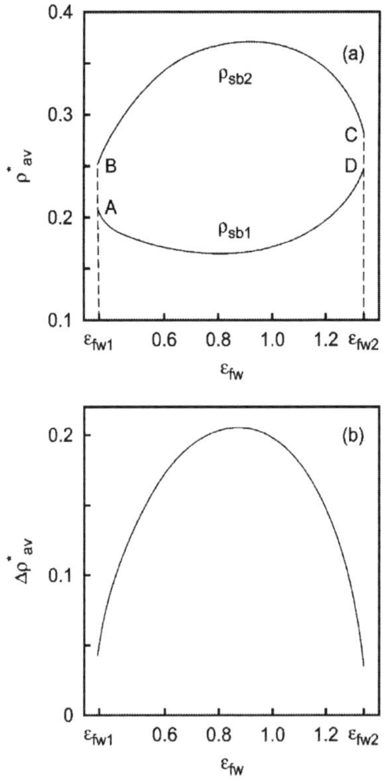

FIGURE 2. (a) Dependence of the critical densities ρ^*_{sb1} and ρ^*_{sb2} on ε_{fw} for $T = 104$ K. Symmetry breaking occurs only for those $\rho^*_{av} - \varepsilon_{fw}$ pairs which are inside the area *ABCD*. The critical values of ε_{fw} are equal to $\varepsilon_{fw1} = 0.34$ and $\varepsilon_{fw2} = 1.38$. (b) Dependence of the interval $\Delta\rho^*_{av} = \rho^*_{sb2} - \rho^*_{sb1}$ on ε_{fw} for $T = 104$ K.

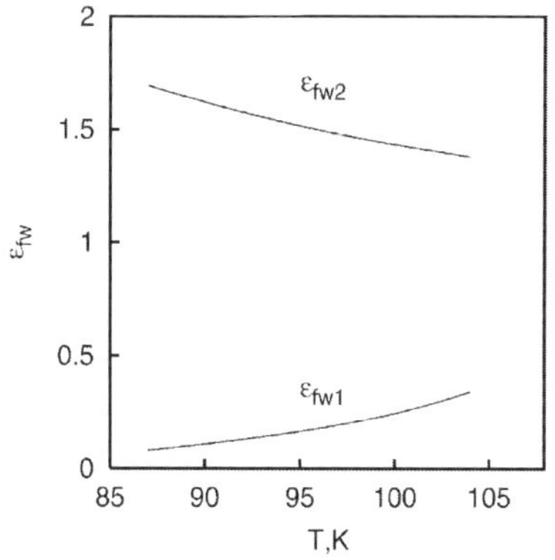

FIGURE 3. Temperature dependence of the critical values ε_{fw1} and ε_{fw2}. At fixed temperature, symmetry breaking occurs only if $\varepsilon_{fw1} < \varepsilon_{fw} < \varepsilon_{fw2}$.

to an asymmetrical density profile, the Euler—Lagrange equation for the same set of parameters provides also a symmetrical profile which has, however, a higher free energy than the asymmetrical one. For illustration, the free energies of the asymmetrical profiles of Figure 4a, b are presented in Table 1 along with the free energies of the related symmetrical profiles.

At fixed values of ρ^*_{av} and T the shape of the stable profile depends on ε_{fw}. As an example, the profiles for $\rho^*_{av} = 0.2319$ and $\varepsilon_{fw} = 0.013$, 0.639, 1.534, and 1.789 are plotted in Figure for $T = 87$ K.

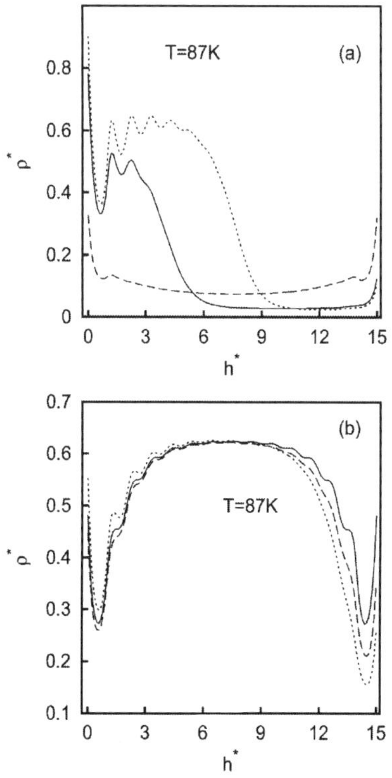

FIGURE 4. Stable density profiles for $\varepsilon_{fw} = 0.639$ and $T = 87$ K. In this case $\rho^*_{sb1} = 0.1051, \rho^*_{sb2} = 0.5241$. (a) Profiles for $\rho^*_{av} = 0.1005$ (dashed line), 0.1546 (solid line), and 0.3092 (dotted line). The first one is symmetrical about the middle of the slit, whereas the other two are asymmetrical. (b) Profiles for $\rho^*_{av} = 0.5194$ (dotted line), 0.5225 (dashed line), and 0.5411 (solid line). The last profile is symmetrical about the middle of the slit, the other two are asymmetrical.

TABLE 1
Values (per Unit Area of One Wall) of the Helmholtz Free Energies F_{sym} and F_{asym} of the Metastable (Symmetric Density Profile) and Stable (Asymmetric Density Profile) States of the Fluid in a Slit with $h_m^* = 15$, Respectively $h_m^* = 15$, for Various Average Densities $\rho_{av1}^* < \rho_{av}^* < \rho_{av2}^*$

ρ_{av}^*	F_{sym}	F_{asym}
0.1546	−25.76	−25.98
0.3092	−52.06	−52.26
0.5194	−88.21	−88.22
0.5225	−88.75	−88.76

[a] The temperature $T = 87$ K and $\varepsilon_{fw} = 0.639$. The free energies are given in k_BT/σ_{ff}^2 units.

In this case, SB occurs for $0.08 < \varepsilon_{fw} < 1.693$. Consequently, the profiles plotted for $\varepsilon_{fw} = 0.013$ and $\varepsilon_{fw} = 1.789$ are symmetrical and those for $\varepsilon_{fw} = 0.639$ and $\varepsilon_{fw} = 1.534$ asymmetrical. For the smallest value of $\varepsilon_{fw} = 0.013$, the fluid density near the walls is gas-like $\left[\rho^*(0) = \rho^*(h_m) \simeq 0.022\right]$ and in the middle of the slit it is liquid-like $\left[\rho^*(h_m/2) \simeq 0.58\right]$. Owing to the weak fluid—wall interactions, there are no visible density oscillations near the walls. For the largest value $\varepsilon_{fw} = 1.4$, the stable density profile is also symmetrical. The fluid density has the largest value at the walls $\left[\rho^*(0) = \rho^*(h_m) \simeq 4.46\right]$ and the smallest at the middle of the slit $\left[\rho^*(h_m/2) \simeq 0.021\right]$. There are density oscillations near the walls which have a large amplitude and penetrate into the slit up to two molecular diameters. The fluid density in the asymmetrical profiles has the largest, liquid-like, value near the left wall, which increases with increasing ε_{fw}. The depth of penetration of the oscillations into the slit increases with ε_{fw} and for $\varepsilon_{fw} = 1.534$ becomes about four molecular diameters.

4. DISCUSSION

As shown in previous papers,[10,11] the symmetry breaking of the fluid density profile across a closed slit with identical walls has several critical-like features. First, SB occurs only if the temperature T is smaller than a critical temperature T_{sb} which for argon between solid carbon dioxide walls is

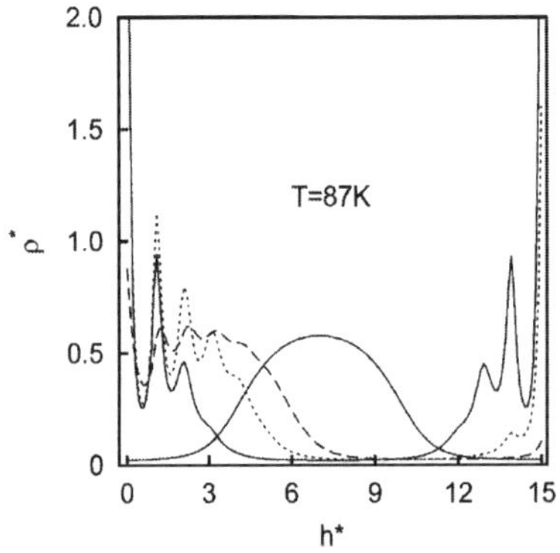

FIGURE 5. Stable density profiles for $\rho_{av}^* = 0.2319$, $T = 87$ K and various values of ε_{fw}. The symmetrical profiles (solid lines) are for $\varepsilon_{fw} = 0.013$ (has a maximum in the middle of the slit) and $\varepsilon_{fw} = 1.789$. The asymmetrical profiles are for $\varepsilon_{fw} = 0.639$ (dashed line) and $\varepsilon_{fw} = 1.534$ (dotted line). The critical values of ε_{fw} for $\rho_{av}^* = 0.2319$, and $T = 87$ K are $\varepsilon_{fw1} = 0.080$ and $\varepsilon_{fw2} = 1.693$.

equal to $T_{sb} \simeq 106$ K. Second, at selected $T < T_{sb}$, SB occurs only for those fluid average densities ρ_{av} in the slit that are between two critical values, ρ_{av1} and ρ_{av2} ($\rho_{av1} < \rho_{av2}$). Both ρ_{av1} and ρ_{av2} are temperature dependent.

In the present paper, one more critical-like feature of SB in a slit was noted. It was shown that, at a selected temperature, SB occurs only if the strength ϵ_{fw} of the fluid—wall interaction is located between two critical values ϵ_{fw1} and ϵ_{fw2}. The value of ϵ_{fw} $\left(\epsilon_{fw1} < \epsilon_{fw} < \epsilon_{fw2}\right)$ affects the critical densities ρ_{av1} and p_{av2} and the interval $\Delta\rho_{av} = \rho_{av2} - \rho_{av1}$ in which SB occurs.

It was also shown that, at a selected value of ρ_{av}, ϵ_{fw1} and ϵ_{fw2} depend on temperature and the interval $\Delta\epsilon_{fw} = \epsilon_{fw2} - \epsilon_{fw1}$ in which SB occurs decreases with increasing temperature.

The results obtained in the present and previous papers (refs 10 and 11) are based on the assumption that the density distribution is homogeneous in x and y directions. This assumption allowed to reduce the dimensionality of the problem to an one-dimensional one and to obtain numerical solutions which revealed the existence of SB in the h direction. It is, however, possible the closed system to undergo symmetry breaking also in x and y directions through the formation of liquid-like drops on the walls or liquid-like bridges between the walls. Such a possibility was identified in refs 16 and 17 for cylindrical pores in the axial direction. If SB takes place across as well as along the slit simultaneously, then the state with the lowest free energy could be inhomogeneous in all three directions. The complete answer to this question involves the difficult minimization of the free energy with respect to all three coordinates (x, y, h). The results of such calculations should be compared to those obtained in the present paper to verify when the free energy is smaller.

The occurrence of SB in h direction is supported by recent Monte Carlo simulations of argon between parallel identical walls of solid carbon dioxide ($\varepsilon_{fw} = 1.277$), obtained by Errington[18] for a slit with $h^*_m = 20$ at $T = 89.82$ K. In Figure 6, three snapshots representing the equilibrium states of three closed systems are presented for the average densities $\rho^*_{av,1} = 0.123$ (Figure 6a), $\rho^*_{av,2} = 0.246$ (Figure 6b), and $\rho^*_{av,3} = 0.556$ (Figure 6c). (Although oscillations of the density near the wall do exist, they are not visible in these figures.) The critical values ρ^*_{sb1} and ρ^*_{sb2} provided by our theory for these systems are 0.08 and 0.48, respectively, and the density profiles calculated for the average densities used in Monte Carlo simulations are plotted in Figure 7. The states presented in Figure 6a, b belong to the range of average densities where SB should exist according to the present theory (see Figure 7a, b). One of those states (Figure 6a) shows, in agreement with the calculations, an SB in the density distribution across the slit. Another one (Figure 6b) has a density distribution less asymmetrical than the predicted one.

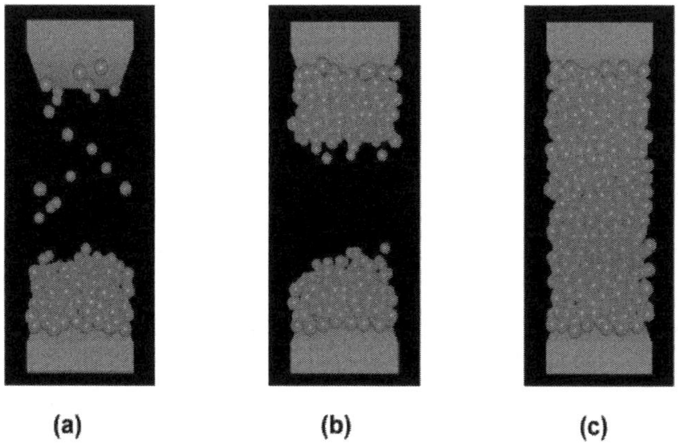

FIGURE 6. Monte Carlo images of the equilibrium states of argon between walls of solid carbon dioxide ($\varepsilon_{fw} = 1.277$). The width of the slit $h^*_m = 20$ and $T = 89.82$ K. The average densities ρ^*_{av} are (a) 0.123, (b) 0.246, and (c) 0.556.

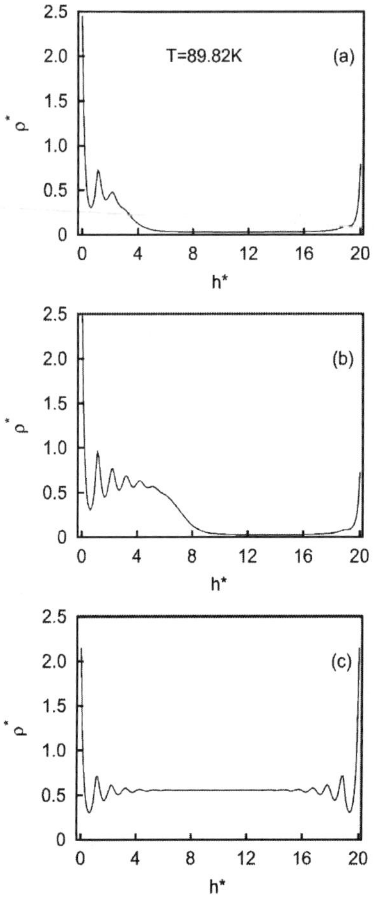

FIGURE 7. Stable density profiles calculated for argon between walls of solid carbon dioxide ($\varepsilon_{fw} = 1.277$). $h^*_m = 20$, $T = 89.82$ K. The average densities ρ^*_{av} are (a) 0.123, (b) 0.246, and (c) 0.556.

The third (Figure 6c) has a symmetrical distribution corresponding to the average density $\rho^*_{av3} > \rho^*_{av2}$, in agreement with the theory (Figure 7c).

ACKNOWLEDGMENT

The authors are grateful to Dr. J. Errington for permission to use the results of his Monte Carlo simulations prior to their publication.

APPENDIX A

The total fluid—walls interaction $U_s(h)$ in eq 3 is provided by the Lennard-Jones 3–9 potential, obtained by integrating ϕ_{fw} over the volume of the walls, by assuming constant densities of the walls.[15] This potential has the form

$$U_S(h) = \psi(h) + \psi(h_m - h) \tag{A.1}$$

where

$$\psi(h)=\frac{2\pi}{3}\epsilon_{\mathrm{fw}}\rho_w\sigma_{\mathrm{fw}}^3\left[\frac{2}{15}\left(\frac{\sigma_{\mathrm{fw}}}{\sigma_{\mathrm{fw}}+h}\right)^9-\left(\frac{\sigma_{\mathrm{fw}}}{\sigma_{\mathrm{fw}}+h}\right)^3\right] \tag{A.2}$$

Taking into account that the density $\rho(\mathbf{r})$ depends only on the distance h, $U_\mathrm{f}(h)$ can be written after integration over the lateral coordinates in the form

$$U_f(h)=\int_0^{h_\mathrm{m}}\mathrm{d}h'\rho(\mathrm{h}')K(h-h') \tag{A.3}$$

where

$$K(h-h')=2\pi\epsilon_{\mathrm{ff}}\,\sigma_{\mathrm{ff}}^2\begin{cases}\frac{2}{5}\left(\frac{\sigma_{\mathrm{ff}}}{(h-h')}\right)^{10}-\left(\frac{\sigma_{\mathrm{ff}}}{(h-h')}\right)^4, & |h-h'|>\sigma_{\mathrm{ff}}, \quad \text{(A.4)}\\ -\frac{3}{5}, & |h-h'|\le\sigma_{\mathrm{ff}} \quad \text{(A.5)}\end{cases}$$

The derivative $\left(\partial\Delta\Psi_{\mathrm{hs}}\left[\bar{\rho}(h')\right]/\partial\bar{\rho}(h')\right)$ in eq 5 is given by

$$\frac{\partial\Delta\Psi_{\mathrm{hs}}\left[\bar{\rho}(h')\right]}{\partial\bar{\rho}(h')}=K_\mathrm{B}T\frac{\pi}{6}\sigma_{\mathrm{ff}}^3\frac{4-2\eta_{\bar{\rho}}}{\left(1-\eta_{\bar{\rho}}\right)^2},\ \eta_{\bar{\rho}}=\frac{\pi}{6}\bar{\rho}(h')\sigma_{\mathrm{ff}}^3 \tag{A.6}$$

APPENDIX B

Using eq 1 for the Helmholtz free energy, one obtains the following expression for the second variation δ^2F with respect to small perturbations $\delta\rho^*(z)$ and $\delta\rho^*(z')$:

$$\frac{\delta^2F}{k_\mathrm{B}T/\sigma_{\mathrm{ff}}^2}=\int_0^{z_\mathrm{m}}\frac{\mathrm{d}z}{\rho^*(z)}\delta\rho^*(z)+\int_0^{z_\mathrm{m}}\mathrm{d}z\delta\rho^*(z)\int_0^{z_\mathrm{m}}\mathrm{d}z'X(z,z')\delta\rho^*(z') \tag{B.1}$$

where the dimensionless variable z stands for h/σ_{ff} and $\rho^*(z)=\rho(z)\sigma_{\mathrm{ff}}^3$. In eq B.1

$$X(z,z')=f_0\left[\bar{\rho}^*(z)\right]\varphi_1(z,z')+K_1(z-z')+2\pi\sum_{k=0}^{2}X_k(z,z') \tag{B.2}$$

where

$$X_k(z,z')=\frac{1}{\alpha(z')}f_k\left[\bar{\rho}^*(z')J_k(z-z')+\right.$$

$$\int_0^{h_m}\mathrm{d}z''\frac{\rho(z'')}{\alpha(z'')}\frac{\partial f_k\left[\bar{\rho}^*(z'')\right]}{\partial\bar{\rho}^*(z'')}J_k(z-z'')\varphi_1(z'',z')-$$

$$\int_0^{h_m} dz'' \frac{\rho(z'')}{\alpha^2(z'')} f_k\left[\bar{\rho}^*(z'')\right] J_k(z-z'')\varphi_2(z'',z') \tag{B.3}$$

$$\alpha(z) = 1 - \rho_1^*(z) - 2\bar{\rho}^*(z)\bar{\rho}_2^*(z) \tag{B.4}$$

$$f_k\left[\bar{\rho}^*(z)\right] = \frac{\pi}{6}\frac{4-\dfrac{\pi}{3}\bar{\rho}^*(z)}{\left[1-\dfrac{\pi}{6}\bar{\rho}^*(z)\right]^3}\left[\bar{\rho}^*(z)\right]^k \tag{B.5}$$

$$\varphi_1(z-z') = 2\pi\sum_{k=0}^{2}\frac{\partial\bar{\rho}^*(z)}{\partial\bar{\rho}_k^*(z)}J_k(z-z') \tag{B.6}$$

with

$$\varphi_2(z,z') = 2\pi\left[J_1(z-z') + 2\bar{\rho}^*(z)J_2(z-z')\right] - 2\bar{\rho}_2^*(z)\varphi_1(z,z') \tag{B.7}$$

$$K_1(z-z') = \frac{2\pi\epsilon_{\mathrm{ff}}}{k_{\mathrm{B}}T}\begin{cases}\dfrac{2}{5}\left(\dfrac{1}{z-z'}\right)^{10} - \left(\dfrac{1}{z-z'}\right)^4, & |z-z'|>1,\\ -\dfrac{3}{5'} & |z-z'|\le 1\end{cases}$$

$$J_0(z-z') = \begin{cases}\dfrac{3}{8\pi}\left[1-(z-z')^2\right], & |z-z'|\le 1,\\ 0, & |z-z'|>1\end{cases}$$

$$J_1(z-z') =$$

$$\begin{cases}A - \dfrac{a_0}{2}(z-z') - \dfrac{b_1}{3}|z-z'|^3 - \dfrac{b_2}{4}|z-z'|^4, & |z-z'|\le 1,\\ B - d_1|h-h'| - \dfrac{c_0}{2}(z-z')^2 - \dfrac{\mathrm{m}_1}{3}|z-z'|^3 - \dfrac{m_2}{4}|z-z'|^4, & 1<|z-z'|\le 2,\\ 0, & |z-z'|>2\end{cases}$$

$$J_2(z-z') =$$

$$\begin{cases}\dfrac{5\pi}{144}\left[1-3(z-z')^2 + 4|z-z|^3 - \dfrac{5}{4}(z-z')^4\right], & |z-z'|\le 1,\\ 0, & |z-z'|>1\end{cases}$$

In the above expressions $a_0 = 0.475$, $b_l = -0.648$, $b_2 = -0.113$, $c_0 = -0.924$, $d_1 = 0.288$, $m_1 = 0.764$, $m_2 = -0.187$, $A = (1/2)a_0 + (1/3)b_1 + (1/4)b_2 + d_l + (3/2)c_0 + (7/3)m_1 + (15/4)m_2$, and $B = 2d_1 + 2c_0 + (8/3)m_1 + 4m_2$.

REFERENCES

(1) Tarazona, P.; Marconi, U. M. B.; Evans, R. *Mol. Phys.* **1987**, *60*, 573.
(2) Magda, J. J.; Tirrel, M.; Davis, H. T. *J. Chem. Phys.* **1985**, *83*, 1888.
(3) Evans, R.; Marconi, U. M. B.; Tarazona, P. *J. Chem. Phys.* **1986**, *84*, 2376.
(4) Evans, *R.* J. Phys.: Condens. Matter **1990**, 2, *8989.*
(5) Sarman, S. *J. Chem. Phys.* **1990**, *92*, 4447.
(6) Lastoskie, C.; Gubbins, K. E.; Quirke, N. *Langmuir* **1993**, *9*, 2693.
(7) Maciolek, A.; Evans, R.; Wilding, N. B. *J. Chem. Phys.* **2003**, *119*, 8663.
(8) Löwen, H. *Phys. Rep.* **1994**, *237*, 249.
(9) Merkel, M.; Lowen, H. *Phys. Rev. E* **1996**, *54*, 6623.
(10) Berim, G. O.; Ruckenstein, E. *J. Phys. Chem. B* **2007**, *111*, 2514 (Section 2.1 of this volume).
(11) Berim, G. O.; Ruckenstein, E. *J. Chem. Phys.* **2007**, *126*, 124503 (Section 2.2 of this volume).
(12) Nilson, R. H.; Griffiths, S. K. *J. Chem. Phys.* **1999**, *111*, 4281.
(13) Tarazona, P. *Phys. Rev. A* **1985**, *31*, 2672.
(14) Evans, R.; Tarazona, P. *Phys. Rev. A* **1983**, *28*, 1864.
(15) Ebner, C.; Saam, W. F. *Phys. Rev. Lett.* **1977**, *38*, 1486.
(16) Vishnyakov, A.; Neimark, A. V. *J. Chem. Phys.* **2003**, *119*, 9755.
(17) Ustinov, E. A.; Do, D. D. *J. Chem. Phys.* **2004**, *120*, 9769.
(18) Errington, J. Personal communication.

2.4 Two-Dimensional Symmetry Breaking of Fluid Density Distribution in Closed Nanoslits*

Gersh Berim and Eli Ruckenstein†
Department of Chemical and Biological Engineering, State University of New York at Buffalo, Buffalo, New York 14260, USA

Corresponding Author

† E-mail: feaeliru@acsu.buffalo.edu.
Phone: (716)645-1179. Fax: (716)645-3822.

(Received 17 September 2007; accepted 1 November 2007; published online 9 January 2008)

I. INTRODUCTION

A fluid in a long nanoslit between two planar identical solid walls is exposed to an external potential due to fluid-solid interactions, which is symmetrical about the middle of the slit in the direction across the slit (perpendicular to the walls). For this reason, one usually expects the fluid density distribution (FDD) to possess the same symmetry. However, in Refs. 1 and 2, it was shown that in a closed slit this symmetry can be broken in some ranges of the parameters characterizing the system (fluid average density, strength of the fluid-solid interactions, and temperature) and that the stable FDD can become asymmetrical about the middle of the slit. The existence of such a symmetry breaking, which was unknown previously, is in agreement with the Monte Carlo (MC) simulations.[3]

Similarly, in a long (infinite) slit the fluid-solid interaction potential possesses translational symmetry in lateral (parallel to the walls) directions suggesting uniformity of the FDD along the walls. The same symmetry is also inherent for long cylindrical pores in the axial direction. For the latter system, it was shown using canonical ensemble MC simulations,[4] the Laplace theory of capillarity and the Derjaguin model,[5] and the nonlocal density functional theory[6] that the translational symmetry in the axial direction of the FDD can be broken through the formation of bumps, bubbles, and bridges on and between the solid surfaces. These results were used to explain the Everett-Haynes[7] scenario of bridging through the formation of a bump on the film adsorbed on the solid surface, which grows with increasing fluid average density and transforms into a bridge connecting the solid surfaces. They also explain the capillary condensation in open pores as a morphological transition from a low average density state (thin film adsorbed on the solid surface) to a high average density state (liquid completely filling the pore), both states being uniform in the lateral direction.[5] During such a transition, the bridges and bumps play the roles of unstable nuclei of the

* *J. Chem. Phys.* 128, 024704 (2008). Republished with permission.

new liquid phase. Because the states of interest (bridges and bumps) are not present in open systems which are described by the grand canonical ensemble, but occur in closed systems which are described by the canonical ensemble, MC simulations in canonical ensemble were used in Ref. 4 to obtain information about them. In addition, the canonical ensemble version of the density functional theory (CEDFT) was used to describe the unstable states in open systems. This approach allowed one to calculate the work of formation of droplets or bubbles in a metastable configuration on a solid substrate[8] and inside slit pores.[9] It was also used to describe the backward part of the adsorption isoterm in long (infinite) cylindrical as well as slitlike pores.[10–13] Only uniform states in the axial and lateral directions were considered in the latter papers. The nonuniform states in the axial direction were examined on the basis of CEDFT in Ref. 6 for the cylindrical pore. Additional information regarding the bridge formation and capillary condensation in open systems with various geometries can be found in Refs. 14–17.

The results of MC simulations in Ref. 4 were dependent on the size of the simulation box in the axial direction. For a small box with a length of $10\sigma_{ff}$, where σ_{ff} is the fluid molecular diameter, only a uniform FDD in the axial direction was identified and no symmetry breaking occurred. The latter occurred for a box with a length of $30\sigma_{ff}$ as bumps or bridges.

In the present paper, we examine using a nonlocal CEDFT (Refs. 18 and 19) the FDD symmetry breaking in the lateral direction for a planar slit by first formulating the problem in a way similar to that used in MC simulations. In particular, the assumption that the solution of the Euler-Lagrange equation for the FDD is periodical with a period L_x in one of the lateral directions, referred below as the x direction, is employed. This assumption is equivalent to the periodic boundary conditions widely used in MC simulations[4,12,20] and, therefore, allows us to compare (at least qualitatively) the analytical results based on CEDFT to those obtained via MC simulations. In the other lateral direction, the y direction, the FDD is considered uniform.

The paper is organized as follows. In Sec. II, the model and the iteration procedure for the solution of the Euler-Lagrange equation for FDD are described in details. In Sec. III, the stable and metastable FDDs are identified for intermediate ($L_x = 30\sigma_{ff}$), small ($L_x = 10\sigma_{ff}$), and large ($L_x = 240\sigma_{ff}$) values of the selected period length and for various fluid average densities. The FDDs include bridge- and bumplike states, as well as uniform states in the x direction. In addition, the L_x dependence of the free energy of the stable state is calculated. Finally, the results are discussed in Sec. IV. Two main conclusions were drawn. First, that no symmetry breaking in the x direction occurs at small $L_x \leqslant 14\sigma_{ff}$ for any average density. For $L_x \geqslant 14\sigma_{ff}$ there is always a range of fluid average densities for which a bridge- or bumplike solution is stable, i.e., symmetry breaking occurs. For $L_x \leqslant 14\sigma_{ff}$, symmetry breaking from a symmetric to a nonsymmetric profile can occur across the slit. These results are in agreement with those of the MC simulations. The second conclusion follows from the monotonically decreasing free energy of the stable state with increasing L_x, which was found in the presence of symmetry breaking. The solution of the Euler-Lagrange equation for the stable state in such a case is a single finite size liquid bridge in the entire slit.

II. BACKGROUND

The system under consideration is presented schematically in Fig. 1. It consists of a one component fluid (argon) confined between two parallel structureless identical solid walls W_1 and W_2 separated by a distance H, which is much smaller than the distance between the side walls which close the slit. For this reason, the end effects can be considered negligible. If symmetry breaking occurs in the x and y directions, the fluid density $\rho(\mathbf{r})$ at the point $\mathbf{r}(x, y, h)$ becomes a function of all three coordinates x, y, and h. However, in the present paper, we restrict ourselves to a density distribution that is uniform in the y direction and nonuniform in the x and h directions, i.e., $\rho(\boldsymbol{r}) \equiv \rho(x, h)$. Such an approximation has been often used, for example, in the wetting theory when the so-called cylindrical drop was considered on a solid surface.[21]

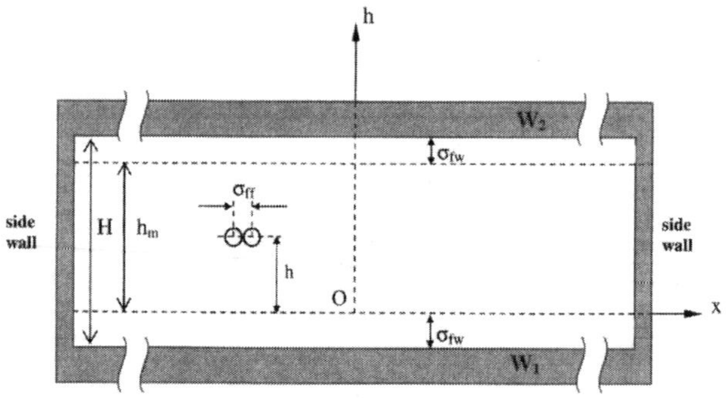

FIG. 1. Closed slit of width H between two identical solid walls W_1 and W_2. σ_{ff} and σ_{fw} are the hard core diameters of the fluid-fluid and fluid-wall interactions, respectively. The y axis is perpendicular to the plane of the figure.

The total Helmholtz free energy $F[\rho(\mathbf{r})]$ is expressed as the sum of the ideal gas free energy $F_{\text{id}}[\rho(\mathbf{r})]$, the excess free energy $F_{\text{ex}}[\rho(\mathbf{r})]$, and the free energy $F_{\text{fw}}[\rho(\mathbf{r})]$ due to the interactions between fluid and walls. The ideal gas free energy has the form

$$F_{\text{id}}\left[\rho(\mathbf{r})\right]=k_BT\int d\mathbf{r}\rho(\mathbf{r})\left\{\log\left[\Lambda^3\rho(\mathbf{r})\right]-1\right\}, \tag{1}$$

here $\Lambda = h_P/(2\pi mk_BT)^{1/2}$ is the thermal de Broglie wavelength, k_B and h_P are the Boltzmann and Planck constants, respectively, T is the absolute temperature, and m is the molecular mass of argon. The excess free energy is composed of a contribution from a reference system of hard spheres $F_{\text{hs}}[\rho(\mathbf{r})]$ and a contribution $F_{\text{attr}}[\rho(\mathbf{r})]$ due to the attractive interactions between the fluid molecules. The former contribution expressed in the smoothed density approximation of Tarazona *et al.*[19] has the form

$$F_{\text{hs}}\left[\rho(\mathbf{r})\right]=\int d\mathbf{r}\rho(\mathbf{r})\Delta\Psi_{\text{hs}}(\mathbf{r}), \tag{2}$$

where

$$\Delta\Psi_{\text{hs}}\left[\rho(\mathbf{r})\right]=k_BT\eta_{\bar{\rho}}\frac{4-3\eta_{\bar{\rho}}}{\left(1-\eta_{\bar{\rho}}\right)^2}, \tag{3}$$

$\eta_{\bar{\rho}} = 1/6\pi\bar{\rho}(\mathbf{r})\sigma_{ff}^3$ is the packing fraction of the fluid molecules, σ_{ff} is the fluid hard core diameter, and $\bar{\rho}(\mathbf{r})$ is the smoothed density defined as

$$\bar{\rho}(\mathbf{r})=\int d\mathbf{r}'\rho(\mathbf{r}')W\left(|\mathbf{r}-\mathbf{r}'|\right). \tag{4}$$

The weighting function $W(|\mathbf{r}-\mathbf{r}'|)$ is selected in the form[22]

$$W\left(|\mathbf{r}-\mathbf{r}'|\right)=\begin{cases}\dfrac{3}{\pi\sigma_{ff}^3}\left(1-\dfrac{r}{\sigma_{ff}}\right), & r\leqslant\sigma_{ff}\\ 0, & r>\sigma_{ff},\end{cases}$$

where $r = |\mathbf{r} - \mathbf{r}'|$.

The contribution to the excess free energy due to the attraction between the fluid-fluid molecules is calculated in the mean-field approximation

$$F_{\text{attr}}\left[\rho(\mathbf{r})\right]=\frac{1}{2}\iint d\mathbf{r}d\mathbf{r}'\rho(\mathbf{r})\rho(\mathbf{r}')\phi_{ff}\left(|\mathbf{r}-\mathbf{r}'|\right), \tag{5}$$

where

$$\phi_{ff}\left(|\mathbf{r}-\mathbf{r}'|\right)=\begin{cases}4\epsilon_{ff}\left[\left(\dfrac{\sigma_{ff}}{r}\right)^{12}-\left(\dfrac{\sigma_{ff}}{r}\right)^{6}\right], & r\geqslant\sigma_{ff}\\ 0, & r<\sigma_{ff}\end{cases} \tag{6}$$

The same kind of potential is selected for the fluid-wall interactions [the energy parameter ϵ_{ff} and hard core diameter σ_{ff} in Eq. (6) are changed to ϵ_{fw} and σ_{fw}, respectively, and r is the distance between a molecule of fluid and those of the solid walls].

The contribution of the fluid-walls interactions, described by the potential $U_{\text{fw}}(\mathbf{r})$, to the free energy is given by

$$F_{\text{fw}}\left[\rho(\mathbf{r})\right]=\int d\mathbf{r}\rho(\mathbf{r})U_{\text{fw}}(\mathbf{r}). \tag{7}$$

Because of the uniformity of the walls, the potential $U_{\text{fw}}(\mathbf{r})$ in Eq. (7) depends only on the distance h from the lower wall (see Fig. 1) and can be written in the form

$$U_{\text{fw}}(h)=\psi(h)+\psi(h_m-h), \tag{8}$$

where

$$\psi(h)=\frac{2\pi}{3}\epsilon_{\text{fw}}\rho_w\left[\frac{2}{15}\left(\frac{\sigma_{\text{fw}}}{\sigma_{\text{fw}}+h}\right)^{9}-\left(\frac{\sigma_{\text{fw}}}{\sigma_{\text{fw}}+h}\right)^{3}\right] \tag{9}$$

and ρ_w is the density of the walls.

The Euler-Lagrange equation for the fluid density distribution $\rho(x, h)$ in a closed slit, obtained by minimizing the Helmholtz free energy

$$F\left[\rho(\mathbf{r})\right]=F_{\text{id}}\left[\rho(\mathbf{r})\right]+F_{\text{ex}}\left[\rho(\mathbf{r})\right]+F_{\text{fw}}\left[\rho(\mathbf{r})\right] \tag{10}$$

has the following form:

$$k_BT\log\left(\Lambda^3\rho\right)+\Delta\Psi_{\text{hs}}(x,h)+\overline{\Delta\Psi'_{\text{hs}}}(x,h)+U_{ff}(x,h)+U_{\text{fw}}(h)=\lambda \tag{11}$$

where

$$U_{ff}(x,h)=\iint dx'dh'\rho(x',h')\phi_{ff,y}\left(|x-x'|,|h-h'|\right), \tag{12}$$

$$\overline{\Delta\Psi'_{\text{hs}}}(x,h)=\iint dx'dh'\rho(x',h')W_y\left(|x-x'|,|h-h'|\right)$$
$$\times\frac{\partial}{\partial\bar{\rho}}\Delta\Psi_{\text{hs}}(\bar{\rho})\bigg|_{\bar{\rho}=\bar{\rho}(x',h')}, \tag{13}$$

$\phi_{ff,y}(|x-x'|,|h-h'|)$ and $W_y(|x-x'|,|h-h'|)$ are obtained by the integrations of the potential $\phi_{ff}(|\mathbf{r}-\mathbf{r}'|)$ and weighted function $W(|\mathbf{r}-\mathbf{r}'|)$ with respect to the y direction, respectively. In Eq. (11), λ is a Lagrange multiplier arising because of the constraint of fixed average density of the fluid in the slit, which has the form

$$\rho_{\mathrm{av}} = \frac{1}{V}\int_V d\mathbf{r}\rho(\mathbf{r}), \tag{14}$$

where V is the volume of the slit.

Formally, Eq. (11) coincides with that for an open system (without side walls) in contact with a reservoir where the chemical potential $\mu = \lambda$. However, the Lagrange multiplier λ in closed systems is not known in advance. The relation between λ and the average density ρ_{av} is provided by the constraint Eq. (14) and has the form

$$\lambda = -k_B T \log\left[\frac{1}{\rho_{\mathrm{av}} V \Lambda^3}\int_V d\mathbf{r} e^{\mathcal{Q}(x,h)}\right], \tag{15}$$

where

$$\mathcal{Q}(x,h) = -\frac{1}{k_B T}\Big[\Delta\Psi_{\mathrm{hs}}(x,h) + \overline{\Delta\Psi'_{\mathrm{hs}}}(x,h)$$

$$+U_{ff}(x,h) + U_{\mathrm{fw}}(h)\Big] \tag{16}$$

By eliminating λ between Eqs. (11) and (15), one obtains an integral equation for the density profile $\rho(x, h)$.

Because of the large (infinite) size of the slit in the x direction, it is practically impossible to apply the iteration procedure for the solution of the Euler-Lagrange equation without an additional assumption about the x dependence of the density $\rho(x, h)$. In this paper, we assume that $\rho(x, h)$ is a periodic function of x with a period L_x that allows us to perform iterations over a finite domain of x and h coordinates. This assumption is equivalent to the periodic boundary condition commonly used in MC simulations of confined fluids.[4,12,20] The same assumption was employed in Ref. 6 where the single value, $L_x = 20\sigma_{ff}$, was considered. In contrast with those simulations, where the period (length of the simulation box) rarely exceeded 30 molecular diameters, the period used in this paper is selected up to several hundred molecular diameters. By considering numerous increasing values of L_x and analyzing the changes in the density profile, one can extract information about the fluid density distribution in a long (infinite) closed slit.

Under the periodicity assumption, Eq. (15) becomes

$$\lambda = -k_B T \log\left[\frac{1}{\rho_{\mathrm{av}} L_x h_m \Lambda^3}\int_\Omega dx dh e^{\mathcal{Q}(x,h)}\right], \tag{17}$$

where $\int_\Omega dx dh e^{\mathcal{Q}(x,h)} = \int_0^{L_x} dx \int_0^{h_m} dh e^{\mathcal{Q}(x,h)}$.

To calculate the term $U_{ff}(x, h)$ defined by Eq. (12), a cutoff at a distance equal to four molecular diameters σ_{ff} for the range of Lennard-Jones attraction was employed. The increase of this distance up to $10\sigma_{ff}$ changed the results by less than 1%.

To avoid the divergence of the iteration procedure, the input density profile $\rho_i^{\mathrm{in}}(x,h)$ for the $(i + 1)$ th iteration $\rho_{i+1}(x, h)$, generated by the Euler-Lagrange equation, was constructed as follows:[18]

$$\rho_i^{\mathrm{in}}(x,h) = (1-\alpha)\rho_{i-1}^{\mathrm{in}}(x,h) + \alpha\rho_i(x,h), \tag{18}$$

where $\rho_i(x, h)$ is the ith iteration and the constant $\alpha = 0.1$. The Lagrange multiplier can be obtained from Eq. (17) using the resulting FDD. As a measure of the precision of the iterations, the dimensionless quantity

$$\delta = \int_{\Omega} dxdh\left[\rho_{i+1}(x,h) - \rho_i^{\text{in}}(x,h)\right]^2 \Big/ \left(\int_{\Omega} dxdh\rho_i(x,h)\right)^2$$

was introduced. The iterations were carried out on a two dimensional grid with a spacing equal to $0.1\sigma_{ff}$ until δ became smaller than $\epsilon = 10^{-7}$. The additional increase in precision did not lead to appreciable changes in the density profile.

All obtained density profiles were tested with respect to their stability by examining the changes of the free energy of the system to arbitrarily small perturbations of the density distribution. To perform such a calculation, the tested profile was randomly disturbed at all grid points by less than 1%. Then all the densities after perturbation were changed by the same value to satisfy the constraint of constant fluid average density. Finally, the free energy per unit area of one of the walls was calculated and compared with the initial one. The test was repeated several times with different random perturbations. In all considered cases, the free energy of the perturbed system was higher than that of the unperturbed one. While not completely rigorous, such a test indicated that all obtained states correspond to local minima of the free energy and, hence, they are stable or metastable. The state with the lower free energy is considered to be the stable one.

For any selected period L_x and average density of the fluid, several initial guesses $\rho_0^{\text{in}}(x,h)$ for the FDD $\rho(x, h)$ were employed. In general, each of them provided a different solution of the Euler-Lagrange equation. All guessed FDDs were constructed in the interval of $0 \leqslant x \leqslant L_x$ such as to satisfy the periodicity condition $[\rho_0^{\text{in}}(0,h) = \rho_0^{\text{in}}(L_x,h)]$ and to provide the continuity of the derivative with respect to $x[(\partial/\partial x)\rho(x,h)|_{x=0} = (\partial/\partial x)\rho(x,h)|_{x=L_x}]$. Typical initial guesses employed in the calculations are shown in Fig. 2.

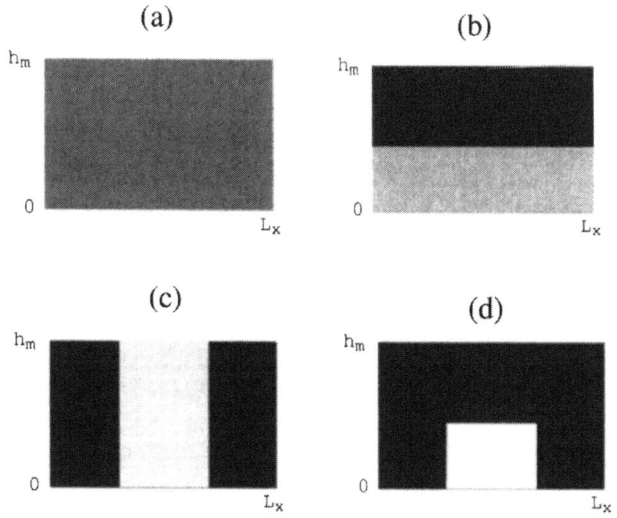

FIG. 2. Typical initial fluid density distributions (FDDs) employed in iterations. The lighter areas correspond to higher fluid density. FDD (a) always leads to a solution of the Euler-Lagrange equation uniform in the x direction and symmetrical in the h direction; FDD (b) can lead to a solution uniform in the x direction and asymmetrical in the h direction; FDDs (c) and (d) can lead to bridgelike and bumplike solutions, respectively.

First, we considered in detail the typical solutions of the Euler-Lagrange equation for $L_x = 30\sigma_{ff}$ and identified the stable and metastable states for various values of the average density. By extracting such information for other values of L_x, from $L_x = 10\sigma_{ff}$ to $L_x = 240\sigma_{ff}$, the free energy of the stable state of the system was represented as a function of L_x and its asymptotic behavior at large values of L_x provided some information about the infinite slit.

III. RESULTS

Argon was selected as the fluid and the solid carbon dioxide as the walls. The interaction parameters and wall density are equal to [23] $\epsilon_{ff}/k_B = 119.76\text{K}$, $\epsilon_{fw}/k_B = 153.0\text{K}$, $\sigma_{ff} = 3.405\text{Å}$, $\rho_w = 1.91\times10^{28}\,\text{m}^{-3}$. The width of the slit was taken as $h_m = 10\sigma_{ff}$. If not mentioned otherwise, the temperature was $T = 87$ K.

A. Density distribution for $L_x = 30\sigma_{ff}$

For a fixed average fluid density, the solution of the Euler-Lagrange equation depends on the initial guess selected for the iterations (see Fig. 2). The number of different solutions depends on the average density ρ_{av}. For small average fluid densities, say $\rho_{av}^* = 0.01$ ($\rho^* \equiv \rho\sigma_{ff}^3$ is the dimensionless number density of the fluid), all initial guesses led to the same solution, namely, uniform distribution along the x axis and symmetrical in the direction perpendicular to the walls with respect to the middle of the slit (US solution). In this case, the walls are covered by a liquid film uniform in the lateral directions and the remaining of the slit is filled by a low density vapor. The free energy per unit area of one of the walls and the corresponding value of the Lagrange multiplier of that US-solution are listed in Table 1. For $\rho_{av}^* = 0.12$, a second, bumplike solution, appears which is shown schematically in Fig. 3(a). The bumplike FDD is nonuniform along the x axis and is nonsymmetrical about the middle of the slit in the h direction. This solution constitutes a symmetry breaking in two dimensions. A density profile for a bumplike solution in the h direction taken for $x = L_x/2$ is presented in Fig. 3(b). The oscillations of the density near the wall are similar to those that occur in a bulk fluid in contact with a solid wall.[22,23] The free energy of a bumplike FDD is slightly smaller than that of a US-solution, hence, the former FDD is stable and the latter one metastable. Note that a single bump solution is twofold degenerate because the bump can be located either on the lower or on the upper wall.

At $\rho_{av}^* = 0.14$, two new solutions were obtained in addition to those mentioned above. One of them, the UA-solution, is uniform in the x direction and asymmetrical about the middle of the slit, and is a result of the symmetry breaking in the h direction described in detail in Refs. 1 and 2. The density profile for this solution is presented in Fig. 4. Another, bridgelike solution, which is nonuniform along the x axis but symmetrical about the middle plane between walls, is shown in Fig. 5.

Comparing the free energies of the four FDDs that occur for $\rho_{av}^* = 0.14$ (see Table 1), one can see that in this case the bumplike FDD [Fig. 3(a)] has the smallest free energy and is therefore the stable one, and the other FDDs are metastable. However, the differences between their free energies are very small. If the energy barriers (which are unknown) between those states are sufficiently high, the system can survive for a long time in any of the metastable state. If those barriers are low, the thermal fluctuation can drive the system through all of these states.

The configuration of the stable state changes at $\rho_{av}^* = 0.17$, where a bridgelike FDD is the stable one, and the bumplike FDD and the other ones are metastable. The values of the free energies for a similar case with $\rho_{av}^* = 0.2$ are listed in Table 1. The metastable bumplike solution disappears at $\rho_{av}^* = 0.22$ and the UA solution disappears at $\rho_{av}^* \simeq 0.5$. The bridgelike FDD remains stable up to $\rho_{av}^* = 0.58$, where the fluid fills completely the slit and the FDD becomes uniform in the x direction and symmetrical about the middle plane between walls (US solution).

TABLE 1
Free energies (*F*) per unit area of one of the walls and the values of Lagrange multiplier (λ) for various solutions of the Euler-Lagrange equation. The stable solutions are distinguished by bold font. $L_x = 30\sigma_{ff}$.

FDD	$F\sigma^2_{ff}/k_BT$	λ/k_BT
	$\rho^*_{av} = 0.01$	
US solution	**–1.268**	**–12.729**
	$\rho^*_{av} = 0.14$	
US solution	–14.860	–11.231
UA solution	–14.866	–11.302
Bridgelike	–14.851	–11.472
Bumplike	**–14.888**	**–11.351**
	$\rho^*_{av} = 0.2$	
US solution	–21.000	–11.297
UA solution	–21.067	–11.414
Bridgelike	**–21.163**	**–11.550**
Bumplike	–21.102	–11.387

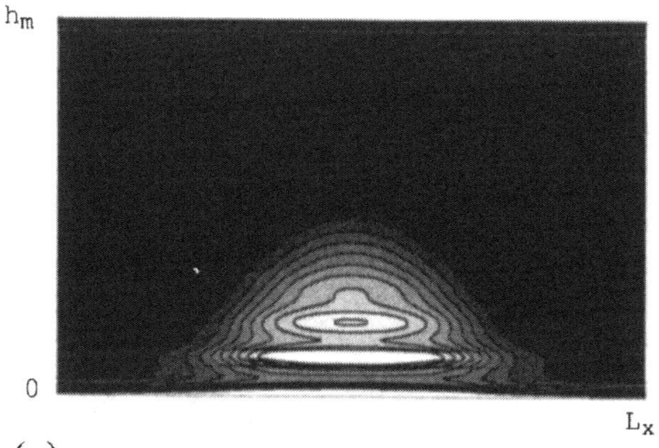

(a)

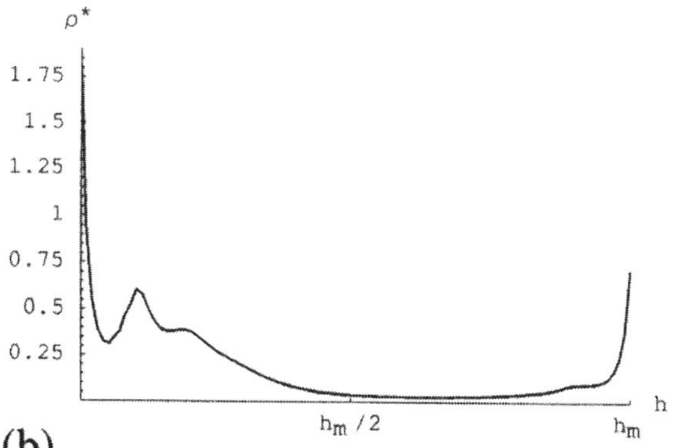

(b)

FIG. 3. (a) Bumplike fluid density distribution. The lighter areas correspond to higher fluid density. (b) Density profile in the *h* direction at $x = L_x/2$. $L_x = 30\sigma_{ff}$ and $\rho_{av} = 0.12$.

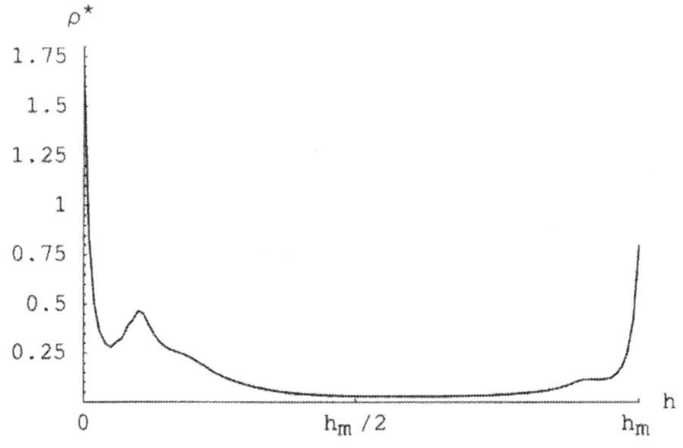

FIG. 4. Typical density profile of a UA solution uniform in the x direction and asymmetrical in the h direction. $L_x = 30\sigma_{ff}$ and $\rho_{av} = 0.14$.

B. Density distribution for $L_x = 10\sigma_{ff}$

In this case, only three solutions of the Euler-Lagrange equation could be identified. A US-solution, which is present at all average densities. A second, UA solution, was found in the range of $0.14 \leqslant \rho_{av}{}^* \leqslant 0.58$. This range of existence of the UA solution coincides with that for $L_x = 30\sigma_{ff}$. This is a consequence of the uniformity of the UA solution in the x direction, that means independence of this FDD of L_x. The third, bridgelike solution, was found along with a UA solution in the range of average densities of $0.38 \leqslant \rho_{av}{}^* \leqslant 0.42$. When the UA solution occurred, it had always the smallest free energy, i.e., it was stable, all the other ones being metastable. No bumplike FDD was found for $L_x = 10\sigma_{ff}$.

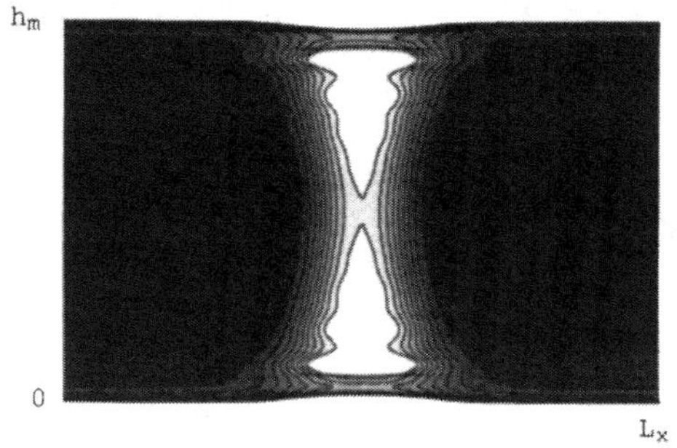

FIG. 5. Bridgelike fluid density distribution. The lighter areas correspond to higher fluid density.

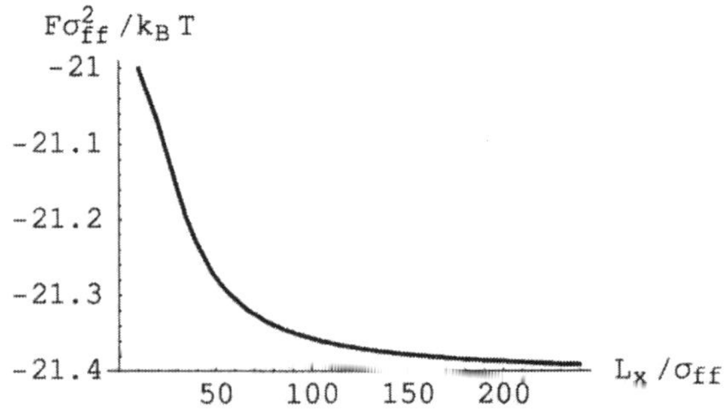

FIG. 6. Dependence of the free energy of the stable state on the length L_x of the period of the fluid density distribution.

C. Density Distribution for $L_x = 240\sigma_{ff}$

Due to their uniformity in the x direction, the US and UA solutions coincide in this case with those obtained for $L_x = 30\sigma_{ff}$. The bridgelike solution appears at the average density $\rho_{av}^* = 0.06$. It remains stable for average densities $\rho_{av}^* < 0.58$. For $\rho_{av}^* \geqslant 0.58$ only the US solution is stable. The UA solution is always metastable, and the bumplike solution could not be identified.

D. L_x Dependence of the Density Distribution at Fixed Average Density

Having finding of the stable FDD for a very long (infinite) closed slit as the goal, we calculated the density distributions when L_x was changed from $10\sigma_{ff}$ to $240\sigma_{ff}$ at the fixed average density $\rho_{nv}^* = 0.2$. This moderate value of density was selected because it is expected to provide a variety of solutions of the Euler-Lagrange equation. For any selected L_x from that interval, all the FDDs were calculated and the stable ones were identified. For $10\sigma_{ff} \leqslant L_x \leqslant 20\sigma_{ff}$ the UA solution (Fig. 4) was the stable FDD. The bumplike FDD [Fig. 3(a)] was stable for $20\sigma_{ff} < L_x < 24\sigma_{ff}$ and the bridgelike FDD was stable for $L_x \geqslant 24\sigma_{ff}$. The free energy of the stable state is plotted as a function of L_x in Fig. 6. This figure indicates an asymptotic behavior of the free energy with increasing L_x. From this behavior, one can conclude that for a given average density a single bridge of finite length occurs as a unique stable state in a long closed slit. For gaslike average densities $\rho_{av}^* \leqslant 0.01$ as well as for liquidlike densities $\rho_{av}^* \geqslant 0.6$, the stable state for any L_x is the US solution of the Euler-Lagrange equation.

IV. DISCUSSION

The results obtained in this paper involve the assumption of periodicity of the FDD in the x direction, which is equivalent to the cyclic boundary conditions widely used in MC simulations. Assuming specific values of L_x, we determined the stable and metastable solutions and the values of the free energy for each of them. For small values of L_x ($L_x \leqslant 14$) and depending on the average density, the stable state is either the US or the UA solution. There is a range of ρ_{av}^* in which a bridgelike solution also occurs but it was always metastable. Similarly, symmetrical and asymmetrical with respect to the middle of the slit FDDs were observed in MC experiment[3] for a simulation box with $L_x = Ly \simeq 10\sigma_{ff}$. Argon between parallel identical walls ($h_m = 20\sigma_{ff}$) of solid carbon dioxide at $T = 89.82$ K was considered. In Fig. 7, three snapshots representing the equilibrium states of that

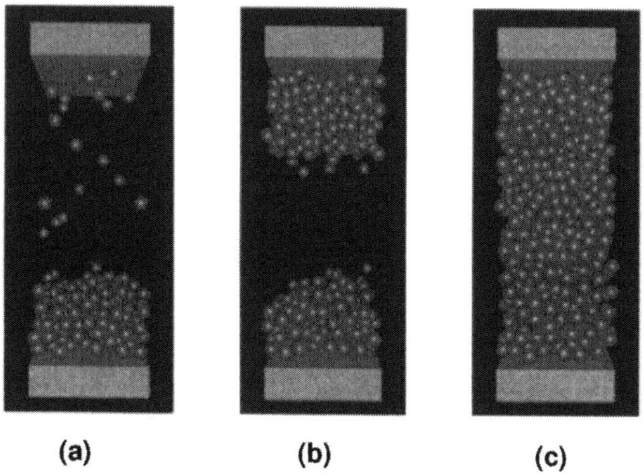

FIG. 7. Monte Carlo images of equilibrium states of argon between walls of solid carbon dioxide. The width of the slit $h_m^* = 20$ and $T = 89.82$ K. The average densities ρ_{av}^* are (a) 0.123, (b) 0.246, and (c) 0.556.

closed system are presented for the average densities $\rho_{av,1} = 0.123$ [Fig. 7(a)], $\rho_{av,2} = 0.246$ [Fig. 7(b)], and $\rho_{av,3} = 0.556$ [Fig. 7(c)]. (While oscillations of the density near the wall do exist, they are not visible in these figures.) Calculations performed on the basis of the present theory predicted for this case a stable UA solution in the range $0.08 < \rho_{av}^* < 0.48$. The density profiles along the *h* direction calculated for the average densities used in Monte Carlo simulations are plotted in Fig. 8. The states presented in Figs. 7(a) and 7(b) belong to the range of average densities where the UA solution is stable according to the present theory [see Figs. 8(a) and 8(b)]. One of those states [Figure 7(a)] exhibits, in agreement with the calculations, a symmetry breaking in the density distribution across the slit. Another one [Fig. 7(b)] has a density distribution less asymmetrical than the predicted one. The third [Fig. 7(c)] has a symmetrical distribution corresponding to the average density $\rho_{av,3} > 0.48$, which is in agreement with the theory [Fig. 8(c)].

Even though we consider another system, the above theoretical results also suggest an explanation why no symmetry breaking in the *x* direction was observed in the MC simulations[4] for a

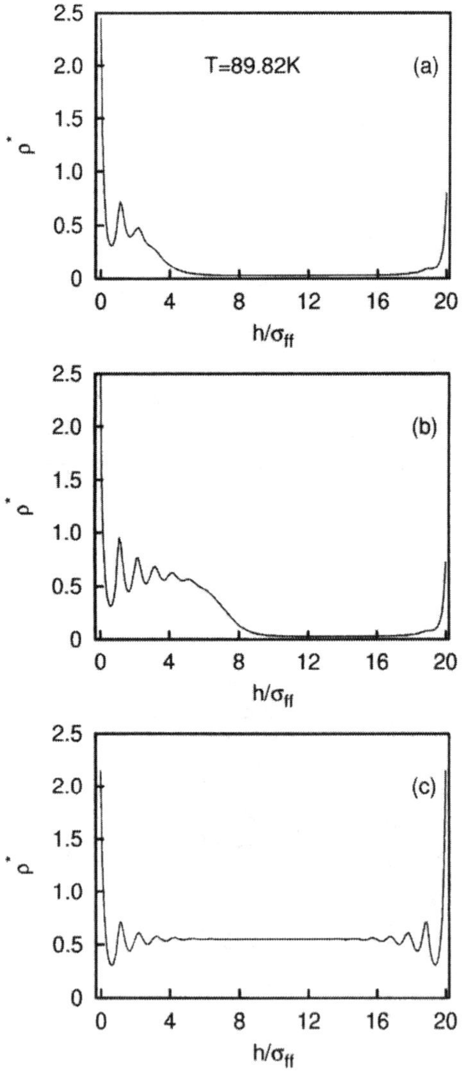

FIG. 8. Stable density profiles calculated for $h_m^* = 20$ and $T = 89.82$ K. The average densities ρ_{av}^* are (a) 0.123, (b) 0.246, and (c) 0.556.

cylindrical pore of length $10\sigma_{ff}$. For larger values of L_x, for instance, $L_x = 30\sigma_{ff}$, the stable FDD in a planar slit can be described, depending on ρ_{av}^*, by US-, bump-, and bridgelike solutions. A similar conclusion on the basis of CEDFT was drawn for a cylindrical pore in Ref. 6, where the period of FDD was selected equal to $20\sigma_{ff}$. In the MC simulations of a fluid located in narrow cylindrical pores, stable bump- and bridgelike solutions were observed, depending on the average density.[4]

The analysis of the density distribution obtained for an enough long slit ($L_x = 240\sigma_{ff}$) has shown that for small ($\rho_{av}^* < 0.06$) and high ($\rho_{av}^* \geqslant 0.58$) average densities the stable state is represented by the US-solution of the Euler-Lagrange equation, whereas for $0.06 \leqslant \rho_{av}^* < 0.58$ the stable state is the bridge solution. Because the considered value of L_x is much larger than the width of the slit h_m, one can extrapolate this conclusion to a very long (infinite) slit.

The assumption of periodicity of FDD in the x direction used in the present paper allowed one to find the stable and metastable states of the system. According to this assumption, the bump- and bridgelike FDDs are periodic sequences of bumps or bridges along the x direction. If the stable FDD is indeed periodical, then its period L_x should minimize the free energy of the system considered as a function of L_x. However, the analysis of this dependence reveals the absence of such a minimum. In the case of FDDs which are uniform in the x direction (US or UA solutions), the free energy does not depend on L_x. For FDDs nonuniform in the x direction (bridge- and bumplike solutions), the calculated free energy monotonously decreases with increasing L_x (see Fig. 6), the bridgelike FDD remaining the only stable state. This suggests that in the presence of symmetry breaking in the x direction the real stable state which exists in a long (infinite) closed slit and, presumably, in the long closed cylindrical pore, is a single bridge of finite length over the entire slit. In other words, this result suggests that, in a rigorous sense, the only reasonable assumption for L_x is $L_x = \infty$. However, the periodical solutions allow us to make a comparison with the Monte Carlo simulations.

As previously shown in Refs. 1 and 2, where uniformity of the system in all lateral (parallel to the walls) directions was supposed, there is a symmetry breaking across the slit which for a given temperature occurs in a specific range of average fluid densities. The solutions of the Euler-Lagrange equation obtained in Refs. 1 and 2 belongs to the UA solutions considered in the present paper. Under the assumptions used here (uniformity of the system in the y direction, periodicity of the density distribution in the x direction), the results obtained provide some new details regarding the symmetry breaking in a closed nanoslit. The main conclusion from our calculations is that symmetry breaking in such a system occurs simultaneously both in the x and h directions, the former being manifested as a bump- or bridgelike solution.

REFERENCES

1. G. O. Berim and E. Ruckenstein, J. Phys. Chem. B **111**, 2514 (2007) (Section 2.1 of this volume).
2. G. O. Berim and E. Ruckenstein, J. Chem. Phys. **126**, 124503 (2007) (Section 2.2 of this volume).
3. J. Errington (private communication).
4. A. Vishnyakov and A. V. Neimark, J. Chem. Phys. **119**, 9755 (2003).
5. K. G. Kornev, I. K. Shingareva, and A. V. Neimark, Adv. Colloid Interface Sci. **96**, 143 (2002).
6. E. A. Ustinov and D. D. Do, J. Phys. Chem. B **109**, 11653 (2005).
7. D. H. Everett and J. M. Haynes, J. Colloid Interface Sci. **38**, 125 (1972).
8. V. Talanquer and D. W. Oxtoby, J. Chem. Phys. **104**, 1483 (1996).
9. V. Talanquer and D. W. Oxtoby, J. Chem. Phys. **114**, 2793 (2001).
10. A. V. Neimark, P. I. Ravikovitch, and A. Vishnyakov, J. Phys.: Condens. Matter **15**, 347 (2003).
11. P. I. Ravikovitch, A. Vishnyakov, and A. V. Neimark, Phys. Rev. E **64**, 011602 (2001).
12. A. V. Neimark, P. I. Ravikovitch, and A. Vishnyakov, Phys. Rev. E **65**, 031505 (2002).
13. E. A. Ustinov and D. D. Do, J. Chem. Phys. **120**, 9769 (2004).
14. H. T. Dobbs, G. A. Darbellay, and Y. M. Yeomans, Europhys. Lett. **18**, 439 (1992).
15. H. T. Dobbs and Y. M. Yeomans, J. Phys.: Condens. Matter **4**, 10133 (1992).
16. H. T. Dobbs and Y. M. Yeomans, Mol. Phys. **80**, 877 (1993).
17. A. O. Parry, C. Rascon, N. B. Wilding, and R. Evans, Phys. Rev. Lett. **98**, 226101 (2007).

18. P. Tarazona, Phys. Rev. A **31**, 2672 (1985).
19. P. Tarazona, U. M. B. Marconi, and R. Evans, Mol. Phys. **60**, 573 (1987).
20. J. Mittal, J. R. Errington, and T. M. Truskett, J. Chem. Phys. **126**, 244708 (2007).
21. E. K. Yeh, J. Newman, and C. J. Radke, Colloids Surf., A **156**, 137 (1999).
22. R. H. Nilson and S. K. Griffiths, J. Chem. Phys. **111**, 4281 (1999).
23. R. Evans and P. Tarazona, Phys. Rev. A **28**, 1864 (1983).

2.5 Symmetry Breaking in Binary Mixtures in Closed Nanoslits*

Gersh Berim and Eli Ruckenstein†
Department of Chemical and Biological Engineering, State University of New York at Buffalo, Buffalo, New York 14260, USA

Corresponding Author

† E-mail: feaeliru@acsu.buffalo.edu.
Phone: (716) 645-1179. Fax: (716) 645-3822.

(Received 22 January 2008; accepted 12 March 2008; published online 7 April 2008)

I. INTRODUCTION

It was shown in Refs. 1–3 on the basis of a nonlocal density functional theory (DFT) that the stable states of a one-component fluid in a long closed nanoslit between two planar identical solid walls can differ considerably from those in an open slit that communicates with a reservoir. While in the open slit the fluid density distribution (FDD) is almost always uniform in the lateral (parallel to the walls) directions and symmetrical in the direction perpendicular to the wall surface about the middle plane between walls (referred below as the middle plane),[4–6] the FDD in a closed slit can exhibit symmetry breaking (SB) in one or in both of those directions. (To our best knowledge, the only example for a SB in open systems was found in Ref. 7, where two parallel planar slits in contact with a reservoir were considered.) In the direction across the slit, SB occurs as a density profile asymmetrical about the middle plane,[1,2] while in the lateral directions, it occurs as bump- or bridgelike density distributions.[3]

SB across a slit between parallel walls was also found by Monte Carlo (MC) and molecular dynamics simulations in Refs. 8 and 9 for systems consisting of constant numbers of fluid molecules. Bump- and bridgelike states were identified for cylindrical pores in canonical ensemble MC simulation in Ref. 10. Such behaviors have been analytically predicted on the basis of the Laplace theory of capillarity combined with the Derjaguin model,[11] as well as on the basis of a nonlocal DFT.[12]

SB states occur for specific values of the parameters of the fluid-fluid and fluid-solid interaction potentials, temperature, as well as average density of the fluid inside the slit.[1–3] For a selected fluid and walls and at a given temperature, SB can occur only in a range of fluid average density.[1] (As an extreme case, SB can be absent for any fluid average density.) For a selected average density and temperature, SB occurs only in a range of values of the fluid-fluid and fluid-wall interaction parameters.[1]

Physically, the SB states (bumps and bridges) can also be considered as unstable nuclei of a liquid phase formed during capillary condensation in open pores.[10,13–15] This interpretation allowed one to calculate the formation of droplets or bubbles on a solid substrate[14] and inside a slitlike pore.[15] It was also used to describe the backward part of the adsorption isotherm in a long (infinite) cylindrical as well as slitlike pores.[12,13,16,17]

Theoretical considerations based either on classical ideas involving the surface tension[18] or MC simulations[19] of binary mixtures confined in cylindrical pores have indicated the possibility of phase separation (analogous to SB) in the axial direction of the pore.

* *J. Chem. Phys.* 128, 134713 (2008). Republished with permission.

In the present paper, the SB states of a two-component mixture confined in a closed nanoslit between identical walls composed of solid carbon dioxide are examined by using a nonlocal canonical ensemble DFT. Under the conditions considered, component 1 taken alone does not possess SB states, whereas component 2 (argon) taken alone exhibits all the above mentioned SB states. We examine SB in such systems by formulating the problem in a way similar to that used in MC simulations. In particular, the assumption that the solution of the Euler—Lagrange equation for the FDD is periodical with a period L_x in one of the lateral directions, referred below as the x direction, is employed. In the other lateral direction, the y direction, the FDD is considered to be uniform

Note that although there are numerous publications regarding binary mixtures (see, for example, Refs. 20–25 and references therein), all of them were devoted to open systems. The asymmetrical states in the h direction and the nonuniform states in the x direction considered in some of them[24,25] occurred not because of SB but because of the absence of symmetry in the interaction potential between the walls and fluid due either to nonidentical[24] or to nonuniform[25] walls.

II. BASIC EQUATIONS

The system under consideration consists of a two-component fluid of spherical molecules of components 1 (C1) and 2 (C2) interacting via a truncated Lennard–Jones potential:

$$U_{ij}(r)\begin{cases}\infty, & r\leqslant\sigma_{ij},\\ 4\epsilon_{ij}\left[\left(\dfrac{\sigma_{ij}}{r}\right)^{12}\left(\dfrac{\sigma_{ij}}{r}\right)^{6}\right], & \sigma_{ij}<r\leqslant r_{ij,\mathrm{cut}},\\ 0,\quad r> & r_{ij,\mathrm{cut}},\end{cases} \tag{1}$$

where i and j take the values 1 and 2 corresponding to components 1 and 2, r is the distance between the centers of a pair of molecules, and $r_{ij,\mathrm{cut}} = 2.5\sigma_{ij}$ is the cutoff distance. The constants σ_{ij} and ϵ_{ij} are the hard core diameters and energy parameters, respectively. The fluid is confined in a closed slit (see Fig. 1) between two parallel structureless identical solid walls W_1 and W_2 separated by

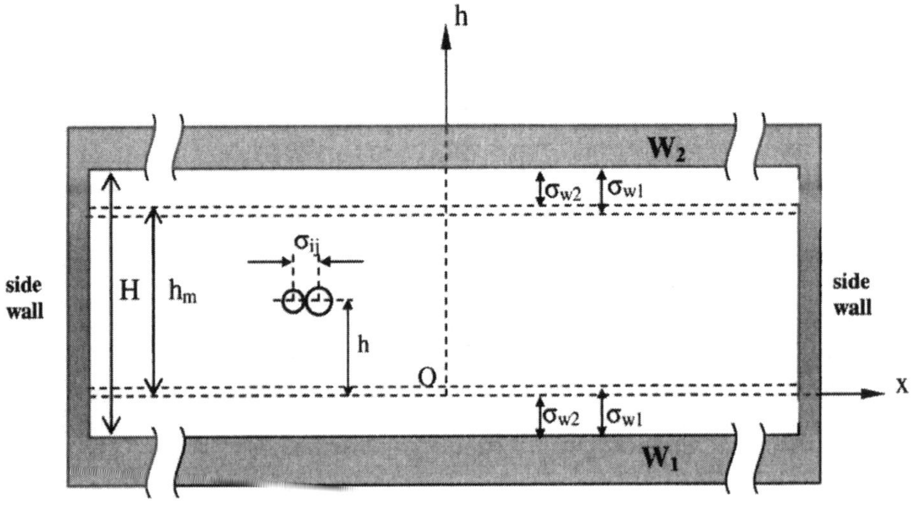

FIG. 1. Closed slit of width H between two identical solid walls W_l and W_2. σ_{ij} and σ_{wi} (i, j = 1,2) are the hard core diameters of component i-component j and wall-component i interactions, respectively. The y axis is perpendicular to the plane of the figure.

a distance H, which is much smaller than the distance between the side walls that close the slit. For this reason, the end effects can be considered negligible.

The solid-fluid interaction of a molecule of component i is described by the potential

$$\psi_i(h) = \frac{2\pi}{3}\epsilon_{wi}\rho_w\left[\frac{2}{15}\left(\frac{\sigma_{wi}}{\sigma_{wi}+h}\right)^9 - \left(\frac{\sigma_{wi}}{\sigma_{wi}+h}\right)^3\right], \tag{2}$$

which represents the interaction via a Lennard—Jones potential (without a cutoff distance) between the fluid molecules and the solid. Here, σ_{wi} and ϵ_{wi} (i = 1,2) are the hard core diameters and the energy parameters for the fluid-solid interaction, ρ_w is the density of the walls, and h is a distance measured from the lower wall, as shown in Fig. 1.

The total external potential U_{wi} exerted on a fluid molecule of component i at distance h is calculated as the superposition of $\psi_i(h)$ for the two walls:

$$U_{wi}(h) = \psi_i(h) + \psi_i(h_m - h), \tag{3}$$

where h and h_m are defined in Fig. 1.

Due to the possible SB of the FDD that can occur in a closed volume, the fluid density $\rho_i(\mathbf{r})$ of component i can be a function of all three coordinates x, y, and h. However, as previously, we restrict ourselves to a density distribution that is uniform in the y direction but can be nonuniform in the x and h directions, i.e., $\rho_i(\mathbf{r}) = \rho_i(x, h)$.

The density distributions $\rho_i(\mathbf{r})$ (i = 1,2) are calculated by using a density functional approach formulated by Rosenfeld.[26] As shown in Refs. 27–30, this theory, developed originally for open systems (grand canonical ensemble), can also be applied to closed systems (canonical ensemble). The only restriction, which is fulfilled in the present calculations, is that the number of molecules must be sufficiently large. A detailed discussion on the applicability of the canonical ensemble DFT can be found in Ref. 16.

The total Helmholtz free energy $F_{\mathbf{tot}}[\rho_1(\mathbf{r}), \rho_2(\mathbf{r})]$ can be expressed as the sum of an ideal gas free energy $F_{\mathbf{id}}[\rho_1(\mathbf{r}),\rho_2(\mathbf{r})]$, an excess free energy $F[\rho_1(\mathbf{r}),\rho_2(\mathbf{r})]$, and a free energy $F_{fw}[\rho_1(\mathbf{r})] + F_{fw}[\rho_2(\mathbf{r})]$ due to the interactions between the fluid and walls. The ideal gas free energy has the form

$$F_{\mathrm{id}}\left[\rho_1(\mathbf{r}),\rho_2(\mathbf{r})\right] = k_B T\sum_{i=1,2}\int d\mathbf{r}\rho_i(\mathbf{r})\left\{\log\left[\Lambda_i^3\rho_i(\mathbf{r})\right]-1\right\}, \tag{4}$$

where $\Lambda_i = h_p/(2\pi m_i k_B T)^{1/2}$ is the thermal de Broglie wavelength of the molecules of component i, k_B and h_P are the Boltzmann and Planck constants, respectively, T is the absolute temperature, and m_i is the molecular mass of component i. The excess free energy is composed of a contribution from a reference system of hard spheres, $\Phi[\rho_1(\mathbf{r}), \rho_2(\mathbf{r})]$, and a contribution, $F_{\mathrm{attr}}[\rho_1(\mathbf{r}), \rho_2(\mathbf{r})]$, due to the attractive interactions between the fluid molecules. The former contribution, expressed in Rosenfeld's approximation,[26] has the form

$$\Phi\left[\rho_1(\mathbf{r}),\rho_2(\mathbf{r})\right]/\mathrm{k}_B T = -n_0\log(1-n_3) + \frac{n_1 n_2 - \mathbf{n}_1\mathbf{n}_2}{4\pi(1-n_2)}$$

$$+\frac{n_2^2(n_2^2 - \mathbf{n}_1\mathbf{n}_2)}{24\pi(1-n_3)^2}, \tag{5}$$

where n_α ($\alpha = 0,1,2,3$) and $\mathbf{n}_\alpha$ ($\alpha = 1,2$) are averaged densities given by

$$n_\alpha(\mathbf{r}) = \sum_{i=l}^{2} \int d\mathbf{r}' \rho_i(\mathbf{r}') \omega_i^{(\alpha)}(r - r'), \; \alpha = 0,1,2,3, \tag{6}$$

$$\mathbf{n}_\alpha(\mathbf{r}) = \sum_{i=l}^{2} \int d\mathbf{r}' \rho_i(\mathbf{r}') \bar{\omega}_i^{(\alpha)}(r - r'), \; \alpha = 1,2. \tag{7}$$

The scalar, $\omega_i^{(\alpha)}(\mathbf{r} - \mathbf{r}')$, and vector, $\bar{\omega}_i^{(\alpha)}(\mathbf{r} - \mathbf{r}')$, weight functions are provided in Ref. 20.

The contribution to the excess free energy due to the attraction between the fluid molecules, calculated in the mean-field approximation, is given by

$$F_{\text{attr}}\left[\rho_1(\mathbf{r})\rho_2(\mathbf{r})\right] = \frac{1}{2} \sum_{i,j=1}^{2} \iint d\mathbf{r}\, d\mathbf{r}' \rho_i(\mathbf{r}) \rho_j(\mathbf{r}') \times U_{ij}(|\mathbf{r} - \mathbf{r}'|)$$

$$\left[\text{with } U_{ij}(|\mathbf{r} - \mathbf{r}'|) = 0 \text{ for } |\mathbf{r} - \mathbf{r}'| \leqslant \sigma_{ij}\right]. \tag{8}$$

The contribution of the fluid-wall interactions is given by

$$F_{fw}\left[\rho_i(\mathbf{r})\right] = \int d\mathbf{r}\, \rho_i(\mathbf{r}) U_{wi}(\mathbf{r}). \tag{9}$$

The Euler-Lagrange equations for the FDDs $\rho_i(x, h)$ (i = 1,2) in a closed slit, obtained by minimizing the Helmholtz free energy

$$F_{\text{tot}}\left[\rho_1(\mathbf{r}), \rho_2(\mathbf{r})\right] = F_{\text{id}}\left[\rho_1(\mathbf{r}), \rho_2(\mathbf{r})\right] + \Phi\left[\rho_1(\mathbf{r}), \rho_2(\mathbf{r})\right] + F_{\text{attr}}\left[\rho_1(\mathbf{r}), \rho_2(\mathbf{r})\right] + \sum_{i=l}^{2} F_{fw}\left[\rho_i(\mathbf{r})\right] \tag{10}$$

has the following form:

$$\log\left[\Lambda_i^3 \rho_i(\mathbf{r})\right] - Q_i(\mathbf{r}) = \lambda_i / k_B T \;\; (i = 1,2), \tag{11}$$

where

$$k_B T Q_i(\mathbf{r}) = -\sum_{\alpha=0}^{3} \int d\mathbf{r}' \frac{\partial \Phi}{\partial n_\alpha(\mathbf{r}')} \omega_i^{(\alpha)}(\mathbf{r} - \mathbf{r}') - \sum_{\alpha=1}^{2} \int d\mathbf{r}' \frac{\partial \Phi}{\partial n_\alpha(\mathbf{r}')} \bar{\omega}_i^{(\alpha)}(\mathbf{r} - \mathbf{r}') - \int d\mathbf{r}' [\rho_i(\mathbf{r}') U_{ii}(|\mathbf{r} - \mathbf{r}'|) + \rho_j(\mathbf{r}') \times U_{ij}(|\mathbf{r} - \mathbf{r}'|) - U_{wi}(\mathbf{r}) \;\; (j \neq i), \tag{12}$$

and λ is a Lagrange multiplier arising because of the constant of fixed average density of component i in the slit:

$$\rho_{i,\mathrm{av}} = \frac{1}{V_i}\int_{V_i} d\mathbf{r}\rho_i(\mathbf{r}), \tag{13}$$

where V_i is the volume of the slit accessible to component i. $V_1(V_2)$ is located between planes parallel to the walls and separated from the walls by the distance $\sigma_{w1}(\sigma_{w2})$ (see Fig. 1). Using Eqs. (11) and (13), the Lagrange multipliers can be rewritten in the form

$$\lambda_i = -k_B T \log\left[\frac{1}{\rho_{i,\mathrm{av}} V_i \Lambda_i^3}\int_{V_i} d\mathbf{r} e^{Q_i(r)}\right]. \tag{14}$$

By eliminating λ_i between Eqs. (11) and (14) and performing the integration in Eq. (12) with respect to the y direction, one obtains two integral equations for the FDDs $\rho_1(x, h)$ and $\rho_2(x, h)$ that can be solved by numerical iterations. Because of the large (infinite) size of the slit in the x direction, an additional assumption is required to make the problem solvable. Similar to our previous calculations, we assume that $\rho_1(x, h)$ and $\rho_2(x, h)$ are periodic functions of x with a period L_x that allows one to perform iterations over finite domains of the x and h coordinates. This assumption has the same meaning as the periodic boundary condition commonly used in MC simulations of confined fluids.[10,13,19,31] In contrast to the above mentioned simulations, where the period (length of the simulation box) rarely exceeded 30 molecular diameters, the period used in this paper is up to 120 molecular diameters. By considering numerous values of L_x and analyzing the changes in the density profile, one can obtain information about the real FDD in a long (infinite) closed slit.

The details of the numerical procedure are similar to those employed in Ref. 3 and can be found in that paper.

First, we consider the solutions of the Euler-Lagrange equation for $L_x = 30$ and identify the stable and metastable states for various values of the average density. By extracting such information for numerous values of L_x, from $L_x = 10$ to $L_x = 120$, the free energy of the stable state of the system could be represented as a function of L_x and its asymptotic behavior at large values of L_x provided information about the infinite slit.

III. RESULTS

In the computations, the solid carbon dioxide was selected for the walls and argon was selected as component 2 (C2). The interaction parameters and wall density are[6] $\epsilon_{22}/k_B = 119.76$ K, $\epsilon_{w2}/k_B = 153.0$ K, $\sigma_{22} = 3.405$ Å, $\sigma_{w2} = 3.727$ Å, and $\rho_w = 1.91 \times 10^{28}$ m^{-3}. The parameters for component 1 (C1) and the cross parameters were selected as follows: $\epsilon_{11} = \epsilon_{22}/4$, $\epsilon_{w1} = \epsilon_{w2}$, $\sigma_{11} = 1.2\sigma_{22}$, $\sigma_{w1} = 1.2\sigma_{w2}$, $\epsilon_{12} = \sqrt{\epsilon_{11}\epsilon_{22}} = \epsilon_{22}/2$, and $\sigma_{12} = (\sigma_{11} + \sigma_{22})/2 = 1.1\sigma_{22}$. The width of the slit was taken as $h_m = 10\sigma_{22}$ and the temperature was $T = 87$ K. To calculate the de Broglie wavelength, the mass m_1 of a molecule of C1 was taken as $m_1 = (\sigma_{11}/\sigma_{22})^3 m_2$, where m_2 is the mass of the argon molecule. All densities will be expressed in dimensionless form, $\rho_i^* = \sigma_{22}^3$, in units of the arbitrarily chosen density $\rho^* = 0.0116$.

A. Density distributions for $L_x = 30\sigma_{22}$

For fixed average fluid densities $\rho_{1,\mathrm{av}}$ and $\rho_{2,\mathrm{av}}$, the Euler-Lagrange equation can have several solutions that can be obtained by using various initial guesses for the iteration procedure. Four guesses, presented in Fig. 2, were employed as initial density distributions for each of the components.

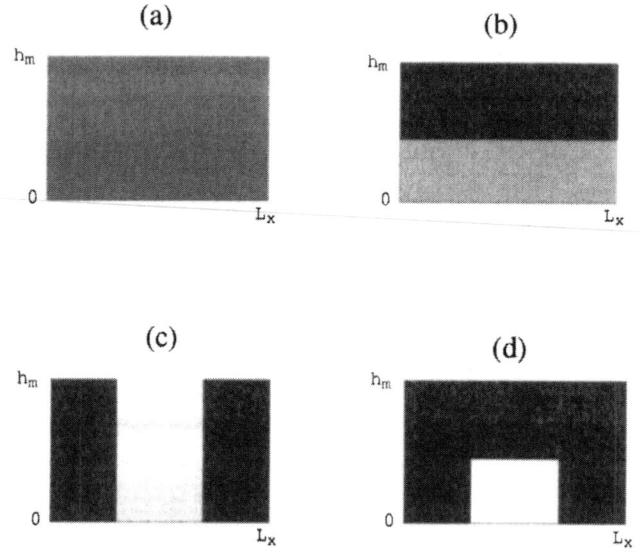

FIG. 2. Typical initial FDDs of each component employed in iterations. The lighter color areas correspond to higher fluid density. (a) FDD always leads to a solution of the Euler-Lagrange equation uniform in the x direction and symmetrical in the h direction; (b) FDD can lead to a solution uniform in the x direction and asymmetrical in the h direction; [(c) and (d)] FDDs can lead to bridgelike and bumplike solutions, respectively.

The number of different solutions of the Euler-Lagrange equation depends on the average densities $\rho_{i,\text{av}}$. Some characteristic examples of those solutions are presented below.

First, let us consider the features of the FDDs for a fixed average density of component $1\left(\rho^*_{i,\text{av}}=3\right)$ and various values of $\rho^*_{i,\text{av}}$. For $\rho^*_{i,\text{av}}<11$, all the initial guesses led to uniform distributions along the x axis and symmetrical in the direction perpendicular to the walls with respect to the middle plane (referred below as US solution). In this case the walls are covered by liquidlike layers of mixed C1 and C2 and the remainder of the slit is filled by a low density mixture. The densities $\rho^*_1(\mathbf{r})$ and $\rho^*_2(\mathbf{r})$ and the local composition of the mixture, defined as $X(\mathbf{r})=\rho^*_1(\mathbf{r})/\left[\rho^*_1(\mathbf{r})+\rho^*_2(\mathbf{r})\right]$, depend in this case on the h coordinate only. $\rho^*_2(h)$ and $X(h)$ are plotted in Figs. 3(a) and 3(b), respectively, for some average densities $\rho^*_{2,\text{av}}$. The density distribution of C1 is almost independent of $\rho^*_{2,\text{av}}$ and is similar to $\rho^*_2(h)$ for $\rho^*_{2,\text{av}}=3$. Note that the ranges of $0<h<(\sigma_{w1}-\sigma)$ and $h_m-(\sigma_{w1}-\sigma_{w2})<h<h_m$ are not accessible to C1 molecules because of their larger hard core diameter compared to that of C2. One can see from Fig. 3(b) that the composition of the fluid in the region of the slit accessible to both components considerably changes across the slit. Close to the wall, the mixture is much richer in C1 than in the middle of the slit, the latter being close to the average composition $X_{\text{av}}=\rho_{1,\text{av}}/(\rho_{1,\text{av}}+\rho_{2,\text{av}})$.

At $\rho^*_{2,\text{av}}=11$, two additional solutions appear. The first, the UA solution, is uniform in the x direction and asymmetric in the h direction. It is a result of the SB described in detail in Refs. 1 and 2. The density profiles of C2 and C1 are presented in Figs. 4(a) and 4(b), respectively. Comparing with the US solution, C1 and C2 are redistributed across the slit in opposite ways: C1 has the largest density near $W_2(h\simeq h_m)$, whereas the density of C2 has the largest value near $W_1(h\simeq 0)$. The density oscillations of the more dense C2 near the wall $W_1(h\simeq 0)$ are similar to those occurring in a bulk fluid in contact with a solid wall.[6,32] The oscillations of the FDD of the less dense C1 are almost absent in the considered case. The UA solution indicates a specific demixing that occurs across a closed narrow slit [see Fig. 4(c) where the composition of the mixture across the slit is plotted]. Second, a bumplike solution is schematically presented in Fig. 5. The density distributions of C1 [Fig. 5(b)] and C2 [Fig. 5(a)] inside the bump are different. C2 is present in the bump core

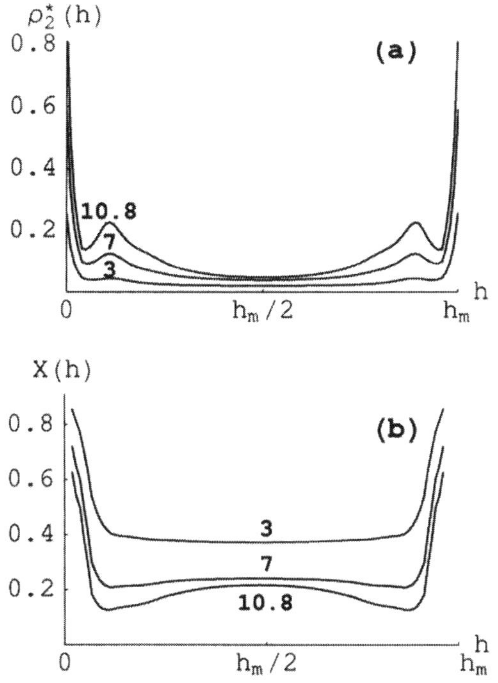

FIG. 3. Mixtures with $\rho^*_{1,\text{av}} = 3$ and various values of $\rho^*_{2,\text{av}}$. $L_x = 30\sigma_{22}$ and T = 87 K. (a) Density profiles of the uniform in the x direction and symmetrical in the h direction US solution of the Euler-Lagrange equation for component 2. The numbers on the curves indicate the values of $\rho^*_{2,\text{av}}$. (b) h dependence of composition X for the same cases as in panel (a).

[light color area in Fig. 5(a)], whereas C1 is located outside the bump. The bumplike FDD is nonuniform along the x axis and nonsymmetrical about the middle plane in the h direction. This solution constitutes a SB in two, x and h, directions. The density profiles across the bump in the h direction resemble the profiles for the UA solution presented in Fig. 4. The density profiles of C1 and C2 in the h direction taken outside the bump are symmetrical and similar to those presented in Fig. 3(a). The free energies of the UA solution and bumplike one are smaller than the free energy of the US solution, the free energy of the bumplike solution being the smallest one (see Table 1). Hence, the latter solution is stable for the considered values of $\rho^*_{1,\text{av}}\left(i = 1, 2\right)$. Note that the bumplike solution presented in Fig. 5 is twofold degenerate because the bump can be located either on the lower or on the upper wall.

At $\rho^*_{2,\text{av}} = 11.8$, a new, bridgelike solution was obtained in addition to those mentioned above [see Figs. 6(a) and 6(b)]. C2 forms a dense bridge between the walls [see Fig. 6(a)] with C1 displaced outside the bridge [Fig. 6(b)]. The bridge consists of a liquidlike core [light color area in Fig. 6(a)] and an interfacial area between the core and vaporlike phase of C2. The bridgelike FDD is symmetrical about the middle plane but is nonuniform along the x axis. This nonuniformity constitutes a SB in the x direction. The composition of the mixture depends on both the h and x coordinates. The h dependence of the composition inside ($x = 0.5L_x$), on the boundary ($x = 0.4L_x$), and outside ($x = 0.2L_x$) the bridge in the direction across the slit is presented in Fig. 6(c). One can see from this figure that for any x, the fraction of C1 decreases quickly near the walls and remains almost constant for the remaining part of the slit. In Fig. 6(d), the x dependence of the composition is presented for various values of h (indicated on the curves). Outside the bridge ($x < L_x/4$, $x > 3L_x/4$), at any given distance h from the wall, this fraction does not depend on x and has values much larger than those inside the bridge ($x \sim L_x/2$). Estimations have shown that the bridge presented in Fig. 6(a) contains

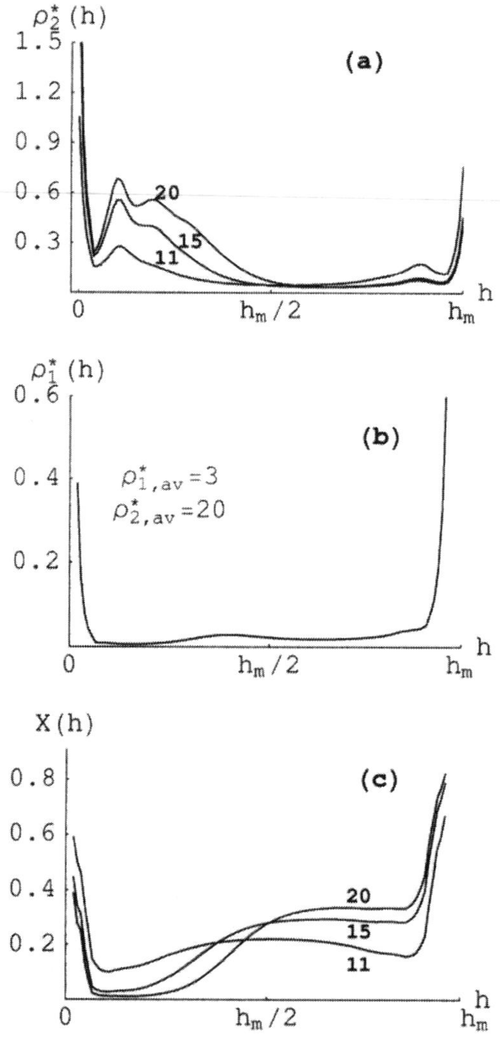

FIG. 4. Mixtures with $\rho^*_{1,\mathrm{av}}=3$ and various values of $\rho^*_{2,\mathrm{av}}.L_x=30\sigma_{22}$ and $T=87$ K. (a) Density profiles of the uniform in the x direction and asymmetrical in the h direction UA solution of the Euler-Lagrange equation for component 2. The numbers on the curves indicate the values of $\rho^*_{2,\mathrm{av}}$. (b) h dependence of ρ_1^* for $\rho^*_{1,\mathrm{av}}=3$ an $\rho^*_{2,\mathrm{av}}=20$. (c) h dependence of composition X for the same cases as in panel (a).

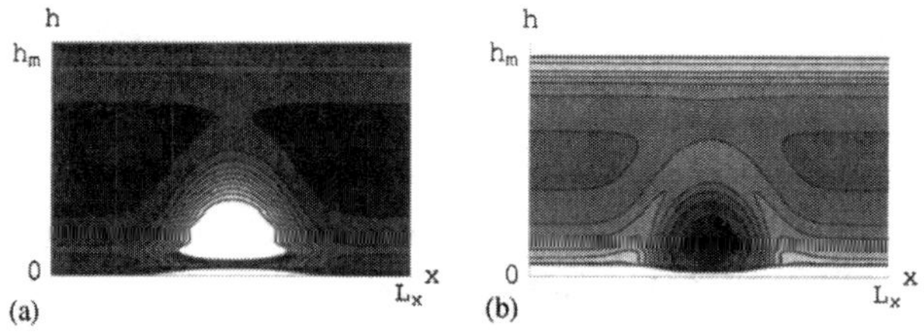

FIG. 5. Bumplike (a) FDD of component 2 and (b) FDD of component 1. The lighter color areas correspond to higher fluid densities. Mixture with $\rho^*_{1,\mathrm{av}}=3$ and $\rho^*_{2,\mathrm{av}}=11$. $L_x=30\sigma_{22}$ and $T=87$ K.

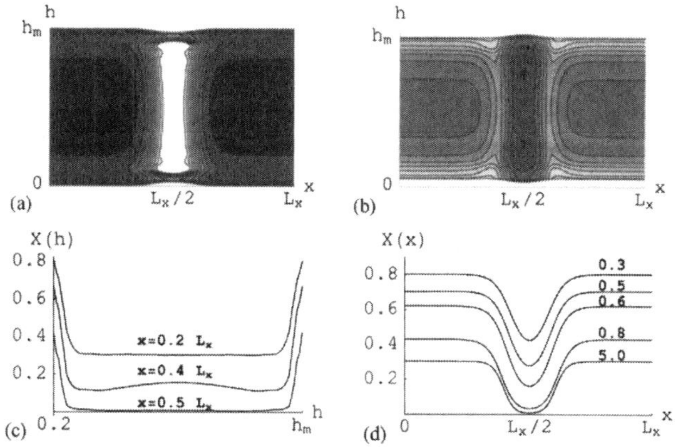

FIG. 6. Mixture with $\rho^*_{1,\text{av}} = 3$ and $\rho^*_{2,\text{av}} = 11.8$. $L_x = 30\ \sigma_{22}$ and $T = 87$ K. (a) Bridgelike FDD of component 2 and (b) FDD of component 1. The lighter color areas correspond to higher fluid densities. (c) h dependence of the mixture composition taken at $x = 0.5L_x$ (center of the bridge), $x = 0.4L_x$ (boundary of the bridge), and $x = 0.2L_x$ (outside of the bridge). (d) x dependence of the mixture composition taken at values of h noted on each curve in units of σ_{22}.

about 66% of the total amount of C2 in the slit and about 20% of C1. Hence, the bridgelike solution involves a demixing of the C1 and C2 in the x direction, in contrast to the UA solution that involves demixing across the slit (in the h direction).

The free energy of the bridgelike solution is smaller than those of all other solutions (see Table 1); therefore, this FDD represents the stable state of the system. However, the differences between the free energies of all possible states are very small. The lifetimes of the system in those states depend on the energy barriers between them.

TABLE 1
Free Energies (*F*) per unit area of one of the walls for various solutions of the Euler-Lagrange equation. The stable solutions are distinguished by the bold font. $L_x = 30\sigma_{22}$

FDD		$F\sigma^2_{22}/k_BT$
	$\rho^*_{1,\text{av}} = 3, \rho^*_{2,\text{av}} = \mathbf{10.8}$	
US solution		**−19.84**
	$\rho^*_{1,\text{av}} = 3, \rho^*_{2,\text{av}} = \mathbf{11.0}$	
US solution		−20.102
UA solution		−20.104
Bumplike		**−20.160**
	$\rho^*_{1,\text{av}} = 3, \rho^*_{2,\text{av}} = \mathbf{14.0}$	
US solution		−23.95
UA solution		−24.00
Bridgelike		**−24.13**
Bumplike		−24.10

By further increasing $\rho^*_{2,av}$, the number of possible solutions decreases. The bumplike solution disappears at $\rho^*_{2,av} \simeq 16$ and the UA solution disappears at $\rho^*_{2,av} \simeq 41$. The bridgelike FDD remains stable until the average density of C2 increases up to $\rho^*_{2,av} \simeq 45$, above which only the US solution exists.

To study the influence of C1 on the confined mixture we considered the possible solutions of the Euler-Lagrange equation for various average densities $\rho^*_{1,av}$ at $\rho^*_{2,av} = 14$. In the range $0 < \rho^*_{1,av} < 5$, all the above listed solutions are present. For $0 < \rho^*_{1,av} \leqslant 0.9$, the bumplike solution is the stable one, all other ones being metastable. For $0.9 < \rho^*_{1,av} < 5$, the stable solution is the bridgelike one. For $\rho^*_{1,av} > 5$, the bumplike solution does not exist, and for $\rho^*_{1,av} > 7.5$, the same happens with the UA solution, the bridgelike solution remaining the stable one and the US solution being metastable. At $\rho^*_{1,av} > \rho^*_{1,0} = 10.5$, only the uniform US solution still exists. The average density $\rho^*_{av} = \rho^*_{1,av} + \rho^*_{2,av}$ of the mixture at $\rho^*_{1,av} = \rho^*_{1,0}$ is equal to $\rho^*_{av} = 24.5$. C2 taken alone at the latter average density has a bridgelike solution as the stable state, the US and UA solutions being metastable.

The absence at $\rho^*_{1,av} > 10.5$ of a bridgelike stable state as well as bumplike and UA solutions in the case of a mixture can be understood by remembering that in a single component fluid the existence of UA bump and bridgelike solutions at a given temperature depends on the strength of fluidfluid interactions and occurs if the latter are larger than a certain critical value.[1] C1 taken alone does not exists in stable UA-bump- or bridgelike states because of the weakness of the intermolecular interactions considered. A small amount of C1 in a mixture with C2 does not appreciably change the behavior of the mixture which is determined by C2. However, the increase of the average density $\rho_{1,av}$ of C1 reduces the interactions between the C2 molecules because of the increased average distance between them. For this reason, the effective C2-C2 intermolecular interaction is no longer sufficiently large to generate UA-, bump-, and bridgelike states. Hence, the composition X_{av} plays an important role in the existence of SB states at a given total average density of the mixture. As an example, let us consider the average density of the mixture equal to $\rho^*_{av} = 14$. In this case, we found that for $0 < X_{av} < 0.179$, all four states exist, the bumplike being the stable one. For $X_{av} > 0.179$, the bridge like solution disappears, and the bumplike solution remains the stable one. For $X_{av} > 0.250$, only the US solution exists. No stable UA or bridgelike solutions were identified for the average density $\rho^*_{av} = 14$ of the mixture.

B. Density Distribution for $L_x = 10\sigma_{22}$

For $\rho^*_{1,av} = 3$ and various values of $\rho^*_{2,av}$, only two solutions of the Euler-Lagrange equation were identified for both components of the mixture. A US solution was found for all considered $\rho^*_{2,av}$ and a UA solution was found for $11 < \rho^*_{2,av} \leqslant 41$. When the latter exists, it constitutes the stable one. The above range of existence of the UA solution coincides with that for $L_x = 30\sigma_{22}$ because the uniformity of the UA solution in the x direction means independence of this FDD of L_x. No bridge or bumplike FDDs were found for $L_x = 10\sigma_{22}$.

C. L_x Dependence of the Density Distributions at Fixed Average Densities

Having as a goal the finding of the stable FDD of a binary mixture in a very long (infinite) closed slit, we calculated the FDDs $\rho^*_1(\mathbf{r})$ and $\rho^*_2(\mathbf{r})$ when L_x was changed from $10\sigma_{22}$ to $120\sigma_{22}$ at the fixed average densities $\rho^*_{1,av} = 3, \rho^*_{2,av} = 26$. These values of the densities were selected because they provide a variety of solutions of the Euler-Lagrange equations. For any selected L_x from that interval, all possible FDDs were calculated and the stable one was identified. For $10\sigma_{22} \leqslant L_x \leqslant 16\sigma_{22}$, the UA solution (Fig. 4) was the stable FDD. The bridgelike FDD [Fig. 6(a)] was stable for $L_x > 16\sigma_{22}$. No stable bumplike states were found in this case.

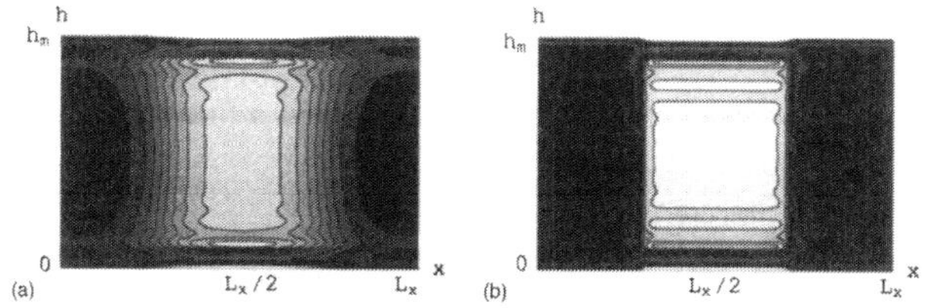

FIG. 7. Mixture with $\rho^*_{1,av} = 3, \rho^*_{2,av} = 26$, and $T = 87$ K. Bridgelike FDD of component 2 for (a) $L_x = 16\sigma_{22}$ and (b) $L_x = 120\sigma_{22}$. The lighter color areas correspond to higher fluid densities.

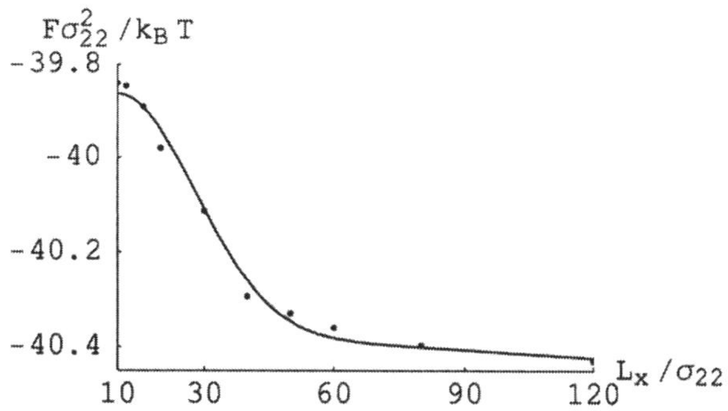

FIG. 8. Mixture with $\rho^*_{1,av} = 3, \rho^*_{2,av} = 26$, and $T = 87$ K. The dependence of the free energy of the stable state on the length L_x of the period of the FDD. The points represent the results of the calculations.

In the considered range of L_x, the width ΔL_{bridge} of the bridge increases almost linearly ($\Delta L_{bridge} \sim 0.5L_x$) and the shape of the bridge changes from that presented in Fig. 7(a) to an almost rectangular shape presented in Fig. 7(b).

The free energy of the stable state is plotted as a function of L_x in Fig. 8. This figure indicates an asymptotic behavior of the free energy with increasing L_x. From this behavior, one can conclude that for selected average densities, a single bridge of finite length consisting mostly of C2, the remainder of the slit being filled with a mixture of C1 and C2, occurs as a unique stable state in a long closed slit.

For gaslike average densities $\rho^*_{1,av}$ and $\rho^*_{2,av} \sim 0.01$, as well as for liquidlike densities $\rho^*_{1,av}$ and $\rho^*_{2,av} \sim 0.6$, the stable state for any L_x is the US solution of the Euler–Lagrange equation.

IV. DISCUSSION

In this paper as well as in a previous one[3], the solution of the Euler-Lagrange equation for FDDs involves the assumption of the FDD periodicity in the x direction that is equivalent to the cyclic boundary conditions used in MC simulations. By assuming specific values of L_x, we determined the possible solutions and the values of the free energy for each of them, identifying the stable state.

For small values of L_x $(L_x \leqslant 16)$ and depending on the average densities of the components, the stable state is represented either by the US or by the UA solution for both components of the mixture.

For larger values of L_x, for instance, $L_x = 30\sigma_{22}$, the stable FDD in a planar slit can be described, depending on $\rho^*_{1,av}$ and $\rho^*_{2,av}$, by a US-, bump-, or bridgelike solution. In the MC simulations of a binary mixture located in narrow cylindrical pores, stable bridgelike solutions were identified in Ref. 19.

The assumption of periodicity of FDD in the x direction used in the present paper allowed one to find the stable and metastable states of the system. According to this assumption, the bump- and bridgelike FDDs are periodic sequences of bumps or bridges along the x direction. If the stable FDD is indeed periodical, then its period L_x should minimize the free energy of the system considered as a function of L_x. However, the analysis of this dependence reveals the absence of such a minimum. In the case of FDDs that are uniform in the x direction (US or UA solutions), the free energy does not depend on L_x. For FDDs that are nonuniform in the x direction (bridge and bumplike solutions), the calculated free energy of the stable state monotonously decreases with increasing L_x (see Fig. 8), the bridgelike FDD remaining the only stable state. This suggests that in the presence of SB in the x direction, the real stable state that exists in a long (infinite) closed slit and, presumably, in a long closed cylindrical pore is a single bridge of finite length over the entire slit. In other words, this result suggests that, in a rigorous sense, the only reasonable assumption for L_x is $L_x = \infty$. However, the periodical solutions allow us to make a comparison with MC simulations.

One of the peculiarities of SB in binary mixtures considered in this paper is that one of the component (C2) generated SB in the density distribution of the other component (C1) in spite of the fact that the latter component taken alone does not have SB states. Vice versa, C1 shrinks the range of existence of SB, which C2 has when taken alone.

As shown in Ref. 10 for a one-component fluid confined in a cylindrical pore, the bridges and bumps found in a canonical ensemble DFT or MC simulations can be considered as the nuclei of a liquid phase during a capillary condensation. While in open systems (grand canonical ensemble) those states are unstable, they become stable in closed systems (canonical ensemble) and can provide information about the nucleation energy barrier, and hence about the nucleation rate. The approach used in the present paper allows one to extend such considerations to the capillary condensation (or wetting) of binary mixtures.

In conclusion, it should be noted that the DFT used in the present paper treats the fluid-fluid interaction in the mean-field approximation. Therefore, it does not account for the fluid density fluctuations in the lateral directions. At finite temperatures, these fluctuations might destroy the SB, thus restoring the uniformity of the density profile. Nevertheless, MC simulations (see Refs. 10 and 19) confirm the existence of SB in the lateral direction. This means that the mean-field approximation provides, at least qualitatively, a correct description of the density distribution when SB in the lateral direction occurs.

REFERENCES

1. G O. Berim and E. Ruckenstein, J. Phys. Chem. B **111**, 2514 (2007); **111**, 12823 (2007) (Sections 2.1 and 2.3 of this volume).
2. G. O. Berim and E. Ruckenstein, J. Chem. Phys. **126**, 124503 (2007) (Section 2.2 of this volume).
3. G. O. Berim and E. Ruckenstein, J. Chem. Phys. **128**, 024704 (2008) (Section 2.4 of this volume).
4. P. Tarazona, Phys. Rev. A **31**, 2672 (1985).
5. P. Tarazona, U. M. B. Marconi, and R. Evans, Mol. Phys. **60**, 573 (1987).
6. R. Evans and P. Tarazona, Phys. Rev. A **28**, 1864 (1983).
7. M. Merkel and H. LOwen, Phys. Rev. E **54**, 6623 (1996).
8. J. Errington, personal communication (2007).
9. J. Z. Tang and J. G. Harris, J. Chem. Phys. **103**, 8201 (1995).
10. A. Vishnyakov and A. V. Neimark, J. Chem. Phys. **119**, 9755 (2003).
11. K. G. Kornev, I. K. Shingareva, and A. V. Neimark, Adv. Colloid Interface Sci. 96, 143 (2002).
12. E. A. Ustinov and D. D. Do, J. Chem. Phys. 120, 9769 (2004).
13. A. V. Neimark, P. I. Ravikovitch, and A. Vishnyakov, Phys. Rev. E **65**, 031505 (2002).
14. V. Talanquer and D. W. Oxtoby, J. Chem. Phys. **104**, 1483 (1996).
15. V. Talanquer and D. W. Oxtoby, J. Chem. Phys. **114**, 2793 (2001).
16. A. V. Neimark, P. I. Ravikovitch, and A. Vishnyakov, J. Phys.: Condens. Matter **15**, 347 (2003).
17. P. I. Ravikovitch, A. Vishnyakov, and A. V. Neimark, Phys. Rev. E **64**, 011602 (2001).
18. A. J. Liu, D. J. Durian, E. Herbolzheimer, and S. A. Safran, Phys. Rev. Lett. **65**, 1897 (1990).
19. Z. Zhang and A. Chakrabarti, Phys. Rev. E **50**, R4290 (1994).
20. R. Roth and S. Dietrich, Phys. Rev. E **62**, 6926 (2000).
21. B. Yan and X. Yang, Ind. Eng. Chem. Res. **43**, 6577 (2004).

22. O. Pizio, A. Patrykiejew, and S. Sokolowski, Mol. Phys. **99**, 57 (2001).
23. K. Bucior, A. Patrykiejew, O. Pizio, and S. Sokolowski, J. Colloid Interface Sci. **259**, 209 (2003).
24. K. Bucior, Colloids Surf., A **243**, 105 (2004).
25. C. J. Hemming and G. N. Patey, J. Phys. Chem. B **110**, 3764 (2006).
26. Y. Rosenfeld, Phys. Rev. Lett. **63**, 980 (1989).
27. A. Gonzalez, J. A. White, F. L. Roman, and S. Velasco, Phys. Rev. Lett. **79**, 2466 (1997).
28. A. Gonzalez, J. A. White, F. L. Roman, and R. Evans, J. Chem. Phys. **109**, 3637 (1998).
29. J. A. White and S. Velasco, Phys. Rev. E **62**, 4427 (2000).
30. J. A. White, A. Gonzalez, F. L. Roman, and S. Velasco, Phys. Rev. Lett. 84, 1220 (2000).
31. J. Mittal, J. R. Errington, and T. M. Truskett, J. Chem. Phys. **126**, 244708 (2007).
32. R. H. Nilson and S. K. Griffiths, J. Chem. Phys. **111**, 4281 (1999).

2.6 Symmetry Breaking of the Density Distribution of a Quantum Fluid in a Nanoslit*

Gersh Berim and Eli Ruckenstein†
Department of Chemical and Biological Engineering, State University of New York at Buffalo, Buffalo, New York 14260, USA

† E-Mail: feaeliru@buffalo.edu.
Phone: (716)645-1179. Fax: (716)645-3822.

(Received 10 August 2009; accepted 18 October 2009; published online 11 November 2009)

I. INTRODUCTION

The symmetry breaking (SB) in the fluid density distribution (FDD) of a fluid confined in a slit between two parallel identical solid walls manifests through the existence of a stable asymmetric FDD, which provides a lower free energy of the system than the symmetric FDD. Such asymmetric FDDs were obtained first via molecular dynamics simulations of simple fluids,[1–3] as well as through a density functional theory.[4] In the latter case the density profile was a stable solution of the Euler–Lagrange equation describing the equilibrium state of a fluid in the slit. Near one of the walls this profile was similar to that for a bulk liquid in contact with a solid wall, whereas near the other wall it was similar to the profile for a bulk vapor in contact with a wall. It should be noted that Refs. 1–4 were concerned with the wetting and drying transitions and that the existence of an asymmetric profile was not associated with the phenomenon of SB. The asymmetric profile obtained in Ref. 4 was for structured walls; for smooth walls the profile was symmetric. The first study focused on SB was published in 1996 by Merkel and Löven[5] who demonstrated that in an open system consisting of two parallel identical monolayers of a two-dimensional liquid coupled by an interlayer potential, SB occurs at a critical strength of the interlayer interaction. The fluid in one layer had a liquidlike density, while in the other a gaslike density. In Refs. 6 and 7 the existence of spontaneous SB was proven for a fluid (argon) confined in a closed nanoslit with a width of about 5 nm and with walls made of solid carbon dioxide. It was assumed that each wall generates a hard core repulsion at some distance h_r from the wall, which was taken to be equal to the parameter σ_{fs} of the Lennard-Jones fluid-solid interaction potential. It was shown that the occurrence of SB is affected by the temperature and the average density ρ_{av} of the fluid. SB occurred only for $T < T_{sb}$, where the critical temperature T_{sb} was found to be 106 K. At each temperature below T_{sb}, SB occurred when ρ_{av} had values between ρ_{sb1} and ρ_{sb2} ($\rho_{sb1} < \rho_{av} < \rho_{sb2}$), ρ_{sb1} and ρ_{sb2} being temperature dependent critical densities. In Ref. 8, the dependence of SB on the Lennard-Jones energy parameter ϵ_{fs} of the fluid-solid interactions was investigated. The range of existence of SB decreased with increasing ϵ_{fs} and disappeared for sufficiently large values of ϵ_{fs}. In Ref. 9 the possibility of a two-dimensional SB in closed nanoslits was also examined.

* *J. Chem. Phys.* 131, 184707 (2009). Republished with permission.

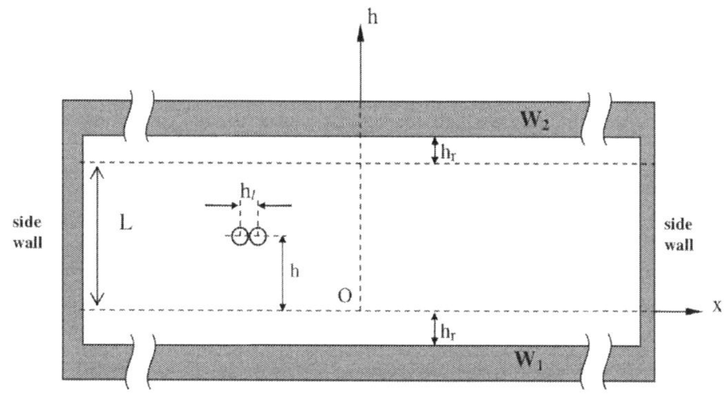

FIG. 1. A closed slit of width $L + 2h_r$ between two identical solid walls W_1 and W_2. h_r is the distance up to which the wall generates hard core repulsion and h_l is the screening distance of the fluid-fluid interactions. L is the effective width of the slit accessible to the fluid molecules. The y-axis is perpendicular to the plane of the figure. The size of the slit in the x and y directions is large for the end effects to be negligible.

Recently, it was shown[10] that in a closed slit SB is affected by the value of the distance h_r at which the solid wall generates a hard core repulsion. The change in h_r in the range $\sigma_{fs}/2 < h_r < \sigma_{fs}$ performed in Ref. 10 revealed that for argon in contact with solid carbon dioxide, SB occurs only for $h_r > 0.847\sigma_{fs}$ Qualitatively, the effect of h_r on SB was explained in Ref. 10 as the result of the increase in the fluid-solid attraction with decreasing h_r. For $h_r \simeq \sigma_{fs}$ of the free energy of the asymmetric profile at some ρ_{av} becomes smaller than that of the symmetric one and hence the former profile is the stable one. When h_r decreases, the liquid-solid attraction at the two solid walls increases. Consequently, the free energy of the symmetric profile decreases more rapidly than that of the asymmetric one which involves a liquid-solid interface at one of the solid walls and a vapor-solid interface at the other one. As a result, the free energy of the symmetric profile becomes smaller than that of the asymmetric one and SB disappears. The same arguments can explain the dependence of the SB on the parameter ϵ_{fs}.

The examination of the equilibrium between a quantum fluid (^{4}He) and a solid substrate, which was performed mostly at $T = 0$ K has shown that many of its behaviors (e.g., the layering, wetting, and prewetting transitions) are similar to those of the classical fluids.[11–14] The effect of temperature was taken into account in Ref. 15, where the contact angles θ of a ^{4}He drop on the surfaces of some alkali metals (Cs, Rb, and K) were calculated. It was observed that near the λ-point of ^{4}He ($T_\lambda = 2.17$ K) at which the phase transition from normal to superfluid state takes place, the liquid-vapor surface tension had a small but observable kink. Because of the unique character of quantum liquids it is of interest to examine the SB in such cases and to verify whether the quantum effects affect this phenomenon. To do this, a nonlocal density functional theory formulated in Ref. 13 for $T = 0$ K and generalized to $T \neq 0$ K in Ref. 15 will be employed. Cesium will be considered as the solid substrate.

II. THE SYSTEM

In the present paper we consider a fluid confined between two parallel structureless identical solid walls W_1 and W_2 separated by a distance $L + 2h_r$ (see Fig. 1), where h_r is the distance at which the walls generate a hard core repulsion and L is the effective width of the slit accessible to the ^{4}He molecules. The slit is considered to be closed, the distance between the sidewalls which close the slit being much larger than the width of the slit for the end effects to be negligible.

The interactions between the molecules of ^{4}He are described by the Lennard-Jones potential $\phi_{ff}(|\mathbf{r}-\mathbf{r}'|) \equiv \epsilon_{ff}\bar{\phi}_{ff}(|\mathbf{r}-\mathbf{r}'|)$, where $\epsilon_{ff}/k_B = 10.22\text{K}$ (Ref. 13) (k_B being the Boltzmann constant) and $\bar{\phi}_{ff}(|\mathbf{r}-\mathbf{r}'|)$ is a dimensionless quantity given by

$$\phi_{ff}\left(\left|\mathbf{r}-\mathbf{r}'\right|\right)=\begin{cases}4\left[\left(\frac{\sigma_{ff}}{r}\right)^{12}-\left(\frac{\sigma_{ff}}{r}\right)^{6}\right], & r\geq h_l\\ \infty, & r<h_l.\end{cases} \tag{1}$$

In Eq. (1), the molecular diameter $\sigma_{ff} = 2.556$ Å, h_l is a screening distance which depends on temperature,[13,15] the coordinates $\mathbf{r}$ and $\mathbf{r}'$ provide the locations of the interacting fluid molecules, $r = \left|\mathbf{r}-\mathbf{r}'\right|$. The dependence of h_l on temperature was taken from Ref. 15.

To describe the interaction between a molecule of fluid and the solid walls, two potentials will be employed. First, the standard 9–3 Lennard-Jones potential, which is provided by the expression

$$U_{\mathrm{LJ}}\left(\mathbf{r}\right)=U_{\mathrm{LJ}}\left(h\right)\equiv V_{\mathrm{LJ}}\bar{U}_{\mathrm{LJ}}\left(h\right), \tag{2}$$

where $V_{\mathrm{LJ}}=\left(2\pi/3\right)\epsilon_{\mathrm{fs}}\rho_s\sigma_{\mathrm{fs}}^3$, ρ_s being the wall density, and σ_{fs} and ϵ_{fs} parameters in the Lennard-Jones potential for the fluid-solid interactions, and the dimensionless quantity $\bar{U}_{\mathrm{LJ}}\left(h\right)$ is given by

$$\bar{U}_{\mathrm{LJ}}\left(h\right)=\psi_{\mathrm{LJ}}\left(h\right)+\psi_{\mathrm{LJ}}\left(L-h\right) \tag{3}$$

with

$$\psi_{\mathrm{LJ}}\left(h\right)=\frac{2}{15}\left(\frac{\sigma_{\mathrm{fs}}}{h_r+h}\right)^{9}-\left(\frac{\sigma_{\mathrm{fs}}}{h_r+h}\right)^{3},\ \left(h\geq 0\right). \tag{4}$$

(Note that the origin of the h-axis is taken at a distance h_r from the wall W_1.) The potential Eq. (2) written for a single wall was used in Ref. 13 in another equivalent form to examine the adsorption of ^{4}He on the surface of cesium. Using for the number density of Cs the value $\rho_s = 8.74 \times 10^{27}\ \mathrm{m}^{-3}$, the parameters of the potential $U_{\mathrm{LJ}}(\mathbf{r})$ can be recalculated from those of Ref. 13 to obtain $V_{\mathrm{LJ}}/k_B = 4.18$ K, $\sigma_{\mathrm{fs}} = 5.44$ Å, and $\epsilon_{\mathrm{fs}}/k_B = 1.42$ K.

Another potential, $U_{\mathrm{CCZ}}(h)$, which was employed for the description of the interaction between ^{4}He and alkali metals by Chizmeshya *et al.*,[16] has the following form:

$$U_{CCZ}\left(h\right)=V_{CCZ}\left(1+\alpha h\right)\exp\left(-\alpha h\right)$$
$$-f_2\left(\beta\left(h\right)\left(h-h_{\mathrm{vdW}}\right)\right)\frac{C_{\mathrm{vdw}}}{\left(h-h_{\mathrm{vdw}}\right)^3}, \tag{5}$$

where

$$f_2\left(x\right)=1-\left(1+x+\frac{x^2}{2}\right)\exp\left(-x\right), \tag{6}$$

and

$$\beta\left(h\right)=\frac{\alpha^2 h}{1+\alpha h}. \tag{7}$$

The parameters of this potential for Cs in the units used in the present paper are[16] $V_{\mathrm{CCZ}}/k_B = 1636.83$ K, $\alpha\sigma_{ff} = 4.52735$, $C_{\mathrm{vdw}}/(\sigma_{ff}^3 k_B) = 59.5863\,\mathrm{K}$, and $h_{\mathrm{vdw}}/\sigma_{ff} = 0.0749209$.

Because of the uniformity of the walls, the fluid-solid potential and FDD $\rho(\mathbf{r})$ inside the slit is a function of h only $[\rho(\mathbf{r}) \equiv \rho(h)]$.

III. DENSITY FUNCTIONAL AND EULER-LAGRANGE EQUATION

In the present paper, the density functional for ^{4}He is selected in the form

$$\Omega_{\mathrm{He}}[\rho]=\int[f-\mu\rho(\mathbf{r})]d\mathbf{r}, \tag{8}$$

suggested in Ref. 15. In Eq. (8), μ is the chemical potential and f is the free-energy density of ^{4}He

$$f=\frac{\hbar^2}{2M}\left(\nabla\sqrt{\rho}\right)^2+c_4 f_{id}+\frac{1}{2}\int d\mathbf{r}'\rho(\mathbf{r}')\phi_{ff}(|\mathbf{r}-\mathbf{r}'|)$$
$$+\frac{c_2}{2}\rho(\mathbf{r})[\bar{\rho}(\mathbf{r})]^2+\frac{c_3}{3}\rho(\mathbf{r})[\bar{\rho}(\mathbf{r})]^3-\frac{\hbar^2}{4M}\alpha_s$$
$$\times\int d\mathbf{r}'F(|\mathbf{r}-\mathbf{r}'|)\left(1-\frac{\rho(\mathbf{r})}{\rho_{0s}}\right)\nabla\rho(\mathbf{r})\nabla\rho(\mathbf{r}')\left(1-\frac{\rho(\mathbf{r}')}{\rho_{0s}}\right), \tag{9}$$

where $\hbar$ is the Planck constant and M is the mass of a molecule of ^{4}He. The first and second terms in the above expression account for the quantum kinetic energy and the free-energy density of an ideal Bose gas, respectively. The third term represents the contribution of the fluid-fluid interactions. The terms that contain c_2 and c_3 provide the hard core He-He repulsion and are important at high densities. They contain the density $\bar{\rho}(\mathbf{r})$ which is obtained by averaging the real density $\rho(\mathbf{r})$ over a sphere of radius h_l (the screening distance for fluid-fluid interactions) at the constant weight $3(4\pi h_l^3)^{-1}$. The last term in Eq. (9) is a nonlocal correction of the kinetic energy.[13,15] In that term, $\alpha_s = 54.31\,\text{Å}, \rho_{0s}\sigma_{ff}^3 = 0.6679$, and $F(|\mathbf{r}-\mathbf{r}'|) \equiv F(r) = \pi^{-3/2}l^{-3}\exp(-r^2/l^2)$ with $l = 1$ Å.[13] The coefficients c_2, c_3, and c_4 depend on temperature. Their values were taken from Ref. 15.

The ideal-gas contribution f_{id} in Eq. (9) is[15]

$$f_{id}=\rho k_B T\log(z)-\frac{k_B T}{\Lambda^3}g_{5/2}(z), \tag{10}$$

where $\Lambda=\sqrt{2\pi\hbar^2/Mk_BT}$ is the thermal de Broglie wave-length. The fugacity z is given by

$$z=\begin{cases}1, & \rho\Lambda^3\geq g_{3/2}(1)\\ z_0, & \rho\Lambda^3<g_{3/2}(1),\end{cases} \tag{11}$$

where $g_n(z)\equiv\sum_{l=1}^{\infty}z^l/l^n$ and z_0 is the root of the equation $\rho\Lambda^3=g_{3/2}(z)$.

Taking into account the fluid-solid interaction, Eq. (2), the total free-energy functional is given by

$$\Omega[\rho]=\Omega_{\mathrm{He}}[\rho]+\int_0^L dh\rho(h)U_{\mathrm{fs}}(h), \tag{12}$$

where $U_{\mathrm{fs}}(h)$ represents either $U_{\mathrm{LJ}}(h)$ or $U_{\mathrm{CCZ}}(h)$ potential. The equilibrium fluid density distribution $\rho(\mathbf{r}) \equiv \rho(h)$ can be found from the Euler-Lagrange equation

$$\left\{-\frac{\hbar^2}{2M}\frac{d^2}{dh^2}+U(\rho,h)\right\}\sqrt{\rho(h)}=\mu\sqrt{\rho(h)}, \tag{13}$$

which can be obtained by minimizing the grand-canonical potential $\Omega[\rho]$ with respect to $\rho \equiv \rho(h)$. In Eq. (13), $U(\rho, h)$ is an effective potential caused by the fluid-fluid and fluid-solid interactions.

The latter potential is examined in some details in Appendix. Note that for the closed slit μ has the meaning of a Lagrange multiplier, which can be determined from the constraint of fixed average fluid density $\rho_{av} = (1 / V)\int_V \rho(\mathbf{r})d\mathbf{r}$ in the slit.

Equation (13) was solved by standard numerical iterations[7] using a grid with 40 grid points per molecular diameter. For a selected average density, an initial guess profile as a homogeneous distribution $\rho(h) = \rho_{av}$ of the fluid density was employed. In such a case, the iterations always converged to a symmetric profile which constitutes one of the solutions of the Euler-Lagrange equation. To determine if an asymmetric solution also exists, an asymmetric profile was selected as initial guess for a system with identical walls. It was constructed for a selected average density ρ_{av} as a straight line between $\rho(0) = \rho_0$ and $\rho(L) = \rho_1$, where $\rho_0 > \rho_1$. Depending on the selected value of ρ_{av} the iterations converged to a symmetric or asymmetric profile. When the profile was symmetric it coincided with that obtained for a homogeneous initial guess. If the profile was asymmetric, it provided a second solution of the Euler-Lagrange equation. The profile corresponding to the smaller free energy provides the stable solution.

IV. RESULTS

A. Lennard-Jones Fluid-Solid Potential

In this section, SB is examined for walls that generate hard core repulsions at distances $h_r = 0.4$, 0.2, and $0.1\sigma_{fs}$. Let us consider first the case $h_r = 0.4\sigma_{fs}$.

The fluid density profile obtained as solution of Eq. (13) for identical walls and an initial guess considered as a uniform density profile, is symmetric about the middle of the slit and has a shape dependent on the average density of the fluid and temperature. Typical profiles for small and large values of ρ^*_{av} $\left(\rho^* = \rho\sigma^3_{ff}\right)$ are presented in Fig. 2. The profiles in Fig. 2(a) are plotted for various ρ^*_{av} (marked near the curves) for a slit with $L = 60\sigma_{ff}$ and for $T = 2.0$ K. For a very small average density $\rho^*_{av} = 0.01$, the fluid density distribution is almost homogeneous with a small increase near the walls. For the larger average density $\rho^*_{av} = 0.02$, several layers of liquid-like density appear near the wall. The location of these layers is indicated by the maxima of the density in the profile, which has liquid-like values $\rho^* \sim 0.20$ in the vicinity of the walls. The density becomes gaslike $\rho^* \sim 0.001$ for distances from the wall of the order of several molecular diameters. When ρ^*_{av} increases, the oscillations of the density become larger and the region with gaslike fluid density decreases in size. For relatively high densities, $\rho^*_{av} \sim 0.20$, the whole slit is filled with a liquidlike fluid, and the oscillations of the fluid density become significant at larger distances from the walls. The fluid density in the layers near the walls increases with increasing ρ^*_{av}. In Fig. 2(b) profiles corresponding to $\rho^*_{av} = 0.10$ are presented at various temperatures. The amplitude of the oscillations of the density near the walls decreases and the density in the central part of the slit increases with increasing temperature. This is an obvious consequence of the penetration of the ^{4}He molecules from the walls into the internal part of the slit due to the increase in their mobility with increasing temperature.

Asymmetric density profiles were obtained using the procedure described in Sec. III. Several asymmetric profiles are shown in Fig. 3(a) for $T = 2.0$ K and various values of ρ^*_{av}. When ρ^*_{av} increases, the number as well as the amplitudes of the oscillations near the left wall increase and the region where the fluid has a gaslike density decreases moving toward the right wall. Note that the asymmetric solutions are twofold degenerate. The profiles obtained from those shown in Fig. 3(a) by reflection about the middle of the slit are also solutions of the Euler–Lagrange equation.

In Fig. 3(b) two profiles, one asymmetric and another symmetric, obtained for $\rho^*_{av} = 0.10$ and $T = 2.0$ K are presented as an example of multiple solutions of the Euler–Lagrange equation. It was found that the asymmetric density profile always provides a slightly smaller Helmholtz free energy than the symmetric one and, hence, the former represents a stable state. (The free energies corresponding to symmetric and asymmetric profiles for $T = 2.0$ K at various ρ^*_{av} are presented in Table 1). This indicates that, indeed, SB occurs in closed systems.

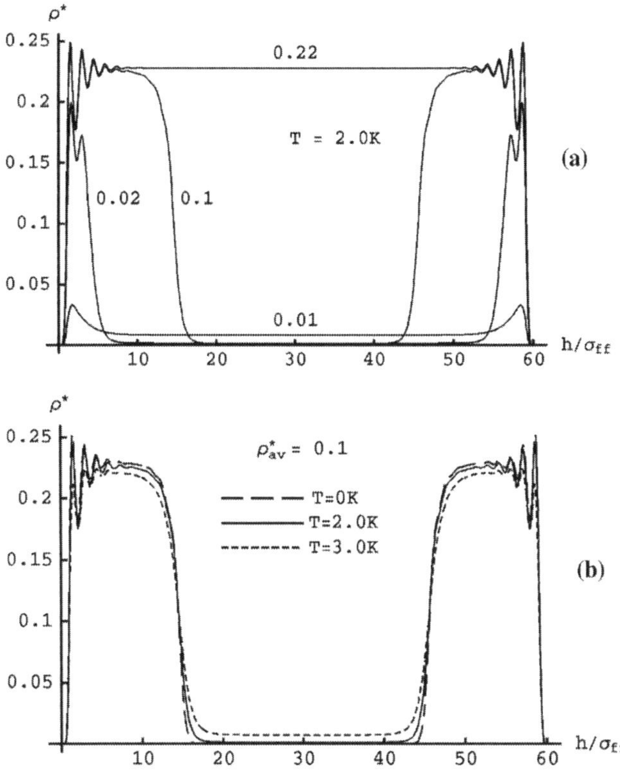

FIG. 2. (a) Symmetric fluid density profiles plotted at various average densities $\rho^*_{\rm av} = \rho_{\rm av}\sigma^3_{ff}$ indicated by the numbers at each of the curves. $T = 2.0$ K and $\rho^* = \rho\sigma^3_{ff}$. (b) Symmetric profiles at various temperatures for $\rho^*_{av} = 0.1$.

To characterize quantitatively the asymmetry of the density profiles, one can use the parameter[7]

$$\Delta_N = \frac{1}{2N_s}\int_0^L dh\left|\rho(h) - \rho(L-h)\right|. \tag{14}$$

This parameter is equal to zero when the profile is symmetric about the middle of the slit and $0 < \Delta_N < 1$ in the opposite case. In the extremely asymmetric case where all molecules are located on one side of the plane of symmetry, $\Delta_N = 1$. The dependence of Δ_N on the average density $\rho^*_{\rm av}$ is presented in Fig. 4 for $T = 0$, 2.0, and 3.0 K. Two features should be emphasized.

First, for each of the considered temperatures there is a range of average densities $\rho^*_{\rm sb1} < \rho^*_{\rm av} < \rho^*_{\rm sb2}$ for which Δ_N has significant values and is almost zero outside of that range.

Near $\rho^*_{\rm sb1}$ the parameter Δ_N as function of $\rho^*_{\rm av}$ changes rapidly and therefore those values of $\rho^*_{\rm av}$ can be considered as critical ones. The change in Δ_N is not so rapid when $\rho^*_{\rm av} \to \rho^*_{\rm sb2}$. Second, the critical value $\rho^*_{\rm sb1}$ increases, and $\rho^*_{\rm sb2}$ decreases with increasing temperature. The difference $\rho^*_{\rm sb2}-\rho^*_{\rm sb1}$, which provides the range of existence of the asymmetric solution and hence of the symmetry breaking, decreases with increasing temperature.

The results for $h_{\rm r} = 0.2\sigma_{\rm fs}$ and $h_{\rm r} = 0.1\sigma_{\rm fs}$ coincide qualitatively with those obtained for $h_{\rm r} = 0.4\sigma_{\rm fs}$. At the same average density, the asymmetry parameter Δ_N changes little with changing h_r. For instance, at $\rho^*_{\rm av} = 0.03$ and $T = 2.0$ K this parameter has the values 0.9539, 0.9545, and 0.9549 for $h_r = 0.4$, $h_r = 0.2$, and $h_r = 0.1\sigma_{\rm fs}$, respectively.

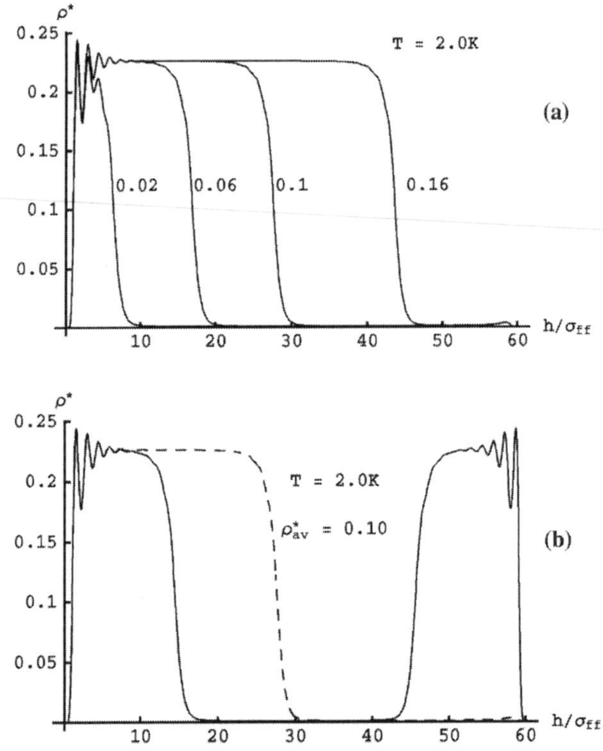

FIG. 3. **(a)** Stable asymmetric fluid density profiles plotted at various average densities $\rho^*_{av} = \rho_{av}\sigma^3_{ff}$ indicated by the numbers at each of the curves. $T = 2.0$ K and $\rho^* = \rho\sigma^3_{ff}$. (b) Metastable symmetric density profile (solid line) and stable asymmetric density profile (dashed line) calculated for $T = 2.0$ K and $\rho^*_{av} = 0.10$.

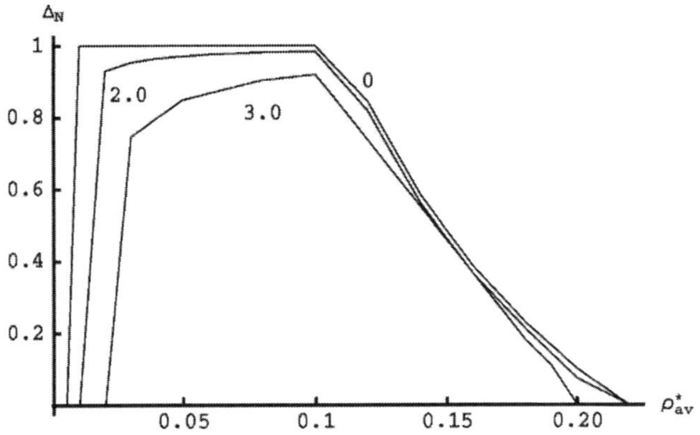

FIG. 4. The dependence of the parameter Δ_N, characterizing the asymmetic density distribution across a closed slit on the average density $\rho^*_{av}\sigma^3_{ff}$ of the fluid inside the slit for three different temperatures in kelvins indicated by the numbers near each of the curves.

B. CCZ Potential of Fluid-Solid Interactions

For this potential, defined by Eq. (5), the smallest distance of molecules of fluid from the wall is given by the parameter h_{vdW}. The potential is constructed in such a way to avoid divergence for $h \to h_{vdw}$. For this reason, one should not introduce a hard core repulsion generated by the walls. Hence this potential is a good example to test if SB occurs for $h_r = 0$.

TABLE 1
Free energies (per unit area of one of the walls) of the symmetric and asymmetric density profiles for a slit with a width $L = 60\sigma_{ff}$. $T = 2.0$ K and $h_r = 0.4\sigma_{ff}$

	Free energy (J / m²)	
$\rho_{av}\sigma^3_{ff}$	Symmetric profile	Asymmetric profile
0.02	−0.002 261 1	−0.002 443 4
0.05	−0.006 119 0	−0.006 268 5
0.10	−0.012 484	−0.012 607
0.12	−0.015 021	−0.015 140
0.14	−0.017 557	−0.017 675
0.18	−0.022 337	−0.022 749
0.19	−0.023 669	−0.024 023

An analysis similar to that carried out for the Lennard-Jones potential has shown that SB occurs for a quantum fluid even for $h_r = 0$ but has some specific features different from those of the Lennard-Jones potential. First, for a Chizmeshya-Cole-Zaremba (CCZ)-potential a stable asymmetric density profile exists only for small average densities in the slit. For example, at $T = 2.0$ K, this range is $0.005 < \rho^*_{av} < 0.02$. In Fig. 5, three stable asymmetric density profiles at $T = 2.0$ K are presented for $\rho^*_{av} = 0.007$, 0.01, and 0.014. As for the Lennard-Jones potential, the number of oscillations near the wall and their amplitude increase with increasing average density.

For $\rho^*_{av} \geq 0.02$, the asymmetric solution still exists and is similar to that presented in Fig. 3(a). However the corresponding FDD in this case has a higher free energy than the symmetric solution and, therefore, is metastable. This feature constitutes the main difference between the SB for the CCZ potential when compared with the SB for the Lennard-Jones one. In the latter case, when an asymmetric profile exists, it has always a lower free energy than the symmetric one.

In Fig. 6 the asymmetric profiles are plotted for $\rho^*_{av} = 0.10$ for the CCZ (dashed line) and Lennard-Jones (solid line) potentials. One can see that the density of the fluid near the wall is larger for the CCZ potential than for the Lennard-Jones one because the former potential has a higher attractive strength than the latter. All other features, such as the number and location of the density oscillations and the behavior far from the surface are the same for both potentials.

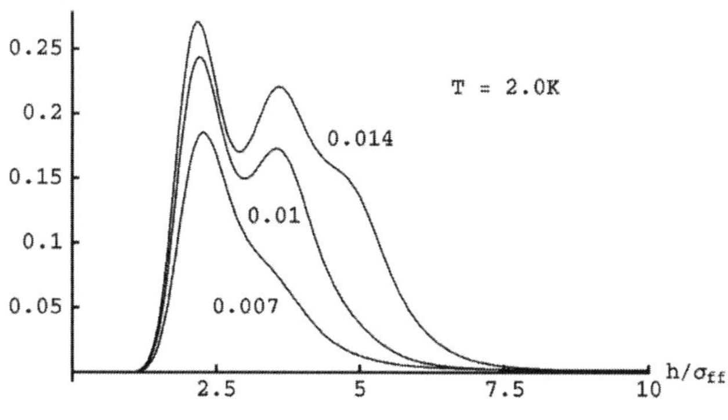

FIG. 5. Stable asymmetric fluid density profiles for the CCZ potential of the fluid-solid interactions. The profiles are plotted at various average densities $\rho^*_{av} = \rho_{av}\sigma^3_{ff}$ indicated by the numbers at each of the curves. Only the left-hand side of the profiles is presented. The fluid density in the right hand side has approximately the same value as at $h/\sigma_{ff} = 10$. $T = 2.0$ K, and $\rho^* = \rho\sigma^3_{ff}$.

V. DISCUSSION

Let us compare the results obtained in the present paper regarding the SB for quantum fluids (^4He) to those obtained previously for a classical fluid[6–8,10] (Ar) in contact with solid walls made of cesium or solid carbon dioxide, respectively.

In spite of the different natures of the fluids and of the corresponding Euler-Lagrange equations, there is a qualitative similarity of SB for both fluids. First, in both cases SB occurs in a range of average densities of the fluid in the slit which decreases with increasing temperature. As an example of such ranges, in Fig. 7(a) the asymmetry parameter Δ_N is plotted for quantum (solid line) and classical (dashed line) fluids as functions of the dimensionless density $\rho^*_{av} = \rho_{av}\sigma^3_{ff}$, where $\sigma_{ff} = 2.556$ Å for ^4He and 3.405 Å for Ar. One can see that the ranges of ρ^*_{av} of SB existence are very different as a consequence of differences in bulk liquid densities ($\rho^*_b = 0.226$ for ^4He at $T = 2.0$ K and $\rho^*_b = 0.664$ for Ar at $T = 87$ K). However if one plots these dependencies as function of ρ^*_{av}/ρ^*_b where $\rho^*_b$is the corresponding density of the bulk liquid then the ranges of SB become comparable [see Fig. 7(b)].

For classical fluids, it was shown[7,10] that SB does not occur at temperatures larger than a certain critical temperature T_{sb}, which for argon is 106 K. In the present calculations we could not identify the temperature T_{sb} for ^4He because no information about the temperature dependence of the coefficients c_i ($i = 2$, 3, and 4) of the density functional, Eq. (9), for $T > 3.0$ K is available. However the above noted tendency of decrease in the range of occurrence of the SB with increasing temperature indicates the existence of such a temperature for ^4He as well. For $T \leq 3.0$ K SB occurs at any temperature.

The main difference between the SB in classical and quantum fluids is related to the dependence of SB on the parameter h_r, the distance at which the wall generates hard core repulsion. For classical fluids there is a critical value $h_{r,c} \simeq 0.847\sigma_{fs}$ below which ($h_r < h_{r,c}$) SB does not occur.[10] As argued in Ref. 10, the decrease in h_r from $h_r = \sigma_{fs}$ is accompanied by an increase in the effective strength of attraction of the fluid by the wall. As a result, the symmetric profile becomes energetically favorable and the SB disappears. In the present paper it was shown that in quantum fluids SB occurs for all considered values of h_r (up to $h_r = 0.1\sigma_{fs}$ for the Lennard-Jones potential). Even for the CCZ-potential, which provides stronger attraction of the fluid by the wall than the Lennard-Jones potential, SB exists for small values of ρ^*_{av}. In this case there is a range of ρ^*_{av} in which an asymmetric profile exists as a metastable solution of the Euler-Lagrange equation. Note that for the Lennard-Jones potential when an asymmetric profile occurs, it constitutes the stable solution.

It is of interest to compare SB in the nonsuperfluid ($T > T_\lambda$) and superfluid ($T < T_\lambda$) states of ^4He. To make such a comparison, one should take into account that in the nanoslit T_λ can be different from its value $T_\lambda = 2.17$ K founded for bulk ^4He. Estimation of this change is a problem

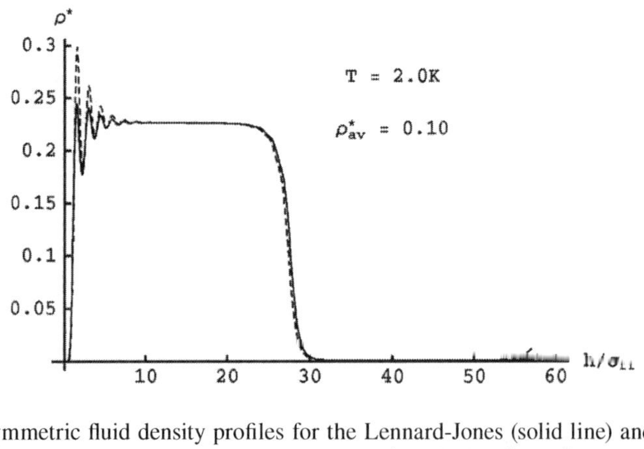

FIG. 6. Stable asymmetric fluid density profiles for the Lennard-Jones (solid line) and CCZ (dashed line) potentials of fluid-solid interactions plotted at $T = 2.0$ K for $\rho^*_{av} = 0.10$. ($\rho^* = \rho\sigma^3_{ff}$).

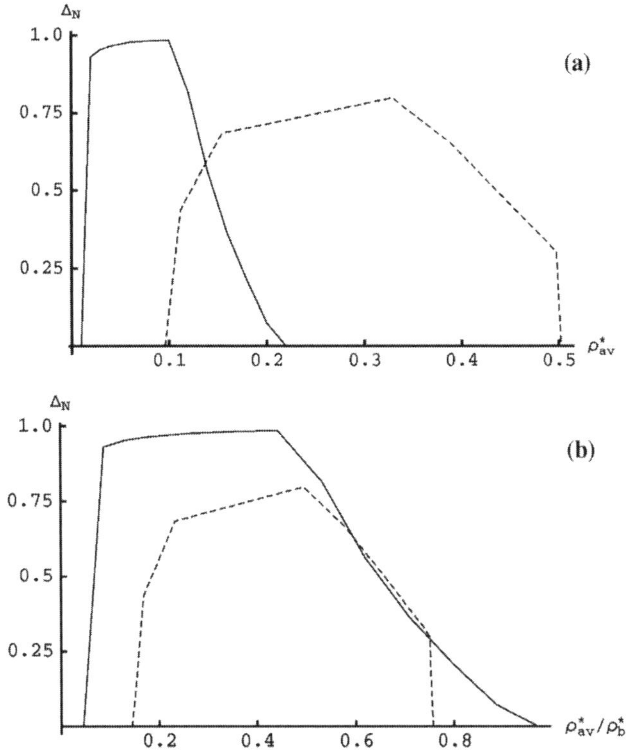

FIG. 7. (a) Anisotropy parameter Δ_N for quantum (solid line) and classical (dashed line) fluids at $T = 2.0$ K and $T = 87$ K, respectively. Δ_N is plotted as function of the dimenisonless average density $\rho^*_{av} = \rho_{av}\sigma^3_{ff}$, where $\sigma_{ff} = 2.556$ Å for the quantum and $\sigma_{ff} = 3.405$ Å for the classical fluid. (b) Δ_N as function of ρ^*_{av}/ρ^*_b, where ρ^*_b is the dimensionless density of the bulk liquid ($\rho^*_b = 0.226$ for the quantum and $\rho^*_b = 0.664$ for the classical fluids).

in itself and is not the goal of the present paper. Assuming, however, that the change in T_λ is not large, one can use for comparison the data obtained for $T = 0$ K (superfluid state) and $T = 3.0$ K (nonsuperfluid state). It turned out, that in both cases SB has the same features regarding the effects of h_r, ϵ_{fs} and ρ_{av}. Hence, one can conclude, that the transition to superfluidity does not affect the characteristics of SB.

ACKNOWLEDGMENTS

We are grateful to Professor Francesco Ancilotto (University of Padova, Italy) for providing us with the details of the numerical procedure used in Ref. 15.

APPENDIX: EULER-LAGRANGE EQUATION FOR THE FLUID DENSITY DISTRIBUTION

For a closed slit between identical walls filled with a fluid at fixed average density $\rho_{av} = (1/V)\int_V \rho(\mathbf{r})d\mathbf{r}$, the Euler-Lagrange equation for the fluid density distribution $\rho(h)$ can be obtained through the minimization of the free-energy functional of the system given by Eq. (12). One thus obtains the following equation:

$$\left\{-\frac{\hbar^2}{2M}\frac{d^2}{dh^2} + U(\rho, h)\right\}\sqrt{\rho(h)} = \mu\sqrt{\rho(h)}, \tag{A1}$$

where μ is a Lagrange multiplier and the effective potential $U(\rho, h)$ is given by

$$U(\rho,h)=U_{\text{fs}}(h)+k_BT\log(z)+U_{ff}(h)+\varphi_2(h)+\varphi_3(h)$$
$$-\frac{\hbar^2}{4M}\alpha_s\left[\varphi_{G,1}(h)+\varphi_{G,2}(h)\right]. \tag{A2}$$

In Eq. (A2), the first two terms represent the fluid-solid interactions [$U_{\text{fs}}(h)$ represents either $U_{\text{LJ}}(h)$ or $U_{\text{CCZ}}(h)$ potentials] and ideal-gas contributions to the total free energy, respectively, and the quantity z in the second term is defined by Eq. (11). The third term is due to the fluid-fluid interactions

$$U_{ff}(h)=\int dh'\rho(h')\phi_{ff,h}(|h-h'|), \tag{A3}$$

where $\phi_{ff,h}(|h-h'|)$ is obtained by integrating the potential $\phi_{ff}(|\mathbf{r}-\mathbf{r}'|)$ with respect to the x-and y-coordinates. The functions $\varphi_n(h)(n=2,3)$ provide the hard core He-He repulsion and are given by

$$\varphi_n(h)=\frac{c_n}{n}\left\{\left[\bar{\rho}(h)\right]^n+2\pi n\int dh'\rho(h')\left[\bar{\rho}(h')\right]^{n-1}\mathrm{X}_0(|h-h'|)\right\}, \tag{A4}$$

where

$$\mathrm{X}_0(|h-h'|)=\begin{cases}\dfrac{3}{8\pi h_l^3}\left[h_l^2-(h-h')^2\right], & |h-h'|<h_l\\ 0, & |h-h'|\geq h_l.\end{cases} \tag{A5}$$

The last term of Eq. (A2) represents a nonlocal correction of the quantum kinetic energy in which the functions $\varphi_{G,1}(h)$ and $\varphi_{G,2}(h)$ are given by

$$\varphi_{G,1}(h)=-\frac{2}{\rho_{0s}}\mathrm{X}_1(h)\frac{d}{dh}\rho(h), \tag{A6}$$

$$\varphi_{G,2}(h)=-2\frac{d}{dh}\left\{\left[1-\frac{\rho(h)}{\rho_{0s}}\right]\mathrm{X}_1(h)\right\}, \tag{A7}$$

where

$$\mathrm{X}_1(h)=\int dh'\left[1-\frac{\rho(h)}{\rho_{0s}}\right]G(|h-h'|)\frac{d}{dh'}\rho(h'),$$

$$G(|h-h'|)=\frac{1}{\sqrt{\pi}\,l}\exp\left[-(h-h')^2/l^2\right],$$

and $l = 1$Å.[13]

REFERENCES

1. J. H. Sikkenk, J. O. Indekeu, J. M. J. van Leeuwen, and E. O. Vossnack, Phys. Rev. Lett. **59**, 98 (1987).
2. M. J. P. Nijmeijer, C. Bruin, A. F. Bakker, and J. M. J. van Leeuwen, Phys. Rev. A **42**, 6052 (1990).
3. T. Ingebrigtsen and S. Toxvaerd, J. Phys. Chem. C **111**, 8518 (2007).
4. E. Velasco and P. Tarazona, J. Chem. Phys. **91**, 7916 (1989).
5. M. Merkel and H. Löwen, Phys. Rev. E **54**, 6623 (1996).

6. G. O. Berim and E. Ruckenstein, J. Phys. Chem. B **111**, 2514 (2007) (Section 2.1 of this volume).
7. G. O. Berim and E. Ruckenstein, J. Chem. Phys. **126**, 124503 (2007) (Section 2.2 of this volume).
8. G. O. Berim and E. Ruckenstein, J. Phys. Chem. B **111**, 12823 (2007) (Section 2.3 of this volume).
9. G. O. Berim and E. Ruckenstein, J. Chem. Phys. **128**, 024704 (2008) (Section 2.4 of this volume).
10. L. Szybisz and S. A. Sartarelli, J. Chem. Phys. **128**, 124702 (2008).
11. E. Cheng, M. W. Cole, W. F. Saam, and J. Treiner, Phys. Rev. Lett. **67**, 1007 (1991).
12. E. Cheng, M. W. Cole, W. F. Saam, and J. Treiner, Phys. Rev. B **46**, 13967 (1992).
13. F. Dalfovo, A. Lastri, L. Pricaupenko, S. Stingari, and J. Treiner, Phys. Rev. B **52**, 1193 (1995).
14. L. Szybisz, Phys. Rev. B **62**, 12381 (2000).
15. F. Ancilotto, F. Faccin, and F. Toigo, Phys. Rev. B **62**, 17035 (2000).
16. A. Chizmeshya, M. W. Cole, and E. Zaremba, J. Low Temp. Phys. **110**, 677 (1998).

2.7 Symmetry Breaking in Confined Fluids*

Eli Ruckenstein† *and Gersh Berim*
Department of Chemical and Biological Engineering, State University of New York at Buffalo, Buffalo, New York 14260, United States

Corresponding Author

† E-mail: feaeliru@buffalo.edu.
Phone: +1 716 645 1179. Fax: +1 716 645 3822.

1. INTRODUCTION

The phenomenon of symmetry breaking (SB) is well known in physics (see the recent reviews Refs. [1–3]), chemistry [4–8], biology [9–11], fluid mechanics [12–14], etc. In general, SB occurs when the symmetry of the stable state of the system is lower than the symmetry of the interaction potentials of the system and, as a consequence, of the equations describing this system. Both continuous translational symmetry inherent to unconstrained fluids in gaseous or liquid states and discrete (e.g. reflectional or left-right) symmetries can be broken. As specific examples of breaking the continuous translational symmetry one can mention the homogeneous nucleation of a liquid or solid phase from a uniform vapor or liquid phase, respectively [15–20], the freezing transition in the absence of an external potential [21–24], and the nucleation of bumps and bridges in capillary condensation in pores [25,26].

Discrete symmetry breaking manifests itself in such phenomena as spontaneous magnetization of ferromagnetic materials in the absence of an external magnetic field at a temperature smaller than the Curie temperature, appearance in solutions of stable asymmetrical configurations of molecules which are symmetrical in the gaseous phase [27], uniaxial to biaxial transition in wetting and drying films of nematic fluids in contact with hard substrates [28,29], reflection and chiral symmetry breaking [30–38], appearance of a stable asymmetrical fluid density distribution in slits with identical walls [39–45].

Even these few examples show that the investigation of symmetry breaking in physical systems can have fundamental and practical implications. In this paper the recent advances in the theoretical consideration of SB in confined fluid are reviewed. Note that the theoretical description of symmetry breaking in fluids is based in most cases either on molecular dynamics and Monte Carlo simulations [27,28,31,46–50] or on various versions of the density functional theory [22,29,39,40,44].

The main attention is given here to the application of the density functional theory to the description of SB in fluids confined in nanoslit between parallel identical solid walls.

A fluid in such a slit is exposed to an external potential $U_s(\mathbf{r})$ due to the fluid-solid interactions which possesses a reflectional symmetry about the middle of the slit. Therefore it is expected the fluid density distribution to possess the same symmetry. However, the following qualitative arguments based on a macroscopic theory of wetting suggest that a symmetry breaking of the density profile can occur.

* *Advances in Colloid and Interface Science*. 154. (2010) 56–76. Republished with permission.

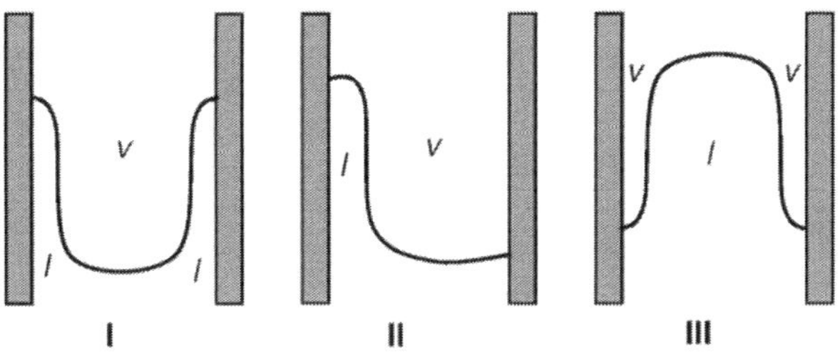

FIG. 1. (Reprinted from Ref. [41] with permission) Schematic presentation of the three possible density profiles in the slit. Profile I corresponds to the state where both walls are covered by thin films of fluid of liquid-like density, the remaining of the slit being filled by a vapor-like fluid. In contrast, in the state with profile III layers of vapor-like fluid are present near the walls, the remaining of the slit being filled by a liquid-like fluid. Profile II corresponds to the state with asymmetric distribution of molecules (a mirror asymmetric profile is also possible).

Let us consider the three possible density profiles shown in Fig. 1. In the first profile, both walls of the slit are covered by a liquid-like density fluid, and a vapor-like density fluid is present between them. In the second, a film of liquid-like density covers one of the walls (the left or right one) and the remaining of the slit is occupied by a vaporlike fluid. Finally, in the third profile the walls are covered by a vaporlike fluid and the remaining of the slit is filled with a liquid-like fluid. Denoting by γ_{ls}, γ_{vs}, and γ_{lv} the liquid-solid, vapor-solid, and liquid-vapor surface tensions, respectively, which satisfy the Young equation $\gamma_{vs} - \gamma_{ls} = \gamma_{lv} \cos\theta$, where θ is the contact angle, one can calculate the differences $\Delta F_{2,1}$ and $\Delta F_{2,3}$ between the free energy of the asymmetric profile II and the free energies of the symmetric profiles I and III, respectively

$$\Delta F_{2,1} = \gamma_{lv}\left(cos\theta - 1\right) \leq 0,$$

$$\Delta F_{2,3} = -\gamma_{lv}\left(cos\theta + 1\right) \leq 0. \tag{1}$$

Eqs. (1) show that the free energy for the asymmetric profile is smaller than those for the symmetric ones, with the exception of the cases for which $\theta = 0°$ or $\theta = 180°$ when they are equal. While the above macroscopic considerations are not strictly valid for a nanoslit, they suggest that even for identical walls an asymmetric profile might be the stable one.

In Section 2, a short description of the application of the DFT to the examination of the fluid density distribution in confined fluids and procedures to identify symmetry breaking density profiles is provided. The reflectional symmetry breaking in open slit-like systems is examined in Section 3. New results obtained recently regarding the symmetry breaking of the fluid density distribution in an open slit-like systems, closed slits for one-and two-component classical fluids and for one component quantum fluid are considered in Sections 4 and 6, respectively. In Section 5, results obtained via simulations are discussed and compared to those obtained by DFT.

2. APPLICATION OF THE DFT TO THE DESCRIPTION OF THE DENSITY DISTRIBUTION OF CONFINED FLUIDS

2.1. Density functional and the Euler–Lagrange equation for the fluid density distribution

The use of DFT to the study of confined fluids has a long history. It started with the papers of Ebner and Saam [51] and Evans and Tarazona [52–54], who considered a bulk fluid in contact with a solid wall; it was followed with various fluids subjected to various kinds of confinements (cavities, pores, slits, etc). We refer the readers to Refs. [21,55–69] where the traditional problems, such as melting and freezing, adsorption of fluids on solid surfaces, wetting, phase transitions and phase separations in single fluids and binary mixtures, capillary condensation, etc were reviewed. In all mentioned studies, the system (fluid plus solid) was considered open, i.e. in contact with an infinite reservoir of fluid molecules at temperature T and chemical potential μ. The equilibrium density distribution (density profile) $\rho(\mathbf{r})$ was obtained in this case by minimizing the free energy functional $\Omega\left[\rho(\mathbf{r})\right]$ which has the form

$$\Omega\left[\rho(\mathbf{r})\right] = F_{HS} + \frac{1}{2}\int_V d\mathbf{r}d\mathbf{r}'\rho(\mathbf{r})\rho(\mathbf{r}')\phi_{ff}\left(|\mathbf{r}-\mathbf{r}'|\right)$$

$$+\int_V d\mathbf{r}\rho(\mathbf{r})U_s(\mathbf{r}) - \mu\int_V d\mathbf{r}\rho(\mathbf{r}) \tag{2}$$

where $\phi_{ff}(|\mathbf{r}-\mathbf{r}'|)$ is the potential of the fluid-fluid interactions and V is the volume occupied by fluid. The first term, F_{HS}, in Eq. (2) is the Helmholtz free energy of a reference system which for classical fluids is considered to be a system of hard spheres. The second term represents the contribution of the attractive part of fluid-fluid interactions to the total free energy (this term is calculated in the mean-field approximation). The function $U_s(\mathbf{r})$ is the external potential field generated by the solid in contact with the fluid. The sum of the first two terms in the right hand side of Eq. (2) constitutes the intrinsic (calculated in the absence of an external field) Helmholtz free energy of the system.

The Helmholtz free energy of the reference system of hard spheres can be written as

$$F_{HS} = \int_V f_{id}(\rho)dV + \int_V \Delta\Psi_{hs}(\bar{\rho}) + dV \tag{3}$$

where

$$f_{id}(\rho) = k_B T\rho\left[\log\left(\Lambda^3\rho\right)\right] - 1 \tag{4}$$

is the free energy density of an ideal gas, and $\Delta\Psi_{hs}(\bar{\rho})$ is the excess free energy density of a bulk hard sphere fluid provided by an equation of state. In Eq. (3), $\Lambda = h_P/\left(2\pi m k_B T\right)^{1/2}$ is the thermal deBroglie wavelength, m is the molecular mass of the fluid, h_P and k_B are the Planck and Boltzmann constants, respectively, T is the absolute temperature, $\rho \equiv \rho(\mathbf{r})$, is the fluid density, and $\bar{\rho} \equiv \bar{\rho}(\mathbf{r})$ is a smoothed (weighted) density, obtained by averaging the fluid density $\rho(\mathbf{r})$ over a sphere with a diameter equal to the hard core diameter σ_{ff}.

The excess free energy density, $\Delta\Psi_{hs}(\bar{\rho})$, and the density $\bar{\rho}(\mathbf{r})$ are treated in various ways which provide two major versions of the theory.

In the first version, $\Delta\Psi_{hs}(\bar{\rho})$ is provided by an expression [54,70], obtained using the Carnahan–Starling equation of state [71]

$$\Delta\Psi_{hs}(\bar{\rho}) = k_B T \eta_{\bar{\rho}} \frac{4 - 3\eta_{\bar{\rho}}}{\left(1-\eta_{\bar{\rho}}\right)^2} \tag{5}$$

where $\eta_{\bar{\rho}} = (1/6)\pi\bar{\rho}(\mathbf{r})\sigma_{ff}^3$ is the packing fraction of the fluid molecules. The density $\bar{\rho}(\mathbf{r})$ is taken either equal to the local density $\rho(\mathbf{r})$ (local density approximation) [52] or is calculated by averaging $\rho(\mathbf{r})$ with weighting functions chosen to provide the best possible description of the bulk fluid (nonlocal density approximation [54]).

In the most successful, Tarazona's version, the weighting function depends on a weighted density [54].

Another approach to calculate the excess fluid repulsion was developed by Rosenfeld and Kierlik [72–74] who introduced four weighting functions (independent of the local density) chosen such as to exactly reproduce the Percus–Yevick's direct pair correlation function in an uniform hard-sphere fluid. This version of the DFT provides a more accurate description of fluids in very small pores [75,76] and can be easily extended to the description of adsorbed mixtures [77–81].

The Euler–Lagrange equation for the density profile $\rho(\mathbf{r})$ of classical fluids, derived by minimizing the functional Eq. (2) has the following form

$$k_B T \log\left(\Lambda^3 \rho\right) + \frac{\partial \Delta \Psi_{hs}\left(\bar{\rho}(\mathbf{r})\right)}{\partial \rho} + \int_V d\mathbf{r}' \rho(\mathbf{r}') \phi_{ff}\left(|\mathbf{r} - \mathbf{r}'|\right) + U_s(\mathbf{r}) = \mu \tag{6}$$

and can be solved numerically by standard iterations. Note that for quantum fluids the Euler-Lagrange equation has additional terms [82] (see Section 6).

In the absence of confinements and external fields ($U_s(\mathbf{r}) = 0$) the density $\rho(\mathbf{r}) \equiv \rho$ is constant, hence, $\bar{\rho}(\mathbf{r}) = \rho$ and Eq. (6) transforms into one for a bulk fluid

$$k_B T \log\left(\Lambda^3 \rho\right) + \frac{\partial \Delta \Psi_{hs}\left(\bar{\rho}\right)}{\partial \rho} + a\rho = \mu \tag{7}$$

where the constant $a = \int_V d\mathbf{r}' \phi_{ff}\left(|\mathbf{r} - \mathbf{r}'|\right)$.

If the slit is closed, the number of molecules inside is constant and the behavior of the fluid is expected to be very different from that in an open system because in the absence of an external reservoir of molecules only their redistribution inside the closed volume can occur. The behavior of a fluid confined in a closed volume has to be treated in a canonical ensemble. The description of confined fluids in a canonical ensemble was carried out first in Refs. [75,76,83,84].

These papers were concerned with small spherical cavities containing small fixed numbers of molecules, and suitable DFT approaches have been developed which accounted for the large fluctuations of the local density arising due to the small number of molecules. The density profile in a canonical ensemble was expressed in terms of that corresponding to a grand canonical ensemble with an additional term involving the reciprocal of the number of molecules. If this number is macroscopically large, this additional term can be neglected and the canonical version of the density functional coincides with the grand canonical one with an unknown Lagrange multiplier λ used instead of the chemical potential μ. The Lagrange multiplier can be found using the constraint

$$N = \int_V \rho(\mathbf{r}) d\boldsymbol{r} \tag{8}$$

where N is the total number of molecules in the system. Such a canonical ensemble approach was used in Refs. [40–45,59,75,76,83–90].

2.2. Interaction Potentials

In most studies of confined classical fluids, the Lennard-Jones 6–12 potential

$$\phi_{ff}\left(|\mathbf{r} - \mathbf{r}'|\right) = \begin{cases} 4\epsilon_{ff}\left[\left(\frac{\sigma_{ff}}{r}\right)^{12} - \left(\frac{\sigma_{ff}}{r}\right)^{6}\right], & r \leq \sigma_{ff} \\ \infty, & r < \sigma_{ff} \end{cases}, \tag{9}$$

where ϵ_{ff} and σ_{ff}, are the energy parameter and the hard core diameter, respectively, is employed to describe the fluid-fluid interactions. Here **r** and **r′** are the coordinates of the interacting molecules and $r = |\mathbf{r}-\mathbf{r}'|$. Note that in some cases (see e.g. Ref. [91]) the potential $\phi_{ff}(\mathbf{r})$ at $r<\sigma_{ff}$ is taken zero instead of ∞. Despite the dramatic difference, this choice does not affect the total potential energy because in both cases the region where $r<\sigma_{ff}$ is excluded from consideration.

It is usually assumed that a solid wall generates a hard core repulsion at distance h_r from the wall. In this case, the potential $\phi_{fs}(|\mathbf{r}-\mathbf{r}'|)$ describing the interaction between a fluid molecule located at point **r** with a molecule of solid located at **r′** has the form

$$\phi_{fs}(|\mathbf{r}-\mathbf{r}'|)=\begin{cases}4\epsilon_{fs}\left[\left(\frac{\sigma_{fs}}{r}\right)^{12}-\left(\frac{\sigma_{fs}}{r}\right)^{6}\right], & r\geq h_r,\\ \infty, & r<h_r\end{cases} \tag{10}$$

where the parameters σ_{fs} and ϵ_{fs} provide the length scale and the energy parameter of the flui—olid interactions, respectively.

The total potential energy of a fluid molecule due to its interaction with the solid walls depends on the geometry of the system. Let us consider as an example a one component fluid confined in a closed nanoslit between two parallel structureless semiinfinite identical solid walls W_1 and W_2 of density ρ_s, separated by a distance L (see Fig. 2). The distance between the side walls which close the slit is considered to be very large compared to L. As a result, the end effects can be considered negligible. (The closed slit with planar walls can be imagined also as a thin layer of thickness L between a solid sphere of large radius $R\to\infty$ embedded into a spherical cavity of radius $R+L$ in the same solid.) In this case, the total fluid-wall interaction is provided by the Lennard-Jones 3–9 potential, obtained by integrating ϕ_{fs} over the volume of the walls, by assuming constant densities for the walls [51]. This potential depends on the distance h of the molecule from one of the wall (W_1 in the considered example) and has the form

$$U_s(h)=\psi(h)+\psi(h_m-h), \tag{11}$$

where h_m characterizes the size of the part of the slit available for fluid molecules (see Fig. 2) and

$$\psi(h)=\frac{2}{3}\epsilon_{fs}\rho_s\left[\frac{2}{15}\left(\frac{\sigma_{fs}}{h_r+h}\right)^{9}-\left(\frac{\sigma_{fs}}{h_r+h}\right)^{3}\right] \tag{12}$$

The smaller is h_r, the "softer" is the wall and the larger is the repulsion at $h=0$.

2.3. Determination of Symmetry Breaking Solutions of the Euler—Lagrange Equation

The Euler—Lagrange equations describing inhomogeneous fluids are complex nonlinear integral [52,54] or integro-differential [82,92] equations (The latter equations describe quantum fluids.). Because of this, one expects multiple solutions for the density profile (including the symmetry breaking ones) to exist. Indeed, it was found in Ref. [91] that for open slits the equation for the density profile can have two solutions among which the one providing the smallest value of the grand canonical potential provides the stable state of the system. However all those solutions possess reflectional symmetry and no symmetry breaking was detected. Note that for the time being, the only known example of a reflectional symmetry breaking in an open system was demonstrated in Ref. [39] where it was shown that in a system consisting of two parallel identical layers of a two-dimensional liquid separated by some distance and coupled by an interlayer potential, a symmetry breaking can occur for some strengths of the interlayers interaction.

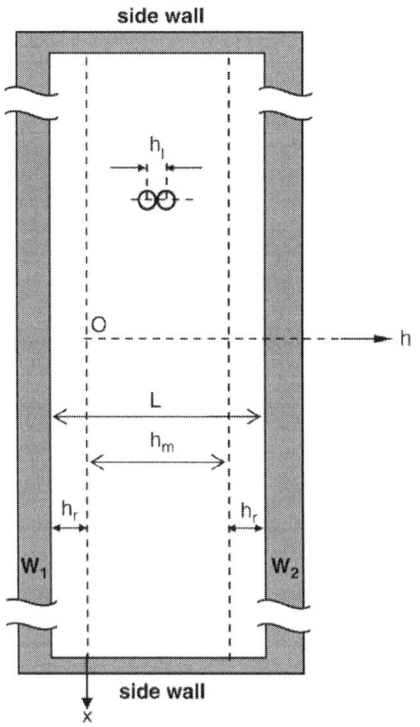

FIG. 2. (Reprinted from Ref. [41] with permission) A closed slit of width $L = h_m + 2\ h_r$ between two identical solid walls W_1 and W_2. h_r is the distance up to which the wall generates a hard core repulsion and h_l is the screening distance of the fluid-fluid interactions which for classical fluids is taken equal to σ_{ff}, h_m is the effective width of the slit accessible to the fluid molecules. The y-axis is perpendicular to the plane of the figure. The size of the slit in the x and y directions is large for the end effects to be negligible.

The fluid in one layer has a liquid-like density, while in the other it has a gas-like density (We will return to this example later).

Let us consider the methods of identifying the symmetry breaking solutions in the example of the closed system presented in Fig. 2. This system possesses a reflectional symmetry in h-direction with respect to the middle of the slit and a translational symmetry in lateral (parallel to the walls) directions. The standard iteration procedure to obtain a numerical solution of Eq. (6) starts with an initial guess $\rho^{in}(\mathbf{r})$ which is used as input for the right hand side of the equation

$$\rho(\mathbf{r}) = \rho_0 exp\left[\frac{Q(\rho(\mathbf{r}))}{k_B T}\right], \tag{13}$$

obtained from Eq. (6). In Eq (13),

$$Q(\rho(\mathbf{r})) = -\frac{\partial \Delta\Psi_{hs}(\bar{\rho}(\mathbf{r}))}{\partial \rho} - \int_V d\mathbf{r}'\rho(\mathbf{r}')\phi_{ff}(|\mathbf{r}-\mathbf{r}'|) - U_s(\mathbf{r}) \tag{14}$$

and $\rho_0 = \Lambda^{-3} exp(\lambda / k_B T)$ (Here, the chemical potential μ in Eq. (6), is changed to the Lagrange multiplier λ). The density profile $\rho(\mathbf{r})$ generated by Eq. (13) from $\rho^{in}(\mathbf{r})$ is usually combined in some way (see for example Refs. [54]) with $\rho^{in}(\mathbf{r})$ to provide new input for Eq. (13). The iteration procedure is repeated until the difference between two successive iterations becomes sufficiently small (see Ref. [54] for details). If $\rho^{in}(\mathbf{r})$ is symmetrical with respect to the middle of the slit, the final solution has the same symmetry for any values of the parameters of Eq. (6).

To find the asymmetrical solution in h-direction, the initial guess, $\rho^{in}(\mathbf{r})$, can be selected asymmetrical about the middle of the slit and uniform in lateral directions, i.e. $\rho^{in}(\mathbf{r}) \equiv \rho^{in}(h)$ [40]. For example, it can be considered as a straight line $\rho = \rho(h)$ in $h - \rho$ plane between the points $\rho_0 = \rho(0)$ and $\rho_1 = \rho(L)$, where the densities $\rho_0 \neq \rho_l$ are adjusted to provide the average density of the initial FDD ρ_{av}. Depending on the values of the parameters of Eq. (6) and the average density of the fluid in the slit, the final solution can be symmetrical or asymmetrical. This approach was used in Refs. [40–42] where it was assumed that the FDD is uniform in the lateral directions i.e. $\rho(\mathbf{r}) \equiv \rho(\mathrm{h})$. When the result was symmetrical, it coincided with that obtained starting from a symmetrical initial guess. Comparing the free energy corresponding to the asymmetrical solution with that corresponding to the symmetrical one, one can identify the stable and metastable solutions of Eq. (6).

A similar procedure can be used if one is looking for symmetry breaking in both h- and one of the lateral directions (say, x-direction). In this case, $\rho(\mathbf{r}) = \rho(x, h)$ and an initial guess can be selected possessing various symmetries. For example, if the latter is taken uniform in the x-direction and symmetrical in the h-direction, the final solution will always have the same symmetry. If $\rho^{in}(x, h)$ is selected nonuniform in the x-direction and asymmetrical in the h-direction, the final solution can have symmetry breaking in both or one of these directions [43,44], or can be uniform in the x-direction and symmetrical in the h-direction [43,44] (no symmetry breaking).

Another possible method to identify an asymmetrical solution of the Euler-Lagrange equation consists in introducing a small perturbation $\Delta U_s(\mathbf{r})$ of the fluid-solid interaction potential which breaks the symmetry of the system [26]. In this case, the solution of the Euler-Lagrange equation will be asymmetrical for any initial guess. Analyzing this solution at vanishing values of the perturbation, one can conclude whether this solution is stable (exists for $\Delta U_s(\mathbf{r}) \to 0$) or not.

3. SYMMETRY BREAKING IN A FLUID CONFINED IN A SLIT-LIKE OPEN SYSTEM

Let us consider, following Ref. [39], two layers (each of area A) of a two-dimensional uniform fluid separated by the distance h_m (see Fig. 3) and coupled by the interlayer potential $V_{12}(r)$ possessing a reflection symmetry with respect to the midplane between the two layers. In the potential $V_{12}(r)$, r denotes the distance between a molecule from layer 1 to another molecule from layer 2. The layers are in contact with a two-dimensional bulk reservoir of molecules exhibiting a liquid—gas phase transition at a temperature dependent chemical potential μ_c. The fluid molecules in the layers interact via a Lennard-Jones potential and three kinds of interlayer potential were considered. Two of these potentials were slightly modified truncated Lennard—Jones potentials and the third potential was taken in the Yukawa form [39].

Assuming homogeneous density distributions in both layers and considering the interlayer interaction in the mean field approximation, the grand canonical potential $\omega(\rho_1, \rho_2)$ of the system per unit area of one of the layer can be written in the form [39]

$$\omega(\rho_1,\rho_2) = \rho_1 f(\rho_1) + \rho_2 f(\rho_2) - \mu(\rho_1 + \rho_2) + \rho_1\rho_2 V_0. \tag{15}$$

where ρ_1 and ρ_2 are the densities of the fluid in the first and second layers, respectively, $f(\rho)$ is the Helmholtz free energy per molecule, and the constant V_0 represents the average interlayer interaction

$$V_0 = 2\pi \int_{h_m}^{\sigma_{ff}} r g_{12}(r) V_{12}(r)\, dr \tag{16}$$

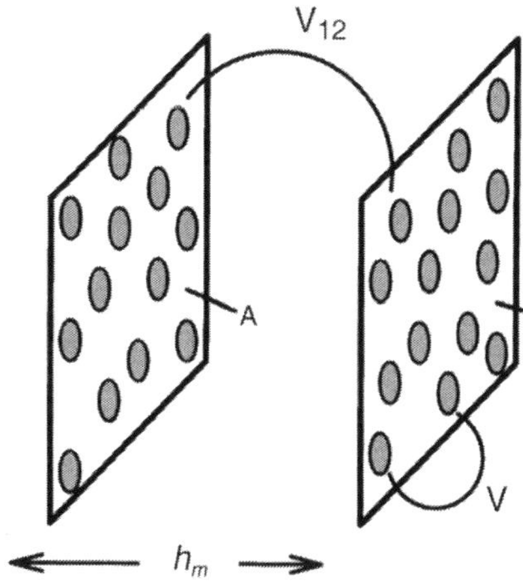

FIG. 3. (Reprinted from Ref. [39] with permission) Two two-dimensional layers of area A are separated by the distance h_m. The molecules in each layer are coupled with the in-plane interaction $V = V(r)$ and interlayer interaction $V_{12} = V_{12}(r)$, r being the distance between fluid molecules located in the same layer or in the different layers, respectively.

$g_{12}(r)$ being the interlayer pair correlation function. At fixed temperature T, area A, and chemical potential μ, the layer densities can be determined by minimizing $\omega(\rho_l, \rho_2)$ with respect to ρ_1 and ρ_2. Numerical minimization on the basis of the density functional theory for two-dimensional fluids [21] has provided the phase diagram of Fig. 4 which contains information about the stability of the three situations, vapor-vapor, liquid-liquid, and vapor-liquid versus $\Delta\mu = \mu - \mu_c$ and V_0. The symmetry-broken situation, where one of the layer is filled with a vapor-like fluid and the other one is filled with a liquid-like fluid, occurs only for $V_0 > 0$ (repulsive interlayer interaction) and $\Delta\mu>0$. If the interlayer interaction is attractive ($V_0 < 0$), the symmetry breaking is absent for any value of μ. There is no symmetry breaking for repulsive interlayer interaction $V_0 > 0$ if the chemical potential μ is smaller then μ_c. The topology of this phase diagram is similar for all considered interlayer potentials.

The results obtained through the density functional approach were verified in Ref. [39] by computer simulations. It turned out that in some cases the reflectional symmetry breaking predicted by

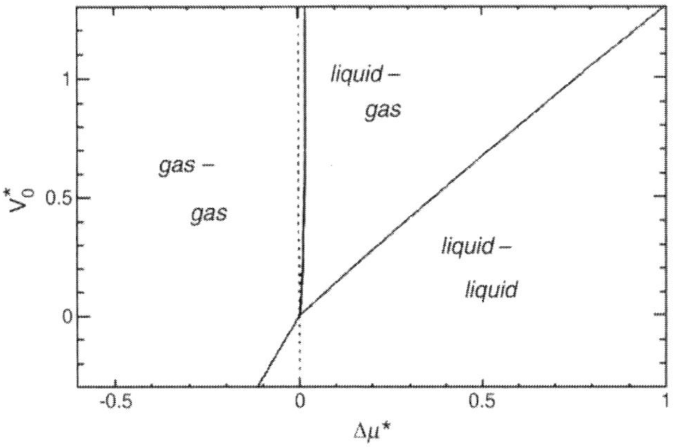

FIG. 4. (Reprinted from Ref. [39] with permission) Phase diagram in the plane of reduced chemical potential difference $\Delta\mu = (\mu - \mu_c)/\epsilon$ and reduced interlayer coupling $V_0^* = V_0/\epsilon$, μ_c and ϵ being the chemical potential at vapor-liquid coexistence and the energy parameter of the interlayer potential, respectively [39].

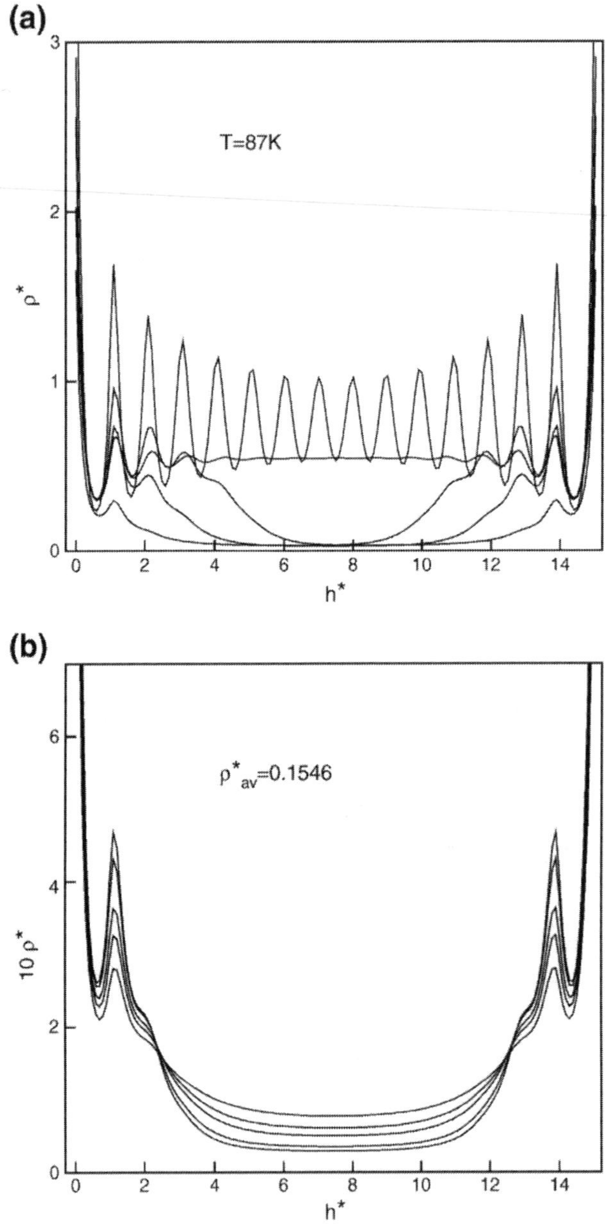

FIG. 5. (Reprinted from Ref. [41] with permission) Density profiles in the closed slit for $h_m = 15$. (a) The symmetric profiles at T = 87 K plotted for $\rho^*_{av} = 0.1159$; 0.2319; 0.3865; 0.5411 and 0.7729. The fluid density at any point inside the slit increases with increasing average density. (b) The symmetric profiles corresponding to the average density $\rho^*_{av} = 0.1546$ for the temperatures $T = 87$, 90, 96, 100, and 106 K. The density in the middle of the slit increases with increasing temperature.

DFT is preempted by a freezing transition in which the molecules in both layers are ordered into triangular lattices. This freezing transition is not incorporated into the density functional approach, hence the theory fails in predicting the correct densities (see for details Ref. [39]).

Despite the apparently unrealistic character of this quasi two-dimensional model, the authors argued the possibility of its experimental realization. They suggest to consider mesoscopic colloidal particles suspended in a solvent and confined between two parallel plates. The van der Waals attraction between the walls and the colloidal particles should lead to two deep symmetric minima near the wall, which can be tuned by changing the solvent. Assuming that the particles are mainly captured in these two minima, one can justify the applicability of the model. Desirable interactions can be provided by adding sterically stabilized suspensions that are additionally low charged with added polymer in the solution [39].

In an experiment with such confined colloids one would notice symmetry-breaking density profiles by watching the particle configurations in real space using e.g. video microscopy. Another possibility to obtain the colloidal density profile is by light scattering [39].

4. SYMMETRY BREAKING IN ONE COMPONENT CLASSICAL FLUIDS IN A CLOSED NANOSLIT

Below, if not mentioned otherwise, the interaction parameters ϵ_{ff} and σ_{ff} are for argon [52]: $\epsilon_{ff} = 119.76\,\text{K}, \sigma_{ff} = 3.405\,\text{Å}$, and the values $\epsilon_{fs}/k_B = 153\,\text{K}$, $\sigma_{fs} = 3.727$ Å and $\rho_s = 1.191\times10^{28}\,m^{-3}$ are those for the solid carbon dioxide interacting with argon [52].

The fluid is confined in the closed slit presented in Fig. 2 with the distance h_r at which a wall generates hard core repulsion equal to σ_{fs}. The dimensionless quantities

$$\rho^* = \rho\sigma_{ff}^3, h^* = h/\sigma_{ff} \tag{17}$$

for the fluid density and distance from the wall, respectively, will be used for the presentation of the results.

4.1 SYMMETRY BREAKING FOR ONE-DIMENSIONAL FLUID DENSITY DISTRIBUTION

Under the assumption that the fluid density distribution (FDD) is uniform in the lateral direction, the density $\rho(\mathbf{r})$ depends only on the single coordinate ($\rho(\mathbf{r}) \equiv \rho(h)$). In this case, the fluid density profile, obtained as solution of Eq. (6) for an initial guess taken as a uniform density profile ($\rho^{in}(h) =$ const), is symmetrical about the middle of the slit and its shape depends, for selected values of the parameters of the model and temperature, on the number N of molecules in the system (or the average density $\rho_{av}^* = N\sigma_{ff}^2/h_m^*$ of the fluid). Typical profiles for small and large values of ρ_{av}^* are presented in Fig. 5a and b. The profiles in Fig. 5a are plotted for various ρ_{av}^* in a slit with $h_m^* = 15$ at $T = 87$ K. For small average densities, the oscillations of the density profiles are significant only near the walls. The density of the fluid in the very vicinity of the walls has liquid-like values ($\rho^* \sim 1$) and becomes vapor-like ($\rho^* \sim 10^{-2}$) at distances from the wall of the order of several molecular diameters (Note, that such profiles are present in open systems far from condensation conditions [70,91]).

When ρ^*_{av} increases, the oscillations become larger and the region with vapor-like fluid density decreases. For relatively high densities ($\rho^*_{av} \sim 0.5411$) the whole slit is filled with a liquid-like fluid and the oscillations of the fluid density become significant everywhere in the slit. The fluid density at the walls increases with increasing ρ^*_{av}.

Note that for a selected temperature the bulk densities ρ_l^* and ρ_v^* of argon calculated using Eq. (7) at $\mu = \mu_c, \mu_c$ being the chemical potential at liquid—vapor coexistence are equal to $\rho_l^* = 0.6633$ and $\rho_v^* = 0.0217$.

In Fig. 5b, the profiles corresponding to $\rho_{av}^* = 0.1546$ are presented at various temperatures. The amplitude of the oscillations of the density near the walls decreases and the density in the central part of the slit increases with increasing temperature. This is the obvious consequence of the penetration of the molecules from the walls into the internal part of the slit.

Asymmetric density profiles were obtained using the two procedures mentioned in a previous section. Several asymmetric profiles are shown in Fig. 6a for $T = 87$ K and various values of ρ_{av}^*. When ρ_{av}^* increases, the number as well as the amplitude of oscillations near the left wall increases and the region where the fluid has a vapor-like density decreases and moves toward the

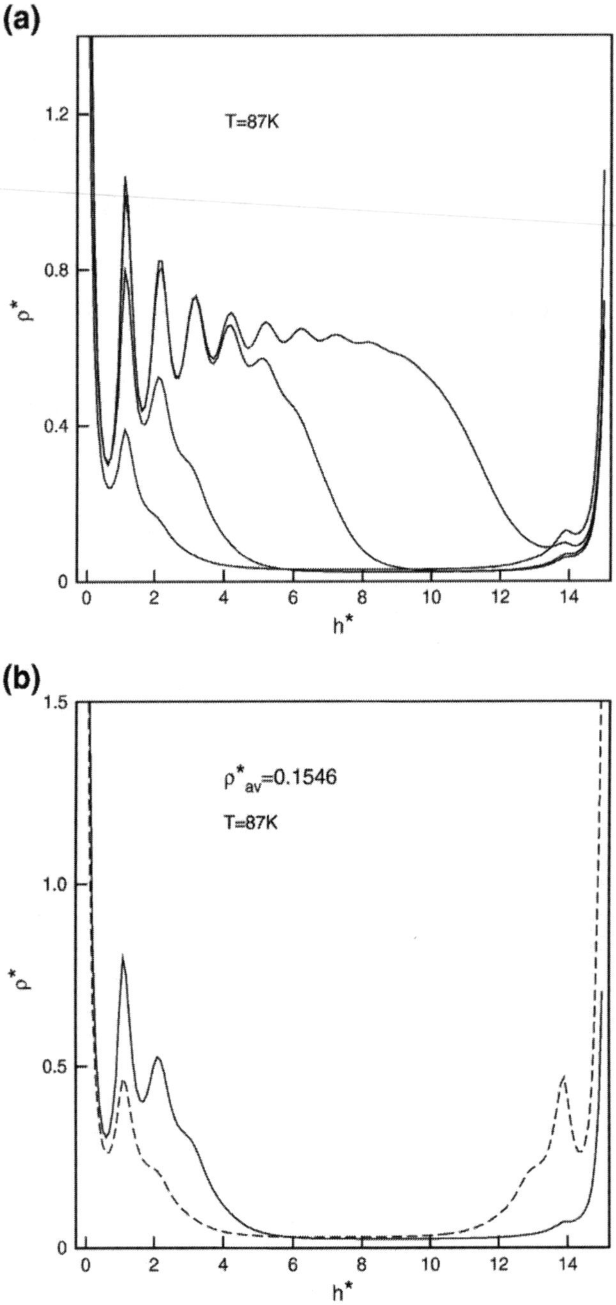

FIG. 6. (Reprinted from Ref. [41] with permission) Density profiles in the closed slit for $h^*_m = 15$. (a) The stable asymmetric profiles at $T = 87$ K plotted for $\rho^*_{av} = 0.1043$; 0.1546; 0.3092; and 0.4978. The fluid density near the left wall increases with increasing average density. (b) The stable (asymmetric profile) and metastable (symmetric profile) solutions of the Euler-Lagrange equation in the closed slit for the average density $\rho^*_{av} = 0.1546$ and temperature $T = 87$ K.

TABLE 1
(Reprinted from Ref. [41] with permission) The values (per unit area of one wall) of the Helmholtz free energies F_{sym} and F_{asym} of the metastable (symmetric density profile) and stable (asymmetric density profile) states of the fluid in the slit with $h_m^* = 15$, respectively for various average densities $\rho_{av1}^* < \rho_{av}^* < \rho_{av2}^*$. The temperature $T = 87$ K. Free energies are given in $k_B T/\sigma_{ff}^2$ units.

ρ_{av}^*	F_{sym}	F_{asym}
0.1546	−26.59	−26.67
0.2319	−39.66	−39.86
0.3092	−52.81	−53.06
0.3865	−66.01	−66.26
0.4638	−79.22	−79.48

right wall. Note, that the asymmetric solutions are two-fold degenerate. The profiles obtained from those shown in Fig. 6a by reflection about the middle of the slit are also solutions of the Euler-Lagrange equation.

In Fig. 6b two profiles, one asymmetric and the other symmetric, obtained for $\rho_{av}^* = 0.1546$ and $T = 87$ K, are presented as an example of multiple solutions of the Euler-Lagrange equation. It was found that the asymmetric density profile always provides a slightly smaller Helmholtz free energy than the symmetrical one (see Table 1) and, hence, the former corresponds to a stable state. This means that, indeed, symmetry breaking occurs in closed systems.

To characterize quantitatively the asymmetry of the density profiles, the parameter

$$\Delta_N = \frac{1}{2N}\int_0^{h_m} dh\left|\rho(h) - \rho(h_m - h)\right| \tag{18}$$

was introduced [41].

This parameter is equal to zero if the profile is symmetric about the middle of the slit and $0 < \Delta_N \leq 1$ in the opposite case. In the extremely asymmetric case when all molecules are located on one side of the plane of symmetry, $\Delta_N = 1$. The dependence of Δ_N on the average density ρ^*_{av} is presented in Fig. 7 for $T = 87$, 96, and 104 K. Two features should be mentioned. First, for each of the considered temperatures there is a range of average densities $\rho_{sb1}^* < \rho_{av}^* < \rho_{sb2}^*$ where Δ_N has significant values and becomes almost zero outside this range. Near ρ_{sb1}^* and ρ_{sb2}^* the parameter Δ_N changes very rapidly with ρ_{av}^* and therefore those values of ρ_{av}^* should be considered as "critical" values. Second, the critical value ρ_{sb1}^* increases, and ρ_{sb2}^* decreases with increasing temperature. The temperature dependencies of ρ_{sb1}^* and ρ_{sb2}^* are presented in Fig. 8. Symmetry breaking occurs only for those $\rho_{av}^* - T$ pairs which are located between the two curves on that figure. The difference $\rho_{sb1}^* - \rho_{sb2}^*$ which provides the range of existence of the asymmetric solution and, hence, of the symmetry breaking, decreases with increasing temperature. For $T = T_{sb1} \simeq 106$ K no symmetry breaking occurs and the profile is symmetrical about the middle of the slit.

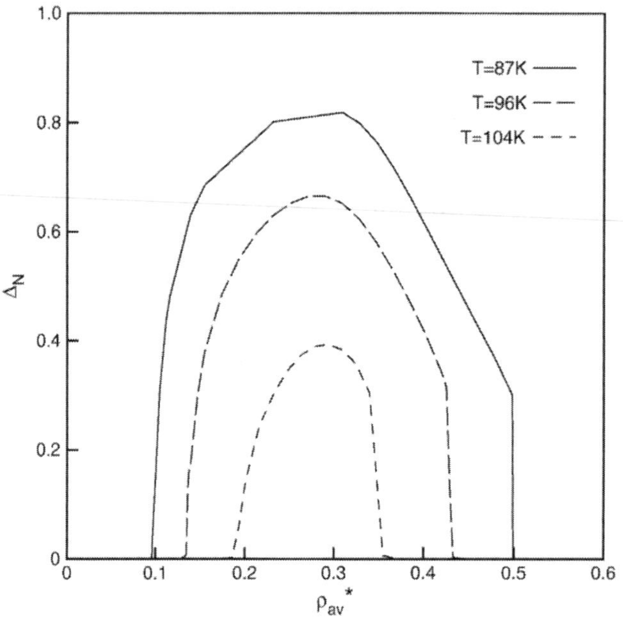

FIG. 7. (Reprinted from Ref. [41] with permission) The dependence of the parameter Δ_N, characterizing the asymmetical density distribution across a closed slit with $h_r = 15$, on the average density ρ^*_{av} of the fluid inside the slit for three different temperatures. For $T \geq 106$ K $\Delta_N = 0$ for all $T \geq 106$K and all density profiles are symmetrical about the middle of the slit.

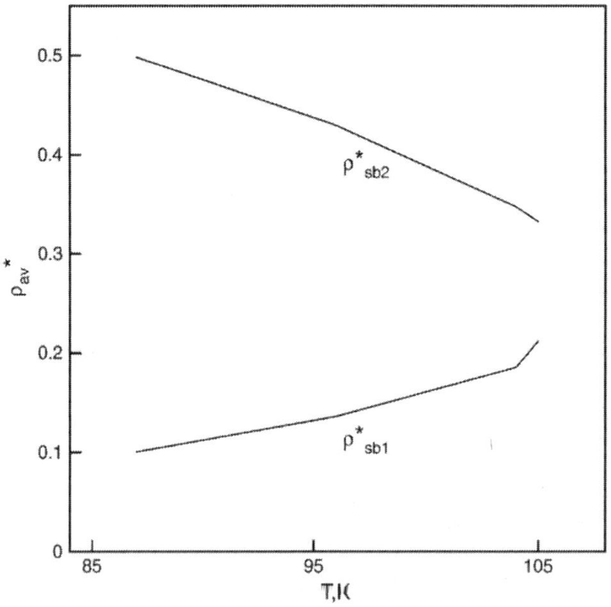

FIG. 8. (Reprinted from Ref. [41] with permission) Temperature dependence of the critical values ρ^*_{sb1} and ρ^*_{sb2} of the average density in a closed slit with $h^*_m = 15$ for $T \geq 87$ K. Symmetry breaking occurs when $\rho^*_{sb1} < \rho^*_{av} < \rho^*_{sb2}$. For temperatures higher than $T = 106$ K no symmetry breaking occurs.

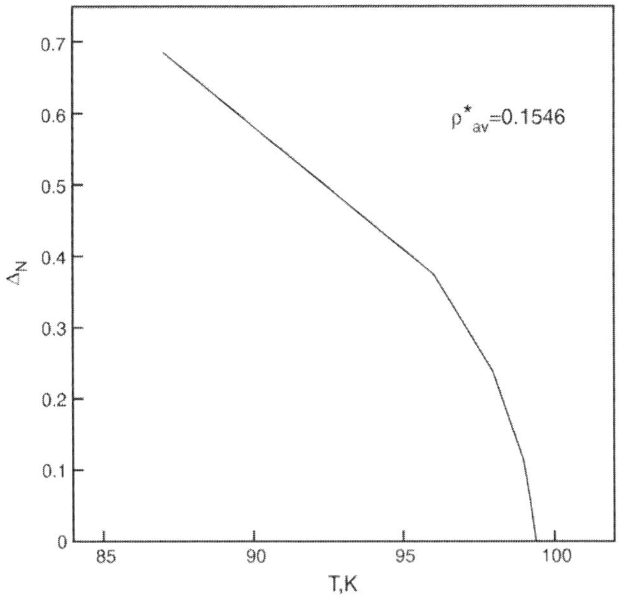

FIG. 9. (Reprinted from Ref. [41] with permission) Temperature dependence of the parameter Δ_N in a closed slit with $h_m^* = 15$ for $T > 87$ K and p = 0.1546. Symmetry breaking occurs (i.e. $\Delta_N > 0$) for $T < 99.4$ K.

At a selected ρ_{av}^*, Δ_N has also a critical-like behavior as function of temperature. As an example, the T-dependence of Δ_N in a slit with $\rho_{av}^* = 0.1546$ is presented in Fig. 9 which shows that there is a very rapid change of Δ_N near $T = T_{sb1} \simeq 99.2$ K.

Note that in Ref. [40] the phenomenon of SB was considered using a local DFT for the Lennard-Jones as well as for the van der Waals potentials for the fluid-solid interactions. The latter potential is provided by Eqs. (13) and (14) without the term $\sim (\sigma_{fs}/(h_r + h))^9$ in Eq. (14). For the Lennard-Jones potential the results are in qualitative agreement with those obtained with a nonlocal DFT. The main quantitative difference is in the value of the critical temperature T_{sb} that is predicted to be equal to 106 K for the nonlocal and 112 K for the local version of the DFT.

For the van der Waals potential, the local DFT provided only a single critical density, ρ_{sb}, such that SB exists for $\rho_{av} < \rho_{sb}$ and is absent for $\rho_{av} < \rho_{sb}$. A critical temperature T_{sb} could not be identified for any ρ_{av}^* in the temperature range 87 K < T< 140 K.

The dependence of the domain of existence of symmetry breaking on the energy parameter ε_{fs} of the Lennard-Jones potential of the fluid-solid interaction was considered in Ref. [42].

In Fig. 10a the dependencies of ρ_{sb1}^* and ρ_{sb2}^* on ε_{fs} are plotted for $T = 104$K. Symmetry breaking occurs only for those $\rho_{av}^* - \varepsilon_{fs}$ pairs which are located inside the area *ABCD*. For $\varepsilon_{fs} < \varepsilon_{fs1}$ or ε_{fs2}, where $\varepsilon_{fs1} = 0.34$ and $\varepsilon_{fs2} = 1.38$ are critical values, no SB occurs and the density profile is symmetric.

The dependence on ε_{fs} of the difference $\Delta\rho_{av}^* = \rho_{sb2}^* - \rho_{sb1}^*$ between the average critical densities in which SB occurs is presented in Fig. 10b which shows that $\Delta\rho_{av}^*$ passes through a maximum $\Delta\rho_{av}^* = 0.1855$ at $\varepsilon_{fs} = 0.986$. In the latter case $\rho_{sb1}^* = 0.1662$ and $\rho_{sb2}^* = 0.3517$.

As shown in Fig. 11, the critical values ε_{fs1} and ε_{fs2} depend on temperature. The former slightly increases and the latter decreases with increasing temperature. As a result, the interval $\Delta\varepsilon_{fs1} = \varepsilon_{fs2} - \varepsilon_{fs1}$ in which SB occurs decreases with increasing temperature.

By increasing the average fluid density ρ_{av}^* at a fixed value of ε_{fs}, the density profile of the stable state of the system first loses its symmetry for ρ_{av}^* larger than ρ_{sb1}^* and then restores it for ρ_{av}^* larger than ρ_{sb2}^*. In Fig. 12a the stable profiles corresponding to $\rho_{av}^* = 0.1005, 0.1546,$ and 0.3092 are presented for $T = 87$ K and $\varepsilon_{fs} = 0.639$ (At that temperature, $\rho_{sb1}^* = 0.1051$, and

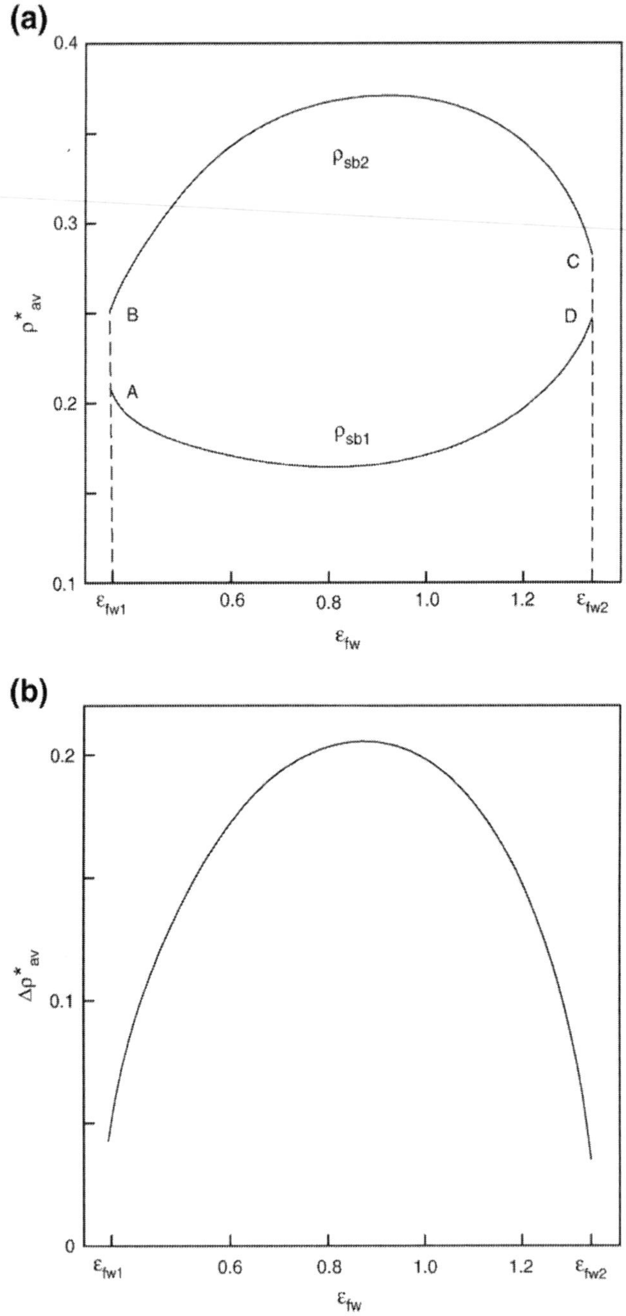

FIG. 10. (Reprinted from Ref. [42] with permission) (a) Dependence of the critical densities ρ^*_{sb1} and ρ^*_{sb2} on ε_{fs} for $T = 104$ K. Symmetry breaking occurs only for those $\rho^*_{av} - \varepsilon_{fs}$ pairs which are inside the area $ABCD$. The critical values of ε_{fs} are $\varepsilon_{fs1} = 0.34$ and $\varepsilon_{fs2} = 1.38$. (b) Dependence of the interval $\Delta\rho^*_{av} = \rho^*_{sb2} - \rho^*_{sb1}$ on ε_{fs} for $T = 104$ K. In both figures subscript fw stands for fs.

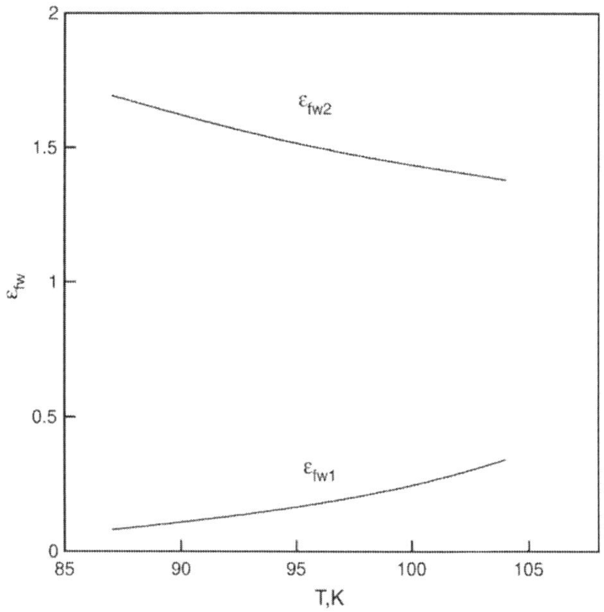

FIG. 11. (Reprinted from Ref. [42] with permission) Temperature dependence of the critical values ε_{fs1} and ε_{fs2}. At fixed temperature, symmetry breaking occurs only if $\varepsilon_{fs1} < \varepsilon_{fs} < \varepsilon_{fs2}$. In the figure, subscript fw stands for fs.

$\rho^*_{sb2} = 0.5241$). For $\rho^*_{av} = 0.1005$ the profile is symmetrical and for $\rho^*_{av} = 0.1546$ and $\rho^*_{av} = 0.3092$ they are asymmetrical. The restoring of the profile symmetry near ρ^*_{sb2} is illustrated in Fig. 12b where the profiles for $\rho^*_{av} = 0.5194, 0.5225$ (asymmetrical), and 0.5411 (symmetrical) are presented. Note that, in addition to an asymmetrical density profile, the Euler-Lagrange equation for the same set of parameters provides also a symmetrical profile which has, however, a higher free energy than the asymmetrical one. For illustration, the free energies of the asymmetrical profiles of Fig. 12a, b are presented in Table 2 along with the free energies of the related symmetrical profiles.

At fixed values of ρ^*_{av} and T the shape of the stable profile depends on ε_{fs}. As an example, the profiles for $\rho^*_{av} = 0.2319$ and $\varepsilon_{fs} = 0.013$; 0.639; 1.534; and 1.789 are plotted in Fig. 13 for $T = 87$ K. In this case SB occurs for $0.08 < \varepsilon_{fs} < 1.693$. Consequently, the profiles plotted for $\varepsilon_{fs} = 0.013$ and $\varepsilon_{fs} = 1.789$ are symmetrical and those for $\varepsilon_{fs} = 0.639$ and $\varepsilon_{fs} = 1.534$ asymmetrical. For the smallest value of $\varepsilon_{fs} = 0.013$, the fluid density near the walls is vapor-like $\left[\rho^*(0) = \rho^*(h_m) \simeq 0.022\right]$ and in the middle of the slit it is liquid-like $\left[\rho^*(h_m)/2 \simeq 0.58\right]$. Owing to the weak fluid-wall interactions, there are no visible density oscillations near the walls. For the largest value $\varepsilon_{fs} = 1.4$, the stable density profile is also symmetrical. The fluid density has the largest value at the walls $\left[\rho^*(0) = \rho^*(h_m) \simeq 4.46\right]$ and the smallest at the middle of the slit $\left[\rho^*(h_m)/2 \simeq 0.021\right]$. There are density oscillations near the walls which have a large amplitude and penetrate into the slit up to two molecular diameters. The fluid density in the asymmetrical profiles has the largest, liquid-like, value near the left wall, which increases with increasing ε_{fs}. The depth of penetration of the oscillations into the slit increases with ε_{fs} and for $\varepsilon_{fs} = 1.534$ becomes about four molecular diameters.

Recently [93] it was shown that the existence of stable asymmetric solutions for the density profiles depends on the position of the hard-wall repulsion generated by the solid wall. (This position is determined by the parameter h_r in potential Eq. (14)). When the location of this hard wall is moved, the strength of the potential near the wall changes and this is assumed to be the reason for the disappearance (appearance) of the SB

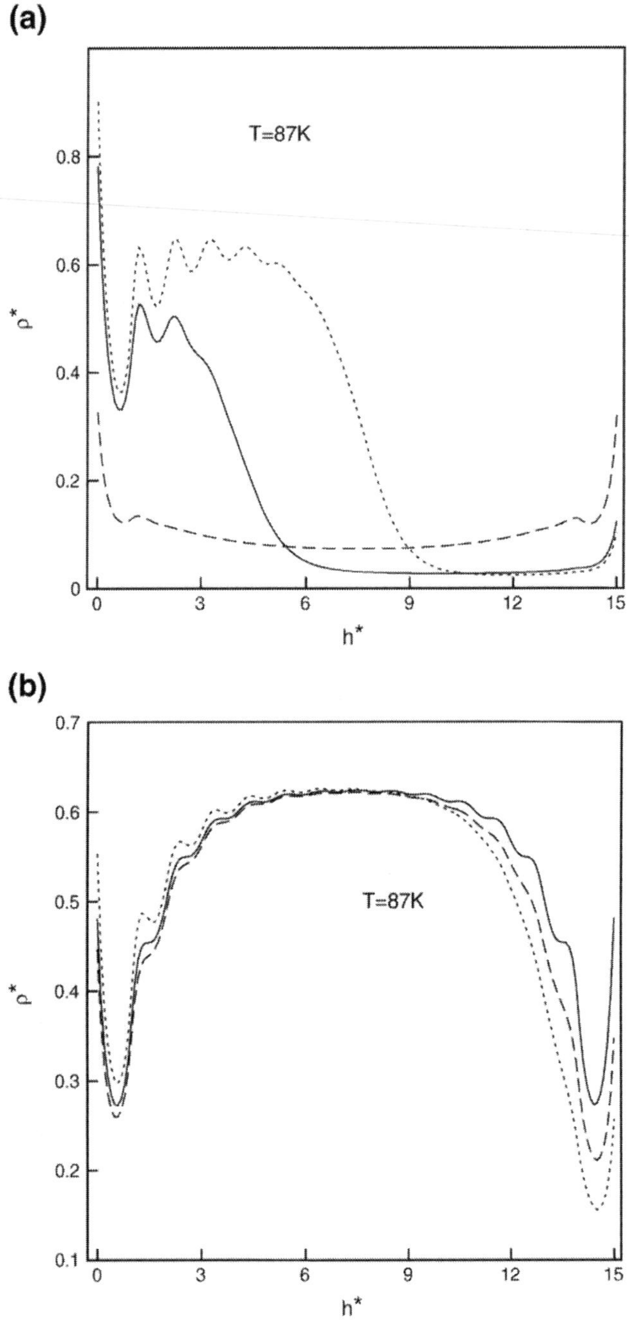

FIG. 12. (Reprinted from Ref. [42] with permission) Stable density profiles for $\varepsilon_{fs} = 0.639$ and $T = 87$ K. In this case $\rho^*_{sb1} = 0.1051$ and $\rho^*_{sb2} = 0.5241$. (a) Profiles for $\rho^*_{av} = 0.1005$ (dashed line), 0.1546 (solid line), and 0.3092 (dotted line). The first one is symmetrical about the middle of the slit, whereas the other two are asymmetrical. (b) Profiles for $\rho^*_{av} = 0.5194$ (dotted line), 0.5225 (dashed line), and 0.5411 (solid line). The last profile is symmetrical about the middle of the slit, the other two are asymmetrical.

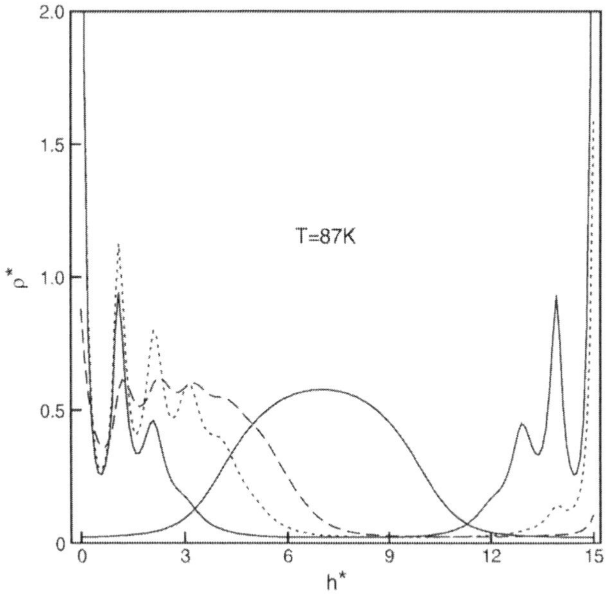

FIG. 13. (Reprinted from Ref. [42] with permission) Stable density profiles for $\rho^*_{av} = 0.2319$, $T = 87$ K and various values of ε_{fs}. The symmetrical profiles (solid lines) are for $\varepsilon_{fs} = 0.013$ (has a maximum in the middle of the slit) and $\varepsilon_{fs} = 1.789$. The asymmetrical profiles are for $\varepsilon_{fs} = 0.639$ (dashed line) and $\varepsilon_{fs} = 1.534$ (dotted line). The critical values of ε_{fs} for $\rho^*_{av} = 0.2319$, and $T = 87$ K are $\varepsilon_{fs1} = 0.080$ and $\varepsilon_{fs2} = 1.693$.

TABLE 2
(Reprinted from Ref. [42] with permission) The values (per unit area of one wall) of the Helmholtz free energies F_{sym} and F_{asym} of the metastable (symmetric density profile) and stable (asymmetric density profile) states of the fluid in a slit with $h^*_m = 15$, respectively for various average densities $\rho^*_{av1} < \rho^*_{av} < \rho^*_{av2}$. The temperature $t = 87$ K and $\varepsilon_{fs} = 0.639$. The free energies are given in $k_B T/\sigma^2_{ff}$ units.

ρ^*_{av}	F_{sym}	F_{asym}
0.1546	−25.76	−25.98
0.3092	−52.06	−52.26
0.5194	−88.21	−88.22
0.5225	−88.75	−88.76

Fig. 14 presents the asymmetry parameter Δ_N as a function of the parameter $v = \sigma_{fs}/h_r$ at the average density $\rho^*_{av} = 0.1932$ and $T = 87$ K. The data indicate that the asymmetric solution disappears for a critical value $v_c \simeq 1.18$. The evolution of the density profiles from asymmetric to symmetric is displayed in Fig. 15. In this plot one can note the diffusion of argon from the neighborhood of the left wall toward the right one. This process continues until a symmetric density profile is formed for $v \simeq 1.2$ [93].

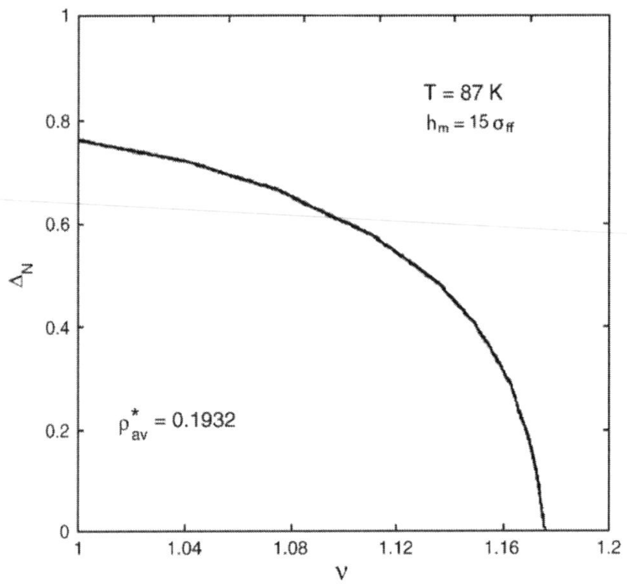

FIG. 14. (Reprinted from Ref. [93] with permission) Asymmetry parameter of density profiles, defined by Eq. (18), obtained for the average density $\rho^*_{av} = 0.1932$ in a slit with $h^*_m = 15$ as a function of the factor $\nu = \sigma_{fs}/h_r$.

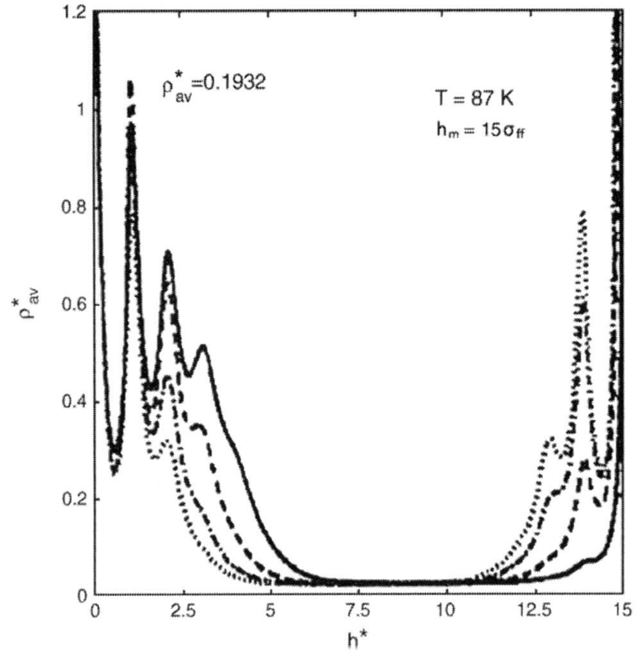

FIG. 15. (Reprinted from Ref. [93] with permission) Evolution of the asymmetric solution for the average density $\rho^*_{av} = 0.1932$ in a slit with $h^*_m = 15$ in terms of the factor ν. The solid curve corresponds to $\nu = 1$, the dashed to 1.136, the dashed-dotted to 1.156, and the dotted to 1.2.

Similar results were obtained for temperatures higher than $T = 87$ K.

It was observed [93] that at given average density and temperature, the SB disappears at that value of v at which a minimum of the fluid-solid potential is reached inside the slit, i.e. the hard-core repulsion is located closer to the wall than in the cases considered in Refs. [40–42].

Symmetry breaking can be understood in terms of the traditional wetting theory. By locating the hard-wall repulsion relatively far from the solid surface diminishes the attraction of the fluid-solid potential exerted on the fluid causing the system to enter in a one-wall-wetting regime where asymmetric profiles were detected by molecular dynamics simulations (see Fig. 2 in Ref. [94]).

By moving the hard-wall repulsion toward the solid wall the attraction increases causing eventually the system to enter in the two-wall-wetting regime characterized by a symmetric density profile (see Fig. 3 of Ref. [94]).

In Ref. [93], along with the parameters of the fluid-solid interaction potential, the effect of the slit width h_m on the domain of existence of the SB is also examined. In Fig. 16 the asymmetry parameter Δ_N is presented for several values of h_m labeled on each curve. These results are obtained for $T = 87$ K and $v = 1$. A comparison of the curves for $h_m^* = 30$ and $h_m^* = 15$ indicates that their shapes do not differ significantly, particularly in the range of large average densities ($0.35 < \rho^*_{\text{av}} < 0.51$) where the values of Δ_N are almost the same. This means, that the capillary effects in the case of a moderately thick slit of $h_m^* = 15$ are not very important. However, for $h_m^* < 15$ Δ_N differs considerable from the case $h_m^* = 30$. Finally, for $h_m^* = 5.5$ the parameter A_N becomes zero for all ρ_{av}^*, indicating that the asymmetric solution disappears [93].

4.2 Symmetry breaking for a two-dimensional fluid density distribution

The translational symmetry of a macroscopically large slit in lateral (parallel to the walls) directions suggests the same symmetry of the fluid density distribution. Possible breaking of this symmetry in one of these directions (x-direction in Fig. 2) was considered in Ref. [43].

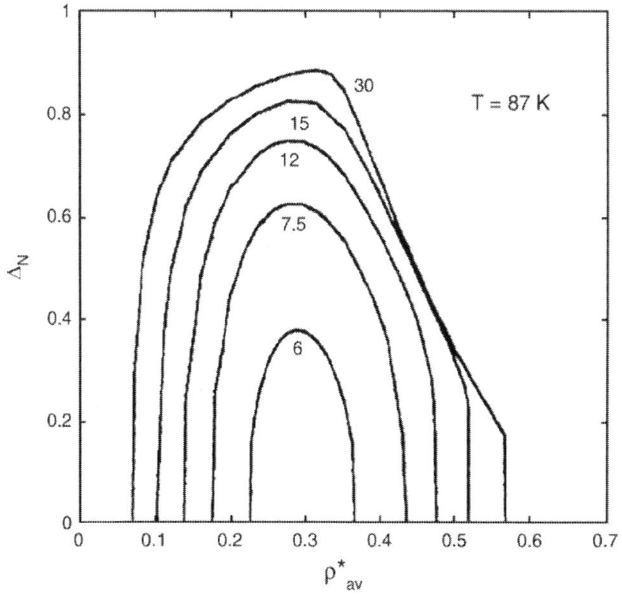

FIG. 16. (Reprinted from Ref. [93] with permission) Asymmetry parameter defined in Eq. (18) as a function of the average density. The successive curves are results obtained for slits with different effective widths labeled with corresponding values of h^*_m. For each width the asymmetric solutions occur in different ranges $\rho^*_{sb1} < \rho^*_{av} < \rho^*_{sb2}$.

The FDD in this case depends on two coordinates $(\rho(\mathbf{r}) \equiv \rho(x,h))$ and the Euler—Lagrange equation for $\rho(x, h)$ becomes two-dimensional, the coordinate x varying from $-\infty$ to $+\infty$. The latter makes the numerical solution of that equation practically impossible if no other restrictions are employed. To simplify the problem, the latter can be reformulated in a manner similar to that used in Monte Carlo simulations, namely, using the assumption that the FDD $\rho(x, h)$ is periodical with a period L_x in the x-direction. This assumption is equivalent to the periodic boundary conditions widely used in MC simulations [26,87,95,96] and, therefore, allows to compare (at least qualitatively) the analytical results based on DFT to those obtained via MC simulations. In the other lateral direction, the y-direction, the FDD is considered uniform. The same assumption was employed in Ref. [97] were a single value, $L_x = 20\sigma_{ff}$, was considered. In contrast to the MC simulations, where the period (length of the simulation box) rarely exceeds thirty molecular diameters, the period used in Ref. [43] is selected up to several hundred molecular diameters. By considering numerous increasing values of L_x and analyzing the changes in the density profile, one can extract information about the fluid density distribution in a long (infinite) closed slit.

In the numerical calculations, the potential Eq. (9) of the fluid-fluid interactions was truncated at the distance $r_{cut} = 2.5\sigma_{ff}$, i.e. it is assumed that $\phi_{ff} = 0$ for $r > r_{cut}$.

For any selected period L_x and average density of the fluid, several initial guesses $\rho^{in}(x, h)$ for the FDD $\rho(x, h)$ were employed. In general, each of them provided a different solution of the Euler-Lagrange equation. All guessed FDDs were constructed in the interval $0 \le x \le L_x$ such as to satisfy the periodicity condition $[\rho^{in}(0,h) = \rho^{in}(L_x, h)]$ and to provide the continuity of the derivative with respect to x $[(\partial/\partial x\, \rho(x,h))|_{x=0} = (\partial/\partial x\, \rho(x,h))|_{x=L_x}]$. Typical initial guesses employed in the calculations are presented in Fig. 17. First, typical solutions of the Euler-Lagrange equation for $L_x = 30$ and $h_m^* = 10$ are considered and stable and metastable states for various values of the average density were identified. By extracting such information for other values of L_x, from $L_x = 10\sigma_{ff}$ to $L_x = 240\sigma_{ff}$, the free energy of the stable state of the system was represented as a function of L_x and its asymptotic behavior at large values of L_x provided information about the infinite slit.

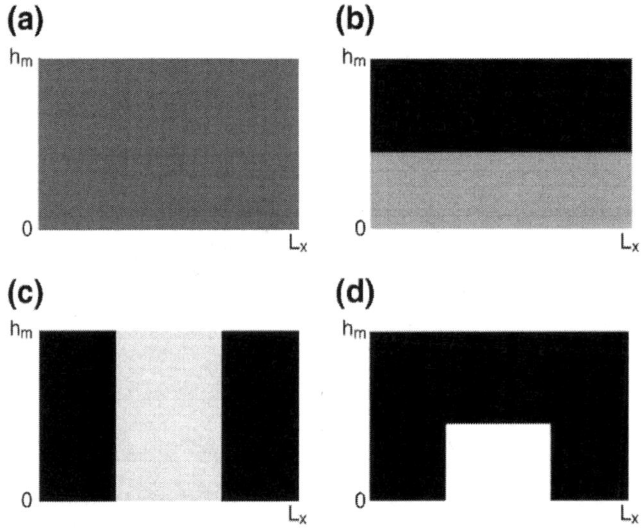

FIG. 17. (Reprinted from Ref. [43] with permission) Typical initial fluid density distributions (FDDs) employed in iterations. The lighter areas correspond to higher fluid density. FDD (a) always leads to a solution of the Euler-Lagrange equation uniform in x-direction and symmetrical in h-direction; FDD (b) can lead to a solution uniform in x-direction and asymmetrical in h-direction; FDDs (c) and (d) can lead to bridge like and bump like solutions, respectively.

Let us consider in more details the possible FDD for $L_x = 30\sigma_{ff}$. In all considered cases, the number of different solutions depends on the average density ρ_{av}. For small average fluid densities, say $\rho_{av} = 0.01$, all initial guesses led to the same solution, namely the uniform distribution along the x-axis and symmetrical in the direction perpendicular to the walls with respect to the middle of the slit (US-solution). In this case the walls are covered by a liquid film uniform in lateral directions and the remaining of the slit is filled by a low density vapor. The free energy per unit area of one of the walls and the corresponding value of the Lagrange multiplier of that US-solution are listed in Table 3. For $\rho_{av} = 0.12$, a second, bump like solution, appears which is presented schematically in Fig. 18a. The bump-like FDD is nonuniform along the x-axis and is non symmetrical about the middle of the slit in the h-direction. This solution constitutes a symmetry breaking in two dimensions. A density profile for a bumplike solution in h-direction taken for $x = L_x/2$ is presented in Fig. 18b. The oscillations of the density near the wall are similar to those that occur in a bulk fluid in contact with a solid wall [52,91].

The free energy of a bump-like FDD is slightly smaller than that of the US-solution, hence the former FDD is stable and the latter one metastable. Note that a single bump solution is two-fold degenerate, because the bump can be located either on the lower or on the upper wall.

At $\rho_{av} = 0.14$ two new solutions were obtained in addition to those mentioned above. One of them, the UA-solution, is uniform in the x-direction and asymmetrical about the middle of the slit, and is a result of the symmetry breaking in the h-direction described in details in Refs. [40,41].

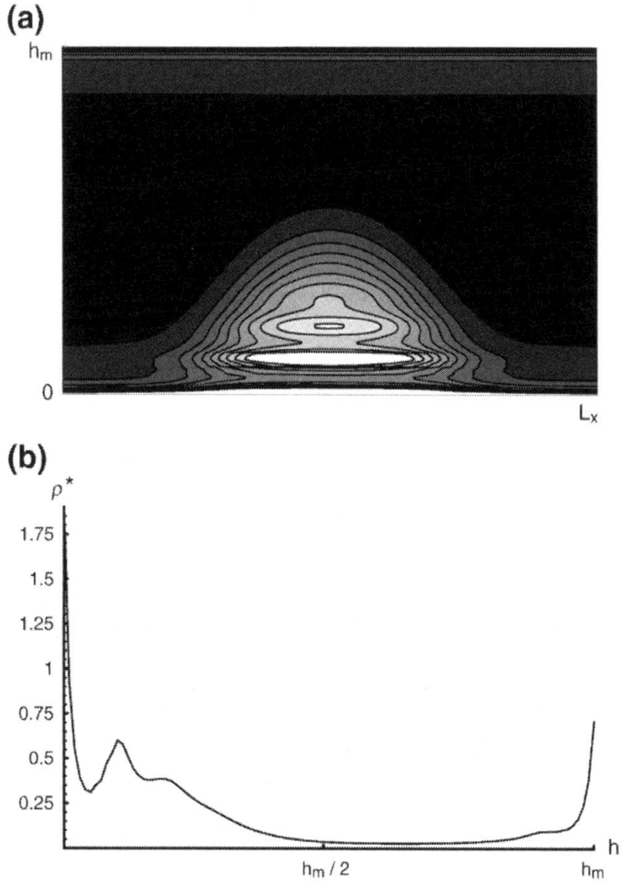

FIG. 18. (Reprinted from Ref. [43] with permission) (a) Bump like fluid density distribution. The lighter areas correspond to higher fluid density. (b) Density profile in h-direction at $x = L_x/2$. $L_x = 30\sigma_{ff}$ and $\rho^*_{av} = 0.12$.

The density profile for this solution is presented in Fig. 19. Another, bridge like solution, which is nonuniform along the x-axis but symmetrical about the middle of the slit, is shown in Fig. 20.

Comparing the free energies of the four FDDs that occur for $\rho_{av} = 0.14$ (see Table 3) one can see that in this case the bump like FDD (Fig. 18a) has the smallest free energy and is therefore the stable one, and the other FDDs are metastable. However, the differences between their free energies are very small. If the energy barriers (which are unknown) between those states are sufficiently high, the system can survive for a long time in any of the metastable state. If those barriers are low, the thermal fluctuations can drive the system through all of these states.

The configuration of the stable state changes at $\rho_{av} = 0.17$ where a bridge like FDD is the stable one, and the bump-like FDD and the other ones are metastable. The values of the free energies for a similar case with $\rho_{av} = 0.2$ are listed in Table 3. The metastable bump-like solution disappears at $\rho_{av} = 0.22$ and the UA-solution disappears at $\rho_{av} \simeq 0.5$.

The bridge like FDD remains stable up to $\rho^*_{av} = 0.58$ where the fluid fills completely the slit and the FDD becomes uniform in the x-direction and symmetrical about the middle of the slit (US-solution).

For smaller L_x ($L_x = 10\sigma_{ff}$), only three solutions of the Euler-Lagrange equation could be identified. A US-solution, which is present at all average densities. A second, UA-solution, was found in the range

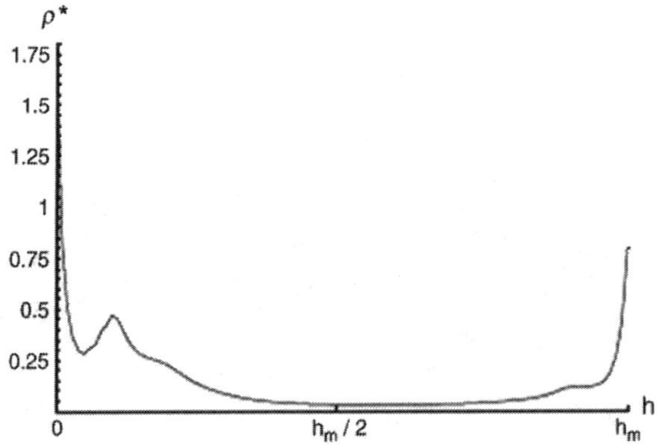

FIG. 19. (Reprinted from Ref. [43] with permission) Typical density profile of a UA-solution uniform in x-direction and asymmetrical in h-direction. $L_x = 30\sigma_{ff}$ and $\rho^*_{av} = 0.14$.

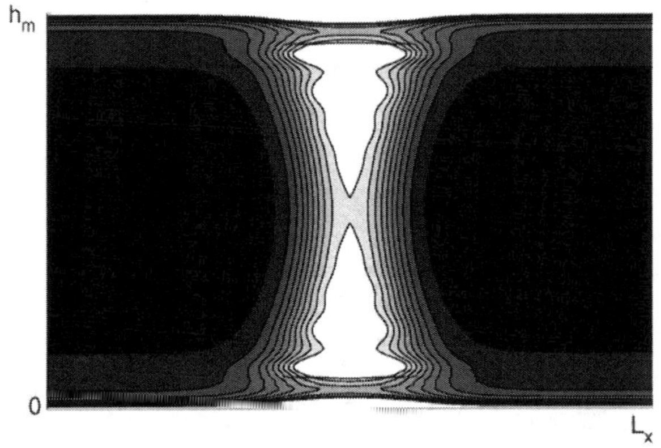

FIG. 20. (Reprinted from Ref. [43] with permission) Bridge like fluid density distribution. The lighter areas correspond to higher fluid density.

TABLE 3
(Reprinted from Ref. [43] with permission) Free energies (F) per unit area of one of the walls and the values of Lagrange multiplier (λ) for various solutions of the Euler-Lagrange equation. The stable solutions are distinguished by bold font. $L_x = 30\sigma_{ff}$.

FDD	$F\sigma_{ff}^2/k_BT$	λ/k_BT
$\rho_{av}^* = 0.01$		
US-solution	**−1.268**	**−12.729**
$\rho_{av}^* = 0.14$		
US-solution	−14.860	−11.231
UA-solution	−14.866	−11.302
Bridge like	−14.851	−11.472
Bump like	**−14.888**	**−11.351**
$\rho_{av}^* = 0.2$		
US-solution	−21.000	−11.297
UA-solution	−21.067	−11.414
Bridge like	**−21.163**	**−11.550**
Bump like	−21.102	−11.387

$0.14 \leq \rho_{av}^* \leq 0.58$. This range of existence of the UA-solution coincides with that for $L_x = 30\sigma_{ff}$. This is a consequence of the uniformity of the UA-solution in the x-direction, which means independence of this FDD of L_x. The third, bridge like solution, was found along with an UA-solution in the range of average densities $0.38 \leq \rho_{av}^* \leq 0.42$. When a UA-solution exists, it has always the smallest free energy, i.e. it is stable, all the other ones being metastable. No bump-like FDD was found for $L_x = 10\sigma_{ff}$.

For the largest considered period $L_x = 240\sigma_{ff}$, the US- and UA-solutions coincide with those obtained for $L_x = 30\sigma_{ff}$ due to their uniformity in the x-direction. The bridge-like solution appears at the average density $\rho_{av} = 0.06$. It remains stable for average densities $\rho_{av} < 0.58$. For $\rho_{av} \geq 0.58$ only the US-solution remains stable. The UA-solution is always metastable, and a bump like solution could not be identified.

Having as a goal the finding of a stable FDD for a very long (infinite) closed slit we calculated the density distributions when L_x was changed from $10\sigma_{ff}$ to $240\sigma_{ff}$ at the fixed average density $\rho_{av} = 0.2$. This moderate value of the density was selected because it is expected to provide a variety of solutions of the Euler-Lagrange equation. For any selected L_x from that interval, all the FDDs were calculated and the stable ones were identified. For $10\sigma_{ff} \leq L_x \leq 20\sigma_{ff}$ the UA-solution (Fig. 19) was the stable FDD one. The bump like FDD (Fig. 18a) was stable for $20\sigma_{ff} \leq L_x \leq 24\sigma_{ff}$ and the bridge like FDD was stable for $L_x \geq 24\sigma_{ff}$. The free energy of the stable state is plotted as a function of L_x in Fig. 21. This figure indicates an asymptotic behavior of the free energy with increasing L_x. From this behavior one can conclude that for a given average density a single bridge of finite length occurs as a unique stable state in a long closed slit. For vaporlike average densities $\rho_{av}\, \rho_{av} \leq 0.01$ as well as for liquid like densities $\rho_{av} \leq 0.6$ the stable state for any L_x is the US-solution of the Euler-Lagrange equation.

4.3 Symmetry Breaking for Confined Mixtures

The behavior of confined mixtures is much more complex than that of a single component fluid. In an open slit, the problem was treated in Refs. [98–103] and the asymmetrical states in the h-direction and the nonuniform states in the x-direction were studied in some of them [102,103].

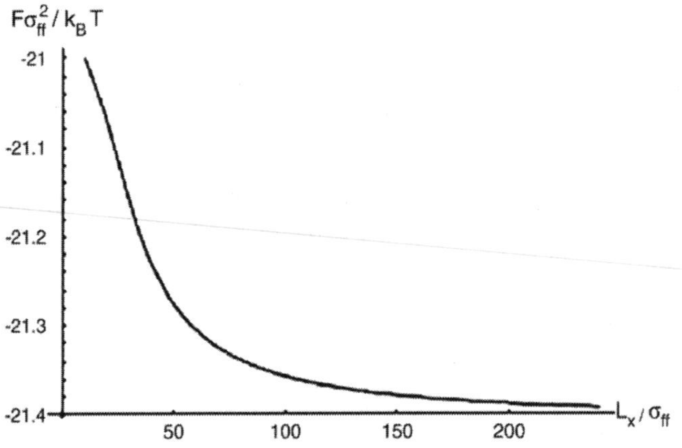

FIG. 21. (Reprinted from Ref. [43] with permission) Dependence of the free energy of the stable state on the length L_x of the period of the fluid density distribution.

However these asymmetrical states occur not because of symmetry breaking, but because of the absence of symmetry in the interaction potential between the walls and fluid, due either to nonidentical [102] or nonuniform [103] walls.

The symmetry breaking in *x*- and *h*-directions in binary mixture confined in a closed slit with identical walls was studied in Ref. [44].

To be more specific, conditions were identified when component 1 taken alone, does not possess SB states, whereas component 2 (argon) taken alone exhibits all SB states described in Sections 1 and 2. As for the one component fluid, the problem was formulated in a way similar to that used in MC simulations. In particular, the assumption that the solution of the Euler-Lagrange equation for the FDD is periodical with a period L_x in the *x*-direction, is employed. In the other lateral direction, the *y*-direction, the FDD is considered uniform, hence $\rho_i(\mathbf{r}) \equiv \rho_i(x, h)$.

The interactions between the molecules of components 1 (C1) and 2 (C2) were described by the truncated Lennard—Jones potential

$$U_{ij}(r) = \begin{cases} \infty, & r \le \sigma_{ij} \\ 4\epsilon_{ij}\left[\left(\dfrac{\sigma_{ij}}{r}\right)^{12} - \left(\dfrac{\sigma_{ij}}{r}\right)^{6}\right], & \sigma_{ij} < r \le r_{ij,cut} \\ 0, & r > r_{ij,cut} \end{cases} \tag{19}$$

where *i* and *j* take the values 1 and 2 corresponding to components 1 and 2, *r* is the distance between the centers of a pair of molecules, and $r_{ij,cut} = 2.5\sigma_{ff}$ is the cutoff distance. The constants σ_{ff} and ϵ_{ij} are the hard core diameters and energy parameters, respectively.

The total external potential U_{si} exerted on a fluid molecule of component *i* at distance *h* from the wall W_1 is calculated as the superposition of the $\psi_i(h)$ for the two walls:

$$U_{si}(h) = \psi_i(h) + \psi_i(h_m - h) \tag{20}$$

where h and h_m are defined in Fig. 22 and $\psi_i(h)$ is given by the expression

$$\psi_i(h) = \frac{2\pi}{3}\epsilon_{si}\rho_s\left[\frac{2}{15}\left(\frac{\sigma_{si}}{\sigma_{si}+h}\right)^9 - \left(\frac{\sigma_{si}}{\sigma_{si}+h}\right)^3\right] \tag{21}$$

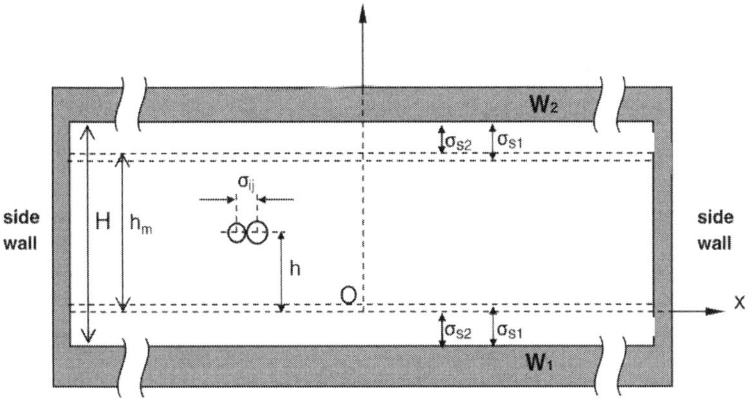

FIG. 22. (Reprinted from Ref. [44] with permission) Closed slit of width H between two identical solid walls W_1 and W_2. σ_{ij} and σ_{si} $(i = 1,2)$ are the hard core diameters of *component i - component j* and wall - *component i* interactions, respectively. The y axis is perpendicular to the plane of the figure.

Here σ_{si} and ϵ_{si} are the hard core diameters and the energy parameters for the fluid-solid interaction, ρ_s is the density of the walls and h is the distance measured from the lower wall as shown in Fig. 22.

The density distributions $\rho_i(\mathbf{r})$ $(i = 1,2)$ are calculated using the density functional approach formulated by Rosenfeld [72].

As shown in Refs. [75,76,83,84], this theory, developed originally for open systems (grand canonical ensemble), can be also applied to closed systems (canonical ensemble). The only restriction, which is fulfilled in the calculations performed in Ref. [44] is for the number of molecules to be sufficiently large. A detailed discussion on the applicability of the canonical ensemble DFT can be found in Ref. [88].

For a binary mixture, the system of two integral equations

$$\rho_i(x,h) = \rho_{i,0} exp\left[\frac{Q_i(\rho_1(x,h),\rho_2(x,h))}{k_B T}\right], (i = 1,2), \tag{22}$$

each similar to Eq. (13) must be solved. In Eq. (22), $\rho_{i,0} = \Lambda_i^{-3} exp(\lambda / k_B T), \Lambda_i$ is the thermal de Broglie wavelength of the molecules of component i. Explicit expressions for the functions $Q_i(\rho_1(x, h), \rho_2(x, h))$ can be found in Ref. [44].

In the computations, the solid carbon dioxide was selected for the walls and argon was selected as component 2 (C2). The parameters for component 1 (C1) and the cross parameters were selected as follows: $\epsilon_{11} = \epsilon_{22}/4$, $\epsilon_{W1} = \epsilon_{W2}$, $\sigma_{11} = 1.2\sigma_{22}$, $\sigma_{W1} = 1.2\sigma_{W2}$, $\epsilon_{12} = \sqrt{\epsilon_{11}\epsilon_{12}} = \epsilon_{22}/2$, $\sigma_{12} = (\sigma_{11} + \sigma_{22})/2 = 1.1\sigma_{22}$. The width of the slit was taken $h_m = 10\sigma_{22}$ and the temperature was $T = 87$ K. To calculate the de Broglie wavelength, the mass m_1 of a molecule of C1 was taken as $m_1 = (\sigma_{11}/\sigma_{22})^3 m_2$, where m_2 is the mass of an argon molecule. All densities in this section are expressed in dimensionless form, $\rho_i^* = \rho_i \sigma_{22}^3$ in units of the arbitrarily chosen density $\rho_{ref}^* = 0.0116$, i.e. the numerical values of the density are obtained by dividing ρ_i^* by ρ_{ref}^*.

4.3.1. Density distributions for $L_x = 30\sigma_{ff}$

For fixed average fluid densities $\rho_{1,av}^*$ and $\rho_{2,av}^*$, the Euler-Lagrange equation can have several solutions which can be obtained using various initial guesses for the iteration procedure. Four guesses, similar to those presented in Fig. 17, were employed as initial density distributions for each of the components. The number of different solutions of the Euler-Lagrange equation depends on the average densities $\rho_{i,av}^*$. Some characteristic examples of those solutions are presented below.

First, let us consider the features of the FDDs for a fixed average density of component 1 $(\rho_{1,av}^* = 3)$ and various values of $\rho_{2,av}^*$ For $\rho_{2,av}^* \leq 11$ all initial guesses led to uniform distributions along the

x axis and symmetrical in the direction perpendicular to the walls with respect to the middle plane (US-solution). In this case, the walls are covered by liquid-like layers of mixed C1 and C2 and the remaining of the slit is filled by a low density mixture. The densities $\rho_1^*(\mathbf{r})$, and the local composition of the mixture, defined as $X(\mathbf{r}) = \rho_1^*(\mathbf{r})/[\rho_1^*(\mathbf{r}) + \rho_2^*(\mathbf{r})]$, depend in this case on the h coordinate only. $\rho_2^*(h)$, and $X(h)$ are plotted in Fig. 23a and b, respectively for some average densities $\rho_{2,av}^*$. The density distribution of C1 is almost independent of $\rho_2^*(h)$ and is similar to $\rho_2^*(h)$ for $\rho_{2,av}^* = 3$. Note that the ranges of $0 < h < (\sigma_{s1} - \sigma_{s2})$ and $h_m - (\sigma_{s1} - \sigma_{s2}) < h < h_m$ are not accessible to C1 molecules because of its larger hard core diameter compared to that of C2. One can see from Fig. 23b that the composition of the fluid in the region of the slit accessible to both components, changes considerably across the slit. Close to the wall the mixture is much richer in C1 than in the middle of the slit, the latter being close to the average composition $X_{av} = \rho_{1,av}^*/(\rho_{1,av}^* + \rho_{2,av}^*)$.

At $\rho_{2,av}^* = 11$, two additional solutions appear. The first, a UA-solution, is uniform in the x-direction and asymmetric in the h-direction. It is a result of the symmetry breaking described in details in Refs. [40,41].

The density profiles of C2 and C1 are presented in Fig. 24a and b, respectively. Compared with the US solution, C1 and C2 are redistributed across the slit in opposite ways: C1 has the largest density near W_2 $(h \simeq h_2)$, whereas the density of C2 has the largest value near W_l $(h \simeq 0)$. The density oscillations of the more dense C2 near the wall W_1 $(h \simeq 0)$ are similar to those occurring in a bulk fluid in contact with a solid wall [52,91].

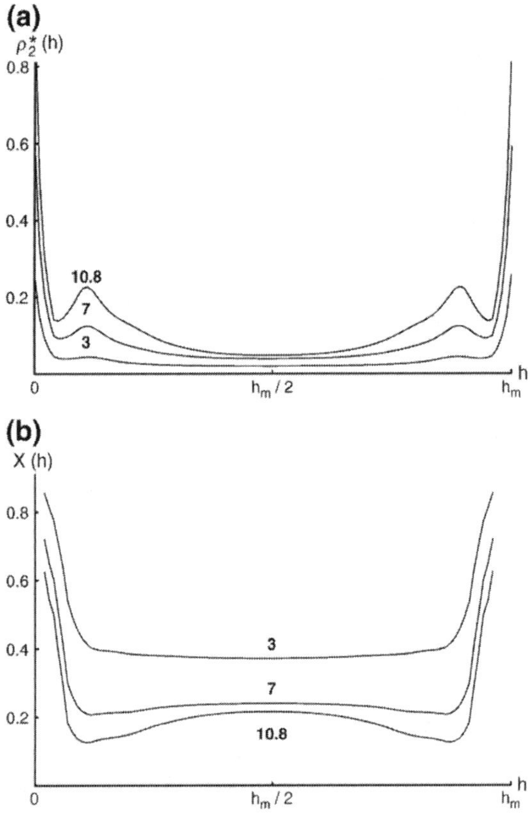

FIG. 23. (Reprinted from Ref. [44] with permission) Mixtures with $\rho_{1,av}^* = 3$ and various values of $\rho_{2,av}^*, L_x = 30\sigma_{22}, \mathrm{T} = 87$ K. (a) Density profiles of the uniform in x-direction and symmetrical in h-direction US-solution of the Euler-Lagrange equation for component 2. The numbers on the curves indicate the values of $\rho_{2,av}^*$. (b) h dependence of composition X for the same cases as in panel (a).

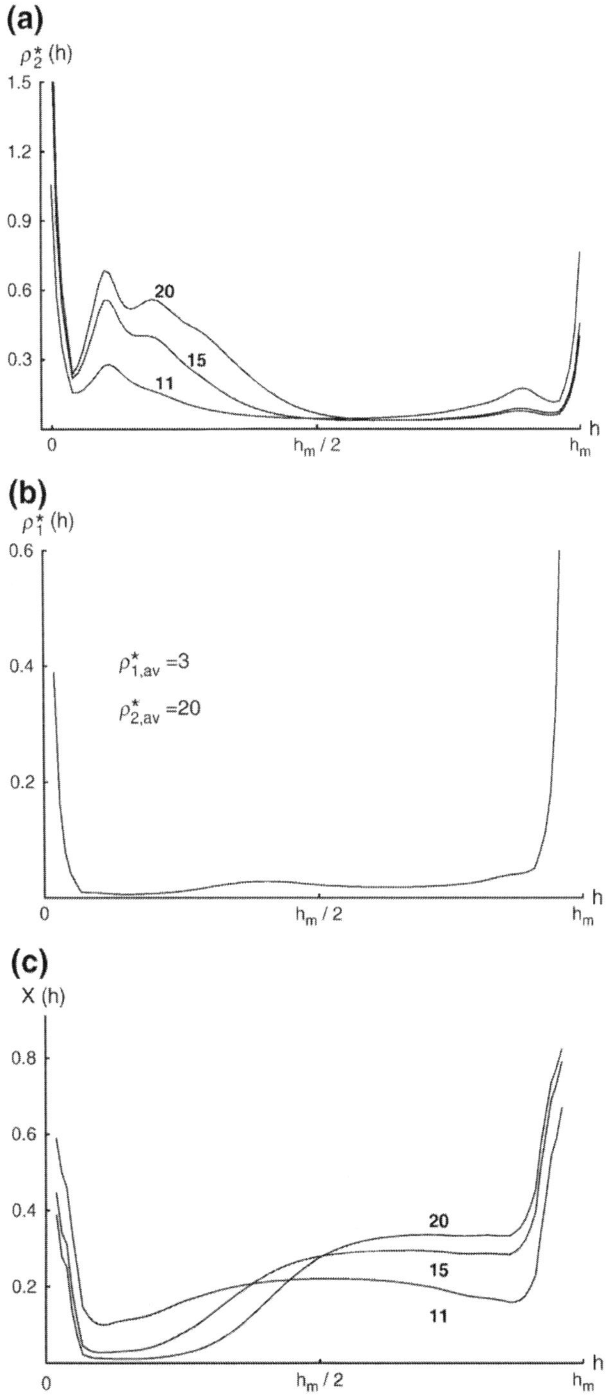

FIG. 24. (Reprinted from Ref. [44] with permission) Mixtures with $\rho_{1,av}^* = 3$ and various values of $\rho_{2,av}^*, L_x = 30\sigma_{22}, T = 87$ K. (a) Density profiles of the uniform in *x*-direction and asymmetrical in *h*-direction UA-solution of the Euler-Lagrange equation for component 2. The numbers on the curves indicate the values of $\rho_{2,av}^*$. (b) *h* dependence of ρ_1^* for $\rho_{1,av}^* = 3$ and $\rho_{2,av}^* = 20$. (c) *h* dependence of composition *X* for the same cases as in panel (a).

The oscillations of the FDD of the less dense C1 are almost absent in the considered case. The UA solution indicates a specific demixing that occurs across a closed narrow slit (see Fig. 24c where the composition of the mixture across the slit is plotted). Second, a bump-like solution is presented schematically in Fig. 25. The density distributions of C1 (Fig. 25b) and C2 (Fig. 25a) inside the bump are different. C2 is present in the bump core (light color area in Fig. 25a), whereas C1 is located outside the bump. The bump-like FDD is nonuniform along the x axis and nonsymmetrical about the middle plane in the h-direction. This solution constitutes a symmetry breaking in two, x and h, directions. The density profiles across the bump in the h-direction resemble the profiles for the UA solution presented in Fig. 24. The density profiles of C1 and C2 in the h-direction taken outside the bump are symmetrical and similar to those presented in Fig. 23a. The free energies of the UA-solution and bump-like one are smaller than the free energy of the US-solution, the free energy of the bump-like solution being the smallest one (see Table 3). Hence, the latter solution is stable for the considered values of $\rho^*_{i,av}$ ($i = 1, 2$). Note that the bump-like solution presented in Fig. 25 is two-fold degenerate, because the bump can be located either on the lower or on the upper wall.

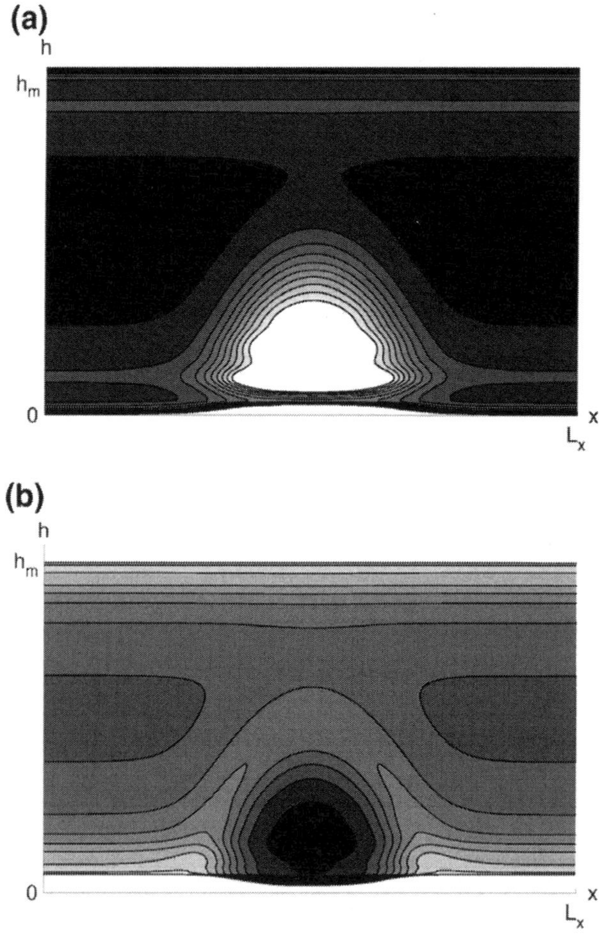

FIG. 25. (Reprinted from Ref. [44] with permission) Bump-like FDD of component 2 (a) and FDD of component 1 (b). The lighter color areas correspond to higher fluid densities. Mixture with $\rho^*_{1,av} = 3$ and $\rho^*_{2,av} = 11$. $L_x = 30\sigma_{22}$, $T = 87$ K.

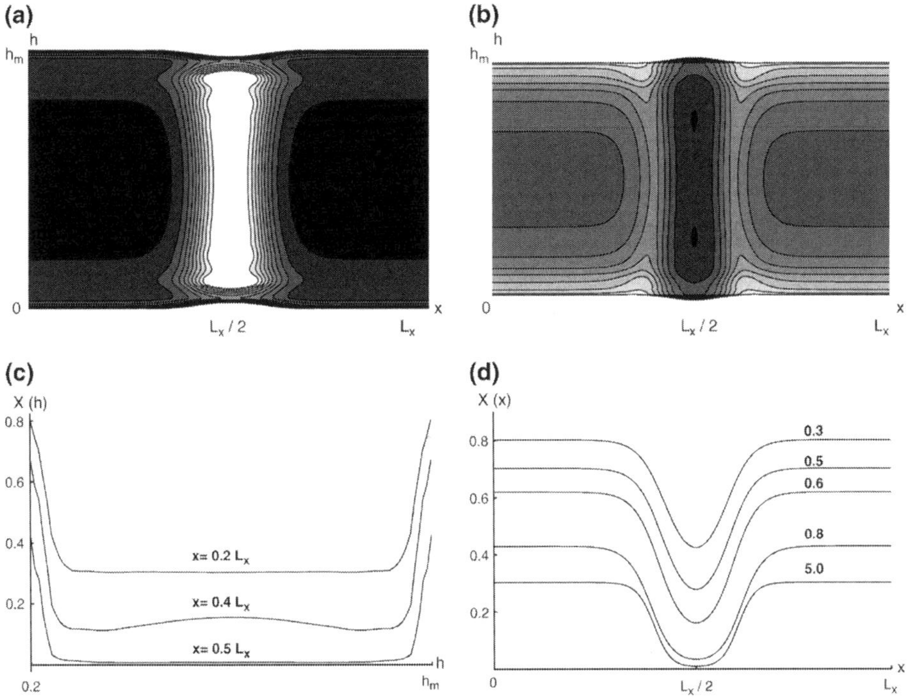

FIG. 26. (Reprinted from Ref. [44] with permission) Mixture with $\rho^*_{1,av} = 3$ and $\rho^*_{1,av} = 11.8$. $L_x = 30\sigma_{\text{ff}}$, $T = 87$ K. Bridge-like FDD of component 2 (a) and FDD of component 1 (b). The lighter color areas correspond to higher fluid densities. (c) h-dependence of the mixture composition taken at $x = 0.5L_x$ (center of the bridge), $x = 0.4L_x$ (boundary of the bridge), and $x = 0.2L_x$ (outside of the bridge). (d) x-dependence of the mixture composition taken for values of h noted on each curve in units of σ_{22}.

At $\rho^*_{2,av} = 11.8$ a new, bridge-like solution is obtained in addition to those mentioned above (see Fig. 26a, b). C2 forms a dense bridge between the walls (see Fig. 26a) with C1 displaced outside the bridge (Fig. 26b). The bridge consists of a liquid-like core (light color area in Fig. 26a) and an interfacial area between the core and a vapor-like phase of C2. The bridge-like FDD is symmetrical about the middle plane but is non-uniform along the x axis. This nonuniformity constitutes a symmetry breaking in the x-direction. The composition of the mixture depends on both, h and x coordinates. The h dependence of the composition inside ($x = 0.5L_x$), on the boundary ($x = 0.4L_x$), and outside ($x = 0.2L_x$) the bridge in the direction across the slit is presented in Fig. 26c. One can see from this figure that for any x the fraction of C1 decreases quickly near the walls and remains almost constant for the remaining part of the slit. In Fig. 26d the x dependence of composition is presented for various values of h (indicated on the curves). Outside the bridge ($x<L_x/4$, $x>3L_x/4$), at any distance h from the wall, this fraction does not depend on x and has values much larger than inside the bridge ($x\sim L_x/2$). Estimations have shown that the bridge presented in Fig. 26a contains about 66% of the total amount of C2 in the slit and about 20% of C1. Hence, the bridge-like solution involves a demixing of the C1 and C2 in the x-direction, in contrast to the UA-solution which involves demixing across the slit (in the h-direction).

The free energy of the bridge-like solution is smaller than those of all other solutions (see Table 3); therefore, this FDD represents the stable state of the system. However the differences between the free energies of all possible states are very small. The life times of the system in those states depend on the energy barriers between them.

By further increasing $\rho^*_{2,av}$, the number of possible solutions decreases. The bump-like solution disappears at $\rho^*_{2,av} \simeq 16$ and the UA-solution disappears at $\rho^*_{2,av} \simeq 41$. The bridge-like FDD remains stable until the average density of C2 increases up to $\rho^*_{2,av} \simeq 45$ above which only the US solution exists.

To study the influence of C1 on the confined mixture we considered the possible solutions of Euler-Lagrange equation for various average densities $\rho^*_{1,av}$ at $\rho^*_{2,av} = 14$. In the range $0 < \rho^*_{1,av} < 5$, all the above listed solutions are present. For $0 < \rho^*_{1,av} \leq 0.9$, the bump-like solution is stable, all other ones being metastable. For $0.9 < \rho^*_{1,av} < 5$, the stable solution is the bridge-like one. For $\rho^*_{1,av} > 5$, a bump-like solution does not exist, and for $\rho^*_{1,av} > 7.5$ the same happens with the UA solution, the bridge-like solution remaining the stable one and the US solution metastable. At $\rho^*_{1,av} > \rho^*_{1,av} = 10.5$ only the uniform US-solution still exists. The average density $\rho^*_{av} = \rho^*_{1,av} + \rho^*_{2,av}$ of the mixture at $\rho^*_{1,av} = \rho^*_{1,0}$ is equal to $\rho^*_{av} = 24.5$. C2 taken alone at the latter average density, has a bridge-like solution as the stable state, the US and UA solutions being metastable.

The absence at $\rho^*_{1,av} > 10.5$ of a bridge-like stable state as well as bump-like and UA solutions in the case of a mixture can be understood remembering that in a single component fluid the existence of UA, bump, and bridge-like solutions at a given temperature depends on the strength of fluid-fluid interactions and occurs if the latter are larger than a critical value [40,43].

C1 taken alone does not exists in stable UA-, bump-, or bridge-like states because of the weakness of the intermolecular interactions considered. A small amount of C1 in a mixture with C2 does not change appreciably the behavior of the mixture, which is determined by C2. However, the increase of the average density $\rho^*_{1,av}$ of C1 reduces the interactions between the C2 molecules because of the increased average distance between them. For this reason, the effective C2-C2 intermolecular interaction is no longer sufficiently large to generate UA-, bump-, and bridge-like states. Hence, the composition X_{av} plays an important role in the existence of SB states at a given total average density of the mixture. As an example let us consider the average density of the mixture equal to $\rho^*_{av} = 14$. In this case we found that for $0 < X_{av} < 0.179$ all four states exist, the bump-like being the stable one. For $X_{av} > 0.179$, the bridge-like solution disappears, and the bump-like solution remains the stable one. For $X_{av} > 0.250$, only the US solution exists. No stable UA or bridge-like solutions were identified for the average density $\rho^*_{av} = 14$ of the mixture.

4.3.2. Density distribution for $L_x = 10\sigma_{22}$

For $\rho^*_{1,av} = 3$ and various values of $\rho^*_{2,av}$ only two solutions of the Euler-Lagrange equation were identified for both components of the mixture. A US-solution was found for all considered $\rho^*_{2,av}$ and a UA-solution for $11 < \rho^*_{2,av} \leq 41$. When the latter exists, it constitutes the stable one. The above range of existence of the UA-solution coincides with that for $L_x = 30\sigma_{22}$ because the uniformity of the UA-solution in the x-direction means independence of this FDD of L_x. No bridge or bump-like FDDs were found for $L_x = 10\sigma_{22}$.

4.3.3. L_x-dependence of the density distributions at fixed average densities

Having as a goal the finding of the stable FDD of a binary mixture in a very long (infinite) closed slit we calculated the FDDs $\rho^*_1(\mathbf{r})$ and $\rho^*_2(\mathbf{r})$ when L_x was changed from $10\sigma_{22}$ to $120\sigma_{22}$ at the fixed average densities $\rho^*_{1,av} = 3, \rho^*_{2,av} = 26$. These values of the densities were selected because they provide a variety of solutions of the Euler-Lagrange equations. For any selected L_x from that interval, all possible FDDs were calculated and the stable one was identified. For $10\sigma_{22} \leq L_x \leq 16\sigma_{22}$ the UA-solution (Fig. 24) is the stable FDD. The bridge-like FDD (Fig. 26a) is stable for $L_x > 16\sigma_{22}$. No stable bump-like state is found in this case.

In the considered range of L_x, the width ΔL_{bridge} of the bridge increases almost linearly ($\Delta L_{bridge} \sim 0.5L_x$) and the shape of the bridge changes from that presented in Fig. 27a to the almost rectangular shape presented in Fig. 27b.

The free energy of the stable state is plotted as a function of L_x in Fig. 28. This figure indicates an asymptotic behavior of the free energy with increasing L_x. From this behavior one can conclude that for selected average densities a single bridge of finite length consisting mostly of C2 (the remaining of the slit being filled with a mixture of C1 and C2) occurs as a unique stable state in a long closed slit.

For vapor-like average densities $\rho^*_{1,av}, \rho^*_{2,av} \sim 0.01$ as well as for liquid-like densities $\rho^*_{1,av}, \rho^*_{2,av} \sim 0.6$, the stable state for any L_x is the US-solution of the Euler-Lagrange equation.

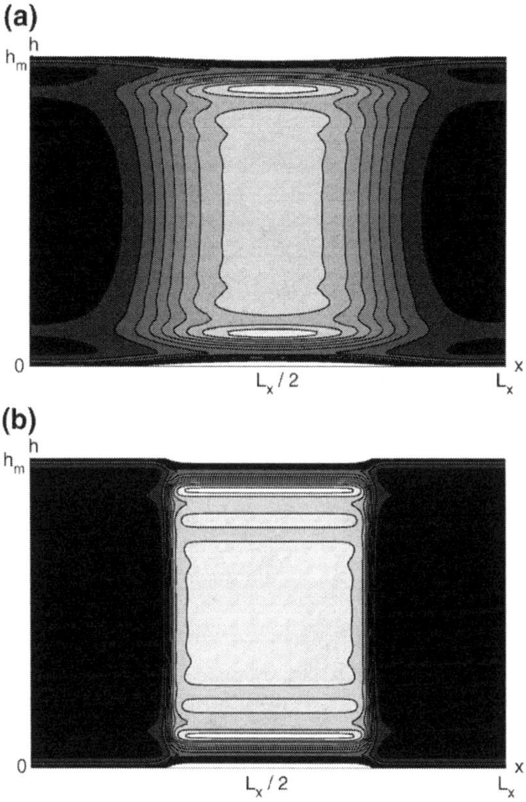

FIG. 27. (Reprinted from Ref. [44] with permission) Mixture with $\rho^*_{1,av} = 3$, $\rho^*_{2,av} = 26$, and $T = 87$ K. Bridge-like FDD of component 2 for $L_x = 16\sigma_{22}$ (a) and $L_x = 120\sigma_{22}$ (b). The lighter color areas correspond to higher fluid densities.

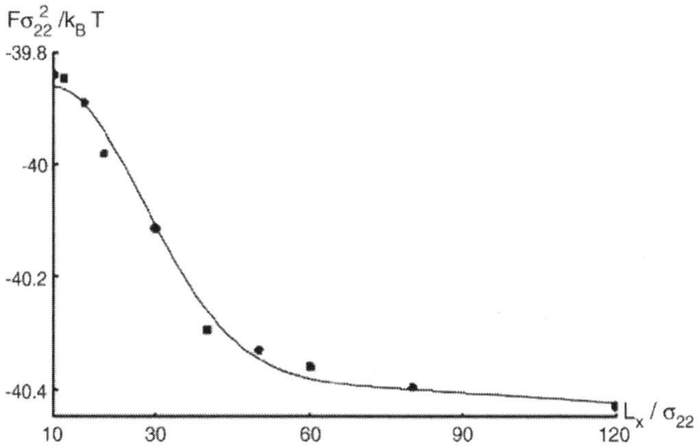

FIG. 28. (Reprinted from Ref. [44] with permission) Mixture with $\rho^*_{1,av} = 3$, $\rho^*_{1,av} = 26$, and $T = 87$ K. Dependence of the free energy of the stable state on the length L_x of the period of the fluid density distribution. Points represent the results of the calculations.

5. OBSERVATION OF ASYMMETRICAL DENSITY PROFILES IN COMPUTER SIMULATIONS

As a separate issue, the symmetry breaking was not considered systematically neither experimentally nor by computer simulations. However, asymmetrical density profiles were observed in the studies of various wetting phenomena in slits [94,96,104–108] and cylindrical pores [26] performed in simulations. The common features of most simulations is a fixed number of fluid molecules in the simulation box and periodic boundary conditions commonly used in both molecular dynamics and Monte Carlo simulations. Some results of computer simulations can be compared (at least qualitatively) with the results obtained via DFT for closed system.

In Ref. [94] the density profiles of a simple Lennard-Jones fluid were examined for various values of the parameter $\varepsilon_{fs} = \epsilon_{fs}/\epsilon_{ff}$ at the temperature $T = 0.9\varepsilon_{fs}/k_B$ (for argon, this temperature corresponds to 107.8 K) in a slit with $h_m \sim 30\sigma_{ff}$. The number of molecules in the slit was varied between 2400 and 3200. It was shown that for $\varepsilon_{fs} \leq 0.20$ and for $\varepsilon_{fs} \leq 0.7$ the stable density profile was symmetrical with respect to the middle of the slit. In the former case, small ε_{fs} it has a shape similar to that presented by the solid line (with a maximum in the middle of the slit) in Fig. 13. In the latter case of comparatively large ε_{fs}, the profile is similar to that presented by the solid line with a minimum in the middle of the slit in Fig. 13. In the range $0.2 \leq \varepsilon_{fs} \leq 0.7$, the shape of the profile was asymmetric and similar to the profiles represented by the dotted and dashed lines in Fig. 13. Such a dependence of the shape of the stable density profile on the energy parameter of the liquid-solid interaction is in qualitative agreement with the predictions of DFT (see Section 4.1). However, the absence of detailed information about the dependence of the density profiles on the average density of the fluid in a slit does not allow one a quantitative comparison of computer simulations with DFT results.

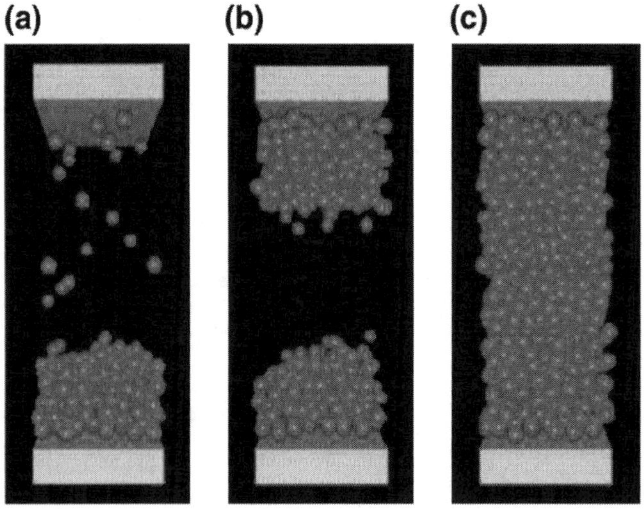

FIG. 29. (Reprinted from Ref. [43] with permission) Monte Carlo images of the equilibrium states of argon between walls of solid carbon dioxide ($\varepsilon_{fs} = 1.277$). The width of the slit $h_m^* = 20$ and $T = 89.82$ K. The average densities ρ_{av}^* are (a) 0.123, (b) 0.246, and (c) 0.556.

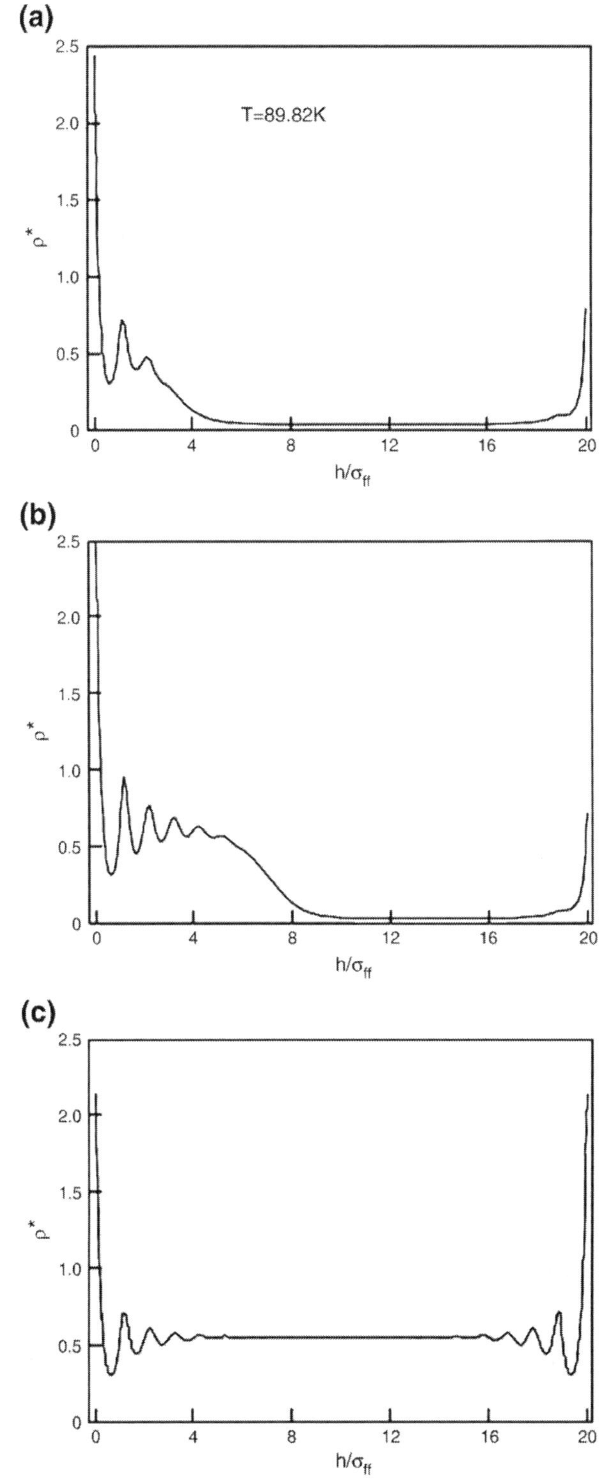

FIG. 30. (Reprinted from Ref. [43] with permission) Stable density profiles calculated for argon between walls of solid carbon dioxide (ε_{fs} =1.277). h_m^* =20, T = 89.82 K. The average densities ρ_{av}^* are (a) 0.123, (b) 0.246, and (c) 0.556.

For instance, the conclusion of nonexistence of asymmetrical profiles for $\varepsilon_{fs} \geq 0.7$ and for $\varepsilon_{fs} \geq 0.2$ [94] may not be valid because such profiles may occur at other number of molecules in the system than that used in simulations. Note that several asymmetrical density profiles similar to those noted in Ref. [94] were obtained also in Ref. [107] where the effect of the wall roughness on the wetting process was examined. However the authors made the simulations only for those values of ε_{fs} which belong to a range in which the asymmetric profile is the stable one and a transition between symmetrical and asymmetrical profiles could not be detected.

The occurrence of SB in h-direction is supported by the recent Monte Carlo simulations of argon between parallel identical walls of solid carbon dioxide $\varepsilon_{fs} = 1.277$), obtained by Errington [96] for a slit with $h_m^* = 20$ at $T = 89.82$ K. In Fig. 29, three snapshots representing the equilibrium states of three closed systems are presented for the average densities $\rho_{av,1}^* = 0.123$ (Fig. 29a), $\rho_{av,2}^* = 0.246$ (Fig. 29b), and $\rho_{av,3}^* = 0.556$ (Fig. 29c) (While oscillations of the density near the wall do exist, they are not visible in these figures.). The critical values ρ_{sb1}^* and ρ_{sb2}^* provided by our DFT theory for these systems are 0.08 and 0.48, respectively, and the density profiles calculated for the average densities used in Monte Carlo simulations are plotted in Fig. 30. The states presented in Fig. 29a, b belong to the range of average densities in which SB exists according to the present theory (see Fig. 30a, b). One of those states, (Fig. 29a) shows, in agreement with the calculations, a SB in the density distribution across the slit. Another one (Fig. 29b) has a density distribution less asymmetrical than the predicted one. The third, (Fig. 29c) has a symmetrical distribution corresponding to the average density $\rho_{av,3}^* > \rho_{sb,2}^*$, in agreement with the DFT (Fig. 30c).

6. SYMMETRY BREAKING FOR QUANTUM FLUID IN A CLOSED NANOSLIT

Quantum fluids possess such unusual properties as superfluidity, extremely high heat conductivity, etc. The theoretical examination of the equilibrium between a quantum fluid (^{4}He) and a solid substrate, which was performed mostly at $T = 0$ K has shown that many of its behaviors (e.g. layering, wetting, and prewetting transitions) are similar to those of classical fluids [92,109–111].

The effect of temperature was taken into account in Ref. [82] where the contact angles θ of a ^{4}He drop on the surfaces of some alkali metals (Cs, Rb, and K) were calculated. It was observed that near the λ-point of ^{4}He ($T_\lambda = 2.17$ K) at which the phase transition from normal to superfluid state takes place, the liquid-vapor surface tension had a small but observable kink. Because of the unique character of quantum liquids one expects the symmetry breaking in them to have specific features. In order to examine these features, in Ref. [45] a nonlocal density functional theory, formulated in Ref. [92] for $T = 0$ K and generalized to $T \neq 0$ K in Ref. [82], is employed. Cesium was considered as the solid substrate.

The interactions between the ^{4}He molecules are described by the Lennard-Jones potential Eq. (9) with $\epsilon_{ff}/k_B = 10.22K$ and $\sigma_{ff} = 2.556$ Å [92] but with the splitting at $r = \sigma_{ff}$ changed to that at $r = h_l$ where h_l depends on temperature [92]. The dependence of h_l on temperature is provided in Ref. [82].

To describe the interaction between a molecule of fluid and the solid walls, two potentials are usually employed. First, the standard 9–3 Lennard-Jones potential given by Eqs. (13) and (14). The potential Eq. (14), written for a single wall, was used in Ref. [92] in another equivalent form to examine the adsorption of ^{4}He on the surface of cesium. Using for the number density of Cs the value $p_s = 8.74 \times 10^{27}\ m^{-3}$, the parameters of the potential Eq. (14) can be recalculated from those of Ref. [92] to obtain $\sigma f_s = 5.44$ Å, and $\epsilon_{ff}/k_B = 1.42$ K.

Another potential, $U_{CCZ}(h)$, which was employed for the description of the interaction between ^{4}He and alkali metals by Chizmeshya et al. [112], has the following form

$$U_{ccz}(h) = V_{ccz}(1+\alpha h)exp(-\alpha h) - f_2(\beta(h)(h-h_{VdW}))\frac{C_{VdW}}{(h-h_{VdW})^3}, \tag{23}$$

where

$$f_2(x)=1-\left(1+x+\frac{x^2}{2}\right)exp(-x), \tag{24}$$

and

$$\beta(h)=\frac{\alpha^2 h}{1+\alpha h} \tag{25}$$

The parameters of this potential for Cs in the units used in the present paper are [112] V_{CCZ}/k_B = 1636.83 K, $\alpha\sigma_{ff}=4.52735$, $C_{VdW}/(\sigma_{ff}^3 k_B)=59.5863K$, and $h_{VdW}/\sigma_{ff}=0.0749209$.

Because of the uniformity of the walls, the fluid-solid potential and fluid density distribution $\rho(\mathbf{r})$ inside the slit is a function of h only $\left[\rho(\boldsymbol{r})\equiv\rho(h)\right]$.

The density functional used for the description of liquid helium is selected in the form suggested in Ref. [82]

$$\Omega_{He}[\rho]=\int\left[f-\mu\rho(\boldsymbol{r})\right]d\mathbf{r} \tag{26}$$

where μ is the chemical potential and f is the free-energy density of ^{4}He

$$\begin{aligned} f &= \frac{\hbar^2}{2M}\left(\nabla\sqrt{\rho}\right)^2+C_4 f_{id} \\ &+\frac{1}{2}\int d\mathbf{r}'\rho(\boldsymbol{r})\rho(\boldsymbol{r}')\phi_{ff}\left(\left|\mathbf{r}-\mathbf{r}'\right|\right) \\ &+\frac{c_2}{2}\rho(\boldsymbol{r})\left[\bar{\rho}(\boldsymbol{r})\right]^2+\frac{c_3}{3}\rho(\boldsymbol{r})\left[\bar{\rho}(\boldsymbol{r})\right]^3 \\ &-\frac{\hbar^2}{4M}\alpha_s\int d\mathbf{r}'F\left(\left|\mathbf{r}-\mathbf{r}'\right|\right)\left(1-\frac{\rho(\boldsymbol{r})}{\rho_{0s}}\right)\nabla\rho(\boldsymbol{r})\nabla\rho(\mathbf{r}')\left(1-\frac{\rho(\boldsymbol{r})'}{\rho_{0s}}\right) \end{aligned} \tag{27}$$

where $\hbar$ is the Planck constant and M is the mass of a molecule of ^{4}He. The first and second terms in the above expression account for the quantum kinetic energy and the free energy density of an ideal Bose gas, respectively. The third term represents the contribution of the fluid-fluid interactions. The terms that contain c_2 and c_3 provide the hard core He-He repulsion and are important at high densities. They contain the density $\rho(\mathbf{r})$ which is obtained by averaging the real density $\rho(\mathbf{r})$ over a sphere of radius h_l (the screening distance for fluid-fluid interactions) at the constant weight $3\left(4\pi h_l^3\right)^{-1}$. The last term in Eq. (27) is a nonlocal correction of the kinetic energy [82,92].

In that term, α_s=54.31Å, $\rho_{0s}\sigma_{ff}^3=0.6679$, and $F(|\mathbf{r}-\mathbf{r}'|)\equiv F(\mathrm{r})=\pi^{-3/2}l^{-3}exp\left(-r^2/l^2\right)$ with l = 1 Å [92].

The coefficients c_2, c_3, and c_4 depend on temperature. Their values were taken from Ref. [82].

The ideal-gas contribution f_{id} in Eq. (27) is [82]

$$f_{id}=\rho k_B T\log z-\frac{k_B T}{\Lambda^3}g_{5/2}(z), \tag{28}$$

where $\Lambda = \sqrt{2\pi\hbar^2/Mk_BT}$ is the thermal de Broglie wavelength. The fugacity z is given by

$$z = \begin{cases} 1, \ \rho\Lambda^3 \geq g_{3/2}(1), \\ z_0, \ \rho\Lambda^3 \geq g_{3/2}(1) \end{cases} \tag{29}$$

where $g_n(z) \equiv \sum_{l=1}^{\infty} z^l/l^n$ and z_0 is the root of the equation $\rho\Lambda^3 = g_{3/2}(z)$.

Taking into account the fluid-solid interaction, $Us(h)$, the total free energy functional is given by

$$\Omega[\rho] = \Omega_{He}[\rho] + \int_0^L dh\rho(h)U_s(h) \tag{30}$$

where $U_s(h)$ represents either the potential given by Eq. (11) or the $U_{CCZ}(h)$ potential. The equilibrium fluid density distribution $\rho(\mathbf{r}) = \rho(h)$ can be found from the Euler-Lagrange equation

$$\left\{-\frac{\hbar^2}{2M}\frac{d^2}{dh^2} + U(\rho, h)\right\}\sqrt{\rho(h)} = \mu\sqrt{\rho(h)} \tag{31}$$

which can be obtained by minimizing the grand-canonical potential $\Omega[\rho]$ with respect to $\rho \equiv \rho(h)$. In Eq. (31), $U(\rho,h)$ is an effective potential caused by the fluid-fluid and fluid-solid interactions. The latter potential is examined in some details in the Appendix of Ref. [44]. Remember that for a closed slit μ has the meaning of a Lagrange multiplier, which can be determined from the constraint of fixed average fluid density $\rho_{av} = 1/V \int_V \rho(\mathbf{r})d\mathbf{r}$ in the slit.

Eq. (31) was solved by standard numerical iterations [41] using a grid with 40 grid points per molecular diameter using the procedure outlined in Section 2.3.

6.1 LENNARD-JONES POTENTIAL FOR FLUID-SOLID INTERACTIONS

Let us consider symmetry breaking in a system with walls that generate hard core repulsions at distances $h_r = 0.4$, 0.2, and $0.1\sigma_{fs}$ focusing on the case $h_r = 0.4\sigma_{fs}$.

Typical symmetrical profiles for small and large values of $\rho^*_{av}\left(\rho^* = \rho\sigma_{ff}^3\right)$ are presented in Fig. 31. The profiles in Fig. 31a are plotted for various ρ^*_{av} (marked near the curves) for a slit with $L = 60\sigma_{ff}$ and for $T = 2.0$ K. For a very small average density $\rho^*_{av} = 0.01$, the fluid density distribution is almost homogeneous with a small increase near the walls. For the larger average density $\rho^*_{av} = 0.02$, several layers of liquid like density appear near the wall. The location of these layers is indicated by the maxima of the density in the profile, which has liquid like values $\rho^* \sim 0.20$ in the vicinity of the walls. The density becomes gas like $\rho^* \sim 0.001$ for distances from the wall of the order of several molecular diameters. When ρ^*_{av} increases, the oscillations of the density become larger and the region with gaslike fluid density decreases in size. For relatively high densities, $\rho^* \sim 0.20$, the whole slit is filled with a liquid like fluid, and the oscillations of the fluid density become significant at larger distances from the walls. The fluid density in the layers near the walls increases with increasing ρ^*_{av}. In Fig. 31b profiles corresponding to $\rho^*_{av} = 0.10$ are presented at various temperatures. The amplitude of the oscillations of the density near the walls decreases and the density in the central part of the slit increases with increasing temperature. This is an obvious consequence of the penetration of the ^{4}He molecules from the walls into the internal part of the slit due to the increase of their mobility with increasing temperature.

Several asymmetric profiles are shown in Fig. 32a for $T = 2.0$ K and various values of ρ^*_{av}. As in the case of classical fluids, the number as well as the amplitudes of the oscillations near the left wall

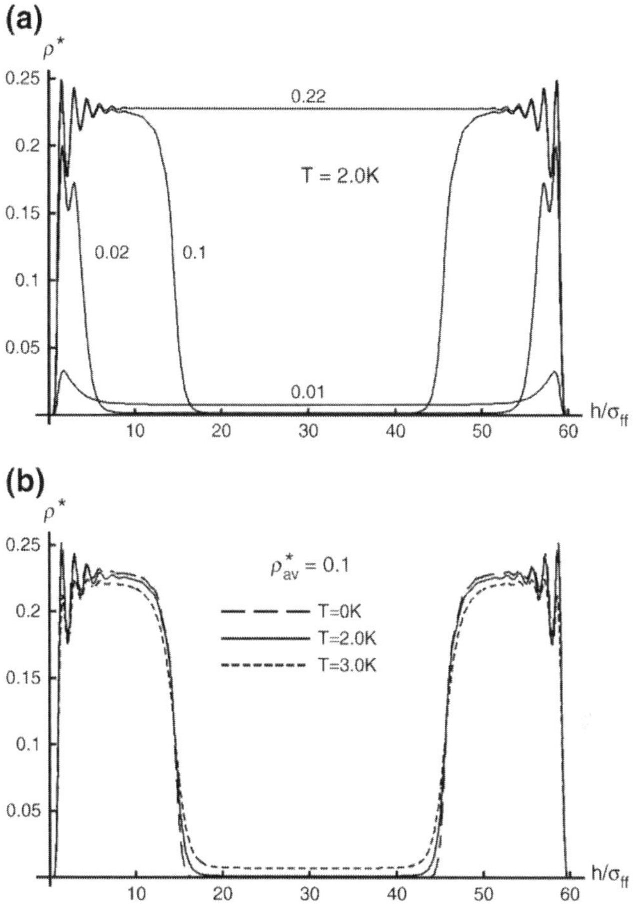

FIG. 31. (Reprinted from Ref. [45] with permission) (a) Symmetric fluid density profiles plotted at various average densities $\rho^*_{av} = \rho^*_{av}\sigma^3_{ff}$ indicated by the numbers for each of the curves. $T = 2.0$ K, $\rho^*_{av} = \rho_{av}\sigma^3_{ff}$ (b) Symmetric profiles at various temperatures for $\rho^*_{av} = 0.1$.

increase with increasing ρ^*_{av} and the region where the fluid has a gas like density decreases moving toward the right wall.

In Fig. 32b two profiles, one asymmetric and another symmetric, obtained for $\rho^*_{av} = 0.10$ and $T = 2.0$ K are presented as an example of multiple solutions of the Euler-Lagrange equation. It was found that the asymmetric density profile always provides a slightly smaller Helmholtz free energy than the symmetric one and, hence, the former represents the stable state (The free energies corresponding to symmetric and asymmetric profiles for $T = 2.0$ K at various ρ^*_{av} are presented in Table 4). This indicates that, indeed, symmetry breaking occurs in closed systems.

The dependence of the parameter Δ_N (Eq. (18)) characterizing the asymmetry of the density profile on the average density ρ^*_{av} is presented in Fig. 33 for $T = 0$, 2.0, and 3.0 K. Two features should be emphasized.

First, for each of the considered temperatures there is a range of average densities $\rho^*_{sb1} < \rho^*_{av} < \rho^*_{sb2}$ for which Δ_N has significant values and is almost zero outside of that range. Near ρ^*_{sb1} the parameter Δ_N as function of ρ^*_{av} changes rapidly and therefore ρ^*_{sb1} can be considered as a critical value. The change of Δ_N is not so rapid when $\rho^*_{av} \to \rho^*_{sb2}$. Second, the critical value ρ^*_{sb1} increases, and ρ^*_{sb2} decreases

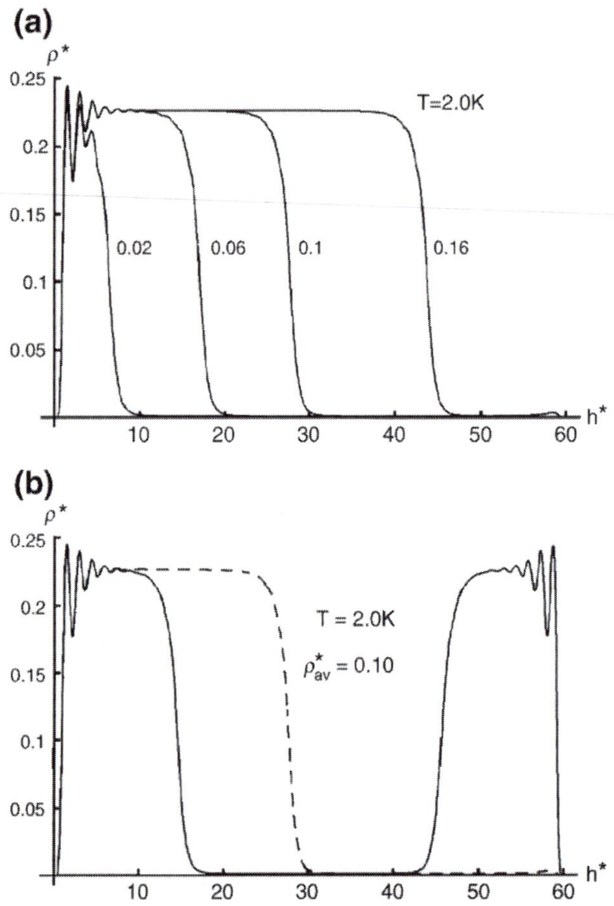

FIG. 32. (Reprinted from Ref. [45] with permission) (a) Stable asymmetric fluid density profiles plotted at various average densities $\rho^*_{av} = \rho_{av}\sigma^3_{ff}$ indicated by the numbers for each of the curves. $T = 2.0$ K, $\rho^* = \rho\sigma^3_{ff}$. (b) Metastable symmetric density profile (solid line) and stable asymmetric density profile (dashed line) calculated for $T = 2.0$ K and $\rho^*_{av} = 0.10$.

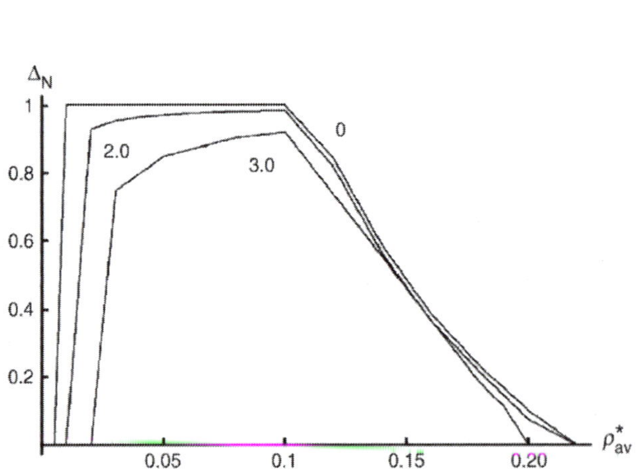

FIG. 33. (Reprinted from Ref. [45] with permission) The dependence of the parameter Δ_N, characterizing the asymmetric density distribution across a closed slit on the average density $\rho^*_{av} = \rho_{av}\sigma^3_{ff}$ of the fluid inside the slit for three different temperatures in K indicated by the numbers near each of the curves.

TABLE 4
(Reprinted from Ref. [45] with permission) Free energies per unit area of the symmetric and asymmetric density profiles for a slit with a width $L = 60\sigma_{ff}$. $T = 2.0$ K, $h_r = 0.4\sigma_{ff}$

	Free energy (J/m^2)	
$\rho^*_{av}\sigma^3_{ff}$	Symmetric profile	Asymmetric profile
0.02	−0.0022611	−0.0024434
0.05	−0.0061190	−0.0062685
0.10	−0.012484	−0.012607
0.12	−0.015021	−0.015140
0.14	−0.017557	−0.017675
0.18	−0.022337	−0.022749
0.19	−0.023669	−0.024023

with increasing temperature. The difference $\rho^*_{sb2} - \rho^*_{sb1}$ which provides the range of existence of the asymmetric solution and, hence, of the symmetry breaking, decreases with increasing temperature.

The results for $h_r = 0.2\sigma_{fs}$ and $h_r = 0.1\sigma_{fs}$ coincide qualitatively with those obtained for $h_r = 0.4\sigma_{fs}$. At the same average density, the asymmetry parameter, Δ_N, changes little with changing h_r. For instance, at $\rho^*_{av} = 0.03$ and $T = 2.0$ K this parameter has the values 0.9539, 0.9545, and 0.9549 for $h_r = 0.4$, $h_r = 0.2$, and $h_r = 0.1\sigma_{fs}$, respectively.

6.2 Zaremba Potential for Fluid-Solid Interactions

For this potential, defined by Eq. (23), the smallest distance of molecules of fluid from the wall is given by the parameter h_{vdW}. The potential is constructed in such a way to avoid divergence for $h \to h_{vdW}$. For this reason, one should not introduce a hard core repulsion generated by the walls. Hence this potential is a good example to test if SB occurs for $h_r = 0$.

An analysis similar to that carried out for the Lennard-Jones potential has shown that SB occurs for a quantum fluid even for $h_r = 0$ but has some specific features different from those of the Lennard-Jones potential. First, for a CCZ-potential a stable asymmetric density profile exists only for small average densities in the slit. For example, at $T = 2.0$K, this range is $0.005 < \rho^*_{av} < 0.02$. In Fig. 34, three stable asymmetric density profiles at $T = 2.0$ K are presented for $\rho^*_{av} = 0.007$, 0.01, and 0.014. As for the Lennard-Jones potential, the number of oscillations near the wall and their amplitude increase with increasing average density.

For $\rho^*_{av} \geq 0.02$, the asymmetric solution still exists and is similar to that presented in Fig. 32a. However the corresponding FDD has in this case a higher free energy than the symmetric solution and, therefore, is metastable. This feature constitutes the main difference between the SB for the CCZ potential when compared to the SB for the Lennard-Jones one. In the latter case, when an asymmetric profile exists, it has always a lower free energy than the symmetric one.

In Fig. 35, the asymmetric profiles are plotted for $\rho^*_{av} = 0.10$ for the CCZ (dashed line) and Lennard-Jones (solid line) potentials. One can see that the density of the fluid near the wall is larger for the CCZ potential than for the Lennard-Jones one because the former potential has a higher attractive strength than the latter. All other features, such as the number and location of the density oscillations and the behavior far from the surface are the same for both potentials.

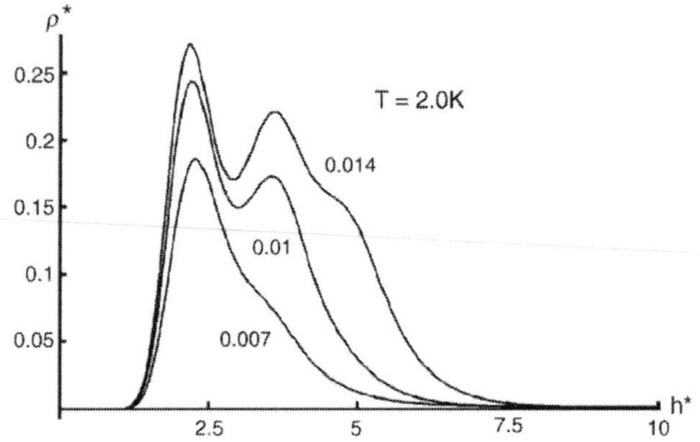

FIG. 34. (Reprinted from Ref. [45] with permission) Stable asymmetric fluid density profiles for the CCZ potential of the fluid-solid interactions. The profiles are plotted at various average densities $\rho^*_{av} = \rho_{av}\sigma^3_{ff}$ indicated by the numbers at each of the curves. Only the left-hand side of the profiles is presented. The fluid density in the right hand side has approximately the same value as at $h^* = 10$. $T = 2.0$ K, $\rho^* = \rho\sigma^3_{ff}$.

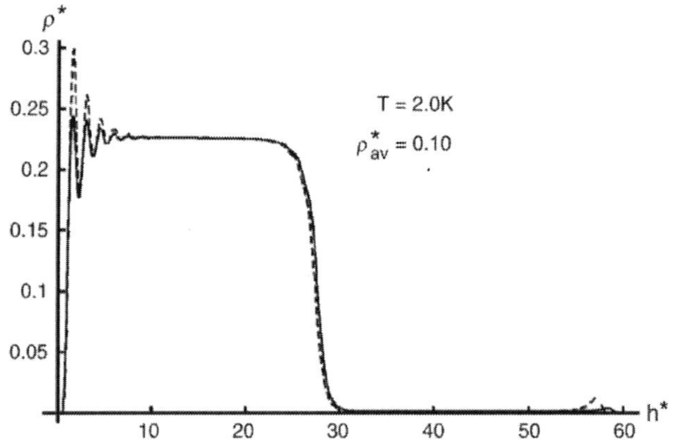

FIG. 35. (Reprinted from Ref. [45] with permission) Stable asymmetric fluid density profiles for the Lennard-Jones (solid line) and CCZ (dashed line) potentials of fluid-solid interactions plotted at $T = 2.0$ K for $\rho^*_{av} = 0.10$. $\rho^* = \rho\sigma^3_{ff}$.

REFERENCES

[1] Frauendorf S. Rev Mod Phys 2001;73:463.
[2] Yannouleas C, Landman U. Rep Prog Phys 2007;70:2067.
[3] Quigg C. Rep Prog Phys 2007;70:1019.
[4] Poniewierski A, Samborski A. Pol J Chem 2001;75:463.
[5] Mikami K, Yamanaka M. Chem Rev 2003;103:3369.
[6] Weissbuch I, Leiserowitz L, Lahav M. Prebiotic chemistry: from simple amphiphiles to protocell models vol. 259; 2005. p. 123.
[7] Terenziani F, Painelli A, Katan C, Charlot M, Blanchard-Desce M. J Am Chem Soc 2006;128:15742.
[8] Sorkin A, Iron MA, Truhlar DG. J Chem Theory Comput 2008;4:307.
[9] Goldanskii VI, Kuzmin VV. Uspekhi Fizicheskikh Nauk 1989;157:3.
[10] Avetisov VA, Goldanskii VI. Uspekhi Fizicheskikh Nauk 1996;166:873.

[11] Palmer AR. Science 2004;306:828.
[12] Crawford JD, Knobloch E. Annu Rev Fluid Mech 1991;23:341.
[13] Shtern V, Hussain F. Annu Rev Fluid Mech 1999;31:537.
[14] Drikakis D, Grinstein F, Youngs D. Prog Aerosp Sci 2005;41:609.
[15] Kondepudi DK, Sabanayagam C. Chem Phys Lett 1994;217:364.
[16] Metaxas D, Weinberg EJ. Phys Rev D 1996;53:836.
[17] Blaak R, Auer S, Frenkel D, Lowen H. J Phys C 2004;16:S3873.
[18] Viedma C. J Cryst Growth 2004;261:118.
[19] Vega DA, Gomez LR. Phys Rev E 2009;79:051607.
[20] Dagnino D, Barberan N, Lewenstein M. Dalibard. J Nat Phys 2009;5:431.
[21] Lowen H. Phys Rep Rev Section Phys Lett 1994;237:249.
[22] Rosenfeld Y. J Phys C 1996;8:L795.
[23] Hernando JA, Blum L. Physica A 1996;232:74.
[24] Singh SL, Singh Y. Epl 2009;88:6.
[25] Kornev K, Shingareva I, Neimark A. Adv Colloid Interface Sci 2002;96:143.
[26] Vishnyakov A, Neimark AV. J Chem Phys 2003;119:9755.
[27] Lebrero M, Bikiel D, Elola M, Estrin D, Roitberg A. J Chem Phys 2002;117:2718.
[28] Phuong N, Schmid F. J Chem Phys 2003;119:1214.
[29] Shundyak K, van Roij R. Europhys Lett 2006;74:1039.
[30] Gross M. Phys Rev E 1998;58:6124.
[31] Pickett GT, Gross M, Okuyama H. Phys Rev Lett 2000;85:3652.
[32] Kohl PB, Patrick DL. J Phys Chem B 2001;105:8203.
[33] Kondepudi DK, Asakura K. Acc Chem Res 2001;34:946.
[34] Gridnev ID. Chem Lett 2006;35:148.
[35] Walba DM. Korblova E, Huang CC, Shao RF, Nakata M, Clark NA. J Am Chem Soc 2006;128:5318.
[36] Palto SP, Mottram NJ, Osipov MA. Phys Rev E 2007;75:061707.
[37] Longa L, Pajak G, Wydro T. Phys Rev E 2009;79:040701.
[38] Guillemot L, Bobrov K. Phys Rev B 2009;79:201406.
[39] Merkel M, Lowen H. Phys Rev E 1996;54:6623.
[40] Berim G, Ruckenstein E. J Phys Chem B 2007;111:2514 (Section 2.1 of this volume).
[41] Berim G, Ruckenstein E. J Chem Phys 2007;126:124503 (Section 2.2 of this volume).
[42] Berim G, Ruckenstein E. J Phem Chem B 2007;111:12823 (Section 2.3 of this volume).
[43] Berim G, Ruckenstein E. J Chem Phys 2008;128:024704 (Section 2.4 of this volume).
[44] Berim G, Ruckenstein E. J Chem Phys 2008;128:134713 (Section 2.5 of this volume).
[45] Berim G, Ruckenstein E. J Chem Phys 2008;131:1. 184707 (Section 2.6 of this volume).
[46] Pederiva F, Ferrante A, Fantoni S, Reatto L. Phys Rev B 1995;52:7564.
[47] Xu JL, Selinger RLB, Selinger JV, Shashidhar R. J Chem Phys 2001;115:4333.
[48] Germano G, Schmid F. J Chem Phys 2005;123:214703.
[49] Rzysko W, Patrykiejew A, Sokolowski S. Phys Rev E 2008;77:061602.
[50] Zhang FS, Lynden-Bell RM. Phys Rev Lett 2003;90:185505.
[51] Ebner C, Saam WF. Phys Rev Lett 1977;38:1486.
[52] Evans R, Tarazona P. Phys Rev A 1983;28:1864.
[53] Tarazona P, Evans R. Mol Phys 1984;52:847.
[54] Tarazona P. Phys Rev A 1985;31:2672–9 ibid 1985; 31: 2672.
[55] Singh Y. Phys Rep Rev Section Phys Lett 1991;207:351.
[56] Sinclair PE, Catlow CRA. J Chem Soc Faraday Trans 1997;93:333.
[57] Bieker T, Dietrich S. Physica A 1998;252:85.
[58] Gelb LD, Gubbins KE, Radhakrishnan R, Sliwinska-Bartkowiak M. Rep Prog Phys 1999;62:1573.
[59] Ravikovitch PI, Vishnyakov A, Neimark AV. Phys Rev E 2001;64:011602.
[60] Truskett TM, Debenedetti PG, Torquato S. J Chem Phys 2001;114:2401.
[61] Do DD, Do HD. Adsorp Sci Technol 2003;21:389.
[62] Brader JM, Evans R, Schmidt M. Mol Phys 2003;101:3349.
[63] Corminboeuf C, Tran F, Weber J. J Mol Struct (Thoechem) 2006;762:1.
[64] Alba-Simionesco C, Coasne B, Dosseh G, Dudziak G, Gubbins KE, Radhakrishnan R, et al. J Phys Condens Matter 2006;18:R15.
[65] Kaplan WD, Kauffmann Y. Annu Rev Mater Res 2006;36:1.
[66] Zhou YH, Lv PH, Wang GC. J Mol Catal A Chem 2006;258:203.
[67] Bruch LW, Diehl RD, Venables JA. Rev Mod Phys 2007;79:1381.

[68] Schoen M. Phys Chem Chem Phys 2008;10:223.
[69] Gelb LD. Mrs Bull 2009;34:592.
[70] Tarazona P, Marconi UMB, Evans R. Mol Phys 1987;60:573.
[71] Carnahan NF, Starling KE. J Chem Phys 1969;51:635.
[72] Rosenfeld Y. Phys Rev Lett 1989;63:980.
[73] Kierlik E, Rosinberg ML. Phys Rev A 1990;42:3382.
[74] Phan S, Kierlik E, Rosinberg ML, Bildstein B, Kahl G. Phys Rev E 1993;48:618.
[75] Gonzalez A, White JA, Roman FL, Velasco S, Evans R. Phys Rev Lett 1997;79:2466.
[76] Gonzalez A, White JA, Roman FL, Evans R. J Chem Phys 1998;109:3637.
[77] Zeng XC, Oxtoby DW, Rosenfeld Y. Phys Rev A 1991;43:2064.
[78] Kierlik E, Rosinberg ML. Phys Rev A 1991;44:5025.
[79] Kierlik E, Rosinberg M, Finn JE, Monson PA. Mol Phys 1992;75:1435.
[80] Rosenfeld Y. J Chem Phys 1993;98:8126.
[81] Rosenfeld Y. Phys Rev Lett 1994;72:3831.
[82] Ancilotto F, Faccin F, Toigo F. Phys Rev B 2000;62:17035.
[83] White JA, Velasco S. Phys Rev E 2000;62:4427.
[84] White JA, Gonzalez A, Roman FL, Velasco S. Phys Rev Lett 2000;84:1220.
[85] Talanquer V, Oxtoby DW. J Chem Phys 1996;104:1483.
[86] Talanquer V, Oxtoby DW. J Chem Phys 2001;114:2793.
[87] Neimark AV, Ravikovitch PI, Vishnyakov A. Phys Rev E 2002;65:031505.
[88] Neimark AV, Ravikovitch PI, Vishnyakov A. J Phys Condens Matter 2003;15:347.
[89] Ustinov EA, Do DD. J Chem Phys 2004;120:9769.
[90] Berim G, Ruckenstein E. J Chem Phys 2006;125:164717.
[91] Nilson RH, Griffiths SK. J Chem Phys 1999;111:4281.
[92] Dalfovo F, Lastri A, Pricaupenko L, Stringari S, Treiner J. Phys Rev B 1995;52:1193.
[93] Szybisz L, Sartarelli SA. J Chem Phys 2008;128:124702.
[94] Nijmeijer MJP, Bruin C, Bakker AF, Vanleeuwen JMJ. Phys Rev A 1990;42:6052.
[95] Frenkel D, Smith B, editors. Understanding molecular simulations. 2nd ed. NY: Academic Press; 2001. p. 664.
[96] Errington J. unpublished.
[97] Ustinov EA, Do DD. J Phys Chem B 2005;109:11653.
[98] Roth R, Dietrich S. Phys Rev E 2000;62:6926.
[99] Yang XN, DingJQ. J Chem Phys 2004;121:7449.
[100] Pizio O, Patrykiejew A, Sokolowski S. Mol Phys 2001;99:57.
[101] Bucior K, Patrykiejew A, Pizio O, Sokolowski S. J Colloid Interface Sci 2003;259:209.
[102] Bucior K. Colloids Surf A 2004;243:105.
[103] Hemming CJ, Patey GN. J Phys Chem B 2006;110:3764.
[104] SikkenkJH, Indekeu JO, Vanleeuwen JMJ, Vossnack EO. Phys Rev Lett 1987;59:98.
[105] Sikkenk JH, Indekeu JO, Vanleeuwen JMJ, Vossnack EO, Bakker AF. J Stat Phys 1988;52:23.
[106] Nijmeijer MJP, Bruin C, Bakker AF, Vanleeuwen JMJ. Physica A 1989;160:166.
[107] Tang JZ, Harris JG. J Chem Phys 1995;103:8201.
[108] Ingebrigtsen T, Toxvaerd S. J Phys Chem C 2007;111:8518.
[109] Cheng E, Cole MW, Saam WF, Treiner J. Phys Rev Lett 1991;67:1007.
[110] Cheng E, Cole MW, Saam WF, Treiner J. Phys Rev B 1992;46:13967.
[111] Szybisz L. Phys Rev B 2000;62:12381.
[112] Chizmeshya A, Cole MW, Zaremba E.J Low Temp Phys 1998;110:677.

3 Macroscopic Drop on a Solid Surface

Eli Ruckenstein and Gersh Berim

INTRODUCTION TO CHAPTER 3

In this chapter, three microscopic approaches to the treatment of macroscopic drops on smooth or rough, planar or curved, solid surfaces, based on fluid–fluid and fluid–solid interaction potentials are discussed.

The first approach (Sec. 3.1) is based on the Born-Green- Yvon integral equations for the density distribution function, and provides an approximate integral equation for the profile of the surface of the drop. From numerical solutions and analytical solutions for limiting cases one concludes that there is a rapid spatial variation of shape near the leading edge and that for large drops the measured macroscopic wetting angle is reached at a distance of about 20 to 40 Å from the leading edge. For very small drops the wetting angle is weakly size dependent. A condition for drop stability is established, which if not satisfied, the liquid will spread over the surface of the solid.

The second approach (Secs. 3.2–3.7) employs the minimization of the total potential energy of a drop by assuming that the drop has a well-defined profile and a constant liquid density in its entire volume with the exception of the monolayer nearest to the surface where the density has a different value. As a result, a differential equation for the drop profile as well as the necessary boundary conditions are derived which involve the parameters of the interaction potentials and do not contain such macroscopic characteristics as the surface tensions. As a consequence, the macroscopic and microscopic contact angles which the drop profile makes with the surface can be calculated. The former angle is obtained via the extrapolation of the circular part of the drop profile valid at some distance from the surface up to the solid surface. The microscopic angle is formed at the intersection of the real profile (which is not circular near the surface) with the surface.

The third approach (Sec. 3.8) is based on the nonlocal density functional theory which allows to determine the liquid-vapor, liquid-solid, and vapor-solid surface tensions. Using these results and the classical Young equation one can find the contact angle for macroscopic drop in wide interval of liquid-solid interactions and temperature.

3.1 The Wetting Angle of Very Small and Large Drops*

Eli Ruckenstein and P. S. Lee
State University of New York at Buffalo, Faculty of Engineering and Applied Sciences, Buffalo, New York 14214, USA

Received 28 May 1975; manuscript received in final form 14 July 1975

1. INTRODUCTION

A droplet of liquid spreads on a solid surface until an equilibruim state is reached. The equilibrium state is characterized by the wetting (contact) angle, α, at the contact line of the liquid and solid phases. Young's equation

$$\gamma^{lg}\cos\alpha = \gamma^{gs} - \gamma^{ls}; \tag{1}$$

developed from thermodynamics, relates this angle to the surface tensions, γ^{ij}, of the interfaces separating the phases i and j ($i, j = l$, g, s refer to the liquid, the gas, and the solid phases respectively).

Recently Miller and Ruckenstein [1] obtained a relationship between the wetting angle and the parameters characterizing the intermolecular interactions by calculating the tangential force component from the interaction potential between the molecules and by observing that at thermodynamic equilibruim this force component must vanish at every point on the liquid–gas interface*.

The advantage of this treatment lies in the direct relation established between the wetting angle and the parameters characterizing the intermolecular interactions; the use of the surface tensions is avoided. However, the treatment completely neglects the effect of the entropy and thus is strictly valid only at 0 K. On the other hand, since the wetting angle is weakly dependent upon temperature, this kind of approach is not unreasonable.

The traditional thermodynamic approach which leads to Young's equation is valid for length scales which are large compared to the range of the interaction forces. The profile of a large droplet near the leading edge in a region whose dimension is smaller than the range of the interaction forces between one atom and all the other atoms, and the profile of a droplet whose size is smaller than the range of the interaction forces, cannot be obtained from traditional thermodynamics. They can be obtained, however, from statistical mechanical considerations.

The goal of the paper is to use this kind of approach to obtain information about the profile of a droplet, in particular near the leading edge, and about the effect of the size of very small droplets upon their profile. It is shown that the experimentally determined macroscopic wetting angle is reached only at a distance of about 20 to 40 Å from the leading edge and that in the region of the leading edge a rapid change in the angle between the liquid–gas interface and the horizontal occurs. Furthermore, for small enough droplets, the angle is size dependent.

---

* *Surface Science*. 52 (1975) 298–310. Republished with permission.

* Thermodynamic considerations [2] have shown that Young's equation is a consequence of the vanishing of the tangential force acting per unit area of the liquid–gas interface.

2. BASIC EQUATIONS

For a non-homogeneous system, Born–Green–Yvon established, on the basis of statistical mechanics, the following integral equation relating the one-body density distribution function $\rho^{(1)}$ and the two-body density distribution function $\rho^{(2)}$ [3,4]:

$$kT\,\nabla_r\rho^{(1)}(\boldsymbol{r})+\iiint_V \mathrm{d}\boldsymbol{r}'\rho^{(2)}(\boldsymbol{r},\boldsymbol{r}')\nabla_r\phi(\boldsymbol{r},\boldsymbol{r}')=0. \tag{2}$$

In eq. (2) k is the Boltzmann constant, T the absolute temperature, $\rho^{(1)}(\boldsymbol{r})\mathrm{d}\boldsymbol{r}$ is the probability that one of the molecules of the system will be found in the volume element $\mathrm{d}\boldsymbol{r}$ at $\boldsymbol{r}$, while $\rho^{(2)}(\boldsymbol{r},\boldsymbol{r}')\mathrm{d}\boldsymbol{r}\,\mathrm{d}\boldsymbol{r}'$ is the probability that one molecule of the system will be found in $\mathrm{d}\boldsymbol{r}$ at $\boldsymbol{r}$ and another in $\mathrm{d}\boldsymbol{r}'$ at $\boldsymbol{r}'$, $\phi(\boldsymbol{r},\boldsymbol{r}')$ is the interaction potential of a molecule at $\boldsymbol{r}$ with that at $\boldsymbol{r}'$ and the volume of integration $\boldsymbol{V}$ includes the complete system.

For a droplet sitting on a solid surface, the integral term in eq. (2) can be split into two terms. One of them expresses the interactions between solid and the molecules of the fluid at $\boldsymbol{r}$, and the other gives the interactions between the molecules of the fluid at $\boldsymbol{r}$ and all the other molecules of the fluid. Writing

$$\rho^{(2)}(\boldsymbol{r},\boldsymbol{r}')=n(\boldsymbol{r})n(\boldsymbol{r}')\,\mathrm{g}(\boldsymbol{r},\boldsymbol{r}'), \tag{3}$$

where $g(\boldsymbol{r},\boldsymbol{r}')$ is the radial distribution function of the fluid, and $\rho^{(1)}(\boldsymbol{r})=n(\boldsymbol{r})$, where n is the number density of the molecules at $\boldsymbol{r}$, one obtains:

$$\begin{aligned}\iiint_V \mathrm{d}\boldsymbol{r}'\rho^{(2)}(\boldsymbol{r},\boldsymbol{r}')\nabla_r\phi(\boldsymbol{r},\boldsymbol{r}') &= n(\boldsymbol{r})\iiint_{V^{\mathrm{s}}}\mathrm{d}\boldsymbol{r}'\mathrm{g}(\boldsymbol{r},\boldsymbol{r}')n(\boldsymbol{r}')\nabla_r\phi(\boldsymbol{r},\boldsymbol{r}')\\ &\quad+n(\boldsymbol{r})\iiint_{V^{\mathrm{f}}}\mathrm{d}\boldsymbol{r}'\mathrm{g}(\boldsymbol{r},\boldsymbol{r}')n(\boldsymbol{r}')\nabla_r\phi(\boldsymbol{r},\boldsymbol{r}'),\end{aligned} \tag{4}$$

where the integration volumes V^{s} and V^{f} refer to those of the solid and fluid respectively.

The integral in the first term of the right-hand side can be interpreted as an external field $\nabla_r\psi(\boldsymbol{r})$ acting upon the fluid. Consequently, eq. (2) becomes

$$kT\,\nabla_r n(\boldsymbol{r})+n(\boldsymbol{r})\nabla_r\psi(\boldsymbol{r})+\iiint_{V^{\mathrm{f}}}\mathrm{d}\boldsymbol{r}'n(\boldsymbol{r})\mathrm{g}(\boldsymbol{r},\boldsymbol{r}')n(\boldsymbol{r}')\nabla_r\phi(\boldsymbol{r},\boldsymbol{r}')=0. \tag{5}$$

Eq. (5) can be formally solved to yield for n the non-linear integral equation:

$$n(\boldsymbol{r})=C\exp\left\{-\frac{1}{kT}\left[\psi(\boldsymbol{r})+\iiint_{V^f}\mathrm{d}\boldsymbol{r}'n(\boldsymbol{r}')\int_\infty^r \mathrm{d}\xi\cdot\mathrm{g}(\xi,\boldsymbol{r}')\nabla_\xi\phi(\xi,\boldsymbol{r}')\right]\right\}, \tag{6}$$

where C is a constant. This equation requires knowledge of the radial distribution function $\mathrm{g}(\xi,\boldsymbol{r}')$ to be solved for $n(\boldsymbol{r})$. Nevertheless, some qualitative information concerning the density distribution near the liquid–gas interface can be obtained. Taking the radial distribution function g as unity, accounting in the integral for the fact that the density in the liquid is much larger than that in the gas and that the effective range of the interaction between two atoms is only a few angstrom, eq. (6) shows that the density decreases from that of the liquid n_l to that of the gas

n_g within a few angstrom. Consequently, as a good approximation, one can use the bulk values n_l and n_g for each of the phases.

On the liquid–gas interface, eq. (5) becomes

$$kT\frac{\partial}{\partial r_t}\ln n(\boldsymbol{r})+\frac{\partial}{\partial r_t}\psi(\boldsymbol{r})+\iiint_{V^f} d\boldsymbol{r}' n(\boldsymbol{r}')\,g(\boldsymbol{r},\boldsymbol{r}')\frac{\partial}{\partial r_t}\phi(\boldsymbol{r},\boldsymbol{r}')=0, \tag{5a}$$

and

$$kT\frac{\partial}{\partial r_n}\ln n(\boldsymbol{r})+\frac{\partial}{\partial r_n}\psi(\boldsymbol{r})+\iiint_{V^f} d\boldsymbol{r}' n(\boldsymbol{r}')\,g(\boldsymbol{r},\boldsymbol{r}')\frac{\partial}{\partial r_n}\phi(\boldsymbol{r},\boldsymbol{r}')=0, \tag{5b}$$

where $\partial/\partial r_t$ indicates a derivative along the tangent to the surface, and $\partial/\partial r_n$ indicates a derivative along the normal to the surface.

Eqs. (5), (5a) and (5b) are the basic equations which determine both the surface profile of a liquid droplet and the density distribution.

For the sake of simplicity the emphasis will be put on the profile of the droplet, assuming that the variations of the density of the liquid along the tangential direction are very small. Eq. (5a) in which the density n is assumed constant is used in the computation because $(\partial/\partial r_t)\ln n(\boldsymbol{r}) \approx 0$ along the interface, while $(\partial/\partial r_n)\ln n(\boldsymbol{r})$ in eq. (5b) has large values near the interface which cannot be predicted easily.

The resulting equation can be rederived in a more simple, intuitive manner, which shows that in this approximation only the interaction energy at 0 K is taken into account and that the entropic effect is neglected.

The interaction potential, ϕ_{ij}, between two molecules of substances *i* and *j* separated by a distance $\boldsymbol{r}$ is taken in the form

$$\phi_{ij}=\begin{cases}-\beta_{ij}/r^q, & r\ge a_{ij},\\ \infty & r<a_{ij},\end{cases} \tag{7}$$

where a_{ij} is a parameter characterizing the size of the repulsive core of the interaction between the molecules of substances *i* and *j*, β_{ij} is a parameter characterizing the interaction strength and the exponent *q* is a constant. The subscripts $i, j = l$, s, g denote the liquid, the solid, and the gas phases respectively.

The force $\boldsymbol{f}_{ij}$ exerted on a molecule of phase *i* by a molecule of phase *j* is given by

$$f_{ij}=-\nabla_i\phi_{ij}=\begin{cases}\infty & r\le a_{ij},\\ -q\dfrac{\beta_{ij}}{r^{q+2}}\left(\boldsymbol{r}_i-\boldsymbol{r}_j\right), & r>a_{ij},\end{cases} \tag{8}$$

where $\boldsymbol{r}_i$ and $\boldsymbol{r}_j$ are the position vectors of the molecules *i* and *j* respectively.

To compute the force acting on a liquid molecule on the surface, Ω^{lg}, of the liquid–gas interface, any arbitrary point on it may be chosen. For a planar solid surface it is convenient, because of symmetry, to choose $Q(\chi_q, 0, z_Q)$ (see fig. 1).

Because of the low density of the gas phase, the interaction of the liquid molecule with the gas molecules is neglected. Furthermore, the effect of gravitation is neglected since the drops are small. Consequently, only the interactions between the molecule of the liquid at Q with the molecules of the solid and with the remaining molecules of the liquid are taken into account.

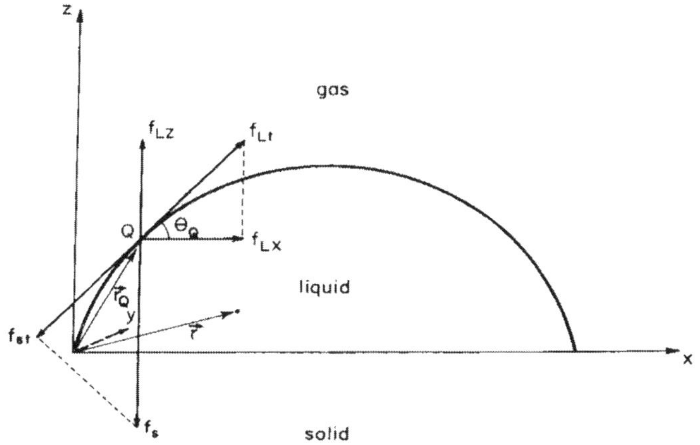

FIG. 1. A liquid drop on a plane solid surface. The point Q lies on x–z plane, and at $\boldsymbol{r}_{\mathrm{Q}} \equiv \left(x_{\mathrm{Q}},0,z_{\mathrm{Q}}\right)$.

For symmetry reasons, the force f_s on a molecule at Q, due to its interaction with molecules of the solid, is normal to the solid surface. Furthermore, f_s has a tangential force component $f_{st} = f_s \sin \theta_Q$ along the liquid–gas interface directed toward the leading edge of the drop, θ_Q being the angle between the tangent to the liquid–gas interface at Q and the x–axis (see fig. 1).

The force f_l acting on a molecule at Q, resulting from its interaction with the liquid molecules, is directed into the liquid phase and has, along the liquid–gas interface, the tangential component $f_{lt} = f_{lx} \cos\theta_Q + f_{lz} \sin\theta_Q$ (fig. 1).

Because, at thermodynamic equilibrium, the net tangential force along the liquid–gas interface has to vanish, one obtains

$$\begin{aligned} & n_s\beta_{ls} \sin\theta_Q \int_{V^s} \mathrm{d}x\,\mathrm{d}y\,\mathrm{d}z \frac{\left(z-z_Q\right)}{\left|\boldsymbol{r}-\boldsymbol{r}_Q\right|^{q+2}} \\ & + n_l\beta_{ll} \int_{V^l} \mathrm{d}x\,\mathrm{d}y\,\mathrm{d}z \frac{(x-x_Q)\cos\theta_Q + (z-z_Q)\sin\theta_Q}{\left|\boldsymbol{r}-\boldsymbol{r}_Q\right|^{q+2}} = 0, \end{aligned} \tag{9}$$

where $\boldsymbol{r} \equiv (x, y, z)$ is the position vector of a molecule, of liquid or solid, $\boldsymbol{r}_Q \equiv (x_q, 0, z_q)$ is the position vector of the point Q, n_l and n_s the number density of the liquid and solid phases respectively, and V^l and V^s mean that the integrations are performed over the volume of the liquid or the solid respectively.

Eq. (9) contains two integrals. The first integral is the force exerted on a liquid molecule at Q resulting from its interaction with the solid molecules, whereas the second integral is the force resulting from its interaction with the liquid molecules.

For concreteness, in what follows, the retardation effect is neglected and q is taken to be 6. For a plane. smooth, and homogeneous solid surface denoted by $z = -a_{ls}$, the first integral leads to

$$\int_{-\infty}^{\infty} \mathrm{d}x \int_{-\infty}^{\infty} \mathrm{d}y \int_{-\infty}^{-a_{ls}} \mathrm{d}z \frac{z-z_Q}{\left|\boldsymbol{r}-\boldsymbol{r}_Q\right|^8} = -\frac{\pi}{12} \frac{1}{\left(z_Q+a_{ls}\right)^4}, \tag{10}$$

where a_{ls} represents the size of the repulsive core of the liquid–solid interaction.

To evaluate the second integral it is convenient to divide the domain of integration into two regions. The first region is bounded by the liquid–gas interface defined by

$$W_1(x,y,z)=0, \tag{11a}$$

and a conical surface defined by

$$W_2(x,y,z)=0, \tag{11b}$$

which has its apex at Q and passes through the liquid–solid contact line. The second region includes the rest of the volume of the liquid and is bounded by the surface defined by (11b) and the plane

$$z=0. \tag{11c}$$

Consequently,

$$\int_{V^l} dx\,dy\,dz \frac{(x-x_Q)\cos\theta_Q+(z-z_Q)\sin\theta_Q}{|\boldsymbol{r}-\boldsymbol{r}_Q|^8}$$

$$=\int\int_{W_2=0}^{W_1=0}\int d\varphi \sin\theta\, d\theta\, R^2\, dR \frac{1}{R^7}(\sin\theta\cos\theta_Q\cos\varphi+\cos\theta\sin\theta_Q)$$

$$+\int\int_{z=0}^{W_2=0}\int d\varphi \sin\theta\, d\theta\, R^2\, dR \frac{\sin\theta\cos\theta_Q\cos\varphi+\cos\theta\sin\theta_Q}{R^7}, \tag{12}$$

where $R\equiv(\boldsymbol{r}-\boldsymbol{r}_Q), R\equiv|\boldsymbol{R}|$, and

$$x-x_Q\equiv R\sin\theta\cos\varphi,\quad y-y_Q=y=R\sin\theta\cos\varphi,\quad z-z_Q\equiv R\cos\theta \tag{13}$$

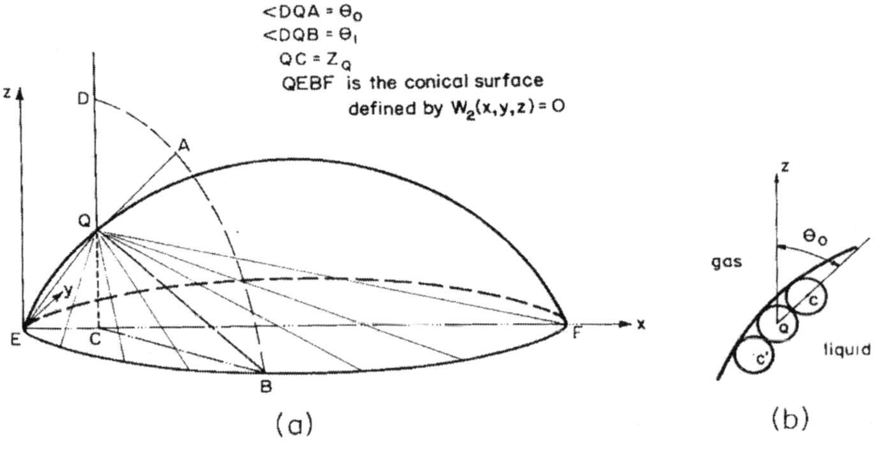

FIG. 2. Definitions of θ_0, θ_1 and the conical surface $W_2(x, y, z)$.

Using the coordinates R, θ and φ as defined by (13), the integration limits of (12) can be found using (11a), (11b) and (11c). Consequently, eq. (12) leads to

$$\int_0^{2\pi} d\varphi \int_{\theta_0}^{\theta_1} \sin\theta\, d\theta(\sin\theta\cos\theta_Q\cos\varphi+\cos\theta\sin\theta_Q)\int_{a_{ll}}^{R_1}\frac{dR}{R^5}$$

$$+\int_0^{2\pi} d\varphi \int_{\theta_1}^{\pi} \sin\theta\, d\theta\,(\sin\theta\cos\theta_Q\cos\varphi+\cos\theta\sin\theta_Q)\int_{a_{ll}}^{-z_Q/\cos\theta}\frac{dR}{R^5}$$

$$=\frac{1}{4a_{ll}^4}\int_0^{2\pi} d\varphi \int_{\theta_0(\varphi,\theta_Q)}^{\pi} \sin\theta\, d\theta\,(\sin\theta\cos\theta_Q\cos\varphi+\cos\theta\sin\theta_Q)$$

$$-\frac{1}{4}\left\{\int_0^{2\pi} d\varphi \int_{\theta_0(\varphi,\theta_Q)}^{\theta_1(\varphi,\theta_Q)} \sin\theta\, d\theta\frac{(\sin\theta\cos\theta_Q\cos\varphi+\cos\theta\sin\theta_Q)}{R_1^4(\theta,\varphi,\boldsymbol{r}_Q)}\right.$$

$$\left.+\int_0^{2\pi} d\varphi \int_{\theta_1(\varphi,\theta_Q)}^{\pi} \sin\theta\, d\theta\frac{(\sin\theta\cos\theta_Q\cos\varphi+\cos\theta\sin\theta_Q)\cos^4\theta}{z_Q^4}\right\}, \quad (14)$$

where $R_1(\theta,\varphi,\boldsymbol{r}_Q)$ is determined from the solution of

$$W_1(x,y,z)=W_1(R_1,\theta,\varphi,\boldsymbol{r}_Q)=0, \quad (11\ a')$$

θ_1 is determined by

$$W_2(x,y,z)=W_2(\theta_1,\varphi,\boldsymbol{r}_Q)=0, \quad (11\ b')$$

and θ_0 by

$$W_1(x,y,z)=W_1(R=a_{ll},\theta_0,\varphi,\boldsymbol{r}_Q)=0, \quad (11d)$$

as shown in fig. 2.

The first integral on the right-hand side of eq. (14) is zero. It will be shown below that the second integral is negligibly small for large droplets and introduces a size effect for small droplets. The third integral is important for both large and small droplets. Consequently, the right hand side of eq. (14) becomes

$$-\frac{1}{4}\int_0^{2\pi} d\varphi\left\{\int_{\theta_0(\varphi,\theta_Q)}^{\theta_1(\varphi,\theta_Q)} \sin\theta\, d\theta(\sin\theta_Q\cos\theta+\cos\theta_Q\sin\theta\cos\varphi)\frac{1}{R_1^4(\theta,\varphi,\theta_Q)}\right.$$

$$\left.+\frac{1}{z_Q^4}\int_{\theta_1(\varphi,\theta_Q)}^{\pi} \sin\theta\, d\theta\cos^4\theta(\sin\theta_Q\cos\theta+\cos\theta_Q\sin\theta\cos\varphi)\right\}. \quad (15)$$

Combining (9), (10), and (15) one obtains

$$\frac{3}{2}n_l\beta_u\int_0^{2\pi}\mathrm{d}\varphi\left\{\int_{\theta_0(\varphi,\theta_Q)}^{\theta_1(\varphi,\theta_Q)}\sin\theta\,\mathrm{d}\theta\frac{\sin\theta\cos\theta_Q+\cos\theta\sin\theta_Q\cos\varphi}{R_1^4(\theta,\varphi,\boldsymbol{r}_Q)}\right.$$

$$\left.+\frac{1}{z_Q^4}\int_{\theta_1(\varphi,\theta_Q)}^{\pi}\sin\theta\,\mathrm{d}\theta\cos^4\theta\left(\sin\theta\cos\theta_Q+\cos\theta\sin\theta_Q\cos\varphi\right)\right\}$$

$$+\frac{\pi}{2}n_s\,\beta_{ls}\frac{\sin\theta_Q}{\left(z_Q+a_{ls}\right)^4}=0, \tag{16}$$

subject to

$$V^l=\int_0^{2\pi}\mathrm{d}\varphi\left\{\int_{\theta_0}^{\theta_1}\sin\theta\,\mathrm{d}\theta\int_0^{R_1}R^2\mathrm{d}R+\int_{\theta_1}^{\pi}\sin\theta\,\mathrm{d}\theta\int_0^{-z_Q/\cos\theta}R^2\mathrm{d}R\right\}$$

= constant, where V^l is the volume of the drop.

Since in eq. (16) the quantity $R_1(\theta,\varphi,\boldsymbol{r}_Q)$ is unknown and has to be determined from the equation itself, eq. (16) is, in fact, an integral equation. This integral equation can be solved for $R_1(\theta,\varphi,\boldsymbol{r}_Q)$, for a given solid plane $z=-a_{ls}$, to determine the profile of the drop.

3. APPROXIMATE SOLUTION OF EQ. (16)

A solution to the integral eq. (16) is difficult to find. It is, however, easier to obtain information about the angle θ_Q near the leading edge. In this region, θ_1 is practically a constant ($\approx \pi/2$) for macroscopic droplets and depends weakly on the shape of the drop for small droplets (but not smaller than about five to ten molecular diameters). Thus, for the leading edge region, the second integral in eq. (16) is practically independent of the shape. The first integral of (16) depends, because of R_1^4, upon the complete shape of the drop. However, the first integral is small compared to the second because for most of the integration range, $R_1 \gg z_Q$. Consequently, any reasonably chosen profile may suffice for the evaluation of the first integral. Large droplets have a spherical shape. For this reason, as a first approximation, for evaluating the integrals, the shape of the droplet is described as a section of a sphere of radius A. Eq. (11 a) which represents the liquid–gas interface, leads to

$$R_1=2A\left(\sin\theta\cos\varphi\sin\theta_Q-\cos\theta\cos\theta_Q\right). \tag{17}$$

Substituting (17) into (16) gives

$$\frac{3}{2}n_l\beta_u\left\{-\int_0^{2\pi}\mathrm{d}\varphi\int_{\theta_0(\varphi,\theta_Q)}^{\pi}\sin\theta\,\mathrm{d}\theta\,(\sin\theta_Q\cos\theta+\cos\theta_Q\sin\theta\cos\varphi)\frac{1}{a_{ll}^4}\right.$$

$$+\frac{1}{z_Q^4}\int_0^{2\pi}\mathrm{d}\varphi\int_{\theta_1(\varphi,\theta_Q)}^{\pi}\sin\theta\,\mathrm{d}\theta\,(\sin\theta_Q\cos\theta+\cos\theta_Q\sin\theta\cos\varphi)\cos^4\theta$$

$$+\frac{1}{(2A)^4}\int_0^{2\pi}\mathrm{d}\varphi\int_{\theta_0(\varphi,\theta_Q)}^{\theta_1(\varphi,\theta_Q)}\sin\theta\,\mathrm{d}\theta\frac{\sin\theta_Q\cos\theta+\cos\theta_Q\sin\theta\cos\varphi}{[\sin\theta\sin\theta_Q\cos\varphi-\cos\theta_Q\cos\theta]^4}\Bigg\}$$

$$+\frac{\pi}{2}n_s\beta_{ls}\frac{\sin\theta_Q}{\left(z_Q+a_{ls}\right)^4}=0, \tag{18}$$

where the angles θ_1 and θ_0 can be obtained from eqs. (11a′), (11b′), (11d) and (18) in the form

$$\cos^2\theta_1=\frac{1}{2\left(\cos^2\theta_Q+\sin^2\theta_Q\cos^2\varphi\right)}\left\{\sin^2\theta_Q\cos^2\varphi+\frac{z_Q}{A}\cos\theta_Q-\sin\theta_Q\cos\varphi\right.$$

$$\left.\times\left[\sin^2\theta_Q\cos^2\varphi+\frac{2z_Q}{A}\cos\theta_Q-\left(\frac{z_Q}{A}\right)^2\right]^{1/2}\right\}, \tag{19a}$$

and

$$\cos\theta_0=\frac{1}{\left(\cos^2\theta_Q+\sin^2\theta_Q\cos^2\varphi\right)}\left\{-\frac{a_{ll}}{2A}\cos\theta_Q\right.$$

$$\left.+\sin\theta_Q\cos\varphi\left[\left(\cos^2\theta_Q+\sin^2\theta_Q\cos^2\varphi\right)-\left(\frac{a_{ll}}{2A}\right)^2\right]^{1/2}\right\}. \tag{19b}$$

Neglecting terms in $(t_0/A)^2$ and those of higher order, eq. (18) gives, upon integration,

$$\frac{1}{t_0^4}\left\{\left[\frac{1}{4}\pi n_l\beta_{ll}\left(\csc^3\theta_Q+\frac{3}{2}\cot\theta_Q+\cot^3\theta_Q\right)-\frac{1}{2}\pi n_s\beta_{ls}K\csc^3\theta_Q\right]\right.$$

$$-\frac{3n_l\beta_{ll}t_0\left[1-\cos\alpha-\frac{1}{2}\cos\alpha\sin^2\alpha\right]^{1/3}}{2\left(\sin\theta_Q\right)\left(\frac{2}{3}\pi V^l\right)^{1/3}}\left[\frac{3}{32}\left(\sin^2\theta_Q\right)\left(1-3\sin^2\theta_Q\right)\right.$$

$$\left.\left.-\frac{1}{8}\left(\cos^2\theta_Q\right)\left(1-\frac{3}{4}\sin^2\theta_Q\right)+\frac{1}{16}\cos\theta_Q+\frac{1}{16}\sin^2\theta_Q\right]\right\}=0, \tag{20a}$$

where

$$t_0=z_Q/\sin\theta_Q\text{ and }K=1/\left(1+\left(a_{ls}/z_Q\right)\right)^4. \tag{20b}$$

4. BEHAVIOR OF LARGE DROPS

Let us first consider a large macroscopic droplet and ask about the angles θ_Q near the leading edge. The smallest value of z_Q for which an angle has meaning corresponds to the second lowest layer of molecules of liquid, hence to $z_Q = a_{ll}$. Because $z_Q / (V^l)^{1/3} \ll 1$, the term containing this ratio can be neglected. Therefore, at the leading edge,

$$\frac{1}{4}\pi n_l \beta_{ll}\left(\csc^3\alpha_0 + \frac{3}{2}\cot\alpha_0 + \cot^3\alpha_0\right) - \frac{1}{2}\pi n_s \beta_{ls} K_0 \csc^2\alpha_0 = 0, \tag{22}$$

where $K_0 \equiv \left(1+\left(a_{ls}/a_{ll}\right)\right)^{-4}$, and α_0 is the equilibrium contact angle at the leading edge. When $a_{ls} \approx a_{ll}$, $K_0 = 1/16$. For values of z_Q larger than about $20a_{ls}$, $K - 1$ is small and eq. (20a) leads to the eq. (23) previously established by Miller and Ruckenstein [1]:

$$2n_s\beta_{ls} / n_l\beta_{ll} = \left(\csc^3\alpha + \frac{3}{2}\cot\alpha + \cot^3\alpha\right) / \csc^3\alpha \equiv F(\alpha). \tag{23}$$

Because $a_{ls} \sim 1$ Å, eqs. (22) and (23) show that, in a region of about $20 \div 40$ Å, the angle θ_Q decreases from α_0 to α. Since the thickness of this region is too small to be detected in a macroscopic experiment, α is the experimentally observed angle.

The rapid change of shape near the leading edge is caused by the repulsive forces. Indeed, the repulsive forces introduce in eq. (10) the repulsive core a_{ls} whose effect appears in the fundamental eq. (20a) through the parameter K. If the ratio z_Q/a_{ls} increases, K increases towards unity. Because the repulsive forces decrease the interaction forces with the solid, and because for $z_Q/a_{ls} \approx 20$ their effect is negligible, the angle near the leading edge changes rapidly from a large angle α_0 to a smaller asymptotic angle α. Fig. 3 shows that α_0 is larger and can be very much larger than α.

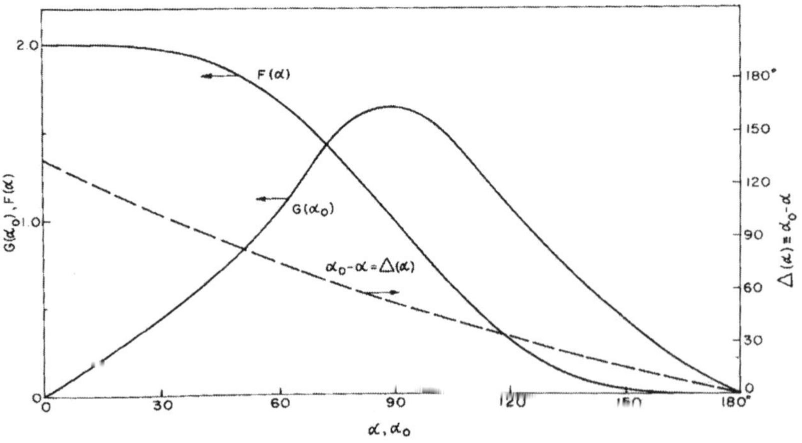

FIG. 3. Dependence of G(α_0), $F(\alpha)$, and $\alpha_0 - \alpha$ on α_0 or α.

5. THE SIZE EFFECT

Now consider the situation where the radius of the droplet is so small that terms of the order of z_q/A are no longer negligible, but terms of the order of $(z_Q/A)^2$ are still small. Assuming that the angle at the leading edge deviates from α_0 by the small amount $\delta\alpha$, eq. (20a) leads to

$$\delta\alpha = -\frac{a_{ll}}{\left(V^l\right)^{1/3}} G\left(\alpha_0\right), \tag{24a}$$

where

$$G\left(\alpha_0\right) = \frac{1}{4}\left(1 - 3\sin^2\alpha_0 + 3\sin^4\alpha_0 - \frac{1}{2}\cos\alpha_0\right)$$

$$\times\left[\frac{2}{3}\pi\left(1 - \cos\alpha_0 - \frac{1}{2}\cos\alpha_0\sin^2\alpha_0\right)\right]^{1/3}$$

$$\times\left\{\left(\sin^2\alpha_0\right)\left[\left(1 - K_0\right)\left(\cot\alpha_0\csc^3\alpha_0 + \frac{1}{2}\cot^2\alpha_0 + \cot^2\alpha_0\csc^2\alpha_0\right) + \frac{1}{2}\right]\right\}^{-1}. \tag{24b}$$

$G(\alpha_0)$ is plotted in fig. 3. The correction $\delta\alpha$ acts in the opposite direction to the repulsive forces tending to decrease the wetting angle. As explained before, the angle α_0 is larger than α because of the repulsive forces. The decrease of the wetting angle when the drop is very small is caused by a decrease of the attractive forces due to the liquid, because a smaller population of liquid molecules than that corresponding to the range of interaction forces is acting.

6. NUMERICAL SOLUTION OF THE INTEGRAL EQUATION (16)

Eq. (16) was solved numerically by an iterative procedure using eq. (17) as the zero order approximation for the profile of the droplet. The results are given in figs. 4 and 5. Fig. 4 shows that near the leading edge there are three numerical solutions. Two of them have to be, however, excluded as unphysical. When the equivalent radius A is very large, the profile of the drop is given by the curve denoted by $A \to \infty$. The profile of a small drop is also included in fig. 4 for $A = 30\ a_{ll}$. Numerical solutions for the profile for various values of β are plotted in fig. 5.

7. DROP STABILITY

Because the function $F(\alpha)$, defined by equation

$$F\left(\alpha\right) \equiv \left(\csc^3\alpha + \frac{3}{2}\cot\alpha + \cot^3\alpha\right) / \csc^3\alpha \tag{25}$$

has to be smaller than 2, eq. (22) shows that if

$$\beta K_0 \equiv n_s\beta_{ls}K_0 / n_l\beta_{ll} > 1, \tag{26}$$

the drop is no longer stable but will spread over the surface of the solid. If $\beta > 1$ and $\beta K_0 < 1$ then the macroscopic droplet will have a flat shape with a rapid variation of angle near the leading edge only (see also fig. 5).

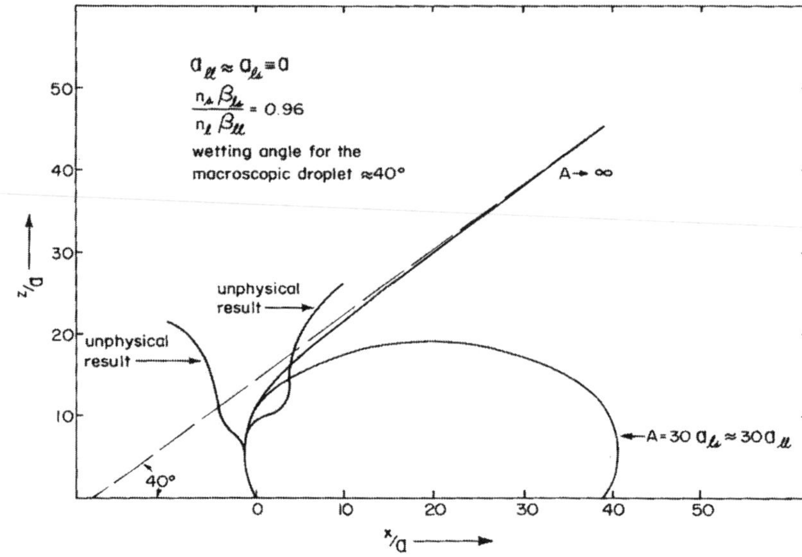

FIG. 4. Numerical solution of eq. (16) when the macroscopic wetting angle α is 40° and $a_{ll} = a_{ls}$. $A \to \infty$ represents the behavior of a large drop, while $A = 30\, a_{ls}$ represents the behavior of a small droplet.

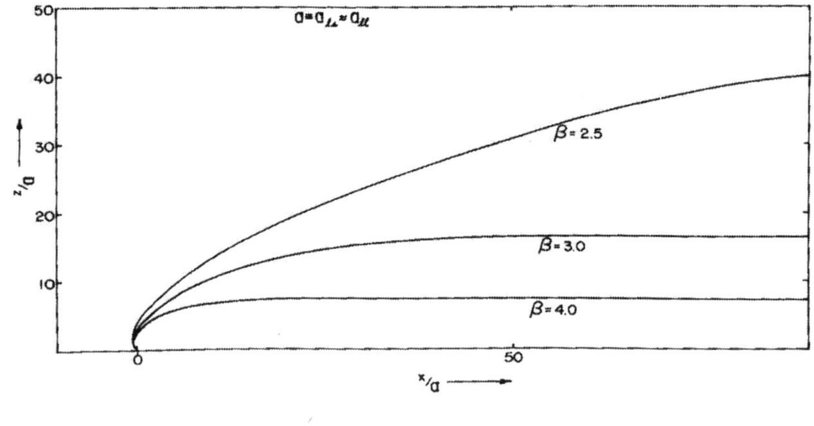

FIG. 5. Numerical solution of eq. (16) for various values of $\beta \equiv n_s\beta_{ls}/n_l\beta_{ll}$.

8. CONCLUSIONS

The main results of the paper can be synthesized as follows:

1. The tangential force condition allows to obtain information about the surface profile of the drops and to relate the wetting angle near the leading edge to parameters characterizing the interaction forces. A size effect occurs for sufficiently small drops.
2. There is a rapid variation of shape very near to the leading edge caused by the repulsive forces.
3. The macroscopic wetting angle is reached at a distance of about 20 to 40 Å from the leading edge.
4. If $K_0\beta > 1$, the droplet is no longer stable but spread over the surface of the solid.

There are at least two important problems for which the results obtained here are of significance; namely the problem of heterogeneous nucleation and the problem of supported metal catalyst. Both the nuclei and the crystallites of supported metals can be very small, smaller than the range of interaction forces, and consequently the traditional thermodynamics can no longer be applied to them.

ACKNOWLEDGEMENT

This work was support in part by NSF.

REFERENCES

[1] C. Miller and E. Ruckenstein, J. Colloid Interface Sc. 48 (1974) 368 (Section 1.1 of this volume).
[2] E. Ruckenstein and P.S. Lee, Surface Sci. 50 (1975) 597 (Section 1.4 of this volume).
[3] J. Yvon, Actualités Scientifiques et Industrielles (Hermann, Paris, 1935).
[4] H.S. Green, Molecular Theory of Fluids (North-Holland, Amsterdam, 1952

3.2 On the Shape and Stability of a Drop on a Solid Surface*

Gersh Berim and Eli Ruckenstein†
Department of Chemical and Biological Engineering,
State University of New York at Buffalo, Buffalo, New York 14260

Corresponding Author

† E-mail: feaeliru@buffalo.edu.
Phone: (716)645-1179, Fax: (716)645-3822.

Received: May 3, 2004; In Final Form: July 7, 2004

I. INTRODUCTION

The study of the shape of a drop on a solid substrate[1–7] was motivated by the existence of numerous cases in which the shape plays a significant role. As examples, one can mention heterogeneous nucleation,[8,9] supported metal catalysts,[10] and the roll-up transition on cylindrical surfaces.[7]

Most of the mentioned studies are based on thermodynamic considerations applied to large droplets and involved such macroscopic parameters as the pressure and the surface and line tensions. The drop was usually considered as a continuum medium of constant density, and the differential equation of the drop profile has been obtained from the minimum of the calculated macroscopic free energy of the drop.[3,5–7] Two main regimes have been considered, which correspond to (i) a bare surface[2,6] and (ii) one covered with a thin liquid film.[5,6] Among the obtained results, let us note an expression for the microcontact angle θ between the drop surface and the bare solid surface at the leading edge of the drop[6]

$$\cos\theta = 1 + \frac{S - P(0)}{\gamma} \tag{1}$$

where $S = \gamma_{SG} - \gamma_{SL} - \gamma$ is the spreading coefficient, γ_{SG} and γ_{SL} are the surface free energies of the solid/gas and solid/liquid interfaces, γ is the liquid/gas surface tension, and $P(0)$ is the potential solid/liquid interaction energy near the bare surface of the solid. This expression provides a stability condition for a drop, which involves the macroscopic parameters of interaction of the liquid with the environment. A drop exists only if $|1 + (S - P(0))/\gamma| \leq 1$.

It is a challenging problem to describe the drop shape and the contact angle from the point of view of a microscopic theory that uses only the potentials of the intermolecular interactions. A first microscopic approach to investigate the droplet shape, which took into account the liquid—liquid and liquid—solid interactions, was developed by Ruckenstein and Lee.[1] Using the condition of zero tangential force on the surface of a drop in equilibrium and realistic interaction potentials, they obtained an integral equation that can provide the shape of a droplet on a bare surface. From the analysis of that equation, the possibilities of a planar shape of a droplet and of a rapid variation of the direction of the tangent to the profile in the close vicinity of the leading edge were predicted. The analysis of the contact angle provided a condition for the droplet stability.

* *J. Phys. Chem.* B 2004, 108, 19330–19338. Republished with permission.

In the present paper, the investigation of the problem on the basis of a microscopic approach is continued. The shape of a drop on a bare planar solid surface is obtained by minimizing the total potential energy of the drop and neglecting for simplicity the effect of the entropy. In section II, the intermolecular interactions are defined and the total potential energy of a drop is calculated in a general form. Instead of an integral equation, a simpler differential equation for the droplet profile, which is similar to the augmented Young–Laplace equation,[6] is derived for cylindrical (section III) and axisymmetrical (section IV) homogeneous droplets. From the general conditions that the apex of the droplet should be located on the symmetry axis of the profile and the leading edge located on the solid surface, the microcontact angle θ between a drop and the solid surface could be determined to be equal to 180°, hence independent of the interaction parameters. It was shown that the change of the profile slope from an acute angle to 180° occurs in the very vicinity of the leading edge of a large drop, a region that is not observable in macroscopic experiments. An observable macroscopic contact angle θ_m was introduced and calculated for a large droplet using a continuation of the circular part of its profile. As expected that angle did depend on the interaction parameters. A condition for the stability of the droplets was found that was derived from the analysis of the existence of a solution of the differential equation for the droplet profile. From this condition, two ranges of values of the interaction parameters were found in which the droplet can have any height or the height y_m is limited by a critical value $y_{m,c}$ and a stable droplet with a height larger than $y_{m,c}$ cannot exist. The critical height $y_{m,c}$ depends on the parameters that chracterize the interaction potentials. One should note that the expressions for the contact angles θ and θ_m were not involved in the determination of the stability condition.

II. FORMULATION OF THE PROBLEM AND BASIC EQUATIONS

A. The Potentials.

A molecule of a liquid drop immersed in its vapor and located on the bare solid surface interacts with the other molecules of the drop with an interaction potential ϕ_{LL}, with the molecules of vapor with a potential ϕ_{LV}, and with the molecules of the solid with a potential ϕ_{LS}. At equilibrium, a drop acquires the shape which for a given volume V minimizes its free energy. Because the contact angle is weakly dependent upon temperature, one can neglect the role of entropy and minimize only the total potential energy of the droplet. Below it will be assumed that the interaction of the drop molecules with the vapor molecules is negligible. The interaction potential, $\phi_{LL}(r)$, between two molecules of liquid the centers of which are separated by a distance r is assumed to be of the simplest, square-well form

$$\phi_{LL}(r)=\begin{cases}-\epsilon_{LL} & r\le\eta\\ 0 & r>\eta\end{cases} \tag{2}$$

where η is the radius of interaction and ϵ_{LL} characterizes the strength of the interaction. Despite its oversimplified form, such a potential was used as a reasonable approximation for the qualitative understanding of various phenomena.[11]

The interaction potential ϕ_{LS} between the molecules of liquid and solid is chosen by combining the London–van der Waals attraction with a rigid core repulsion

$$\phi_{LS}(r)=\begin{cases}-\epsilon_{LS}\left(\dfrac{\sigma}{r}\right)^6 & r\ge\sigma\\ \infty & r<\sigma\end{cases} \tag{3}$$

where σ and $\epsilon_{LS} > 0$ are the size of the repulsive core and the interaction constant, respectively. Below σ is used as a unit of length, and for numerical estimations, the value $\sigma \simeq 3.4\,\text{Å}$ was selected.[12]

B. Liquid—Solid Interaction.

Assuming pairwise additivity, the interaction potential of a molecule of liquid with a semiinfinite solid can be written as

$$\Phi(y) = -\frac{1}{6}\pi\epsilon_{LS}\rho_S\sigma^3\left(\frac{\sigma}{\sigma + y}\right)^3, \quad y \geq 0 \tag{4}$$

where $\sigma + y$ is the distance of the center of a molecule of liquid from the center of the first layer of solid atoms and ρ_S is the (constant) number density of molecules in the solid. The total potential energy, U_{LS}, associated with the liquid—solid interactions is given by

$$U_{LS} = \int_V \rho_L \Phi(y)\,dV \tag{5}$$

where V is the volume of the droplet.

C. Liquid—Liquid Interaction.

Let us assume that the size of the droplet is much larger than the radius η of the range of liquid—liquid interactions. Then a molecule of liquid at a distance larger than η from the droplet surface interacts with all the molecules inside a sphere of radius η and the center in the center of the molecule (interaction sphere). The effective potential of this molecule is proportional to the number of molecules inside the sphere and is equal to

$$\phi_1 = -\frac{4}{3}\pi\eta^3\rho_L\epsilon_{LL} \tag{6}$$

where ρ_L is the number density of molecules in the liquid, which is considered constant.

A molecule at a distance $l < \eta$ from the droplet surface interacts only with those molecules that are present both in the drop and in the interaction sphere described above (Figure 1). Assuming that the radius of curvature of the surface is much larger than η, one can consider the intersection of the interaction sphere with the drop to be planar. (Such an assumption is valid if the drop volume

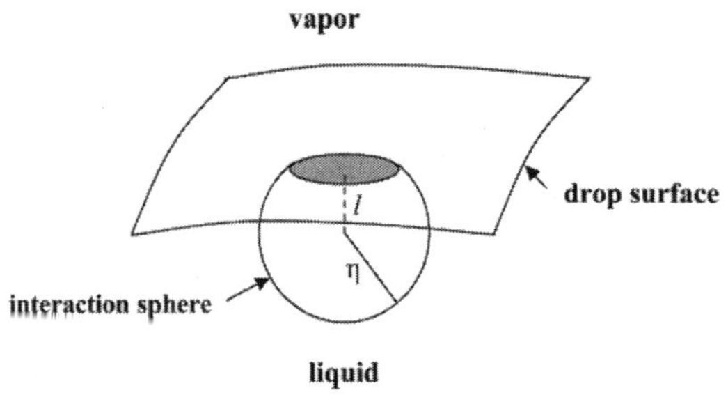

FIGURE 1. Interaction range of a molecule near the surface of a drop ($l < \eta$). If the drop surface has a radius of curvature much larger than η, the intersection of the interaction sphere with the drop can be considered planar.

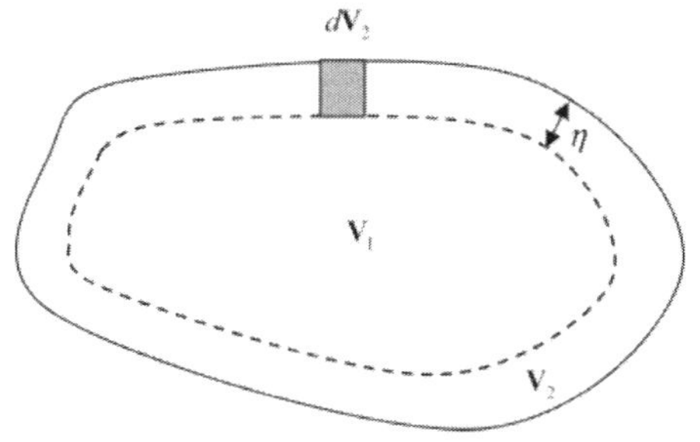

FIGURE 2. Cross-sections of a drop surface (—) and of the surface located at a distance η from the drop surface (– – –). Inside the volume V_1, any molecule interacts with all the molecules inside the interaction sphere with the center in the center of the selected molecule. The elementary volume dV_2 has the base area ds and height η.

V is much larger than the volume of the interaction sphere, $V \gg {}^4/_3\,\pi\,\eta^3$). In this case, the effective potential $\phi_2(l)$ of a molecule near the drop surface is

$$\phi_2(l) = -\pi\epsilon_{LL}\rho_L \int_{-\eta}^{l} \left(\eta^2 - z^2\right) dz = -\frac{2}{3}\pi\eta^3\epsilon_{LL}\rho_L\left[1 + \frac{3}{2}\frac{l}{\eta} - \frac{1}{2}\left(\frac{l}{\eta}\right)^3\right] \quad (l \le \eta) \tag{7}$$

On a schematic representation of a drop having an arbitrary shape (see Figure 2), the dashed line separates the volume V_1 in which each molecule has the effective potential eq 6 from the volume V_2 in which eq 7 is valid. After integration over the volume V_1, one obtains for the total potential energy U_1 of molecules in this volume

$$U_1 = \frac{1}{2}\phi_1\rho_L V_1 = -\frac{2}{3}\pi\eta^3\rho_L^2\epsilon_{LL}V_1 \tag{8}$$

where the factor $^1/_2$ avoids the double counting of the interaction energy during integration.

To calculate the energy U_2 of all molecules in the volume V_2, let us first find the energy dU_2 of the molecules in the cylindrical volume dV_2 with the base area ds and height η normal to the surface (Figure 2)

$$dU_2 = \frac{1}{2} ds\rho_L \int_0^{\eta} \phi_2(l)\,dl = -\frac{13}{24}\pi\eta^4\epsilon_{LL}\rho_L^2\, ds \tag{9}$$

Then the energy U_2 can be written as follows:

$$U_2 = -\frac{13}{24}\pi\eta^4\epsilon_{LL}\rho_L^2\, S \tag{10}$$

where $S = S_{LV} + S_{LS}$, S_{LV} (S_{LS}) being the area of the interface between the droplet and vapor (solid). When the approximation $V_2 = S\eta$ is used for the volume V_2, the total potential energy $U_{LL} = U_1 + U_2$ of the droplet associated with the liquid–liquid interactions can be written as

$$U_{LL} = -\frac{2}{3}\pi\eta^3\rho_L^2\epsilon_{LL}V + \frac{1}{8}\pi\eta^4\rho_L^2\epsilon_{LL}S \tag{11}$$

The two terms in eq 11 represent the bulk and surface energy of the droplet, and the coefficient multiplying the area S provides a microscopic expression for the surface tension γ_L of the liquid

$$\gamma_L = \frac{1}{8}\pi\eta^4 \rho_L^2 \epsilon_{LL} \quad (12)$$

The total droplet energy U_{total} is given by

$$U_{total} = U_{LL} + U_{LS} = -\frac{2}{3}\pi\eta^3 \rho_L^2 \epsilon_{LL} V + \frac{1}{8}\pi\eta^4 \rho_L^2 \epsilon_{LL} S - \frac{\pi}{6}\epsilon_{LS}\rho_S\rho_L\sigma^3 \int_V \left(\frac{\sigma}{\sigma + y}\right)^3 dV \quad (13)$$

One should note that the total potential energy (eq 13) does not include the gravitational and liquid—vapor contributions, which are considered small.

In the following sections, the shape of the drop will be examined for cylindrical and axisymmetrical drops.

III. CYLINDRICAL DROPLET

A. An Equation for the Profile.

Let us consider a cylindrical liquid drop on a smooth bare solid surface XOZ, extended along the z-axis and with a cross-section symmetrical with respect to the y-axis (Figure 3). Usually one assumes[5,6] that the microcontact angle θ, that is, the angle between the drop surface and the solid surface at the leading edge of the drop, is less than 90° and that the droplet has a bell-like shape (Figure 3a). In this case, the equation of the profile can be chosen in the form $y = y(x)$ where $y(x)$ satisfies an equation of the form $f(x, y, (dy/dx), (d^2y/dx^2)) = 0$ for the entire profile. However, if the drop has the shape presented in Figure 3b with $\theta > 90°$, then the above equation is no longer valid due to the lack of uniqueness of the function $y = y(x)$ for $|x| > x_0$, where x_0 is the abscissa of the leading edge of the droplet. To avoid this difficulty and include both shapes (Figure 3 parts a and b), the equation for the droplet profile will be written in the form $x = x(y)$ and only that part of the profile for which $x \geq 0$ will be considered. The volume V of the drop, the areas of the liquid—vapor (S_{LV}), and liquid—solid (S_{LS}) interfaces, and $\int_V (\sigma^3/(\sigma + y)^3)dV$ are given by

$$V = 2\int_0^{y_m} x(y)\, dy$$

$$S_{LV} = 2\int_0^{y_m} \sqrt{1 + x_y^2}\, dy$$

$$S_{LV} = 2x(0)$$

$$\int_V \frac{\sigma^3}{(\sigma + y)^3} dV = 2\int_0^{y_m} x \frac{\sigma^3}{(\sigma + y)^3} dy \quad (14)$$

where y_m is the largest ordinate of the droplet profile (the height of the droplet) and the subscript y denotes the derivative with respect to y. All the quantities in eq 14 are given per unit length of the droplet in the z direction.

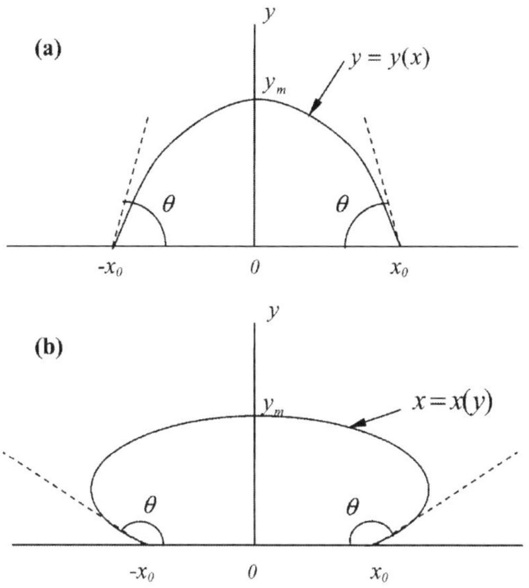

FIGURE 3. Two characteristic profiles of a cylindrical droplet: (a) the contact angle θ at the leading edge of the droplet is less than 90°; (b) the same angle is larger than 90°. The z-axis is normal to the plane of the figure.

With the use of the identity $x(0) = -\int_0^{y_m} x_y dy$, which follows from $x(y_m) = 0$, the total potential energy eq 13 per unit length becomes

$$U_{\text{total}} = \int_0^{y_m} f(y, x, x_y)\, \mathrm{d}y \tag{15}$$

where

$$f(y, x, x_y) = \frac{1}{4}\pi\eta^4 \rho_L^2 \epsilon_{LL}\left[-a_0 x + \sqrt{1 + x_y^2} - x_y - ax\phi(y)\right] \tag{16}$$

and

$$a_0 = \frac{16}{3\eta}, \quad a = \frac{4}{3}\frac{\epsilon_{LS}}{\epsilon_{LL}}\frac{\rho_S}{\rho_L}\left(\frac{\sigma}{\eta}\right)^4, \quad \phi(y) = \frac{\sigma^2}{(\sigma + y)^3} \tag{17}$$

The actual drop profile has to minimize the total potential energy of the drop for a given drop volume. The problem is thus reduced to the minimization of the functional

$$v[x(y)] = \int_0^{y_m} F(y, x, x_y)\, \mathrm{d}y \tag{18}$$

where

$$F(y, x, x_y) = \lambda_c x + \sqrt{1 + x_y^2} - x_y - ax\phi(y) \tag{19}$$

and λ_c is a Lagrange multiplier for the drop volume.

The standard calculus of variations[13] provides the following differential equation for the profile:

$$\frac{x_{yy}}{\left(1+x_y{}^2\right)^{3/2}}+a\phi\left(y\right)-\lambda_c=0 \tag{20}$$

where the first term has the meaning of a curvature $\kappa(y)$ of the profile. This equation has to be complemented with the so-called transversality conditions[13]

$$\frac{\partial}{\partial x_y}F\left(y,x,x_y\right)\Big|_{y=0}=0,\ \left[F\left(y,x,x_y\right)-x_y\frac{\partial}{\partial x_y}F\left(y,x,x_y\right)\Big|_{y=y_m}=0\right] \tag{21}$$

which reflect the fact that the ends of the considered part of the profile ($x \geq 0$) should be located on the x (contact point with the solid) and y (drop apex) axes. In the considered case, these conditions have the forms

$$\left.\frac{x_y}{\sqrt{1+x_y{}^2}}\right|_{y=0}=1 \tag{22}$$

and

$$\left.\frac{1}{\sqrt{1+x_y{}^2}}\right|_{y=y_m}=0 \tag{23}$$

The left-hand side of eq 22 is equal to $-\cos\theta$, where θ is the microcontact angle of the drop on the solid surface (Figure 3b). Therefore condition 22 provides $\theta = 180°$ between the droplet and the solid, independent of the parameters of the model. One should note that at the point of contact the derivative x_y diverges ($x_y|_{y=0}=\infty$). The second transversality condition, eq 23, can be replaced by

$$x_y|_{y=y_m}=\infty \tag{24}$$

or, equivalently, by $y_x|_{x=0}=0$, and is a consequence of the symmetry of the drop profile with respect to the y-axis.

Denoting $p(y) = x_y$, one can transform eq 20 into a first-order equation

$$\frac{p_y}{\left(1+p^2\right)^{3/2}}=\lambda_c-a\phi\left(y\right) \tag{25}$$

with the solution

$$p=\frac{\Psi\left(y\right)}{\sqrt{1-\Psi^2\left(y\right)}} \tag{26}$$

where

$$\Psi\left(y\right)=\lambda_c y+\frac{a\sigma^2}{2\left(\sigma+y\right)^2}+C \tag{27}$$

and C is a constant. Substituting $p = x_y$ into eq 26, one can obtain a final general solution for $x(y)$ in the form

$$x\left(y\right)-x\left(0\right)=\int_0^y\frac{\Psi\left(y\right)}{\sqrt{1-\Psi^2\left(y\right)}}dy \tag{28}$$

The point of contact $x(0) = x_0$ is given by

$$x(0) = -\int_0^{y_m} \frac{\Psi(y)}{\sqrt{1-\Psi^2(y)}} \mathrm{d}y \tag{29}$$

which follows from eq 28 and the condition $x(y_m) = 0$.

The solution 28 contains three unknown quantities, namely, the constant of integration C, the Lagrange multiplier λ_c, and the height of the droplet y_m, which have to be found using suitable constraints. First of these constraints is provided by the transversality condition 22, which is equivalent to the equation $\Psi(0) = 1$ and which gives for C the value $C = 1-a/2$. As a result, function $\Psi(y)$ becomes

$$\Psi(y) = 1 - \frac{a}{2} + \lambda_c y + \frac{a\sigma^2}{2(\sigma + y)^2} \tag{30}$$

The second transversality condition, provided by eq 24, and eq 26 together with the definition of p lead to $\Psi(y_m) = \pm 1$. To select the correct sign, one should take into account that in the limit $a \to 0$, that is, in the absence of liquid—solid interactions, the profile has to be circular. To recover such a result from eq 26, one has to select $\Psi(y_m) = -1$. Obviously, the same sign is also valid when $a \neq 0$, and this provides a relation between y_m and λ_c

$$\lambda_c = \frac{1}{y_m}\left[-2 + a/2 - \frac{a\sigma^2}{2(\sigma + y_m)^2}\right] \tag{31}$$

which follows from eq 30 at $y = y_m$. Finally, one has the condition of fixed volume ($V = 2\int_0^{y_m} x\,\mathrm{d}y =$ const), which after integration by parts and employing eq 26 and condition $x(y_m) = 0$ can be rewritten in the form

$$V = -2\int_0^{y_m} \frac{y\Psi(y)}{\sqrt{1-\Psi^2(y)}} \mathrm{d}y \tag{32}$$

Equation 32 together with eqs 30 and 31 allows one to find the unknown parameters λ_c, y_m and $x(0)$ and hence to find the shape of the droplet. Even though the detailed profile can be obtained only numerically, eqs 20 and 28 allow one to extract some general conclusions. Thus, eq 20 shows that the curvature of the profile is equal to $\kappa(y) = \lambda_c - a\phi(y)$, where $\phi(y)$ is given by eq 17. The latter function vanishes for $y \gg \sigma$. Consequently, for $y/\sigma \gg 1$, the curvature of the profile is almost constant and equal to λ_c, that is, the shape of that part of the drop can be represented by a circle of radius $R \simeq |1/\lambda c|$. The question how large is this part can be answered only on the basis of the complete solution of eq 20, and this issue will be examined in the next section.

Equation 28 has a solution only if $|\Psi(y)| < 1$ for $0 < y < y_m$. Because, as noted above, $\Psi(y)$ acquires for $y = y_m$ its smallest value equal to -1, this function has to decrease in the vicinity of y_m and its derivative

$$\frac{\mathrm{d}\Psi(y)}{\mathrm{d}y} = \lambda_c - \frac{a\sigma^2}{(\sigma + y)^3} \tag{33}$$

has to be negative at $y = y_m$. Therefore,

$$\lambda_c - \frac{a\sigma^2}{(\sigma + y_m)^3} < 0 \tag{34}$$

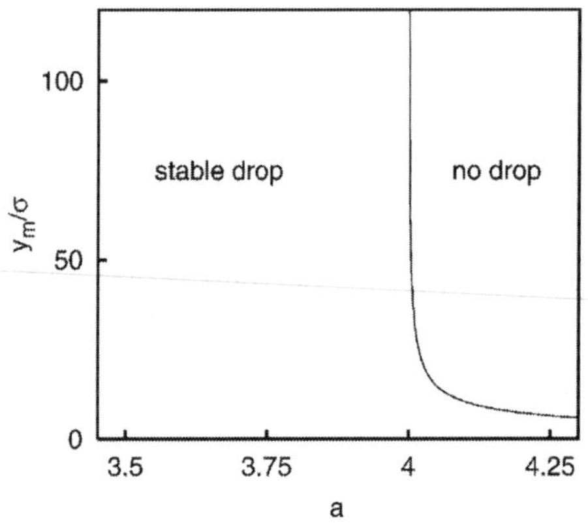

FIGURE 4. Domains of drop stability in the a–(y_m/σ) plane. The solid line provides the critical height $y_{m,c}$ as a function of a. The critical height $y_{m,c}$ increases as a decreases and approaches infinity when $a \to 4$.

The latter inequality together with the explicit expression for λ_c given by eq 31 provides the following inequality between the parameter a and the droplet height y_m:

$$a < 4\left(\frac{\sigma}{y_m}\right)^2 \frac{(1 + y_m/\sigma)^3}{3 + y_m/\sigma} \tag{35}$$

Condition 35 for the existence of a solution to eq 20 for the drop profile consitutes a necessary stability condition for the droplet, which will be examined in more detail in section III.B.

Eq 20 has the same form as the augmented Young–Laplace equation[6]

$$\frac{y_{xx}}{\left(1 + y_x^{\,2}\right)^{3/2}} + \frac{\Pi(y)}{\gamma_L} = -P_c/\gamma_L \tag{36}$$

for the drop profile given as a function of $x[y = y(x)]$, which is applicable to drop shapes as in Figure 3a. Here γ_L is the bulk liquid/gas surface tension, $\Pi(y)$ is the disjoining pressure, and p_c is the capillary pressure. Because the first terms of eqs 36 and 20 represent the profile curvature, one can establish the folowing correspondence between the macroscopic parameters of eq 36 and the microscopic parameters of eq 20: $P_c/\gamma_L = -\lambda_c$; $\Pi(y)/\gamma_L = a\phi(y)$.

B. Stability of the Drop.

As mentioned above, the equation for the droplet profile has a solution only if the parameter a and the droplet height y_m satisfy inequality 35. The examination of that inequality shows that for $a > 4$ a solution exists only when the height of the droplet, y_m, is smaller than a critical value $y_{m,c}$ provided by the equation

$$\left(\frac{\sigma}{y_{m,c}}\right)^2 \frac{(1 + y_{m,c}/\sigma)^3}{3 + y_{m,c}/\sigma} = \frac{a}{4} \tag{37}$$

The dependence of $y_{m,c}$ on a is given by the solid line of Figure 4. The solid line separates two regions; in one, at the right, droplets cannot exist, while in the other, at the left, droplets do exist. If $a \to 4$, $y_{m,c}$ grows to infinity. For $a < 4$, droplets of any height can exist.

C. Shape of the Drop.

A typical profile of a cylindrical droplet for $a < 4$ ($a = 2.5$) is presented in Figure 5 for two different drop volumes, a relatively small volume ($V/\sigma^3 = 183.5$, $y_m/\sigma = 10$) and a relatively large one ($V/\sigma^3 = 1.9 \times 10^4$, $y_m/\sigma = 100$). In both cases, the microcontact angle is equal to 180°, and near the leading edge, there is a rapid variation of the profile curvature as well as of the angle between the tangent to the profile and the solid surface (see insert in Figure 5b).

The traditional approximation of the profile of a drop on a plane surface is to consider it as a part of a circle.[4,9] If the height of the drop (y_m) and the radius R of the circle are known, then one can easily calculate the drop volume and the areas of the interfaces. If the approximation of the drop profile by a circle is valid, then the radius of curvature will be the same at any (or almost any) point of the profile. To verify its validity in the framework of the present microscopic approach, we calculated the radius of curvature, $R^{curv}(y)$, of the profile as a function of the distance y from the solid surface for several values of the droplet height y_m, using the expression

$$R^{curv}(y) \equiv \left|\frac{1}{\kappa(y)}\right| = \left|\lambda_c - \frac{a\sigma^2}{(\sigma + y)^3}\right|^{-1} \tag{38}$$

which was obtained from the definition of the radius of curvature and eq 20. In Figure 6, this dependence is presented for several values of y_m by plotting the normalized coordinate $R^{curv}(y)/R_m^{curv}$ vs y/y_m, where $R_m^{curv} \equiv R^{curv}(y_m)$ is the largest value of the curvature radius, which the profile has at the drop apex ($y = y_m$, $x = 0$). The parameter a was selected smaller than 4 ($a = 3$) to allow for the existence of droplets of any size. One can see that for small droplets ($y_m/\sigma = 10$), there is no range in which the radius of curvature is constant. For such droplets, the profile cannot be approximated by a circle. However, for large drops, $y_m/\sigma = 1000$, the radius of curvature for $y > 0.1y_m$ is almost equal to its largest value $\left(R_m^{curv}/\sigma = 2000.0\right)$ and the profile can be considered as a part of a circle. Numerical analysis shows that for $a < 4$ the radius R of that circle is proportional to the height of the droplet y_m ($R = ky_m$). The proportionality coefficient k is a function of a and, for example, for $a = 3$, $k = 2$. The approximation by a circle is valid starting from $y_m/\sigma \simeq 100$.

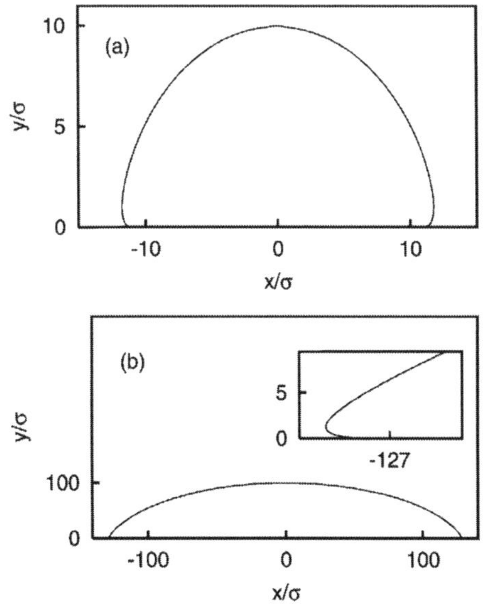

FIGURE 5. Profiles of droplets of different volumes for $a = 2.5$: (a) small droplet with $V/\sigma^3 \simeq 183.5$, $y_m/\sigma = 10$; (b) large droplet with $V/\sigma^3 \simeq 1.9 \times 10^4$, $y_m/\sigma = 100$. On both figures, the shape of the drop near the leading edge is similar to that shown in the insert.

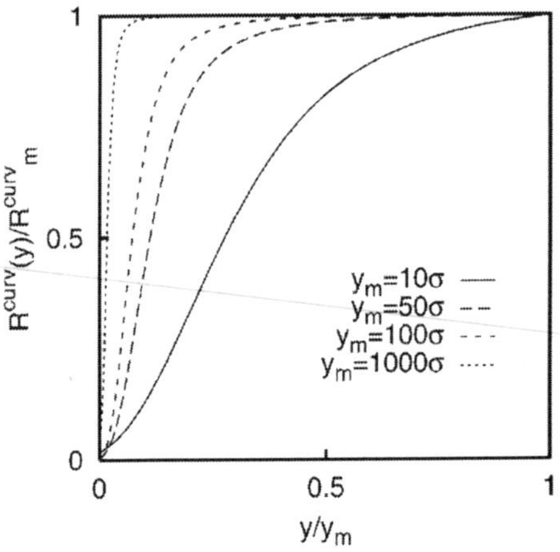

FIGURE 6. Radius of the profile curvature for droplets of various sizes for $a < 4$ ($a = 3$). For a relatively large droplet ($y_m/\sigma = 100$), the curvature radius is almost constant for $y > 0.1y_m$. R_m^{curv}/σ is equal to 18.7, 99.7, 199.8, and 2000.0 for y_m/σ equal to 10, 50, 100, and 1000, respectively.

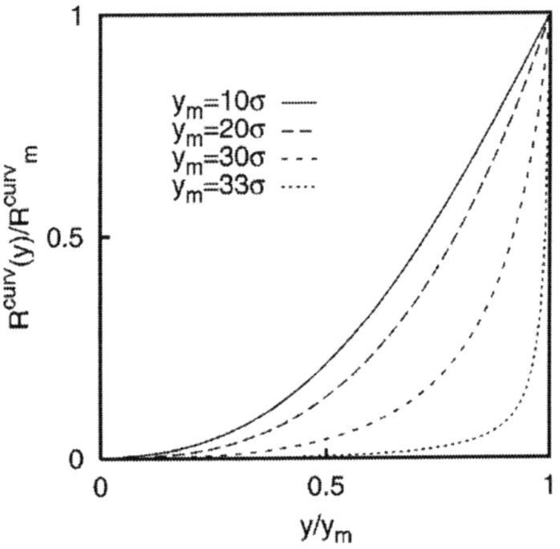

FIGURE 7. The radius of the profile curvature for droplets of various sizes for $a > 4$ ($a = 4.01$). R_m^{curv}/σ is equal to 240.0, 2437.0, 2.67×10^3, and 3.26×10^5 for y_m/σ equal to 10, 20, 30, and 33, respectively.

In Figure 7, a similar plot, but for $a = 4.01$, is presented. For this value of a, the height of the drop cannot exceed the critical value $y_{m,c}/\sigma = 33.34$, and there is no range in which $R^{curv}(y)$ of the profile is constant. Therefore for this case, the approximation by a circle is not at all valid. For a height y_m of the droplet close to the critical one ($y_m/\sigma = 33$), the curvature radius for $y > 0.9y_m$ is much larger than y_m $\left(R_m^{curv}/y_m \approx 10^4\right)$, and the width of the droplet, $2x(0)$, is much larger than the height $\left(2x(0)/y_m \simeq 100\right)$. The shape of the droplet can be considered to be planar in this range.

D. Contact Angle.

As shown in section III.A, the microcontact angle θ of a drop on the bare surface at the leading edge is always equal to 180°, independent of the intermolecular interactions. This is in apparent contradiction with the experimental results, which provide for θ a variety of values (see, for example, ref 14). To resolve this contradiction, one should note that the measurements of the contact angles were performed for large drops ($y_m > 1\ \mu$m) for which the main part of the drop profile can be considered circular and the part near the surface of the solid (which is similar to that shown in the insert of Figure 5b), where very rapid curvature changes occur, is extremely small. For such a large drop, it is reasonable to define a macroscopic contact angle, which can be measured in macroscopic experiments, as the angle between the continuation of the circular part of the profile until its intersection with the solid surface (see Figure 9a). In this case, the value θ_m of this angle is given by the simple formula

$$\cos\theta_m = 1 - \frac{y_m}{R_m^{curv}} \tag{39}$$

where R_m^{curv} is the curvature radius at the drop apex and y_m is the height of the drop. The curvature radius R_m^{curv} can be obtained from eq 38, where λ_c is given by eq 31. One can see from Figure 4 that stable large droplets can exist only for $a \le 4$. In this case

$$R_m^{curv} = 2y_m\left[4 - a + a\frac{1+3y_m/\sigma}{\left(1+y_m/\sigma\right)^3}\right]^{-1} \tag{40}$$

and eq 39 provides for cos θ_m the expression

$$\cos\theta_m = \frac{1}{2}\left[a - 2 + a\frac{1+3y_m/\sigma}{\left(1+y_m/\sigma\right)^3}\right] \tag{41}$$

For large droplet, y_m, $\gg \sigma$ and eq 41 acquires the very simple form

$$\cos\theta_m \simeq \frac{1}{2}(a-2) \tag{42}$$

which connects the macroscopic contact angle to the microscopic parameters involved in the definition of a (see eq 17). It follows from eq 42 that the macroscopic contact angle exists only for $0 \le a \le 4$. The latter inequality provides for a the same range of values in which a macroscopic droplet can exist as those of section III.B on the stability of droplets. Therefore the simple eq 42 provides a stability condition for large droplets, which is equivalent to that obtained from the condition (eq 35) of existence of a solution of the differential equation for the drop profile. According to eq 42, a large droplet cannot exist on a bare solid surface for $a > 4$. Once artificially created, it has to spread over the surface to form a planar droplet with a height less than the critical value $y_{m,c}$, similar to those presented in Figure 8.

Note that expressions 41 and 42 can be used if the domain of constant curvature radius of the profile (h_1 in Figure 9a) is much larger than that in which a rapid variation in the profile curvature

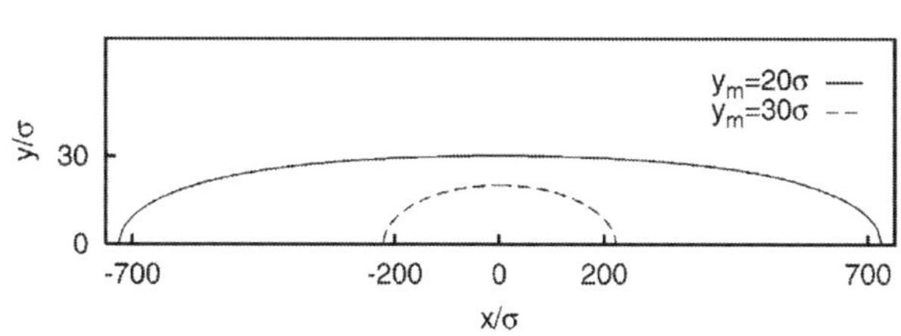

FIGURE 8. The droplet shape for ***a*** > 4 (a = 4.01) for two values of the droplet height. For the height $y_m/\sigma = 30$ close to the critical value $y_{m,c}/\sigma = 33.34$, the shape is planar, while for smaller heights ($y_m/\sigma = 20$), it has a nonplanar profile.

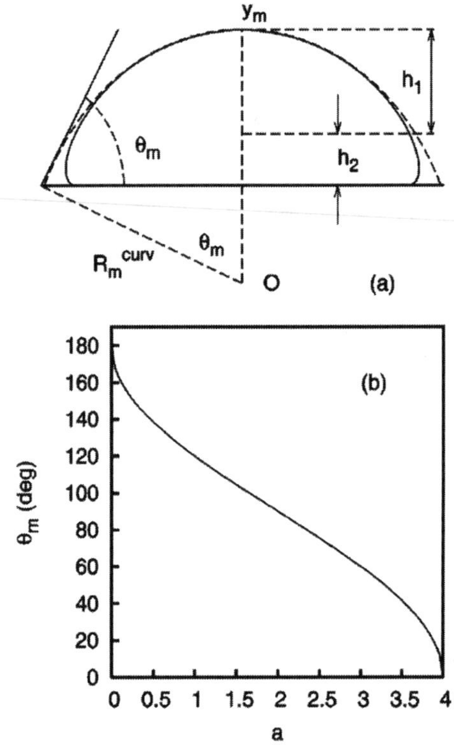

FIGURE 9. Panel a provides the definition of a macroscopic contact angle θ_m. Here h_1 is the range of constant profile curvature corresponding to the curvature radius R_m^{curv}, y_m is the droplet height, h_2 is the range of rapid change of the profile curvature. It is assumed that $h_2/h_1 \leq 0.1$. Panel b shows the a-dependence of the macroscopic contact angle given by eq 42.

radius occurs (h_2 in Figure 9a). In the numerical estimations, a radius of curvature was considered constant in the upper part of the profile if its variation within that part did not exceed 5%. The ratio (h_2/h_1) was calculated for several values of a, and the results are presented in Figure 10. One can see that starting from relatively small heights ($y_m = 150\sigma$ for $a = 1.5$ and $y_m = 400\sigma$ for $a = 3$), the above ratio does not exceed 0.1. The larger the droplet height, the smaller is the relative range of rapid variation of the curvature, the size of the range of rapid variation being microscopically small. For example, for a drop with a height $y_m = 400\sigma$ (~120 nm), this range is equal approximately to 9.9 nm.

One can see from Figure 9b that the macroscopic contact angle increases with decreasing a. To relate this angle to thermodynamic characteristics of the system, one can use the expression for the liquid surface tension γ_L given by eq 12 and the definition of a given by eq 17. Then, θ_m is given by the equation

$$\cos\theta_m \simeq \frac{1}{2}\left(\frac{K}{\gamma_L} - 2\right) \tag{43}$$

where $K = (\pi/6)\epsilon_{LS}\rho_L\rho_S\sigma^4$. From this equation, it follows that for a given solid cos θ_m decreases with increasing γ_L, that is, the macroscopic contact angle increases with increasing liquid surface tension. This conclusion is in agreement with experiment (ref 14). One can see from Figures 6 and 7 that the curvature of a small droplet with $y_m < 100\ \sigma$ varies rapidly along the profile. In this case, it is not possible to define a macroscopic contact angle as it was possible for a large droplet.

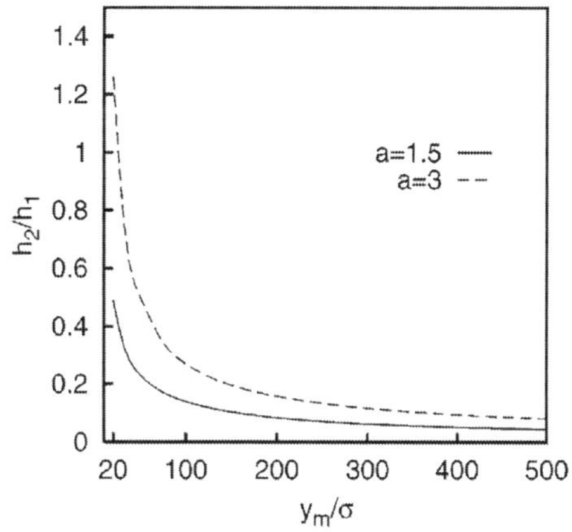

FIGURE 10. The dependence of the ratio h_2/h_1 characterizing the quality of approximation of a drop profile by a circle on the droplet height for $a = 1.5$ and $a = 3$.

IV. AXISYMMETRICAL DROPLET

This section considers a drop that has a rotational symmetry with respect to the y-axis on a solid surface XOZ. The derivation of the Euler equation for the profile is similar to that for a cylindrical droplet. Instead of eq 14, the following expressions for the volume V and the areas of the interfaces have to be employed

$$V = \pi \int_0^{y_m} x^2 \, dy$$

$$S_{LV} = 2\pi \int_0^{y_m} x\sqrt{1+x_y^2} \, dy$$

$$S_{LS} = \pi x^2(0) = -2\pi \int_0^{y_m} x x_y \, dy$$

$$\int_V \frac{\sigma^3}{(\sigma+y)^3} dV = \pi \int_0^{y_m} x^2 \frac{\sigma^3}{(\sigma+y)^3} dy$$

The total potential energy can be written in the form of eq 15 with

$$f(y,x,x_y) = \frac{\pi^2}{4}\eta^4 \rho_L^2 \epsilon_{LL} \left[-\frac{a_0}{2}x^2 + x\left(\sqrt{1+x_y^2} - x_y\right) - \frac{a}{2}x^2\phi(y) \right] \tag{44}$$

where $\phi(y)$ is given by eq 17. The Euler equation for the profile becomes

$$\frac{x_{yy}}{\left(1+x_y^2\right)^{3/2}} - \frac{1}{x\sqrt{1+x_y^2}} + a\phi(y) - 2\lambda_a = 0 \tag{45}$$

where λ_a is a Lagrange multiplier for the drop volume. As for the cylindrical droplet, the transversality conditions 21 can be written in the form

$$x_y|_{y=0} = \infty, \quad x_y|_{y=y_m} = \infty, \tag{46}$$

Again, the first condition provides the value $\theta = 180°$ for the microcontact angle, and the second one reflects the fact that the tangent at the drop apex is parallel to the plane of the solid. However, eq 45 cannot be reduced to a first-order differential equation, as for the cylindrical droplet. As a result, it is not possible to formulate analytically, as in section III.A, the necessary condition for the existence of a solution and calculate the drop profile. All the results below were obtained numerically.

A. Stability of the Axisymmetrical Drop

As for the cylindrical droplet, there is a critical value, a_c, of the parameter a such that for $a \le a_c$ droplets of any height can exist but for $a > a_c$ the height cannot exceed the (critical) value $y_{m,c}$. Within the precision of the numerical calculations, the value of a_c for axisymmetrical droplets is the same as that for the cylindrical ones, that is, $a_c = 4$. The a-dependence of the critical droplet height $y_{m,c}$ differs, however, somewhat from that for a cylindrical one. One can see from Figure 11 that $y_{m,c}$ for the axisymmetrical droplet is larger than that for the cylindrical one, particularly near the critical value of a.

B. Shape of the Axisymmetrical Drop and Contact Angles.

The numerical solution of eq 45 provides the shape of the drop, which is qualitatively similar to that of the cylindrical drop. The cylindrical and axisymmetrical droplets will be compared either for the same height, for the same volume per unit liquid/solid contact area, or for the same volume per unit length of the contact line.

In Figure 12, the profile of an axisymmetrical droplet with the height $y_m/\sigma = 27.0$ and volume $V/\sigma^3 \approx 2 \times 10^5$ (solid curve) is compared with three cylindrical droplets, each of them satisfying one of the above conditions. Only the profiles for $x > 0$ are presented. The difference between the profiles with the same height is small. The profiles have almost the same macroscopic contact angles, $\theta_m = 43.7°$ for the axisymmetrical drop and $\theta_m = 42.0°$ for the cylindrical one. Note that the numerical values of the free terms, $-\lambda_c$ and $-2\lambda_a$, in eqs 20 and 45 for the cylindrical and axisymmetrical

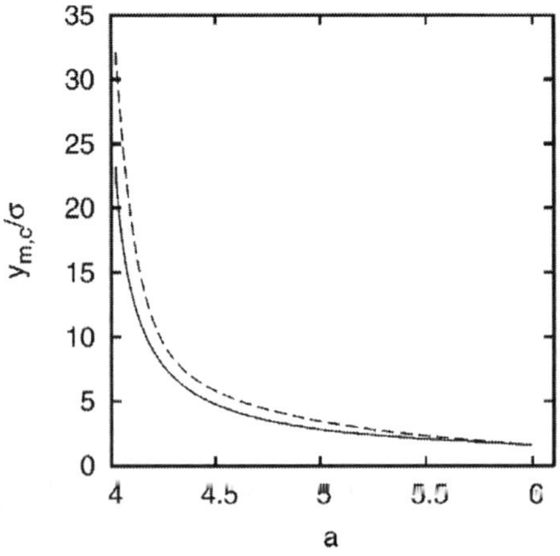

FIGURE 11. The critical height, $y_{m,c}$, of the axisymmetrical (– – –) and cylindrical (—) droplets as function of a for $a > 4$.

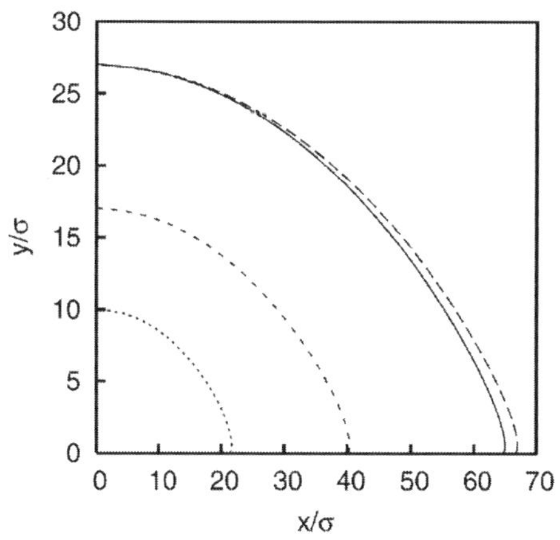

FIGURE 12. Profile of the axisymmetrical drop (—) and profiles of three cylindrical drops having the same height (– – –), the same volume/(contact area) ratio (•••), and volume/(contact line length) ratio (---) as the axisymmetric one.

profiles, respectively, differ by a factor of $1.86 \simeq 2$, the latter one being the larger. If the height of the drops grows, the above ratio becomes closer to 2. In addition, the difference between the two profiles becomes smaller, and the axisymmetrical profile can be represented by the much simpler eq 20 for the cylindrical drop. One should note that in the considered case the characteristic volume of the cylindrical droplet (per unit length) is equal to $V_{\rm cyl}/\sigma^3 \simeq 2600$, while the volume of the axisymmetrical one, $V_{\rm axis}/\sigma^3 \simeq 2\times10^5$, is much larger.

In Figure 12, the profiles of cylindrical drops that have the same volume/(contact area) or volume/(length of the contact line) ratios as the axisymmetrical drop are presented by dashed and dotted lines, respectively. These profiles are lower than that of the axisymmetrical drop but have approximately the same macroscopic contact angles, $\theta_{\rm m} = 44.8°$ and $\theta_{\rm m} = 42.7°$, respectively, as the axisymmetrical drop ($\theta_{\rm m} = 43.7°$).

As for the cylindrical droplet, for $a > 4$, the shape of the axisymmetrical drop becomes planar as the drop height approaches the critical value $y_{\rm m,c}$.

V. DISCUSSION AND CONCLUSION

The analysis provided in sections III and IV for the cylindrical and axisymmetrical drops revealed that their main characteristics are similar. For this reason, the discussion below, if not otherwise mentioned, concerns both kinds of drops.

The interesting result that follows from the microscopic considerations regarding the drop profile is the existence of the stability condition provided by eq 35. This condition, which so far was not noted in the literature, connects the drop stability to the value of a single parameter $a = \frac{4}{3}\left(\epsilon_{\rm LS}/\epsilon_{\rm LL}\right)\left(\rho_{\rm S}/\rho_{\rm L}\right)\left(\sigma/\eta\right)^4$, which depends on the microscopic characteristics of the intermolecular interactions. For $a > 4$, the droplet is stable if its height is smaller than a critical height $y_{\rm m,c}$, which depends on a. The critical height increases when a decreases and approaches infinity when $a \rightarrow 4$. For $a \leq 4$, droplets of any height are stable.

One should note that the obtained stability condition does not involve the contact angle θ at the leading edge (microcontact angle), which is constant ($\theta = 180°$). This fact distinguishes it from the stability conditions discussed in refs 1, 4, and 6 and based on the existence of a solution for the contact angle.

Let us note that an unstable state occurs only when the parameter a exceeds the value $a = 4$. Among the quantities contained in this parameter, the most important one is the ratio (σ/η) of the hard core repulsion radius σ of the solid—liquid interactions and the radius η of the liquid—liquid interactions, because $a \approx (\sigma/\eta)^4$. Only when the ratio σ/η is not too small the ratios $\epsilon_{SL}/\epsilon_{LL}$ and ρ_S/ρ_L, which are also involved in the definition of a, can lead to values of a of the order of 4 required for the drops to become unstable. To evaluate η, let us compare the total potential energy ϕ_1 of a molecule of liquid, given by eq 6, with that provided by the potential

$$\phi'_{LL}(r) = \begin{cases} -\epsilon'_{LL}\left(\frac{\sigma_{LL}}{r}\right)^6 & r \geq \sigma_{LL} \\ \infty & r < \sigma_{LL} \end{cases} \tag{47}$$

which is frequently used to describe the intermolecular interactions in liquids. The latter energy of a molecule in the bulk of the liquid is given by

$$\phi'_1 = \int_0^{2\pi} d\phi \int_0^{\pi} d\theta \int_{\sigma_{LL}}^{\infty} \phi'_{LL}(r)\rho_L r^2 \sin\theta \, dr =$$

$$-\frac{4}{3}\pi\sigma_{LL}{}^3\epsilon'_{LL}\rho_L \tag{48}$$

Comparing eqs 48 and 6 and considering $\epsilon_{LL} = \epsilon'_{LL}$, one can conclude that $\eta \simeq \sigma_{LL}$. Because the hard core radius σ of liquid—solid interactions can be larger than σ_{LL} (see refs 12, 15, and 16), the ratio σ/η can be larger than unity and cases in which a can acquire values larger than 4 can exist.

The droplet profile depends on the droplet size and on the value of the parameter a. By examining the curvature radius, $R^{curv}(y)$, of the droplet profile, one can see (Figure 5) that for large droplets and $0 < a \leq 4$, $R^{curv}(y)$ is almost constant for all points located at distances y from the solid surface larger than $0.1y_m$, where y_m is the height of the droplet. In such cases, the profile has almost a circular shape. However, for small droplets, $R^{curv}(y)$ changes considerably along the profile and the approximation by a circle is no longer valid.

An interesting transformation of the drop profile occurs for $a > 4$ when the height of the drop approaches the critical value $y_{m,c}$. In this case, the width of the drop becomes much larger than its height and goes to infinity as $y_m \rightarrow y_{m,c}$, and the droplet acquires a pancake shape (see Figure 8) with an almost planar upper boundary.

The microcontact angle of the droplet on the solid surface is always equal to 180°, regardless the values of the other parameters. However, the macroscopic contact angle, defined using a continuation of the circular part of a large droplet, depends on the parameters of intermolecular interactions and densities of the liquid and solid substances. This dependence, as shown in section III.D is in qualitative agreement with experiment.

ACKNOWLEDGMENT

The authors are grateful to Prof. C. J. Radke for providing details regarding the numerical calculations in papers of ref 6. This work was supported by the National Science Foundation.

REFERENCES AND NOTES

(1) Ruckenstein, E.; Lee, P. S. *Surf. Sci.* **1975**, *52*, 298 (Section 3.1 of this volume)
(2) de Gennes, P. G. *Rev. Mod. Phys.* **1985**, *57*, 827.
(3) Brochard-Wyart, F.; di Meglio, J. M.; Quéré, D.; de Gennes, P. M. *Langmuir* **1991**, *7*, 335.
(4) Widom, B. *J. Phys. Chem.* **1995**, *99*, 2803.
(5) Solomentsev, Y.; White, L. R. *J. Colloid Interface Sci.* **1999**, *218*, 122.

(6) Yeh, E. K.; Newman, J.; Radke, C. J. *Colloids Surf.* **1999**, *156*, 137; **1999**, *156*, 525.
(7) McHale, G.; Newton, M. I. *Colloids Surf.* **2002**, *206*, 79.
(8) Zettlemoyer, A. C. In *Nucleation*; Dekker: New York, 1969.
(9) Nowakowski, B.; Ruckenstein, E. *J. Chem. Phys.* **1992**, *96*, 2313.
(10) Ruckenstein, E. In *Metal-support interactions in catalysis, sintering, and redispersion;* Stevenson, S. A., Dumesis, J. A., Baker, R. T. K., Ruckenstein, E., Eds.; Van Nostrand Reinhold: New York, 1987.
(11) Abraham, F. *Homogeneous Nucleation Theory;* Academic: New York, 1974.
(12) Lee, S. H.; Rossky, P. J. *J. Chem. Phys.* **1994**, *100*, 3334.
(13) Elsgolc, L. E. *Calculus of variations;* Pergamon Press: Addison-Wesley: New York, 1962.
(14) Spelt, J. K.; Li, D.; Neumann, A. W. In *Modern Approaches to Wettability;* Loeb, G. I., Ed.; Plenum Press: New York, 1992; p 101.
(15) Stillinger, F. H.; Rahman, A. *J. Chem. Phys.* **1974**, *60*, 1545.
(16) Martoňàk, R.; Colombo, L.; Molteni, C.; Parrinello, M. *J. Chem. Phys.* **2002**, *117*, 11329.

3.3 Microcontact and Macrocontact Angles and the Drop Stability on a Bare Surface*

Gersh Berim and Eli Ruckenstein†
Department of Chemical and Biological Engineering, State University of New York at Buffalo, Buffalo, New York 14260

Corresponding Author

† Electronic mail: feaeliru@buffalo.edu.
Phone: (716)645-1179, Fax: (716)645-3822.

Received: August 10, 2004

I. INTRODUCTION

The microscopic theory that describes a liquid drop on a bare solid substrate, which was developed in refs 1–3, is based on a continuum picture implying that the drop is spatially homogeneous and can be characterized by a constant number density. Equations for the drop profile were obtained either by considering the mechanical equilibrium of the molecules on the drop surface[2,3] or through the variational minimization of the total potential energy of the drop.[1] In the framework of the latter approach, it was shown[1] that the height h_m of the drop at its apex can have, even in the absence of gravity, any value only if a parameter a, which is a combination of the interaction parameters characterizing the interactions between the molecules of the liquid as well as between the molecules of the solid and liquid, is less than a critical value, a_c. For $a > a_c$, the droplet height, h_m, is limited by a critical value, h_c, which depends on a. Were a droplet with a height larger than h_c to be artificially created, it would transform into a droplet with a new height less than h_c. If h_m is close to h_c, the droplet will have a planar shape.

Among the results, the theory of ref 1 provided a constant value $\theta_0 = 180°$ for the microcontact angle, which the drop profile makes with the solid surface. Whereas the macroscopic (apparent) contact angle, θ_m, which was defined for sufficiently large drops using a continuation of the circular part of the drop profile, was dependent on the intermolecular interactions and is accessible experimentally, the angle θ_0 was constant. A possible reason for such a behavior of the microcontact angle might be the inadequacy of the continuum picture for the description of the drop. Indeed, the assumption of drop homogeneity may be not valid, especially in the vicinity of the solid surface where the interactions between the molecules of liquid and solid are both short- and long-range. Hence, near the solid surface, a layer of liquid molecules can appear that is beyond the continuum picture.

In this paper, the microscopic approach of refs 1–3 is modified to include the characteristics of the liquid near the solid surface. It is shown that in this case the microcontact angle θ_0 at the solid

* *J. Phys. Chem.* B 2004, 108, 19339-19347. Republished with permission.

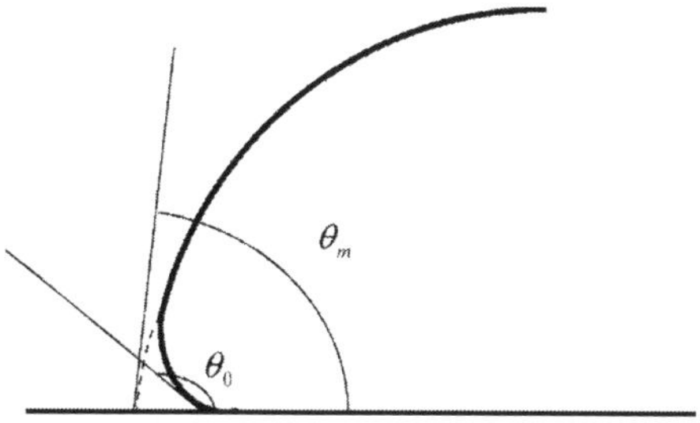

FIGURE 1. Microcontact angle θ_0 and macroscopic (apparent) contact angle θ_m. The latter angle is defined using a continuation of the circular part of the drop profile.

surface becomes a function of the parameters of the intermolecular interactions. The dependence of the microcontact angle on the parameters of intermolecular interactions affects both the shape and the stability of the drop. For example, in the new picture the above-mentioned quantities, a_c and h_c, become dependent on the microcontact angle, θ_0.

It was also shown previously[1–3] that the profile of a liquid drop on a solid substrate exhibits a rapid variation of curvature in a small region (~ 10–100 Å) near the surface, due to the rapid variation of the interactions between molecules of liquid and solid. Because of its small size, this region and, in particular, the microcontact angle θ_0 are practically undetectable in macroscopic experiments, while the macroscopic contact angle, which represents an extrapolation of the circular part of the drop profile to the solid surface, is a measurable quantity (see Figure 1).

In section II, the interaction potentials between the molecules of liquid and solid are introduced and a general form of the total potential energy is derived by taking into account the presence of a special liquid layer near the solid surface. The application of these expressions to a cylindrical droplet is considered in section III and that for an axisymmetrical droplet in section IV.

II. BASIC EQUATION

A. Potentials.

Assuming that the interaction of the drop molecules with the molecules of the surrounding vapor is negligible, two potentials, $\phi_{LL}(r)$ and $\phi_{LS}(r)$ are considered, which describe, respectively, the interaction of a molecule of a liquid drop with the other molecules of the drop and that with the molecules of the solid (r is the distance between the centers of the interacting molecules). For a variety of liquids, the potential $\phi_{LL}(r)$ can be considered to be a combination between a London—van der Waals (LvdW) attraction and a rigid core repulsion[4]

$$\tilde{\phi}_{LL}(r) = \begin{cases} -\epsilon'_{LS}\left(\dfrac{\sigma'_{LL}}{r}\right)^6 & r \geq \sigma'_{LL} \\ \infty & r < \sigma'_{LL} \end{cases} \tag{1}$$

where σ'_{LL} and $\epsilon'_{LL} > 0$ are the size of the repulsive core and an interaction constant, respectively. The calculation of the potential energy of a liquid drop based on potential eq 1 is a complicated problem. To avoid difficulties, a more simple square-well potential

$$\phi_{LL}(r) = \begin{cases} -\epsilon_{LL} & r \leq \eta \\ 0 & r > \eta \end{cases} \tag{2}$$

will be used, where η is the radius of the so-called interaction sphere[1] and ϵ_{LL} characterizes the strength of the interactions. Despite its oversimplified form, such a potential is a reasonable approximation for the qualitative understanding of various phenomena.[5] With appropriate choices for the parameters ϵ_{LL} and η, the potential eqs 2 and 1 can provide equal values for at least some of the physical characteristics of the system. As an example, let us compare the total potential energy of a molecule in the bulk of a homogeneous liquid calculated using potential eq 1 to that obtained using potential eq 2. The first potential energy, ϕ_1', is given by

$$\phi_1' = \int_0^{2\pi} d\phi \int_0^{\pi} d\theta \sin\theta \int_{\sigma_{LL}'}^{\infty} \phi_{LL}'(r)\rho_L r^2 dr = -\frac{4}{3}\pi\sigma_{LL}'^3\epsilon_{LL}'\rho_L \tag{3}$$

where ρ_L is the liquid density. The second potential energy, ϕ_1, is proportional to the volume of the interaction sphere and is equal to

$$\phi_1 = -\frac{4}{3}\pi\eta^3\epsilon_{LL}\rho_L \tag{4}$$

Comparing eqs 3 and 4, one arrives at an expression connecting the parameters of the two potentials

$$\frac{\epsilon_{LL}}{\epsilon_{LL}'} = \left(\frac{\sigma_{LL}'}{\eta}\right)^3 \tag{5}$$

In particular, one of the possible choices of the parameters ϵ_{LL} and η is $\epsilon_{LL} = \epsilon_{LL}'$ and $\eta = \sigma_{LL}'$.

The interaction potential $\phi_{LS}(r)$ will be represented as the sum of two components, $\phi_{LS}^{l}(r)$ and $\phi_{LS}^{s}(r)$, related to the long-range and the short-range (acid—base) interactions between the liquid and solid. The long-range potential is chosen in the form

$$\phi_{LS}^{l}(r) = \begin{cases} \epsilon_{LS}\left[k_\phi\left(\frac{\sigma_{LS}}{r}\right)^{12} - \left(\frac{\sigma_{LS}}{r}\right)^{6}\right] & r \geq \sigma_{LS} \\ \infty & r < \sigma_{LS} \end{cases} \tag{6}$$

where $k_\phi = 0$ or 1, σ_{ls} and $\epsilon_{LS} > 0$ are, respectively, the size of the repulsive core and the interaction constant of the liquid—solid interactions. For $k_\phi = 0$, one recovers the LvdW potential and for $k_\phi = 1$ and $r \geq \sigma_{LS}$ the Lennard-Jones (LJ) potential. In what follows, σ_{LS} is used as unit of length. In real systems, σ_{LS} is of the order of several angstroms.[6]

Assuming pairwise additivity, the potential of a liquid molecule interacting with a semi-infinite solid through the long-range interactions can be written as

$$\Phi_{LS}^{l}(h) = \frac{\pi}{6}\epsilon_{LS}\rho_S\sigma_{LS}^3\left[\frac{2}{15}k_\phi\left(\frac{\sigma_{LS}}{h+\sigma_{LS}}\right)^9 - \left(\frac{\sigma_{LS}}{h+\sigma_{LS}}\right)^3\right], h \geq 0 \tag{7}$$

where ρ_S is the constant number density of molecules in the solid and $h + \sigma_{LS}$ is the distance of the center of the molecule of liquid from the first layer of solid atoms ($h = 0$ corresponds to the centers of the layer of liquid molecules on the solid surface).

The short-range potential, $\phi_{LS}^{s}(r)$, accounts for the acid—base interactions, which can be attractive as well as repulsive.[7] It is assumed that $\phi_{LS}^{s}(r)$ decays exponentially and can be considered zero at distances larger than $l_0 \approx 1$ nm.[7,8] It will be supposed that the potential energy $\Phi_{LS}^{s}(h)$ of a liquid molecule due to the short-range interactions has the value $-\epsilon_{LS}^{s}$ for the molecules of the first layer of liquid near the solid surface ($h = 0$) and is equal to zero for all other molecules. The distance of the first layer of liquid molecules from the solid surface is assumed to be equal to the hard core radius, σ_{LS}, of the potential eq 6. As a result, the potential $\Phi_{LS}^{s}(h)$ is given by

$$\Phi_{LS}^{s}(h) = \begin{cases} -\epsilon_{LS}^{s} & h = 0 \\ 0 & h > 0 \end{cases} \tag{8}$$

Note that the interaction constant ϵ_{LS}^{s} can be positive or negative depending on the system. The microscopic calculation of ϵ_{LS}^{s} is a separate issue.

The potential eqs 7 and 8 have to be combined to provide the total potential $\Phi_{LS}(h)$ of a molecule of liquid interacting with a semi-infinite solid

$$\Phi_{LS}(h) = \begin{cases} \Phi_{LS}^{l}(0) + \Phi_{LS}^{s}(0) & h = 0 \\ \Phi_{LS}^{l}(h) & h > 0 \end{cases} \tag{9}$$

The most frequently used model for the theoretical description of liquid drops is a continuum model, which considers the liquid as a continuous medium characterized by a macroscopic number density, ρ_L, the latter being, in general, a function of the space coordinates. Such a description implies that all interactions change slowly for distances comparable with the intermolecular distances in the liquid. This assumption is not valid in the very close vicinity of the solid surface where the liquid—solid interactions (long- as well as short-range) change rapidly with the moving away from the solid surface. For example, the potential $\Phi_{LS}^{l}(h)$ (for $k_\phi = 0$ and $\sigma_{LS} \simeq \sigma'_{LL}$) is 8-fold smaller for the molecules of the second layer of liquid than for the molecules of the first layer. Due to the strong liquid—solid interactions, the liquid molecules nearest to the solid surface are arranged in a layer "stuck" to the solid surface, which will be assumed to be a monolayer. The rest of the drop is assumed to be a continuous medium of constant density (see Figure 2).

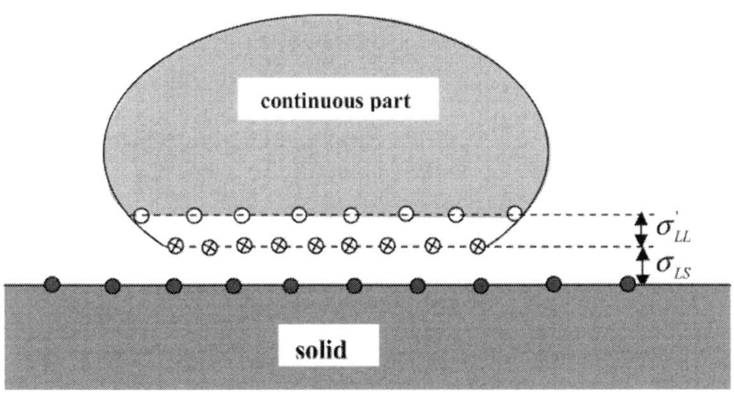

FIGURE 2. The cross-section of a drop on a solid surface. The filled and crossed circles represent the molecules of the solid surface and the molecules of the first (nearest to the solid surface) layer of liquid molecules, respectively. Beginning with the second layer (open circles), the liquid is considered as a continuous medium.

B. Potential Energy of a Drop.

The total potential energy of a drop (the gravity is neglected)

$$U_{\text{total}} = U_{LL} + U_{LS} \tag{10}$$

consists of a potential energy U_{LL} due to the interactions of the molecules of liquid between themselves and a potential energy U_{LS} due to the interactions between the molecules of liquid and the molecules of solid. The presence of a liquid monolayer on the solid surface will be taken into account during the calculation of U_{LS}. In the calculation of U_{LL}, the deviation from the continuum model is neglected.

The energy U_{LL} for a constant number density ρ_L and the interaction potential eq 2 was calculated in ref 1, and the result has the form

$$U_{LL} = -\frac{2}{3}\pi\eta^3\rho_L{}^2\epsilon_{LL}V + \frac{1}{8}\pi\eta^4\rho_L{}^2\epsilon_{LL}S \tag{11}$$

where V and $S = S_{LV} + S_{LS}$ are the total volume and surface of the drop, respectively, the latter including the areas of the liquid—vapor (S_{LV}) and liquid—solid (S_{LS}) interfaces. The two terms in the right-hand side of eq 11 represent the bulk and liquid/vapor surface energy of the droplet. The coefficient multiplying the area S can be considered as a microscopic expression for the surface tension γ_{lv} of the liquid/vapor interface ($\gamma_{lv} = 1/8\,\pi\eta^4\rho_L{}^2\epsilon_{LL}$).

To calculate the interaction energy U_{LS}, it will be supposed that the molecules of liquid near the planar solid surface form a monolayer with a constant surface number density, ρ_{LS}, separated from the solid surface by the distance σ_{LS}. The molecules of that layer are affected by both long-range and short-range interactions with the solid, which are represented by potential eq 9 for $h = 0$. The continuous part of the drop is separated from the first layer by the distance $\sigma'_{LL} \approx \eta$ and its molecules do not exhibit short-range interactions.

Under the above assumptions, the energy U_{ls} is given by

$$U_{LS} = \rho_{LS}\Phi_{LS}(0)S_{LS} + \int_{V_{\text{cont}}} \rho_L\Phi^1_{LS}(h)\mathrm{d}V \tag{12}$$

where S_{LS} is the contact area between the droplet and the solid, and V_{cont} denotes the volume of the continuous part of the droplet. The latter can be represented as the difference between the drop volume V and the volume V_g of the gap between the first layer and the continuous part of the droplet, which is approximately equal to $V_g = \sigma'_{LL}S_{LL}$,, that is, $V_{\text{cont}} = V - S_{LS}\sigma'_{LL}$. Then the second term in the right-hand side of eq 12 can be written as

$$\int_{V_{\text{cont}}} \rho_L\Phi^1_{LS}(h)\mathrm{d}V = \int_V \rho_L\Phi^1_{LS}(h)\mathrm{d}V - \int_{V_s} \rho_L\Phi^1_{LS}(h)\mathrm{d}V \simeq \int_V \rho_L\Phi^1_{LS}(h)\mathrm{d}V - \rho_L\sigma'_{LL}\Phi^1_{LS}(\sigma'_{LL})S_{LS} \tag{13}$$

Consequently,

$$U_{LS} \simeq \int_V \rho_L\Phi^1_{LS}(h)\mathrm{d}V + \left[\rho_{LS}\Phi_{LS}(0) - \rho_L\sigma'_{LL}\Phi^1_{LS}(\sigma'_{LL})\right]S_{LS} \tag{14}$$

Equations 11 and 14 provide a basis for the analysis of the drop characteristics (shape, stability, etc.) using the minimization of the total potential energy, eq 10. Below, these characteristics will be examined separately for cylindrical and axisymmetrical droplets.

III. CYLINDRICAL DROPLET

A. Equations for the Profile and the Microcontact Angle.

Let us assume that a cylindrical liquid drop is extended along the y-axis and has a cross-section symmetrical with respect to the h-axis. Depending on the value of the microcontact angle, the drop profile can have the shapes presented in Figure 3. It is usual practice[9–11] to describe the drop profile through the classical augmented Young—Laplace differential equation, which for a cylindrical droplet has the form

$$\frac{h_{xx}}{\left(1+h_x^{\,2}\right)^{3/2}}+\Pi\left(h\right)/\gamma_{\mathrm{lv}}=-p_{\mathrm{c}}/\gamma_{\mathrm{lv}} \tag{15}$$

where $h \equiv h(x)$ is the drop thickness, the subscript of h denotes the derivative with respect to the corresponding variable, γ_{lv} is the bulk liquid/vapor surface tension, $\Pi(h)$ is the disjoining pressure originating from the liquid—solid and liquid—liquid interactions, and p_{c} is the capillary pressure. The disjoining pressure can be obtained from the potential interaction energy, $P(h)$, that is $\Pi(h) = -(\mathrm{d}/\mathrm{d}h)P(h)$. However, eq 15 describes only the profiles for which $\theta_0 \leq 90°$ (Figure 3a), because for $\theta_0 > 90°$ the function $h(x)$ is not unique for $|x| > x_0$, where x_0 is the abscissa of the leading edge of the droplet. To include both shapes (Figure 3, parts a and b) into consideration, the equation of the droplet profile is chosen in the form $x = x(h)$, and only the part of the profile for which $x \geq 0$ will be considered further. With such a choice, one can write (all quantities are calculated per unit length of the droplet in the y-direction)

$$V=2\int_0^{h_{\mathrm{m}}} x\,\mathrm{d}h,\quad S_{\mathrm{LV}}=2\int_0^{h_{\mathrm{m}}}\sqrt{1+x_{\mathrm{h}}^{\,2}}\,\mathrm{d}h$$

$$S_{\mathrm{LS}}=2x\left(0\right)=-2\int_0^{h_{\mathrm{m}}} x_h\,\mathrm{d}h$$

$$\int_V \rho_{\mathrm{L}}\Phi_{\mathrm{LS}}^{1}\left(h\right)\mathrm{d}V=2\rho_{\mathrm{L}}\int_0^{h_{\mathrm{m}}}\Phi_{\mathrm{LS}}^{1}\left(h\right)x\,\mathrm{d}h \tag{16}$$

where the subscript of x denotes the derivative with respect to the corresponding variable, h_m is the droplet height, and $x\left(0\right)=-\int_0^{h_{\mathrm{m}}} x_h\,\mathrm{d}h$ follows from the obvious expression $x(h_m) = 0$. The total potential energy, eq 10, per unit length is given by

$$U_{\mathrm{total}}=\int_0^{h_{\mathrm{m}}} f\left(h,x,x_h\right)\mathrm{d}h \tag{17}$$

where

$$f\left(h,x,x_h\right)=\frac{1}{4}\pi\eta^4\rho_{\mathrm{L}}^{\,2}\epsilon_{\mathrm{LL}}\left\{\left[-a_0+a_1\Phi_{\mathrm{LS}}^{1}\left(h\right)\right]x+\sqrt{1+x_h^{\,2}}-bx_h\right\} \tag{18}$$

$$a_0 = \frac{16}{3\eta}, \; a_1 = \frac{8}{\pi\eta^4 \rho_L \epsilon_{LL}} \tag{19}$$

and

$$b = 1 + \frac{8\sigma'_{LL}}{\pi\rho_L \eta^4 \epsilon_{LL}} \left[\frac{\rho_{LS}}{\rho_L \sigma'_{LL}} \Phi_{LS}(0) - \Phi^1_{LS}(\sigma'_{LL}) \right] \tag{20}$$

In eq 20, according to eqs 9, 8, and 7

$$\Phi_{LS}(0) = \frac{\pi}{6} \epsilon_{LS} \rho_S \sigma_{LS}^3 \left(\frac{2}{15} k_\phi - 1 \right) - \epsilon^s_{LS} \tag{21}$$

and

$$\Phi^1_{LS}(\sigma'_{LL}) = \frac{\pi}{6} \epsilon_{LS} \rho_S \sigma_{LS}^3 \left[\frac{2}{15} k_\phi \left(\frac{\sigma_{LS}}{\sigma'_{LL} + \sigma_{LS}} \right)^9 - \left(\frac{\sigma_{LS}}{\sigma'_{LL} + \sigma_{LS}} \right)^3 \right] \tag{22}$$

If one uses as a constraint the given volume V of the drop, then the equation of the drop profile can be obtained by the variational minimization of the functional

$$\int_0^{h_m} f(h, x, x_h) \, dh + 2\lambda \int_0^{h_m} x \, dh \tag{23}$$

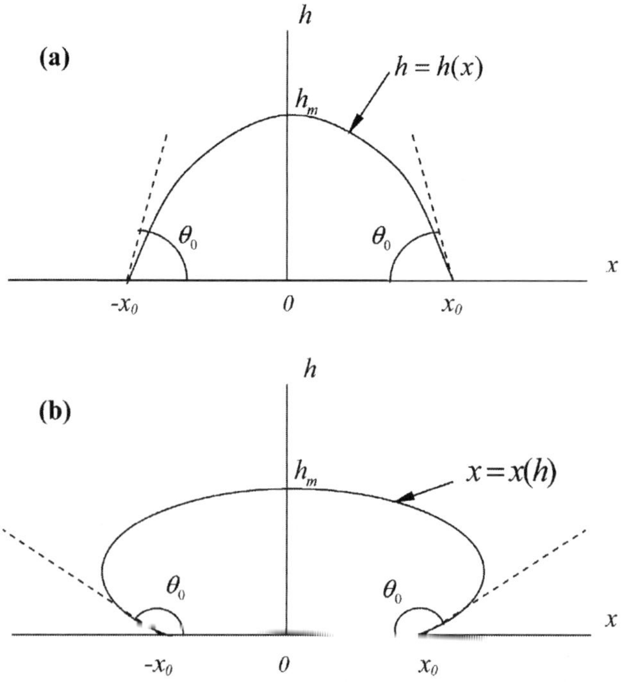

FIGURE 3. Two characteristic profiles of a cylindrical droplet: (a) microcontact angle θ_0 smaller than 90°; (b) θ_0 larger than 90°. The y-axis is normal to the plane of the figure.

where λ is a Lagrange multiplier for the drop volume. The functional eq 23 can be rewritten in the following more convenient form

$$\int_0^{h_m} F(h,x,x_h)\,\mathrm{d}h \tag{24}$$

where

$$F(h,x,x_h) = \left[\lambda_c + a_1\Phi^1_{LS}(h)\right]x + \sqrt{1+x_h^2} - bx_h$$

$$\lambda_c = 8\lambda / \left(\pi\eta^4 \rho_L^2 \epsilon_{LL}\right) - a_0 \tag{25}$$

The latter functional is obtained by dividing the functional eq 23 with the nonzero coefficient $\pi\eta^4 \rho_L^2 \epsilon_{LL}/4$.

Using the standard calculus of variations,[12] one can obtain the following differential equation for the profile:

$$\frac{x_{hh}}{\left(1+x_h^2\right)^{3/2}} - a_1\Phi^1_{LS}(h) = \lambda_c \tag{26}$$

which coincides in form with the augmented Young—Laplace equation (eq 15) for cylindrical droplets. The first terms of eqs 26 and 15 represent the curvature $\kappa(h)$ of the profile, which in the considered case can be rewritten as

$$\kappa(h) = \lambda_c + a_1\Phi^1_{LS}(h) \tag{27}$$

with a curvature radius given by

$$R^{\text{curv}}(h) = \frac{1}{|\kappa(h)|} \tag{28}$$

One can establish the following correspondence between the disjoining [$\Pi(h)$] and the capillary (p_c) pressures in eq 15 and the terms involved in eq 26:

$$\Pi(h)/\gamma_{lv} = -a_1\Phi^1_{LS}(h);\ \lambda_c = -p_c/\gamma_{lv} \tag{29}$$

Because the apex of the droplet and the contact point of the profile with the solid should be located on the h- and x-axes, respectively, one can also write the so-called transversality conditions[12]

$$\left.\frac{\partial}{\partial x_h}F(h,x,x_h)\right|_{h=0} = 0,\ \left.\left[F(h,x,x_h) - x_h\frac{\partial}{\partial x_h}F(h,x,x_h)\right]\right|_{h=h_m} = 0 \tag{30}$$

which provide the following two boundary conditions for eq 26

$$\left.\frac{1}{\sqrt{1+x_h^2}}\right|_{h=h_m} = 0 \tag{31}$$

$$\left.\frac{x_h}{\sqrt{1+x_h^2}}\right|_{h=0} = b \tag{32}$$

The first condition, eq 31, or, equivalently, $x_h|_{h=h_m} = \infty \left(h_x|_{x=0} = 0\right)$, reflects the fact that the tangent to the drop profile at its apex should be parallel to the solid surface. The second condition, eq 32, provides an expression for the microcontact angle θ_0 of the drop on the solid surface (Figure 1) through the equation

$$\cos\theta_0 = -b \tag{33}$$

which follows from eq 32 because $x_h|_{h=0} = -\cos\theta_0$. It should be noted that the microcontact angle is obtained from pure microscopic considerations, without involving such thermodynamic quantities as the surface tensions.

The substitution $p = x_h$ transforms eq 26 into the first-order equation

$$\frac{p_h}{\left(1+p^2\right)^{3/2}} = \lambda_c + a_1\Phi^1_{LS}(h) \tag{34}$$

which has the solution

$$p = \frac{\Psi(h)}{\sqrt{1-\Psi^2(h)}} \tag{35}$$

where

$$\Psi(h) = \frac{x_h}{\sqrt{1+x_h^{\,2}}} = \lambda_c h + D(h) + C \tag{36}$$

$$D(h) = a_1\int \Phi^1_{LS}(h)\,\mathrm{d}h \tag{37}$$

and C is a constant of integration. After substitution of x_h instead of p into eq 35, the following general solution for $x(h)$ can be obtained[1]

$$x(h) - x_0 = \int_0^h \frac{\Psi(h)}{\sqrt{1-\Psi^2(h)}}\,\mathrm{d}h \tag{38}$$

where $x_0 = x(0)$ (Figure 3) is a second integration constant. It can be determined from the condition for the drop apex to be located on the h-axis $[x(h_m) = 0]$, which leads to the expression

$$x_0 = -\int_0^{h_m} \frac{\Psi(h)}{\sqrt{1-\Psi^2(h)}}\,\mathrm{d}h \tag{39}$$

The first integration constant, C, can be obtained using the boundary condition eq 32, which is equivalent to the equation $\Psi(0) = b = -\cos\theta_0$. The latter equation provides for C the value $C = b - D(0)$, and the function $\Psi(h)$ becomes

$$\Psi(h) = b + \lambda_c h + D(h) - D(0) \tag{40}$$

After the determination of x_0 and C, the solution, eq 38, still contains two unknown parameters, λ_c and h_m. A connection between them is provided by the boundary condition eq 31, which leads to the equation $\Psi(h_m) = \pm 1$. The sign can be selected by observing that in the absence of liquid—solid interactions a drop must have a spherical shape and contact the solid surface only at $x_0 = 0$. Such a solution can be obtained[1] if $\Psi(h_m) = -1$, which provides the following equation for λ_c:

$$\lambda_c = -\frac{1}{h_m}\left[1 - \cos\theta_0 + D(h_m) - D(0)\right] \tag{41}$$

The last necessary condition for the determination of all unknown parameters is provided by the equation for the drop volume, $V = 2\int_0^{h_m} x\,dh$. Integrating by parts and taking into account that $x(h_m) = 0$, one can rewrite the latter equation in the form $V = -2\int_0^{h_m} hx_h\,dh$, which after substitution of expression eq 35 for $x_h = p$ becomes

$$V = -2\int_0^{h_m} \frac{h\Psi(h)}{\sqrt{1-\Psi^2(h)}}dh \tag{42}$$

B. Existence of a Cylindrical Drop on a Bare Surface.

The first necessary condition for the existence of a drop on a bare surface follows from eq 33 and has the simple form $|b| \leq 1$. It establishes [through eq 20] a restriction for the values of the microscopic parameters for which a drop can be formed on a solid surface. If $|b| > 1$, the drop will spread completely over the surface of the solid. It should be noted that this restriction is very different from the traditional one. The latter implies $|\cos\theta_m| > 1$.

Another necessary condition results from the fact that the integral in the right-hand side of eq 38 exists only if the function $\Psi(h)$ satisfies the inequality $|\Psi(h)| < 1$ for $0 < h < h_m$. Because, as shown above, $\Psi(h_m) = -1$, the function $\Psi(h)$ has to be a decreasing one in the vicinity of $h = h_m$, that is, its derivative at that point must be negative

$$\left.\frac{d\Psi(h)}{dh}\right|_{h=h_m} < 0 \tag{43}$$

This inequality provides a second necessary condition for the existence of a drop on a bare surface.

When expression 40 is used for $\Psi(h)$ and expression 37 is used for $D(h)$, condition 43 becomes

$$\lambda_c + a_1\Phi^1_{LS}(h_m) < 0 \tag{44}$$

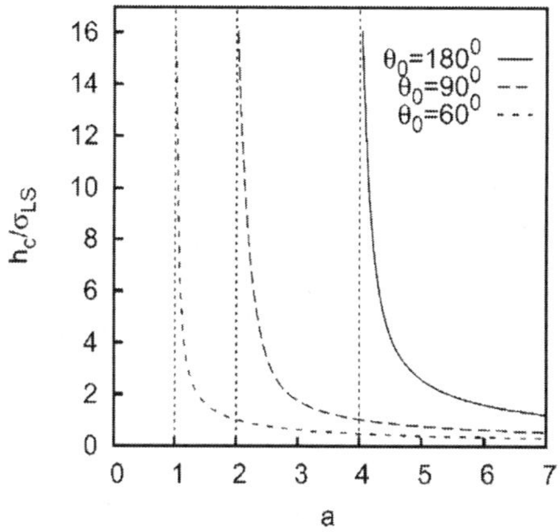

FIGURE 4. a-dependence of the critical height h_c for various values of the microcontact angle from the domain of limited height. The vertical asymptotes for θ_0 equal to 60°, 90°, and 180° are given by equations $a = 1$, $a = 2$, and $a = 4$, respectively.

which is equivalent to the condition for the drop profile to be convex at the drop apex. Together with eq 41 for λ_c, inequality 44 acquires the form

$$-1+\cos\theta_0 - D(h_m) + D(0) + a_1\Phi^1_{LS}(h_m) < 0 \quad (45)$$

and provides a restriction for the drop height, h_m, the specific form of which depends on the form of potential $\Phi^1_{LS}(h)$.

B.1. London—van der Waals Potential. In this case, the potential $\Phi^1_{LS}(h)$ is given by eq 7 with $k_\phi = 0$, and inequality 45 can be rewritten as

$$a < A_{LW}(h_m, \theta_0) \quad (46)$$

where

$$a = \frac{4}{3}\frac{\epsilon_{LS}}{\epsilon_{LL}}\frac{\rho_S}{\rho_L}\left(\frac{\sigma_{LS}}{\eta}\right)^4 \quad (47)$$

and

$$A_{LW}(h) = 2(1-\cos\theta_0)\left(\frac{\sigma_{LS}}{h}\right)^2 \frac{(1+h/\sigma_{LS})^3}{3+h/\sigma_{LS}} \quad (48)$$

The analysis of inequality 46 shows that it has a solution for any pair (θ_0, a) of values of the parameters θ_0 and a, where $0 < \theta_0 \leq 180°$ and $a > 0$. However, for some pairs the droplet height h_m can have any value, while for other pairs, it is limited by a critical value h_c given by equation

$$A_{LW}(h_c, \theta_0) = a \quad (49)$$

The domain of pairs of the first kind lies in the plane $\theta_0 - a$ below a critical curve, the equation of which can be obtained from eq 49 in the limit $h_c \to \infty$

$$a = 2(1-\cos\theta_0) \quad (50)$$

The domain of the pairs of the second kind lies above that curve. In the latter domain

$$a > 2(1-\cos\theta_0) \quad (51)$$

The above-mentioned domains and the critical curve (solid line) for the LvdW potential are presented in Figure 5.

The a-dependence of h_c provided by eq 49 is presented in Figure 4 for several values of the microcontact angle. For each value of θ_0, there is a critical value a_c of a, near which the height h_c grows rapidly. Obviously, the point (θ_0, a_c) lies on the critical curve [eq 50].

B.2. Lennard-Jones Potential. In this case, inequality 45 acquires the form of inequality 46, where $A_{LW}(h,\theta_0)$ is replaced by

$$A_{LJ}(h,\theta_0) = 60(1-\cos\theta_0)/\left[29 - 30(1+h)^{-2} - 60h(1+h)^{-3} + (1+h)^{-8} + 8h(1+h)^{-9}\right] \quad (52)$$

The equation for the critical curve, which separates the domains of a and θ_0 in which the height h_m of the drop is limited or unlimited, has the form

$$a = \frac{60}{29}(1-\cos\theta_0) \simeq 2.07(1-\cos\theta_0) \quad (53)$$

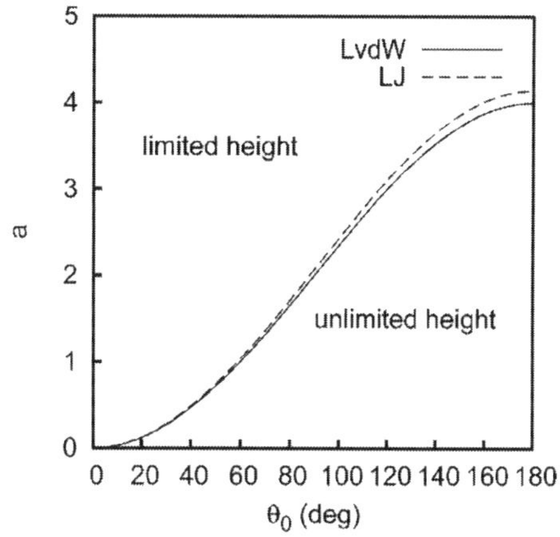

FIGURE 5. The critical curves in the θ_0–a plane for the LvdW (solid line) and LJ (dashed line) potentials. The domain of limited height for each potential lies above the corresponding critical curve, and the domain of unlimited height lies below it. The points on the critical curves corresponding to $a = 0.8$ have the coordinates θ_0 equal to $\theta_{0,1} = 53.1301°$ and $\theta_{0,2} = 52.1691°$ for the LvdW and LJ potentials, respectively.

which is similar to that for the LvdW potential [eq 50]. This curve is presented in Figure 5 by the dashed line. The difference between the critical curves for the LvdW and LJ potentials is small, the value of a for the LJ potential at the same θ_0 being larger than that for the LvdW potential. In the domain in which h_m is limited for both potentials, the value of h_c is larger for the same a and θ_0 for the LJ potential than for the LvdW potential (see Figure 6 plotted for $\theta_0 = 60°$).

C. Shape of the Drop.

For the LvdW potential, typical profiles of cylindrical droplets with a constant volume $V/\sigma_{LS}^3 = 4.21 \times 10^4$, but various microcontact angles are presented in Figure 7a for $a = 0.8$ and $x \geq 0$. For this value of a, the drop can have any height for microcontact angles larger that $\theta_{0,1} = 53.1301°$. If $\theta_0 < \theta_{0,1}$, the height is limited. Figure 7a shows that for $\theta_0 > \theta_{0,1}$ the drop height decreases and the half-width x_0 of the contact area of the drop with the solid increases with decreasing microcontact

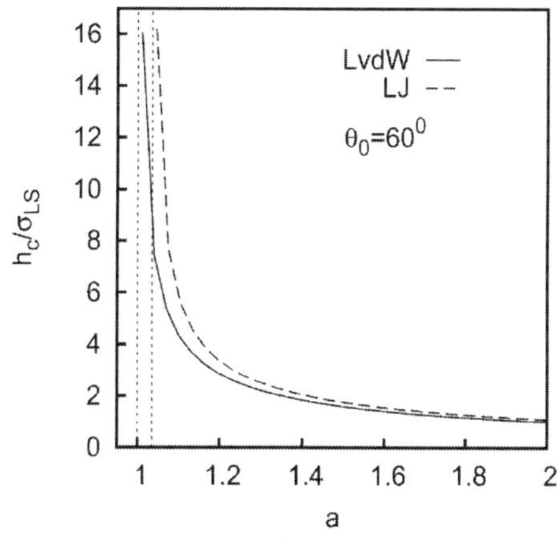

FIGURE 6. a-dependence of the critical height h_c for the LvdW (solid line) and LJ (dashed line) potentials and for a microcontact angle $\theta_0 = 60°$. The critical values of a are in this case $a_{c,1} = 1$ and $a_{c,2} = 1.035$ for the LvdW and LJ potentials, respectively.

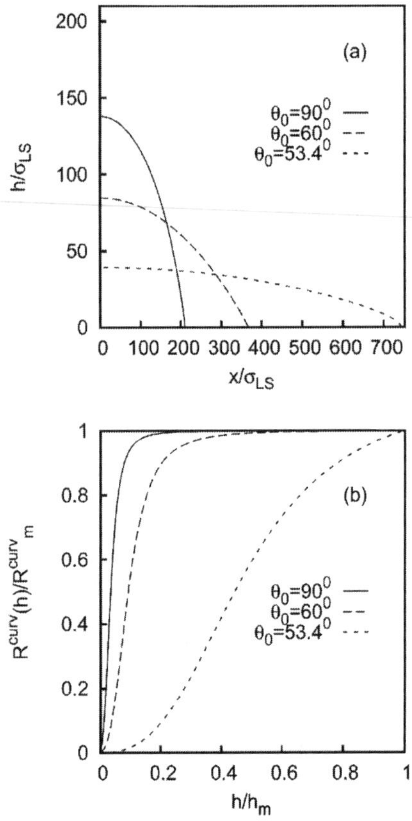

FIGURE 7. Panel a shows dependence of the drop shape on the microcontact angle for $a = 0.8$. All drops have the same volume $V/\sigma_{ls}^3 \simeq 4.2 \times 10^4$. Panel b shows normalized h-dependence of the curvature radius $R^{curv}(h)$ of the profile for a droplet of volume $V/\sigma_{ls}^3 \simeq 4.2 \times 10^4$ for various microcontact angles. R_m^{curv}/σ_{LS} is equal to 230.3, 845.0, and 8773.0 for θ_0 equal to 90°, 60°, and 53.4°, respectively. R_m^{curv} and h_m are the curvature and height at the apex of the droplet.

angle. The closer θ_0 is to $\theta_{0,1}$, the larger is x_0, and the drops become almost planar. (This planar drop is not included in Figure 7a because of the very large value of x_0).

As noted earlier in ref 1, the shape of the drop can be often considered circular starting from some distance h_2 from the solid surface, that is, for $h_2 < h < h_m$, the curvature radius is almost constant (see Figure 8). In those cases, the circular part of the profile decreases with decreasing microcontact angle. To illustrate this point, which cannot be clearly seen from Figure 7a, the h-dependence of the curvature radius defined by eq 28 is presented in Figure 7b by plotting the normalized values $R^{curv}(h)/R_m^{curv}$ against h/h_m, where $R_m^{curv} \equiv R^{curv}(h_m)$ is the largest value of the curvature radius, which the profile has at the drop apex. One can see that, indeed, the relative range in which $R^{curv}(h)$ can be considered constant is smaller for smaller microcontact angles.

For large drops, the largest value R_m^{curv} of the curvature radius is a linear function of the drop height

$$R_m^{curv} = h_m\left(1 - \cos\theta_0 - a/2\right)^{-1} \tag{54}$$

where $a/2 < 1 - \cos\theta_0$. Equation 54 was obtained by combining eq 28 with eqs 27, 41, and 7 and by considering the limit $h_m/\sigma_{LS} \gg 1$.

Equation 54 can be used to calculate the macroscopic contact angle θ_m, which can be defined (see Figure 8) as the angle between the solid surface and the continuation of the circular part of the

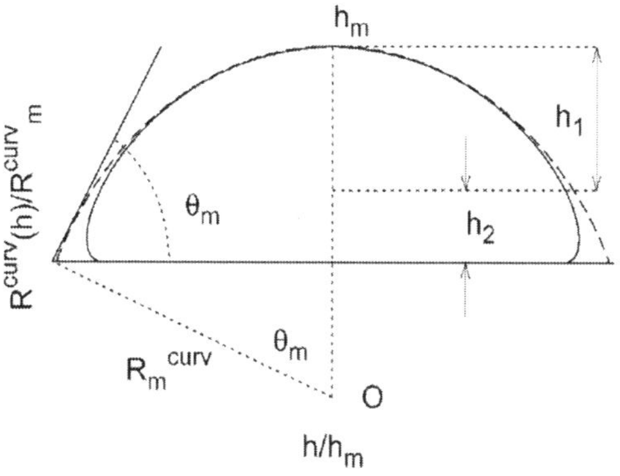

FIGURE 8. Definition of the macroscopic contact angle θ_m. The dashed line corresponds to the hypothetical circular profile of radius R_m^{curv}, h_1 is the range in which the circular profile almost coincides with the drop profile (solid line), and h_2 is the range of rapid change of the drop profile curvature. The ratio h_2/h_1 has to be small for the macroscopic contact angle to be meaningful.

profile until its intersection with the surface of the solid. Such a definition is reasonable if the range h_2 of rapid change of the drop curvature is much smaller than the range h_1 where the curvature is almost constant. One can easily see from Figure 8 that θ_m is given by the equation

$$\cos\theta_m = 1 - \frac{h_m}{R_m^{curv}} \tag{55}$$

which, by combination of eqs 54 and 55, becomes

$$\cos\theta_m = \frac{a}{2} + \cos\theta_0 \tag{56}$$

The latter formula provides a connection between the micro (θ_0) and macrocontact angles (θ_m); it shows that θ_m is always less than θ_0. Indeed, for the profiles presented in Figure 7a, θ_m is equal to 138°, 75.5°, and 41.4° for θ_0 equal to 180°, 90°, and 53.4°, respectively.

If in the above example the microcontact angle θ_0 would acquire values smaller than $\theta_{0,1}$ = 53.1301°, then the droplet height would become limited because the pair (θ_0, a) will move into the domain of limited height. The largest possible height h_c depends on θ_0 and decreases rapidly with decreasing θ_0. Indeed, h_c/σ_{LS} = 869.6 for θ_0 = 53.13° and h_c/σ_{LS} = 24.4 for θ_0 = 53.0°. If the drop height at a given θ_0 approaches h_c, the drop becomes planar and its volume grows indefinitely, as in the case θ_0 = 180° and a = 4 considered in ref 1.

For a LJ potential, the shape of the drop has exactly the same peculiarities as for the LvdW potential. However, the corresponding profiles for the same values of a and θ_0 differ in details. Figure 9 presents the profiles of drops of volumes equal to $V/\sigma_{ls}^3 \simeq 2.35 \times 10^5$ for both potentials, a = 0.8, and different values of θ_0 in the domain of unlimited height. If the point (a, θ_0) is not close to the critical curves separating the two domains, then the difference between the profiles is not large (the curves of Figure 9 for θ_0 = 60°). However, the difference becomes considerable if that point is in the vicinity of the critical curves (the curves for θ_0 = 53.2°). The reason for such a difference is that, for a = 0.8, the smallest possible value of θ_0 from the domain of unlimited droplet height is equal to $\theta_{0,2}$ = 52.1691° for the LJ potential and to $\theta_{0,1}$ = 53.1301° > $\theta_{0,2}$ for the LvdW potential. As already mentioned, the height of the profiles decreases when θ_0 approaches the corresponding smallest value ($\theta_{0,1}$ or $\theta_{0,2}$) at constant a. The closer is θ_0 to $\theta_{0,1}$ ($\theta_{0,2}$), the higher is the rate of decrease. Because $\theta_{0,2}$ < $\theta_{0,1}$ the selected θ_0 is closer to $\theta_{0,1}$ than to $\theta_{0,2}$. Therefore the height of the profile decreases more rapidly for the LvdW potential than for the LJ potential, and as a consequence, the droplet height for

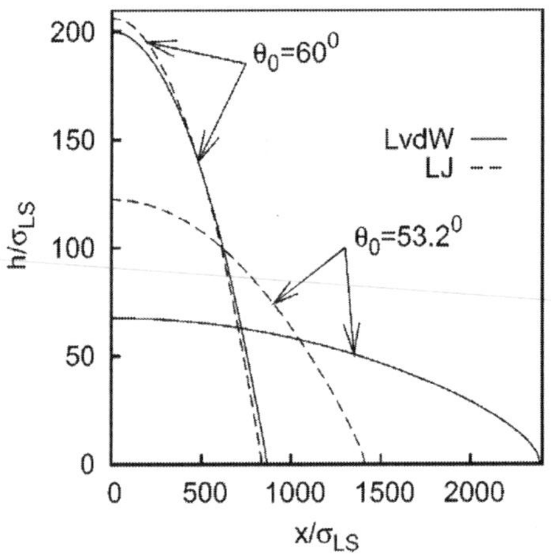

FIGURE 9. The profiles of a drop of volume $V/\sigma_{ls}^3 = 2.35 \times 10^5$ for the LvdW (solid lines) and LJ (dashed lines) potentials and $a = 0.8$ for two microcontact angles. All the profiles belong to the domain of unlimited height. The values of θ_0 that separate the domains of limited and unlimited heights for $a = 0.8$ are $\theta_{0,1} = 53.1301°$ and $\theta_{0,2} = 52.1691°$ for the LvdW and LJ potentials, respectively.

the LvdW potential becomes smaller than that for the LJ potential. The physical explanation is that the repulsive part of the LJ potential reduces the effect of the attractive forces between the solid and liquid molecules. This reduction is more effective for small droplet heights (the case of $\theta_0 = 53.2°$) and almost negligible for larger heights ($\theta_0 = 60°$ in our example).

The same features could be observed when the profiles with the same height for LvdW and LJ potentials were compared. Near the critical line, the difference between profiles is very large (Figure 10a), it decreases with increasing θ_0 and becomes almost negligible far from the critical curves (Figure 10b, c).

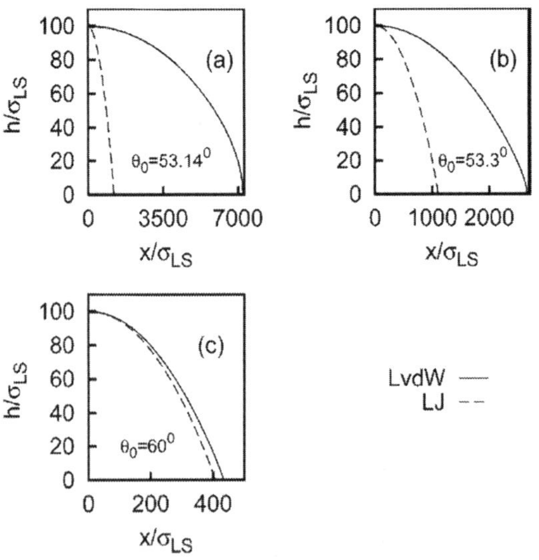

FIGURE 10. Profiles of cylindrical drops of the same height of the apexes for $a = 0.8$ and various values of the microcontact angle in the domain of unlimited height for the LvdW (solid lines) and LJ (dashed lines) potentials.

IV. AXISYMMETRICAL DROPLET

If the drop is axisymmetrical with respect to the h-axis, the volume V and the areas of the interfaces are

$$V = \pi \int_0^{h_m} x^2\,\mathrm{d}h,\; S_{LV} = 2\pi \int_0^{h_m} x\sqrt{1+x_h^{\,2}}\,\mathrm{d}h$$

$$S_{LS} = \pi x_0^{\,2} = -2\pi \int_0^{h_m} x x_h \mathrm{d}h,$$

$$\int_V \rho_L \Phi_{LS}^1(h)\mathrm{d}V = \pi\rho_L \int_0^{h_m} x^2 \Phi_{LS}^1(h)\mathrm{d}V \tag{57}$$

The total potential energy can be written in the form of eq 18 with

$$f(h,x,x_h) = \frac{\pi^2}{4}\eta^4 \rho_L^{\,2} \epsilon_{LL}\left[-\frac{a_0}{2}x^2 + x\left(\sqrt{1+x_h^{\,2}} - bx_h\right) - \frac{a_1}{2}x^2\Phi_{LS}^1(h)\right] \tag{58}$$

where the coefficients a_0, a_1, and b and the function $\Phi_{LS}^1(h)$ are given by expressions 19, 20, and 7, respectively. In this case, the Euler equation for the profile has the form

$$\frac{x_{hh}}{\left(1+x_h^{\,2}\right)^{3/2}} - \frac{1}{x\sqrt{1+x_h^{\,2}}} + a_1\Phi_{LS}^1(h) - 2\lambda_a = 0 \tag{59}$$

where λ_a is a Lagrange multiplier. The boundary conditions can be written in the same form as those for cylindrical droplets [eqs 31 and 32] or, equivalently, $x_h|_{h=h_m} = \infty \left(h_x|_{x=0} = 0\right)$ and $\cos\theta_0 = -b$, where θ_0 is the microcontact angle. Equation 59 cannot be reduced to a first-order differential equation, as for the cylindrical droplets, and therefore all the results presented below have been obtained numerically.

A. Stability of the Axisymmetrical Drop.

As for the cylindrical drops (Figure 5), there is a critical curve in the plane θ_0–a that divides the plane into two domains. In one of them (below the critical curve), the height of the drop can have any value, whereas in the other (above the critical curve), the height is limited. Within the precision of the numerical calculations for each potential, LvdW and LJ, the critical curves for axisymmetrical droplets coincide with those for the cylindrical ones. For example, for $a = 0.8$, the critical values of θ_0 for the LvdW potential are equal to $\theta_{c,\,cyl} = 53.1301°$ and $\theta_{c,\,a} = 53.1295°$ for the cylindrical and axisymmetrical drops, respectively. For the LJ potential, these values are $\theta_{c,\,cyl} = 52.1691°$ and $\theta_{c,\,a} = 52.1690°$, respectively. In the domain of limited heights, the largest possible droplet height h_c for the same a and θ_0 is larger for the axisymmetrical droplet than for the cylindrical one. For example, for $a = 0.8$ and $\theta_0 = 52.1689°$, $h_{c,\,a} = 1441.9$ and $h_{c,\,cyl} = 688.1$ for the axisymmetrical and cylindrical droplets, respectively.

B. Shape of the Axisymmetrical Drop.

The numerical solution of eq 59 provides the shape of the drop, which is qualitatively similar to that of the cylindrical drop. However, there are quantitative differences, which increase when the pair values (a, θ_0) become closer to the critical curve.

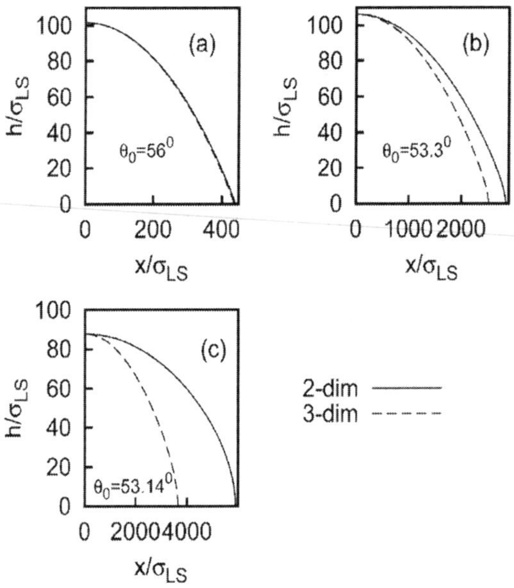

FIGURE 11. Profiles of the cylindrical (solid lines) and axisymmetrical (dashed lines) droplets of the same height of the apexes for various microcontact angles from the domain of unlimited height for the LvdW potential.

In Figure 11, the profiles of cylindrical and axisymmetrical droplets with the same heights are compared for $a = 0.8$ and various values of θ_0 (the LvdW potential was employed in the calculations) in the domain of unlimited height. Only the profiles for $x \geq 0$ are presented. Far from the critical curve (profiles for $\theta_0 = 60°$), the profiles are almost identical except of a small region near the surface. The macroscopic contact angle for both profiles is equal approximately to 25.9° and remains almost constant when the height of the droplet increases. Therefore, the profile of an axisymmetrical drop far from the critical curve can be represented by the profile of a cylindrical drop, which is much easier to calculate. When θ_0 decreases, the difference between the profiles becomes more pronounced. One can see in the curves for $\theta_0 = 53.3°$ and $\theta_0 = 53.14°$ that the width of the cylindrical drop profile is larger than that of the axisymmetrical one and that both profiles become more planar (the width becoming much larger than the height), when the angle θ_0 decreases. The macroscopic contact angle decreases and becomes very small with decreasing θ_0. For instance, for $\theta_0 = 53.14°$, $\theta_{m,\,cyl} = 1.37°$ and $\theta_{m,\,a} = 2.47°$, and for $\theta_0 = 60°$, $\theta_{m,\,cyl} \simeq \theta_{m,\,a} \simeq 26.0°$.

For the LJ potential, all the characteristics of the axisymmetrical droplets are qualitatively similar to those for the LvdW potential.

V. DISCUSSION

The analysis of cylindrical and axisymmetrical droplets provided in sections III and IV revealed that their main characteristics are similar. The same similarity was observed with respect to the interaction potential (London—van der Waals or Lennard—Jones potentials). Therefore, the discussion that follows, if not otherwise mentioned, concerns both kinds of drops as well as both kinds of interaction potentials.

The consideration of a monolayer of liquid near a solid surface provided a more realistic description of the characteristics of a drop. The microcontact angle θ_0, which was constant ($\theta_0 = 180°$) in the previous continuum description, ref 1, acquired now a dependence on the microscopic parameters of the model [see eqs 33 and 20], becoming the second parameter that affects the drop features. The first parameter, a, given by eq 47, is also a function of the microscopic parameters of the model (strength of intermolecular interactions, densities of liquid and solid phases, hard core and interaction radius).

Although the microcontact angle is practically undetectable, it influences strongly such drop characteristics as stability, shape, and macroscopic contact angle. One of the consequences of the variability of θ_0 is that instead of a single value a_c of the parameter a, which separates the one-dimensional domain ($a < a_c$) of drops of any height from those ($a > a_c$) in which the height of the drop was limited, there are two two-dimensional domains (Figure 5) in the θ_0–a plane, in which the drops possess those features. Those domains are separated by a critical curve given by eqs 50 and 53. The critical droplet height h_c in the domain of limited height is a function of a and θ_0 and rapidly grows when the point (θ_0, a) approaches the critical curve.

The effect of the microcontact angle θ_0 upon the shape of the drop was illustrated with an example of a drop of a given constant volume. In this case, the relative part of the profile, which can be considered circular, decreases with decreasing θ_0 (Figure 7b). In the domain of limited droplet height, the shape of the drop is planar when its height is close to h_c.

Two stability conditions can be formulated on the basis of the developed theory. The first and strongest condition is that the absolute value of parameter b, eq 20, which is equal to $-\cos\theta_0$, must be not larger than unity. The fulfillment of this condition depends exclusively on the values of the parameters of microscopic liquid—liquid and liquid—solid interactions, which determine the magnitude of b. If $|b| > 1$, no drop can exist on the bare surface, and if created, it has to completely spread on the surface.

The second, weaker stability condition is valid only in the domain of limited droplet height. According to this condition, the droplet height h_m has to be smaller than a critical height h_c, and drops with $h_m > h_c$ cannot exist. If such a drop is created, it has to acquire a shape with a height smaller than h_c.

It seems natural to add to the above two conditions a third one, based on the eq 56 for the macroscopic contact angle θ_m, which also depends on a and θ_0. It follows from that equation that θ_m does not exist if the value $(a/2) + \cos\theta_0$ is larger than unity. This can happen even for small $a > 0$ when the microcontact angle is small and $\cos\theta_0$ is close to unity. However, the nonexistence of the macroscopic contact angle in the considered case does not mean the nonexistence of the drop. Indeed, the inequality $(a/2) + \cos\theta_0 > 1$, which is equivalent to the inequality 51, means only that the pair (θ_0, a) belongs to the domain of limited droplet height in which the large droplets have a planar shape.

The interaction potentials employed provided quantitative differences between the drop profiles, which became more pronounced when the values of the parameters a and θ_0 were near the critical curve. In the latter case, the differences between cylindrical and axisymmetrical droplets were also larger.

ACKNOWLEDGMENT

This work was supported by the National Science Foundation.

REFERENCES AND NOTES

(1) Berim, G. O.; Ruckenstein, E. *J.Phys. Chem B*, **2004**, *108*, 19330 (Section 3.2 of this volume).
(2) Miller, C. A.; Ruckenstein, E. *J. Colloid Interface Sci.* **1974**, *48*, 368 (Section 1.1 of this volume).
(3) Ruckenstein, E.; Lee, P. S. *Surf Sci.* **1975**, 52, 298 (Section 3.1 of this volume).
(4) Nowakowski, B.; Ruckenstein, E. *J. Phys. Chem.* **1992**, *96*, 2313.
(5) Abraham, F. In *Homogeneous Nucleation Theory*; Academic: New York, 1974.

(6) Lee, S. H.; Rossky, P. J. *J. Phys. Chem.* **1994**, *100*, 3334.
(7) Neimark, A. V. *J. Adhes. Sci. Technol.* **1999**, *13*, 1137.
(8) Sharma, A. *Langmuir* **1993**, 9, 3580.
(9) Yeh, E. K.; Newman, J.; Radke, C. J. *Colloids Surf* **1999**, *156,* 137; **1999**, *156*, 525.
(10) Solomentsev, Y.; White, L. R. *J. Colloid Interface Sci.* **1999**, *218*, 122.
(11) McHale, G.; Newton, M. I. *Colloids Surf.* **2002**, *206*, 79.
(12) Elsgolc, L. E. *Calculus of Variations*; Pergamon Press, Addison-Wesley: New York, 1962.

3.4 Microscopic Treatment of a Barrel Drop on Fibers and Nanofibers*

Gersh Berim and Eli Ruckenstein†
Department of Chemical and Biological Engineering,
State University of New York at Buffalo, Buffalo, NY 14260, USA

† Corresponding author. Fax: +1(716)645-3822.

E-mail address: feaeliru@buffalo.edu (Eli Ruckenstein).

Received 29 October 2004; accepted 11 February 2005

Available online 22 March 2005

1. INTRODUCTION

The theoretical understandings of the wetting of cylindrical surfaces (fibers) are important for the textile industry, fabrication of composite materials, detergency, adhesion, and so on. The discovery of nanotubes stimulated the interest in the behavior of small droplets on fibers, which can provide information about the properties of the fiber surface. The modern methods of observation of droplet shapes allowed to obtain drop profiles at the nanometer scale and to evaluate the contact angles using a variety of methods (see, for example, Refs. [3-6]). For these reasons a careful description of the profile, especially near the surface of the fiber where the strength and variability of the interaction potential between the molecules of the fiber and those of the liquid are strong, is useful.

As a rule, the shape of a drop on a fiber was described in the framework of a macroscopic approach, based on thermodynamic considerations of a drop at equilibrium and involving such macroscopic quantities as surface tensions, pressure, etc. Two specific drop shapes, barrel and clam-shell, were usually considered [5,7–9]. The first of them has an axial symmetry, and the corresponding differential equation for the drop profile allowed order reduction and, hence, an analytical solution in quadrature. The second shape has symmetry only about a plane containing the fiber axis and the drop apex. In this case only a numerical solution was possible. Below, only the barrel drop will be considered. For that case Carroll [7] derived and solved the differential equation for the profile using the supposition of constancy of the Laplace pressure across the entire drop profile. As one of the boundary conditions the (unknown) value of the angle θ which the profile makes with the surface of the fiber was employed, assuming that this angle is the conventional wetting angle. In this way relations between this angle and measurable characteristics of the barrel droplet (drop height and length, volume, contact area) were found, which allowed its determination from the latter characteristics [5,7]. In addition, the stability of a droplet on a fiber with respect to the roll-up transition from a barrel to a clam-shell configuration was examined and a condition involving θ, the height of the droplet, and the radius of the fiber was derived [10]. Note that Carroll's considerations do not involve explicitly the liquid-liquid and liquid-solid interactions, and involve the extrapolation of the macroscopic equation till the

* *Journal of Colloid and Interface Science* 286 (2005) 681–695. Republished with permission.

leading edge of the drop. As a consequence, the typical profile provided by Carroll's solution has a slow curvature variation near the fiber surface, and the angle θ is identified as the macroscopic wetting angle (or, in other nomenclature, apparent contact angle) measured experimentally.

A completely microscopic approach concerning the shape and stability of drops on a planar surface was developed by Ruckenstein et al. in Refs. [1,2,11,12]. Equations for the drop profile were obtained either by considering the mechanical equilibrium of the molecules on the drop surface [12], or through a variational minimization of the total potential energy of the drop [1,2].

It was shown that the profile of the drop near the solid surface (on an interval of several nanometers) has a rapid change in curvature and that the actual angle which the profile makes with the solid surface cannot be determined from macroscopic experiments. However, this angle, which was called microcontact angle and was denoted θ_0, plays an important role in the theory because it serves as a boundary condition to the differential equation for the drop profile and its value provides information about the microscopic parameters of the model. The measurable, apparent, contact angle θ, can be found by extrapolating near the solid surface a part of the profile with slow varying curvature up to the solid surface. Obviously, the latter angle does not coincide with θ_0. Nevertheless, the theory provides an equation connecting both angles from which one of them can be found if the value of the other is known.

It was found that the microcontact angle has the constant value ($\theta_0 = 180°$) if a continuous picture for the liquid is used (Ref. [1]), or is dependent on the constants of the liquid–liquid and liquid-solid interactions if the structure of the liquid in the very vicinity of the solid surface is taken into account (Ref. [2]). Even when the angle θ_0 is constant, the angle θ can have different values dependent on the drop size and the values of the interaction parameters.

For large droplets the part of the profile with a slow varying curvature can be considered circular (gravity is neglected). In that case the apparent contact angle θ coincides with the angle θ_m which the circular profile with the curvature equal to the curvature of the drop profile at the drop apex makes with the fiber surface. In general, $\theta \neq \theta_0$.

In the framework of the microscopic approach it was also shown [1] that the height h_m of the drop at its apex can have, even in the absence of gravity, any value only if a parameter a, which is a combination of the interaction parameters characterizing the interactions between the molecules of the liquid as well as between the molecules of the solid and liquid, is less than a critical value a_c. For $a > a_c$ the droplet height h_m is limited by a critical value h_c which depends on a. Were a droplet with a height larger than h_c artificially created, it would transform into a droplet with a new height less than h_c. If h_m is close to h_c, the droplet will have a planar shape. In Ref. [2] the dependence of the critical values a_c and h_c on the microcontact angle θ_0 was established.

Note that the interactions of a drop with a fiber were previously included in the theory by Brochard [13] and Neimark [8] using the Derjaguin disjoining pressure [14], which was assumed independent of the curvature of the fiber, and by Bauer and Dietrich [9] who used the Lennard–Jones potential to describe the liquid–fiber interactions in the framework of the density functional theory.

In the present paper the microscopic approach of Refs. [1,2,11,12] was applied to drops on a fiber. The intermolecular interactions are assumed of the form of London–van der Waals potentials (Sections 2.1 and 2.2 of the present paper), and the interaction potential of the liquid molecule with an infinitely long fiber is calculated in Section 2.2. Based on those potentials the total potential energy of a drop on a fiber is calculated in Section 2.3. In Section 3 and Appendixes A and B an equation for the profile is derived by the minimization of the total potential energy of a drop and solved in quadratures. It is also shown that an expression for the microcontact angle follows from the minimization procedure in a natural way. On the basis of the obtained solution for the profile the stability conditions of a barrel drop are identified in Section 4. It turns out that these conditions are more complicated than for a drop on a planar surface, due to the finite curvature of the substrate. The same complication appears with respect to the shape of the drop profile (Section 5) which depends on the droplet size, on the contact angle θ_0, as well as on the radius of the fiber. In Section 6 expressions which can be used to evaluate the contact angle θ_0 from experimental observations are derived.

2. POTENTIALS

2.1 LIQUID-LIQUID INTERACTION

The potential $\tilde{\phi}_{LL}$ of the interaction between the molecules of the liquid can be chosen in the form of the London- van der Waals potential with a rigid core repulsion

$$\tilde{\phi}_{LL}(r)=\begin{cases}-\epsilon'_{LL}\left(\dfrac{\sigma'_{LL}}{r}\right)^6, & r\geqslant\sigma'_{LL},\\ \infty, & r<\sigma'_{LL},\end{cases}\tag{1}$$

where r is the distance between the centers of the interacting molecules, and $\sigma'_{LL}>0$ and $\epsilon'_{LL}>0$ are the size of the repulsive core and an interaction constant, respectively. If no simplifying assumptions are made concerning the drop shape, the calculation of the potential energy based on Eq. (1) is a complicated problem due to the long-range nature of the latter potential. To avoid such difficulties we consider instead of Eq. (1) the potential

$$\phi_{LL}(r)=\begin{cases}0, & r>\eta,\\ -\epsilon_{LL}\left(\dfrac{\sigma_{LL}}{r}\right)^6, & \eta\geqslant r\geqslant\sigma_{LL},\\ \infty & r<\sigma_{LL},\end{cases}\tag{2}$$

where the range of interactions is diminished to a sphere of radius η (interaction sphere). Because the potential of Eq. (1) decreases rapidly with r, the approximate Eq. (2) seems reasonable. To evaluate ϵ_{LL} and σ_{LL} in terms of ϵ'_{LL} and σ'_{LL}, let us require the equality of the total potential energies ϕ'_l and ϕ_l of a molecule in the bulk of a homogeneous liquid, calculated using potentials (1) and (2), respectively. Assuming pairwise additivity, the first energy is given by

$$\phi'_l=\int_0^{2\pi}\mathrm{d}\phi\int_0^{\pi}\mathrm{d}\theta\sin\theta\int_{\sigma'_{LL}}^{\infty}\phi'_{LL}(r)\rho_L r^2\mathrm{d}r$$

$$=-\frac{4}{3}\pi\sigma'^3_{LL}\epsilon'_{LL}\rho_L,\tag{3}$$

where ρ_L is the liquid density. The second potential energy, ϕ_l, is given by $\phi_l=\rho_L\int_{V_{is}}\phi_{LL}(r)\mathrm{d}V$ where ρ_L is the number density of molecules in the liquid, which is considered constant and the integration is carried out over the entire volume V_{is} of the interaction sphere. Due to the spherical symmetry of the potential $\phi_{LL}(r)$, the calculation of ϕ_l is straightforward and one obtains

$$\phi_l=-\frac{4\pi}{3}\epsilon_{LL}\sigma^3_{LL}\rho_L\left(1-\Delta^3\right),\tag{4}$$

where $\Delta=\sigma_{LL}/\eta$. From the equality $\phi_l=\phi'_l$ one arrives at the following expression connecting the parameters of the two potentials:

$$\frac{\epsilon_{LL}}{\epsilon'_{LL}}=\left(\frac{\sigma'_{LL}}{\sigma_{LL}}\right)^3\left(1-\Delta^3\right)^{-1}\tag{5}$$

If, for example, $\eta=3\sigma_{LL}$, the following particular choices for the parameters ϵ_{LL} and σ_{LL} can be obtained: $\sigma_{LL}=\sigma'_{LL}$, $\epsilon_{LL}={}^{27}\!/_{26}\epsilon'_{LL}$.

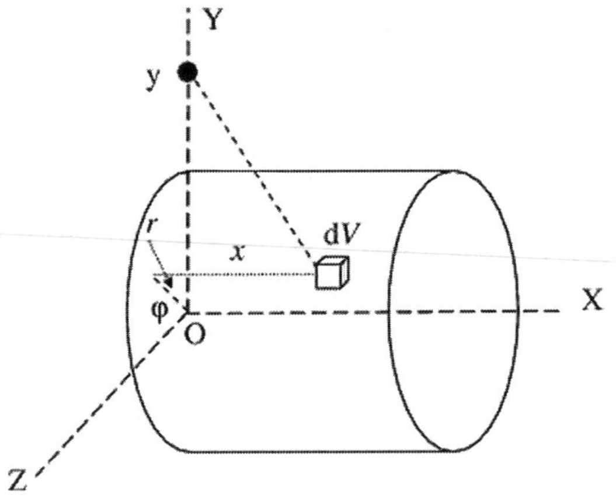

FIG. 1. Coordinate system used for the calculation of the potential $\Phi^l_{LS}(y)$ of the interaction of a liquid molecule with a fiber.

2.2. Liquid-solid interaction

It is supposed that the interaction potential between a molecule of liquid and a molecule of solid is equal to the sum of two components, $\phi^l_{LS}(r)$ and $\phi^s_{LS}(r)$, related to the long-range and the short-range (acid-base) interactions. The long-range potential is chosen in a form similar to $\tilde{\phi}_{LL}(r)$ (Eq. (1)),

$$\phi^l_{LS}(r) = \begin{cases} -\epsilon_{LS}\left(\frac{\sigma_{LS}}{r}\right)^6, & r \geqslant \sigma_{LS}, \\ \infty, & r < \sigma_{LS}, \end{cases} \tag{6}$$

where $\sigma_{LS} > 0$ and $\epsilon_{LS} > 0$ are the size of the repulsive core and the interaction constant of the liquid–solid interactions, respectively. In what follows σ_{LS} is used as unit of length. In real systems σ_{LS} is of the order 3–4 Å [15]; we selected $\sigma_{LS} = 3$Å.

To calculate the potential of a liquid molecule interacting with a fiber through the long-range interactions let us consider the fiber as an infinite cylindrical solid body of radius R_f with a constant number density ρ_S. If the molecule of liquid is located at a distance $y > R_f$ from the fiber axis X then its potential energy $d\Phi^l_{LS}(y)$ due to its interactions with the molecules of solid contained in the element of volume dV (see Fig. 1) is given by

$$d\Phi^l_{LS}(y) = -\frac{\epsilon_{LS}\rho_S\sigma^6_{LS}}{\left(r^2 + x^2 + y^2 - 2yr\sin\phi\right)^3} dV = -\frac{\epsilon_{LS}\rho_S\sigma^6_{LS}}{\left(r^2 + x^2 + y^2 - 2yr\sin\phi\right)^3} r\,d\phi\,dr\,dx. \tag{7}$$

The total potential $\Phi^l_{LS}(y) = \int_V d\Phi^l_{LS}(y)$ after integration over the volume of the fiber has the form

$$\Phi^l_{LS}(y) = \frac{\pi\epsilon_{LS}\rho_S\sigma^6_{LS}}{24(R_f + y)(R_f^2 - y^2)^2} \times \left\{(7R_f^2 + y^2)\left[E\left(\frac{\pi}{4}, -\frac{4R_f y}{(R_f - y)^2}\right) + E\left(\frac{3\pi}{4}, \frac{4R_f y}{(R_f - y)^2}\right)\right]\right.$$

$$-(R_f+y)^2\left[F\left(\frac{\pi}{4},-\frac{4R_fy}{(R_f-y)^2}\right)\right.$$

$$\left.\left.+F\left(\frac{3\pi}{4},-\frac{4R_fy}{(R_f-y)^2}\right)\right]\right\}, \tag{8}$$

where $F(\phi,m)$ and $E(\phi,m)$ are the elliptic integrals of the first and second kind, respectively [16]. In the limiting case $R_f\to\infty$ the potential $\Phi_{\mathrm{LS}}^{l}(y)$ coincides with that of a molecule of liquid interacting with a semi-infinite solid.

Potential (8) is too complicated for further calculations and it is reasonable to approximate it by a more simple function. Let us represent the distance y as $y=R_f+\sigma_{\mathrm{LS}}+h$ where the length h will be specified later. The potential $\Phi_{\mathrm{LS}}^{l}(y)$ becomes a function of h, which for simplicity will be denoted $\Phi_{\mathrm{LS}}^{l}(h)$. For $R_f\gg\sigma_{\mathrm{LS}}$ the potential $\Phi_{\mathrm{LS}}^{l}(h)$ can be approximated with high accuracy by the power law function,

$$\Phi_{\mathrm{LS}}^{l}(h)=-\frac{\pi}{6}\epsilon_{\mathrm{LS}}\sigma_{\mathrm{LS}}^{3}\rho_{\mathrm{S}}\frac{k}{(1+h/\sigma_{\mathrm{LS}})^{\nu}}, \tag{9}$$

where k and ν depend on R_f/σ_{LS} and are independent of h. The parameters k and ν can be found by comparing $\Phi_{\mathrm{LS}}^{l}(h)$ (Eq. (9)) with the exact expression, Eq. (8). Taking for example $R_f=100\sigma_{\mathrm{LS}}\simeq 30\,\mathrm{nm}$, one obtains $\nu=3.0107$ and $k=0.9926$. In this case the difference between the exact and the approximate potentials in the region $h<5\sigma_{\mathrm{LS}}$ where the solid-liquid interaction is strong, did not exceed 1.7%. Note that for $R_f\to\infty$ one has $\nu\to 3$ and $k\to 1$, and potential (9) coincides with that for the interaction of a liquid molecule with a semi-infinite solid [12].

It is assumed that the short-range potential $\Phi_{\mathrm{LS}}^{\mathrm{S}}(r)$ which accounts for the acid-base interactions decays rapidly and can be considered zero at distances larger than 1 nm $\sim 3\sigma_{\mathrm{LS}}$ [8,17]. Consequently it will be supposed that the potential energy $\Phi_{\mathrm{LS}}^{\mathrm{S}}(y)$ (or, equivalently, $\Phi_{\mathrm{LS}}^{\mathrm{S}}(h)$) of a liquid molecule due to its short-range interactions with a fiber has the value $-\epsilon_{\mathrm{LS}}^{S}$ for the molecules located at a distance of the order of σ_{LS} $(h=0)$ near the fiber surface and is equal to zero for all the other molecules. As a result, the potential $\Phi_{\mathrm{LS}}^{s}(h)$ is given by

$$\Phi_{\mathrm{LS}}^{s}(h)=\begin{cases}-\epsilon_{\mathrm{LS}}^{s}, & h=0,\\ 0, & h>0.\end{cases} \tag{10}$$

The interaction constant $\epsilon_{\mathrm{LS}}^{s}$ can be positive or negative [8].

Combining potentials (8) and (10), one obtains the total potential $\Phi_{\mathrm{LS}}(h)$ of a molecule of liquid interacting with a fiber:

$$\Phi_{LS}(h)=\begin{cases}\Phi_{LS}^{l}(0)+\Phi_{LS}^{s}(0), & h=0,\\ \Phi_{LS}^{l}(h), & h>0.\end{cases} \tag{11}$$

2.3. The Potential Energy of a Drop

To calculate the potential energy of a drop one has to specify a model for the liquid. The most frequently used one considers the liquid as a continuous medium characterized by a macroscopic number density ρ_{L} [8,12,13]. Such a continuous picture is not valid in the very vicinity of the fiber surface where the interactions (long- as well as short-range) of the liquid molecules with the fiber are strong and change rapidly with the distance from the fiber surface. Due to these interactions, the liquid molecules nearest to the fiber surface are rearranged in a layer "sticked" to the fiber surface, which will be assumed to be a monolayer separated from the surface by a distance σ_{LS} The rest of the drop is separated from the first layer by a distance of the order of σ_{LL} and is assumed to be a continuous medium of constant density (see Fig. 2). The total potential energy of a drop,

$$U_{drop} = U_{LL} + U_{LS}, \tag{12}$$

consists of a potential energy U_{LL} due to the interactions of the molecules of liquid between themselves and a potential energy U_{LS} due to the interactions between the molecules of liquid and those of the fiber. The interaction between liquid and vapor is neglected. The presence of a liquid monolayer on the fiber surface will be taken into account during the calculation of U_{LS}. In the calculation of U_{LL} the deviation from the continuum model is neglected.

Let us assume that the size of the droplet is much larger than the radius η of the range of the liquid-liquid interactions. Then a molecule of liquid located at a distance larger than η from the droplet surface interacts with all the molecules inside a sphere of radius η with the center in the center of the molecule (interaction sphere). The potential energy of this molecule is given by Eq. (4). A molecule at a distance $l < \eta$ from the droplet surface interacts only with those molecules which are present both in the drop and in the interaction sphere described above (Fig. 3). Assuming that the principal radii of curvature of the drop surface is much larger than η, one can consider that the intersection of the interaction sphere with the drop is planar. (Such an assumption is valid if the volume V of the drop is much larger than the volume of the interaction sphere, $V \gg \frac{4}{3}\pi\eta^3$.) In this case the effective potential $\phi_2(l)$ of a molecule near the drop surface consists of two parts,

$$\phi_2(l) = \phi_{2,1}(l) + \phi_{2,2}(l),$$

where $\phi_{2,1}(l)$ $(\phi_{2,2}(l))$ is the potential energy of a central molecule due to its interaction with the molecules inside the volume Ω_1 (Ω_2) of the interaction sphere (see Fig. 4). The first potential is given by the integral

$$\phi_{2,1}(l) = -\epsilon_{LL}\sigma_{LL}^6 \int_0^{2\pi} d\phi \int_0^{\theta_l} \sin\theta \, d\theta \int_{\sigma_{LL}}^{l/\cos\theta} \frac{dr}{r^4}, \tag{13}$$

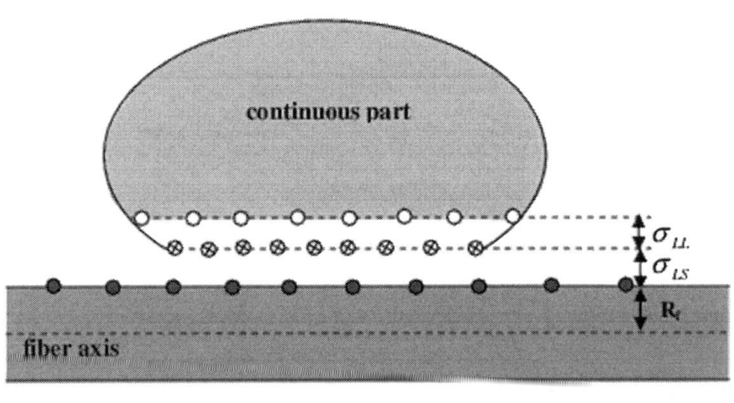

FIG. 2. The cross section of a drop on a fiber. The filled and crossed circles represent the molecules on the fiber surface and the molecules of the first (nearest to the solid surface) layer of liquid molecules, respectively. Beginning with the second layer (open circle), the liquid is considered as a continuous medium

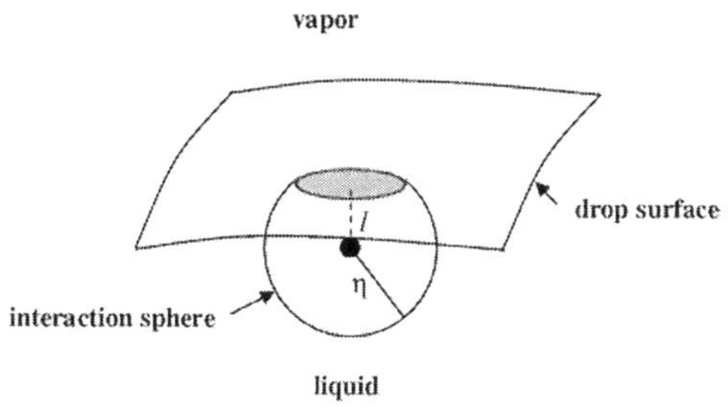

FIG. 3. Interaction range of a molecule near the surface of a drop. If the principal radii of curvature of the drop surface are much larger than the radius η of the interaction sphere, the intersection of the interaction sphere with the drop can be considered planar. The filled circle represents the hard core σ_{LL} of the liquid-liquid interaction.

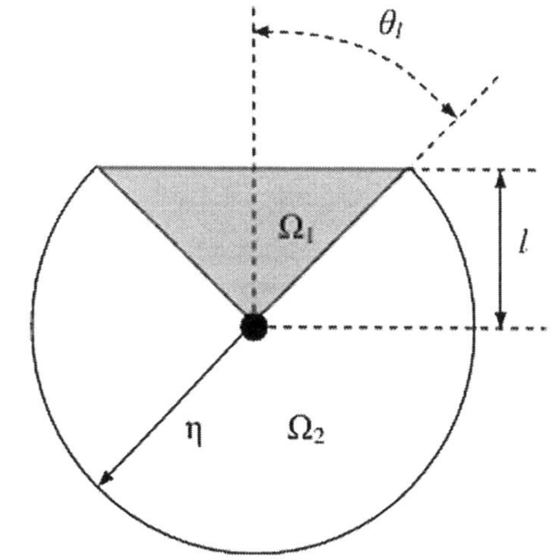

FIG. 4. Part of the interaction sphere of a molecule located at a distance $l < \eta$ from the drop surface, which contributes to its potential energy resulting from the interaction with other molecules of liquid. For convenience of the calculation, the total volume of that part is divided into two volumes, Ω_2 and Ω_2.

Where the angle $\theta_l = \arccos\frac{l}{\eta}$ is defined in Fig. 4. The potential $\phi_{2,2}(l)$ is given by

$$\phi_{2,2}(l) = -\epsilon_{LL}\sigma_{LL}^6\rho_L \int_0^{2\pi} d\phi \int_{\theta_l}^{\pi} \sin\theta \, d\theta \int_{\sigma_{LL}}^{\eta} \frac{dr}{r^4}. \tag{14}$$

Consequently, the potential $\phi_2(l)$ acquires the form

$$\phi_2(l) = -\frac{2\pi}{3}\epsilon_{LL}\sigma_{LL}^3\rho_L\left[2 - \frac{\sigma_{LL}^3}{4l^3} - \Delta^3\left(1 + \frac{3l}{4\eta}\right)\right]. \tag{15}$$

The molecules for which $l \leqslant \sigma_{LL}$ will be considered to be located on the surface of the drop. The potential energy ϕ_3 of such a molecule is given by

$$\phi_3 = \frac{1}{2}\phi_1 = -\frac{2\pi}{3}\epsilon_{LL}\sigma_{LL}^3\rho_L\left(1 - \Delta^3\right). \tag{16}$$

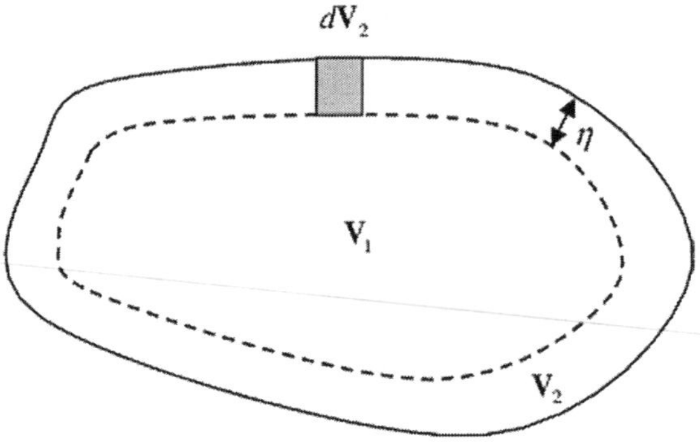

FIG. 5. Cross section of a drop surface (solid line) and of the surface located at a distance η from the drop surface (dashed line). Inside the volume V_1, any molecule interacts with all the molecules inside the interaction sphere with the center in the center of the selected molecule. The elementary volume dV_2 has the base area dS and height η.

In a schematic representation of a drop having an arbitrary shape (see Fig. 5) the dashed line separates the volume V_1, in which each molecule has Eq. (4) as effective potential, from the volume V_2, in which Eqs. (15) and (16) are valid. After integration over the volume V_1, one obtains for the total potential energy U_1 of the molecules in this volume the following expression,

$$U_1 = \frac{1}{2}\phi_1 \rho_L V_1 = -\frac{2\pi}{3}\epsilon_{LL}\sigma_{LL}^3\rho_L^2\left(1-\Delta^3\right)V_1, \tag{17}$$

where the factor 1/2 avoids the double counting of the interaction energy during integration. To calculate the energy U_2 of all the molecules in the volume V_2, let us first find the energy dU_2 of the molecules in the cylindrical volume dV_2 with the base dS and the height η normal to the surface (Fig. 5),

$$dU_2 = \frac{1}{2}\, dS\rho_L \int_{\sigma_{LL}}^{\eta} \phi_2(l)dl + \frac{1}{2}dS\rho_{LS}\phi_3, \tag{18}$$

where ρ_{LS} is the surface density of the molecules on the drop surface. Using Eqs. (15) and (16), the energy U_2 can be written as follows,

$$U_2 = -\frac{2\pi}{3}\epsilon_{LL}\sigma_{LL}^3\rho_L\left[\rho_L\eta K + (\rho_{LS}/2)\left(1-\Delta^3\right)\right]S, \tag{19}$$

where $K = 1 - \frac{17}{16}\Delta - \frac{5}{8}\Delta^3 + \frac{1}{2}\Delta^4 + \frac{3}{16}\Delta^5$ and $S = S_{LV} + S_{LS}, S_{LV}\left(S_{LS}\right)$ being the area of the interface between the droplet and vapor (fiber). Using for the volume V_2 the approximation $V_2 = S\eta$, the total potential energy $U_{LL} = U_1 + U_2$ of the drop, associated with the liquid-liquid interactions, can be written as

$$U_{LL} = -K_\upsilon V + K_S S, \tag{20}$$

where

$$K_\upsilon = \frac{2\pi}{3}\epsilon_{LL}\sigma_{LL}^3\rho_L^2\left(1-\Delta^3\right), K_s = \frac{2\pi}{3}\epsilon_{LL}\sigma_{LL}^4\rho_L^2\left[\frac{17}{16} - \frac{3}{8}\Delta^2 - \frac{1}{2}\Delta^3 - \frac{3}{16}\Delta^4 - \frac{\rho_{LS}/2}{\sigma_{LL}\rho_L}\left(1-\Delta^3\right)\right]. \tag{21}$$

Equation (21) implies that the surface densities of the liquid at the liquid-vapor and liquid-solid interfaces are comparable.

To calculate the interaction energy U_{LS} it will be supposed that the molecules of liquid near the fiber surface form a monolayer with a constant surface number density ρ_{LS}, separated from the fiber surface by the distance σ_{LS}. The molecules of that layer are affected by both long-range and short-range interactions with the fiber which are represented by potential (11) for $h = 0$. The continuous part of the drop is separated from the first layer by a distance equal to the repulsive core radius σ_{LL} and its molecules do not exhibit short range interactions.

Under the above assumptions, the energy U_{LS} is given by

$$U_{LS} = \rho_{LS}\Phi_{LS}(0)S_{LS} + \int_{V_{cont}} \rho_L \Phi_{LS}^l(h)\mathrm{d}V, \tag{22}$$

where S_{LS} is the contact area between the droplet and the fiber, and V_{cont} denotes the volume of the continuous part of the droplet. The latter can be represented as the difference between the drop volume V and the volume V_g of the gap between the first layer and the continuous part of the droplet which is approximately equal to $V_g = \sigma_{LL}S_{LS}$, i.e., $V_{cont} = V - S_{LS}\sigma_{LL}$. Then the second term on the right-hand side of Eq. (22) can be written as

$$\int_{V_{cont}} \rho_L \Phi_{LS}^l(h)\mathrm{d}V$$

$$= \int_V \rho_L \Phi_{LS}^l(h)\mathrm{d}V - \int_{V_g} \rho_L \Phi_{LS}^l(h)\mathrm{d}V \simeq \int_V \rho_L \Phi_{LS}^l(h)\mathrm{d}V - \rho_L\sigma_{LL}\Phi_{LS}^l(\sigma_{LS})S_{LS}.$$

Consequently,

$$U_{LS} \simeq \int_V \rho_{LS}\Phi_{LS}^l(h)\mathrm{d}V$$

$$+\left[\rho_{LS}\Phi_{LS}(0) - \rho_L\sigma_{LL}\Phi_{LS}^l(\sigma_{LL})S_{LS}\right]. \tag{23}$$

The total potential energy of the system, U_{total}, also contains the vapor (gas)-solid potential energy U_{vs}. For low density vapors, only the adsorbed layer ($h = 0$) contributes appreciably to U_{vs} which is therefore given by $U_{VS} = \rho_{VS}(S_{solid} - S_{LS})\Phi_{VS}^l(0)$, where $\Phi_{VS}^l(0)$ is the interaction potential of a vapor molecule with the solid, ρ_{VS} is the surface density of the vapor, and S_{solid} is the surface area of the solid. Consequently, U_{total} is given by

$$U_{total} = U_{LL} + U_{LS} + U_{VS}. \tag{24}$$

Equations (20), (23), and (24) provide a basis for the analysis of the drop characteristics (shape, stability, etc.) using the minimization of the total potential energy.

3. EQUATION FOR THE BARREL DROP PROFILE

In Fig. 6 the upper half of a profile of a barrel drop on a fiber in the plane passing through the axis of the fiber is presented together with the coordinate system used in calculations. The x axis passes through the molecules of the liquid nearest to the surface of the fiber, represented as crossed circles in Fig. 2. The surface of a barrel drop is obtained by rotation of the profile about the axis of the fiber which is parallel to the x-axis and located at distance $R = R_f + \sigma_{LS}$ below it. We consider a drop on a bare surface, therefore the leading edge of the drop is located just on the fiber surface. The profile in Fig. 6a has a microcontact angle θ_0 larger than 90°, while for the profile in Fig. 6b, $\theta_0 < 90°$. In both cases the macroscopic contact angle θ is smaller than 90°. Usually [8,9], the drop profile is described by an equation of the form (here and below the gravity is neglected)

$$\frac{h_{xx}}{\left(1+h_x^2\right)^{3/2}}-\frac{1}{(d+h)\sqrt{1+h_x^2}}-\Pi(h)/\gamma=-P_{\text{cap}}/\gamma, \tag{25}$$

where $h = h(x)$ is the profile equation, the subscript of h denotes the derivative with respect to the corresponding variable, $d + h$ is the distance of the profile point from the fiber axis, γ is the liquid/vapor surface tension, $\Pi(h)$ is the disjoining pressure originating from the liquid-solid and liquid-liquid interactions, and P_{cap} is the capillary pressure. However, Eq. (25) describes only the profiles when $\theta_0 \leqslant 90°$ (Fig. 6b), because for $\theta_0 > 90°$ the function $h(x)$ is not unique for $|x| > x_0$, where $x_0 \equiv x(0)$ is the abscissa of the leading edge of the droplet. To include both shapes (Figs. 6a and 6b), the equation of the droplet profile is chosen in the present paper in the form $x = x(h)$, and only the part of the profile for which $x \geqslant 0$ will be examined further.

The equation of the drop profile can be derived using the variational minimization of the total potential energy U_{total} of the drop given by Eq. (24), assuming the constancy of the drop volume. The details are given in Appendix A. The equation obtained for the profile has the form

$$\frac{x_{hh}}{\left(1+x_h^2\right)^{3/2}}+\frac{x_h}{(R+h)\left(1+x_h^2\right)^{1/2}}-a_1\Phi_{\text{LS}}^{l}(h)=\lambda_a, \tag{26}$$

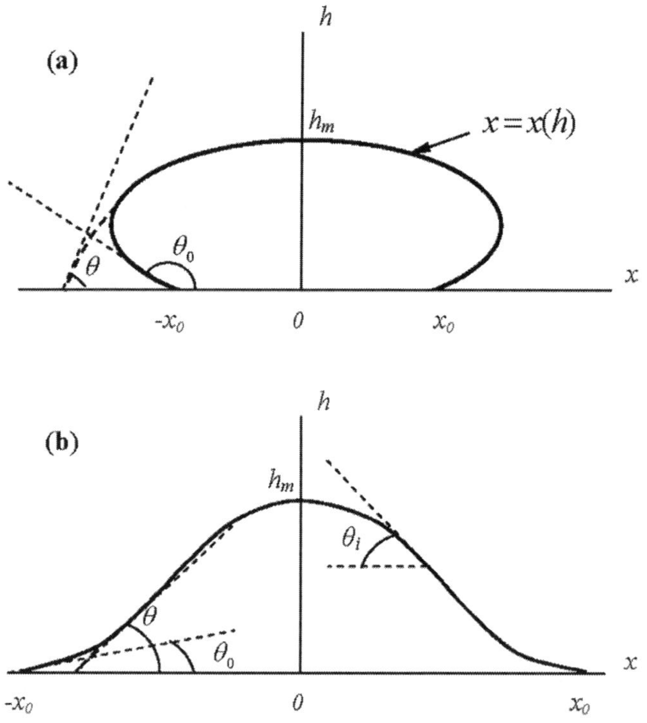

FIG. 6. Two characteristic profiles (solid lines) of a barrel drop in a plane containing the fiber axis: (a) the microcontact angle θ_0 is larger than 90°; (b) θ_0 is smaller than 90°. The dashed curves are the continuations of the upper parts of the profiles. The tangents to those parts at the points of intersection with the fiber surface make with x-axis the apparent contact angles θ. In part (b), θ_i is the inflection angle (the angle between fiber axis and a tangent to the drop profile at the inflection point).

where $a_1 = \rho_L/K_s$ and λ_a is a constant dependent on the drop volume. The first term of Eq. (26) represents the principal curvature of the drop surface in the plane of the drop profile and the second term represents the one in the perpendicular plane. The same meaning can be attributed to the first two terms of Eqs. (25) which is valid for $\theta_0 < 90°$. Therefore, in the latter case one can establish the following correspondence between the disjoining $\left[\Pi(h)\right]$ and the capillary (P_{cap}) pressures of Eq. (25) and the terms involved in Eq. (26):

$$\Pi(h)/\gamma = -a_1\Phi_{LS}^{l\prime}(h), \quad \lambda_a = -P_{cap}/\gamma. \tag{27}$$

The boundary conditions for Eq. (26) can be written in the form

$$x\big|_{h=h_m} = 0, \tag{28}$$

$$x_h\big|_{h=0} = -\cot\theta_0, \tag{29}$$

where h_m is the droplet height and θ_0 is the microcontact angle of the drop with the solid surface. The variational procedure allows to connect θ_0 to the microscopic parameters of the interaction potentials using the so-called transversality conditions which account for the fact that the apex of the droplet and the contact point of the profile with the fiber should be located on the h- and x-axes, respectively. The result is (see Appendix A)

$$\cos\theta_0 = -b, \tag{30}$$

where

$$b = 1 + \frac{1}{K_S}\left[\rho_{LS}\Phi_{LS}(0) - \rho_L\sigma_{LL}\Phi_{LS}^{l\prime}(\sigma_{LL}) - \rho_{VS}\Phi_{VS}^{l\prime}(0)\right]. \tag{31}$$

Equation (30) plays in the microscopic approach the role of Young equation for the contact angle in the macroscopic theory. It should be noted that the expression for the microcontact angle θ_0 was obtained from pure microscopic considerations, which do not involve such thermodynamic quantities as the surface tensions. The angle θ_0 depends on the constants of the intermolecular interactions as well as on the fiber radius R_f. The latter dependence arises from the R_f-dependence of the parameters k and v in the potential $\Phi_{LS}^{l\prime}(h)$ (see Section 2.2).

The solution for Eq. (26) is (see Appendix B)

$$x(h) - x_0 = \int_0^h \frac{\Psi(h)}{\sqrt{1-\Psi^2(h)}}\,dh, \tag{32}$$

where $x_0 \equiv x(0)$ (Fig. 6),

$$\Psi(h) = \frac{1}{R+h}\left\{-R\cos\theta_0 + \frac{\lambda_a}{2}h(h+2R) - a\left[D(h) - D(0)\right]\right\}, \tag{33}$$

$$a = \frac{1}{4K_1}\frac{\epsilon_{LS}\rho_S}{\epsilon_{LL}\rho_L}\left(\frac{\sigma_{LS}}{\sigma_{LL}}\right)^4, \tag{34}$$

$$K_1 = \frac{17}{16} - \frac{3}{8}\Delta^2 - \frac{1}{2}\Delta^3 - \frac{3}{16}\Delta^4 - \frac{\rho_{LS}/2}{\sigma_{LL}\rho_L}\left(1-\Delta^3\right), \tag{35}$$

$$D(h) = -\frac{a_1}{a}\int \Phi_{LS}^{l}(h)(R+h)\mathrm{d}h. \tag{36}$$

Explicit expressions for x_0, λ_a, and $D(h)$ are provided in Appendix B. To find the three unknown parameters x_0, λ_a, and h_m one needs in addition to Eqs. (28), (29) one more condition. The latter is provided by the equation for the drop volume $V = 4\pi\int_0^{h_m}(R+h)x\,dh$, where $R = R_f + \sigma_{LS}$. Integrating by parts and taking into account that $x(h_m) = 0$, one can rewrite the latter equation in the form $V = -2\pi R^2 x(0) - 2\pi\int_0^{h_m}(R+h)^2 x_h$ which, after substituting expression (B.11) for x_h, becomes

$$V = -2\pi R^2 x_0 - 2\pi\int_0^{h_m}\frac{(R+h)^2\Psi(h)}{\sqrt{1-\Psi^2(h)}}\mathrm{d}h. \tag{37}$$

4. EXISTENCE OF A BARREL DROP ON A BARE SURFACE OF A FIBER

The first necessary condition for the existence of a drop on a bare surface follows from Eq. (30), and has the simple form, $|b| \leqslant 1$. It establishes (through Eq. (31)) a restriction for the values of the microscopic parameters for which a barrel drop can be formed on a surface. If $|b| > 1$, a barrel drop cannot exist either because it becomes a clam-shell drop or it has spread completely as a monolayer over the surface of the fiber. It should be noted that the spreading condition $|b| > 1$ (or, equivalently, $|\cos\theta_0| > 1$) differs from the traditional one, which implies $|\cos\theta| > 1$. One should note that $\cos\theta_0$, and hence the stability condition, depends on the fiber radius.

Another general necessary condition results from the fact that the integral on the right-hand side of Eq. (32) has meaning only if the function $\Psi(h)$ satisfies the inequality

$$|\Psi(h)| < 1 \tag{38}$$

for $0 < h < h_m$. Because, as shown in Appendix B, $\Psi(h_m) = -1$ the function $\Psi(h)$ has to be a decreasing one in the vicinity of $h = h_m$, i.e., its derivative at that point must be negative:

$$\left.\frac{d\Psi(h)}{dh}\right|_{h=h_m} < 0. \tag{39}$$

Using Eq. (33) for $\Psi(h)$, Eq. (36) for $D(h)$, and Eq. (B.14) for λ_a, inequality (39) can be rewritten in the form

$$a < A_{LW}(h_m, \theta_0), \tag{40}$$

where

$$A_{LW}(h,\theta_0) = \frac{\pi}{6}\epsilon_{LS}\rho_S\sigma_{LS}^4\left\{-\frac{\cos\theta_0 R}{(R+h)^2}\right.$$

$$\left.+\frac{h+(1-\cos\theta_0)}{h(h+2R)}\frac{R(R+h)^2+R^2}{(R+h)^2}\right\}\times\left[\Phi_{LS}^{l}(h) - 2\frac{D(h)-D(0)}{h(h+2R)}\right]^{-1}. \tag{41}$$

The analysis of inequality (40) shows that it has a solution for h_m for any pair (Θ_0, a) of values of the parameters θ_0 and a, where $0 < \theta_0 \leqslant 180°$ and $a > 0$. However, for some pairs the droplet height h_m can have any value, while for others it can have only a restricted set of values. To find in the plane

$\theta_0 - a$ the domains of the pairs (θ_0, a) of the first and second kinds let us consider first the curve in the plane a–h_m given by the equation

$$a = A_{LW}\left(h_m, \theta_0\right) \tag{42}$$

for a fixed value of θ_0. A typical plot of that curve is presented in Fig. 7a for $\theta_0 = 45°$ by the solid line. The necessary condition for drop existence (Eq. (40)) is satisfied by all points below the curve, provided by Eq. (42), which passes through a minimum $a = a_c$ at $h = h_{m,c}$. However the more general necessary condition, Eq. (38), is not satisfied by all those points. If $h_m > h_{m,c}$, Eq. (38) is valid only if a is smaller than some value $a_m(h_m, \theta_0)$, which can be obtained using inequality (38). Note that $a_c = a_m(h_{m,c}, \theta_0)$. In Fig. 7a the numerically found curve $a = a_m(h_m, \theta_0)$ is shown as a dashed line for $\theta_0 = 45°$. For large h_m the function $a_m(h_m, \theta_0)$ acquires the value $a_{c,1} = a_m(\infty, \theta_0)$. As a result, a drop of any height can exist for $a < a_c$. If a specific value a_0 of the parameter a lies between a_c and a_{cl} $(a_c < a_0 < a_{cl})$, the drop can have a height smaller than some value $h_{m,1}$ but larger than another value, $h_{m,2}$. Those values can be found as intersection points between the horizontal line $a = a_0$ with the critical curves $a = A_{LW}(h_m, \theta_0)$ for $h_m < h_{m,c}$ and $a = a_m(h_m, \theta_0)$ for $h_m > h_{m,c}$. For example, for $R_f = 100\sigma_{LS}$ and $\theta_0 = 45°$ one has $a_c = 0.611272$ and $a_{cl} = 0.692162$. For $a = 0.65$ the limiting heights are $h_{m,1} = 4.3729\sigma_{LS}$ and $h_{m,2} = 160.8161\sigma_{LS}$. For $a_0 > a_{cl}$ the possible drop heights are restricted only by the value of $h_{m,1}$.

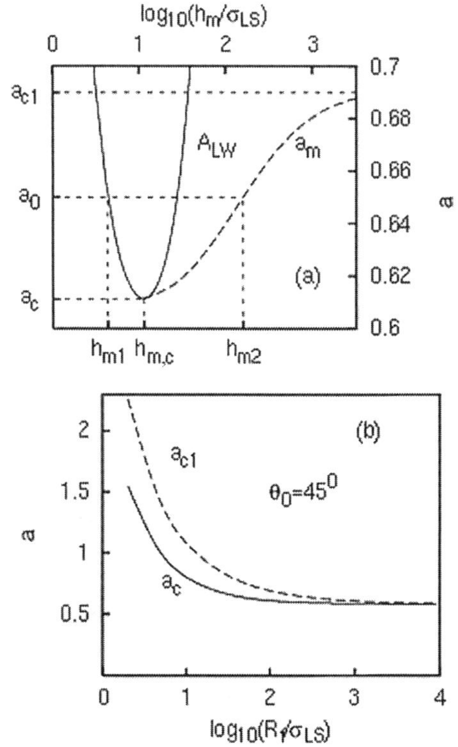

FIG. 7. (a) The curves $a = A_{LW}(h_m, \theta_0)$ (solid line) and $a = a_m(h_m, \theta_0)$ (dashed line) which follow from the necessary conditions of drop existence. For $a \leqslant a_c$ the height h_m of a drop can acquire any value. For $a_c < a \leqslant a_{cl}$, h_m is restricted to intervals $h_m \geqslant h_{m,2}$ or $h_m \leqslant h_{m1}$. If $a > a_{cl}$, the height of the drop cannot be larger than h_{m1}. The curves were plotted for $\theta_0 = 45°$. (b) Dependence of the critical values a_c and a_{cl} on the fiber radius R_f. Microcontact angle is equal to $\theta_0 = 45°$.

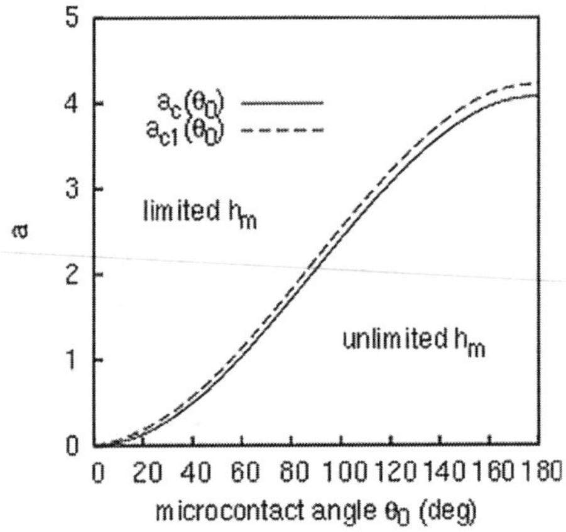

FIG. 8. Dependence of the critical values a_c and a_{cl} of parameter a on the microcontact angle θ_0 for $R_f = 100\sigma_{LS} \sim 30$ nm. The solid line ($a = a_c(\theta_0)$) divides the θ_0–a plane into domains of unlimited (below the solid line) and limited (above solid line) heights.

The specific features of the critical curve in the h_m-a plane depend on the radius R_f of the fiber as well as on the value of the microcontact angle θ_0. As R_f increases, both a_c and a_{cl} as well as the difference between them become smaller (see Fig. 7b), and for $R_f \to \infty$ the critical curve takes a form similar to that for a planar solid surface [1]. The curve has no extrema and approaches the value $a_c = 0.5857$ as $h_m \to \infty$.

By performing a similar analysis for other values of θ_0, one can plot the θ_0 dependencies of the critical values of a_c and a_{cl} in the θ_0-a plane (see Fig. 8). The curve $a = a_c(\theta_0)$ separates the domains of restricted and unrestricted droplet heights for $R_f = 100\sigma_{LS}$. The pairs (θ_0,a) for which the droplet height h_m can have any value lie below that curve. The domain between the curves $a = a_c(\theta_0)$ and $a = a_{cl}(\theta_0)$ corresponds to the case when the drop height can take the values $h_m < h_{m,1}$ or $h_m > h_{m,2}$ ($h_{m2} > h_{m,1}$), and the domain above the curve $a = a_{cl}(\theta_0)$ corresponds to the case when the drop height can take only a value $h_m < h_{m,1}$.

5. SHAPE OF THE DROP PROFILE

In this section the shape of the drop profile will be examined for various values of the microcontact angle θ_0 and unlimited (below the solid line) and limited (above solid line) heights. the parameter a which characterizes the liquid-liquid and liquid-solid interactions. The specific features of the drop profile depend on several factors, including the distance of the point (θ_0, a) from the critical line and the drop volume. Taking as an example a fiber of radius $R_f = 100\sigma_{LS}$, the typical profiles of a small droplet of volume $V = 8.2\times10^6\sigma_{LS}^3 \simeq 2.2\times10^5$ nm^3 are presented in Fig. 9 for $a = 0.611227$ and various values of the microcontact angle θ_0 in the domain of unlimited heights. The selected value of a is close to the critical value $a_c = 0.611272$ for $\theta_0 = 45°$. Therefore, the plot of the profile (solid line) in Fig. 9a corresponds to a point (θ_0, a) in the very vicinity of the critical line $a = a_c(\theta_0)$. In this case the ratio of the droplet height ($h_m \simeq 6$ nm) to the droplet length ($2x_0 \simeq 270$ nm) is small ($h_m/2x_0 \simeq 0.022$) and a considerable part of the profile can be considered planar. The dashed curve represents the circular profile with a radius equal to the curvature radius of the profile at the drop apex, and the dotted line represents Carroll's solution [7] for the drop profile with the same height h_m and contact angle $\theta = 45°$. The latter two curves considerably differ from the drop profile obtained microscopically. When θ_0 increases the height/length ratio increases and the approximation of the drop profile by a circular one with the curvature radius equal to that at the apex of the drop becomes better as long as $\theta_0 < 90°$. In all cases Carroll's solution for the same h_m and $\theta = \theta_0$ provides a drop with smaller length and volume than the microscopic solution. The above features are illustrated in Figs. 9b and 9c where the three mentioned profiles (ours, Carroll's solutions, and the circular approximation) are plotted for $\theta_0 = 75°$ and $\theta_0 = 135°$. One should note that for an appropriate choice of the contact angle θ Carroll's solution provides a profile close to the microscopic one except near the fiber surface. For example, the microscopic profile for $\theta_0 = 45°$ almost coincides with Carroll's solution obtained for $\theta = 6°$.

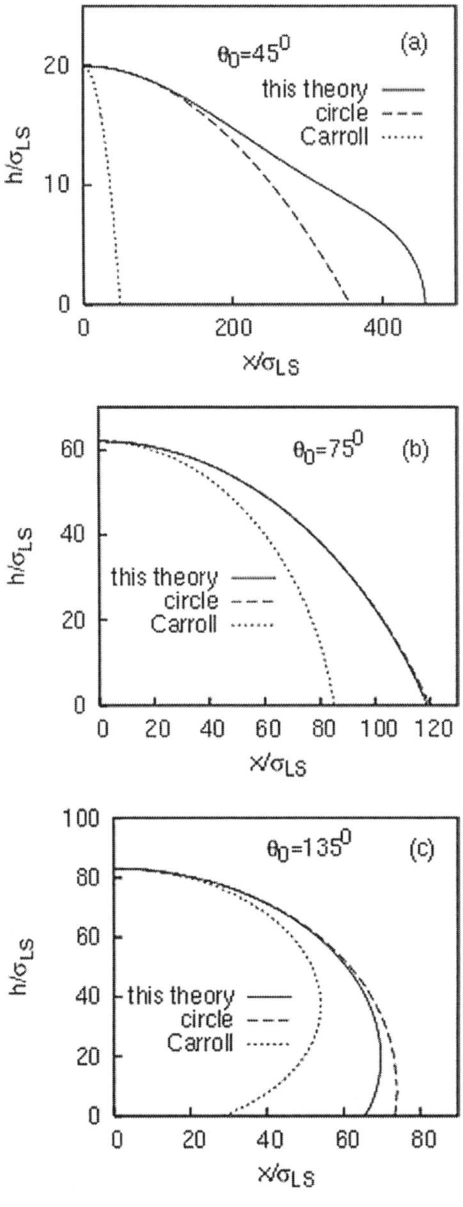

FIG. 9. Profiles of a small drop of constant volume $\left(V = 8.2 \times 10^6 \sigma_{LS}^3 \simeq 2.5 \times 10^2 \text{ nm}^3\right)$ for several values of the microcontact angle θ_0 (solid lines) and $a = 0.6112272$. The dashed line in each figure represents a circular profile with the same height h_m as the original one and with a radius equal to curvature radius of the original profile at the drop apex. The dotted line in each figure is Carroll's solution with the apparent contact angle $\theta = \theta_0$ and with the height of profile equal to h_m. All the profiles correspond to the points (θ_0, a) in the domain of unlimited height, the point (θ_0, a) for profile in part (a) being located in the very vicinity of the critical line.

The profile of a large droplet $V = 6.6 \times 10^8 \sigma_{LS}^3 \simeq 1.8 \times 10^7 \, nm^3$ is presented in Fig. 10. In no case this profile can be considered planar. For $\theta_0 = 45°$ it has an inflection point (Fig. 10a) that disappears for larger values of θ_0 (Figs. 10b, 10c).

For the large droplet, the approximation of the profile by a circle is a better one than for small droplets. To even a larger extent this is valid for Carroll's solution as well, as one can see from Figs. 10a–10c.

If the pair (θ_0, a) lies in that part of the domain of restricted heights that is located between the critical curves (see Fig. 8), for instance $a = 0.65$ and $\theta_0 = 45°$, then the droplet height can take only the values $h_m < h_{m,1} = 4.3729\sigma_{LS}$ or $h_m > h_{m,2} = 160.8161\sigma_{LS}$. If the drop height has a value close to $h_{m,2}$ (for example, $h_m = 161.08\sigma_{LS}$), the drop acquires the shape of a manchon (annulus) [13] which represents the precursor of a wetting film on a cylindrical surface (Fig. 11a). The precision of our

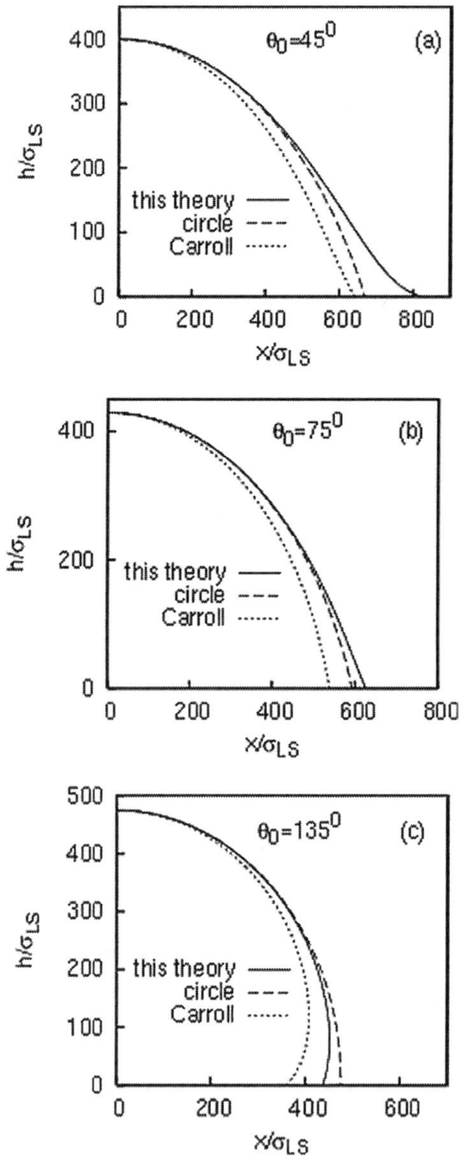

FIG. 10. Profiles of a relatively large drop of constant volume $V = 6.6 \times 10^8 \sigma_{LS}^3 \simeq 2.0 \times 10^7 \, mn^3$ for several values of the microcontact angle θ_0 (solid lines) and $a = 0.612272$. All other notations are the same as in Fig. 9.

numerical calculations does not allow to examine droplets with heights closer to $h_{m,2}$ than that used in the above example. However, the analysis of Eqs. (B.13) and (37) shows that the drop length and volume increase indefinitely when h_m approaches $h_{m,2}$. Therefore, the manchon-like part of the drop is expected to grow and a film to appear on the bare fiber surface.

For larger values of the droplet heights the manchon size decreases and the droplet profile looks like one in the range of unlimited sizes (Fig. 11b).

In all the examples displayed in Figs. 11a and 11b the circular approximation is reasonable only for the upper part of the droplet. Carroll's solution is again unsatisfactory in those cases.

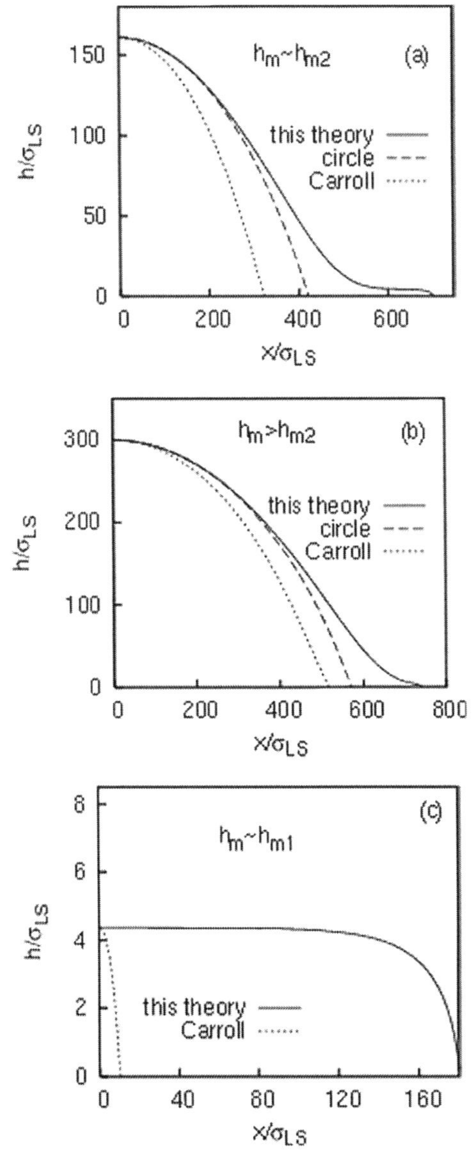

FIG. 11. Drop profiles in the domain of limited height when the point (θ_0, a) is located between the critical lines shown in Fig. 8. (In present case, $a = 0.65$, $\theta_0 = 45°$.) (a) The height $h_m = 161.08\sigma_{LS}$ of the drop is close to $h_{m2} = 160.8161\sigma_{LS}$; (b) h_m is sufficiently large compared to h_{m2}; (c) the height $h_m = 4.372\sigma_{LS}$ is close to $h_{m1} = 4.37294\sigma_{LS}$. The dashed line represents the circular approximation of the profile and the dotted line represents Carroll's solution for the drop with the same height and $\theta_m = \theta_0$.

In the other limiting situation, $h_m \simeq h_{m,1}$, the height/length ratio is very small ($h_m/2x_0 \simeq 0.011$) and the droplet has a planar shape (Fig. 11c) for the description of which neither the circular approximation, nor Carroll's solution, is appropriate.

When the pair (θ_0, a) lies above the critical line $a = a_{cl}(\theta_0)$ (Fig. 8), the droplet height can acquire only the values $h_m < h_{m,1}$ and the droplet has a planar shape similar to that shown in Fig. 11c when $h_m \to h_{m,1}$. It should be noted that the value of $h_{m,1}$ in the cases considered above is small ($h_{m,1} < 5$ Å) and therefore the applicability of the theory becomes questionable because the thickness of the drop has to be much larger than the radius of the interaction sphere in the liquid (see Section 2.3).

6. CONTACT ANGLE

To evaluate the contact angle θ_0 using the experimental data regarding the drop profile it is necessary to derive expressions connecting θ_0 to such observable quantities, as for example the droplet height, length, volume [7], or position of an inflection point [5]. The macroscopic wetting angle θ_m (which is the only one which can be determined in the conventional macroscopic experiments) can be evaluated by ignoring the rapid variation of curvature in the region near the leading edge (where the interactions between solid and liquid are strong) and considering the remaining profile as part of a circle of radius R_m^{curv} equal to the curvature radius at the apex of the drop. Consequently [1]

$$\cos\theta_m = 1 - \frac{h_m}{R_m^{curv}}. \tag{43}$$

Obviously, θ_m does not coincide with θ_0.

In this section, expressions based on the microscopic approach will be established, which connect the microcontact angle θ_0 to the geometrical parameters of the droplet profile. Note that together with θ_0 the theory contains one more parameter, a, which is generally unknown and has to be also found from the experimental data. In fact, the values of a and θ_0 can be obtained in several ways using various measurable quantities. Below we consider some of these ways separately. A detailed comparison of the theory with specific experiments will make this paper too long and for this reason will be made in a separate publication.

6.1. Calculation of θ_0 Using the Height, Length and Volume of the Drop

If the height h_m, length $L = 2x_0$, and volume V of the drop are known, then the parameter a and the microcontact angle θ_0 can be calculated from Eqs. (B.13) and (37). The function $\Psi(h)$ is linear with respect to a and $\cos\theta_0$, and can be written in the form

$$\Psi(h) = \Psi_0(h)\cos\theta_0 + \Psi_a(h)a + \Psi_1(h), \tag{44}$$

where

$$\Psi_0(h) = -\frac{h(h+2R)}{h+R}\frac{h_m+R}{h_m(h_m+2R)}, \Psi_a(h) = \frac{h(h+2R)}{h+R}\frac{D(h_m)-D(0)}{h_m(h_m+2R)} - \frac{D(h)-D(0)}{h+R}, \tag{45}$$

$$\Psi_1(h) = \frac{R}{R+h}\left[\frac{h(h+2R)}{h_m(h_m+R)} - 1\right]. \tag{46}$$

However, the integrals in Eqs. (B.13) and (37) cannot be calculated analytically and therefore a and $\cos\theta_0$ can be calculated only numerically.

6.2. Calculation of θ_0 from the Height of the Drop and the Curvature Radius at the Drop Apex

For relatively large drops the profile near the drop apex can be approximated by a circle with a radius equal to the curvature radius $R_m^{\text{curv}} \equiv 1/\left|\kappa\left(h_m\right)\right|$ at the apex, which can be obtained together with the droplet height h_m from experiment. Using Eqs. (B.2) and (33) the curvature $\kappa(h)$ can be rewritten in a form in which a and $\cos\theta_0$ are expressed explicitly,

$$\kappa(h) = \kappa_0(h)\cos\theta_0 + \kappa_0(h)a + \kappa_1(h), \tag{47}$$

where

$$\kappa_0(h) = \frac{2R}{h_m(h_m+2R)} - \frac{R}{(R+h)^2}\left[\frac{h(h+2R)}{h_m(h_m+2R)} - 1\right], \quad \kappa_a(h) = 2\frac{D(h_m)-D(0)}{h_m(h_m+2R)}\left[1-\frac{h(h+2R)}{2(h+R)^2}\right]$$

$$+\frac{D(h)-D(0)}{(h+R)^2} - \frac{k}{\sigma_{\text{LS}}\left(1+h/\sigma_{\text{LS}}\right)^{\nu}}, \quad \kappa_1(h) = -\frac{2(R+h_m)}{h_m(h_m+2R)}\left[1-\frac{h(h+2R)}{2(h+R)^2}\right], \tag{48}$$

and $D(h)$is given by Eq. (B.8). Then at $h = h_m$ one has

$$\kappa_0(h_m)\cos\theta_0 + \kappa_a(h_m)a + \kappa_1(h_m) = -\frac{1}{R_m^{\text{curv}}}, \tag{49}$$

where the minus sign on the right-hand side reflects the convex shape ($\kappa(h_m) < 0$) of the profile at the drop apex. Assuming that θ_0 is independent of the droplet size one can measure h_m and R_m^{curv} for two droplets of different size on the same fiber and obtain a system of two equations of the same type as Eq. (49) which can provide the values of a and $\cos\theta_0$.

The equation relating the contact angle θ_m (circular approximation) to the microcontact angle θ_0 can be obtained by combining Eq. (43) with Eq. (49):

$$\cos\theta_m = 1 + h_m\left[\kappa_0(h_m)\cos\theta_0 + \kappa_a(h_m)a + \kappa_1(h_m)\right]. \tag{50}$$

For $R \to \infty$ the above expression transforms to the much more simple one,

$$\cos\theta_m = \cos\theta_0 + \frac{a}{2}, \tag{51}$$

which coincides with the corresponding formula for a cylindrical droplet on a planar surface [1].

6.3. Connection of θ_0 with an Infletion Point

One can see from section 5 that a typical profile of a large drop corresponding to a contact angle $\theta_0 < 90°$ contains at least one inflection point at which the curvature $\kappa(h)$ changes sign. If the thickness of the drop at the inflection point is denoted h_{inf}, then $\kappa(h_{\text{inf}}) = 0$ and instead of Eq. (49) one can use the more simple expression

$$\kappa_0\left(h_{inf}\right)\cos\theta_0+\kappa_a\left(h_{inf}\right)a+\kappa_1\left(h_{inf}\right)=0, \tag{52}$$

where h_{inf} is provided by experiment. Again, by determining h_{inf} for two different drops on the same fiber one can obtain two equations which will allow to determine θ_0 and a. Another way to find θ_0 and a is to combine Eqs. (49) and (52).

Note that the use of the inflection point of the drop profile for the determination of the angle at the leading edge was discussed first by McHale [5] who employed Carroll's solution [7] for the drop profile.

7. DISCUSSION AND SUMMARY

Unlike the previous macroscopic considerations of the drop equilibrium on a solid cylinder (fiber) [5,7–10,13], the theory developed in the present paper uses a microscopic representation of the solid-liquid and liquid–liquid interactions followed by the minimization of the potential energy to obtain a differential equation for the drop profile. The potential of the interaction of a liquid molecule with an infinitely long fiber is obtained analytically in terms of elliptic integrals. An approximation of this potential by a much more simple function is suggested, which is valid in the range in which the liquid-solid interaction is strong. The obtained equation for the drop profile contains terms representing the principal curvatures of the drop surface as well as a term representing the contribution of the liquid–solid interactions. In addition, the minimization procedure provides an expression for the microcontact angle between the drop profile and the fiber axis, which contains only the microscopic parameters of the model. Due to the axial symmetry of the problem it was possible to find a solution of the equation for the profile by quadratures.

The analysis of the obtained equation in the absence of gravity reveals several new features which were not noted in the previous macroscopic studies [5,7–9,13]. The drop stability depends on the value of a parameter a, Eq. (34), which incorporates the microscopic parameters of liquid-liquid and solid-liquid interactions as well as the liquid and the solid number densities. For each value of the microcontact angle θ_0 two critical values, a_c and a_{cl} ($a_c < a_{cl}$), of a exist which specify three ranges of possible drop heights. In the first range, with $a < a_c$, the drop can have any height h_m, length, and volume (we do not consider here the possibility of a roll-up transition which according to Carroll (Ref. [10]) takes place at small droplet sizes). In the second range, where $a > a_{cl}$, the drop height is limited by some value h_{ml} ($h_m < h_{ml}$). The third range is defined by the inequalities $a_c < a < a_{cl}$. In the latter case two limiting heights, h_{ml} and h_{m2} ($h_{ml} < h_{m2}$), occur such that the height of the drop can have values only in the intervals $h_m < h_{ml}$ and $h_m > h_{m2}$. Despite the fact that the droplet height in the second and third ranges described above is limited, the droplet length and volume become very large for $h_m \to h_{ml}$ or $h_m \to h_{m2}$.

One should note that the difference between a_c and a_{cl} decreases with increasing fiber radius R_f and that for $R_f \to \infty$, $a_c = a_{cl}$ and only two ranges of stability, the first and the second, occur. The last limit coincides with the case of a cylindrical droplet on a planar surface, considered in Refs. [1,2].

The loci of the points (θ_0, a_c) and (θ_0,a_{cl}) (critical lines) divide the θ_0–a plane into three domains, shown in Fig. 8, each of them corresponding to one of the stability ranges discussed above. The shape of the drop profile depends on the drop size and the values of the microcontact angle θ_0 and parameter a. If the point (θ_0, a) belongs to the domain of unlimited heights (below the curve $a = a_c(\theta_0)$ in Fig. 8) and lies in the very vicinity of the critical line then the small droplets have a shape close to a planar one. However, the large droplets do not possess such shape. If the point (θ_0, a) lies far from the critical line and the microcontact angle has a value less than 90° then the drop shape is close to a circular one. The circular approximation is less satisfactory for $\theta_0 > 90°$. In all cases Carroll's solution [7] for the same R_f, h_m, and $\theta = \theta_0$ provides a profile with a smaller length than predicted by the microscopic theory. The discrepancy is larger for small droplets than for the large ones. Changing the value of the contact angle θ in Carroll's solution it is possible to adjust the two profiles. For example, in the case illustrated in Fig. 9a ($\theta_0 = 45°$) the two profiles almost coincide if one selects $\theta = 6°$. The large difference between the two angles is not surprising because the first of them appears as a result of a rapidly changed curvature of the drop profile near the fiber surface due to the microscopic liquid-solid and liquid-liquid interactions, while the second was calculated using a macroscopic approach. Actually, the angle θ should be compared with the angle between the fiber axis and the continuation of that part of the profile which is far enough from the surface to be unaffected by the strong liquid–solid interactions. Calculations performed for the profile presented in Fig. 9a using the continuation of the interval $200 < x/\sigma_{LS} < 350$ provided a value for the apparent contact angle equal to 2.3° which is much closer to Carroll's result, but nevertheless does not coincide with it. However, the circular approximation which represents a continuation of the profile near the drop apex provided a contact angle $\theta_m = 6.4°$ which almost coincides with Carroll's result.

When the point (θ_0, a) is located in the domain of limited droplet height corresponding to the second range of drop stability (above the curve $a = a_{cl}(\theta_0)$ in Fig. 8), the drops become planar when h_m is close to h_{m1} (see Fig. 11c). For the corresponding profile the height/length ratio is approximately 0.025. One of the consequences is that were a droplet with a height larger than h_{m1} artificially created, it would transform into a planar droplet with a new height close to h_{m1}. Because this height is very small it would be equivalent to spreading of the initial droplet into a film on the fiber.

Another interesting feature of the droplet profile one can find in the domain of limited heights corresponding to the third range of drop stability (between the critical curves in Fig. 8). If the drop height is close to h_{m2}, the drop has a long film-like part (manchon) which increases when h_m becomes closer to h_{m2}.

The obtained solution for the profile provides also several expressions which can be used to evaluate the parameter a and microcontact angle θ_0 using experimental data regarding the drop height, length, volume and position of the inflection point (if it exists) or the inflection angle.

The present theory is based on the minimization of the potential energy of a drop possessing a barrel-like shape. In reality, a drop with the same volume but with another shape (clam-shell, for example) may have a smaller potential energy, and for this reason, the latter type of the droplet will be present on the fiber instead of the barrel. This issue is related to the so-called roll-up transition and was considered by Carroll [7] and McHale [19] on the basis of a macroscopic approach. To consider this problem microscopically one has to calculate the potential energy of a clam-shell drop and compare with one for barrel drop, but this is out of the scope of the present paper.

APPENDIX A. EQUATION FOR THE DROP PROFILE—DERIVATION

To minimize the total potential energy of the barrel drop given by Eq. (24) one needs expressions for the drop volume V, the surfaces S_{LV} and S_{LS}, and the integral $\int_V \rho_L \Phi^l_{LS}(y)\mathrm{d}V$ on the right-hand side of Eq. (23). Because of the axial symmetry of the barrel drop, the calculation of all those quantities is straightforward,

$$V = 4\pi\int_0^{h_m}(R+h)x\,\mathrm{d}h,\ S_{LV} = 4\pi\int_0^{h_m}(R+h)\sqrt{1+x_h^2}\mathrm{d}h,\ S_{LS} = 4\pi R x_0 = -4\pi R\int_0^{h_m}x_h\mathrm{d}h, \tag{A.1}$$

$$\int_V \rho_L\Phi^l_{LS}(y)\mathrm{d}V = 4\pi\rho_L\int_0^{h_m}\Phi^l_{LS}(h)(R+h)x\mathrm{d}h,$$

where $R = R_f + \sigma_{LS}$, $x_h \equiv dx/dh$, and h_m is the droplet height at its apex. The equality $x_0 = -\int_0^{h_m} x_h \mathrm{d}h = -x(h)|_0^{h_m}$ follows from the obvious fact that $x(h_m) = 0$. The total potential energy is given by

$$U_{\text{total}} = \int_0^{h_m} f(h, x, x_h)\mathrm{d}h, \tag{A.2}$$

where

$$f(h,x,x_h)=4\pi K_s\left\{\left[-a_0+a_1\Phi^l_{\rm LS}(h)\right](R+h)x\right.$$
$$\left.+(R+h)\sqrt{1+x_h^2}-bRx_h\right\}, \tag{A.3}$$

$$a_0=\frac{K_\upsilon}{K_S},\quad a_1=\frac{\rho_{\rm L}}{K_S} \tag{A.4}$$

and

$$b=1+\frac{1}{K_S}\left[\rho_{\rm LS}\Phi_{\rm LS}(0)-\rho_{\rm L}\sigma_{\rm LL}\Phi^l_{\rm LS}(\sigma_{\rm LL})-\sigma_{\rm VS}\Phi^l_{\rm VS}(0)\right]. \tag{A.5}$$

In Eq. (A.5), according to Eqs. (9), (10), and (11),

$$\Phi_{\rm LS}(0)=-\frac{\pi}{6}\epsilon_{\rm LS}\rho_{\rm S}\sigma^3_{\rm LS}k-\epsilon^s_{\rm LS} \tag{A.6}$$

and

$$\Phi^l_{\rm LS}(\sigma_{\rm LL})=-\frac{\pi}{6}\epsilon_{\rm LS}\rho_{\rm S}\sigma^3_{\rm LS}\frac{k}{\left(1+\sigma_{\rm LL}/\sigma_{\rm LS}\right)^{\nu}}. \tag{A.7}$$

Because the potentials $\Phi_{\rm LS}(h)$ and $\Phi^l_{\rm LS}(h)$ depend on the radius of the fiber, R_f, b is R_f-dependent.

If one uses as a constraint the given volume V of the drop, then the equation of the drop profile can be obtained by the variational minimization of the functional,

$$\int_0^{h_m} f(h,x,x_h)\,{\rm d}h+4\pi\lambda\int_0^{h_m}(R+h)x\,{\rm d}h, \tag{A.8}$$

where λ is a Lagrange multiplier for the drop volume. The functional in Eq. (A.8) can be rewritten in the following more convenient form,

$$\int_0^{h_m} F(h,x,x_h)\,{\rm d}h, \tag{A.9}$$

where

$$F(h,x,x_h)=\left[\lambda_a+a_1\Phi^l_{\rm LS}(h)\right](R+h)x+(R+h)\sqrt{1+x_h^2}-bRx_h,$$
$$\lambda_a=\frac{\lambda}{K_s}-a_0. \tag{A.10}$$

The latter functional is obtained by dividing the functional of Eq. (A.8) with the nonzero coefficient $4\pi K_s$.

Using the standard calculus of variations [18], the following differential equation for the profile could be obtained:

$$\frac{x_{hh}}{\left(1+x_h^2\right)^{3/2}}+\frac{x_h}{(R+h)\left(1+x_h^2\right)^{1/2}}-a_1\Phi^{l}_{\mathrm{LS}}(h)=\lambda_a. \tag{A.11}$$

Because the apex of the droplet and the contact point of the profile with the fiber should be located on the h- and x- axes, respectively (Fig. 6), one can also write the so-called transversality conditions [18]

$$\left.\frac{\partial}{\partial x_h}F(h,x,x_h)\right|_{h=0}=0,\quad \left[F(h,x,x_h)-x_h\frac{\partial}{\partial x_h}F(h,x,x_h)\right]\Bigg|_{h=h_m}=0 \tag{A.12}$$

which provide the following two conditions for the derivative x_h:

$$\left.\frac{1}{\sqrt{1+x_h^2}}\right|_{h=h_m}=0, \tag{A.13}$$

$$\left.\frac{x_h}{\sqrt{1+x_h^2}}\right|_{h=0}=b. \tag{A.14}$$

The first condition, Eq. (A.13), or, equivalently, $x_h\mid_{h=h_m}=\infty$ ($h_x\,|_{x=0}=0$), reflects the fact that the tangent to the drop profile at its apex should be parallel to the solid surface. The second condition, Eq. (A.14), provides an expression for the microcontact angle θ_0 of the drop on the solid surface (Fig. 6) through the equation

$$\cos\theta_0=-b \tag{A.15}$$

which follows from Eq. (A.14) because $x_h\mid_{h=0}=-\cot\theta_0$.

APPENDIX B. EQUATION FOR THE DROP PROFILE—SOLUTION

To transform Eq. (26) into a differential equation of the first order, let us introduce the function

$$\Psi(h)=\frac{x_h}{\sqrt{1+x_h^2}}, \tag{B.1}$$

the derivative of which provides the curvature κ (h) of the profile in the plane containing the fiber axis

$$\frac{d\Psi(h)}{dh}=\frac{x_{hh}}{\left(1+x_h^2\right)^{3/2}}=\kappa(\mathrm{h}). \tag{B.2}$$

Geometrically, $\Psi(h)=-\cos\theta_h$, where θ_h is the angle which a tangent to the drop profile in the point (x, h) makes with the negative direction of the x-axis. Then Eq. (26) becomes

$$\frac{d\Psi(h)}{dh}+\frac{\Psi(h)}{R+h}-a_1\Phi_{\mathrm{LS}}^{l}(h)=\lambda_a \tag{B.3}$$

and has the general solution

$$\Psi(h)=\frac{\lambda_a}{2}(R+h)-a\frac{D(h)}{R+h}+\frac{C}{R+h}, \tag{B.4}$$

where C is a constant of integration,

$$a=\frac{1}{4K_1}\frac{\epsilon_{\mathrm{LS}}\rho_{\mathrm{S}}}{\epsilon_{\mathrm{LL}}\rho_{\mathrm{L}}}\left(\frac{\sigma_{\mathrm{LS}}}{\sigma_{\mathrm{LL}}}\right)^4, \tag{B.5}$$

$$K_1=\frac{17}{16}-\frac{3}{8}\Delta^2-\frac{1}{2}\Delta^3-\frac{3}{16}\Delta^4-\frac{\rho_{\mathrm{LS}}/2}{\sigma_{\mathrm{LL}}\rho_{\mathrm{L}}}\left(1-\Delta^3\right), \tag{B.6}$$

and

$$D(h)=-\frac{a_1}{a}\int\Phi_{\mathrm{LS}}^{l}(h)(R+h)\mathrm{d}h. \tag{B.7}$$

Using Eq. (9) for the potential $\Phi_{\mathrm{LS}}^{l}(h)$, Eq. (B.7) becomes

$$D(h)=\frac{k\sigma_{\mathrm{LS}}}{(2-\nu)(1+h/\sigma_{\mathrm{LS}})^{\nu-2}}$$

$$+\frac{kR_f}{(1-\nu)(1+h/\sigma_{\mathrm{LS}})^{\nu-1}}. \tag{B.8}$$

Combining the boundary condition (A.14) with Eqs. (B.1) and (A.15) leads to $\Psi(0)=-\cos\theta_0$, which provides the following expression for C:

$$C=-R\cos\theta_0-\frac{\lambda_a}{2}R^2+aD(0). \tag{B.9}$$

Consequently, the function $\Psi(\mathrm{h})$ becomes

$$\Psi(h)=\frac{1}{R+h}\left\{-R\cos\theta_0+\frac{\lambda_a}{2}h(h+2R)\right.$$

$$\left.-a\left[D(h)-D(0)\right]\right\}. \tag{B.10}$$

Equation (B.1) can be rewritten as

$$x_h=\frac{\Psi(h)}{\sqrt{1-\Psi^2(h)}} \tag{B.11}$$

which has the obvious solution

$$x(h) - x_0 = \int_0^h \frac{\Psi(h)}{\sqrt{1-\Psi^2(h)}} dh, \tag{B.12}$$

where $x_0 \equiv x(0)$ (Fig. 6) is a second integration constant which can be determined from the condition that the drop apex to be located on the h-axis ($x(h_m) = 0$). One thus obtains the expression

$$x_0 = -\int_0^h \frac{\Psi(h)}{\sqrt{1-\Psi^2(h)}} dh. \tag{B.13}$$

After the determination of x_0 and C, the solution (B.12) still contains two unknown parameters, λ_a and h_m. A connection between them is provided by the boundary condition Eq. (A.13), which together with Eq. (B.11) leads to the equation $\Psi(h_m) = \pm 1$. The correct sign of Ψ (h_m) can be selected by observing that near the drop apex the derivative x_h must be negative. According to Eq. (B.1), the function $\Psi(h)$ has the same sign and therefore, $\Psi(h_m) = -1$. The latter condition provides the following equation for λ_a:

$$\lambda_a = -\frac{2}{h_m(h_m + 2R)} \times \left\{(1-\cos\theta_0)R + h_m - a\left[D(h_m) - D(0)\right]\right\}. \tag{B.14}$$

REFERENCES

[1] G.O. Berim, E. Ruckenstein, J. Phys. Chem. B 108 (2004) 19330 (Section 3.2 of this volume).
[2] G.O. Berim, E. Ruckenstein, J. Phys. Chem. B 108 (2004) 19339 (Section 3.3 of this volume).
[3] T.S. Meiron, A. Marmur, I.S. Saguy, J. Colloid Interface Sci. 274 (2004) 637.
[4] Y. Gu, D. Li, J. Colloid Interface Sci. 206 (1998) 288.
[5] G. McHale, N.A. Käb, M.I. Newton, S.M. Rowan, J. Colloid Interface Sci. 186 (1997) 453.
[6] A.W. Neumann, R.J. Good, in: R.J. Good, R.R. Stromberg (Eds.), Surface and Colloid Science, vol. 11, Plenum, New York, 1979, p. 31.
[7] B.J. Carroll, J. Colloid Interface Sci. 57 (1976) 488.
[8] A.V. Neimark, J. Adhesion Sci. Technol. 13 (1999) 1137.
[9] C. Bauer, S. Dietrich, Phys. Rev. E 62 (2000) 2428.
[10] B.J. Carroll, Langmuir 2 (1986) 248.
[11] C.A. Miller, E. Ruckenstein, J. Colloid Interface Sci. 48 (1974) 368.
[12] E. Ruckenstein, P.S. Lee, Surf. Sci. 52 (1975) 298 (Section 3.1 of this volume).
[13] F. Brochard, J. Chem. Phys. 84 (1986) 637.
[14] B.V Derjaguin, Acta Phys. Chim. USSR 12 (1940) 181, J. Colloid Interface Sci. 49 (1974) 249.
[15] S.H. Lee, P.J. Rossky, J. Chem. Phys. 100 (1994) 3334.
[16] M. Abramovitz, I. Stegun (Eds.), Handbook of Mathematic Functions with Formulas, Graphs, and Mathematical Tables, Dover, New York, 1972.
[17] A Sharma, Langmuir 9 (1993) 3580.
[18] L.E. Elsgolc, Calculus of Variations, Pergamon/Addison Wesley, New York, 1962.
[19] G. McHale, M.I. Newton, Colloids Surf. A 206 (2002) 79.

3.5 Cylindrical Droplet on Nanofibers

*A Step toward the Clam-Shell Drop Description**

Gersh Berim and Eli Ruckenstein[†]
Department of Chemical and Biological Engineering, State University of New York at Buffalo, Buffalo, New York 14260

Corresponding Author

[†] E-mail: feaeliru@buffalo.edu;
Phone: (716)645-1179 Fax: (716)645-3822.

Received: January 3, 2005; In Final Form: April 19, 2005

I. INTRODUCTION

A liquid droplet placed on a cylindrical surface of a solid (fiber) and immersed in a vapor (gas) can acquire either an axisymmetrical (barrel) or a nonaxisymmetrical (clam-shell) configuration (Figure 1). Experiment[1] showed that the configuration that a given drop takes depends on the drop volume and the conventionally macroscpically determined contact angle between the drop and the fiber. For larger volumes and smaller contact angles, the drop tends to acquire a barrel-like shape. With decreasing volume, that axisymmetrical configuration is transformed through the so-called roll-up transition[2] to a clam-shell configuration. Full theoretical consideration of that transition requires a comparison of the free energies of the two configurations for the same drop volume. An equation for the barrel drop surface was established in numerous papers on the basis of a macroscopic approach[3–7] that involved such quantities as the surface tensions and the disjoining pressure, as well as on the basis of microscopic theories[8,9] based on liquid—liquid and liquid—solid interaction potentials. The obtained solutions provided equations of the surface in explicit form through elliptic integrals[3] or quadratures[7–9] and allowed one to calculate the energy of any given barrel droplet. Carroll[1] has employed his solution for the barrel-drop profile[3] for an indirect theoretical treatment of the roll-up transition by analyzing the stability to small perturbations of its shape. If the profile is not stable, the barrel drop is in a metastable state and can acquire a clam-shell shape. The conditions for existence of a barrel drop were also obtained in ref 8 using a microscopic theory.

Much less is known about the shape of the clam-shell droplet, for which no analytical results could be obtained because of the complexity of the problem. The only known calculations were carried out by the finite element method[10,11] based on the classical idea of the constancy of the Laplace excess pressure along the drop surface. The necessity of a large-scale computation makes that method inconvinient and any analytical solution highly desirable.

In this paper, the characteristics of a clam-shell droplet profile in a plane that includes the drop apex and is normal to the fiber axis are examined. The profile is approximated by that of a cylindrical drop on a fiber, which can be described as an infinite stripe with the same profile in any plane

* *J. Phys. Chem. B*, Vol. 109, No. 25, 2005. Republished with permission.

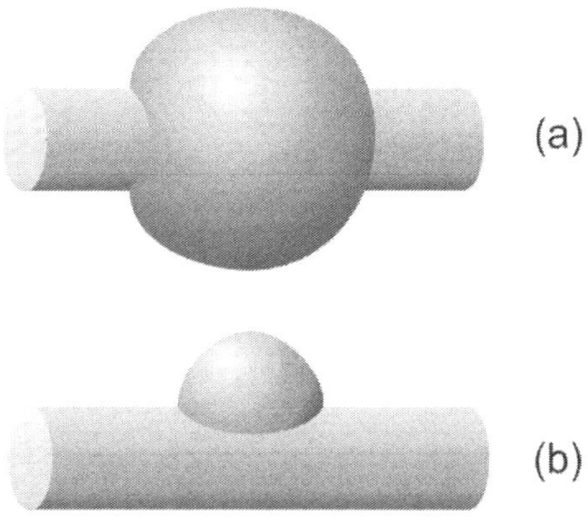

FIGURE 1. Axisymmetric (a) and clam-shell (b) configurations of a liquid drop on a fiber.

normal to the fiber axis. Such a drop can be considered as a clam shell with an infinite dimension in the direction of the fiber axis. One can suppose that the profiles of the cylindrical and clam-shell drops are close to each other provided that both droplets have equal heights. As an argument in favor of such a conclusion, one can mention that the same similarity occurs for the profiles of cylindrical and axisymmetrical drops of equal heights on a planar surface.[12,13] Consideration of the cylindrical drop on a fiber can also provide information about the roll-up transition. Although a cylindrical drop can be used to approximate the clam-shell one, a film with the same volume can be used to approximate the barrel drop. By calculating and comparing the potential energies of those two configurations of the liquid on the fiber, one can find the transition between the two.

Note that the Laplace pressure is no longer constant if one takes into account the liquid—liquid and liquid—solid interactions. Those interactions can be described in the framework of a macroscopic approach by introducing the disjoining pressure[14] that depends on the distance of the liquid molecules from the solid surface. In this case, it is the sum of the disjoining and Laplace pressures that is constant along the drop surface.

Another, microscopic, approach that accounts for the inter-molecular interactions was developed by Ruckenstein et al.[15,16] and applied to the calculation of the droplet shapes on a planar solid surface[12,13,15,16] and of barrel droplets on fibers.[8] The microscopic theory provides in a natural way, without additional assumptions, the microscopic contact angle (θ_0) between the actual drop profile and the solid surface, which depends on the interaction potentials. The theory reveals a rapid change of the curvature of the drop profile in the nanometric range near the solid surface, where the interactions between liquid and solid are strong. This leads to a difference between the microcontact angle (θ_0) between the drop profile and the solid surface and the measurable macroscopic contact angle (θ_m) determined with the wetting angle instrument. The latter represents in reality the angle between the extrapolation of the profile somewhat farther from the leading edge (outside the range of strong interactions between the liquid and the solid) and the surface. θ_m can be evaluated by considering the profile as a part of a circle with the curvature equal to that at the apex of the drop profile.[12,13] Let us note that the distinction between the microscopic and macroscopic contact angles was initially made in ref 16. The microscopic approach is applied in the present paper to the description of a cylindrical droplet on a fiber.

In Section II, the interaction potentials are introduced, and the total potential energy of a system is calculated in Section III. Using the variational minimization of that energy, an equation for the profile is derived in Section IV and solved by quadrature, and an expression for the microcontact angle θ_0 is obtained. Section V is devoted to the finding of the existence conditions of a cylindrical drop on a fiber, which are further used in Section VI to examine the drop shape. The features of the roll-up transition between the cylindrical and the film-like configurations are discussed in Section VII.

II. POTENTIALS OF INTERMOLECULAR INTERACTIONS

A. Interaction between the Molecules of the Liquid.

The potential $\phi_{LL}(r)$ of the interaction between the molecules of the liquid is chosen in the form

$$\phi_{LL}(r)=\begin{cases} 0, & r>\eta, \\ -\epsilon_{LL}\left(\dfrac{\sigma_{LL}}{r}\right)^6, & \eta\geq r\geq \sigma_{LL}, \\ \infty, & r<\sigma_{LL}, \end{cases} \tag{1}$$

where r is the distance between the centers of the interacting molecules, $\sigma_{LL} > 0$ and $\epsilon_{LL} > 0$ are the size of the repulsive core and an interaction constant, respectively, and η (a radius of the interaction sphere) is the largest value of r for which the interaction is considered as nonzero. The potential eq 1 is a simplified version of the London-van der Waals potential with a rigid core repulsion

$$\tilde{\phi}_{LL}(r)=\begin{cases} -\epsilon'_{LL}\left(\dfrac{\sigma'_{LL}}{r}\right)^6, & r\geq \sigma'_{LL} \\ \infty, & r^{<}\sigma'_{LL} \end{cases} \tag{2}$$

which is difficult to use in the variational calculation of the drop shape because of the long-range nature of the potential. Because the London-van der Waals potential decreases rapidly with r, the approximate eq 1 seems reasonable. In ref 8, the relation between ϵ_{LL}, σ_{LL}, and η on one hand and ϵ'_{LL} and σ'_{LL} on the other hand was obtained using the requirement of equality of the total potential energies ϕ_1 and ϕ'_1 of a molecule in the bulk of a homogeneous liquid, calculated using the potentials eq 1 and eq 2, respectively. That relation has the form

$$\frac{\epsilon_{LL}}{\epsilon'_{LL}}=\left(\frac{\sigma'_{LL}}{\sigma_{LL}}\right)^3\left(1-\Delta^3\right)^{-1} \tag{3}$$

where $\Delta = \sigma_{LL}/\eta$. If, for example, $\eta = 3\sigma_{LL}$, then the following particular choices for parameters ϵ_{LL} and σ_{LL} are obtained: $\sigma_{LL} = \sigma'_{LL}$, $\epsilon_{LL} = (27/26)\epsilon'_{LL}$.

B. Interactions between the Molecules of Liquid and Solid, and Vapor and Solid.

It is supposed that the interaction potential between a molecule of liquid and a molecule of solid is equal to the sum of two components, $\phi^{l}_{LS}(r)$ and, $\phi^{s}_{LS}(r)$, related to the long-range and the short-range interactions. The long-range potential is chosen in a form similar to $\tilde{\phi}_{LL}(r)$

$$\phi^{l}_{LS}(r)=\begin{cases} -\epsilon_{LS}\left(\dfrac{\sigma_{LL}}{r}\right)^6, & r\geq \sigma_{LS}, \\ \infty, & r<\sigma_{LS}, \end{cases} \tag{4}$$

where $\sigma_{LS} > 0$ and $\epsilon_{LS} > 0$ are the size of the repulsive core and the interaction constant of the liquid—solid interactions, respectively. In what follows, σ_{LS} is used as the unit of length. In real systems, σ_{LS} is of the order of 3 to 4 Å[17] and for definiteness we selected $\sigma_{LS} = 3$ Å.

The total potential $\Phi_{LS}^{l}(y)$ of a liquid molecule interacting with a fiber through the long-range interactions was calculated in ref 8, where the fiber was considered as an infinite cylindrical solid body of radius R_f with a constant number density (ρ_s) of atoms. The result is

$$\Phi_{LS}^{l}(y) = \frac{\pi \epsilon_{LS} \rho_S \sigma_{LS}^{6}}{24(R_f + y)(R_f^2 - y^2)^2} \left\{ (7R_f^2 + y^2) \times \right.$$

$$\left[E\left(\frac{\pi}{4}, -\frac{4R_f y}{(R_f - y)^2} \right) + E\left(\frac{3\pi}{4}, -\frac{4R_f y}{(R_f - y)^2} \right) \right] -$$

$$\left. (R_f + y)^2 \left[F\left(\frac{\pi}{4}, -\frac{4R_f y}{(R_f - y)^2} \right) + F\left(\frac{3\pi}{4}, -\frac{4R_f y}{(R_f - y)^2} \right) \right] \right\} \tag{5}$$

where $y > R_f$ is the distance of a molecule from the fiber axis, and $F(\phi, m)$ and $E(\phi, m)$ are the elliptic integrals of the first and second kind, respectively.[18] In the limiting case $R_f \to \infty$ the potential $\Phi_{LS}^{l}(y)$ coincides with that of a molecule of liquid interacting with a semi-infinite solid.

Potential eq 5 decreases rapidly within several hard core radii from the fiber surface and for $R_f \gg \sigma_{LS}$ can be approximated with high accuracy by the power law function

$$\Phi_{LS}^{l}(h) = -\frac{\pi}{6} \epsilon_{LS} \sigma_{LS}^{3} \rho_S \frac{k}{\left(1 + \frac{h}{\sigma_{LS}}\right)^{v}} \tag{6}$$

where $h = y - R_f - \sigma_{LS}$ and k and v are constants (for given R_f).[8] For example, for $R_f = 100\ \sigma_{LS} \simeq 30$ nm, one has $v = 3.01712$, and $k = 0.9926$.[8]

The short-range interactions between liquid and solid molecules, such as the acid—base ones, decrease rapidly (exponentially) after distances of the order of 1 nm from the liquid—solid interface.[7,19] Hence, the total potential energy $\Phi_{LS}^{s}(h)$ of the liquid molecule due to those interactions can be approximated by the potential[8]

$$\Phi_{LS}^{s}(h) = \begin{cases} -\epsilon_{LS}^{s}, & h = 0, \\ 0, & h > 0 \end{cases} \tag{7}$$

where ϵ_{LS}^{s} is a constant.

Combining the potentials of eqs 6 and 7, one obtains the total potential, $\Phi_{LS}(h)$, of a molecule of liquid interacting with a fiber

$$\Phi_{LS}(h) = \begin{cases} \Phi_{LS}^{l}(0) + \Phi_{LS}^{s}(0), & h = 0, \\ \Phi_{LS}^{l}(h), & h > 0 \end{cases} \tag{8}$$

The interaction between the molecules of vapor (gas) and solid has in most cases only a long-range component, which, for low-density vapors, is determined only by the molecules adsorbed on the solid surface ($h = 0$). The potential $\Phi_{VS}(h)$ of that interaction can be written in the form

$$\Phi_{VS}(h) = \begin{cases} \Phi_{VS}^{l}(0), & h = 0 \\ 0, & h > 0 \end{cases} \tag{9}$$

where the potential $\Phi_{VS}^{l}(0)$ has the same form as $\Phi_{LS}^{l}(0)$ given by eq 6, where ϵ_{LS} and σ_{LS} have to be replaced by ϵ_{VS} and σ_{VS}, respectively.

III. POTENTIAL ENERGY OF THE SYSTEM

If a vapor (gas) that surrounds a drop has a low density, one can neglect the vapor—vapor and vapor—liquid potential energies in the calculation of the total potential energy (U_{total}) of the system and take into account only the interaction of the adsorbed vapor molecules with the solid. The interaction energy of the molecules of the solid between themselves will be considered constant and will also be excluded from consideration. The possible changes of the latter energy caused by the restructuring of the surface molecules of the solid due to their interaction with the molecules of liquid or vapor can be included into the corresponding solid—liquid and solid—vapor interactions. Under those assumptions, U_{total} can be written in the form

$$U_{\text{total}} = U_{\text{LL}} + U_{\text{LS}} + U_{\text{VS}} \tag{10}$$

where U_{LL} and U_{LS} are the potential energies due to the interactions of the molecules of liquid between themselves and with those of the fiber, respectively, and U_{VS} is a potential energy due to the interactions between the molecules of the vapor and the fiber.

A. The Potential Energy of a Drop.

The model used most frequently for a liquid considers the liquid as a continuous medium characterized by a constant macroscopic number density (ρ_l) of molecules.[5,7,16] Such a continuous picture is not valid in the very vicinity of the fiber surface where the interactions of the liquid molecules with the fiber are the strongest. Because of these interactions, the liquid molecules closest to the fiber surface are rearranged in a layer "sticked" to the fiber surface, which will be assumed to be a monolayer separated from the surface by a distance σ_{LS}. The remainder of the drop is separated from the first layer by a distance, σ_{LL}, and is assumed to be a continuous medium of constant density (see Figure 2).

Assuming that the size of the droplet is much larger than the radius η of the range of the liquid—liquid interactions, the energy U_{LL} can be represented in the form[8]

$$U_{\text{LL}} = -K_{\text{V}}V + K_{\text{s}}S_{\text{LV}} + K_{\text{s}}'S_{\text{LS}} \tag{11}$$

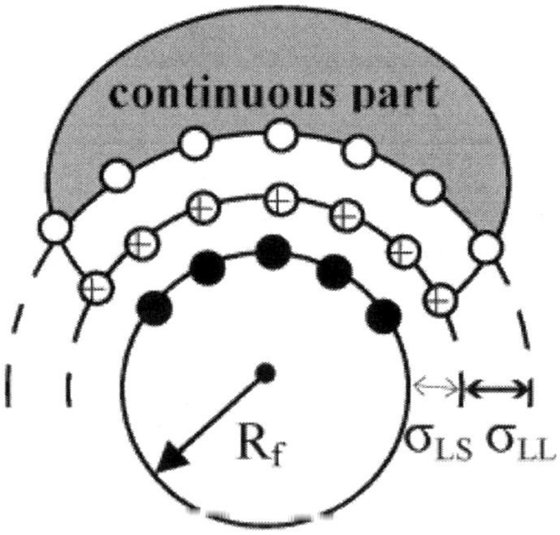

FIGURE 2. Cross-section of a cylindrical drop on a fiber. The filled and crossed circles represent the molecules on the fiber surface and the molecules of the first (closest to the solid surface) layer of liquid molecules, respectively. Beginning with the second layer (open circle) the liquid is considered as a continuous medium.

where V, S_{LV}, and S_{LS} are the total volume of the droplet and the surfaces of the liquid—vapor and liquid—solid interfaces, respectively

$$K_v = \frac{2\pi}{3}\epsilon_{LL}\sigma_{LL}^3\rho_L^2\left(1-\Delta^3\right)$$

$$K_s = \frac{2\pi}{3}\epsilon_{LL}\sigma_{LL}^4\rho_L^2\left[\frac{17}{16}-\frac{3}{8}\Delta^2-\frac{1}{2}\Delta^3-\frac{3}{16}\Delta^4-\frac{\rho_{LV}/2}{\sigma_{LL}\rho_L}\left(1-\Delta^3\right)\right]$$

$$K_s' = \frac{2\pi}{3}\epsilon_{LL}\sigma_{LL}^4\rho_L^2\left[\frac{17}{16}-\frac{3}{8}\Delta^2-\frac{1}{2}\Delta^3-\frac{3}{16}\Delta^4-\frac{\rho_{LS}/2}{\sigma_{LL}\rho_L}\left(1-\Delta^3\right)\right] \quad (12)$$

and ρ_{lv} and ρ_{ls} are the surface densities of the liquid at the liquid—vapor and liquid—solid interfaces, respectively.

To calculate the interaction energy, U_{LS}, let us note that the molecules closest to the solid layer are affected by both long-range and short-range interactions with the fiber, which are represented by the potential eq 8 for $h = 0$. The molecules of the continuous part of the drop do not exhibit short-range interactions.

Under the above assumptions, the energy (U_{LS}) is given by

$$U_{LS} = \rho_{LS}\Phi_{LS}(0)S_{LS} + \int_{V_{cont}} \rho_L\Phi_{LS}^l(h)\mathrm{d}V \quad (13)$$

where S_{LS} is the contact area between the droplet and the fiber, and V_{cont} denotes the volume of the continuous part of the droplet. The latter can be written as the difference between the drop volume (V) and the volume (V_g) of the gap between the first layer and the continuous part of the droplet that is approximately equal to $V_g = \sigma_{LL}S_{LS}$, that is, $V_{cont} = V - S_{LS}\sigma_{LL}$. Then the second term in the right-hand side of eq 13 becomes

$$\int_{V_{cont}} \rho_L\Phi_{LS}^l(h)\mathrm{d}V = \int_V \rho_L\Phi_{LS}^l(h)\mathrm{d}V - \int_{V_g} \rho_L\Phi_{LS}^l(h)\mathrm{d}V \simeq \int_V \rho_L\Phi_{LS}^l(h)\mathrm{d}V - \rho_L\sigma_{LL}\Phi_{LS}^l(\sigma_{LL})S_{LS} \quad (14)$$

Consequently

$$U_{LS} \simeq \int_V \rho_L\Phi_{LS}^l(h)\mathrm{d}V + \left[\rho_{LS}\Phi_{LS}(0) - \rho_L\sigma_{LL}\Phi_{LS}^l(\sigma_{LL})\right]S_{LS} \quad (15)$$

B. Potential Energy of the Vapor—Solid Interactions.

As already mentioned in Section II.B, only the contribution of the adsorbed vapor molecules to the vapor—solid potential energy will be taken into account. Using the potential eq 9 and denoting the surface density of the vapor molecules on the solid surface ρ_{VS}, we can write the potential energy (U_{VS}) of the vapor—solid interactions in the form

$$U_{VS} = \rho_{VS}\left(S_{solid} - S_{LS}\right)\Phi^{l}_{LS}(0) \tag{16}$$

where S_{solid} is the total area of the solid surface and S_{LS} is the area of the liquid—solid interface.

Using eqs 11, 15, and 16 and omitting the constant term proportional to S_{total} in the last equation, the following expression for the total potential energy of the system is obtained

$$U_{total} = -K_v V + K_s S_{LV} + \int_V \rho_L \Phi^{l}_{LS}(h)\mathrm{d}V + K_s b S_{LS} \tag{17}$$

where

$$b = \frac{K'_s}{K_s} + \frac{1}{K_s}\left[\rho_{LS}\Phi_{LS}(0) - \rho_L \sigma_{LL}\Phi^{l}_{LS}(\sigma_{LL}) - \rho_{VS}\Phi^{l}_{VS}(0)\right] \tag{18}$$

Equation 17 provides the basis for the analysis of the drop characteristics (existence, shape, etc.)

IV. EQUATION FOR THE PROFILE OF A CYLINDRICAL DROP

In Figure 3 two possible profiles of a cylindrical drop on a fiber in the plane perpendicular to the axis of the fiber are presented together with the polar coordinate system used in the calculations. The surface of the cylindrical drop is obtained by translating the profile along the axis of the fiber. We consider a drop on a bare surface; thefore, the leading edge of the drop is located just on the fiber surface.

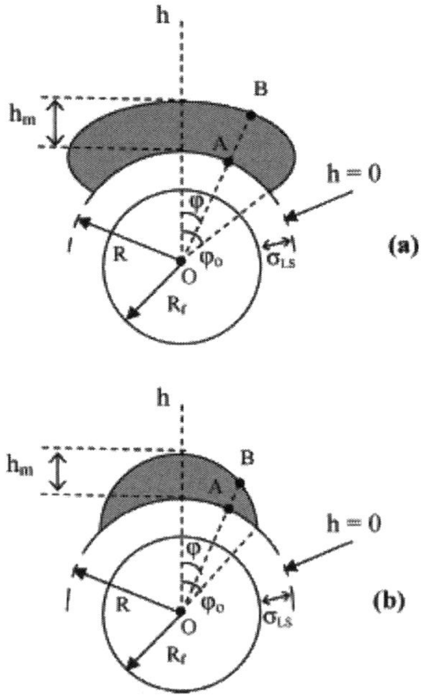

FIGURE 3. Characteristic profiles of a cylindrical drop in a plane normal to the fiber axis: (a) microcontact angle θ_0 is larger than 90°; (b) θ_0 is smaller than 90°.

To provide a unique description for both shapes (Figure 3a and b), the equation of the drop profile is chosen in the present paper in the form $\phi = \phi\,(h)$, and only the part of the profile for which $\phi \geq 0$ will be examined further.

The equation of the drop profile can be derived using the variational minimization of the total potential energy (U_{total}) of the drop defined by eq 17 assuming the constancy of the drop volume. The derivation details are given in Appendix A. The equation for the profile has the form

$$\frac{(R+h)\phi_{hh}+2\phi_h+(R+h)^2\phi_h^3}{\left[1+(R+h)^2\phi_h^2\right]^{3/2}}-a_1\Phi_{\mathrm{LS}}^1(h)=\lambda_{\mathrm{c}} \tag{19}$$

where the subscript h means differentiation with respect to h, $R = R_f + \sigma_{\mathrm{LS}}$, $a_1 = \rho_{\mathrm{L}}/K_{\mathrm{s}}$, and λ_{c} is a constant dependent on the drop volume. The first term of eq 19 represents the curvature $\kappa(h)$ of the drop profile in polar coordinates.

The boundary conditions for eq 19 have the form

$$\phi\,|_{h=h_m}=0 \tag{20}$$

$$\phi_h|_{h=0}=-\frac{1}{R}\cot\theta_0 \tag{21}$$

where h_{m} is the droplet height at its apex (Figure 3) and θ_0 is the microcontact angle (θ_0 is defined as the angle between the tangents to the profiles of the fiber and profile of the drop at the point of contact). Note that the height h_{m} is not given and can be found using the known value of the drop volume.

Eq 21 follows from the fact that in polar coordinates the angle a that the curve $\phi = \phi(r)$ makes with the circle $r = R$ is given by the expression

$$\tan\alpha=-\frac{1}{R\phi_r}\quad\left(\phi_r\equiv\frac{\mathrm{d}\phi}{\mathrm{d}r}\right) \tag{22}$$

and that in our case $\phi_r = \phi_h$. The expression for angle θ_0 is obtained from the so-called transversality conditions,[20] which allows one to connect θ_0 to the microscopic parameters of the interaction potentials. The result is (see Appendix A)

$$\cos\theta_0=-b \tag{23}$$

where b is provided by eq 18

It should be noted that the above expression for the microcontact angle (θ_0) was obtained from pure microscopic considerations. Angle θ_0 depends on the parameters of the intermolecular interactions and on the fiber radius, R_{f}.

The solution of eq 19 for the boundary conditions eqs 20 and 21 is (see Appendix B)

$$\phi(h)-\phi_0=\int_0^h\frac{\Psi(h)}{(R+h)\sqrt{1-\Psi^2(h)}}\,\mathrm{d}h \tag{24}$$

where $\phi_0 = \phi(0)$ (Figure 3)

$$\Psi(h)=\frac{1}{R+h}\left\{-R\cos\theta_0+\frac{\lambda_c}{2}h(h+2R)-a\left[D(h)-D(0)\right]\right\} \tag{25}$$

$$a = \frac{1}{4K_1}\frac{\epsilon_{LS}\rho_S}{\epsilon_{LL}\rho_L}\left(\frac{\sigma_{LS}}{\sigma_{LL}}\right)^4 \tag{26}$$

$$K_1 = \frac{17}{16} - \frac{3}{8}\Delta^2 - \frac{1}{2}\Delta^3 - \frac{3}{16}\Delta^4 - \frac{\rho_{LV}/2}{\sigma_{LL}\rho_L}\left(1-\Delta^3\right) \tag{27}$$

and

$$D(h) = -\frac{a_1}{a}\int \Phi_{LS}^{l}(h)(R+h)\mathrm{d}h \tag{28}$$

The explicit expressions for constants ϕ_0 and λ_c and the function $D(h)$ are provided in Appendix B. To find all of the unknown parameters (ϕ_0, λ_c, and h_m) it is necessary to add to eqs 20 and 21 the equation for the drop volume $V = 2\int_0^{h_m}(R+h)\phi\mathrm{d}h$ substituting $\phi(h)$ given by eq 24 and changing the order of integration one obtains

$$V = -\int_0^{h_m}\frac{\left(h^2+2Rh\right)\Psi(h)}{\sqrt{1-\Psi^2(h)}}\mathrm{d}h \tag{29}$$

V. EXISTENCE OF A CYLINDRICAL DROP ON A BARE SURFACE OF A FIBER

The first necessary condition for the existence of a drop on a bare surface follows from eq 23 and has the simple form $|b| \leq 1$. It establishes (through eq 18) a restriction for the values of the microscopic parameters for which a cylindrical drop can be formed on a surface. If $|b| > 1$, a cylindrical drop cannot exist. It should be noted that the restriction $|b| < 1$ (or, equivalently, $|\cos\theta_0| < 1$) differs from the traditional one, which implies $|\cos\theta_m| < 1$.

A second general necessary condition results from the fact that the integral on the right-hand side of eq 24 has meaning only if the function $\Psi(h)$ satisfies the inequality

$$|\Psi(h)| < 1, \ (0 < h < h_m) \tag{30}$$

The function $\Psi(h)$ depends on the fiber radius (R_f), the drop height (h_m), the microcontact angle (θ_0), and microparameter a and satisfies the conditions $\Psi(h_m) = -1$, $\Psi(0) = -\cos\theta_0$. For any given values of the parameters, $\Psi(h)$ (eq 25) can have one of the three behaviors shown in Figure 4a. However, because only the first satisfies restriction eq 30, a cylindrical droplet will exist only for those values of the parameters for which $\Psi(h)$ exhibits the first behavior.

To examine the consequences of the second necessary condition in more detail, let us first consider the case in which R_f (= $100\sigma_{LS}$) and θ_0 (= 45°) are fixed and only h_m and a are variables. The analysis of the behavior of function $\Psi(h)$ shows that inequality 30 is valid only in a domain of the $h_m - a$ plane below a critical curve $a = a_m(h_m)$ (see Figure 4b), which exhibits a minimum $a = a_c = 0.613$ at $h_m = h_{m,c} = 11.2\sigma_{LS}$ and reaches the asymptotic value $a = a_{c1} = 0.692$ for $h_m \to \infty$. If $a < a_c$, the drop can have any height. If a specific value a_0 of the parameter a lies between a_c and a_{c1} (e.g., $a_0 = 0.65$), then the second necessary condition is satisfied by h_m values smaller than $h_{m,1} = 1.11\sigma_{LS}$ or larger than $h_{m,2} = 147.77\sigma_{LS}$. The analysis of the shape of the drop profile (section VI) shows that for $h_m > h_{m,2}$ the angular length of the droplet (the angle ϕ_0 of Figure 3) is larger than 180° and therefore those shapes do not have any physical meaning. As a result, the actual critical curve in the

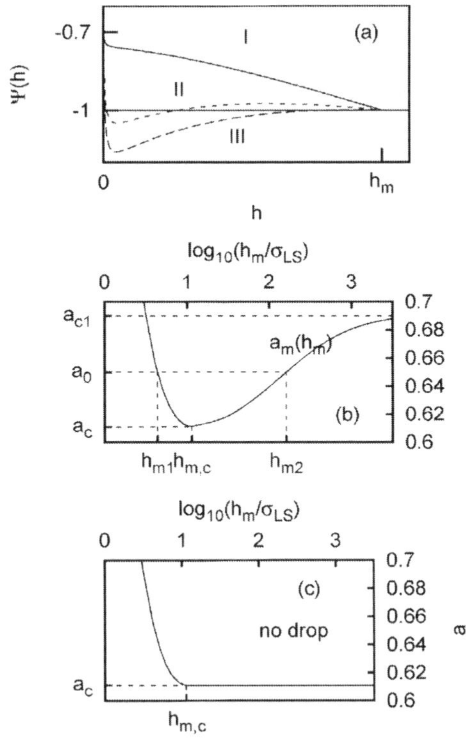

FIGURE 4. (a) Three types of behavior of the function $\Psi(h)$ given by eq 25. A cylindrical drop on the fiber exists only for those values of the parameters that provide the first type of behavior. (b) The critical curve $a = a_m(h_m)$ that divides the $h_m - a$ plane into two domains for which the second necessary condition given by inequality 30 is not fulfilled (above the curve) and fulfilled (below the curve). (c) The final critical curve $a_c(h_m)$.

$h_m - a$ plane has the form presented in Figure 4c. The critical value (a_c) depends on the radius (R_f) of the fiber and on the microcontact angle (θ_0). As R_f increases, a_c decreases and for $R_f \to \infty$ (planar surface) acquires the value $a_c = 0.5857$.

By performing similar analyses for other values of θ_0, one can plot a_c against θ_0 in a $\theta_0 - a$ plane (see Figure 5). The curve $a = a_c(\theta_0)$ separates the domains of restricted and unrestricted droplet

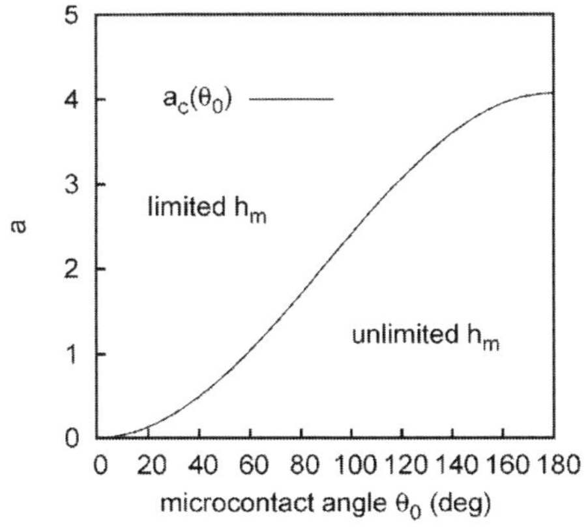

FIGURE 5. Dependence of the critical values a_c and a_{c1} of the parameter a on the microcontact angle θ_0 for $R_f = 100\sigma_{LS} \approx 30$ nm. The solid line ($a = a_c(\theta_0)$) divides the $\theta_0 - a$ plane into domains of unlimited (below the solid line) and limited (above solid line) heights.

heights for $R_f = 100\sigma_{LS}$. The pairs (θ_0, a) for which the droplet height h_m can have any value lie below that curve. In the domain located above the curve, the drop height can take only values $h_m < h_{m,l}$.

VI. SHAPE OF THE DROP PROFILE AND THE CONTACT ANGLE

In this section, the shape of the drop profile will be examined for various values of microcontact angle θ_0 and parameter a, which characterizes the liquid—liquid and liquid—solid interactions. Because of the strong interaction between solid and liquid near the liquid—solid interface, the curvature and slope of the drop change rapidly with the distance from the solid. For a drop on a planar surface, it was shown[12,13] that the main change of the curvature occurs in a range of about 10 to $40\sigma_{LS} \simeq 3$ to 12 nm. At larger distances from the solid, the curvature of the profile remains almost constant. The slope of the profile also changes rapidly in that nanometric range. The microcontact angle (θ_0) cannot be determined with the conventional equipment employed for wetting angle measurements which cannot "see" what happens in the very vicinity of the solid surface. The experimentally measured angle (θ_m) is in reality an extrapolation of the profile slope from a macroscopic distance from the leading edge up to the solid surface. A useful approximation of the drop profile away from the solid surface is a circular shape with a radius equal to the curvature radius of the profile at the droplet apex. For a drop on a planar solid, the angles θ_0 and θ_m are represented in Figure 6a. For a drop on a fiber, the angles θ_0 and θ_m can be defined in a similar way (see Figures 6b and c). To calculate angle θ_m for the latter case one can use the expression

$$\cos\theta_m = \frac{R_m^2 + R^2 - d^2}{2R_m R} \tag{31}$$

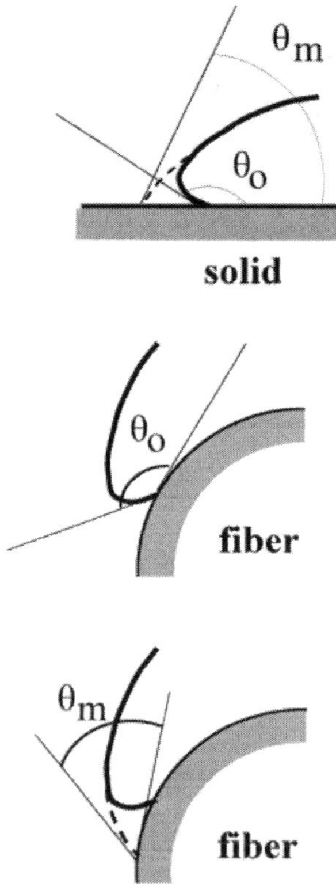

FIGURE 6. Microscopic, θ_0, and macroscopic, θ_m, contact angles for the drop profile on a planar solid (a) and fiber (b, c). The circular approximation for the profile is represented by the dashed lines.

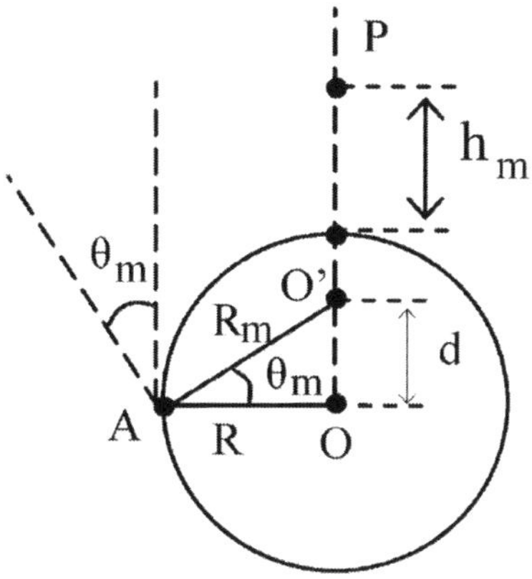

FIGURE 7. Auxiliary plot for the calculation of macroscopic contact angle θ_m. O and O' are, respectively, the centers of a fiber of radius R and a circle with a radius R_m equal to the curvature radius at the drop apex P. The latter circle is not shown in the figure.

where $R_m = 1/\kappa(h_m)|$ is the curvature radius at the drop apex, and $d = R - R_m + h_m$. (Equation 31 follows from the application of the cosine theorem to the triangle OAO' of Figure 7.) The curvature $\kappa(h)$ can be found from eq 19 and has the form

$$\kappa(h) = \lambda_c + a_1\Phi^1_{LS}(h) \tag{32}$$

where λ_c is given by B.14 and $\Phi^1_{LS}(h)$ by eq 6. One can thus obtain a relation between $\cos\theta_m$ and $\cos\theta_0$. For a drop for which $\sigma_{LS} \ll h_m \ll R$, one obtains the simple expression

$$\cos\theta_m = \left(1 + \frac{h_m}{2R}\right)\cos\theta_0 + \frac{ak}{\nu - 1} - \frac{h_m}{2R}\left(1 - \frac{ak}{\nu - 1}\right) \tag{33}$$

which for $R \to \infty$ leads to the equation

$$\cos\theta_m = \cos\theta_0 + \frac{a}{2} \tag{34}$$

obtained previously for a drop on a planar surface.[12,13]

The specific features of the cylindrical drop profile on the fiber depend on several factors, including the distance of the point (θ_0, a) from the critical line and the drop volume. Taking as an example a fiber of radius $R_f = 100\sigma_{LS}$, the typical profiles of a small droplet of volume $V = 3.63\times10^3\sigma_{LS}^2 \simeq 1.0\times10^2$ nm^2 per unit length of the drop are presented in Figure 8 for $a = 0.61$ and various values of the microcontact angle θ_0 from the domain of unlimited heights. The selected value of a is close to the critical value $a_c = 0.613$ for $\theta_0 = 45°$. Therefore, the plot of the profile (solid line) in Figure 8a corresponds to a point (θ_0, a) in the very vicinity of the critical line $a = a_c(\theta_0)$ (see Figure 5). The dashed curve represents the circular approximation for the drop profile. In the case considered, the circle has no intersection with the fiber and most of the droplet profile and the profile of the fiber can be considered as approximately concentric circles. This case is similar to that of a planar droplet on a planar solid surface,[12,13] which occurs when the point (θ_0, a) is located near the critical line. For this reason, the above drop on the fiber can be considered as a quasi planar one.

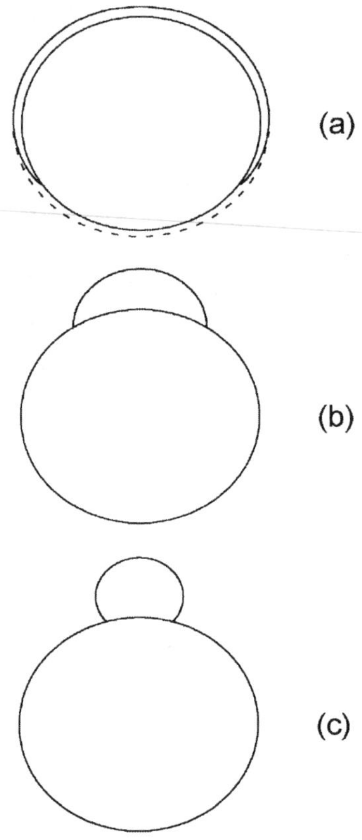

FIGURE 8. Profiles of a small drop of volume $V = 3.63\times10^3\sigma_{LS}^2 \simeq 1.0\times10^2$ nm² per unit length for several values of the microcontact angle θ_0 (solid lines) and a = 0.612: (a) θ_0 = 45°; (b) θ_0 = 75°; (c) θ_0 = 135°. The dashed line on figure a represents a circular profile with the same height (h_m) as the original one and with a radius equal to the curvature radius of the original profile at the drop apex. The same approximations for figures b and c almost coincide with the profiles of the drops. There is a difference at the leading edge that cannot be detected by eyes. All of the profiles correspond to the points (θ_0, a) in the domain of unlimited height, the point (θ_0, a) for profile a being located in the very vicinity of the critical line.

When θ_0 increases, the angular length ϕ_0 of the droplet becomes smaller, the droplet height increases, and the approximation of the drop profile by a circular one becomes better. As a consequence, the approximate and exact curves in Figure 8b and c become indistinguishable by eye. However, in the very vicinity of the fiber surface, the circular approximation does not coincide with the exact solution and therefore the microscopic and macroscopic contact angles have different values. For instance, for the profiles shown in Figure 8 b and c for θ_0 = 75 and 135° the angles (θ_m) are 55.9 and 113.9°, respectively. For a similar droplet on a planar surface the angles (θ_m) were 55 and 113.7°, respectively, almost the same as on the fiber. The profiles of a larger droplet $\left(V = 4.1\times10^4\sigma_{LS}^2 \simeq 1.1\times10^3\,\text{nm}^2\right)$ are presented in Figure 9. These profiles cannot be considered as quasi planar. The macroscopic contact angles (θ_m) are practically the same as those for the small droplet.

When the pair (θ_0, a) lies above the critical line $a = a_c(\theta_0)$ (Figure 5) the droplet height can acquire only values $h_m < h_{m,1}$ and the droplet has a quasi planar shape when $h_m \to h_{m,1}$. It should be noted that $h_{m,1}$ in the cases considered above is small ($h_{m,1}$ < 5 Å), and therefore the applicability of the theory is questionable.

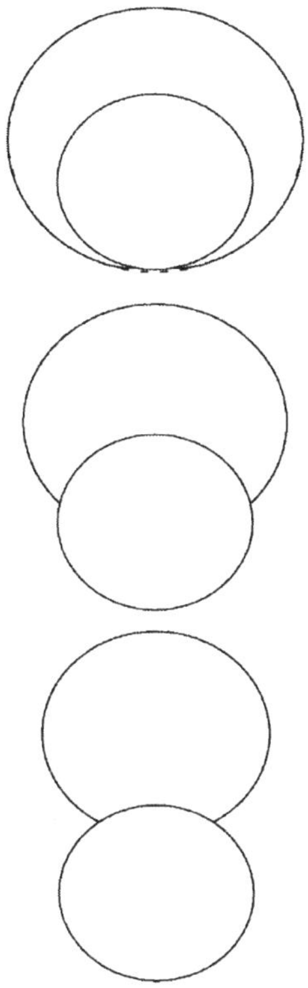

FIGURE 9. Profiles of a relatively large drop of a volume per unit length $V = 4.1\times10^4\sigma_{LS}^2 \simeq 1.1\times10^3$ nm^2 for several values of microcontact angle θ_0 (solid lines) and $a = 0.612$: (a) $\theta_0 = 45°$; (b) $\theta_0 = 75°$; (c) $\theta_0 = 135°$. All of the other notations are the same as those in Figure 8.

VII. ROLL-UP TRANSITION

For the complete examination of the existence conditions for a cylindrical drop on a fiber, it is necessary to take into account its possible transformation to a film that covers the fiber. The latter transformation takes place if the potential energy for a film is smaller than that for a cylindrical droplet, provided that both the film and the droplet have the same volume (V) per unit length.

The total potential energy, $U_{\text{total}}^{\text{cyl}}$, of a cylindrical droplet is provided by eq 17, where the drop volume and areas of the liquid—solid and liquid—vapor interfaces can be calculated using eq A.1 from Appendix A.

For a film, the total potential energy can be calculated in the same way as for a cylindrical drop and the result is

$$U_{\text{total}}^{\text{f}} = -K_v V_{\text{f}} + K_s S_{\text{LV}}^{\text{f}} + \int_V \rho_L \Phi_{\text{LS}}^{1}(h)\,\text{d}V + K_s b S_{\text{LS}}^{\text{f}} \qquad (35)$$

where V_{f}, S_{LS}^{f}, and S_{LS}^{f} are the volume and the areas of liquid—vapor and liquid—solid interfaces of the film, respectively. Simple geometrical considerations give

$$S_{\text{LS}}^{\text{f}} = 2\pi\left(h_{\text{f}} + 2R\right)$$

$$S_{\text{LS}}^{\text{f}} = 2\pi R$$

$$\int_{V_f} \rho_L \Phi_{LS}^1(h) dV = 2\pi\rho_L \int_0^{h_f} (R+h)\Phi_{LS}^1(h) dh \tag{36}$$

In eq 36, h_f is the film thickness which for a volume $V_f = V$ is given by

$$h_f = \sqrt{R^2 + \frac{V}{\pi}} - R$$

Employing eqs 17 and 35, the difference of potential energies $U_{total}^{f} - U_{total}^{cyl} \equiv \Delta U$ can be represented in the form

$$\frac{1}{2K_s}\Delta U = \pi(h_f + R) - \int_0^{h_m} \sqrt{1+(R+h)^2 \phi_h^2}\, dh + Rb(\pi - \phi_0)$$

$$-a\left\{\pi\left[D(h_f) - D(0)\right] - \int_0^{h_m} \frac{k(R+h)}{\left(1+\frac{h}{\sigma_{LS}}\right)^{\nu}} \phi dh\right\} \tag{37}$$

where the functions $\phi \equiv \phi(h)$ and D(h) are provided by eqs 24 and B.8, respectively. In Figure 10 the difference ΔU is plotted as a function of the droplet volume for $\theta_0 = 45°$ for various values of the parameter a from the domain of unlimited heights ($a < a_c = 0.613$) and $R = 100\sigma_{LS}$. For $a > a_{lim} \simeq 0.536$, ΔU changes sign at a volume Vc, which depends on a. Therefore, for $V > V_c$, U_{total}^{f} is smaller than U_{total}^{cyl} and the film configuration is more stable than the cylindrical droplet. If $a < a_{lim}$ the potential energy of the cylindrical drop is smaller than that of the film for any value of the volume, that is, the cylindrical configuration is always stable.

Experimental data[1] show that the critical volume (V_c) of a drop at which the roll-up transition occurs increases with increasing macroscopic contact angle. In Figure 11 the θ_0– dependence of the critical volume is presented for two values of a and for a fiber with $R_f = 20\sigma_{LS}$. (The θ_m–dependence is qualitatively similar). For each a, the roll-up transition takes place only in a narrow range of

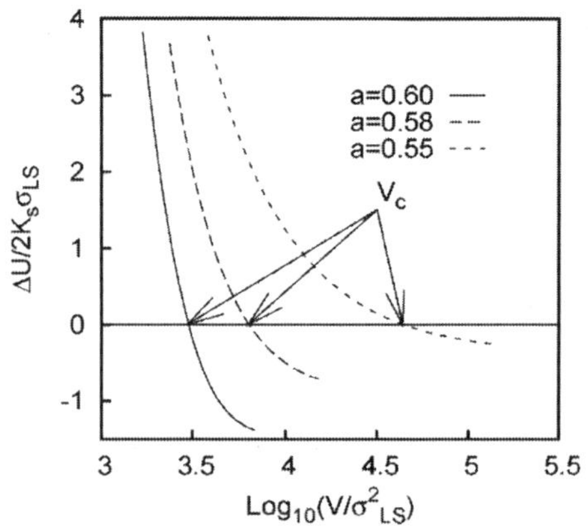

FIGURE 10. The difference (ΔU) between the potential energies of the system for a film and a cylindrical drop of the same volume (V) as a function of V for several values of parameter a. The microcontact angle is equal to $\theta_0 = 45°$. V_c represents the critical volume at which the roll-up transition occurs.

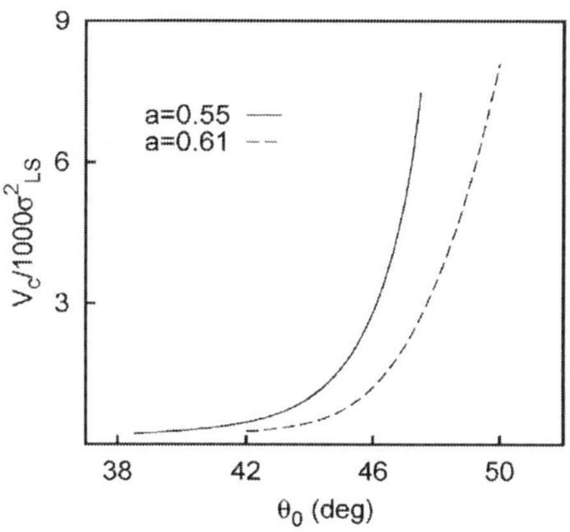

FIGURE 11. Dependence of the critical volume (V_c) on microcontact angle θ_0 for several values of parameter a.

microcontact angles ($\theta_{0,1} < \theta_0 < \theta_{0,2}$) from the domain of unlimited heights. Outside this region, the cylindrical droplets are energetically preferable at any drop size. In the range $\theta_{0,1} < \theta_0 < \theta_{0,2}$, the volume ($V_c$) increases rapidly with increasing θ_0. Such a behavior coincides qualitatively with the experimental observations.[11] However, experiment provides a much wider (but also restricted) range of contact angles where the roll-up transition occurs. The possible reason for such a difference is that the experimental data were obtained for finite-size drops (barrel and clam-shell), whereas the theory considers infinite sizes for both the barrel and the clam-shell drops. The smaller the value of the parameter a, the smaller is the angle θ_0 for which the roll-up transition occurs. Note that the range of existence of the roll-up transition increases with decreasing R_f. For example, for $R_f = 20\sigma_{LS}$ and $a = 0.61$ the difference $\theta_{0,2} - \theta_{0,1}$ is 8.0°, whereas for $R_f = 100\sigma_{LS}$ it is 4.2°.

VIII. DISCUSSION AND SUMMARY

In this paper, a microscopic approach is applied to the description of a cylindrical drop on a fiber and to the transition from a drop to a film covering the fiber (roll-up transition). Considering the cylindrical drop as a limiting case of a clam-shell one, and the film as a limiting case of a barrel drop, the obtained analytical results can be related to the barrel-drop—clam-shell drop transition (roll-up transition), which was treated before only numerically in the framework of a macroscopic approach. The analysis of the equation for the drop profile reveals that the drop existence conditions depend on the value of microparameter a, eq 26, which incorporates the microscopic parameters of liquid—liquid, solid—liquid, and solid—gas interactions as well as the liquid and the solid densities. For each value of microcontact angle θ_0, a critical value, a_c, exists which separates two ranges of possible drop heights. In the first range $a < a_c$, and the drop can have any height (h_m) and volume. In the second range $a > a_c$, and the drop height is upper bounded by some value, h_{m1}.

The locus of the points (θ_0, a_c) (critical line) divides the $\theta_0 - a$ plane into two domains shown in Figure 5, each of them corresponding to one of the ranges mentioned above. The shape of the drop profile for a given fiber radius (R_f) depends on the drop size, microcontact angle θ_0, and parameter a. If the point (θ_0, a) belongs to the domain of unlimited heights (below the curve $a = a_c(\theta_0)$ in Figure 5) and lies in the very vicinity of the critical line, then the small droplets have a quasi planar shape, that is, most of the drop profile is located on a circle concentric with the profile of the fiber. However, the larger droplets do not possess such a shape (compare Figures 8a and 9a). If the point (θ_0, a) lies far from the critical line, then the drop shape is close to a circular one. In all cases, the values of the macrocontact (θ_m) and microcontact (θ_0) angles do not coincide. The difference between θ_0 and angle θ_m (the angle determined with the conventional wetting angle instrument) is a consequence of the strong interactions between the solid and liquid near the leading edge of the drop. Near the leading edge there is a very strong variation of the drop curvature, whereas somewhat farther from the leading edge, where the interactions with the solid are not as strong, the curvature becomes almost constant. The wetting angle instrument can determine angles in the macroscopic range and not in the microscopic range near the leading edge. For this reason, the measured wetting angle is an extrapolation of the slope of the drop from some distance from the leading edge to the solid surface.

When the point (θ_0, a) is located in the domain of limited droplet heights corresponding to the second range of drop existence (above the curve $a = a_c(\theta_0)$ in Figure 5) the drops become quasi planar when h_m is close to h_{m1}.

Analysis of the roll-up transition shows that for given values of a and R_f it occurs only if the microcontact angle lies in a narrow interval $\theta_{0,1} < \theta_0 < \theta_{0,2}$, $\theta_{0,1}$ and $\theta_{0,2}$ being functions of R_f and a, and the drop volume is larger than some critical value, V_c. Note that the point ($\theta_{0,1}$, a) is located near the critical curve in Figure 5, and the difference between $\theta_{0,2}$ and $\theta_{0,1}$ increases with decreasing R_f. If the microcontact angle θ_0 lies outside the interval ($\theta_{0,1}$, $\theta_{0,2}$), then the roll-up transition does not occur for any drop volume.

ACKNOWLEDGMENT.

We are grateful to Ilya G.Berim for his help in numerical calculations.

APPENDIX A. DERIVATION OF THE EQUATION FOR THE DROP PROFILE

Expressions for drop volume V, surfaces S_{LV} and S_{LS}, and the integral $\int_V \rho_L \Phi^l_{LS}(y)dV$, which are necessary to find the total potential energy, U_{total}, have the form (all quantities are given per unit length of the drop)

$$V = 2\int_0^{h_m} (R+h)\phi dh, S_{LV} =$$

$$2\int_0^{h_m} \sqrt{1+(R+h)^2 \phi_h^2} dh, S_{LS} = 2R\phi_0 =$$

$$-2R\int_0^{h_m} \phi_h dh, \int_V \rho_L \Phi^l_{LS}(y) dV =$$

$$4\pi\rho_L \int_0^{h_m} \Phi^l_{LS}(h)(R+h)\phi dh, \quad (A.1)$$

where $R = R_f + \sigma_{LS}$, $\phi_h \equiv d\phi/dh$, and h_m is the droplet height. The angular length, ϕ_0, of the drop

$$\phi_0 = -\int_0^{h_m} \phi_h dh = -\phi(h)\vert_0^{h_m} = \phi(0) \quad (A.2)$$

follows from the obvious fact that $\phi(h_m) = 0$ (Figure 3). The total potential energy is given by

$$U_{total} = \int_0^{h_m} f(h,\phi,\phi_h) dh \quad (A.3)$$

where

$$f(h,\phi,\phi_h) = 2K_s\left\{\left[-a_0 + a_1\Phi^1_{LS}(h)\right](R+h)\phi + \right.$$

$$\left.\sqrt{1+(R+h)^2\phi_h^2} - bR\phi_h\right\} \quad (A.4)$$

$$a_0 = \frac{K_v}{K_s},\ a_1 = \frac{\rho_L}{K_s} \quad (A.5)$$

and

$$b = \frac{K_s'}{K_s} + \frac{1}{K_s}\left[\rho_{LS}\Phi_{LS}(0) - \rho_L\sigma_{LL}\Phi^1_{LS}(\sigma_{LL}) - \rho_{VS}\Phi^1_{LS}(0)\right] \quad (A.6)$$

From eqs 6, 7, and 8 one has

$$\Phi_{LS}(0) = -\frac{\pi}{6}\epsilon_{LS}\rho_S\sigma^3_{LS}k - \epsilon^s_{LS} \qquad \Phi^1_{VS}(0) = -\frac{\pi}{6}\epsilon_{VS}\rho_S\sigma^3_{VS}k \quad (A.7)$$

and

$$\Phi^1_{LS}(\sigma_{LL}) = -\frac{\pi}{6}\epsilon_{LS}\rho_S\sigma^3_{LS}\frac{k}{\left(1+\frac{\sigma_{LL}}{\sigma_{LS}}\right)^v} \quad (A.8)$$

Because of the dependence of potentials $\Phi_{LS}(h)$, Φ^1_{LS} (h), and $\Phi^1_{VS}(0)$ on the radius of the fiber, R_f, b is also R_f-dependent.

In the case of constant drop volume, the equation of the drop profile can be obtained by the variational minimization of the functional

$$\int_0^{h_m} f(h,x,x_h)\mathrm{d}h + 2\lambda\int_0^{h_m}(R+h)\phi\,\mathrm{d}h \quad (A.9)$$

where λ is a Lagrange multiplier for the drop volume. The latter functional can be rewritten in the following more convenient form

$$\int_0^{h_m} F(h,\phi,\phi_h)\mathrm{d}h \quad (A.10)$$

where

$$F(h,\phi,\phi_h) = \left[\lambda_c + a_1\Phi^1_{LS}(h)\right](R+h)\phi +$$

$$\sqrt{1+(R+h)^2\phi^2{}_h} - bR\phi_h \quad (A.11)$$

and

$$\lambda_c = \frac{\lambda}{K_s} - a_0 \quad (A.12)$$

The functional $F(h,\phi,\phi_h)$ in A.11 is obtained by dividing the functional of A.9 with the nonzero coefficient $2K_s$.

Employing the standard calculus of variations,[20] the following differential equation for the profile is obtained

$$\frac{(R+h)\phi_{hh}+2\phi_h+(R+h)^2\phi_h^3}{\left[1+(R+h)^2\phi_h^2\right]^{3/2}}-a_1\Phi_{\mathrm{LS}}^1(h)=\lambda_c \tag{A.13}$$

Because the apex of the droplet and the contact point of the profile with the fiber should be located at $h=h_m$ and $h=0$, respectively (Figure 3), one can also write the so-called transversality conditions[20]

$$\left.\frac{\partial}{\partial\phi_h}F(h,\phi,\phi_h)\right|_{h=0}=0$$

$$\left.\left[F(h,\phi,\phi_h)-\phi_h\frac{\partial}{\partial\phi_h}F(h,\phi,\phi_h)\right]\right|_{h=h_m}=0 \tag{A.14}$$

which can be rewritten as

$$\left.\frac{1}{\sqrt{1+(R+h)^2\phi_h^2}}\right|_{h=h_m}=0 \tag{A.15}$$

$$\left.\frac{\phi_h}{\sqrt{1+R^2\phi_h^2}}\right|_{h=0}=\frac{b}{R} \tag{A.16}$$

The first condition, A.15, or, equivalently, $\phi_h|_{h=h_m}=\infty\ (h_\phi|_{\phi=0}=0)$, reflects the fact that the tangent to the drop profile at its apex should be normal to the polar radius at the apex. The second condition together with eq 21 provides the value of microcontact angle θ_0

$$\cos\theta_0=-b \tag{A.17}$$

APPENDIX B. SOLUTION OF THE EQUATION FOR THE DROP PROFILE

Let us introduce the function

$$\Psi(h)=\frac{(R+h)\phi_h}{\sqrt{1+(R+h)^2\phi_h^2}} \tag{B.1}$$

which is related to the profile curvature $\kappa(h)$ through the equation

$$\frac{d\Psi(h)}{dh}+\frac{\Psi(h)}{R+h}=\kappa(h) \tag{B.2}$$

Then, eq 19 becomes a first order differential equation

$$\frac{d\Psi(h)}{dh}+\frac{\Psi(h)}{R+h}-a_1\Phi^1_{LS}(h)=\lambda_c \tag{B.3}$$

which has the general solution

$$\Psi(h)=\frac{\lambda_c}{2}(R+h)-a\frac{D(h)}{R+h}+\frac{C}{R+h} \tag{B.4}$$

where C is a constant of integration

$$a=\frac{1}{4K_1}\frac{\epsilon_{LS}\rho_S}{\epsilon_{LL}\rho_L}\left(\frac{\sigma_{LS}}{\sigma_{LL}}\right)^4 \tag{B.5}$$

$$K_1=\frac{17}{16}-\frac{3}{8}\Delta^2-\frac{1}{2}\Delta^3-\frac{3}{16}\Delta^4-\frac{\rho_{LV}/2}{\sigma_{LL}\rho_L}\left(1-\Delta^3\right) \tag{B.6}$$

and

$$D(h)=-\frac{a_1}{a}\int\Phi^1_{LS}(h)(R+h)dh \tag{B.7}$$

Using eq 6 for the potential $\Phi^1_{LS}(h)$, eq 28 becomes

$$D(h)=\frac{k\sigma_{LS}}{(2-\nu)\left(1+\frac{h}{\sigma_{LS}}\right)^{\nu-2}}+\frac{kR_f}{(1-\nu)\left(1+\frac{h}{\sigma_{LS}}\right)^{\nu-1}} \tag{B.8}$$

Combining the boundary condition A.16 with eqs B.1 and A.17 leads to $\Psi(0)=-\cos\theta_0$ which provides the following expression for C

$$C=-R\cos\theta_0-\frac{\lambda_c}{2}R^2+aD(0) \tag{B.9}$$

Hence, function $\Psi(h)$ becomes

$$\Psi(h)=\frac{1}{R+h}\left\{-R\cos\theta_0+\frac{\lambda_c}{2}h(h+2R)-a\left[D(h)-D(0)\right]\right\} \tag{B.10}$$

Equation B.1 can be rewritten as

$$\phi_h=\frac{\Psi(h)}{(R+h)\sqrt{1-\Psi^2(h)}} \tag{B.11}$$

which has the obvious solution

$$\phi(h)-\phi_0=\int_0^h\frac{\Psi(h)}{(R+h)\sqrt{1-\Psi^2(h)}}dh \tag{B.12}$$

where the constant of integration (ϕ_0) can be determined from the condition for the drop apex to be located on the polar axis of the coordinate system [$\phi(h_m) = 0$]. One thus obtains the expression

$$\phi_0 = -\int_0^{h_m} \frac{\Psi(h)}{(R+h)\sqrt{1-\Psi^2(h)}} dh \tag{B.13}$$

After the determination of ϕ_0 and C, eq B.12 still contains two unknown parameters, λ_c and h_m. A connection between them is provided by the boundary condition eq A.15, which together with eq B.11 leads to $\Psi(h_m) = \pm 1$. The correct sign of $\Psi(h_m)$ can be selected by observing that near the drop apex the derivative ϕ_h must be negative because $\phi(h)$ should increase with decreasing h. According to eq B.1 the functions $\Psi(h)$ and $\phi(h)$ have the same sign and therefore $\Psi(h_m) = -1$. The latter condition provides the following expression for λ_c

$$\lambda_c = -\frac{2}{h_m(h_m + 2R)} \times$$

$$\left\{(1-\cos\theta_0)R + h_m - a\left[D(h_m) - D(0)\right]\right\} \tag{B.14}$$

REFERENCES AND NOTES

(1) Carroll, B. J. *Langmuir* **1986**, *2*, 248.
(2) Adam, N. K. *J. Soc. Dyers Colour.* **1937**, *53*, 122.
(3) Carroll, B. J. *J.* Colloid *Interface Sci.* **1976**, *57*, 488.
(4) Carroll, B. J. *J. Colloid Interface Sci.* **1984**, *97*, 195.
(5) Brochard, F. *J. Chem. Phys.* **1986**, *84*, 637.
(6) McHale, G.; Käb, N. A.; Newton, M. I.; Rowan, S. M. *J. Colloid Interface Sci.* **1997**, *186*, 453.
(7) Neimark, A. V. *J. Adhes. Sci. Technol.* **1999**, *13*, 1137.
(8) Berim, G. O.; Ruckenstein, E. *J. Colloid Interface Sci.* **2005**, *286*. 681 (Section 3.4 of this volume).
(9) Bauer, C.; Dietrich, S. *Phys. Rev. E.* **2000**, *62*, 2428.
(10) McHale, G.; Newton, M. I.; Carroll, B. J. *Oil Gas Sci. Technol.* **2000**, *56*, 47.
(11) McHale, G.; Newton, M. I. *Colloids Surf.* **2002**, *206*, 79.
(12) Berim, G. O.; Ruckenstein, E. *J. Phys. Chem. B* **2004**, *108,* 19330 (Section 3.2 of this volume).
(13) Berim, G. O.; Ruckenstein, E. *J. Phys. Chem. B* **2004**, *108,* 19339 (Section 3.3 of this volume).
(14) (a) Derjaguin, B. V. *Acta Phys. Chem.* **1940**, *12*, 181. (b) Derjaguin, B. V. *J. Colloid Interface Sci.* **1974**, *49*, 249.
(15) Miller, C. A.; Ruckenstein, E. *J. Colloid Interface Sci.* **1974**, *48*. 368 (Section 1.1 of this volume).
(16) Ruckenstein, E.; Lee, P. S. *Surf. Sci.* **1975**, *52*, 298 (Section 3.1 of this volume).
(17) Lee, S. H.; Rossky, P. J. *J. Phys. Chem.* **1994**, *100*, 3334.
(18) *Handbook of Mathematic Functions with Formulas, Graphs, and Mathematical Tables;* Abramovitz, M., Stegun, I., Eds.; Dover: New York, 1972.
(19) Sharma, A. *Langmuir* **1993**, *9,* 3580.
(20) Elsgolc, L. E. In *Calculus of Variations*; Pergamon Press: Addison-Wesley: New York, 1962 *J. Phys. Chem. B*, Vol. 109, No. 25, 2005

3.6 Microscopic Interpretation of the Dependence of the Contact Angle on Roughness*

Gersh Berim and Eli Ruckenstein†
Department of Chemical and Biological Engineering, State University of New York at Buffalo, Buffalo, New York 14260

Corresponding Author

† E-mail: feaeliru@buffalo.edu.
Phone: (716)645-1179, Fax: (716) 645-3822.

Received March 12, 2005. In Final Form: May 25, 2005

INTRODUCTION

The increase of the hydrophobicity of a hydrophobic solid surface by increasing the roughness was the concern of numerous experimental and theoretical investigations (see the review paper[2] and the references therein). The basic treatment of that phenomenon was developed by Wenzel[3] and Cassie and Baxter[4] (CB) who suggested phenomenological equations connecting the macroscopic contact angles θ_m and $\theta_{m,0}$ which the drop profile makes with the rough and the smooth solid surfaces, respectively.

The Wenzel equation has the form

$$\cos\theta_m = r\cos\theta_{m,0} \tag{1}$$

where the roughness r is defined as the ratio of the true area of the solid surface to its projected area. In this limiting case (Wenzel regime), one considers that the liquid penetrates into the space between asperities and that there is no gas beneath the drop.

The CB equation is given by

$$\cos\theta_m = -1 + f(1+\cos\theta_{m,0}) \tag{2}$$

where f is the fraction of the area of the top of the asperities with respect to the projected area. In this limiting case, one considers that the drop is located on the top of the asperities and that gas is trapped between the liquid and the solid surface (heterogeneous wetting or CB regime). In both expressions, the angle $\theta_{m,0}$ is identified with the Young contact angle given by the equation

$$\cos\,\theta_{m,0} = \frac{\gamma_{sv}-\gamma_{sl}}{\gamma_{lv}} \tag{3}$$

where γ_{lv}, γ_{sl}, and γ_{sv} are the surface tensions for liquid–gas, solid–liquid, and solid–gas interfaces, respectively. One should note that the Wenzel and CB equations are linear with respect to the parameters r and f which characterize the roughness of the surface. The theoretical versions of

* *Langmuir*. Vol. 21, No. 17, 2005–7745. Republished with permission.

the Wenzel and CB equations developed in refs 5–8 on the basis of a macroscopic approach which involves the surface tensions are also linear with respect to r and f.

Despite the large number of experimental papers regarding the wetting of rough surfaces, the linearity predicted by the Wenzel and CB equations could not be verified for a long time, probably because of the difficulties in controlling the surface roughness. In the past decade, surfaces with regular arrangements of asperities (e.g., cylindrical or square pillars) could be prepared. By changing the sizes of the asperities and the distances between them one could control the roughness of the surface and thus one could check the validity of eqs 1 and 2. A most complete experiment in this direction was carried out recently by Krupenkin et al.[1] In their study, five chemically identical surfaces with various roughnesses were prepared, and the contact angle θ_m of four liquids was determined on each surface. A drop of the liquid with the highest surface tension was located on the top of the asperities (CB regime), and it was concluded that the cosine of its contact angle had a linear dependence on roughness. For the other three liquids with lower surface tensions, it was shown that they penetrate into the space between asperities (Wenzel regime), and therefore, their contact angles should follow the Wenzel law eq 1. However, the latter experiments[1] provided a nonlinear dependence of $\cos\theta_m$ on r. As a possible explanation for such a behavior, the authors suggested to include the line tension which is expected to affect the contact angle. However, a more detailed examination of the contribution of the line tension[6,7] still provided a linear dependence of $\cos\theta_m$ on r.

The above discrepancy between theory and experiment determined us to attempt a more detailed theoretical consideration of the wetting of rough surfaces. In the present paper, the problem is treated on the basis of a microscopic approach which was developed by Ruckenstein et al.[9,10] and applied recently to planar[11,12] and curved[13] surfaces. In that approach, the interactions of the liquid and gas molecules between themselves and with the molecules of solid were taken into account explicitely using appropriate microscopic potentials. A long range (London—van der Waals) as well as a short range (e.g., acid—base) interaction were included in the calculations. The minimization of the total potential energy of the system at equilibrium has provided a differential equation for the drop profile as well as the value of the microcontact angle θ_0 which the profile makes with the solid surface. The latter angle plays a central role in the theory instead of the Young contact angle $\theta_{m,0}$ of the classical macroscopic theory, which in reality does not provide the angle at the leading edge but only the angle between the solid surface and the continuation till the surface of the drop profile near the solid surface (outside the range of strong interactions between the liquid and the solid; see Figure 1). The values of θ_0 and θ_m for large drops which have almost circular profiles are connected by the simple relation (see refs 11 and 12)

$$\cos\theta_m = \cos\theta_0 + \frac{a}{2} \tag{4}$$

where a depends on the microscopic parameters of the interaction potentials employed.

It was shown[11] that if only the long-range interactions between the molecules of the liquid and solid are taken into account the microcontact angle θ_0 is always equal to 180°. It can acquire any value if one considers that the liquid layer near the solid surface (assumed for simplicity a monolayer) is structured by the solid and if short range interactions between that layer and the solid

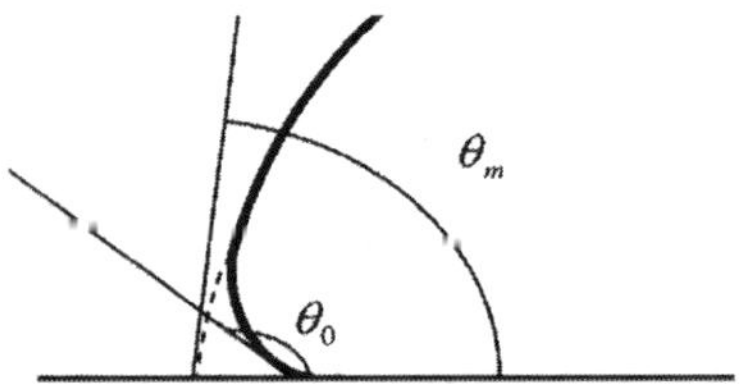

FIGURE 1. Macroscopic, θ_m, and microscopic, θ_0, contact angles.

additionally occur. The application of the microscopic approach of refs 11 and 12 to the study of the surfaces with regular arrangement of asperities is straightforward. However, to explain the nonlinear dependence of $\cos\theta_m$ on roughness an additional assumption about the liquid molecules at the edges of the asperities must be made.

In section 2, the interaction potentials between the molecules of liquid, gas, and solid are defined, and in section 3, the models for the rough surfaces and for the liquid drop are presented. On the basis of those potentials and models, the total potential energy of the drop is calculated in section 4 for Wenzel (section 4.1) and Cassie—Baxter (section 4.2) regimes. The differential equations for the drop profile in each regime are derived and solved in section 5. In section 6, the macroscopic contact angle θ_m is calculated and compared with experiment.

2. INTERACTION POTENTIALS

The considered system consists of a liquid drop embeded in an insoluble gas and residing on a rough solid surface. All three media are considered uniform and having the bulk number densities ρ_L, ρ_G, and ρ_S, respectively. The interaction potentials of the liquid molecules between themselves and with those of the solid are selected, respectively, in the form

$$\phi_{LL}(r_{ll}) = \begin{bmatrix} 0, & r_{ll} > \eta, \\ -\epsilon_{LL}\left(\dfrac{\sigma_{LL}}{r_{ll}}\right)^6, & \eta \geq r_{ll} \geq \sigma_{LL}, \\ \infty, & r_{ll} < \sigma_{LL}, \end{bmatrix} \tag{5}$$

and

$$\phi^l_{LS}(r_{ls}) = \begin{bmatrix} -\epsilon_{LS}\left(\dfrac{\sigma_{LS}}{r_{ls}}\right)^6, & r_{ls} \geq \sigma_{LS}, \\ \infty, & r_{ls} < \sigma_{LS}, \end{bmatrix} \tag{6}$$

where σ_{LL} and σ_{LS} are hard core radii, ϵ_{LL} and ϵ_{LS} are interaction constants, r_{ll} and r_{ls} are the distances between the centers of the interacting molecules (liquid—liquid and liquid—solid), and η in eq 5 is the distance (radius of interaction sphere) after which the liquid—liquid interactions are considered negligible. The hard core radius σ_{LS}, which will be used as unit of length, has a typical value of about 3 Å.[14] The potential $\phi^l_{LS}(r_{ls})$ for the liquid—solid interactions is a long-range one, whereas $\phi_{LL}(r_{ll})$ is a short-range potential. The potential energy of a molecule of liquid interacting with the bulk solid via the potential $\phi^l_{LS}(r_{ls})$ is given by[12]

$$\Phi^l_{LS}(h) = -\frac{\pi}{6}\epsilon_{LS}\sigma_{LS}{}^3\rho_S\frac{1}{(1+h/\sigma_{LS})^3} \tag{7}$$

where h is the distance of the molecule from the solid surface.

In addition to the above potentials, a short-range potential, which has a nonzero value only in the very vicinity of the solid surface (at a distance smaller than 1 nm[15]) and decays exponentially with the distance, will be also taken into account. This potential will be selected in the form[12]

$$\Phi^l_{LS}(h) = \begin{bmatrix} -\epsilon^s_{LS} & h = 0, \\ 0, & h > 0 \end{bmatrix} \tag{8}$$

where ϵ^s_{LS} is an interaction constant.

The interaction between the molecules of gas and solid has in most cases only a long-range component. For a low-density gas, only the molecules adsorbed on the solid surface ($h = 0$) provide a noticeable contribution to the total potential energy. For those molecules, the potential Φ_{GS} of gas—solid interaction can be written in the form

$$\Phi_{GS} = \Phi^{l}_{GS}(0) \tag{9}$$

where the potential $\Phi^{l}_{GS}(0)$ has the same form as $\Phi^{l}_{LS}(0)$ given by eq 7 where ϵ_{LS} and σ_{LS} have to be replaced by ϵ_{GS} and σ_{GS} respectively.

The gas—gas and gas—liquid interactions as well as gravity will be neglected.

3. MODELS FOR THE SOLID SURFACE AND THE LIQUID DROP

In the experimental work of Krupenkin et al.,[1] both the solid and the liquid have a complex structure. A solid substrate (silicon) consisting of a bulk solid and pillars was coated with thin layers of oxide (50 nm) and fluorocarbon (20 nm). The latter layer was added to generate a hydrophobic surface. 30, 40, and 50 volume percent methanol—water mixtures were used to form drops. Such a complex structure provides additional difficulties in calculating the interaction energy between the molecules of liquid and solid. For simplicity, a homogeneous solid and a one component liquid have been considered. Consequently, all microscopic parameters used, have to be considered as mean values.

3.1. Solid Surface

A rough surface is usually characterized by the roughness r and the area fraction of the top of the asperities f. For natural surfaces with random roughnesses the parameters r and f can be determined only approximately from experimental measurements (see, e.g., refs 16 and 17). For surfaces with a regular roughness, r and f can be expressed through the geometric parameters of the asperities. For cylindrical pillars arranged as shown in Figure 2, one has

$$f = \frac{\pi R^2}{l^2}, \quad r = 1 + \frac{2h_p}{R} f \tag{10}$$

where h_P is the pillar height, R is their radius and l is the distance between pillars. From eq 10, it follows that in the considered cases the parameters r and f are connected by linear relations and any of them can be used as a characteristic for roughness.

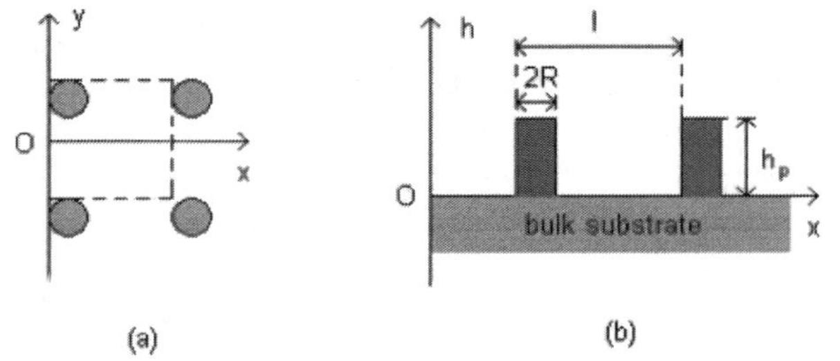

FIGURE 2. Top (a) and side (b) views of a rough solid surface. The asperities are assumed to be cylinders with a radius R and a height h_P. l is the distance between cylinders.

The total area of the solid consists of that between pillars, the area of the top of the pillars, and the side area of the pillars. The fractions t_{bp}, t_{pt} and t_{sp} of these areas with respect to the projected area are given, respectively, by

$$t_{bp} = 1 - f,\ t_{pt} = f,\ t_{sp} = r - 1 + f \tag{11}$$

In the experimental work[1] mentioned above, cylindrical pillars were used with a radius of 0.175μm height $h_P = 7\mu$m and an interpillar distance between 1 and 4 μm. Consequently, the parameter f was varied between 0.01 and 0.11, and r had values between 1.8 and 8.7. It was supposed that the pillars constitute the only source of roughness. The surfaces between pillars as well as those of the pillars are considered to be smooth.

3.2. Liquid Drop

The liquid drops used by Krupenkin et al.[1] in their experiments were comparatively large and had a characteristic radius of the contact area larger than 1 mm. Therefore, the number of pillars beneath the drop was large. The contact line of the drop with the solid surface was most likely located partly on the top of the pillars and partly on the interpillar surface. Such a complicated geometry makes extremely difficult to perform calculations regarding the drop profile. We will consider below two-dimensional (cylindrical) drops with the profiles presented in Figure 3, panels a and b. The drop has at its apex the height h_m, its profile is symmetric about the h axis and the drop is extended indefinitely along the y- axis that is normal to the plane of the figure. The profile in Figure 3a is typical for the Wenzel regime and the profile in Figure 3b is typical for the Cassie—Baxter regime. In Figure 3c, the reference drop on a smooth planar solid surface is shown, which differs from the drop in the Figure 3a by the absence of asperities. It was shown[11,12] that the profiles of large cylindrical and axisymmetrical drops on a planar surface are close to each other. For this reason, one can assume that they are also similar on a rough surface. The equation for that part of the profile which corresponds to the external part ABC of the liquid—gas interface for $x \geq 0$ will be selected in the form $x = x(h)$ to include into consideration the cases with $\theta_0 > 90°$.[12] (Note that the traditional form, $h = h(x)$, of the profile equation is not applicable to the latter cases because near the drop's edges the profile is not a unique function of x).

The small circles in Figure 3, panels a and b, represent the first layer (monolayer) of liquid molecules which are assumed to be structured by the solid and can be subjected both to long and short-range interactions with the solid. The remaining of the drop separated from the first layer by a distance σ_{LL} (hard core diameter of liquid—liquid interactions) will be considered as a uniform single component medium characterized by the number density ρ_L. It will be also supposed that small portions of insoluble gas can be accumulated during the formation of the drop between the drop and the solid surface near the edges of the asperities and that this accumulation does not allow the liquid to wet completely the solid surface. This partial unwetting is assumed to be sufficiently small for not to affect the potential energy of the drop due to the long-range interactions, but sufficiently large to affect the energy of the short range interactions. Because of the presence of the gas only a fraction of the solid surface beneath the drop is covered by liquid. (This fraction provides the liquid—solid interface.) To characterize that fraction two functions are introduced. The first, $k_s(f)$, represents the fraction of the solid surface between pillars, which is covered by liquid, and the second, $k_{sp}(f)$, represents the fraction of the surface of the pillars covered by liquid. Correspondingly, $1 - k_s(f)$, and $1 - k_{sp}(f)$ represent fractions of the solid surface covered by adsorbed gas molecules. As a result, the liquid—gas interface consists of an external part ABC of the drop surface and that part beneath the drop which is in contact with the trapped gas. The surface number densities ρ_{LS} and ρ_{LG} of the liquid molecules at the liquid—solid and liquid—gas interfaces, respectively, are generally not equal but very likely comparable.

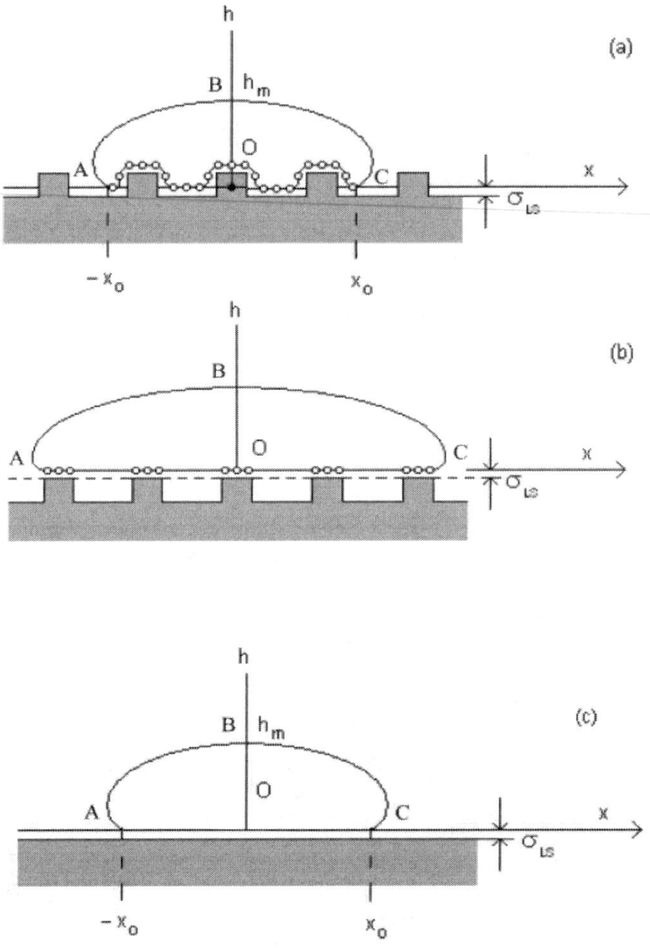

FIGURE 3. Two possible regimes for a cylindrical drop on a rough surface: (a) The liquid penetrates completely into the space between asperities (Wenzel regime). (b) The drop resides on the top of the asperities (Cassie—Baxter regime). (c) The reference drop on the smooth planar solid surface. The open circles constitute the surface layer of molecules of liquid, nearest to the solid, h_m is the drop height, σ_{LS} is the hard core diameter for liquid—solid interaction, and x_0 is the half width of the drop.

One should note that the functions $k_s(f)$ and $k_{sp}(f)$ are unknown and have to be determined by comparing with the experimental data regarding observable quantities (contact angles, area of contact, etc.).

4. POTENTIAL ENERGY OF THE SYSTEM

Using the potentials 5, 7, 8, and 9, the total potential energy U_{total} of the drop can be calculated as the sum of the potential energies, U_{LL}, U_{LS}, and U_{GS} due to the liquid—liquid, liquid—solid, and gas—solid interactions, respectively

$$U_{\text{total}} = U_{LL} + U_{LS} + U_{GS} \tag{12}$$

The first energy, U_{LL}, is given by the sum of volume and surface contributions[13]

$$U_{LL} = -K_v V + K_s S_{GS} + K_s' S_{LS} \tag{13}$$

where V, S_{LG}, and S_{LS} are the total volume of the drop and the surfaces of the liquid—gas and liquid—solid interfaces, respectively and $\Delta = \sigma_{LL}/\eta$. Note, that eq 13 was derived using the

$$K_v = \frac{2\pi}{3}\epsilon_{LL}\sigma_{LL}{}^3\rho_L{}^2\left(1-\Delta^3\right)$$

$$K_s = \frac{2\pi}{3}\epsilon_{LL}\sigma_{LL}{}^4\rho_L{}^2\left[\frac{17}{16}-\frac{3}{8}\Delta^2-\frac{1}{2}\Delta^3-\frac{3}{16}\Delta^4-\frac{\rho_{LG}/2}{\sigma_{LL}\rho_L}\left(1-\Delta^3\right)\right] \quad (14)$$

$$K_s' = \frac{2\pi}{3}\epsilon_{LL}\sigma_{LL}{}^4\rho_L{}^2\left[\frac{17}{16}-\frac{3}{8}\Delta^2-\frac{1}{2}\Delta^3-\frac{3}{16}\Delta^4-\frac{\rho_{LS}/2}{\sigma_{LL}\rho_L}\left(1-\Delta^3\right)\right]$$

assumption that the size of the drop is much larger than the radius η of the interaction sphere.

The energy U_{LS} can be represented as the sum of four terms

$$U_{LS} = U_{LB}^{l} + U_{LP}^{l} + U_{LB}^{s} + U_{LP}^{s} \quad (15)$$

where U_{LB}^{l} and U_{LP}^{l} are, respectively, the total potential energies due to the long-range interaction (potential eq 6) of the molecules of liquid with the bulk part of the solid and with the pillars, and U_{LB}^{s} and U_{LP}^{s} are the corresponding potential energies of interactions due to the short-range interactions.

The potential energy U_{GS} of gas—solid interactions is proportional to the surface number density ρ_{GS} of the gas molecules on the solid surface and the area of contact between gas and solid

$$U_{GS} = \rho_{GS}\left(S_{solid} - S_{LS}\right)\Phi_{GS}^{l}(0) \quad (16)$$

where S_{solid} is the total area of the solid surface.

Below, the total energy U_{total}, given by eq 12, will be calculated separately for the Wenzel regime (Figure 3a) and for the CB regime (Figure 3b).

4.1. Wenzel Regime

In the Wenzel regime, the volume of the drop is equal to the difference beween the volume V_0 of a reference drop (Figure 3c)

$$V_0 = 2\int_0^{h_m} x\,\mathrm{d}h \quad (17)$$

and the volume of all pillars covered by the liquid. Here and below all the quantities are given per unit length of the drop. The volume of a single pillar is equal to $\pi R^2 h_p$ and their number is

$$N_p = \frac{2f}{\pi R^2}x_0 \quad (18)$$

where x_0 is the half width of the cylindrical drop (Figure 3a). As a result the drop volume V which appears in eq 13 is given by

$$V = 2\int_0^{h_m}\left(x + h_p f x_h\right)\mathrm{d}h \quad (19)$$

where $x_h \equiv dx/dh$ and the identity $x_0 = -\int_0^{h_m} x_h dh$ which follows from the fact that $x(h_m) = 0$ is employed.

Similarly, the areas of the liquid—gas and liquid—solid interfaces are given by

$$S_{LG} = 2\int_0^{h_m} [\sqrt{1+x_h^2} - \phi_1(f)x_h]dh \tag{20}$$

$$S_{LS} = -2\phi_2(f)\int_0^{h_m} x_h \, dh \tag{21}$$

where

$$\begin{aligned} \phi_1(f) &= r - k_s(f)t_{bp} - k_{sp}(f)t_{sp} \\ \phi_2(f) &= k_s(f)t_{bp} - k_{sp}(f)t_{sp} \end{aligned} \tag{22}$$

and r is the roughness of the solid surface. As a result, the energy of the liquid—liquid interactions, U_{LL}, is given by

$$\begin{aligned} U_{LL} = 2\int_0^{h_m} dh\Big\{-K_v x + K_s\sqrt{1+x_h^2} - \big[K_s\phi_1(f) + \\ K_s' \phi_2(f)\big]x_h\Big\} \end{aligned} \tag{23}$$

In the calculations of the energies U_{LB}^1 and U_{LP}^1 with the potential eq 7, the contributions of the molecules of the first layer of liquid and the molecules of the continuous part have to be calculated separately because of the different number densities of those molecules.

The energy U_{LB}^1 can be obtained as the difference between the energy U_{ref}^1 of a reference drop and that of "liquid pillars" (pillars which look like the solid pillars but consist of a liquid with the number density ρ_L). The energy of a "liquid pillar", U_1, is given by

$$U_1 = \int_{V_p} \rho_L \Phi_{LS}^1(h) dV \tag{24}$$

where V_p is the volume of a pillar. The total (negative) contribution of these "liquid pillars" is proportional to the number of pillars immersed in the liquid (eq 18). The first energy was calculated in ref 12 (see eq 14 in that paper).

Finally, one can write for the energy U_{LB}^1 the expression

$$\begin{aligned} U_{LB}^1 = 2\int_0^{h_m} \rho_L \Psi_{LS}^1(h)dh + 2x_0\big[\rho_{LS}\Phi_{LS}^1(0) - \\ \rho_L\sigma_{LL}\Phi_{LS}^1(\sigma_{LL}) - \frac{f}{\pi R^2}U_1\Big] \end{aligned} \tag{25}$$

The first two terms in the parentesis describe the contribution of the long-range liquid—solid interactions of the molecules of the surface layer of the liquid molecules (see Figure 3) to U_{LB}^{l} (see ref 12 for details). To provide a rigorous expression for the energy U_{LP}^{l}, one has to take into account the shape and dimensions of pillars. For liquid molecules interacting with a cylindrical solid, the calculations were carried out in ref 13. It was shown that near the surface of the solid the difference between a cylindrical and a planar semiinfinite solid became important only when the radius of the cylinder was less than $100\sigma_{LS} \sim 30$ nm. Because in the experiments examined here the radii of the pillars were much larger (R = 175 nm), the interaction of the liquid molecules located near the pillars with the pillars will be considered given by the interactions with a planar semiinfinite solid. Furthermore, only the contribution of those molecules located at distances smaller than $t_p \sim 1\,\mu$m from the surface of the pillar will be taken into account. If the distance l between pillars is larger than t_p then the potential energy associated with a single pillar is independent of the presence of other pillars and the energy U_{LP}^{l} is proportional to the number of pillars N_p given by eq 18. Under those conditions one has

$$U_{LP}^{l} = 2\rho_L\left(1+\frac{2h}{R}\right)fx_0\int_0^{t_p}\Phi_{LS}^{l}(h)dh \tag{26}$$

where the integral on the right-hand side is independent of the drop dimensions and roughness.

According to the model of a liquid drop presented in Section 3.2, the energy U_{LB}^{s} can be rewritten in the form

$$U_{LB}^{s} = 2\epsilon_{LS}^{s}\rho_{LS}k_s(f)t_{bp}\int_0^{h_m}x_h\mathrm{d}h,\ 0 \le k_s(f) \le 1 \tag{27}$$

In a similar way the energy U_{LP}^{s} can be rewritten as

$$U_{LP}^{s} = 2\epsilon_{LS}^{s}\rho_{LS}k_{sp}(f)t_{sp}\int_0^{h_m}x_h\,dh \tag{28}$$

Substituting eqs 25–28 into eq 15, one obtains for the Wenzel regime

$$U_{LS} = 2\rho_L\int_0^{h_m}\Phi_{LS}^{l}(h)xdh - 2b_{LS}\int_0^{h_m}x_hdh \tag{29}$$

where

$$\begin{aligned} b_{LS} &= \rho_{LS}\Phi_{LS}^{l}(0) - \rho_L\sigma_{LL}\Phi_{LS}^{l}(\sigma_{LL}) \\ &+\rho_L f\left(1+\frac{2h_p}{R}\right)\int_0^{t_p}\Phi_{LS}^{l}(h)\,\mathrm{d}h - \frac{f}{\pi R^2}U_1 - \epsilon_{LS}^{s}\rho_{LS}k_s(f)t_{bp} \\ &-\epsilon_{LS}^{s}\rho_{LS}k_{sp}(f)\left(1+\frac{2h_p}{R}\right) \end{aligned} \tag{30}$$

The potential energy of gas—solid interactions can be calculated using eqs 16 and 21. Ommiting the constant term in eq 16 proportional to S_{solid} one arrives at

$$U_{\text{GS}} = 2\rho_{\text{GS}} k_{\text{s}}(f)\Phi^{1}_{\text{GS}}(0)\int_0^{h_{\text{m}}} x_{\text{h}}\,dh - 2b_{\text{LS}}\int_0^{h_{\text{m}}} x_{\text{h}}\,dh \tag{31}$$

The total potential energy per unit drop length in the Wenzel regime, $U^{\text{w}}_{\text{total}}$, is given by the sum of the energies given by eqs 23, 29, and 31. After adding the Lagrange term λV associated with the condition of constant drop volume, the following functional $\upsilon_{\text{w}}[x]$ for the potential energy, which has to be minimized, is obtained:

$$\upsilon_{\text{w}}[x] = \int_0^{h_{\text{m}}} F_{\text{w}}(h, x, x_{\text{h}})\,dh \tag{32}$$

where

$$F_{\text{w}}(h, x, x_{\text{h}}) = \left[\lambda_{\text{w}} + a_1\Phi^{1}_{\text{LS}}(h)\right]x + \sqrt{1 + x_{\text{h}}^{2}} - b_{\text{w}} x_{\text{h}}$$

$$\lambda_{\text{w}} = \frac{\lambda}{K_{\text{s}}} - a_0, \qquad a_0 = \frac{K_{\text{v}}}{K_{\text{s}}},\quad a_1 = \frac{\rho_{\text{L}}}{K_{\text{s}}}$$

$$b_{\text{w}} = -\lambda_{\text{w}} h_{\text{p}} f + \phi_1(f) + \frac{1}{K_{\text{s}}}\left\{\left[K_{\text{s}}' - \rho_{\text{GS}}\Phi^{1}_{\text{GS}}(0)\right]\phi_2(f) + b_{\text{LS}}\right\} \tag{33}$$

4.2. CASSIE—BAXTER REGIME

In the CB regime, the shape of the drop surface in the selected coordinates (Figure 3b) is similar to that of the reference drop. The drop volume is provided by the right-hand-side of eq 17 and the surfaces S_{LS} and S_{LG} are given by

$$S_{\text{LG}} = 2\int_0^{h_{\text{m}}} dh\left\{\sqrt{1 + x_{\text{h}}^{2}} - \left[1 - fk_{\text{sp}}(f)\right]x_{\text{h}}\right\}$$

$$S_{\text{LS}} = -2fk_{\text{sp}}(f)\int_0^{h_{\text{m}}} x_{\text{h}}\,dh \tag{34}$$

Consequently, the energy of the liquid—liquid interactions, U_{LL}, is given by the equation

$$U_{\text{LL}} = 2K_{\text{s}}\int_0^{h_{\text{m}}} dh\left\{-a_0 x + \sqrt{1 + x_{\text{h}}^{2}} - \left[1 - \left(1 - \frac{K_{\text{s}}'}{K_{\text{s}}}\right)fk_{\text{sp}}(f)\right]x_{\text{h}}\right\} \tag{35}$$

The potential of the liquid—solid interaction in the CB regime consists of two parts, $\tilde{\Phi}^{1}_{\text{LS}}\big|_{\text{bulk}}$ and $\tilde{\Phi}^{1}_{\text{LS}}\big|_{\text{pillars}}$, representing the interaction of a liquid molecule with the bulk solid and with the pillars, respectively. The first potential depends only on the distance $h_{\text{p}} + h$ of the molecule from the surface of the bulk solid and has a form similar to eq 7

$$\tilde{\Phi}^{1}_{\text{LS}}\big|_{\text{bulk}} = -\frac{\pi}{6}\epsilon_{\text{LS}}\rho_{\text{S}}\sigma_{\text{LS}}^{3}\frac{1}{\left[1 + (h_{\text{p}} + h)/\sigma_{\text{LS}}\right]^{3}} \tag{36}$$

The calculation of the second potential, $\tilde{\Phi}^{1}_{LS}\big|_{pillars}$, is a difficult problem because of the finite size of the asperities and the dependence of the resulting potential on the location of the molecule in the horizontal plane. To simplify the calculation, this potential will be approximated by a potential of interaction with a uniform solid layer which has a thickness h_p equal to the height of the pillars and the number of molecules equal to the total number of molecules in all pillars. The number density of that layer, $\rho_{s,1}$, is equal to $\rho_{S,1} = f\rho_S$. If h_p is much larger than the hard core radius σ_{LS}, then the interaction with this layer can be considered as the interaction with a bulk solid (provided by eq 7) and the potential $\tilde{\Phi}^{1}_{LS}\big|_{pillars}$ is given by

$$\tilde{\Phi}^{1}_{LS}\Big|_{pillars} = \frac{\pi}{6} f\epsilon_{LS}\rho_S\sigma_{LS}{}^3 \frac{1}{\left(1+h/\sigma_{LS}\right)^3} \tag{37}$$

For $h_p \gg \sigma_{LS}$, $\tilde{\Phi}^{1}_{LS}\big|_{pillars}$ is much larger than $\tilde{\Phi}^{1}_{LS}\big|_{bulk}$ even for small f and the total potential of the liquid—solid interaction in the CB regime, $\tilde{\Phi}^{1}_{LS}(h)$, has the form

$$\tilde{\Phi}^{1}_{LS}(h) = -\frac{\pi}{6} f\epsilon_{LS}\rho_S\sigma_{LS}{}^3 \frac{1}{\left(1+h/\sigma_{LS}\right)^3} \tag{38}$$

Using the potential eq 38, the potential energy, U^{1}_{LS}, due to the long-range interactions is given by

$$U^{1}_{LS} = 2\int_0^{h_m} \rho_L \tilde{\Phi}^{1}_{LS}(h)\,\mathrm{d}h + 2\left[\rho_{LS}\tilde{\Phi}^{1}_{LS}(0) - \rho_L\sigma_{LL}\tilde{\Phi}^{1}_{LS}(\sigma_{LL})\right] f x_0 \tag{39}$$

The energy U^{s}_{LS} due to the short-range interactions is proportional to the contact area between the first layer of liquid and the top of the pillars. Because the latter area is equal to $2x_0 f$ one can write

$$U^{s}_{LS} = -2\epsilon^{s}_{LS}\rho_{LS}k_{sp}(f)\,f x_0 \tag{40}$$

where the coefficient $k_{sp}(f)$ has the same meaning as in eq 27.

The energy of the gas—solid interaction in the considered case is given by

$$U_{GS} = 2\rho_{GS}\left[1 - f k_{sp}(f)\right]\Phi^{1}_{GS}(0)\int_0^{h_m} x_h\,\mathrm{d}h \tag{41}$$

Combining eqs 39, 40, and 41, one obtains for the functional $\upsilon^{cb}[x]$ of the potential energy in the CB-regime the expression

$$\upsilon^{cb}[x] = \int_0^{h_m} F_{cb}(h, x, x_h)\,\mathrm{d}h \tag{42}$$

where

$$F_{cb}(h, x, x_h) = \left[\lambda_c + a_1\tilde{\Phi}^{1}_{LS}(h)\right]x + \sqrt{1+x_h{}^2} - b_c x_h$$

$$\lambda_c = \frac{\lambda}{K_s} - a_0, \quad b_c = 1 - \frac{\rho_{GS}}{K_s}\Phi^{1}_{GS}(0)\alpha f \tag{43}$$

$$\alpha K_s = \left[K_s - K_s' - \rho_{GS}\Phi^{1}_{GS}(0) + \rho_{LS}\epsilon^{s}_{LS}\right]k_{sp}(f) - \left[\rho_{LS}\tilde{\Phi}^{1}_{LS}(h) - \rho_L\sigma_{LL}\tilde{\Phi}^{1}_{LS}(\sigma_{LL})\right]$$

5. EQUATION FOR THE DROP PROFILE AND ITS SOLUTION

The functionals eqs 33 and 42 have the same form as the functional considered in ref 12. Their minimization provides the differential equation for the drop profile. For the Wenzel regime, this equation has the form

$$\frac{x_{\mathrm{hh}}}{\left(1+x_{\mathrm{h}}^{2}\right)^{3/2}}-a_{1}\Phi_{\mathrm{LS}}^{1}\left(h\right)-\lambda_{\mathrm{w}}=0 \tag{44}$$

where the first term represents the curvature $\kappa(y)$ of the profile. For the CB regime, the coefficient λ_w and the potential $\Phi_{\mathrm{LS}}^{1}\left(h\right)$ in eq 44 and in its solution have to be replaced by λ_c given by eq 43 and $\tilde{\Phi}_{\mathrm{LS}}^{1}\left(h\right)$ given by eq 38. The boundary conditions are

$$x\big|_{\mathrm{h=h_m}}=0 \tag{45}$$

and

$$x_{\mathrm{h}}\big|_{\mathrm{h=0}}=-\cot\,\theta_{0} \tag{46}$$

where θ_0 is the microcontact angle of the drop with the solid surface. The variational procedure allows to relate θ_0 to the microscopic parameters of the interaction potentials using the so-called transversality conditions[18] which account for the fact that the apex of the drop and the contact point of the profile with the solid should be located on the *h* and *x* axes (Figure 3), respectively. The result is[13]

$$\cos\,\theta_{0}=-b_{\mathrm{w}} \tag{47}$$

Unlike refs 11 and 12, where a drop on a smooth solid surface was considered, the parameter b_w and, correspondingly, the microcontact angle θ_0 depends on the Lagrange multiplier λ_w (see eq 33) and therefore on the volume of the drop. However, as it will be shown later, for large droplets ($h_m \gg h_p$), this dependence is negligible.

The solution of eq 44 was obtained in the form[11]

$$x\left(h\right)-x_{0}=\int_{0}^{h}\frac{\Psi\left(h\right)}{\sqrt{1-\Psi^{2}\left(h\right)}}\mathrm{d}h \tag{48}$$

where

$$\Psi\left(h\right)=b_{\mathrm{w}}+\lambda_{\mathrm{w}}h+D\left(h\right)-D\left(0\right) \tag{49}$$

$$D\left(h\right)=a_{1}\int\Phi_{\mathrm{LS}}^{1}\left(h\right)\mathrm{d}h \tag{50}$$

The point of contact $x_0 = x\left(0\right)$ is given by

$$x_{0}=-\int_{0}^{h_{\mathrm{m}}}\frac{\Psi\left(h\right)}{\sqrt{1-\Psi^{2}\left(h\right)}}\mathrm{d}h \tag{51}$$

which follows from eq 48 and the condition $x\left(h_{\mathrm{m}}\right)=0$.

In what follows, it will by assumed that the surface number densities of liquid molecules on the liquid—gas and liquid—solid interfaces are comparable $\left(\rho_{\mathrm{LG}}=\rho_{\mathrm{LS}}\right)$ and that the surface number density of gas molecules on the solid surface, ρ_{GS}, can be neglected compared with that of the liquid molecules, ρ_{LS}. Under these assumptions, an analysis similar to that performed in refs 11 and 12 provided the following expression for the Lagrange multiplier λ_w

$$\lambda_w = -\frac{2}{h_m - h_p f}\left\{r + \frac{1}{2}\left[D(h_m) - D(0)\right] + \frac{b_{LS}}{2K_s}\right\} + \frac{b_{LS}}{K_s} \tag{52}$$

If $h_m \gg h_p$ (large drop), evaluations have shown that the term containing λ_w in eq 33 can be neglected. Then eqs 33, 47, and 52 provide an expression for cos θ_0 independent of the droplet height (volume)

$$\cos\theta_0 = b_{w,0} + f b_{w,1} - b_{w,2} k_s(f) t_{bf} \tag{53}$$

where

$$b_{w,0} = -1 + a_l \sigma_{LL}\left[\Phi_{LS}^1(\sigma_{LL}) - \delta_l \Phi_{LS}^1(0)\right]$$

$$b_{w,1} = a_l \int_{t_p}^{h_p} \Phi_{LS}^1(h)\,dh - \frac{2h_p}{R}[1 + a_l \int_0^{t_p} \Phi_{LS}^1(h)\,dh - b_{w,2} k_{s,p}(f)] + b_{w,2} \tag{54}$$

$$b_{w,2} = a_l \epsilon_{LS}^{s} \frac{\rho_{LS}}{\rho_L}$$

and $\delta_l = \rho_{LS}/\sigma_{LL}\rho_L$. The dependence of $\cos\theta_0$ on the roughness r is provided through the parameter f which is connected to r through eq 10.

6. MACROSCOPIC CONTACT ANGLE AND COMPARISON WITH EXPERIMENT

As previously shown,[11,12] the profile of a large drop at distances larger than several hard core radii σ_{LS} from the solid surface can be approximated by a circle with a radius equal to the curvature radius of the profile at the drop apex. Considering the angle which the circle makes with the solid surface as the macroscopic contact angle θ_m, a relation between θ_m and θ_0 was established in the form of eq 4. Due to the similarity between the profile equations for smooth and rough surfaces the same equation

$$\cos\theta_m = \cos\theta_0 + \frac{a}{2} \tag{55}$$

can be also considered valid in the latter case for both the Wenzel and the CB regimes. However, the parameter a for rough surfaces has the forms

$$a_w = \frac{1}{4K_1}\frac{\epsilon_{LS}\rho_S}{\epsilon_{LS}\rho_L}\left(\frac{\sigma_{LS}}{\sigma_{LL}}\right)^4 \quad \text{(Wenzel regime) (56)}$$

The coefficient K_1 in eq 56 is given by

$$K_1 = \frac{17}{16} - \frac{3}{8}\Delta^3 - \frac{1}{2}\Delta^3 - \frac{\rho_{LG}}{2\sigma_{LL}\rho_L}\left(1 - \Delta^3\right)$$

$$a_{cb} = f a_w \quad \text{(CB regime) (57)}$$

In the experiments reported by Krupenkin et al.,[1] the solid surface was decorated with cylindrical pillars having a radius $R = 0.175\ \mu$m and a height $h_p = 7\mu$m, arranged as shown in Figure 2 a. The area fraction f covered by the pillars depends on the distance l between them through eq 10.

Four surfaces were prepared with various f values in the range $0 \le f \le 0.11$. The corresponding values for the roughness r were in the range $1 \le r \le 8.7$. The f dependence of the macroscopic contact angle θ_m was determined for four liquids with various surface tensions. The liquid with the largest surface tension did not penetrate into the space between pillars (CB regime), and for the three others, with smaller surface tensions, a complete penetration was observed (Wenzel regime).

Assuming that the hard core radii σ_{LS} and σ_{LL} are of the order of 3 Å and the scale t_p in which the interaction between a molecule of liquid and the pillars is nonnegligible is about $1 \mu m$, the following inequality

$$\sigma_{LS}, \sigma_{LL} \ll t_p \ll h_p \tag{58}$$

can be written for the experimental data of Krupenkin et al.[1] This inequalities provide the estimations

$$a_l \int_0^{t_p} \Phi_{LS}^1(h) dh \simeq -\frac{a}{2}$$
$$a_l \int_{t_p}^{t_p} \Phi_{LS}^1(h) dh \simeq 0 \tag{59}$$

where $a = a_w$ for the Wenzel regime and $a = a_{cb}$ for the CB regime. By considering the radius η of the interaction sphere of the liquid—liquid interactions equal to $5\sigma_{LL}(\Delta = 1/5$ in eqs 14), and employing eqs 55, 53, 54, and 59, the following equation for the macroscopic contact angle in the Wenzel regime is obtained

$$\cos\theta_m^w = d_{w,0} + f d_{w,1} + d_{w,2}\left[\frac{2h_p}{R} f k_{s,p}(f) + t_{bp} k_s(f)\right] \tag{60}$$

where

$$d_{w,0} = -1 + \frac{\delta_0^4 \delta_2}{1 - n_1\delta_1}\left[1 - \frac{2\delta_0^2}{(1+\delta_0)^3}\right] + \frac{2\delta_0^3 \delta_1 \delta_2}{1 - n_1\delta_1}$$
$$d_{w,1} = -\frac{2h_p}{R}\left(1 - \frac{\delta_0^4 \delta_2}{1 - n_1\delta_1}\right) + \frac{\delta_1\delta_3}{1 - n_1\delta_1} \tag{61}$$
$$d_{w,2} = \frac{\delta_1\delta_3}{1 - n_1\delta_1}, \; n_1 = 0.9437$$

and

$$\delta_0 = \frac{\sigma_{LS}}{\sigma_{LL}}, \; \delta_1 = \frac{\rho_{LS}}{\rho_L \sigma_{LL}}, \; \delta_2 = 0.119 \frac{\epsilon_{LS}\rho_S}{\epsilon_{LL}\rho_L},$$
$$\delta_3 = 0.454 \frac{\epsilon_{LS}^S}{\epsilon_{LL}} \frac{1}{\rho_L \sigma_{LL}^3} \tag{62}$$

are dimensionless parameters. Similarly, for the CB regime

$$\cos\theta_m^c = -1 + f d_{c,1} \tag{63}$$

where

$$d_{c,1} = \frac{\delta_0{}^4\delta_2}{1-n_1\delta_1}\left\{1+\frac{1}{\delta_0}\left[4k_{s,p}(f)\frac{\delta_3}{1-n_1\delta_1}+\delta_1-\frac{1}{8}\right]\right\} \tag{64}$$

Using the known values for the densities of water, methanol, and silicon, as well as estimations of the microscopic parameters σ_{LL}, σ_{LS}, ϵ_{LL}, and ϵ_{LS},[14] one can roughly estimate the constants δ_0 and δ_2

$$0.56 \le \delta_0 \le 0.85, \quad 0.21 \le \delta_2 \le 0.62$$

Considering the surface number density as $\rho_{LS} \simeq \rho_L\sigma_{LL}$ one obtains that $\delta_1 \sim 1$. The most difficult task is to estimate the parameter δ_3 because of insufficient information about the energy ϵ_{LS}^{s} of short-range interaction. Assuming that ϵ_{LS}^{s} has the same order of magnitude as ϵ_{LL} one can conclude that $0.46 \le \delta_3 \le 0.6$.

Applying eq 60 to a smooth surface ($f = 0, \theta_m^w = \theta_{m,0}$), one finds that

$$\cos\,\theta_{m,0} = d_{w,0} + d_{w,2}k_s(0) \tag{66}$$

Using the latter equation and eqs 61 and 64, eqs 60 and 63 for θ_m^w and $\cos\theta_m^c$ can be rewritten, respectively, in the form

$$\cos\theta_m^w = \frac{1}{2\delta_1+3/4}\left(2\delta_1-\frac{1}{4}+r\right)\cos\theta_{m,0} + fd_{w,3} + d_{w,2}t_{bp}k_s(f) \tag{67}$$

$$\cos\,\theta_m^c = -1 + fd_{c,2}\left(1-d_{w,2}+\cos\theta_{m,0}\right) \tag{68}$$

where

$$d_{w,3} = -\frac{2h_p}{R}\left[1-\frac{1-d_{w,2}k_s(0)}{2\delta_1+3/4}-k_{sp}(f)\frac{\delta_1\delta_3}{1-n_1\delta_1}\right]+\frac{\delta_1\delta_3}{1-n_1\delta_1}$$

and

$$d_{c,2} = 1+\frac{1}{\delta_0}\left[4k_{sp}(f)\frac{\delta_3}{1-n_1\delta_1}+\delta_1-\frac{1}{8}\right] \tag{69}$$

Equations 67 and 68 have more complicated forms than the traditional eqs 1 and 2. Along with the terms resembling those of the traditional equations, they contain extra terms proportional to the fractional area f of the top of the asperities or to the fractional area, $1-f$, of the solid surface between the asperities. In addition, two functions, $k_s(f)$ and $k_{sp}(f)$, characterizing the partial wetting are involved. Even when $k_s(f) = k_{sp}(f) \equiv 1$ (there is no gas beneath the drop) differences with respect to the traditional cases still persist. Note that $\cos\theta_m^c$ is a linear function of the roughness if $k_{sp}(f)$ does not depend on f. Linearity of $\cos\theta_m^w$ with respect to the roughness occurs only if both $k_s(f)$ and $k_{sp}(f)$ become independent of f.

To describe the nonlinear dependence of $\cos\theta_m$ on roughness some assumption have to be made about the functions $k_s(f)$ and $k_{sp}(f)$. The first of them, $k_s(f)$, provides the fraction of molecules of the first layer of liquid on the solid surface which are not displaced by the gas

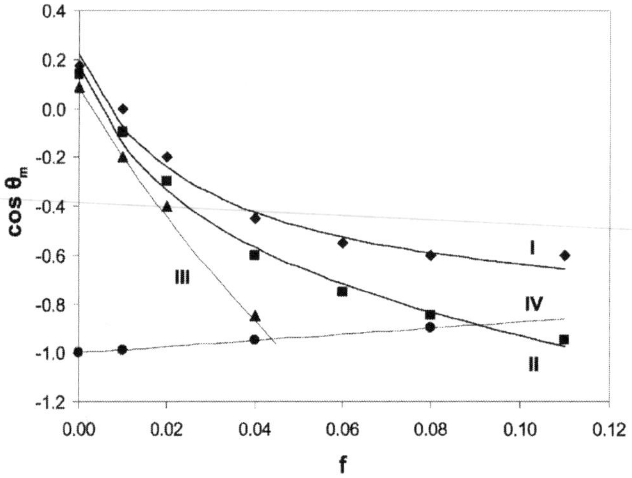

FIGURE 4. Experimental results of Krupenkin et al.[1] (filled points) and theoretical curves given by eq 60 for curves I, II, and III and eq 63 for curve IV. The coefficients of those equations are listed in Table 1. The surface tensions of the employed liquids increased with increasing curve index.

accumulated at the edges of the asperities and are subjected to short-range interactions. The function $k_s(f)$ has to be equal to unity for a smooth surface ($f = 0$). When the roughness increases $k_s(f)$ is expected to decrease. For those reasons, this function will be represented by the simple espression

$$k_s(f) = \frac{1}{1 + n_s f} \tag{70}$$

where n_s is a constant.

The second function, $k_{sp}(f)$ provides the fraction of molecules which are not displaced from the surface of a pillar. It is reasonable to suppose that this fraction has a weak dependence on the number of pillars and, therefore, $k_{sp}(f)$ will be selected to be a constant

$$k_{sp}(f) = n_{sp} \tag{71}$$

where n_{sp} is a constant. The constants n_s and n_{sp} have to be determined from comparison with experiment. In Figure 4, the theoretical curves for $\cos\theta_m^w$ and $\cos\theta_m^c$ which have the best fit with experimental data are presented together with the experiments of Krupenkin et al.[1] The

TABLE 1
Values of the Parameters $\delta_i (i = 0–3), n_s$, and n_{sp} in eqs 60 and 63 Providing the Best Fit between Experimental and Theoretical Results

Experiment	δ_0	δ_1	δ_2	δ_3	n_s	n_{sp}
I	0.65	0.71	0.12	0.48	38	0.9
II	0.65	0.70	0.12	0.47	40	0.9
III	0.65	0.69	0.21	0.40	20	0.9
IV	0.73	0.70	0.41	0.48		0.9

theoretical curves are labeled in the order of increasing surface tension of the employed liquids. Curves I, II, and III correspond to the Wenzel regime with $\cos\theta_m^w$ nonlinearly dependent on roughness, and curve IV to the CB regime, with $\cos\theta_m^c$ linearly dependent on f.

The values of the parameters n_s and n_{sp}, and δ_i $(i = 0,1,2,$ and $3)$ are listed in Table 1 for each curve. It is interesting to note that the values of the parameters are close in all experiments. The obtained values for δ_i can be, in principle, employed as a source of additional information about the microscopic interactions for the liquids involved in the experiments.

7. DISCUSSION

The main purpose of the present paper is to show that the nonlinear dependence of the macroscopic contact angle of a liquid drop on the roughness in the Wenzel regime can be caused by the displacement of some liquid molecules of the liguid layer nearest to the solid surface. Such a displacement can occur at the edges of the asperities because of the accumulation of microscopic amounts of gas during the wetting process. The derived microscopic eqs 60 and 63 for the macroscopic contact angles in Wenzel and CB regimes differ from the corresponding traditional equations. As another possible source for a nonlinear dependence of θ_m^w on f one can mention the interaction between the molecules of the liquid with the asperities. In the present calculations this interaction was taken into account only when the distance l between the pillars was larger than the characteristic length t_p $(t_p \sim 1\,\mu\text{m})$ of the range of the interactions between the molecules of the liquid and the pillars (see derivation of eq 26). In that case each pillar contributed separately to the total energy U_{LP}^1 and therefore $U_{LP}^1 \sim N_p$, where N_p is the number of pillars (eq 18). If $l < t_p$, the energy associated with a single pillar becomes dependent on N_p and the total energy becomes nonlinear with respect to $N_p$and, consequently, with respect to f. This dependence will introduce additional terms in eq 30 for the coefficient b_{LS} and finally in the expression of $\cos\theta_m^w$.

ACKNOWLEDGMENT

The authors are grateful to Ilya G.Berim for his valuable contribution to this work.

REFERENCES

(1) Krupenkin, T. N.; Taylor, J. A.; Schneider, T. M.; Yang, S. *Langmuir* **2004**, *20*, 3824.
(2) Findenegg, G. H.; Herminghaus, S. *Curr. Opin. Colloid Interface Sci.* **1997**, *2*, 301.
(3) Wenzel, R. N. *Ind. Eng. Chem.* **1936**, *28*, 988; *J. Phys. Colloid Chem.* **1949**, *53*, 1466.
(4) Cassie, A. B. D.; Baxter, S. *Trans. Faraday Soc.* **1944**, *40*, 546. Cassie, A. B. D. *Discuss. Faraday Soc.* **1948**, *3*, 11.
(5) Marmur, A. *Langmuir* **1996,** *12*, 5704.
(6) Drelich, J. *Colloid Surf. A* **1996,** *116*, 43.
(7) Swain, P. S.; Lipovsky, R. *Langmuir* **1998,** *14*, 6772.
(8) Bico J.; Marzolin, C.; Quere, D. *Europhys. Lett.* **1999**, *47*, 220.
(9) Miller, C. A.; Ruckenstein, E. *J. Colloid Interface Sci.* **1974**, *48*, 368.
(10) Ruckenstein, E.; Lee, P. S. *Surf. Sci.* **1975**, *52*, 298.
(11) Berim, G. O.; Ruckenstein, E. *J. Phys. Chem. B* **2004**, *108*, 19330 (Section 3.2 of this volume).
(12) Berim, G. O.; Ruckenstein, E. *J. Phys. Chem.* **2004**, *108*, 19339 (Section 3.3 of this volume).
(13) Berim, G. O.; Ruckenstein, E. *J. Colloid. Interface Sci.* **2005**, *286*, 681 (Section 3.4 of this volume).

(14) Lee, S.H.;Rossky, P.J.J. Chem. Phys. 1994, 100, 3334.Stillinger, F. H.; Rahman, A. J. J. Chem. Phys. 1974, 60, 1545. Molteni, C.; Martonak, R.; Parrinello, M. J. Chem. Phys. 2001, 114, 5358. Martonak, R.; Colombo, L.; Molteni, C.; Parrinello, M. J. Chem. Phys. 2002, 117, 11329.
(15) Neimark, A. V. J. Adhesion Sci. Technol. 1999, 13, 1137.
(16) Meiron, T. S.; Marmur, A.; Saguy, I. S. J. Colloid Interface Sci. 2004, 274, 637.
(17) Thomas, R. Rough surfaces; Imperial College: London, 1999.
(18) Elsgolc, L. E. *Calculus of variations*; Pergamon Press:NewYork, 1962.

3.7 Nanodroplets on a Planar Solid Surface

*Temperature, Pressure, and Size Dependence of Their Density and Contact Angles**

Gersh Berim and Eli Ruckenstein†
Department of Chemical and Bioogical Engineering,
State University of New York at Buffalo, Buffalo, New York 14260

Corresponding Author

†E-mail: feaeliru@buffalo.edu.
Phone: (716)645-1179, Fax: (716)645-3822.

Received August 17, 2005. In Final Form: November 11, 2005

1. INTRODUCTION

Most of the theories that describe a liquid drop on a solid surface consider the drop to be a uniform body with a density independent of its size and temperature. The differential equation for the drop profile has the form of the Young—Laplace equation without[1] or with[2] a term due to the disjoining pressure. The boundary conditions for that equation are provided by the drop volume (or drop height) and the macroscopic (apparent) contact angle, θ_m, which the profile makes with the solid surface. (Actually, θ_m is experimentally measured as the angle between the solid surface and the extrapolation of the drop profile from the region where the interaction with the solid becomes negligible.) The angle θ_m is considered to be given by the Young equation

$$\cos\theta_m = \frac{\gamma_{sg} - \gamma_{sl}}{\gamma_{lg}} \tag{1}$$

where γ_{lg}, and γ_{sg} are the liquid—gas, solid—liquid, and solid—gas surface tensions, respectively, and are independent of the drop volume and, consequently (because of the assumption of uniform density), the number of molecules in the drop. The temperature dependence of θ_m arises from the temperature dependence of the surface tensions involved in eq 1.

A microscopic theory developed in refs 3 and 4 provided a differential equation for the drop profile as well as a value for the true contact angle θ_0 that the actual profile makes with the solid surface. The derivations were based on the minimization of the total potential energy of the system, calculated microscopically, and for this reason, the results were independent of temperature. The assumption of constant density and hence of a constant volume of a drop containing a given number of molecules was also employed in the theory. Obviously, it is important to overcome these restrictions and develop a theory that takes the temperature and nonconstancy of the density into consideration. Note that macroscopically large amounts of fluids, composed of a liquid in thermodynamic equilibrium with its vapor, have

* *Langmuir*, Vol. 22, No. 3, 2006. Republished with permission.

been investigated theoretically for a long time (e.g., refs 5–10) using the density functional approach that considers the contribution of the repulsive forces between molecules to the free energy to be a hard sphere free energy and treats the attractive forces in the mean field approximation.

Two kinds of problems were examined. In those of the first kind, the densities of the liquid and vapor were considered uniform and equal to those corresponding to the bulk fluids. The shape of the liquid—vapor interface was assumed, and only its position had to be found. Usually, simple shapes (planar, cylindrical, or spherical) that made the problems solvable were considered.[7,9] For example, in ref 7 a liquid film on a cylindrical solid surface embedded into its vapor was considered, and the thickness of that film, which defined the position of the liquid—vapor interface, was determined as a function of the parameters of the problem. In the second kind of problems, the shape and position of the interface were considered given, and the variation of the density in the vicinity of the interfaces was calculated[5,6] by considering that far from the interface the densities were equal to their bulk values.

However, neither the profile at the leading edge of a large drop and of a small drop in contact with its vapor nor the shape of the interface and the drop and vapor densities can be considered given. All of these characteristics have to be found from the simultaneous minimization of the free energy of the system with respect to the drop and vapor densities and the shape of the liquid—vapor interface. As a first step toward the solution of this extremely complicated problem, a liquid drop on a solid surface, embedded in an inert, insoluble gas is examined microscopically using the density functional approach. This approach includes the entropic contribution to the free energy and therefore provides the possibility to account for temperature-dependent effects. The system is considered to be at a given temperature and pressure, and the vapor pressure of the liquid is considered negligible. Generally, the density in the small droplets is not uniform, and the dependence of the density on the position coordinates must be taken into account. However, for the sake of simplicity we will consider for the time being the density to be uniform and thus, at a given number of molecules in the drop, dependent only on the drop volume.

In section 2, the description of the system under consideration is presented, and the main expressions for the calculation of the system free energy are provided. Differential equations for the drop profile and equilibrium number density of the drop are derived and solved numerically in section 3. Section 4 is devoted to the examination of the drop characteristics in various cases.

2. BACKGROUND

2.1. The System

Let us consider a solid substrate with a constant number density (below referred to as density) ρ_S and a smooth planar surface, a liquid drop containing a constant number, N_l, of molecules that resides on the solid surface, and an inert, insoluble gas with a constant density ρ_G and pressure p surrounding the drop at temperature T. (Figure 1). Some gas molecules are adsorbed on the solid

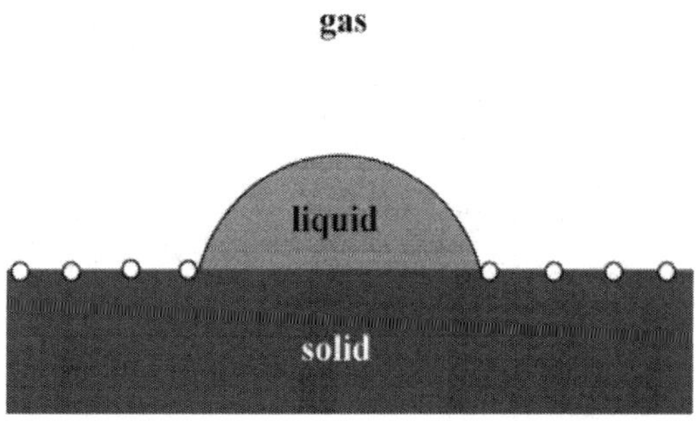

FIGURE 1. System under consideration. The open circles represent gas molecules adsorbed on the solid surface.

surface generating a surface density ρ_{GS} that is also considered constant. It will be assumed that the chemical and physical characteristics of the substrate are not changed because of the contact with the gas and liquid molecules.

If the vapor pressure of the liquid is negligible and the volumes and molar masses of the gas and solid substrate are much larger than those of the liquid drop, then the gas and substrate can be considered to be a heat bath for the drop. The drop is subjected to a constant gas pressure and to interactions with the solid, and the liquid molecules interact between themselves. The quantities of interest are the density of the liquid and the shape of the drop profile, which have to be found by minimizing the free energy of the system. Because the number of molecules in the drop, the temperature, and the external pressure of the system are constant, one should use for its thermodynamic description the Gibbs free energy $G(N, p, T)$, which is connected to the Helmholtz free energy $F(N, V, T)$ through the relation

$$G(N, p, T) = F(N, V_d, T) + pV_d \tag{2}$$

where p is the external pressure and V_d is the volume of the drop. The free energy due to the adsorbed gas will be considered equal to its potential energy because of the interactions with the solid and will be added to the free energy of the drop.

2.2. Potentials

It is supposed that the interaction potentials of the liquid molecules between themselves and with those of the solid have, respectively, the form

$$\phi_{LL}(r_{ll}) = \begin{cases} 0, & r_{ll} > \eta \\ -\epsilon_{LL}\left(\dfrac{\sigma_{LL}}{r_{ll}}\right)^6, & \eta \geq r_{ll} \geq \sigma_{LL}, \\ \infty, & r_{ll} < \sigma_{LL} \end{cases} \tag{3}$$

and

$$\phi_{LS}^1(r_{ls}) = \begin{cases} -\epsilon_{LS}\left(\dfrac{\sigma_{LS}}{r_{ls}}\right)^6, & r_{ls} \geq \sigma_{LS}, \\ \infty, & r_{ls} < \sigma_{LS} \end{cases} \tag{4}$$

where σ_{LL} and σ_{LS} are hard core radii, ϵ_{LL} and ϵ_{LS} are interaction constants, r_{ll} and r_{ls} are the distances between the centers of the interacting liquid and liquid and solid molecules, respectively, and η in eq 3 is the distance (radius of interaction sphere) after which the liquid—liquid interactions are considered to be negligible. The length η imposes a restriction on the drop size that has to be much larger than η. The potential energy of a molecule of liquid interacting with the bulk solid via the molecular interaction potential $\Phi_{LS}^1(r_{ls})$ is given by[4]

$$\Phi_{LS}^1(h) = -\frac{\pi}{6}\epsilon_{LS}\sigma_{LS}^3\rho_S \frac{1}{(1 + h/\sigma_{LS})^3} \tag{5}$$

where $h + \sigma_{LS}$ is the distance of the molecule from the solid surface.

In addition to the above potentials, a short-range potential, which has a nonzero value only in the vicinity of the solid surface (at a distance smaller than 1 nm[11]) and decays exponentially with distance, will be also taken into account. This potential will be approximated by[4]

$$\Phi_{LS}^{s}(h)=\begin{cases}-\epsilon_{LS}^{s}, & h=0,\\ 0, & h>0\end{cases} \tag{6}$$

where ϵ_{LS}^{s} is an interaction constant.

The interaction between the molecules of gas and solid has in most cases only a long-range component. For a low-density gas, only the molecules adsorbed on the solid surface ($h = 0$) provide a noticeable contribution to the total potential energy. For those molecules, the potential Φ_{GS} of gas-solid interaction can be written in the form

$$\Phi_{GS} = -\Phi_{GS}^{l}(0) \tag{7}$$

where the potential $\Phi_{GS}^{l}(0)$ has the same form as $\Phi_{LS}^{l}(0)$ given by eq 5 where ϵ_{LS} and σ_{LS} have to be replaced by ϵ_{GS} and σ_{GS}, respectively.

The gas—gas, gas—solid, and gas—liquid interactions as well as gravity will be neglected.

2.3. Free Energy of the System

For the thermodynamic description of a drop on a solid surface, the Gibbs free energy (eq 2) will be used, where the Helmholtz free energy $F(N, V_d, T)$ is given by the expression[7]

$$F(N,V_d,T)=\int_{V_d} d^3r f_H(\rho,T)+U_{total} \tag{8}$$

where V_d is the volume of the drop, $\rho \equiv \rho(\mathbf{r})$ is its density at a point $\mathbf{r}$ of the system, U_{total} is the total potential energy of the system, and $f_H(\rho, T)$ is the free-energy density of a homogeneous system of hard spheres of diameter σ_{LL}. The latter function is provided by the Carnahan—Starling expression[12] in the form presented in ref 7

$$f_H(\rho,T)=k_BT\rho\left[\log(\eta_\rho)-1+\eta_\rho\frac{4-3\eta_\rho}{(1-\eta_\rho)^2}\right] \tag{9}$$

where $\eta_\rho = 1/6\pi\rho\sigma_{LL}^{3}$ is the packing fraction of liquid molecules. Equation 8 for the free energy was used to describe a liquid as a film[7] or as a large droplet[9] in thermodynamic equlibrium with its vapor. Because of the complexity of the calculation of the density $\rho(\mathbf{r})$, two kinds of approximations were usually employed. First, the densities of liquid and gas phases were considered uniform and equal to their bulk densities at the given temperature and chemical potential (sharp-kink approximation), and the shape and position of the interface were determined.[7,9] second, the shape and position of the interface were given, and the density variation near the interface was calculated[5,6] assuming that far from the interface the density of the fluid was equal to the corresponding bulk value.

For the small drops that are considered in the present article, neither the density nor the shape can be considered given quantities. Hence, a more general case than those examined in refs 5–7 and 9 had to be solved. For its solution, the sharp-kink approximation for the density will be used (i.e., the density ρ will be considered uniform inside the drop ($\rho(\mathbf{r}) = \rho_L$)). The value of ρ_L will not be assumed to be equal to the bulk density in a liquid but will be determined via optimization simultaneously with the shape of the drop. Because of the constraint of a constant number of molecules in the drop, the change in density causes a change in the drop volume and affects the angle that the drop makes with the solid surface.

The total potential energy U_{total} of the system can be calculated as the sum of the potential energies U_{LL}, U_{LS}, and U_{GS} due to the liquid—liquid, liquid—solid, and absorbed gas—solid interactions, respectively.

$$U_{total} = U_{LL} + U_{LS} + U_{GS} \tag{10}$$

To calculate all of the energies involved in eq 10, some assumptions about the drop structure must be made. In this respect, the model of a liquid drop used in refs 4 and 13 will be employed. In that model, it was assumed that the liquid molecules closest to the solid surface form a monolayer a distance σ_{LS} from the solid (Figure 2), which is characterized by a surface density ρ_{LS} and is subjected to short- and long-range interactions with the solid. The remainder of the drop, separated from the first layer by a distance σ_{LL}, is treated in the framework of the sharp-kink approximation[7] as a continuous medium with a uniform volume density ρ_L that interacts only through long-range forces with the solid. The liquid molecules at the liquid—gas interface are considered to have a surface density ρ_{LG}.

Using this model, the energy U_{LL} can be represented as the sum of volume and surface contributions[13]

$$U_{LL} = \left(-K_v V_d + K_s S_{LG} + K_s' S_{LS}\right)\rho_L^2 \tag{11}$$

where S_{LG} and S_{LS} are the surfaces of the liquid—gas and liquid—solid interfaces, respectively,

$$K_v = \frac{2\pi}{3}\epsilon_{LL}\sigma_{LL}^3\left(1-\Delta^3\right)$$

$$K_s = \frac{2\pi}{3}\epsilon_{LL}\sigma_{LL}^4\left[\frac{17}{16} - \frac{3}{8}\Delta^2 - \frac{1}{2}\Delta^3 - \frac{3}{16}\Delta^4 - \frac{\rho_{LG}}{2\sigma_{LL}\rho_L}\left(1-\Delta^3\right)\right]$$

$$K_s' = \frac{2\pi}{3}\epsilon_{LL}\sigma_{LL}^4\left[\frac{17}{16} - \frac{3}{8}\Delta^2 - \frac{1}{2}\Delta^3 - \frac{3}{16}\Delta^4 - \frac{\rho_{LS}}{2\sigma_{LL}\rho_L}\left(1-\Delta^3\right)\right] \tag{12}$$

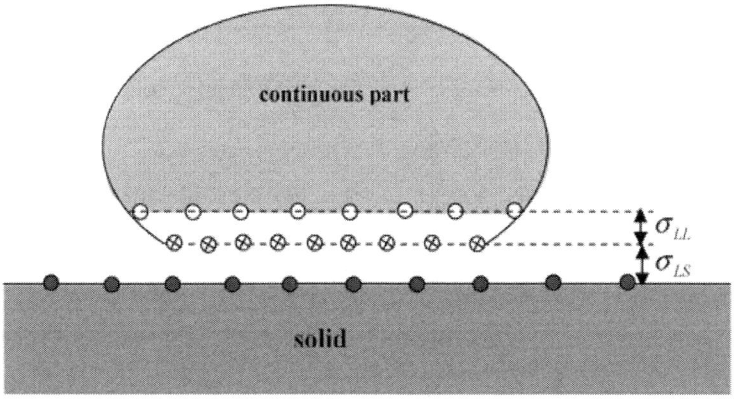

FIGURE 2. Cross section of a drop on a solid surface. The filled and crossed circles represent the molecules of the solid surface and those of the first (closest to the solid surface) layer of liquid, respectively. Beginning with the second layer (open circles), the liquid is considered to be a continuous medium.

and $\Delta = \sigma_{LL}/\eta$. It should be noted that eq 11 implies that the size of the drop is much larger than the radius η of the interaction sphere.

The energy U_{LS} can be represented as the sum of two terms

$$U_{LS} = U_{LS}^{l} + U_{LS}^{s} \tag{13}$$

where U_{LS}^{l} and U_{LS}^{s} are, respectively, the total potential energies of the drop due to the long- and short-range interactions (potentials eqs 5 and 6) of the molecules of liquid with the solid. The energy U_{LS}^{l} is given by the expression

$$U_{LS}^{l} = \rho_{LS}\Phi_{LS}^{l}(0)S_{LS} + \int_{V_{cont}} \rho_{L}\Phi_{LS}^{l}(h)\,dV \tag{14}$$

where the first and second terms represent, respectively, the potential energy of the first layer and of the continuous part of the drop (of volume V_{cont}) due to the interaction potential (eq 5). The volume V_{cont} represents the difference between the drop volume V_d and the volume V_g of the gap between the first layer and the continuous part of the droplet (Figure 2), which is approximately equal to $V_g = \sigma_{LL}S_{LL}$, hence $V_{cont} = V_d - S_{LS}\sigma_{LL}$ and the second term on the right-hand side of eq 14 can be rewritten as

$$\begin{aligned}\int_{V_{cont}} \rho_{L}\Phi_{LS}^{l}(h)\,dV &= \int_{V_d} \rho_{L}\Phi_{LS}^{l}(h)\,dV - \int_{V_g} \rho_{L}\Phi_{LS}^{l}(h)\,dV \\ &\simeq \int_{V_d} \rho_{L}\Phi_{LS}^{l}(h)\,dV - \rho_{L}\sigma_{LL}\Phi_{LS}^{l}(\sigma_{LL})S_{LS}\end{aligned} \tag{15}$$

The energy U_{LS}^{s} has the form

$$U_{LS}^{s} = \rho_{LS}\Phi_{LS}^{S}(0)S_{LS} \tag{16}$$

Consequently,

$$U_{LS} \simeq \int_{V_d} \rho_{L}\Phi_{LS}^{l}(h)\,dV + \left\{\rho_{LS}\left[\Phi_{LS}^{l}(0) + \Phi_{LS}^{l}(0)\right] - \rho_{L}\sigma_{LL}\Phi_{LS}^{l}(\sigma_{LL})S_{LS}\right\} \tag{17}$$

The potential energy U_{GS} of the absorbed gas–solid interactions is proportional to the surface density ρ_{GS} of the gas molecules on the solid surface and the area of contact between gas and solid.

$$U_{GS} = \rho_{GS}S_{solid}\Phi_{GS}^{l}(0) - \rho_{GS}S_{LS}\Phi_{LS}^{l}(0) \tag{18}$$

where S_{solid} is the total area of the solid surface, which is a constant. Because the density ρ_{GS} is considered below to be constant, the first term in eq 18 can be omitted.

Below, a cylindrical drop extended along the y axis is considered in the system of coordinates presented in Figure 3. For such a drop, the volume V_d, the surfaces S_{LG} and S_{LS}, and the integral $\int V_{cont}\rho\Psi_{LS}^{l}(h)\,dV$ are given by the expressions

$$V_d = 2\int_0^{h_m} x\,dh \qquad S_{LV} = 2\int_0^{h_m} \sqrt{1 + x_h^2}$$

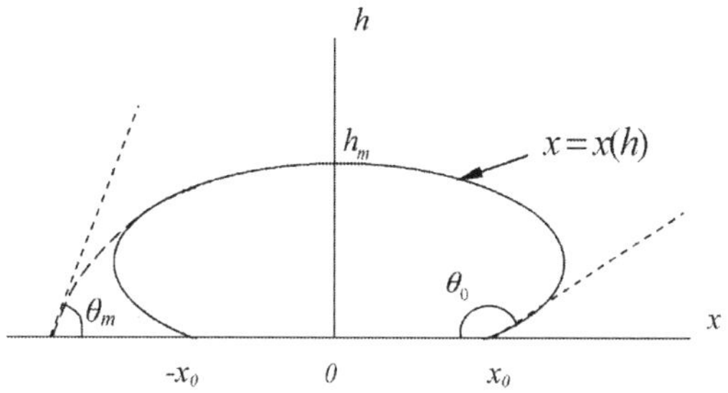

FIGURE 3. Coordinate system used for the description of the cylindrical drop profile. The drop is extended in the direction normal to the plane of the figure. θ_0 and θ_m are the microscopic (true) and the macroscopic (apparent) contact angle, respectively.

$$S_{\mathrm{LS}} = 2x(0) = -2\int_0^{h_{\mathrm{m}}} x_h \mathrm{d}h, \quad \int_{V_{\mathrm{d}}} \rho_{\mathrm{L}} \Phi_{\mathrm{LS}}^{\mathrm{l}}(h) \mathrm{d}V = 2\rho_{\mathrm{L}} \int_0^{h_{\mathrm{m}}} \Phi_{\mathrm{LS}}^{\mathrm{l}}(h) x \, \mathrm{d}h \tag{19}$$

where $x \equiv x(h)$ represents the drop profile, x_{h} denotes the derivative of x with respect to h, h_{m} is the droplet height at the drop apex, and the equality

$$x(0) = -\int_0^{h_{\mathrm{m}}} x_h \mathrm{d}h$$

follows from the obvious expression $x(h_{\mathrm{m}}) = 0$. In eq 19 and below, all quantities are expressed per unit length of the drop.

Using eqs 8, 10, 11, 17, 18, and 19, the Gibbs free energy (eq 2) can be rewritten as a sum of three terms

$$G(N, p, T) = G_{\mathrm{V}}(N, p, T) + G_{\mathrm{S}}(N, p, T) + G_{\mathrm{E}}(N, p, T) \tag{20}$$

where

$$G_{\mathrm{V}}(N, p, T) = \left[-K_{\mathrm{v}}\rho_{\mathrm{L}}^2 + f_{\mathrm{H}}(\rho_{\mathrm{L}}; T) + p\right] V_{\mathrm{d}} \tag{21}$$

$$G_{\mathrm{S}}(N, p, T) = K_{\mathrm{s}}\rho_{\mathrm{L}}^2 \left(S_{\mathrm{LG}} + bS_{\mathrm{LS}}\right) \tag{22}$$

$$G_{\mathrm{E}}(N, p, T) = 2\int_0^{h_{\mathrm{m}}} \rho_{\mathrm{L}} \Phi_{\mathrm{LS}}^{\mathrm{l}}(h) x \mathrm{d}h \tag{23}$$

S_{LG} and S_{LS} are given by eqs 19, and

$$b = \frac{1}{K_{\mathrm{S}}\rho_{\mathrm{L}}^2} \left\{ K'_{\mathrm{S}}\rho_{\mathrm{L}}^2 + \rho_{\mathrm{LS}}\left[\Phi_{\mathrm{LS}}^{\mathrm{l}}(0) + \Phi_{\mathrm{LS}}^{\mathrm{l}}(0)\right] - \rho_{\mathrm{GS}}\Phi_{\mathrm{GS}}^{\mathrm{l}}(0) + \rho_{\mathrm{L}}\sigma_{\mathrm{LL}}\Phi_{\mathrm{LS}}^{\mathrm{l}}(\sigma_{\mathrm{LL}}) \right\} \tag{24}$$

Note that $G_{\mathrm{V}}(N, p, T)$ contains all temperature-dependent terms of $G(N, p, T)$.

If the volume of the drop is fixed, then the first term, $G_{\mathrm{V}}(N, p, T)$, does not depend on the shape of the drop, whereas the second and the third ones are shape–dependent. This difference provides the specific roles of these contributions to the drop properties, which will be examined in the next section.

3. EQUATIONS FOR THE PROFILE AND DENSITY ρ_L AND THEIR SOLUTION

3.1. Derivation of the Basic Equations

The Gibbs free energy (eq 20) has to be minimized with respect to the drop profile $x(h)$ and number density ρ_L under the constraint of a constant number of molecules per unit length of the drop. The latter constraint leads to the expression

$$N_l = \rho_L V_d = 2\rho_L \int_0^{h_m} x\,dh \tag{25}$$

Adding the Lagrange term $-\mu N_l$ to the Gibbs free energy, one arrives at the minimization of the functional

$$\upsilon[x, \rho_L] = 2\int_0^{h_m} F(h, x, x_h; \rho_L)\,dh \tag{26}$$

where

$$F = (h, x, x_h; \rho_L) = [-K_V\rho_L^2 + f_H(\rho_L; T) + p - \mu\rho_L + \rho_L\Phi_{LS}^l(h)x + K_s\rho_L^2\left(\sqrt{1+x_h^2} - bx_h\right) \tag{27}$$

and ρ_L and $x \equiv x(h)$ are independent variables. The Lagrange multiplier μ can be found using eq 25. To find the extremum of the functional $\upsilon[x, \rho_L]$, one has to solve the following system of equations

$$\delta\upsilon[x, \rho_L] = 0 \tag{28}$$

$$\frac{\partial\upsilon[x, \rho_L]}{\partial\rho_L} = 0 \tag{29}$$

where $\delta\upsilon[x, \rho_L]$ represents the variation of $\upsilon[x, \rho_L]$ with respect to the drop profile. The first of the above equations provides the Euler—Lagrange differential equation for the drop profile at a fixed density. The second one provides the minimim of $\upsilon[x, \rho_L]$ with respect to the density ρ_L for a fixed profile. Performing standard variational procedures, one arrives at the following explicit forms of eqs 28 and 29

$$\frac{x_{hh}}{\left(1+x_h^2\right)^{3/2}} - a_1(\rho_L)\Phi_{LS}^l(h) = \frac{1}{K_S\rho_L^2}\left[-\mu\rho_L - K_v\rho_L^2 + f_H(\rho_L; T) + p\right] \tag{30}$$

$$\left[-\mu - 2K_v\rho_L + \frac{\partial}{\partial\rho_L}f_H(\rho_L; T)\right]\frac{N_l}{\rho_L} + 2\int_0^{h_m}\Phi_{LS}^l(h)x\,dh + S_{LG}\frac{\partial}{\partial\rho_L}\left(\rho_L^2 K_S\right)$$

$$+\left[\frac{\partial}{\partial\rho_L}\left(\rho_L^2 K_S'\right) + \sigma_{LL}\Phi_{LS}^l(\sigma_{LL}) + \left[\Phi_{LS}(0) + \Phi_{LS}^s(0)\right]\frac{\partial}{\partial\rho_L}\rho_{LS}\right]S_{LS} = 0 \tag{31}$$

where $a_1(\rho_L) = 1/\rho_L K_s$.

Equations 30 and 31 together with the constraint (eq 25) form a closed set of equations for the profile of the drop $x(h)$, its density ρ_L, and Lagrange multiplier μ. The boundary conditions for eq 30 can be written in the forms

$$x\big|_{h=h_m} = 0 \tag{32}$$

and

$$x_h\big|_{h=0} = -\cot\theta_0 \tag{33}$$

where θ_0 is the true contact angle of the drop with the solid surface.

Because neither h_m nor θ_0 is given, additional information is needed to find the complete solution of eq 30. This information is provided by the transversality conditions,[14] which are consequences of the facts that the apex of the drop and the contact line of the drop on the solid should be located on the h and x axes, respectively. These conditions have the form[14]

$$\frac{\partial}{\partial x_h} F\left(h, x, x_h; \rho_L\right)\Big|_{h=0} = 0$$

$$\left[F\left(h, x, x_h; \rho_L\right) - x_h \frac{\partial}{\partial x_h} F\left(h, x, x_h; \rho_L\right)\right]\Big|_{h=h_m} = 0 \tag{34}$$

which provide the following relations regarding the profile:

$$\left.\frac{1}{\sqrt{1+x_h^2}}\right|_{h=h_m} = 0, \tag{35}$$

$$\left.\frac{x_h}{\sqrt{1+x_h^2}}\right|_{h=0} = b \tag{36}$$

The first relation, eq 35, or equivalently $x_h|_{h=hm} = \infty$ ($h_x|_{x=0} = 0$), reflects the fact that the tangent to the drop profile at its apex should be parallel to the solid surface. Because $x_h|_{h=0} = -\cot\theta_0$, the second relation, eq 36, provides the following expression for the contact angle θ_0:

$$\cos\theta_0 = -b \tag{37}$$

Quantity b in eq 37 is provided by eq 24 and is independent of the geometrical characteristics of the drop.

Note that in the particular case of constant density and, consequently, constant drop volume, eq 31 no longer appears, and the right-hand side of eq 30 becomes independent of the drop profile and can be found from the constraint of the constant number of molecules. In this case, eq 30 coincides with the equation for the drop profile obtained in refs 3 and 4 where only the potential energy of intermolecular interactions was taken into account. In particular, this means that the results of refs 3 and 4 are applicable at any temperature if the drop volume is constant.

3.2 Selection of the Parameters of Microscopic Interactions

Specific values of most microscopic parameters that are needed for the numerical solution of the obtained equation were selected to be consistent with those obtained experimentally or used in numerical simulations of analogous systems.[15] Because sometimes the published information differs considerably, several sets of parameters were used to analyze the role of specific parameters. Some of the parameters, namely, ϵ_{LL}, ρ_S, and η, were selected to be the same in all sets. The energy parameter of the liquid—liquid interaction, ϵ_{LL}, which is related to the critical temperature of the liquid through the equation[15]

TABLE 1
Two Sets of Parameters of Microscopic Interactions Used for Calculations of Drop Properties

set	σ_{LS}, nm	ϵ_{LS}, J	ϵ^{s}_{LS}, J	ρ_{LS}, m^{-2}	ρ_{LG}, m^{-2}
1	0.2	1.48×10^{-21}	1.4×10^{-22}	1.8×10^{18}	$6 \times 10^{-11}\rho_L$
2	0.2	1.1×10^{-23}	1.0×10^{-22}	$4.2 \times 10^{17} + 6 \times 10^{-11}\rho_L$	$6 \times 10^{-11}\rho_L$

$$\frac{K_B T_c}{\epsilon_{LL}} = 1.30 \tag{38}$$

was selected to correspond to T_c = 513 K (the critical temperature of methanol) and has the value $\epsilon_{LL} = 1.58 \times 10^{-20}$ J.

The density of the solid substrate, ρ_S, was taken to be equal to $\rho_S = 5 \times 10^{28}$ m^{-3}, corresponding to that of silicon. The radius of the interaction sphere in the liquid was selected to be $\eta = 5\sigma_{LL}$, and hence $\Delta = \sigma_{LL}/\eta = 0.2$. If should be noted that energies calculated for this value of Δ with eq 11 differ by less than a few percent from those in which η is considered to be very large.

Several values for the other microscopic parameters were used in the calculations. Thus, hard core diameters σ_{LL} and σ_{LS} were selected in the range 1 ÷ 3.5Å, and the values of ϵ_{LS}, and ϵ^{s}_{LS}, in the range $\epsilon_{LL}/100 < (\epsilon_{LS}, \epsilon^{s}_{LS}) < 10\epsilon_{LL}$.

It was difficult to estimate the surface densities ρ_{LG}, ρ_{LS}, and ρ_{GS} because of the absence of experimental data. The surface density ρ_{LG} of the liquid molecules on the liquid–gas interface was taken to be proportional to the density of the liquid in the form $\rho_{LG} = k\rho_L\sigma_{LL}$ using several values for the coefficient k ($0 < k < 2$). The analogous density ρ_L for the liquid—solid interface was taken to be either constant ($\rho_{LS} \simeq 10^{19}$ m^{-2}) or as the sum of a constant contribution (4.2×10^{17} m^{-2}) and a term proportional to the liquid density ρ_L. The density of the gas molecules adsorbed on the solid surface was taken to be constant and equal to $\rho_{LS}/2$. The hard core diameter, σ_{GS}, and the energy parameter, ϵ_{GS}, of the gas—solid interactions were taken to be equal to those of the liquid—solid interactions.

Continuing to consider methanol, which has a melting temperature of T_m = 176 K, to be the liquid that composes the drop, the temperature range for calculation was restricted to values larger than 176 K.

The drop behavior depends on the specific values of the parameters. Only two sets of values were selected, which are listed in Table 1. (The nonchangeable parameters ϵ_{LL} and σ_{LL} are not included it the Table). The main distinction between the two sets consists of different values for the liquid—solid interaction parameters that are smaller in set 2 than in set 1.

The number of molecules in the drop is selected in the interval $10^{12} \leq N_l \leq 10^{20}$. With such a choice, the height h_m and the width of the drop $2x_0$ (Figure 3) vary from several nanometers for $N_l = 10^{12}$ to several micrometers for $N_l = 10^{20}$. The pressure in the system is changed between 1 and 10^4 atm.

3.3. Solution of the Equations for the Drop Profile

3.3.1. Qualitative Consideration

Equations 30, 31, and 25 for the drop profile, density, and Lagrange multiplier, respectively, can be solved only numerically. Before performing numerical computations, it seems reasonable to analyze the general features of the Gibbs free energy to acquire some qualitative information about the possible number and location of its extrema. As shown in section 2.3, eq 20 for the Gibbs free energy $G(N, p, T)$ can be decomposed into the sum of three terms. The first of them, $G_V(N, p, T)$, given by eq 21, is proportional to the drop volume V_d and at fixed V_d is independent of the drop shape. The second, $G_S(N, p, T)$, accounts for contributions to the total free energy due to the liquid—gas and liquid—solid interfaces, whereas the third, $G_E(N, p, T)$, provides the contribution due to the interaction between liquid and solid. The latter two terms depend on the shape of the drop even when the drop volume is fixed.

Because the considerations that follow (section 3.3.2) have shown that the qualitative features of $G(N, p, T)$ are similar to those of $G_V(N, p, T)$, it is reasonable for the sake of simplicity to examine first $G_V(N, p, T)$. Typical plots of $G_V(N, p, T)$ at various temperatures are presented in Figure 4 for the set 1 parameters (Table 1) and $p = 1$ atm. For temperatures below a certain value $T_w \simeq 378.8$ K, the function $G_V(N, p, T)$ has two minima—one at $\rho_L = \rho_1$ which is located in the range of relatively low densities around $10^{25}\,m^{-3}$ corresponding to a less dense phase, and the other one, at $\rho_L = \rho_2$, which is located in the range around $\rho_L \approx 10^{28}\,m^{-3}$ corresponding to a dense phase. The positions and the depths of the minima at the selected pressure depend on temperature and are separated by a local maximum at $\rho_L = \rho_m$. At temperatures lower than $T_{w,1} \simeq 209.8$ K, say $T = 176$ K, the minimum at $\rho_L = \rho_2 \simeq 3.49 \times 10^{28}$ m^{-3} corresponds to a stable state, and the minimum at $\rho_L = \rho_2 \simeq 4.17 \times 10^{25}$ m^{-3} corresponds to a metastable one (Figure 4a). At $T = T_{w,1}$, both minima have the same depth; at $T_w > T > T_{w,1}$ (e.g., $T = 300$ K), the minimum at $\rho_L = \rho_1$ corresponds to a stable state, and at $\rho_L = \rho_2$, it corresponds to a metastable one (Figure 4b). When the temperature increases further and reaches the value T_w, the minimum at $\rho_L = \rho_2$ is no longer present, and only a stable state corresponding to $\rho_L = \rho_1$ is present (Figure 4c).

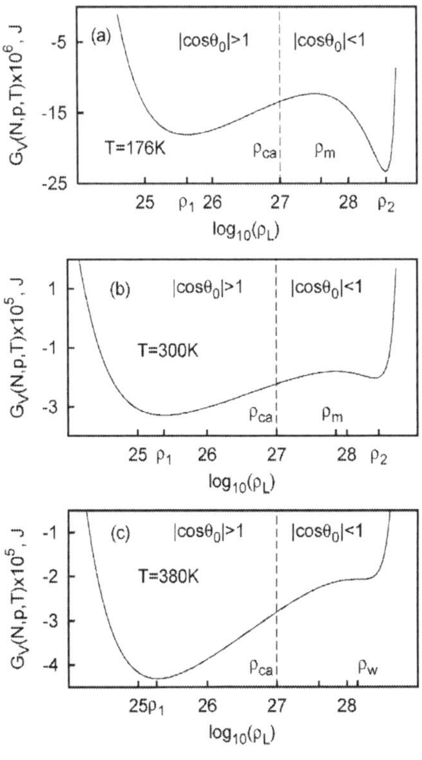

FIGURE 4. Typical plots of the Gibbs free-energy contribution proportional to the drop volume, $G_V(N, p, T)$. (a) The temperature is equal to the melting point $T_m = 176$ K. The stable state corresponds to the density $\rho_L = \rho_2$, and the metastable one, to the density $\rho_L = \rho_1$ (b) At higher temperatures ($T = 300$ K), the stable and metastable states exchange their locations. (c) At the temperature $T = T_w \simeq 380$ K, the minimum at $\rho_L = \rho_2$ disappears, and the only stable state corresponds to $\rho_L = \rho_1$. At densities $\rho_L < \rho_{ca}$, $|\cos\theta_0| > 1$ and a drop can no longer exist.

Note that the temperatures T_w and $T_{w,1}$ increase as the pressure increases. Their values for various pressures are listed in Table 2.

In Table 3, the densities ρ_1, ρ_m, and ρ_2 are listed at various temperatures and pressures. At the temperatures $T = 176$ and 300 K, only one minimum of the Gibbs free energy (corresponding to the high-density state with $\rho_L = \rho_2$) remains when the pressure becomes sufficently large.

From the above considerations, it follows that at $T > T_{w,1}$ the stable state is that which has the density ρ_1. However, that density is approximately the same as that of the inert gas. (For example, the density of nitrogen at atmospheric pressure and $T = 273$ K is equal to 4.12×10^{25} m^{-3}.) Obviously, a drop with such a density cannot be stable. However, the state with $\rho_L = \rho_2$ is stable for $T < T_{w,1}$ and metastable for $T_{w,1} < T < T_w$. The metastable state can, however, have a long life when the potential barier is sufficiently high.

TABLE 2
Temperatures $T_{w,1}$ and T_w Calculated on the Basis of the Volume Contribution $G_V(N, p, T)$ to the Gibbs Free Energy at Various Values of the External Pressure[a]

p, atm	$T_{w,1}$, K	T_w, K
1	209.8	378.8
10	267.4	380.6
20	392.7	382.7
100	377.6	400.8

[a] The microscopic parameters for set 1 (Table 1) were used in calculations.

TABLE 3
Densities ρ_1, ρ_m and ρ_2[a] **Calculated on the Basis of the Volume Contribution $G_V(N, p, T)$ to the Gibbs Free Energy at Various Temperatures and External Pressures**[b]

T, K	p, atm	$\rho 1 \times 10^{-25}$	$\rho_m \times 10^{-27}$	$\rho_2 \times 10^{-28}$
176	1	4.17	3.23	3.49416
	10	48.49	2.78	3.49537
	20			3.49671
	100			3.50728
300	1	2.42	7.01	2.50276
	10	25.16	6.74	2.50675
	20	52.71	6.42	2.51113
	100			2.54477
450	1	1.61		
	10	16.33		
	20	33.16		
	100	191.58		

[a] See Figure 4 for their description.

[b] The empty places in the last three columns indicate the absence of the corresponding extremum. The microscopic parameters for set 1 (Table 1) were used in calculations.

Two additional factors affect the above picture of drop stability. First, as shown later in section 3.3.2, for small drops the surface contribution to the Gibbs free energy, given by eq 22, becomes comparable with the volume contribution (eq 21), and the positions and depths of the minima are changed. The other factor is the true contact angle θ_0 that the drop makes with the solid surface. As one can see from eqs 37 and 24, this angle depends on ρ_L, and for some densities, $|\cos\theta_0|$ can become larger than unity. This means that for those densities a drop cannot be formed. This feature imposes a natural restriction to the existence of a drop on the solid surface, and the temperature at which this happens can be considered to be the wetting transition temperature. For the case considered in Figure 4, a drop cannot exist for densities smaller than $\rho_{ca} \approx 10^{27}$ m^{-3}. Note that the value of ρ_{ca} is provided by the equation $|\cos\theta_0| = 1$ and can be obtained independently of the solution for the drop profile.

It should be noted that neither the locations of the minima of $G_V(N, p, T)$ nor the densities ρ_1, ρ_m, and ρ_2 depend on the number of molecules in the drop.

3.3.2. Numerical Solution

Regarding the drop profile and density, let us note that after extracting μ from eq 31 and substituting into eq 30 the following integro-differential equation that does not contain μ can be obtained

$$\frac{x_{hh}}{\left(1+x_h^2\right)^{3/2}} - a_1\left(\rho_L\right)\Phi_{LS}^{l}\left(h\right) = \lambda_t\left(\rho_L; x\right) \tag{39}$$

where

$$\begin{aligned} K_s\rho_L^2\lambda_1\left(\rho_L;x\right) &= p + K_v\rho_L^2 - k_B T_{\rho_L}\left[1+\frac{2\eta_\rho\left(2-\eta_\rho\right)}{\left(1-\eta_\rho\right)^3}\right] - \\ &\frac{2}{N_1}\int_0^{h_m}\Phi_{LS}^{l}\left(h\right)x\,\mathrm{d}h - \frac{S_{LG}}{N_1}\frac{\partial}{\partial\rho_L}\left(\rho_L^2 K_s\right) - \left[\frac{\partial}{\partial\rho_L}\left(\rho_L^2 K_s'\right) + \right. \\ &\left. \sigma_{LL}\Phi_{LS}^{l}\left(\sigma_{LL}\right) + \left[\Phi_{LS}\left(0\right)+\Phi_{LS}^{s}\left(0\right)\right]\frac{\partial}{\partial\rho_L}\rho_{LS}\right]\frac{S_{LS}}{N_1} \end{aligned} \tag{40}$$

Because eq 39 is of second order, its solution contains two unknown constants that can be determined using the boundary conditions eqs 32 and 33. However, the latter conditions contain two other unknown quantities, namely, the height of the drop h_m and the contact angle θ_0. Furthermore, eq 39 contains the density ρ_L, which is also unknown. Therefore, three extra conditions are needed for the determination of h_m, θ_0, and ρ_L. As noted in section 3.1, these additional conditions are provided by the transversality conditions, eqs 35 and 36, and by the constraint, eq 25, of the constant number of molecules.

A trial solution $x = x'(h)$ for eq 39 for any selected $\rho_L > \rho_{ca}$ can be attempted as a solution of the equation

$$\frac{x_{hh}}{\left(1+x_h^2\right)^{3/2}} - a_1\left(\rho_L\right)\Phi_{LS}^{l}\left(h\right) = \lambda_c\left(\rho_L\right) \tag{41}$$

for the same boundary and transversality conditions as for eq 39 and for the constraint of constant volume $V_d = N_l/\rho_L$. In eq 41, $\lambda_c(\rho_L)$ is considered to be an unknown parameter independent of x and h. The left-hand side of eq 41 is the same as in eq 39, whereas the right-hand side is a constant for a selected value of ρ_L. Because $\lambda_c(\rho_L)$ is a constant, eq 39 can be solved in quadrature (Appendix A), and the parameters of the trial profile $x'(h)$, such as its height h_m and the value of $\lambda_c(\rho_L)$, can be determined. Using the profile $x'(h)$, the right-hand side $\lambda_t(\rho_L; x)$ of eq 39 can be calculated and compared with $\lambda_c(\rho_L)$. If $\lambda_c(\rho_L) = \lambda_t(\rho_L; x'(h))$, then the density ρ_L and the profile equation $x = x'(h)$ constitute the solution of eq 39. This observation suggests the following procedure for the numerical solution of eq 39. For selected values of N_1 and of the parameters characterizing the microscopic interactions, the right-hand sides $\lambda_c(\rho_L)$ and $\lambda_t(\rho_L; x'(h))$ of eqs 41 and 39 have to be calculated as functions of ρ_L for $\rho_L > \rho_{ca}$ The intersection of the curves $\lambda_c(\rho_L)$ and $\lambda_t(\rho_L; x'(h))$ provides the equilibrium density and equilibrium profile.

Three typical plots of $\lambda_c(\rho_L)$ and $\lambda_t(\rho_L; x(h))$ versus ρ_L corresponding to various temperatures are presented in Figure 5 (panels a–c) for cases in which $\rho_1 < \rho_{ca} < \rho_m$. (The definitions of ρ_1, ρ_{ca} and ρ_m are provided in Figure 4a.) In this case, the equation for the drop profile does not have a solution if the drop density is less than ρ_{ca} and therefore $|\cos \theta_0|$ is greater than unity. For this reason, the drop with a density of $\rho_L = \rho_1 < \rho_{ca}$ cannot be formed. In Figure 5a, the plots of $\lambda_c(\rho_L)$ and $\lambda_t(\rho_L; x'(h))$ against ρ_L have two intersection points, hence eq 39 has two solutions. The thermodynamic analysis of those solutions shows that one of them corresponds to a local maximum at $\rho_L = \rho_m$ of the Gibbs free energy whereas the other corresponds to a minimum at $\rho_L = \rho_2$. The latter solution is the physically meaningful one. As the temperature increases, the intersection points become closer, and at a temperature T_w, they collaps in an inflection point (Figures 5b and 4c). For $T > T_w$, there is no solution (Figure 5c). When $\rho_{ca} < \rho_2$ the plots of $\lambda_c(\rho_L)$and $\lambda_t(\rho_L; x(h))$ can have three intersection points (Figure 5d) that provide the densities ρ_1, ρ_2 and ρ_m. At sufficently high temperatures, the intersection

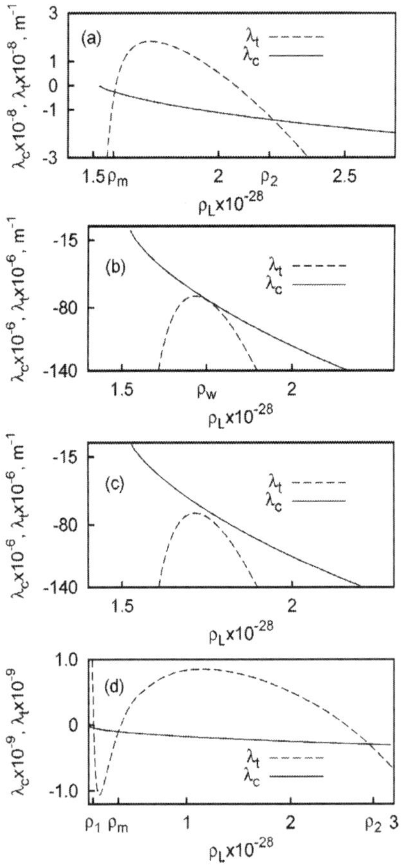

FIGURE 5. Plots of density dependencies of $\lambda_t \equiv \lambda_t(\rho_L; x)$ and $\lambda_c \equiv \lambda_c(\rho_L)$ from eqs 39 and 41, respectively. The densities at which $\lambda_c = \lambda_t$ correspond to extrema of the Gibbs free energy shown in Figure 4. (a) The case with two solutions, one at $\rho_L = \rho_m$, corresponding to a local maximum of the free energy and another at $\rho_L = \rho_2$, corresponding to a minimum (Figure 4a). (b) The case with one solution, corresponding to the inflection point in Figure 4b. (c) The case with no solution (Figure 4c). Panels a–c correspond to the case where $\rho_1 < \rho_{ca} < \rho_m$. (d) The case where three solutions are possible, corresponding to $\rho_{ca} < \rho_1$.

points corresponding to ρ_2 and ρ_m no longer exist, and the only solution for the drop profile is that which corresponds to the densiy ρ_1. However, as already noted, this low density state cannot be stable.

An alternative way to find the density and the profile of the drop is to examine the ρ_1 dependence of the Gibbs free energy $G(N, p, T)$, eq 20, for selected values of N_l and pressure p, as it was carried out for $G_V(N, p, T)$ in section 3.3.1. For any selected ρ_L, the drop characteristics that are necessary for calculating $G(N, p, T)$ can be found from eq 41 by taking into account that the drop volume V_d is given by $V_d = N_l/\rho_L$. The true profile corresponds to that value of ρ_L for which $G(N, p, T)$ has a minimum.

One should note that for large drops with $N_l \geq 10^{16}$ the total free energy $G(N, p, T)$ almost coincides with $G_V(N, p, T)$. For small drops with $N_l < 10^{14}$, there is a difference between $G(N, p, T)$ and $G_V(N, p, T)$. For illustration, in Figure 6 the behavior of these functions in the vicinity of $\rho_L = \rho_2$ is plotted for the set 1 parameters from Table 1 and $T = 200$ K and $p = 1$ atm. In all cases, the contribution of $G_E(N, p, T)$ to the total Gibbs free energy is much smaller than those of $G_V(N, p, T)$ and

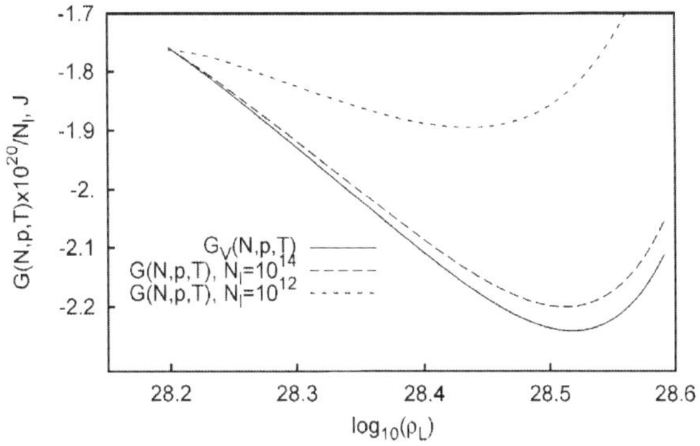

FIGURE 6. Density dependence of $G(N, p, T)$ and $G_V(N, p, T)$ calculated per molecule of liquid for various numbers N_l of molecules per unit length of the drop at $T = 200$ K. For $N_l \geq 10^{18}$, the difference between $G(N, p, T)$ and $G_V(N, p, T)$ becomes visually indistinguishable. Note that $G_V(N, p, T)$ per molecule is independent of ρ_L.

TABLE 4
Temperature T_w (K) as a Function of the Droplet Size and External Pressure for Set 1 (Table 1) for the Microscopic Parameters

	p, atm		
N_l	1	10	100
10^{12}	215.5	216.4	225.7
10^{14}	355.0	356.3	370.0
10^{20}	377.4	380.0	394.4

TABLE 5
Temperature T_w (K) as Function of the Droplet Size and External Pressure for Set 2 (Table 1) for the Microscopic Parameters

	p, atm		
N_l	1	10	100
10^{12}	309.7	311.9	337.7
10^{14}	371.8	373.6	394.1
10^{20}	378.9	380.7	400.8

$G_S(N, p, T)$. For example, for $N_l = 10^{12}$ the function $G_E(N, p, T)$ is on the order of 10^{-11} J, whereas $G_V(N, p, T)$ and $G_S(N, p, T)$ are on the order of 10^{-8} and 10^{-9}J, respectively. This is why the difference between $G(N, p, T)$ and $G_V(N, p, T)$ provides the surface contribution to the total Gibbs free energy, which increases as the drop size decreases.

Below, various drop characteristics that are dependent on the number of molecules it contains, temperature, pressure, and interaction parameters are examined.

4. DROPS UNDER VARIOUS CONDITIONS

The analysis that follows is restricted to the two sets of microscopic parameters listed in Table 1. Section 4.1 contains results regarding the temperature T_w, and the temperature, size, and pressure dependencies of the equilibrium drop density ρ_L and contact angle θ_0 are considered in sections 4.2 and 4.3, respectively.

4.1. Temperature T_w.

The temperature T_w at which the state corresponding to the density ρ_2 in Figure 4a becomes unstable depends on the size of the droplet, the external pressure, and the microscopic parameters. The values of T_w for the selected sets of parameters are listed in Tables 4 and 5. In all cases, T_w is smaller than the critical temperature $T_c = 513$ K. At a given size of the drop, T_w increases with increasing pressure. Similarly, at a given external pressure, T_w increases with increasing number of molecules in the drop. For large drops ($N_l = 10^{20}$), T_w is the same for both sets of microscopic parameters and coincides with the values for T_w obtained on the basis of the function $G_V(N, p, T)$ (Table 2). For small drops, the stronger liquid–solid interactions of set 1 provide smaller values of T_w compared to the weaker interactions provided by set 2.

4.2. Equilibrium Drop Density.

For set 1 of the microscopic parameters (Table 1), the density ρ_{ca} has the value 1.524×10^{28} m^{-3}, which is much larger than the density $\rho_l \approx 10^{25}$ m^{-3} at which the second minimum of the free energy occurs (Figure 4a). In this case for any drop size and external pressure, the equilibrium density $\rho_L = \rho_2$ decreases with increasing temperature from the melting point $T_m = 176$ K up to the temperature T_w. A typical temperature dependence of ρ_L is presented in Figure 7 for $p = 1$ atm. In a wide range of temperature far enough from T_w, the decrease in density follows a linear law

$$\rho_L = \rho_{L,m}\left(1-\alpha_T \Delta T\right) \tag{42}$$

where $\rho_{L,m}$ is the density at the melting temperature T_m, $\Delta T = T - T_m$, and α_T is the thermal expansion coefficient. The latter is almost independent of N_l for $N_l > 10^{16}$ and equal to $\alpha_T = 0.00233$ K^{-1}. However, for a smaller number of molecules, α_T depends on N_l. Thus, for $N_l = 10^{14}$, $\alpha_T = 0.00245$ K^{-1}, and for $N_l = 10^{12}$, $\alpha_T = 0.0048$ K^{-1}. An increase in the external pressure results in a weak decrease in α_T. For example, for $p = 100$ atm, $\alpha_T = 0.00225$ K^{-1}, for $N_l = 10^{20}$, $\alpha_T = 0.00234$ K^{-1}, for $N_l = 10^{14}$, and $\alpha_T = 0.0045$ K^{-1} for $N_l = 10^{12}$. The change in the microscopic parameters to those of set 2 changes the value of the density ρ_{ca} considerably, which becomes equal to $\rho_{ca} = 5.25 \times 10^{26}$ m^{-3}. However, the temperature behavior of the density ρ_L is in this case the same as in eq 42 with almost the same coefficient α_T.

The volume of the drop increases with increasing temperature according to the expression

$$V_d = V_{d,m}\left(1+\alpha_T \Delta T\right) \tag{43}$$

where $V_{d,m}$ is the volume of the drop at the temperature T_m. The drop height changes approximately from 5 nm to 50 μm when the number of molecules per unit length of the drop changes from 10^{12} to 10^{20}.

Close to temperature T_w, the decrease in ρ_L becomes nonlinear, and the absolute rate of decrease increases when the temperature approaches T_w (Figure 7b).

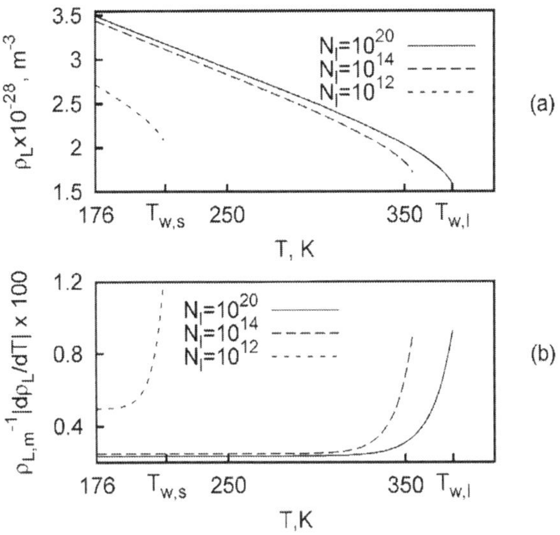

FIGURE 7. (a) Temperature dependence of ρ_L for large ($N_l \simeq 10^{20}$) and small ($N_l \simeq 10^{14}$, 10^{12}) droplets at $p = 1$ atm and set 1 of the parameters of microscopic interactions from Table 1. $T_{w,s} = 215.5$ K and $T_{w,l} = 377.4$ K. (b) Temperature dependence of the rate of decrease of the density ρ_L with increasing temperature for the same conditions. Both dependencies are qualitatively the same at other pressures and set 2 of the parameters.

The density $\rho_{L,m}$ in eq 42 is almost constant for $N_l > 10^{15}$ but is subjected to a rapid change when N_l decreases from 10^{15} to 10^{12} (Figure 8). Such behavior could be a consequence of the smallness of the drop but could also be a result of the nonapplicability of the mean-field approximation to small values of N_l. Indeed, if one considers a small part of a cylindrical drop that has a length equal to the drop width (i.e., several nanometers), then for $N_l = 10^{12} \div 10^{13}$ the number of molecules it contains ($10^3 \div 10^5$) is perhaps too small to employ a mean-field theory.

The pressure dependence of the equilibrium density $\rho_L = \rho_2$ is presented in Figure 9 for large ($N_l \approx 10^{20}$) and small ($N_l \approx 10^{12} \div 10^{14}$) droplets and $T = 200$ K. In all cases, ρ_L increases with increasing p and N_l. For $N_l > 10^{14}$, there is almost no difference between the densities of the drops having different sizes.

4.3. Contact Angle

The contact angle θ_0 provided by eqs 37 and 24 depends on the density ρ_L and hence on the temperature and N_l. The dependencies of θ_0 on temperature and N_l are very sensitive to the values of the parameters of the microscopic interactions. For example, for set 2 of the parameters and any number of molecules in the drop, θ_0 remains almost constant when the temperature increases from T_m to T_w. In that case, θ_0 decreases from 167.0° at $T = T_m$ to 160.6° at $T = T_w$ for $N_l = 10^{12}$ and from 168.0 to 161.8° for $N_l = 10^{20}$. For set 1 of the parameters, the dependence is much more pronounced as one can see from Figure 10 (solid lines) for $p = 1$ atm. For small drops with $N_l = 10^{12}$, the microscopic contact angle decreases from 85.3° at $T = T_m$ to 61.5° at $T = T_{w,s} \simeq 215$ K. For a large drop with $N_l = 10^{20}$, the value of θ_0 at $T_{w,l} \simeq 378$ K is much smaller ($\theta_0 \simeq 15°$) than the value $\theta_0 = 101°$ at $T = T_m$.

Similar behavior is exhibited by the apparent contact angle θ_m (dashed lines in Figure 10), which can be defined as the angle between the solid surface and the continuation of the circular part of the profile until its intersection with the surface of the solid (Figure 3). The angle θ_m can be calculated using the expression[4]

$$\cos\theta_m = 1 - \frac{h_m}{R_m^{curv}} \tag{44}$$

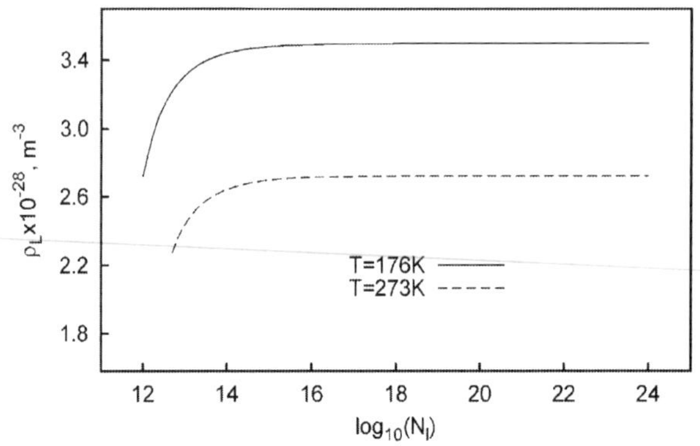

FIGURE 8. Size dependencies of the drop density at $T = 176$ and 273 K and pressure $p = 1$ atm for set 1 of the parameters of microscopic interactions from Table 1. (For set 2, the dependence is qualitatively the same as for set 1.)

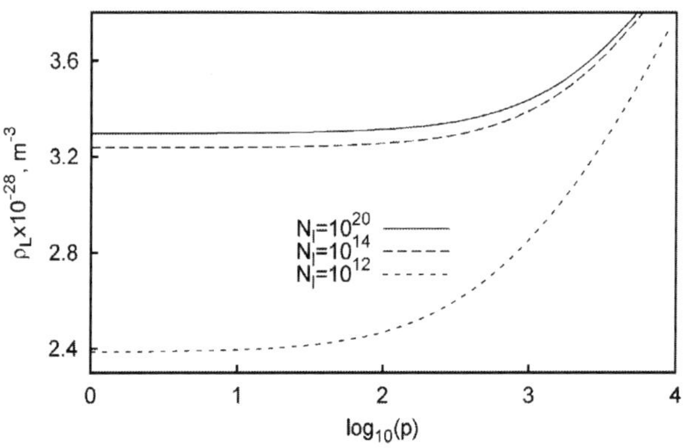

FIGURE 9. Pressure dependence of the drop density at $T = 200$ K for various numbers of molecules in the drop. The dependence is qualitatively the same for both sets 1 and 2 of the parameters of microscopic interactions (Table 1).

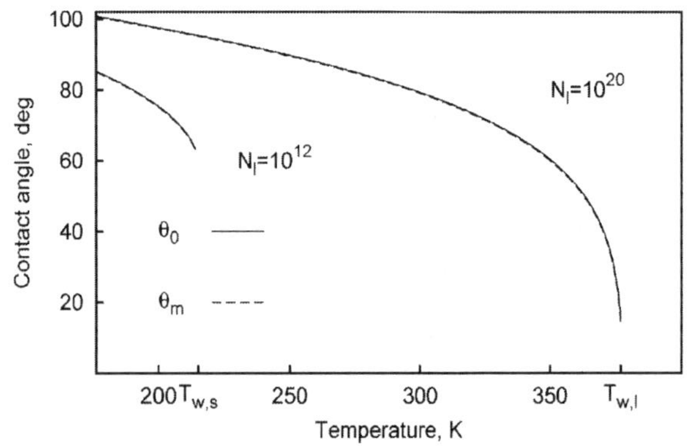

FIGURE 10. Temperature dependence of the true wetting angle, θ_0, and apparent wetting angle, θ_m, for large ($N_l \simeq 10^{20}$) and small ($N_l \simeq 10^{12}$) droplets at the pressure $p = 1$ atm and set 1 of the parameters of microscopic interactions from Table 1. (For set 2, the dependence is qualitatively the same as for set 1.) The temperatures $T_{w,s}$ and $T_{w,l}$ are equal to 215.5 and 377.4 K, respectively.

where h_m is the height of the drop and R_m^{curv} is the curvature radius at the drop apex. The small difference between θ_0 and θ_m observed in this case indicates a small deviation of the drop profile from the circular shape.

As already noted, a drop cannot exist above T_w, the temperature at which it spreads completely over the surface. At $T = T_w$, θ_0 is a discontinuous function of temperature (Figure 10), changing from a nonzero to a zero value. For a better understanding of what happens, let us consider qualitatively the density dependence of the Gibbs free energy at temperatures close to T_w. For the set 1 parameters and $T < T_w$, the Gibbs free energy as a function of ρ_L exhibits the behavior shown in Figures 4b and c. The drop is in a metastable state at the density $\rho = \rho_2$, which is far enough from the density $\rho = \rho_{ca}$ at which the contact angle θ_0 becomes zero. As the temperature increases, ρ_2 decreases, and the density ρ_m at which the Gibbs free energy has a local maximum increases. At $T = T_w$, ρ_m and ρ_2 become equal and acquire the value $\rho_w > \rho_{ca}$, which does not correspond to any stable or metastable state. The state of the system becomes unstable, and the density decreases and reaches the value ρ_{ca} at which a drop cannot exist because cos θ_0 becomes larger than unity. For large drops, the difference between ρ_w and ρ_{ca} is less than for small drops; therefore, the equilibrium contact angle for large drops near $T_{w,\,l}$ is less than for small drop near $T_{w,\,s}$.

The results of sections 4.2 and 4.3 have shown that the drop size influences the drop characteristics (e.g., density, contact angle) only for $N_l \leq 10^{14}$. This is due to the increase in the relative contribution of the term $G_S(N, p, T)$ to the total free energy. Estimations have shown that for N_l larger than 10^{14} this term is much smaller than $G_V(N, p, T)$. However, for $N_l \leq 10^{14}$, $G_V(N, p, T)$ and $G_S(N, p, T)$ become comparable.

4.4. Drop Shape

It was shown in refs 3, 4, and 13 that the shape of the drop depends on the drop volume and contact angle. In the case of fixed density, which was considered in the cited papers, the drop volume for the given number of molecules in the drop was constant. The magnitude of the contact angle depends on the drop density and the parameters characterizing the intermolecular interactions and is independent of temperature. In the case examined in the present article, the drop density depends on temperature, and for this reason, so do the drop volume and contact angle. For illustration, in Figure 11 the profile of the drop with $N_l = 10^{20}$ is presented for two temperatures, $T_1 = 293$ K and $T_2 = 350$ K. For $T_1 = 293$ K, the drop density is equal to $\rho_L = 2.56 \times 10^{28}$ m^{-3}, the volume per unit length of the drop is equal to $V = 3.91 \times 10^{-9}$m^3/m, the contact angle $\theta_0 = 86.6°$, and the drop height is equal to $h_m = 4.68 \times 10^{-5}$ m. For $T_2 = 350$ K, these quantities are $\rho_L = 2.03 \times 10^{28}$ m^{-3}, $V = 4.93 \times 10^{-9}$ m^3/m, $\theta_0 = 60.3°$, and $h_m = 4.53 \times 10^{-5}$ m. As one can

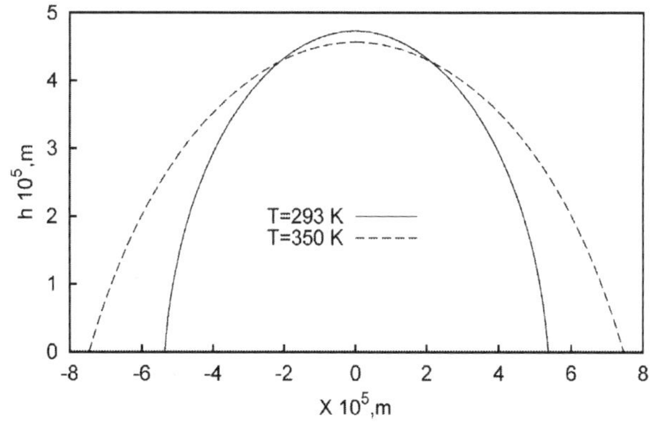

FIGURE 11. Profiles of the drop with $N_l = 10^{20}$ molecules per unit length for two temperatures. Interaction parameters correspond to set 1 (Table 1).

see from Figure 11, the increase in volume with increasing temperature is due to the increase in the drop width, whereas the drop height slightly decreases. This is accompanied by a decrease in the contact angle and hence an increase in wettability with increasing temperature.

5. CONCLUSIONS

The equilibrium description of a small drop on a solid, embedded in an inert gas at constant pressure has required the simultaneous minimization of the Gibbs free energy $G(N, p, T)$ of the system with respect to the density of the liquid in the drop and shape of the drop profile. The minimization procedure provides integro-differential equations for the drop profile and the density that have to be solved using suitable boundary conditions, transversality conditions, and the constraint of a constant number of molecules in the drop. It turns out that below a certain temperature T_w, which depends on the drop size, external pressure, and on the parameters of microscopic interactions, the drop can be in a stable or in a metastable state, one of them corresponding to a gas phase and the other to a liquid phase. At $T = T_{w,1} < T_w$, the stable and metastable states interchange their locations. Above T_w, a drop cannot be formed on a solid surface because the drop spreads completely (wets the surface). However, the transition to this state has to be distinguished from the usually considered[10,16] wetting transition. In the latter, the true contact angle is a continuous function of temperature, which at and above the transition point is equal to zero. The derivative of the true contact angle θ_0 with respect to temperature has a discontinuity at $T = T_w$. In the cases examined in the present article, the contact angle θ_0 has a nonzero value at $T = T_w$. Considering that its value above T_w is equal to zero, one arrives at the conclusion that both the value of θ_0 and its temperature derivative have discontinuities at $T = T_w$. The difference between the two kinds of transitions results from the difference between the physical conditions under which they take place.

The obtained numerical solution for the drop profile has shown an unusually strong dependence of some of the drop characteristics on the drop size. For the considered values of the microscopic parameters, the temperature interval in which a drop still exists on the surface is much larger for larger drops (containing more than 10^{14} molecules per unit length) than for small ones with $N_l \simeq 10^{12}$ (compare Tables 4 and 5). The size of the drop strongly affects the drop density. For large drops, the density is almost independent of size. However, our calculations showed that the density rapidly decreases when N_l decreases from 10^{14} to 10^{12}. The true contact angle that the drop makes with the surface also depends on its size and decreases with decreasing size (Figure 10). Note that for $N_l > 10^{20}$ all of the above quantities become practically independent of N_l (i.e., the case $N_l > 10^{20}$ represents an infinite-sized drop).

The density of the drop decreases with increasing temperature and follows in a wide range of temperature a linear law, which becomes nonlinear near the limiting temperature T_w. The density is almost independent of pressure for $p < 100$ atm and increases when the pressure is further increased.

If the density of the liquid is given and consequently the volume of the drop is fixed, then the equation for the drop profile reduces to the one obtained in refs 3 and 4 on the basis of the minimization of the total potential energy of the intermolecular interactions.

APPENDIX A. SOLUTION OF THE EQUATION FOR THE DROP PROFILE

To transform eq 41 into a first-order differential equation, let us introduce the function

$$\Psi(h) = \frac{x_h}{\sqrt{1 + x_h^2}} \tag{A 1}$$

which has the derivative

$$\frac{d\Psi(h)}{dh} = \frac{x_{hh}}{\left(1 + x_h^2\right)^{3/2}} \tag{A.2}$$

Then, eq 41 becomes

$$\frac{d\Psi(h)}{dh} - a_1(\rho_L)\Phi_{LS}^1(h) = \lambda_c(\rho_L) \tag{A.3}$$

and has the general solution

$$\Psi(h) = \lambda_c(\rho_L)h + D(h) + C \tag{A.4}$$

where

$$D(h) = a_1(\rho_L)\int \Phi_{LS}^1(h)\ dh \tag{A.5}$$

and C is a constant of integration. Because eq 1 can be rewritten as

$$x_h = \frac{\Psi(h)}{\sqrt{1 - \Psi^2(h)}} \tag{A.6}$$

one can find the following general solution for $x(h)$[3]

$$x(h) - x_0 = \int_0^h \frac{\Psi(h)}{\sqrt{1 - \Psi^2(h)}} dh \tag{A.7}$$

where $x_0 \equiv x(0)$ (Figure 3) is a second integration constant. x_0 can be determined from the condition for the drop apex to be located on the h axis $[x(h_m) = 0]$, which leads to the expression

$$x_0 = -\int_0^{h_m} \frac{\Psi(h)}{\sqrt{1 - \Psi^2(h)}} dh \tag{A.8}$$

The first integration constant, C, can be obtained using eqs 36 and 37, which provide the expression $\Psi(0) = b = -\cos\theta_0$. The latter expression leads for C to $C = b - D(0)$, and the function $\Psi(h)$ becomes

$$\Psi(h) = b + \lambda_c(\rho_L)h + D(h) - D(0) \tag{A.9}$$

After x_0 and C are determined, eq 7 still contains two unknown quantities, $\lambda_c(\rho_L)$ and h_m. A connection between them can be found from the fact that the tangent to the drop profile at the drop apex should be parallel to the x axis. Consequently,

$$x_h\,|_{h=h_m} = -\infty \tag{A.10}$$

which together with eq 6 leads to $\Psi(h_m) = -1$. As a result, one obtains for $\lambda_c(\rho_L)$ the expression

$$\lambda_c(\rho_L) = -\frac{1}{h_m}\left[1 + b + D(h_m) - D(0)\right] \tag{A.11}$$

The last required information is provided by the equation for the drop volume. $V_d = 2\int_0^{h_m} x\,dh$ ($V_d = N_1/\rho_L$) Integrating by parts and taking into account that $x(h_m) = 0$, one can rewrite the latter equation in the form $V_d = -2\int_0^{h_m} hx_h\,dh$, which after substituting eq 6 for x_h becomes

$$V_d = -2\int_0^{h_m} \frac{h\Psi(h)}{\sqrt{1-\Psi^2(h)}}\,dh \quad \text{(A.12)}$$

If the volume of the drop is known, then one can find numerically from the latter equation the drop height h_m.

REFERENCES

(1) Carroll, B. J. J. *Cooid Interface Sci.* **1976**, *57*, 488. Carroll, B. J. J. *Cooid Interface Sci.* **1984**, *97*, 195.
(2) Yeh, E. K.; Newman, J.; Radke, C. J. *Colloids Surf* **1999**, *156*, 137. Yeh, E. K.; Newman, J.; Radke, C. J. *Colloids Surf.* **1999**, *156*, 525. Solomentsev, Y.; White, L. R. J. *Colloid Interface Sci.* **1999**, *218*, 122.
(3) Berim, G. O.; Ruckenstein, E. J. *Phys. Chem. B* **2004**, *108*, 19330 (Section 3.2 of this volume).
(4) Berim, G. O.; Ruckenstein, E. J. *Phys. Chem. B* **2004**, *108*, 19339 (Section 3.3 of this volume).
(5) Tarazona, P.; Evans R. *Mol. Phys.* **1983**, *48*, 799.
(6) van Giessen, A. E.; Bukman, D. J.; Widom, B. *J. Colloid Interface Sci.* **1997**, *192*, 257.
(7) Bieker, T.; Dietrich, S. *Physica A* **1998**, *252*, 85.
(8) Lee, D. J.; Telo da Gama, M. M.; Gubbins, K. E. *J. Chem. Phys.* **1986**, *85*, 490.
(9) Bauer, C.; Dietrich, S. *Eur. Phys. J. B* **1999**, *10*, 767.
(10) Dietrich, S. In *Phase Transitions and Critical Phenomena*; Domb, C., Lebowitz, J. L., Eds.; Academic: London, 1988; Vol. 12, p 1.
(11) Neimark, A. V. *J. Adhes. Sci. Technol.* **1999**, *13*, 1137.
(12) Carnahan, N. F.; Starling, K. E. *J. Chem. Phys.* **1969**, *51*, 635.
(13) Berim, G. O.; Ruckenstein, E. *J. Colloid Interface Sci.* **2005**, *286*, 681 (Section 3.4 of this volume).
(14) Elsgolc, L. E. In *Calculus of Variations*; Pergamon Press: Addison–Wesley: New York, 1962.
(15) Hirschfelder, J. O.; Curtis, C. F.; Bird, R. B. *Molecular Theory of Gases and Liquids*; John Wiley & Sons: New York, 1967.
(16) Bonn, D.; Ross, D. *Rep. Prog. Phys.* **2001**, *64*, 1085.

3.8 Dependence of the Macroscopic Contact Angle on the Liquid-Solid Interaction Parameters and Temperature*

Gersh Berim and Eli Ruckenstein†
Department of Chemical and Biological Engineering, State University of New York at Buffalo, Buffalo, New York 14260, USA
Corresponding Author
†E-mail: feaeliru@buffalo.edu.
Phone: (716) 645-1179, Fax: (716) 645-3822

(Received 16 February 2009; accepted 22 April 2009; published online 13 May 2009)

I. INTRODUCTION

In a paper, Ref. 1, the simple expression

$$\theta - \theta_0 = -C\delta^{-\nu}\left(\varepsilon - \varepsilon_0\right) \tag{1}$$

representing the contact angle θ of a nanodrop of a Lennard-Jones fluid on smooth homogeneous solid surfaces as a function of the temperature T and the energy parameter ϵ_{fs} of the Lennard-Jones potential of the fluid-solid interactions was suggested. In this equation $\delta = \epsilon_{ff}/k_BT$, ($\epsilon_{ff}$ being the energy parameter of the Lennard-Jones fluid-fluid interactions, and k_B the Boltzmann constant), $\varepsilon = \epsilon_{fs}/\epsilon_{ff}$, $C>0$ and ε_0 are quantities dependent on the hard core parameter σ_{fs} of the Lennard-Jones fluid-solid interactions, and θ_0 and ν are universal constants independent of T and σ_{fs}. Equation (1) was derived by calculating θ from the drop profile obtained on the basis of a nonlocal density functional theory (DFT) in wide ranges of temperature, ε and σ_{fs}. The DFT involves only the microscopic parameters of the interaction potentials and does not require the use of macroscopic quantities such as the surface tensions. One of the consequences of DFT analysis is the existence of a (quasi)universal contact angle, $\theta_0 \simeq 112°$. If a nanodrop makes with the surface a contact angle $\theta > \theta_0$ (this occurs for $\varepsilon < \varepsilon_0$), then θ increases with increasing temperature. Vice versa, if on a given surface $\theta < \theta_0$ ($\varepsilon > \varepsilon_0$) then θ decreases with increasing temperature.

The drops considered in Ref. 1 had sizes of the order of 10 nm. For such small drops, the sizes of the liquid-vapor and liquid-solid interfaces become comparable with the size of the drop and the definition and calculation of the contact angles cannot be made in a unique way. In this case, the drop profile is obtained from the fluid density distribution as a line along which the fluid density has a constant value corresponding to that of an equimolar dividing surface located at some distance from the solid surface.[2–4] This profile has a nonsmooth shape near the solid surface

* *J. Chem. Phys.* 130, 184712 (2009). Republished with permission.

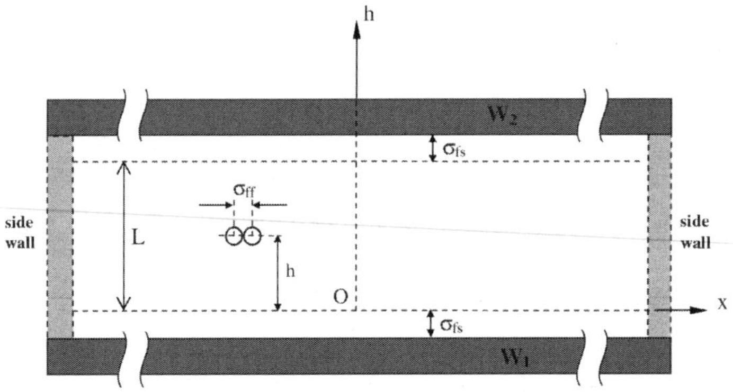

FIG. 1. A slit of width $L + 2\sigma_{fs}$ between two identical solid walls W_1 and W_2. σ_{ff} and σ_{fs} are the hard core diameters of the fluid-fluid and fluid-solid interactions, respectively. The *y*-axis is perpendicular to the plane of the figure. The open slit (without side walls) is infinite in the *x* and *y* directions. The closed slit has the size in the *x* and *y* directions large for the end effects to be negligible.

because of the oscillations of the fluid density and the contact angle cannot be uniquely defined. Is Eq. (1) also valid for macroscopic drops? Because of the computational restrictions caused by the large number of molecules contained in macrodrops, it is not practically possible to apply DFT to the latter case. This difficulty can be overcome by taking into account that DFT can be used to calculate, under some reasonable approximations, the liquid-vapor γ_{lv}, solid-liquid (γ_{sl}), and solid-vapor γ_{sv} surface tensions.[5,6] The contact angle θ can then be determined using the Young equation

$$\gamma_{lv}\cos\theta = \gamma_{sv} - \gamma_{sl}, \tag{2}$$

which is valid for macroscopic drops.

In this paper we performed such calculations for a one component fluid (argon) and show that, indeed, Eq. (1) with slightly different parameters θ_0, ε_0, and v remains still valid for macroscopic drops. The procedure for calculating the surface tensions is outlined in Sec. III and the results are presented in Sec. IV.

II. THE SYSTEM

To calculate the surface tensions, we consider a fluid confined between two parallel structureless identical solid walls W_1 and W_2 separated at a sufficiently large distance $L + 2\sigma_{fs}$ (see Fig. 1). The slit is considered to be either connected to a reservoir of bulk fluid at a temperature T and chemical potential $\mu = \mu_c$, where μ_c is the chemical potential at liquid-vapor coexistence, or to be closed. In the latter case, the distance between the side walls which close the slit should be much larger than the width of the slit for the end effects to be negligible.

The interactions between the fluid molecules are described by the Lennard-Jones potential $\phi_{ff}(|\mathbf{r}-\mathbf{r}'|) \equiv \epsilon_{ff}\bar{\phi}_{ff}(|\mathbf{r}-\mathbf{r}'|)$, where $\bar{\phi}_{ff}(|\mathbf{r}-\mathbf{r}'|)$ is a dimensionless quantity given by

$$\bar{\phi}_{ff}(|\mathbf{r}-\mathbf{r}'|) = \begin{cases} 4\left[\left(\dfrac{\sigma_{ff}}{r}\right)^{12} - \left(\dfrac{\sigma_{ff}}{r}\right)^{6}\right], & r \geq \sigma_{ff} \\ \infty, & r < \sigma_{ff}. \end{cases} \tag{3}$$

In Eq. (3), the coordinates **r** and **r**′ provide the locations of the interacting fluid molecules, $r = |\mathbf{r} - \mathbf{r}'|$, and σ_{ff} and ϵ_{ff} are the fluid hard core diameter and the energy parameter, respectively.

The same kind of potential is selected for the interactions between the fluid and solid molecules (the energy parameter ϵ_{ff} and the hard core diameter σ_{ff} being replaced by ϵ_{fs} and σ_{fs}, respectively, r in Eq. (3) being the distance between a molecule of fluid and that of the solid). Because of walls uniformity, the total potential $U_{fs}(\mathbf{r})$ of the interaction between a fluid molecule with the walls depends only on the distance h from one of the walls (see Fig. 1) and can be written in the form

$$U_{fs}\left(h\right) = \epsilon_{fs}\bar{U}_{fs}\left(h\right), \tag{4}$$

where the dimensionless quantity $\bar{U}_{fs}\left(h\right)$ is given by

$$\bar{U}_{fs}\left(h\right) = \psi\left(h\right) + \psi\left(L-h\right) \tag{5}$$

and

$$\psi\left(h\right) = \frac{2\pi}{3}\rho_s\sigma_{fs}^3\left[\frac{2}{15}\left(\frac{\sigma_{fs}}{\sigma_{fs}+h}\right)^9 - \left(\frac{\sigma_{fs}}{\sigma_{fs}+h}\right)^3\right]. \tag{6}$$

As a consequence, the fluid density distribution $\rho(\mathbf{r})$ inside the slit is a function of h only ($\rho(\mathbf{r}) \equiv \rho(h)$).

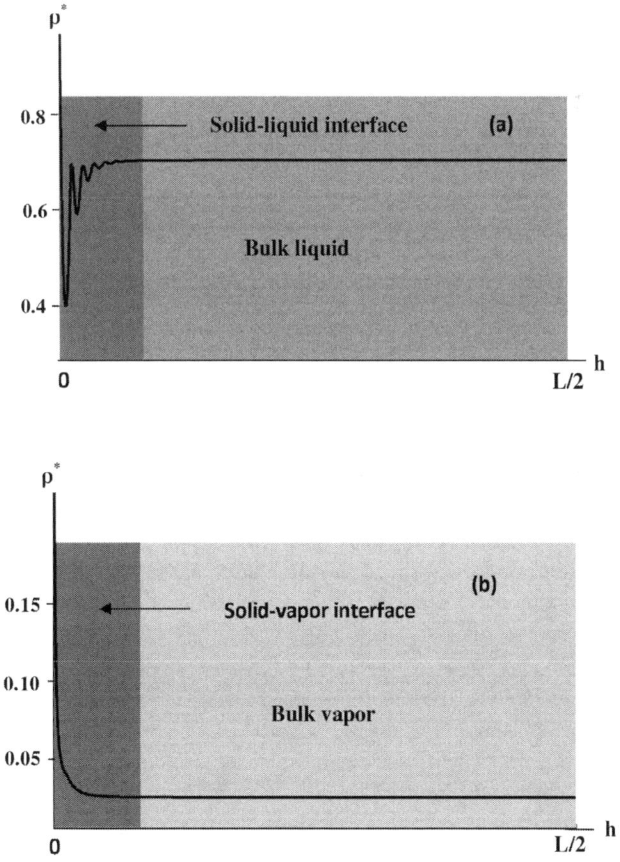

FIG. 2. (a) Schematic presentation of a symmetric fluid density distribution across the open slit, containing a solid-liquid interface. The solid line is the density profile provided by the DFT calculations. Only one-half of the slit with $L = 90\sigma_{ff}$ is presented, and $\rho^* = \rho\sigma_{ff}^3$. (b) The same for the fluid density distribution containing a solid-vapor interface.

Because the main purpose of the present paper is to examine the dependence of the contact angle of a macrodrop on a solid surface on the parameters of the fluid-solid interactions ($\sigma_{fs}, \epsilon_{fs}$), it is not necessary to select those parameters for specific substrates. As a reference substrate, we consider the solid carbon dioxide which has a density $\rho_s = 1.91 \times 10^{28}$ m^{-3}.[7]

III. GENERAL PROCEDURES FOR CALCULATING THE SURFACE TENSIONS

To determine γ_{sl}, and γ_{sv} an open slit was considered. In such a slit the fluid density distribution $\rho(h)$ is symmetric with respect to the middle of the slit.[8,9] The effective width of the slit, L, is selected large enough to provide a bulklike density of the fluid in the middle region of the slit (light shaded area in Fig. 2). The density distribution $\rho(h)$ can be obtained as a solution of the Euler—Lagrange equation derived by minimizing the grand canonical potential of the system. This equation is provided in the Appendix [Eq. (A7)] together with its derivation on the basis of DFT.

After determining $\rho(h)$, the grand canonical potential of the system, Ω_{sf}, and grand canonical potential Ω_f^b of a bulk fluid having the volume of the slit can be calculated as follows:

$$\Omega_{sf} = F_H\left[\rho(h)\right] - \mu_c \int_0^l \rho(h)\,dh, \tag{7}$$

$$\Omega_f^b = F_H\left[\rho_f^b\right] - \mu_c \rho_f^b L. \tag{8}$$

Here the subscript f stands either for the liquid ($f = l$) or for the vapor ($f = \upsilon$), $F_H[\rho(h)]$ and $F_H[\rho_f^b]$ (provided in the Appendix) are the Helmholtz free energies of the fluid in the entire slit for the real density distribution and the homogeneous bulk density distribution, respectively, ρ_f^b is the density of the bulk fluid. Both potentials are calculated per unit area of one of the walls. The surface tensions γ_{sl} and $\gamma_{s\upsilon}$ can be obtained from the expression

$$\gamma_{sf} = \left(\Omega_{sf} - \Omega_f^b\right)/2, \; \left(f = l, \upsilon\right). \tag{9}$$

To illustrate the calculation procedure in some details, let us examine as an example the case of $T = 85\text{K}$, $\sigma_{fs} = 3.1\text{Å}$, $\epsilon_{fs} = 0.6\epsilon_{ff}\,(\varepsilon = 0.6)$. At a selected temperature, the dimensionless chemical potential $\bar{\mu} \equiv \mu/k_B T$ of the bulk fluid corresponding to vapor-liquid coexistence is $\bar{\mu}_c = -11.4852$ and the dimensionless densities ρ_l^* and $\rho_\upsilon^* (\rho^* \equiv \rho\sigma_{ff}^3)$ of the liquid and vapor are $\rho_l^* = 0.6789$ and $\rho_\upsilon^* = 0.01868$, respectively. Under such conditions, two density distributions of the fluid in the slit with $L = 90\sigma_{ff}$ can be obtained from the Euler—Lagrange equation. First, a liquidlike solution is shown in Fig. 2(a). In the middle region of the slit (light shaded area) the fluid density coincides almost perfectly with the density of the bulk liquid at coexistence (the difference between those densities is less than 0.1%). By calculating the corresponding grand canonical potentials one can obtain for γ_{sl} the value $\gamma_{sl} = 0.003\,695$ N/m. The second solution, shown in Fig. 2(b) has the fluid density in the middle region of the slit equal to the bulk vapor density and provides $\gamma_{s\upsilon} = -0.0002259$ N/m.

A similar approach with some modifications can be used to calculate $\gamma_{l\upsilon}$. In this case one must have a density distribution in the slit, that contains the interface between a bulklike liquid and a bulklike vapor phase. The corresponding density distribution is shown schematically in Fig. 3. However, in an open system, the numerical solution corresponding to such a distribution is unstable due to the uncertainty of the position of the interface.[5,6] To overcome this difficulty, one must correct the position of the interface to remain fixed at the same place during calculations.[5,6] In the present paper we use another approach for calculating $\gamma_{l\upsilon}$, that is based on the existence in a closed system of a stable symmetry breaking solution of the Euler—Lagrange equation[10,11] which

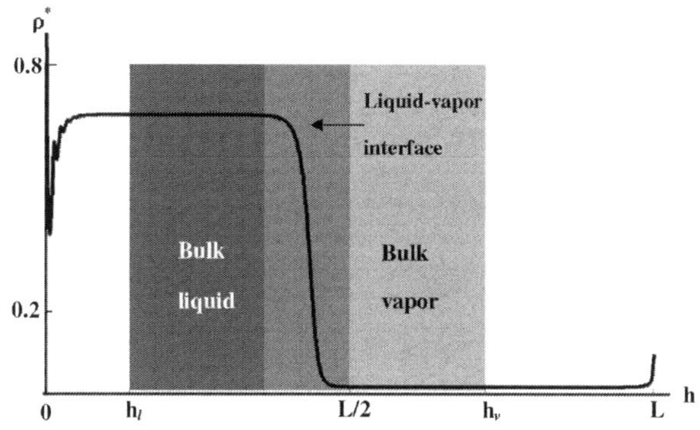

FIG. 3. Schematic presentation of an asymmetric fluid density distribution across the closed slit ($L = 90\sigma_{ff}$), containing a liquid-vapor interface. The solid line is the density profile provided by the DFT calculations. h_l and h_υ provide the external boundaries for the liquidlike and vaporlike bulk fluids, and $\rho^* = \rho\sigma_{ff}^3$.

provides an asymmetric density profile. For closed systems, the number of molecules is constant and the Lagrange multiplier which is determined by this number has the dimensions of a chemical potential. By varying the number of molecules in a closed slit for the Lagrange multiplier to acquire the value of the chemical potential in the bulk at coexistence, one can obtain the proper stable asymmetric density profile. After such a profile is found, $\gamma_{l\upsilon}$ can be calculated using the expression

$$\gamma_{l\upsilon} = F_H'\left[\rho(h)\right] - F_H'\left[\rho_f^b\right], \tag{10}$$

where $F_H'[\rho(h)]$ and $F_H'[\rho_f^b]$ are the Helmholtz free energies per unit area of one of the walls over the filled regions (i.e., between $h = h_l$ and $h = h_\upsilon$) in Fig. 3 calculated with the actual density profile and for the homogeneous density distribution with $\rho = \rho_l^b$ (or $\rho = \rho_\upsilon^b$). Note that the value of $F_H'[\rho_f^b]$ is the same for $\rho_f^b = \rho_l^b$ and $\rho_f^b = \rho_\upsilon^b$.

For the example considered above, $\gamma_{l\upsilon} = 0.006591$ N/m. The corresponding value of the contact angle calculated with Young's equation is $\theta = 126.5°$.

IV. RESULTS

For the system described in Sec. II the calculations of all surface tensions and contact angles were performed at four temperatures, $T = 85$, 90, 95, and 100 K and for ε varying from 0.2 to 1.8. We selected σ_{ff} as the unit of length and introduced two independent dimensionless parameters, $\sigma = \sigma_{fs}/\sigma_{ff}$ for the substrate, and the parameter $\delta = \epsilon_{ff}/k_BT$ characterizing the effect of temperature.

A. Surface Tensions and Contact Angle for $\sigma_{fs} = 3.5$ Å ($\sigma = 1.028$)

1. ε-dependence at various temperatures

A typical dependence of γ_{sl} and $\gamma_{s\upsilon}$ on ε is presented in Fig. 4 for $T = 85$ K (solid lines) and $T = 95$ K (dashed lines). In all cases, both surface tensions decrease with increasing ε, one of them, $\gamma_{s\upsilon}$ being always negative. The solid-liquid surface tension, γ_{sl}, is positive at smaller ε and becomes negative at larger ones. The intersection of the curves for γ_{sl} and $\gamma_{s\upsilon}$ provides the value of ε at which the contact angle $\theta = 90°$. This value decreases as the temperature increases. The liquid-vapor surface tension does not depend on ε and has the values of 0.006 600, 0.005 253, 0.003 987, and 0.002 893 N/m for $T = 85$, 90, 95, and 100 K, respectively.

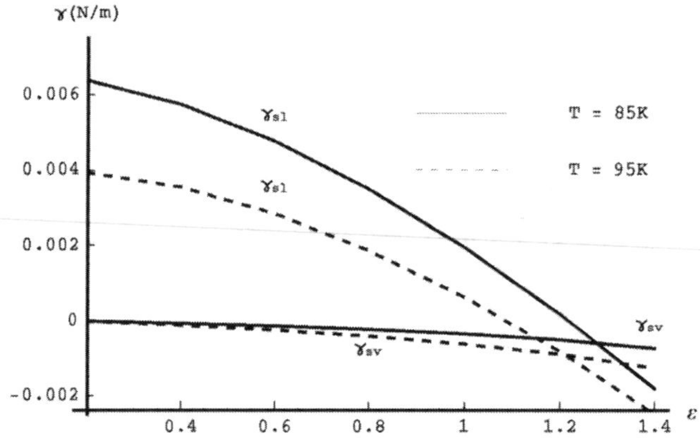

FIG. 4. ε-dependence of the solid-liquid and solid-vapor surface tensions for T = 85 K (solid line) and T = 95 K (dashed lines). The liquid-vapor surface tension is equal to 6.60×10^{-3} N/m for T = 85 K and 4.00×10^{-3} N/m for T = 95 K.

The dependence of θ on ε is presented in Fig. 5 for T = 85 K by the solid line. This dependence is almost linear in the range of $0.4< \varepsilon < 1.2$, with small deviations from linearity for ε <0.4 and ε> 1.2. For comparison, the analogous dependence for nanodrops under the same conditions is presented in Fig. 5 as a dashed line. One can see that for ε <0.6 the contact angle for nanodrops is larger than for macrodrops, while for ε >0.6 it is smaller. Below, only the linear parts of the dependencies of θ on ε, which can be represented by the equation

$$\theta = k\varepsilon + b, \tag{11}$$

will be considered. For the case presented in Fig. 5, k = –70.1°, and b = 180.4°.

The values of k and b for other temperatures (other δ) are listed in Table 1. The slopes k of the lines and the values of δ satisfy the simple approximate relation

$$k\delta^{\nu} = -C, \tag{12}$$

where ν = 1.65 and C = 122.3°. In Fig. 6 the plots of θ versus ε are presented for the four considered values of δ. The rectangle in this figure indicates the area in which the intersection

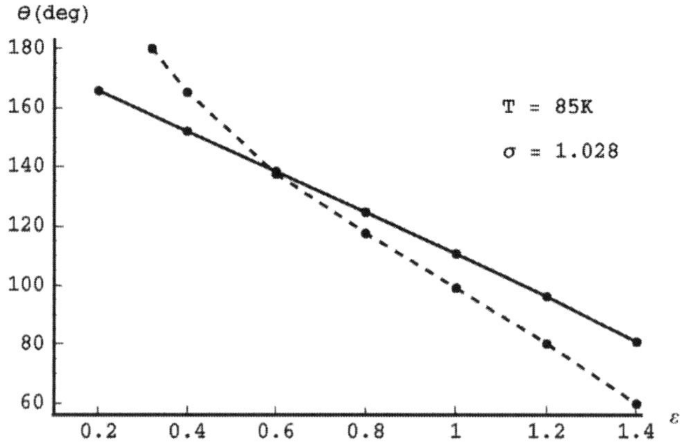

FIG. 5. The solid line represents the ε-dependence of the macroscopic contact angle for T = 85 K, σ = 1.028 (σ_{fs} = 3.5 Å). The points represent the results of the DFT calculations. The dashed line represents the results obtained for nanodrops (Ref. 1).

TABLE 1
Parameters of Eq. (11) for various temperatures for $\sigma = 1.028(\sigma_{fs} = 3.5\text{Å})$.

	T	k	b
δ	(k)	(deg)	(deg)
1.409	85	–70.1	180.4
1.331	90	–75.3	184.5
1.261	95	–83.0	191.1
1.198	100	–91.5	197.8

points of the four straight lines are located. The bold point on the plot has the coordinates $\varepsilon = \varepsilon_0 = 0.82$ and $\theta = \theta_0 = 123.3°$ and represents the mean location of all intersection points. Since the latter points are near to each other, one can consider that all lines intersect at the same point (ε_0, θ_0) and combine all the straight lines presented in Fig. 6 into a single line given by the equation

$$\theta - \theta_0 = -C\delta^{-\nu}\left(\varepsilon - \varepsilon_0\right), \tag{13}$$

the form of which coincides with that for nanodrops [Eq. (1)]. In Eq. (13), θ_0, ε_0, and C are constants provided above and the parameter δ contains the temperature. The parameters θ_0, ε_0, and C are slightly different from those for nanodrops. For the latter, under the same conditions, $\nu = 2.64$, $\theta_0 = 112.2°$, $\varepsilon_0 = 0.86$, and $C = 232.7°$.[1] Note that the relative difference between the values of θ obtained through computations of the surface tensions based on DFT and those obtained from Eq. (13) does not exceed 4% for the range of $0.6< \varepsilon <1.4$ and 1% for the range of $0.6<\varepsilon<1.1$. Under the employed conditions (constant σ_{fs} and ρ_s), Eqs. (11)–(13) provide a complete description of the fluid-solid pairs for which the parameters δ and ε are in the ranges noted above.

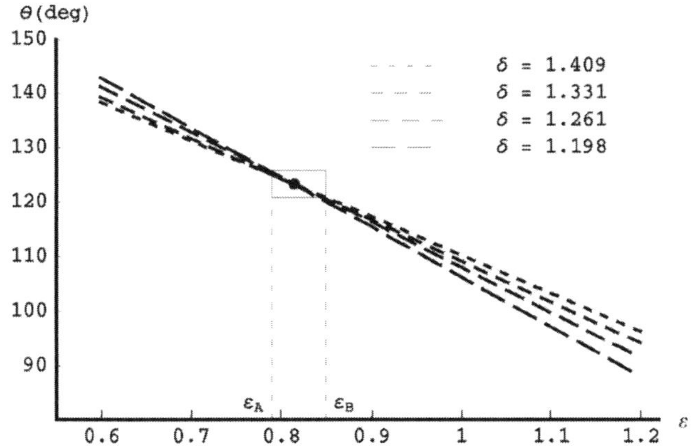

FIG. 6. ε-dependence of the macroscopic contact angle for $\delta = 1.409$, 1.331, 1.261, and 1.198 ($T = 85$, 90, 95, and 100 K, respectively), and $\sigma = 1.028$ ($\sigma_{fs} = 3.5$ Å). The bold point has the coordinates $\varepsilon_0 = 0.82$ and $\theta_0 = 123.3°$ which are equal to the mean values of the coordinates of all intersection points of the presented straight lines. All those points are located inside the drawn rectangular area. $\varepsilon_A = 0.79$, $\varepsilon_B = 0.85$.

TABLE 2
Parameters of Eq. (11) for various σ and temperatures.

$\sigma = \sigma_{fs}/\sigma_{ff}$	σ_{fs} (Å)	T (K)	k (deg)	b (deg)
		85	–88.1	179.2
1.095	3.727	90	–94.9	183.2
		95	–105.6	190.3
		100	–115.2	195.5
		85	–56.3	181.1
0.969	3.3	90	–60.2	185.0
		95	–66.4	192.0
		100	–71.6	197.1
		85	–45.8	183.4
0.910	3.1	90	–49.0	187.8
		95	–54.6	196.3
		100	–59.4	202.6

2. Temperature dependence of the contact angle at constant ε

Using Eq. (13) one can easily find the temperature dependence of the contact angle for selected values of ε.

For $\varepsilon > \varepsilon_0$, the contact angle decreases and for $\varepsilon < \varepsilon_0$ it increases with increasing temperature. The rate $d\theta/dT$ of change in the contact angle with temperature is given by

$$\frac{d\theta}{dT} = \frac{\nu k_B}{\epsilon_{ff}} C\left(\varepsilon - \varepsilon_0\right)\left(\frac{k_B T}{\epsilon_{ff}}\right)^{\nu-1}. \tag{14}$$

Because $\nu > 1$, the absolute value of $d\theta/dT$ increases with increasing temperature. For $\varepsilon = \varepsilon_0$ the contact angle does not depend on temperature and is always equal to θ_0.

Taking $T_0 = 92.5$ K as a reference temperature, the variation ΔT of the temperature considered in the present paper can be considered to be small ($\Delta T/T_0 \ll 1$); therefore, Eq. (13) can be linearized with respect to ΔT

$$\theta(T) = \theta(T_0) + \alpha(T_0)\Delta T, \tag{15}$$

where

$$\theta(T_0) = \theta_0 - C\left(\varepsilon - \varepsilon_0\right)\left(\frac{k_B T_0}{\epsilon_{ff}}\right)^{\nu}, \tag{16}$$

and

$$\alpha(T_0) = -\frac{C_\nu}{T_0}\left(\varepsilon - \varepsilon_0\right)\left(\frac{k_B T_0}{\epsilon_{ff}}\right)^{\nu}. \tag{17}$$

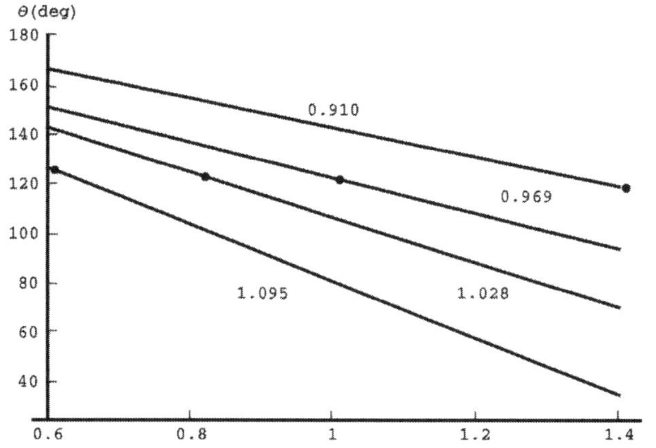

FIG. 7. The ε-dependence of macroscopic contact angle provided by Eq. (13) for $\delta \equiv \epsilon_{ff}/k_BT = 1.198$ ($T = 100$ K) and various values of $\sigma \equiv \sigma_{fs}/\sigma_{ff}$ indicated on each line. The parameters of Eq. (13) are listed in Table 3. The bold points on each of the lines have the coordinates (ε_0, θ_0).

For the case considered, $\alpha(T_0) = -1.40(\varepsilon - \varepsilon_0)$ The above analysis of the temperature dependence of the contact angle involves the existence of a unique intersection point (at $\varepsilon = \varepsilon_0$) of all straight lines that represent the ε-dependence of θ at various temperatures. Even if this is not absolutely true, the ranges of ε in which θ monotonously increases (decreases) with the temperature are still present. However, they are shifted to $\varepsilon<\varepsilon_A$ ($\varepsilon>\varepsilon_B$), respectively (see Fig. 6).

B. Contact Angles for $\sigma \equiv \sigma_{fs}/\sigma_{ff}$ = 1.095, 0.969, 0.910

The results for the contact angle obtained for σ_{fs} = 3.727 Å, σ_{fs} = 3.3 Å, and σ_{fs} = 3.1 Å (σ = 1.095, 0.969, and 0.910, respectively) coincide qualitatively with those obtained in Sec. IV A for σ_{fs} = 3.5 Å (σ = 1.028). In the range $0.6 < \varepsilon < 1.4$, θ versus ε can be represented by straight lines, Eq. (11). The values of the parameters k and b are listed in Table 2.

For each σ, the straight lines at various temperatures can be combined in a single line of the form of Eq. (13) with the parameters v, C, and ε_0 dependent on σ. Such single lines are plotted in Fig. 7 for four values of σ, and the values of the parameters v, C, θ_0, and ε_0 are listed in Table 3. An important feature is that the values of θ_0 are almost the same for all considered values of σ. Assuming that for other values of σ the parameters v, C, and ε_0 can be found by interpolating between the obtained results, one can consider that Eq. (13) provides a general curve describing the contact angle for fluids and solids which have intermolecular interactions of the Lennard-Jones type.

V. DISCUSSION AND CONCLUSION

In this paper, the theoretical results obtained previously in Ref. 1 for the contact angle of nanodrops on smooth solid surfaces are confirmed for macroscopic drops on such surfaces. As previously, the fluid-fluid and fluid-solid interactions were selected in the form of Lennard-Jones potential with hard core repulsion. For a nanodrop, the contact angle was extracted directly from the two-dimensional density distribution of a fluid confined in a closed slit.[1] For macroscopic drops, the surface tensions, γ_{sl}, and γ_{sv} were determined from one-dimensional density distributions of a fluid in an open slit in contact with a reservoir of bulk fluid under liquid-vapor

TABLE 3
Parameters of Eq. (13) for various values of $\sigma = \sigma_{fs}/\sigma_{ff}$.

σ	v	θ_0 (deg)	ε_0	C (deg)
1.095	1.68	126.0	0.61	155.6
1.028	1.65	123.3	0.82	122.3
0.969	1.54	122.2	1.01	93.8
0.910	1.64	119.0	1.41	79.6

coexistence conditions. The third surface tension, γ_{lv}, was calculated on the basis of an asymmetrical fluid density distribution in a closed slit. The contact angle of the macroscopic drop was then calculated using the Young equation.

The main conclusion is that for macroscopic drops as well as for nanodrops the contact angle, characterizing the wetting properties of a fluid on a solid surface, has some general characteristics with respect to its dependence on temperature and the parameters that characterize the fluid-solid intermolecular interactions, assumed to be of the Lennard-Jones type. At various temperatures, the macroscopic contact angle has a linear dependence on the dimensionless energy parameter $\varepsilon = \epsilon_{fs}/\epsilon_{ff}$, where ϵ_{fs} and ϵ_{ff} are the fluid-solid and fluid-fluid energy parameters in the Lennard-Jones interaction potentials, respectively. The straight lines representing these dependencies at various temperatures intersect almost in a single point (ε_0, θ_0) where ε_0 depends on σ_{fs} and θ_0 is almost a universal constant. We use the word "almost" because θ_0 varies between 122.5° ± 3.5° (see Table 3). The contact angle θ_0 for $\varepsilon = \varepsilon_0$ is independent of temperature. For $\varepsilon < \varepsilon_0$, the contact angle monotonously increases and for $\varepsilon > \varepsilon_0$ monotonously decreases with increasing temperature. These results can qualitatively explain the different temperature behaviors of the contact angle observed experimentally.[12–16]

The temperature independent value θ_0 of the contact angle corresponding to ε_0 is independent of $\sigma = \sigma_{fs}/\sigma_{ff}$ and is therefore a general characteristic of a macroscopic drop on a solid surface. The existence of such a point, if confirmed, can lead to additional classification of materials with respect to the temperature dependence of the contact angle (or, more general, with respect to the temperature dependence of wetting properties) complementing the usual classification of materials into hydrophobic and hydrophilic. For $\theta < \theta_0$, the interactions of the liquid with the solid are stronger ($\varepsilon > \varepsilon_0$); whereas for $\theta > \theta_0$ the interactions are weaker ($\varepsilon < \varepsilon_0$).

As well known, the short range interactions such a polar ones can affect the contact angle including the angle θ_0. It will be interesting to examine this problem in the future.

APPENDIX: DERIVATION OF THE EULER-LAGRANGE EQUATION FOR THE FLUID DENSITY DISTRIBUTION

For an open slit between identical walls in contact with a bulk fluid at fixed chemical potential μ, the Euler-Lagrange equation for fluid density distribution $\rho(h)$ can be obtained through the minimization of the grand canonical potential of the system

$$\Omega = F_H\left[\rho(h)\right] - \mu N, \tag{A1}$$

where $F[\rho(h)]$ is the Helmholtz free energy functional, μ is the chemical potential of the bulk fluid and N is the number of molecules in the system, which depends on μ. (All quantities are considered per unit area of one of the walls). For the Helmholtz free energy one can write the expression

$$F_H\left[\rho(h)\right] = F_{\rm id}\left[\rho(h)\right] + F_{\rm ex}\left[\rho(h)\right] + \frac{\epsilon_{ff}}{2}\iint d\mathbf{r}d\mathbf{r}'\rho(h)\rho(h')\bar{\phi}_{ff}\left(\left|\mathbf{r}-\mathbf{r}'\right|\right) + \epsilon_{fs}\int dh\rho(h)\bar{U}_{fs}(h), \tag{A2}$$

where $F_{\rm id}[\rho(h)]$ and $F_{\rm ex}[\rho(h)]$ are the ideal gas and excess contributions, respectively, of a reference hard sphere fluid, $\bar{\phi}_{ff}(|\mathbf{r}-\mathbf{r}'|)$ is given by Eq. (3), $\mathbf{r} = \{x, y, h\}$ determines the location of a fluid molecule, and $\bar{U}_{fs}(h)$ is the dimensionless fluid-solid intermolecular potential Eq. (5). The third term in Eq. (A2) represents the contribution of the fluid-fluid interaction treated in the mean-field approximation. In explicit form

$$F_{\rm id}\left[\rho(h)\right] = k_B T\int dh\rho(h)\left\{\log\left[\Lambda^3\rho(h)\right]-1\right\}, \tag{A3}$$

$$F_{\rm ex}\left[\rho(h)\right] = \int dh\rho(h)\Delta\Psi_{hs}(h), \tag{A4}$$

where $\Lambda = h_P/(2\pi m k_B T)^{1/2}$ is the thermal de Broglie wavelength, h_P and k_B are the Planck and Boltzmann constants, respectively, m is the mass of a fluid molecule,

$$\Delta\Psi_{hs}(h) = k_B T\eta_{\bar{\rho}}\frac{4-3\eta_{\bar{\rho}}}{\left(1-\eta_{\bar{\rho}}\right)^2}. \tag{A5}$$

In Eq. (A5), $\eta_{\bar{\rho}} = 1/6\pi\bar{\rho}(h)\sigma_{ff}^3$ is the packing fraction of the fluid molecules, σ_{ff} is the fluid hard core diameter, and $\bar{\rho}(h)$ is the smoothed density defined as

$$\bar{\rho}(h) = \int d\mathbf{r}'\rho(h')W\left(\left|\mathbf{r}-\mathbf{r}'\right|\right). \tag{A6}$$

The weighting function $W(|\mathbf{r}-\mathbf{r}'|)$ is selected in the form[9]

$$W\left(\left|\mathbf{r}-\mathbf{r}'\right|\right) = \begin{cases} \dfrac{3}{\pi\sigma_{ff}^3}\left(- \dfrac{r}{\sigma_{ff}}\right), & r \le \sigma_{ff} \\ 0, & r > \sigma_{ff}, \end{cases}$$

where $r = |\mathbf{r}-\mathbf{r}'|$.

The Euler—Lagrange equation for the fluid density distribution $\rho(h)$, obtained by minimizing the Helmholtz free energy, has the following form:

$$\log\left[\Lambda^3\rho(h)\right] + Q(h) + \frac{\epsilon_{ff}}{k_B T}U_{ff}(h) + \frac{\epsilon_{fs}}{k_B T}\bar{U}_{fs}(h) = \frac{\mu}{k_B T}. \tag{A7}$$

The functions $Q(h)$, $U_{ff}(h)$, and $\bar{U}_{fs}(h)$ represent a reference hard spheres system, fluid-fluid interactions, and fluid-solid interactions, respectively. All these functions are independent of temperature and of the parameters ϵ_{ff} and ϵ_{fs} and are given by the expressions

$$Q(h) = -\left[\Delta\Psi_{hs}(h) + \overline{\Delta\Psi'_{hs}(h)}\right], \tag{A8}$$

Where

$$\overline{\Delta\Psi'_{hs}}(h)=\int dh'\rho(h')W_y(|h-h'|)\frac{\partial}{\partial\bar{\rho}}\Delta\Psi_{hs}(\bar{\rho})\Big|_{\bar{\rho}=\bar{\rho}(h')}. \tag{A9}$$

$W_y(|h-h'|)$ is obtained by integrating the weighted function $W(|\mathbf{r}-\mathbf{r}'|)$ with respect to the x and y directions.

The function $U_{ff}(h)$, which represents in Eq. (A7) the fluid-fluid interactions is given by

$$U_{ff}(h)=\int dh'\rho(h')\phi_{ff,h}(|h-h'|), \tag{A10}$$

where $\phi_{ff,h}(|h-h'|)$ is obtained by integrating the potential $\phi_{ff,h}(|\mathbf{r}-\mathbf{r}'|)$ with respect to the x and y directions.

For a closed slit the Euler-Lagrange equation has the same form as Eq. (A7). However, in this case μ is not the chemical potential but a Lagrange multiplier which should be determined from the constraint of fixed average density (or fixed N) in the slit

$$\rho_{\text{av}}=\frac{1}{V}\int_V d\mathbf{r}\rho(\mathbf{r}), \tag{A11}$$

where V is the volume of the system. This constraint leads to the following expression for μ:

$$\mu=-k_BT\log\left\{\frac{1}{\rho_{av}V\Lambda^3}\int_V d\mathbf{r}\exp\left[-Q(h)-\frac{\epsilon_{ff}}{k_BT}U_{ff}(h)-\frac{\epsilon_{fs}}{k_BT}\bar{U}_{fs}(h)\right]\right\}. \tag{A12}$$

By eliminating μ between Eqs. (A7) and (A12), one obtains an integral equation for the FDD $\rho(h)$, which can be solved by iteration. The details of the iteration procedure can be found in Ref. 4.

Note that in Eqs. (A2)-(A4) the integration is performed over h from $h_1=0$ to $h_2=L$ for the open slit, and from $h-h_l$ to $h=h_\upsilon$ for the closed slit. In the latter case h_l and $h_l=h_\upsilon$ provide the boundaries for the liquidlike and vaporlike bulk fluids, respectively (see Fig. 3).

REFERENCES

[1] G. O. Berim and E. Ruckenstein, J. Chem. Phys. **130**, 044709 (2009) (Section 4.1 of this volume).
[2] M. J. de Ruijter, T. D. Blake, and J. De Coninck, Langmuir **15**, 7836 (1999).
[3] N. Giovambattista, P. G. Debenedetti, and P. J. Rossky, J. Phys. Chem. B **111**, 9581 (2007).
[4] G. O. Berim and E. Ruckenstein, J. Chem. Phys. **129**, 014708 (2008); **129**, 114709 (2008) (Sections 4.4 and 4.11 of this volume).
[5] E. van Giessen, D. J. Bukman, and B. Widom, J. Colloid Interface Sci. **192**, 257 (1997).
[6] F. Ancilotto, F. Faccin, and F. Toigo, Phys. Rev. B **62**, 17035 (2000). [7]R. Evans and P. Tarazona, Phys. Rev. A **28**, 1864 (1983).
[7] P. Tarazona, Phys. Rev. A **31**, 2672 (1985).
[8] R. H. Nilson and S. K. Griffiths, J. Chem. Phys. **111**, 4281 (1999).
[9] G. O. Berim and E.Ruckenstein, J. Phys. Chem. B **111**, 2514 (2007) (Section 2.1 of this volume).
[10] G. O. Berim and E.Ruckenstein, J. Chem. Phys. **126**, 124503 (2007) (Section 2.2 of this volume).
[11] E. Rolley and C. J. Guthmann, Low Temp. Phys. **108**, 1 (1997).
[12] C. J. Budziak, E. I. Vargha-Butler, and A. W. Neumann, J. Appl. Polym. Sci. **42**, 1959 (1991)
[13] A. Suzuki and Y. Koboki, Jpn. J. Appl. Phys., Part 1 **38**, 2910 (1999).
[14] H. Y. She and B. E.Sleep, Water Resour. Res. **34**, 2587, DOI: 10.1029/98WR01199 (1998).
[15] J. D. Bernardin, I. Mudawar, C. B. Walsh, and E. I. Franses, Int. J. Heat Mass Transfer **40**, 1017 (1997).

4 Nanodrops on the Solid Surface

Contact Angle, Sticking Force

Eli Ruckenstein and Gersh Berim

INTRODUCTION TO CHAPTER 4

In this chapter, a liquid drop on a smooth or rough planar solid surface was examined on the basis of a microscopic nonlocal density functional theory in canonical ensemble. The main idea was to consider drop-like nonuniform density distribution in a closed volume of fluid in contact with the solid and from this distribution identify and analyze the drop profile. The obtained results for the contact angle θ which the drop profile makes with the solid surface are compared with the predictions of classical theories for macroscopic drops and similarities and differences between them are analyzed.

In the first part of the chapter, the nanodroplets on the smooth planar solid surface were considered. A simple linear dependence of the contact angle on the fluid-solid interaction parameter ε_{fs} was found for various temperatures, hard core fluid-solid parameter σ_{fs}, and average fluid density of the system (Sec. 4.1) and a simple expression is suggested which represents all the results in a unified form. The results obtained in Sec. 4.1 for nanodrops and in Sec. 3.7 for macrodrops are summarized in Sec. 4.2. In Sec. 4.3, the theory is extended to a magnetic fluid in contact with a solid surface. The dependence of the contact angle on the parameters involved in the magnetic interactions between the molecules of fluid and between the molecules of fluid and an external magnetic field is calculated. The obtained results are in qualitative agreement with the experimental data on the contact angle of magnetic drops on a solid surface available in the literature

In the second part of the chapter (Secs 4.4–4.11), the nanodrops on the rough surfaces are examined. Considerable attention is given to the comparison of the obtained results with the predictions of classical theory of wetting. The latter theory predicts two major regimes, Cassie-Baxter regime and Wenzel regime. In Cassie-Baxter regime, the contact angle θ increases with increasing roughness for both hydrophobic and hydrophilic surfaces. In the Wenzel regime, θ increases with increasing roughness for hydrophobic surfaces and θ decreases for hydrophilic surfaces. The results obtained in this chapter revealed a lot of cases in which the dependence of the contact angle on roughness for nanodrops does not match that of classical theory. For example, it is shown that (i) in contrast to the commonly used Wenzel formula, even an extremely small roughness can considerably increase the contact angle for the drop on hydrophilic surfaces (Secs. 4.4, 4.5), (ii) For hydrophobic substrates the trend of the contact angle to increase with increasing roughness is not universally valid, but depends on the fluid–pillar interactions, pillar height, interpillar distance, as well as on the size of the drop (Sec. 4.6), (iii) the conclusion of classical theory that only the interfacial area near the leading edges of the drop on physically smooth but chemically rough solid surfaces affects the contact angle and that most of the contact area has no effect is not always valid for nanodrops and cases are identified in which the above conclusion is not valid (Sec. 4.7).

In Secs. 4.8 and 4.9 the influence on the nanodrop of the so called hidden roughness is examined. The hidden roughness occurs when the rough surface is covered either by a solid material having one smooth surface (Sec. 4.8) or by a lubricating liquid (Sec.4.9). For both cases, the contact angles at the leading edges of the drop are determined as functions of drop size and horizontal perturbative force. The main conclusion is that the hidden roughness has a similar effect on the nanodrop features as the traditionally considered physical and chemical roughness.

The specific feature of the nanodrop is that it can be located inside the microscopic defects on the solid surface, e.g. inside a nanocavity. In Sec. 4.10, the dependence of the contact angles of nanodrops in nanocavities on their sizes is calculated. It is argued that this dependence might affect strongly, for instance, the rate of heterogeneous nucleation on rough surfaces, which is usually calculated under the assumption of constant contact angle.

Another important characteristic of a drop on a solid rough surface is the sticking force which opposes the drop motion along an inclined surface, In Sec. 4.11, a nanodrop on a vertical rough solid surface is examined in the presence of gravity. It is shown that the macroscopically derived equation for a drop in equilibrium on an inclined surface is also applicable to nanodrops. The liquid-vapor surface tension involved in this equation was calculated for various specific cases, and the values obtained are of the same order of magnitude as those obtained in macroscopic experiments.

In Sec. 4.12, two approaches are examined, which can be used to determine the drop profile from the fluid density distributions (FDDs) obtained on the basis of microscopic theories. The first approach is based on the sharp–kink interface approximation in which the density of the liquid inside and the density of the vapor outside the drop are constant with the exception of the surface layer of the drop where the density is different from the above ones. In this case, the drop profile was calculated by minimizing the total potential energy of the system. The second approach is based on a nonuniform FDD obtained by either the density functional theory or molecular dynamics simulations. Several procedures are discussed to determine the drop profile from such an FDD, which does not contain sharp interfaces. The procedure which provides a drop profile which is more reasonable than the other ones is identified.

In the final section of this chapter (Sec. 4.13), the results of Ch.3 and Secs 4.1, 4.4, and 4.7 are summarized.

4.1 Simple Expression for the Dependence of the Nanodrop Contact Angle on Liquid-Solid Interactions and Temperature*

Gersh Berim and Eli Ruckenstein†
Department of Chemical and Biological Engineering, State University of New York at Buffalo, Buffalo, New York 14260, USA

Corresponding Author

† E-mail: feaeliru@buffalo.edu.
Phone: (716)-645-1179, Fax: (716)-645-3822.

(Received 7 August 2008; accepted 17 December 2008; published online 29 January 2009)

I. INTRODUCTION

The knowledge of the surface properties of various materials and their temperature dependence allows one to utilize them in a more efficient way. One of the methods frequently used to obtain information about the surface properties is via the contact angle which a drop makes with a solid surface. It can be measured using various methods[1–4] or can be calculated.[5–10] In the latter case, one calculates theoretically the surface tensions γ_{lv}, γ_{ls}, and γ_{vs} of the liquid-vapor, liquid-solid, and vapor solid interfaces, respectively, considering planar interfaces. Further, the Young equation

$$\gamma_{lv}\cos\theta = \gamma_{sv} - \gamma_{sl} \tag{1}$$

is used to obtain the contact angle. Wetting angles for nanodroplets can be also obtained directly from molecular dynamics simulations.[11–14] The contact angle is a result of a balance between the fluid-fluid and fluid-solid interactions. When the latter is sufficiently large, the liquid wets better the surface and the contact angle tends to become smaller (or even zero). The existing experimental data provide examples of different temperature behaviors of the contact angle for different materials. For example, an almost linear decrease in θ, with $d\theta/dT = -0.13$ deg/K, was reported in Ref. 5 for water on naphthalene. A nonlinear decrease in the contact angle of water on polymer surfaces with increasing temperature was observed in Ref. 15. The opposite behavior, an increase in θ with temperature was found in Ref. 16 for water on perchloroethylene, in Ref. 3 for glycerol, ethylene glycol, and diethylene glycol on elastomers, and in Ref. 9 for mixtures of ^{4}He and ^{3}He on Cs.

* *The Journal of Chemical Physics* 130, 044709 (2009). Republished with permission.

There are several theoretical approaches which can provide the dependence of the contact angle, which a drop makes with a solid surface, on temperature. The simplest one is based on the Young equation, Eq. (1), and can be used when theoretical methods of calculation of the surface tensions are available.

Another approach is based on the Young—Laplace equation for the drop profile,[17,18] which for a cylindrical drop (drop infinite in one of the direction on the surface), has the simple form

$$\frac{\gamma_{\mathrm{lv}} h_{xx}}{\left(1+h_x^2\right)^{3/2}} + \Pi\left(h\right) = -p_c, \tag{2}$$

where $h \equiv h(x)$ is the drop profile, $\Pi(h)$ is the disjoining pressure, p_c is the capillary pressure, and $h_x \equiv dh/dx$. The quantities $\Pi(h)$, p_c, and γ_{lv} depend on temperature and this dependence must be known to find the dependence of θ on temperature.

The third, the most general one, is the microscopic approach. It was developed in the framework of the grand canonical density functional theory (DFT) (Refs. 19–21) and provided a qualitative and often quantitative description of various phase transitions (wetting, prewetting, layering, etc.), as well as the fluid density distribution (FDD) near the solid surface. In a modified form,[22–26] which involved a canonical ensemble instead of the grand canonical one, DFT was used to examine the droplike aggregates on solids, including the drop profile and the contact angle which were obtained from the FDD.[26]

Most of the above cited papers were devoted to specific fluids on specific solids but no systematic theoretical analysis of the dependence of the contact angle on the parameters of the model was performed. In our previous papers[26] we calculated the contact angles of a nanodrop of argon on various substrates for various values of the energy parameter ϵ_{fs} of the Lennard-Jones fluid-solid interactions at a single temperature. Surprisingly, a simple, almost linear, dependence of θ on ϵ_{fs} was found. This observation stimulated us to examine the dependence of the wetting angle on smooth solid surfaces on the parameters involved in the Lennard-Jones potentials for fluid-fluid and fluid-solid interactions and on temperature. First, the fluid density distribution was obtained by minimizing the Helmholtz free energy of the system and using the approach developed in Refs. 22 and 26. Second, the profile of the drop was extracted from the obtained FDD by assuming an equimolar dividing surface. Finally, the contact angle θ which this profile makes with the solid surface was determined for various values of the parameters involved in the interaction potentials and various temperatures. Among the results obtained was the observation that there is a critical value ϵ_{fs}^0 of ϵ_{fs} that divides the materials for which θ increases from those in which θ decreases with increasing temperature. For $\epsilon_{\mathrm{fs}} = \epsilon_{\mathrm{fs}}^0$ the corresponding wetting angle θ_0 becomes independent of temperature. The critical value θ_0 of θ divides the surfaces for which $\epsilon_{\mathrm{fs}} > \epsilon_{\mathrm{fs}}^0$ from those for which $\epsilon_{\mathrm{fs}} > \epsilon_{\mathrm{fs}}^0$.

Because the largest size of the system for which computations could be carried out during reasonable times was restricted to several tenths of nanometers, the calculations were performed only for nanodrops.

II. BACKGROUND

The system under consideration has the finite dimensions L_x and L_h in the x-and h-directions, respectively, and infinite dimension in the y direction, and contains a one-component fluid of fixed average density ρ_{av} (see Fig. 1). Considering that no symmetry breaking occurs in the y-direction, one can assume that the FDD $\rho(\mathbf{r})$ in such a system is independent of y, i.e., $\rho(\mathbf{r}) \equiv \rho(x, h)$. The periodic boundary condition commonly used in Monte Carlo and molecular dynamics simulations of confined fluids[27–30] is employed in the x-direction and the upper boundary of the box is treated as a hard wall (without attractive interaction with the fluid molecules).

The interaction between fluid molecules is described by the Lennard-Jones potential $\phi_{\mathrm{ff}}\left(|\mathbf{r}-\mathbf{r}'|\right) \equiv \epsilon_{\mathrm{ff}}\bar{\phi}_{\mathrm{ff}}\left(|\mathbf{r}-\mathbf{r}'|\right)$, where the dimensionless $\bar{\phi}_{\mathrm{ff}}\left(|\mathbf{r}-\mathbf{r}'|\right)$ is given by

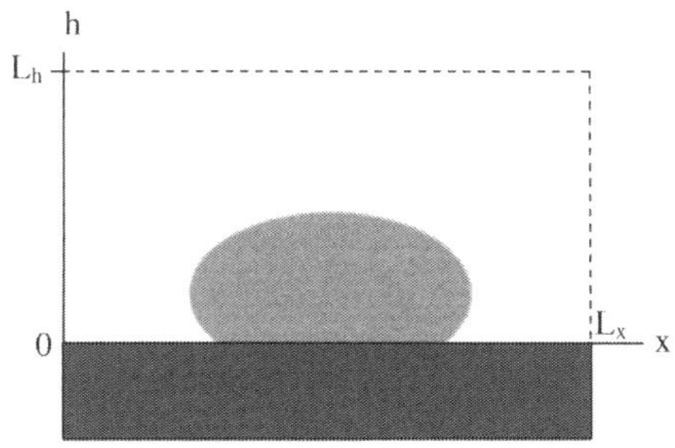

FIG. 1. Schematic representation of a cylindrical drop (drop uniform in the y-direction which is perpendicular to the plane of the figure) on a smooth surface.

$$\bar{\phi}_{ff}\left(\left|\mathbf{r}-\mathbf{r}'\right|\right)=\begin{cases}4\left[\left(\frac{\sigma_{ff}}{r}\right)^{12}-\left(\frac{\sigma_{ff}}{r}\right)^{6}\right], & r\geq\sigma_{ff}\\ \infty, & r<\sigma_{ff},\end{cases} \tag{3}$$

where the coordinates $\mathbf{r}$ and $\mathbf{r}'$ provide the locations of the interacting fluid molecules, $r = |\mathbf{r} - \mathbf{r}'|$, and σ_{ff} and ϵ_{ff} are the fluid hard core diameter and the energy parameter, respectively.

The same kind of potential is selected for the fluid-solid interactions (the energy parameter ϵ_{ff} and the hard core diameter σ_{ff} in Eq. (3) being replaced by ϵ_{fs} and σ_{fs}, respectively, r being the distance between a molecule of the fluid and that of the solid). Under the constraint of constant average density, which is equivalent to the constraint of constant number of molecules in the system, the Helmholtz free energy was minimized with respect to the fluid density distribution. The latter free energy, $F[\rho(\mathbf{r})]$, can be written in the form

$$\begin{aligned}F\left[\rho(\mathbf{r})\right]&=F_{id}\left[\rho(\mathbf{r})\right]+F_{ex}\left[\rho(\mathbf{r})\right]\\&+\frac{\epsilon_{ff}}{2}\int\int d\mathbf{r}d\mathbf{r}'\rho(\mathbf{r})\rho(\mathbf{r}')\bar{\phi}_{ff}\left(\left|\mathbf{r}-\mathbf{r}'\right|\right)\\&+\epsilon_{fs}\int d\mathbf{r}\rho(\mathbf{r})U_{fs}(\mathbf{r}),\end{aligned} \tag{4}$$

where $F_{id}[\rho(\mathbf{r})]$ and $F_{ex}[\rho(\mathbf{r})]$ are the ideal gas and excess contributions, respectively, of a reference hard sphere fluid and $U_{fs}(\mathbf{r})$ is the dimensionless fluid-solid intermolecular potential. The third term in Eq. (4) represents the contribution of the fluid-fluid interaction treated in the mean-field approximation. Explicit expressions for $F_{id}[\rho(\mathbf{r})]$ and $F_{ex}[\rho(\mathbf{r})]$ and $U_{fs}(\mathbf{r})$ can be found in the Appendix.

The Euler—Lagrange equation for the fluid density distribution $\rho(x, h)$, obtained by minimizing the Helmholtz free energy, has the following form:

$$\begin{aligned}\log\left[\Lambda^3\rho(x,h)\right]+Q(x,h)+\frac{\epsilon_{ff}}{k_BT}U_{ff}(x,h)+\frac{\epsilon_{fs}}{k_BT}U_{fs}(h)\\=\frac{\lambda}{k_BT},\end{aligned} \tag{5}$$

where $\Lambda = h_P/(2\pi m k_B T)^{1/2}$ is the thermal de Broglie wavelength, k_B and h_P are the Boltzmann and Planck constants, respectively, T is the absolute temperature, m is the mass of a fluid molecule, and λ is a Lagrange multiplier arising because of the constraint of fixed average density of the fluid. The functions $Q(x, h)$, $U_{ff}(x, h)$, and $U_{fs}(h)$ represent the reference hard spheres system, fluid-fluid interactions, and fluid-solid interactions, respectively. All these functions are independent of temperature and of the parameters ϵ_{ff} and ϵ_{fs}. Their explicit forms are given in the Appendix.

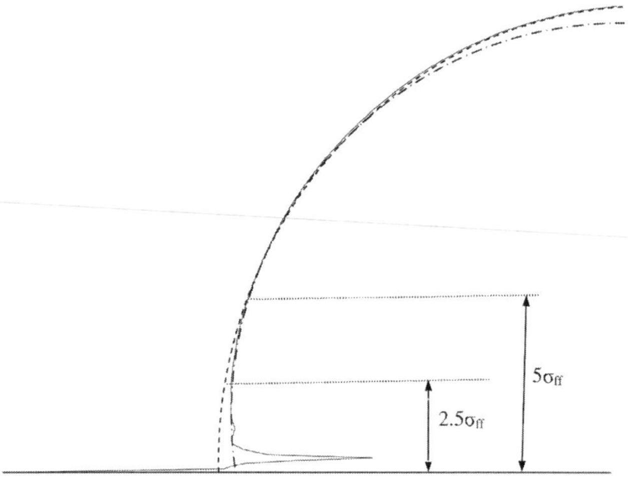

FIG. 2. Example of a drop profile (solid line, only the left-hand side part of the profile is presented) extracted from the fluid density distribution using the procedure described in the text. The profile is approximated by two circles. The first one (dashed line) has the best fit with the upper part ($h > 5\sigma_{ff}$) of the profile. The second circle (dashed-dotted line) has the best fit with the profile in the range $2.5\sigma_{ff} < h < 5\sigma_{ff}$.

The constraint of fixed average density has the form

$$\rho_{av} = \frac{1}{V}\int_V d\mathbf{r}\rho(\mathbf{r}), \tag{6}$$

where V is the volume of the system and leads to the following expression for λ

$$\lambda = -k_B T \log\left\{\frac{1}{\rho_{av} V \Lambda^3}\int_V d\mathbf{r} \times \exp\left[-\mathcal{Q}(x,h) - \frac{\epsilon_{ff}}{k_B T}U_{ff}(x,h) - \frac{\epsilon_{ff}}{k_B T}U_{fs}(h)\right]\right\}. \tag{7}$$

By eliminating λ between Eqs. (5) and (7), one obtains an integral equation for the FDD $\rho(x, h)$, which was solved by iteration and provided FDD $\rho(x, h)$. The details of the iteration procedure can be found in Ref. 26.

To extract the drop profile from the obtained FDD and determine the contact angle, we use a procedure similar to that employed by us in Ref. 26. This procedure consists in determining the drop profile $h = h(x)$ inside the vapor-liquid interface as the line passing through the points where the local density $\rho(x, h)$ has a constant value ρ_{div} and finding the angle which this profile makes with the solid surface.[14,31] The usual way to calculate ρ_{div} is to consider that it is provided by an equimolar dividing surface at some distance h from the solid surface (see Refs. 14 and 31). In the present calculations this distance was $5\sigma_{ff}$. It should be noted that for $h > 2.5\sigma_{ff}$ ρ_{div} is almost independent of h. To avoid the uncertainties due to the strong fluid density oscillations near the solid surface, only that part of the obtained profile with $h > 2.5\sigma_{ff}$ was taken into account and approximated by two circles (see Fig. 2). The first (the dashed line in Fig. 2) has the best fit with the upper part ($h > 5\sigma_{ff}$) of the drop profile, whereas the second (dashed-dotted line of Fig. 2) has the best fit with the lower part ($2.5\sigma_{ff} < h < 5\sigma_{ff}$) of the profile.

These circles, when extended up to the solid surface, make with that surface the angles θ_u and θ_l, respectively, which can be easily determined and can be considered as two complementary contact angles. The definition of the contact angle θ_u is similar to that used in the goniometric measurements of the contact angles of macroscopic drops.[1] The angle θ_l is similar to the microcontact angle that is formed near a solid surface by macroscopic drops.[32] Below, the dependence of both angles with respect to the liquid-solid interactions and temperature is examined. The procedure usually used for macrodrops, which involves the Young equation [Eq. (1)] and the macroscopic surface tensions for planar interfaces is not applicable to nanodrops because the thicknesses of the interfaces are comparable in these cases with the size of the drop.

III. RESULTS

Because the main purpose of the present paper is to examine the dependence of the contact angle of a nanodrop on a solid surface on the parameters of the intermolecular interactions (σ_{ff}, ϵ_{ff}, σ_{fs}, and, ϵ_{fs}), it is not necessary to select those parameters for specific fluids and substrates. However, to have a specific reference fluid, we select σ_{ff} and ϵ_{ff} as for argon ($\epsilon_{ff}/k_B = 119.76$ K, $\sigma_{ff} = 3.405$ Å) and take the density of the solid equal to that of solid carbon dioxide ($\rho_s = 1.91 \times 10^{28}$ m^{-3}).[20] We selected σ_{ff} as the unit of length and introduced two independent dimensionless parameters, $\sigma = \sigma_{fs}/\sigma_{ff}$ and $\varepsilon = \epsilon_{fs}/\epsilon_{ff}$ characterizing the substrate and fluid, and the parameter $\delta = \epsilon_{ff}/k_B T$ characterizing the effect of temperature. The calculations of the contact angles were performed for four temperatures, $T = 85$, 90, 95, and 100 K ($\delta = 1.409$, 1.331, 1.261, and 1.198, respectively) and for ε varying from 0.3 to 1.6. If not mentioned otherwise, the width of the system L_x was taken $L_x = 50\sigma_{ff}$ and the height $L_h = 30\sigma_{ff} = 2\sigma_{fs}$. The dimensionless average density $\rho^*_{av} = \rho_{av}\sigma^3_{ff}$ was considered in the range $0.1 \le \rho^*_{av} \le 0.25$. In Secs. IIIA 1, IIIA 2, and IIIB the case $\rho^*_{av} = 0.1$ will be presented in detail. The results for other values of ρ^*_{av} will be outlined in Sec. III C.

A. CONTACT ANGLE FOR $\sigma_{fs} = 3.5$ Å ($\sigma = 1.028$)

1. ε-dependence at various temperatures

In Figs. 3(a) and 3(c) typical fluid density distributions for a relatively weak fluid-solid interaction ($\varepsilon = 0.6$) and for a stronger one ($\varepsilon = 1.4$), obtained by solving Eq. (5), are presented schematically for $T = 85$ K together with the corresponding drop profiles [Figs. 3(b) and 3(d)] calculated as described in Sec. II. The circular extension of the upper part of the first profile provides the angle $\theta_u \simeq 137°$ with the solid surface and the second profile provides $\theta_u \simeq 60°$. Similar extensions of the lower parts of the profile provide the angles $\theta_l \simeq 147°$ and $\theta_l \simeq 61°$, respectively. A typical dependence of θ_u on ε is shown in Fig. 4 for $T = 85$ K ($\delta = 1.409$). This dependence is almost linear in the range

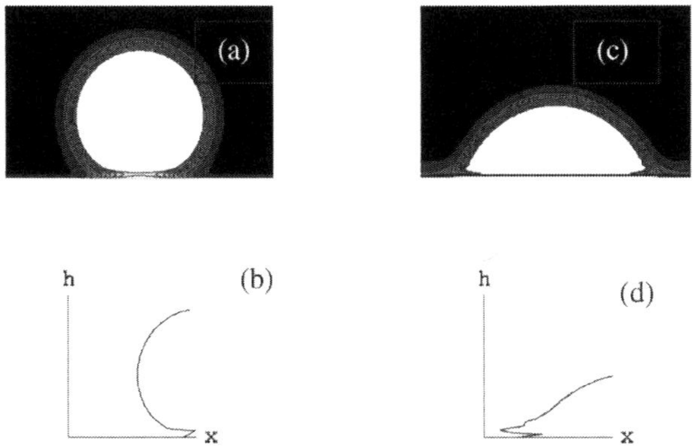

FIG. 3. [(a) and (b)] Schematic presentations of the fluid density distribution and the left-hand side of the symmetrical drop profile, respectively, for $T = 85$ K, $\varepsilon = 0.6$, $\rho^*_{div} = 0.371$, and $\rho^*_{av} = 0.1$. [(c) and (d)] The same as in panels (a) and (b), respectively, but for $\varepsilon = 1.4$ and $\rho^*_{div} = 0.356$. The lighter areas in panels (a) and (c) correspond to larger fluid density.

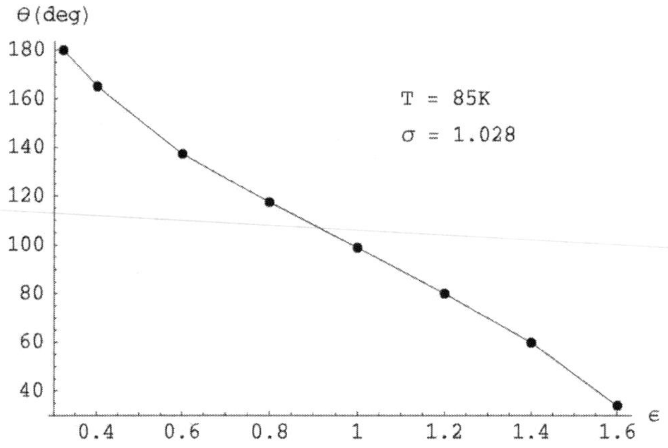

FIG. 4. ε-dependence of the contact angle for $T = 85$ K, $\sigma = 1.028$ ($\sigma_{fs} = 3.5$ Å). The points represent the results of the DFT calculations and the solid line is a guide for eyes.

$0.6 < \varepsilon < 1.4$, with small deviations from linearity for $\varepsilon < 0.6$ and $\varepsilon > 1.4$. The dependence of θ_l on ε is similar. Only the linear parts of these dependencies, which can be represented by the equation

$$\theta = k\varepsilon + b, \tag{8}$$

where the subscripts u and l for θ, k, and b are omitted will be considered below. For the case presented in Fig. 4, $k = k_u = -96.0°$ and $b = b_u = 194.8°$. The values of k and b for other temperatures (other δ) are listed in Table 1 for both θ_u and θ_l. The slopes k of the lines and the values of δ satisfy approximately the simple relation

$$k\delta^{\nu} = -C, \tag{9}$$

where $\nu = 2.64$ and $C = 232.7°$ for θ_u and $\nu = 3.23$, and $C = 317.22°$ for θ_l. In Fig. 5 plots of θ_u versus ε are presented for the four considered values of δ. The rectangle in this figure indicates the area in which the intersection points of the four straight lines are located. The bold point on the plot has the coordinates $\varepsilon = \varepsilon_{u,0} = 0.86$ and $\theta = \theta_{u,0} = 112.2°$ and represents the mean location of all intersection points. Since the latter points are near to each other, one can consider that all lines intersect at the same point (ε_0, θ_0) and combine all the straight lines presented in Fig. 5 into a single line given by

$$\theta - \theta_0 = -C\delta^{-\nu}\left(\varepsilon - \varepsilon_0\right). \tag{10}$$

TABLE 1
Parameters of Eq. (8) for various temperatures, for $\sigma = 1.028$ ($\sigma_{fs} = 3.5$ Å) and $\rho^*_{av} = 0.1$.

δ	T (K)	k_u (deg)	b_u (deg)	k_l (deg)	b_l (deg)
1.409	85	−96.0	194.8	−108.1	211.7
1.331	90	−107.0	203.8	−119.0	222.8
1.261	95	−121.8	216.5	−147.5	249.4
1.198	100	−149.8	241.0	−184.8	281.2

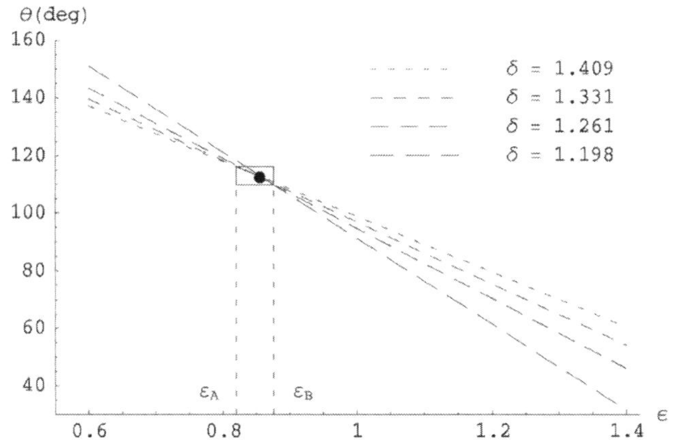

FIG. 5. ε-dependence of the contact angle for $\delta = 1.409$, 1.331, 1.261, and 1.198 ($T = 85$, 90, 95, and 100 K, respectively) and $\sigma = 1.028$ ($\sigma_{\text{fs}} = 3.5$ Å). The bold point has the coordinates $\varepsilon_0 = 0.86$ and $\theta_0 = 112.4°$ which are equal to the mean values of the coordinates of all intersection points of the presented straight lines. All those points are inside the drawn rectangular area. $\varepsilon_A = 0.82$, $\varepsilon_C = 0.88$.

Because this equation is valid for both θ_u and θ_l, the corresponding subscripts of θ, θ_0, C, ν, and ε_0 are omitted and each of the angles θ_u and θ_l will be referred to below as contact angle. The subscripts will be used to specify if θ is θ_u or θ_l. In Eq. (10), θ_0, ε_0, and C are constants provided above and the parameter δ contains the temperature. Note that the relative difference between the values of θ obtained through computations based on DFT and those obtained from Eq. (10) does not exceed 5% for the range $0.6 < \varepsilon < 1.4$ and 1% for the range $0.6 < \varepsilon < 1.1$. Under the employed conditions (constants σ_{fs} and ρ_s), Eqs. (8)–(10) provide a complete description of the fluid-solid pairs for which the parameters δ and ε are in the ranges noted above.

2. Temperature dependence of the contact angle at constant ε

Using Eq. (10) one can easily find the temperature dependence of the contact angle at fixed values of ε. For $\varepsilon > \varepsilon_0$, the contact angle increases and for $\varepsilon < \varepsilon_0$ it decreases with increasing temperature. The rate $d\theta/dT$ of change in the contact angle with temperature is given by

$$\frac{d\theta}{dT} = -\frac{\nu k_B}{\epsilon_{\text{ff}}} C\left(\varepsilon - \varepsilon_0\right)\left(\frac{k_B T}{\epsilon_{\text{ff}}}\right)^{\nu-1} \tag{11}$$

Because $\nu > 1$, the absolute value of $d\theta/dT$ increases with increasing temperature.

An interesting implication of Eq. (10) is that for $\varepsilon = \varepsilon_0$ the contact angle does not depend on temperature and is always equal to θ_0.

Taking $T_0 = 92.5$ K as a reference temperature, the variation ΔT of the temperature considered in the present paper can be considered to be small $\left(\Delta T / T_0 \ll 1\right)$; therefore, Eq. (10) can be linearized with respect to ΔT,

$$\theta(T) = \theta(T_0) + \alpha(T_0)\Delta T, \tag{12}$$

where

$$\theta(T_0) = \theta_0 - C\left(\varepsilon - \varepsilon_0\right)\left(\frac{k_B T_0}{\epsilon_{\text{ff}}}\right)^{\nu} \tag{13}$$

TABLE 2
Parameters of Eq. (8) for various σ and temperatures for $\rho^*_{av} = 0.1$.

$\sigma = \sigma_{fs}/\sigma_{ff}$	σ_{fs} (Å)	T (K)	k_u (deg)	b_u (deg)	k_l (deg)	b_l (deg)
1.095	3.727	85	–117.4	190.9	–141.3	215.1
		90	–126.8	197.5	–159.7	229.3
		95	–143.5	208.0	–184.6	250.3
		100	–171.0	226.8	–207.2	265.6
0.969	3.3	85	–77.8	195.4	–88.7	212.7
		90	–85.0	201.9	–99.8	226.6
		95	–98.0	216.8	–120.5	251.4
		100	–119.0	239.6	–140.0	272.4
0.910	3.1	85	–59.5	191.9	–73.2	215.6
		90	–65.5	199.6	–81.5	228.6
		95	–73.5	210.4	–92.6	246.2
		100	–90.2	234.2	–120.6	285.6

and

$$\alpha(T_0) = -\frac{C\nu}{T_0}(\varepsilon - \varepsilon_0)\left(\frac{k_B T_0}{\epsilon_{ff}}\right)^{\nu}. \tag{14}$$

For θ_u, $\alpha(T_0) = -3.36(\varepsilon - \varepsilon_{u,0})$, and for θ_l, $\alpha(T_0) = -4.88(\varepsilon - \varepsilon_{l,0})$.

The above analysis of the temperature dependence of the contact angle involves the existence of a unique intersection point of all straight lines that represent the ε-dependence of θ at various temperatures. Even if this is not absolutely correct, there are ranges of monotonous increase (decrease) in θ with temperature. They are displaced to $\varepsilon < \varepsilon_A$ ($\varepsilon > \varepsilon_B$) accordingly (see Fig. 5).

B. Contact Angles for $\sigma = 1.095$, 0.969, and 0.910

The results for the contact angle obtained for $\sigma_{fs} = 3.727$ Å, $\sigma_{fs} = 3.3$ Å, and $\sigma_{fs} = 3.1$ Å ($\sigma = 1.095$, 0.969, and 0.910, respectively) coincide qualitatively with those obtained in Sec. III A for $\sigma_{fs} = 3.5$ Å ($\sigma = 1.028$). In the range $0.6 < \varepsilon < 1.4$, θ versus ϵ is represented by straight lines in Eq. (8). The values of the parameters along with the average value θ_0 of the contact angles corresponding to the intersection points of these straight lines are listed in Table 2. An interesting feature is that the values of $\theta_{u,0}$ and $\theta_{l,0}$ are almost constant for all considered values of σ.

For each σ, the straight lines at various temperatures can be combined in a single one of the form of Eq. (10) with the parameters ν, C, and ε_0 dependent on σ. Such single lines are plotted in Fig. 6 for four values of σ, and the values of the parameters ν, C, and ϵ_0 are listed in Table 3 for both θ_u and θ_l. The values of the coordinates ε and θ of the bold points in Fig. 6 are ε_0 and θ_0, respectively. Assuming that for other values of σ the parameters ν, C, and ε_0 can be found by interpolating between the obtained results, one can consider Eq. (10) as a general curve describing the contact angle for wide ranges of fluids and solids which have intermolecular interactions of the Lennard-Jones type.

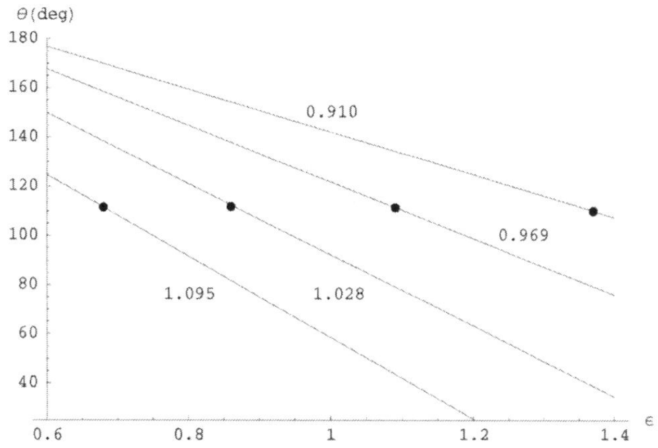

FIG. 6. The ε-dependence of the contact angle $\theta = \theta_u$ provided by Eq. (10) for $\delta = \varepsilon_{ff}/k_BT = 1.198$ ($T = 100$ K) and various values of $\sigma \equiv \sigma_{fs}/\sigma_{ff}$ indicated on each line. The parameters of Eq. (10) are listed in Table III. The bold points on each of the lines have the coordinates (ε_0, θ_0).

TABLE 3
Parameters of Eq. (10) for various values of $\sigma = \sigma_{fs}/\sigma_{ff}$ and $\rho_{av}^* = 0.1$.

σ	ν_u	$\theta_{u,0}$ (deg)	$\varepsilon_{u,0}$	C_u (deg)	ν_l	$\theta_{l,0}$ (deg)	$\varepsilon_{l,0}$	C_l (deg)
1.095	2.25	112.4	0.68	248.5	2.39	106.6	0.77	319.3
1.028	2.62	112.2	0.86	231.0	3.23	112.0	0.93	317.2
0.969	2.55	112.1	1.06	182.1	2.92	110.0	1.17	240.8
0.910	2.45	110.7	1.36	135.0	2.87	105.1	1.52	191.0

C. Contact Angle at Various Average Fluid Densities

In order to verify the generality of Eq. (10), calculations of the contact angle were also performed for $\rho_{av}^* = 0.2$ and all the above considered values of σ as well as for $\rho_{av}^* = 0.15$, 0.2, and 0.25 and $\sigma = 1.028$. The results obtained are listed in Tables 4 and 5 which suggest that there are indeed some general features in the dependence of the contact angles of nanodroplets on temperature and the fluid-solid energy parameter in the Lennard-Jones interaction potential. Table 5 shows that the calculated contact angles depend weakly on the average density of the fluid in the system. This can be caused by the weak change in the density of the vapor phase in the close systems considered by us when ρ_{av}^* is changed.

TABLE 4
Parameters of Eq. (10) for various values of $\sigma = \sigma_{fs}/\sigma_{ff}$ and $\rho_{av}^* = 0.2$.

σ	ν_u	$\theta_{u,0}$ (deg)	$\varepsilon_{u,0}$	C_u (deg)	ν_l	$\theta_{l,0}$ (deg)	$\varepsilon_{l,0}$	C_l (deg)
1.095	2.28	120.6	0.60	257.0	2.69	104.4	0.79	332.0
1.028	2.03	120.6	0.76	183.9	2.57	105.8	1.00	242.1
0.969	2.04	119.8	0.92	144.0	3.22	106.4	1.20	244.2
0.910	1.9	119.2	1.18	111.7	2.75	101.5	1.58	175.7

TABLE 5
Parameters of Eq. (10) for various values of average fluid density and $\sigma = 1.028$ (σ_{fs} 3.5Å).

ρ^*_{av}	v_u	$\theta_{u,0}$ (deg)	$\varepsilon_{u,0}$	C_u (deg)	v_l	$\theta_{l,0}$ (deg)	$\varepsilon_{l,0}$	C_l (deg)
0.1	2.62	112.2	0.86	231.0	3.23	112.0	0.93	317.2
0.15	2.19	119.2	0.73	188.1	2.53	109.0	0.96	244.8
0.20	2.03	120.6	0.76	183.9	2.57	105.8	1.00	242.1
0.25	1.87	124.1	0.68	173.3	2.47	106.8	0.99	230.7

IV. DISCUSSION AND CONCLUSION

The main conclusion is that for nanodrops the contact angle, characterizing the wetting properties of a fluid on a solid surface, has some general characteristics with respect to its dependence on temperature and the parameters that characterize the fluid-solid intermolecular interactions, assuming that the latter are also of the Lennard-Jones type. At various temperatures, the contact angle has a linear dependence on the dimensionless energy parameter $\varepsilon = \epsilon_{fs}/\epsilon_{ff}$, where ϵ_{fs} and ϵ_{ff} are the fluid-solid and fluid-fluid energy parameters, respectively. The straight lines representing these dependencies at various temperatures intersect almost in a single point (ε_0, θ_0), where ε_0 depends on σ_{fs} and θ_0 is almost a universal constant. The contact angle θ_0 for $\varepsilon = \varepsilon_0$ is independent of temperature. For $\varepsilon < \varepsilon_0$, the contact angle monotonously increases and for $\varepsilon > \varepsilon_0$ monotonously decreases with increasing temperature. Assuming that the obtained result can be extrapolated to macroscopic drops, this result can qualitatively explain the different temperature behaviors of the contact angle observed in experiments.[2–4,7,16]

The temperature independent value θ_0 of the contact angle corresponding to ε_0 is independent of $\sigma = \sigma_{fs}/\sigma_{ff}$ and is therefore a general characteristic of a nanodrop on a solid surface. For $\theta < \theta_0$, the interactions of the liquid with the solid are stronger ($\varepsilon > \varepsilon_0$); whereas for $\theta > \theta_0$ the interactions are weaker ($\varepsilon < \varepsilon_0$).

The general features of the contact angle identified in the present paper can be compared to the results obtained in Refs. 11–13 via molecular dynamics simulations for truncated Lennard-Jones interactions. Although the available data[11–13] do not provide enough information for a full comparison, they are in qualitative agreement with our results. Indeed, in both treatments θ decreases with increasing σ_{fs}, ϵ_{ff}, or the size of the droplet. It should be noted that in Refs. 12 and 13 it was reported that $\cos\theta$ has a linear dependence on ϵ_{fs} (in contrast to the linear dependence of θ on ϵ_{fs} obtained in the present paper). However, the replotting of the data presented in Refs. 12 and 13 has shown that the linear fit of θ on ϵ_{fs} is as good as that for $\cos\theta$. The quantitative differences between the values of θ can be explained by the difference in dimensions of the drops considered in Refs. 11–13 (three dimensional drops) and in the present paper (two dimensional drops).

APPENDIX: CONTRIBUTIONS TO THE FREE ENERGY OF THE SYSTEM

The contributions to the free energy listed in Sec. II can be expressed as follows:

$$F_{id}\left[\rho(\mathbf{r})\right] = k_B T \int d\mathbf{r}\rho(\mathbf{r})\left\{\log\left[\Lambda^3\rho(\mathbf{r})\right]-1\right\}, \tag{A1}$$

$$F_{ex}\left[\rho(\mathbf{r})\right] = \int d\mathbf{r}\rho(\mathbf{r})\Delta\Psi_{hs}(\mathbf{r}), \tag{A2}$$

where

$$\Delta\Psi_{\text{hs}}(\mathbf{r}) = k_B T \eta_\rho \frac{4-3\eta_\rho}{(1-\eta_\rho)^2}, \tag{A3}$$

$\eta_\rho = \frac{1}{6}\pi\bar{\rho}(\mathbf{r})\sigma_{\text{ff}}^3$ is the packing fraction of the fluid molecules, is the fluid hard core diameter, and $\bar{\rho}(\mathbf{r})$ is the smoothed density defined as

$$\bar{\rho}(\mathbf{r}) = \int d\mathbf{r}' \rho(\mathbf{r}') W(|\mathbf{r}-\mathbf{r}'|). \tag{A4}$$

The weighting function $W(|\mathbf{r}-\mathbf{r}'|)$ is selected in the form[21]

$$W(|\mathbf{r}-\mathbf{r}'|) = \begin{cases} \dfrac{3}{\pi\sigma_{\text{ff}}^3}\left(1-\dfrac{r}{\sigma_{\text{ff}}}\right), & r \le \sigma_{\text{ff}} \\ 0, & r > \sigma_{\text{ff}}, \end{cases}$$

where $r = |\mathbf{r}-\mathbf{r}'|$.

The potential $U_{\text{fs}}(\mathbf{r})$ generated by the solid depends on the coordinate h only and is given by the equation

$$U_{\text{fs}}(h) = \frac{2\pi}{3}\epsilon_{\text{fs}}\rho_s\sigma_{\text{fs}}^3\left[\frac{2}{15}\left(\frac{\sigma_{\text{fs}}}{\sigma_{\text{fs}}+h}\right)^9 - \left(\frac{\sigma_{\text{fs}}}{\sigma_{\text{fs}}+h}\right)^3\right]. \tag{A5}$$

The function $Q(x, h)$ in Eq. (5) is given by

$$Q(x,h) = -\left[\Delta\Psi_{\text{hs}}(x,h) + \overline{\Delta\Psi'_{\text{hs}}}(x,h)\right], \tag{A6}$$

where

$$\overline{\Delta\Psi'_{\text{hs}}}(x,h) = \iint dx'dh'\rho(x',h')$$

$$\times W_y(|x-x'|,|h-h'|)\frac{\partial}{\partial\bar{\rho}}\Delta\Psi_{\text{hs}}(\bar{\rho})|_{\bar{\rho}=\bar{\rho}(x',h')}, \tag{A7}$$

$W_y(|x-x'|,|h-h'|)$ is obtained by the integration of the weighted function $W(|\mathbf{r}-\mathbf{r}'|)$ with respect to the y-direction, from $-\infty$ to ∞.

The function $U_{\text{ff}}(x, h)$, which represents in Eq. (5) the fluid-fluid interactions is given by

$$U_{\text{ff}}(x,h), \iint dx'dh'\rho(x',h')\phi_{\text{ff},y}(|x-x'|,|h-h'|) \tag{A8}$$

where $\phi_{\text{ff},y}(|x-x'|,|h-h'|)$ is obtained by the integrations of the potential $\phi_{\text{ff}}(|\mathbf{r}-\mathbf{r}'|)$ with respect to the y-direction, from $-\infty$ to $+\infty$.

When evaluating this function, a cutoff at a distance equal to four molecular diameters σ_{ff} for the range of Lennard-Jones attraction was employed. The increase in this distance up to $10\sigma_{\text{ff}}$ changed the results by less than 1%.

REFERENCES

1. R. E. Johnson, Jr. and R. H. Dettre, Surf. Colloid Sci. **2**, 85 (1969).
2. E. Rolley and C. J. Guthmann, Low Temp. Phys. **108**, 1 (1997).
3. C. J. Budziak, E. I. Vargha-Butler, and A. W. Neumann, J. Appl. Polym. Sci. **42**, 1959 (1991).
4. A. Suzuki and Y. Koboki, Jpn. J. Appl. Phys., Part 1 **38**, 2910 (1999).
5. J. B. Jones and A. W. Adamson, J. Phys. Chem. **72**, 646 (1968).
6. S. H. Yuk and M. S. Jhon, J. Colloid Interface Sci. **116**, 25 (1987).
7. J. D. Bernardin, I. Mudawar, C. B. Walsh, and E. I. Franses, Int. J. Heat Mass Transfer **40**, 1017 (1997).
8. J. Klier, P. Stefanyi, and A. F. G. Wyatt, Phys. Rev. Lett. **75**, 3709 (1995).
9. J. Klier and A. F. G. Wyatt, Phys. Rev. B **54**, 7350 (1996).
10. F. Ancilotto, F. Faccin, and F. Toigo, Phys. Rev. B **62**, 17035 (2000).
11. S. Matsumoto, S. Maruyama, and H. Saruwatari, Proceedings of the ASME/JSME Thermal Engineering Joint Conference, Maui, HI, 1995 (unpublished), Vol. 2, p. 557.
12. S. Maruyama, T. Kurashige, S. Matsumoto, Y. Yamaguchi, and T. Kimura, Microscale Thermophys. Eng. **2**, 49 (1998).
13. M. J. De Ruijter, T. D. Blake, and J. De Coninck, Langmuir **15**, 7836 (1999).
14. N. Giovambattista, P. G. Debenedetti, and P. J. Rossky, J. Phys. Chem. B **111**, 9581 (2007).
15. F. D. Petke and B. R. Ray, J. Colloid Interface Sci. **31**, 216 (1969).
16. H. Y. She and B. E. Sleep, Water Resour. Res. **34**, 2587 (1998).
17. D. H. Everett and J. M. Haynes, J. Colloid Interface Sci. **38**, 125 (1972).
18. J. R. Philip, J. Chem. Phys. **66**, 5069 (1977).
19. P. Tarazona, Phys. Rev. A **31**, 2672 (1985).
20. R. Evans and P. Tarazona, Phys. Rev. A **28**, 1864 (1983).
21. R. H. Nilson and S. K. Griffiths, J. Chem. Phys. **111**, 4281 (1999).
22. A. V. Neimark, P. I. Ravikovitch, and A. Vishnyakov, J. Phys.: Condens. Matter **15**, 347 (2003).
23. E. A. Ustinov and D. D. Do, J. Phys. Chem. B **109**, 11653 (2005).
24. V. Talanquer and D. W. Oxtoby, J. Chem. Phys. **104**, 1483 (1996).
25. V. Talanquer and D. W. Oxtoby, J. Chem. Phys. **114**, 2793 (2001).
26. G. O. Berim and E. Ruckenstein, J. Chem. Phys. **129**, 014708 (2008); G. O. Berim and E. Ruckenstein, ibid. **129**, 114709 (2008) (Sections 4.4 and 4.11 of this volume).
27. D. Frenkel and B. Smit, *Understanding Molecular Simulation* (Academic, New York, 1966).
28. A. Vishnyakov and A. V. Neimark, J. Chem. Phys. **119**, 9755 (2003).
29. A. V. Neimark, P. I. Ravikovitch, and A. Vishnyakov, Phys. Rev. E **65**, 031505 (2002).
30. J. Mittal, J. R. Errington, and T. M. Truskett, J. Chem. Phys. **126**, 244708 (2007).
31. F. Porcheron and P. A. Monson, Langmuir **22**, 1595 (2006).
32. E. Ruckenstein and P. S. Lee, Surf. Sci. **52**, 298 (1975) (Section 3.1 of this volume).

4.2 Universality in the Dependence of the Drop Contact Angle on Liquid-Solid Interactions and Temperature Obtained by the Density Functional Theory*

Gersh Berim and Eli Ruckenstein†
Department of Chemical and Biological Engineering,
State University of New York at Buffalo, Buffalo, NY 14260, USA

Corresponding Author
† E-mail: feaeliru@buffalo.edu

Received 02 June 2011 / Received in final form 15
June 2011 Published online 30 August 2011

1 INTRODUCTION

The density functional theory (DFT) developed in Refs. [1–5] provides the necessary tools for the theoretical description of bulk fluids and confined fluids. The major advantage of DFT is that it is based on fluid-fluid and fluid-solid microscopic interactions and does not involve macroscopic characteristics such as the surface tensions. The latter can be also calculated on the basis of interaction potentials using DFT. As examples of DFT applications, one can mention the accurate descriptions of adsorption isotherms of simple fluids in contact with smooth or rough solid surfaces [6–9], of wetting and layering transitions at solid surfaces [1,9–11], of capillary condensation in slits and cylindrical pores [12–15]. In addition, the properties of nanodrops on smooth and rough solid surfaces could be examined [16–20]. The DFT can also predict subtle details regarding the fluid behavior in the presence of a surface. For instance, it can predict the prewetting (thick-thin film) transition [1] as well as symmetry breaking of the fluid density distribution in slits [21–24].

In the present paper, DFT is used to derive simple expressions for the dependence of the contact angles which nano- and macro drops make with the solid surface on liquid-solid interaction parameters and temperature. Evidence that such dependencies do exist was provided by Monte Carlo simulations of Lennard-Jones fluids in Refs. [25,26], where it was shown that the cosine of the contact angle of a nanodrop has a linear dependence on the energy parameter ϵ_{fs} of the

* *Eur. Phys. J. Special Topics* 197, 163–178 (2011). Republished with permission.

fluid-solid interactions. A similar conclusion was drawn in Ref. [18] where a simple, almost linear, dependence of the contact angle on ϵ_{fs} was found for a nanodrop of argon on various substrates.

Depending on the size of the drop, there are two different ways for calculating the contact angles. For a macroscopic drop, the contact angle can be determined by calculating the surface tensions γ_{lv}, γ_{ls}, γ_{vs} of the liquid-vapor, liquid-solid, and vapor solid interfaces, respectively, and the further usage of the Young equation

$$\gamma_{lv} \cos\theta = \gamma_{sv} - \gamma_{sl}. \tag{1}$$

Several ways of calculating the surface tensions on the basis of DFT were suggested in Refs. [27–29].

Note that the Young equation, Eq. (1), is applicable only to macroscopic drops for which the thickness of the liquid-vapor interface is much smaller than the size of the drop, which in this case can be considered of uniform density.

For nanodrops, the thickness of the liquid-vapor interface is comparable with the drop size and, in addition, considerable variation of the fluid density inside the drop takes place. In this case the drop profile and the contact angle can be extracted from the fluid density distribution (FDD) obtained on the basis of DFT using the approach suggested in Refs. [18,26,30,31].

2 BACKGROUND

2.1 THE SYSTEM

Let us consider a fluid confined between two parallel structureless solid walls W_1 and W_2 separated at a sufficiently large distance $L + 2\sigma_{fs}$, where σ_{fs} is a length parameter of the fluid-solid interaction defined below. Three different specific systems shown in Figure 1 will be considered. Note that all of them are infinite in the y direction which is normal to the plane of the figure.

In the first two systems presented in Figure 1a, the slit is considered to be either connected (no side walls) to a reservoir of bulk fluid at a temperature T and chemical potential μ (system I), or to be closed (system II). In the latter case, the distance between the side walls which close the slit should be much larger than the width of the slit for the end effects to be negligible. The walls W_l and W_2 are assumed to be identical and the fluid density distribution (FDD) to be uniform in the x and y directions, i.e. $\rho(\mathbf{r}) \equiv \rho(h)$. These systems will be considered for calculating the surface tensions γ_{ls}, γ_{vs}, and γ_{lv}.

The third system (system III) has the finite dimensions L_x and L_h in the x and h directions, respectively (see Figure 1b). For this system, the wall W_2 is considered neutral (the interaction with the fluid molecules being via hard core repulsion only) and the FDD in the x direction is considered to be nonuniform but having a fixed average density ρ_{av} (closed system). Assuming periodic boundary conditions, nanodrop-like FDDs are obtained and the contact angles are calculated [18].

The interactions between the fluid molecules in the system are described by the Lennard-Jones potential $\phi_{ff}\left(|\mathbf{r}-\mathbf{r}'|\right) \equiv \epsilon_{ff}\bar{\phi}_{ff}\left(|\mathbf{r}-\mathbf{r}'|\right)$, where $\bar{\phi}_{ff}\left(|\mathbf{r}-\mathbf{r}'|\right)$ is a dimensionless quantity given by

$$\bar{\phi}_{ff}\left(|\mathbf{r}-\mathbf{r}'|\right) = \begin{cases} 4\left[\left(\dfrac{\sigma_{ff}}{r}\right)^{12} - \left(\dfrac{\sigma_{ff}}{r}\right)^{6}\right], & r \geq \sigma_{ff} \\ \infty, & r < \sigma_{ff}. \end{cases} \tag{2}$$

In Eq. (2), the coordinates $\mathbf{r}$ and $\mathbf{r}'$ provide the locations of the interacting fluid molecules, $r = |\mathbf{r} - \mathbf{r}'|$, and σ_{ff} and ϵ_{ff} are the fluid hard core diameter and the energy parameter, respectively.

Because of the walls uniformity, the total potential $U_1(\mathbf{r})$ of the interaction between a fluid molecule and wall W_1 depends only on the distance h from the wall and can be written in the form

$$U_1(h) = \epsilon_{fs}\psi(h), \tag{3}$$

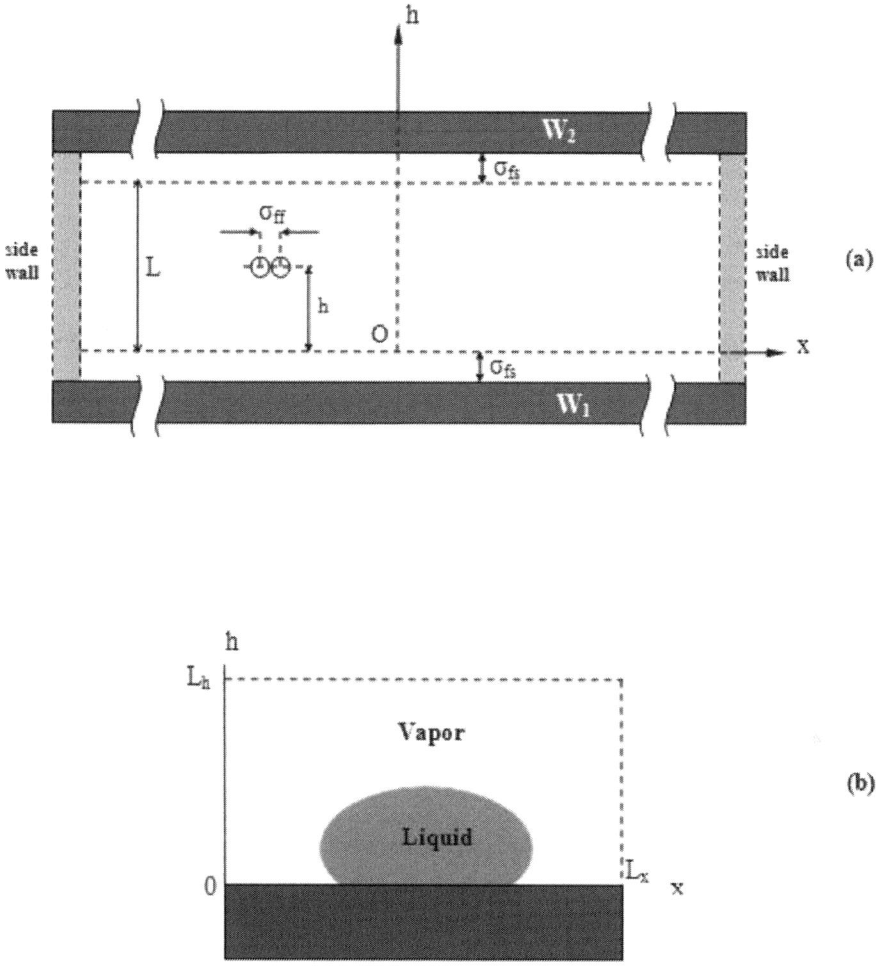

FIG. 1. (a) A slit of width $L + 2\sigma_{fs}$ between two identical solid walls W_1 and W_2. σ_{ff} and σ_{fs} are the hard core diameters of the fluid-fluid and fluid-solid interactions, respectively. The closed slit has the sizes in the x and y directions large for the end effects to be negligible. (Reprinted from Ref. [29] with permission) (b) Schematic representation of a closed system consisting of a cylindrical drop (drop uniform in the y-direction which is normal to the plane of the figure) on a smooth surface in contact with the vapor of the same fluid. In both panels, the y-axis is normal to the plane of the figure. The fluid density distribution is uniform in x- and y-directions. (Reprinted from Ref. [32] with permission.)

where the dimensionless quantity $\psi(h)$ is given by

$$\psi(h) = \frac{2\pi}{3}\rho_s\sigma_{fs}^3\left[\frac{2}{15}\left(\frac{\sigma_{fs}}{\sigma_{fs}+h}\right)^9 - \left(\frac{\sigma_{fs}}{\sigma_{fs}+h}\right)^9\right]. \tag{4}$$

In systems I and II (see Figure la), both walls W_1 and W_2 generate attractive potentials and the net potential is provided by the expression

$$U_{fs}(h) = \epsilon_{fs}\left[\psi(h) + \psi(L-h)\right]. \tag{5}$$

Because the main purpose of the present paper is to examine the dependence of the contact angle of a nanodrop on a solid surface on the parameters of the intermolecular interactions (σ_{ff}, ϵ_{ff}, σ_{fs}, ϵ_{fs}), it is not necessary to select those parameters for specific fluids and substrates. However, to have a specific reference fluid, we select σ_{ff} and ϵ_{ff} as for argon ($\epsilon_{ff}/k_B = 119.76$K, $\sigma_{ff} = 3.405$ Å) and select the density of the solid equal to that of solid carbon dioxide ($\rho_s = 1.91 \times 10^{28}$ m^{-3}) [1]. We selected σ_{ff} as the unit of length and introduced two independent dimensionless parameters, $\sigma = \sigma_{fs}/\sigma_{ff}$ and $\varepsilon = \epsilon_{fs}/\epsilon_{ff}$ characterizing the substrate and the fluid, and the parameter $\delta = \epsilon_{ff}/k_B T$ characterizing the effect of temperature. The calculations of the contact angles were performed for four temperatures, T = 85, 90, 95, and 100 K (δ = 1.409, 1.331, 1.261, 1.198, respectively) and for ε varying between 0.3 and 1.6. For numerical calculations, the width L_x of the system was taken $L_x = 50\sigma_{ff}$ and the height $L = L_h = 30\sigma_{ff} + 2\sigma_{fs}$. The dimensionless average density $\rho_{av}^* = \rho_{av}\sigma_{ff}^3$ was considered in the range $0.1 \le \rho_{av}^* \le 0.25$.

2.2 Euler-Lagrange equation for fluid density distribution

The Helmholtz free energy $F[\rho(\mathbf{r})]$ of the system can be written in the form

$$F\left[\rho(\mathbf{r})\right] = F_{id}\left[\rho(\mathbf{r})\right] + F_{ex}\left[\rho(\mathbf{r})\right] + \frac{\epsilon_{ff}}{2}\iint d\mathbf{r}d\mathbf{r}'\rho(\mathbf{r})\rho(\mathbf{r}')\bar{\phi}_{ff}\left(|\mathbf{r}-\mathbf{r}'|\right) + \epsilon_{fs}\int d\mathbf{r}\rho(\mathbf{r})U_{fs}(\mathbf{r}) \tag{6}$$

where $F_{id}[\rho(\mathbf{r})]$ and $F_{ex}[\rho(\mathbf{r})]$ are the ideal gas contribution and excess contribution due to nonideality, respectively, of a hard sphere fluid. The third term in Eq. (6) represents the contribution of the fluid-fluid interaction treated in the mean-field approximation. Explicit expressions for $F_{id}[\rho(\mathbf{r})]$ and $F_{ex}[\rho(\mathbf{r})]$ can be found in the Appendix. The potential $U_{fs}(\mathbf{r})$ generated by the solid walls is given by Eq. (5) for systems I and II and by Eq. (3) for system III.

In closed systems (I and II), the Euler-Lagrange equation for the fluid density distribution $\rho(\mathbf{r})$ can be obtained by minimizing the Helmholtz free energy Eq. (6) under the constraint of fixed average density (fixed number of molecules in the system) which has the form

$$\rho_{av} = \frac{1}{V}\int_V d\mathbf{r}\rho(\mathbf{r}), \tag{7}$$

where V is the volume of the system.

In an open system (system III), the Euler-Lagrange equation follows from the minimization of the grand canonical potential of the system

$$\Omega_{sf} = F_H\left[\rho(h)\right] - \mu\int_0^L \rho(h)dh, \tag{8}$$

where μ is the chemical potential of the bulk fluid.

In both cases, the Euler-Lagrange equation has the form

$$\log\left[\Lambda^3\rho(\mathbf{r})\right] + Q(\mathbf{r}) + \frac{\epsilon_{ff}}{k_B T}U_{ff}(\mathbf{r}) + \frac{\epsilon_{fs}}{k_B T}U_{fs}(\mathbf{r}) = \frac{\mu}{k_B T}, \tag{9}$$

where $\Lambda = h_P/(2\pi m k_B T)^{1/2}$ is the thermal de Broglie wavelength, k_B and h_P are the Boltzmann and Planck constants, respectively, T is the absolute temperature, and m is the mass of a fluid molecule. The functions $Q(\mathbf{r})$, $U_{ff}(\mathbf{r})$, and $U_{fs}(\mathbf{r})$ represent a hard spheres system, the fluid-fluid interactions, and the fluid-solid interactions, respectively, and are provided in the Appendix.

Here we note that for an open system the parameter μ, in Eq. (9) is the chemical potential of the bulk fluid, whereas for a closed system μ is a Lagrange multiplier which is not known in advance but can be determined using the constraint of fixed average density, Eq. (7). One thus obtains

$$\mu = -k_B T \log\left\{\frac{1}{\rho_{av} V \Lambda^3}\int d\mathbf{r} \exp\left[-Q(\mathbf{r}) - \frac{\epsilon_{ff}}{k_B T} U_{ff}(\mathbf{r}) - \frac{\epsilon_{fs}}{k_B T} U_{fs}(\mathbf{r})\right]\right\}. \tag{10}$$

By eliminating μ between Eqs. (9) and (10), one obtains a closed equation for the density profile $\rho(\mathbf{r})$. For both open and closed systems, the Euler-Lagrange equation represents a nonlinear integral equation which can be solved by iterations. The details regarding the iteration procedure can be found in Ref. [18].

2.3 Calculation of the contact angle for a nanodrop in a closed system

Because the liquid-vapor interface for a nanodrop has a thickness comparable to that of the drop, the contact angle for a nanodrop is not clearly defined. A convenient way to calculate this angle is to determine the profile inside the vapor-liquid interface corresponding to a constant local density ρ_{div} and to find the angle which this profile makes with the solid surface [30,31]. The FDD inside the drop has a complex structure in the vicinity of the liquid-solid and liquid-vapor interfaces. In particular, near the solid surface there are strong oscillations of density due to the ordering of the fluid molecules which form several liquid layers of various densities [3,4]. For this reason the profiles corresponding to liquid like values of ρ_{div} have a complicated shape near the solid surface. As an illustration, in Figure 2, three typical profiles corresponding to vapor like $\left(\rho^*_{div} = 0.061, \rho^* = \rho\sigma^3_{ff}\right)$ and liquid like $\rho^*_{div} = 0.361$ and 0.451) densities are presented for a drop on a smooth solid surface. Evidently, these profiles provide different angles with the surface and this leads to uncertainty in the definition of the contact angle. The common procedure to specify the value of ρ^*_{div} is to consider that it is provided by an equimolar dividing surface (see Refs. [30,31]). Considering the profile corresponding to this value of ρ_{div} as the drop profile, its smooth part located outside the region of strong density oscillations ($h > 3\sigma_{ff}$) was approximated by a circle and extended up to the solid surface. The angle which this circle makes with the surface was considered as the apparent contact angle.

2.4 General procedures for calculating the surface tensions

To determine γ_{sl}, and γ_{sv}, an open slit (system I) was considered. In such a slit the fluid density distribution $\rho(h)$ is symmetric with respect to the middle of the slit. The effective width of the slit, L, is selected large enough to provide a bulk-like density of the fluid in the middle region of the slit (light shaded area in Figure 3). The density distribution $\rho(h)$ can be obtained as the solution of the Euler-Lagrange equation, Eq. (9), for the chemical potential $\mu = \mu_c$, where μ_c is the chemical potential for bulk liquid-vapor coexistence.

After determining $\rho(h)$, the grand canonical potential of the system, Ω_{sf}, can be calculated using Eq. (8), and the grand canonical potential Ω^b_f of a bulk fluid having the volume of the slit is given by the expression

$$\Omega^b_f = F_H\left[\rho^b_f\right] - \mu_c \rho^b_f L. \tag{11}$$

In Eq. (11) the subscript f stands either for the liquid ($f = l$) or for the vapor ($f = v$), $F_H\left[\rho^b_f\right]$ is the Helmholtz free energy for a homogeneous bulk density distribution, and ρ^b_f is the density of the bulk fluid. All extensive quantities are calculated per unit area of one of the walls. The surface tensions γ_{sl} and γ_{sv} can be obtained using the expression

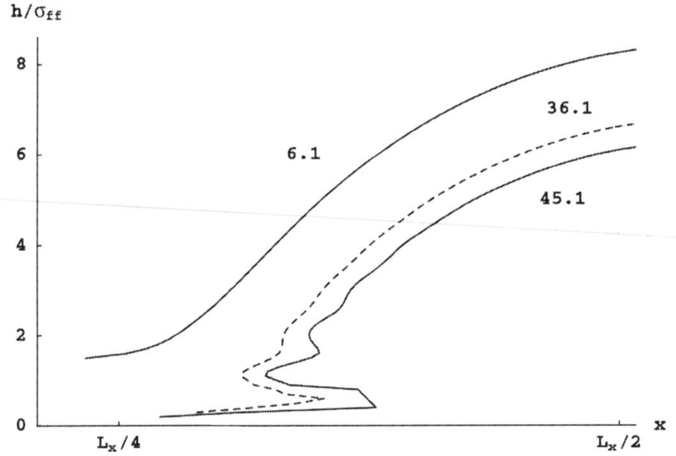

FIG. 2. (Reprinted from Ref. [18] with permission.) Drop profiles for various dividing surfaces corresponding to $\rho^*_{div} = 0.061; 0.361;$ and 0.451. The average density $\rho^*_{av} = 0.1$. The numbers on curves indicate the value of $\rho^*_{div} \times 10^2$.

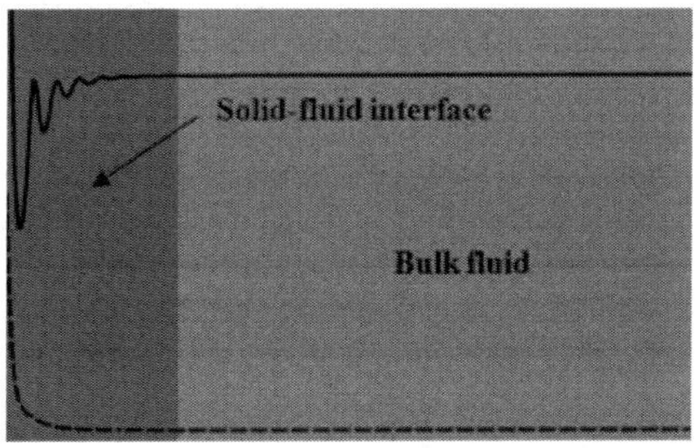

FIG. 3. (Reprinted from Ref. [20] with permission.) Examples of fluid density distributions in the open slit containing a solid-fluid interface. Solid and dashed lines represent distributions which are liquid-like and vapor-like in the bulk phase, respectively. Only one-half of the slit is presented.

$$\gamma_{sf} = \left(\Omega_{sf} - \Omega_f^b\right)/2, \left(f = l, \upsilon\right), \tag{12}$$

where the factor 1/2 accounts for the two solid-fluid interfaces in the system.

A similar approach with some modifications can be used to calculate $\gamma_{l\upsilon}$. In the latter case one must obtain a density distribution in a slit that contains the interface between a bulk-like liquid and a bulk-like vapor. One of the possible density distribution is presented schematically in Figure 4. Such a distribution can exist in the slit in the absence of the fluid-solid interactions ($\epsilon_{fs} = 0$). However, in an open system, the numerical solution for this distribution is unstable due to the uncertainty in the position of the interface [27]. To overcome this difficulty, one must correct the position of the interface such as to remain fixed in the same place during calculations [27], or to fix the number of molecules in the system [28]. The latter procedure is equivalent to the consideration of a closed system and it is used in the present paper. For a closed system (system II) the size of the slit is selected sufficiently large to allow the Lagrange multiplier μ to take a value close to the chemical potential μ_c for the vapor-liquid coexistence. Using the obtained density profile, $\gamma_{l\upsilon}$ can be calculated using the expression [28]

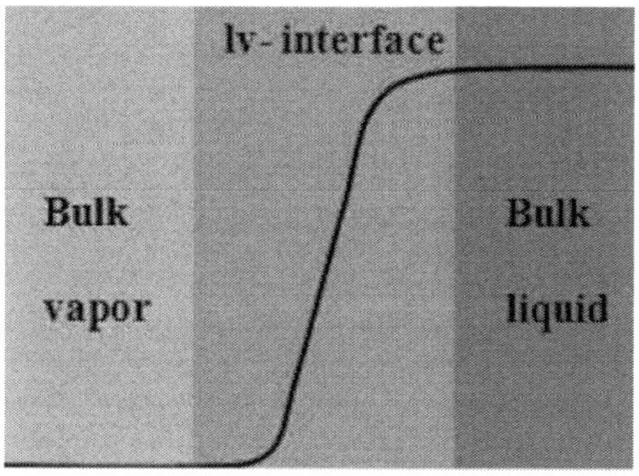

FIG. 4. (Reprinted from Ref. [20] with permission.) Symmetric fluid density distribution in the open slit with neutral (only hard core repulsion) walls, containing two liquid-vapor interfaces. Only one-half of the slit containing one of the interfaces is presented.

$$\gamma_{lv} = \left(\Omega_{lv} - \Omega^b\right)/2, \tag{13}$$

where Ω_{lv} is the grand canonical potential of the system containing two liquid-vapor interfaces and Ω^b is that for a homogeneous fluid distribution.

3 RESULTS

Because the general features regarding the contact angle dependence on the parameters of interaction potentials and temperature are similar for nano and macrodrops, below they will be presented in details for nanodrops only. The specific features for macrodrops will be briefly discussed in Sec. 3.5. In Secs. 3.1–3.3 the average fluid density was selected equal to $\rho^*_{av} = 0.1$. The results for other values of ρ^*_{av} are presented in Sec. 3.4.

3.1 ε-DEPENDENCE OF A NANODROP CONTACT ANGLE AT $\sigma = 1.028$ ($\sigma_{fs} = 3.5$ Å)

In Figures 5a, b typical fluid density distributions for a relatively weak fluid solid-interaction ($\varepsilon = 0.6$) and for a stronger one ($\varepsilon = 1.4$), obtained by solving Eq. (9), are presented schematically for $T = 85$ K together with the corresponding drop profiles (Figures 5c, d) calculated as described in Sec. 2.3. The circular extension of the upper part of the first profile provides the angle $\theta \simeq 137°$ with the solid surface and of the second profile provides $\theta \simeq 60°$. A typical dependence of θ on ε is presented in Figure 6 for $T = 85$ K ($\delta = 1.409$). This dependence is almost linear in the range $0.6 < \varepsilon < 1.4$, with small deviations from linearity for $\varepsilon < 0.6$ and $\varepsilon > 1.4$. The linear part of this dependence can be represented by the equation

$$\theta = k\varepsilon + \mathrm{b}, \tag{14}$$

For the case presented in Figure 6, $k = -96.0°$, and $b = 194.8°$. The values of k and b for other temperatures (other δ) are listed in Table 1.

The slopes k of the lines and the values of δ satisfy approximately the simple relation

$$\mathrm{k}\delta^{\nu} = -\mathrm{C}, \tag{15}$$

where $\nu = 2.64$ and $C = 232.7°$. In Figure 7 the plots of θ vs ε are presented for the four considered values of δ. All the intersection points of the four straight lines are located inside the rectangle shown in this figure. The bold point on the plot represents the mean location of all intersection points and has the coordinates $\varepsilon = \varepsilon_0 = 0.86$ and $\theta = \theta_0 = 112.2°$. Since the intersection points are

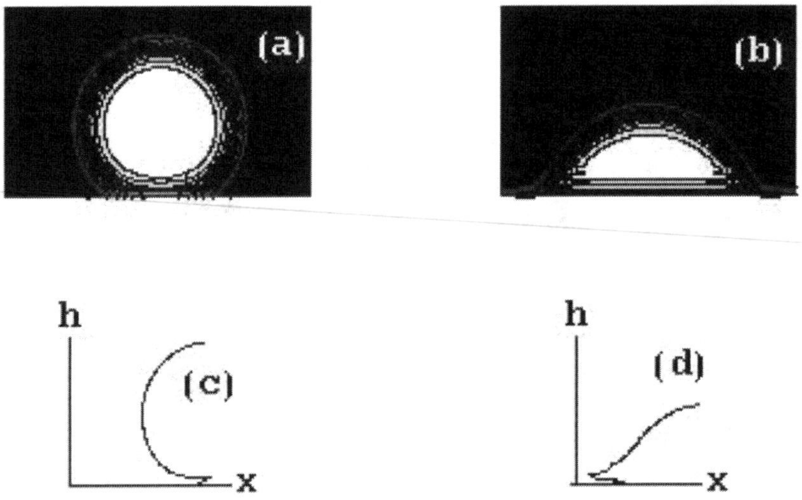

FIG. 5. (Reprinted from Ref. [32] with permission.) (a) and (c) Fluid density distribution and the left hand side of the symmetrical drop profile, respectively, for $T = 85, \varepsilon = 0.6, \rho^*_{div} = 0.371, \rho^*_{av} = 0.1$ (b) and (d) The same as in panels (a) and (c), respectively, but for $\varepsilon = 1.4$, and $\rho^*_{div} = 0.356$. The lighter areas in panels (a) and (b) correspond to larger fluid densities.

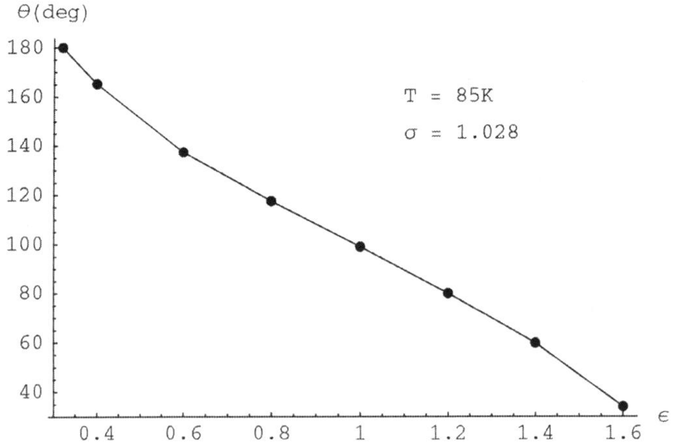

FIG. 6. (Reprinted from Ref. [32] with permission.) ε-dependence of the contact angle for $T = 85$K, $\sigma = 1.028$ ($\sigma_{fs} = 3.5$Å).

TABLE 1
Parameters of Eq. (14) for nanodrops at various temperatures for $\sigma = 1.028$ ($\sigma_{fs} = 3.5$Å) and $\rho^*_{av} = 0.1$

δ	T(K)	k (deg)	b (deg)
1.409	85	−96.0	194.8
1.331	90	−107.0	203.8
1.261	95	−121.8	216.5
1.198	100	−149.8	241.0

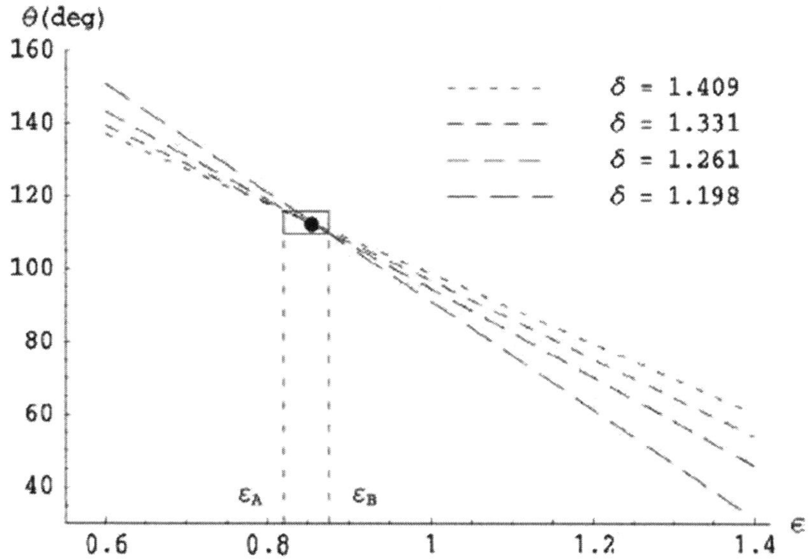

FIG. 7. (Reprinted from Ref. [32] with permission.) ε-dependence of the contact angle for δ = 1.409, 1.331, 1.261, and 1.198 (T = 85K, 90K, 95K, and 100K, respectively) and σ = 1.028 (σ_{fs} = 3.5 Å). The bold point has the coordinates ε_0 = 0.86 and θ_0 = 112.4° which are equal to the mean values of the coordinates of all intersection points of the presented straight lines. All those points are inside the drawn rectangular area. ε_A = 0.82, ε_B = 0.88.

near to each other, one can consider that all lines intersect in the same point (ε_0, θ_0) and combine all the straight lines presented in Figure 7 into a single line given by

$$\theta - \theta_0 = -C\delta^{-\nu}\left(\varepsilon - \varepsilon_0\right). \tag{16}$$

In Eq. (16), θ_0, ε_0, and C are constants provided above and the parameter δ contains the temperature. Note, that the difference between the values of θ obtained through computations based on DFT and those obtained from Eq. (16) does not exceed 5% for the range $0.6 < \varepsilon < 1.4$ and 1% for the range $0.6 < \varepsilon < 1.1$. Under the employed conditions (constant σ_{fs} and constant ρ_s), Eqs. (14) – (16), provide a complete description of the fluid-solid pairs for which the parameters δ and ε are in the ranges noted above.

3.2 Temperature Dependence of the Nanodrop Contact Angle at Constant ε for σ_{fs} = 3.5 Å

The temperature dependence of the contact angle at fixed values of ε can be easily found using Eq. (16) where $\delta = \epsilon_{ff}/k_BT$. One can see from Figure 7, which represents this equation at various temperatures, that for $\varepsilon < \varepsilon_0$ the contact angle increases and for $\varepsilon > \varepsilon_0$ it decreases with increasing temperature (decreasing δ). The rate $\frac{d\theta}{dT}$ of change of the contact angle with temperature is provided by the equation

$$\frac{d\theta}{dT} = -\frac{\nu k_B}{\epsilon_{ff}} C\left(\varepsilon - \varepsilon_0\right)\left(\frac{k_B T}{\epsilon_{ff}}\right)^{\nu-1}. \tag{17}$$

Because $\nu > 1$, the absolute value of $\frac{d\theta}{dT}$ increases with increasing temperature.

An interesting implication of Eq. (16) is that for $\varepsilon = \varepsilon_0$ the contact angle does not depend on temperature and is always equal to θ_0. This observation suggests that for a given fluid one can find (or manufacture) a surface with wetting properties independent of temperature in some temperature range.

Considering $T_0 = 92.5$ K as a reference temperature, the variation $\Delta T = T - T_0$ of the temperature in the examined range 85 K $\leq T \leq$100 K can be considered to be small ($\Delta T/T_0 \ll 1$); therefore, Eq. (16) can be linearized with respect to ΔT

$$\theta(T) = \theta(T_0) + \alpha(T_0)\Delta T, \tag{18}$$

where

$$\theta(T_0) = \theta_0 - C(\varepsilon - \varepsilon_0)\left(\frac{k_B T_0}{\epsilon_{ff}}\right)^{v}, \tag{19}$$

and

$$\alpha(T_0) = -\frac{Cv}{T_0}(\varepsilon - \varepsilon_0)\left(\frac{k_B T_0}{\epsilon_{ff}}\right)^{v}. \tag{20}$$

The above analysis of the temperature dependence of the contact angle involves the existence of a unique intersection point of all straight lines that represent the ε-dependence of θ at various temperatures. Even if this is not absolutely true, there are ranges of monotonous increase (decrease) of θ with temperature. They are located at $\varepsilon < \varepsilon_A$ ($\varepsilon > \varepsilon_B$), respectively (see Figure 7).

3.3 Nanodrop Contact Angles for σ = 1.095, 0.969, and 0.910

The results for the contact angle obtained for $\sigma_{fs} = 3.727$ Å, $\sigma_{fs} = 3.3$ Å, and $\sigma_{fs} = 3.1$ Å ($\sigma = 1.095$, 0.969, and 0.910, respectively) [32] are similar to those discussed in Sec. 3.1 for $\sigma_{fs} = 3.5$ Å ($\sigma = 1.028$). In the range $0.6 < \varepsilon < 1.4$, the plot of θ vs ε is represented by straight lines, Eq. (14). The values of the parameters k and b are listed in Table 2.

TABLE 2
Parameters of Eq. (14) for nanodrops for various σ and temperatures for $\rho^*_{av} = 0.1$.

$\sigma = \sigma_{fs}/\sigma_{ff}$	σ_{fs}(Å)	T(K)	k (deg)	b (deg)
		85	−117.4	190.9
1.095	3.727	90	−126.8	197.5
		95	−143.5	208.0
		100	−171.0	226.8
		85	−77.8	195.4
0.969	3.3	90	−85.0	201.9
		95	−98.0	216.8
		100	−119.0	239.6
		85	−59.5	191.9
0.910	3.1	90	−65.5	199.6
		95	−73.5	210.4
		100	−90.2	234.2

TABLE 3
Parameters of Eqs. (15) and (16) for nanodrops for various values of $\sigma = \sigma_{fs}/\sigma_{ff}$ and $\rho_{av}^* = 0.1$.

σ	ν	θ_0 (deg)	ε_0	C (deg)
1.095	2.25	112.4	0.68	248.5
1.028	2.62	112.2	0.86	231.0
0.969	2.55	112.1	1.06	182.1
0.910	2.45	110.7	1.36	135.0

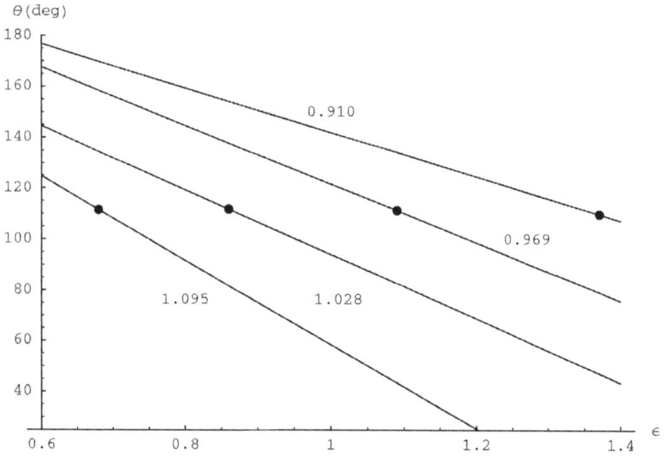

FIG. 8. (Reprinted from Ref. [32] with permission.) The ε-dependence of the contact angle θ provided by Eq. (16) for $\delta \equiv \epsilon_{ff}/k_B T = 1.198$ ($T = 100$ K) and various values of $\sigma \equiv \sigma_{fs}/\sigma_{ff}$ indicated on each line. The parameters of Eq. (16) are listed in Table 3. The bold points on each of the lines have the coordinates $(\varepsilon_0, \theta_0)$.

For each σ, the straight lines at various temperatures can be combined in a single one of the form of Eq. (16) with the parameters ν, C, and ε_0 dependent on σ. Such single lines are plotted in Figure 8 for each considered value of σ, and the values of the parameters ν, C, and ε_0 are listed in Table 3.

The values of the coordinates ε and θ of the bold points in Figure 8 are ε_0 and θ_0, respectively. For other values of σ, the parameters ν, C, and ε_0 can be found by interpolating between the obtained results. Consequently, one can consider Eq. (16) as a universal equation describing the contact angle for wide ranges of fluids and solids which have intermolecular interactions of the Lennard-Jones type.

3.4 Nanodrop Contact Angle at Various Average Fluid Density

In order to verify the generality of Eq. (16), calculations of the contact angle were also performed for $\rho_{av}^* = 0.2$ and all the above considered values of σ as well as for $\rho_{av}^* = 0.15, 0.2$, and 0.25 and $\sigma = 1.028$. The results obtained are listed in Tables 4 and 5 which suggest that there are indeed some general features in the dependence of the contact angles of nanodrops on temperature and the fluid-solid energy parameter in the Lennard-Jones interaction potential. In particular, Table 5 shows that the calculated contact angles depend weakly on the average density of the fluid in the system. This occurs because of the weak change in the density of the vapor phase in the considered close systems when ρ_{av}^* is changed.

TABLE 4
Parameters of Eqs. (15) and (16) for nanodrops for various values of $\sigma = \sigma_{fs}/\sigma_{ff}$ and $\rho_{av}^{*} = 0.2$.

σ	ν	θ_0 (deg)	ϵ_0	C (deg)
1.095	2.28	120.6	0.60	257.0
1.028	2.03	120.6	0.76	183.9
0.969	2.04	119.8	0.92	144.0
0.910	1.9	119.2	1.18	111.7

TABLE 5
Parameters of Eqs. (15) and (16) for nanodrops for various values of the average fluid density and σ = 1.028 (σ_{fs} = 3.5Å).

ρ_{av}^{*}	ν	θ_0 (deg)	ϵ_0	C (deg)
0.1	2.62	112.2	0.86	231.0
0.15	2.19	119.2	0.73	188.1
0.20	2.03	120.6	0.76	183.9
0.25	1.87	124.1	0.68	173.3

3.5 Dependence of the Macroscopic Contact Angle on Liquid-Solid Interactions and Temperature

For macroscopic drops, the DFT provides the values of the surface tensions on the basis of the procedure described in Sec. 2.4. Typical dependencies of γ_{sl} and γ_{sv} on ε are presented in Figure 9 for $T = 85$ K (solid lines) and $T = 95$K (dashed lines). In all cases, both surface tensions decrease with increasing ε, one of them, γ_{sv} remaining always negative. The solid-liquid surface tension γ_{sl} changes sign when ε increases; it is positive at smaller e but becomes negative at larger ones. The intersection of the curves for γ_{sl} and γ_{sv} provides the value of ε at which the contact angle $\theta = 90°$. This value decreases as the temperature increases. The liquid-vapor surface tension does not depend on ε and has the values 0.006600 N/m, 0.005253 N/m, 0.003987 N/m, and 0.002893 N/m for $T = 85$ K, 90 K, 95 K, and 100 K, respectively. Note that the negative values of the surface tensions are likely due to the neglect of the modification of the surface of the solid by the liquid and vapor, issue which we are now investigating.

The analysis of the macroscopic contact angle calculated with Eq. (1) using the obtained values of the surface tensions has shown that its behavior as function of ε and temperature is described by the same equations as those derived for nanodrops (Eqs. (14)–(17), (19), and (20)). However, the parameters of this equations for macroscopic drops are somewhat different from those for nanodrops. For comparison, the values of k and b of Eq. (14) and those of ν, θ_0, ε_0, and C of Eqs. (15) and (16) for macrodrops are presented in Tables 6 and 7, respectively.

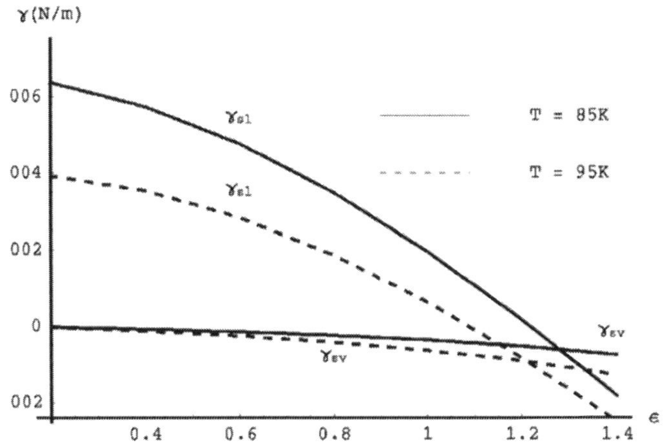

FIG.9. (Reprinted from Ref. [29] with permission.) ε-dependence of the solid-liquid and solid-vapor surface tensions for $T = 85$ K (solid lines) and $T = 95$ K (dashed lines). The liquid-vapor surface tension is equal to 6.60×10^{-3} N/m for $T = 85$ K and 3.99×10^{-3} N/m for $T = 95$K.

TABLE 6
Parameters of Eq. (14) for nanodrops for various σ and temperatures for $\rho_{av}^{*} = 0.1$.

$\sigma = \sigma_{fs}/\sigma_{ff}$	σ_{fs}(Å)	T(K)	k (deg)	b (deg)
		85	–70.1	180.4
1.028	3.5	90	–75.3	184.5
		95	–83.0	191.1
		100	–91.5	197.8
		85	–88.1	179.2
1.095	3.727	90	–94.9	183.2
		95	–105.6	190.3
		100	–115.2	195.5
		85	–56.3	181.1
0.969	3.3	90	–60.2	185.0
		95	–66.4	192.0
		100	–71.6	197.1
		85	–45.8	183.4
0.910	3.1	90	–49.0	187.8
		95	–54.6	196.3
		100	–59.4	202.6

TABLE 7
Parameters of Eqs. (15) and (16) for macrodrops for various values of $\sigma = \sigma_{fs}/\sigma_{ff}$.

σ	ν	θ_0 (deg)	ϵ_0	C (deg)
1.095	1.68	126.0	0.61	155.6
1.028	1.65	123.3	0.82	122.3
0.969	1.54	122.2	1.01	93.8
0.910	1.64	119.0	1.41	79.6

3.6 Conclusions

The main conclusion which can be drawn for nano- and macrodrops on a planar smooth solid surface is that the contact angle, characterizing the wetting properties of a fluid, has some general characteristics with respect to its dependence on temperature and the parameters of the fluid-solid intermolecular interactions, assuming that the latter interactions are of the Lennard-Jones type. At various temperatures, the contact angle has a linear dependence on the dimensionless energy parameter $\varepsilon = \epsilon_{fs}/\epsilon_{ff}$, where ϵ_{fs} and ϵ_{ff} are the fluid-solid and fluid-fluid energy parameters, respectively. The straight lines representing these dependencies at various temperatures intersect almost in a single point (ε_0,θ_0) where ε_0 depends on σ_{fs}, and θ_0 is an almost universal constant. We use the word "almost" because θ_0 varies between 122.5° ± 3.5°, and between 117.4° ± 6.7° for macro and nanodrops, respectively. The contact angle θ_0 for $\varepsilon = \varepsilon_0$ is independent of temperature. For $\varepsilon < \varepsilon_0$ $(\theta > \theta_0)$, the contact angle monotonously increases and for $\varepsilon > \varepsilon_0$ $(\theta < \theta_0)$ monotonously decreases with increasing temperature. This result can qualitatively explain the different temperature behaviors of the contact angle observed in macroscopic experiments [33–39]. For example, an almost linear decrease of θ, with $\frac{d\theta}{dT} = -0.13°/K$, was reported in Ref. [34] for water on naphthalene. A nonlinear decrease of the contact angle of water on polymer surfaces with increasing temperature was observed in Ref. [35]. The opposite behavior, an increase of θ with temperature was found in Ref. [36] for water on perchloroethylene, in Ref. [37] for glycerol, ethylene glycol, and diethylene glycol on elastomers, and in Ref. [38] for mixtures of ^{4}He and ^{3}He on Cs.

The temperature independent value θ_0 of the contact angle corresponding to ε_0 is independent of $\sigma = \sigma_{fs}/\sigma_{ff}$ and is therefore a general characteristic of a nanodrop on a solid surface. For $\theta < \theta_0$, the interactions of the liquid with the solid are stronger $(\varepsilon > \varepsilon_0)$; whereas for $\theta > \theta_0$ the interactions are weaker $(\varepsilon < \varepsilon_0)$. The existence of such a point, if confirmed, may lead to a new classification of materials in hydrophobic and hydrophilic. The conventional classification into hydrophobic and hydrophilic arbitrarely selects $\theta = 90°$ as the separator between hydrophilic and hydrophobic surfaces, while the suggested classification based on θ_0 as the separator has a microscopic origin.

The general features of the contact angle identified for nanodrops can be compared with the results obtained in Refs. [25,26] via molecular dynamics simulations for truncated Lennard-Jones interactions. Although the available data [25,26] do not provide enough information for a full comparison, they are in qualitative agreement with our results. Indeed, in both treatments θ decreases with increasing σ_{fs}, ϵ_{fs}, or the size of the droplet. It should be noted that in Refs. [25,26] it was reported that $\cos\theta$ has a linear dependence on ϵ_{fs} (in contrast to the linear dependence of θ on ϵ_{fs} obtained by us). However, the replotting of the data presented in Refs. [25,26] has shown that the linear fit of θ on ϵ_{fs} is as good as that for $\cos\theta$. The quantitative differences between the values of θ can be explained by the difference in the dimensionality of the drops considered in Refs. [25,26] (three dimensional drops) and in Refs. [29,32] (two dimensional drops).

4 APPENDIX: CONTRIBUTIONS TO THE FREE ENERGY OF THE SYSTEM

The contributions to the free energy listed in Sec. 2.2 can be expressed as follows.

$$F_{id}\left[\rho(\mathbf{r})\right] = k_B T \int d\mathbf{r}\rho(\mathbf{r})\left\{\log\left[\Lambda^3\rho(\mathbf{r})\right]-1\right\} \tag{A.1}$$

$$F_{ex}\left[\rho(\mathbf{r})\right] = \int d\mathbf{r}\rho(\mathbf{r})\Delta\Psi_{hs}(\mathbf{r},) \tag{A.2}$$

where

$$\Delta\Psi_{hs}(\mathbf{r}) = k_B T \eta_{\bar{\rho}} \frac{4-3\eta_{\bar{\rho}}}{(1-\eta_{\bar{\rho}})^2}, \tag{A.3}$$

$\eta_\rho = \frac{1}{6}\pi\bar{\rho}(\mathrm{r})\sigma_{ff}^3$ is the packing fraction of the fluid molecules, σ_{ff} is the fluid hard core diameter, and $\bar{\rho}(\mathbf{r})$ is the smoothed density defined as

$$\bar{\rho}(\mathbf{r}) = \int d\mathbf{r}'\rho(\mathbf{r}')W(\mathbf{r}-\mathbf{r}'). \tag{A.4}$$

The weighting function $W(|\mathbf{r} - \mathbf{r}'|)$ is selected in the form [40]

$$Q(x,h) = -\left[\Delta\Psi_{hs}(x,h) + \overline{\Delta\Psi'}_{hs}(x,h)\right],$$

where $\mathrm{r} = |\mathbf{r} - \mathbf{r}'|$.

The function $Q(\mathbf{r})$ in Eq. (9) depends in general on coordinates x and h and has the form

$$Q(x,h) = -\left[\Delta\Psi_{hs}(x,h) + \overline{\Delta\Psi'}_{hs}(x,h)\right], \tag{A.5}$$

where

$$\overline{\Delta\Psi'}_{hs}(x,h) = \iint dx'dh\rho(x',h')W_y\left(|x-x'|,|h-h'|\right)\frac{\partial}{\partial\bar{\rho}}\Delta\Psi_{hs}(\bar{\rho})|_{\rho=\rho(x',h')}, \tag{A.6}$$

Wy $(|x - x'|, |h - h'|)$ is obtained by the integration of the weighted function $W(|\mathbf{r} - \mathbf{r}'|)$ with respect to the y-direction, from $-\infty$ to $+\infty$.

The function $U_{ff}(x, h)$, which represents in Eq. (9) the fluid-fluid interactions is given by

$$U_{ff}(x,h) = \iint dx'dh'\rho(x',h')\phi_{ff,y}\left(|x-x'|,|h-h'|\right), \tag{A.7}$$

where $\phi_{ff,y}$ $(|x - x'|, |h - h'|)$ is obtained by the integration of the potential $\phi_{ff}(|\mathbf{r} - \mathbf{r}'|)$ with respect to the y-direction, from $-\infty$ to $+\infty$. When evaluating this function, a cutoff at a distance equal to four molecular diameters σ_{ff} for the range of Lennard-Jones attraction was employed. The increase of this distance up to $10\sigma_{ff}$ changed the results by less than one percent.

REFERENCES

1. R. Evans, P. Tarazona, Phys. Rev. A **28**, 1864 (1983)
2. P. Tarazona, R. Evans, Mol. Phys. **52**, 847 (1984)
3. P. Tarazona, Phys. Rev. A **31**, 2672 (1985)
4. P. Tarazona, Phys. Rev. A **32**, 3148 (1985)
5. P. Tarazona, U.M.B. Marconi, R. Evans, Mol. Phys. **60**, 573 (1987)
6. E.A. Ustinov, J. Chem. Phys. **132**, 194703 (2010)
7. P. Bryk, D. Henderson, S. Sokolowski, Langmuir **15**, 6026 (1999)
8. P.I. Ravikovitch, S.C. Odomhnaill, A.V. Neimark, et al., Langmuir **11**, 4765 (1995)
9. P.B. Balbuena, K.E. Gubbins, Langmuir **9**, 1801 (1993)
10. P. Tarazona, R. Evans, Mol. Phys. **48**, 799 (1983)
11. Y. Singh, Physics Reports-Review Section of Physics Letters **207**, 351 (1991).
12. M. Schmidt, A. Fortini, M. Dijkstra, J. Phys.-Cond. Matter **15**, S3411 (2003)
13. A.V. Neimark, P.I. Ravikovitch, Micropor. Mesopor. Mater. **44**, 697 (2001)
14. P. Rocken, A. Somoza, P. Tarazona, et al., J. Chem. Phys. **108**, 8689 (1998)
15. R. Evans, U.M.B. Marconi, P. Tarazona, J. Chem. Phys. **84**, 2376 (1986)
16. V. Talanquer, D.W. Oxtoby, J. Chem. Phys. **104**, 1483 (1996)
17. V. Talanquer, D.W. Oxtoby, J. Chem. Phys. **119**, 9121 (2003)
18. G.O. Berim, E. Ruckenstein, J. Chem. Phys. **129**, 014708 (2008) (Section 4.4 of this volume).
19. G.O. Berim, E. Ruckenstein, J. Chem. Phys. **129**, 114709 (2008) (Section 4.11 of this volume).

20. E. Ruckenstein, G.O. Berim, Adv. Colloid and Interface Sci. **157**, 1 (2010) (Section 4.13 of this volume).
21. M. Merkel, H. Lowen, Phys. Rev. E **54**, 6623 (1996)
22. G.O. Berim, E. Ruckenstein, J. Phys. Chem. B **111**, 2514 (2007) (Section 2.1 of this volume).
23. G.O. Berim, E. Ruckenstein, J. Chem. Phys. **126**, 124503 (2007) (Section 2.2 of this volume).
24. L. Szybisz, S.A. Sartarelli, J. Chem. Phys. **128**, 124702 (2008)
25. S. Maruyama, T. Kurashige, S. Matsumoto, et al., Microscale Thermophysical Engineering **2**, 49 (1998)
26. M.J. de Ruijter, T.D. Blake, J. De Coninck, Langmuir **15**, 7836 (1999)
27. A.E. vanGiessen, D.J. Bukman, B. Widom, J. Coll. Interf. Sci. **192**, 257 (1997)
28. F. Ancilotto, F. Faccin, F. Toigo, Phys. Rev. B **62**, 17035 (2000)
29. G.O. Berim, E. Ruckenstein, J. Chem. Phys. **130**, 184712 (2009) (Section 3.8 of this volume).
30. F. Porcheron, P.A. Monson, Langmuir **22**, 1595 (2006)
31. N. Giovambattista, P.G. Debenedetti, P.J. Rossky, J. Phys. Chem. B **111**, 9581 (2007)
32. G.O. Berim, E. Ruckenstein, J. Chem. Phys. **130**, 044709 (2009) (Section 4.1 of this volume).
33. E. Rolley, C. Guthmann, J. Low Temp. Phys. **108**, 1 (1997)
34. J.B. Jones, A.W. Adamson, J. Phys. Chem. **72**, 646 (1968)
35. F.D. Petke, B.R. Ray, J. Coll. Interf. Sci. **31**, 216 (1969)
36. H.Y. She, B.E. Sleep, Water Resources Research **34**, 2587 (1998)
37. C.J. Budziak, E.I. Varghabutler, A.W. Neumann, J. Appl. Polym. Sci. **42**, 1959 (1991)
38. J. Klier, A.F.G. Wyatt, Phys. Rev. B **54**, 7350 (1996)
39. J.D. Bernardin, I. Mudawar, C.B. Walsh, et al., Int. J. Heat Mass Transfer **40**, 1017 (1997)
40. R.H. Nilson, S.K. Griffiths, J. Chem. Phys. **111**, 4281 (1999)

4.3 Nanodrop of an Ising Magnetic Fluid on a Solid Surface*

Gersh Berim and Eli Ruckenstein†
Department of Chemical and Biological Engineering, State University of New York at Buffalo, Buffalo, New York 14260, United States

Corresponding Author

† E-mail: feaeliru@buffalo.edu.
Phone: (716) 645–1179. Fax: (716) 645–3822.

1. INTRODUCTION

During the past decade, increasing interest in the wetting of solid surfaces by magnetic fluids and its control by an external magnetic field has taken place.[1–11] This was a result of their use in various applications,[12–17] for instance, in the synthesis of carbon nanotubes,[15] magnetic resonance imaging,[16] the preparation of polymer particles for drug transport in living organisms,[17] and so forth.

The magnetic fluids constitute a particular case of the so-called dipolar fluids (i.e., fluids with molecules that possess electric or magnetic dipole moments). In magnetic fluids, the dipole moment either can be induced by an external magnetic field (superparamagnetic fluids) or can exist independently of its presence (ferrofluids).

Dipolar fluids have been investigated both experimentally and theoretically for a long time (see the reviews[18,19] and the references therein) using various approaches. Most of the theoretical contributions have been carried out for bulk fluids with an emphasis on the long–range ordering that occurs in such systems. The interfacial properties of dipolar fluids and their behavior in confined environments have attracted less attention. For instance, the vapor—liquid interface was examined in refs 20–27, and the magnetic fluids in slitlike pores were considered in refs 28–30 using Monte Carlo simulations[29,30] and density functional theory.[28] In ref 31, the shape of the free surface of a magnetic fluid confined between parallel plates was examined using a modified Young—Laplace equation.

In refs 1, 3, and 32, ferrofluid drops on solid surfaces were studied experimentally in the presence of a permanent magnet. It was shown that the contact angle decreases linearly with increasing external magnetic field and that the drop changes its shape from circular to puddlelike[3] because its base diameter increases and its drop height decreases. Another interesting observation was made in ref 1, where it was found that for a magnetic fluid the contact angle decreases rapidly with increasing surfactant concentration, reaching an asymptotic value.

The magnetic dipole—dipole interaction that determines the behavior of magnetic fluids is long–range and depends on the orientation of the intermolecular axis with respect to the direction of the magnetic moments, ranging from attractive (tail–to–tail orientation of the dipoles) to repulsive (side–by–side orientation). To simplify the problem, the so–called Ising fluid[33–37] in which the interaction between the magnetic moments of the fluid molecules does not depend on the orientation of the intermolecular axis and involves only one component of the magnetic moment

* *Langmuir* 2011, 27, 8753–8760. Republished with permission.

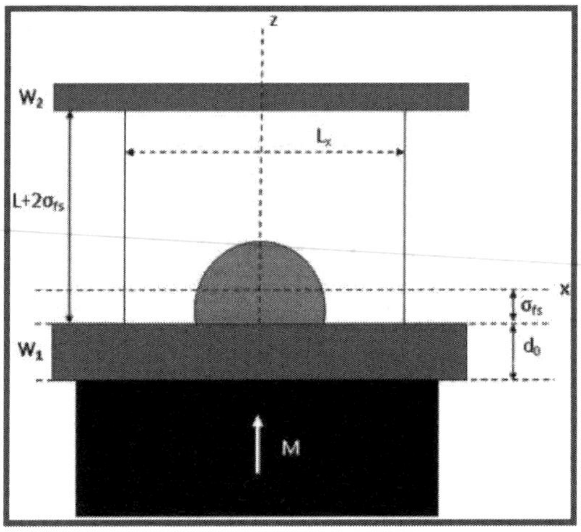

FIGURE 1. Schematic presentation of the system. The fluid is located between two solid walls, with the fluid density distribution being periodic in the x direction with period L_x. The molecules of wall W_1 interact with those of the fluid via the Lennard–Jones potential with hard core repulsion, whereas wall W_2 generates only hard core repulsion. σ_{fs} is the hard core diameter of the fluid—solid interaction. M is the magnetic moment of a permanent magnet that is located under wall W_1.

can be considered. Whereas in general this interaction can be long–range, the common practice is to restrict it to several neighbors. In spite of the model simplicity, the obtained results demonstrate[33–37] that the Ising fluids possess features qualitatively similar to those of real dipolar magnetic fluids. In addition, one can mention the successful application of the Ising model to the description of the solid magnetic materials. (See, for example, refs 38 and 39.)

In this article, a nanodrop of an Ising fluid on a solid surface is examined using density functional theory (DFT). This theory provides an appropriate tool for extracting information regarding nonuniform fluids. As examples of applications of DFT to magnetic fluids, one can mention refs 28 and 40–45, where confined as well as bulk fluids have been of concern. Below, only nanodrops are considered because the computational time rapidly increases with the size of the system.

The presentation of the model is provided in sections 2.1 and 2.2. Details regarding DFT and the procedure used to obtain the drop profile from a continuous fluid density distribution are provided in sections 2.3 and 2.4 and in the Appendix. The results obtained for various external magnetic fields and magnetic interaction constants are presented in section 3.

2. MODEL

2.1. System

The considered system (Figure 1) consists of a one–component fluid of spherical molecules interacting via the Lennard–Jones potential coupled with a hard core repulsion, which is confined between two parallel structureless smooth, solid walls W_1 and W_2. The fluid is subjected to an external magnetic field that can be uniform or nonuniform, the latter being generated by a permanent magnet located under the lower wall (Figure 1). It is assumed that the fluid molecules possess a magnetic dipole moment m of constant magnitude and that they interact with an external magnetic field and between themselves. Walls W_1 and W_2 are separated by a sufficiently large distance $L + 2\sigma_{fs}$, where σ_{fs} is a length parameter (hard core diameter) involved in the fluid—solid interactions that are defined below. For simplicity, a 2D case (cylindrical drop) is considered (i.e., the fluid density distribution (FDD) is considered to depend only on the x and z coordinates and to be uniform in the y direction, with the y axis being normal to the plane of the Figure). It was shown in ref 46 that the cylindrical drop of a nonmagnetic fluid has a profile close to that of an axisymmetric drop usually employed in experiments. It is expected that the same observation will be valid for magnetic fluids. As in molecular dynamics simulations, the FDD is assumed to be periodic in the lateral direction with a period of L_x, with the average density, ρ_{av}, taken over that period being constant.

The lower solid wall, W_1, has a thickness of d_0 and generates an external potential that acts on the fluid molecules. The thickness d_0 is selected to be large enough for this potential to be considered to be caused by a semiinfinite solid. The upper wall, W_2, serves exclusively as a boundary of the system and has no influence on the drop formation on the lower wall. For this reason, W_2 is selected as a hard wall with no attractive interactions with fluid molecules. This avoids the formation of a liquid phase (film or drop) on the upper wall and thus eliminates its influence on what happens on the lower wall.

A permanent magnet has a constant magnetic moment M pointed in the positive direction of the z axis and produces a nonuniform magnetic field.

In the present study, characteristic sizes L and L_x (Figure 1) of the system are on the order of a hundred nanometers (i.e., much smaller than the width of the magnet that is assumed to be macroscopic in size).

2.2. Interaction Potentials

The nonmagnetic interactions between the fluid molecules of the system are described by the potential $\varphi_{ff}\left(|\mathrm{r}-\mathrm{r}'|\right) \equiv \epsilon_{ff}\bar{\varphi}_{ff}\left(|\mathrm{r}-\mathrm{r}'|\right)$ where ϵ_{ff} is an energy parameter and $\bar{\varphi}_{ff}\left(|\mathrm{r}-\mathrm{r}'|\right)$ represents the following dimensionless quantity:

$$\bar{\varphi}_{ff}\left(|\mathbf{r}-\mathbf{r}'|\right)=\begin{cases}4\left[\left(\dfrac{\sigma_{ff}}{r}\right)^{12}-\left(\dfrac{\sigma_{ff}}{r}\right)^{6}\right], & r\geq\sigma_{ff}\\ \infty, & r<\sigma_{ff}\end{cases} \tag{1}$$

In eq 1, the coordinates $\mathbf{r}$ and $\mathbf{r}'$ provide the locations of the interacting fluid molecules, $r=|\mathbf{r}-\mathrm{r}'|$, and σ_{ff} is the fluid hard core diameter. Potential $\bar{\varphi}_{ff}\left(|\mathrm{r}-\mathbf{r}'|\right)$ in eq 1 combines a hard core repulsion at distance r between molecules smaller than σ_{ff} with a Lennard-Jones-type attraction for $r\geq\sigma_{ff}$.

The potential for the fluid—wall W_1 interactions is selected in the same form as $\varphi_{ff}(|\mathbf{r}-\mathbf{r}'|)$, with parameters ϵ_{ff} and σ_{ff} being replaced by ϵ_{fs} and σ_{fs}, respectively. Because of the walls' uniformity, the total potential $U_1(r)$ of interaction between a fluid molecule and wall W_1 depends only on the distance z from the wall, and for a sufficiently large thickness, d_0 can be written in the form

$$U_1(\mathbf{r})=\epsilon_{fs}U_{fs}\,(\mathbf{r}) \tag{2}$$

where the dimensionless quantity $U_{fs}\left(\mathrm{r}\right)\equiv U_{fs}\left(z\right)$ is given by

$$U_{fs}\left(z\right)=\frac{2\pi}{3}\rho_s\sigma_{fs}^3\left[\frac{2}{15}\left(\frac{\sigma_{fs}}{\sigma_{fs}+z}\right)^9-\left(\frac{\sigma_{fs}}{\sigma_{fs}+z}\right)^3\right] \tag{3}$$

The magnetic moment $\mathbf{m}$ associated with a fluid molecule is assumed to have a single nonzero component, mS_z, along the vertical z axis, where $S_z=\pm1$ (the Ising model). Therefore, the potential energy due to the magnetic interaction of a molecule of fluid, located at the point (x,z) with an external magnetic field can be written as

$$U_{mf}\left(x,z\right)=-mS_z\left(x,z\right)B_z\left(x,z\right) \tag{4}$$

where $B_z(x,z)$ is the z component of the field at point (x,z).

The z component of the magnetic field generated by a permanent magnet can be represented in the form (ref 47 Ch. 8) of $B_z(x,z) = (\mu_0 / 2\pi)(\mathrm{M}\cos^2\Theta / \mathrm{r}^3)$, where M is the magnitude of the magnetic moment of the magnet, μ_0 is the permeability constant, and r and Θ are the distance from the magnet and the inclination angle of point (x,z), respectively. Because of the small size of the system compared to the width of the magnet, the angle Θ is small and therefore

$$B_z(x,z) = \frac{\mu_0}{2\pi}\frac{M}{(d_0 + \sigma_{\mathrm{fs}} + z)^3} \tag{5}$$

and the other two components of the magnetic field of the magnet can be taken to be zero because they are proportional to sin Θ. In addition to the nonuniform magnetic field in eq 5, the effect of a uniform magnetic field parallel to the z axis will also be considered. In such a case, $B_z(x,z) \equiv \mathrm{B} = \mathrm{const.}$

In the Ising model, the magnetic interaction between two fluid molecules located at points **r** and **r′** is taken in the form of

$$\mathrm{U}(|\mathrm{r}-\mathrm{r}'|) = -J(|\mathrm{r}-\mathrm{r}'|)S_z(\mathrm{r})S_z(\mathrm{r}') \tag{6}$$

where the energy parameter $J(|\mathrm{r}-\mathrm{r}'|) \geq 0$ is, in general, a decreasing function of $|\mathrm{r}-\mathrm{r}'|$ between molecules. This parameter is selection in the form of

$$J(|\mathbf{r}-\mathbf{r}'|) = \begin{cases} J_0, & \sigma_{\mathrm{ff}} \leq |\mathbf{r}-\mathbf{r}'| \leq R_{\mathrm{cut}} \\ 0, & |\mathbf{r}-\mathbf{r}'| > R_{\mathrm{cut}} \end{cases} \tag{7}$$

where $J_0 \geq 0$ is a constant and R_{cut} is a cutoff radius for the Ising interaction. The interaction potential of a fluid molecule, located at point r, with all other molecules can be written in the form of

$$U_{\mathrm{mm}}(\mathbf{r}) = -mS_z(\mathbf{r})B_{\mathrm{mm}}(\mathbf{r}) \tag{8}$$

where

$$B_{\mathrm{mm}}(\mathbf{r}) = \frac{1}{m}\int_{V'} J(|\mathbf{r}-\mathrm{r}'|)S_z(\mathrm{r}')\rho(\mathrm{r}')\mathrm{d}V' \tag{9}$$

with V' being the volume occupied by the fluid for $|\mathbf{r} - \mathbf{r}'| \geq \sigma_{\mathrm{ff}}$. After integrating over y from $-\infty$ to $+\infty$, one obtains

$$B_{\mathrm{mm}}(x,z) = \frac{J_0}{m}\int_{A_1} \mathrm{d}x'\,dz'\rho(x',z')S_z(x',z')I_1(x,z;x',z') + \frac{J_0}{m}\int_{A_2} \mathrm{d}x'\,dz'\rho(x',z')S_z(x',z')I_2(x,z;x',z') \tag{10}$$

where A_1 is the area in the $x-z$ plane between concentric circles of radii σ_{ff} and R_{cut} with the center at point (x,z), A_2 is the area restricted by the circle of radius σ_{ff} with the center at (x,z), $I_1(x,z;x',z') = 2\left(R_{\mathrm{cut}}{}^2 - (x-x')^2 - (z-z')^2\right)^{1/2}$ and $I_2(x,z;x',z') = I_1(x,z;x',z') - 2(\sigma_{\mathrm{ff}}^2 - (x-x')^2 - (z-z')^2)^{1/2}$. As a consequence, each magnetic moment can be considered to be subjected to a total effective field

$$B_{\mathrm{eff}}(\mathbf{r}) = B_z(\mathbf{r}) + B_{\mathrm{mm}}(\mathbf{r}) \tag{11}$$

where $B_z(\mathbf{r})$ and $B_{\mathrm{mm}}(\mathbf{r})$ are the external magnetic field and the effective field due to the fluid—fluid magnetic interactions.

In the mean-field approximation, the variable $S_z(x',z')$ in eq 10 can be replaced by its equilibrium value, $\langle S_z(x',z')\rangle$ in the effective field $B_{\text{eff}}(x',z')$. The magnetic potential energy of a molecule in this field is given by $-mS_z(x',z')B_{\text{eff}}(x',z')$. One can therefore write (see, for example, ref 48)

$$\langle S_z(x',z')\rangle = \frac{1}{Z}\sum_{S_z=\pm 1} S_z \exp\left[\frac{mS_zB_{\text{eff}}(x',z')}{k_BT}\right]$$

$$= \frac{2}{Z}\sinh\left[\frac{mB_{\text{eff}}(x',z')}{k_BT}\right] \tag{12}$$

where k_B is the Boltzmann constant, T is the absolute temperature, and Z is the partition function

$$Z = \sum_{S_z=\pm 1} \exp\left[\frac{mS_zB_{\text{eff}}(x',z')}{k_BT}\right]$$

$$= 2\cosh\left[\frac{mB_{\text{eff}}(x',z')}{k_BT}\right] \tag{13}$$

Consequently,

$$\langle S_z(x',z')\rangle = \tanh\left[\frac{mB_{\text{eff}}(x',z')}{k_BT}\right] \tag{14}$$

Introducing the dimensionless quantities $B^*_{\text{mm}} = mB_{\text{mm}}(x,z)/k_BT, B^*_{\text{eff}} = mB_{\text{eff}}(x,z)/k_BT$, and $J^*_0 = J_0/k_BT$, one can write

$$B^*_{\text{mm}}(x,z) = \frac{J^*_0}{m}\int_{A1} dx'dz'\rho(x',z')\tanh\left[B^*_{\text{eff}}(x',z')\right]I_1(x,z;x',z')$$

$$+\frac{J^*_0}{m}\int_{A2} dx'dz'\rho(x',z')\tanh\left[B^*_{\text{eff}}(x',z')\right]I_2(x,z;x',z') \tag{15}$$

Because

$$B^*_{\text{eff}}(x,z) = B^*_z(x,z) + B^*_{\text{mm}}(x,z) \tag{16}$$

where $B^*_z(x,z) = mB_z(x,z)/k_BT$ is the external field, the integral equation, eq 15, contains two unknown functions, namely, $B^*_{\text{mm}}(x,z)$ and FDD $\rho(x,z)$. The additional necessary equation is obtained below via the minimization of the Helmholtz free energy of the system.

2.3. Density Functional Theory

The Helmholtz free energy density $F[\rho(r)]$ of the considered system can be written in the form

$$F[\rho(\mathbf{r})] = F_f[\rho(\mathbf{r})] + F_m[\rho(\mathbf{r})] \tag{17}$$

where $F_f[\rho(\mathbf{r})]$ and $F_m[\rho(\mathbf{r})]$ are contributions provided by the nonmagnetic and magnetic interactions, respectively. The former contribution is selected in the form suggested by Tarazona[49]

$$F_{\mathrm{f}}\left[\rho(\mathbf{r})\right]=F_{\mathrm{id}}\left[\rho(\mathbf{r})\right]+F_{\mathrm{ex}}\left[\rho(\mathbf{r})\right]+\frac{\epsilon_{\mathrm{ff}}}{2}\iint d\mathbf{r}\,d\mathbf{r}'\rho(\mathbf{r})\rho(\mathbf{r}')\bar{\varphi}_{\mathrm{ff}}\left(|\mathbf{r}-\mathbf{r}'|\right)+\epsilon_{\mathrm{fs}}\int d\mathbf{r}\rho(\mathbf{r})U_{\mathrm{fs}}(\mathbf{r}) \tag{18}$$

where $F_{\mathrm{id}}[\rho(\mathbf{r})]$ and $F_{\mathrm{ex}}[\rho(\mathbf{r})]$ are the ideal gas and the excess free energy densities, respectively, of a reference hard sphere fluid. The third term in eq 18 represents the contribution of the fluid—fluid interactions treated in the mean–field approximation. Explicit expressions for $F_{\mathrm{id}}[\rho(\mathbf{r})]$ and $F_{\mathrm{ex}}[\rho(\mathbf{r})]$ can be found in the Appendix. The potential $U_{\mathrm{fs}}(\mathbf{r})$ generated by the wall W_1 is provided by eq 3.

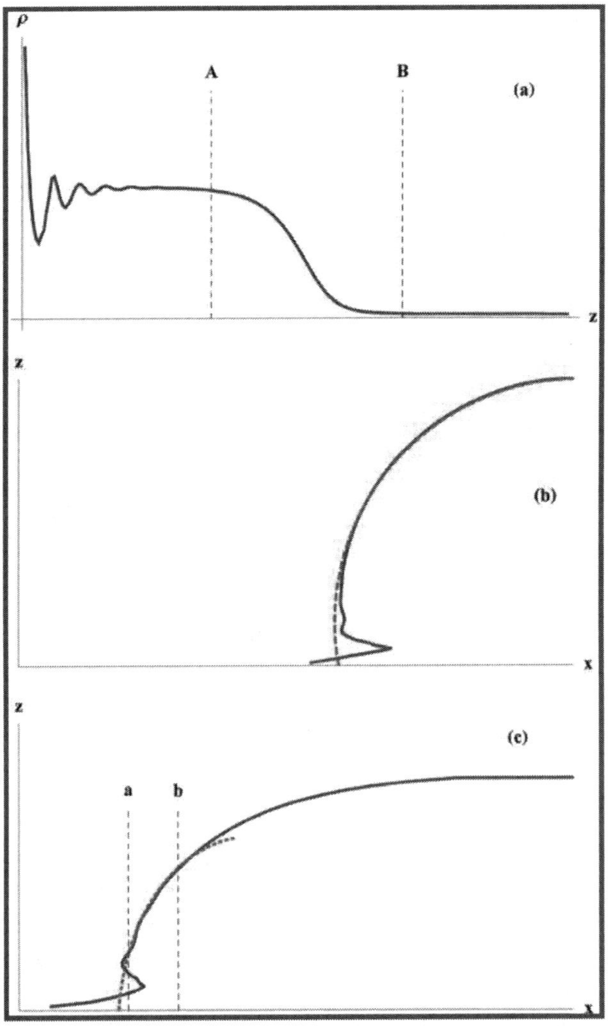

FIGURE 2. (a) Example of a fluid density distribution along the symmetry axis of the drop. Range AB of this distribution corresponds to a liquid—vapor interface. (b) Weak magnetic field. Example of a drop profile (solid line, only the left–hand part of the profile is presented) extracted from the fluid density distribution using the procedure described in the text. The profile is approximated by a circle (dotted line) that has the best fit with the upper part of the profile. (c) Example of a drop profile for a strong magnetic field. The lower part of the profile is approximated by a circle (dotted line) that has the best fit with the ab range of the profile.

The contribution $F_m[\rho(\mathbf{r})]$ to the total free energy density due to the magnetic interactions can be found by combining the standard expression $F_H = -k_B T \log Z$ (free energy per molecule) with eq 13. One thus obtains

$$F_m\left[\rho(\mathbf{r})\right] = -k_B T \rho(\mathbf{r}) \log\left[2\cosh B_{\text{eff}}^*(\mathbf{r})\right] \tag{19}$$

where the factor $\rho(\mathbf{r})$ accounts for the number of molecules at point r and the effective field $B_{\text{eff}}^*(r)$ depends on $\rho(\mathbf{r})$ according to eqs 15 and 16.

In a system with a fixed average density ρ_{av}, the Euler—Lagrange equation for the fluid density distribution $\rho(\mathbf{r})$ can be obtained by minimizing the Helmholtz free energy $F_h = \int_V dVF[\rho(\mathbf{r})]$ under the constraint

$$\rho_{av} = \frac{1}{V}\int_V d\mathbf{r}\rho(\mathbf{r}) \tag{20}$$

As a result, one obtains

$$\log\left[\Lambda^3\rho(\mathbf{r})\right] + Q_{hs}(\mathbf{r}) + Q_m(\mathbf{r}) + \epsilon_{ff}^* U_{ff}(\mathbf{r}) + \epsilon_{fs}^* U_{fs}(\mathbf{r}) = \lambda^* \tag{21}$$

where $\epsilon^* = \epsilon/k_B T$, $\Lambda = h_p/(2\pi M_{mol} k_B T)^{1/2}$ is the thermal de Broglie wavelength, h_P is the Planck constant, M_{mol} is the mass of a fluid molecule, $\lambda^* = \lambda/k_B T$, and λ is a Lagrange multiplier. The functions $Q_{hs}(\mathbf{r}), Q_m(\mathbf{r}), U_{ff}(\mathbf{r})$, and $U_{fs}(\mathbf{r})$ represent the hard sphere system, the magnetic interactions, the fluid—fluid nonmagnetic interactions, and the fluid—solid nonmagnetic interactions, respectively. The first three are provided in the Appendix, and the last one is given by eq 3.

The two integral equations, eqs 15 and 21, allow one to calculate via iterations both $B_{mm}^*(x,z)$ and $\rho(x,z)$. For nonmagnetic fluids, the iteration procedure of solving the Euler—Lagrange equation was described in ref 50. The only difference between the present procedure and that used in ref 50 is that at each iteration step the effective field $B_{\text{eff}}^*(r)$ which is involved in the term $Q_m(\mathbf{r})$ of eq 21, has to be determined as the solution of eq 15 with the $\rho(\mathbf{r})$ used in the iteration step.

2.4. Calculation of the Contact Angle

To calculate the contact angle, the drop profile was first extracted from the obtained continuous 2D FDD as the curve along which the density $\rho(x,z)$ is constant. The value ρ_{div} of the constant was selected as the density on an equimolar dividing surface in a 1D FDD along the z axis (vertical axis of symmetry). An example of such an FDD is presented in Figure 2a. In the vicinity of the liquid—solid interface, there are strong fluid density oscillations resulting from the ordering of the fluid molecules that form several liquid layers of various densities. At larger distances from the solid, the density becomes almost constant, acquiring liquidlike values, and then it quickly decreases across the liquid—vapor interface, approaching vaporlike values. By considering the AB range (Figure 2a) of the fluid density distribution that contains the vapor—liquid interface, the location of the equimolar dividing surface and the corresponding density ρ_{div} were determined. Note that a similar approach to calculating the contact angle was used in refs 52–54, where the FDDs were obtained by molecular dynamics simulations.

Typical examples of drop profiles obtained as described above for nanodrops of magnetic fluids are presented schematically in Figures 2b, c. For the profile presented in Figure 2b that corresponds to weak magnetic interactions, its upper part was approximated by a circle that was extended up to the surface (dotted line in Figure 2b). The angle θ that this circle makes with the surface was considered to be the actual contact angle. For the profile presented in Figure 2c for strong magnetic interactions, the upper part of the profile is almost planar. In this case, the circular approximation of the profile was applied to the lower portion (ab in Figure 2c) of the profile that is located outside the range of strong density oscillations (dotted line in Figure 2c).

Note that the change in the location of the dividing surface and the corresponding change in ρ_{div} weakly affect the calculated contact angle.[50]

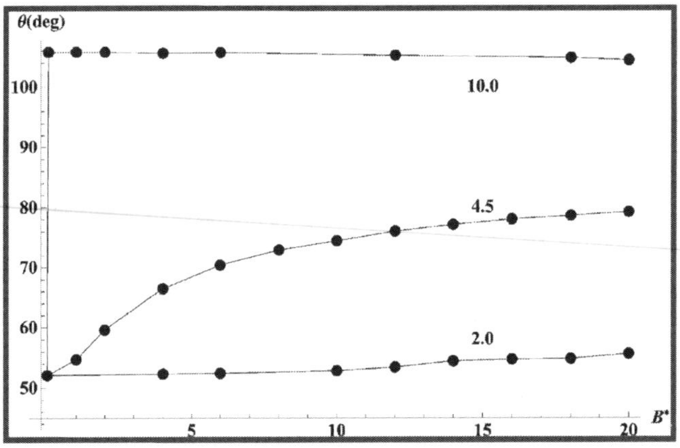

FIGURE 3. Dependence of the contact angle on the external uniform magnetic field for various values of J_0^* indicated on each curve and $\epsilon_{fs}^* = 1.800$. The points represent the results of the calculation of the contact angle. The lines are guides for the eye.

3. RESULTS

Below, the dimensionless energy parameter of the fluid—fluid nonmagnetic interaction potential is selected: $\epsilon_{ff}^* \equiv \epsilon_{ff}/k_BT = 1.409$. The parameters of the fluid—solid interaction potential are selected: $\sigma_{fs} = 1.095\sigma_{ff}$ and $\epsilon_{fs}^* = 1.800, 1.409$, and 1.127. Several values for ϵ_{fs}^* were used to estimate the effect of magnetic interactions on various kinds of substrates. Dimensionless parameters $J_{0,}^*$ $B_M^* = \mu_0 Mm/2\pi\sigma_{ff}^3 k_BT$, and $B^* = mB/k_BT$ characterizing the magnetic fluid—fluid interactions, the interactions of a permanent magnet with a magnetic fluid, and the interaction of an external uniform magnetic field B with the magnetic fluid were changed over wide ranges to estimate their effect on the contact angle of the drop. The dimensions of the system were selected as $L = 20\sigma_{ff}$, $L_x = 60\sigma_{ff}$, and $d_0 = 100\sigma_{ff}$, and the cutoff radius for Ising interactions was taken to be $R_{cut} = 2\sigma_{ff}$.

In section 3.1, the simplest case of a drop of magnetic fluid in a uniform magnetic field B^* is considered, whereas in section 3.2 the external field is considered to be generated by a permanent magnet.

3.1. CONTACT ANGLE AS A FUNCTION OF A UNIFORM EXTERNAL MAGNETIC FIELD

If the external magnetic field is parallel to the z axis and is uniform, then field $B_z^*(x,z)$ does not depend on the coordinates $\left(B_z^*(x,z) \equiv B^* = \text{const}\right)$. In Figure 3, the contact angle dependence on the external uniform magnetic field B* is presented for $J_0^* = 2.0, 4.5$, and $\epsilon_{fs}^* = 1.800$. In all cases, the contact angle at $B^* = 0$ is $\theta \simeq 52^\circ$, a value that coincides with the contact angle of a nonmagnetic drop on the same surface. For $J_0^* = 0$, the contact angle does not depend on B^*. This result is expected because in the absence of magnetic fluid—fluid interactions the magnetic and nonmagnetic subsystems are independent and the effective magnetic field B_{eff}, given by eq 16, is constant, $B_{eff}^* = B^*$. The external magnetic field generates an ordering of the magnetic moments of the molecules along the z axis, which is independent of the molecules' locations and does not affect in any way the FDD $\rho(r)$. Mathematically, this occurs because the function $Q_m(r)$ in the Euler—Lagrange equation, representing the magnetic fluid—fluid interactions, is constant for $J_0^* = 0$, and can be included in Lagrange multiplier λ*.

For $J_0^* \neq 0$, the contact angle increases with increasing B^*, reaching an asymptotic value that increases as the energy parameter J_0^* increases (Figure 3). This Figure also indicates that for the same $B^* \neq 0$ the contact angle increases with increasing J_0^*. To examine the latter feature in more detail, the dependence of θ on J_0^* for the weak external field $B^* = 0.01$ is presented in Figure 4a.

The contact angle is almost constant for $0 < J_0^* < 4.73$ but then rapidly increases with increasing J_0^*. To interpret these behaviors one should take into account that the potential of the fluid—fluid magnetic interactions, eq 6, generates an additional attractive force between fluid molecules. Indeed, in the mean– field approximation this potential has the form

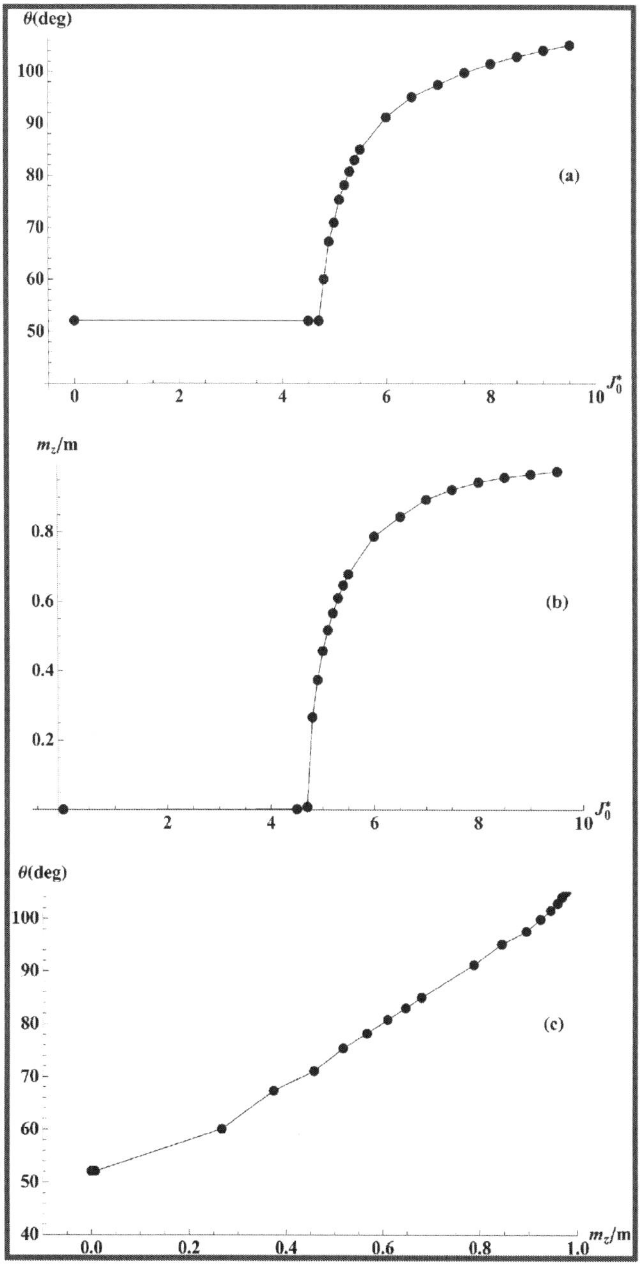

FIGURE 4. (a) Dependence of the contact angle on the energy parameter J_0^* of the magnetic fluid—fluid interactions. (b) Average magnetic moment of the molecules in the drop as a function of J_0^*. (c) Dependence of the contact angle on the average magnetic moment of the molecules in the drop. For all panels, B* = 0.01 and $\epsilon_{fs}^* = 1.800$. The points represent results of calculations of the contact angle. The lines are guides for the eye.

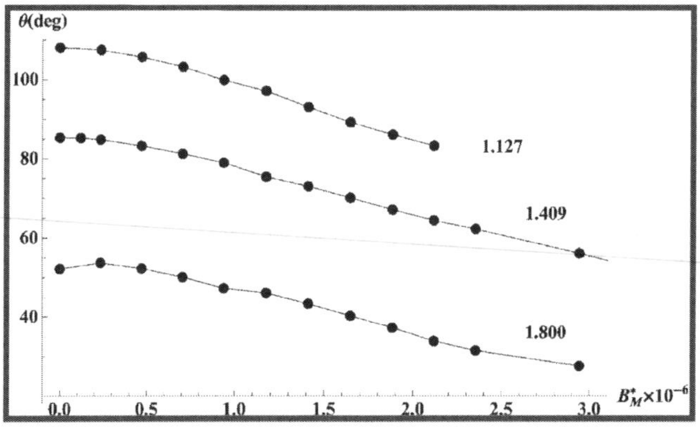

FIGURE 5. Dependence of the contact angle on the magnetic field of a permanent magnet for various values of the energy parameter of the fluid—solid interaction ϵ_{fs}^* indicated on each curve and $J_0^* = 0$. The points represent the results of calculations of the contact angle. The lines are guides for the eye.

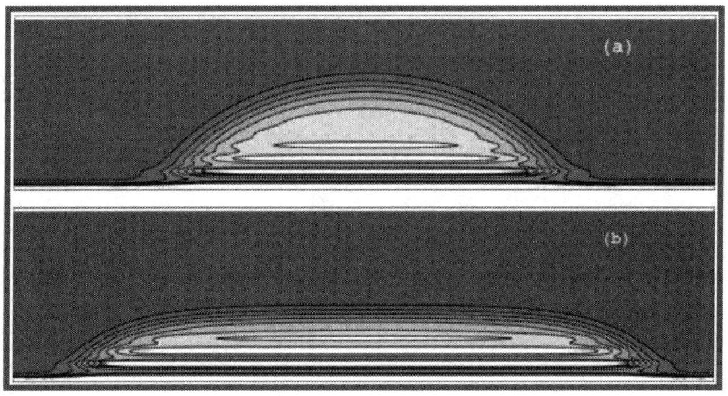

FIGURE 6. Fluid density distributions for (a) weak and (b) strong permanent magnets. The lighter areas correspond to higher fluid densities.

$$U(|\mathbf{r}-\mathbf{r}'|) = -J(|\mathbf{r}-\mathbf{r}'|)\langle S_z(\mathbf{r})\rangle\langle S_z(\mathbf{r}')\rangle \tag{22}$$

where the averages $\langle S_z(\mathbf{r})\rangle$ and $\langle S_z(\mathbf{r}')\rangle$ have in the considered case$\left(J_0^* > 0\right)$ the same sign and $J(|\mathbf{r} - \mathbf{r}'|) > 0$ decreases with increasing distance between molecules. This leads to an increase of the potential $U(|\mathbf{r} - \mathbf{r}'|)$ with increasing distance between molecules that is equivalent to an attractive interaction between them. The attractive force increases with increasing J_0^* and increasing z component, $m\langle S_z(\mathbf{r})\rangle$, of the average molecular magnetic moment. The latter quantity increases with increasing external magnetic field B^* and increasing J_0^*. As shown in ref 51, the larger the ratio $\epsilon_{ff}^* / \epsilon_{fs}^*$, the larger the contact angle θ. For this reason, θ increases with increasing B^* or J_0^*.

The dependence of m_z, the z component of the average of the magnetic moment of the molecules of the drop, on J_0^* is similar to that of the contact angle (Figure 4b). Figure 4b reveals the existence of a magnetic (quasi–)phase transition from a disordered to an ordered state of the magnetic moments of the fluid molecules that occurs at $J_0^* \simeq 4.73$. Note that such a transition was not identified in real magnetic fluids where the magnetic interactions are of the dipole—dipole (not Ising) type.

It is interesting that the dependence of the contact angle on m_z is almost linear (Figure 4c).

3.2. Contact angle in the presence of a permanent magnet

In this section, the contact angle for a drop of magnetic fluid on a solid surface is examined as a function of the dimensionless parameter B_M^*, representing the interaction of a fluid molecule with a nonuniform magnetic field generated by a permanent magnet. In Figure 5, such a dependence is presented for various values of the energy parameter ϵ_{fs}^* of the fluid—solid (nonmagnetic) interactions. The results plotted in this Figure were obtained for $J_0^* = 0$ and hence for a zero contribution of the magnetic fluid—fluid interactions.

One can see that for all ϵ_{fs}^*, θ decreases with increasing B_M^*, with the dependence being nonlinear for small B_M^* and almost linear for large B_M^*. As expected, at constant B_M^* the contact angle increases with decreasing ϵ_{fs}^*.

A strong permanent magnet affects not only the contact angle but also the shape of the drop. In Figure 6, nanodrops of the same volume are presented for $B_M^* = 1.0\times10^5$ and 2.0×10^6 for $\epsilon_{fs}^* = 1.800$. In the first case, the shape of the drop outside the range of density oscillations near the solid surface is close to circular, whereas in the second case, the shape of the upper portion of the drop is almost planar.

The decrease in θ with the increasing nonuniform magnetic field of a permanent magnet seems to contradict its behavior for increasing uniform magnetic fields, presented in section 3.1. However, one should take into account that in contrast to the uniform field, the nonuniform field generates an attractive force on the drop in the negative z direction that is similar to gravity but can be much stronger. Such a force can deform the shape of the drop and thus can decrease the contact angle. Both effects- the linear decrease in the contact angle and the change in the drop shape—have been observed in the experiments of ref 3. Consequently, one can use nonuniform magnetic fields to control the drop geometry.

For $J_0^* \neq 0$, the contact angle dependence on B_M^* is more complicated. In Figure 7, the B_M^* dependence of θ is presented for $J_0^* = 0.5, 2.5$, and 5.0 and $\epsilon_{fs}^* = 1.800$. The contact angle first increases, attains a maximum, and then decreases with increasing B_M^*. Such behavior occurs because of two competitive mechanisms. First, as noted in section 3.1, for $J_0^* \neq 0$ the magnetization increases rapidly with increasing external magnetic field. This leads to an increase in the magnetic attraction between fluid molecules, which can be interpreted as an increase in the energy parameter ϵ_{ff} in the Lennard–Jones potential. As a result,[51] angle θ increases with increasing magnetic field. Second, the increase in B_M^* results in an increase in the attraction of the fluid molecules by the magnet. This attraction has to be added to the Lennard–Jones attraction between fluid and solid and can be interpreted as an increase in energy parameter ϵ_{fs}. This leads[51] to the decrease in the contact angle θ (Figure 5). For small B_M^*, the first mechanism dominates, whereas for larger B_M^*, when m is close to saturation, the second mechanism becomes responsible for the decrease in the contact angle with increasing B_M^*.

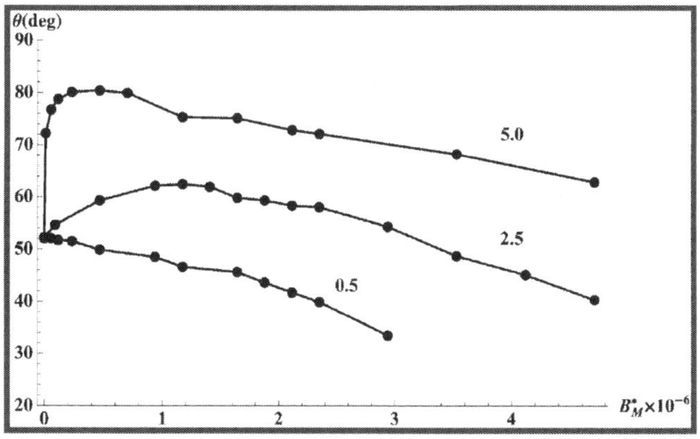

FIGURE 7. Dependence of the contact angle on the strength of a permanent magnet for various values of J_0^* indicated on each curve. The points represent the results of calculations of the contact angle. The lines are guides for the eye.

4. DISCUSSION

The considered Ising magnetic fluid model describes from a microscopic point of view the equilibrium of a nanodroplet on a solid surface in the presence of an external magnetic field (uniform or nonuniform). It is based on the density functional theory for simple fluids but involves an additional term describing the contribution of the magnetic interactions between the molecules of fluid and between the molecules of fluid and the magnetic field. In spite of the oversimplified description of the magnetic interactions, the theory provides qualitative agreement with some of the experimental data available in the literature. Nguyen et al.[3] observed a linear decrease in contact angle θ with increasing strength of a permanent magnet. Similar behavior is predicted by the present theory (Figures 5) for small magnetic interactions between fluid molecules. The system examined by Nguyen et al. differs from that described by our model; consequently, only qualitative agreement is expected. Indeed, the present theory predicts the nonlinear dependence of the contact angle for small B_M^*, which becomes linear for large B_M^*. The experimental data are represented by the linear dependence in the entire range of the magnetic field.

Another feature of a real drop of magnetic fluid on a solid surface is the change in the shape of the drop profile with increasing strength of the permanent magnet. As shown by Nguyen et al.[3] for large magnetic fields produced by a permanent magnet, the upper portion of the drop profile departs from the circular shape and becomes almost planar. This is also predicted by the present theory (cf. Figure 6a, b).

In ref 1, a decrease in the contact angle with increasing concentration of surfactant in the fluid was observed. In real magnetic fluids, the surfactant covers the surface of magnetic particles, and this decreases the fluid—fluid magnetic interactions that are represented by J_0^*. One can see from Figures 3 and 7 that the contact angle decreases with decreasing J_0^*, in agreement with the observations in ref 1.

In conclusion, the approximation of the real magnetic moment by a single component in the z direction (as employed in the Ising model) is reasonable in the presence of an external magnetic field in that direction. In this case, the field aligns the magnetic moments in the z direction and their x and y components become negligible.

It is, however, hard to estimate the consequences of another approximation employed in the considered Ising fluid, namely, the independence of the magnetic fluid—fluid interaction of the direction of the vector linking the interacting molecules. For Ising magnetic fluids, the dipole—dipole interaction potential $U_{1,2}$ of two interacting molecules has the form $U_{1,2} \approx (1-3\cos^2\Theta)\mu_{z,1}\mu_{z,2}$, where Θ is the inclination angle of the vector connecting these molecules. For the side–by–side location of the magnetic moments, $U_{1,2}$ has minima for the antiparallel orientation of $\mu_{z,1}$ and $\mu_{z,2}$. In contrast, for the tail-to-tail location, $U_{1,2}$ has minima for the parallel orientation of the magnetic moments. For this reason, it is expected that if the external field is sufficiently weak and cannot force all magnetic moments to become aligned in the z direction then the tail-to-tail magnetic moments become ordered ferromagnetically, in contrast to those located side-by-side that will be ordered antiferromagnetically. As a consequence, sharp increases in magnetization (as in Figure 4b) and in the contact angle (as in Figures 3 and 4a) may not occur. The actual behavior of magnetization and the contact angle as functions of model parameters can be obtained only using more appropriate models.

APPENDIX: CONTRIBUTIONS TO THE FREE ENERGY OF THE SYSTEM

The contributions to the free energy listed in section 2.3 can be expressed as follows

$$F_{\mathrm{id}}\left[\rho(\mathbf{r})\right] = k_B T \int d\mathbf{r}\rho(\mathbf{r})\left\{\log\left[\Lambda^3\rho(\mathbf{r})\right]-1\right\} \tag{A.1}$$

$$F_{\mathrm{id}}\left[\rho(\mathbf{r})\right] = \int d\mathbf{r}\rho(\mathbf{r})\Delta\Psi_{\mathrm{hs}}(\mathbf{r}) \tag{A.2}$$

where

$$\Delta\Psi_{hs}(\mathbf{r}) = k_B T \eta_\rho \frac{4-3\eta_\rho}{(1-\eta_\rho)^2} \tag{A.3}$$

$\eta_\rho = (1/6)\pi\bar{\rho}(r)\sigma_{ff}^3$ is the packing fraction of the fluid molecules, and $\bar{\rho}(\mathbf{r})$ is the smoothed density defined as

$$\bar{\rho}(\mathbf{r}) = \int d\mathbf{r}' \rho(\mathbf{r}') W(|\mathbf{r}-\mathbf{r}'|) \tag{A.4}$$

The weighting function $W(|\mathbf{r}-\mathbf{r}'|)$ is selected in the form[55]

$$W(|\mathbf{r}-\mathbf{r}'|) = \begin{cases} \dfrac{3}{\pi\sigma_{ff}^3}\left(1-\dfrac{r}{\sigma_{ff}}\right), & r \le \sigma_{ff} \\ 0, & r > \sigma_{ff} \end{cases}$$

where $r = |\mathbf{r}-\mathbf{r}'|$.

Functions $Q_{hs}(\mathbf{r})$ and $Q_m(\mathbf{r})$ in eq 21 depend in general on coordinates x and z and have the form

$$Q_{hs}(x,z) = -\left[\Delta\Psi_{hs}(x,z) + \overline{\Delta\Psi'}_{hs}(x,z)\right] \tag{A.5}$$

where

$$\overline{\Delta\Psi'}_{hs}(x,z) = \iint dx'dz' \rho(x',z') W_y(|x-x'|,$$

$$|z-z'|) \frac{\partial}{\partial\bar{\rho}} \Delta\Psi_{hs}(\bar{\rho})|_{\rho=\rho(x',z')} \tag{A.6}$$

$W_y(|x-x'|,|z-z'|)$ is obtained by integrating weighted function $W(|\mathrm{r}-\mathrm{r}'|)$ with respect to the y direction from $-\infty$ to $+\infty$

$$Q_m(x,z) = -\log\left[2\cosh B_{ff}^*(\mathbf{r})\right]$$

$$-J_0^*\tanh\left[B_{ff}^*(x,z)\right]\int_{A1} dx'dz'\rho(x',z')\tanh\left[B_{eff}^*(x',z')I_1(x,z;x'z')\right.$$

$$-J_0^*\tanh\left[B_{ff}^*(x,z)\right]\int_{A2} dx'dz'\rho(x',z')\tanh\left[B_{eff}^*(x',z')I_2(x,z;x'z')\right. \tag{A.7}$$

where functions I_1 and I_2 are defined in section 2.2.

Function $U_{ff}(x,z)$, which accounts in eq 21 for the fluid—fluid interactions, is given by

$$U_{ff}(x,z) = \iint dx'dz'\rho(x',z')\bar{\varphi}_{ff,y}(|x-x'|, |z-z'|) \tag{A.8}$$

Where $\bar{\varphi}_{ff,y}(|x-x'|,|z-z'|)$ is obtained by integrating potential $\bar{\varphi}_{ff}(|\mathbf{r}-\mathbf{r}'|)$, eq 1, with respect to the y direction from $-\infty$ to $+\infty$. When evaluating this function, a cutoff at a distance equal to four molecular diameters σ_{ff} for the range of Lennard–Jones attraction was employed. The increase in this distance up to $10\sigma_{ff}$ changed the results by less than 1%.

REFERENCES

1) Asakura, H.; Nakajima, A.; Sakai, M.; Suzuki, S.; Kameshima, Y.; Okada, K. *Appl. Surf. Sci* **2007**, *253*, 3098–3102.
2) Chen, C. Y.; Cheng, Z. Y. *Phys. Fluids* **2008**, *20*, 054105.
3) Nguyen, N. T.; Zhu, G. P.; Chua, Y. C.; Phan, V. N.; Tan, S. H. *Langmuir* **2010**, *26*, 12553–12559.
4) Egatz–Gomez, A.; Melle, S.; Garcia, A. A.; Lindsay, S. A.; Marquez, M.; Dominguez–Garcia, P.; Rubio, M. A.; Picraux, S. T.; Taraci, J. L.; Clement, T.; Yang, D.; Hayes, M. A.; Gust, D. *Appl. Phps. Lett.* **2006**, *89*, 034106.
5) Garcia, A. A.; Egatz–Gomez, A.; Lindsay, S. A.; Dominguez–Garcia, P.; Melle, S.; Marquez, M.; Rubio, M. A.; Picraux, S. T.; Yang, D. Q.; Aella, P.; Hayes, M. A.; Gust, D.; Loyprasert, S.; Vazquez–Alvarez, T.; Wang, J. *J. Magn.Magn. Mater.* **2007**, 311, 238–243.
6) Pipper, J.; Inoue, M.; Ng, L. F. P.; Neuzil, P.; Zhang, Y.; Novak, L. *Nat. Med.* **2007**, *13*, 1259–1263.
7) Long, Z. C.; Shetty, A. M.; Solomon, M. J.; Larson, R. G. *Lab Chip* **2009**, *9*, 1567–1575.
8) Nguyen, N. T.; Ng, K. M.; Huang, X. Y. *Appl. Phys. Lett.* **2006**, *89*, 052509.
9) Beyzavi, A.; Nguyen, N. T. *J. Micromech. Microeng.* **2010**, *20*, 015018.
10) Sun, Y.; Nguyen, N. T.; Kwok, Y. C. *Anal. Chem.* **2008**, *80*, 6127–6130.
11) Plaza, R. C.; de Vicente, J.; Gomez–Lopera, S.; Delgado, A. V. *J. Colloid Interface Sci.* **2001**, *242*, 306–313.
12) Williams, R. A. Colloid and Surface Engineering: Applications in the Process Industries; Butterworth–Heinemann Ltd.: Oxford, U.K., 1992.
13) McKay, R., Ed. *Technological Applications of Dispersions;* Dekker: New York, 1994.
14) Pugh, R., Bergstrom, L., Eds. Surface and Colloid Chemistry in Advanced Ceramics Processing; Dekker: New York, 1994.
15) Cho, Y. S., Choi, G. S.; Kim, D. J.; Kim, H. J.; Yoon, S. G. *Met. Mater. Int.* **2003**, *9*, 427–431.
16) Iannone, A.; Magin, R. L.; Walczak, T.; Federico, M.; Swartz, H. M.; Tomasi, A.; Vannini, V. *Magn. Reson. Med.* **1991,** *22,* 435–442.
17) Pouliquen, D.; Chouly, C. *Medical and Biotechnology Applications:* The MML Series; Citus Books: London, 1999; Vol. 2.
18) Huke, B.; Lucke, M. *Rep. Prog. Phys.* **2004**, *67*, 1731–1768.
19) Klapp, S. H. L. *J. Phys.: Condens. Matter* **2005**, *17*, R525–R550.
20) Szalai, I.; Chan, K. Y.; Tang, Y. W. *Mol. Phys.* **2003**, *101*, 1819–1828.
21) Boda, D.; Winkelmann, J. C.; Liszi, J.; Szalai, I. *Mol. Phys.* **1996**, *87*, 601–624.
22) Shelley, J. C.; Patey, G. N.; Levesque, D.; Weis, J.J. *Phys. Rev. E* **1999**, *59*, 3065–3070.
23) Tavares, J. M.; da Gama, M. M. T.; Osipov, M. A. *Phys. Rev. E* **1997**, *56*, R6252–R6255.
24) Ng, K. C.; Valleau, J. P.; Torrie, G. M.; Patey, G. N. *Mol. Phys.* **1979**, *38*, 781–788.
25) Vanleeuwen, M. E.; Smit, B.; Hendriks, E. M. *Mol. Phys.* **1993**, *78*, 271–283.
26) Smit, B.; Williams, C. P.; Hendriks, E. M.; Deleeuw, S. W. *Mol. Phys.* **1989**, *68*, 765–769.
27) Shilov, V. P.*J.Magn. Magn. Mater.* **2006**, *302*, 495–502.
28) Gramzow, M.; Klapp, S. H. L. *Phys.Rev. E* **2007**, *75*, 011605.
29) Jordanovic, J.; Klapp, S. H. L. *Phys. Rev. Lett.* **2008**, *101*, 038302.
30) Trasca, R. A.; Klapp, S. H. L. *J. Chem. Phys.* **2008**, *129*, 084702.
31) Polevikov, V.; Tobiska, L. *J. Magn. Magn. Mater.* **2005**, *289*, 379–381.
32) Zhang, J. H.; Cheng, Z. J.; Zheng, Y. M.; Jiang, L. *Appl. Phys. Lett.* **2009**, *94*, 144104.
33) Ferreira, A. L.; Korneta, W. *Phys. Rev. E* **1998**, *57*, 3107–3114.
34) Nijmeijer, M. J. P.; Parola, A.; Reatto, L. *Phys. Rev. E* **1998**, *57*, 465–474.
35) Omelyan, I. P.; Mryglod, I. M.; Folk, R.; Fenz, W. *Phys. Rev. E* **2004**, *69*, 061506.
36) Fenz, W.; Folk, R.; Mryglod, I. M.; Omelyan, I. P. *Phys. Rev. E* **2007**, *75*, 061504.
37) Omelyan, I. P.; Folk, R.; Mryglod, I. M.; Fenz, W. *J. Chem. Phys.* **2007**, *126*, 124702.
38) Jongh de, L. J.; Miedema, A. R. *Adv. Phys.* **1974**, *23*, 1–260.
39) Steiner, M; Villain, J.; Windsor, C. G. *Adv. Phys.* **1976**, *25*, 87–209.
40) Teixeira, P. I.; Dagama, M. M. T. *J. phys.; Condens. Matter* **1991**, *3*, 111–125.
41) Groh, B.; Dietrich, S. *Phys. Rev. E* **1994**, 50, 3814–3833.
42) Groh, B.; Dietrich, S. *Phys. Rev. Lett.* **1994**, *72*, 2422–2425.
43) Groh, B.; Dietrich, S. *Phys. Rev. E* **1997**, *55*, 2892–2901.
44) Groh, B.; Dietrich, S. *Phys. Rev. Lett.* **1997**, *79*, 749–752.
45) Groh, B.; Dietrich, S. *Phys. Rev. E* **1998**, *57*, 4535–4546. 20.
46) Berim, G. O.; Ruckenstein, E. *J. Phys. Chem.* **2004**, *108*, 19330–19338 (Section 3.2 of this volume).

47) Franklin, J. *Classical Electromagnetism*; Pearson Addison Wesley: San Francisco, **2005**.
48) Aharoni, A. *Introduction to the Theory of Ferromagnetism;* Clarendon Press: Oxford, U.K., 1998.
49) Tarazona, P. *Phys. Rev. A* **1985**, *31*, 2672–2679.
50) Berim, G.; Ruckenstein, E. *J. Chem. Phys.* **2008**, *129*, 014708 (Section 4.4 of this volume).
51) Berim, G. O.; Ruckenstein, E. *J. Chem. Phys.* **2009**, *130*, 044709 (Section 4.1 of this volume).
52) Daub, C. D.; Wang, J. H.; Kudesia, S.; Bratko, D.; Luzar, A. Faraday Discuss. **2010**, *146*, 67–77.
53) Giovambattista, N.; Debenedetti, P. G.; Rossky, P. J. *J. Phys. Chem. B* **2007**, *111*, 9581–9587.
54) de Ruijter, M. J.; Blake, T. D.; De Coninck, J. *Langmuir* **1999**, *15*, 7836–7847.
55) Nilson, R. H.; Griffiths, S. K. *J. Chem. Phys.* **1999**, *111*, 4281–4290.

4.4 Nanodrop on a Nanorough Solid Surface

*Density Functional Theory Considerations**

Gersh Berim and Eli Ruckenstein†
Department of Chemical and Biological Engineering, State University of New York at Buffalo, Buffalo, New York 14260, USA

Corresponding Author
†E-mail: feaeliru@buffalo.edu.
Tel.: (716)645-1179, Fax: (716)645-3822.

(Received 21 April 2008; accepted 4 June 2008; published online 7 July 2008)

I. INTRODUCTION

The theoretical description of a liquid drop on the surface of a solid has a long history. It started more than two hundred years ago with Young's equation[1]

$$\gamma_{lv} \cos\theta = \gamma_{sv} - \gamma_{sl}, \tag{1}$$

where γ_{sv}, γ_{sl}, and γ_{lv} are the solid-vapor, solid-liquid, and liquid-vapor surface tensions, respectively, and θ is the contact angle which the drop profile makes with an ideal (without any defects) solid surface. Young's equation relates an observable quantity such as the contact angle to some macroscopic characteristics of the fluid and solid such as the surface tensions that incorporate the effects of fluid-solid and fluid-fluid interactions as well as that of the temperature.

The theory was improved later by developing a Young—Laplace[2] and an augmented Young—Laplace[3] equation for the drop profile, for which Young's equation [Eq. (1)] provides a boundary condition. For a cylindrical drop (drop which is infinite in one of the direction on the surface), the latter equation has the simple form

$$\frac{h_{xx}}{\left(1+h_x^2\right)^{3/2}} + \Pi\left(h\right)\gamma_{lv} = -p_c/\gamma_{lv}, \tag{2}$$

where $h \equiv h\left(x\right)$ is the drop profile, $\Pi(h)$ is the disjoining pressure, p_c is the capillary pressure, and $h_x \equiv dh/dx$. Far from the solid surface, this equation provides a circular drop profile, the extension of which up to the surface provides the contact angle θ. However in the vicinity of the solid surface, the shape of the drop is not a circular one. Therefore an additional microcontact angle can be considered along with the apparent one.[4]

* *The Journal of Chemical Physics* 129, 014708 (2008). Republished with permission.

For a rough solid surface, Young's equation [Eq. (1)] does not provide a correct description of the contact angle because in this case the surface tensions γ_{sv} and γ_{sl} are not clearly defined. For chemically inhomogeneous smooth solids such as those composed of two types, A and B, of homogeneous patches, the surface generates a nonuniform potential that affects the drop shape and the value of the contact angle. The apparent contact angle θ of a macroscopic drop is usually provided in this case by the Cassie—Baxter formula[5]

$$\cos\theta = f_A \cos\theta_A + f_B \cos\theta_B, \tag{3}$$

where θ_A and θ_B are the contact angles on the corresponding homogeneous surfaces, which can be found from Eq. (1), and f_A and f_B ($f_A+f_B=1$) are the surface area fractions of the A and B patches, respectively.

For chemically homogeneous but physically rough (corrugated) surfaces, the apparent contact angle θ is described by the Wenzel formula[6]

$$\cos\theta = r\cos\theta_{\text{flat}}, \tag{4}$$

where the roughness r is the ratio of the true area of the solid to its planar projection and θ_{flat} is the contact angle for a smooth flat surface.

While widely used, the applicability of the Cassie—Baxter and Wenzel generalizations of Young's equation have been extensively discussed during the past years.[7–13] It was argued that in the general case, the surface area fractions f_A, f_B and roughness r should be considered as local parameters, dependent on the position on the surface. In this case the surface tensions become position dependent and hence the macroscopic approach on which Eq. (3) is based becomes unsatisfactory. Consequently, a microscopic approach seems to be the only one that can provide the contact angle in terms of the intermolecular interactions without involving macroscopic parameters such as the surface tensions. Such considerations were used in Ref. 14 where microscopic drops of the order of several hundreds of molecular diameters on physically rough surfaces were examined using a lattice model. By considering superhydrophobic surfaces, a large difference was found between the calculated contact angles and those predicted by the Cassie-Baxter and Wenzel expressions. The former overestimated the angle θ compared with those provided by the microscopic theory, whereas the latter underestimated the contact angle. The lattice model used in Ref. 14 accounted only for the interactions between the nearest-neighbor fluid molecules and treated the free energy of the system in the local density approximation that employs the free energy density of a homogeneous system to calculate the free energy of an inhomogeneous one. However, the local density approximation is restricted to weak external fields. When applied to strong liquid-solid interactions, this theory cannot reveal the complex oscillatory structure of the fluid density distribution (FDD) near the solid surface provided by numerical experiments.[15,16] Because the contact angle was calculated in Ref. 14 on the basis of a FDD in the very vicinity (four molecular diameters apart) of the solid surface, its value might have been affected by the approximate FDD employed.

A local density functional analysis of a drop on a smooth solid surface was performed earlier in Refs. 17 and 18, in the context of the nucleation theory, but no wetting features were of concern there. Several droplike aggregations such as bumps in cylindrical pores and planar slits[19,20] or bulges on an inhomogeneous solid surface[21] were also identified.

Some microscopic approaches are based on Monte Carlo (MC) or molecular dynamics (MD) simulations,[22,23] which involve periodic boundary conditions. Another limitation is the size of the simulation box, which rarely exceeds several tens of molecular diameters. However, the results of MC and MD simulations are often the only source of "experimental" information at the nanoscale and therefore the theories have to be adjusted to the settings of MC and MD experiments.

In the present paper we consider liquid drops on chemically or physically rough solid surfaces on the basis of a nonlocal density functional theory (DFT),[24,25] which provides a more accurate description of FDD near a solid surface. In the nonlocal approach, the local density is replaced in one of the terms of the free energy [see Eq. (A3) in the Appendix] by a coarse grained density. The latter can be imagined to represent the mean density around a molecule in a volume dependent on the range of interactions. For the sake of simplicity, two-dimensional (cylindrical) drops and roughness of infinite length in one of the lateral directions on the surface were considered. The typical sizes of the drops in the remaining directions did not exceed 10–12 nm. However these small aggregates possess the main drop characteristics, such as a core with an almost constant fluid density and an interfacial area between the core and the surrounding of low fluid density (vapor). The sizes of the roughness were selected in the range of 1–5 nm. From the calculated density distributions, contact angles were determined as functions of roughness, and the accuracy of the Wenzel and Cassie-Baxter equations was examined.

II. BACKGROUND

Let us consider a system of finite dimensions L_x and L_h in the x- and h-directions, respectively, and infinite dimension in the y-direction, which contains a one-component fluid of fixed average density ρ_{av} (see Fig. 1). The roughnesses (patches or pillars) on the solid surface are uniform in the y-direction. Consequently, it is assumed that the FDD $\rho(\mathbf{r})$ in such a system is uniform in the y-direction and nonuniform in the x- and h-directions, i.e., $\rho(\mathbf{r}) \equiv \rho(x,h)$. The periodic boundary conditions commonly used in MC and MD simulations of confined fluids[26–28] are employed in the x-direction and the upper boundary of the box is treated as a hard wall (without attractive interaction with the fluid molecules).

The interaction between fluid molecules is described by the Lennard-Jones potential

$$\phi_{ff}\left(|\mathbf{r}-\mathbf{r}'|\right) = \begin{cases} 4\epsilon_{ff}\left[\left(\dfrac{\sigma_{ff}}{r}\right)^{12} - \left(\dfrac{\sigma_{ff}}{r}\right)^{6}\right], & r \geq \sigma_{ff} \\ \infty, & r < \sigma_{ff}, \end{cases} \quad (5)$$

where the coordinates $\mathbf{r}$ and $\mathbf{r}'$ provide the locations of the fluid molecules, $r = |\mathbf{r}-\mathbf{r}'|$ [should not be confused with the roughness defined in Eq. (4)] and σ_{ff} and ϵ_{ff} are the fluid hard core diameter and energy parameter, respectively.

The same kind of potential is selected for the fluid-solid interactions [the energy parameter ϵ_{ff} and hard core diameter σ_{ff} in Eq. (5) being changed to ϵ_{fs} and σ_{fs}, respectively, r being the distance between a molecule of fluid and that of the solid walls].

The total Helmholtz free energy $F[\rho(\mathbf{r})]$ is expressed as the sum of an ideal gas free energy, $F_{id}[\rho(\mathbf{r})]$, a free energy $F_{hs}[\rho(\mathbf{r})]$ of a reference system of hard spheres, a free energy $F_{attr}[\rho(\mathbf{r})]$ due to

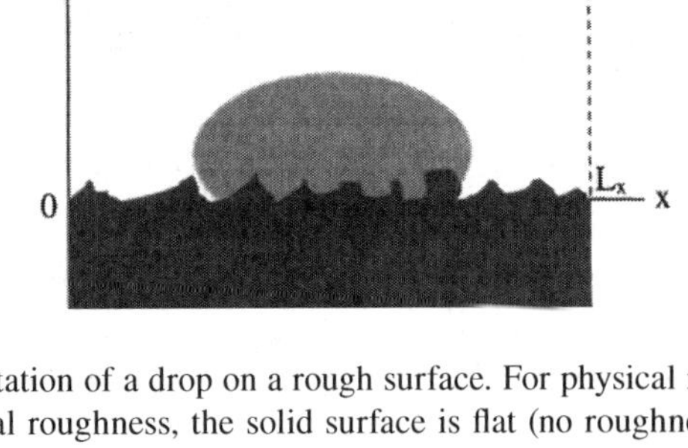

FIG. 1. Schematic representation of a drop on a rough surface. For physical roughness, the solid for $h < 0$ is homogeneous. For chemical roughness, the solid surface is flat (no roughness) but the solid for $h < 0$ is inhomogeneous.

the attractive interactions between the fluid molecules (in the mean field approximation), and a free energy $F_{fs}[\rho(\mathbf{r})]$ due to the interactions between the fluid and solid. Explicit expressions for these contributions are provided in the Appendix. Here we specify only the contribution $F_{fs}[\rho(\mathbf{r})]$, which is given by the expression

$$F_{fs}\left[\rho(\mathbf{r})\right] = \int_V d\mathbf{r}\rho(\mathbf{r})U_{fs}(\mathbf{r}), \tag{6}$$

where V is the volume of the system and $U_{fs}(\mathbf{r})$ is the potential generated by the solid. This potential is equal to

$$U_{fs}(\mathbf{r}) = \int_{V_s} \rho_s(\mathbf{r}')\phi_{fs}(|\mathbf{r}-\mathbf{r}'|)d\mathbf{r}', \tag{7}$$

where V_s is the volume of the solid and $\rho_s(\mathbf{r}')$ is the density of the solid that, in general, depends on the coordinates.

The Euler—Lagrange equation for the FDD $\rho(x, h)$ obtained by minimizing the Helmholtz free energy has the following form:

$$\log\left[\Lambda^3\rho(x,h)\right] - Q(x.h) = \frac{\lambda}{k_B T}, \tag{8}$$

where the function $Q(x, h)$ is given in Appendix [Eq. (A6)], $\Lambda = h_P/(2\pi m k_B T)^{1/2}$ is the thermal de Broglie wavelength, k_B and h_P are the Boltzmann and Planck constants, respectively, T is the absolute temperature, m is the mass of a fluid molecule, and λ is a Lagrange multiplier arising because of the constraint of fixed average density of the fluid. This constraint has the form

$$\rho_{\text{av}} = \frac{1}{V}\int_V d\mathbf{r}\rho(\mathbf{r}) \tag{9}$$

and leads to the following expression for λ:

$$\lambda = -k_B T \log\left[\frac{1}{\rho_{\text{av}} V \Lambda^3}\int_V d\mathbf{r} e^{Q(x,h)}\right]. \tag{10}$$

By eliminating λ between Eqs. (8) and (10), one obtains an integral equation for the FDD $\rho(x, h)$, which was solved by iteration.

To avoid the divergence of the iteration procedure, the input density profile $\rho_i^{\text{in}}(x,h)$ for the $(i+1)$th iteration $\rho_{i+1}(x, h)$, generated by the Euler-Lagrange equation, was selected as follows:[24]

$$\rho_i^{\text{in}}(x,h) = (1-\gamma)\rho_{i-1}^{\text{in}}(x,h) + \gamma\rho_i(x,h), \tag{11}$$

where $\rho_i(x, h)$ is the ith iteration and the constant γ=0.1. As a measure of the precision of the iterations, the dimension-less quantity

$$\delta = \frac{\lambda_{i+1} - \lambda_i}{\lambda_i}$$

was employed, where λ_i is the Lagrange multiplier calculated from Eq. (10) using FDD $\rho_i(x, h)$. The iterations were carried out on a two-dimensional grid with a spacing equal to $0.1\sigma_{ff}$ until δ became smaller than 10^{-7}. Of course, the substitution of the ith iteration $\rho_i(x, h)$ into the Euler-Lagrange equation [Eq. (8)] does not provide a constant value for λ, as it would be expected if $\rho_i(x, h)$ represented its exact solution. Due to the approximate nature of $\rho_i(x, h)$, the Lagrange multiplier calculated in this way depends on coordinates $\left[\lambda_i = \lambda_i(x,h)\right]$ To characterize how well the Euler—Lagrange equation is satisfied at each point of the grid, we introduce the relative deviation

$$\Delta\lambda(x,h) = \left|\frac{\lambda_f(x,h) - \lambda_f}{\lambda_f}\right|,$$

where λ_f and $\lambda_f(x,h)$ are the final values of the Lagrange multiplier at the end of iterations calculated with Eq. (10) (which represents the constraint of fixed average density) and Eq. (8) (which represents the Euler-Lagrange equation), respectively. In our calculations, the maximum value of $\Delta\lambda(x,h)$ at the end of iteration did not exceed 10^{-4} at the vapor-liquid interface and was less than 10^{-6} outside this interface. An increase in the precision of the iteration procedure up to δ=10^{-9} decreased the above values to 10^{-6} and 10^{-8}, respectively, but did not lead to any appreciable changes in the FDD.

A. Chemical Roughness

We model the chemically rough surface as a sequence of two kinds of plates, A and B, of finite thickness d_j (j=A, B), which are infinite in the y-direction and semiinfinite in the h-direction (Fig. 2). The system is assumed to be symmetrical with respect to the y-h plane. Due to different chemical natures, those plates interact differently with the fluid molecules. The external field generated by a single plate at point $M(x, y,$ h) depends on the distance h of this point from the surface and on the distance from the vertical plane dividing the plate into two equal parts. For example, plate A at the left of Fig. 2 generates the field described by the potential.

$$U_{sf,A}(x,h) = \rho_{s,A}\int_{V_A} dx'dy'dh'\phi_{sf,A}\left(\left|\mathbf{r}-\mathbf{r}'\right|\right) = \rho_{s,A}\int_{-d_A/2}^{d_A/2} dh'\int_{-\infty}^{0} dh'\int_{-\infty}^{\infty} dy'\phi_{sf,A}\left(\left|\mathbf{r}-\mathbf{r}'\right|\right), \quad (12)$$

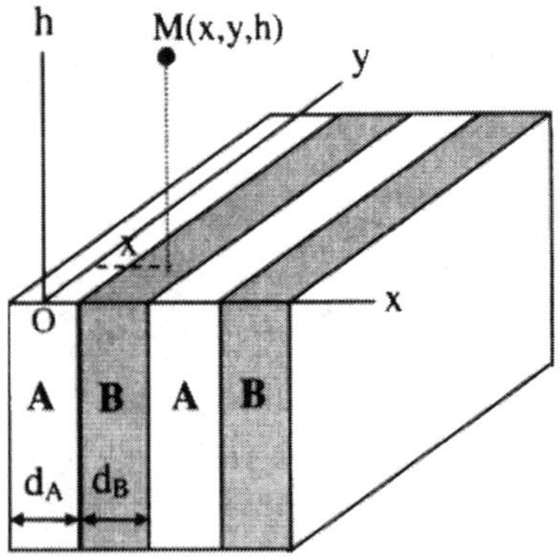

FIG. 2. Schematic representation of a chemically rough smooth surface in the x-y plane considered in the present paper. Plates A and B are infinite in the y-direction, semi-infinite in the h-direction, and have different values of the energy interaction parameters $\epsilon_{fs,A}$ and $\epsilon_{fs,B}$. The solid is symmetrical with respect to the y-h plane.

where V_A is the volume of plate A, $\mathbf{r} = \{x, y, h\}$, $\mathbf{r}' = \{x', y', h'\}$, $\rho_{s,A}$ is the number density of plate A, and $\phi_{sf,A}(|\mathbf{r}-\mathbf{r}'|)$ has the form of the potential Eq. (5) with parameters $\epsilon_{fs,A}$ and $\sigma_{fs,A}$ replacing ϵ_{ff} and σ_{ff}, respectively. In a similar way one can calculate the contribution to the total potential from any plate. The resulting potential is a periodic function of x with a period equal to $d = d_A + d_B$. Assuming that the difference between plates A and B consists only in the difference of the energy parameters, $\epsilon_{fs,A}$ and $\epsilon_{fs,B}$, and $\rho_{s,A} = \rho_{s,B} \equiv \rho_s$, one can represent the total potential $U_{sf}(x,h)$ in the form

$$U_{sf}(x,h) = \sum_{k=-\infty}^{+\infty} U_{sf,A}(x-kd,h)$$

$$+\sum_{k=-\infty}^{+\infty} U_{sf,B}\left(x-\left(k+\frac{1}{2}\right)d,h\right), \tag{13}$$

where k=0 indicates plates A and B at the left end of Fig. 2 and $k > 0$ ($k < 0$) indicates plates A and B located at $x > 0$ ($x < 0$). In the numerical calculations limits $-\infty$ and $+\infty$ for the sums over k were replaced by k_{min}<0 and k_{max}>0, respectively, which were selected to satisfy the condition $(k_{max} - k_{min})d \geqslant 2L_x$. Due to this selection, the boundaries of the system, located at $x = \pm L_x/2$, were far enough from the left and right boundaries of the solid to avoid the distortion of the potential $U_{fs}(x, h)$ due to the finite size of the solid considered in the calculations. Far from the solid surface ($h > 5\sigma_{fs}$) the total potential can be approximated by an oscillatory function $U_{sf}(x,h) \sim \sin((2\pi/d)x)$, but near the surface the shape of the potential is more complex. At any distance from the solid surface, the local minima (maxima) of $U_{sf}(x,h)$ are located right above the middle of the A-plate (B-plate).

B. Physical Roughness

Because of the random nature of the physical roughness (in shape, size, and location), it is difficult to develop an adequate model for rough surfaces. One of the frequently employed approximations is to decorate the planar surface with a regular array of identical protrusions such as pillars[14] or grooves.[29] In such cases it is possible to calculate the potential of pillar-fluid interaction and analyze the dependence of the drop parameters on roughness. In the present paper we consider a surface decorated with an infinite number of identical parallelepipeds (denoted below as pillars) of width d_p, height h_p, and infinite length in the y-direction, as shown in Fig. 3. The roughness of such a surface is provided by the expression

$$r = 1 + \frac{2h_p}{d_p + \Delta d_p}, \tag{14}$$

where Δd_p is the distance between pillars.

The contribution $U_{fp}(x, h)$ of a pillar located at $x = 0$ to the total fluid-solid potential is given by

$$U_{fp}(x,h) = \rho_p \int_{V_p} dx'dy'dh'\phi_{fs,p}(|\mathbf{r}-\mathbf{r}'|) = \rho_p \int_{-d_p/2}^{d_p/2} dx' \int_0^{h_p} dh' \int_{-\infty}^{\infty} dy'\phi_{fs,p}(|\mathbf{r}-\mathbf{r}'|), \tag{15}$$

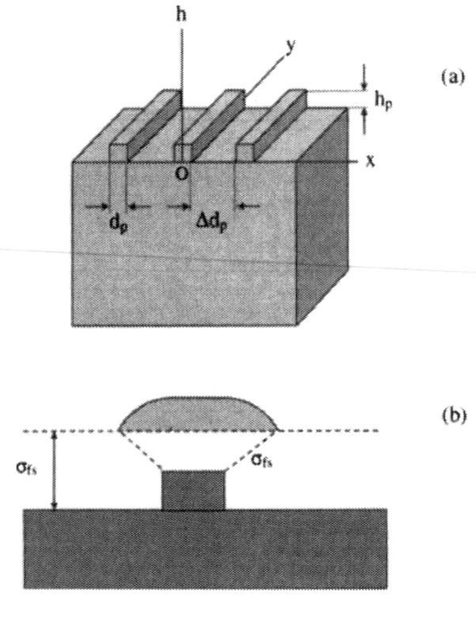

FIG. 3. (a) Pillars on a flat surface. (b) Additional volume (the shaded area above the horizontal dashed line) inaccessible to fluid molecules due to the existence of a pillar.

where $\mathbf{r}$ and $\mathbf{r}'$ provide the locations of the fluid and pillar molecules, respectively, ρ_p is the number density of molecules in the pillars, V_p is the pillar volume, and $\phi_{fs,p}\left(\left|\mathbf{r}-\mathbf{r}'\right|\right)$ is the Lennard-Jones potential [Eq. (5)] in which $\epsilon_{fs,p}$ and $\sigma_{fs,p}$, which characterize the pillars, replace ϵ_{ff} and σ_{ff}, respectively. In general, the energy parameter $\epsilon_{fs,\,p}$ involved in the potential $\phi_{fs,p}\left(\left|\mathbf{r}-\mathbf{r}'\right|\right)$ can be different from that involved in the interaction potential $\phi_{fs}\left(\left|\mathbf{r}-\mathbf{r}'\right|\right)$ between fluid and the part of the solid with $h < 0$ (Fig. 1).

For a selected array of pillars, their contribution to the total external potential is

$$U_{fp,\mathrm{tot}}\left(x,h\right)=\sum_{k=-\infty}^{\infty}U_{fp}\left(x-x_k,h\right), \tag{16}$$

where x_k is the location of the center of the kth pillar. For numerical calculations the limits of k were selected as for chemical roughness (Sec. II A).

Note that the hard core repulsion between fluid and solid molecules decreases somewhat the volume accessible to the fluid molecules. For a planar solid surface, the inaccessible volume is located between the surface and a plane parallel to the latter at the distance σ_{fs} from it. The presence of a pillar on the solid surface increases the inaccessible volume, and the shape of the additional part is different from that of the pillar. As an example, this additional excluded volume is presented schematically in Fig. 3(b) for $h_p < \sigma_{fs}$.

C. Calculation of the Contact Angle

A convenient way to calculate the contact angle, which is not clearly defined for nanodrops, is to determine the profile inside the vapor-liquid interface corresponding to a constant local density $\rho_{l/v}$ and to find the angle which this profile makes with the solid surface.[14,23] The FDD inside the drop on the solid surface has a complex structure in the vicinity of the liquid-solid and liquid-vapor interfaces. In particular, near the solid surface there are strong oscillations of density due to the ordering

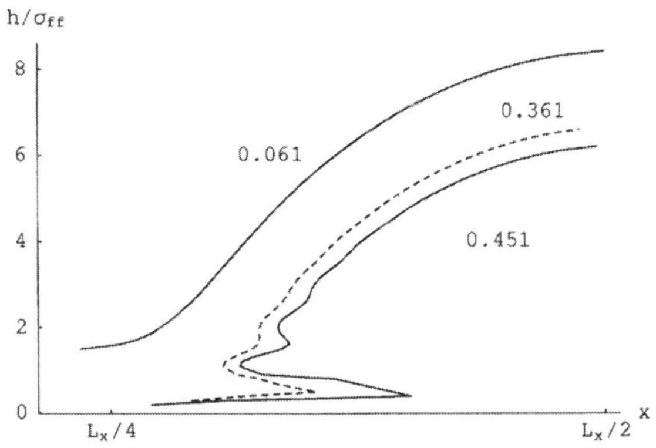

FIG. 4. Drop profiles for various dividing surfaces corresponding to $\rho^*_{\text{div}} = 0.061, 0.361$, and 0.451. The average density $\rho^*_{\text{av}} = 0.1$. The numbers on curves indicate the value of ρ^*_{div}.

of the fluid molecules that form several liquid layers of various densities.[24] As an illustration, in Fig. 4 three typical profiles corresponding to the densities $\rho^*_{\text{div}} = 0.061, 0.361$ and $0.451\left(\rho^* = \rho\sigma^3_{ff}\right)$ are presented for a drop on the smooth solid surface considered later in the paper. The profile for $\rho^*_{\text{div}} = 0.061$ corresponds to a vaporlike shell around the drop that ends on a fluid layer on the solid; the other two represent liquidlike fluids. These profiles provide different angles with the surface and this leads to uncertainty in the definition of the contact angle. The usual procedure to calculate the profile of the interface is to consider that it is provided by an equimolar dividing surface (see Refs. 14 and 23). Our calculations have shown that except the region corresponding to $h < 3\,\sigma_{ff}$, the density ρ_{div} is almost constant under the conditions employed $\rho^*_{\text{div}} \simeq 0.371$. Considering the profile corresponding to ρ_{div} as the drop profile, its part with $h > 3\,\sigma_{ff}$ was approximated by a circle and extended up to the solid surface. The angle which this circle makes with the surface was considered as the apparent contact angle. For a surface decorated with identical pillars, the contact angle was calculated with respect to the plane containing the top of the pillars.

III. RESULTS

In the cases considered, argon was selected as the fluid with the interaction parameters $\epsilon_{ff}/k_B = 119.76\,\text{K}$ and $\sigma_{ff} = 3.405\,\text{Å}$, and the temperature was selected $T = 87.3\,\text{K}$. As reference solid, the solid carbon dioxide was considered with $\epsilon^0_{fs}/k_B = 153.0\,\text{K}$, $\sigma_{fs} = 3.727\,\text{Å}$, and $\rho_s = 1.91\times10^{28}\,\text{m}^{-3}$.[30] The other solids considered below differ from the above one only through the value of the energy parameter, all the other parameters remaining unchanged. If not mentioned otherwise, the width of the system L_x was taken $L_x = 60\,\sigma_{ff}$ and the height $L_h = 20\sigma_{ff} + 2\sigma_{fs}$.

A. Nanodrop on a Chemically Rough Surface

Let us consider first a nanodrop on a chemically inhomogeneous solid surface as described in Sec. II A. The widths of plates A and B were selected equal to $3\sigma_{ff}$ and six values for the ratio $\alpha_{BA} \equiv \epsilon_{fs,B}/\epsilon_{fs,A}$ were considered (α_{BA} = 1, 0.8, 0.6, 0.4, 0.2, and 0). (Case α_{BA} = 1 corresponds to a homogeneous surface and case $\alpha_{BA} = 0$ represents a plate B with $\epsilon_{fs,B} \ll \epsilon_{fs,A}$). In all the cases considered the widths of the liquid-vapor and liquid-solid interfaces are comparable with the droplet size and the liquid density distribution has an oscillatory behavior near the latter interface.

A typical drop on a homogeneous surface (α_{BA} =1) is presented in Fig. 5(a) for $\rho^*_{\text{av}} = 0.1$. The lighter areas correspond to higher densities of the fluid. The x-dependence of the fluid density at various distances h from the surface is presented in Fig. 5(b). The density as a function of x is almost constant in the middle region of the drop. In the very vicinity of the solid surface ($h = 0.1\,\sigma_{ff}$), it has

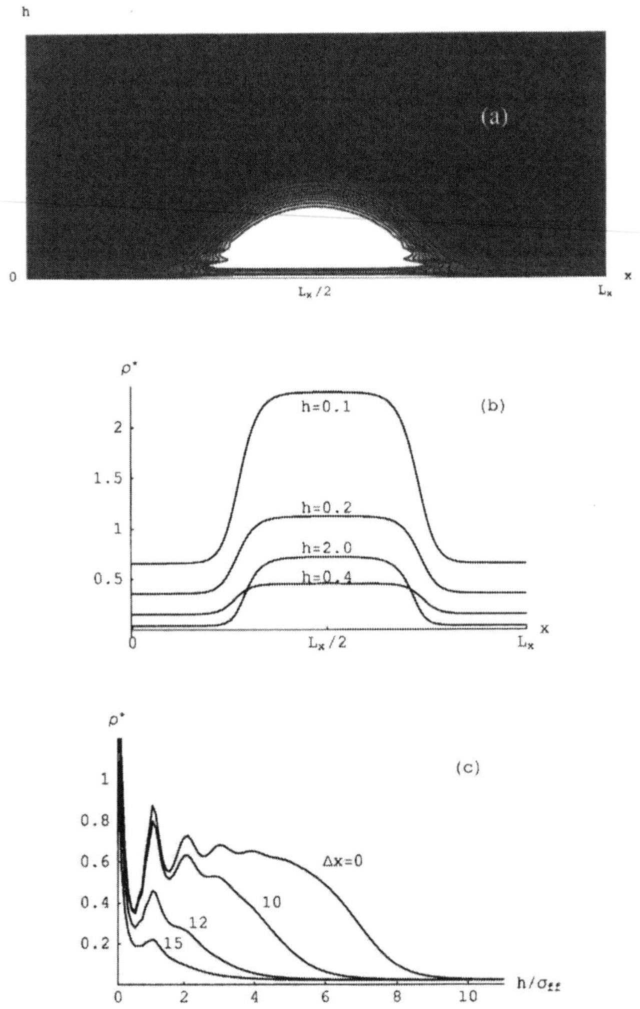

FIG. 5. $\rho_{av}^{*} = 0.1$. (a) Typical drop on a homogeneous surface. The lighter areas correspond to higher fluid densities. (b) The x-dependence of the fluid density at various distances h from the surface. The values of h are indicated on the curves in units of σ_{ff}. (c) The h-dependence of the fluid density at various distances Δx from the plane of symmetry of the drop. The values of Δx are indicated on the curves in units of σ_{ff}.

a solidlike value due to the structuring of the fluid[19] but decreases quickly to liquidlike values for $h \geqslant 0.2\sigma_{ff}$. The h-dependence of the fluid density at various distances Δx from the plane of symmetry of the drop is presented in Fig. 5(c).

Using the approach described in Sec. II C, the contact angle for the drop shown in Fig. 5(a) was found to be equal to $\theta = 48.5^{\circ}\left(\rho_{div}^{*} = 0.371\right)$. For the drops on the other homogeneous solid surfaces with various values of the fluid-solid energy parameter, the contact angles are listed in the second column of Table 1.

The chemical inhomogeneity affects the drop shape and, consequently, the contact angle. In the presence of two kinds of plates (Fig. 2), the Euler-Lagrange equation [Eq. (8)] has generally two solutions for a given average fluid density. The first solution, $D1$, describes a drop with the apex located at the local minimum of the fluid-solid interaction potential. The apex of the drop described by the second solution ($D2$) is located at the local maximum of the potential. Note that in both cases the leading edge of the drop is located on a B-plate, the location being closer to that of the two neighboring A-plates, which is covered by the drop. Depending on the average density ρ_{div}^{*}, one of

TABLE 1
Contact angles of a drop on various chemically rough surfaces. θ_B is the contact angle of a drop on the homogeneous surface of a solid of type *B* (Fig. 2) calculated by DFT with $\varepsilon_{fs,B} = \alpha_{BA}\varepsilon_{fs,A}$, θ_r is the contact angle on chemically rough surface, and $\theta_{r,\text{Cassie}}$ is the contact angle calculated using Cassie—Baxter equation [Eq. (3)].

$\alpha_{BA} = e_{fs,B}/\varepsilon_{fs,A}$	θ_B (deg)	θ_r (deg)	$\theta_{r,\text{Cassie}}$ (deg)
0.8	81.4	68.4	66.0
0.6	109.7	86.3	80.6
0.4	139.9	102.7	92.9
0.2	161.5	112.0	98.2
0.0	180	119.8	99.7

these drops, *D*1 or *D*2, has a smaller free energy and is stable, the other one being metastable. For example, in the range of fluid average densities $0.1 \leqslant \rho^*_{\text{av}} \leqslant 0.26$ considered in this paper and for $\alpha_{BA} = 0.2$, the drop *D*2 is stable for $0.14 \leqslant \rho^*_{\text{av}} \leqslant 0.19$ and the drop *D*1 is stable for all the other values of ρ^*_{av}. As expected, the contact angle of a drop on a rough surface depends on the drop volume (or on ρ^*_{av}). For the considered example, the ρ^*_{av}-dependence of θ for a stable drop is presented in Fig. 6(a). Figure 6(b) presents the ρ^*_{av}-dependence of the area *S* of the solid surface covered by the drop. One can observe that the rapid decrease of the contact angle at $\rho^*_{\text{av}} \simeq 0.135$ and $\rho^*_{\text{av}} \simeq 0.195$ can be correlated with the rapid increase of *S* at the same values of ρ^*_{av}. In contrast to *S*, the height of the stable drop increases smoothly with ρ^*_{av}. Because when ρ^*_{av}. increases the leading edge of the stable drop is displaced along the periodic potential generated by the solid surface, one expects the contact angle to be a periodic function of the drop volume (or of ρ^*_{av}). The same values of θ at $\rho^*_{\text{av}} = 0.1$ and $\rho^*_{\text{av}} = 0.26$ in Fig. 6(a) provide an indication in that direction.

Let us now consider the effect of the chemical roughness on the contact angle for $\rho^*_{\text{av}} = 0.1$. Figure 7 presents the FDDs for α_{BA} = 0.8, 0.4, 0.2, 0. The contact angles for those drops are listed in the third column of Table 1 and show that they increase with decreasing α_{BA}, i.e., the chemical roughness increases the hydrophobicity of the solid surface.

To compare our results to those predicted by the macroscopic treatment, we calculated the contact angle for chemically rough surfaces using the Cassie—Baxter equation [Eq. (3)]. (In the considered case $f_A = f_B = 1/2$.) The contact angles θ calculated with that equation are listed in the last column of Table 1. For all considered values of α_{BA}, the Cassie—Baxter equation provides a smaller contact angle than the DFT theory, the difference being greater for smaller values of α_{BA}, i.e., for larger differences between the plates *A* and *B*.

B. Droplet on a Physically Rough Surface

The pillars used in the calculations had the width $d_p = 2\sigma_{ff}$ and the distance between them $\Delta d_p = 4\sigma_{ff}$. The height of the pillars h_p was varied between $h_p = 0$ (no pillars) and $h_p = 10\sigma_{ff}$. Surfaces that are hydrophobic or hydrophilic when free of pillars are considered. The difference between them was achieved by changing only the energy parameter ϵ_{fs} in the Lennard-Jones potential of the fluid-solid interactions, keeping all other parameters the same. The contact angle of a drop on a smooth surface as function of $\alpha = \epsilon_{fs}/\epsilon^0_{fs}$ is presented in Fig. 8. One should note the almost linear decrease of θ with increasing α. From this dependence one can conclude that for $\epsilon_{fs} < 0.75\epsilon^0_{fs}$ the surface can be considered as hydrophobic ($\theta > 90°$), and for $\epsilon_{fs} > 0.75\epsilon^0_{fs}$ as hydrophilic ($\theta < 90°$).

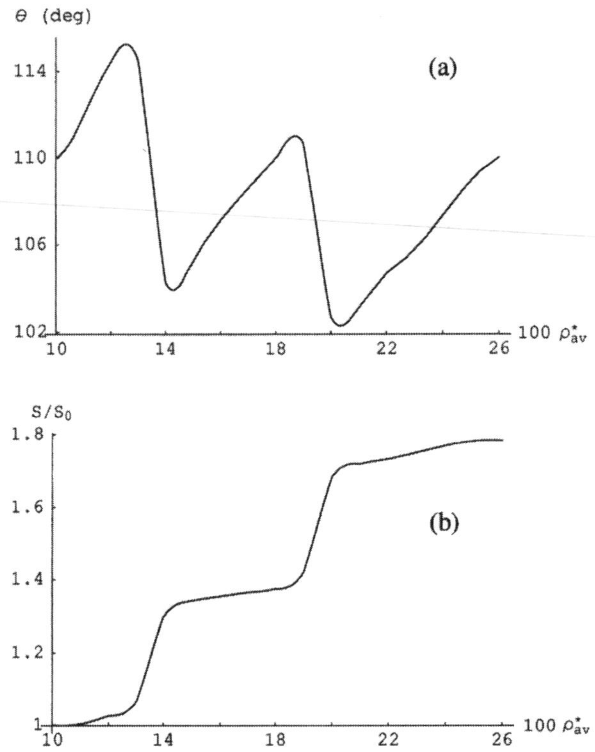

FIG. 6. $\rho^*_{av} = 0.1, \alpha_{BA} = 0.2$. (a) ρ^*_{av}-dependence of the contact angle of a drop on a chemically rough surface. (b) ρ^*_{av}-dependence of the area of the solid surface covered by the drop. S_0 is this area at $\rho^*_{av} = 0.1$.

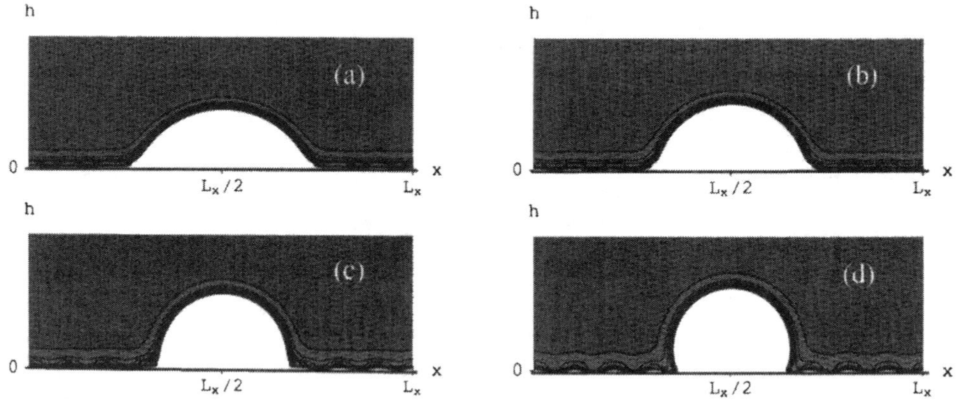

FIG. 7. Drop on chemically rough surfaces for various values of $\alpha_{BA} = \epsilon_{fs,B}/\epsilon_{fs,A}$. (a) $\alpha_{BA} = 0.8$, (b) $\alpha_{BA} = 0.4$, (c) $\alpha_{BA} = 0.2$, and (d) $\alpha_{BA} = 0$.

1. Droplet on a rough hydrophobic surface

Let us first consider the case of a superhydrophobic surface with $\epsilon_{fs} = 0.4\epsilon^0_{fs}$, which is additionally decorated with pillars. The FDDs for various heights of the pillars and $\rho^*_{av} = 0.1$ are presented in Fig. 9. For small pillar heights $(h_p \leqslant 4\sigma_{ff})$ the liquid penetrates between the pillars under the drop (Wenzel regime). For larger heights $(h_p \geqslant 6\sigma_{ff})$ most of the space between pillars is occupied by a low density (vaporlike) fluid and the drop is located on the tops of the pillars (Cassie—Baxter regime). The FDD in the x-direction between pillars under the drop is presented in Fig. 10 for

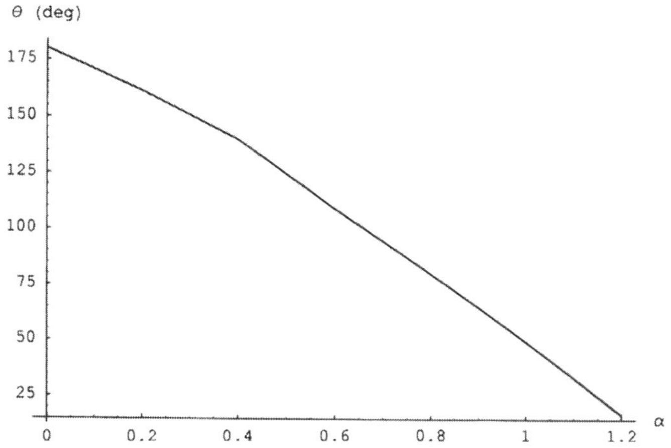

FIG. 8. Contact angle θ of a drop on a flat homogeneous surface as function of $\alpha = \epsilon_{fs}/\epsilon_{fs}^0$.

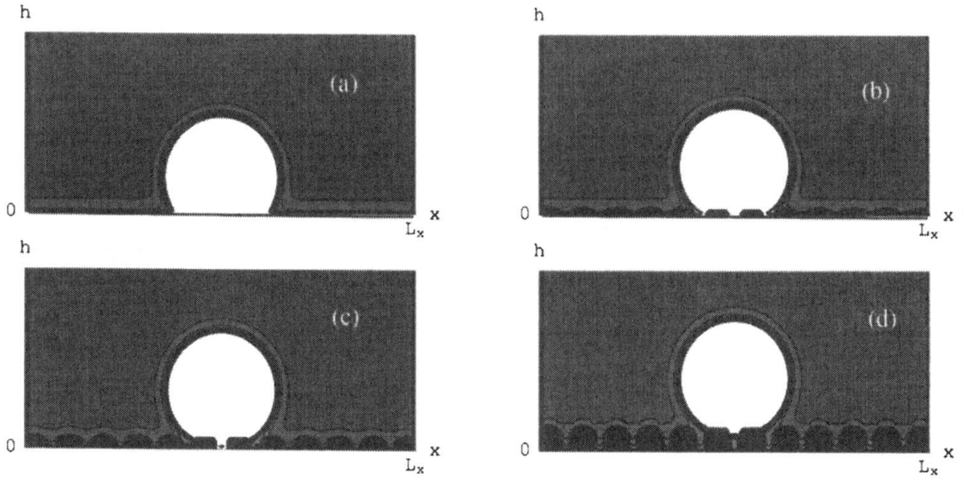

FIG. 9. Drops on superhydrophobic surfaces with $\epsilon_{fs} = 0.4\epsilon_{fs}^0$ and various pillar heights. The lighter areas correspond to higher fluid densities. (a) $h_p = 0$, (b) $h_p = 2\sigma_{ff}$, (c) $h_p = 4\sigma_{ff}$, and (d) $h_p = 8\sigma_{ff}$.

$h_p = 3\sigma_{ff}$ (Wenzel regime) and $h_p = 8\sigma_{ff}$ (Cassie—Baxter regime). Near the bottom and top of the pillars the density distributions in these two cases are almost the same. However, for $h_p = 8\sigma_{ff}$ a considerable part of the space $(0 < h < 0.7h_p = 5.6\sigma_{ff})$ is occupied by a low-density vaporlike fluid represented in Fig. 10(b) by the curve for $h = 5.6\sigma_{ff}$. In both cases the surface of the pillars is covered by a thin layer of fluid molecules. The dependence of the cosine of the contact angle on the pillar height h_p is presented in Fig. 11. Note that for the selected geometry of the pillar location (pillar width $d_p = 2\sigma_{ff}$ and distance between pillars $\Delta d_p = 4\sigma_{ff}$), the roughness defined by Eq. (14) is a linear function of the pillar height $(r = 1 + h_p/3)$. One can see from Fig. 11 that for small pillar heights $(h_p \leqslant \sigma_{ff})$ the Wenzel formula and DFT results agree qualitatively, the contact angle provided by Wenzel formula being larger than that calculated by DFT. However, according to the Wenzel formula, the contact angle reaches its maximum possible value θ=180° ($\cos\theta = -1$) at $h_p \sim \sigma_{ff}$, whereas the contact angle provided by the DFT at $h_p = \sigma_{ff}$ is $\theta \sim 158°(\cos\theta = -0.927)$. By increasing the pillar height, the DFT provided increasing contact angles that reached at $h_p \sim 2\sigma_{ff}$ the asymptotic value of $\theta_{\max} \sim 167°$

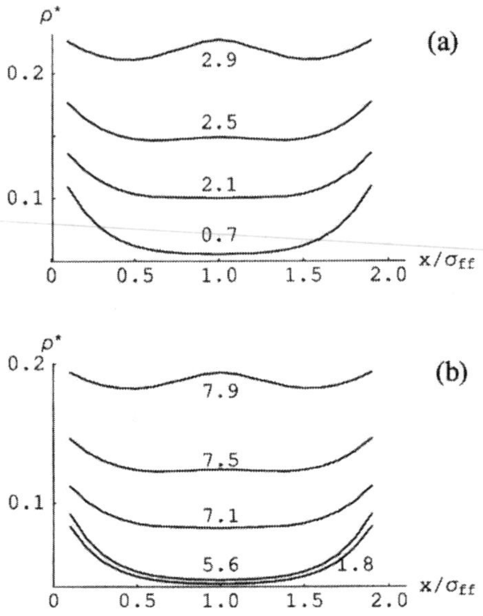

FIG. 10. x dependence of the fluid density between pillars. $\rho^*_{\mathrm{av}} = 0.1$. (a) $h_p = 3\sigma_{ff}$ and (b) $h_p = 8\sigma_{ff}$. The numbers on the curves provide the distance from the solid surface in units of σ_{ff}.

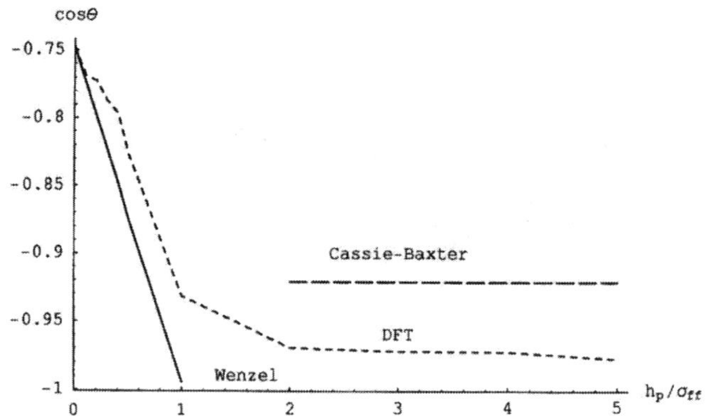

FIG. 11. Dependence of cos θ on the pillar height for $\rho^*_{\mathrm{av}} = 0.1$. The solid line represents the Wenzel formula and the dashed line represents the DFT calculations. The long-dashed line represents the contact angle obtained for $h > 2\ \sigma_{ff}$ using the Cassie—Baxter formula [Eq. (3)].

(cos $\theta = -0.974$). The difference in the behavior of the contact angle can be qualitatively explained by a transition from the Wenzel to the Cassie—Baxter regime. To apply the Cassie—Baxter formula [Eq. (3)], for $h_p > 2\ \sigma_{ff}$, θ_A was taken equal to the contact angle $\theta = 139.9°$ (cos $\theta = -0.76$) which the drop makes with the smooth surface, and θ_B was taken 180° (negligible interaction with the solid). For the selected geometry of the pillars, $f_A = \frac{1}{3}$ and $f_B = \frac{2}{3}$. As a result, Eq. (3) provides for the contact angle of the drop on the pillars the value θ=157.2° (cos $\theta = -0.92$), which is smaller than the asymptotic value $\theta = 167°$ provided by the DFT for the Cassie—Baxter regime. Note that in contrast to the nonlocal DFT, the lattice model employed in Ref. 14 provided smaller contact angles for the Cassie—Baxter regime and larger ones for the Wenzel regime than those provided by Eqs. (3) and (4), respectively.

2. Droplet on a rough hydrophilic surface

As a first example of a hydrophilic surface, let us consider the solid carbon dioxide surface $\left(\epsilon_{fs}=\epsilon_{fs}^{0}, \epsilon_{fs}^{0}/k_B=153.0\mathrm{K}\right)$. The FDDs for $\rho_{\mathrm{av}}^{*}=0.2$ and various pillar heights are presented in Fig. 12. Unexpectedly, in contrast to the predictions of the Wenzel formula, the presence of pillars does not increase the hydrophilicity (i.e., a decrease of the contact angle) but provides an increase in θ. One should also note the decrease of the drop dimensions with increasing height of the pillars. This decrease can be explained by the increase of the amount of fluid accumulated between pillars. Because the total number of molecules in the system is fixed, this accumulation leads to a decrease in the number of molecules remaining to form a drop, hence, to a decrease in the size of the drop on the pillars.

For stronger fluid-solid interactions $\left(\epsilon_{fs}=1.4\epsilon_{fs}^{0}\right)$, no drop is present on the smooth surface that is covered by a liquidlike film (consequently, cos $\theta = 1$). However, the presence of pillars changes completely the behavior. While the short pillars with $h_p < 2\sigma_{ff}$ cause only periodical nonuniformity in the film in the x-direction [see Fig. 13(a)], the larger pillars generate drops on the rough surface. However, the size of the drop decreases with increasing pillar height as for a hydrophobic surface [Figs. 13(b) and 13(c)]. At some pillar height ($h_p \simeq 9\sigma_{ff}$ in the considered case) the drop disappears

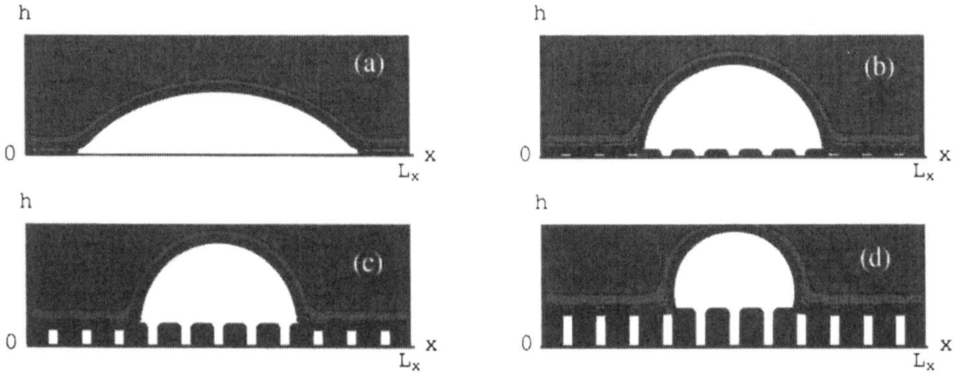

FIG. 12. Drop on a hydrophilic surface with $\epsilon_{fs}=\epsilon_{fs}^{0}$, various pillars heights, and $\rho_{\mathrm{av}}^{*}=0.2$. The lighter areas correspond to higher fluid densities. (a) $h_p=0$, (b) $h_p=3\sigma_{ff}$, (c) $h_p=6\sigma_{ff}$, and (d) $h_p=8\sigma_{ff}$.

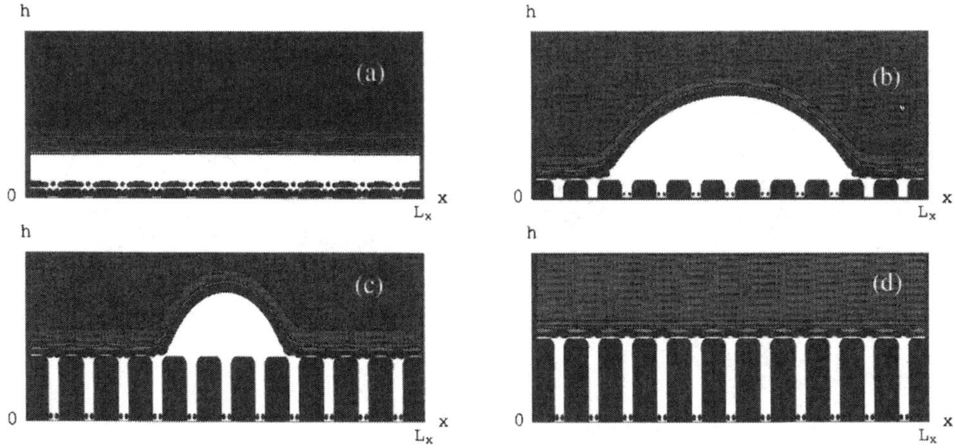

FIG. 13. Drops on hydrophilic surfaces for $\epsilon_{fs}=1.4\epsilon_{fs}^{0}$ and various pillar heights and $\rho_{\mathrm{av}}^{*}=0.2$. The lighter areas correspond to higher fluid densities. (a) $h_p=0$, (b) $h_p=4\sigma_{ff}$, (c) $h_p=8\sigma_{ff}$, and (d) $h_p=10\sigma_{ff}$.

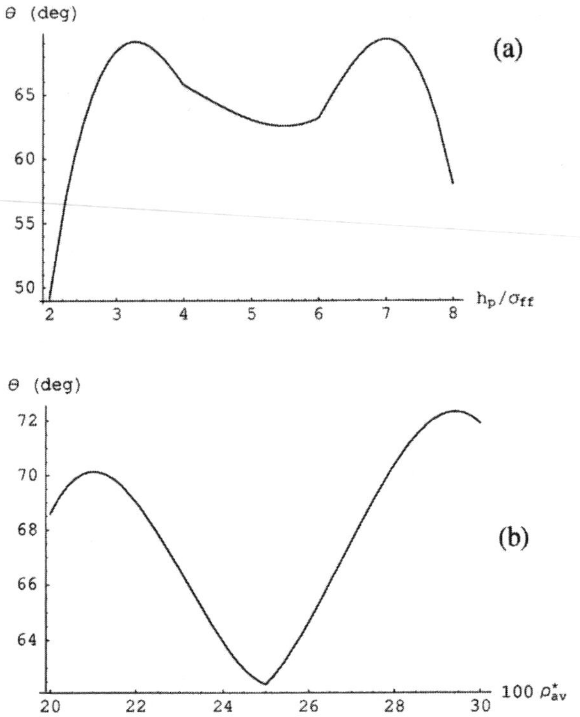

FIG. 14. Dependence of the contact angle of the drop on a hydrophilic surface for $\epsilon_{fs} = 1.4\epsilon^0_{fs}$ (a) on the pillar height for $\rho^*_{av} = 0.2$ and (b) on the average density ρ^*_{av} for $h_p = 7\sigma_{ff}$.

completely because the fluid fills the space between pillars and forms a film above the top of the pillars [Fig. 13(d)], in which the density varies periodically in the x-direction.

The calculated contact angle changes nonmonotonously with increasing pillar height [see Fig. 14(a)] and does not follow the Wenzel formula [Eq. (4)]. Indeed, in the considered case cos $\theta_{\text{flat}} = 1$ in Eq. (4) and the latter formula provides for cos θ values larger than unity for all pillar heights, i.e., a drop cannot be formed on the surface. Note that at a fixed pillar height, the contact angle calculated by DFT depends slightly on the volume of the drop, which can be changed by changing the average density ρ^*_{av}. This dependence is nonmonotonous [Fig. 14(b)]. The possible reason for the nonmonotonous dependence of the contact angle on the pillar height and on the drop volume is that this angle depends on the position of the leading edge of the nanodrop with respect to the pillar surface. With increasing (decreasing) size of the drop, the leading edge periodically changes its position and can be located either between pillars or on their tops being thus exposed to different fluid-solid interactions. This behavior is similar to the ρ^*_{av}-dependence of the contact angle on a chemically rough surface considered in Sec. III A.

IV. CONCLUSION

In the present paper a two-dimensional version of the nonlocal DFT was used to analyze the behavior of a nanodrop of argon on a chemically or physically rough surface with roughness not exceeding 5 nm. In the case of chemical roughness, which originates from the chemically inhomogeneous substrate, the dependence of the contact angle on the substrate composition agrees qualitatively with the predictions of the Cassie—Baxter formula developed for macroscopic drops. However, the contact angles for the nanodrops calculated by DFT is larger than that provided by the Cassie—Baxter formula [Eq. (3)]. The difference between those two contact angles increases with increasing chemical inhomogeneity of the substrate. For physical roughness, two kinds of surfaces, hydrophobic and hydrophilic when free of pillars, were considered. For a hydrophobic surface, the calculated contact angle increases with increasing roughness. This behavior is in qualitative agreement with the Wenzel formula given by Eq. (4). Quantitatively, the values of θ provided by Eq. (4) are larger than those obtained on the basis of the microscopic theory. For a hydrophilic surface, the dependence of the contact angle on roughness disagrees qualitatively with the Wenzel formula, according to which the contact angle should always decrease as the roughness increases. One of the consequences of this rule is the impossibility of formation of a drop on a rough surface of a solid if it is not formed on the smooth surface of the same solid. In contrast, the present calculations based on DFT predict that such a possibility exists as well as the possibility of increase of the contact angle with increasing roughness. However, there are ranges of roughness in which the contact angle decreases with increasing roughness. This feature was not observed with macroscopic drops. Another specific feature of a nanodrop on a rough surface is the volume dependence of the contact angle, which can be of the order of 10% of its magnitude. Such a behavior can not be explained neither by Wenzel nor by Cassie—Baxter formulas.

APPENDIX: FREE ENERGY CONTRIBUTIONS

The contributions to the free energy listed in Sec. II can be represented as follows:

$$F_{\text{id}}\left[\rho(\mathbf{r})\right] = k_B T \int d\mathbf{r}\rho(\mathbf{r})\left\{\log\left[\Lambda^3\rho(\mathbf{r})\right]-1\right\}, \tag{A1}$$

$$F_{\text{hs}}\left[\rho(\mathbf{r})\right] = \int d\mathbf{r}\rho(\mathbf{r})\Delta\Psi_{\text{hs}}(\mathbf{r}), \tag{A2}$$

where

$$\Delta\Psi_{\text{hs}}(\mathbf{r}) = k_B T \eta_\rho \frac{4-3\eta_\rho}{\left(1-\eta_\rho\right)^2}, \tag{A3}$$

$\eta_\rho = \frac{1}{6}\pi\bar{\rho}(\mathbf{r})\sigma_{ff}^3$ is the packing fraction of the fluid molecules, σ_{ff} is the fluid hard core diameter, and $\bar{\rho}(\mathbf{r})$ is the smoothed density defined as

$$\bar{\rho}(\mathbf{r}) = \int d\mathbf{r}'\rho(\mathbf{r}')W\left(|\mathbf{r}-\mathbf{r}'|\right). \tag{A4}$$

The weighting function $W\left(|\mathbf{r}-\mathbf{r}'|\right)$ is selected in the form[25]

$$W\left(|\mathbf{r}-\mathbf{r}'|\right) = \begin{cases} \dfrac{3}{\pi\sigma_{ff}^3}\left(1-\dfrac{r}{\sigma_{ff}}\right), & r \leqslant \sigma_{ff} \\ 0, & r > \sigma_{ff}, \end{cases}$$

where $r = |\mathbf{r}-\mathbf{r}'|$.

The contribution to the excess free energy due to the attraction between the fluid-fluid molecules is calculated in the mean-field approximation

$$F_{\text{attr}}\left[\rho(\mathbf{r})\right] = \frac{1}{2}\iint d\mathbf{r}d\mathbf{r}'\rho(\mathbf{r})\rho(\mathbf{r}')\phi_{ff}(\mathbf{r}-\mathbf{r}'). \tag{A5}$$

The function $Q(x, h)$ in Eq. (8) is given by

$$Q(x,h) = -\frac{1}{k_B T}\Big[\Delta\Psi_{\text{hs}}(x,h) + \overline{\Delta\Psi'}_{\text{hs}}(x,h) + U_{ff}(x,h) + U_{fs}(x,h)\Big] \tag{A6}$$

where

$$U_{ff}(x,h) = \iint dx'dh'\rho(x',h')\phi_{ff,y}\left(|x-x'|,|h-h'|\right), \tag{A7}$$

$$\overline{\Delta\Psi'}_{\text{hs}}(x,h) = \iint dx'dh'\rho(x',h')W_y\left(|x-x'|,|h-h'|\right) \times \frac{\partial}{\partial\bar{\rho}}\Delta\Psi_{\text{hs}}(\bar{\rho})_{\bar{\rho}=\bar{\rho}(x',h')}. \tag{A8}$$

$\phi_{ff,y}\left(|x-x'|,|h-h'|\right)$ and $W_y\left(|x-x'|,|h-h'|\right)$ are obtained by the integrations of the potential $\phi_{ff}\left(|\mathbf{r}-\mathbf{r}'|\right)$ and weighted function $W\left(|\mathbf{r}-\mathbf{r}'|\right)$ with respect to the y-direction from $-\infty$ to $+\infty$, respectively.

When calculating the term $U_{ff}(x, h)$ of the Euler—Lagrange equation arising due to the long-range fluid-fluid interactions, a cutoff at a distance equal to four molecular diameters σ_{ff} for the range of Lennard-Jones attraction was employed. The increase of this distance up to $10\sigma_{ff}$ changed the results by less than 1%.

REFERENCES

1. T. Young, Philos. Trans. R. Soc. London **95**, 65 (1805).
2. D. H. Everett and J. M. Haynes, J. Colloid Interface Sci. **38**, 125 (1972).
3. J. R. Philip, J. Chem. Phys. **66**, 5069 (1977).
4. E. Ruckenstein and P. S. Lee, Surf. Sci. **52**, 298 (1975) (Section 3.1 of this volume).
5. A. B. D. Cassie and S. Baxter, Trans. Faraday Soc. **40**, 546 (1944).
6. R. N. Wenzel, Ind. Eng. Chem. **28**, 988 (1936).
7. P. S. Swain and R. Lipowsky, Langmuir **14**, 6772 (1998).
8. L. Gao and Th. J. McCarthy, Langmuir **23**, 3762 (2007).
9. M. V. Panchagnula and S. Vedantam, Langmuir **23**, 13242 (2007).
10. N. Anantharaju, M. V. Panchagnula, S. Vedantam, and S. Neti, Langmuir **23**, 11673 (2007).
11. M. Nosonovsky, Langmuir **23**, 9919 (2007).
12. L. Gao and Th. J. McCarthy, Langmuir **23**, 13243 (2007).
13. G. McHale, Langmuir **23**, 8200 (2007).
14. F. Porcheron and P. A. Monson, Langmuir **22**, 1595 (2006).
15. W. G. Hoorer and F. H. Ree, J. Chem. Phys. **49**, 3602 (1968).
16. J. P. Noworyta, D. Henderson, S. Sokolowski, and K.-Y. Chan, Mol. Phys. **95**, 415 (1998).
17. V. Talanquer and D. W. Oxtoby, J. Chem. Phys. **104**, 1483 (1996).
18. V. Talanquer and D. W. Oxtoby, J. Chem. Phys. **114**, 2793 (2001).
19. E. A. Ustinov and D. D. Do, J. Phys. Chem. B **109**, 11653 (2005).
20. G. O. Berim and E. Ruckenstein, J. Chem. Phys. **128**, 024704 (2008) (Section 2.4 of this volume).
21. F. Porcheron, P. A. Monson, and M. Schoen, Phys. Rev. E 73, 041603 (2006)
22. M. Lundgren, N. L. Allan, T. Cosgrove, and N. George, Langmuir **19**, 7127 (2003).
23. N. Giovambattista, P. G. Debenedetti, and P. J. Rossky, J. Phys. Chem. B **111**, 9581 (2007).
24. P. Tarazona, Phys. Rev. A **31**, 2672 (1985).

25. R. H. Nilson and S. K. Griffiths, J. Chem. Phys. **111**, 4281 (1999).
26. A. Vishnyakov and A. V. Neimark, J. Chem. Phys. **119**, 9755 (2003).
27. A. V. Neimark, P. I. Ravikovitch, and A. Vishnyakov, Phys. Rev. E **65**, 031505 (2002).
28. J. Mittal, J. R. Errington, and T. M. Truskett, J. Chem. Phys. **126**, 244708 (2007).
29. R. Kannan and D. Sivakumar, Colloids Surf., A **317**, 694 (2008).
30. R. Evans and P. Tarazona, Phys. Rev. A **28**, 1864 (1983).

4.5 Nanodrop on a Nanorough Hydrophilic Solid Surface

Contact Angle Dependence on the Size, Arrangement, and Composition of the Pillars[*]

Gersh Berim and Eli Ruckenstein[†]

Department of Chemical and Biological Engineering, State University of New York at Buffalo, Buffalo, NY 14260, United States

Corresponding Author

[†]E-mail: feaeliru@buffalo.edu Corresponding author. Fax: +1 716 645 3822.

1. INTRODUCTION

The progress of modern nanotechnologies allows one to fabricate surfaces with prescribed structure, for instance, decorated with asperities of various shapes, sizes, and compositions [1–6]. Such modifications of the surface lead to considerable changes of its wetting properties. In particular, if the fluid penetrates between asperities (so called Wenzel regime), the contact angle θ_W which a liquid drop makes with a rough surface is provided by the Wenzel equation [7]

$$\cos\theta_w = r\cos\theta \tag{1}$$

where θ is the contact angle for a smooth surface, and r is the roughness defined as the ratio between the real solid surface and the projected one. Because, obviously, r is greater than unity, θ_W is greater than θ and increases with increasing roughness for a hydrophobic smooth surface ($\theta > 90°$). However, θ_W is smaller than θ and decreases with increasing roughness r for a hydrophilic ($\theta < 90°$) surface.

Wenzel formula, Eq. (1), was successfully tested experimentally for macroscopic drops. For nanodrops, there is theoretical evidence, obtained by the density functional theory [8] and Monte Carlo simulations [9], which even on a hydrophilic surface the contact angle increases with increasing roughness. In order to answer to the question of how specific modifications of a hydrophilic surface affect its wetting properties, in the present paper, we examine nanodrops on such a surface in more details than in the previous papers [8,9]. In particular, by changing the distance between pillars, the most effective configuration is identified which provides the largest increase in the contact angle. Pillars composed of materials which differ from the substrate are also examined. By changing the strength of the interaction between the molecules of pillars and the molecules of fluid, the effect of the gradient of the pillar fluid interaction on the contact angle θ is examined.

[*] *Journal of Colloid and Interface Science* 359 (2011) 304-310. Republished with permission.

2. GENERAL CONSIDERATIONS

2.1. The system

The considered system, which involves a one-component fluid (argon) of fixed average density ρ_{av} in contact with the surface of a semiinfinite substrate (solid carbon dioxide) decorated with pillars of height h_p, width d_p and distance between pillars Δ_p, is presented in Fig. 1. All distances between surfaces are measured between the centers of the molecules of their first layers.

The system has the finite lengths L_x and L_h in the x- and h-directions, respectively, and an infinite dimension in the y-direction, which is perpendicular to the plane of the figure. The pillars are infinite in the y-direction and are composed of a material, which can differ from the solid CO2. All pillars have rectangular shape with the same width $d_p = 1.81\sigma_{ff}$ (σ_{ff} is the hard core diameter of the fluid-fluid interactions) and are located equidistant from each other. The fluid density distribution (FDD) $\rho(\mathbf{r})$ in this system is considered uniform in the y-direction and nonuniform in the x- and h-directions, hence $\rho(\mathbf{r}) \equiv \rho(x, h)$. A periodic boundary condition is assumed in the x-direction, and the upper boundary of the system is treated as a hard wall without attractive interactions between the wall and the fluid molecules. The fluid is exposed to an external potential due to the fluid-substrate and fluid-pillars interactions. The temperature is fixed at $T = 87.0$ K.

Due to the regular geometrical location of the pillars, the rough surface generates a potential which for identical pillars is periodic in the x-direction.

The origin of the coordinate system is located at a distance from the solid substrate equal to one hard core diameter σ_{fs} of the fluid-solid interaction.

2.2. Interaction potentials

The interaction potentials between the molecules of fluid and between the molecules of fluid and those of the solid are selected in the Lennard-Jones form with hard core repulsion

$$\phi_\alpha\left(|\mathbf{r}-\mathbf{r}'|\right) = \begin{cases} 4\varepsilon_\alpha\left[\left(\dfrac{\sigma_\alpha}{r}\right)^{12} - \left(\dfrac{\sigma_\alpha}{r}\right)^{6}\right], & r \geq \sigma_\alpha \\ \infty & r < \sigma_\alpha \end{cases} \tag{2}$$

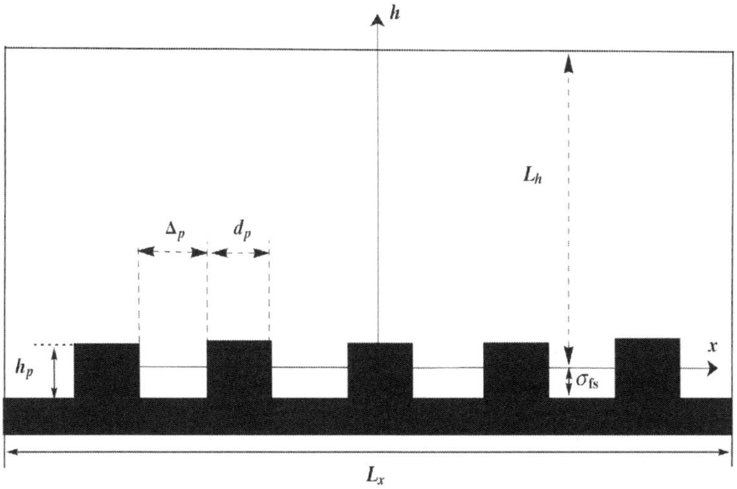

FIG. 1. Schematic representation of the considered system. σ_{fs} is the hard core diameter of the fluid-solid interactions. All distances between surfaces are measured between the centers of the molecules forming the first layers.

where the subscript α should be *ff*, *fs*, and *fp* for the fluid-fluid, fluid-substrate, and fluid-pillars interactions, respectively, ϵ_{ff}, ϵ_{fs}, and ϵ_{fp}, are energy parameters, σ_{ff}, σ_{fs}, and σ_{fp}, are hard core diameters, $\mathbf{r}$ and $\mathbf{r}'$ provide the locations of the interacting molecules, and $r = |\mathbf{r} - \mathbf{r}'|$. For argon, $\epsilon_{ff}/k_B = 119.76$ K, $\sigma_{ff} = 3.405$ Å, where k_B is the Boltzmann constant; for the solid carbon dioxide which is used as substrate, $\epsilon_{fs}/k_B = 153.0$ K, $\sigma_{fs} = 3.727$ Å and the number density $\rho_s = 1.91 \times 10^{28}$ m^{-3} [10].

Because of the geometry of the system, the external potential U_{sf} generated by the solid (substrate plus pillars) at a point $M(x, y, h)$ depends on the distance $h + \sigma_{fs}$ of this point from the surface of the substrate and the position x of point M along the surface, and is independent of y, i.e. $U_{sf} \equiv U_{sf}(x, h)$. The potential U_{sf} can be obtained by integrating the Lennard-Jones potential (Eq. (2)) for the fluid-substrate and fluid-pillars interactions over the entire volume of the solid and can be written as

$$U_{fs}(\mathbf{r}) = \int_{V_s} \rho_s(\mathbf{r}')\phi_{fs}(|\mathbf{r}-\mathbf{r}'|)d\mathbf{r}' + \int_{V_p} \rho_p(\mathbf{r}')\phi_{fp}(|\mathbf{r}-\mathbf{r}'|)d\mathbf{r}' \tag{3}$$

where V_s is the volume of the substrate, V_p is the volume occupied by the pillars, $\rho_s(\mathbf{r}')$ and $\rho_p(\mathbf{r}')$ are the densities of the substrate and pillars, respectively, which, in general, depend on the coordinates. For a uniform substrate, $\rho_s(\mathbf{r}') \equiv \rho_s$, and the first integral in Eq. (3) can be calculated analytically

$$\int_{V_s} \rho_s(\mathbf{r}')\phi_{fs}(|\mathbf{r}-\mathbf{r}'|)d\mathbf{r}' = \frac{2\pi}{3}\epsilon_{fs}\rho_s\sigma_{fs}^3\left[\frac{2}{15}\left(\frac{\sigma_{fs}}{\sigma_{fs}+h}\right)^9 - \left(\frac{\sigma_{fs}}{\sigma_{fs}+h}\right)^3\right] \tag{4}$$

The second integral in Eq. (3) was calculated numerically. Note that for identical pillars, U_{sf} is a periodic function of x with a period equal to $d_p + \Delta_p$.

Below, two kinds of arrangements of the pillars on the surface of the substrate are considered. In the first ($S1$), the pillars are identical but composed of a material, which differs from that of the substrate by the energy parameter ($\epsilon_{fp} \neq \epsilon_{fs}$). For the second ($S2$), the pillars have the same location as for $S1$ but interact differently with the fluid (the energy parameter ϵ_{fp} is different for different pillars). For the surfaces $S1$ and $S2$, the net potentials $\Phi(x, h)$ to which a molecule of fluid is exposed are presented schematically in Figs. 2a and b as functions of x for $\Delta_p = 5.69\sigma_{ff}$, $h_p = 2\,\sigma_{ff}$, and $h = h_p$.

For the first surface (with $\epsilon_{fp} = \epsilon_{fs}$), the total potential possesses multiple wells with the same depth separated by potential barriers of equal height. For the second kind of surface, $\epsilon_{fp} = 0.2n\epsilon_{fs}$, n being the sequential number of the pillar, and the depth of the potential well increases in the positive direction of the x-axis.

The form of the potential presented in Fig. 2b suggests that farther the drop is from the left hand side of the system, the smaller is its free energy. In this case, a drop can exist only in metastable states ensured by the potential barriers between the minima. If the heights of these barriers are not large enough, then drops in metastable states cannot be formed.

2.3. The Euler-Lagrange Equation for the Fluid Density Distribution and Calculation of the Contact Angle

The equation for the fluid density distribution (FDD) $\rho(x, h)$ is obtained through the minimization of the total Helmholtz free energy $F[\rho(\mathbf{r})]$, which can be expressed as the sum of an ideal gas free energy, $F_{id}[\rho(\mathbf{r})]$, an excess free energy $F_{hs}[\rho(\mathbf{r})]$ (with respect to the ideal gas) of a reference system of hard spheres, a free energy $F_{attr}[\rho(\mathbf{r})]$ due to the attractive interactions between the fluid molecules (in the mean field approximation), a free energy $F_{fs}[\rho(\mathbf{r})]$ due to the interactions between the fluid and solid (substrate plus pillars). Explicit expressions for these contributions, which can be obtained in standard ways [11], are provided in Ref. [8].

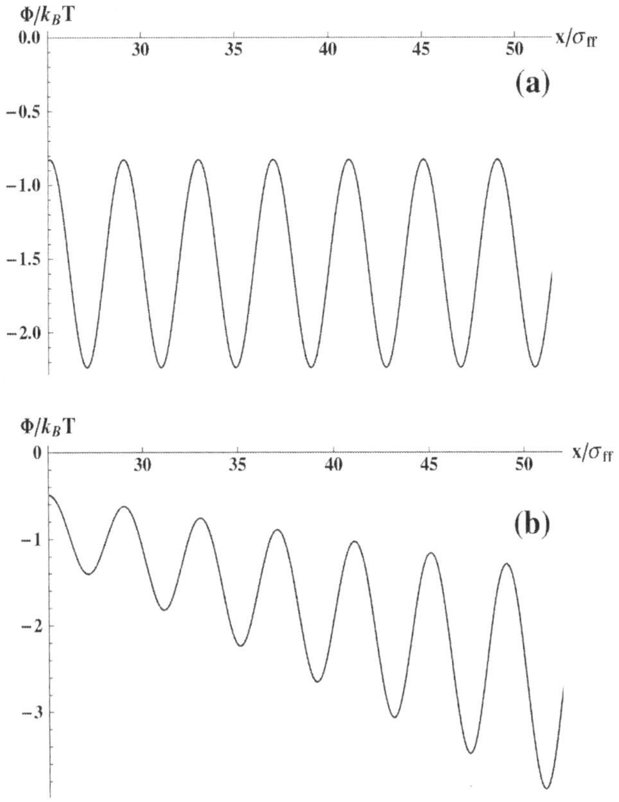

FIG. 2. Net potential $\Phi \equiv \Phi(x)$ due to the interaction of a fluid molecule, located in the first fluid layer above the pillars, with the substrate and pillars for $h_p = 2\sigma_{ff}$ and $\Delta_p = 5.69\sigma_{ff}$. (a) The case of identical pillars with $\epsilon_{fp} = \epsilon_{fs}$. (b) The case of nonidentical pillars with $\epsilon_{fp} = 0.2n\epsilon_{fs}$, n being the sequential number of the pillar. Here, x represents the distance from the left hand side of the system.

The Euler-Lagrange equation for $\rho(x, h)$ can be written in the following form [8]

$$\log\left[\Lambda^3\rho(x,h)\right] - Q(x,h) = \frac{\lambda}{k_B T} \tag{5}$$

where the function $Q(x, h)$, which is a functional of $\rho(x, h)$, is given in Ref. [8], $\Lambda = h_P/(2\pi m k_B T)^{1/2}$ is the thermal de Broglie wavelength, h_P is the Planck constant, T is the absolute temperature, m is the mass of a fluid molecule, and λ is a Lagrange multiplier arising because of the constraint of fixed average density of the fluid. This constraint has the form

$$\rho_{av} = \frac{1}{V}\int_V d\mathbf{r}\rho(\mathbf{r}) \tag{6}$$

and leads to the following expression for λ [8]:

$$\lambda = -k_B T \log\left[\frac{1}{\rho_{av} V \Lambda^3}\int_V d\mathbf{r} e^{Q(x,y)}\right] \tag{7}$$

where V is the volume occupied by the fluid. By eliminating λ between Eqs. (5) and (7), one obtains an integral equation for the FDD $\rho(x, h)$, which can be solved by iterations. The details of the iteration procedure have been presented in Ref. [8].

The drop profile was extracted from the obtained continuous FDD as the curve along which the density $\rho(\mathbf{r})$ is constant. The value ρ_{div} of the constant was selected as the density on an equimolar dividing surface in an one-dimensional FDD $\rho(x, h_0)$ obtained from $\rho(x, h)$ at some distance $h = h_0$ from the top of the pillars. The latter distance was selected $h_0 = 4\sigma_{ff}$ because at this distance the oscillations of the fluid density near the tops of the pillars are small (see Ref. [8] for more details). After the drop profile was obtained, its upper portion was considered as part of a circle, which was extrapolated up to the horizontal plane passing over the tops of the pillars. The angle that this circle makes with that plane was considered to be the contact angle. Note, that a similar procedure for calculating the contact angle was used in Refs. [12–14] for drops on smooth surface.

3. RESULTS: DROP ON A SURFACE *S1*

If otherwise mentioned, the calculations in this section have been carried out for a fixed number N_{mol} of molecules per unit length in the *y*-direction of the system ($N_{mol}\sigma_{ff} = 90$).

3.1. Dependence of the Contact Angle on the Pillars Height and Distance between Pillars

Let us first consider the dependence of the contact angle of a drop on the pillar height for various values of the energy parameter ϵ_{fp} characterizing the interaction of the pillars with the fluid. In Fig. 3, this dependence is presented for a distance between pillars $\Delta_p = 5.69\sigma_{ff}$. The lower scale on the horizontal axis provides the roughness of the surface, *r*, defined as the ratio of the real solid surface to the projected one, which for the considered geometry of the surface is given by

$$r = 1 + \frac{2h_p}{d_p + \Delta_p} \tag{8}$$

The smallest height h_p of the pillars was selected $0.2\sigma_{ff}$ (such a pillar can be considered as a small defect of the substrate), and $\epsilon_{fp}/\epsilon_{fs}$ was taken 0.25, 1.0, and 1.4. In all considered cases, the contact angle increases with increasing pillar height (increasing roughness) and with decreasing pillar-fluid energy parameter. Each height dependence suggests the existence of an asymptotic value of the contact angle (different for different ϵ_{fp}). This is expected because the substrate potential decreases rapidly with increasing pillar height, and the shape of the drop becomes dominated by the interaction between the drop and pillars. The latter interaction increases rapidly in magnitude with increasing h_p approaching an asymptotic value. For illustration purposes, the dependence of the potentials generated by the substrate and pillars at a distance σ_{fs} above the top of the pillars (the location of the first liquid layer of molecules above the pillars) on the pillar height is presented in Fig. 4.

It should be noted that the presence of even very small pillars on the substrate surface considerably affects the contact angle. For instance, in the absence of pillars, the contact angle of the nanodrop on the smooth surface of the considered substrate is $\theta_{pl} \simeq 52°$. However, for pillars of height $h_p = 0.2\sigma_{ff}$ with $\epsilon_{fp} = \epsilon_{fs}$, $\theta \simeq 67°$.

The h_p-dependence of the contact angle for other distances Δ_p between pillars is qualitatively the same as that presented in Fig. 3. They are plotted in Fig. 5 for $\epsilon_{fp}/\epsilon_{fs} = 1$ and $\Delta_p = 2.19\sigma_{ff}$, $2.99\sigma_{ff}$, and $4.19\sigma_{ff}$.

Typical dependencies of the contact angle on the distance Δ_p between pillars and on the roughness at constant height of the pillars are presented in Fig. 6a and b for $h_p = \sigma_{ff}$ and $h_p = 2\sigma_{ff}$, respectively, for various $\epsilon_{fp}/\epsilon_{fs}$. For all considered cases, this dependence attains a maximum for $\Delta_{p,m} = 2.19\sigma_{ff}$. For $\Delta_p < \Delta_{p,m}$, the contact angle decreases with decreasing Δ_p, approaching for small Δ_p (not shown in the figure) the value $\theta_{pl} \simeq 52°$ of the contact angle on an ideal planar surface.

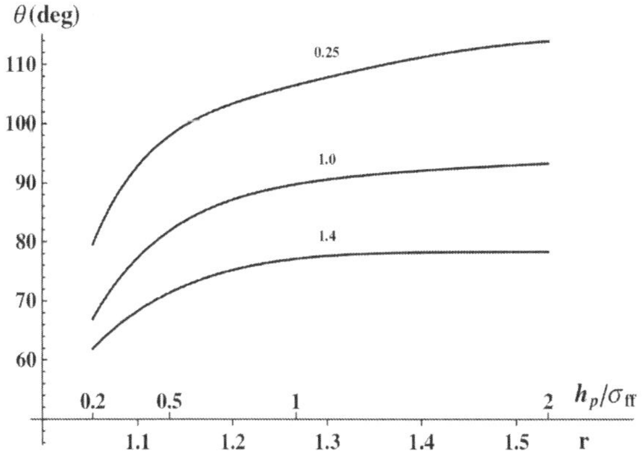

FIG. 3. Dependence of the contact angle θ on the roughness r and pillar height h_p for various ratios $\epsilon_{fp}/\epsilon_{fs}$ indicated on each curve. The distance between pillars $\Delta_p = 5.69\sigma_{ff}$.

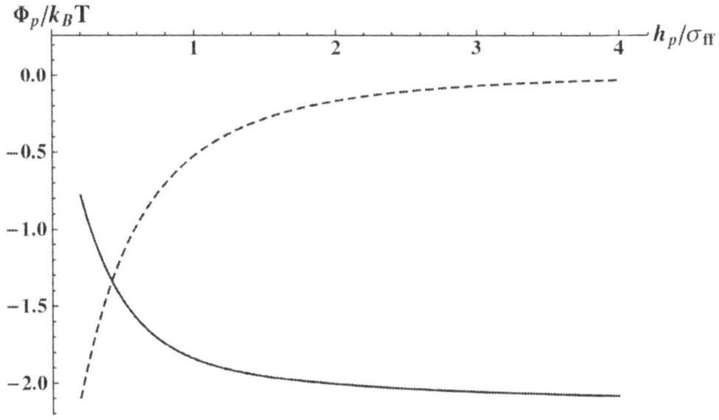

FIG. 4. Potential $\Phi_p \equiv \Phi_p(h_p)$ generated by a pillar at a point located in the first fluid layer above the pillars as function of pillars height h_p (solid line). For comparison, the potential generated by the substrate at the same point is presented by the dashed line. In this figure, $\epsilon_{fp}/\epsilon_{fs} = 1$.

4. RESULTS: DROP ON A SURFACE *S*2

Let us consider now a surface decorated with nonidentical pillars separated by the same distance Δ_p. The difference between pillars arises because the energy parameter $\epsilon_{fp} \equiv \epsilon_{fp}(n)$ is considered a function of the sequential number of the pillar. All other parameters of the pillars (width, height) were kept as for the surface *S*1. A linear dependence

$$\epsilon_{fp}(n) = \epsilon_{fp}^{0} + (n-1)\Delta\epsilon, \quad (n = 1,\ldots,N_P) \tag{9}$$

where N_p is the total number of pillars, was selected for the energy parameter ϵ_{fp} with $\epsilon_{fp}^{0} = 0.2\epsilon_{fs}$ and various $\Delta\epsilon$. The value $n = 1$ is assigned to the leftmost pillar. The interpillar distance was selected $\Delta_p = 5.69\sigma_{ff}$ ($N_p = 9$).

In previous studies based on classical thermodynamics [15–19], and in the microscopic approaches [8,20], it was shown that a drop on an inhomogeneous surface can be present in several metastable states separated by potential barriers. To identify some of such states for the present system, three initial guesses for the fluid density distribution, required in the iteration procedure employed for solving the Euler-Lagrange equation, were used. In the first, the fluid density was assumed uniform and liquid like inside a rectangular area symmetric with respect to the central pillar, and uniform

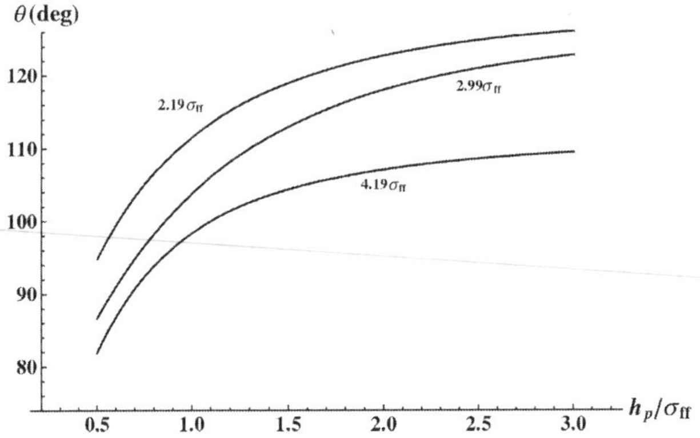

FIG. 5. Dependence of the contact angle θ on the pillar height h_p for various distances Δ_p between pillars indicated on each curve. In this figure, $\epsilon_{fp}/\epsilon_{fs} = 1$.

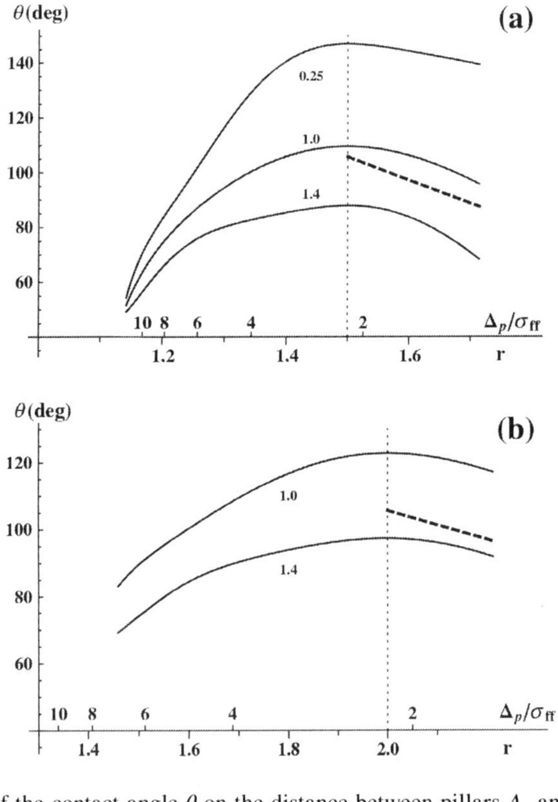

FIG. 6. Dependence of the contact angle θ on the distance between pillars Δ_p and on the roughness r for various $\epsilon_{fp}/\epsilon_{fs}$ indicated on each curve. (a) $h_p = \sigma_{ff}$, (b) $h_p = 2\sigma_{ff}$. The dashed lines in both figures represent the results provided by the Cassie-Baxter formula (see discussion in Section 5) for $\epsilon_{fp}/\epsilon_{fs} = 1$. The dotted vertical lines indicate the locations of the maxima of the curves.

but vapor like outside that area. The area of the rectangle and the fluid densities were adjusted to provide the selected number of molecules in the system ($N_{mol}\sigma_{ff} = 90$). In the other two guesses, the rectangles were taken symmetric with respect to the pillars next to the central one.

As expected, the results of the calculations depend on the height of the pillars, the distance between them and the gradient of the pillar-fluid interaction, which is characterized by the parameter $\Delta\epsilon$. In Fig. 7, the FDDs for $\Delta\epsilon = 0.2\epsilon_{fs}$ are presented for $h_p = 3\sigma_{ff}$ and $\Delta_p = 5.69\sigma_{ff}$. Obviously,

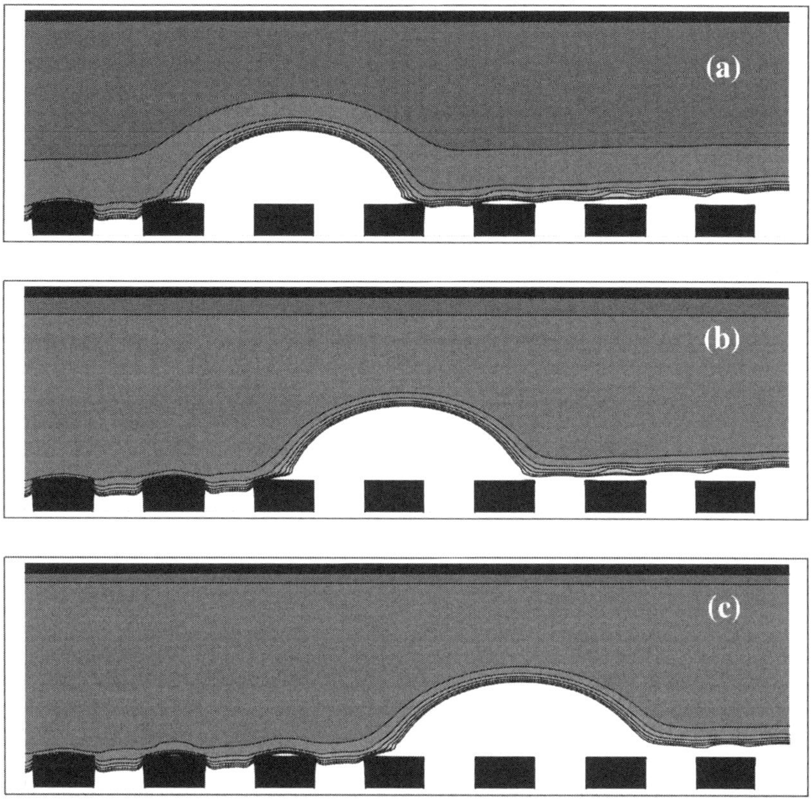

FIG. 7. Three metastable FDDs obtained for $\epsilon_{fp} = \epsilon_{fp}^0 + (n-1)\Delta\epsilon$ where n = 1,..., 9 is the sequential number of the pillar, $\epsilon_{fp}^0 = 0.2\epsilon_{fs}$ $\Delta\epsilon = 0.2\epsilon_{fs}$ and the height of the pillars $h_p = 3\sigma_{ff}$. FDDs in panels (a), (b), (c) were obtained using the initial guesses symmetrical with respect to different pillars. Only seven of the nine pillars of the system are present.

TABLE 1

θ_f, right hand side (front) and θ_r, left hand side (rear) contact angles for the drops presented in Fig. 7 and the free energy F of the system per unit area of the substrate. The height of pillars $h_p = 2\sigma_{ff}$, distance between pillars $\Delta_p = 5.69\sigma_{ff}$, and $\Delta\epsilon = 0.2\epsilon_{fs}$.

	Drop (a)	Drop (b)	Drop (c)
θ_f(deg)	96.3	90.8	84.8
θ_r(deg)	95.9	90.4	84.5
$\theta_f - \theta_r$(deg)	0.4	0.4	0.3
$F\sigma_{ff}^2/k_BT$	−958.6	−959.7	−961.0

these FDDs are asymmetric due to the gradient of the pillar-fluid interaction, which generates a force that acts on the drop in the horizontal direction. For this reason, the contact angles on the right (θ_f) and left (θ_r) hand sides of the drop profile (the front and rear contact angles, respectively) are slightly different. These contact angles along with the free energies per unit area of the substrate are listed in Tables 1 and 2 for each of the three drops for $h_p = 2\sigma_{ff}$ and $h_p = 3\sigma_{ff}$, respectively.

TABLE 2
θ_f, right hand side (front) and θ_r, left hand side (rear) contact angles for the drops located as in Fig. 7 but for the height of pillars $h_p = 3\sigma_{ff}$ and the free energy F of the system per unit area of the substrate. The distance between pillars $\Delta_p = 5.69\sigma_{ff}$ and $\Delta\epsilon = 0.2\ \epsilon_{fs}$.

	Drop (a)	Drop (b)	Drop (c)
θ_f(deg)	90.1	85.0	79.6
θ_r(deg)	87.7	83.3	78.0
$\theta_f - \theta_r$(deg)	2.4	1.7	1.6
$F\sigma_{ff}^2/k_BT$	–961.5	–962.9	–964.4

The free energies listed in the tables represent local minima of the Helmholtz free energy of the system, which become deeper in the direction of the gradient of the pillar-fluid interaction.

As well known, the difference between the front and rear contact angles determines the magnitude of the force which acts on the drop in the direction parallel to the surface of the substrate [18–20]. The larger the difference between θ_f and θ_r, the larger is the force exerted on the drop. If this force becomes larger than the sticking force caused by the free energy barriers between the minima of the free energy, the drop will not survive in any of the metastable states. In the present case, the difference in angles is small and slightly increases with increasing pillars height (compare, for example, the data in Table 1 with those in Table 2). If the gradient of the energy parameter of the pillar-fluid interaction becomes sufficiently large, for instance, if $\Delta\epsilon = 0.3\epsilon_{fs}$ instead of $\Delta\epsilon = 0.2\epsilon_{fs}$, the calculations revealed that no drops in metastable states can be formed. This means that the magnitude of the force that acts on the drop becomes larger than the sticking force.

5. DISCUSSION

The common view regarding the behavior of a macroscopic liquid drop on a rough hydrophilic surface is based on the idea that the surface energy of a wet hydrophilic solid is smaller than that of a dry solid. For this reason, the solid surface beneath the drop prefers to be completely wet (the liquid to penetrate between the pillars). In this case, the contact angle which the drop profile makes with the rough solid is given by the Wenzel formula, Eq. (1). Note that the theoretical justification of the Wenzel equation involves the concepts of surface tensions, which are meaningful only for macroscopic interfaces.

The calculations carried out in the present paper and in Ref. [8] for a nanodrop of argon on the basis of DFT as well as the molecular dynamics simulations performed in Ref. [9] for a water nanodrop have shown that for a nanodrop on a hydrophilic surface the contact angle increases as the roughness r increases. Hence, there is a qualitative difference between the results predicted by Eq. (1) for macrodrops and those predicted for nanodrops. Indeed, for macrodrops on a hydrophilic surface ($\theta < 90°$), the wetting angle decreases with increasing roughness. This difference can be attributed to the inapplicability of the macroscopic concepts of surface tensions to the nanoscale and consequently, the inapplicability of Wenzel equation to nanodrops.

Our DFT analysis also has shown that the Δ_p-dependence, as well as the r-dependence of the contact angle at constant h_p is nonmonotonous (see Fig. 6). For all considered pillars heights, the maximum of θ is reached when the distance between pillars becomes equal to $2.19\sigma_{ff} = 2\sigma_{fs}$, the smallest distance that allows to locate a fluid molecule between pillars. In such a case, the space between pillars remains empty. In the nomenclature of the macroscopic treatment of wetting, this corresponds to a transition from the Wenzel to the Cassie-Baxter regime. In the latter regime, the drop is located only on the top of the pillars and the fluid does not penetrate between pillars. According to the Cassie-Baxter theory [21], the contact angle is a function of f_p (the ratio between the surface of the tops of the pillars and the projected solid surface)

$$\cos\theta_{\text{CB}} = -1 + f_p\left(\cos\theta + 1\right) \tag{10}$$

where for the considered cases $f_p = d_p/(d_p + \Delta_p) = d_p/2h_p(r-1)$. Eq. (10) describes qualitatively the decrease in the contact angle for $\Delta_p < \Delta_{p,\,m}$ (see Fig. 6). However, there is a quantitative difference between the results provided by the Cassie-Baxter formula and the DFT, which increases with increasing height of the pillars.

Note that in Ref. [8] and numerical simulations [9], the height of the pillars was selected to be larger or equal to two molecular diameters. In the present calculations, much smaller heights, starting with $0.2\sigma_{ff} \simeq 0.68$ Å, were considered. The results have shown that for the smallest pillars considered, the contact angle is appreciably larger than that for a smooth surface. Because it is unlikely for real surfaces to be completely free of in homogeneities of the order of 1 Å, the present results for nanodroplets suggest that the contact angle θ measured on such a surface which is considered smooth differs from the contact angle θ_{pl} on an ideal smooth surface. For example, calculations performed for pillars with $h_p = 0.2\sigma_{ff} \simeq 0.7$ Å, $\epsilon_{fp} = \epsilon_{fs}$, and $\Delta_p = 5.69\sigma_{ff}$ provided a contact angle $\theta \simeq 67°$ which is very different from that of an ideally smooth surface, $\theta_{pl} \simeq 52°$. Can this observation be extended to macrodrops?

Our results show that the roughness r is not the only parameter which determines the contact angle of a nanodrop on a rough surface. For instance, let us compare the contact angles of a nanodrop on two different surfaces of the same roughness $r = 1.50$. The two surfaces selected for comparison are decorated with pillars for which $h_p = \sigma_{ff}$ and $2\sigma_{ff}$, $\Delta_p = 2.19\sigma_{ff}$ and $6.31\sigma_{ff}$, respectively, and $\epsilon_{fp}/\epsilon_{fs} = 1$. Fig. 6a and b shows that for the selected roughness r the contact angles are very different, namely $\theta \simeq 109°$ and $\theta \simeq 89°$, respectively. These results are in contradiction with the Wenzel formula, Eq. (1), which predicts a unique dependence of the apparent contact angle on roughness. However, as argued in Refs. [6,22,23], the roughness r in Eq. (1) should be calculated as an average over a large number of pillars located in the vicinity of the contact line. This observation provides an additional reason for the inapplicability of Wenzel equation to nanodrops (the first and main reason is its use of the macroscopic concepts of surface tensions). Indeed, the size of such drops is comparable (or much smaller) than the characteristic size of the roughness, and the contact line is located only on a few pillars. The difference between the contact angles of the two considered surfaces with the same r is a result of the different sizes and arrangement of the pillars, which generate very different potentials near the solid surface. Therefore, the nanodrops on such surfaces are subjected to different potentials and hence have different contact angles.

For identical equidistant pillars, drops of the same volume located at equivalent positions with respect to the pillars provide the same free energy of the system and are stable. When a gradient of the pillar-fluid interaction is present, the potential of the pillar- fluid interaction as function of x has minima which become deeper with increasing x (see Fig. 2b). For this reason, the free energy of the system decreases if the location of the drop is displaced in the direction of increasing pillar-fluid interaction. The numerical analysis of the Euler-Lagrange equation, Eq. (5), reveals two possible scenarios. In the first, Eq. (5) has no solutions and drops are not formed. This occurs when the gradient of the pillar-fluid interaction is relatively large ($\Delta\epsilon > 0.25\epsilon_{fs}$ in Eq. (9)).

In the second case ($\Delta\epsilon < 0.25\epsilon_{fs}$), several solutions of the Euler- Lagrange equation describing drops at different locations, as shown in Fig. 7, do exist. All the drops are metastable, and their characteristics are listed in Tables 1 and 2. These metastable states are separated by free energy barriers, which cannot be estimated within the statistical thermodynamics approach employed. (Only a kinetic approach can provide the values of the barriers.) The shape of the

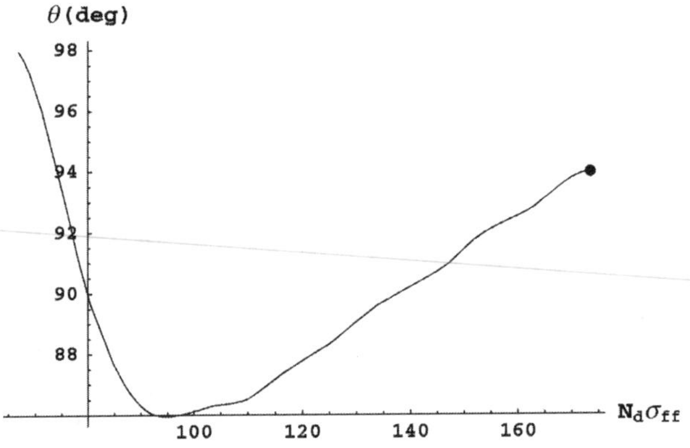

FIG. 8. Dependence of the contact angle θ on the number of molecules, N_d, in the drop for $h_p = \sigma_{ff}$, $\Delta_p = 2\sigma_{ff}$, and $\epsilon_{fp} = \epsilon_{fs}$. The points represent results of calculations, the solid line is a guide for eyes.

drops is slightly asymmetric, the difference between the front and rear contact angles being however small (see Tables 1 and 2).

The disappearance of the metastable states, when the parameter $\Delta\epsilon$ increases above a certain value, can be understood in terms of a sticking force caused by the interaction of the fluid with the solid. The gradient of the pillar-surface interaction generates a force which tends to displace the drop in the direction of increasing pillar-fluid interaction. The larger this gradient, the larger is the force. If this force becomes greater than the sticking force, the drop overcomes the free energy barriers between the metastable states of the system and moves along the surface (phenomenon similar to the Marangoni effect). Neither equilibrium nor metastable states are possible in such a case. If the force acting on the drop due to the gradient of pillar-fluid interaction is smaller than the sticking force, the drop can remain in metastable equilibrium in one of the possible metastable states.

Note that the size, arrangement, and composition of pillars are not the only characteristics that affect the contact angle of a nanodrop on a rough surface. For instance, in Fig. 8, the contact angle as function of the number of molecules in the drop (drop size) is presented for $h_p = \sigma_{ff}$ and $\Delta_p = 2\sigma_{ff}$. The analysis of the location of the drop contact line as a function of the size of the drop has shown that the contact angle increases with increasing size of the drop when the contact line is located on the pillars and decreases when it is located between pillars. This observation suggests a quasiperiodic behavior of θ as a function of the number of molecules, N_d, in the drop and can be attributed to the well-known phenomenon of contact angle hysteresis on a rough surface.

REFERENCES

[1] S. Shibuichi, T. Onda, N. Satoh, K. Tsujii, J. Phys. Chem. 100 (1996) 19512.
[2] T. Onda, S. Shibuichi, N. Satoh, K. Tsujii, Langmuir 12 (1996) 2125.
[3] Y. Song, R.P. Nair, M. Zou, Y.Q. Wang, Nano Res. 2 (2009) 143.
[4] H. Wang, M. Zou, R. Wei, Thin Solid Films 518 (2009) 1571.
[5] M. Reyssat, F. Pardo, D. Quere, EPL 87 (2009) 36003.
[6] Y. Kwon, S. Choi, N. Anantharaju, J. Lee, M.V. Panchagnula, N.A. Patankar, Langmuir 26 (2010) 17528.
[7] R.N. Wenzel, Ind. Eng. Chem. 28 (1936) 988.
[8] G.O. Berim, E. Ruckenstein, J. Chem. Phys. 129 (2008) 014708 (Section 4.4 of this volume).
[9] C.D. Daub, J.H. Wang, S. Kudesia, D. Bratko, A. Luzar, Faraday Discuss. 146 (2010) 67.
[10] R. Evans, P. Tarazona, Phys. Rev. A 28 (1983) 1864.
[11] P. Tarazona, Phys. Rev. A 31 (1985) 2672.
[12] F. Porcheron, P.A. Monson, M. Schoen, Phys. Rev. E 73 (2006) 041603.
[13] N. Giovambattista, P.G. Debenedetti, P.J. Rossky, J. Phys. Chem. B 111 (2007) 9581.
[14] M.J. de Ruijter, T.D. Blake, J. de Coninck, Langmuir 15 (1999) 7836.
[15] R.E. Johnson Jr., R.H. Dettre, Surf. Colloid Sci. 2 (1969) 85.

[16] A. Marmur, Colloids Surf. A 136 (1998) 209.
[17] A. Marmur, Langmuir 20 (2004) 3517.
[18] B. Krasovitski, A. Marmur, Langmuir 21 (2005) 3881.
[19] D. Quere, M. Reyssat, Philos. Trans. Roy. Soc. A 366 (2008) 1539.
[20] G. O. Berim, E. Ruckenstein, J. Chem. Phys. 129 (2008) 114709 (Section 4.11 of this volume).
[21] A. B.D. Cassie, S. Baxter, Trans. Faraday Soc. 40 (1944) 546.
[22] M. V. Panchagnula, S. Vedantam, Langmuir 23 (2007) 13242.
[23] A. Marmur, E. Bittoun, Langmuir 25 (2009) 1277.

4.6 Contact Angle of a Nanodrop on a Nanorough Solid Surface*

Gersh Berim and Eli Ruckenstein†
Department of Chemical and Biological Engineering, State University of New York at Buffalo, Buffalo, New York 14260, USA.

Corresponding Author

† E-mail: feaeliru@buffalo.edu;
Fax: +1 (716)645–3822; Tel: +1 (716)645-1179

1. INTRODUCTION

For more than two centuries, the behavior of a macroscopic liquid drop on a solid surface was the object of great practical and theoretical interests. The three basic equations employed are the Young equation[1]

$$\gamma_{\mathrm{lv}} \cos\theta = \gamma_{\mathrm{sv}} - \gamma_{\mathrm{sl}} \tag{1}$$

for smooth surfaces as well as the Wenzel equation[2]

$$\cos\theta_{\mathrm{W}} = r\cos\theta \tag{2}$$

and Cassie-Baxter equation[3]

$$\cos\theta_{\mathrm{CB}} = f\cos\theta + f - 1 \tag{3}$$

for rough surfaces. In the above equations, θ, θ_{W}, and θ_{CB} are the contact angles on a smooth surface, a rough surface in which the liquid penetrates into the space between asperities (Wenzel regime), and a rough surface in which the space beneath the drop is filled with air (Cassie-Baxter regime). In addition, γ_{lv}, γ_{sv}, and γ_{sl} are the liquid-vapor, solid-vapor, and solid-liquid surface tensions, respectively, r is the roughness defined as the ratio between the actual area of the rough surface and the projected area of this surface onto a horizontal smooth one, and f is the roughness defined as the ratio between the top area of asperities and the area of the surface free of asperities.

Because the surface tensions as well as r and f do not depend on the drop size and shape, eqn (1)–(3) predict that the contact angle of a macroscopic drop is independent of the drop size. Another feature is related to the dependence of the contact angle on the roughness of the surface. For both Wenzel and Cassie-Baxter regimes, the contact angle increases with increasing roughness if the solid substrate is hydrophobic ($\theta > 90°$) and decreases if it is hydrophilic ($\theta < 90°$) (below, the terms hydrophobic and hydrophilic will be used for any surface-fluid pairs, not only for the surface-water one).

Even though in most cases eqn (1)–(3) provide correct descriptions of the experimental data, some cases were identified for which the above mentioned characteristics are not valid. First, it was observed that for small spherical drops (microdrops) on a smooth surface the contact angle depends

* *Nanoscale*, 2015, 7, 3088–3099. Republished with permission.

on the radius R of the contact line between drop and surface. In this case, the Young equation has to be modified by introducing a term involving the ratio between the line tension τ and radius R. The modified equation has the form[4]

$$\gamma_{lv}\cos\theta = \gamma_{sv} - \gamma_{sl} - \frac{\tau}{R}. \tag{4}$$

Eqn (4) coincides with eqn (1) in the limit $R \rightarrow \infty$ (large drop). However, for a cylindrical (two-dimensional) drop, *i.e.* a drop which is elongated in a single direction, the last term in eqn (4) should not occur because the contact line between the drop and solid is a straight line ($R = \infty$). Therefore, θ for a cylindrical drop does not depend on the drop size.

The situation becomes more complex when the drop size becomes of the order of a few nanometers. In this case, the thickness of the liquid-fluid interface between the core of the drop and surrounding vapors becomes comparable with the drop size and the macroscopic concept of surface tension is no longer valid. For this reason, eqn (1)–(3) are not applicable to such drops and microscopic considerations are required to describe a nanodrop on a solid surface. Comparatively few studies of nanodrops on nanorough surfaces have been carried out.[5–11] In ref. 5,10, the calculations were performed using a lattice DFT approach, which is appropriate for three-dimensional nanodrops with heights of up to about $100\sigma_{ff}$, where σ_{ff} is the diameter of a fluid molecule. The off lattice density functional theory (DFT) approach was employed in ref. 6, 8, 11. Because of a large increase of the computational time for large nanodrops the heights of the considered cylindrical nanodrops were restricted to the range $10\sigma_{ff}$–$20\sigma_{ff}$. In ref. 7,9 molecular dynamic simulations were used for drops with heights of several nanometers.

As expected, in many cases the behavior of the contact angles of nanodrops on nanorough surfaces was different from that described by eqn (1)–(3). As an example, one can mention the qualitative disagreement between the r-dependence of angle θ for a nanodrop on a nanorough hydrophilic surface provided by the Wenzel equation and that provided by molecular dynamics simulation[7] and DFT.[8] In the latter cases, θ increases with increasing roughness, whereas eqn (2) predicts the opposite behavior.

Porcheron *et al.*[5] and Malanoski *et al.*,[10] have noted that on a hydrophobic smooth surface the contact angle increases with increasing height of the drop from about $10\sigma_{ff}$ to $120\sigma_{ff}$, after which it remains constant. That behavior of θ was attributed to the line tension. A nonmonotonous dependence of θ on the size of the cylindrical drop on a rough hydrophilic surface was also noted.[6]

Surprisingly, the behavior of the contact angle of a nanodrop on a nanorough hydrophobic surface agrees qualitatively with the predictions of eqn (1)–(3). In particular, in ref. 5–7,10 was noted that on such surfaces angle θ increases with increasing roughness, in agreement with eqn (2) and (3). However, in the cases presented in ref. 5–7,10, the material of the pillars was selected the same as that of the substrate. Is this conclusion also valid when the interactions between fluid and substrate and between fluid and pillars are different? In this paper it is shown that in the latter cases the contact angle of a nanodrop on a rough hydrophobic substrate depends on roughness differently from the predictions provided by the Wenzel and Cassie-Baxter equations. Along with hydrophobic substrates, the hydrophilic ones are also considered.

2. BACKGROUND

2.1. The system and interaction potentials

The considered system (see Fig. 1) consists of a one-component fluid of fixed average density ρ_{av} inside a rectangular box which has finite dimensions L_x and $L_h + \sigma_{fs}$ in the horizontal (x) and vertical (h) directions, respectively, and infinite dimension in the y-direction perpendicular to the plane of the figure (the length σ_{fs} is the hard core diameter of the fluidsubstrate interaction).

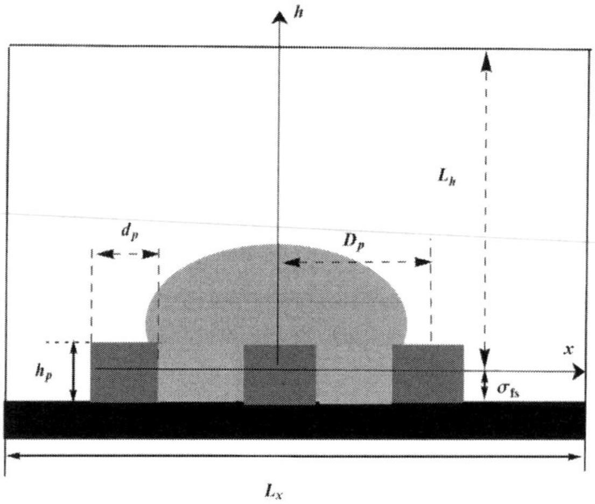

FIG. 1. Schematic representation of the considered system. The distances between surfaces are measured between the centers of the molecules forming the first layers.

The system is in contact with a rough solid surface which is composed of a semiinfinite uniform substrate having a smooth surface decorated with evenly distributed rectangular pillars which model the roughness. The material of the pillars can be different from that of the substrate. The pillars have heights h_p, widths d_p, and distance between their centers D_p. The parameters of all materials as well as the size of the pillars will be specified below. At the distance $L_h + \sigma_{fs}$ from the substrate, the system is limited by a hard wall which has no attractive interactions with the fluid molecules. In the horizontal direction along the x axis, a periodic boundary condition is assumed. It will be assumed that the fluid density distribution (FDD) $\rho_f(\mathbf{r})$ in the system is uniform in the y-direction and non-uniform in the x- and h-directions *i.e.* $\rho_f(\mathbf{r}) = \rho_f(x, h)$.

The interactions between fluid molecules, between fluid molecules and molecules of the substrate and those of the pillars are provided by the Lennard-Jones potential with hard core repulsion:

$$\phi_\alpha(r) = \begin{cases} 4\varepsilon_\alpha \left[\left(\frac{\sigma_\alpha}{r} \right)^{12} - \left(\frac{\sigma_\alpha}{r} \right)^{6} \right], & r \geq \sigma_\alpha \\ & r < \sigma_\alpha \\ \infty, \quad r < \sigma_\alpha \end{cases} \tag{5}$$

where $r = |\mathbf{r} - \mathbf{r}'|$, $\mathbf{r}$ and $\mathbf{r}'$ provide the locations of the interacting molecules, the subscripts α are ff, fs, and fp for fluid-fluid, fluid-substrate, and fluid-pillar interactions, respectively, ε_{ff}, ε_{fs}, and ε_{fp} are energy parameters, and σ_{ff}, σ_{fs}, and σ_{fp} are hard core diameters.

For the fluid-fluid interaction, the energy parameter and hard core diameter were selected as for argon ($\varepsilon_{ff}/k_B = 119.76$ K, $\sigma_{ff} = 3.405$ Å),[12] where k_B is the Boltzmann constant. The substrate-fluid interaction ε_{fs} was selected $0.6\varepsilon_{ff}$ and $1.28\varepsilon_{ff}$ for hydrophobic and hydrophilic substrates, respectively, with the hard core diameter $\sigma_{fs} = 3.727$ Å the same for both substrates (The contact angles for a nanodrop on smooth hydrophobic and hydrophilic substrates were 121° and 44°, respectively[13]). For fluid-pillar interaction the hard core diameter was selected as for the fluid-substrate one ($\sigma_f = \sigma_{fs}$), but the energy parameter ε_{fp} was varied. The number densities of the substrate and pillars were taken the same $\rho_s = 1.91 \times 10^{28}$ m^{-3}. The temperature was selected $T = 87.0$ K $= 0.66\, T_c$, where $T_c = 131.6$ K is the critical temperature of the bulk fluid.[12]

The external potential U_{fs} generated by the solid (substrate plus pillars) does not depend on the coordinate y, *i.e.* $U_{fs} \equiv U_{fs}(x, h)$. This potential can be calculated by integrating the Lennard-Jones potential (eqn (5)) for the fluid-substrate and fluid-pillar interactions over the volume of the substrate and pillars, respectively, and can be written in the form

$$U_{\mathrm{fs}}(\mathbf{r})=\int_{V_s}\rho_s(\mathbf{r}')\phi_{\mathrm{fs}}(|\mathbf{r}-\mathbf{r}'|)\mathrm{d}\mathbf{r}'+\int_{V_p}\rho_p(\mathbf{r}')\phi_{\mathrm{fp}}(|\mathbf{r}-\mathbf{r}'|)\mathrm{d}\mathbf{r}' \tag{6}$$

where V_s and V_p are the volumes occupied by the substrate and pillars, respectively, $\rho_s(\mathbf{r}')$ and $\rho_p(\mathbf{r}')$ are the densities of the substrate and pillars, respectively, which, in general, depend on coordinates. For a uniform substrate ($\rho_s(\mathbf{r}') \equiv \rho_s = \text{const}$) the first integral in eqn (6) has the form

$$\int_{V_s}\rho_s(\mathbf{r}')\phi_{\mathrm{fs}}(|\mathbf{r}-\mathbf{r}'|)\mathrm{d}\mathbf{r}'=\frac{2\pi}{3}\varepsilon_{\mathrm{f}\rho_s}\sigma_{\mathrm{fs}}^3\left[\frac{2}{15}\left(\frac{\sigma_{\mathrm{fs}}}{\sigma_{\mathrm{fs}}+h}\right)^9-\left(\frac{\sigma_{\mathrm{fs}}}{\sigma_{\mathrm{fs}}+h}\right)^3\right]. \tag{7}$$

The integration of the second part of eqn (6) can be carried out analytically only over the y-coordinate. The integration over the x- and h-coordinates was performed numerically. Note that the origin of the coordinate system was selected at the distance σ_{fs} from the substrate surface (see Fig. 1).

2.2. Basic equation of DFT and its solution

The basic equation for the FDD $\rho_f(\mathbf{r})$ was obtained by minimizing the Helmholtz free energy of the system, $F[\rho_f(\mathbf{r})]$. An explicit expressions for $F[\rho_f(\mathbf{r})]$ is provided in Appendix A.

As a result of minimization, one obtains the following Euler-Lagrange equation[6]

$$\log\left[\Lambda^3\rho_f(x,h)\right]-Q_f(x,h)=\frac{\lambda}{k_BT} \tag{8}$$

where the function $Q_f(x, h)$, which is a functional of $\rho_f(x, h)$, is provided in Appendix A, $\Lambda = h_P/(2\pi m k_B T)^{1/2}$ is the thermal de Broglie wavelength, h_P is the Planck constant, m is the mass of a fluid molecule, and λ is a Lagrange multiplier which accounts for the constraint of fixed average density of the fluid. This constraint provides the equation

$$\rho_{\mathrm{av}}=\frac{1}{V}\int_V \mathrm{d}\mathbf{r}\rho_f(\mathbf{r}) \tag{9}$$

where V is the volume occupied by the fluid. Eqn (9) leads to the following expression for λ[6]

$$\lambda=-k_BT\log\left[\frac{1}{\rho_{\mathrm{av}}V\Lambda^3}\int_V \mathrm{d}\mathbf{r}e^{Q_f(x,\ h)}\right] \tag{10}$$

which after combining with eqn (9) provides an integral Euler-Lagrange equation for the FDD $\rho_f(x, h)$. The latter equation can be solved numerically by iterations. The details of the iteration procedure are presented in Appendix A. Here we emphasize only the choice of the initial fluid density distribution (initial guess) used in the calculations. Due to the symmetry of the system, it is natural to assume that a nanodrop on a rough surface can exist only in two states which differ by the drop position with respect to the pillars (see Fig. 2 where those states are presented schematically). One of them (D1) is symmetrical with respect to a vertical plane passing through the middle of the pillar, the second one (D2) is symmetrical with respect to a vertical plane located midway between pillars. For this reason, the initial density distribution was usually selected as a cylindrical (two-dimensional) rectangular drop of a reasonable density elongated in the y-direction which possesses the symmetry of one of the mentioned above drops. Note that the location of the initial guess at any other position increases the time of calculation but results, nevertheless, in one of the above mentioned two solutions.

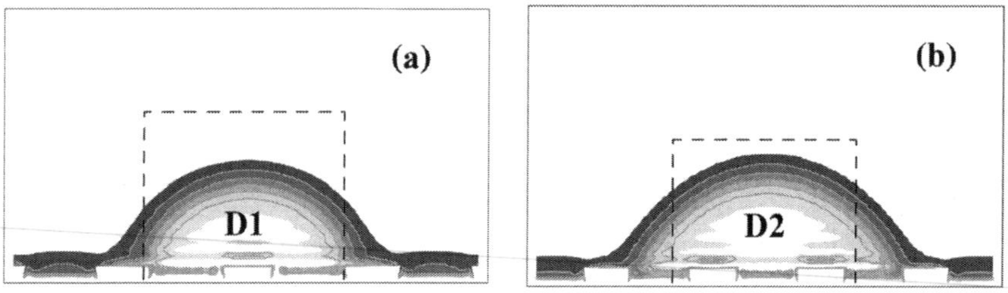

FIG. 2. Schematic representation of two possible solutions of the Euler-Lagrange equation. The dashed rectangles represent an initial guess for the iteration procedure. In the figure, the pillars have white color.

As expected, depending on the selected parameters of the system, there are several possible outcomes of the iterations. In the first case, the drop does not form. This occurs when the amount of fluid in the system is too small to ensure a drop existence, or when the solid-fluid interaction is so strong that the fluid molecules completely wet the surface. In the second case, two initial guesses provide finally the same drop (D1 or D2). In the last, third case, the initial guesses mentioned above converge to different drops, the free energy of the system being different for the two drops. The drop corresponding to the smaller free energy is considered as stable, the other one as metastable.

When using a numerical procedure to solve the Euler- Lagrange equation, one always encounter the question whether the selected precision can ensure that the obtained solution is not an intermediate step of the iteration procedure, which, in reality, leads to another solution. We do not have an exact answer to this question. In Appendix B a qualitative approach is presented for handling this uncertain situation.

In conclusion of this section, let us note that the selected temperature, $T = 87.0$ K, is far enough from the critical temperature, T_c of the considered fluid ($T/T_c = 0.66$) for the thermal fluctuations to be small. For this reason, the use of DFT, which is based on a mean-field approximation for the fluid-fluid interactions, is justified.

2.3. Calculation of the Contact Angle

Because of the nonuniformity of the fluid in nanodrops, the drop profile needed to calculate the contact angle, is not clearly defined. In this paper, a simple procedure is used in which the drop profile is given by the line in the vapor-liquid interface which corresponds to a local constant density ρ_{div}. The procedure how to determine ρ_{div} was described in ref. 14. An example of a drop profile obtained in this way is presented in Fig. 3. On this figure and all similar figures below, the lighter areas correspond to higher fluid densities.

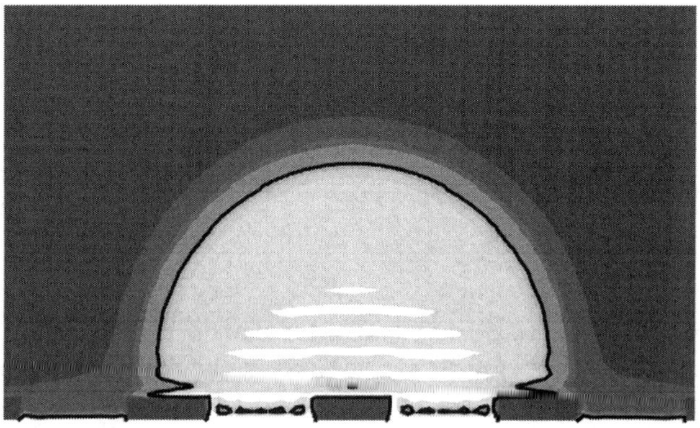

FIG. 3. Example of the drop profile (solid line) obtained by the method described in ref. 14. The dividing density $\rho_{div}\sigma_{ff}^3 = 0.375$ in this example. The lighter areas correspond to higher densities.

After the drop profile was obtained, its upper part was approximated by a circle and extrapolated until it intersected the solid. The angle that this circle makes with the solid was considered as the contact angle. This definition of the contact angle does not account for the profile of the drop in the very vicinity of the solid. Close to the solid, the fluid density in the drop has an oscillatory behavior and changes considerable. This makes impossible the use of this region for meaningful contact angle calculations.

Note that in some cases the drop profile has the shape of a closed loop which does not intersect the solid surface. In this case, the contact angle was considered to be 180°. Using the obtained drop profile one can calculate the number of molecules in the drop N_d (per unit length) which will be used as a characteristic of the drop size.

3. RESULTS AND DISCUSSION

The characteristics of nanodrops (size, contact angle) were examined for 12 rough surfaces which differ in the distance between pillars and their height. Depending on those parameters, the surfaces have different roughnesses, *r*, which were calculated using the equation[8]

$$r = 1 + \frac{2h_p}{D_p} \tag{11}$$

The notations which will be used to identify the surfaces are presented in Table 1 along with their roughnesses. In most calculations, the total number N_{tot} of molecules of fluid in the system per unit length in the *y*-direction was selected between $N_{tot}\sigma_{ff} = 80$ and $N_{tot}\sigma_{ff} = 180$. The parameter ε_{fp} of the fluid-pillar interaction was varied between $\varepsilon_{fp} = 0.20\varepsilon_{fs}$ and $\varepsilon_{fp} = 6.0\varepsilon_{fs}$ for a hydrophobic substrate and between $\varepsilon_{fp} = 0.05\varepsilon_{fs}$ and $2.0\varepsilon_{fs}$ for a hydrophilic one. In all considered cases, the fluid penetrated into the space between pillars (the Wenzel regime).

In section 3.1 it is shown that, for some values of the interaction parameters and size of the drop, the system can be in a stable or metastable state. The dependence of the contact angle of a nanodrop on a rough surface on the drop size is examined in section 3.2, its dependence on the distance between pillars and their height is presented in section 3.3. Section 3.4 is concerned with the effect of fluid-pillar interactions.

3.1. STABLE AND METASTABLE SOLUTIONS

As already mentioned, two solutions of the Euler-Lagrange equation were identified in most cases. One of these solutions corresponds to a global minimum in the Helmholtz free energy (stable state), and the other one corresponds to a local minimum (metastable state). The formalism employed in

TABLE 1
Notations for selected surfaces and their roughnesses

	h_p/σ_{ff}		
D_p/σ_{ff}	1	2	3
14.0	S_{11} $r = 1.143$	S_{12} $r = 1.286$	S_{13} $r = 1.429$
9.3	S_{21} $r = 1.215$	S_{22} $r = 1.430$	S_{23} $r = 1.645$
7.5	S_{31} $r = 1.267$	S_{32} $r = 1.533$	S_{33} $r = 1.800$
4.8	S_{41} $r = 1.417$	S_{42} $r = 1.833$	S_{43} $r = 2.250$

the present paper allows one to calculate the free energies of those minima and discriminate between stable and metastable states. However it does not allow to obtain the height of the energy barrier, which separates the stable from the metastable state. To make the considerations more general, it will be assumed that the metastable states have enough long lifetimes and for this reason they will be examined along with the stable states.

Because the Euler-Lagrange equation provides solutions corresponding to the extremum of the free energy (minimum or maximum) there is, in principle, the possibility for the state with greater free energy to correspond to a local maximum of the free energy which is, in fact, not metastable but unstable. Another possibility is that the numerically obtained solution is only an intermediate step in the sequence of iterations which eventually lead to a stable solution if a higher calculation precision would have been used. Those issues are examined in Appendix B.

The stability and metastability of the drops on rough surfaces depend on a number of factors and it is difficult to formulate general rules for their identification. Some specific results for drops D1 and D2 on a hydrophobic substrate with pillars of height $h_p = \sigma_{ff}$ and $\varepsilon_{fp} = \varepsilon_{fs}$ are presented in Table 2 for $N_{tot}\sigma_{ff}$ between 80 and 180.

In the examples listed in Table 2, both drops D1 and D2 are presented for all considered N_{tot}. However for other choices of the fluid-pillar energy parameter ε_{fp}, there are cases in which only drop D1 exists and drop D2 is unstable. This happens, for example, for the hydrophilic surface S_{21} with $N_{tot} \leq 90$ and $\varepsilon_{fp}/\varepsilon_{fs} \leq 0.75$.

One can see from Table 2, that on surfaces S_{21}, S_{31}, and S_{41}, the state of the drop (stable or metastable) depends on the number of molecules in the system and as a consequence, on the size of the drop. A possible reason for such a behavior is the change in the location of the leading edges of the drops with respect to pillars. In Fig. 4, the drops D1 and D2 are presented for surface S_{21} and $N_{tot}\sigma_{ff} = 100$ (first row) and 150 (second row). In the first case, the leading edges of both drops are located on the pillars, with the free energy of drop D2 smaller (stable drop) than that of drop D1 (metastable drop). For the greater N_{tot}, the leading edges of D1 are located between pillars where the attraction by the substrate is larger, whereas those for D2 remain on the pillars. In this case, the free energy of D1 is smaller than that of D2 and D1 is stable while D2 metastable. For surface S_{11}, for which the distance between pillars is larger than that for S_{21}, and S_{31}, the leading edges of D2 are

TABLE 2
The dependence of the state of drops D1 and D2 on rough hydrophobic substrates on the total number N_{tot} of fluid molecules per unit length of the system. In the considered examples the substrate is decorated with pillars of the same material ($\varepsilon_{fp} = \varepsilon_{fs}$); $h_p = \sigma_{ff}$. In the bistable states, the stable and metastable states have the same free energies

Surface	$N_{tot}\sigma_{ff}$	D1	D2
S_{11}	80–180	Metastable	Stable
S_{21}	80–140	Metastable	Stable
	150–180	Stable	Metastable
S_{31}	80–110	Metastable	Stable
	120–180	Stable	Metastable
S_{41}	80–100	Metastable	Stable
	110	Bistable	Bistable
	120–180	Stable	Metastable

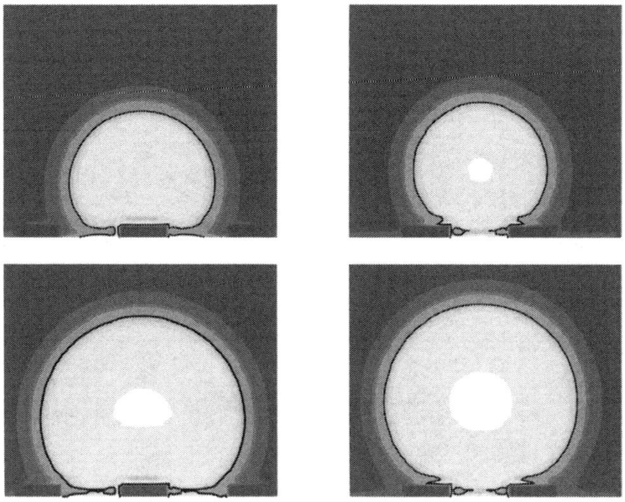

FIG. 4. Drops D1 (first column) and D2 (second column) for the systems with $N_{tot}\sigma_{ff} = 100$ (first row) and $N_{tot}\sigma_{ff} = 150$ (second row) on the hydrophobic surface S_{21}. $\varepsilon_{fp}/\varepsilon_{fs} = 1$.

between pillars for all considered values of N_{tot}. The leading edges of D1 are located either on the pillars or between them. In the latter case, the contact area of D1 with the substrate was always less than that for D2. For this reason, D1 is always metastable and D2 stable.

3.2. Size dependence of θ on rough surfaces

In this section, the dependence of the contact angle of a nanodrop on a nanorough surface is examined as function of the size of the nanodrop expressed via the number N_d of fluid molecules in the drop per unit length in the y-direction. This number was calculated using FDD and the drop profile extracted from FDD (see section 2.3).

In Fig. 5, the size dependence of the contact angle of the nanodrop on four hydrophobic surfaces is presented for drops D1 (Fig. 5a) and D2 (Fig. 5b). The size of the drop was varied between $N_d\sigma_{ff} = 40$ and $N_d\sigma_{ff} = 150$.

For drop D1, the contact angles on surfaces S_{11} and S_{41} decrease monotonously with increasing size of the drop, whereas for the drops on surfaces S_{21} and S_{31} this dependence passes through a minimum after which the contact angle only slightly increases with increasing drop size. For drop D2, the contact angles on surfaces S_{11} and S_{21} increase monotonously and for the surface S_{41} it decreases with increasing size of the drop. For surface S_{31}, the change in contact angle with N_d is small. For both nanodrops, the contact angles are not constant, hence in contradiction with the predictions of the traditional macroscopic theory. Note that the line tension cannot explain the dependence of θ on the drop size because the leading edges of the drop have infinite radius of curvature and the last term in eqn (4) should not be present.

All the peculiarities of contact angle behavior can be qualitatively explained considering the location of the leading edges for different drop sizes. As an example, in Fig. 6 drops D1 and D2 of different sizes are present on the hydrophobic surface S_{21}. In the considered range of drop sizes, drop D2 has its leading edges pinned to the edges of neighboring pillars. This pinning prevents the motion of the drop leading edges on the surface of the pillars and causes a monotonous increase of the contact angle with increasing size of the nanodrop (see the curve for S_{21} in Fig. 5b). It is expected, that this increase will continue until the leading edges of the drop detach from the pillars edges and move on the surface of a pillar toward the next edge of the pillar. For this reason, the width of the drop base increases and the contact angle decreases. The change from the increase of θ to its decrease means that there is a local maximum in the dependence of θ on the drop size and suggests that the shape of the curve for S_{21} in Fig. 5b should

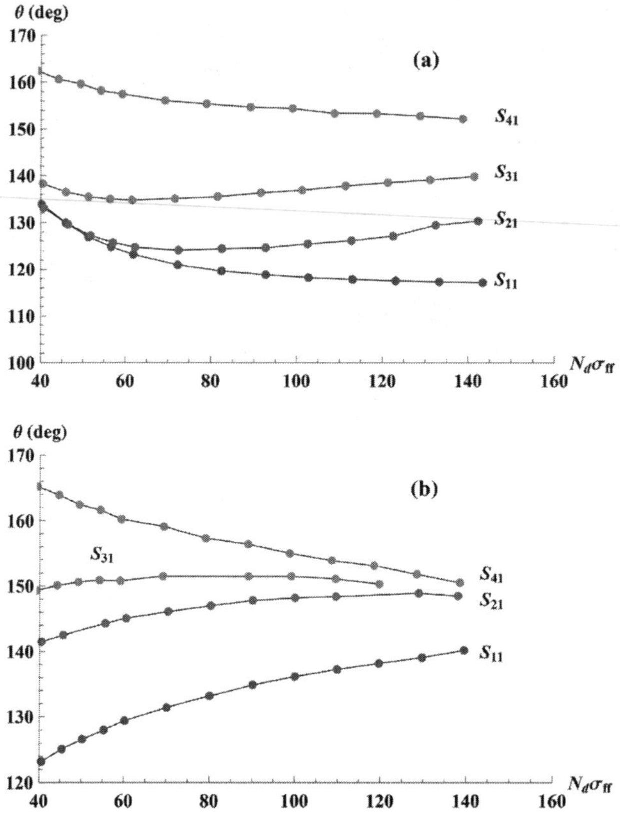

FIG. 5. Size dependence of the contact angle of a nanodrop on a rough hydrophobic substrate for $h_p = \sigma_{ff}$ for drops D1 (a) and D2 (b) when $\varepsilon_{fp} = \varepsilon_{fs}$. The points represent results of calculations, the curves are guides for eye.

be convex. Note that the suggested maximum is not present in Fig. 5b because the required drop size was not achieved in the calculations due to the small size of the system.

For the smallest size of drop D1, its leading edges are pinned to the edges of the pillar on which the drop is sitting. When the size of the drop increases, the drop leading edges will change their position by moving on the surface of the substrate where the attractive drop-substrate interactions are stronger. This leads to the decrease of the contact angle with increasing drop size until the leading edges become located close to the neighboring pillars and are pinned to them. This is followed by a contact angle increase with increasing drop size as it was in the case of drop D2. Hence, the size dependence of θ for drop D1 should have a minimum (see the curve for S_{21} in Fig. 5a) and the curve for S_{21} in Fig. 5a is convex downward. Note that the changes in contact angle in the used example are small and cannot by detected by eye in Fig. 6.

The size dependence of the contact angle on a rough hydrophilic substrate presented in Fig. 7 demonstrates a more complex behavior. For drop D1 (Fig. 7a), the contact angles on surfaces S_{21} and S_{31} first increase with increasing size, then rapidly decrease after which they increase again. As for hydrophobic surfaces, this behavior is related to the change in the location of the leading edges of the drop profile on the surface. Before θ decreases, the leading edges are located on the pillars (as shown in Fig. 8a) and are pinned to the edges of the pillars.

As long as the size of the drop increases up to a critical value, the contact angle increases because the pinning prevents the drop leading edges to move to the space between pillars. At the critical size, the leading edges separate from pillars edges and become located between pillars (Fig. 8b). In this case, the contact angle becomes considerably smaller than in the previous case thus explaining the kink-like behavior.

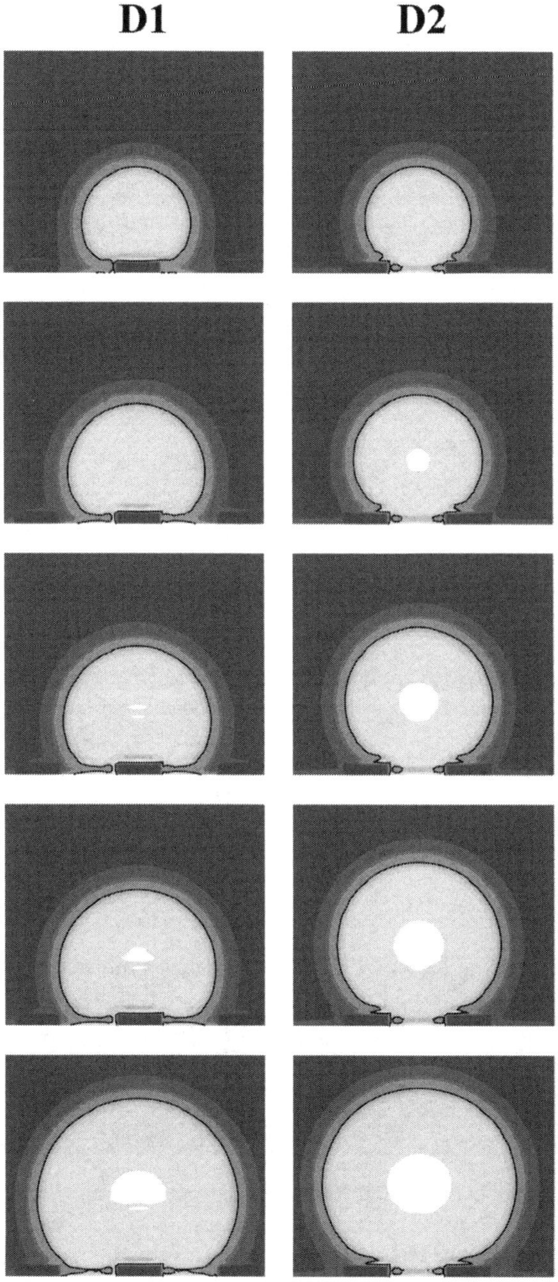

FIG. 6. Drops D1 and D2 of various sizes on the hydrophobic surface S_{21} with $\varepsilon_{fp} = \varepsilon_{fs} = 0.6\varepsilon_{ff}$.

A similar behavior of the contact angle occurs for drop D2 on surface S_{11} (see Fig. 7b).

For drop D1 on the hydrophilic surfaces S_{11} and S_{41}, as well as for drop D2 on surfaces S_{12}, S_{31}, and S_{41} the leading edges of the drop are located either on the pillars or between pillars for all considered drop sizes. For this reason, kink-like changes in the contact angle dependence on N_d do not occur for those surfaces. Nevertheless, one can expect kinks for larger drop sizes because of changes in the location of the leading edges. However the sizes of these systems do not allow to perform calculations for large nanodrops to identify those kinks.

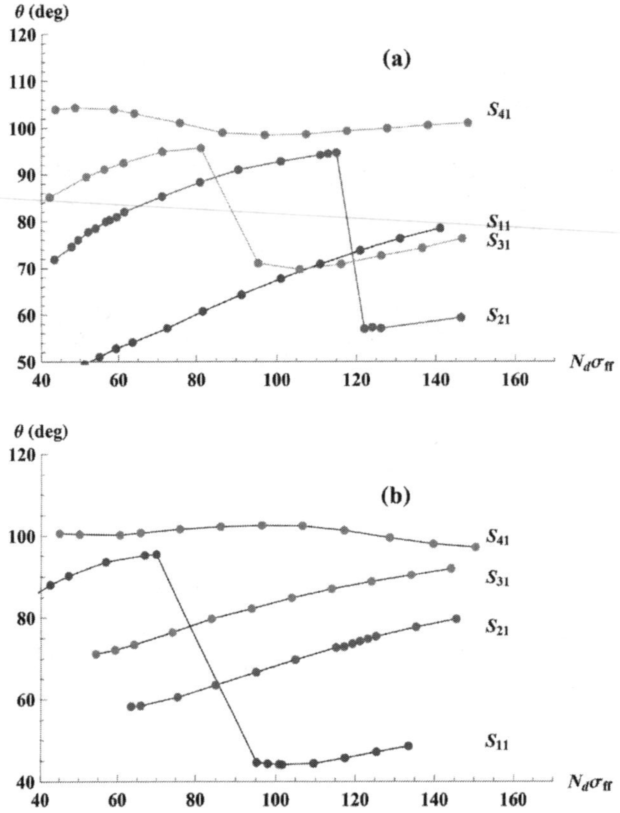

FIG. 7. Size dependence of the contact angle of a nanodrop on a rough hydrophilic substrate with $h_p = \sigma_{ff}$ for drops D1 (a) and D2 (b) when $\varepsilon_{fp} = \varepsilon_{fs}$. The points represent results of calculations, the lines are the guides for eye.

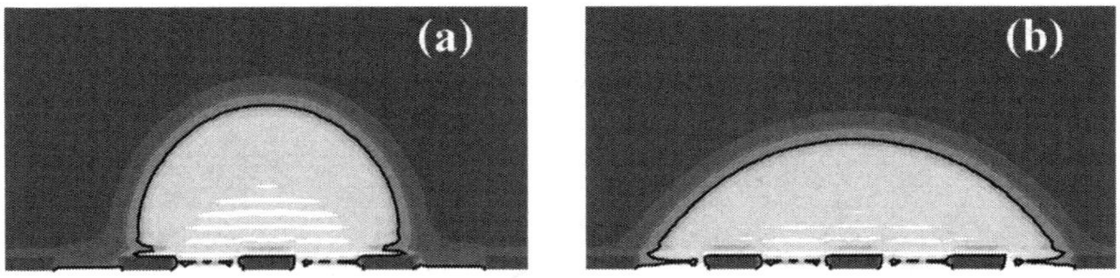

FIG. 8. Two characteristic shapes of nanodrop D1 on a rough hydrophilic surface.

Note that the absence of such kinks for hydrophobic substrates (Fig. 5) can be explained by the lower drop-substrate and drop-pillar interactions compared with the hydrophilic surfaces. As a result, the pinning force is smaller than for hydrophilic surfaces and the change in the contact angle is much smoother when the drop leading edges move from pillars to the space between them.

The dependence of the contact angle on drop size was also examined in connection with the phenomenon of contact angle hysteresis in ref. 15 by minimizing the free energy of the system, (the latter involving the surface tension), and in ref. 9 by molecular dynamic simulations. In ref. 15, the sizes of the considered two-dimensional drops were much larger than the size of the pillars; in ref. 9, where only three dimensional drops were considered, the sizes of the droplets were comparable to those of the pillars and of the order of a few nanometers. In both papers, kink-like changes in the contact angle with increasing drop volume were identified in some cases that agree with our results obtained by DFT.

Note in conclusion that for both hydrophilic and hydrophobic surfaces the size dependence of the contact angle should be quasiperiodic because the drop leading edges sequentially move along the upper surface of the pillars and along the surface of the substrate, the contact angle being determined by the location of the leading edges of the drop.

3.3. Dependence of the Contact Angle on the Distance between Pillars and on Their Height

Examples regarding the dependence of the contact angle of a nanodrop on the distance D_p between pillars for drop D1 on rough hydrophobic substrate are presented in Fig. 9 for surfaces with various pillar heights, h_p, and strength ε_{fp} of the fluid-pillar interaction. The roughnesses of the considered surfaces depend both on D_p and h_p and are provided above the horizontal axis.

In each case, the contact angle first decreases with increasing roughness (decreasing D_p) in contradiction with Wenzel equation, but then increases, in qualitative agreement with that equation.

Even though D_p-dependence of θ in all considered cases is similar, there are some differences which should be emphasized. For example, for $h_p/\sigma_{ff} = 1$ and $h_p/\sigma_{ff} = 2$, the D_p-dependencies of θ for $\varepsilon_{fp}/\varepsilon_{fs} = 3.5$ have minima at different roughnesses 1.27 and 1.43, respectively (compare the dashed lines in Fig. 9(a) and (b)). To explain this observation, let us examine the changes of the drop profiles of drop D1 with changes in the distance between pillars (see Fig. 10) for both cases. Comparing the drop profiles for $h_p/\sigma_{ff} = 1$ and $h_p/\sigma_{ff} = 2$ one can see that in both cases the

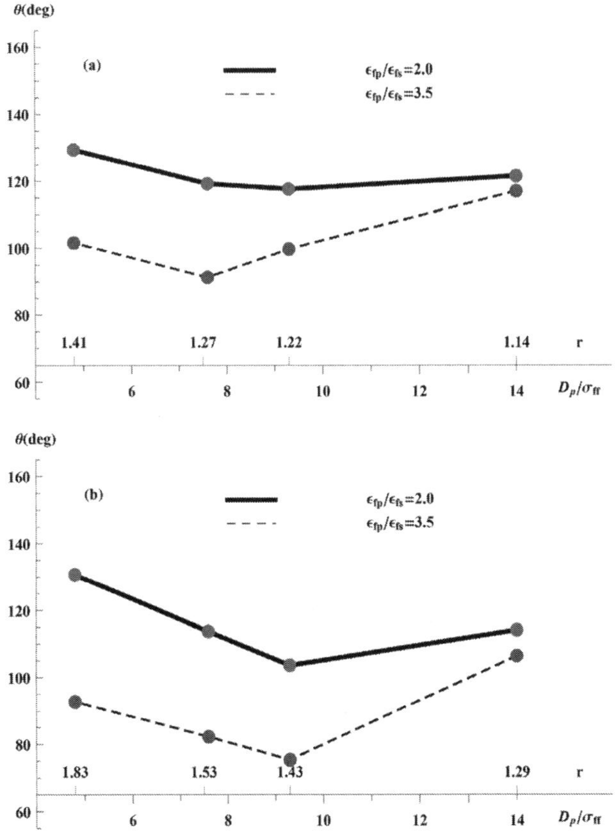

FIG. 9. Dependence of the contact angle of drop D1 on a rough hydrophobic substrate on the distance between pillars for $\varepsilon_{fp}/\varepsilon_{fs} = 2.0$, and 3.5 for pillars heights $h_p/\sigma_{ff} = 1$ (a) and $h_p/\sigma_{ff} = 2$ (b). In all cases $N_{tot}\sigma_{ff} = 90$. The points represent results of calculations, the lines are guides for eye.

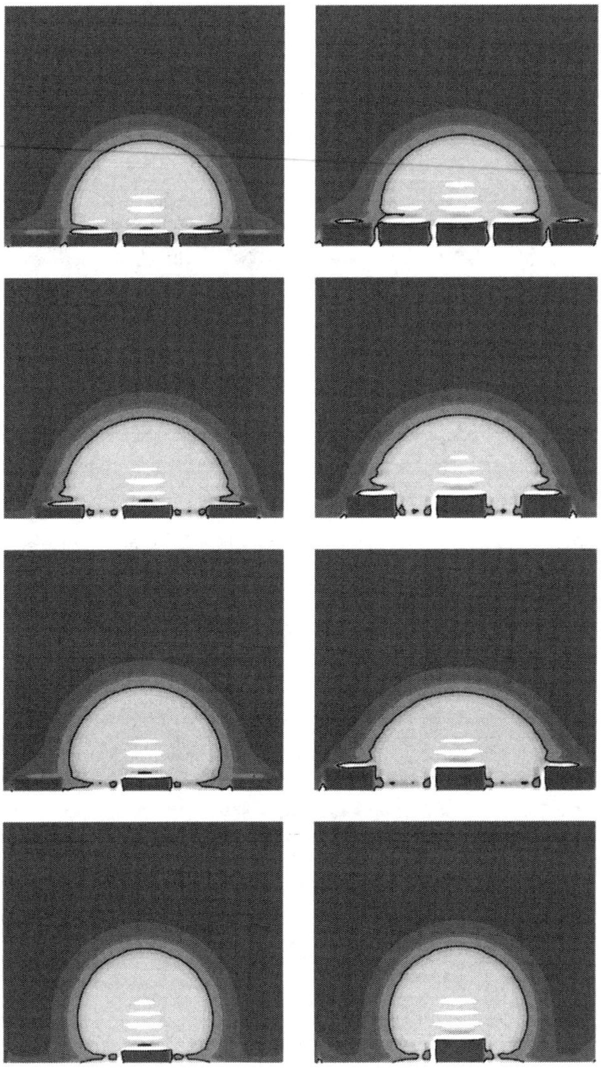

FIG. 10. Drop on various hydrophobic surfaces decorated with pillars of height $h_p/\sigma_{ff} = 1$ (first column) and $h_p/\sigma_{ff} = 2$ (second column). The distances between pillars are $4.8\sigma_{ff}$ (first row), $7.5\sigma_{ff}$ (second row), $9.3\sigma_{ff}$ (third row), and $14\sigma_{ff}$ (forth row). $\varepsilon_{fp}/\varepsilon_{fs} = 3.5$ for all cases.

contact angle decreases with increasing D_p as long as the leading edges of the drop remain on the pillars. The contact angle acquires a minimum when D_p approaches the critical value D_{pc} at which the leading edges of the drop "jump" from pillars into the area between pillars and the contact angle increases. For $h_p/\sigma_{ff} = 1$, the distance D_{pc} is in the interval $7.5 < D_{pc}/\sigma_{ff} < 9.3$. For $h_p/\sigma_{ff} = 2$, D_{pc} is greater ($9.3 < D_{pc}/\sigma_{ff} < 14$) because the attraction between the leading edges of the drop and the pillars is greater for the larger pillars. For this reason, θ exhibits minima at various D_p. From the above consideration, it is clear that the location of the minimum in the D_p-dependence of the contact angle depends on the size of the drop and on the strength of fluid-pillar interaction.

Note that the deviation in the contact angle behavior from that predicted by Wenzel equation was noted previously for a nanodrop on a hydrophilic surface.[7,8,11] It was, in particular, shown[8] that the contact angle of a nanodrop on a hydrophilic surface increases with increasing roughness, passes through a maximum and then decreases. The latter behavior is in qualitative agreement with eqn (2) and (3).

For a hydrophobic substrate, the dependence of θ on the height of the pillars is presented in Fig. 11 for various distances between pillars, D_p, and energy parameters ε_{fp}.

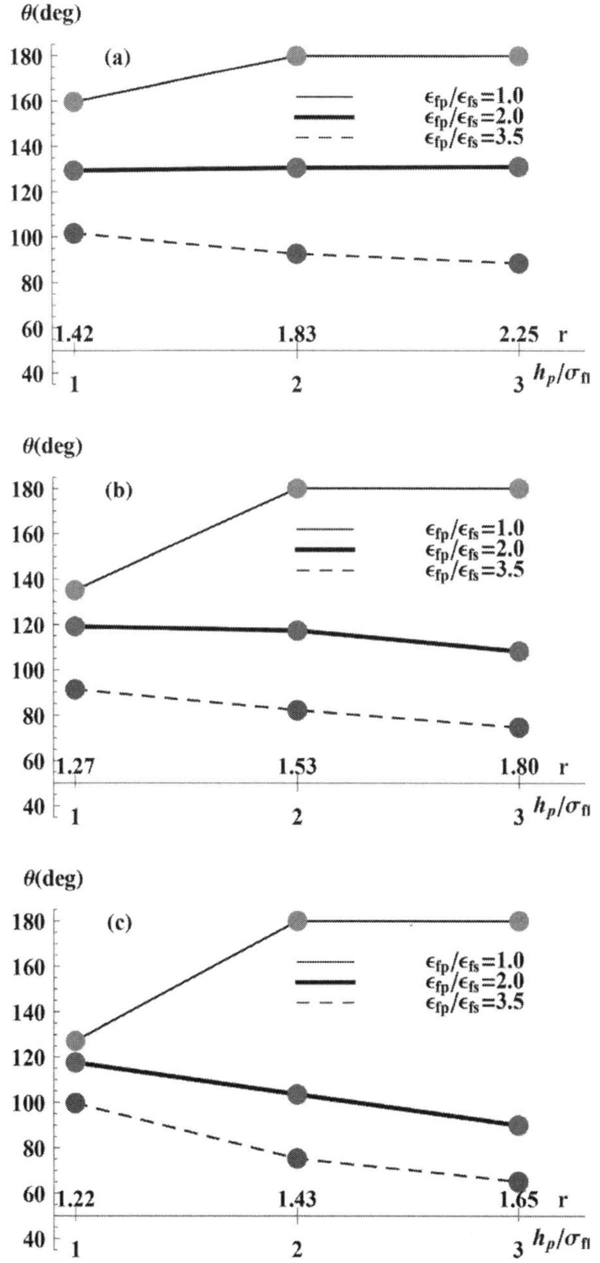

FIG. 11. Dependence of the contact angle of drop D1 on a rough hydrophobic substrate on the height of pillars for $\varepsilon_{fp}/\varepsilon_{fs} = 1.0$, 2.0, and 3.5 for several surfaces. The distances between pillars are $4.8\sigma_{ff}$ (a), $7.5\sigma_{ff}$ (b), and $9.3\sigma_{ff}$ (c). The roughness r is provided above the horizontal axis. The points represent results of calculations, the lines are guides for eye. In all cases, $N_{tot}\sigma_{ff} = 90$.

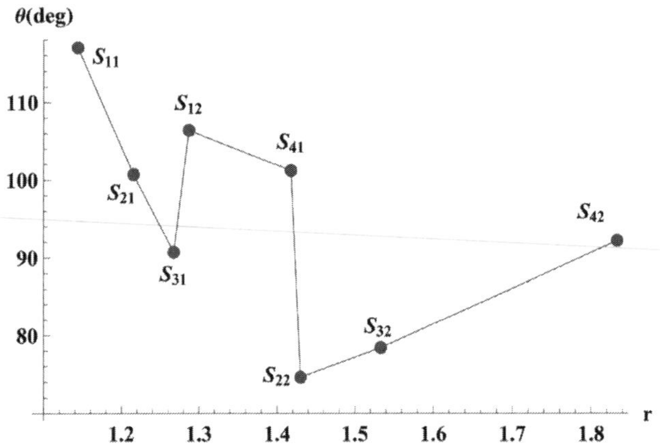

FIG. 12. Dependence of the contact angle on roughness for a drop on a rough hydrophobic substrate for $\varepsilon_{fp}/\varepsilon_{fs} = 3.5$. The points represent results of calculations, the lines are guides for eye. The labels at each point indicate the surface used in calculations. In all cases, $N_{tot}\sigma_{ff} = 90$.

For most of the considered cases, the contact angle decreases with increasing h_p (increasing roughness), again in contradiction with eqn (2) and (3). Such a monotonous behavior of θ as function of r differs from that in Fig. 9 in which the change in roughness was caused by the change in the distance between pillars.

As expected, the larger ε_{fp}, the smaller is the contact angle, *i.e.* the surface becomes less hydrophobic with increasing ε_{fp}. For the largest ε_{fp} presented in Fig. 11 ($\varepsilon_{fp}/\varepsilon_{fs} = 3.5$) the surface becomes even hydrophilic ($\theta < 90°$) even though the substrate is hydrophobic. Note that for small distances between pillars (*e.g.* $D_p/\sigma_{ff} = 4.8$ in Fig. 11), the change in contact angle with the change in their height is much smaller than for the larger values of D_p. This feature of the contact angle behavior can be understood by taking into account that for small values of D_p the pillars are close to each other and form a quasi smooth surface. The interaction of this surface with the fluid molecules is only slightly dependent on the height of the pillars.

All the above considerations indicate that the roughness as defined by eqn (11) cannot be considered the appropriate unique characteristics of the wetting of a surface. To provide additional evidence, the contact angle is plotted in Fig. 12 as a function of roughness for $\varepsilon_{fp}/\varepsilon_{fs} = 3.5$. One can see that different surfaces (*e.g.* S_{31} and S_{12}), for which the contact angles are very different (about 91° and 106°, respectively) exhibit approximately the same roughness (1.27 and 1.29, respectively). This is also true for surfaces S_{41} and S_{22}. For this reason, θ behaves irregularly as function of r in the range $r \simeq 1.26$ and $r = 1.44$ (Fig. 12).

3.4. Dependence of the contact angle on fluid-pillar interaction

Fig. 13 presents the dependence of the contact angle on the strength ε_{fp} of the fluid-pillar interaction for a hydrophobic substrate decorated with pillars of height $h_p = \sigma_{ff}$.

As expected, θ decreases with increasing ε_{fp} for both drops, D1 and D2. In both cases, the change in contact angle when ε_{fp} is changed from $\varepsilon_{fp}/\varepsilon_{fs} = 0.2$ to $\varepsilon_{fp}/\varepsilon_{fs} = 6.0$ is the smallest for surface S_{11}, the change for D1 being smaller than for D2 ($\Delta\theta_1 \simeq 15°$ and $\Delta\theta_2 \simeq 55°$, respectively). Such a difference between $\Delta\theta_1$ and $\Delta\theta_2$ occurs because in the first case (D1) the leading edges of the drop are located on the substrate and the pillars are beneath the drop (see Fig. 6, left column). For drop D2, the leading edges of the drop are on the pillars and for this reason the pillars affect the contact angle stronger than for D1, especially for larger $\varepsilon_{fp}/\varepsilon_{fs}$. When the distance between pillars is the smallest (surface S_{41}), the number of pillars beneath the drop increases and this causes higher changes in θ with changing ε_{fp}.

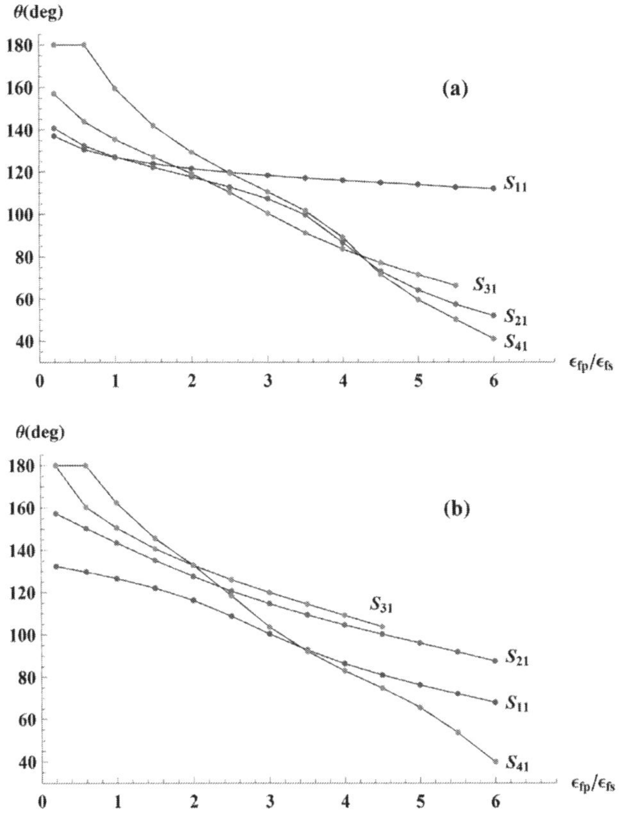

FIG. 13. Dependence of the contact angles of drops D1 (a) and D2 (b) on the ratio $\varepsilon_{fp}/\varepsilon_{fs}$ for various surfaces in the case of hydrophobic substrates. The points are results of calculations, the lines are guides for eyes. In all cases, $N_{tot}\sigma_{ff} = 90$.

4. CONCLUSIONS

In contrast to macroscopic drops for which the contact angles on smooth and rough surfaces are governed by eqn (1)–(3) and for which the dependence on drop size and roughness of the surface have some universal features, the contact angle of nanodrops on nanorough surfaces exhibits non-universalities both with respect to the drop size and roughness.[5–10]

In this paper it is shown that depending on the size of the drop as well as on the size and position of the pillars, there is even less universality in the contact angle behavior than it was found previously.

First, the contact angle depends on drop size and such a dependence is absent for macrodrops. In ref. 10 it was shown that on smooth as well as on rough hydrophobic surfaces the contact angle increases with increasing drop size. Because the drop is three-dimensional, one can assign this behavior to the existence of line tension (see eqn (4)). In our calculations in which a two-dimensional drop of much smaller size is considered, both an increase and a decrease in the contact angle with increasing drop size was identified. This behavior cannot be related to the line tension because the latter plays no role for two-dimensional drops. In addition, kink-like dependence of the contact angle on drop size was found for rough hydrophilic substrates, which occur when the leading edges of the drop change their location with respect to the pillars. One can expect that such kinks can repeat with increasing drop size when the leading edges of the drop change their location from on the pillars to between pillars.

In ref. 5–7,10, the increase of θ with increasing roughness of a hydrophobic surface was found to occur in all considered cases, in agreement with the behavior of θ for macroscopic drops. Note that in those studies, the condition $\varepsilon_{fp} = \varepsilon_{fs}$ was employed. However, for $\varepsilon_{fp} > \varepsilon_{fs}$, we have found cases when the contact angle first decreases with increasing roughness and then increases, as well as cases when the contact angle monotonously decreases with increasing roughness.

On the basis of the previous results and those obtained in the present paper, the conclusion is that the roughness r alone cannot explain all the peculiarities of the contact angle behavior of nanodrops on rough surfaces. Only a microscopic theory can provide adequate results for any values of the interaction parameters.

In this paper, only the Wenzel regime was considered. To obtain a stable Cassie-Baxter type drop, surfaces with higher hydrophobicity or pillars with larger heights should be examined.

5. APPENDIX A: FREE ENERGY CONTRIBUTIONS AND SOLUTION OF THE EULER-LAGRANGE EQUATION

Following ref. 16,17, the total Helmholtz free energy $F[\rho_f(\mathbf{r})]$ of a considered system can be represented as the sum of four contributions. The first one is the ideal gas free energy

$$F_{id}\left[\rho_f(\mathbf{r})\right] = k_B T \int d\mathbf{r}\rho_f(\mathbf{r})\left\{\log\left[\Lambda^3\rho_f(\mathbf{r})\right]-1\right\}, \tag{12}$$

where $\Lambda = h_P/(2\pi m k_B T)^{1/2}$ is the thermal de Broglie wavelength, h_P the Planck constant, and m the mass of a fluid molecule. The second contribution is the free energy of a reference system of hard spheres

$$F_{hs}\left[\rho_f(\mathbf{r})\right] = \int d\mathbf{r}\rho_f(\mathbf{r})\Delta\Psi_{hs}(\mathbf{r}) \tag{13}$$

where

$$\Delta\Psi_{hs}(\mathbf{r}) = k_B T \eta_{\bar{\rho}_f}\frac{4-3\eta_{\bar{\rho}_f}}{\left(1-\eta_{\bar{\rho}_f}\right)^2} \tag{14}$$

$\eta_{\bar{\rho}_f} = \frac{1}{6}\pi\bar{\rho}_f(\mathbf{r})\sigma_{ff^3}$ being the packing fraction of the fluid molecules and $\bar{\rho}_f(\mathbf{r})$ is a smoothed density defined as

$$\bar{\rho}_f(\mathbf{r}) = \int d\mathbf{r}'\rho_f W\left(\left|\mathbf{r}-\mathbf{r}'\right|\right). \tag{15}$$

The weighting function $W(|\mathbf{r} - \mathbf{r}'|)$ in eqn (15) is selected in the form[18]

$$W\left(\left|\mathbf{r}-\mathbf{r}'\right|\right) = \begin{cases} \frac{3}{\pi\sigma_{ff}^3}\left(1-\frac{r}{\sigma_{ff}}\right), & r \le \sigma_{ff} \\ 0, & r > \sigma_{ff} \end{cases}$$

where $r = |\mathbf{r} - \mathbf{r}'|$. The third contribution is the excess free energy due to fluid-fluid attractive interaction which is accounted for in the mean-field approximation

$$F_{attr}\left[\rho_f(\mathbf{r})\right] = \frac{1}{2}\iint d\mathbf{r}d\mathbf{r}'\rho_f(\mathbf{r})\rho_f(\mathbf{r}')\phi_{ff}\left(\left|\mathbf{r}-\mathbf{r}'\right|\right) \tag{16}$$

where $\phi_{ff}(|\mathbf{r}- \mathbf{r}'|)$ is the Lennard-Jones potential of the fluid-fluid interactions provided by eqn (5).

The last contribution is due to the interaction between fluid and solid molecules

$$F_{fs}\left[\rho_f(\mathbf{r})\right]=\int_V d\mathbf{r}\rho_f(\mathbf{r})U_{fs}(\mathbf{r}) \tag{17}$$

where V is the volume occupied by the fluid, and $U_{fs}(\mathbf{r})$ is provided by eqn (6).

The minimization of the Helmholtz free energy with respect to the fluid density distribution $\rho_f(x, h)$ leads to the following Euler-Lagrange equation for $\rho_f(x, h)$.

$$\log\left[\Lambda^3\rho_f(x,\ h)\right]-Q_f(x,h)=\frac{\lambda}{k_BT}. \tag{18}$$

In eqn (19), λ is a Lagrange multiplier and the function $Q_f(x, h)$ is given by

$$Q_f(x,h)=-\frac{1}{k_BT}\left[\Delta\Psi_{hs}(x,h)+\overline{\Delta\Psi'_{hs}}(x,h)+U_{ff}(x,h)+U_{fs}(x,h)\right] \tag{19}$$

where

$$U_{ff}(x,h)=\iint dx'dh'\rho_f(x',h')\phi_{ff,y}(|x-x'|,|h-h'|), \tag{20}$$

$$\overline{\Delta\Psi'_{hs}}(x,h)=\iint dx'dh'\rho_f(x',h')W_y(|x-x'|,|h-h'|)\frac{\partial}{\partial\bar{\rho}}\Delta\Psi_{hs}(\bar{\rho})|_{\rho=\rho_j(x',h')}. \tag{21}$$

The functions $\phi_{ff,y}(|x-x'|,|h-h'|)$ and $W_y(|x-x'|,|h-h'|)$ are obtained by integrating the potential $\phi_{ff}(|r-r'|)$ and the weighted function $W(|r-r'|)$ with respect to y from $-\infty$ to $+\infty$, respectively.

When calculating $U_{ff}(x, h)$ which is due to the long-range fluid-fluid interactions, a cutoff at a distance equal to four molecular diameters σ_{ff} for the range of Lennard-Jones attraction was employed. The precision of the iterations was characterized by the dimensionless quantity

$$\delta=\int_V dxdh\left[\rho_{f,i+1}(x,h)-\rho_{f,i}^{in}(x,h)\right]^2\bigg/\left(\int_V dxdh\rho_{f,i}(x,h)\right)^2$$

where $\rho_{f,i}^{in}(x,h)$ is the input density profile for the $(i+1)$-th iteration $\rho_{f,i+1}(x, h)$, generated by the Euler-Lagrange equation. The iterations were carried out on a two dimensional grid with a spacing $0.1\ \sigma_{ff}$ until δ became smaller than 10^{-7}.

6. APPENDIX B

The obvious way to proof that the solution of the Euler-Lagrange equation (let say, for drop D1), obtained with a selected precision, is not an intermediate step in the iteration procedure that eventually will provide the other solution (drop D2) is to increase the precision of the calculation. However for the two-dimensional case considered in the present paper this way is impractical because it involves a considerable increase in the calculation time, which even for the selected precision $\delta = 10^{-7}$ lasts about seven hours. To obtain at least a qualitative answer to the above question, the following procedure was developed which uses an intrinsic stability criterion for a system which has a minimum in its free energy. First, the obtained density profile $\rho_f(x, h)$ was disturbed by shifting it in some direction (let say, to the left) by a few grid intervals. This new density distribution, $\rho'_f(x, h)$, which now is asymmetric with respect to the initial plane of symmetry, is used as a new initial guess for the Euler-Lagrange equation. If the nonshifted FDD $\rho_f(x, h)$ provides the

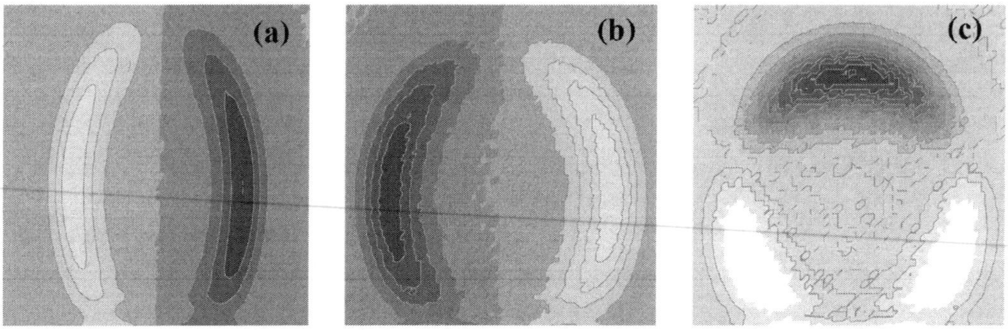

FIG. 14. (a), (b), and (c) Possible contour plots of the differences between two consecutive iterations. Dark areas correspond to negative values of the differences and light areas correspond to positive ones.

minimum (local or global) in the free energy, the iteration procedure which started with $\rho'_f(x, h)$ should converge to $\rho_f(x, h)$. However the absence of symmetry in $\rho'_f(x, h)$ leads to an extremely slow convergence of the iteration procedure. To obtain the decision more rapidly, track was kept of the difference $\Delta\rho'_{f,i}(x, h) = \rho'_{f,i+1}(x, h) - \rho'_{f,i}(x, h)$ between the density distributions, $\rho'_{f,i}(x, h)$ and $\rho'_{f,i+1}(x, h)$, provided by two successive iterations started with the initial guess $\rho'_f(x, h)$. After 50–100 iterations this difference as function of x and h has, generally, one of the two shapes presented in Fig. 14a and b, where the light (dark) areas represent positive (negative) values of $\Delta\rho_{f,i}(x, h)$. Fig. 14a and b indicate the tendency of the density distribution $\rho_{f,i}(x, h)$ to "move" in the direction of the positive part of $\Delta\rho_{f,i}(x, h)$, i.e. to the left for Fig. 14a and to the right for Fig. 14b.

If $\Delta\rho_{f,i}(x, h)$ had the shape shown in Fig. 14a, the unshifted solution was considered unstable and was disregarded. In the opposite case, when $\Delta\rho_{f,i}(x, h)$ had the shape shown in Fig. 14b, the unshifted solution was considered stable or metastable.

Note that if the initial guess for the iteration procedure is selected at the location of the stable or metastable solution of Euler-Lagrange equation, the difference $\Delta\rho_{f,i}$(x, h) has the symmetric shape shown in Fig. 14c.

REFERENCES

1 T. Young, *Philos. Trans. R. Soc. London*, 1805, **95**, 65–87.
2 R. N. Wenzel, *Ind. Eng. Chem.*, 1936, **28**, 988–994.
3 A. B. D. Cassie and S. Baxter, *Trans. Faraday Soc.*, 1944, **40**, 0546–0550.
4 B. A. Pethica, *Rep. Prog. Appl. Chem.*, 1961, **46**, 14;B. A. Pethica, *J. Colloid Interface Sci.*, 1977, **62**, 567–569.
5 F. Porcheron, P. A. Monson and M. Schoen, *Phys. Rev. E: Stat. Phys., Plasmas, Fluids, Relat. Interdiscip. Top.*, 2006, **73**, 041603.
6 G. O. Berim and E. Ruckenstein, *J. Chem. Phys.*, 2008, **129**, 014708 (Section 4.4 of this volume).
7 C. D. Daub, J. H. Wang, S. Kudesia, D. Bratko and A. Luzar, *Faraday Discuss.*, 2010, **146**, 67–77.
8 G. O. Berim and E. Ruckenstein, *J. Colloid Interface Sci.*, 2011, **359**, 304–310 (Section 4.5 of this volume).
9 T. Koishi, K. Yasuoka, S. Fujikawa and X. C. Zeng, *ACS Nano*, 2011, **5**, 6834–6842.
10 A. P. Malanoski, B. J. Johnson and J. S. Erickson, *Nanoscale*, 2014, **6**, 5260–5269.
11 A. Malijevsky, *J. Chem. Phys.*, 2014, **141**, 184703.
12 R. Evans and P. Tarazona, *Phys. Rev. A*, 1983, **28**, 1864–1868.
13 G. O. Berim and E. Ruckenstein, *J. Chem. Phys.*, 2009, **130**, 044709 (Section 4.1 of this volume).
14 G. O. Berim and E. Ruckenstein, *J. Chem. Phys.*, 2008, **129**, 114709 (Section 4.11 of this volume).
15 H. Kusumaatmaja and J. M. Yeomas, *Langmuir*, 2007, **23**, 6019–6032.
16 P. Tarazona, *Phys. Rev. A*, 1985, **31**, 2672–2679.
17 P. Tarazona, U. M. B. Marconi and R. Evans, *Mol. Phys.*, 1987, **60**, 573–595.
18 R. H. Nilson and S. K. Griffiths, *J. Chem. Phys.*, 1999, **111**, 4281–4290.

4.7 Contact Angles of Nanodrops on Chemically Rough Surfaces*

Gersh Berim and Eli Ruckenstein†
Department of Chemical and Biological Engineering,
State University of New York at Buffalo, Buffalo, New York 14260

Corresponding Author

† E-mail: feaeliru@buffalo.edu.
Phone: 716-645-1179. Fax: 716-645-3822.
Received March 9, 2009. Revised Manuscript Received April 14, 2009

1. INTRODUCTION

In a recent paper, Gao and McCarthy[1] provided experimental evidence that the contact angle θ of a liquid drop on a chemically rough solid surface is determined by the interactions near the three-phase contact line, with the contribution of the remaining solid-liquid interfacial area inside the drop having no effect. They attempted to represent the data using the standard Cassie—Baxter equation[2]

$$\cos\theta_{\mathrm{CB}} = f_A \cos\theta_A + f_B \cos\theta_B \tag{1}$$

where θ_A and θ_B are the contact angles which the drop makes with the uniform surfaces A and B, and f_A and f_B are the area fractions of A and B surfaces beneath the drop. They found that there is no agreement and concluded that this equation is incorrect in general. In the discussion that followed in the literature (see refs 3–7), various arguments were brought pro and contra that conclusion. The main conclusion of refs 3–5 was that the area fractions f_A and f_B involved in the Cassie—Baxter equation are local quantities calculated in the vicinity of the contact line. However, the considerations of Whyman et al.[6] regarding the equilibrium conditions of a drop on a surface, based on the minimization of the global free energy of the drop, revealed that the entire contact area between fluid and solid contributes to the value of the contact angle. Finally, Marmur and Bittoun[7] have shown that the Cassie-Baxter as well as the Wentzel equation for a physically rough surface[8] are valid when the drop size is much larger (about 3 orders of magnitude larger) than the wavelength of the chemical heterogeneity or physical roughness. In the experiments of Gao and McCarthy, the above condition is not satisfied; consequently, the Cassie-Baxter equation is not applicable. All the above mentioned considerations were made in the framework of classical thermodynamics, which involves the surface tensions and the Young equation for the contact angle of the drop on smooth surfaces.

Because of the inapplicability of the Cassie—Baxter equation (eq 1) to small drops (nanodrops) on rough surfaces, it is reasonable to employ another tool for the treatment of such a case. In refs 10 and 11, a nanodrop on a rough surface was treated on the basis of a density functional theory,[9] which provided a rigorous microscopic approach to the problem. Such statistical mechanical considerations involve only the potentials of the fluid-fluid and fluid-solid interactions and are more

* *Langmuir* 2009, 25(16), 9285–9289. Republished with permission.

appropriate for the examination of inhomogeneous surfaces than the classical approach that uses spatially dependent surface tensions, quantities which, in fact, are not clearly defined.

In the present paper, we use a nonlocal density functional theory (DFT) to examine nanodrops on chemically inhomogeneous surfaces which are similar to those used in the experiment of ref 1. We restrict the calculations to nanodrops because the macrodrops contain a too large number of molecules to allow us to carry out with the available computers the extensive numerical calculations required by DFT.

The system under consideration is presented in Figure 1 and consists of a box of infinite dimension in the y-direction (normal to the plane of the figure) and finite dimensions L_x and L_h in the x- and h-directions, respectively. The box is filled with a one-component fluid of constant average density ρ_{av} that is in contact with a chemically rough solid surface. The surface is composed of a sequence of two kinds of plates, A and B, which are made of different materials and interact differently with the fluid molecules. The plates can have, generally, different thicknesses and are infinite in the y-direction and semi-infinite in the h-direction. Consequently, the fluid density distribution (FDD) $\rho(\mathbf{r})$ is uniform in the y-direction and nonuniform in the x- and h-directions, that is, $\rho(\mathbf{r}) \equiv \rho(x, h)$. Periodic boundary conditions are employed in the x-direction, and the upper boundary of the box is treated as a hard wall without attractive interaction with the fluid molecules. The suitability of such a two-dimensional model for a three-dimensional drop was examined in ref 7.

The interaction between the fluid molecules is described by the Lennard-Jones potential $\phi_{ff}(r) = 4\epsilon_{ff}\left[(\sigma_{ff}/r)^{12} - (\sigma_{ff}/r)^{6}\right]$ for $r \geq \sigma_{ff}$ and $\phi_{ff}(r) = \infty$ for $r < \sigma_{ff}$, where $r \equiv |\mathbf{r} - \mathbf{r}'|$; the coordinates $\mathbf{r}$ and $\mathbf{r}'$ provide the locations of the fluid molecules, and σ_{ff} and ϵ_{ff} are the fluid hard core diameter and energy parameter, respectively.

The same kind of potential is selected for the fluid—solid interactions, with the energy parameter ϵ_{ff} and hard core diameter σ_{ff} in $\phi_{ff}(r)$ being replaced by ϵ_{fs} and σ_{fs}, respectively, and r being the distance between a molecule of fluid and that of the solid wall.

For such a system, the Euler—Lagrange equation for the FDD $\rho(x, h)$ has the form

$$\log\left[\Lambda^3 \rho(x, h) - Q(x, h) = \frac{\lambda}{k_B T}\right] \tag{2}$$

where the first term represents the ideal gas contribution, $Q(x, h)$ is a functional of $\rho(x,h)$, $\Lambda = \left(h_P / 2\pi m k_B T^{1/2}\right)$ is the thermal de Broglie wavelength, k_B and h_P are the Boltzmann and Planck constants, respectively, T is the absolute temperature, m is the mass of a fluid molecule, and λ is a Lagrange multiplier arising because of the constraint of fixed average density of the fluid. The functional $Q(x, h)$ accounts for a reference hard sphere system of fluid molecules, the fluid—solid

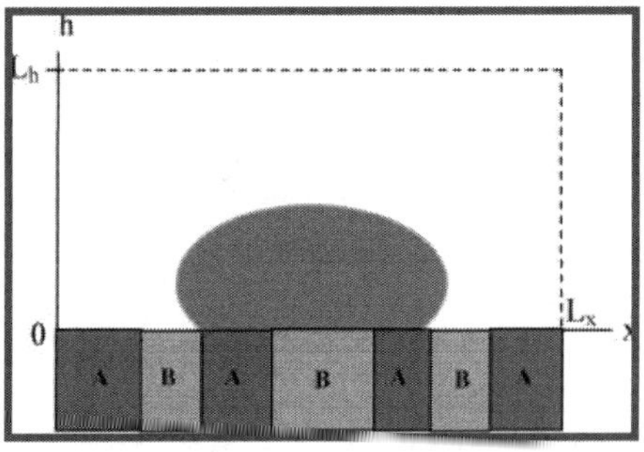

FIGURE 1. Schematic representation of a drop on a chemically rough surface in the x–h plane. Plates A and B are infinite in the y-direction, semi-infinite in the h-direction and have different values of the energy interaction parameters $\epsilon_{fs,A}$ and $\epsilon_{fs,B}$ (the y-axis is normal to the plane of the figure).

interactions, and the fluid—fluid interactions treated in the mean-field approximation. The explicit form of $Q(x, h)$ and details of the derivation of eq 2 can be found in ref 10.

The constraint of fixed average density leads to

$$\rho_{av} = \frac{1}{V}\int_V d\mathbf{r}\,\rho(\mathbf{r}) \tag{3}$$

where V is the fixed volume of the system and provides the following expression for λ

$$\lambda = -k_B T \log\left[\frac{1}{\rho_{av} V \Lambda^3}\int_V d\mathbf{r}\, e^{Q(x,h)}\right] \tag{4}$$

By eliminating λ between eqs 2 and 4, one obtains an integral equation for the FDD $\rho(x, h)$, which can be solved by iterations. The details of the iteration procedure are provided in ref 10.

In Figure 2, an example of the fluid density distribution obtained as a solution of the Euler—Lagrange equation is presented. The lighter areas correspond to higher fluid densities. Because of the absence of a sharp vapor—liquid interface, the contact angle for nanodrops is not clearly defined. To extract it from the given FDD, we used an approach employed in refs 10 and 12. First, the drop profile $h = h(x)$ is determined inside the vapor—liquid interface as the line passing through the points where the local density $\rho(x, h)$ has a constant value ρ_{div}. The latter quantity is considered equal to the fluid density for an equimolar dividing surface at some distance h_0 from the solid surface (see refs 12 and 13) In the present calculations, this distance was selected to be $5\sigma_{ff}$. The determined profile is presented in Figure 2 as a white solid line. After the profile is determined, its part for $h > 2.5\sigma_{ff}$ was approximated by a circle and extended up to the solid surface. The contact angle was taken equal to that made by the circle with the solid surface. (Note that such a definition of the contact angle is similar to that used in the goniometric measurements of the contact angles of macroscopic drops.[14]) The intersections of the circle with the solid surface provided the locations of the contact lines, and the coordinates of intersections were used to evaluate the fractions of the contact area beneath the drop occupied by the A- and B-plates.

To estimate the importance of the liquid—solid interfacial area inside the drop to the value of the contact angle, several specific cases will be examined below. In all of them, the leading edges of the drop are considered to be located on A-plates, at a fixed interaction energy parameter between them and the fluid. As for the B-plates, either we change the fraction of the surface area beneath the drop occupied by them by changing the width of the B-plates or we change the energy parameter $\epsilon_{fs,B}$ characterizing the interaction of the B-plates with the fluid molecules. In this way, situations are generated in which the conditions in the vicinity of the contact line are unchanged and those in the remaining solid liquid interface beneath the drop are varied.

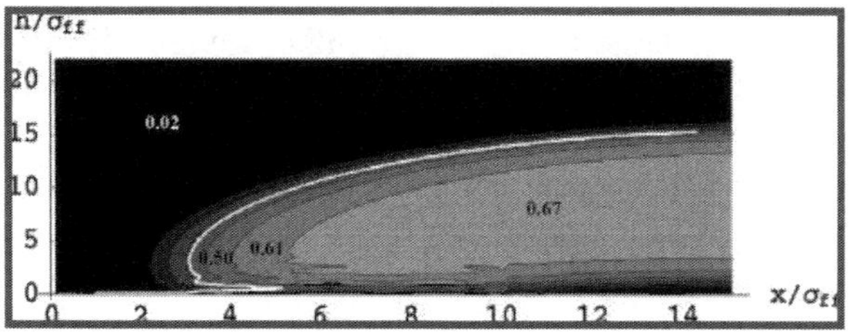

FIGURE 2. Example of the fluid density distribution in the system (only its left-hand side is presented). The lighter areas correspond to higher fluid densities. The numbers indicate the values of the dimensionless fluid density $\rho\sigma_{LL}^3$. The white line represents the drop profile extracted for $\rho_{div}\sigma_{ff}^3 = 0.383$ as described in the text.

In all the considered cases, argon was selected as the fluid with the parameters of the Lennard—Jones interaction potential $\epsilon_{ff}/k_B = 119.76\,K$, $\sigma_{ff} = 3.405\,Å$. For both kinds of plates, the hard core diameters and the number densities were taken the same $\left(\sigma_{fs} = 3.27\,Å,\ \rho_s = 1.91 \times 10^{28} m^{-3}\right)$. The B-plates differ from the A-plates by the energy parameter $\epsilon_{fs,B} \neq \epsilon_{fs,A}$ only. More details of the considered cases and the results obtained are presented in the next sections.

2. CASE 1: HYDROPHILIC AREA INSIDE A HYDROPHOBIC SURFACE

In this case, presented in Figure 3, the solid—liquid interface consists of three parts, with a plate B located completely beneath the drop. The energy parameter $\epsilon_{fs,A}$ of the A-plates is selected to be $\epsilon_{fs}/k_B = 91.8$ K. The surface of the A-plate is slightly hydrophobic with a contact angle of $\theta_A = 100.6°$. The energy parameter $\epsilon_{fs,B}$ of the B-plates is selected to be $\epsilon_{fs,B}/k_B = 183.6$ K and corresponds to a hydrophilic surface with a wetting angle $\theta_B \simeq 0°$. The thickness d_B of the B-plate was changed from $6\sigma_{ff}$ to $26\sigma_{ff}$. For $T = 85$ K and $\rho_{av}\sigma_{ff}^3 = 0.16$, the width of the contact area of the drop was varied from $25\sigma_{ff}$ for the smallest thickness of the B-plate to $29\sigma_{ff}$ for the largest one.

The contact angles calculated on the basis of DFT are listed in Table 1 in the fourth column. The third column provides the distance Δl between the leading edge of the drop and the nearest boundary of a B-plate (see Figure 3). The contact angle θ which the drop makes with the surface is constant ($\theta = 100.6°$) for $d_B \leq 14\sigma_{ff}$ $\left(f_B \leq 0.560, \Delta l > 5\sigma_{ff}\right)$ and decreases to 85.6° with increasing d_B up to $d_B = 26\sigma_{ff}$ $\left(f_B \simeq 0.88\right)$. For comparison purposes, the contact angle θ_{CB} calculated with eq 1 is provided in the last column of Table 1. This angle changes from 84.3° for $d_B = 6\sigma_{ff}$ to 30.5° for $d_B = 26\sigma_{ff}$.

3. CASE 2: ULTRAHYDROPHOBIC AREA INSIDE A HYDROPHOBIC SURFACE

In this case, the surface has the same structure as in the previous one (see Figure 3), but the B-plate is ultrahydrophobic with $\epsilon_{fs,B}/k_B = 0K$ and $\theta_B = 180°$. (Such a case can be imagined by assuming the presence of trapped air instead of a B-plate). The A-plates are as those considered in section 2.

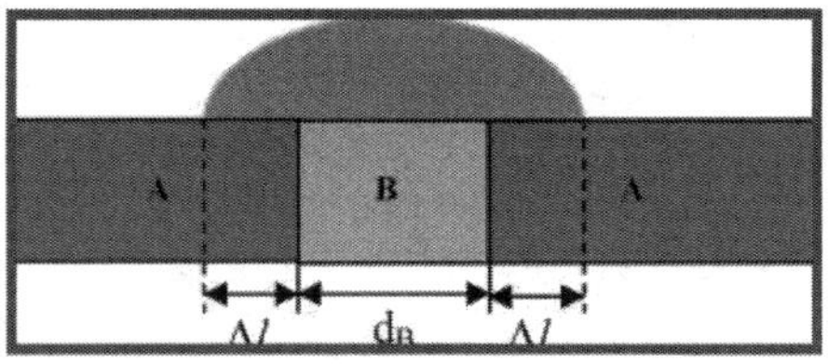

FIGURE 3. Schematic representation of a hydrophilic area (B-plate) inside a hydrophobic surface (A-plate)

TABLE 1
Contact Angles for a Hydrophilic Area Inside a Hydrophobic Surface

d_B/σ_{ff}	f_B	$\Delta l/\sigma_{ff}$	θ (deg)	θ_{CB} (deg)
6	0.239	9.5	100.6	84.3
10	0.401	7.4	100.6	73.1
14	0.560	5.4	100.6	61.4
18	0.710	3.7	99.2	49.0
22	0.821	2.4	94.2	38.0
26	0.883	1.8	85.6	30.5

The width of the contact area of the drop with the solid surface is about $26\sigma_{ff}$. The results obtained are listed in Table 2. A drop, containing a *B*-plate beneath it, can exist only if the width d_B of the *B*-plate is smaller than $17.2\sigma_{ff}$. For $d_B > 17.2\sigma_{ff}$, the Euler—Lagrange equation, eq 2, does not have a stable or metastable solution with the leading edge located on *A*-plates.

The contact angle θ which the drop makes with the surface varies between 100.6° and 103.6° as d_B increases from 0 to $17.2\sigma_{ff}$. (The contact angle θ_{CB} calculated with eq 1 changes from 100.6° to 137.6°.) A visible change of the contact angle begins when the distance Δl between the leading edge and the nearest boundary of a *B*-plate becomes smaller than $5\sigma_{ff}$.

4. CASE 3: A SURFACE WITH PERIODIC CHEMICAL ROUGHNESS

In this case, the drop is located on a surface composed of plates *A* and *B* of equal thicknesses *d*, with the leading edges of the drop being located on the *A*-plates (Figure 4). The energy parameter $\epsilon_{fs,A} = 153$ K and $\epsilon_{fs,B}$ varies from $\epsilon_{fs,B} = 0$K to $\epsilon_{fs,B} = 0.8\epsilon_{fs,A}$. The thickness of the plates is $d = 3\sigma_{ff}$. The surface generates a periodic potential with a wavelength equal to $2d$ ($6\sigma_{ff}$). The details regarding the calculations of the interaction potential can be found in ref 10.

As shown previously,[11] a nanodrop on the surface presented in Figure 4 can be symmetrical with respect to the middle of plate *A* (drop D1, Figure 4a) or plate *B* (drop D2, Figure 4b). For the same fluid average density in the system, one of these drops is stable (i.e., corresponds to a global minimum of the free energy) and the other one is metastable (corresponds to a local minimum of the free energy). Depending on specific conditions (chemical nature and sizes of the plates, temperature,

TABLE 2
Contact Angles for an Ultrahydrophobic Area Inside a Hydrophobic Surface

d_B/σ_{ff}	f_B	$\Delta l/\sigma_{ff}$	θ (deg)	θ_{CB} (deg)
10	0.389	7.8	100.6	120.0
16	0.620	4.8	101.6	133.6
16.4	0.640	4.6	102.0	134.9
16.8	0.660	4.3	102.0	136.3
17.2	0.680	4.0	103.6	137.6

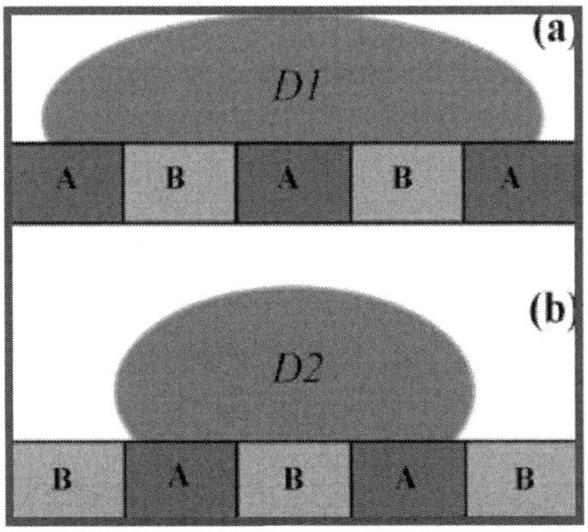

FIGURE 4. Two characteristic states of a nanodrop on a surface with periodic chemical roughness. The drop D1 (D2) is symmetrical with respect to the vertical plane passing through the middle of plate *A* (*B*).

fluid average density), the leading edges of the drop can be located on either *A*- or *B*-plates. Only those drops which have the leading edges on *A*-plates were selected by us for analysis. In those cases, the variation of $\epsilon_{fs,B}$ changes mainly the contribution to the free energy of that part of the area beneath the drop which is at some distance from the leading edges, with the contribution of the area near the leading edges remaining almost the same. Changing the energy parameter $\epsilon_{fs,B}$ of the liquid—solid interactions for the *B*-plates but keeping that of *A*-plates fixed, one can model the above situations. To check the role of the wavelength of the inhomogeneity, one can also change the thickness *d* of the plates.

In Table 3, the contact angles θ extracted from the calculated fluid density distribution are listed for drops D1 and D2 together with the contact angle θ_{CB} calculated with eq 1. The contact angle $\theta_A = 49.3°$ and the values of θ_B and f_B are provided in the second and third columns of Table 3, respectively.

The first conclusion, which can be drawn from the data presented in Table 3, is that the change of the strength of interactions inside the internal area beneath the drop does affect the contact angle θ of nanodrops on rough surfaces. Such an effect occurs for both drops D1 and D2. The contact angle θ for drop D1 changes from 66.7° to 117.6° when $\delta \equiv \epsilon_{fs,B}/\epsilon_{fs,A}$ varies from $\delta = 0$ to $\delta = 0.8$. For the drop D2, θ varies from 80.7° to 103.3° when δ varies from $\delta = 0.6$ to $\delta = 0.2$. The change of the contact angle takes place along with that of the distance Δl between the leading edge of the drop (located on an *A*-plate) and the closest boundary between the *A*- and *B*-plates (see Table 3).

A similar analysis was carried out for a surface with a larger width of the plates, $d = 5\sigma_{\mathrm{ff}}$. The corresponding results are listed in Table 4. Drop D1 is stable for $\delta = 0.8$ but metastable at other values of δ, whereas drop D2 is stable at all listed values of δ.

5. DISCUSSION

The solid lines in Figure 5 present the contact angles obtained via DFT as functions of the fraction of *B*-plate beneath the drop for the cases considered in sections 2 and 3 (cases 1 and 2, respectively). For $f_B < 0.6$, the contact angle calculated via DFT is almost independent of f_{B} and remains the same independent of the nature of the *B*-plate (hydrophilic or hydrophobic) present beneath the drop.

TABLE 3
Contact Angles for a Surface with Periodic Chemical Roughness ($d_B = 3\sigma_{ff}$)

$\epsilon_{fs,B}/\epsilon_{fs,A}$	θ_B (deg)	f_B	$\Delta l/\sigma_{ff}$	θ (deg)	θ_{CB} (deg)
		Drop D1			
0.8	79.3	0.500	1.5	66.7	65.3
0.2	180	0.409	0.1	109.8	91.4
0.0	180	0.455	0.8	117.6	95.1
		Drop D2			
0.6	107.2	0.43	0.2	80.7	75.8
0.4	136.8	0.48	1.2	90.3	90.6
0.2	180	0.56	2.5	103.3	105.8

TABLE 4
Contact Angles for a Surface with Periodic Chemical Roughness ($d_B = 5\sigma_{ff}$)

$\epsilon_{fs,B}/\epsilon_{fs,A}$	θ_B (deg)	f_B	$\Delta l/\sigma_{ff}$	θ (deg)	θ_{CB} (deg)
		Drop D1			
0.8	79.3	0.412	0.35	65.0	62.6
0.6	107.2	0.441	1.2	71.5	76.5
0.4	136.8	0.468	1.8	77.2	89.7
0.2	180	0.504	2.5	84.4	100.4
		Drop D2			
0.2	180	0.661	0.1	109.3	123.9
0.0	180	0.637	0.6	114.5	127.1

The distance Δl between the leading edges of the drop and the B-plates is in this case larger than $5\sigma_{ff}$. For the hydrophilic B-plate, the results are in agreement with the experimental observations of Gao and Mccarthy[1] which have shown that for a hydrophilic spot in a hydrophobic field (analogous to case 1 in our considerations) the contact angle is almost independent of f_B for $f_B < 0.64$. (Note that in their experiments surfaces with $f_B > 0.64$ and hydrophobic surfaces have not been considered.)

For $f_B > 0.6$ ($\Delta l < 5\sigma_{ff}$), our DFT results for cases 1 and 2 exhibit a change in the contact angle. One of the reasons for that change is that for $f_B > 0.6$ the distance between the leading edges of the drop and the B-plates becomes very small, and hence, the local conditions near the leading edges are considerably changed compared to those on an uniform surface. From a microscopic point of view, those conditions are determined solely by the potential $U_{fs}(x,h)$ of the fluid—solid interactions.

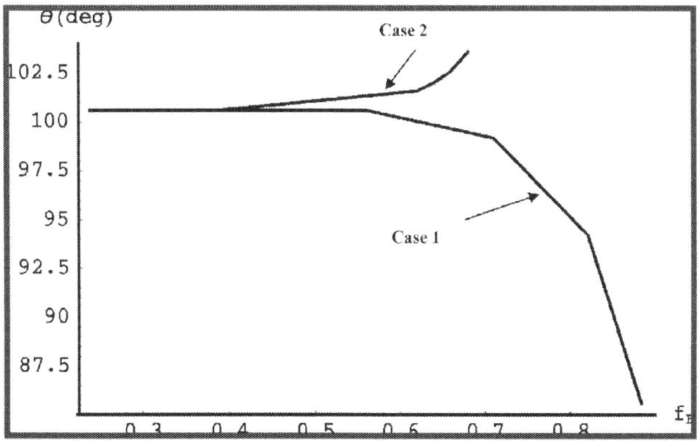

FIGURE 5. Dependence of the contact angles calculated via DFT on the fraction of the surface of B-type beneath the drop for a hydrophilic area inside a hydrophobic surface (case 1) and for an ultrahydrophobic area inside a hydrophobic surface (case 2).

In Figure 6, the change of this potential in the vicinity of the boundary between the A- and B-plates is plotted as a function of x for semi-infinite A-plates and $d_B = 10\sigma_{ff}$. The plateaus at the right and left had sides of the B-plate and one in the middle of the B-plate provide potentials $U_{fs}(x,h)$ approximately equal to those for uniform infinite A- and B-plates, respectively. The transition between those values occurs within a distance of about $8\sigma_{ff}$ and begins at a distance of $5\sigma_{ff}$ when approaching the boundary between A- and B-plates from the A-plate side. Just at the latter distance begins the change of the calculated contact angle (see Tables 1 and 2).

The results for cases 1 and 2 support the suggestion that the contact angle is determined by conditions near the leading edges of the drop and that the remaining area of the solid—liquid interface is irrelevant.

The results obtained for case 3 seem to be in contradiction with the above conclusions. Indeed, whereas the fraction of B-plates beneath the drop is approximately equal to $f_B = 0.5$, the change in the energy parameter $\epsilon_{fs,B}$ from $\epsilon_{fs,B} = 0$ to $\epsilon_{fs,B} = 0.8\epsilon_{fs,A}$ results in large changes of

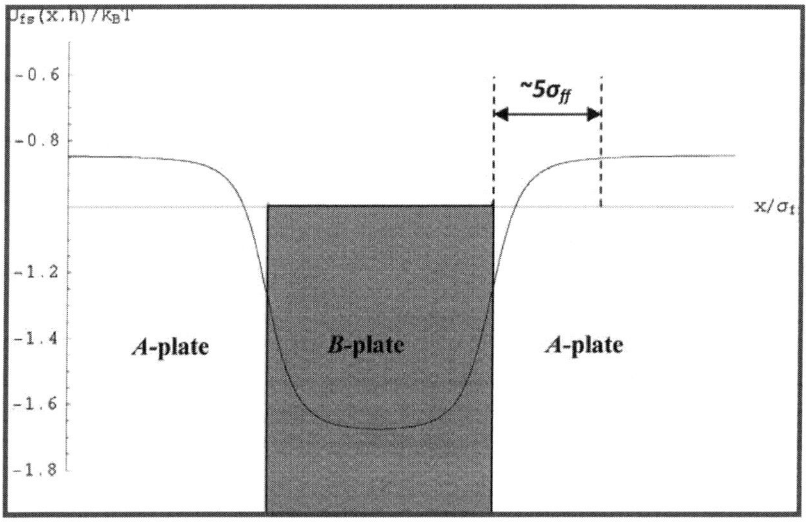

FIGURE 6. Potential of the fluid-solid interaction as function of x for the surface considered in section 3 (case 1) with $d_B = 10\sigma_{ff}$, calculated at a distance from the solid surface $h = 0.5\sigma_{ff}$.

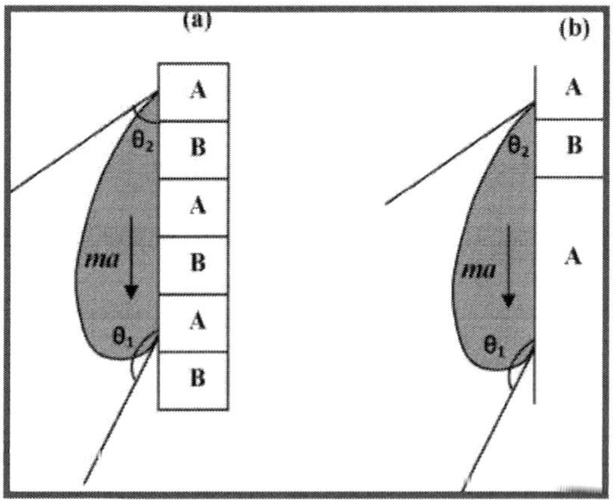

FIGURE 7. Schematic representation of a nanodrop of mass m on a chemically rough vertical surface in the presence of gravity. The considered surfaces correspond to case 3 [panel (a)] and case 1 [panel (b)].

the contact angle θ. However, this can be explained if one takes into account that the leading edges of the drop which are located on A-plates are very close to the neighboring B-plates (see Tables 3 and 4) and, therefore, are strongly affected by the interactions with those plates. Even a microscopic displacement of the order of one molecular diameter of the leading edge due to changes in $\epsilon_{fs,B}$ can provide large changes in the contact angle.

Note that, from a thermodynamic point of view, the role of the local conditions in the value of the contact angle was examined in detail by Marmur and Bittoun.[7] Along with a qualitative analysis of the origin of the contact angle changes with changing of the location of the contact line, it was shown that a drop of a fixed (macroscopic) volume on a rough surface can have multiple metastable states possessing different contact angles (contact angle hysteresis). For the nanodrops considered in the present paper, which have very small volumes, either the metastable states are absent or their number is small. Among the surfaces examined in this paper, only those composed of periodic sequences of A- and B-plates (case 3) provided a metastable drop in addition to the stable one. For this reason, it is not possible to estimate the contact angle hysteresis from the analysis of the metastable states.

However, the contact angle hysteresis for nanodrops can be detected in another way. Let us consider a drop on a vertical surface identical to that examined in section 4 (case 3) in the presence of gravity (see Figure 7a). As was shown previously in ref 11, the gravitational acceleration a has to be taken enormously large ($a \sim 10^{10}$ m/s^2) to affect the shape of the nanodrop. In this case, the angles θ_1 and θ_2 can be considered as estimates for the advanced and receding contact angles of a nanodrop on a rough surface. Increasing the acceleration a up to the critical value a_c at which the drop loses its mechanical equilibrium, one can find the largest (smallest) values of θ_1 (θ_2) which provide the contact angle hysteresis. For example, for the case $d_B = d_A = 3\sigma_{ff}$ examined in ref 11, the contact angle changes between 76.5° and 128.4° for drop D1 and between 102.6° and 136.2° for drop D2. Similar calculations performed for a surface corresponding to case 1 (see Figure 7b) with $d_B = 6\sigma_{ff}$ provide a range of values between 58.7° and 117.6°.

Note that the value of the contact angle θ_2 for the surface presented in Figure 7b depends on the width d_B of plate B, that is, on the fraction of surface area beneath the drop occupied by the B-plate. For example, for $d_B = 6\sigma_{ff}$ ($f_B = 0.23$) and $d_B = 18\sigma_{ff}$ ($f_B = 0.64$), this angle has the values 93.5° and 81.7°, respectively, for a gravitational acceleration $a = 7.3 \times 10^{10}$ m/s^2.

REFERENCES

(1) Gao, L.; McCarthy, T. *Langmuir* **2007**, *23*, 3762.
(2) Cassie, A. B. D.; Baxter, S. *Trans, Faraday Soc.* **1944**, *40*, 546.
(3) McHale, G. *Langmuir* **2007**, *23*, 8200.
(4) Nosonovsky, M. *Langmuir* **2007**, *23*, 9919.
(5) Panchagnula, M. V.; Vedantam, S. *Langmuir* **2007**, *23*, 13242.
(6) Whyman, G.; Bormashenko, E.; Stein, T. *Chem, Phys, Lett.* **2008**, *450*, 355.
(7) Marmur, A.; Bittoun, E. *Langmuir* **2009**, *25*, 1277.
(8) Wenzel, R. N. *Ind, Eng, chem.* **1936**, *28*, 988.
(9) Tarazona, P. *Phys. Rev. A* **1985**, *31*, 2672.
(10) Berim, G. O.; Ruckenstein, E. *J. Chem. Phys.* **2008**, *129*, 014708 (Section 4.4 of this volume).
(11) Berim, G. O.; Ruckenstein, E. *J. Chem. Phys.* **2008**, *129*, 114709 (Section 4.11 of this volume).
(12) Giovambattista, N.; Debenedetti, P. G.; Rossky, P. J. *J. phys, chem. B* **2007**, *111*, 9581.
(13) Porcheron, F.; Monson, P. A. *Langmuir* **2006**, *22*, 1595.
(14) Johnson, R. E.; Dettre, R. H. Surf, Colloid, Sci. **1969**, *2*, 85.

4.8 Nanodrop on a Smooth Solid Surface with Hidden Roughness

*Density Functional Theory Considerations**

*Gersh Berim and Eli Ruckenstein**

Received 29th January 2015, Accepted 26th March 2015

1. INTRODUCTION

For a long time, a liquid drop on the surface of a solid was the object of intense experimental and theoretical investigations and numerous results were obtained using various methods. Particular attention was given to rough surfaces because of the large effect which roughness has on the wetting of a solid substrate. Two kinds of roughnesses were considered. One of them is due to the asperities present on the surface of a homogeneous solid substrate (physical roughness). The second, chemical roughness, occurs when the substrate has a smooth surface with nonuniform chemical composition. For both types of roughnesses, the contact angle of the drop on the rough surface is usually greater than on the smooth one, *i.e.* the roughness increases the hydrophobicity.[1,2] Another important feature of a rough surface, which is absent for a smooth one, is the appearance of the sticking (pinning) of the drop-solid contact line to the solid surface due to the direct contact of this line with the asperities (see *e.g.* ref. 3).

In the present paper, a new type of roughness, which will be called as hidden roughness, is considered using a microscopic approach based on the density functional theory (DFT). The system which possesses such a roughness, can be imagined, for instance, as a uniform substrate decorated with asperities (SDA) and covered by a layer of a second solid material (SSM) which has a smooth surface but can have a nonuniform density. The drop on the smooth surface of SSM is subjected to the interaction potentials of SDA and SSM, which determine its properties. The contact angle of the drop is expected to depend on the details of the fluid–solid interactions and nonuniformity of SSM. Another expectation is that if an external force acts on the drop parallel to the surface, the SSM and SDA will generate a sticking force which maintains the equilibrium of the drop on the surface. The presence of only one kind of liquid in the system (that of the drop) allows one to use for its study the one-component DFT developed in ref. 4.

As an example of a real system possessing hidden roughness, one can mention the recently examined systems involving the slipping of a drop on inclined surfaces.[5–12] The main idea consisted in filling the space between asperities of a textured surface with a lubricating liquid which adheres to the substrate, and place a drop of the test liquid on the surface of the lubricating liquid (see Fig. 1). In Fig. 1a, $\mathbf{F}_{g,\tau}$ is the component of gravity along the surface and $\mathbf{F}_{st}$ is the sticking force which provides the mechanical equilibrium of the drop. The angles θ_1 and θ_2 in Fig. 1b are the contact angles of the advancing and receding edges of the drop. The drop is in mechanical equilibrium

* *Nanoscale*, 2015, 7, 7873–7884. Republished with permission.

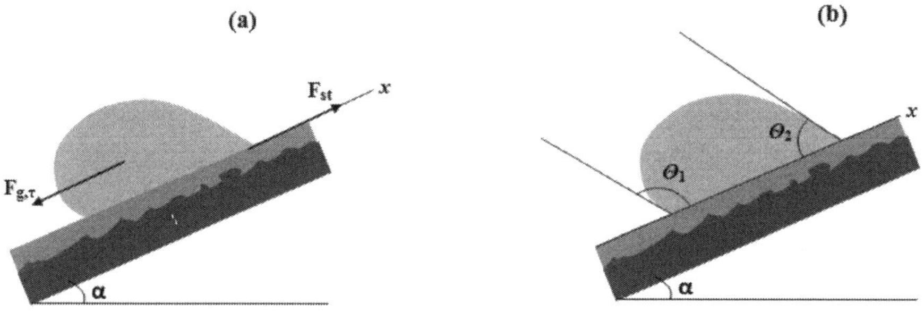

FIG. 1 (a) Schematic representation of the drop on an inclined rough substrate (black area) covered with a lubricating liquid. $F_{g,\tau}$ is the component of gravity parallel to the surface of the lubricating liquid and F_{st} is the sticking force which maintains the drop equilibrium. (b) Contact angles at advancing (θ_1) and receding (θ_2) leading edges of the drop.

until $\alpha \leq \alpha_c$, α_c being a critical angle. For $\alpha > \alpha_c$ the drop slips or rolls along the surface. For $\alpha = \alpha_c$ the sticking force acquires the critical value $F_{st,c}$ and angles θ_1 and θ_2 become θ_a (advancing contact angle) and θ_r (receding contact angle), respectively. When the test and lubricating liquids are immiscible the surface of the lubricating liquid is smooth and the drop can slip along the surface at very small inclinations ($\alpha \sim 3°$). The name SLIPS (slippery liquid-infused porous surface) was given to such a system.[5]

The main difference between SLIPS and the system considered in this paper is that in the former case the drop of the test fluid is in contact with another, lubricating, liquid but not with SDA. Because of the presence of two different fluids, the one-component DFT can not be directly applied to such a system and more sophisticated and time-consuming versions of DFT must be used. In spite of this, one can mimic SLIPS using particular choices for SSM such as the nonuniform density distribution and the interaction parameters with the drop. In this case the one-component DFT can provide, to some extent, a microscopic insight for SLIPS.

Note that the DFT approach has the advantage that it does not involve any phenomenological parameters and accounts explicitly for the microscopic details of the fluid–fluid and fluid–solid interactions. This allows to consider nanosystems to which macroscopic concepts, such as surface tension, are not applicable. As a disadvantage of DFT one can mention the extremely large computational time necessary to obtain results for macroscopic drops that makes such calculations impractical. However, there are cases in which one can extract some information about macroscopic drops from results obtained for nanodrops. An example is the cylindrical drop, *i.e.* a very (infinitely) long drop, which is frequently used in considerations regarding wetting phenomena (see ref. 13 and references therein). The contact line of such a drop is a straight line, which does not depend on the drop size. If the interactions of the drop molecules with those of the SDA and SSM decrease sufficiently rapid with increasing distance between molecules one can expect the critical sticking force acting on the drop at the contact line to be the same for macro- and nanodrops if the two contact lines have identical locations on the surface with respect to the horizontal profile of the fluid-solid interaction potential. The critical sticking force for a nanodrop can be calculated using DFT in a way described in ref. 14 and this value is used for the sticking force of a macrodrop.

The two goals of the present paper are (i) to examine the new kind of roughness and its influence on the contact angle and sticking force for a drop located on the surface of such a solid, and (ii) to estimate the sticking force for macroscopic drops using results from nanodrops. Because the force of gravity is extremely small for nanodrops, it cannot break their mechanical equilibrium even when the drops are on a vertical surface with $\alpha = 90°$.[14] To obtain information about the sticking force, we will consider a nanodrop on a horizontal surface and apply a perturbative horizontal force f_τ on each molecule of the drop. By changing f_τ, a critical value $f_{\tau,c}$ can be found such that for $f_\tau > f_{\tau,c}$

the Euler–Lagrange equation of DFT, which provides the equilibrium state of the system, has no drop-like solution. Then the critical sticking force $F_{st,c}$ is provided by $F_{st,c} = N_d \times f_{\tau,c}$, where N_d is the number of molecules in the drop, which can be found from the solution of the Euler-Lagrange equation of DFT. For the sake of generality, in the present paper no specific fluids, liquids, and surfaces are considered, but the parameters of intermolecular interactions are selected from reasonable ranges.

Note that in experiments involving inclined surfaces, the magnitude of the critical sticking force $F_{st,c}$ on a drop of mass M can be easily found using the equation

$$F_{st,c} = Mg\sin\alpha_c. \tag{1}$$

2. GENERAL CONSIDERATIONS

2.1. THE SYSTEM

2.1.1. Geometry.

The considered system consists of three components which are presented in Fig. 2. The first component, SDA, (black area in Fig. 2) is a semiinfinite substrate of constant density ρ_s decorated with regular arrays of pillars of height h_p, width d_p and distance between pillars Δ_p. All distances between surfaces are measured between the centers of the molecules of their first layers. The pillars are infinite in the y-direction (normal to the plane of the figure) and composed of the same material as the substrate. The second solid material (SSM) (lighter area) fills the space between pillars and forms a layer of thickness h_l above them which has a smooth upper surface. The density of SSM is considered nonconstant and will be provided in section 2.1.4. The third component is a test fluid (TF) which forms a drop on the surface of SSM. The drop is in thermodynamic equilibrium with its vapor. Note that the density distribution of SSM is considered independent of the density distribution of TF.

The upper boundary of the system (not shown in Fig. 2), located at distance h_u from the surface of SSM, is treated as a hard wall. Because of the low density of TF outside the drop, the influence of the

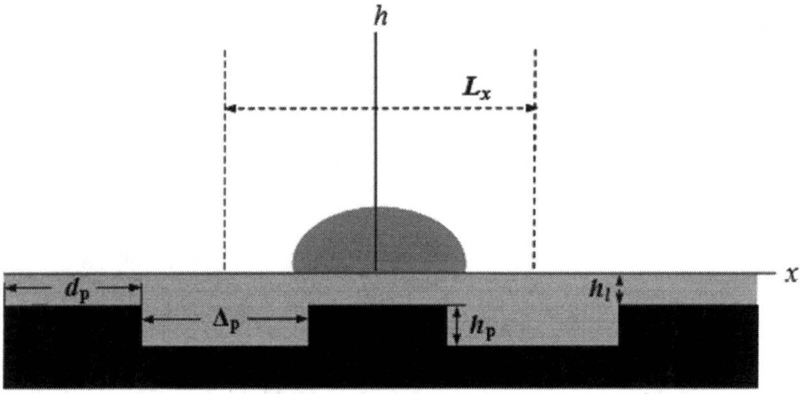

FIG. 2 Schematic representation of the considered system which consists of a solid (substrate and pillars) of constant density (black area) covered by a second solid material (SSM) (light area), and the drop of the test fluid on the smooth surface of SSM. The lengths d_p, h_p, and Δ_p are the pillars width, pillars height, and distance between pillars, respectively, h_l is the thickness of SSM above the pillars, $L_x = d_p + \Delta_p$ is the width of the unit cell used in the calculations. All distances between surfaces are measured between the centers of the molecules forming the first layers of the corresponding surfaces. The x-axis passes through the centers of molecules of the test fluid located at the bottom surface of the drop.

upper boundary on the state of the system can be neglected. In the x-direction, the system is considered periodic with period L_x, the number of molecules of TF per L_x being constant (closed system).

The density distribution (FDD) of TF, $\rho_f(\mathbf{r})$, in this system is considered uniform in the y-direction (cylindrical drop) and non-uniform in the x- and h-directions, hence $\rho_f(\mathbf{r}) \equiv \rho_f(x, h)$. The test fluid is exposed to an external potential due to the TF-SDA and TF-SSM interactions.

2.1.2. Interaction potentials.

The interaction potentials between the molecules of TF and those of TF, of SSM and SDA are selected in the Lennard–Jones form with hard core repulsion

$$\phi_\alpha\left(|\mathbf{r}-\mathbf{r}'|\right)=\begin{cases}4\varepsilon_\alpha\left[\left(\dfrac{\sigma_\alpha}{r}\right)^{12}-\left(\dfrac{\sigma_\alpha}{r}\right)^{6}\right], & r\geq\sigma_\alpha\\ \infty & r<\sigma_\alpha\end{cases} \tag{2}$$

where the subscript α should be ff, fm, and fs for the TF-TF, TF-SSM, and TF-SDA interactions, respectively, ε_{ff}, ε_{fm}, and ε_{fs} are energy parameters, σ_{ff}, σ_{fs}, and σ_{fm} are hard core diameters of the corresponding interaction potentials, $\mathbf{r}$ and $\mathbf{r}'$ provide the locations of the interacting molecules, and $r = |\mathbf{r} - \mathbf{r}'|$.

Because of the geometry of the system, the external potential $U_{fs}(\mathbf{r})$ generated by the substrate and pillars depends on x and h and is independent of y, *i.e.* $U_{fs}(\mathbf{r}) \equiv U_{fs}(x, h)$. Due to the ordered geometrical location of the pillars and uniformity of the substrate, this potential is periodic in the x-direction with period L_x. The potential $U_{fs}(\mathbf{r})$ can be obtained by integrating the Lennard–Jones potential (eqn (2)) for the TF-SDA interactions over the entire volume of SDA and can be written as

$$U_{fs}(\mathbf{r})=\int_{V_s}\rho_s(\mathbf{r}')\phi_{fs}\left(|\mathbf{r}-\mathbf{r}'|\right)d\mathbf{r}'+\int_{V_P}\rho_s(\mathbf{r}')\phi_{fs}\left(|\mathbf{r}-\mathbf{r}'|\right)d\mathbf{r}' \tag{3}$$

where V_s is the volume occupied by the substrate, V_p is the volume occupied by the pillars, $\rho_s(\mathbf{r}')$ is the density of SDA. For a uniform SDA, $\rho_s(\mathbf{r}') \equiv \rho_s$, and the first integral in eqn (3) can be calculated analytically

$$\int_{V_s}\rho_s(\mathbf{r}')\phi_{fs}\left(|\mathbf{r}-\mathbf{r}'|\right)d\mathbf{r}'=\frac{2\pi}{3}\varepsilon_{fs}\rho_s\sigma_{fs}^3\Psi\left(\sigma_{fs},h_l+h_p+h\right). \tag{4}$$

where

$$\Psi(\sigma,H)=\frac{2}{15}\left(\frac{\sigma}{\sigma+H}\right)^9-\left(\frac{\sigma}{\sigma+H}\right)^3.$$

In the second integral in eqn (3), the integration with respect to y could be performed analytically. Integration with respect of x- and h-coordinate was carried out numerically.

The contribution $U_{fm}(\mathbf{r})$ of SSM to the total potential is provided by equation

$$U_{fm}(\mathbf{r})\equiv U_{fm}(x,h)=\int_{V_m}\rho_m(\mathbf{r}')\phi_{fm}\left(|\mathbf{r}-\mathbf{r}'|\right)d\mathbf{r}', \tag{5}$$

where V_m is the volume occupied by SSM, and depends on the density distribution $\rho_m(\mathbf{r}) \equiv \rho_m(x, h)$. The choice of $\rho_m(x, h)$ is explained in section 2.1.4. As for the case of TF-SDA potential, the integration with respect to y in eqn (5) could be performed analytically and with respect to x- and h-coordinates, numerically.

A perturbative horizontal force f_τ acting on a molecule of TF in the negative direction of the x-axis generates the potential

$$U_e(\mathbf{r}) \equiv U_e(x) = f_\tau x \tag{6}$$

which is zero in origin.

Finally, the net potential has the form

$$U_{net}(x,h) = U_{fs}(x,h) + U_{fm}(x,h) + U_e(x). \tag{7}$$

2.1.3. Parameters of the substrate and pillars, and test fluid.

Below, the lengths will be provided in units of TF-TF hard core diameter, σ_{ff}. The following geometrical characteristics of the system were selected as constants: $h_p = 4$, $d_p = 2$, $\Delta_p = 6$, and $h_u = 20$. The small size of asperities was selected to decrease the computational time. Comparable or even smaller sizes are often used in molecular dynamics simulations (see e.g. ref. 15). Three thicknesses of SSM were considered. They are characterized by the thicknesses of that part of SSM which is above the pillars and were selected as $h_l = 1$ (system S1), $h_l = 3$ (system S2), $h_l = 10$ (system S3). The parameters of the interaction potential for TF-TF interactions were selected as for argon:[16] $\sigma_{ff} = 3.405$ Å, $\varepsilon_{ff}/k_B = 119.76$ K, where k_B is the Boltzmann constant. For TF-SDA and TF-SSM interactions the energy parameters were selected as $\varepsilon_{fs}/k_B = 145$ K and $\varepsilon_{fm}/k_B = 135.8$ K, respectively. The reason for such a selection will be explained in section 2.1.4. The hard core diameters for those interaction potentials were considered equal ($\sigma_{fs} = \sigma_{fm} = \sigma_{ff}$). The temperature was selected $T = 87.3$ K, that is far from the critical temperature of bulk TF ($T_c = 131.6$ K).[16] For this reason one can neglect the critical density fluctuations and the mean-field DFT can be employed to describe the selected system. The number density of SDA was selected $\rho_s = 1.91 \times 10^{28}$ m^{-3} and the mass of a molecule of TF $m_f = 6.63 \times 10^{-26}$ kg.

2.1.4. Selection of the second solid material (SSM).

As mentioned in the Introduction, the system considered in the present paper can mimic SLIPS by using a particular choice of SSM instead of the lubricating liquid (LF) of SLIPS. To make this choice, two auxiliary systems were considered. The goal was (i) to select an LF which is immiscible with TF and (ii) to determine the density distribution of that LF in contact with a rough solid and use it as the density distribution of SSM.

To select a suitable LF, the system consisting of a smooth substrate in contact with a mixture of LF and TF was considered first using the density functional approach formulated by Rosenfeld[17] for binary mixtures. The substrate and TF were the same as those described in section 2.1.3. The applied calculational procedure is similar to that used in ref. 18 where a binary mixture in contact with a uniform solid was considered. However, in the present calculations the system is considered connected to the reservoir of TF at a chemical potential equal $-12.5k_BT$ (open system). Analyzing the density distributions of TF and LF for various values of the parameters of LF (σ_{ll}, ε_{ll}, and ε_{ls}) the latter quantities were selected as follows: $\sigma_{ll} = \sigma_{ff}$, $\varepsilon_{ll}/k_B = 194$ K, and $\varepsilon_{ls}/k_B = 220$ K. This choice of parameters provides an example of a suitable LF. Finally, the same parameters were used in a second auxiliary system, consisting of a rough solid and LF, to find the density distribution of LF in contact with a rough surface using the one-component DFT.[4] The obtained density distribution (FDD), $\rho_l(x, h)$, is presented in Fig. 3 for the system S1. In Fig. 3a, the FDD in the x-direction is provided for the range between pillars. One can see that the space between pillars is filled with the "liquid-like" LF. The FDD in the x-direction above the pillars is presented in Fig. 3b. As expected, the amplitude of the density oscillations decreases with increasing distance from SDA. The FDD in the vertical direction is presented in Fig. 3c along the line passing through the midway between pillars (solid line) and the line passing in the middle of a pillar. Finally, the part of this distribution

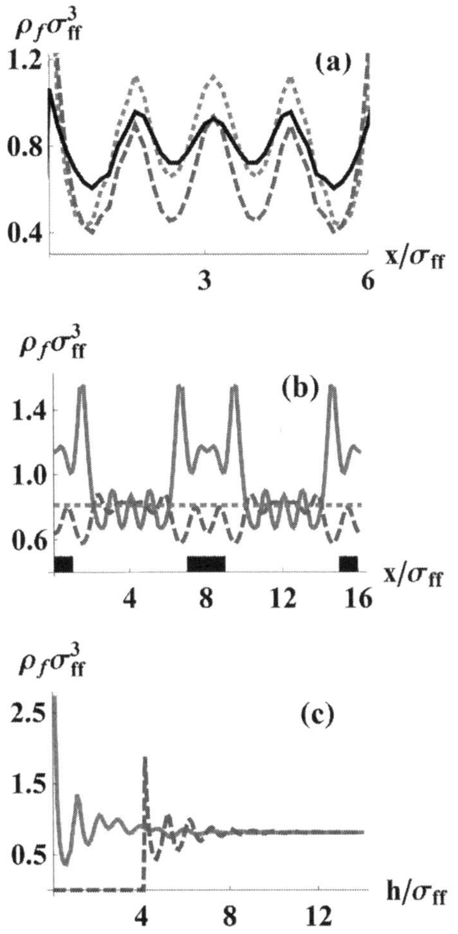

FIG. 3 System S1. (a) Density distribution of LF between the solid pillars at various distances h from the surface of the substrate: $h = 1$ (dotted line), $h = 2.5$ (dashed line), and $h = 4$ (solid line); (b) density distribution of LF above the solid pillars at various distances h from their upper surfaces: $h = 0.3$ (solid line), $h = 1$ (dashed line), and $h = 10$ (dotted line). The rectangles in the bottom part show the locations of the solid pillars; (c) density distribution of LF along the vertical lines passing through the midway between pillars (solid line) and through the middle of the pillars (dashed line).

for a thickness equal to that of SSM was extracted and used to calculate the potential $U_{fm}(x, h)$ with eqn (5), where $\rho_m(x, h) = \rho_l(x, h)$.

Another feature which should characterize SSM is its wetting temperature T_w, which should be higher than the selected $T = 87.3$ K because only in this case the drop is stable on the surface. To find T_w, the open system consisting of uniform SSM in contact with TF was considered and the wetting temperature estimated using the method described in ref. 19. It turned out that the wetting temperature of considered system is greater than 100 K and hence, the selected temperature is suitable.

2.2. The Euler-Lagrange equation for the fluid density distribution

To find the FDD of the test fluid (TF), $\rho_f(x, h)$, the weighted density approximation (WDA) version of the DFT suggested first in ref. 20 and developed further in ref. 4 was used. This version of DFT accounts for the non-local, short-ranged correlations in the fluid and provides the correct structure of fluids near solid walls. The equation for the FDD of TF, $\rho_f(x, h)$, is obtained through the minimization of the total Helmholtz free energy $F[\rho_f(\mathbf{r})]$ of the system under the constraint of a constant number of molecules, *i.e.* in a canonical ensemble (see Appendix). Such a procedure was developed in ref. 21–24 and provided the convergence of the numerical iterations to drop-like solutions. The Euler–Lagrange equation can be represented in the form

$$\log\left[\Lambda^3 \rho_f(x,h)\right] - Q_f(x,h) = \frac{\lambda}{k_B T} \tag{8}$$

where the function $Q_f(x, h)$, which is a functional of $\rho_f(x, h)$, is provided in Appendix (eqn (19)), $\Lambda = h_p/(2\pi m_f k_B T)^{1/2}$ is the thermal de Broglie wavelength, h_p is the Planck constant, and λ is a Lagrange multiplier arising because of the constraint of fixed average density of the fluid. This constraint has the form

$$\rho_{f,\,av} = \frac{1}{V}\int_V d\mathbf{r}\rho_f(\mathbf{r}) \tag{9}$$

where V is the volume occupied by TF, and leads to the following expression for λ

$$\lambda = -k_B T \log\left[\frac{1}{\rho_{f,\,av} V' \Lambda^3}\int_{V'} d\mathbf{r} e^{Q_f(x,h)}\right]. \tag{10}$$

In the above equation, $V' = L_x h_u$ is the volume per unit length in y-direction occupied by TF. By eliminating λ between eqn (8) and (10), one obtains an integral equation for the FDD $\rho_f(x, h)$, which can be solved by iterations. When calculating the density distribution $\rho_f(x, h)$ of TF, the net potential is provided by the sum of $U_{fs}(x, h)$, $U_{fm}(x, h)$, and $U_e(x, h)$ the latter two being calculated with eqn (5) and (6).

The main details of the iteration procedure are provided in ref. 14 and 25 and in the Appendix where a calculation tactic, that considerably reduce the time to find the numerical solution of the Euler–Lagrange equation for the density distribution of the test fluid, is presented.

The procedure described in the present section, was also employed to find the FDD of the lubricating fluid in contact with a rough solid. The latter FDD is used to model the nonuniform SSM (see section 2.1.4).

2.3. Calculation of the Drop Profile

Because of the relatively large width of the vapor–liquid interface detected in molecular dynamics experiments as well as in DFT calculations,[25,26] the profile of the nanodrops is not clearly defined. There are several approaches to extract the profile from a known FDD (see ref. 26 and 27 as examples). In this paper we use the simplest one employed in ref. 27 and determine the profile inside the vapor–liquid interface as that corresponding to a constant local density ρ_{div} which can be defined as the density for an equimolar dividing surface of a horizontal FDD $\rho(x, h_0)$ at some distance h_0 from the surface, by considering this FDD as that of a planar vapor–liquid interface.[25] To select the most appropriate value of h_0, one should note, that for $h < 3\sigma_{ff}$ the fluid density distribution $\rho_f(x, h)$ inside the drop has an oscillatory behavior as a function of h (because the fluid molecules form several liquid layers of various densities)[28] and as a function of x (due to the roughness of the surface). For these reasons, it is not clear how to determine ρ_{div} for $h_0 < 3\sigma_{ff}$. For $h_0 > 3\sigma_{ff}$ those oscillations become smaller and FDD $\rho_f(x, h_0)$ ($h_0 > 3\sigma_{ff}$) exhibits a clearly observable interface between high and low density phases, for which one can easily find ρ_{div} and define a dividing surface for the selected h_0. Hence, it is reasonable to select a value of h_0 larger than $3\sigma_{ff}$.

After the profile is determined, other characteristics of the drop, such as the advancing and receding contact angles and the number of molecules in the drop can be calculated.[25]

3. RESULTS

3.1. POTENTIALS OF INTERMOLECULAR INTERACTIONS

In Fig. 4, the potential $U_{fm}(x, h)$ generated in the system S1 by a nonuniform SSM at distance σ_{ff} from the SSM surface ($h = 0$) is presented (by the solid line) as function of x. The nonuniform density $\rho_m(x, h)$ of SSM was calculated as described in section 2.2 and $U_{fm}(x, h)$ was obtained using eqn (5). To estimate the role of nonuniformity of SSM, the potential generated by a uniform SSM with the same average density as the nonuniform SSM $\left(\rho_{av}\sigma_{ff}^3 \simeq 0.82\right)$ is presented as a dashed line. As expected, both potentials are periodic with respect to x with a period $L_x = \Delta_p + d_p$. The periodicity of the potential generated by the uniform SSM is due to the presence of "upside down pillars" of SSM regularly located between the pillars on the substrate, whereas the layer above them produces a uniform potential. For the nonuniform SSM considered below, the nonuniformity of the potential is caused both by the pillars of SSM and by the nonuniformity of the density distribution, $\rho_m(x, h)$. One can see in Fig. 4, that the amplitude of the changes of $U_{fm}(x, h)$ is slightly larger for a nonuniform SSM than for a uniform one, while their magnitudes differ very little. In both cases, the minima of $U_{fm}(x, h)$ as function of x are located on the vertical lines of symmetry of the pillars of SSM.

The potential $U_{fs}(x, h)$, generated by the substrate and pillars located on its surface, behaves similarly to $U_{fm}(x, h)$, the minima of the potential being displaced by $(\Delta_p + d_p)/2$ with respect to those of $U_{fm}(x, h)$. The magnitude of U_{fs} at $h = 0$ is much smaller than that of U_{fm} due to the larger distance from the substrate.

In Fig. 5, the net potential $U_{net}(x, h)$ provided by eqn (7), is presented as function of x for the system S1. In this example, the horizontal force f_τ in the negative x-direction on a single molecule of TF was 9.7×10^{-15} N. The total potential possesses multiple minima separated by potential barriers; due to the presence of the perturbative force f_τ the local minima of the potential gradually increase with increasing x.

Because of the periodicity of the potentials $U_{fs}(x, h)$ and $U_{fm}(x, h)$ in the x direction, any drop, in the absence of the horizontal perturbative force f_τ, has the same free energy when it is displaced along the surface over an integer number of periods. However, in the presence of that force, the potential energy of the drop decreases if the drop is displaced in the negative x-direction. This means that any drop on a surface is metastable.

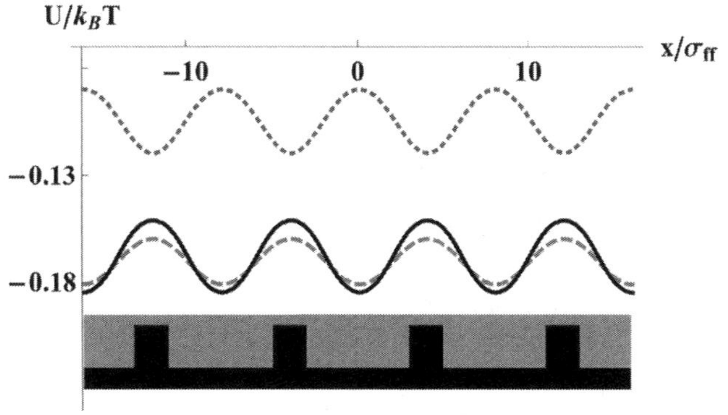

FIG. 4 TF-SSM interaction potential in the system S1 for a nonuniform (solid line) and uniform (dashed line) SSM. The potential is calculated at distance σ_{ff} from the surface of SSM. The dotted line is the net potential of substrate and pillars. For clarity, it is multiplied by a factor of 10. The insert beneath the plots indicates the locations of the pillars.

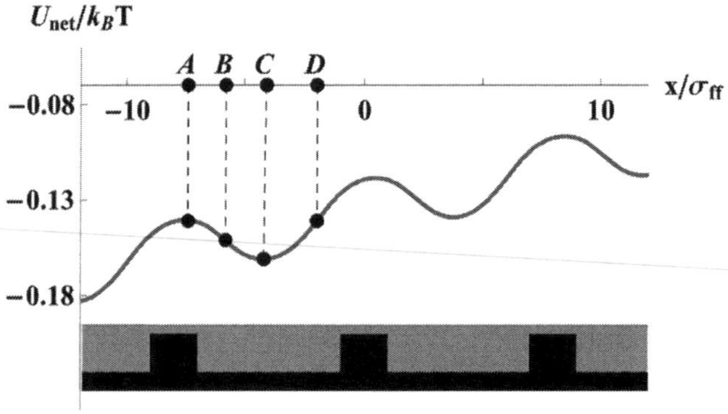

FIG. 5 Net potential as function of x for $h = 0$ in system S1. The extrema of the potential have labels A and C and the inflection points are marked with B and D.

3.2. Drop profile and contact angle

In Fig. 6, a typical drop in the system S1 is presented in the absence of the external perturbative force ($f_\tau = 0$). The size of the drop (number of molecules per unit length in y-direction) $N_d = 2.82 \times 10^{11}$ m^{-1}. The drop profile, determined as described in section 2.3, is provided by the solid line. The densities given on the legend bar above the figure are provided in dimensionless form as $\rho\sigma_{ff}^3$. The drop is symmetrical with respect to the vertical line passing midway between solid pillars. Note that another solution, which is symmetrical with respect of the middle of the solid pillars, was also identified; however the free energy of this solution is greater than that of the drop from Fig. 6. However, the analysis of its stability by the method described in Appendix, did not provide unambiguous proof of whether this solution is metastable or unstable. The lack of evidence may be due to the very small height of the potential barrier (if the drop is, actually, metastable) or because the extremum of the free energy obtained by solving the Euler–Lagrange equation is a maximum (unstable drop). Because of this, only the stable drops (with smallest free energies) will be presented below.

In the absence of the external force ($f_\tau = 0$), the left and right parts of the drop profile make the same angles with the surface ($\theta_1 = \theta_2 = \theta$). The magnitude of θ depends, in particular, on the size

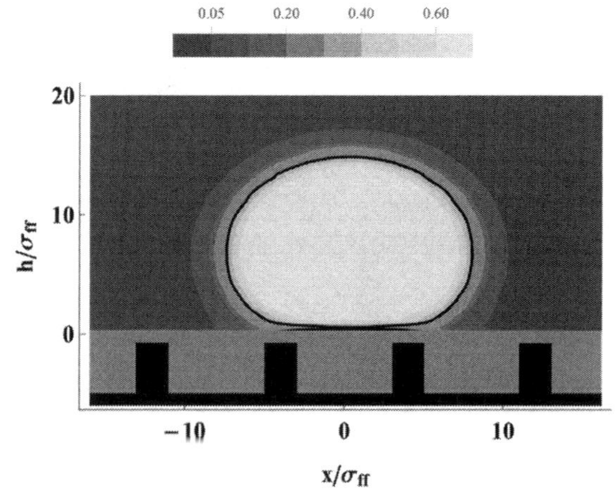

FIG. 6 Example of a drop on the surface of SSM for the system S1. The solid line presents the drop profile. The magnitude of the fluid density is provided in dimensionless form as $\rho\sigma_{ff}^3$.

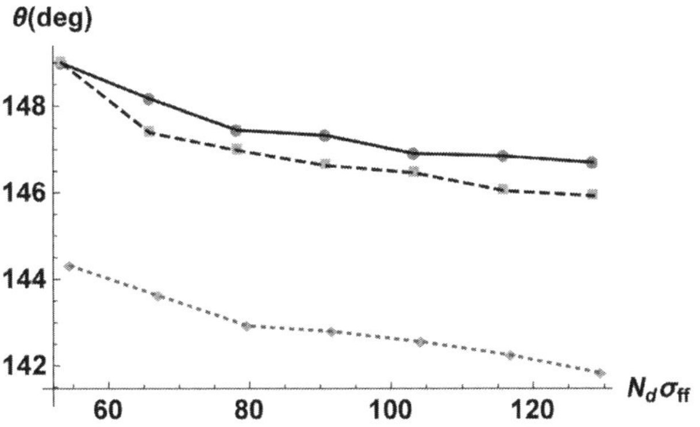

FIG. 7 Calculated dependence of θ on the drop size in the absence of external force ($f_\tau = 0$) for systems S1 (diamonds), S2 (squares), and S3 (points). The lines are guides for eye.

of the drop. In Fig. 7, this dependence is presented for all three considered systems. In all cases, the contact angle decreases with increasing size of the drop. For the same size, the drop in the system S1 has the smallest and in S3, the largest contact angles.

The size dependence of θ for nanodrops was examined earlier in several studies using DFT[25,29,30] or molecular dynamics simulations[31] but till now no universal explanation of this dependence was found.

One of the possible explanations involves the concept of line tension introduced in ref. 32 which leads (for three dimensional drops) to an additional term in the Young equation containing the reciprocal of the radius of the contact line between drop and solid surface. However, for the cylindrical drop considered in the present paper, the line tension can not explain the size dependence of θ because the contact line is a straight line ($R = \infty$) and the additional term in Young equation disappears. Even more, the Young equation itself and its modifications suggested in ref. 33 and 34 are not applicable to nanodrops because they are based on macroscopic considerations involving surface tensions, the latter quantities are not clearly defined at the nanoscale. A possible explanation of this dependence for the considered case will be provided in Discussion.

In the presence of an external perturbative force ($f_\tau \neq 0$) in the negative x direction, $\theta_1 \neq \theta_2$ and both angles depend on the size of the drop. An example of this dependence is presented in Fig. 8 for the system S2 and $f_\tau = 2.92 \times 10^{-16}$ N. The N_d dependence is close to linear and both angles decrease with increasing N_d.

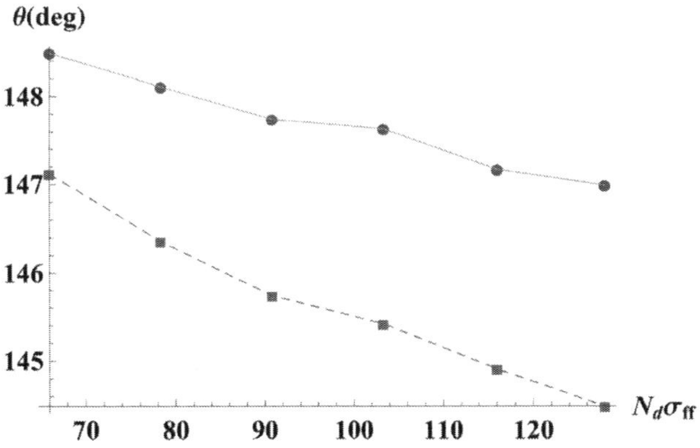

FIG. 8 System S2. Calculated dependence of θ_1 (points) and θ_2 (squares) on the number N_d of molecules of the test fluid in the drop for $f_\tau = 2.92 \times 10^{-16}$ N. The lines are guides for eye.

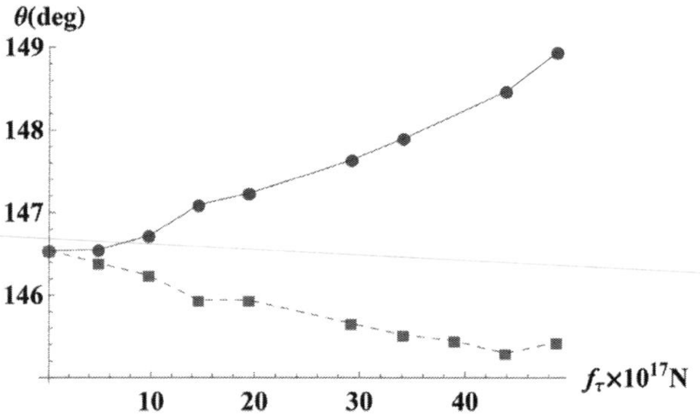

FIG. 9 Calculated dependence of θ_1 (points) and θ_2 (squares) on f_τ. The lines are guides for eye.

At constant size of the drop, the angles θ_1 and θ_2 depend on f_τ. This dependence is presented in Fig. 9 for the system S2. For all values of f_τ, the number of molecules in the drop per unit length is the same ($N_d \simeq 2.85 \times 10^{11}$). The angle θ_1 increases and θ_2 decreases with increasing f_τ. As a consequence, the difference between these angles increases with increasing external force.

To estimate the influence of SSM inhomogeneity on the contact angle, the latter was calculated for the same drop on the surface of nonuniform and uniform SSM. At $f_\tau = 0$, the contact angle of the drop containing 2.82×10^{11} molecules per unit length is 150.62° and 149.73° for the nonuniform and uniform SSM, respectively. The contact angle on nonuniform SSM is slightly larger than that on the uniform SSM, in agreement with general ideas about the influence of roughness on the contact angle.

3.3. Sticking force

The shape of the net potential allows one to explain qualitatively the microscopic origin of the sticking force acting on a drop of TF, which arises due to the interaction of the fluid molecules with a smooth nonuniform SSM that covers the rough solid. The basic idea is that the main contributions to the sticking force comes from the forces imposed on the molecules located at both leading edges of the drop.[35–37] When the leading edge (advancing or receding) of the drop on the SSM surface is located between points A and C of Fig. 5,[†] the potential $U_{net}(\mathbf{r})$ generates a force on the molecules of the drop at the leading edge in the positive direction of the x axis, which opposes the motion of the drop in the negative x-direction along the surface due to the external force $\mathbf{F}_{ext}$. Only a large enough perturbative force can overcome the potential barrier and set the drop in motion. The magnitudes of the components of the sticking force at any of the two leading edges are proportional to the x-components of the gradient of the net potential at the location of the leading edges with respect to the potential profile. Each of them increases when the leading edge is displaced in the direction from point C to point B (see Fig. 5) and decreases when it is displaced from point B to point A.

Tables 1 and 2 list the values of the critical sticking forces $F_{st,\,c}$ and contact angles obtained for systems S1, S2, and S3 for several situations. In addition, the weight of the nanodrop and the contact angle hysteresis $\Delta\theta = \theta_{a,\,c} - \theta_{r,\,c}$ are also presented. Comparing the results listed in Table 1, one can mention, first, that for the drop of approximately the same size the magnitude of the sticking force decreases with increasing thickness of the SSM above the pillars. This is an obvious consequence of the decreasing influence of the inhomogeneity of the SDA and SSM in the x-direction because of the increasing distance of the smooth surface from the pillars on the surface of the substrate.

The second interesting observation is that increasing of the size of the drop can lead to opposite results for the sticking force. For example (see Table 2), when the drop size increases from $N_d\sigma_{ff} = 66.0$ to $N_d\sigma_{ff} = 96.9$ the sticking force increases from 131 μN m^{-1} to 140 μN m^{-1}. However a further increase

[†] "Between points A and C" means all similar ranges of the potential curve.

TABLE 1
Critical sticking force for the drop containing about 2.84 × 10¹¹ molecules per unit length for all considered systems. $\theta_{a,c}$ and $\theta_{r,c}$ are the advancing and receding contact angles, respectively, Mg is the weight of a drop per unit length, $\Delta\theta$ is the contact angle hysteresis

System	$F_{st,c}$ (μN m^{-1})	$\theta_{a,c}$ (°)	$\theta_{r,c}$ (°)	Mg (pN m^{-1})	$\Delta\theta$ (°)	$N_d\sigma_{ff}$
S1	151	149.66	146.01	0.185	3.65	96.8
S2	140	148.90	145.43	0.185	3.47	96.9
S3	99.3	143.80	141.64	0.188	2.16	98.5

TABLE 2
Critical sticking force for the drops containing various numbers of molecules per unit length for the system S2. $f_{st,c}$ is the critical sticking force per molecule. Other notations are the same as in Table 1

$N_d\sigma_{ff}$	$F_{st,c}$ (μN m^{-1})	$f_{st,c} \times 10^{14}$ N	$\theta_{a,c}$ (°)	$\theta_{r,c}$ (°)	Mg(pN m^{-1})	$\Delta\theta$ (°)
66.0	131	4.46	149.20	147.30	0.126	1.90
96.9	140	4.77	148.90	145.43	0.185	3.47
127.1	99.3	2.44	143.80	141.64	0.188	2.16

of the size to $N_d\sigma_{ff} = 127.1$ leads to the decrease of the sticking force to 99.3 μN m^{-1}. To explain such a behavior of $F_{st,c}$, one should note that the drop size affects the locations of the leading edges of the drop. In Fig. 10 those locations are presented schematically with respect to the net potential generated by SSM and SDA. The points present the location of the leading edges of the drop with $N_d\sigma_{ff} = 66.0$, diamonds and squares correspond to drops with $N_d\sigma_{ff} = 96.9$ and $N_d\sigma_{ff} = 127.1$, respectively. One can see that both leading edges of the drop with $N_d\sigma_{ff} = 96.9$ are located close to the points with the largest slope of the potential curve, *i.e.* the sticking force is close to the maximum possible. The leading edges of the drop with $N_d\sigma_{ff} = 127.1$ are located close to the points with the smallest slope of the potential curve. As a consequence, the sticking force for this drop is smaller than for the previous one. For the drop with $N_d\sigma_{ff} = 66.0$, the advancing leading edge is located close to the point with the smallest slope

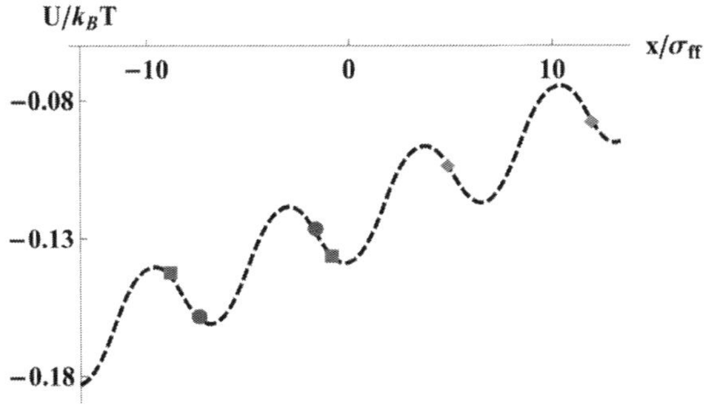

FIG. 10 System S2. Location of the leading edges, with respect to the net potential, of critical drops for $N_d\sigma_{ff} = 66.0$ (points), 96.9 (diamonds), and 127.1 (squares).

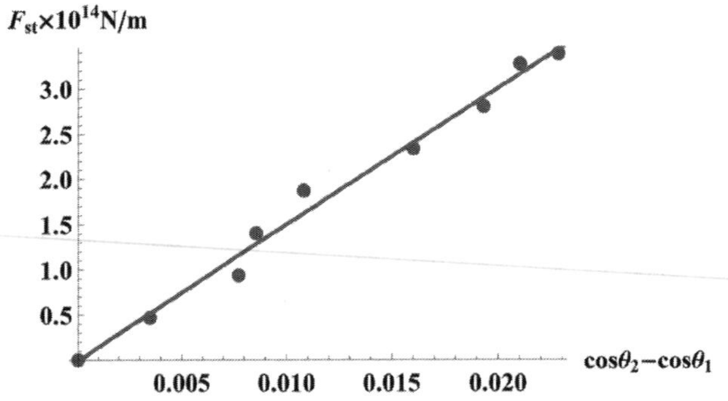

FIG. 11 System S3. Dependence of the sticking force on the difference cos θ_2 – cos θ_1. Points are results of calculations, line is the best linear fit. θ_1 and θ_2 are defined in Fig. 1.

and the receding leading edge close to the point with the largest slope. The magnitude of the sticking force has in this case an intermediate value when compared with two above considered drops.

It is interesting to compare the results obtained for the sticking force for nanodrops with the predictions made on the basis of the traditional macroscopic approach. The latter provides for the sticking force the equation

$$F_{st} = C\gamma_{lv}\left(\cos\theta_2 - \cos\theta_1\right) \tag{11}$$

where θ_1 and θ_2 are defined in Fig. 1, γ_{lv} is the liquid–vapor surface tension and C is a constant dependent on the geometry of the drop.[38,39] In Fig. 11, the dependence of the sticking force on cos θ_2 – cos θ_1 is presented for the system S3. One can see that the F_{st} calculated on the basis of DFT can be reasonable approximated by eqn (11).

Comparing the values of $F_{st,\,c}$ and Mg listed in Tables 1 and 2 one can see that the weight of a nanodrop is much smaller that the sticking force acting on it. This means that the gravity alone cannot overcome the sticking force even for $\alpha = 90°$ (vertical wall). However, the results obtained for nanodrops allow to estimate an upper bound for the critical inclination angle, α_c, for macroscopic cylindrical drops on an inclined surface. As noted in the Introduction, this can be achieved because the leading edges of macro- and nanodrops are identical for cylindrical drops. For this reason, the sticking force on a cylindrical drop of any size depends only on the location of the leading edges with respect to the interaction potential profile. Considering nanodrops of different sizes on the surface of the lubricating liquid, which have various locations of the leading edges, the largest sticking force $F_{st,\,c}$ on the nanodrops can be obtained and associated with the largest possible sticking force acting on a macrodrop. If the latter has a mass M_d, then the largest inclination angle α_c at which the macrodrop is at rest can be calculated from eqn (1) as sin $\alpha_c = F_{st,\,c}/M_d$ g. For example, let us consider in the system S3 a cylindrical drop of liquid argon of volume ν = 1 μL cm^{-1}. Such a drop contains about 2 × 10^{21} molecules m^{-1} and has a weight $M_d g \simeq 1.4$ mN m^{-1}. For estimation of the inclination angle α_c, the largest calculated sticking force $F_{st,\,max}$ = 99.3 μN m^{-1} will be selected. In this case, α_c is equal to about 4.1°, *i.e.* close to the inclination angles observed in SLIPS experiments. The largest inclination angle α_c will decrease (increase) with increasing (decreasing) mass of the macrodrop.

4. DISCUSSION

The results obtained in the present paper show that the hidden roughness has qualitatively the same influence on the behavior of a nanodrop as the physical and chemical roughnesses of the solid surface. Namely, the hidden roughness increases the hydrophobicity of the solid and generates the sticking force.

The microscopic picture of a drop on the surface in the presence of roughness developed in ref. 14 and 25 and in the present paper shows, in particular, that the sticking force and contact angle depend on different characteristics of the interaction potential. While for the same geometrical characteristics of the solid, the potential contributes mainly to the contact angle θ,[13] the sticking force is affected mainly by the slope of the horizontal profile of the interaction potential at the locations of the leading edges of the drop. If this slope is large, the sticking force can be large even for superhydrophobic surfaces. For small variation of the interaction potential in the horizontal direction, the sticking force can be extremely small. The difference between the potential profiles may be the reason of the difference between the adhesive properties of the surfaces of the rose petal and lotus leaf.[40] Both surfaces are superhydrophobic ($\theta \geq 150°$), but the first one possesses a large contact angle hysteresis and a sticking force that prevents slipping even at high inclination angles, while the second has a small contact angle hysteresis and a small sticking force. The liquid drop on the lotus leaf can easily roll down at very small inclination angles α. Detailed analysis which is needed to check this suggestion is not of concern in the present paper.

The nonuniformity in the horizontal direction of the interaction potential due to the hidden roughness and the appearance of the sticking force it causes can qualitatively explain the dependence of the contact angle θ on the size of the drop. Indeed, the presence of the sticking force opposes (to some extent) the expansion of the drop base with increasing drop size which leads to the increase of θ. On the other hand, the displacements of the leading edges with respect to the profile of the interaction potential can lead to the decrease of the sticking force and, as a consequence, to the increase of the drop base and decrease of the contact angle. As a result, the contact angle can decrease or increase with increasing size of the drop. The actual change of θ can be found only by calculations.

The method used in the present paper to find the conditions of breaking equilibrium of a drop in the presence of the external perturbative force can not provide an answer what kind of motion, rolling or slipping, will have the drop after unpinning from the surface. Intuitively, rolling is expected to occur when the drop shape is close to circular (contact angle is close to 180°), while slipping can occur even on a hydrophilic surface (the contact angle is smaller than 90°). Such expectations were confirmed in ref. 41 on the basis of the solution of the Navier–Stokes and Cahn–Hillard equations. The drops considered in the present paper had an equilibrium contact angle of about 150°. One can assume that such a drop will slip along the surface, because it has considerable contact area with the solid (see Fig. 6).

The model considered in the present paper suggests that the sticking force on a drop on the SLIPS can be caused, at least in part, by the hidden roughness. Full microscopic insight to that problem can be obtained by using the density functional theory for binary mixtures which is based on the microscopic considerations of fluid–fluid and fluid–solid interactions. Such a theory, called as Fundamental Measure Theory (FMT), was developed by Rosenfeld.[17] However, the application of the FMT to the above problem is extremely difficult for computational reasons and up to our best knowledge, no attempts have been done in this directions. The SLIPS were examined theoretically only on the basis of the traditional thermodynamics involving the macroscopic concepts of surface tensions.[5,42,43]

5. APPENDIX: FREE ENERGY CONTRIBUTIONS AND SOLUTION OF THE EULER–LAGRANGE EQUATION

The total Helmholtz free energy $F[\rho_f(\mathbf{r})]$ of a fluid under the external potential generated by a solid is expressed as the sum of an ideal gas free energy, $F_{id}[\rho_f(\mathbf{r})]$, a free energy $F_{hs}[\rho_f(\mathbf{r})]$ of a reference system of hard spheres, a free energy $F_{attr}[\rho_f(\mathbf{r})]$ due to the attractive interactions between fluid molecules (in the mean field approximation), and a free energy $F_{fs}[\rho_f(\mathbf{r})]$ due to the interactions between fluid and solid. These contributions to the free energy can be represented as follows[28,44]

$$F_{id}\left[\rho_f(\mathbf{r})\right] = k_B T \int d\mathbf{r}\rho_f(\mathbf{r})\left\{\log\left[\Lambda^3\rho_f(\mathbf{r})\right]-1\right\}, \tag{12}$$

$$F_{hs}\left[\rho_f(\mathbf{r})\right]=\int d\mathbf{r}\rho_f(\mathbf{r})\Delta\Psi_{hs}(\mathbf{r}) \tag{13}$$

where $\Lambda = h_p/(2\pi m k_B T)^{1/2}$ is the thermal de Broglie wavelength, h_p and k_B are the Planck and Boltzmann constants, respectively, T is the absolute temperature, m is the mass of a fluid molecule,

$$\Delta\Psi_{hs}(\mathbf{r})=k_B T\eta_{\bar{\rho}_f}\frac{4-3\eta_{\bar{\rho}_f}}{\left(1-\eta_{\bar{\rho}_f}\right)^2} \tag{14}$$

$\eta_{\bar{\rho}_f}=\frac{1}{6}\pi\bar{\rho}_f(\mathbf{r})\sigma_{ff}^3$ is the packing fraction of the fluid molecules, σ_{ff} is the fluid hard core diameter, and $\bar{\rho}(\mathbf{r})$ is the smoothed density defined as

$$\bar{\rho}(\mathbf{r})=\int d\mathbf{r}'\rho(\mathbf{r}')W\left(|\mathbf{r}-\mathbf{r}'|\right) \tag{15}$$

The weighting function $W(|\mathbf{r}-\mathbf{r}'|)$ is selected in the form[45]

$$W\left(|\mathbf{r}-\mathbf{r}'|\right)=\begin{cases}\dfrac{3}{\pi\sigma_{ff}^3}\left(1-\dfrac{r}{\sigma_{ff}}\right), & r\le\sigma_{ff}\\ 0, & r>\sigma_{ff}\end{cases}$$

where $r = |\mathbf{r}-\mathbf{r}'|$.

The contribution to the excess free energy due to the attraction between the fluid–fluid molecules is calculated in the mean-field approximation

$$F_{attr}\left[\rho_f(\mathbf{r})\right]=\frac{1}{2}\iint d\mathbf{r}d\mathbf{r}'\rho_f(\mathbf{r})\rho_f(\mathbf{r}')\phi_{ff}\left(|\mathbf{r}-\mathbf{r}'|\right) \tag{16}$$

where $\phi_{ff}\left(|\mathbf{r}-\mathbf{r}'|\right)$ is the Lennard–Jones potential of the fluid–fluid interactions.

The last contribution, $F_{fs}[\rho_f(\mathbf{r})]$, is given by the expression

$$F_{fs}\left[\rho_f(\mathbf{r})\right]=\int_V d\mathbf{r}\rho_f(\mathbf{r})\left[U_{fs}(\mathbf{r})+U_{fm}(\mathbf{r})+U_e(\mathbf{r})\right] \tag{17}$$

where V is the volume occupied by the fluid, and $U_{fs}(\mathbf{r})$, $U_{fm}(\mathbf{r})$, and $U_e(\mathbf{r})$ are provided by eqn (3), (5) and (6).

The Euler–Lagrange equation for the fluid density distribution $\rho_f(x, h)$ obtained by minimizing the Helmholtz free energy can be represented in the following general form

$$\log\left[\Lambda^3\rho_f(x,h)\right]-Q_f(x,h)=\frac{\lambda}{k_B T} \tag{18}$$

where λ is a Lagrange multiplier and the function $Q_f(x, h)$ is given by

$$Q_f(x,h)=-\frac{1}{k_B T}\left[\Delta\Psi_{hs}(x,h)+\overline{\Delta\Psi'_{hs}}(x,h)+U_{ff}(x,h)\right.$$
$$\left.+U_{fs}(x,h)+U_{fm}(x,h)+U_e(x)\right] \tag{19}$$

where

$$U_{ff}(x,h)=\iint dx'dh'\rho_f(x',h')\phi_{ff,y}\left(|x-x'|,|h-h'|\right), \tag{20}$$

$$\overline{\Delta\Psi'}_{\mathrm{hs}}(x,h) = \iint \mathrm{d}x'\mathrm{d}h' \rho_{\mathrm{f}}(x',h') W_y(|x-x'|,|h-h'|)$$

$$\times \frac{\partial}{\partial\bar{\rho}} \Delta\Psi_{\mathrm{hs}}(\bar{\rho})|_{\bar{\rho}=\rho_{\mathrm{f}}(x',h')}, \tag{21}$$

$\phi_{\mathrm{ff},\,y}(|x-x'|,|h-h'|)$ and $W_y(|x-x'|,|h-h'|)$ are obtained by integrating the potential $\phi_{\mathrm{ff}}(|\mathbf{r}-\mathbf{r}'|)$ and the weighted function $W(|\mathbf{r}-\mathbf{r}'|)$ with respect to y from $-\infty$ to $+\infty$.

When calculating $U_{\mathrm{ff}}(x, h)$ of the Euler–Lagrange equation arising due to the long-range fluid–fluid interactions, a cutoff at a distance equal to four molecular diameters σ_{ff} for the range of Lennard–Jones attraction was employed.

The general iteration procedure used in this paper is explained in ref. 14 and 25. Here we will discuss only the selection of the initial guess which constitutes an important part of the calculations. As shown in ref. 14 and 25, the location of a stable drop on a nonuniform surface depends on the properties of the surface and the size of the drop. When the initial guess is selected arbitrarily, usually as a "rectangular" drop at an arbitrary location on the surface,[13] the iterations transform its location and shape toward the location and shape of the stable drop. The required number of iterations depends on how close is the initial guess to the location of the solution of the Euler–Lagrange equation and how quick is the transformation of the intermediate density distribution during iterations. For the case considered in the present paper, the convergence of the iteration procedure is extremely slow and, for this reason, the choice of the initial guess has a critical importance in finding the solution in a reasonable time. For the selection of an initial FDD a special approach was developed which is based on the following observation.

Let us suppose that the iterations start with a rectangular initial guess located arbitrarily on the surface of SSM (see Fig. 12a) and that one keeps track of the difference $\Delta\rho_{\mathrm{f},\,i}(x, h) = \rho_{\mathrm{f},i+1}(x, h) - \rho_{\mathrm{f},i}(x, h)$ between the density distributions, $\rho_{\mathrm{f},i}(x, h)$ and $\rho_{\mathrm{f},i+1}(x, h)$, provided by two consecutive iterations. Then,

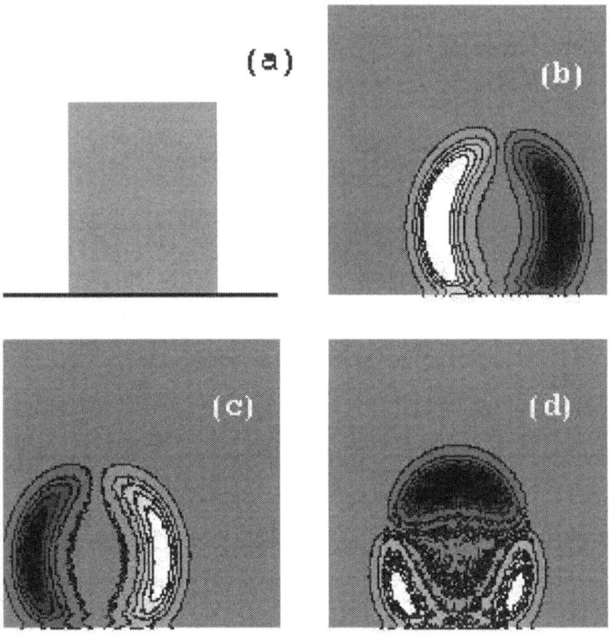

FIG. 12 (a) Initial guess for the iteration procedure. (b), (c), and (d) Possible graphs of the differences between two consecutive iterations. Dark areas correspond to negative values of this differences and light areas correspond to positive values.

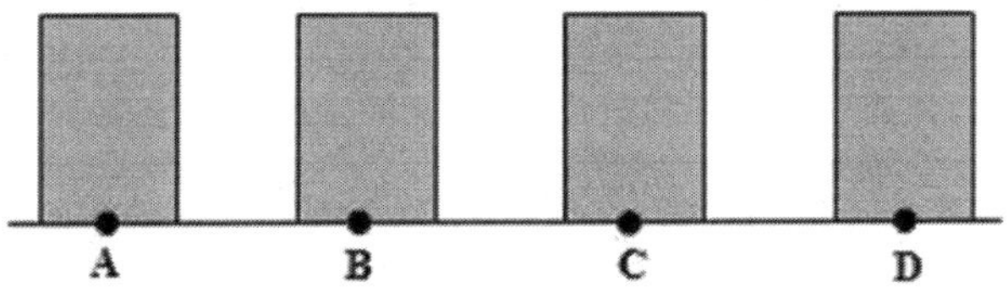

FIG. 13 Four starting initial guesses. The meaning of points A, B, C, and D is provided in Fig. 5.

after 50–100 iterations this difference as function of x and h has, generally, one of the three shapes presented in Fig. 12b–d where the light (dark) areas represent positive (negative) values of $\Delta\rho_{f,i}(x, h)$. Fig. 12b and c indicate the tendency of the density distribution $\rho_{f,i}(x, h)$ to "move" in the direction of the positive part of $\Delta\rho_{f,i}(x, h)$, *i.e.* to the left for Fig. 12b and to the right for Fig. 12c. Fig. 12d indicates the case in which the initial guess is selected almost at the location of the solution of Euler–Lagrange equation.

On the basis of the above observation, we first selected (see Fig. 13) four initial guesses of the same rectangular shape at the positions of the minimum, maximum, and of the two inflection points of the net potential $U_{net}(\mathbf{r})$ (points A, C, B, and D in Fig. 5).

After several iterations (about 100), the differences $\Delta\rho_{f,i}(x, h)$ for each initial guess were analyzed. The solution is located between the points where the distributions of $\Delta\rho_{f,i}(x, h)$ move towards each other. To determine the location of the solution more precisely, the described procedure was applied to the interval between those specific points found in the previous part of the calculations.

The analysis of $\Delta\rho_{f,i}(x, h)$ provides information whether the obtained solution is stable or unstable. In more details, this issue is discussed in Appendix B of ref. 46.

To avoid the divergence of the iteration procedure, the input density profile $\rho_{f,i}^{in}(x,h)$ for the $(i + 1)$-th iteration $\rho_{f,i+1}(x, h)$, generated by the Euler–Lagrange equation, was selected as follows[28]

$$\rho_{f,i}^{in}(x,h) = (1-\gamma)\rho_{f,i-1}^{in}(x,h) + \gamma\rho_{f,i}(x,h) \tag{22}$$

where $\rho_{f,i}(x, h)$ is the i-th iteration and the constant $\gamma = 0.01$. As a measure of the precision of the iterations the dimensionless quantity

$$\delta = \int_V dx dh \left[\rho_{f,i+1}(x,h) - \rho_{f,i}^{in}(x,h)\right]^2 \Big/ \left(\int_V dx dh \rho_{f,i}(x,h)\right)^2$$

was introduced. The iterations were carried out on a two dimensional grid with a spacing equal to $0.1\sigma_{ff}$ until δ became smaller than 10^{-7}.

REFERENCES

1. R. N. Wenzel, *Ind. Eng. Chem.*, 1936, **28**, 988–994.
2. A. B. D. Cassie and S. Baxter, *Trans. Faraday Soc.*, 1944, **40**, 0546–0550.
3. M. A. Sarshar, W. Xu and C.-H. Choi, in *Adhesion and Adhesives: Fundamental and Applied Aspects: Advances in Contact Angle, Wettability and Adhesion,* ed. K. L. Mittal, John Wiley & Sons, Somerset, NJ, USA, 2013, pp. 3–18.
4. P. Tarazona and R. Evans, *Mol. Phys.*, 1984, **52**, 847–857.
5. T. S. Wong, S. H. Kang, S. K. Y. Tang, E. J. Smythe, B. D. Hatton, A. Grinthal and J. Aizenberg, *Nature,* 2011, **477**, 443–447.
6. W. Ma, Y. Higaki, H. Otsuka and A. Takahara, *Chem. Commun.*, 2013, **49**, 597–599.
7. A.K. Epstein, T. S. Wong, R. A. Belisle, E. M. Boggs and J. Aizenberg, *Proc. Natl. Acad. Sci. U. S. A.*, 2012, **109**, 13182–13187.

8. P. W. Wilson, W. Z. Lu, H. J. Xu, P. Kim, M. J. Kreder, J. Alvarenga and J. Aizenberg, *Phys. Chem. Chem. Phys.,* 2013, **15**, 581–585.
9. J. D. Smith, R. Dhiman, S. Anand, E. Reza-Garduno, R. E. Cohen, G. H. McKinley and K. K. Varanasi, *Soft Matter,* 2013, **9**, 1772–1780.
10. R. Qiu, Q. Zhang, P. Wang, L. Jiang, J. Hou, W. Guo and H. Zhang, *Colloids Surf., A,* 2014, **453**, 132–141.
11. L. Xiao, J. Li, S. Mieszkin, A. Di Fino, A. S. Clare, M. E. Callow, J. A. Callow, M. Grunze, A. Rosenhahn and P. A. Levkin, *ACS Appl. Mater. Interfaces,* 2013, 5, 10074–10080.
12. J. Li, T. Kleintschek, A. Rieder, Y. Cheng, T. Baumbach, U. Obst, T. Schwartz and P. A. Levki, *ACS Appl. Mater. Interfaces*, 2013, **5**, 6704–6711.
13. E. Ruckenstein and G. O. Berim, *Adv. Colloid Interface Sci.,* 2010, **157**, 1–33 (Section 4.13 of this volume).
14. G. O. Berim and E. Ruckenstein, *J. Chem. Phys.*, 2008, **129**, 114709 (Section 4.11 of this volume).
15. C. D. Daub, J. H. Wang, S. Kudesia, D. Bratko and A. Luzar, *Faraday Discuss.*, 2010, **146**, 67–77.
16. R. Evans and P. Tarazona, *Phys. Rev. A,* 1983, **28**, 1864–1868.
17. Y. Rosenfeld, *Phys. Rev. Lett.,* 1989, **63**, 980–983.
18. G. O. Berim and E. Ruckenstein, *J. Chem. Phys.*, 2008, **128**, 134713 (Section 2.5 of this volume).
19. F. Ancilotto and F. Toigo, *J. Chem. Phys.*, 2000, **112**, 4768–4772.
20. S. Nordholm, M. Johnson and B. C. Freasier, *Aust. J. Phys.,* 1980, **33**, 2139.
21. A. González, J. A. White, F. L. Román and S. Velasco, *Phys. Rev. Lett.,* 1997, **79**, 2466–2469.
22. A. González, J. A. White, F. L. Román and R. Evans, *J. Chem. Phys.*, 1998, **109**, 3637–3650.
23. J. A. White and S. Velasco, *Phys. Rev. E: Stat. Phys., Plasmas, Fluids, Relat. Interdiscip. Top.,* 2000, **62**, 4427–4430.
24. J. A. White, A. González, F. L. Román and S. Velasco, *Phys. Rev. Lett.,* 2000, **84**, 1220–1223.
25. G. O. Berim and E. Ruckenstein, *J. Chem. Phys.*, 2008, **129**, 014708 (Section 4.4 of this volume).
26. M. J. de Ruijter, T. D. Blake and J. De Coninck, *Langmuir,* 1999, **15**, 7836–7847.
27. N. Giovambattista, P. G. Debenedetti and P. J. Rossky, *J. Phys. Chem. B,* 2007, **111**, 9581–9587.
28. P. Tarazona, *Phys. Rev. A,* 1985, **31**, 2672–2679.
29. F. Porcheron, P. A. Monson and M. Schoen, *Phys. Rev. E: Stat. Phys., Plasmas, Fluids, Relat. Interdiscip. Top.,* 2006, **73**, 041603.
30. A. P. Malanoski, B. J. Johnson and J. S. Erickson, *Nanoscale,* 2014, **6**, 5260–5269.
31. G. Scocchi, D. Sergi, C. DAngelo and A. Ortona, *Phys. Rev. E: Stat. Phys., Plasmas, Fluids, Relat. Interdiscip. Top.,* 2011, **84**, 061602.
32. B. A. Pethica, *Rep. Prog. Appl. Chem.,* 1961, **46**, 14; B. A. Pethica, *J. Colloid Interface Sci.,* 1977, **62**, 567–569.
33. R. David and A. W. Neumann, *Langmuir,* 2007, **23**, 11999–12002.
34. L. Schimmele and S. Dietrich, *Eur. Phys. J. E,* 2009, **30**, 427–430.
35. L. Gao and T. McCarthy, *Langmuir,* 2007, **23**, 3762–3765.
36. G. McHale, *Langmuir,* 2007, **23**, 8200–8205.
37. M. Nosonovsky, *Langmuir,* 2007, **23**, 9919–9920.
38. C. G. L. Furmige, *J. Colloid Sci.,* 1962, **17**, 309.
39. M. Callies and D. Quéré, *Soft Matter,* 2005, **1**, 55.
40. *Green Tribology: Biomimetics, Energy Conservation and Sustainability,* ed. M. Nosonovsky and B. Bhushan, Springer, 2012, ch. 2.
41. S. P. Thampi, R. Adhikari and R. Govidarajan, *Langmuir,* **29**, 3339–3346.
42. V. Hejazi and M. Nosonovsky, *Langmuir,* 2012, **28**, 2173–2180.
43. V. Hejazi and M. Nosonovsky, *Colloid Polym. Sci.,* 2013, **291**, 329–338.
44. P. Tarazona, U. M. B. Marconi and R. Evans, *Mol. Phys.,* 1987, **60**, 573–595.
45. R. N. Nilson and S. K. Griffiths, *J. Chem. Phys.,* 1999, **111**, 4281–4290.
46. G. O. Berim and E. Ruckenstein, *Nanoscale,* 2015, **7**, 3088–3099 (Section 4.6 of this volume).

4.9 A Nanodrop on the Surface of a Lubricating Liquid Covering a Rough Solid Surface*†

*Gersh Berim and Eli Ruckenstein**

Department of Chemical and Biological Engineering, State University of New York, at Buffalo, Buffalo, New York 14260, USA.

Corresponding Author

*E-mail: feaeliru@buffalo.edu;
Fax: +(716)645-3822; Tel: +(716)645-1210

1 INTRODUCTION

In paper,[1] a nanodrop located on the smooth surface of a solid possessing hidden roughness was considered on the basis of the density functional theory (DFT). One of its goals was to use a microscopic approach to describe (qualitatively) the experimental results[2–9] regarding the slipping of a nanodrop on a SLIPS (slippery liquid-infused porous surface). The main feature of SLIPS, which is interesting for practical applications, is the extremely small inclination angle at which the drop starts to slip along the surface of the lubricating liquid. This angle remains small both for large $(\theta > 90°)$ and small $(\theta < 90°)$ contact angle θ of the drop. The lubricating fluid, that covers in those experiments the rough solid surface, was modeled in ref. 1 by a nonuniform second solid material possessing a smooth surface which generated, together with the rough substrate, an interaction potential acting on the drop of the test fluid (TF). Even though this approximation changes considerably the original system, it led to reasonable values for the slipping angle (the inclination angle of the surface when slipping begins).

In the present paper, a more realistic model consisting of a lubricating fluid (LF) that covers the rough surface of a substrate and a nanodrop of TF located onto LF is examined using the two-component DFT. A perturbative external force F_τ acting on the molecules of TF is added in the horizontal direction. This force is balanced by a sticking force F_{st} which maintains the drop equilibrium. Schematically, the considered case is sketched in Fig. 1 where, for simplicity, the nonuniformities of TF and LF are not indicated.

Such a model which, to our best knowledge was never considered microscopically, has several new features compared with the traditional one-component liquid drop on a solid surface. First, the latter is surrounded by a vapor of the same fluid which has a much smaller density compared to that of the drop. For this reason, the vapor-liquid interaction plays a minor role in the formation of the drop. In the present model, the environment of the drop includes a low density mixture of TF and LF around the upper part of the drop (which has a neg-ligible role in the formation of the drop) and

* *Nanoscale*, 2015, 7, 15701–15710. Republished with permission.
† Electronic supplementary information (ESI) available. See DOI: 10.1039/c5nr04654h

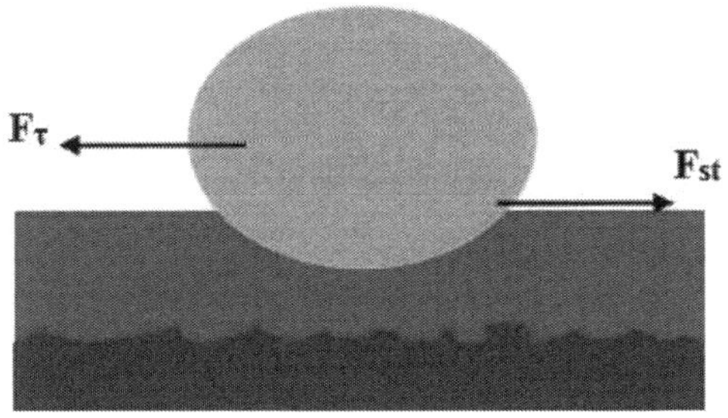

FIG. 1 Schematic representation of a drop of the test fluid (light gray area), lubricating fluid (darker gray area), and rough substrate (black area). F_τ is the horizontal perturbative force and F_{st} is the sticking force which maintains the drop equilibrium.

a dense LF beneath the drop. The interaction between the latter part of LF and TF is responsible for the drop shape and location. In this case, the bottom part of the drop profile is not planar and this rises questions about the definition of the contact angle between the nanodrop and LF and the location of the leading edges of the nanodrop. (Note, that the drop has no direct contact with the solid and interacts with it indirectly *via* the LF.) Another feature is that the contact between the nanodrop and LF occurs in the region of the LF-TF interface, i.e. in the region where the densities of LF and TF are nonuniform both in the vertical and horizontal directions with respect to the solid substrate. For this reason and because the thickness of LF-TF interface is comparable with the size of the drop, additional difficulties arise in the definition of the contact angle.

The goal of the present paper is to examine microscopically the new kind of systems using reasonable definitions of the quantities which characterize the drop of TF located on LF. In section 2, the system is defined in more details including the interaction potentials and the basic equations for the fluid density distributions. In section 3, the drop profile is calculated, contact angles are defined and the results of the calculation of the contact angle and sticking force are provided along with their discussion.

2 THE SYSTEM AND BASIC EQUATIONS

2.1 THE SYSTEM

2.1.1 Geometry.

The three components involved in our considerations are presented in Fig. 2. The first component is the rough substrate of constant density ρ_s, its roughness being modeled by regular arrays of pillars of height h_p, width d_p and distance between pillars Δ_p. All distances between surfaces are measured between the centers of the molecules of their first layer. The pillars are infinite in the y-direction (normal to the plane of the figure). The second component, the lubricating fluid (LF), forms a nonuniform layer of liquid-like density on the surface of the substrate. Outside this layer, the LF is present in small amounts as a mixture with TF. The last component is the test fluid (TF) which forms a two-dimensional (cylindrical) drop extended in the y-direction. Outside the drop, TF is present in small concentrations as a mixture with LF. Evidently, a small amount of LF is also present in the drop, and a small amount of TF is present in the LF layer.

The upper boundary of the system is located at the distance h_u from the upper surface of the pillars and is treated as a hard wall (not shown in Fig. 2). Because of the low densities of TF and LF close to the upper boundary, the influence of the latter on the state of the system is neglected. The system is periodic in the x-direction with period L_x, with the number of molecules of LF and TF per L_x constant (closed system).

It is supposed that there is no symmetry breaking in the y-direction and that the density distributions (FDDs) of LF ($\rho_1(\mathbf{r})$) and TF ($\rho_2(\mathbf{r})$) are uniform in this direction and non-uniform in the x- and h-directions, hence $\rho_i(\mathbf{r}) \equiv \rho_i(x,h)(i=1,2)$.

2.1.2 Interaction potentials.

It is supposed that the spherical molecules of LF and TF interact *via* a truncated Lennard-Jones potential

$$\phi_{ij}(r) = \begin{Bmatrix} \infty, & r \le \sigma_{ij} \\ 4\varepsilon_{ij}\left[\left(\frac{\sigma_{ij}}{r}\right)^{12} - \left(\frac{\sigma_{ij}}{r}\right)^{6}\right], & \sigma_{ij} < r \le r_{ij,\mathrm{cut}} \\ 0, & r > r_{ij,\mathrm{cut}} \end{Bmatrix} \tag{1}$$

where i and j take the values 1 and 2 for LF and TF, respectively, r is the distance between the centers of a pair of interacting molecules, and $r_{ij,\,\mathrm{cut}} = 3\sigma_{ij}$ is the cutoff distance. In eqn (1), σ_{ij} and ε_{ij} are the hard core diameters and energy parameters, respectively. The molecules of TF and LF interact also with the rough solid (substrate plus pillars) of constant density which is considered as the source of an external potential which has the same form as eqn (1) with no upper bound for the radius of the fluid-solid interactions. For these interactions, the hard core diameters σ_{ij} and interaction parameters ε_{ij} should be replaced with σ_{is}, and ε_{is}, respectively, the subscript s indicating the solid.

It is assumed also that each molecule of TF is exposed to a constant external perturbative force f_τ which acts in the negative direction of the x-axis.

As a result, the net external potential, $U_{\mathrm{net,\,ext}}(x,\,h)$ has the form

$$U_{\mathrm{net,\,ext}}(x,h) = U_{1\mathrm{s}}(x,h) + U_{2\mathrm{s}}(x,h) + U_{\mathrm{e}}(x) \tag{2}$$

where $U_{is}(x,h)$ $(i=1,2)$ are potentials due to LF-solid and TF-solid interactions, respectively and $U_e(x) = F_\tau x$ is the potential due to the external perturbative force. For more details, one can see the ESI. †Note that at equilibrium, the magnitude of the sticking force $F_{\mathrm{st}} = f_\tau N_{\mathrm{d}}$ where N_{d} is the number of molecule of TF in the drop. (Both F_{st} and N_{d} are calculated per unit length of the drop in y-direction.)

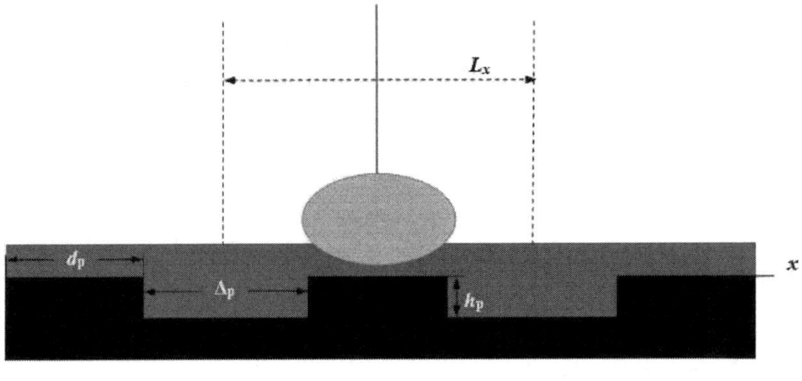

FIG. 2 Schematic representation of the considered system which consists of a solid (substrate and pillars) of constant density (black area) covered by the lubricating fluid (lighter area), and the drop of the test fluid (light area). The lengths $d_{\mathrm{p}}, h_{\mathrm{p}}$ and Δ_{p} are the pillars width, height, and distance between pillars, respectively, $L_x = d_{\mathrm{p}} + \Delta_{\mathrm{p}}$ is the width of the unit cell used in the calculations. All distances between surfaces are measured between the centers of the molecules forming the first layers of the corresponding surfaces. The x-axis passes through the centers of the molecules of the lubricating fluid located next to the upper surface of pillars.

2.2.3 Selection of the parameters.

Below, all lengths will be provided in units of TF-TF hard core diameter, $\sigma \equiv \sigma_{22}$. In this units, the geometrical characteristics of the system were selected as follows: $h_p = 3$, $d_p = 2$, $\Delta_p = 6$, and $h_p = 16$.

The hard core diameters of all intermolecular interactions were considered equal ($\sigma_{11} = \sigma_{22} = \sigma_{12} = \sigma_{1s} = \sigma_{2s} \equiv \sigma = 3.405\text{Å}$).

The energy parameters of the potentials of the fluid-fluid interactions have the values $\varepsilon_{11}/k_B = \varepsilon_{22}/k_B \equiv \varepsilon/k_B = 119.76\text{K}$, and $\varepsilon_{12} = 0.5\varepsilon$, where k_B is the Boltzmann constant. For the fluid-solid interactions the energy parameters were selected as $\varepsilon_{1s} = 4\varepsilon$ and $\varepsilon_{2s} = 0$, respectively. The TF-solid interaction is neglected because of the large distance between the drop and solid due to the presence of LF. The selected interaction parameters ε_{11}, ε_{22}, and ε_{12} provided the immiscibility of the two components, hence the existence of a drop of TF. In addition, the value employed for the energy parameter ε_{1s} allows the lubricating fluid to wet the solid completely forming a liquid layer that covers the pillars. The number density of the solid was taken $\rho_s = 1.92 \times 10^{28}\ \text{m}^{-3}$ and the masses of the molecules of TF and LF were $m_1 = m_2 = 6.63 \times 10^{26}\text{kg}$. The temperature was selected $T = 87.3$ K.

2.2 The Euler-Lagrange equations for the fluid density distributions

The density distributions $\rho_i(\mathbf{r})\,(i = 1,2)$ are calculated using the density functional approach formulated by Rosenfeld.[10] As shown in ref. 11–14, the density functional theory, developed originally for open systems (grand canonical ensemble), can be also applied to closed systems (canonical ensemble). The only restriction, which is fulfilled by the present calculations, is the number of molecules to be sufficiently large. A detailed discussion regarding the applicability of the canonical ensemble DFT can be found in ref. 15.

The total Helmholtz free energy $F_{tot}[\rho_1(\mathbf{r}), \rho_2(\mathbf{r})]$ can be expressed as the sum of an ideal gas free energy, $F_{id}[\rho_1(\mathbf{r}), \rho_2(\mathbf{r})]$, an excess free energy $F_{ex}[\rho_1(\mathbf{r}), \rho_2(\mathbf{r})]$, and a free energy $F_{1s}[\rho_1(\mathbf{r}), \rho_2(\mathbf{r})]$ due to the interactions between fluids and solid. Minimization of the $F_{tot}[\rho_1(\mathbf{r}), \rho_2(\mathbf{r})]$ leads to a system of two Euler-Lagrange equations for the FDDs $\rho_i(x,h)\,(i = 1,2)$

$$\log\left[\Lambda_i^3 \rho_i(\mathbf{r})\right] - Q_i(\mathbf{r}) = \lambda_i / k_B T, \, (i = 1,2) \tag{3}$$

where $\Lambda_i = h_p/(2\pi m_i k_B T)^{1/2}$ is the thermal de Broglie wavelength of the molecules of component i, h_p is the Planck constant, T is the absolute temperature, m_i is the molecular mass of component i, $Q_i(\mathbf{r})$ is a functional of the densities $\rho_1(\mathbf{r})$ and $\rho_2(\mathbf{r})$ and λ_i is a Lagrange multiplier arising because of the constraint of fixed average density of component i in the system

$$\rho_{i,av} = \frac{1}{V_i} \int_{V_i} d\mathbf{r} \rho_i(\mathbf{r}) \tag{4}$$

where V_i is the volume of the system accessible to component i.

Using eqn (3) and (4), the Lagrange multipliers can be rewritten in the form

$$\lambda_i = -K_B T \log\left[\frac{1}{\rho_{i,av} V_i \Lambda_i^3} \int_{V_i} d\mathbf{r} e^{Q_i(\mathbf{r})}\right] \tag{5}$$

By eliminating λ_i between eqn (3) and (5), one obtains two integral equations for the FDDs $\rho_1(x,h)$ and $\rho_2(x,h)$ which can be solved by numerical iterations.

The details of the derivation of the Euler-Lagrange equation and of the numerical procedure used for their solutions are provided in the ESI.†

3 RESULTS AND DISCUSSION

Below, five specific cases, $C_m\ (m=1,\ldots,5)$, are considered which differ from each other by the average number densities $\rho_{i,\mathrm{av}}\ (i=1,2)$ and, as a consequence, by the numbers of molecules of TF and LF in the system. In Table 1, the dimensionless values, $\rho_{i,\mathrm{av}}\sigma^3\ (i=1,2)$, are listed for all five cases. The results obtained for each of these cases will be used below to analyze the characteristics of the drop for various thicknesses of LF and various sizes of the drop.

3.1 Density distributions of the test and lubricating fluids and the drop profile

In Fig. 3, various FDDs of TF and LF are presented for the case C_3 and for $f_\tau = 1.36 \times 10^{-15}$ N. Because of the nonzero horizontal perturbative force, the drop location is displaced to the left of the middle of the pillar. For this reason, LF and TF densities at the left hand side of the drop is slightly (not detectable by eye) different from those on the right hand side, so the drop is slightly asymmetric. Fig. 3a, presents the two-dimensional FDD of TF calculated using eqn (3). The white dashed solid line in this figure represents the profile of a drop which was obtained using a procedure similar to that for a drop on a solid surface (see *e.g.* ref. 1). According to that procedure, the drop profile is defined as a line along which the local density of TF $\rho_2(x,h)$ is constant and equal to the TF density at the location of the equimolar dividing surface of a horizontal FDD $\rho_2(x,h_0)$ at some distance h_0 from the upper surface of the pillars, by considering this FDD as that of a planar vapor-liquid interface. In the present paper, the horizontal FDD was taken along the line passing through the location of the maximum TF density (cross in Fig. 3a). In this case, the density $\rho_{2,\mathrm{div}}$ of TF on the dividing surface is $\rho_{2,\mathrm{div}}\sigma^3 = 0.336$. After the profile is determined, other characteristics of the drop, such as, for example, the number of molecules in the drop (drop size) can be calculated.[16]

Several examples of one-dimensional density distributions of TF along the horizontal and vertical lines are presented in Fig. 3b and c, respectively, in the region of the drop location. The points on each curve represent the locations of the drop profile for the corresponding one-dimensional FDD. All presented FDDs are slightly asymmetrical with respect to the vertical line passing through their maximum. For the horizontal FDDs, this asymmetry is due to the presence of pillars and external force f_τ, and for vertical FDDs to the increasing distance from the solid.

Similar density distributions for LF are displayed in Fig. 3d–f. One can see that LF penetrates into the drop, its density decreasing with increasing distance from the drop surface. However, the fraction of LF molecules inside the drop is about only 4.7% of the total number of molecules in the drop.

This fact justifies the neglect of the action of the external per-turbative force on the molecules of the lubricating fluid.

Let us note that in all considered cases only the stable nanodrops were analyzed. In the absence of an external force, those drops are centered on the pillars. The drops which are centered between neighboring pillars were metastable (had greater free energy than the stable ones) $f_\tau = 0$ and unstable at $f_\tau \neq 0$. Other solutions of Euler-Lagrange equations were not found.

TABLE 1
Average number densities $\rho_{1,\mathrm{av}}$ and $\rho_{2,\mathrm{av}}$ of LF and TF, respectively for five considered cases

Case	C_1	C_2	C_3	C_4	C_5
$\rho_{1,\mathrm{av}}\sigma^3$	0.175	0.211	0.211	0.211	0.250
$\rho_{2,\mathrm{av}}\sigma^3$	0.075	0.075	0.105	0.120	0.075

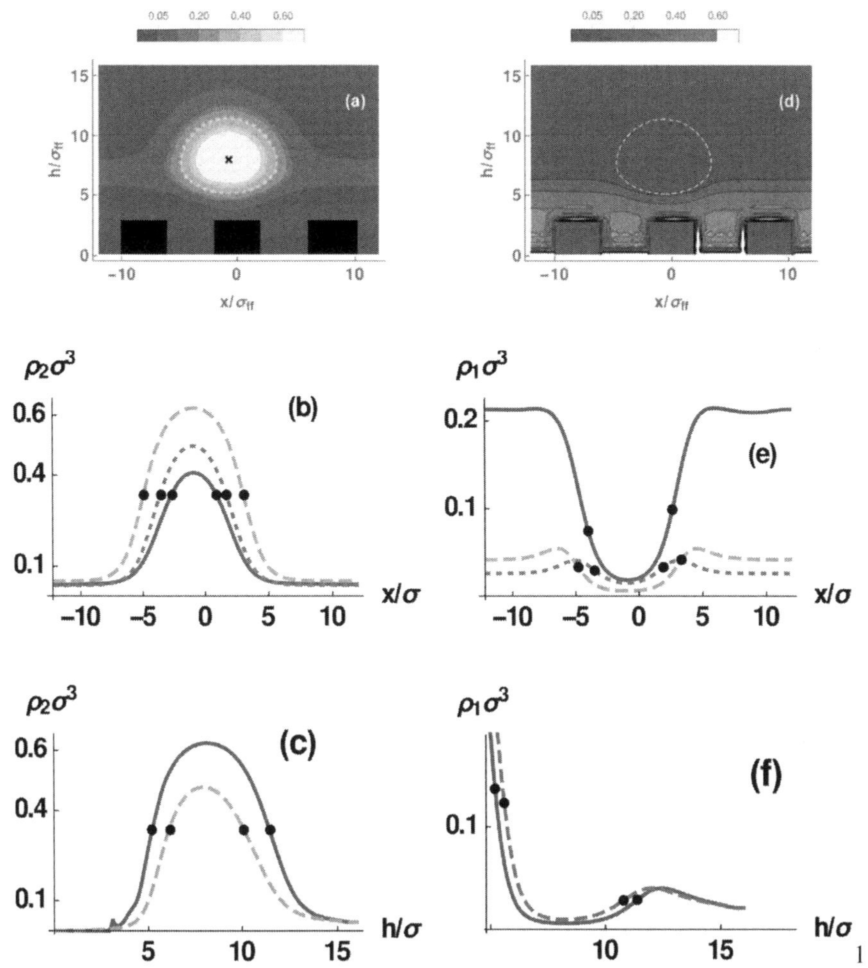

FIG. 3 Various FDDs of TF and LF for the case C_3 and $f_\tau = 1.36 \times 10^{-15}$N. (a) and (d) Two-dimensional density distributions of TF and LF, respectively. The cross in panel (a) indicates the location of the maximum density of TF inside the drop. The white dashed lines indicate the drop profile; (b) TF density distributions along the horizontal lines at the distances 2.5 (solid line), 5.0 (dashed), and 8.0 (dotted) from the upper surface of the pillars. (c) TF density distributions along the vertical lines passing through the point of maximum density of TF (solid line) and midway between pillars (dashed line). Curves in panels (e) and (f) represent similar FDDs as in (b) and (c) for LF.

3.2 Contact angle

Due to the nonplanar shape of the bottom part of the drop and, as a result, the absence of a fixed reference surface (such as, *e.g.*, the surface of the solid), the definition of leading edges of the drop and the contact angle which the drop makes with the reference surface is not straightforward and unique. To make a reasonable choice, let us consider for simplicity the specific case of a nanodrop on the surface of a lubricating fluid that covers a smooth solid surface. The density distribution of the test fluid in such a system is presented in Fig. 4. Below, several possibilities are examined regarding the reference line with which the profile (or its extension) intersects to form an angle which can be considered as the contact one.

Definition 1: Similar to the drop profile, the surface of the lubricating fluid can be considered as the line along which the density of LF is a constant equal to the density of LF on the dividing surface

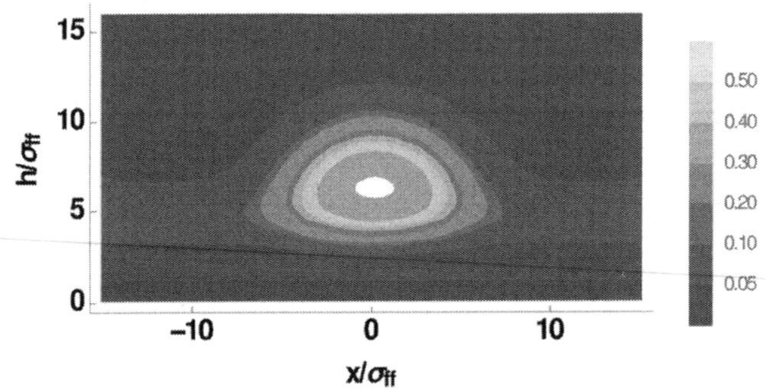

FIG. 4 Dimensionless density distribution $(\rho(x,h)\sigma^3)$ of the test fluid for a smooth solid surface at $\rho_{1,av}\sigma^3 = 0.211$, $\rho_{2,av}\sigma^3 = 0.120$, and $\varepsilon_{12}/\varepsilon_{22} = 0.6$. The black line in this figure represents the drop profile obtained by the procedure described in section 1.

of a particular one-dimensional density distribution of LF. The latter distribution is considered to be along the vertical line passing through a point inside the drop where the density of the test fluid has a maximum. Because of the oscillating behavior of the LF density close to the wall (see Fig. 5a), a special procedure to find the dividing surface which separates the vapor and liquid phases, should be developed. In the present case, we select the part of the FDD between points A and B (see Fig. 5a) and replace the actual FDD with one presented in Fig. 5b. (At points A and B, the density of LF has liquid-like and vapor-like values, respectively.) This FDD is used to find the location of the equimolar dividing surface and the density of LF at that location.

In Fig. 6, the drop profile and the defined surface of LF (solid curve beneath the drop) are presented. One can see that the drop profile and the surface of LF do not intersect and this can be interpreted that between the drop and the surface of LF there is a thin layer of a low density mixture of TF and LF. The reference line was selected as the horizontal line which far from the drop coincides with the surface of the lubricating fluid. This line is presented as the dashed one in Fig. 6. After the reference line is introduced, one can define two types of contact angles which characterize the nanodrop. The first, θ_m, which can be called microcontact angle,[17] is defined using the part of the drop profile in Fig. 6 in the vicinity of the reference line (see the left hand side of Fig. 6). The second, the apparent contact angle θ_a, is defined using a circular extension of the upper part of the drop profile (see right hand side of Fig. 6).

Definition 2: Similar to Definition 1, the surface of the lubricating fluid can be considered as a line of constant density. However, this density is selected as the density of LF at the location of the lowest point of the profile. The micro- and apparent contact angles obtained according to this definition are presented in Fig. 7.

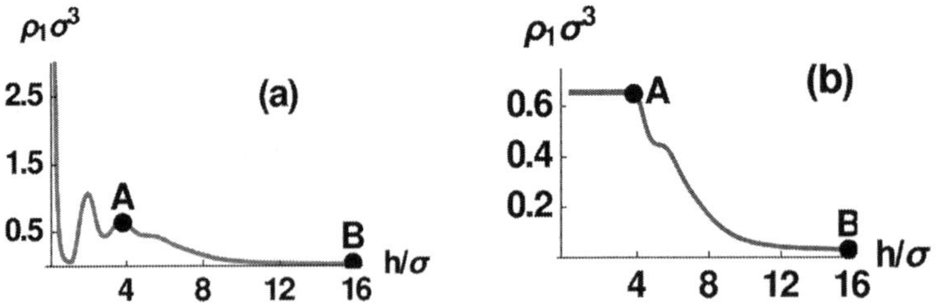

FIG. 5 (a) Density distribution of the lubricating fluid along the vertical line passing through the point of maximum density of the test fluid in the drop. At points A and B, the density of LF has liquid-like and vapor-like values, respectively. (b) Fictitious density distribution of the lubricating fluid which was used to find the equimolar dividing surface.

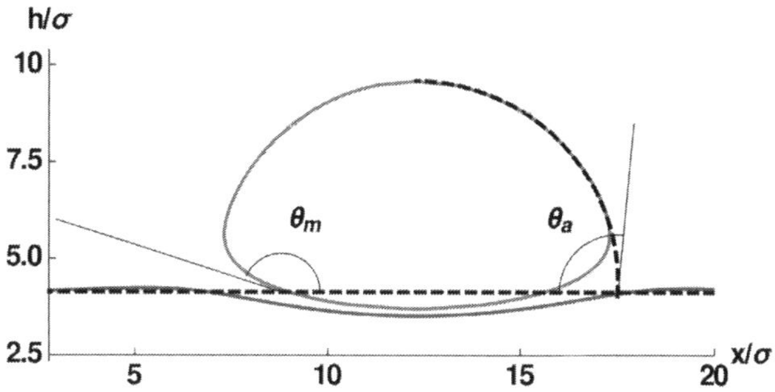

FIG. 6 Microscopic (θ_m) and apparent (θ_a) contact angles according to Definition 1. The surface of the lubricating fluid (LF) (the solid curve beneath the drop) is obtained using the LF density on the dividing surface for one-dimensional density distribution of LF along the vertical line passing through the center of the drop (point of maximum density of the test fluid).

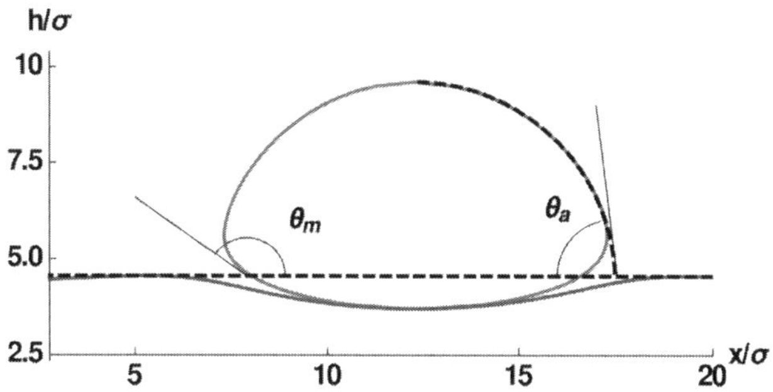

FIG. 7 Microscopic (θ_m) and apparent (θ_a) contact angles according to Definition 2. The surface of the lubricating fluid (LF) (the solid curve beneath the drop) is obtained using the LF density at the lowest point of the drop profile.

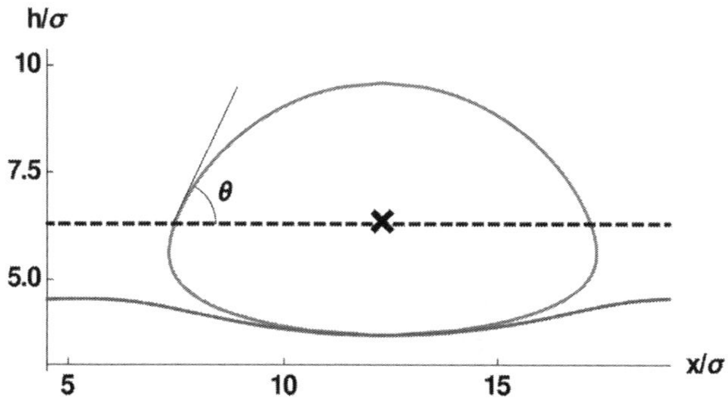

FIG. 8 Contact angle θ according to Definition 3. The cross indicates the point where the density of TF has a maximum. The surface of the lubricating fluid (LF) (the solid curve beneath the drop) does not participate in forming θ.

Definition 3: Another possible definition of the contact angles is to consider that the reference line passes through a characteristic point of the drop. As such a point, one can select, for example, the point inside the drop where the density of TF has a maximum (see Fig. 8). In this case, the microscopic and apparent contact angles practically coincide $\left(\theta_m = \theta_a = \theta\right)$.

TABLE 2
Microscopic (Θ_m) and apparent (θ_a) contact angles calculating according to various definitions for smooth solid surface and $\rho_{1,av} = 0.211, \rho_{2,av} = 0.120$, and various values of ε_{12}. (The latter is provided in units of ε_{22}.)

	Definition 1		Definition 2		Definition 3
$\varepsilon_{12}/\varepsilon_{22}$	θ_a	θ_m	θ_a	θ_m	θ
0.5	120.1	169.6	110.3	142.5	79.9
0.55	106.9	163.3	100.9	144.2	75.6
0.6	93.3	160.1	90.0	148.2	70.3
0.65	82.6	176.0	75.6	156.1	60.1

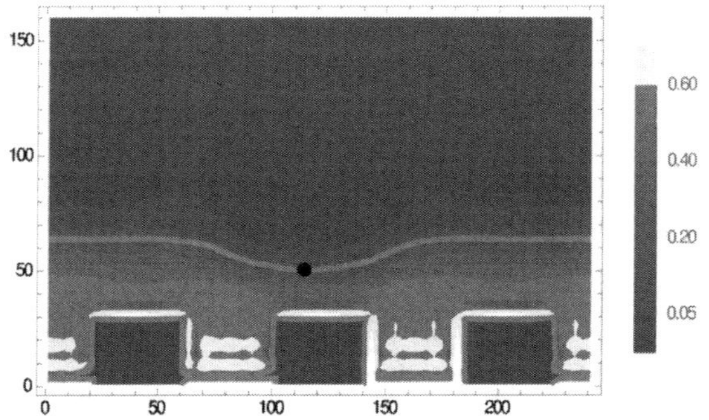

FIG. 9 FDD of LF for the case C_3 and $f_\tau = 1.36 \times 10^{-15}$N. The dot indicates location of the lowest point of the drop profile and the thick line represent the line of constant density of LF passing through this point. The magnitude of the fluid density is provided in dimensionless form by $\rho_l(x,h)\sigma^3$.

For a smooth solid surface, the contact angles obtained for a nanodrop in the system with $\rho_{1,av}\sigma^3 = 0.211$, $\rho_{2,av}\sigma^3 = 0.120$ and various values of ε_{12} are provided in Table 2. One can see that different definitions provide very different values of the contact angles.

Even though the first definition appears to be the most natural one, we will select below Definition 2. The reason is that the practical realization of Definition 1 is not straightforward because of the uncertainty in the selection of point A on the actual density profile (see Fig. 5) that leads to uncertainty in determining the location of the equimolar dividing surface. Definition 3 does not take explicitly into account the presence of LF and for this reason seems to be inappropriate.

It is obvious that the microcontact angle depends stronger on the TF-LF interactions than the apparent one. For this reason, below only the microcontact angle will be used.

Let us consider the application of Definition 2 to a rough surface using as an example case C_3. For this case, FDD of LF is presented in Fig. 9.

The dot in this figure provides the location of the lowest point of the drop profile and the thick line represents the line of constant density of LF passing through this point. The part of this line far from the drop is almost straight and its extension will be used as the reference line for the definition of the leading edges and advancing (θ_1) and receding (θ_2) contact angles in the way illustrated in Fig. 10. In the latter figure, the reference line is represented by the horizontal dashed line and points A and R indicate the locations of the advancing and receding leading edges of the drop.

Following the definition of contact angle, the dependence of θ_1 and θ_2 on the size of the drop and on the magnitude of the perturbative force were calculated.

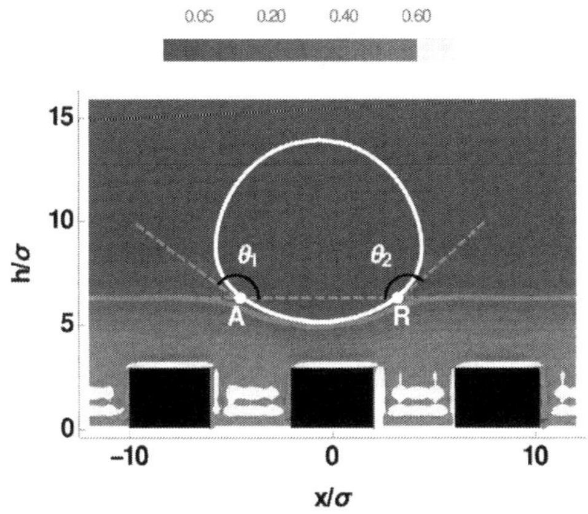

FIG. 10 FDD of LF for the case C_3 and $f_\tau = 1.36 \times 10^{-15}$N and profile of the drop of TF (closed white line) along with the reference line (horizontal dashed line) defined as noted in the text. θ_1 and θ_2 are considered as advancing and receding contact angles and points A and R as locations of the leading edges of the drop.

TABLE 3
The values of the contact angles for different drop sizes at $f_\tau = 0$ and $f_\tau = 1.36 \times 10^{-15}$ N ($\rho_{1,av}\sigma^3 = 0.211$)

	$f_\tau = 0$		$f_\tau = 1.36 \times 10^{-15}$ N	
$N_d\sigma$	θ_A (°)	θ_R (°)	θ_A (°)	θ_R (°)
8.8	127.1		133.5	131.9
19.8	122.5		123.5	121.8
24.9	121.8		122.1	120.0

In Table 3, the contact angles and sizes of the drop are presented for cases $f_\tau = 0$, $f_\tau = 1.36 \times 10^{-35}$ N and $\rho_{1,av}\sigma^3 = 0.211$. In all cases, the magnitudes of the contact angles decrease with increasing drop size. In previous papers[1,18] in which a drop in contact with a rough surface was considered, such a dependence of the contact angle was explained by the change of the location of the leading edges of the drop with respect to the potential of the liquid-solid interactions. In the system considered in the present paper, the drop has no direct contact with the solid and the molecules of TF do not interact with the solid $(\varepsilon_{2,S} = 0.)$ The solid affects the drop indirectly through its interaction with LF, the latter interacting with the molecules of TF. The largest contribution to this interaction comes from the molecules of LF closest to the drop. The larger the density of those molecules, the greater is the net TF-LF interaction.

In Fig. 11, three drop profiles corresponding to drops of different sizes at $f_\tau = 0$ are presented along with the locations of the leading edges of the drop defined as described above. The thick line represents the behavior of the potential (given in arbitrary units) generated by the rough solid in the vicinity of the drop. The magnitude of the potential has its largest value located above the middle of the pillar (note that the potential is negative). The calculations show that the LF density behaves similar to the potential and, hence, the behavior of LF-drop interaction is also similar to that of the potential. With increasing size of the drop, the leading edges are displaced to the region of weaker horizontal interaction forces (which are defined by the horizontal component of the gradient of the interaction potential and are represented by the arrows in the figure) between TF and LF. The decrease of those forces favors the increase of the bottom area of the drop and the decrease of the contact angle.

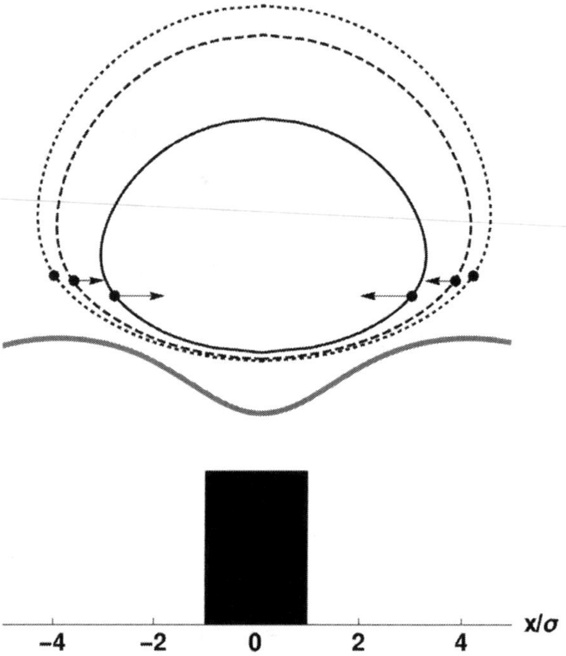

FIG. 11 Drop profiles (closed lines) and profile of the potential (in arbitrary units) generated by a rough solid (thick line). The solid, dashed, and dotted profiles correspond to drops with $N_d = 8.8$, 19.8 and 24.9, respectively. The dots and arrows indicate the locations of the leading edges of the drop and the forces on them due to the interaction with the solid, respectively.

In Fig. 12, the dependencies of the advancing and receding contact angles on the magnitude of the perturbative force f_τ is presented for cases C_2, C_3, and C_4. As expected, in all cases, the difference between θ_1 and θ_2 increases with increasing f_τ. For larger drops (cases C_3 and C_4), the receding angle θ_2 has the tendency to decrease with increasing f_τ while the advancing angle θ_1 has the tendency to increase. For smaller drops, the dependence of the contact angles on f_τ is not monotonous.

3.3 Sticking Force

As discussed in ref. 1, where a drop on the smooth surface of a second solid material (SSM) covering a rough solid was examined, the microscopic origin of the sticking force acting on the drop of TF is the external potential generated by the rough solid, which is nonuniform both in the horizontal and vertical directions. In the case considered in the present paper, the role of SSM is played by a lubricating fluid which interacts with the rough solid and with TF. The former interaction is responsible for the nonuniformity of LF which, in turn, generates a nonuniform potential acting on the drop. (One should note, that the interaction of TF with the solid in not taken into account in the present paper because it is negligible compared to that between LF and TF.)

If, in the presence of an external perturbative force f_τ, the drop remains in mechanical equilibrium, the sticking force per molecule, $f_{st} \equiv F_{st}/N_d$, is equal to f_τ ($f_{st} = f_\tau$). The largest value of f_τ at which the drop remains in equilibrium (the solution of the Euler-Lagrange equations eqn (3) does exist) is equal to the so called critical sticking force $f_{st,c}$ per molecule of the drop. Several values of the force $f_{st,c}$ along with the net critical sticking force per unit length in the y-direction $F_{st,c}$ are provided in Table 4.

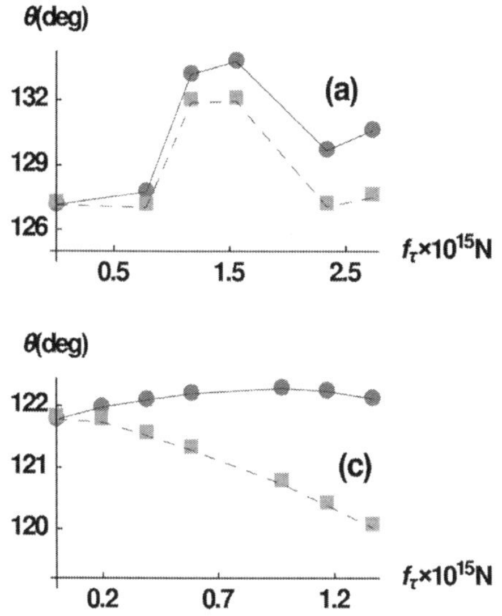

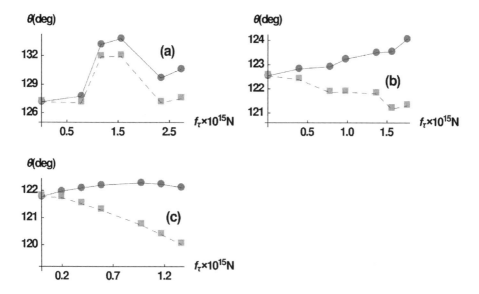

FIG. 12 Calculated advancing (points) and receding (squares) contact angles as functions of external perturbative force for cases C_2 (a), C_3 (b), and C_4 (c). The lines are guides for eye.

Fig. 13 represents the calculated sticking force as function of $\delta \equiv \cos\theta_2 - \cos\theta_1$ for different sizes of the drop and the same thickness of LF (cases C_2 (points), C_3 (squares), and C_4 (diamonds)). The lines are the best linear fits passing through the origin. The results are in agreement with the expression for a drop on a planar solid surface based on classical thermodynamics

$$F_{st} = C\gamma_{lv}\left(\cos\,\theta_2 - \cos\,\theta_1\right) \tag{6}$$

where C is a constant dependent on the drop shape, and γ_{lv} is the liquid-vapor surface tension[19,20] which predicts a linear dependence of the sticking force on δ. (Note, that our previous study of a nanodrop on a rough surface in the absence of a lubricating fluid[1,21] also provided a linear dependence between those two quantities.) Because for a macroscopic drop γ_{lv} as well as C (for a cylindrical drop) are constant the slope of the line does not depend on the size of the drop, and the sticking force is independent of the drop size. However for a nanodrop, one can see from Fig. 13, that the slope of the dependence of F_{st} on δ decreases with increasing drop size (see Table 4 for drop sizes). In terms of macroscopic thermodynamics, such a behavior may be caused by the dependence

TABLE 4

Critical net sticking force, $F_{st,c}$, per unit length in y-direction and critical sticking force per moleculer, $f_{st,c}$, for considered cases

Case	$N_d\sigma$	$F_{st,c}\times 10^5\,Nm^{-1}$	$f_{st,c}\times 10^{15}\,N$
C_1	8.04	9.66	4.09
C_2	8.94	7.17	2.72
C_3	19.89	10.22	1.75
C_4	24.85	9.93	1.36
C_5	9.88	3.94	1.36

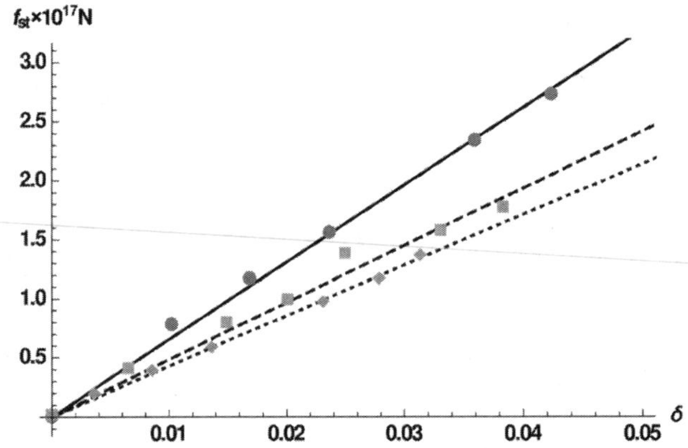

FIG. 13 Dependence of the sticking force on $\delta \equiv \cos\theta_2 - \cos\theta_1$ for the cases C_2 (solid line), C_3 (dashed), C_4 (dotted). The points, squares, and diamonds present results of calculations. The lines are the best linear fits passing through the origin.

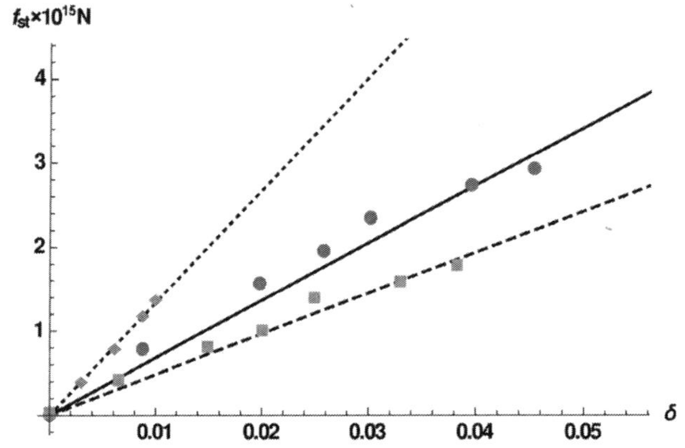

FIG. 14 Dependence of the sticking force on of $\delta \equiv \cos\theta_2 - \cos\theta_1$ for the cases C_1 (solid line), C_3 (dashed), C_5 (dotted). The points, squares, and diamonds present results of calculations. The lines are the best linear fits passing through the origin. The thickness of the lubricating fluid is the smallest for case C_1 and largest for C_5.

of γ_{lv} on the distance from the upper surface of the pillars because of the dependence of the density of LF vapor on this distance.

In Fig. 14 the best linear fits for the sticking force as function of δ are presented for different thicknesses of LF, which correlate with $\rho_{1,av}\sigma^3$ and for the same drop size $(N_d\sigma = 8.8)$ (Cases C_1, C_3, and C_5). In this case, the increase of the thickness of the LF layer results first in a decrease of the slope of F_{st} *vs.* δ dependence (transition from case C_1 (solid line) to case C_3 (dashed line) which is followed by the increase of this slope (transition from case C_3 to case C_5 (dotted line).

4 CONCLUSION

In the present paper, a nanodrop on the surface of a lubricating fluid covering a rough solid surface is considered from a microscopic point of view on the basis of the two-component density functional theory in the presence of an external horizontal perturbative force. For such a system, the comparable values of the thickness of TF-LF interface and of the size of a nanodrop of TF on the surface of LF leads to uncertainty in the definition of the contact angles (advancing and receding) characterizing the nanodrop. Analyzing several possible definitions of the drop profile and of the surface of the lubricating fluid, the most plausible one was selected and the contact angles (advancing and receding) were calculated for various sizes of the drop as functions of the perturbative force. For smaller sizes, this dependence has a maximum, while for larger sizes of the nanodrop it is monotonous. This suggests, that there is a critical size of the drop at which the dependence of contact angles on the external force transforms from non-monotonous to monotonous.

The sticking force which maintains the drop equilibrium is equal in magnitude and opposite in direction to the horizontal external force acting on the drop. It has a linear dependence on $\cos\theta_2 - \cos\theta_1$ which is similar to that for a macrodrop. The maximum possible sticking force (critical sticking force) was calculated as the largest external force at which the Euler-Lagrange equation has a solution.

It is natural to compare the obtained results with those obtained in ref. 1 for a drop on a smooth surface of a solid material covering a rough surface. However because of the substantial difference of the considered systems and interaction parameters, a quantitative comparison is impossible. For this reason, we restrict to a qualitative comparison of the results obtained in ref. 1 and in the present paper.

Two main differences between the drop shapes and composition can be mentioned. A drop of a one-component fluid on the surface of a solid has a flat contact area with the solid and the fluid density distribution in the vertical direction is oscillatory close to the surface of the solid (see, for example, Fig. 6 in ref. 1 and Fig. 5 in ref. 16). In contrast, a drop of a test fluid on the surface of a lubricating fluid has a nonplanar contact area and a nonoscillating behavior of the TF density distribution in the vertical direction (see Fig. 3c, d and 4 of the present paper). Both differences occur because of a layering effect which plays a major role in the region of fluid-solid contact and is weak at a fluid-fluid interface.

REFERENCES

1 G. O. Berim and E. Ruckenstein, *Nanoscale,* 2015, 7, 7873–7884 (Section 4.8 of this volume).
2 T. S. Wong, S. H. Kang, S. K. Y. Tang, E. J. Smythe, B. D. Hatton, A. Grinthal and J. Aizenberg, *Nature,* 2011, **477**, 443–447.
3 W. Ma, Y. Higaki, H. Otsuka and A. Takahara, *Chem. Commun.,* 2013, **49**, 597–599.
4 A. K. Epstein, T. S. Wong, R. A. Belisle, E. M. Boggs and J. Aizenberg, *Proc. Natl. Acad. Sci. U. S. A.,* 2012, **109**, 13182–13187.
5 P. W. Wilson, W. Z. Lu, H. J. Xu, P. Kim, M. J. Kreder, J. Alvarenga and J. Aizenberg, *Phys. Chem. Chem. Phys.,* 2013, **15**, 581–585.
6 J. D. Smith, R. Dhiman, S. Anand, E. Reza-Garduno, R. E. Cohen, G. H. McKinley and K. K. Varanasi, *Soft Matter,* 2013, **9**, 1772–1780.
7 R. Qiu, Q. Zhang, P. Wang, L. Jiang, J. Hou, W. Guo and H. Zhang, *Colloids Surf.,* A, 2014, **453**, 132–141.
8 L. Xiao, J. Li, S. Mieszkin, A. Di Fino, A. S. Clare, M. E. Callow, J. A. Callow, M. Grunze, A. Rosenhahn and P. A. Levkin, *ACS Appl. Mater. Interfaces,* 2013, **5**, 10074–10080.
9 J. Li, T. Kleintschek, A. Rieder, Y. Cheng, T. Baumbach, U. Obst, T. Schwartz and P. A. Levki, *ACS Appl. Mater. Interfaces,* 2013, **5**, 6704–6711.
10 Y. Rosenfeld, *Phys. Rev. Lett.,* 1989, **63**, 980–983.
11 A. González, J. A. White, F. L. Román and S. Velasco, *Phys. Rev. Lett.,* 1997, **79**, 2466–2469.
12 A. González, J. A. White, F. L. Roman and R. Evans, *J. Chem. Phys.,* 1998, **109**, 3637–3650.
13 J. A. White and S. Velasco, *Phys. Rev. E: Stat. Phys., Plasmas, Flu ids, Relat. Interdiscip. Top.,* 2000, **62**, 4427–4430.
14 J. A. White, A. Gonzalez, F. L. Roman and S. Velasco, *Phys. Rev. Lett.,* 2000, **84**, 1220–1223.
15 A.V. Neimark, P. I. Ravikovitch and A. Vishnyakov, *J. Phys.: Condens. Matter,* 2003, **15**, 347.
16 G. O. Berim and E. Ruckenstein, *J. Chem. Phys.,* 2008, **129**, 014708 (Section 4.4 of this volume).
17 A. Miller and E. Ruckenstein, *J. Colloid Interface Sci.,* 1974, **48**, 368; E. Ruckenstein and P. S. Lee, *Surf. Sci.,* 1975, **52**, 298. (Sections 1.1 and 3.1 of this volume)
18 G. O. Berim and E. Ruckenstein, *Nanoscale,* 2015, 7, 3088–3099 (Section 4.6 of this volume).
19 G. L. Furmige, *J. Colloid Sci.,* 1962, **17**, 309.
20 M. Callies and D. Quéré, *Soft Matter,* 2005, **1**, 55.
21 G. O. Berim and E. Ruckenstein, *J. Chem. Phys.,* 2008, **129**, 114709 (Section 4.11 of this volume).

4.10 Size Dependence of the Contact Angle of a Nanodrop in a Nanocavity

*Density Functional Theory Considerations**

Gersh Berim† *and Eli Ruckenstein*
Department of Chemical and Biological Engineering, State University of New York at Buffalo, Buffalo, New York 14260, USA
Corresponding Author
† gberim@buffalo.edu

(Received 28 June 2010; revised manuscript received 22 December 2010; published 10 February 2011)

I. INTRODUCTION

When a small liquid drop is in contact with a smooth homogeneous solid substrate it acquires the shape of a spherical cup, which creates with the solid surface a constant contact angle θ_Y provided by the Young equation $\cos\theta_Y = (\gamma_{sv} - \gamma_{ls})/\gamma_{lv}$, where γ_{sv}, γ_{ls}, and γ_{lv} are the surface tensions for substrate-vapor, substrate-liquid, and liquid-vapor interfaces, respectively. The angle θ_Y was also considered constant when the heterogeneous nucleation of a liquid from vapor was treated by considering the nucleus of the new phase as a uniform drop [1–4]. In the framework of the classical nucleation theory [1–3], this assumption essentially simplifies the calculation of the free-energy barrier for nucleation, which is a function of the unique contact angle given by the Young equation. In the kinetic theory of nucleation [5,6], the constancy of θ_Y simplifies the calculation of the evaporation and condensation rates, which are involved in the rate of nucleation.

However, as well known (see, for example, Refs.[7–12]), for small drops with sizes of several microns on a solid surface, the contact angle θ depends on the radius r of the contact line and hence θ depends on the size of the drop. In the classical theory of wetting, this dependence is accounted for by introducing the line tension τ in a modified Young equation $\cos\theta = \cos\theta_Y - \tau/\gamma_{lv} r$, where $1/r$ is the curvature of the contact line [7–12]. However, when the size of the drop is of the order of a few nanometers (as it is for a nucleus), the thicknesses of the fluid-liquid and fluid-solid interfaces become comparable to the size of the drop, and the classical concepts of surface tensions defined for bulk phases are not applicable. In such cases, the contact angle should be determined differently, on the basis of a microscopic theory. One such theory, the density functional theory (DFT) in a canonical ensemble, was successfully applied to the description of nanodrops on smooth planar solid surfaces as well as to droplike objects (bumps, bridges between walls of nanoslits, etc.) [13–18]. In this paper,

* *Physical Review* E **83**, 021603 (2011). Republished with permission.

DFT is employed to examine a nanodrop in a cavity of a solid surface, focusing on the dependence of the contact angle that the nanodrop makes with the wall of the cavity on the size of the drop. For the sake of simplicity, two-dimensional (cylindrical) cavities and drops will be considered.

II. BACKGROUND

Let us consider a system of finite dimensions L_x and L_h in the x and h directions, respectively, and infinite dimension in the y direction, which contains a one-component fluid of fixed average density ρ_{av} in contact with a solid surface (see Fig. 1 where the y axis is normal to the plane of the figure). The solid contains a cylindrical cavity (groove) of radius R and depth d and is considered to have a uniform density ρ_s. The fluid density distribution (FDD) $\rho(\mathbf{r})$ in such a system is uniform in the y direction and nonuniform in the x and h directions, i.e., $\rho(\mathbf{r}) \equiv \rho(x, h)$. A periodic boundary condition is employed in the x direction, and the upper boundary of the box is treated as a hard wall.

The interactions between the fluid molecules and the fluid molecules and the solid substrate are considered of the Lennard-Jones type with a hard core repulsion $\phi(|\mathbf{r}-\mathbf{r}'|) = 4\epsilon[(\frac{\sigma}{r})^{12}-(\frac{\sigma}{r})^{6}]$ for $r \geqslant \sigma$ and $\phi(|\mathbf{r}-\mathbf{r}'|) = \infty$ for $r < \sigma$, where the coordinates $\mathbf{r}$ and $\mathbf{r}'$ provide the locations of the interacting molecules, $r = |\mathbf{r} - \mathbf{r}'|$, and σ and ϵ are the hard core diameter and the energy parameter, respectively. In the following, the notations ϵ_{ff}, σ_{ff} and ϵ_{fs}, σ_{fs} are used for fluid-fluid and fluid-solid interactions, respectively.

The total Helmholtz free energy $F[\rho(\mathbf{r})]$ of the fluid in the external potential generated by the solid, the Euler-Lagrange equation for FDD, and an outline of the numerical procedure used to solve the Euler-Lagrange equation are provided in Refs. [15] and in the Appendix.

A typical example of a two-dimensional FDD is presented in Fig. 2(a). To determine the profile of a drop on the basis of the obtained FDD, a procedure similar to that used in Refs. [15] was employed. First, the one-dimensional FDD along the h axis (vertical axis of symmetry) was determined from the two-dimensional FDD obtained by solving the Euler-Lagrange equation. An example of such a FDD is presented in Fig. 2(b). In the vicinity of the liquid-solid interface, there are strong density oscillations due to the ordering of the fluid molecules, which form several liquid layers of various densities. At larger distances from the solid, the density becomes almost constant, acquiring liquidlike values and then quickly decreasing across the liquid-vapor interface and

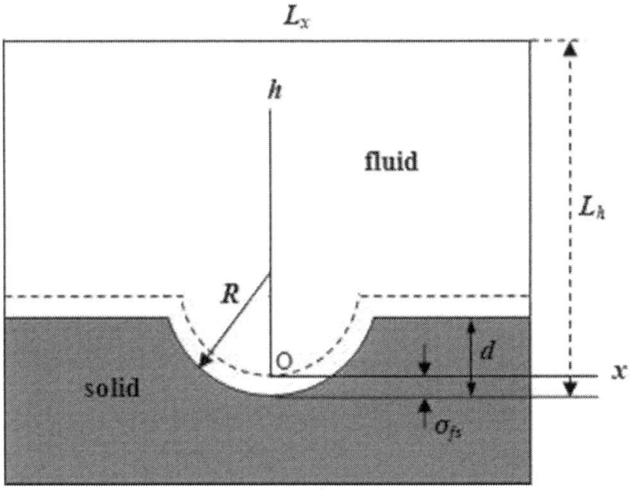

FIG. 1. Schematic presentation of the considered system, which is infinite in the y direction (normal to the plane of the figure) and is periodic with period L_x in the x direction. The dashed line indicates the location of the centers of the fluid molecules in the first (nearest to the solid) layer. Other notations are presented in the text.

approaching vaporlike values. Considering the AB range [see Fig. 2(b)] of the fluid density distribution, which contains the vapor-liquid interface, the location of the equimolar dividing surface and the corresponding density ρ_{div} were determined. The drop profile was extracted from the two-dimensional FDD as a curve along which the local fluid density has the value ρ_{div}. Note that a similar approach for calculating the contact angle was used in Refs. [19–21], where the FDDs were obtained by molecular dynamics simulations.

An example of a drop profile obtained as described above is presented schematically in Fig. 2(c) by the dotted line. Because of the complex structure of this profile near the solid surface, its upper part was approximated by a circle and extended up to the surface of the cavity. The angle θ that this circle makes with the surface of the cavity was considered as the actual contact angle. This procedure is similar to that used in the experimental determination of the contact angle. Note that the change of the location of the dividing surface and the corresponding change of ρ_{div} weakly affect the calculated contact angle. For instance, a change of ρ_{div} by 10% with respect to that obtained for the equimolar dividing surface results in changes of the contact angle of about 0.5%. The size of the cylindrical drop is characterized by the number of molecules N_d it contains per unit length along the y direction of the drop. This number is provided by the expression $N_d = \int_s \rho(x, h)dxdh$, where S is the area between the cavity and the circle that approximates the drop profile. For convenience, the dimensionless quantity $N_d^* \equiv N_d \sigma_{ff}$, which typically has an order of magnitude of 10^2, will be used in the following. This quantity provides the number of molecules in a part of the cylindrical drop of length σ_{ff}.

Note that a droplike solution of the Euler-Lagrange equation can be obtained only if the average fluid density in the system is larger than $\rho_{\text{av},d}$ which depends on the interaction potentials, the geometry of the system, and temperature. For $\rho_{\text{av}} < \rho_{\text{av},d}$, the fluid forms a film. Typically, $\rho_{\text{av},d}\sigma_{ff}^3$ is between 0.04 and 0.05. The obtained droplike solutions are stable in closed systems but unstable in the open ones; similar features were noted for drops on a planar surface in Refs. [13,14] and for bridges and bubbles in cylindrical pores in Ref. [17].

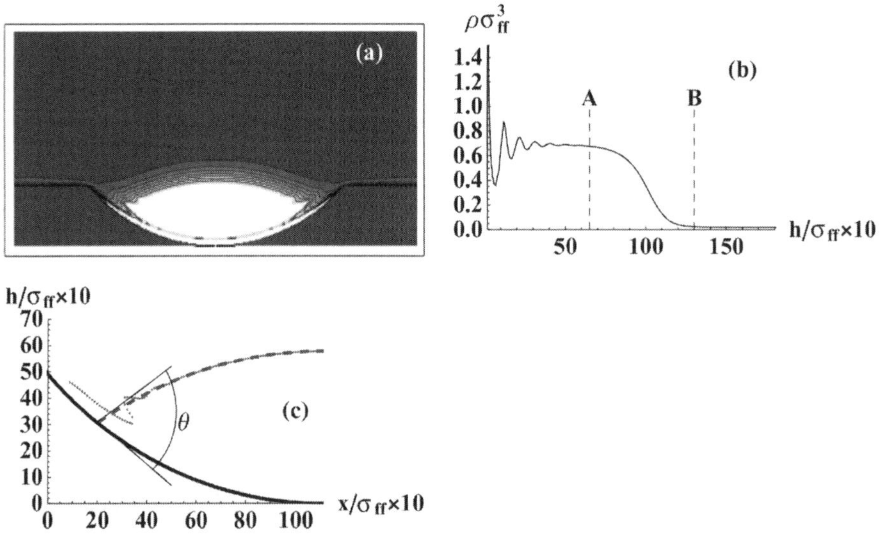

FIG. 2. (Color online) (a) Example of a two-dimensional fluid density distribution (FDD) in a cavity of radius $R = 15\sigma_{ff}$ and depth $d = 5\sigma_{ff}$. The lighter areas correspond to higher fluid densities. (b) One-dimensional FDD along the vertical axis of symmetry of the two-dimensional FDD presented in panel (a). The region AB contains the liquid-vapor interface. (c) Half of the drop profile (dotted line) extracted from the FDD of panel (a). The solid line is the surface of the cavity, and the dashed line represents the circular approximation of the upper part of the drop profile. θ is the contact angle.

III. RESULTS

In this study, argon was selected as the fluid; it has the interaction parameters ϵ_{ff}/k_B = 119.76 K and σ_{ff} = 3.405 Å. The temperatures T = 85 and 95 K were selected. As a supporting solid, the solid carbon dioxide was considered with $\epsilon_{ff} = \epsilon^0_{fs}\left(\epsilon^0_{fs}/k_B = 153.0\,\mathrm{K}\right)$ and $\sigma_{ff} = 3.727$ Å [22]. To examine the effect of the fluid-solid interactions on the FDD in the cavity, other solids were considered, which differ from CO_2 only via the energy parameter ϵ_{fs}; all the other parameters remained unchanged. The width of the system L_x was taken as $L_x = 60\sigma_{ff}$ and the height as $L_h = 20\sigma_{ff} + 2\sigma_{fs}$. Two cavity radii $R_1 = 15\sigma_{ff}$ and $R_2 = 11\sigma_{ff}$ were selected with cavity depths $d_1 = 5\sigma_{ff}$ and $d_2 = 10\sigma_{ff}$, respectively. For comparison purposes, drops on a planar surface were also considered for the same parameters of the fluid-solid interactions.

The size dependence of the contact angle at T = 85 K is presented in Fig. 3 for various cavities and energy parameters ϵ_{fs}. The solid and dashed lines are for cavities with $R = 15\sigma_{ff}$ and $11\sigma_{fs}$, respectively. The curves for different ϵ_{fs} in this figure are plotted for various drop sizes. The upper bounds $N_{d,\max}$ of these sizes correspond to the largest drops with the leading edge located inside the cavity. The lower bounds $N_{d,\min}$ correspond to the smallest drops that can be formed in the cavity. From Fig. 3, one can see that, in all cases, there is a considerable variation $\Delta\theta$ of the contact angle with the drop size. For example, for $\epsilon_{ff} = \epsilon^0_{fs}$ and $R = 15\sigma_{\mathrm{ff}}$, the contact angle increases monotonously from $\theta = 66.3°$ to $116.5°$ ($\Delta\theta = 50.2°$) as N^*_d increases from 20.9 to 127.4, respectively. (The average fluid density in this case changes from $\rho_{\mathrm{av}}\sigma^3_{ff} = 0.042$ to $\rho_{\mathrm{av}}\sigma^3_{ff} = 0.124$.) Note that a nanodrop on a planar surface of the same solid makes with the surface a contact angle $\theta_{\mathrm{pl}} \simeq 52°$, the change of the latter with the size being of the order of only a few degrees. The value of θ_{pl} is close to the value of θ for the smallest of the considered drops in the cavity. This is expected because the smallest drop in the cavity can be approximately considered as located on a planar surface.

For $\epsilon_{ff} = 0.39\epsilon^0_{fs}$ and $R = 15\sigma_{ff}$, the contact angle has a nonmonotonous dependence on the drop size with a much smaller variation $\Delta\theta$ ($\Delta\theta \sim 10°$) (Fig. 3). Such a behavior is the obvious consequence of the weaker fluid-solid interactions. The contact angle θ_{pl} for this case is 137°, which is close to the values of θ for drops in the cavity.

For an intermediate value of the fluid-solid interactions $\left(\epsilon_{fs} = 0.63\epsilon^0_{fs} \text{ and } R = 15\sigma_{ff}, \theta_{pl} \sim 104°\right)$ the contact angle changes from $\theta = 110.5°$ to $\theta = 138.3°$ ($\Delta\theta = 27.8°$) when N^*_d changes from 38.9 to 147.3 (Fig. 3). For the cavity with $R = 11\sigma_{ff}$, the size dependence of the contact angle represented by the dashed curves of Fig. 3 is qualitatively the same as for the cavity with the larger radius ($R = 15\sigma_{ff}$). Comparing the behavior of the contact angle as function of the drop size N_d for cavities with $R = 15\sigma_{ff}$ and $R = 11\sigma_{ff}$, one can see that, for small N_d, the contact angle is smaller for the cavity with the smaller radius. However, this inequality is inverted as N_d increases (see Fig. 3). The

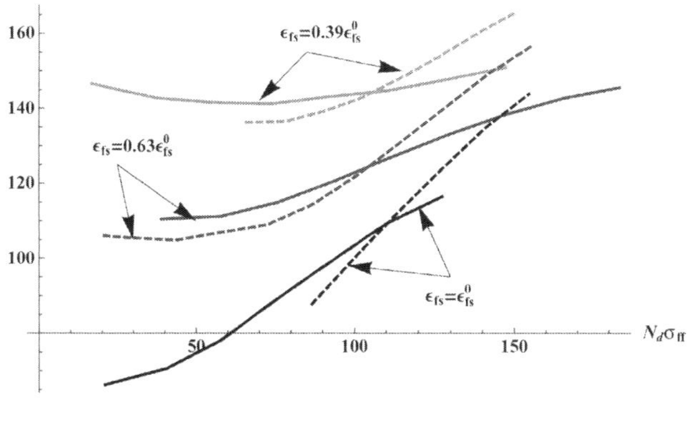

FIG. 3. (Color online) Size dependence of the contact angle in a cavity with $R = 15\sigma_{ff}$ (solid lines) and $R = 11\sigma_{ff}$ (dashed lines) for various values of the energy parameter ϵ_{fs} of the fluid-solid interactions.

inversion occurs for $N_d^* \simeq 10^2$. A summary of the results is provided in Table 1 where, instead of $\Delta\theta$, the average rate $\overline{\Delta\theta} \equiv \Delta\theta/\left(N_{d,max}^* - N_{d,min}^*\right)$ of the variation of $\Delta\theta$ with the change of the number of molecules in the drop is presented. As one can see from Table 1, for a selected cavity, $\overline{\Delta\theta}$ decreases with decreasing ϵ_{fs} but increases with increasing temperature. For the same ϵ_{fs}, $\overline{\Delta\theta}$ increases with decreasing cavity radius.

It is of interest to note that the dependence of the contact angle of a drop on the cavity radius leads, at least in some cases, to a transition from nonwetting ($\theta > 90°$) to wetting ($\theta < 90°$) with decreasing cavity radius (increasing curvature of the surface). See, for example, the case with $\epsilon_{ff} = \epsilon_{fs}^0$ and $N_d^* < 90$ in Fig. 3. The opposite transition from wetting to nonwetting situations with increasing curvatures of the solid was predicted in Ref. [23], where a drop on a cylindrical fiber was considered.

It is instructive to examine the dependence of the contact angle that a drop of given size makes with the cavity on the cavity radius. The results of such calculations for a drop with $N_d^* = 130$ in a cavity of depth $5\sigma_{ff}$ are presented in Table 2 for various substrates (various $\epsilon_{ff}/\epsilon_{fs}^0$). One can see that the contact angle decreases with increasing cavity radius in the direction of the contact angle θ_{pl} of a drop on a planar surface, with the rate of change being the largest for smaller radii. However, even for the largest considered cavity radius $R = 90\sigma_{ff}$, the contact angle does not reach θ_{pl}. In our calculations, the consideration of cavity radii larger than $R = 90\sigma_{ff}$ was not possible because of the small size of the system selected to provide reasonable calculation times. Nevertheless, the results are compatible with the expected contact angle dependence on R.

IV. DISCUSSION

The obtained results show that the contact angle that a nanodrop makes with the surface of a cavity depends on the size N_d of the drop. This dependence becomes more pronounced with increasing ϵ_{fs}, increasing temperature and decreasing cavity radius R. The dependence of the contact angle on the drop size can considerably affect the nucleation rate calculated both by the classical [1] and kinetic [5,6] nucleation theories. In both approaches, the size dependence of the contact angle affects the size dependence of the fluid-liquid and fluid-solid contact areas, which contribute to the free energy of nucleus and hence can change the size of the critical cluster and, as a consequence, the rate of nucleation. In the kinetic theory, in addition, the nonuniformity of the fluid density distribution inside the drop, which can affect the evaporation rate of molecules from the nucleus, should be taken into account.

Because of computational restrictions, the radii of the cavities and the sizes of the drops considered in this paper do not exceed 6–7 nm. Such sizes are much smaller than those for which one

TABLE 1
Summary of Results Obtained for The Size Dependence of the Contact Angle Θ for a Nanodrop in a Nanocavity. All Notations are Explained in the Text.

		T = 85K			T = 95K		
R	$\varepsilon_{fs}/\varepsilon_{fs}^0$	$N_{d,min}\sigma_{ff}$	$N_{d,max}\sigma_{ff}$	$\overline{\Delta\theta}$	$N_{d,min}\sigma_{ff}$	$N_{d,max}\sigma_{ff}$	$\overline{\Delta\theta}$
	1	20.9	127.4	0.471	38.5	91.6	0.683
$15\sigma_{ff}$	0.63	38.9	147.3	0.257	36.7	131.4	0.309
	0.39	16.6	147.2	0.074	35.3	112.9	0.084
	1	86.7	154.6	0.825	77.7	135.5	0.951
$11\sigma_{ff}$	0.63	20.6	156.3	0.376	66.3	140.1	0.633
	0.39	65.9	150.4	0.348	65.2	114.4	0.459

TABLE 2
The Contact Angle for a Drop with $N_{nd}^* = 130$ on Various Substrates and for Various Cavity Radii.

	$\varepsilon_{fs}/\varepsilon_{fs}^0$		
R/σ_{ff}	**0.39**	**0.63**	**1.0**
15	147.7°	133.0°	117.0°
25	144.5°	115.3°	75.8°
45	144.2°	113.9°	71.2°
90	142.3°	114.1°	69.1°
planar surface	137.0°	104.0°	52.0°

expects the classical theory of wetting to be valid. As shown in Refs. [16,24], where the capillary condensation in spherical pores was examined using the classical and the density functional approaches, these approaches provided similar results when the cavity radius was larger than 100 nm.

Note that, for nanodrops, in contrast to macrodrops, the contact angle on a smooth solid surface depends slightly on the size of the drop. For example, at $\epsilon_{fs} = 0.39\epsilon_{fs}^0$, this angle changes from $\theta_1 = 139.4°$ to $\theta_2 = 132.8°$ when the size of the base of the drop changes from $2.6\sigma_{ff} \simeq 9$ Å to $7.4\sigma_{ff} \simeq 25$ Å, and the number of molecules in the drop is changed from $N_d^* \simeq 25$ to $N_d^* \simeq 190$. However, the variation of the contact angle on the smooth surface is much smaller than that in the cavity (compare $\Delta\theta \simeq 7°$ for a planar surface with $\Delta\theta \simeq 35°$ for a cavity with radius $R = 11\sigma_{ff}$).

Note, in conclusion, that the results for the size dependence of the contact angle were obtained here for a cylindrical drop in a cylindrical cavity. However, as previously shown [25] for drops on a planar surface, the difference between the drop profiles of cylindrical and axisymmetrical ones is not large. Therefore, one can expect the size dependence of the contact angle obtained in this paper to remain valid for drops in spherical cavities as well.

APPENDIX: FREE-ENERGY CONTRIBUTIONS AND SOLUTION OF THE EULER-LAGRANGE EQUATION

The total Helmholtz free energy $F[\rho(\mathbf{r})]$ of a fluid in the external potential generated by a solid is expressed as the sum of an ideal gas free energy $F_{id}[\rho(\mathbf{r})]$, a free energy $F_{hs}[\rho(\mathbf{r})]$ of a reference system of hard spheres, a free energy $F_{attr}[\rho(\mathbf{r})]$ due to the attractive interactions between the fluid molecules (in the mean-field approximation), and a free energy $F_{fs}[\rho(\mathbf{r})]$ due to the interactions between the fluid and solid. These contributions to the free energy can be represented as follows [26,27]:

$$F_{id}\left[\rho(\mathbf{r})\right] = k_B T \int d\mathbf{r}\,\rho(\mathbf{r})\left\{\log\left[\Lambda^3\rho(\mathbf{r})\right]-1\right\}, \tag{A1}$$

$$F_{hs}\left[\rho(\mathbf{r})\right] = \int d\mathbf{r}\rho(\mathbf{r})\Delta\Psi_{hs}(\mathbf{r}), \tag{A2}$$

where $\Lambda = h_P/(2\pi m k_B T)^{1/2}$ is the thermal de Broglie wavelength, h_P and k_B are the Planck and Boltzmann constants, respectively, T is the absolute temperature, m is the mass of a fluid molecule,

$$\Delta\Psi_{hs}(\mathbf{r}) = k_B T \eta_{\bar{\rho}} \frac{4-3\eta_{\bar{\rho}}}{\left(1-\eta_{\bar{\rho}}\right)^2}, \tag{A3}$$

$\eta_{\bar{\rho}} = \frac{1}{6}\pi\bar{\rho}(\mathbf{r})\sigma_{ff}^3$ is the packing fraction of the fluid molecules, σ_{ff} is the fluid hard core diameter, and $\bar{\rho}(\mathbf{r})$ is the smoothed density defined as

$$\bar{\rho}(\mathbf{r}) = \int d\mathbf{r}'\rho(\mathbf{r}')W(|\mathbf{r}-\mathbf{r}'|) \tag{A4}$$

The weighting function $W(|\mathbf{r} - \mathbf{r}'|)$ is selected in the form [28]

$$W(|\mathbf{r}-\mathbf{r}'|) = \begin{cases} \dfrac{3}{\pi\sigma_{ff}^3}\left(1-\dfrac{r}{\sigma_{ff}}\right), & r \leqslant \sigma_{ff} \\ 0, & r > \sigma_{ff} \end{cases}$$

where $r = |\mathbf{r} - \mathbf{r}'|$.

The contribution to the excess free energy due to the attraction between the fluid-fluid molecules is calculated in the mean-field approximation

$$F_{attr}[\rho(\mathbf{r})] = \frac{1}{2}\iint d\mathbf{r}\, d\mathbf{r}'\rho(\mathbf{r})\rho(\mathbf{r}')\phi_{ff}(|\mathbf{r}-\mathbf{r}'|), \tag{A5}$$

where $\phi_{ff}(|\mathbf{r}-\mathbf{r}'|)$ is the potential of the fluid-fluid interactions.

The last contribution $F_{fs}[\rho(\mathbf{r})]$ is given by the expression

$$F_{fs}[\rho(\mathbf{r})] = \int_V d\mathbf{r}\,\rho(\mathbf{r})U_{fs}(\mathbf{r}), \tag{A6}$$

where V is the volume occupied by the fluid and $U_{fs}(\mathbf{r})$ is the net potential generated by the solid. This potential is given by

$$U_{fs}(\mathbf{r}) = \int_{V_s} \rho_s(\mathbf{r}')\phi_{fs}(|\mathbf{r}-\mathbf{r}'|)d\mathbf{r}', \tag{A7}$$

where $\phi_{fs}(|\mathbf{r}-\mathbf{r}'|)$ is the potential of the fluid-solid interactions, V_s is the volume of the solid, $\rho_s(\mathbf{r}')$ is the density of the solid, which, in general, depends on coordinates. Note that the integration over y and x coordinates in the three-dimensional integral of Eq. (A7) can be carried out in analytical form, whereas the integration over h can be performed only numerically.

The Euler-Lagrange equation for the fluid density distribution $\rho(x, h)$ obtained by minimizing the Helmholtz free energy can be represented in the following general form:

$$\log\left[\Lambda^3\rho(x,h)\right] - Q(x,h) = \frac{\lambda}{k_B T}, \tag{A8}$$

where λ is the Lagrange multiplier and the function $Q(x, h)$ is given by

$$Q(x,h) = -\frac{1}{k_B T}\left[\Delta\Psi_{hs}(x,h) + \overline{\Delta\Psi'_{hs}(x,h)} + U_{ff}(x,h) + U_{fw}(x,h)\right], \tag{A9}$$

where

$$U_{ff}(x,h) = \iint dx'\, dh' \rho(x',h') \phi_{ff,y}\left(|x-x'|,|h-h'|\right), \quad \text{(A10)}$$

$$\overline{\Delta\Psi'}_{hs}(x,h) = \iint dx'\, dh' \rho(x',h') W_y\left(|x-x'|,|h-h'|\right)$$

$$\times \frac{\partial}{\partial \bar{\rho}} \Delta\Psi_{hs}(\bar{\rho})|_{\rho=\rho(x',h')}, \quad \text{(A11)}$$

$\phi_{ff,y}\left(|x-x'|,|h-h'|\right)$ and $W_y\left(|x-x'|,|h-h'|\right)$ are obtained by integrating the potential $\phi_{ff}\left(|\mathbf{r}-\mathbf{r}'|\right)$ and the weighted function $\mathrm{W}\left(|\mathbf{r}-\mathbf{r}'|\right)$ with respect to y from $-\infty$ to $+\infty$.

When calculating the term $U_{ff}(x, h)$ of the Euler-Lagrange equation arising due to the long-range fluid-fluid interactions, a cutoff at a distance equal to four molecular diameters σ_{ff} for the range of Lennard-Jones attraction was employed. The increase of this distance up to $10\sigma_{ff}$ changed the results by less than 1%.

In Eq. (A8), the Lagrange multiplier λ arises because of the constraint of fixed average density of the fluid, which has the form

$$\rho_{av} = \frac{1}{V}\int_V d\mathbf{r}\, \rho(\mathbf{r}) \quad \text{(A12)}$$

and leads to the following expression for λ:

$$\lambda = -k_B T \log\left(\frac{1}{\rho_{av} V \Lambda^3}\int_V d\mathbf{r} e^{Q(x,h)}\right). \quad \text{(A13)}$$

By eliminating λ between Eqs. (A8) and (A13), one obtains an integral equation for the FDD $\rho(x, h)$, which was solved by iteration.

The initial FDD in the iteration procedure (initial guess) was selected to be uniform inside the cavity, with a liquidlike density ($\rho\sigma_{ff}^3 \sim 0.5$). Outside the cavity, it was taken as uniform with a vaporlike density ($\rho\sigma_{ff}^3 \sim 0.04 \div 0.10$). By changing the fluid densities inside and outside the cavity in the initial guess (i.e., by changing the average density in the system), drops of various sizes could be obtained.

To avoid the divergence of the iteration procedure, the input density profile $\rho_i^{in}(x,h)$ for the $(i + 1)$ th iteration $\rho_{i+1}(x, h)$, generated by the Euler-Lagrange equation, was selected as follows [26]:

$$\rho_i^{in}(x,h) = (1-\gamma)\rho_{i-1}^{in}(x,h) + \gamma\rho_i(x,h), \quad \text{(A14)}$$

where $\rho_i(x, h)$ is the i th iteration and the constant $\gamma = 0.1$. As a measure of the precision of the iterations, the dimensionless quantity

$$\delta = \int_V dx\, dh\left[\rho_{i+1}(x,h) - \rho_i^{in}(x,h)\right]^2 \Bigg/ \left(\int_V dx\, dh\, \rho_i(x,h)\right)^2$$

was introduced. The iterations were carried out on a two-dimensional grid with a spacing equal to $0.1\sigma_{ff}$ until δ became smaller than $\epsilon = 10^{-7}$. The additional increase in precision did not lead to appreciable changes in the density profile.

REFERENCES

[1] Q. X. Liu, Y. J. Zhu, G. W. Yang, and Q. B. Yang, J. Mater. Sci. Technol. **24**, 183 (2008).
[2] S. J. Cooper, C. E. Nicholson, and J. Liu, J. Chem. Phys. **129**, 124715 (2008).
[3] M. Qian and J. Ma, J. Chem. Phys. **130**, 214709 (2009).
[4] E. Ruckenstein and G. O. Berim, J. Colloid Interface Sci. **355**, 259 (2011).
[5] B. Nowakowski and E. Ruckenstein, J. Phys. Chem. **96**, 2313 (1992).
[6] E. Ruckenstein and B. Nowakowski, Langmuir **8**, 1470 (1992).
[7] J. Gaydos and A. W. Neumann, J. Colloid Interface Sci. **120**, 76 (1987).
[8] F. Y. H. Lin, D. Li, and A. W. Neumann, J. Colloid Interface Sci. **159**, 86 (1993).
[9] L. Schimmele, M. Napiorkowski, and S. Dietrich, J. Chem. Phys. **127**, 164715 (2007).
[10] A. Checco, P. Guenoun, and J. Daillant, Phys. Rev. Lett. **91**, 186101 (2003).
[11] A. Marmur, J. Colloid Interface Sci. **186**, 462 (1997).
[12] A. Marmur, Colloids Surf. A **136**, 81 (1998).
[13] V. Talanquer and D. W. Oxtoby, J. Chem. Phys. **104**, 1483 (1996).
[14] V. Talanquer and D. W. Oxtoby, J. Chem. Phys. **114**, 2793(2001).
[15] G. O. Berim and E. Ruckenstein, J. Chem. Phys. **129**, 014708 (2008) (Section 4.4 of this volume).
[16] A. V. Neimark, P. I. Ravikovitch, and A. Vishnyakov, Phys. Rev. E **65**, 031505 (2002).
[17] A. Vishnyakov and A. V. Neimark, J. Chem. Phys. **119**, 9755 (2003)
[18] F. Ancilotto, M. Barranco, E. S. Hernandez, A. Hernando, and M. Pi, Phys. Rev. B **79**, 104514 (2009).
[19] C. D. Daub, J. Wang, S. Kudesia, D. Bratko, and A. Luzar, Faraday Discuss. **146**, 67 (2010).
[20] N. Giovambattista, P. G. Debenedetti, and P. J. Rossky, J. Phys. Chem. B **111**, 9581 (2007).
[21] M. J. de Ruijter, T. D. Blake, and J. De Coninck, Langmuir **15**, 7836 (1999).
[22] R. Evans and P. Tarazona, Phys. Rev. A **28**, 1864 (1983).
[23] A. V. Neimark, J. Adhes. Sci. Technol. **13**, 1137 (1999).
[24] P. I. Ravikovitch and A. V. Neimark, Langmuir **18**, 1550 (2002).
[25] G. O. Berim and E. Ruckenstein, J. Phys. Chem. B **108**, 19330 (2004) (Section 3.2 of this volume).
[26] .P. Tarazona, Phys. Rev. A **31**, 2672 (1985).
[27] P. Tarazona *et al.,* Mol. Phys. **60**, 573 (1987).
[28] R. H. Nilson and S. K. Griffiths, J. Chem. Phys. **111**, 4281 (1999).

4.11 Microscopic Calculation of the Sticking Force for Nanodrops on an Inclined Surface*

Gersh Berim and Eli Ruckenstein†
Department of Chemical and Biological Engineering, State University of New York at Buffalo, Buffalo, New York 14260, USA
Corresponding Author
† E-mail: feaeliru@buffalo.edu. Tel: (716)645-1179, FAX: (716)645-3822.

(Received 19 June 2008; accepted 15 August 2008; published online 18 September 2008)

I. INTRODUCTION

The problem of stability of a drop on an inclined solid surface in the presence of gravity is an important one from a theoretical point of view and its solution can be useful in the design of superhydrophobic materials for various applications, such as the manufacturing of windshields, solar cell panels, and self-cleaning surfaces.[1,2] It was understood a long time ago[3] that a drop on an inclined surface can be stable (metastable) only if the latter is physically or chemically heterogeneous and that it slips (or rolls) down on a homogeneous surface. The roughness generates a sticking force which can fix a drop even on a vertical surface.[1] The explanation of this phenomenon on the basis of a macroscopic theory involves the presence of a range of values of the contact angles which a drop can make with a rough surface (contact angle hysteresis).

Most theoretical studies regarding the macroscopic drop on a solid surface in the absence of gravity involve the assumptions that (i) the size of the drop is much larger than the characteristic size of the roughness, (ii) the liquid density is uniform across the drop, and (iii) the drop has a shape close to spherical. These assumptions allow one to consider that the contact angle θ which a drop makes with the surface is a unique characteristic which determines the behavior of a given liquid on a given surface. For horizontal ideal nonrough surfaces, the contact angle of macroscopic drops does not depend either on the size of the drop or on its location on the surface and is determined by the surface tensions γ_{sv}, γ_{sl}, and γ_{lv}, for the solid-vapor, solid-liquid, and liquid-vapor interfaces, respectively, according to the Young equation

$$\gamma_{lv} \cos\theta = \gamma_{sv} - \gamma_{sl}. \tag{1}$$

On nonideal inclined surfaces and in the presence of gravity, two contact angles, the receding (θ_r) and the advancing (θ_a), are considered to occur for a macroscopic drop at the stability limit, which are assumed to be independent of the size of the drop. The difference between the advancing and receding contact angles is caused by the gravity and the sticking force that ensures the mechanical

* *The Journal of Chemical Physics* 129, 114709 (2008). Republished with permission.

equilibrium of the drop and determines the largest possible inclination angle α_c of the surface. The latter angle is provided by the expression[1,4]

$$Ma\sin\alpha_c = C\gamma_{lv}\left(\cos\theta_r - \cos\theta_a\right), \tag{2}$$

where M is the mass of the drop, a is the gravitational acceleration, and the constant C incorporates the geometrical characteristics of the drop. If the inclination angle α of the solid surface is smaller than α_c, the drop is in mechanical equilibrium and the tangential component, f_τ, of the force of gravity is balanced by the sticking force, f_{st}, caused by the interactions in the system and by the presence of roughness. In this case, f_τ can be calculated using the following expression similar to Eq. (2):

$$f_\tau = Ma\sin\alpha = C\gamma_{lv}\left(\cos\theta_2 - \cos\theta_1\right)\left(f_{st} = f_\tau\right), \tag{3}$$

where α is the inclination angle of the surface, $\theta_1 \leqslant \theta_a$ is the contact angle at the lower edge of the drop, and $\theta_2 \geqslant \theta_r$ is the contact angle at the upper edge (Fig. 1).

Some comments should be made regarding the concept of contact angle. It was shown by Ruckenstein and Lee,[5] who analyzed the equilibrium profile of a liquid drop near a horizontal solid surface, that the shape of the drop is not circular and that the true contact angle, which was called microcontact angle, does not coincide with that provided by the Young equation. It was also shown later[6] on the basis of the minimization of the potential energy of a drop, with the assumptions of uniform density and sharp liquid-vapor interface, that the true (microcontact) angle is equal to 180°, independent of the strength of the liquid-solid interactions. Only if the density of the liquid near the solid surface is considered different from that in the bulk of the drop, the microcontact angle becomes different from 180°.[7]

The problem becomes even more intricate if one considers a drop of several nanometers size. Molecular dynamics (MD) and Monte Carlo (MC) simulations,[8–10] as well as calculations based on density functional theories (DFTs),[11–14] have shown that in this case the thickness of the liquid-vapor interface becomes comparable with the dimension of the drop and that there are strong oscillations of density near the solid surface. In this case, it is even difficult to define a contact angle because of the absence of a unique clear definition for the liquid-vapor interface.

Unlike the large number of publications that consider a drop profile on a horizontal surface in the absence of gravity, the presence of gravity and surface inclination was accounted for in relatively few theoretical papers, which employed macroscopic approaches (see, e.g., Refs. 4 and 15–20). One of the reasons is that the shape of a drop in a gravitational field is not spherical and it is extremely difficult to find the real drop profile. Even for a two-dimensional drop, one has to look for a nonstandard procedure to obtain a solution for the drop profile (see, for example, Ref. 20).

A drop on a solid surface can also be examined using a microscopic approach based on the minimization of the free energy of the system. The advantage of such an approach is that it does not require the *a priori* knowledge of macroscopic characteristics such as the contact angles and surface tensions. In addition, Eqs. (1) and (3) themselves should follow from the basic equations of the microscopic theory. A microscopic treatment based on first principles can also provide information

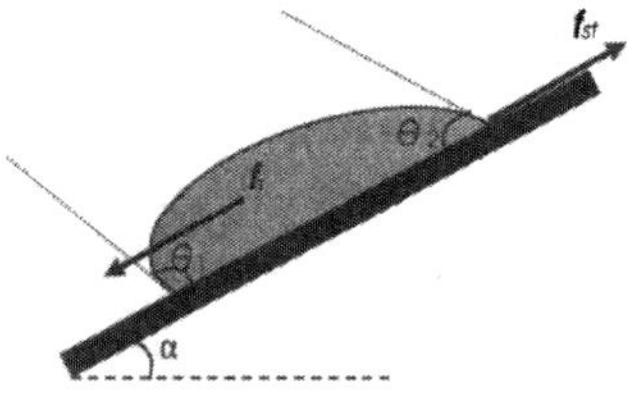

FIG. 1. Liquid drop in mechanical equilibrium on an inclined surface with inclination angle α. f_τ and f_{st} $\left(f_\tau = f_{st}\right)$ are the tangential components of the force of gravity and the sticking force, respectively. The front, θ_1, and rear, θ_2, contact angles for the stability limit are replaced by the advanced and receding contact angles, respectively.

regarding the relation between the sticking force and the parameters of the liquid-liquid and liquid-solid interactions, as well as about the structure of the liquid in the drop.

While widely applied to fluids confined in nanopores,[21–25] the application of the microscopic approach to the description of droplike aggregates has a relatively short history. For example, in Refs. 11–14, 26, and 27 an approach involving the canonical ensemble DFT was used to examine the bumps and bridges in planar and cylindrical nanoslits with smooth walls. A simplified DFT for microscopic drops was developed in Ref. 28. The applicability of the canonical ensemble DFT and the stability of the nonuniform (droplike) fluid density distributions (FDDs) were examined in Refs. 12 and 28–30. In the present paper, liquid drops on chemically or physically rough and inclined solid surfaces are examined using the nonlocal DFT formulated in Refs. 21 and 23. Computational restrictions have not allowed us to examine large three-dimensional drops; therefore, to make the calculations feasible, two-dimensional (cylindrical) nanodrops on a vertical surface were the only ones considered.

This paper is organized as follows. The description of the system and the interaction potentials are provided in Secs. II A and II B. In Sec. II C the Euler–Lagrange equation for the FDD and its solution are outlined. The procedures to obtain the contact angles and to calculate the sticking force from the FDD are examined in Secs. II D and II E, respectively. A drop on a rough horizontal surface is discussed in Sec. III A, and Sec. III B contains the results of the calculations regarding the sticking force and contact angles for a drop on a vertical surface for both chemical and physical roughness. In addition, the applicability of a macroscopic equation to nanodrops is discussed in Sec. III B.

II. GENERAL CONSIDERATIONS

A. The System

The considered system, which involves a one-component fluid of fixed average density ρ_{av} in contact with a rough solid surface inclined at an angle α with respect to the horizontal surface, is presented in Fig. 2. It has the finite lengths L_x and L_h in the x- and h-directions, respectively, and an infinite dimension in the y-direction which is perpendicular to the plane of the figure. The roughness (chemical inhomogeneities or pillars) on the solid surface is uniform in the latter direction. The FDD $\rho(\mathbf{r})$ in such a system is considered uniform in the y-direction and nonuniform in the x- and h-directions, i.e., $\rho(\mathbf{r}) \equiv \rho(x, h)$. A periodic boundary condition is assumed in the x-direction and the upper boundary of the system is treated as a hard wall (without attractive interactions with the fluid molecules). The above conditions are selected to make the treatment consistent with the MC and MD simulations of confined fluids.[9,25,31,32] The fluid is exposed to an external potential due to the fluid-solid interactions and gravity. The temperature is fixed at $T = 87.3$ K.

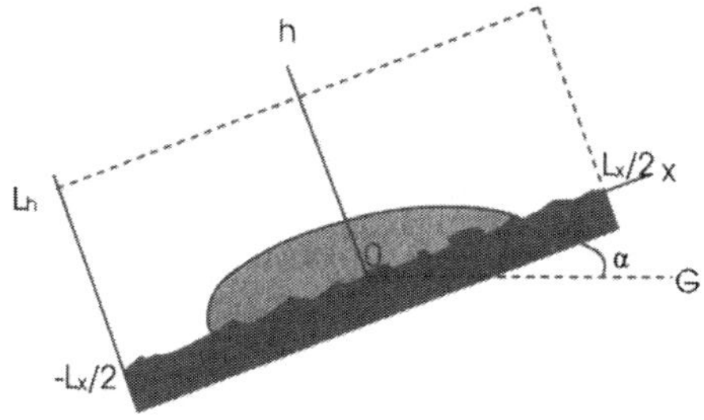

FIG. 2. Schematic representation of a drop on a rough surface with inclination angle α. For the physical roughness, the solid for $h < 0$ is homogeneous. For the chemical roughness, the solid surface is flat (no physical roughness), but the solid for $h < 0$ is inhomogeneous. The y axis is perpendicular to the plane of the figure.

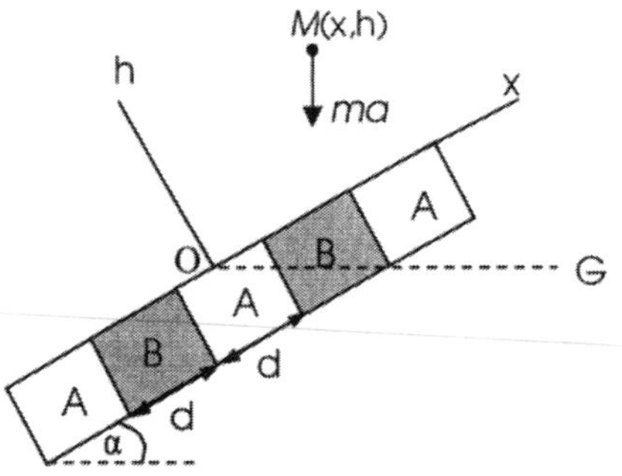

FIG. 3. Schematic representation of a chemically rough inclined planar surface in the *x*-*h*. plane. Plates *A* and *B* are infinite in the *y*-direction, which is perpendicular to the plane of the figure, semi-infinite in the negative *h*-direction, and have different energy interaction parameters $\epsilon_{fs,A}$ and $\epsilon_{fs,B}$. The solid is symmetrical with respect to the *y*-*h* plane. The potential energy due to gravity has zero value at the horizontal *OG*-*y* plane.

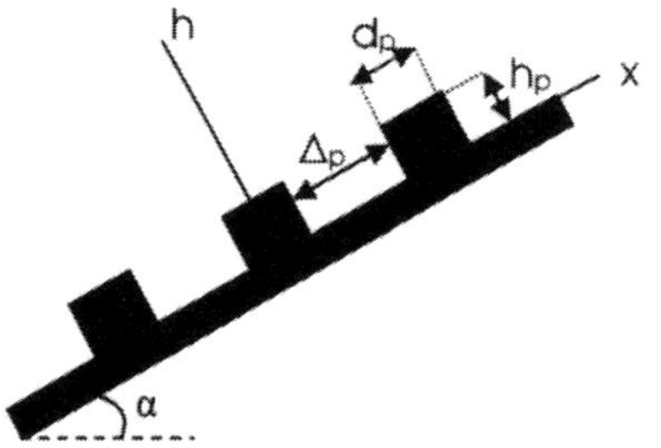

FIG. 4. Schematic representation of a physically rough inclined surface in the *x*-*h* plane. The pillars are infinite in the *y*-direction which is perpendicular to the plane of the figure. The solid is symmetrical with respect to the *y*-*h* plane.

As already mentioned, chemically and physically rough surfaces will be considered. A chemically rough surface is composed of a sequence of two kinds of plates, *A* and B, of equal thickness, *d*, which are made of different materials and interact differently with the fluid molecules. The plates are infinite in the *y*-direction and semi-infinite in the *h*-direction (Fig. 3), and the origin of the reference frame is located in the middle of plate *A*. The surface is assumed to be symmetrical with respect to the *y*-*h* plane.

A physically rough surface is modeled by decorating a planar smooth surface with a regular array of pillars (see Fig. 4). The pillars are assumed to have infinite length in the *y*-direction, a width d_p, a height h_p, and a distance Δ_p between them. Due to the regular geometrical location of plates or pillars, both surfaces (chemically or physically rough) generate potentials which are periodic in the *x*-direction.

In both cases presented in Figs. 3 and 4, the origin of the coordinate system is located at a distance (not shown in the figures) from the main solid body equal to one hard core diameter σ_{fs} of the fluid-solid interaction.

B. Interaction Potentials

The interaction potentials between the molecules of fluid (argon) and between those molecules and the solid (carbon dioxide is considered as one of the solids, but another solid is also considered) are selected in the Lennard-Jones form

$$\phi_b\left(\left|\mathbf{r}-\mathbf{r}'\right|\right)=\begin{cases}4\epsilon_b\left[\left(\frac{\sigma_b}{r}\right)^{12}-\left(\frac{\sigma_b}{r}\right)^{6}\right], & r\geqslant\sigma_b\\ \infty, & r<\sigma_b,\end{cases} \tag{4}$$

where the subscript b should be replaced by ff (fs) for the fluid-fluid (fluid-solid) interactions, ϵ_{ff} and ϵ_{fs} are energy parameters, σ_{ff} and σ_{fs} are hard core diameters, $\mathbf{r}$ and $\mathbf{r}'$ provide the locations of the interacting molecules, and $r = |\mathbf{r} - \mathbf{r}'|$. For argon, $\epsilon_{\text{ff}}/k_B = 119.76$ K and $\sigma_{\text{ff}} = 3.405$ Å, where k_B is the Boltzmann constant; for solid carbon dioxide $\epsilon_{\text{fs}}/k_B = \epsilon_{\text{fs}}^0/k_B = 153.0$ K, $\sigma_{\text{fs}} = 3.727$ Å, and the number density $\rho_s = 1.91 \times 10^{28}$ m^{-3}.[33] The interaction of a molecule of fluid with a semi-infinite solid possessing an ideal surface is given by the potential

$$\psi_{\text{fs}}(h)=\frac{2\pi}{3}\epsilon_{\text{fs}}\rho_s\sigma_{\text{fs}}^3\left[\frac{2}{15}\left(\frac{\sigma_{\text{fs}}}{\sigma_{\text{fs}}+h}\right)^9-\left(\frac{\sigma_{\text{fs}}}{\sigma_{\text{fs}}+h}\right)^3\right], \tag{5}$$

where $h + \sigma_{\text{fs}}$ is the distance of the molecule from the surface.

The gravitational potential is calculated with respect to the horizontal plane OG-y in Fig. 3. If a molecule is located at point $M(x, h)$, its potential energy $U_G(x, h)$ due to gravity is given by the equation

$$U_G(x,h)=ma(x\sin\alpha+h\cos\alpha), \tag{6}$$

where m is the mass of a fluid molecule.

The force of gravity acting on a molecule is extremely small. For this reason, the number of molecules in a drop, which can roll down on a surface under the action of gravity, should be very large. To estimate this number let us consider, for simplicity, a two-dimensional drop on a vertical surface ($\alpha = 90°$) and evaluate with Eq. (3) the sticking force. The constant C in Eq. (3) can be considered equal to the unit of length, and we replace the cosine difference also by unity. Using for the surface tension the value $\gamma_{\text{lv}} \simeq 1.4\times10^{-2}$ N/m for argon at $T = 85$ K,[34,35] one finds that the sticking force (per unit length) is 1.4×10^{-2} N/m. Dividing this force by mg, where $m = 6.6 \times 10^{-26}$ kg is the mass of a molecule of argon, one finds that the number of molecules in a drop should be greater than 2×10^{23} for the drop to be set in motion on the surface. Because the number of molecules in the nanodrops considered in our DFT based calculations was of the order of 10^{11}–10^{12},[14] it is impossible to detect the effect of gravity on such a nanodrop for the real value of a ($a = g$) and to check the validity of Eq. (3) for nanodrops. For the above values of the parameters the nanodrops will be always sticked to the solid surface, and the very small difference between θ_1 and θ_2 could not be calculated because of the limited precision of the numerical procedure. However, one can perform DFT calculations for other values of a (even for unrealistically large) to detect changes in the shape of the drop profile. These values will be taken in the present paper of the order of $a_0 = 10^{10}$ m/s^2. This extremely large value compensates for the small number of molecules which we have to consider for computational restrictions. Still we will refer to this acceleration as gravity.

The external potential U_{sf} generated by a chemically rough surface at a point $M(x, h)$ depends both on the distance h of this point from the surface and the position of point M along the surface, i.e., $U_{\text{sf}} \equiv U_{\text{sf}}(x, h)$. Obviously, U_{sf} is a periodic function of x with a period equal to $2d$. Some details regarding the calculation of $U_{\text{sf}}(x, h)$ were provided in Ref. 14.

Below, two kinds of chemically rough surfaces are considered. In the first ($S1$), the A-plates are made up of solid carbon dioxide $(\epsilon_{\text{fs},A} = \epsilon_{\text{fs}}^0)$ and the B-plates are made up of another (fictitious) material which differs from the solid carbon dioxide only in the value of the energy parameter

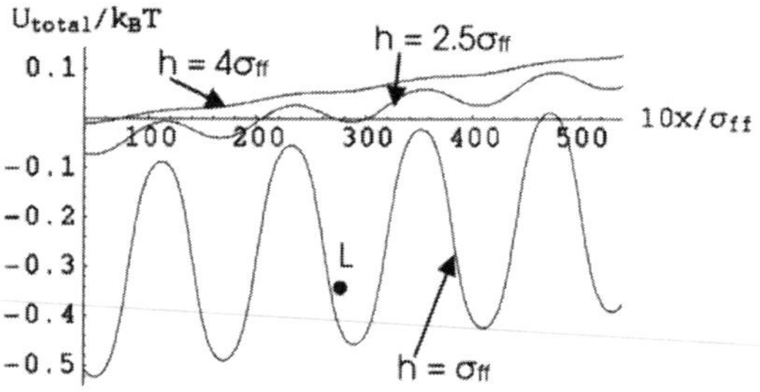

FIG. 5. Dependence of the total external potential, to which a molecule of fluid is subjected, on x for $\alpha = 90°$, $a = 50a_0$, and several values of the distance h from surface $S1$. k_B is the Boltzmann constant and T is the absolute temperature.

$\epsilon_{fs,B}\left(\epsilon_{fs,B} = 0.1\epsilon_{fs}^0\right)$. The thickness of each of the two plates is $d = 6\sigma_{ff}$. For the second surface ($S2$), the energy parameter $\epsilon_{fs,A}$ is selected equal to $\epsilon_{fs,A} = 0.6\epsilon_{fs}^0$, all other parameters remaining those of surface $S1$.

For the surface $S1$, the total potential to which a molecule of fluid is exposed is presented schematically in Fig. 5 as function of x for $\alpha = 90°$ (vertical surface) and several values of the distance h from the surface. The "gravitational" acceleration was taken in this example equal to $14.7a_0$. The total potential for $h < 4\sigma_{ff}$ possesses multiple minima separated by potential barriers, the height of which decreases as h increases. Due to the presence of gravity, the minimum values of the potential gradually increase with increasing x. For $h > 4\sigma_{ff}$, the oscillations due to the surface inhomogeneity become smaller and the total potential increases monotonously as x increases. Qualitatively, the potentials for inclination angles smaller than 90° or for more realistic gravitational accelerations are expected to have similar features. One should note that, at a fixed h, the height of the potential barrier depends on the difference between the energy parameters $\epsilon_{fs,A}$ and $\epsilon_{fs,B}$. The smaller this difference is, the smaller the height of the barrier is; for $\epsilon_{fs,A} = \epsilon_{fs,B}$ there are no minima.

The form of the potential presented in Fig. 5 allows one to explain qualitatively the sticking force which arises due to the interaction of a fluid molecule with the rough solid. Indeed, when the leading edge of the drop on the inclined surface is located near one of the minima of U_{fs}, say, at point L in Fig. 5, this potential generates a force in the positive direction of the x axis which opposes the motion of the drop down (in the negative x-direction) along the inclined surface. Only a large enough gravity can overcome the potential barrier and set the drop in motion. On a smooth surface (without roughness), there is no potential barrier for the drop motion and its mechanical equilibrium on an inclined surface becomes impossible.

Because of the periodicity of the solid-liquid potential in the x-direction, any drop has the same energy when it is displaced along the surface over an integer number of periods. However, the potential energy due to the gravity decreases when the drop moves in the negative x-direction. This means that any sticked state of the drop on an inclined surface is a metastable one.

The above qualitative considerations remain valid for physically rough surfaces. The contribution $U_{fp,\,tot}(x, h)$ of the pillars to the external potential generated by the solid is also a periodic function of x (see Ref. 14 for details).

C. The Euler–Lagrange Equation for the Fluid Density Distribution

The equation for the FDD $\rho(x, h)$ is obtained through the minimization of the total Helmholtz free energy $F[\rho(\mathbf{r})]$, which can be expressed as the sum of an ideal gas free energy $F_{id}[\rho(\mathbf{r})]$, an excess free energy $F_{hs}[\rho(\mathbf{r})]$ (with respect to the ideal gas) of a reference system of hard spheres, a free energy $F_{attr}[\rho(\mathbf{r})]$ due to the attractive interactions between the fluid molecules (in the mean field approximation), a free energy $F_{fs}[\rho(\mathbf{r})]$ due to the interactions between the fluid and solid, and a free energy $F_g[\rho(\mathbf{r})]$ due to the gravity. Explicit expressions for these contributions, which are obtained in the standard way,[21] are provided in the Appendix. Here we consider briefly only the contribution $F_{fs}[\rho(\mathbf{r})]$, which is provided by the expression

$$F_{fs}\left[\rho(\mathbf{r})\right]=\int_V d\mathbf{r}\rho(\mathbf{r})U_{fs}(\mathbf{r}), \tag{7}$$

where V is the volume of the system and $U_{fs}(\mathbf{r})$ is the potential generated by the solid. This potential can be obtained by integrating the Lennard-Jones potential [Eq. (4)] for the fluid-solid interactions over the entire volume V_s of the solid

$$U_{fs}(\mathbf{r})=\int_V \rho_s(\mathbf{r}')\phi_{fs}\left(|\mathbf{r}-\mathbf{r}'|\right)d\mathbf{r}', \tag{8}$$

where $\rho_S(\mathbf{r}')$ is the density of the solid which, in general, depends on the coordinates.

The Euler–Lagrange equation for $\rho(x, h)$ can be written in the following form:[14]

$$\log\left[\Lambda^3\rho(x,h)\right]-Q(x,h)=\frac{\lambda}{k_BT}, \tag{9}$$

where the function $Q(x, h)$, which is a functional of $\rho(x, h)$, is given in the Appendix [Eq. (A7)], $\Lambda = h_P/(2\pi m k_B T)^{1/2}$ is the thermal de Broglie wavelength, k_B and h_P are the Boltzmann and Planck constants, respectively, T is the absolute temperature, m is the mass of a fluid molecule, and λ is a Lagrange multiplier arising because of the constraint of fixed average density of the fluid. This constraint has the form

$$\rho_{av}=\frac{1}{V}\int_V d\mathbf{r}\rho(\mathbf{r}) \tag{10}$$

and leads to the following expression for λ:

$$\lambda=-k_BT\log\left[\frac{1}{\rho_{av}V\Lambda^3}\int_V d\mathbf{r}e^{Q(x,h)}\right]. \tag{11}$$

By eliminating λ between Eqs. (9) and (11), one obtains an integral equation for the FDD $\rho(x, h)$, which can be solved by iterations. The details of the iteration procedure have been discussed in our previous paper.[14]

D. Calculation of the Contact Angle

Because of the relatively large width of the vapor-liquid interface detected in MD experiments as well as in DFT calculations,[8,14] the contact angle for nanodrops is not clearly defined. There are several approaches to define the contact angle from a known FDD. The simplest one used in Refs. 10 and 33 is to determine the profile inside the vapor-liquid interface corresponding to a constant local density ρ_{div} and to find the angle which this profile makes with the solid surface. The density ρ_{div} can be defined as the density for an equimolar dividing surface for a horizontal FDD $\rho(x, h_0)$ at some distance h_0 from the surface by considering this FDD as that for a planar vapor-liquid interface. However, there is no unique way how to select h_0. To find the most appropriate value for h_0, one should note that for $h < 3\sigma_{ff}$ the FDD $\rho(x, h)$ inside the drop has an oscillatory behavior as a function of h (due to the ordering of the fluid molecules which form several liquid layers of various densities)[21] and as function of x (due to the roughness of the surface). For these reasons, it is not clear how to determine ρ_{div} in the range $h_0 < 3\sigma_{ff}$. For $h_0 > 3\sigma_{ff}$ those oscillations become small and FDD $\rho(x, h_0)$ ($h_0 > 3\sigma_{ff}$) exhibits a clearly observable interface between high and low density phases, for which one can easily find ρ_{div} and define a dividing surface at the selected h_0. Hence, it looks reasonable to select a h_0 larger than $3\sigma_{ff}$.

Another more complicated way (see Ref. 8) to define the contact angle is to calculate ρ_{div} as a function of h using the corresponding horizontal FDD $\rho(x, h)$ for all possible distances $h > 0$ and to plot the manifold of points corresponding to the locations of all dividing surfaces. For $h > 3\sigma_{ff}$, this approach provides almost the same drop profile as the previous one, but it is not appropriate for $h < 3\sigma_{ff}$. Therefore, in the present paper we will use the first approach to calculate ρ_{div} by selecting $h_0 \cong 5\sigma_{ff}$.

Even after the drop profile is obtained, there is an additional problem in the identification of the contact angle. Conventionally, the apparent contact angle θ_{ap} is determined by extrapolating the upper part of the profile considered as part of a circle. However, near the surface where the liquid-solid interactions have the largest values, the difference between this circle and the calculated drop profile is large. In this situation, one can make a linear fit of the drop profile in the very vicinity of the surface (at distances smaller than a few molecular diameters)[36] and to consider the angle which this straight line makes with the surface as the contact angle θ. Below, both procedures will be used.

E. Calculation of the Sticking Force

If a drop is in mechanical equilibrium on an inclined surface with an inclination angle α (see Fig. 1), the sticking force f_{st}, which prevents the drop to slide down, is equal to the tangential component f_τ of the force of gravity

$$f_{st} = f_\tau = Ma\sin\alpha, \tag{12}$$

where M is the mass of the drop. By changing α, one can find the critical value α_c such that for $\alpha > \alpha_c$ the drop slides down along the surface. The corresponding critical value $f_{st}^c = Ma\sin\alpha_c$ of f_{st} depends, generally, on the natures of the liquid and solid. The critical sticking force can also be determined by changing the mass of the drop at fixed a and α, say, at α=90° (vertical surface), or changing the acceleration a at fixed M and α. In the first case

$$f_{st}^c = M_c\, a\sin\alpha, \tag{13}$$

and in the second

$$f_{st}^c = Ma_c \sin\alpha, \tag{14}$$

where M_c and a_c are the mass of the drop and the value of a, respectively, which provide the stability limit. For $M > M_c$ or for $a > a_c$ the drop loses its stability on an inclined surface.

It follows from the above considerations that the mass M of the drop calculated from FDD is necessary to determine the sticking force. For macroscopic drops it can be obtained by multiplying the (constant) liquid density by the drop volume. The error due to the existence of density nonhomogeneities at the fluid-liquid and liquid-solid interfaces is in this case negligible. However, for nanodrops, this error can be considerable because the sizes of the interfaces and drop are comparable. In this case, it is convenient to characterize the size of the drop by the total number N_{tot} of molecules it contains. Then the mass of the drop is equal to $M=mN_{tot}$. To calculate N_{tot} for a given drop, the latter was enclosed in a rectangular box with sides tangential to the vapor-liquid interface and N_{tot} was determined by the numerical integration of FDD $\rho(x, h)$ over this box. The presence in such a box of a tiny fraction of molecules which do not belong to the drop has a small effect on N_{tot} because of the small density of the fluid outside the drop.

Evidently, N_{tot} should depend on the average density ρ_{av} of the fluid in the system but the form of this dependence is not clear in advance. The analysis performed in the present paper has shown that in the range of $\rho_{av}^* \geqslant 0.13\left(\rho^* = \rho\sigma_{ff}^3\right)$, N_{tot} is a linear function of ρ_{av}^* As a consequence, the sticking force is also a linear function of ρ_{av}^*. The coefficients of this linear function depend on the case considered and will be specified for each of the particular cases considered.

Finally, a procedure to calculate the critical sticking force f_{st}^c in the framework of the DFT employed in the present paper can be outlined as follows. (i) For a selected ρ_{av}^*, the largest possible value a_c of the acceleration a, which provides a solution of the Euler–Lagrange equation, should be found. (ii) The profile of the drop, the contact angles θ_1 and θ_2, and the total number of molecules should be calculated from the obtained FDD using the procedures described in the previous sections of the paper. (iii) The critical sticking force should be calculated using Eq. (14).

III. RESULTS

A. Drop on a horizontal rough surface

A drop of a given volume on a horizontal chemically rough surface can exist generally in several states which differ by the position of the drop with respect to the inhomogeneities. For example, two characteristic states, *D*1 and *D*2, of a drop on a surface *S*1 are shown in Figs. 6(a) and 6(b) for $\rho_{av}^* = 0.13$ in the absence of gravity (a=0). The plane of symmetry of drop *D*1 coincides with the vertical plane of symmetry of an *A*-plate; on the latter plane are also located the local minima of the solid-liquid potential considered as function of x at various distances h from the solid (see Fig. 5). Similarly, the plane of symmetry of drop *D*2 coincides with the vertical plane of symmetry of a *B*-plate; on the latter plane are also located the local maxima of the solid-liquid potential. The drop *D*1 has a smaller free energy. For a larger average density, $\rho_{av}^* = 0.22$, the above two possible states are shown in Figs. 6(c) and 6(d). In spite of its larger volume, drop *D*1 in Fig. 6(c) has approximately the same height but a larger base area than drop *D*1 in Fig. 6(a). For this reason, the liquid molecules of drop *D*1 of [Fig. 6(c)] cover near the surface more minima than the drop *D*1 of Fig. 6(a).

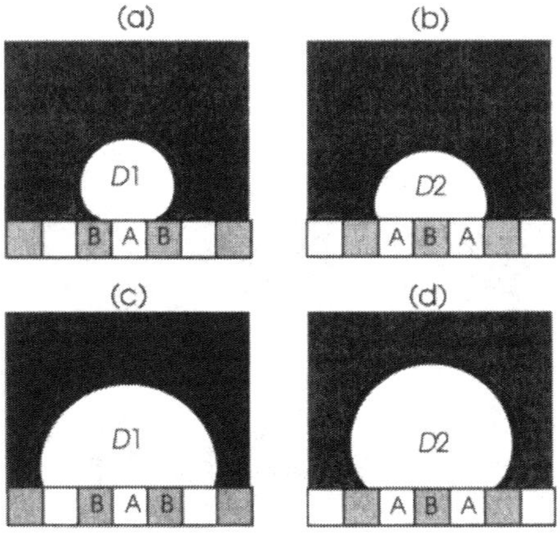

FIG. 6. Two characteristic states (*D*1 and *D*2) of a drop on a chemically rough surface. (a) and (b) $\rho_{av}^* = 0.13$; (c) and (d) $\rho_{av}^* = 0.22$. The lighter areas correspond to higher fluid densities.

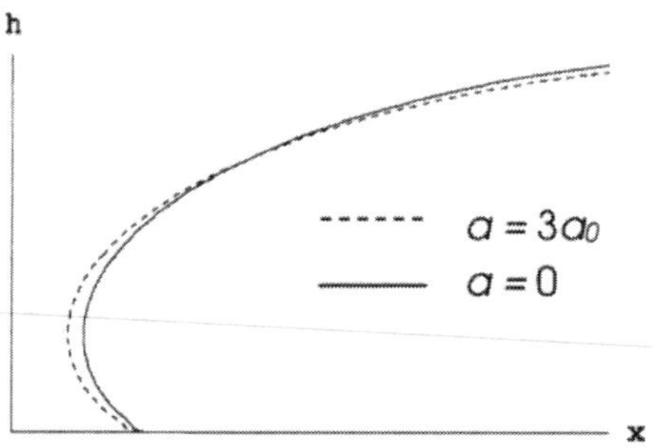

FIG. 7. The profiles of drop $D1$ on surface $S1$ for $\rho^*_{av} = 0.28$ in the absence of gravity (solid line) and for $a = 3a_0$ (dashed line). Only a part of the symmetric profile is presented.

The presence of gravity, even as large as selected in the present calculations, does not change appreciably the shape of the nanodrop and the contact angle. For illustration, the profiles of drop $D1$, obtained as described in Sec. II D for a=0 (solid lines) and a=$3a_0$ (dashed lines), are presented in Fig. 7 for $\rho^*_{av} = 0.28$.

Another feature of the drop on a horizontal rough surface is the volume dependence of the contact angle. As shown in Ref. 14, the contact angle changes periodically as a function of the drop volume (or average density ρ_{av}) between two angles θ_{min} and θ_{max}. For example, for the surface $S1$ and drop $D1$, considered in the present paper, $\theta_{min} = 96.8°$ and $\theta_{max} = 146.2°$.

One should note that even on a smooth surface (without roughness) the calculated contact angle depends slightly on the average density of the fluid in the system. For example, on a smooth surface with $\epsilon_{fs} = 0.6\epsilon^0_{fs}$ the contact angle changes from 100.9° to 98.4° when ρ^*_{av} changes from 0.1 to 0.24. Possible causes for such a behavior are the small difference in the densities of the vapor phase for different ρ^*_{av} and the uncertainties in the procedure to determine the contact angle for nanodrops.

B. A DROP ON A VERTICAL SURFACE

1. Surface S1 $\left(\epsilon_{fs,A} = \epsilon^0_{fs}, \epsilon_{fsB} = 0.1\epsilon^0_{fs}\right)$: Drop D2

The metastable FDDs corresponding to a drop of type $D2$ on a surface $S1$ for an inclination angle of 90° (vertical wall) are presented in Fig. 8 for various average densities and consequently various drop sizes by considering that $a = 3a_0$.

For $\rho^*_{av} = 0.1$, the drop is almost symmetrical [Fig. 8(a)] and the difference between the contact angles θ_1 and θ_2 is small. As the volume increases [Figs. 8(b) and 8(c)] this difference also increases, as one can see from Table 1.

For average densities $\rho^*_{av} > 0.21$ no solutions of the equation for the FDD [Eq. (9)] could be found which correspond to a drop $D2$. Hence, $\rho^*_{av} = 0.21$ provides a critical value for the sticking force

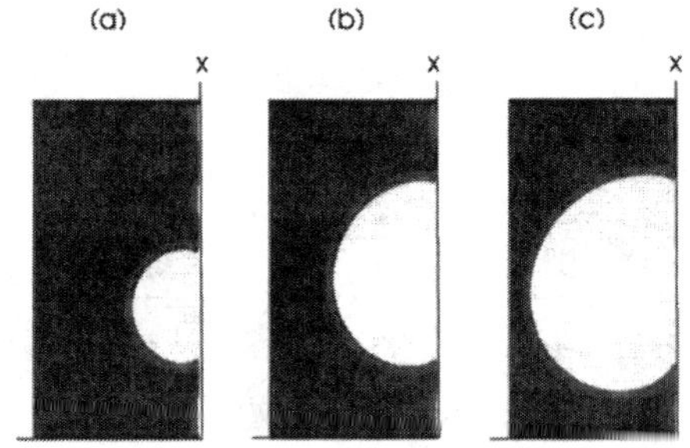

FIG. 8. Drop $D2$ on vertical surface $S1$ for $a = 3a_0$ and various average densities ρ^*_{av}. (a) $\rho^*_{av} = 0.1$, (b) $\rho^*_{av} = 0.16$, and (c) $\rho^*_{av} = 0.21$. The lighter areas correspond to higher fluid densities.

which in this case is equal (per unit length) to 1.67 × 10^{-3} N/m. For $0.1 < \rho^*_{av} =< 0.21$ the sticking force in the considered case can be represented by the linear expression $f_{st} = (91.1\rho^*_{av} - 2.38)\times 10^{-4}$ N/m.

After the determination of the sticking force and of the contact angles θ_1 and θ_2 from the calculated FDD $\rho(x, h)$, one can compare the microscopic results with the predictions of Eq. (3) by calculating the surface tension γ_{lv} from the latter equation. If Eq. (3) provides approximately the same value of γ_{lv} for all values of ρ^*_{av}, one can consider that Eq. (3) remains valid even for nanodrops. In Table 1, the results of the calculations are listed for drop *D*2. The contact angles were determined using the two procedures described in Sec. II D. For relatively small average densities $(\rho^*_{av} \leqslant 0.13)$, there are strong local density oscillations near the solid and, as a result, it is difficult to find a reasonable linear fit for the lower part of the drop profile. For this reason, only the first procedure to determine the contact angle involving the extrapolation of the upper part of the profile was used for $\rho^*_{av} \leqslant 0.13$. One can see from Table 1 that the values of the surface tension obtained using the drop profiles for various drop sizes and the first procedure to find the contact angle do not differ by more than 6% from their mean value γ_{lv} = 7.45 × 10^{-3} N/m and are surprisingly close to the value γ_{lv} = 1.4 × 10^{-2} N/m determined from experiments involving macroscopic interfaces. The values of γ_{lv} obtained when the contact angle was determined by the linear fit of the lower part of the profile differ by more than 12% from their mean value. Note that DFT calculations for a planar liquid-vapor interface provide the value $\gamma_{lv} \cong 1.3 \times 10^{-2}$ N/m.[37]

2. Surface *S*1: drop *D*1

In spite of the fact that the main properties of drop *D*1 on a vertical surface *S*1 are similar to those of drop *D*2, there are some differences which should be mentioned. First, the profile of drop *D*1 is not as smooth near the surface as it is for drop *D*2. For this reason it is impossible to extract the contact angle by employing the linear fit for determining the contact angle. Only the upper part of the profile can be used in this case to determine the contact angle. Second, the area of the base of drop *D*1 is larger than that of drop *D*2 for the same average fluid density. One of the consequences is that the contact angles θ_1 and θ_2 are larger for *D*2 (see Table 2). In the considered range of ρ^*_{av} $(0.13 \leqslant \rho^*_{av} \leqslant 0.34)$, the sticking force can be represented by the linear function $f_{st} = (88.7\rho^*_{av} - 1.797)\times 10^{-4}$ N/m which is slightly different from the similar expression for drop *D*2.

TABLE 1

Drop *D*2 on a vertical chemically rough surface S1. Dependence of the contact angles at the lower (θ_1) and upper (θ_2) contact lines, sticking force f_{st}, and surface tension γ_{iv} on the average density ρ^*_{av}. The contact angles were determined by approximating the upper part of the profile with a circle and extrapolation (column labeled as "circular fit") or by a linear approximation for the profile close to the surface (column labeled as "linear fit").

	θ_1 (deg)		θ_2 (deg)			γ_{iv} × 10^3 N/m	
ρ^*_{av}	Circular fit	Linear fit	Circular fit	Linear fit	f_{st} × 10^4 N/m	Cicular fit	Linear fit
0.10	108.3		102.6		6.7	7.0	
0.13	119.4		111.7		9.5	7.8	
0.16	128.0	124.9	116.8	112.8	12.2	7.4	6.6
0.18	131.5	130.8	118.9	115.4	14.0	7.8	6.2
0.19	133.5	133.7	119.4	116.3	14.9	7.6	6.0
0.20	135.0	136.8	119.7	116.9	15.9	7.5	5.7
0.21	136.2	140.5	119.8	116.9	16.7	7.3	5.2

TABLE 2
Drop *D*1 on a vertical chemically rough surface *S*1. Dependence of the contact angles at the lower (θ_1) and upper (θ_2) contact lines, sticking force f_{st}, and surface tension γ_{iv} on the average density ρ^*_{av}. The contact angles were determined by approximating the upper part of the profile with a circle and extrapoaltion.

ρ^*_{av}	θ_1 (deg)	θ_2 (deg)	$f_{st} \times 10^4$ N/m	$\gamma_{iv} \times 10^3$ N/m
0.13	84.8	76.5	9.8	7.0
0.16	93.8	84.6	12.4	7.7
0.21	107.8	94.8	16.7	7.5
0.26	116.5	100.3	21.3	7.9
0.30	124.6	103.4	24.8	7.4
0.34	128.4	105.3	28.4	7.9

The critical sticking force was calculated using Eq. (14) for α=90°. For $\rho^*_{av} = 0.13$, $a_c = 7.0a_0$ and $f_{st} = 21.3 \times 10^{-4}$ N/m. For $\rho^*_{av} = 0.24$, $a_c = 8.5a_0$ and $f_{st} = 48.2 \times 10^{-4}$ N/m. For $\rho^*_{av} = 0.24$, $a_c = 8.5a_0$ and $f_{st} = 48.2 \times 10^{-4}$ N/m. The values of the surface tension are equal to $\gamma_{lv} = 7.4 \times 10^{-3}$ and $\gamma_{lv} = 6.9 \times 10^{-3}$ N/m, respectively. These values are close to those obtained earlier for drop *D*2.

3. Nanodrops on a surface *S*2 ($\epsilon_{fs,A} = 0.6\epsilon_{fs}$, $\epsilon_{fs,B} = 0.1\epsilon_{fs}$)

In the considered range of average densities $(0.1 \leqslant \rho^*_{av} \leqslant 0.3)$ and for $a = 3a_0$, a drop of type *D*1 could be found only for densities close to $\rho^*_{av} = 0.1$, while a drop of type *D*2 could be found in the range $0.1 \leqslant \rho^*_{av} \leqslant 0.22$. Due to weaker fluid-solid interactions, surface *S*2 is more hydrophobic than *S*1; hence, the contact angles θ_1 and θ_2 for *S*2 are larger than those for *S*1 (for the same average density). For example, for $\rho^*_{av} = 0.16$, $\theta_1 = 128.0°$, $\theta_2 = 116.8°$ for *S*1 and $\theta_1 = 140.1°$, $\theta_2 = 126.1°$ for *S*2.

For $\rho^*_{av} = 0.16$, the critical sticking force for drop *D*2 on *S*2 $f^c_{st} = 1.88 \times 10^{-3}$ N/m and is smaller than that on *S*1 $\left(f^c_{st} = 3.26 \times 10^{-3}\text{ N/m}\right)$. The above results indicate that the sticking force is smaller on a hydrophobic than on a hydrophilic surface.

However, the above observation does not constitute a general rule. Indeed, the sticking force provided by Eq. (3) is proportional to the difference of the cosines of the contact angles θ_1 and θ_2 at the upper and lower contact lines. For the more hydrophobic surface (*S*2), both angles are larger than those on the less hydrophobic surface (*S*1). Nevertheless, the difference of the cosines can be larger or smaller for a hydrophobic surface than for a hydrophilic one. For example, for $\rho^*_{av} = 0.21$, the critical sticking force for drop *D*1 on *S*1 is equal to $f^c_{st} = 1.65 \times 10^{-3}$ N/m but is smaller than that on *S*2 $\left(f^c_{st} = 2.03 \times 10^{-3}\text{ N/m}\right)$ even though surface *S*1 is less hydrophobic than surface *S*2. Another example is presented in Ref. 20 on the basis of macroscopic considerations on a drop on an inclined surface. It was shown[20] that the critical value α_c of the inclination angle can be larger for a hydrophobic than for a hydrophilic surface (for the same volume of the drop). This means that in that particular case the critical sticking force on the hydrophilic was smaller than on the hydrophobic surface.

C. Sticking Force on a Vertical Physically Rough Surface

The pillars used in the present calculations are presented in Fig. 4. They have the width $d_p = 2\sigma_{ff}$ and the distance between them $\Delta_p = 4\sigma_{ff}$. The height of the pillars h_p was selected equal to $h_p = 2.5\sigma_{ff}$ and $h_p = 4\sigma_{ff}$. They are considered composed from the same material as the main solid (carbon dioxide). In the absence of gravity, the contact angle of a drop on a smooth (without pillars) surface is $\theta \simeq 51°$; hence, such a surface is hydrophilic. The presence of pillars increases the hydrophobicity of the surface. For pillars of height $h_p = 2.5\sigma_{ff}$, $\theta = 89.3°$ and for height $h_p = 4\sigma_{ff}$, $\theta = 94.4°$. In the latter case, the rough surface becomes hydrophobic ($\theta > 90°$).

TABLE 3

Drop on a vertical physically rough surface decorated with pillars of height $h_p = 4\sigma_{ff}$. Dependence of the contact angles at the lower (θ_1) and upper (θ_2) contact lines, sticking force f_{st}, and surface tension γ_{lv} on the acceleration *a*. The contact angles were determined by approximating the upper part of the profile with a circle and extrapolation.

a/a_0	θ_1 (deg)	θ_2 (deg)	$f_{st} \times 10^4$ N/m	$f_{lv} \times 10^3$ N/m
2.9	95.6	94.0	2.0	7.0
5.9	96.2	93.0	4.0	7.1
8.8	96.8	91.9	5.9	6.9
11.8	96.9	90.9	7.9	7.6
17.6	98.5	89.1	11.8	7.2
21.2	98.6	88.0	14.2	7.7

The potential of the fluid-solid interactions is a periodic function of x with a wavelength of $d_p + \Delta_p$ which resembles that presented in Fig. 5 for a chemically rough surface. For this reason, the main characteristics of a nanodrop on physically and chemically rough surfaces are similar.

Typical examples of FDDs corresponding to drops on a vertical physically rough surface are presented in Fig. 9(a) for $h_p = 4\sigma_{ff}$ and in Fig. 9(b) for $h_p = 2.5\sigma_{ff}$. In both cases $\rho^*_{av} = 0.1$ and the space between pillars is filled with a fluid of liquid-like density (Wenzel regime). The critical acceleration for the first case is $a_c = 21.2a_0$ and the corresponding critical sticking force is equal to $f^c_{st} = 1.42 \times 10^{-3}$ N/m. The contact angles and the sticking force extracted from FDD as well as the surface tension γ_{lv} calculated with Eq. (3) are listed in Table 3 as functions of the acceleration a. In Table 4 these quantities are listed for $h_p = 2.5\sigma_{ff}$. In this case a drop on a rough surface contains a larger number of molecules than on a surface with pillars of height $h_p = 4\sigma_{ff}$ because in the latter case a larger amount of fluid molecules of the system is located between the pillars [compare Figs. 9(a) and 9(b)]. For this reason, for the same average density $\rho^*_{av} = 0.1$, the critical acceleration is smaller (a_c=10.3a_0) for pillars with $h_p = 2.5\sigma_{ff}$ than for those with $h_p = 4\sigma_{ff}$ ($a_c = 21.2a_0$).

In both examples the calculated values of the surface tension γ_{lv} are in agreement with those calculated for chemically rough surfaces.

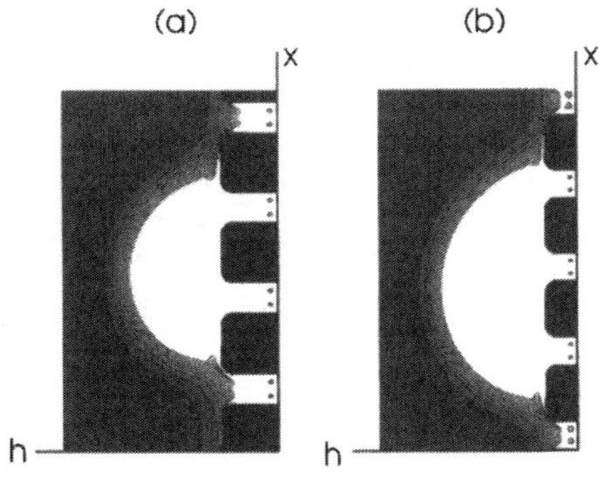

FIG. 9. Drops on physically rough vertical surfaces for $\rho^*_{av} = 0.1$. (a) $h_p = 4\sigma_{ff}$ and (b) $h_p = 2.5\sigma_{ff}$. The lighter areas correspond to higher fluid densities.

TABLE 4
Drop on a vertical physically rough surface decorated with pillars of height $h_p = 2.5\sigma_{ff}$. Dependence of the contact angles at the lower (θ_1) and upper (θ_2) contact lines, sticking force f_{st}, and surface tension γ_{iv} on the acceleration *a*. The contact angles were determined by approximating the upper part of the profile with a circle and extrapolation.

a/a_a	θ_1 (deg)	θ_2 (deg)	$f_{st} \times 10^4$ N/m	$f_{iv} \times 10^3$ N/m
2.9	89.5	86.8	3.5	7.1
5.9	89.8	84.1	7.1	7.1
8.8	90.1	81.3	10.6	6.9
9.7	90.5	81.2	11.6	7.2
10.3	91.2	81.3	12.3	7.2

IV. CONCLUSION

In the present paper the equilibrium of a two-dimensional nanodrop on a vertical rough solid surface is examined on the basis of a nonlocal DFT which involves only the microscopic parameters of the intermolecular interactions. The obtained FDD provides complete information about a drop, namely, the asymmetric drop profile, the contact angles at the upper and lower contact lines, and the liquid density distribution inside the drop. Using this information, the sticking force, which maintains the mechanical equilibrium of the drops of various sizes, is calculated for chemically heterogeneous surfaces and for surfaces decorated with pillars.

Assuming that the traditional equation [Eq. (3)], which provides the sticking force for a macroscopic drop, can be extended to nanodrops and determining the contact angles, the liquid-vapor surface tension γ_{lv} was calculated for a number of specific cases. The obtained values of γ_{lv} are in agreement with each other and with the value determined in macroscopic experiments. The differences between them can be explained by the approximations made during the determining of the contact angles from the FDDs. From these results one can conclude that Eq. (3) and its particular case, Eq. (2) derived for macroscopic drops, are still valid for nanodrops. However, the calculated value of γ_{lv} for nanodrops is approximately twice as small as the experimental one, obtained from macroscopic experiments.

APPENDIX: FREE ENERGY OF THE SYSTEM

The contributions to the free energy listed in Sec. II C can be represented as follows.

$$F_{id}\left[\rho(\mathbf{r})\right] = k_B T \int d\mathbf{r}\rho(\mathbf{r})\left\{\log\left[\Lambda^3\rho(\mathbf{r})\right] - 1\right\}, \tag{A1}$$

$$F_{hs}\left[\rho(\mathbf{r})\right] = \int d\mathbf{r}\rho(\mathbf{r})\Delta\Psi_{hs}(\mathbf{r}), \tag{A2}$$

where

$$\Delta\Psi_{hs}(\mathbf{r}) = k_B T \eta_{\bar{\rho}} \frac{4 - 3\eta_{\bar{\rho}}}{\left(1 - \eta_{\bar{\rho}}\right)^2}, \tag{A3}$$

$\eta_\rho = \frac{1}{6}\pi\bar{\rho}(\mathbf{r})\sigma_{ff}^3$ is the packing fraction of the fluid molecules, σ_{ff} is the fluid hard core diameter, and $\bar{\rho}(\mathbf{r})$ is the smoothed density defined as

$$\bar{\rho}(\mathbf{r}) = \int d\mathbf{r}'\rho(\mathbf{r}')W(|\mathbf{r}-\mathbf{r}'|). \tag{A4}$$

The weighting function $W(|\mathbf{r}-\mathbf{r}'|)$ is selected in the form[23]

$$W(|\mathbf{r}-\mathbf{r}'|) = \begin{cases} \frac{3}{\pi\sigma_{ff}^3}\left(1-\frac{r}{\sigma_{ff}}\right), & r \leqslant \sigma_{ff} \\ 0, & r > \sigma_{ff}, \end{cases}$$

where $r = |\mathbf{r}-\mathbf{r}'|$.

The contribution to the excess free energy due to the attraction between the fluid-fluid molecules is calculated in the mean-field approximation

$$F_{\text{attr}}[\rho(\mathbf{r})] = \frac{1}{2}\iint d\mathbf{r}d\mathbf{r}'\rho(\mathbf{r})\rho(\mathbf{r}')\phi_{ff}(\mathbf{r}-\mathbf{r}'). \tag{A5}$$

and the contribution of the gravity to the total free energy is

$$F_{\text{g}}[\rho(\mathbf{r})] = ma\int d\mathbf{r}\rho(\mathbf{r})(x\sin\alpha + h\cos\alpha). \tag{A6}$$

The function $Q(x, h)$ in Eq. (9) is given by

$$Q(x,h) = \frac{1}{k_BT}\left[\Delta\Psi_{\text{hs}}(x,h) + \overline{\Delta\Psi'_{\text{hs}}}(x,h) + U_{ff}(x,h) + U_{\text{fs}}(x,h) + U_G(x,h)\right], \tag{A7}$$

where $U_{\text{fs}}(x, h)$ and $U_G(x, h)$ are provided by Eq. (8) and (6), respectively,

$$U_{\text{ff}}(x,h) = \iint dx'dh'\rho(x',h')\phi_{\text{ff},y}(|x-x'|,|h-h'|), \tag{A8}$$

$$\overline{\Delta\Psi'_{hs}}(x,h) = \iint dx'dh'\rho(x',h')W_y(|x-x'|,|h-h'|)$$

$$\times\frac{\partial}{\partial\bar{\rho}}\Delta\Psi_{\text{hs}}(\bar{\rho})|_{\bar{\rho}=\bar{\rho}(x',h')}, \tag{A9}$$

$\phi_{\text{ff},y}(|x-x'|,|h-h'|)$ and $W_y(|x-x'|,|h-h'|)$ are obtained by the integrations of the potential $\phi_{ff}(|\mathbf{r}-\mathbf{r}'|)$ and weighted function $W(|\mathbf{r}-\mathbf{r}'|)$ with respect to the y-direction from $-\infty$ to $+\infty$, respectively.

When calculating the term $U_{\text{ff}}(x, h)$ of the Euler-Lagrange equation arising due to the long-range fluid-fluid interactions, a cutoff at a distance equal to four molecular diameters σ_{ff} for the range of Lennard-Jones attraction was employed. The increase in this distance up to $10\sigma_{\text{ff}}$ changed the results by less than 1%.

REFERENCES

1 M. Callies and D. Quéré, Soft Matter **1**, 55 (2005).
2 D. Quéré and M. Reyssat, Philos. Trans. R. Soc. London, Ser. A **366**, 1539 (2008).
3 R. E. Johnson, Jr. and R. H. Dettre, Surf. Colloid Sci. **2**, 85 (1969).
4 C. G. L. Furmige, J. Colloid Sci. **17**, 309 (1962).
5 E. Ruckenstein and P. S. Lee, Surf. Sci. **52**, 298 (1975) (Section 3.1 of this volume).
6 G. O. Berim and E. Ruckenstein, J. Phys. Chem. **108**, 19330 (2004) (Section 3.2 of this volume).
7 G. O. Berim and E. Ruckenstein, J. Phys. Chem. **108**, 19339 (2004) (Section 3.3 of this volume).
8 M. J. de Ruijter, T. D. Blake, and J. De Coninck, Langmuir **15**, 7836 (1999).
9 Vishnyakov and A. V. Neimark, J. Chem. Phys. **119**, 9755 (2003).
10 N. Giovambattista, P. G. Debenedetti, and P. J. Rossky, J. Phys. Chem. B **111**, 9581 (2007).
11 V. Talanquer and D. W. Oxtoby, J. Chem. Phys. **104**, 1483 (1996).
12 V. Talanquer and D. W. Oxtoby, J. Chem. Phys. **114**, 2793 (2001).
13 B. Husowitz and V. Talanquer, J. Chem. Phys. **122**, 194710 (2005).
14 G. O. Berim and E. Ruckenstein, J. Chem. Phys. **129**, 014708 (2008) (Section 4.4 of this volume).
15 R. A. Brown, F. M. Orr, Jr., and L. E. Scriven, J. Colloid Interface Sci. **73**, 76 (1980).
16 Y. Rotenberg, L. Boruvka, and A. W. Neumann, J. Colloid Interface Sci. **102**, 424 (1984).
17 M. Miwa, A. Nakajima, A. Fujishima, K. Hashimoto, and T. Watanabe, Langmuir **16**, 5754 (2000).
18 P. Roura and J. Fort, Langmuir **18**, 566 (2002).
19 Marmur, Langmuir **20**, 3517 (2004).
20 Krasovitski and A. Marmur, Langmuir **21**, 3881 (2005).
21 P. Tarazona, Phys. Rev. A **31**, 2672 (1985).
22 C. Lastoskie, K. E. Gubbins, and N. Quirke, Langmuir **9**, 2693 (1993).
23 R. H. Nilson and S. K. Griffiths, J. Chem. Phys. **111**, 4281 (1999).
24 P. I. Ravikovitch, A. Vishnyakov, and A. V. Neimark, Phys. Rev. E **64**, 011602 (2001).
25 A.V. Neimark, P. I. Ravikovitch, and A. Vishnyakov, Phys. Rev. E **65**, 031505 (2002).
26 E. A. Ustinov and D. D. Do, J. Phys. Chem. B **109**, 11653 (2005).
27 G. O. Berim and E. Ruckenstein, J. Chem. Phys. **128**, 024704 (2008).
28 L. G. MacDowell, M. Muller, and K. Binder, Colloids Surf., A **206**, 277 (2002).
29 J. A. White, A. Gonzalez, F. L. Roman, and S. Velasco, Phys. Rev. Lett. **84**, 1220 (2000).
30 A.V. Neimark, P. I. Ravikovitch, and A. Vishnyakov, J. Phys.: Condens. Matter **15**, 347 (2003).
31 D. Frenkel and B. Smit, *Understanding Molecular Simulation* (Academic, New York, 1966).
32 J. Mittal, J. R. Errington, and T. M. Truskett, J. Chem. Phys. **126**, 244708 (2007).
33 R. Evans and P. Tarazona, Phys. Rev. A **28**, 1864 (1983).
34 C. E. Upstill and R. Evans, J. Phys. C **10**, 2791 (1977).
35 F. P. Buff and R. A. Lovett, *Simple Dense Fluids,* edited by H. L. Frisch and Z. W. Salsburg (Academic, New York, 1968), p. 17.
36 F. Porcheron and P. A. Monson, Langmuir **22**, 1595 (2006).
37 T. Wadewitz and J. Winkelmann, J. Chem. Phys. **113**, 2447 (2000).

4.12 Calculation of Nanodrop Profile from Fluid Density Distribution*

Gersh Berim and Eli Ruckenstein†
Department of Chemical and Biological Engineering, State University of New York at Buffalo, Buffalo, New York 14260, United States

Corresponding Author

† E-mail: feaeliru@buffalo.edu.
Tel.: +1 716 645 1210; fax: +1 716 645 3822.

1. INTRODUCTION

The great success in the last decades in the experimental investigation of the liquid–vapor interfaces at nanoscale led to the increase in the interest to theoretical description of nanodroplets on smooth and rough surfaces. Traditionally, the main attention was given to the calculation of the drop profile and the contact angle which the drop makes with the solid for various solid–liquid pairs. Because the macroscopic concept of surface tension is not clearly defined for nanodrops, the classical Young equation

$$\gamma_{lv} \cos\theta = \gamma_{vs} - \gamma_{ls} \tag{1}$$

where γ_{lv}, γ_{ls}, and γ_{vs} are the liquid–vapor, liquid–solid, and vapor–solid surface tensions, respectively, cannot be used to represent the contact angle θ. Therefore, microscopic approaches based on the interaction potentials between the molecules (fluids and solid) are most appropriate for the calculation of the drop profile which is needed to obtain the contact angle. Another specific feature of profile calculations for nanodrops is the comparable size of the nanodrop and of the range of strong fluid–solid interactions. For this reason, even though the gravity can be neglected for nanodrops, only the upper part of their profile can be approximated by a circle, whereas a considerable part of the profile has a much more complicated shape. In the present paper, a review of the procedures which can be used to calculate the drop profile is provided and a new procedure is suggested.

The first microscopic approach, which is denoted below as the sharp–kink interface approach, was suggested in Ref. [1] and modified later in Refs. [2]–[5]. The equation of the drop profile was obtained directly using the minimization of the total potential energy of the fluid–solid system due to the interactions between all the molecules belonging to the system. The assumptions used in the calculations are presented in Section 2.

The second approach uses various versions of the non–local density functional theory (DFT) [6]–[8] and provides the density distribution of the considered fluid from which the profile of the drop can be extracted. Note that the procedure of extraction of the drop profile from the fluid density distribution obtained by DFT is not unique and different procedures lead to different drop profiles and, as a consequence, to different values of the contact angles. The choice of the most appropriate

* *Advances in Colloid and Interface Science* 231 (2016) 15–22. Republished with permission.

one is a special problem which will be discussed below in Section 3. For simplicity, only drops with infinite length in one direction (cylindrical drops) will be considered.

The fluid density distribution (FDD) can be determined also by molecular dynamics simulations (see e.g. Refs. [9]–[14]).Because those FDDs are similar to the FDDs obtained by the DFT approach, the simulations methods themselves will not be examined separately in the present paper.

2. SHARP–KINK INTERFACE APPROACH

In this approach, the change of fluid density at the interfaces between a liquid drop and the surrounding vapor and solid is considered to occur discontinuously from a liquid like value to a vapor like value (sharp kink approximation [15]). The liquid density ρ_L in the drop is assumed to be uniform everywhere with the exception of the two monolayers at the liquid–solid and liquid–vapor interfaces where the liquid density is equal to ρ_{LS} and ρ_{LV}, respectively. The introduction of those monolayers accounts for the inhomogeneity of the fluid density at the liquid–solid and liquid–vapor interfaces.

It is natural to assume that the vapor around the drop has the constant density ρ_V which is much smaller than ρ_L. For this reason, the contribution of liquid–vapor interactions to the total potential energy of the system can be neglected. The solid substrate is considered homogeneous with a density ρ_S. The values of ρ_L, ρ_{LV}, ρ_{LS}, and ρ_V which should be selected before calculating the drop profile, define the fluid density distribution in the system.

The potential $\phi_{LL}(r)$ of interaction between the liquid molecules is selected in the form of the London–van der Waals potential with a hard core repulsion

$$\phi_{LL}(r)=\begin{cases}-\epsilon_{LL}\left(\dfrac{\sigma_{LL}}{r}\right)^6, & r\geq\sigma_{LL}\\ \infty, & r<\sigma_{LL}\end{cases} \tag{2}$$

where r is the distance between the centers of the interacting molecules, $\sigma_{LL}>0$ and $\epsilon_{LL}>0$ are the size of the repulsive core (hard core diameter) and the energy parameter, respectively. In the calculations, the cutoff diameter $\eta>\sigma_{LL}$ was used ($\phi_{LL}(r)=0$ for all $r>\eta$).

The interaction potential between a molecule of liquid and a molecule of solid is selected as the sum of long–range $\left(\phi_{LS}^{l}(r)\right)$ and short–range $\left(\phi_{LS}^{s}(r)\right)$ interactions. The former potential is selected in the form

$$\phi_{LS}^{l}(r)=\begin{cases}\epsilon_{LS}\left[k_\phi\left(\dfrac{\sigma_{LS}}{r}\right)^{12}-\left(\dfrac{\sigma_{LS}}{r}\right)^6\right], & r\geq\sigma_{LS},\\ \infty, & r<\sigma_{LS},\end{cases} \tag{3}$$

where $k_\phi=0$ or 1, σ_{LS} and $\epsilon_{LS}>0$ are, respectively, the size of the repulsive core and the energy parameter of the liquid–solid interactions. For $k_\phi=0$ one recovers the London van der Waals potential and for $k_\phi=1$ the Lennard–Jones (LJ) potential. In real systems σ_{LL} and σ_{LS} are of the order of several angstroms [16].

The potential energy $\Phi_{LS}^{l}(h)$ of a liquid molecule interacting with a semi–infinite solid, possessing a planar surface, through the long– range interactions can be written as

$$\Phi_{LS}^{l}(h)=\frac{\pi}{6}\epsilon_{LS}\rho_S\sigma_{LS}^3\left[\frac{2}{15}k_\phi\left(\frac{\sigma_{LS}}{h+\sigma_{LS}}\right)^9-\left(\frac{\sigma_{LS}}{h+\sigma_{LS}}\right)^3\right], h\geq 0, \tag{4}$$

where $h + \sigma_{LS}$ is the distance measured from the center of a molecule of liquid to the center of the first layer of solid atoms ($h = 0$ corresponds to the center of the layer of liquid molecules on the solid surface).

The short–range potential $\phi_{LS}^{s}(r)$ which accounts, for example, for the acid–base interactions is assumed to be effective only for the molecules in the first liquid layer near the solid surface. As a result, the potential energy $\Phi_{LS}^{s}(h)$ is given by

$$\Phi_{LS}^{s}(h) = \begin{cases} -\epsilon_{LS}^{s}, & h = 0, \\ 0, & h > 0 \end{cases} \tag{5}$$

where ϵ_{LS}^{s} is the energy parameter for the short–range interaction.

Combining the potentials Eqs. (4) and 5, one obtains the total potential $\Phi_{LS}(h)$ of a molecule of liquid interacting with the solid

$$\Phi_{LS}(h) = \begin{cases} \Phi_{LS}^{l}(0) + \Phi_{LS}^{s}(0), & h = 0, \\ \Phi_{LS}^{l}(h), & h > 0. \end{cases} \tag{6}$$

The interaction between the molecules of vapor and solid is supposed to has only a long–range component. For low–density vapors, only the molecules of a monolayer adsorbed on the solid surface ($h = 0$) are taken into account. As a consequence, the potential $\Phi_{VS}(h)$ of that interaction can be written in the form

$$\Phi_{VS}(h) = \begin{cases} \Phi_{VS}^{l}(0), & h = 0, \\ 0, & h > 0 \end{cases} \tag{7}$$

where the potential $\Phi_{VS}^{l}(h)$ has the same form as $\Phi_{LS}^{l}(h)$ given by Eq. (4), where ϵ_{LS} and σ_{LS} have to be replaced by ϵ_{VS} and σ_{VS}, respectively.

Neglecting the vapor–vapor and vapor–liquid potential energies in the calculation of the total potential energy U_{total} of the system and taking into account only the interaction of the adsorbed vapor molecules with the solid, U_{total} can be written in the form

$$U_{total} = U_{LL} + U_{LS} + U_{VS} \tag{8}$$

where U_{LL} and U_{LS} are the potential energies due to the interactions of the molecules of liquid between themselves and with those of the solid, respectively, and U_{VS} is a potential energy due to the interactions between the molecules of vapor and solid. Details of the calculation of U_{total} can be found in Ref. [5]. Note that calculations of U_{total} were performed only for two–dimensional FDDs which possess either translational symmetry along one of the coordinate axes (cylindrical drop) or axial symmetry about a vertical axis (axisymmetrical drop). The drop profile in those cases can be calculated as a curve in a plane perpendicular to the axis of the cylindrical drop or in a plane containing the axis of symmetry of an axisymmetrical drop.

Minimizing the total potential energy with respect to the drop profile and solving the resulting differential equation, one can obtain (for selected parameters of the interaction potentials and drop volume) the profile's equation in analytical form. A typical drop profile is presented in Fig. 1. In this figure, θ_0 is the so called microscopic contact angle, i.e. the angle which the drop profile makes with a planar surface at their intersection. The value of θ_0 depends on all involved fluid densities and parameters of interaction potentials [2]. Another angle, θ_m, is an apparent (macroscopic) contact angle between the surface and a circular profile obtained by the extension of the upper part of the drop. There is a large difference between the macroscopic and microscopic contact angles which

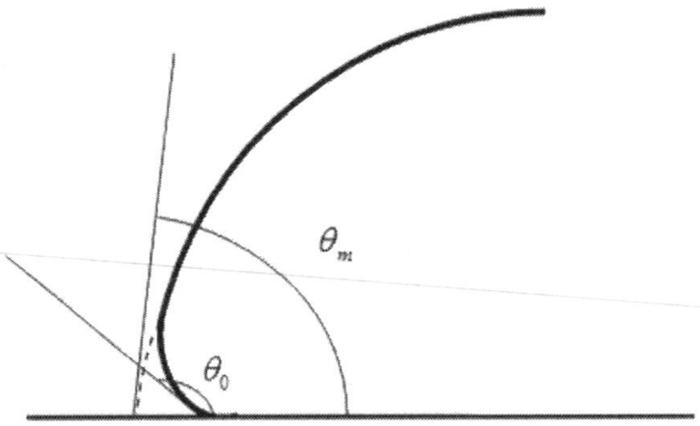

FIG. 1. Microscopic, θ_0 and macroscopic, θ_m contact angles. The angle θ_m is defined by extrapolating the upper (circular) part of the drop profile up to the surface. Only the left hand side of the profile is presented.

occurs due to the rapid change of the fluid–solid interactions close to the solid surface and, because of this, the strong deviation of the drop profile from the circular shape. The region of strong variation of the fluid–solid interaction is very small (of the order of 10–30Å). This fact makes the experimental determination of θ_0 difficult (Another reason why the microscopic contact angle is not a suitable quantity to characterize the drop is presented below in Section 3.). From the above considerations, it follows that the upper part of the drop profile is the most important in calculating the macroscopic contact angle, θ_m and that part should be determined with the highest possible precision.

The advantage of the sharp–kink approximation is that the drop profile is provided by the solution of a differential equation as a unique curve which is determined by the interaction parameters and the size of the drop. After the profile is obtained, the contact angle, θ_m can be calculated.

An obvious disadvantage of this approach is that, in reality, the liquid–vapor interface is not kink–like and has a size of the order of 10 Å in which the fluid density continuously changes from liquid– to vapor–like. For macrodrops much larger than the thickness of the liquid–vapor interface, the latter can be considered as kink–like and, for this case, the sharp–kink approximation is acceptable. However, for nanodrops this approximation is not satisfactory and other microscopic approaches should be employed. Those approaches were developed on the basis of molecular dynamics simulation (Refs. [9]–[14]) as well as on the basis of density functional theory (Refs. [17]–[20]). Both approaches provide the fluid density distribution (FDD) of one or several fluids, which are dependent on the fluid–solid interactions. The approach based on DFT is discussed in the next section.

3. DENSITY FUNCTIONAL APPROACH

In contrast to the theory described in the previous section, the density functional theory (DFT) is based on the minimization of the free energy of the system, which is calculated (using some approximations) from the first principles of statistical mechanics. The interaction between fluid molecules is described by the Lennard–Jones potential

$$\phi_{ff}\left(|\mathbf{r}-\mathbf{r}'|\right)=\begin{cases}4\epsilon_{ff}\left[\left(\dfrac{\sigma_{ff}}{r}\right)^{12}-\left(\dfrac{\sigma_{ff}}{r}\right)^{6}\right], & r\geq\sigma_{ff}\\ \infty, & r<\sigma_{ff}\end{cases} \tag{9}$$

where the coordinates $\mathbf{r}$ and $\mathbf{r}'$ provide the locations of the fluid molecules, $r = |\mathbf{r} - \mathbf{r}'|$, and σ_{ff} and ϵ_{ff} are the fluid hard core diameter and the energy parameter, respectively.

The same kind of potential is selected for the fluid–solid interactions (the energy parameter ϵ_{ff} and hard core diameter σ_{ff} in Eq. (9) being replaced by ϵ_{fs} and σ_{fs}, respectively, r being the distance between a molecule of fluid and that of the solid.)

The total Helmholtz free energy $F_H[\rho(\mathbf{r})]$ is the sum of an ideal gas free energy, $F_{id}[\rho(\mathbf{r})]$, a free energy $F_{hs}[\rho(\mathbf{r})]$, of a reference system of hard spheres, a free energy $F_{attr}[\rho(\mathbf{r})]$ due to the attractive interactions between the fluid molecules (in the mean field approximation), and a free energy $F_{fs}[\rho(\mathbf{r})]$ due to the interactions between fluid and solid molecules. The explicit forms for those contributions areprovided in Refs. [7,20]. It is worth mentioning that the form of $F_{hs}[\rho(\mathbf{r})]$ for the one–component fluid (see for example Ref. [7]) is different from that used for multicomponent fluids. In the latter case, the form suggested by Rosenfeld (see Ref. [8]) was employed.

Note that drop–like solutions for the fluid density distribution (FDD) can be found only if one employs the canonical ensemble version of DFT describing closed systems with fixed number of molecules (fixed average density). The latter constraint has the form

$$\rho_{av} = \frac{1}{V}\int_V d\mathbf{r}\rho(\mathbf{r}) \tag{10}$$

where V is the volume occupied by the fluid. Minimization of the total free energy of the system with respect to the density distribution $\rho(\mathbf{r})$ for the constraint given by Eq. (10) provides the integral Euler–Lagrange equation for $\rho(\mathbf{r})$ which can be solved by iterations (see Ref. [20] for details).

In Fig. 2, three distributions obtained with DFT are presented for cylindrical drops.

The first one, FDD1, presented in Fig. 2a corresponds to a very weak fluid–solid interaction compared with the fluid–fluid one ($\epsilon_{fs} = 0.001\epsilon_{ff}$). In this case, the FDD has a clearly defined axis of symmetry at which the fluid density has a maximum. The location of this axis is marked by the cross in Fig. 2a. The second FDD, FDD2, presented in Fig. 2b, is for nonnegligible fluid–solid interactions $\left(\epsilon_{fs} \simeq \epsilon_{ff}\right)$ and a smooth solid surface. The FDD2 is much more complex than FDD1. In particular, the fluid density has an oscillatory behavior close to the solid surface. The third FDD, FDD3, presented in Fig. 2c represents an FDD of a mixture of two fluids. One of them, the lubricating fluid (not presented in the figure), covers the surface of a smooth solid while the main part of the other fluid (test fluid) is located above the first one. All presented FDDs have $x = 0$ as plane of symmetry. The FDD2 is taken from Ref. [21] where the interaction parameters and temperature were selected as $\epsilon_{ff}/k_B = 119.76$, $\epsilon_{fs} = 1.1\epsilon_{ff}, \sigma_{ff} = \sigma_{ss} \equiv \sigma = 3.405\text{Å}, T = 87.3\text{K}$, The third distribution (Fig. 2c) is taken from Ref. [22] for the following values of the parameters $\epsilon_{11}/k_B = \epsilon_{22}/k_B = 119.76\text{K}, \epsilon_{12} = 0.5\epsilon_{11}, \epsilon_{1s} = 4\epsilon_{11}, \epsilon_{2s} = 0, \sigma_{11} = \sigma_{22} = \sigma_{12} = \sigma_{1s} = \sigma_{2s} = \sigma, T = 87.3\text{K}$, where subindices 1 and 2 label the two components of the fluid covering the solid surface.

To extract from such FDDs the drop profile which is needed to obtain the shape and size of the drop various procedures was employed. To illustrate them, only two–dimensional FDDs will be used below even though some of the previously obtained FDDs were three–dimensional.

The first procedure (P1) was used by Ruijter et al. [9] for an FDD obtained by molecular dynamics simulations. The FDD was considered as a collection of horizontal (parallel to the solid surface) layers of the same thickness (of the order of several molecular diameters) at various distances h_e from the solid. For each of the layers, a one–dimensional FDD along a straight line parallel to the x–axis was considered. At any h_e, the part of the FDD at the left of the vertical plane of symmetry has a shape which can be approximated by the sigmoidal function

$$\rho(x) = \frac{1}{2}(\rho_l + \rho_v) + \frac{1}{2}(\rho_l + \rho_v)\tanh\left(\frac{2(x - x_e)}{d}\right) \tag{11}$$

where ρ_l is the largest (liquid–like) and ρ_v the smallest (vapor–like) densities of the considered FDD, d is the thickness of the liquid–vapor interface, and x_e is the location of the equimolar dividing surface. For illustration, the plot of this function is presented in Fig. 3.

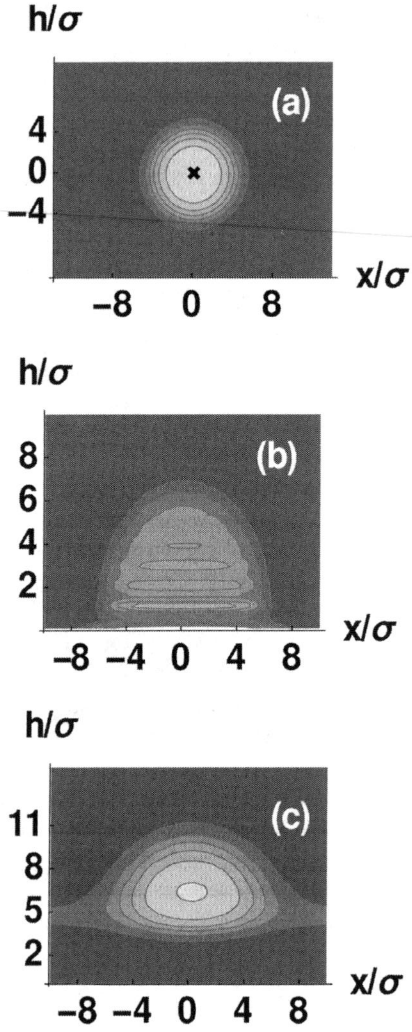

FIG. 2. Two–dimensional FDD for (a) negligible interactions between the solid and fluid. The cross in this figure marks the location of the axis of symmetry which is normal to the plane of the figure; (b) one–component fluid in contact with a smooth solid; (c) the test fluid which is separated from the smooth solid by a lubricating thin film. (The FDD of the lubricating fluid is not shown here). In all panels, the surface of the solid is at the bottom of the figure. The lighter areas correspond to higher densities.

The point of coordinates (x_e, h_e) and another one symmetrical to the first with respect to the plane of symmetry of the two dimensional FDD were considered as points of the drop profile. Repeating this procedure for each of the layers, the locations of the dividing surfaces were collected and used as the points of the drop profile. The fluid density ρ_{div} at those points is equal to $\rho_{div} = (\rho_l + \rho_v)/2$. Note that for large drops and layers which are not close to the top of the drop, the densities ρ_l and ρ_v were close to their bulk values of the considered fluid at the selected temperature. For small drops, ρ_l can be different from the density of the bulk liquid.

A more simple procedure (P2) which was used first in Ref. [13] is based on the consideration of the drop profile as the curve along which the fluid density is constant. This constant density is defined either as the density at the location of the equimolar dividing surface for a one–dimensional FDD considered for a specially selected horizontal layer (reference layer) [20], or it is selected arbitrarily.

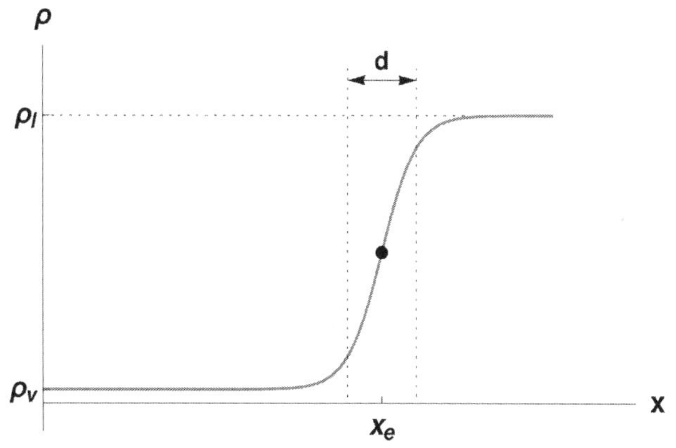

FIG. 3. Sigmoidal function provided by Eq.(11), used for approximation of the fluid density distribution on liquid–vapor interface. Coordinate x_e indicates location of the equimolar dividing surface.

For example, Giovambattista et al. [13] selected this density to be 0.20 g/cm^3 by considering a water drop on the solid surface. The drop profile thus obtained was used to calculate the contact angle which the drop makes with the solid surface and the latter angle was considered to be a measure of the hydrophobicity of the surface. In Ref. [17] where a DFT was developed using a lattice model for a fluid in contact with a solid, the drop profile was obtained by selecting $\rho_{div}\sigma^3 = 0.5$, where σ is the hard core diameter of the fluid–fluid interaction potential. (In the lattice model, $\rho_l\sigma^3 = 1$ and $\rho_v = 0$, consequently the density on the drop surface is equal to the arithmetic average of those two densities). This procedure was used also later in Refs. [18,19].

In the present paper we compare the drop profiles obtained by the above two procedures using as examples the three twodimensional (cylindrical) FDDs presented in Fig. 2. In addition, a new procedure (P3) is suggested which provides a more reasonable profile of the nanodrop.

Let us first consider the simplest case (FDD1 of Fig. 2a) which possesses an axis of symmetry normal to the plane of the figure and hence has a circular liquid–vapor interface.

At the axis of symmetry, FDD1 has a maximum density equal to $\rho_{f,\,max}\,\sigma^3 = 0.689$ that is somewhat smaller than the bulk density of the considered fluid ($\rho_{bulk}\,\sigma^3 = 0.727$). To apply procedure P1, the one–dimensional FDDs along the horizontal layers at various distances h_l from the axis of symmetry of FDD1 have been calculated. Fig. 4a, presents examples of such one–dimensional FDDs. One can see that the shape of all one–dimensional FDDs remains the same, the range of liquid–like densities being small, and there are no obvious plateaus in the upper parts of any FDD, which represents an essential part of the sigmoidal function. In addition, the maximum density ρ_l of the one–dimensional FDDs decreases with increasing h_l and becomes considerably different from the liquid–like density for $h_l/\sigma \geq 3.5$. To find the locations of the equimolar dividing surfaces for each density distribution, only their parts on the left hand side of the plane of symmetry $x = 0$ were taken into account and approximated by Eq. (11). For example, the location of the equimolar dividing surface for $h_l = 0$ is indicated by the point on the left hand side of the curve for $h_l = 0$. The point on the right hand side of that curve is symmetrical with the first one about the plane of symmetry. Note, that ρ_{div} decreases with increasing h_l. The set of all points of the locations of equimolar dividing surfaces provide the drop profile obtained with P1. This profile is presented by the dashed line in Fig. 4b.

The profiles in Fig. 4b provided by the solid lines are obtained with procedure P2 using the densities ρ_{div} for various reference layers. All drop profiles corresponding to different h_l (i.e. to different ρ_{div}) have circular shapes with the centers located on the symmetry axis of FDD1. The radii of the circles increase with increasing h_l.

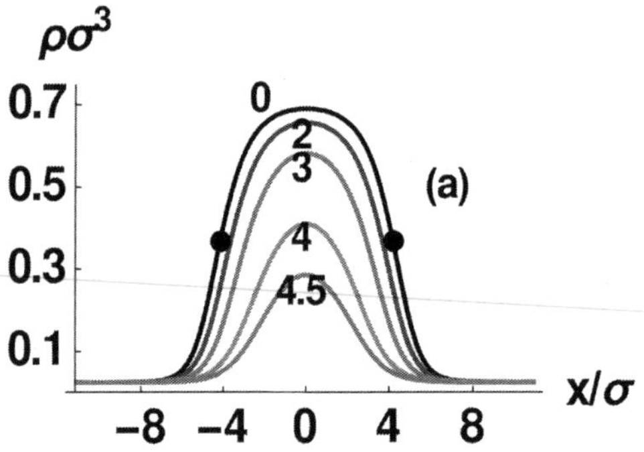

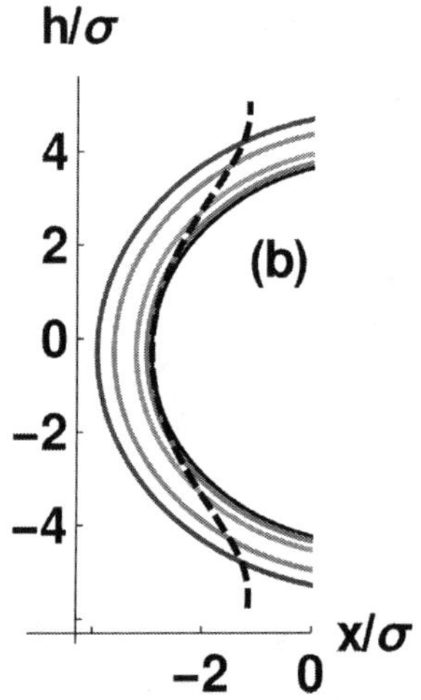

FIG. 4. (a) One–dimensional FDDs along the horizontal layers located at various distances h_l (noted for each of the curves) from the axis of symmetry of FDD1 presented in Fig. 2. The points on the curve for $h_l = 0$ (the thick solid line) indicate the location of the equimolar dividing surfaces. (b) Left hand side of the drop profiles obtained with P2 (solid lines) for h_l/σ =0;3;4; and 4.5 (radius of the drop increases with increasing h_l). The dashed line represents the profile obtained using P1. The profile presented by the thick solid line corresponds to $h_l = 0$.

To select the most suitable drop profile from those obtained with P2, one should take into account that the corresponding one-dimensional FDDs must have the densities ρ_l close to ρ_{bulk}. Such an FDD is that for $h_l = 0$, which is taken along the layer normal to the circles presented in Fig. 4b and passes through the symmetry axis. The corresponding drop profile, which is considered to be the true one, is represented in Fig. 4b by the thick solid line. Note that the FDDs with $h_l/\sigma > 3.5$ do not pass through the region of liquid–like density and therefore cannot provide the true vapor–liquid interface. This leads to a large difference between the drop profile obtained by P1 (dashed line) and P2 (thick solid line) for $h_l/\sigma > 3.5$.

To diversify the possible procedures to calculate the drop profile, let us consider a one–dimensional FDD along a radial line passing through the point of maximum fluid density (cross in Fig. 2a) of FDD1. Because this point, in the considered case, is located on the axis of symmetry of FDD1, that FDD will be the same as that presented in Fig. 4a for $h_l = 0$. It is normal to the drop surface and provides the true drop profile. We will use this observation below when the new procedure for calculating the drop profile is presented.

In the considered example of FDD1, the drop profile at any choice of h_l has no intersection with the solid surface. Because of this, no contact angle can be defined for such a drop.

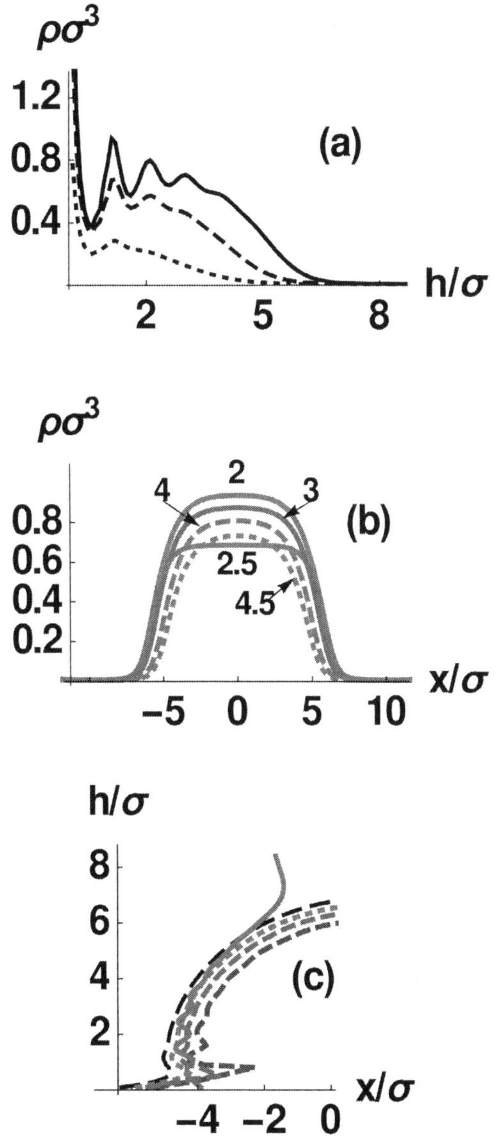

FIG. 5. (a) Examples of one–dimensional FDDs extracted from FDD2 along the vertical lines passing through the points with $x/\sigma = 0$ (solid line), −1 (dashed line), and −2 (dotted line). (b) Examples of one–dimensional FDDs extracted from FDD2 along the horizontal lines at distances from the solid surface noted for each line in units of σ. (c) Profiles of the drop obtained using procedures P1 (solid line) and P2 (dashed lines).

Now let us consider the FDD for a one–component fluid interacting with a solid (FDD2). An example of such an FDD is presented in Fig. 2b. Obviously, this distribution has no axis of symmetry. Another important feature of FDD2 is that, close to the solid surface, the density changes nonmonotonously and can acquire values greater than the bulk density of the considered fluid (see Fig. 5a where the one–dimensional FDDs along several vertical lines are presented).

Even though the gravity is negligible for nanodrops and is not considered here, one cannot expect that the drop profile can be approximated by a single circle because of the attraction between molecules of fluid and solid which is very strong close to the solid surface. However, it is reasonable to assume that the upper part of the nanodrop profile can be circular due to the quick decrease of

TABLE 1
Densities ρ_l and ρ_{div} for various reference layers for FDD2 and macroscopic contact angle for corresponding drop profiles obtained with procedure P2 as functions of h_l.

h_l/σ	2	2.5	3	4	4.5
$\rho_l\sigma^3$	0.935	0.687	0.834	0.808	0.734
$P_{div}\sigma^3$	0.517	0.357	0.484	0.443	0.400
θ_m (deg)	105.0	95.7	103.1	98.9	96.6

fluid–solid interactions with increasing distance from the surface. The one–dimensional horizontal FDDs at various distances h_e from the solid surface are similar to those in Fig. 4a. However they have more pronounced plateaus in the upper part (see Fig. 5b where those FDDs are presented for h_e/σ =2; 2.5; 3; 4; and 4.5), and the maximum density ρ_l as well as dividing densities ρ_{div} are changed nonmonotonous with changing h_e (see Table 1).

The shape of the drop profile obtained with procedure P1 is presented in Fig. 5c by the solid line. Close to the solid surface, the profile has oscillations which disappear in the upper part of the profile. In addition, the upper part has an unrealistic shape for the same reasons as it was for FDD1.

The application of procedure P2 to FDD2 is more complicated than its application to FDD1 because of the effect of the solid surface. In particular, if the horizontal reference layer is selected close to that surface, the value of ρ_l which is associated with the density of the liquid inside the drop can be considerably larger or smaller than ρ_{bulk}. This leads to considerable difference between the drop profiles (see Fig. 5c where various drop profiles obtained with P2 using the one-dimensional FDDs shown in Fig. 5b are presented by dashed lines). The contact angles for drop profiles presented in Fig. 5c are presented in Table 1 in the bottom line. All those angles are macroscopic contact angles i.e. obtained for the circular profile which makes a best fit with the upper part of the drop. The difference between those angles can be about 10°, i.e. is not small. Often [20], one selects as the most suitable reference layer the layer at a distance from the solid surface where the density oscillations due to the presence of the solid surface become small ($h_e/\sigma \geq 3$). For such a choice, the upper part of the profile is almost circular due to the decreasing influence of the solid. However even for this choice the density ρ_l of the nanodrop can be very different from ρ_{bulk} for large h_e and one should pay special attention to avoid this. In the considered example, an appropriate reference layer can be selected at $h_l = 4.5\sigma$. For this choice, the contact angle $\theta \simeq 96.6°$.

Note that oscillations of the drop profile close to the solid surface make the microcontact angle practically useless. For this reason, the knowledge of the upper part of the profile becomes most important because that part allows to determine the apparent (macroscopic) contact angle. To make the extraction of the upper part of the drop profile less sensitive to the choice of the reference layer in procedure P2 and, at the same time, to avoid the unrealistic shape which the drop profile obtained with P1 has at the top part, we suggest a new procedure, P3, which is described below. In this procedure, the one–dimensional FDDs, needed to find the locations of the dividing surfaces, will be taken along the radial directions with the center at the point of maximum density of the two–dimensional FDD2 (the point with $x = h = 0$ in Fig. 2b). The complication that arises with this approach is that for the lines which are close to the solid surface there is no visible plateau in the one–dimensional FDD as one can see from Fig. 6 where an example of such an FDD is presented.

To find the location of the equimolar dividing surface we selected that part of the FDD where the fluid density varies between the vapor– and liquid–like ones (see the rectangle area in Fig. 6) and find the best fit of this part of FDD with the sigmoidal curve given by Eq. (11). Then the

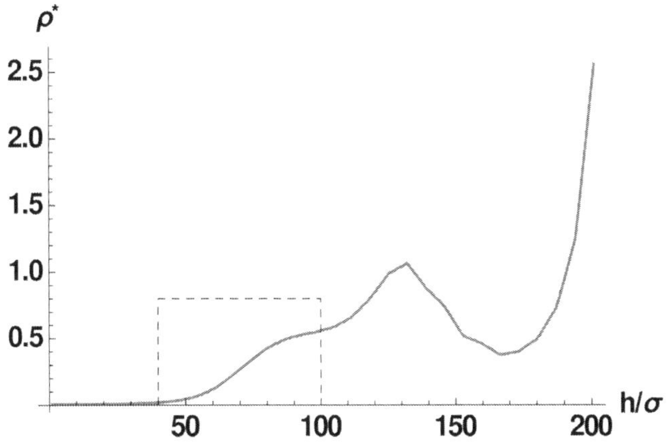

FIG. 6. One–dimensional FDD extracted from a two–dimensional FDD along the radial line at small inclination from the horizontal axis. The rectangle shows the range in which the fluid density changes from vapor– to liquid–like values. $\rho^* = \rho\sigma^3$.

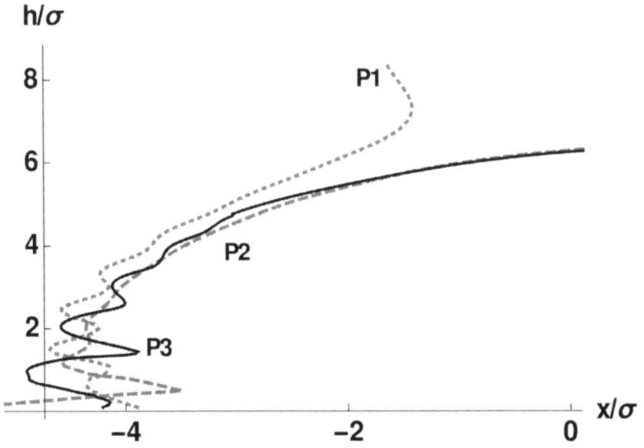

FIG. 7. Profiles obtained by procedures P1 (dotted line), P2 (dashed line), and P3 (solidline) for FDD2.

obtained function is used to calculate ρ_{div}. Note that the plateau in FDD becomes longer when the angle between the radial lines and the horizontal axis increases. As a consequence, the location of the equimolar dividing surface is better justified, especially for the upper part of the profile. In Fig. 7, the final drop profile obtained using P3 (thick solid line) is presented along with the profiles obtained by P1 (dotted line) and P2 (dashed line). The reference layer for P2 was selected at $h_e = 4.5\sigma$. One can see that the lower parts of all profiles have unusual shapes caused by the oscillatory behavior of the fluid density close to the solid surface. For the upper part of the profile ($h > 4.5\sigma$), which is far from the solid, procedures P2 and P3 provide close results. However, the contact angle for the profile obtained by P3 ($\theta = 84.6°$) differs from that obtained by P2 ($\theta = 96.6°$) by more than 10° that is considerable larger than the uncertainty of the modern experimental techniques.

The last example considered in this paper is related to a drop of the test fluid on the surface of a lubricating liquid that covers the solid surface. The FDD of the test fluid for such a case (FDD3) is presented in Fig. 2c. The three drop profiles extracted from that distribution by procedures P1, P2, and P3 are presented in Fig. 8 by dashed, solid, and dotted lines, respectively. The calculations

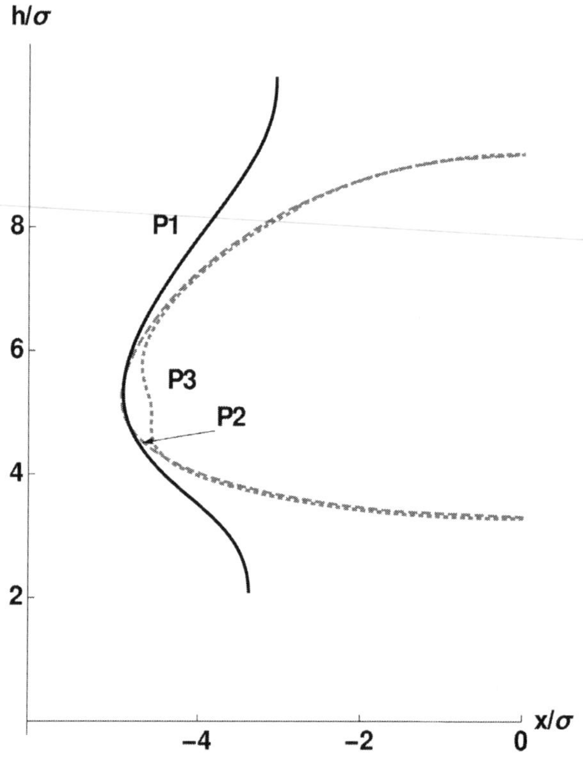

FIG. 8. Profiles of the drop obtained using procedures P1 (solid line) and P2 (dashed lines), and P3 (dotted line).

with procedure P2 involve the horizontal reference layer passing through the point of maximum density of FDD3. As in the two previously considered cases, procedure P1 provides unsatisfactory results in the upper part of the drop profile. Procedures P2 and P3 provide profiles which are close to each other. However, at the left and right sides of the lower part of the profile obtained by P3, there are deviations from the profile generated by P2, which can be explained by the existence of a layering effect which for the considered system is much weaker than for a fluid in contact with a solid surface and is not detected by procedure P2.

4. DISCUSSION

From the considered examples one can conclude that, for nanodrops, procedure P1 has limited applicability because it cannot provide an adequate drop profile for the upper part of the nanodrop. Note that for comparatively large drops this region represents only a few percents of the drop height while for a nanodrop it can represent more than 25% of the drop height. In addition, the upper part of the profile plays a major role in the calculation of the macroscopic contact angle which is the only meaningful one for a nanodrop. In turn, procedure P2 has uncertainty in selecting of the appropriate reference layer used to define the fluid density at the drop surface, which leads to considerable uncertainty in the contact angle (see Table 1). In both procedures, P1 and P2, the horizontal layers generally are not normal to the surface of the drop. For a planar liquid–vapor interface this is not important because the location of the equimolar dividing surface provided by the sigmoidal function does not depend on the direction of the reference layer [9]. However, for a nanodrop the liquid–vapor interface has a large curvature and the location of the equimolar dividing surface may depend on how large is the deviation of the reference layer from the normal to the interface. This is another disadvantage of P1 and P2 which is not present for P3. For the latter procedure, the reference layer is close to the normal to the liquid–vapor interface, particularly, for the upper part of the drop profile. Hence, procedure P3, even though more complicated than P1 and P2, seems to be more appropriate for calculating the profile of nanodrops. Note that the difference in contact angles for profiles obtained by P2 and P3 can be large enough and exceed the experimental uncertainties in contact angle measurements.

It is worth to note that procedures P1 and P2 are based on ideas similar to those used in Sobel procedure (edge detection) in the image analysis to find the boundary of objects [23]. Similarly, procedure P3 has some analogy with Cann procedure for edge detection [24]. The important difference is that the procedures used in image analysis require the knowledge of the threshold between vapor and liquid phases which is not given in advance (especially for nanodrops where the fluid density inside the drop can be different from the bulk value). For this reason, the usage of image analysis technique is not practical in the considered case and is not discussed in the present paper.

REFERENCES

1) C. Miller, E. Ruckenstein, The origin of flow during wetting of solids, J Coll Interface Sci 48 (3) (1974) 368–373 (Section 1.1 of this volume).
2) G.O. Berim, E. Ruckenstein, Microcontact and macrocontact angles and the drop stability on a bare surface, J Phys Chem 108 (50)(2004) 19339–19347 (Section 3.3 of this volume).
3) G.O. Berim, E. Ruckenstein, Microscopic treatment of a barrel drop on fibers and nanofibers, J Coll Interface Sci 286 (2) (2005) 681–695 (Section 3.4 of this volume).
4) Ruckenstein. E., G.O. Berim, Cylindrical droplet on nanofibers: a step toward the clam–shell drop description., J Phys Chem 109 (25) (2005) 12515–12524 (Section 3.5 of this volume).
5) G.O. Berim, E. Ruckenstein, Microscopic interpretation of the dependence of the contactangleon roughness, Langmuir 21 (17) (2005) 7743–7751 (Section 3.6 of this volume).
6) Evans. R., P. Tarazona, A simple density functional theory for inhomogeneous liquids–wetting by gas at a solid liquid interface, Mol Phys 52 (4) (1984) 847–857.
7) P. Tarazona, Free energy density functional for hard spheres., Phys Rev A 31 (4) (1985) 2672–2679.
8) Y. Rosenfeld, Free–energy model for the inhomogeneous hard–sphere fluid mixture and density–functional theory of freezing., Phys. Rev. Lett. 63 (9) (1989) 980–983.
9) M.J de Ruijter, T.D. Blake, J. De Coninck, Dynamic wetting studied by molecular modeling simulations of droplet spreading, Langmuir 15 (22) (1999) 7836–7847.
10) T. Werder, J.H. Walther, R.L. Jaffe, T. Halicioglu, P.J. Koumoutsakos, On the water–carbon interaction for use in molecular dynamics simulations of graphite and carbon nanotubes, Phys Chem B 107 (6) (2003) 1345–1352.
11) T. Koishi, K. Yasuoka, T. Ebisuzaki, S. Yoo, X.C. Zeng, Large–scale molecular– dynamics simulation of nanoscale hydrophobic interaction and nanobubble formation, J Chem Phys 123 (20) (2005) 204707
12) C. Yang, U. Tartaglino, B.N.J. Persson, Phys. Rev. Lett 97(11) (2006) 116103
13) N. Giovambattista, P.G. Debenedetti, P.J. Rossky, Effect of surface polarity on water contact angle and interfacial hydration structure, J Phys Chem B111 (32) (2007) 9581–9587.
14) S. Becker, M. Herbert, H.M. Urbassek, M. Horsch, H. Hasse, Contact angle of sessile drops in Lennard–Jones systems, Langmuir 30 (45) (2014) 13606–13614.
15) T. Bieker, S. Dietrich, Wetting of curved surfaces, Physica A 252 (1-2) (1998) 85–137.
16) S.H. Lee, P.J. Rossky, A comparison of the structure and dynamics of liquid water at hydrophobic and hydrophilic surfaces—a molecular–dynamics simulation study, J. Chem. Phys 100 (4) (1994) 3334–3345.
17) F. Porcheron, Monson PA. Mean–field theory of liquid droplets on roughened solid surfaces: application to superhydrophobicity, Langmuir 22 (4) (2006) 1595–1601.
18) A.P. Malanoski, B.J. Johnson, J.S. Erickson, Contact angles on surfaces using mean field theory: nanodroplets vs. nanoroughness, Nanoscale 6 (10) (2014) 5260–5269.
19) A. Malijevsky, Does surface roughness amplify wetting?, J Chem Phys 141 (18) (2014) 184703.
20) G.O. Berim, E. Ruckenstein, Nanodrop on a nanorough solid surface: density functional theory considerations, J Chem Phys 129 (1) (2008) 014708 (Section 4.4 of this volume). Contact angles of nanodrops on chemically rough surfaces. Langmuir 2009; 25 (16): 9285–9289 (Section 4.7 of this volume).

21) G.O. Berim, E. Ruckenstein, Simple expression for the dependence of the nanodrop contact angle on liquid–solid interactions and temperature., J Chem Phys 130 (4) (2009) 044709 (Section 4.1 of this volume).
22) G.O. Berim, E. Ruckenstein, A nanodrop on the surface of a lubricating liquid covering a rough solid surface, Nanoscale 7 (38) (2015) 7873–7884 (Section 4.8 of this volume).
23) Hart. P., R. Duda, in Pattern classification and scene analysis, Wiley (1973) 271–272.
24) J.F. Canny, A computational approach to edge detection, IEEE Trans. Pattern Anal. Mach. Intell. (1986) 679–698.

4.13 Microscopic Description of a Drop on a Solid Surface*

Eli Ruckenstein† *and Gersh Berim*

Department of Chemical and Biological Engineering, State University of New York at Buffalo, Buffalo, New York 14260, United States

Corresponding Author

† E-mail: feaeliru@buffalo.edu.
Tel.: +1 716 645 1179; fax: +1 716 645 3822.

1. INTRODUCTION

The understanding of the properties of a liquid drop on a solid surface is of importance in numerous applications (dyeing, adhesion, detergency, manufacturing of windshields, solar cell panels, and self cleaning surfaces, to name only a few) as well as in developing a general theoretical basis for the description of a fluid in contact with a solid surface.

The theoretical description of a liquid drop on a solid surface has a long history. It started more than two hundred years ago with Young's equation [1]

$$\gamma_{lv}\cos\theta = \gamma_{sv} - \gamma_{sl} \tag{1}$$

where γ_{lv}, γ_{sv} and γ_{sl} are the liquid—vapor, solid—vapor, and solid—liquid surface tensions, respectively and θ is the contact angle which the drop profile makes with an ideal (without any defects) solid surface.

Young's equation relates an observable quantity (contact angle) to macroscopic characteristics of the fluid and solid (surface tensions) which incorporate the effects of fluid—solid and fluid—fluid interactions as well as that of the temperature.

The theory was further developed by introducing the Young-Laplace and the augmented Young-Laplace equations (see e.g. Refs. [2–4] for the drop profile, for which the Young equation, Eq. (1), provides a boundary condition. For a cylindrical drop (drop which is infinite in one of the direction on the surface), the latter equation has the simple form

$$\frac{h_{xx}}{\left(1+h_x^2\right)^{3/2}} + \Pi\left(h\right)/\gamma_{lv} = -p_c/\gamma_{lv} \tag{2}$$

where $h \equiv h(x)$ is the drop profile (h provides the inclination of the profile above the surface), $\Pi(h)$ is the disjoining pressure [5,6], p_c is the capillary pressure and $h_x \equiv \frac{dh}{dx}$. Far from the solid surface $\Pi(h) = 0$ and this equation provides a circular drop profile, the extension of which up to the surface provides the macroscopic (apparent) contact angle θ_m which is a measurable quantity in macroscopic experiments. Note that the angle θ in Young equation (Eq. (1)) can be identified as θ_m. In the vicinity of the solid surface the profile of a liquid drop on a solid substrate exhibits a rapid variation of

* *Advances in Colloid and Interface Science* 157 (2010) 1-33. Republished with permission.

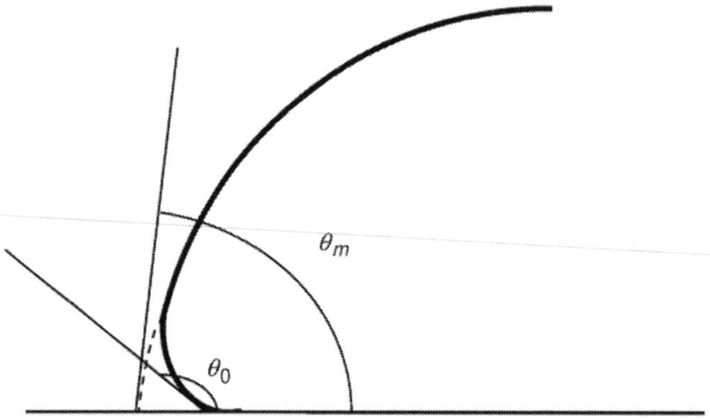

FIG. 1. (Reprinted from Ref. [13] with permission.) Microcontact angle θ_0 and macroscopic (apparent) contact angle θ_m. The latter angle is defined by extrapolating the circular part of the drop profile up to the surface.

curvature in a small region (~ 10–30 Å) near the surface, due to the rapid variation of the interactions between the molecules of liquid and those of the solid and the shape of the drop is not circular. Therefore an additional microcontact angle θ_0 can be considered along with the apparent angle θ_m [7,8] (see Fig. 1). Because of its small size, this region and, in particular, the microcontact angle θ_0 are practically undetectable by macroscopic experiments

For a rough solid surface, the surface tensions γ_{sv} and γ_{sl} are not clearly defined and Young's equation (Eq. (1)) does not provide the proper description of the contact angle. In this case the surface generates a nonuniform potential which affects the drop shape and the contact angle. For chemically inhomogeneous smooth solids such as, for instant, those composed of two types, *A* and *B*, of homogeneous patches, the apparent contact angle θ of a macroscopic drop is usually provided by the semi-empirical Casse—Baxter formula [9]

$$\cos\theta = f_A \cos\theta_A + f_B \cos\theta_B \tag{3}$$

where θ_A and θ_B are the contact angles on the corresponding homogeneous surfaces, which can be found from Eq. (1), and f_A and f_B ($f_B = 1 - f_A$) are the surface area fractions of the *A* and *B* patches, respectively.

For chemically homogeneous but physically rough (corrugated) surfaces, the apparent contact angle θ is described by the semi-empirical Wenzel formula [10]

$$\cos\theta = \mathrm{r}\cos\theta_{\mathrm{flat}} \tag{4}$$

where the roughness *r* is the ratio between the true area of the solid to its planar projection, and θ_{flat} is the contact angle for the smooth flat surface. If the smooth surface is hydrophobic (θ_{flat}>90°, cos θ_{flat}<0), the contact angle θ increases with increasing roughness. For a hydrophilic smooth surface (θ_{flat}<90°, cosθ_{flat}>0), the angle θ decreases with increasing roughness. Consequently, the roughness increases both the hydrophobicity and the hydrophilicity of the surface.

Eqs. (1)–(4) constitute the basis for the thermodynamic treatment of a macroscopic drop on a surface. Their simple form has made them for a long time a most popular tool for interpreting numerous experimental data and for the theoretical analysis of the wetting phenomena. However one can mention two main shortages of these equations. (i) When a thermodynamic approach is used, one assumes that the fluid is a continuous medium of constant density. As it was proven experimentally [11] and theoretically [12] this assumption is not valid in the vicinity of the solid surface where the fluid has an inhomogeneous density due to packing. (ii) When the size of the drop is very small (of the order of several tens of nanometers), the thicknesses of the liquid—vapor and fluid—solid interfaces become comparable with the size of the drop and the concepts of surface tensions become not clearly defined even for smooth surfaces.

For these reasons, the descriptions of the drop profile and of the contact angle from the point of view of a microscopic theory based only on intermolecular interactions become challenging problems. The two recent approaches of such a theory were presented in Refs. [8,13–21]. The first one [8,13–16] is based on the minimization of the potential energy of a drop due to fluid-fluid and fluid-solid interactions by neglecting the effect of entropy. This approach reproduces the augmented Young—Laplace equation (Eq. (2)) where the phenomenological quantities $\Pi(h)/\gamma_{lv}$ and p_c/γ_{lv} are expressed via the microscopic parameters involved in the interaction potentials. Along with Eq. (2), this approach provides the appropriate boundary conditions for this equation which follow from the minimization procedure [13]. Solutions of the derived differential equation for the corresponding boundary conditions exist only in some ranges of the interaction parameters. If the values of these parameters are outside these ranges, drops do not form on the surface (spread upon the surface).

The second microscopic approach to the problem of a drop on a solid surface [17–21] is based on the density functional theory (DFT) which was originally developed to describe inhomogeneous bulk fluids in contact with solid surfaces [12,22–24]. DFT is a statistical mechanics theory which involves interaction potentials and, in contrast to the first microscopic theory, accounts for the effect of temperature. The adaptation of the DFT to the description of drops on solid surfaces was made in Refs. [17,18,25,26], where methods for calculating drop profiles and contact angles which these profiles make with the surface were developed. The DFT theory has been also applied to the description of drops on rough and inclined surfaces [17,18].

Because the generality of the DFT is restricted by calculation difficulties DFT is applicable directly only to drops with sizes of several tens nanometers. Even though such droplets are never encountered in macroscopic experiments they play a role in microfluidic devices operating with ultrasmall aggregates of molecules. In an indirect way, by allowing to calculate the surface tensions and using the Young equation, the DFT can provide information regarding the contact angle of macrodrops.

The present paper reviews the above mentioned two approaches. The review of the first approach involves the consideration of a drop on a smooth planar surface (Sections 2.8 and 2.9), on a smooth fiber (Sections 2.10 and 2.11), and on physically rough planar surfaces (Section 2.12).

The review of the second approach is devoted to nanodrops and, in some cases, to macrodrops on various kinds of surfaces. The universal character of the dependence of the contact angle of a nanodrop and macrodrop on a smooth surface on the fluid-solid interaction parameters and temperature is considered in Section 3.2. Nanodrops on physically and chemically rough surfaces are examined in Sections 3.3 and 3.4, respectively. The sticking force which maintains the mechanical equilibrium of a nanodrop on an inclined surface is considered in Section 3.5.

2. APPROACH BASED ON THE MINIMIZATION OF POTENTIAL ENERGY OF A DROP

In this approach developed in Refs. [7,8,13–16] the equation of the macroscopic drop profile is derived by minimizing the total potential energy of a drop due to the interactions between the molecules of fluid themselves and between the molecules of fluid and those of solid. To simplify the calculations, several assumptions were made which are described below.

2.1. MAIN ASSUMPTIONS

Let us consider a liquid drop on a solid surface, which is in contact either with the vapor of the same fluid or with a neutral gas (air). For convenience, the fluid outside the drop will be referred below as vapor. The liquid in the drop is considered uniform, i.e. it has the same density ρ_L everywhere across the drop with the exception of two regions. In the first region, located in the very vicinity of the solid surface, the interactions of the liquid molecules with the solid are the strongest and for this reason the liquid molecules nearest to the solid surface are rearranged in a layer "sticked" to the surface, which will be assumed to be a monolayer separated from the surface by the distance σ_{LS}. This monolayer is considered to have a surface density ρ_{LS}. The remaining of the drop is separated from the first layer by the distance σ_{LL}. This assumption accounts for the existing of inhomogeneous density distribution in the fluid near the solid surface.

In the second region, the liquid-vapor interface, the density of the fluid changes continuously from ρ_L to the vapor density ρ_V. However this continuous change is often neglected by assuming instead a discontinuous change of the fluid density from ρ_L to ρ_V across the drop profile (sharp kink approximation [27]). To account to some extent for the change of liquid density on the drop boundary, it is assumed that an outward monomolecular layer characterized by the surface density ρ_{LV} is present. By assuming that the vapor has a density much smaller than the density of the liquid, the liquid—vapor interactions can be neglected. The solid substrate is considered homogeneous with a density ρ_S and with a planar or cylindrical surface which can be smooth or rough.

In general, the drop surface can have a complicated shape, especially if the drop is located on a nonplanar solid surface. This can make the calculation of the potential energy more difficult. To simplify the problem, two kinds of drops which are two-dimensional are selected for analysis; cylindrical and axisymmetrical drops on a planar surface presented schematically in Fig. 2; barrel and cylindrical drops on a fiber presented in Fig. 3. Note that in both cases the cylindrical drop is infinite in the y-direction. Because of the geometry of the system, all extensive quantities (potential energy of the drop, drop volume, area of the interface) for cylindrical drops will be calculated per unit length of the drop in the *y*-direction.

2.2. Potentials of intermolecular interactions

2.2.1. Interaction between the molecules of liquid

The potential $\phi_{LL}(r)$ of interaction between liquid molecules is selected in the form

$$\phi_{LL}(r)=\begin{cases} 0, & r>\eta, \\ -\epsilon_{LL}\left(\dfrac{\sigma_{LL}}{r}\right)^6, & \eta\geq r\geq\sigma_{LL}, \\ \infty, & r<\sigma_{LL}, \end{cases} \tag{5}$$

where r is the distance between the centers of the interacting molecules, σ_{LL} and $\epsilon_{LL}>0$ are the size of the repulsive core (hard core diameter) and the energy parameter, respectively, and η (the radius of the interaction sphere) is the largest value of r for which the interaction is considered as non-zero. The potential Eq. (5) constitutes a simplified version of the London—van der Waals (LvdW) potential with a hard core repulsion

$$\tilde{\phi}_{LL}(r)=\begin{cases} -\epsilon'_{LL}\left(\dfrac{\sigma'_{LL}}{r}\right)^6, & r\geq\sigma'_{LL}, \\ \infty, & r<\sigma'_{LL}, \end{cases} \tag{6}$$

which is difficult to use in the variational calculation of the drop shape due to the long-range nature of the potential. Because the London—van der Waals potential decreases rapidly with r, the approximate Eq. (5) seems reasonable. Note that a similar approximation is usually used in Monte Carlo (MC) and molecular dynamics (MD) simulations [28]. In Ref. [14] the relation between ϵ_{LL}, σ_{LL} and

η on one hand and ϵ'_{LL} and σ'_{LL} on the other one, was obtained using the requirement of equality of the total potential energies of a molecule in the bulk of a homogeneous liquid, calculated using the potentials Eqs. (5) and (6), respectively. That relation has the form

$$\frac{\epsilon_{LL}}{\epsilon'_{LL}} = \left(\frac{\sigma'_{LL}}{\sigma_{LL}}\right)^3 \left(1-\Delta^3\right)^{-1} \tag{7}$$

where $\Delta = \sigma_{LL}/\eta$. If, for example, $\eta = 3\sigma_{LL}$, the following particular choices for the parameters ϵ_{ll} and σ_{LL} are obtained: $\sigma_{LL} = \sigma'_{LL}$, $\epsilon_{LL} = \frac{27}{26}\epsilon'_{LL}$.

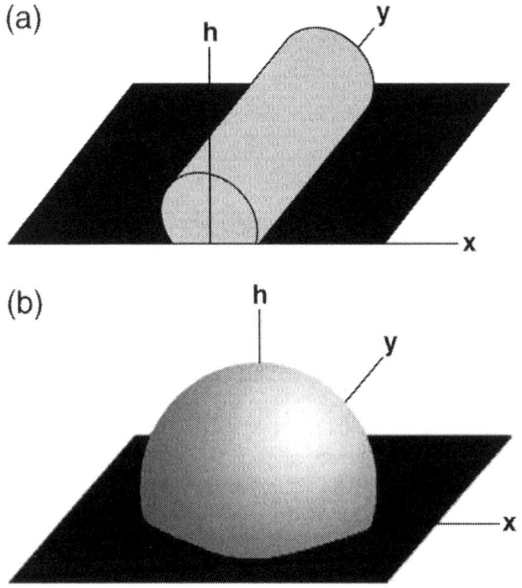

FIG. 2. Cylindrical (a) and axisymmetrical (b) liquid drop on a planar surface.

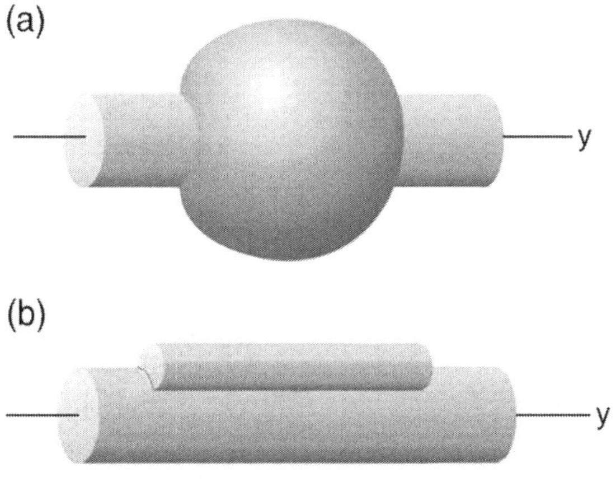

FIG. 3. Axisymmetric (a) and cylindrical (b) liquid drops on a fiber. Panel (a) is reprinted from Ref. [15] with permission.

2.2.2. Interactions between the molecules of fluid and solid

It is supposed that the interaction potential between a molecule of liquid and a molecule of solid is the sum of two components, $\phi_{LS}^{l}(r)$ and $\phi_{LS}^{s}(r)$ which represent the long-range and the short-range interactions, respectively. The former potential is chosen in the form

$$\phi_{LS}^{l}(r)=\begin{cases}\epsilon_{LS}\left[k_{\phi}\left(\dfrac{\sigma_{LS}}{r}\right)^{12}-\left(\dfrac{\sigma_{LS}}{r}\right)^{6}\right], & r\geq\sigma_{LS},\\ \infty, & r<\sigma_{LS},\end{cases} \tag{8}$$

where $k_{\phi}=0$ or 1, σ_{LS} and $\epsilon_{LS}>0$ are, respectively, the size of the repulsive core and the energy parameter of the liquid—solid interactions. For $k_{\phi}=0$ one recovers the LvdW potential and for $k_{\phi}=1$ the Lennard—Johns (LJ) potential. In real systems σ_{LL} and σ_{LS} are of the order of several angstroms [29].

The potential energy of a liquid molecule interacting with a planar semi-infinite solid via long-range interactions can be written as

$$\Phi_{LS}^{l}(h)=\frac{\pi}{6}\epsilon_{LS}\rho_{S}\sigma_{LS}^{3}\left[\frac{2}{15}k_{\phi}\left(\frac{\sigma_{LS}}{h+\sigma_{LS}}\right)^{9}-\left(\frac{\sigma_{LS}}{h+\sigma_{LS}}\right)^{3}\right], h\geq 0, \tag{9}$$

where $h+\sigma_{LS}$ is the distance measured from the center of the molecule of liquid to the center of the first layer of solid atoms ($h=0$ corresponds to the center of the layer of liquid molecules on the solid surface).

The total potential $\Phi_{LS}^{l}(h)$ of a liquid molecule interacting with a fiber via long-range interactions was calculated in Ref. [14], where the fiber was considered as an infinite cylindrical solid body of radius R_f. The result is

$$\begin{aligned}\Phi_{LS}^{l}(y)=&\frac{\pi\epsilon_{LS}\rho_{S}\sigma_{LS}^{6}}{24(R_f+y)(R_f^2-y^2)^2}\left\{(7R_f^2+y^2)\left[E\left(\frac{\pi}{4},-\frac{4R_f y}{(R_f-y)^2}\right)+E\left(\frac{3}{4},-\frac{4R_f y}{(R_f-y)^2}\right)\right]\right.\\ &\left.-(R_f+y)^2\left[F\left(\frac{\pi}{4},-\frac{4R_f y}{(R_f-y)^2}\right)+F\left(\frac{3\pi}{4},-\frac{4R_f y}{(R_f-y)^2}\right)\right]\right\}\end{aligned} \tag{10}$$

where $y>R_f$ is the distance of a molecule from the fiber axis, and $F(\phi, m)$ and $E(\phi, m)$ are elliptic integrals of the first and second kind, respectively [30]. In the limiting case $R_f\rightarrow\infty$ the potential $\Phi_{LS}^{l}(y)$ coincides with that of a molecule of liquid interacting with a semi-infinite solid given by Eq. (9). Potential Eq. (10) decreases rapidly within several hard core radii from the fiber surface and for $R_f\gg\sigma_{LS}$ can be approximated with high accuracy by the power law function

$$\Phi_{LS}^{l}(h)=-\frac{\pi}{6}\epsilon_{LS}\sigma_{LS}^{3}\rho_{S}\frac{k}{(1+h/\sigma_{LS})^{\nu}} \tag{11}$$

where $h=y-R_f-\sigma_{LS}$ and k and ν are constants (for given R_f) [14]. For example, for $R_f=100\sigma_{LS}\simeq 30$ nm, one has $\nu=3.01712$, and $k=0.9926$ [14].

The short-range potential $\phi_{LS}^{s}(r)$ accounts, for example, for the acid-base interactions which can be attractive as well as repulsive [31]. It is assumed that $\phi_{LS}^{s}(r)$ decays exponentially and can be considered zero at distances larger than $l_0\sim 1$ nm [31,32]. It will be supposed that the potential energy

$\Phi_{LS}^{s}(h)$ of a liquid molecule due to the short-range interactions has the value $-\epsilon_{LS}^{s}$ for the molecules of the first layer of liquid near the solid surface ($h = 0$) and is equal to zero for all other molecules. The distance of the first layer of liquid molecules from the solid surface is assumed to be equal to the hard core radius σ_{LS} of the potential Eq. (8). As a result, the potential $\Phi_{LS}^{s}(h)$ is given by

$$\Phi_{LS}^{s}(h) = \begin{cases} -\epsilon_{LS}^{s}, & h = 0, \\ 0, & h > 0. \end{cases} \tag{12}$$

Note that depending on the system the interaction constant ϵ_{LS}^{s} can be positive or negative. The microscopic calculation of ϵ_{LS}^{s} constitutes a separate issue.

Combining the potentials, one obtains the total potential $\Phi_{LS}(h)$ of a molecule of liquid interacting with a solid

$$\Phi_{LS}(h) = \begin{cases} \Phi_{LS}^{l}(0) + \Phi_{LS}^{s}(0), & h = 0, \\ \Phi_{LS}^{l}(0) & h > 0. \end{cases} \tag{13}$$

In most cases, the interaction between the molecules of vapor and solid has only a long-range component, which, for low-density vapors, involves only a monolayer of molecules adsorbed on the solid surface ($h = 0$). The potential $\Phi_{VS}(h)$ of that interaction can be written in the form

$$\Phi_{VS}(h) = \begin{cases} \Phi_{VS}^{l}(0), & h = 0, \\ 0, & h > 0 \end{cases} \tag{14}$$

where the potential $\Phi_{VS}^{l}(h)$ has the same form as $\Phi_{LS}^{l}(h)$ given by Eq. (9) and (11), where ϵ_{LS} and σ_{LS} have to be replaced by ϵ_{VS} and σ_{VS}, respectively.

2.3. Potential energy of the system

If the vapor which surrounds the drop has a low density, one can neglect the vapor—vapor and vapor—liquid potential energies in the calculation of the total potential energy U_{total} of the system and take into account only the interaction of the adsorbed vapor molecules with the solid. The interaction energy of the molecules of the solid between themselves will be considered constant and this will exclude it from consideration. Under those assumptions, U_{total} can be written in the form

$$U_{total} = U_{LL} + U_{LS} + U_{VS} \tag{15}$$

where U_{LL} and U_{LS} are the potential energies due to the interactions of the molecules of liquid between themselves and with those of the solid, respectively, and U_{VS} is a potential energy due to the interactions between the molecules of the vapor and the solid.

2.3.1. Potential energy of the liquid—liquid interactions

To calculate U_{LL}, let us divide the drop into three parts shown in Fig. 4 [13]. In the internal part V_I of the drop, a molecule of fluid interacts with all molecules inside the interaction sphere of radius η and hence its total potential energy does not depend on the location of the molecule and is equal to $\phi_1 = -\frac{4\pi}{3}\epsilon_{LL}\sigma_{LL}^{3}\rho_L(1-\Delta^3)$ where $\Delta = \sigma_{LL}/\eta$. Therefore, the contribution of the molecules located in the internal part of the drop is proportional to the volume V_1 of that part.

In the external parts (V_2 and V_3) of the drop, the liquid molecules are located at distances smaller than η from the drop surface as illustrated in Fig. 5. Assuming that the radius of curvature of the surface is much larger than η, one can consider the intersection of the interaction sphere with the drop to be planar. (Such an assumption is valid if the drop volume V is much larger than the volume of the interaction sphere $\left(V \gg \frac{4}{3}\pi\eta^3\right)$. Calculating the potential energy of a drop as function of distance l from the drop surface and integrating over the volumes $V_2 \simeq \eta S_{LV}$ and $V_3 \simeq \eta S_{LV}$ one can obtain the contributions to the total potential energy of the drop, which are proportional to S_{LV} and S_{LS}, where S_{LV} (S_{LS}) is the area of the liquid—vapor (liquid—solid) interface. As a result, the energy U_{LL} can be represented in the form [15]

$$U_{LL} = -K_v V + K_v S_{LV} + K_s' S_{LS} \tag{16}$$

where

$$K_v = \frac{2\pi}{3}\epsilon_{LL}\sigma_{LL}^3\rho_L^2\left(1-\Delta^3\right), \tag{17}$$

$$K_s = \frac{2\pi}{3}\epsilon_{LL}\sigma_{LL}^4\rho_L^2\left[\frac{17}{16} - \frac{3}{8}\Delta^2 - \frac{1}{2}\Delta^3 - \frac{3}{16}\Delta^4 - \frac{\rho_{LV}}{2\sigma_{LL}\rho_L}\left(1-\Delta^3\right)\right],$$

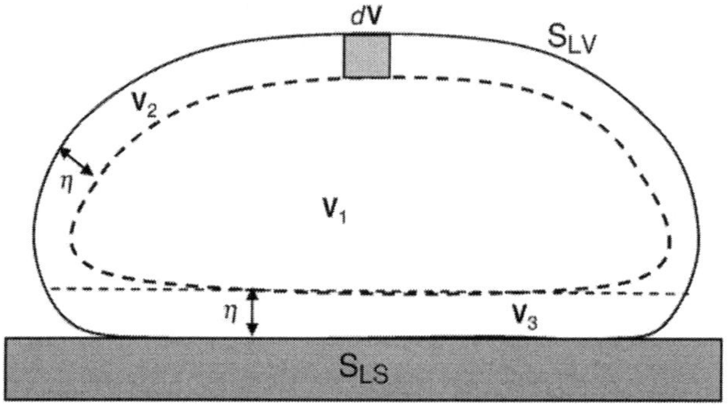

FIG. 4. Cross-section of a drop surface (solid line) and of the surface located at a distance η from the drop surface (dashed line). Inside the volume V_1, any molecule interacts with all the molecules inside the interaction sphere with the center in the center of the selected molecule. The elementary volume dV has the base area dS and height η.

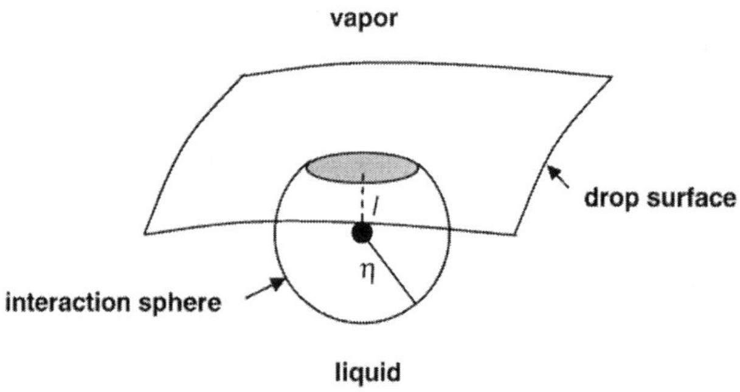

FIG. 5. (Reprinted from Ref. [8] with permission.) Interaction range of a molecule near the surface of a drop. If the principal radii of curvature of the drop surface are much larger than the radius η of the interaction sphere, the intersection of the interaction sphere with the drop can be considered planar. The filled circle represents the hard core σ_{LL} of the liquid—liquid interaction.

$$K_s' = \frac{2\pi}{3}\epsilon_{LL}\sigma_{LL}^4\rho_L^2\left[\frac{17}{16}-\frac{3}{8}\Delta^2-\frac{1}{2}\Delta^3-\frac{3}{16}\Delta^4-\frac{\rho_{LS}}{2\sigma_{LL}\rho_L}\left(1-\Delta^3\right)\right]$$

and ρ_{LV} and ρ_{LS} are the surface densities of the liquid at the liquid-vapor and liquid solid interfaces, respectively. Note that the coefficients K_v, K_s, and K_s' are independent of the geometrical characteristics of the drop.

2.3.2. Potential energy of the liquid—solid interactions

Calculating the interaction energy U_{LS} in Eq. (15), one should take into account that the molecules of the layer nearest to the solid are affected by both long-range and short-range interactions with the solid, the latter being represented by the potential Eq. (12). The molecules of the continuous part of the drop ($h > \sigma_{LS} + \sigma_{LL}$) do not exhibit short-range interactions Eq. (12).

Under the above assumptions, the energy U_{LS} is given by

$$U_{LS} = \rho_{LS}\Phi_{LS}(0)S_{LS} + \int_{V_{cont}} \rho_L\Phi_{LS}^l(h)dV \tag{18}$$

where V_{cont} denotes the volume of the continuous part of the drop. The latter can be written as the difference between the drop volume V and the volume V_g of the gap between the first layer and the continuous part of the drop which is approximately equal to $V_g = \sigma_{LL}S_{LS}$, i.e., $V_{cont} = V - S_{LS}\sigma_{LL}$. Then the second term in the right hand side of Eq. (18) becomes [14]

$$\int_{V_{cont}} \rho_L\Phi_{LS}^l(h)dV = \int_V \rho_L\Phi_{LS}^l(h)dV - \int_{V_g} \rho_L\Phi_{LS}^l(h)dV$$

$$\simeq \int_V \rho_L\Phi_{LS}^l(h)dV - \rho_L\sigma_{LL}\Phi_{LS}^l(\sigma_{LL})S_{LS}. \tag{19}$$

Consequently,

$$U_{LS} \simeq \int_V \rho_L\Phi_{LS}^l(h)dV + \left[\rho_{LS}\Phi_{LS}(0) - \rho_L\sigma_{LL}\Phi_{LS}^l(\sigma_{LL})\right]S_{LS}. \tag{20}$$

2.3.3. Potential energy of the vapor—solid interactions

As already mentioned in Section 2.2.2, only the contribution of the adsorbed vapor molecules to the vapor—solid potential energy will be taken into account. Using the potential Eq. (14) and denoting the surface density of the vapor molecules on the solid surface by ρ_{VS}, the potential energy U_{vs} of the vapor—solid interactions can be written in the form

$$U_{VS} = \rho_{VS}\left(S_{\text{solid}} - S_{LS}\right)\Phi_{VS}^l(0) \tag{21}$$

where S_{solid} is the total area of the solid surface and S_{LS} is the area of the liquid—solid interface.

Using Eqs. (16), (20), and (21) and omitting in the last equation the constant term proportional to S_{solid}, the following expression for the total potential energy of the system is obtained

$$U_{total} = -K_vV + K_sS_{LV} + \int_V \rho_L\Phi_{LS}^l(h)dV + K_sbS_{LS} \tag{22}$$

where

$$b = \frac{K_s'}{K_s} + \frac{1}{K_s}\left[\rho_{LS}\Phi_{LS}(0) - \rho_{LS}\sigma_{LL}\Phi_{LS}^{l}(\sigma_{LL}) - \rho_{L}\sigma_{LL}\Phi_{LS}^{l}(\sigma_{LL}) - \rho_{VS}\Phi_{VS}^{l}(0)\right]. \tag{23}$$

Eq. (22) provides the basis for the analysis of the drop characteristics (shape, stability,...).

2.4. Equation for the Drop Profile

In general, the profile of a drop on a planar surface and of a barrel drop on a fiber can have the shapes shown in Fig. 6a and b. The equation of the profile presented in Fig. 6a can be chosen in the form $h = h(x)$. However, if the drop profile has the shape presented in Fig. 6b with $\theta_0 > 90°$, then the above equation is no longer valid because of the lack of uniqueness of the function $h = h(x)$ for $|x| > x_0$, where x_0 is the abscissa of the leading edge of the drop. To avoid this difficulty and include both shapes, (Fig. 6a and b), one can write the equation for the drop profile in the form $x = x(h)$ [8,13] and consider only that part of the profile for which $x \geq 0$. The part for which $x < 0$ is symmetrical with that for $x \geq 0$ about the h-axis.

In Fig. 7 two possible profiles of a cylindrical drop on a fiber in the plane perpendicular to the axis of the fiber are presented together with the polar coordinate system (h, ϕ) used in the calculations. The surface of the cylindrical drop is obtained by translating the profile along the axis of the fiber. On a smooth surface the leading edge of the drop is located just on the fiber surface. To provide a unique description for both shapes (Fig. 7a and b), the drop profile equation can be expressed in the form $\phi = \phi(h)$, [15] restricting the considerations to the part of the profile for which $\phi \geq 0$.

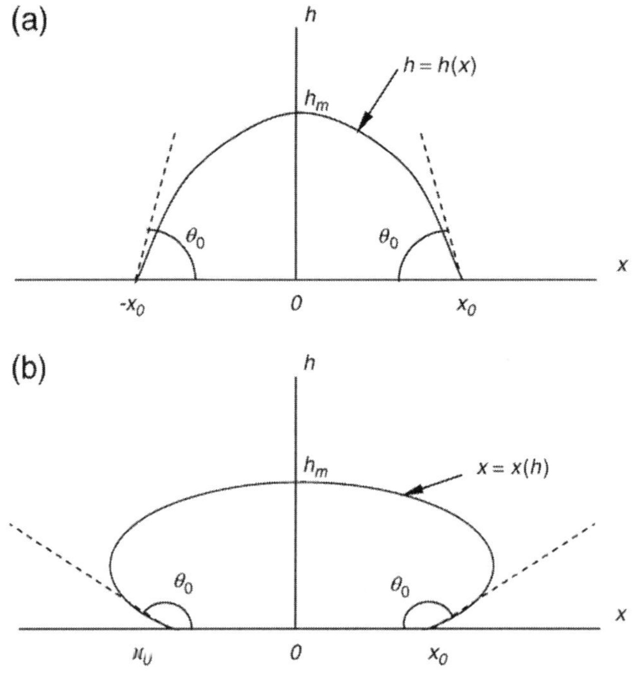

FIG. 6. (Reprinted from Ref. [13] with permission.) Two characteristic profiles of cylindrical and axisymmetrical drops: (a) the contact angle θ_0 at the leading edge of the droplet is less than 90°, (b) the same angle is larger than 90°. The y-axis is normal to the plane of the figure.

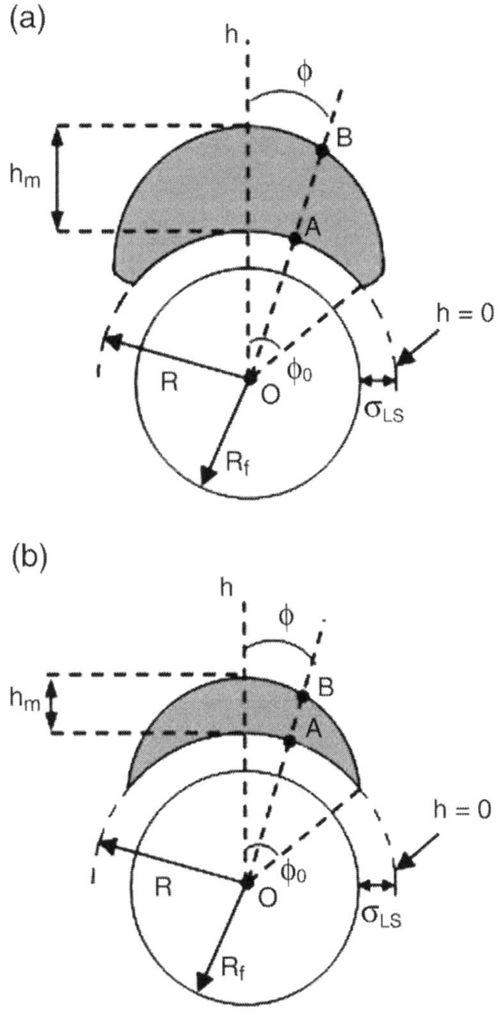

FIG. 7. (Reprinted from Ref. [15] with permission.) Characteristic profiles of a cylindrical drop in a plane normal to the fiber axis: (a) the microcontact angle θ_0 is larger than 90°; (b) θ_0 is smaller than 90°.

In the derivation of an equation for the drop profile, the constraint of constant volume V of the drop (or constant number of molecules $N = \rho_L V$ in the drop) is usually used [8,13–16]. Using such a constraint, the equation for the drop profile can be obtained by minimizing the functional

$$F = \left(-a_0 + a_1\lambda\right)V + S_{LV} + a_1 \int_V \Phi^l_{LS}\left(h\right)dV + bS_{LS} \tag{24}$$

with respect to the function $h = h(x)$ (or $\phi = \phi(h)$) [8,13–16]. In Eq. (24), λ is a Lagrange multiplier, $a_0 = K_l/K_s$, $a_l = \rho_L/K_s$.

The function $h = h(x)$ (or $\phi = \phi(h)$) appears in the functional Eq. (24) through the volume of the drop V and the areas of liquid—vapor (S_{LS}) and fluid—solid (S_{LV}) interfaces. The explicit form of the equation of drop profile depends on the geometry of the problem, namely, on the shape of the drop and shape of the solid surface.

For example [8,13], for a cylindrical drop on a planar surface the drop volume V, surface areas S_{LV} and S_{LS}, and $\int_V \Phi^l_{LS}\left(h\right)\mathrm{d}h$ are given by the expressions

$$V = 2\int_0^{h_m} x dh,\ S_{LV} = 2\int_0^{h_m} \sqrt{1+x_h^2}\, dh \tag{25}$$

$$S_{LS} = 2x(0) = -2\int_0^{h_m} x_h dh,\ \int_V \Phi_{LS}^l(h) dV = 2\int_0^{h_m} \Phi_{LS}^l(h) x dh$$

where the subscript to x denotes the derivative with respect to the corresponding variable, h_m is the drop height and the identity $x(0) = -\int_0^{h_m} x_h \mathrm{dh}$ follows from the obvious expression $x(h_m) = 0$.

The functional Eq. (24) can be rewritten in the following more convenient form

$$F = \int_0^{h_m} F(h, x, x_h) dh, \tag{26}$$

where

$$F(h, x, x_h) = \left[\lambda_c + a_1\Phi_{LS}^l(h)\right]x + \sqrt{1+x_h^2} - bx_h, \tag{27}$$

$$\lambda_c = a_1\lambda - a_0.$$

Using the standard calculus of variations [33], the following differential equation for the profile could be obtained

$$\frac{x_{hh}}{\left(1+x_h^2\right)^{3/2}} - a_1\Phi_{LS}^l(h) = \lambda_c \tag{28}$$

which coincides in form with the augmented Young-Laplace Eq. (2) for cylindrical drops. The first term of Eq. (28) represents the curvature $\kappa(h)$ of the profile, which in the considered case can be rewritten as

$$\kappa(h) = \lambda_c + a_1\Phi_{LS}^l(h) \tag{29}$$

with a curvature radius given by

$$R^{\text{curv}}(h) = \frac{1}{\kappa(h)}. \tag{30}$$

If the drop is axisymmetrical with respect to the h-axis, the Euler–Lagrange equation for the profile has the form [13]

$$\frac{x_{hh}}{\left(1+x_h^2\right)^{3/2}} - \frac{1}{x\sqrt{1+x_h^2}} + a_1\Phi_{LS}^l(h) - 2\lambda_a = 0 \tag{31}$$

where $\lambda_a = (a_1\lambda - a_0)/2$.

For a barrel drop on a fiber, the equation for the drop profile has the from [14]

$$\frac{x_{hh}}{\left(1+x_h^2\right)^{3/2}} + \frac{x_h}{(R+h)\left(1+x_h^2\right)^{1/2}} - a_1\Phi_{LS}^l(h) = \lambda_b \tag{32}$$

where $\lambda_b = a_1\lambda - a_0$, $R = R_f + \sigma_{LS}$, and R_f is the fiber radius.

Finally, the equation for the profile of a cylindrical drop on a fiber has the form [15]

$$\frac{(R+h)\phi_{hh}+2\phi_h+(R+h)^2\phi_h^3}{\left[1+(R+h)^2\phi_h^2\right]^{3/2}} - a_1\Phi_{LS}^{l}(h) = \lambda_c \tag{33}$$

where $\lambda_c = a_1\lambda - a_0$.

2.5. Boundary conditions for the equation for the drop profile

In the considered variational problem the height of the drop h_m, or the half width x_0 (or the angular half width ϕ_0) are not known in advance and hence cannot be used as boundary conditions. However, the boundary conditions can be extracted from the so called transversality conditions [33] which are consequences of the facts that the apex of the drop and the contact point of the profile with the solid should be located on the h- and x-axes, respectively. For drops on a planar surface and for a barrel drop on a fiber, such conditions can be written in the following general form

$$\left.\frac{\partial}{\partial x_h}F(h,x,x_h)\right|_{h=0} = 0, \left.\left[F\left(h,x,x_h - x_h\frac{\partial}{\partial_{xh}}F(h,x,x_h)\right)\right]\right|_{h=h_m} = 0 \tag{34}$$

and provide the following two boundary conditions for Eqs. (28), (31), and (32)

$$\left.\frac{1}{\sqrt{1+x_h^2}}\right|_{h=h_m} = 0, \tag{35}$$

$$\left.\frac{x_h}{\sqrt{1+x_h^2}}\right|_{h=0} = b. \tag{36}$$

The first condition, Eq. (35), or, equivalently,

$$x_h\big|_{h=h_m} = \infty \left(h_x\big|_{x=0} = 0\right), \tag{37}$$

reflects the fact that the tangent to the drop profile at its apex should be parallel to the solid surface. The second condition, Eq. (36), provides the microcontact angle θ_0 of the drop on the solid surface (Fig. 1) through the equation

$$\cos\theta_0 = -b \tag{38}$$

which follows from Eq. (36) because $x_h|_{h=0} = -\cot\theta_0$. It should be noted, that the microcontact angle is obtained from pure microscopic considerations, without involving such thermodynamic quantities as the surface tensions.

For an axisymmetrical drop on a planar surface, the boundary conditions coincide to those for a cylindrical drop.

For a cylindrical drop on a fiber, the transversality conditions provide the following boundary conditions [14]

$$\left.\frac{1}{\sqrt{1+(R+h)^2\phi_h^2}}\right|_{h=h_m}=0, \tag{39}$$

$$\left.\frac{\phi_h}{\sqrt{1+R^2\phi_h^2}}\right|_{h=0}=\frac{b}{R} \tag{40}$$

The first condition, Eq. (39), or, equivalently, $\phi_h\big|_{h=h_m}=\infty\ \left(h_\phi\big|_{\phi=0}=0\right)$, reflects the fact that the tangent to the drop profile at its apex should be normal to the polar radius at the apex. The second condition together with the equation

$$\phi_h|_{h=0}=-\frac{1}{R}\cot\theta_0, \tag{41}$$

provides an equation for the microcontact angle θ_0 which is identical with Eq. (38).

Note that Eq. (41) follows from the fact that in polar coordinates the angle α which the curve $\phi=\phi(r)$ makes with the circle $r=R$ is given by the expression

$$\tan\alpha=-\frac{1}{R\phi_r},\ \left(\phi_r\equiv\frac{d\phi}{dr}\right) \tag{42}$$

and that in our case $\phi_r=\phi_h$.

2.6. Solution of the Equation for the Drop Profile

Eqs. (28) and (33) for a cylindrical drop on a planar surface and on a fiber, respectively, and Eq. (32) for a barrel drop on a fiber allow to obtain exact solutions for the drop profile [8,13–15]. For illustration, the solution of Eq. (28) for the profile of a cylindrical drop on a planar surface is considered below in some details. The solutions of other equations can be found in Refs. [8,13–15].

The substitution $p=x_h$ transforms Eq. (28) into the first order differential equation

$$\frac{p_h}{\left(1+p^2\right)^{3/2}}=\lambda_c+a_1\Phi_{LS}^l(h) \tag{43}$$

which has the solution

$$p=\frac{\Psi(h)}{\sqrt{1-\Psi^2(h)}} \tag{44}$$

where

$$\Psi(h)=\lambda_c h+D(h)+C, \tag{45}$$

$$D(h)-a_1\int\Phi_{la}^l(h)dh\left(a_1=\frac{\rho_L}{K_S}\right) \tag{46}$$

and C is a constant of integration. After the substitution of x_h for p into Eq. (44) the following general solution for $x(h)$ is obtained [8]

$$x(h) - x_0 = \int_0^h \frac{\Psi(h)}{\sqrt{1-\Psi^2(h)}} dh \tag{47}$$

where $x_0 \equiv x(0)$ (Fig. 6) is a second integration constant. For a profile symmetrical about the h-axis, x_0 can be determined from the condition for the drop apex to be located on the h-axis [$x(h_m) = 0$], that leads to the expression

$$x_0 = -\int_0^{h_m} \frac{\Psi(h)}{\sqrt{1-\Psi^2(h)}} dh. \tag{48}$$

The first integration constant, C, can be obtained using the boundary condition Eq. (36), which is equivalent to the equation $\Psi(0) = b = -\cos\theta_0$. The latter equation provides for C the value $C = b - D(0)$, and the function $\Psi(h)$ becomes

$$\Psi(h) = b + \lambda_c h + D(h) - D(0). \tag{49}$$

After the determination of x_0 and C, the solution Eq. (47) still contains two unknown parameters, namely λ_c and h_m. A connection between them is provided by the boundary condition Eq. (35), which leads to the equation $\Psi(h_m) = \pm 1$. The sign can be selected by observing that in the absence of liquid-solid interactions a drop must have a spherical shape and contact the solid surface only at $x_0 = 0$. Such a solution can be obtained [8] if $\Psi(h_m) = -1$, which provides the following expression for λ_c

$$\lambda_c = -\frac{1}{h_m}\left[1 - \cos\theta_0 + D(h_m) - D(0)\right]. \tag{50}$$

As a consequence, the function $\Psi(h)$ acquires the form

$$\Psi(h) = -1 + \left(1 - \frac{h}{h_m}\right)\left[1 - \cos\theta_0 - D(0)\right] + D(h) - \frac{h}{h_m} D(h_m). \tag{51}$$

The last necessary condition for the determination of all unknown parameters is provided by the equation for the drop volume $V = 2\int_0^{hm} \mathrm{xdh}$. Integrating by parts and taking into account that $x(h_m) = 0$, one can rewrite the latter equation in the form $V = -2\int_0^{h_m} h x_h dh$, which after substituting the expression Eq. (44) for $x_h = p$ becomes

$$V = -2\int_0^{h_m} \frac{h\Psi(h)}{\sqrt{1-\Psi^2(h)}} dh. \tag{52}$$

The obtained analytical solution provides the possibility to examine the conditions for drop formation on the solid surface as well as the drop shape. This analysis is provided in the next sections.

2.7. Existence of a cylindrical drop on a planar surface

The first necessary condition for the existence of a drop on a bare surface follows from Eq. (38), and has the simple form

$$|b| \le 1. \tag{53}$$

It provides (through Eq. (23)) a restriction for the values of the microscopic parameters for which a drop can be formed on a solid surface. If $|b| > 1$, $\left(\text{i.e.} |\cos\theta_0| > 1\right)$ the drop will spread completely over the surface of the solid. It should be noted that this restriction is very different from the traditional one. The latter implies $|\cos\theta_{\text{m}}| > 1$ as a condition of nonexistence of a drop on the surface.

The second necessary condition results from the fact that the integral in the right hand side of Eq. (47) has meaning only if the function $\Psi(h)$ satisfies the inequality

$$\left|\Psi(h)\right| < 1 \text{ for } 0 < h < h_m. \tag{54}$$

In general, function $\Psi(h)$ depends on the parameters of the interactions, the geometry of the system, and the drop height h_m (or, equivalently, drop volume) and satisfies the conditions $\Psi(h_m) = -1$, $\Psi(0) = -\cos\theta_0$. Analysis has shown that for any given values of the parameters, $\Psi(h)$ can have the behaviors presented in Fig. 8. However, because only the first one marked with *I* in Fig. 8 satisfies the restriction provided by Eq. (54), a cylindrical drop can form only for those values of the parameters for which $\Psi(h)$ has such a behavior.

To examine in more details the consequences of the second necessary condition let us note that the function $D(h)$ given by Eqs. (46) and (9), can be represented in the form

$$D(h) = \frac{a}{2}\left[-\frac{k_\phi}{30}\left(\frac{\sigma_{LS}}{h+\sigma_{\text{LS}}}\right)^8 + \left(\frac{\sigma_{LS}}{h+\sigma_{\text{LS}}}\right)^2\right] \tag{55}$$

where

$$a = \frac{\overset{*}{\rho}_s S_{LS}^4 \overset{*}{\epsilon}_{LS}}{4\overset{*}{\rho}_L \varphi_S(\Delta)}, \tag{56}$$

$\overset{*}{\rho} = \rho\sigma_{\text{LL}}^3$, $\overset{*}{\epsilon}_{LS} = \epsilon_{LS}/\epsilon_{LL}$, $S_{LS} = \sigma_{LS}/\sigma_{LL}$, and

$$\varphi_s(\Delta) = \frac{17}{16} - \frac{3}{8}\Delta^2 - \frac{1}{2}\Delta^3 - \frac{3}{16}\Delta^3 - \frac{\rho_{LS}}{2\sigma_{LL}\rho_L}\left(1-\Delta^3\right). \tag{57}$$

The dimensionless constant a (Eq. (56)) incorporates the densities of liquid and solid as well as the parameters of fluid-fluid and long range fluid—solid interactions.

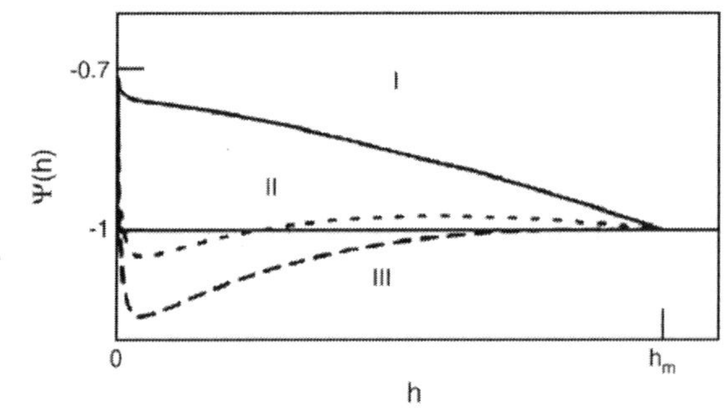

FIG. 8. (Reprinted from Ref. [15] with permission.) Three types of behavior of the function $\Psi(h)$ given by Eq. (49). A cylindrical drop on the fiber exists only for those values of the parameters which provide the first type of behavior.

The parameters of the short range fluid-solid interaction are present through the potential $\Phi_{LS}(0)$ in Eq. (23) which determines the microcontact angle θ_0 (see Eq. (38)). For this reason, θ_0 and a can be considered as two independent variables which affect the solution of the equation for the drop profile. The third independent parameter is the drop volume V, or, equivalently, the drop height h_m which are connected through Eq. (52). Below, the formation of a drop on a planar solid surface is examined with respect to those three parameters (a, θ_0, and h_m).

Let us first consider the case in which the microcontact angle θ_0 is fixed ($\theta_0 = 60°$) and only h_m and a are variables. The analysis of the behavior of the function $\Psi(h)$ shows that inequality Eq. (54) is valid only in a domain of the h_m–a plane below the critical curve presented in Fig. 9a. For the Lennard-Jones potential ($k_\phi = 1$), the latter curve asymptotically approaches the value $a_{c1} = 1.04$ as $h_m \to \infty$. This means that for $a > a_{c1}$ a drop on the surface has for a selected θ_0 a restricted upper bound height. The latter bound monotonously decreases with increasing a approaching zero. Considering the height $h = 5\sigma_{LS}$ as the smallest height which a drop can have on the solid surface one can determine a second specific value, a_{c2}, of the parameter a above which a drop does not exist. For the considered example, $a_{c2} \simeq 1.13$.

By performing similar analysis for other values of θ_0 one can plot a_{c1} and a_{c2} against θ_0 in a θ_0–a plane (see Fig. 9b). The pairs (θ_0, a) for which the drop height h_m can have any value are located below the solid curve. Above the dashed line, a drop does not form, and in the domain between those curves, h_m can take only restricted values which are smaller than an upper bound. The latter should be calculated for each pair (θ_0, a) separately. For $a < a_{c1}$, the drop can have any height.

A similar analysis performed for a fixed value of a ($a = 0.98$) and various microcontact angles θ_0 and drop heights h_m provides the domain of drop existence in the θ_0–h_m plane shown in Fig. 9c. For $\theta_0 < 57.9°$ the drop can have a restricted height. For $\theta_0 < 55.6°$, the possible size of the drop becomes unreasonably small ($h_m < 5\sigma_{LS}$); therefore one can conclude that for $\theta_0 < 55.6°$ a drop does not form on the surface, but for $\theta_0 > 57.9°$ the drop can have any height.

2.8. Shape of a Cylindrical Drop

The drop shape which follows from the solution of Eq. (28) depends on the type of potential used for the fluid-solid interactions and on the values of the parameters of all interaction potentials. However some conclusions can be drawn from the general form of Eq. (28) and its solution provided by Eqs. (47)–(52). Let us rewrite Eq. (29) for the drop curvature using Eq. (50) for λ_c in the form

$$\kappa(h) = -\frac{1}{h_m}\left[1 - \cos\theta_0 + D(h_m) - D(0)\right] + a_1\Phi_{LS}^{1}(h). \tag{58}$$

In Fig. 10 a typical h-dependence of the curvature is presented for a drop with $h_m = 85\sigma_{LS}$ and $a = 0.8$ for a Lennard-Jones potential ($k_\phi = 1$ in Eq. (9)). One can see that $\kappa(h)$ rapidly decreases as h increases from $h = 0$ to $h \sim 5\sigma_{LS}$ and remains almost constant for $h \simeq 5\sigma_{LS}$ acquiring a value approximately equal to that at the drop apex ($h = h_m$). This means that for $h \geq 5\sigma_{LS}$ the drop profile can be considered as part of a circle. Such a feature of the drop profile is expected because of the rapid decrease of the fluid—solid interactions with increasing distance from the surface. For large h_m, the radius of the circle, R_m^{curv}, is a linear function of the drop height

$$R_m^{curv} \equiv \frac{1}{\kappa(h)} = h_m\left[1 - \cos\theta_0 - \frac{a}{2}\left(1 - k_\phi/30\right)\right]^{-1}. \tag{59}$$

The latter equation was obtained by combining Eq. (58) with Eqs. (55) and (9) and by considering the limit $h_m/\sigma_{LS} \gg 1$.

Near the solid surface, the curvature of the drop profile is not constant and the profile cannot be approximated by a circle.

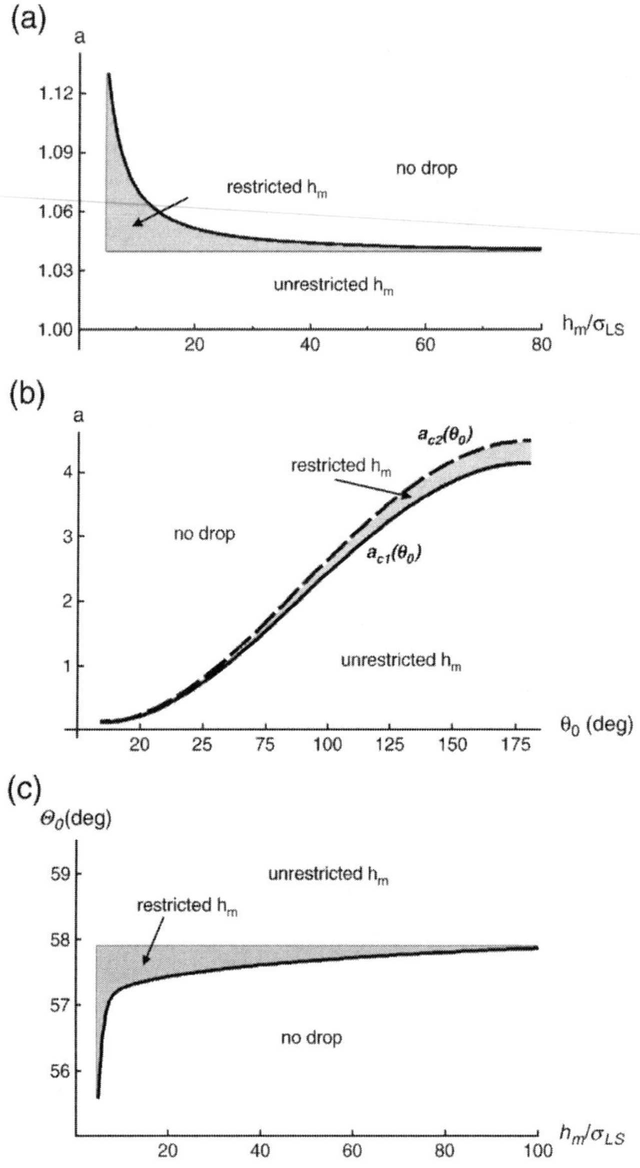

FIG. 9. All curves are plotted for Lennard-Jones potential of fluid—solid interactions. (a) The critical curve (solid line) in the h_m–a plane for $\theta_0 = 60°$. No drops exist on the surface for the points (h_m, a) which lie above this curve. For $a<a_{c1} = 1.04$, the drop on the surface can have any height. (b) The critical curves $a = a_{c1}(\theta_0)$ and $a = a_{c2}(\theta_0)$ in the θ_0–a plane. (c) The critical curve in the h_m–θ_0 plane for $a = 0.98$.

Eq. (59) can be used to calculate the macroscopic contact angle θ_m, which is defined (see Fig. 11) as the angle between the solid surface and the continuation of the circular part of the profile until its intersection with the surface of the solid. Such a definition is reasonable if the range h_2 in Fig. 11 of rapid change of the drop curvature is much smaller than the range h_1 in which the curvature is almost constant. One can easily see from Fig. 11 that θ_m is given by the equation

$$\cos\theta_m = 1 - \frac{h_m}{R_m^{curv}} \tag{60}$$

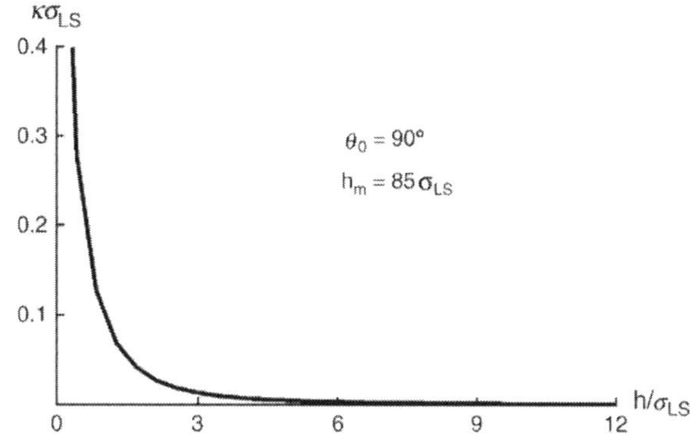

FIG. 10. Curvature κ of the drop profile as function of h for the drop with $h_m = 85\sigma_{LL}$. The microcontact angle $\theta_0 = 90°$.

which combined with Eq. (59) becomes

$$\cos\theta_m = \frac{a}{2}\left(1 - k_\phi / 30\right) + \cos\theta_0. \tag{61}$$

The latter formula provides a connection between the micro- (θ_0) and macro- (θ_m) contact angles of a large drop.

For a LvdW potential, typical profiles of cylindrical drops with a constant volume $V / \sigma_{LS}^3 = 4.21 \times 10^4$, but various microcontact angles are presented in Fig. 12 for $a = 0.8$ and $x \geq 0$. For this value of a the drop can have any height for microcontact angles larger that $\theta_{0,1} = 53.1301°$. For $\theta_0 < \theta_{0,1}$, the height is limited. Fig. 12 shows that, for $\theta_0 > \theta_{0,1}$, the drop height decreases and the half-width x_0 of the contact area of the drop with the solid increases with decreasing microcontact angle. The closer θ_0 is to $\theta_{0,1}$, the larger is x_0, and the drops become almost planar. (This planar drop is not included in Fig. 12 because of the very large value of x_0).

From Eq. (61) it follows that θ_m is always less than θ_0 because a is a positive quantity and, hence, $\cos\theta_m > \cos\theta_0$.

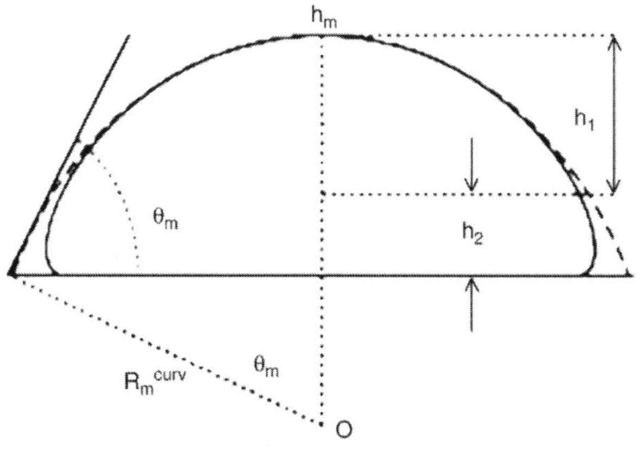

FIG. 11. (Reprinted from Ref. [13] with permission.) Definition of the macroscopic contact angle θ_m. The dashed line corresponds to the hypothetical circular profile of radius R_m^{curv}, h_1 is the range in which the circular profile almost coincides with the drop profile (solid line), and h_2 is the range of rapid change of the drop profile curvature. The ratio h_2/h_1 has to be small for the macroscopic contact angle to be meaningful.

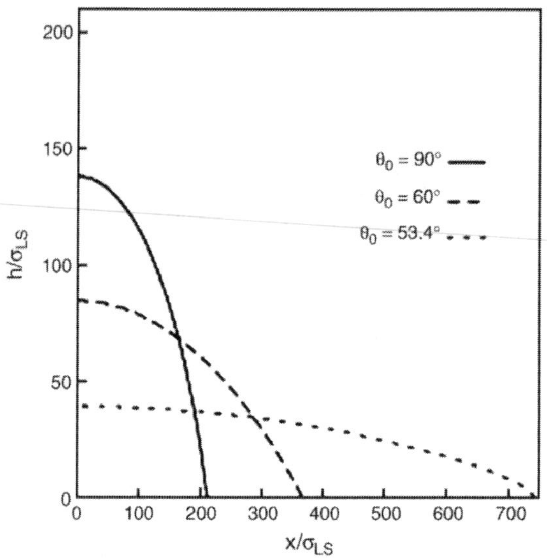

FIG. 12. (Reprinted from Ref. [13] with permission.) Dependence of the drop shape on the microcontact angle for $a = 0.8$. All drops have the same volume $V/\sigma_{LS}^3 \simeq 4.2\times10^4$.

For the LJ potential, the shape of the drop has exactly the same peculiarities as for the LvdW potential. However, the corresponding profiles for the same values of a and θ_0 differ in details. Fig. 13 presents the profiles of the drops of volumes equal to $V/\sigma_{LS}^3 \simeq 2.35\times10^5$, for both potentials, $a = 0.8$ and different values of θ_0, in the domain of unlimited height. If the point (a, θ_0) is not close to the critical curves separating the two domains, then the difference between the profiles is not large (the curves of Fig. 13 for $\theta_0 = 60°$). However, the difference becomes considerable if that point

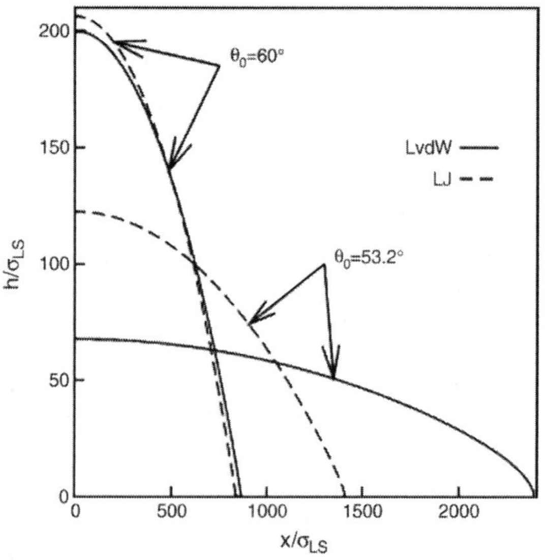

FIG. 13. (Reprinted from Ref. [13] with permission.) The profiles of a drop of volume $V/\sigma_{LS}^3 \simeq 2.35\times10^5$ for the LvdW (solid lines) and LJ (dashed lines) potentials and $a = 0.8$ for two microcontact angles. All the profiles belong to the domain of unlimited height. The values of θ_0 which separate the domains of limited and unlimited heights, for $a = 0.8$, are $\theta_{0,1} = 53.1301°$ and $\theta_{0,2} = 52.1691°$ for the LvdW and LJ potentials, respectively.

is in the vicinity of the critical curves (the curves for $\theta_0 = 53.2°$). The reason for such a difference in behavior is that, for $a = 0.8$, the smallest possible value of θ_0 from the domain of unlimited drop height is $\theta_{0,2} = 52.1691°$ for the LJ potential and $\theta_{0,1} = 53.1301° > \theta_{0,2}$ for the LvdW potential. As already mentioned, the height of the profile decreases when θ_0 approaches the corresponding smallest value ($\theta_{0,1}$ or $\theta_{0,2}$) at constant a. The closer is θ_0 to $\theta_{0,1}$ ($\theta_{0,2}$), the higher is the rate of decrease. Because $\theta_{0,2} < \theta_{0,1}$, the selected $\theta_0 = 53.2°$ is closer to $\theta_{0,1}$ than to $\theta_{0,2}$. Therefore the profile height decreases more rapidly for the LvdW potential than for the LJ potential, and, as a consequence, the drop height for the LvdW potential becomes smaller than that for the LJ potential. The physical explanation of such a behavior is that the repulsive part of the LJ potential reduces the effect of the attractive forces between the solid and liquid molecules. This reduction is more effective for small drop heights (the case of $\theta_0 = 53.2°$), and almost negligible for larger heights ($\theta_0 = 60°$ in our example).

2.9. Axisymmetrical drop on a planar surface

Eq. (31) for the profile of an axisymmetrical drop cannot be reduced to a first order differential equation, as for cylindrical drops, and therefore all the results presented below have been obtained numerically. Below, results obtained in Ref. [13] are presented. Note that for simplicity, the potential of the fluid-fluid interactions $\phi_{LL}(r)$ was taken in that paper in a simplified form (square-well potential)

$$\phi_{LL}(r) = \begin{cases} -\varepsilon_{LL}, & r \leq \eta, \\ 0, & r > \eta, \end{cases} \tag{62}$$

As a consequence, the parameter a (Eq. (56)) which is involved in the calculation of the drop profile is given now by the expression

$$a = \frac{4\epsilon_{LS}\rho_S}{3\epsilon_{LL}\rho_L}\left(\frac{\sigma_{LS}}{\eta}\right)^4. \tag{63}$$

2.9.1. Stability of the axisymmetrical drop

As for cylindrical drops (Fig. 9b), there is a critical curve in the plane θ_0–a which divides the plane into two domains. In one of them (below the critical curve), the height of the drop can have any value, whereas above the critical curve the height is limited. Within the precision of the numerical calculations for each potential, LvdW and LJ, the critical curves for axisymmetrical drops coincide with those for the cylindrical ones when the latter ones are calculated for the potential Eq. (62). For example, for $a = 0.8$, the critical values of θ_0 for the LvdW potential are $\theta_{c,\,cyl} = 53.1301°$ and $\theta_{c,\,a} = 53.1295°$ for the cylindrical and axisymmetrical drops, respectively. For the LJ potential, the critical values are $\theta_{c,\,cyl} = 52.1691°$ and $\theta_{c,\,a} = 52.1690°$, respectively. In the domain of limited heights, the largest possible drop height h_c for the same a and θ_0 is larger for the axisymmetrical drop than for the cylindrical one. For example, for $a = 0.8$ and $\theta_0 = 52.1689°$, $h_{c,a} = 1441.9$ and $h_{c,cyl} = 688.1$ for the axisymmetrical and cylindrical drops, respectively.

2.9.2. Shape of the axisymmetrical drop

The numerical solution of Eq. (31) provides the shape of the drop, which is qualitatively similar to that of the cylindrical drop. However, there are quantitative differences which increase when the pair values (a, θ_0) are closer to the critical curve.

In Fig. 14, the profiles of cylindrical and axisymmetrical drops with the same heights are compared for $a = 0.8$ and various values of θ_0 (the LvdW potential was employed in the calculations) in the domain of unlimited height. Only the profiles for $x \geq 0$ are presented. Far from the critical curve

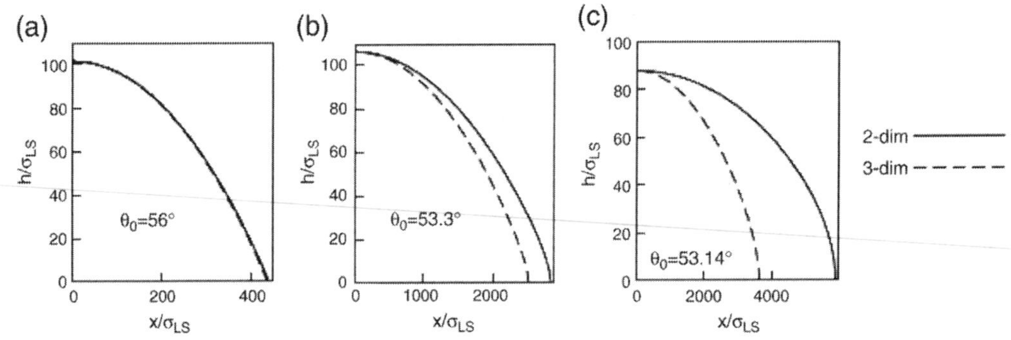

FIG. 14. (Reprinted from Ref. [13] with permission.) Profiles of the cylindrical (solid lines) and axisymmetrical (dashed lines) drops of the same height of the apexes for various microcontact angles from the domain of unlimited height for the LvdW potential.

(profiles for $\theta_0 = 56°$) the profiles are almost identical except of a small region near the surface. The macroscopic contact angle for both profiles is equal approximately to 25.9° and remains almost constant when the height of the drop increases. Therefore, the profile of an axisymmetrical drop far from the critical curve can be represented by that of a cylindrical drop, which is much easier to calculate. When θ_0 decreases, the difference between the profiles becomes more pronounced. From the curves for $\theta_0 = 53.3°$ and $\theta_0 = 53.14°$ one can see that the width of the cylindrical drop profile is larger than that of the axisymmetrical one, and that both profiles become more planar (the width becoming much larger than the height), when the angle θ_0 decreases. The macroscopic contact angle decreases with decreasing θ_0 and becomes sometimes even very small. For instance, for $\theta_0 = 53.14°$, $\theta_{m,cyl} = 1.37°$ and $\theta_{m,\,a} = 2.47°$, and for $\theta_0 = 56°$, $\theta_{m,cyl} \simeq \theta_{m,a} \simeq 26.0°$.

For the LJ potential all the characteristics of the axisymmetrical drops are qualitatively similar to those for the LvdW potential.

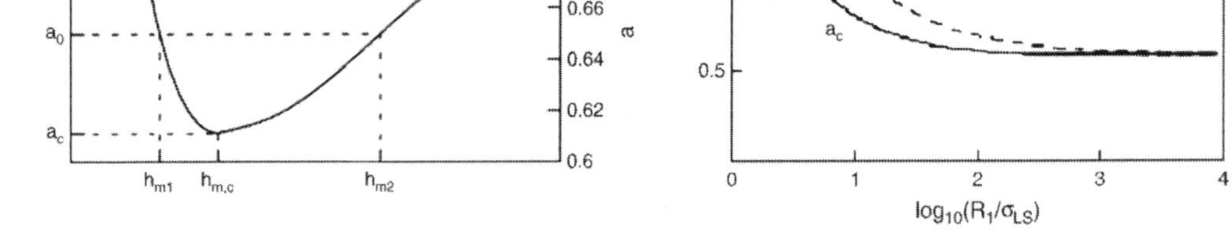

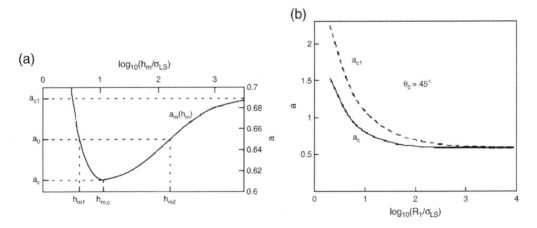

FIG. 15. (a) The critical curve $a = a_m(h_m)$ which follows from the second condition of drop existence Eq. (54) for a barrel drop on a fiber at $\theta_0 = 45°$. For $a \le a_c$ the height h_m of a drop can acquire any value. For $a_c < a \le a_{c1}$, h_m is restricted to intervals $h_m \ge h_{m2}$ and $h_m \le h_{m1}$. If $a > a_{c1}$, the height of the drop cannot be larger than h_{m1}. (b) Dependence of the critical values a_c and a_{cl} on the fiber radius R_f. Microcontact angle is equal to $\theta_0 = 45°$. Panel (b) is reprinted from Ref. [14] with permission.

2.10. Some features of a barrel drop on a fiber

Eq. (32) for the profile of a barrel drop on a fiber of radius R_f can be solved in the same way as Eq. (31) for a cylindrical drop on a planar surface (see Ref. [14] for details). In this case, the analysis of the drop formation involves a new parameter, R_f, which essentially changes the ranges of drop existence with respect to the parameters a, θ_0, and h_m obtained previously. In Fig. 15a typical ranges of barrel drop formation are presented for $R_f = 109\sigma_{LS}$ and $\theta_0 = 45°$ in the a–h_m plane. The curve $a = a_m(h_m)$ which has a minimum $a = a_c$ at $h = h_{m,c}$ and asymptotically approaches $a = a_{c1}$ as $h_m \to \infty$ divides the a–h_m plane into several domains in which the drop has different features. For $a < a_c$, the drop can have any height. If a specific value a_0 of the parameter a is between a_c and a_{c1} ($a_c<a_0<a_{c1}$), the drop can have a height smaller than $h_{m,1}$ or larger than $h_{m,2}$. Those values can be found as the intersection points between the horizontal line $a = a_0$ with the critical curves $a = a_m(h_m, \theta_0)$. For example, for $R_f = 109\sigma_{LS}$ and $\theta_0 = 45°$, one has $a_c = 0.6113$ and $a_{c1} = 0.6922$. For $a = 0.65$ the limiting heights are $h_{m,1} = 4.77\sigma_{LS}$ and $h_{m,2} = 175.29\sigma_{LS}$. For $a_0>a_{c1}$ the possible drop heights are restricted only by $h_{m,1}$.

The specific features of the critical curve in the h_m–a plane depend on the radius R_f of the fiber as well as on the microcontact angle θ_0. As R_f increases, both a_c and a_{c1} as well as the difference between them become smaller (see Fig. 15b) and for $R_f \to \infty$ the critical curve acquires a form similar to that for a planar solid surface [8]. The curve has no extrema and approaches $a_c = 0.5857$ as $h_m \to \infty$.

By performing similar analysis for other values of θ_0, one can plot the θ_0 dependencies of the critical values a_c and a_{c1} in the θ_0–a plane (see Fig. 16). The curve $a = a_c(\theta_0)$ separates the domains of restricted and unrestricted drop heights for $R_f = 100\sigma_{LS}$. The pairs (θ_0, a) for which the drop height h_m can have any value are located below that curve. In the domain between $a = a_c(\theta_0)$ and $a = a_{c1}(\theta_0)$ the drop height can take the values $h_m<h_{m,1}$ and $h_m> h_{m,2}$, whereas in the domain above the curve $a = a_{c1}(\theta_0)$ the drop height can take only values $h_m<h_{m,1}$.

The shape of the barrel drop profile also depends on the location of the pair (θ_0, a) on the a–h_m plane. If the pair (θ_0, a) is located between the critical curves (see Fig. 16), for instance $a = 0.65$,

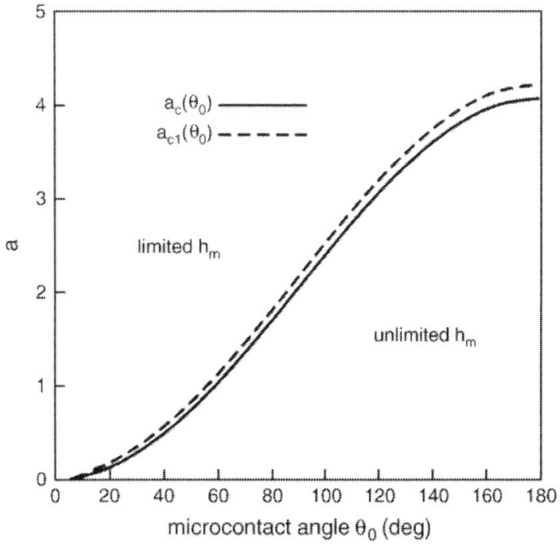

FIG. 16. (Reprinted from Ref. [14] with permission.) Dependence of the critical values a_c and a_{c1} of parameter a on the microcontact angle θ_0 of a barrel drop on a fiber for $R_f = 100\sigma_{LS} \sim 30$ nm. The solid line [$a = a_c(\theta_0)$] divides the θ_0–a plane into domains of unlimited (below the solid line) and of limited (above solid line) heights.

and $\theta_0 = 45°$, then the drop height can take the values $h_m < h_{m,1} = 4.76\sigma_{LS}$ and $h_m > h_{m,2} = 175.29\sigma_{LS}$. If the drop height has a value close to $h_{m,2}$ (for example, $h_m = 175.58\sigma_{LS}$), a part of the drop has the shape of a manchon [34] which constitutes the precursor of a wetting film on a cylindrical surface (Fig. 17a). The precision of our numerical calculations does not allow to examine drops with heights closer to $h_{m,2}$ than that used in the above example. However, the solution of Eq. (32) shows [14] that the drop length and volume increase indefinitely when h_m approaches $h_{m,2}$. Therefore, it is expected the manchon-like part of the drop to grow and a film to cover the bare fiber surface. For larger values of the drop height the size of the manchon decreases and the drop profile looks like those in the range of unlimited sizes (Fig. 17b).

In the examples displayed in Fig. 17a and b the approximation of the drop profile with a circle of radius equal to the curvature radius at the drop apex is shown as a dashed line. The dotted line represent the result obtained using the Carroll's solution [35] for the drop profile with the same height

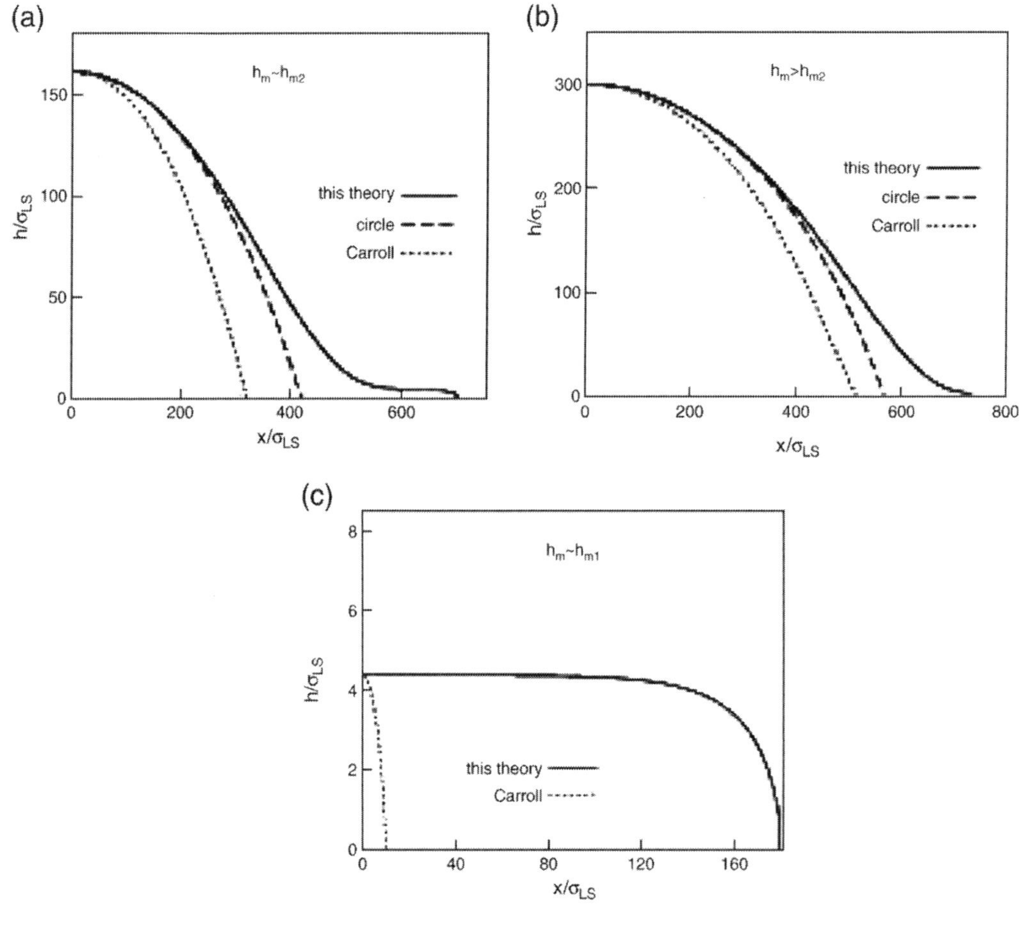

FIG. 17. (Reprinted from Ref. [14] with permission.) Profiles of a drop on a fiber in the domain of limited height when the point (θ_0, a) is located between the critical lines shown in Fig. 16) (In the present case $a = 0.65$, $\theta_0 = 45°$). (a) The height $h_m = 161.08\sigma_{LS}$ of the drop is close to $h_{m2} = 160.8161\sigma_{LS}$; (b) h_m is sufficiently large compared to h_{m2}; (c)The height $h_m = 4.3727\sigma_{LS}$ is close to $h_{m1} = 4.37294\sigma_{LS}$. The dashed line represents the circular approximation of the profile and the dotted-dashed line represents Carroll's solution for the drop with the same height and $\theta_m = \theta_0$.

and contact angle. One can see that the circular approximation is reasonable only for the upper part of the drop. The Carroll's solution is unsatisfactory in those cases.

In the other limiting case, $h_m \simeq h_{m,l}$, the height/length ratio is very small $\left(h_m/2x_0 \simeq 0.011\right)$ and the drop has a planar shape (Fig. 17c) for the description of which neither the circular approximation, nor Carroll's solution, is approriate.

Another difference between drops on a fiber and those on a planar surface is the relation between the microcontact angle θ_0 and the macrocontact angle θ_m for a barrel drop which has in the latter case a more complicated form when compared with Eq. (61) for a drop on a planar surface. For a barrel drop this relation has the form

$$\cos\theta_m = 1 + h_m\left[\kappa_0\left(h_m\right)\cos\theta_0 + \kappa_a\left(h_m\right)a + \kappa_1\left(h_m\right)\right] \tag{64}$$

where

$$\kappa_0\left(h\right) = \frac{2R}{h_m\left(h_m+2R\right)} - \frac{R}{\left(R+h\right)^2}\left[\frac{h\left(h+2R\right)}{h_m\left(h_m+2R\right)} - 1\right], \tag{65}$$

$$\kappa_a\left(h\right) = 2\frac{D\left(h_m\right)-D\left(0\right)}{h_m\left(h_m+2R\right)}\left[1-\frac{h\left(h+2R\right)}{2\left(h+R\right)^2}\right] + \frac{D\left(h\right)-D\left(0\right)}{\left(h+R\right)^2} - \frac{k}{\sigma_{LS}\left(1+h/\sigma_{LS}\right)^{\nu}},$$

$$\kappa_1\left(h\right) = -\frac{2\left(R+h_m\right)}{h_m\left(h_m+2R\right)}\left[1-\frac{h\left(h+2R\right)}{2\left(h+R\right)^2}\right]$$

and

$$D\left(h\right) = -\frac{k\sigma_{LS}}{\left(2-\nu\right)\left(1+h/\sigma_{LS}\right)^{\nu-2}} + \frac{kR}{\left(1-\nu\right)\left(1+h/\sigma_{LS}\right)^{\nu-1}} \tag{66}$$

with k and ν being the parameters of the potential Eq. (11). For $R \to \infty$ Eq. (64) becomes Eq. (61).

2.11. Cylindrical drop on a fiber

2.11.1. Existence, shape, and contact angle

Despite the different shape of cylindrical and barrel drops on a fiber (compare Fig. 3a and b) the main features of their existence, shape, and contact angles are similar (but not identical). In both cases, the second necessary condition of drop existence (Eq. (54)) provides two domains in h_m–a plane which look like those in Fig. 15a. However, for a cylindrical drop the analysis of the shape of the drop profile shows that for $h_m > h_{m,2}$ the angular length of the drop (the angle ϕ_0 in Fig. 7) is larger than 180° and therefore those shapes cannot exist. As a result, the actual critical curve in the h_m–a plane has the form presented schematically in Fig. 18. The critical value a_c depends on the radius R_f of the fiber and on the microcontact angle θ_0. As R_f increases, a_c decreases and for $R_f \to \infty$ (planar surface) acquires the value $a_c = 0.5857$.

In the θ_0–a plane, the curve that separates the domain of unrestricted height from the domain of restricted height is similar to that represented in Fig. 16 by the solid line (see for details Ref. [15]).

Taking as example a fiber of radius $R_f = 100\sigma_{LS}$, typical profiles of a small drop of volume $V = 3.63\times10^3\sigma_{LS}^2 \simeq 1.0\times10^2$ nm^2 per unit length of the drop are presented in Fig. 19 for $a = 0.612$ and various microcontact angles θ_0 from the domain of unlimited heights. The selected value of a is close to the critical value $a_c = 0.613$ for $\theta_0 = 45°$. Therefore, the plot of the profile (solid line) in

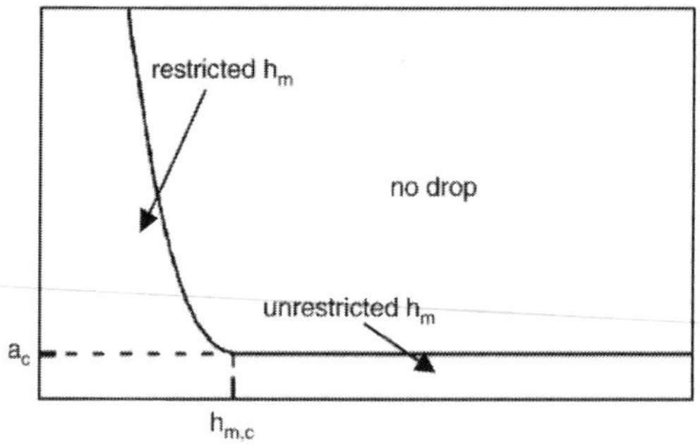

FIG. 18. Schematic presentation of the critical curve (solid line) which divides the h_m–a plane into domains of unrestricted and restricted height of the drop and one where drop does not form.

Fig. 19a corresponds to a point (θ_0, a) in the very vicinity of the critical line $a = a_c(\theta_0)$ (see Fig. 16). The dashed curve represents the circular approximation for the drop profile. In the case considered, the circle has no intersection with the fiber and most of the drop profile and the profile of the fiber can be approximately considered as concentric circles. This case is similar to that of a planar drop on a planar solid surface [8,13] for a point (θ_0, a) which is located near the critical line. For this reason the above drop on the fiber can be considered as a quasi planar one.

When θ_0 increases, the angular length ϕ_0 of the drop becomes smaller, the drop height increases and the approximation of the drop profile by a circular one becomes more accurate. As a consequence the approximate and exact curves in Fig. 19b and c become indistinguishable by eye. However, in the very vicinity of the fiber surface the circular approximation does not coincide with the exact solution and therefore the microscopic and macroscopic contact angles have different values. For instance, for the profiles shown in Fig. 19b, c for $\theta_0 = 75°$ and 135° the angles θ_m are 55.9° and 113.9°, respectively. For a similar drop on a planar surface the angles θ_m were 55°, and 113.7°, respectively, almost the same as on the fiber. The profiles of a larger drop $\left(V = 4.1\times10^4\sigma_{LS}^2 \simeq 1.1\times10^3\text{nm}^2\right)$ are presented in Fig. 20 which shows that they cannot be considered quasi planar. The macroscopic contact angles θ_m are practically the same as for the small drops.

When the pair (θ_0, a) lies above the critical line $a = a_c(\theta_0)$ (Fig. 16) the drop height can acquire only values $h_m < h_{m,1}$ and the drop acquires a quasi planar shape when $h_m \to h_{m,1}$. It should be noted that $h_{m,1}$ in the cases considered above is small ($h_{m,1}$ < 5 Å) and therefore the applicability of the theory is questionable.

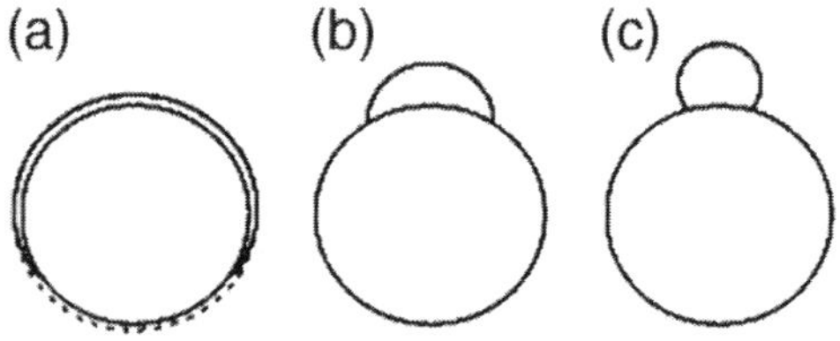

FIG. 19. (Reprinted from Ref. [15] with permission.) Profiles of a small cylindrical drop of volume $V = 3.63\times10^3\sigma_{LS}^2 \simeq 1.0\times10^2\,nm^2$ per unit length for several values of the microcontact angle θ_0 (solid lines) and $a = 0.612$: (a) $\theta_0 = 45°$; (b) $\theta_0 = 75°$; (c) $\theta_0 = 135°$. The dashed line on figure (a) represents a circular profile with the same height h_m as the original one and with a radius equal to the curvature radius of the original profile at the drop apex. The same approximations for figures (b) and (c) almost coincide with the profiles of the drops. There is a difference at the leading edge which cannot be detected by eyes. All profiles correspond to the points (θ_0, a) in the domain of unlimited height, the point (θ_0, a) for profile (a) being located in the very vicinity of the critical line.

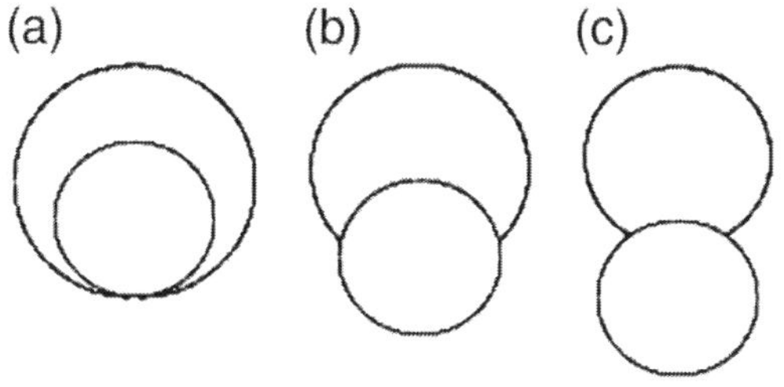

FIG. 20. (Reprinted from Ref. [15] with permission.) Profiles of a relatively large cylindrical drop of a volume per unit length $V = 4.1\times10^4\sigma_{LS}^2 \approx 1.1\times10^3\,\mathrm{nm}^2$ on a fiber with $R_f = 100\sigma_{LS}$ for several values of the microcontact angle θ_0 (solid lines) and $a = 0.612$: (a) $\theta_0 = 45°$; (b) $\theta_0 = 75°$; (c) $\theta_0 = 135°$.

For the drops under consideration the relation between the macro- (θ_m) and microcontact (θ_0) angles is more complicated than for those on the planar surface. As example, such a relation for drops with heights $\sigma_{LS} \ll h_m \ll R$, has the form

$$\cos\theta_m = \left(1 + \frac{h_m}{2R}\right)\cos\theta_0 + \frac{ak}{\nu - 1} - \frac{h_m}{2R}\left(1 - \frac{ak}{\nu - 1}\right). \tag{67}$$

2.11.2. Roll-up transition

Considering the cylindrical drop as a limiting case of a clam-shell one, and the film of constant thickness covering a fiber as the limiting case of a barrel drop, the obtained analytical results can be related to the clam-shell drop-barrel drop transition (roll-up transition [36]) which was treated before only numerically in the framework of a macroscopic approach [35,37,38]. Such a transformation takes place when the potential energy for a film is smaller than that for a cylindrical drop, provided that both the film and the drop have the same volume.

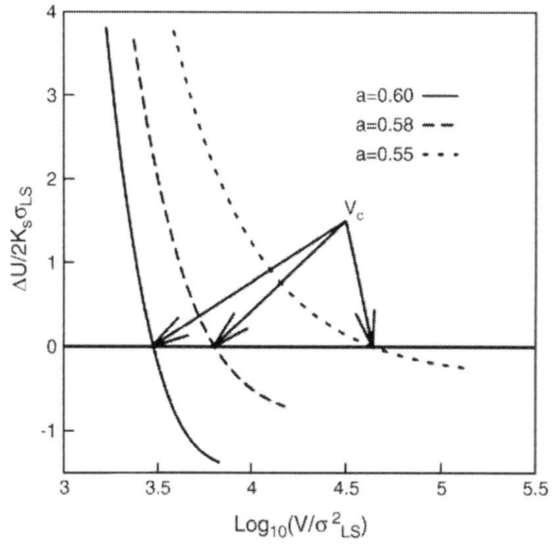

FIG. 21. (Reprinted from Ref. [15] with permission.) The difference ΔU between the potential energies for a film and a cylindrical drop of the same volume V per unit length as a function of V for several values of the parameter a. The microcontact angle is equal to $\theta_0 = 45°$. V_c represents the critical volume at which the roll-up transition occurs.

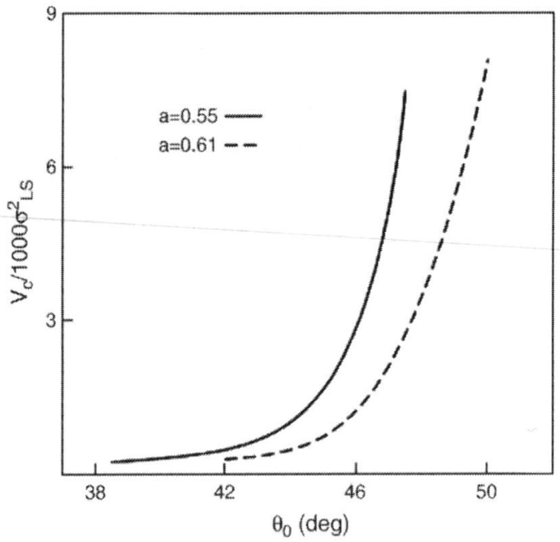

FIG. 22. (Reprinted from Ref. [15] with permission.) Dependence of the critical volume V_c per unit length for roll-up transition on the microcontact angle θ_0 for several values of the parameter a.

In Fig. 21 the difference ΔU between the potential energies of the film U^f_{total} and of the cylindrical drop U^c_{total} is plotted as a function of the drop volume for $\theta_0 = 45°$, for various values of the parameter a from the domain of unlimited heights ($a<a_c = 0.613$), and $R = 100\sigma_{LS}$. For $a > a_{lim} \simeq 0.536$, ΔU changes sign at a volume V_c which depends on a. Therefore, for $V > V_c$, U^f_{total} is smaller than U^c_{total} and the film configuration is more stable than that of the cylindrical drop. If $a < a_{lim}$ the potential energy of the cylindrical drop is smaller than that of the film for any value of the volume, i.e. the cylindrical configuration is always stable.

Experimental data [37] show that the critical volume V_c of a drop at which the roll-up transition occurs increases with increasing macroscopic contact angle. In Fig. 22 the θ_0–dependence of the critical volume is presented for several values of a and for a fiber with $R_f = 20\sigma_{LS}$. (The θ_m–dependence is qualitatively similar). For each value of a the roll-up transition takes place only in a narrow range of microcontact angles ($\theta_{0,1}<\theta_0<\theta_{0,2}$) of the domain of unlimited heights. Outside this region the cylindrical drops are energetically preferred at any drop size. In the range $\theta_{0,1}<\theta_0<\theta_{0,2}$, the volume V_c increases rapidly with increasing θ_0. Such a behavior coincides qualitatively with the experimental observations [38]. However, experiment provides a much wider, (but also restricted) range of contact angles in which the roll-up transition occurs. The possible reason for such a difference is that the experimental data were obtained for finite size drops (barrel and clam-shell), whereas the theory considers infinite sizes for both the barrel and the clam-shell drops which are represented as films and as cylindrical drops, respectively. The smaller the parameter a, the smaller is the angle θ_0 for which the roll-up transition occurs. It should be noted that the range of existence of the roll-up transition increases with decreasing R_f. For example, for $R_f = 20\sigma_{LS}$ and $a = 0.61$ the difference $\theta_{0,2}$-$\theta_{0,1}$ is 8.0°, while for $R_f = 100\sigma_{LS}$ it is 4.2°.

2.12. Dependence of the contact angle on roughness

Despite the large number of experimental papers regarding the wetting of rough surfaces, the linearity predicted by the Cassie-Baxter (CB) and Wenzel equations (Eqs. (3) and (4)) could not be verified for a long time, probably because of the difficulties in controlling the surface roughness. In the last decade, surfaces with regular arrangements of asperities (e.g. cylindrical or square pillars) could be prepared. By changing the sizes of the asperities and the distances between them one could control the roughness of the surface and thus one could check the validity of Eqs. (3) and (4). Most complete experiments in this direction were carried out by Krupenkin et al. [39]. In their study, five chemically identical surfaces with various roughnesses were prepared, and the contact angle θ_m of four liquids was determined on each surface. A drop of the liquid with the highest surface tension was located on the top of the asperities (CB regime) and it was concluded that the cosine of its contact angle had a linear dependence on roughness. For the other three liquids with lower surface tensions, it was shown that they penetrate into the space between asperities (Wenzel regime) and therefore their contact angles should follow the Wenzel law Eq. (4). However, the latter experiments [39] indicated a non-linear dependence of $\cos\theta_m$ on r. As a possible explanation for such a behavior, the authors suggested to include the line tension which is expected to affect the contact angle. However, a more detailed examination of the contribution of the line tension [40,41] still provided a linear dependence of $\cos\theta_m$ on r.

In Ref. [16] this problem was treated on the basis of a microscopic approach described above. The LvdW potential (Eq. (9) with $k_\phi = 0$) was employed to describe the fluid-solid interaction. The vapor-liquid interactions as well as gravity were neglected.

In the experimental work of Krupenkin et al. [39] both the solid and the liquid have a complex structure. The solid substrate (silicon) consisting of a bulk solid and pillars was coated with thin layers of oxide (50 nm) and fluorocarbon (20 nm). The latter layer was added to generate a hydrophobic surface. 30, 40, and 50 vol.% methanol-water mixtures were used to generate drops. Such a complex structure provides additional difficulties in calculating the interaction energy between the molecules of liquid and solid. For simplicity, in our calculations a homogeneous solid and a one component liquid have been considered. Consequently, all the microscopic parameters used, have to be considered as mean values.

To agree with the experimental set up, the roughness was modeled as a regular array of cylindrical pillars of height h_p and radius R separated by the distance l (see Fig. 23). The surfaces between pillars as well as those of the pillars are considered to be smooth. For such a geometry, the roughness r and the area fraction of the top of asperities f can be expressed as

$$f = \frac{\pi R^2}{l^2}, \quad r = 1 + \frac{2h_p}{R} f. \tag{68}$$

From Eq. (68), it follows that in the considered cases the parameters r and f are connected by a linear relation and any of them can be used as a characteristic of roughness.

The liquid drops used by Krupenkin et al. [39] in their experiments were comparatively large and had a characteristic radius of the contact area larger than 1 mm. Therefore, the number of pillars beneath the drop was large. The contact line of the drop with the solid surface was most likely located partly on the top of the pillars and partly on the interpillar surface. Such a complicated

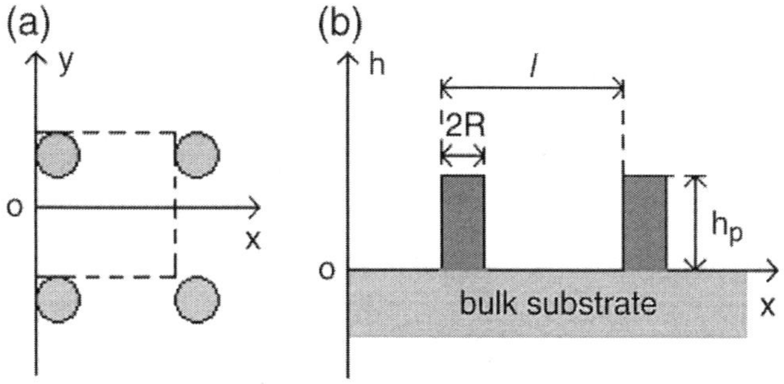

FIG. 23. (Reprinted from Ref. [16] with permission.) The top, (a), and side, (b), views of a rough solid surface. The asperities are assumed to be cylinders with a radius R and a height h_p. l is the distance between cylinders.

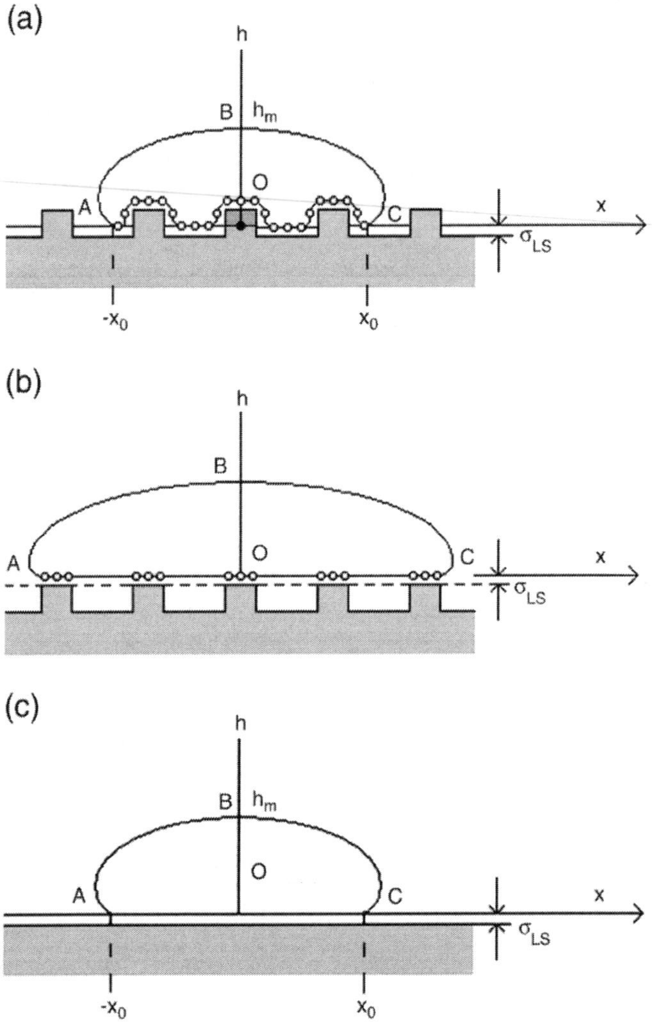

FIG. 24. (Reprinted from Ref. [16] with permission.) Two possible regimes for a cylindrical drop on a rough surface: (a) The liquid penetrates completely into the space between asperities (Wenzel regime); (b) The drop resides on the top of the asperities (Cassie-Baxter regime). (c) The reference drop on a smooth planar solid surface. The open circles constitute the surface monolayer of molecules of liquid, nearest to the solid, h_m is the drop height, σ_{LS} is the hard core diameter for liquid–solid interaction, and x_0 is the half width of the drop.

geometry makes extremely difficult to perform calculations regarding the drop profile. We will consider below two-dimensional (cylindrical) drops with the profiles presented in Fig. 24(a) and (b). The drop profile is symmetric about the h-axis and the drop is extended indefinitely along the y-axis that is normal to the plane of the figure. The profile in Fig. 24(a) is typical for the Wenzel regime and the profile in Fig. 24(b) is typical for the Cassie-Baxter regime. In Fig. 24(c) a reference drop on a smooth planar solid surface is presented. In Section 2.9.2 it was shown [8,13] that the profiles of large cylindrical and axisymmetrical drops on a planar surface are close to each other. For this reason one can assume that they are also similar on a rough surface.

The small circles in Fig. 24 (a) and (b) represent the first layer (monolayer) of liquid molecules which are assumed to be structured by the solid and can be subjected to both long and short range interactions with the solid. The remaining of the drop separated from the first layer by a distance σ_{LL} (hard core diameter of liquid–liquid interactions) will be considered as a uniform single component medium characterized by the number density ρ_L. It will be also supposed that small portions of insoluble gas can be accumulated during the formation of the drop between the drop and the solid surface near the edges of the asperities and that this accumulation does not allow the liquid to wet completely the solid surface. This partial unwetting is assumed to be sufficiently small for not to affect the potential energy of the drop due to the long-range interactions, but sufficiently large to affect the energy of the short range interactions. Because of the presence of the gas only a fraction of the solid surface beneath the drop is covered by liquid. (This fraction provides the liquid-solid interface.) To characterize that fraction two functions are introduced. The first, $k_s(f)$, represents the fraction of the solid surface between pillars, which is covered by liquid, and the second, $k_{sp}(f)$, represents the fraction of the pillars surface covered by liquid. Correspondingly, $1-k_s(f)$ and $1-k_{sp}(f)$ represent fractions of the solid surface covered by the adsorbed gas molecules. As a result, the liquid–gas interface consists of an external part *ABC* of the drop surface and a part beneath the drop which is in contact with the trapped gas. The surface number densities ρ_{LS} and ρ_{LG} of the liquid molecules at the liquid–solid and liquid–gas interfaces, respectively, are generally not equal but very likely comparable.

One should note that the functions $k_s(f)$ and $k_{sp}(f)$ are unknown and have to be determined by comparing with the experimental data regarding observable quantities (contact angles, contact area, etc).

Calculations similar to those performed for a cylindrical drop on a smooth planar surface provide the following results for the macroscopic contact angles θ_m^w (Wenzel regime) and θ_m^c (Cassie-Baxter regime) [14]

$$\cos\theta_m^w = \frac{1}{2\delta_1 + 3/4}\left(2\delta_1 - \frac{1}{4} + r\right)\cos\theta_{m,0} + fd_{w,3} + d_{w,2}t_{bp}k_s(f), \tag{69}$$

$$\cos\theta_m^c = -1 + fd_{c,2}\left(1 - d_{w,2} + \cos\theta_{m,0}\right) \tag{70}$$

where $\theta_{m,0}$ is the contact angle of the drop on a smooth surface given by the equation

$$\cos\theta_{m,0} = d_{w,0} + d_{w,2}k_s(0). \tag{71}$$

In the above equations, $t_{bp} = 1 - f$ and the coefficients δ_1 and $d_{w,i}$ ($i = 0$, 2, and 3) depend on the parameters of the interaction potentials, roughness f, functions $k_s(f)$ and $k_{sp}(f)$, and geometrical parameters of the pillars. Explicit expressions for these coefficients can be found in Ref. [16]. The function $k_s(f)$ is taken in the form $k_s(f) = \frac{1}{1+n_s f}$, where n_s is a constant. This choice was made because the function $k_s(f)$, which provides the fraction of molecules of the first layer of liquid on the solid surface which are not displaced by the gas accumulated at the edges of the asperities and are subjected to short range interactions, has to be equal to unity for a smooth surface ($f = 0$) and expected to decrease with increasing roughness.

The second function, $k_{sp}(f)$, provides the fraction of molecules which are not displaced from the surface of a pillar. It is reasonable to suppose that this fraction has a weak dependence on the number of pillars and, therefore, $k_{sp}(f)$ will be selected to be a constant

$$k_{sp}(f) = n_{sp} \tag{72}$$

where n_{sp} is a constant. The constants n_s and n_{sp} have to be determined from comparison with experiment.

In Fig. 25 the theoretical curves for $\cos\theta_m^w$ and $\cos\theta_m^c$ which have the best fit with experimental data are presented together with the experiments of Krupenkin et al [39]. The theoretical curves are labeled in the order of increasing surface tension of the employed liquids. Curves I, II, and III

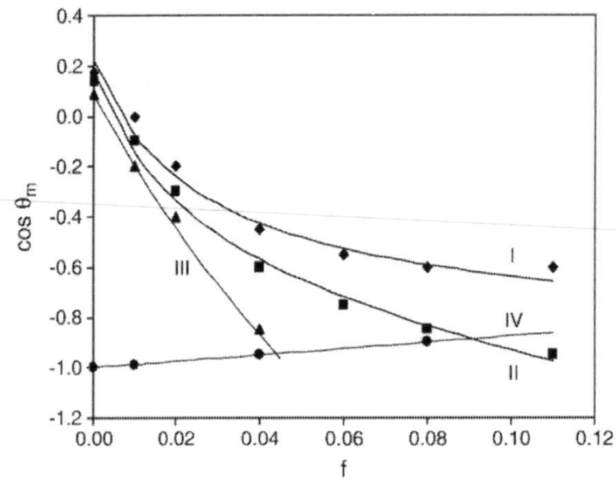

FIG. 25. (Reprinted from Ref. [16] with permission.) The experimental results of Krupenkin et al. [39] (filled points) and theoretical curves given by Eq. (69) for curves I, II, and III and Eq. (70) for curve IV. The surface tensions of the employed liquids increased with increasing curve index.

correspond to the Wenzel regime with $\cos\theta_m^w$ nonlinearly dependent on roughness, and curve IV to the CB regime, with $\cos\theta_m^c$ linearly dependent on f. The values of the parameters of the model, providing the best fit of experimental and theoretical results are given in Table 1 of Ref. [16]. One can see from Fig. 25, that the theory provides a good description of experimental data.

3. DENSITY FUNCTIONAL THEORY APPROACH TO THE DESCRIPTION OF A NANODROP ON A SOLID SURFACE

In this section, a nonlocal density functional theory (DFT) which is applicable to the description of drops on smooth and rough solid surfaces is considered. In contrast to the theory described in Section 2, DFT is based on the minimization of the free energy of the system, which is calculated (using some approximations) from the first principles of statistical mechanics. The basic equations of DFT, the definitions of chemically and physically rough surfaces, and methods of calculating the nanodrop profile and the contact angle are provided in Section 3.1. The quasi-universal dependence of the contact angle on the energy parameter of the fluid—solid interactions and temperature for nano-and macrodrops on a smooth solid surface is presented in Section 3.2. The drops on physically and chemically rough surfaces are considered in Sections 3.3 and 3.4, respectively. In Section 3.5 the sticking force which maintains the drop equilibrium on inclined surfaces is examined.

3.1. Background

3.1.1. Density functional and Euler—Lagrange equation

Even though the density functional theory was employed intensively in the treatment of bulk and confined classical and quantum fluids for more than 30 years, (see e.g. Refs. [12,22,23,42–47]) the number of studies devoted to drops on solid surfaces in not large. The main attention was given to the adsorption from a bulk fluid on solid surfaces and in slits and cylindrical pores and to phase transformations such as layering, prewetting, and wetting in these open systems. The main reason was the absence of drop-like solutions of the DFT equations for the fluid density distribution (FDD) in open systems.

Drop-like solutions for FDD can be, however, found if the canonical ensemble DFT describing closed systems with fixed number of molecules (fixed average density) is employed. Such an approach was employed in Refs. [17,18–21,25,26,48–52]. The applicability of the canonical ensemble DFT was examined in Refs. [53–57]. In computer simulations, the canonical ensemble was used in Refs. [58–63] in which drops and droplike aggregates (bubbles, bumps and bridges) on solid surfaces were considered. In most cases, the considerations were restricted to two-dimensional (cylindrical) drops similar to those described in Section 2.1.

Generally, the DFT description of a cylindrical drop on a planar solid surface involves a system of finite dimensions L_x and L_h in the x- and h-directions, respectively, and infinite in the y direction, which contains a one-component fluid of fixed average density ρ_{av} [17] (see Fig. 26). (The other two-dimensional systems frequently considered in the literature are embedded in a cylinder of radius R and length L [60].) The system is in contact with a smooth or rough solid surface which generates an external potential $U_{sf}(\mathbf{r})$ that acts on the fluid molecules. The solid (planar or cylindrical) is assumed to be homogeneous with a density ρ_s. The asperities (as patches or pillars) of the solid surface are considered uniform in the y-direction. The FDD $\rho(\mathbf{r})$ in such a system is assumed uniform in the y-direction and non-uniform in the x- and h-directions, i.e. $\rho(\mathbf{r}) = \rho(x, h)$. Note that such an assumption neglects the possibility of symmetry breaking in the y-direction which may lead to a transition from a cylindrical (two-dimensional) to a three dimensional drop. This transition is the subject of a separate study.

A periodic boundary condition $\rho(x, h) = \rho(x + L_x, h)$ is employed in the x direction and the upper boundary of the box is treated as a hard wall (without attractive interaction for fluid molecules). The above periodic boundary condition was suggested by Monte Carlo (MC) and molecular dynamics (MD) simulations [28,60,62,64–66] which often constitute the only source of "experimental" information at nanoscale. Therefore the theories had to be adjusted to the MC and MD "experiments" to make possible the comparison of the results.

The interaction between fluid molecules is described by the Lennard-Jones potential

$$\phi_{ff}\left(|\mathrm{r}-\mathrm{r}'|\right) = \begin{cases} 4\epsilon_{ff}\left[\left(\dfrac{\sigma_{ff}}{r}\right)^{12} - \left(\dfrac{\sigma_{ff}}{r}\right)^{6}\right], & r \geq \sigma_{ff} \\ \infty, & r < \sigma_{ff} \end{cases} \tag{73}$$

where the coordinates $\mathbf{r}$ and $\mathbf{r}'$ provide the locations of the fluid molecules, $r = |\mathbf{r}-\mathbf{r}'|$, and σ_{ff} and ϵ_{ff} are the fluid hard core diameter and the energy parameter, respectively.[1] In what follows, argon will be considered as the fluid with $\sigma_{ff} = 3.406$ Å and $\epsilon_{ff}/k_B = 119.76$ K [23].

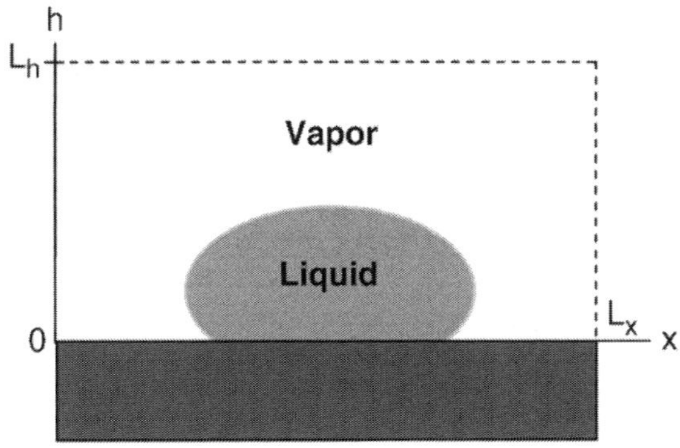

FIG. 26. (Reprinted from Ref. [19] with permission.) Schematic representation of a closed system consisting of a cylindrical drop (drop uniform in the y-direction which is perpendicular to the plane of the figure) on a smooth surface in contact with vapor of the same fluid.

The same kind of potential is selected for the fluid—solid interactions (the energy parameter ϵ_{ff} and hard core diameter σ_{ff} in Eq. (73) being replaced by ϵ_{fs} and σ_{fs}, respectively, r being the distance between a molecule of fluid and that of the solid wall.

The total Helmholtz free energy $F_H[\rho(\mathbf{r})]$ is the sum of an ideal gas free energy, $F_{id}[\rho(\mathbf{r})]$, a free energy $F_{attr}[\rho(\mathbf{r})]$ of a reference system of hard spheres, a free energy $F_{attr}[\rho(\mathbf{r})]$ due to the attractive interactions between the fluid molecules (in the mean field approximation), and a free energy $F_{fs}[\rho(\mathbf{r})]$ due to the interactions between fluid and solid molecules.

[1]To make the description closer to the original works, the subscripts here and below are a little different from those used in Section 2 for similar quantities.

These contributions can be represented as follows:

$$F_{id}\left[\rho(\mathbf{r})\right]=k_BT\int d\mathbf{r}\rho(\mathbf{r})\left\{\log\left[\Lambda^3\rho(\mathbf{r})\right]-1\right\}, \tag{74}$$

where $\Lambda=h_P/\left(2\pi mk_BT\right)^{1/2}$ is the thermal de Broglie wavelength, k_B and h_P are the Boltzmann and Planck constants, respectively, T is the absolute temperature, and m is the mass of a fluid molecule, and

$$F_{hs}\left[\rho(\mathbf{r})\right]=\int d\mathbf{r}\rho(\mathbf{r})\Delta\Psi_{hs}(\mathbf{r}) \tag{75}$$

where [67]

$$\Delta\Psi_{hs}(\mathbf{r})=k_BT\eta_{\bar{\rho}}\frac{4-3\eta_{\bar{\rho}}}{\left(1-\eta_{\bar{\rho}}\right)^2}, \tag{76}$$

$\eta_{\bar{\rho}}=\frac{1}{6}\pi\bar{\rho}(\mathbf{r})\sigma_{ff}^3$ is the packing fraction of the fluid molecules, and $\bar{\rho}(\mathbf{r})$ is the smoothed density defined as

$$\bar{\rho}(\mathbf{r})=\int d\mathbf{r}'\rho(\mathbf{r}')W\left(\left|\mathbf{r}-\mathbf{r}'\right|\right) \tag{77}$$

The weighting function $W(|\mathbf{r}-\mathbf{r}'|)$ is selected of the form [43]

$$W\left(\left|\mathbf{r}-\mathbf{r}'\right|\right)=\begin{cases}\dfrac{3}{\pi\sigma_{ff}^3}\left(1-\dfrac{r}{\sigma_{ff}}\right), & r\le\sigma_{ff}\ ,\\ 0, & r>\sigma_{ff}.\end{cases}$$

This selection of $\bar{\rho}(\mathbf{r})$ is not unique. Another, more complicated expression for $\bar{\rho}(\mathbf{r})$ was suggested in Ref. [12].

The contribution of the fluid—fluid interactions is

$$F_{attr}\left[\rho(\mathbf{r})\right]=\frac{1}{2}\iint d\mathbf{r}d\mathbf{r}'\rho(\mathbf{r}')\phi_{ff}\left(\left|\mathbf{r}-\mathbf{r}'\right|\right) \tag{78}$$

and the contribution $F_{fs}\left[\rho(\mathbf{r})\right]$ is given by

$$F_{fs}\left[\rho(\mathbf{r})\right]=\int_V d\mathbf{r}\rho(\mathbf{r})U_{fs}(\mathbf{r}) \tag{79}$$

where V is the volume of the system and the potential $U_{fs}(\mathbf{r})$ is provided by the expression

$$U_{fs}(\mathbf{r}) = \int_{V_s} \rho_s(\mathbf{r}')\phi_{fs}(|\mathbf{r}-\mathbf{r}'|)d\mathbf{r}' \tag{80}$$

where V_s is the volume of the solid, $\rho_s(\mathbf{r}')$ is the density of the solid which, in general, may depend on the coordinates.

The Euler-Lagrange equation for the fluid density distribution $\rho(x, h)$ obtained by minimizing the Helmholtz free energy has the following form [17]

$$\log\left[\Lambda^3\rho(x,h)\right] - Q(x,h) = \frac{\lambda}{k_B T} \tag{81}$$

where the function $Q(x, h)$ is given by

$$Q(x,h) = -\frac{1}{k_B T}\left[\Delta\Psi_{hs}(x,h) + \overline{\Delta\Psi'}_{hs}(x,h) + U_{ff}(x,h) + U_{fs}(x,h)\right] \tag{82}$$

with

$$U_{ff}(x,h) = \iint dx'dh'\rho(x',h')\phi_{ff,y}(x-x',h-h'), \tag{83}$$

and

$$\overline{\Delta\Psi'}_{hs}(x,h) = \iint dx'dh'\rho(x',h')W_y(|x-x'|,|h-h'|)\frac{\partial}{\partial\bar{\rho}}\Delta\Psi_{hs}(\bar{\rho})\Big|_{\bar{\rho}=\rho(x',h')}. \tag{84}$$

The functions $\phi_{ff,y}(|x'-x'|,|h'-h'|)$ and $W_y(|x'-x'|,|h'-h'|)$ in Eq. (83) and $\phi_{ff}(|\mathbf{r}-\mathbf{r}'|)$ in Eq. (84) are obtained by integrating the potential $W(|\mathbf{r}-\mathbf{r}'|)$ and weighting function $W(|\mathbf{r}-\mathbf{r}'|)$ with respect to the y-direction, from $-\infty$ to $+\infty$.

When calculating the term $U_{ff}(x, h)$ in Eq. (82), a cutoff at a distance equal to four molecular diameters σ_{ff} for the range of Lennard—Jones attraction is employed. The increase of this distance up to $10\sigma_{ff}$ changed the results by less than 1%.

In Eq. (81), λ is a Lagrange multiplier arising because of the constraint of fixed average density of the fluid. This constraint has the form

$$\rho_{av} = \frac{1}{V}\int_V d\mathbf{r}\rho(\mathbf{r}) \tag{85}$$

and leads to the following expression for λ

$$\lambda = -k_B T \log\left[\frac{1}{\rho_{av} V \Lambda^3}\int_V d\mathbf{r} e^{Q(x,h)}\right]. \tag{86}$$

By eliminating λ between Eqs. (81) and (86), one obtains an integral equation for the FDD $\rho(x, h)$, which can be solved by iterations.

The standard iteration procedure starts with an initial guess $\rho^{in}(x, h)$ which is used as input for calculating the right hand side of equation

$$\rho(x,h)=\rho_0\exp\left[\frac{Q(x,h)}{k_BT}\right], \tag{87}$$

obtained from Eq. (81). In Eq. (87), $\rho_0=\Lambda^{-3}\exp(\lambda/k_BT)$. The density distribution $\rho(x, h)$ generated by Eq. (87) from $\rho^{in}(x, h)$ is usually combined in some way (see for example Ref. [12]) with $\rho^{in}(x, h)$ to provide a new input for Eq. (87). The iteration procedure is repeated until the difference between two successive iterations becomes sufficiently small [17]. An initial guess similar to that presented in Fig. 27 was employed in Refs. [17–21]. Because of computational restrictions, the above iteration procedure can be effectively applied only to nanosystems smaller than 100 nm.

3.1.2. Chemical roughness

Chemically rough surfaces are frequently modeled as a sequence of two kinds of plates, A and B, of finite thickness d_A and d_B, respectively, which are infinite in the y-direction and semi-infinite in the h-direction (Fig. 28). The origin of the coordinate system is selected in such a manner for the system to be symmetrical with respect to the y–h plane. The external field generated by a single plate at point $M(x, y, h)$ depends on the distance h of this point from the surface and on the distance from the vertical plane parallel to y–h plane that divides the plate into two equal parts. For example, plate A at the left of Fig. 28 generates a field which can be described by the potential

$$U_{sf,A}(x,h)=\rho_{s,A}\int_{V_A}dx'dy'dh'\phi_{sf,A}\left(\left|\mathrm{r}-\mathrm{r}'\right|\right)=\rho_{s,A}\int_{-d_A/2}^{d_A/2}dx'\int_{-\infty}^{0}dh'\int_{-\infty}^{\infty}dy'\phi_{sf,A}\left(\left|\mathrm{r}-\mathrm{r}'\right|\right), \tag{88}$$

where V_A is the volume of plate A, $\mathrm{r}=\{x,y,h\}$, $\mathrm{r}'=\{x',y',h'\}$, $\rho_{s,A}$ is the density of plate A, $\phi_{sf,A}(|\mathrm{r}-\mathrm{r}'|)$ has the form of the potential Eq. (73) with the parameters $\epsilon_{fs,A}$ and $\sigma_{fs,A}$ replacing ϵ_{ff} and σ_{ff}, respectively. The contribution to the total potential from any plate can be calculated in a similar way. The resulting potential is a periodic function of x with a period equal to $d=d_A+d_B$. Assuming that plates A and B differ only by the energy parameters, $\epsilon_{fs,A}$ and $\epsilon_{fs,B}$ and that $\rho_{s,A}=\rho_{s,B}\equiv\rho_s$, one can represent the total potential $U_{sf}(x, h)$ in the form

$$U_{sf}(x,h)=\sum_{k=-\infty}^{+\infty}U_{sf,A}(x-kd,h)+\sum_{k=-\infty}^{+\infty}U_{sf,B}\left(x-\left(k+\frac{1}{2}\right)d,h\right), \tag{89}$$

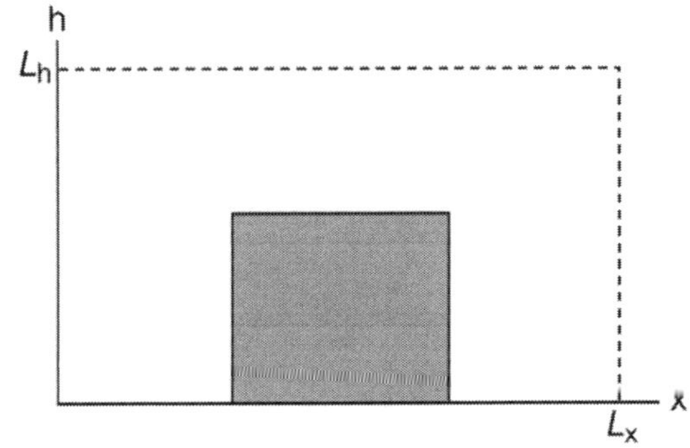

FIG. 27. Initial fluid density distribution of the Euler—Lagrange equation. The densities in filled and blank areas are constant and have the order of magnitude of liquid and vapor phase, respectively.

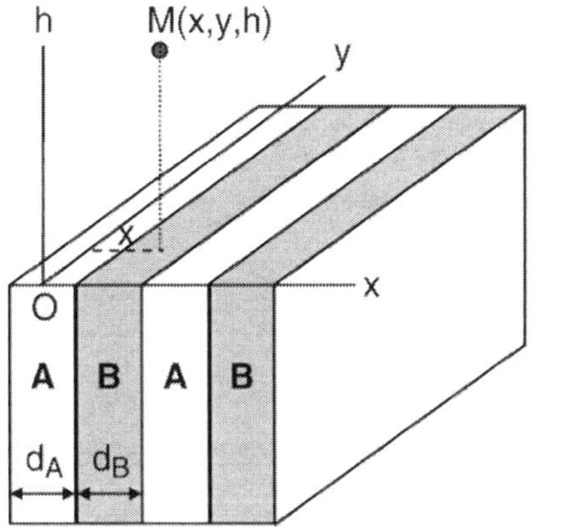

FIG. 28. (Reprinted from Ref. [17] with permission.) Schematic representation of a chemically rough smooth surface in the x–y plane, considered in the present paper. The plates A and B are infinite in the y-direction, semi-infinite in the h-direction and have different values of the energy interaction parameter $\epsilon_{fs,A}$ and $\epsilon_{fs,B}$. The solid is symmetrical with respect to the y–h plane.

where $k = 0$ indicates the plates A and B at the left end of Fig. 28, $k>0$ ($k<0$) indicates the plates A and B located at $x>0$ ($x<0$). In the numerical calculations the limits $-\infty$ and $+\infty$ for the sums over k were replaced by $k_{min} < 0$ and $k_{max} = -k_{min} > 0$, respectively, which were selected to satisfy the condition $(k_{max} - k_{min})d \geq 2L_x$. Due to this choice, the boundaries of the system, located at $x = \pm L_x/2$, were far enough from the left and right boundaries of the solid to avoid the distortion of the potential $U_{fs}(x, h)$ caused by the finite size of the solid considered in the calculations. Far enough from the solid surface, ($h>5\sigma_{fs}$) the total potential can be represented by an oscillatory function $U_{sf}(x,h) \sim \sin\left(\frac{2\pi}{d}x\right)$. Near the surface, the shape of the potential is more complex. Note that at any distance from the solid surface the local minima (maxima) of $U_{sf}(x, h)$ are located just above the middle of the A-plate (B-plate).

3.1.3. Physical roughness

The random nature of the physical roughness (in shape, size and location of asperities on a smooth surface), makes difficult the development of an adequate model for rough surfaces. One of the popular approximations is to consider that the planar surface is decorated with a regular array of identical protrusions such as, for instance, pillars [16,17,50,68,69] or grooves [70–73]. In such cases, the potential of pillar-fluid interaction can be calculated analytically or numerically and the dependence of the drop parameters on roughness can be obtained. In Refs. [17,18] a surface decorated with an infinite number of identical parallelepipeds (denoted below as pillars) of width d_p, height h_p, and infinite length in the y direction, as shown in Fig. 29 was considered. The roughness of such a surface, defined as the ratio of the true area of the solid to its planar projection, is provided by the expression

$$r = 1 + \frac{2h_p}{d_p + \Delta d_p} \tag{90}$$

where Δd_p is the distance between pillars.

The contribution $U_{fp}(x, h)$ of a pillar located at $x = 0$ to the total fluid—solid potential is given by

$$U_{fp}(x,h) = \rho_p \int_{V_p} dx' dy' dh' \phi_{fs,p}\left(\left|\mathrm{r} - \mathrm{r}'\right|\right) = \rho_p \int_{-d_p/2}^{d_p/2} dx' \int_0^{h_p} dh' \int_{-\infty}^{\infty} dy' \phi_{fs,p}\left(\left|\mathrm{r} - \mathrm{r}'\right|\right), \tag{91}$$

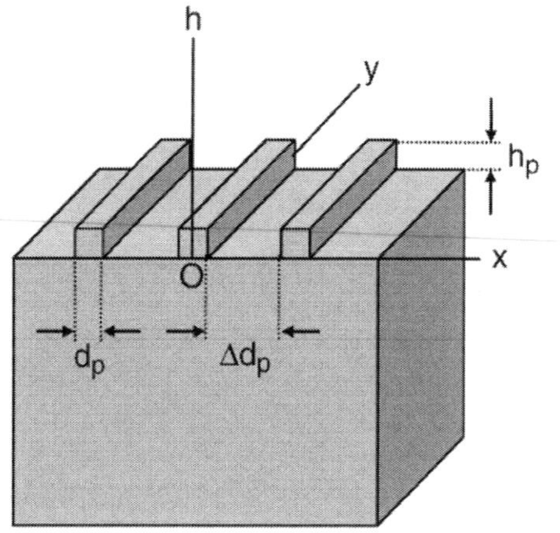

FIG. 29. (Reprinted from Ref. [17] with permission.) Pillars on a flat surface.

where **r** and **r**′ provide the locations of the fluid and pillar molecules, respectively, ρ_p is the density of the pillars, V_p is the pillar volume, $\phi_{fs,p}(|\mathrm{r}-\mathrm{r}'|)$ is the Lennard—Jones potential (Eq. (73)) in which $\epsilon_{fs,p}$ and $\sigma_{fs,p}$, which characterize the pillars, replace ϵ_{ff} and σ_{ff} respectively. In general, the energy parameter $\epsilon_{fs,p}$ involved in the potential $\phi_{fs,p}(|\mathrm{r}-\mathrm{r}'|)$ can be different from that involved in the interaction potential $\phi_{fs,p}(|\mathrm{r}-\mathrm{r}'|)$ between fluid and the bulk part of the solid (the part with $h<0$ in Fig. 26).

The total contribution of pillars to the external potential is given by

$$U_{fp,total}(x,h)=\sum_{k=-\infty}^{\infty}U_{fp}(x-x_k,h), \tag{92}$$

where x_k is the location of the center of the kth pillar. In the numerical calculations the limits of k were selected as for chemical roughness (Section 3.1.2).

3.1.4. Calculation of the contact angle of a nanodrop

Because of the small size of the nanodrops (less than 10 nm) and the relatively large widths (~ 1 nm) of the vapor—liquid interface in the MD experiments as well as DFT calculations [17,59], the procedure usually used to calculate the contact angle for macrodrops, which involves the Young equation (Eq. (1)) and the macroscopic surface tensions for planar interfaces, is not applicable to nanodrops. For the same reason, the contact angle for nanodrops is not clearly defined, because of the inapplicability of the sharp kink approximation, and there are several approaches to extract it from known FDDs [17,50,62]. With small variations, all these approaches are based on determining the drop profile $h = h(x)$ inside the vapor—liquid interface as the line passing through the points where the local density $\rho(x, h)$ has a constant value ρ_{div} and finding the angle which this profile makes with the solid surface. The density ρ_{div} can be defined as the density of an equimolar dividing surface for a horizontal FDD $\rho(x, h_o)$ at some distance h_0 from the surface, by considering this FDD as that for a planar vapor—liquid interface. However, there is no unique way to select h_0. To find the most appropriate value for h_0, one should note, that for $h \lesssim 3\sigma_{ff}$ the fluid density distribution $\rho(x, h)$ inside the drop has an oscillatory behavior as a function of h due to the ordering of the fluid molecules which form several liquid layers of various densities [12]. For these reasons, it is not clear how to determine ρ_{div} in the range $h_0 < 3\sigma_{ff}$. For $h_0 > 3\sigma_{ff}$ those oscillations become small and FDD $\rho(x, h_0)$ ($h_0>\sigma_{ff}$) exhibits a clearly observable interface between high and low density phases, for which one can easily find ρ_{div} and determine the location of a dividing surface at the selected $h0$. Hence, it looks reasonable to select a h_0 larger than $3\sigma_{ff}$.

A more complicated modification of this approach (see Ref. [59]) is to calculate ρ_{div} as a function of h by using the corresponding horizontal FDD $\rho(x, h)$ for each $h>0$ and plotting the manifold of points corresponding to the locations of all dividing surfaces. For $h > 3\sigma_{ff}$, this approach provides almost the same drop profile as the previous one, but it is not appropriate for $h < 3\sigma_{ff}$.

Even after the drop profile is obtained, there is an additional problem in the identification of the contact angle. Conventionally, the contact angle is determined by extrapolating the upper part of the profile considered as part of a circle (dashed line in Fig. 30). However, near the surface (h $<5\sigma_{ff}$) where the liquid-solid interactions have the largest values, the difference between this circle and the calculated drop profile is large. If one attempts to approximate the drop profile by a circle which better fits the actual profile for $h<5\sigma_{ff}$ (dotted-dashed line in Fig. 30) then there is a difference between the circle and the profile in the upper part of the latter. These two circles, when extrapolated up to the solid surface, make with that surface the angles θ_u and θ_l, respectively, which can be easily determined and can be considered as two complementary contact angles. The definition of the contact angle θ_u is similar to that used in the goniometric measurements of the contact angles of macroscopic drops [74] and θ_u will be considered as the macroscopic contact angle in most cases examined below. The angle θ_l is similar to the microcontact angle that is formed near a solid surface by macroscopic drops [7].

In some studies (see e.g. Ref. [50]), one makes a linear fit of the drop profile in the very vicinity of the surface (at distances smaller than a few molecular diameters) and consider the angle which this straight line makes with the surface as the contact angle.

3.1.5. DFT based calculation of the contact angle of a macrodrop

Even though DFT is applicable to the description of nonhomogeneous fluids only at the scale of nanosystems, it can be combined with Young equation (Eq. (1)) for calculating the contact angle of macroscopic drops [42,47,75]. To perform such calculations, the surface tensions γ_{lv}, γ_{sl}, and γ_{sv} are determined via DFT by considering planar interfaces and then the macroscopic contact angle is calculated using the Young equation which is valid for macroscopic drops.

To determine γ_{sl}, and γ_{sv} an open slit between identical walls is considered. In such a slit the FDD depends only on the coordinate h ($\rho(\mathbf{r}) = \rho(h)$) and is symmetric with respect to the middle of the slit [12,43,47]. If the effective width of the slit, L, is selected large enough then the fluid in the middle

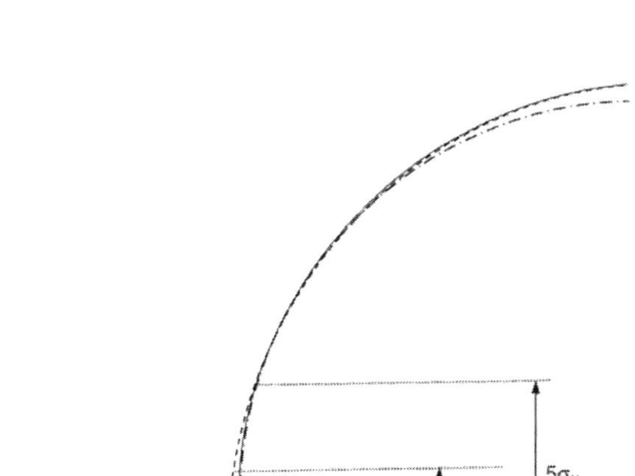

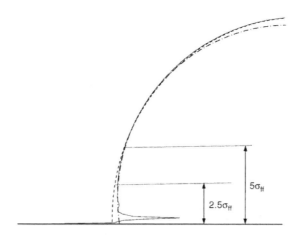

FIG. 30. (Reprinted from Ref. [19] with permission.) Example of a drop profile (solid line, only the left hand side part of the profile is presented) extracted from the fluid density distribution using the procedure described in the text. The profile is approximated by two circles. The first one (dashed line) has the best fit with the upper part (h $> 5\sigma_{ff}$) of the profile. The second circle (dashed-dotted line) has the best fit with the profile in the range $2.5\sigma_{ff}<h<5\sigma_{ff}$.

region of the slit (shaded area in Fig. 31) has a bulklike density. Depending on the selected value of the chemical potential this density can be liquid- or vapor-like. The FDD $\rho(h)$ can be obtained as a solution of the Euler—Lagrange equation derived by minimizing the grand canonical potential of the system. This equation coincides with the Euler—Lagrange equation for a closed slit (Eq. (81)), where the Lagrange multiplier λ is replaced by the chemical potential μ_c, and the constraint of constant average density, Eq. (85), is eliminated.

After determining $\rho(h)$, the grand canonical potential of the system, Ω_{sf}, and the grand canonical potential, Ω_f^b, of the bulk fluid having the volume of the slit can be calculated from

$$\Omega_{sf} = F_H\left[\rho(h)\right] - \mu\int_0^L \rho(h)\,dh, \tag{93}$$

$$\Omega_f^b = F_H\left[\rho_f^b\right] - \mu\rho_f^b L. \tag{94}$$

Here the subscript f stands either for the liquid ($f = l$) or for the vapor ($f = \text{v}$), $F_H[\rho(h)]$ and $F_H\left[\rho_f^b\right]$ are the Helmholtz free energies of the fluid in the entire slit for the real density distribution and the homogeneous bulk density distribution, respectively, and ρ_f^b is the density of the bulk fluid. The calculation of $F_H[\rho(h)]$ is described in Section 3.1.1. Both potentials are calculated per unit area of one of the walls. The surface tensions γ_{sl} and γ_{sv} can be obtained using the expression

$$\gamma_{sf} = \left(\Omega_{sf} - \Omega_f^b\right)/2,\ \left(f = l, \text{v}\right), \tag{95}$$

where the factor 1/2 accounts for the two solid—fluid interfaces in the system.

A similar approach with some modifications can be used to calculate γ_{lv}. In the latter case one must have a density distribution in the slit that contains the interface between a bulk-like liquid and a bulk-like vapor phase. One of the possible density distributions is presented schematically in Fig. 32. Such a distribution can exist in the slit in the absence of the fluid-solid interactions ($\epsilon_{fs} = 0$). However, in an open system, the numerical solution for this distribution is unstable due to the uncertainty in the position of the interface [42]. To overcome this difficulty, one must correct the position of the interface such as to remain fixed in the same place during calculations [42], or to fix the number of molecules in the system [47]. The latter procedure is equivalent to the consideration of a closed system. For such a system the size of the slit should be selected sufficiently large to allow the Lagrange multiplier λ to take a value close to the chemical potential μ_c for the vapor—liquid coexistence. Using the obtained density profile, γ_{lv} can be calculated using the expression [47]

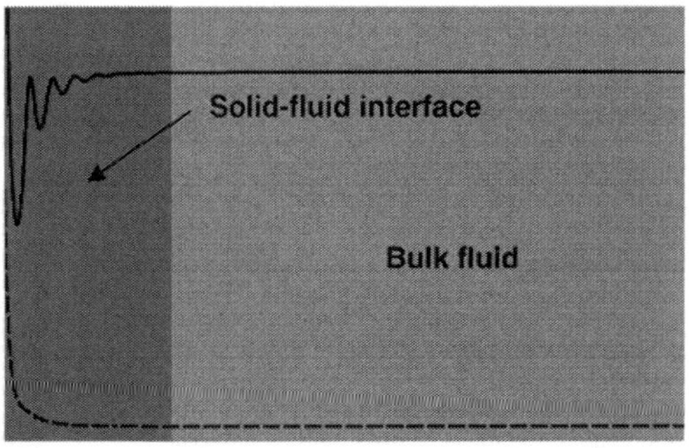

FIG. 31. Schematic presentation of a fluid density distributions across the open slit containing a solid—fluid interfaces. Solid and dashed lines represent distributions which are liquid-like and vapor-like in the bulk phase, respectively. Only one-half of the slit is presented.

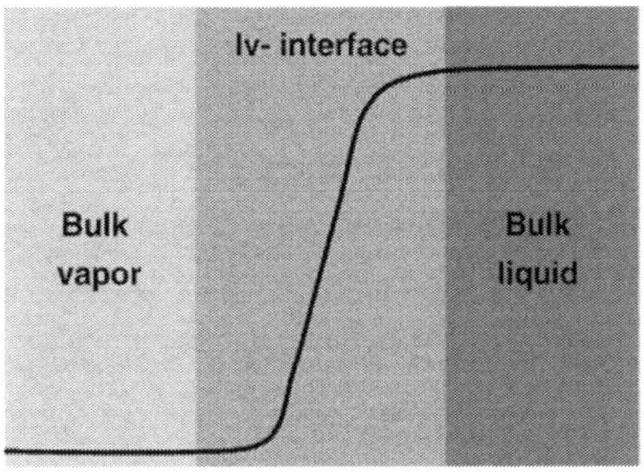

FIG. 32. Schematic presentation of a symmetric fluid density distribution across the open slit with neutral walls, containing liquid–vapor (lv) interface. Only one-half of the slit is presented.

$$\gamma_{lv} = \left(\Omega\left[\rho(h)\right] - \Omega\left[\rho_f^b\right]\right)/2. \tag{96}$$

3.2. A drop on a smooth surface. Dependence of the contact angle on liquid-solid interactions and temperature

A typical drop on a homogeneous surface is presented in Fig. 33a for $\rho_{av}^* = 0.10 \left(\rho^* = \rho\sigma_{ff}^3\right)$ [17]. The lighter areas correspond to higher densities of the fluid. The x-dependence of the fluid density at various distances h from the surface is presented in Fig. 33b. The density as a function of x is almost constant in the middle region of the drop. In the very vicinity of the solid surface (h = $0.1\sigma_{ff}$), it has a solid-like value due to the structuring of the fluid [44] but decreases quickly to liquid-like values for $h \geq 0.2\sigma_{ff}$. The h-dependence of the fluid density at various distances Δx from the plane of symmetry of the drop is presented in Fig. 33c. Using the approach described in Section 3.1.4, the contact angle for the drop shown in Fig. 33a was found to be equal to $\theta = 48.5^\circ \left(\rho_{div}^* = 0.371\right)$.

Most of the DFT studies [17,18,25,26,47] on drops located on solid surfaces were devoted to specific fluids on specific solids and, therefore, were performed for fixed parameters of the interaction potentials. The dependence of the contact angle of argon on various smooth substrates, i.e. for various values of the energy parameter ϵ_{fs} and hard core diameter σ_{fs} of the Lennard-Jones potential was examined in Refs. [17,19] using the procedure described in Section 3.1.4. For the presentation of the results it is convenient to select σ_{ff} as the unit of length and introduce two independent dimensionless parameters, $\sigma = \sigma_{fs}/\sigma_{ff}$ and $\varepsilon = \epsilon_{fs}/\epsilon_{ff}$ characterizing the substrate and fluid, and the parameter $\delta = \epsilon_{ff}/k_B T$ characterizing the effect of temperature. The calculations of the contact angles were performed at four temperatures, T = 85, 90, 95, and 100 K (δ = 1.409, 1.331, 1.261, and 1.198, respectively) and for s varying from 0.3 to 1.6. The values of the parameter σ were selected 1.095, 1.028, 0.969, and 0.910 (σ_{fs} = 3.727, 3.5, 3.3, and 3.1 Å, respectively). The width of the system L_x was taken $L_x = 50\sigma_{ff}$ and the height $L_h = 30\sigma_{ff} + 2\sigma_{fs}$. The dimensionless average density $\rho_{av}^* = \rho_{av}\sigma_{ff}^3$ was considered in the range $0.1 \leq \rho_{av}^* \leq 0.25$. Below, in Sections 3.2.1–3.2.4 the results obtained for nanodrops are presented in details. The results for macroscopic drops are outlined in Section 3.2.5.

3.2.1. s-dependence of nanodrop contact angle at various temperatures and σ = 1.028 (σ_{fs} = 3.5Å)

In Fig. 34a, b typical fluid density distributions for a relatively weak fluid solid—interaction (ε = 0.6) and for a stronger one (ε = 1.4), obtained by solving Eq. (81), are presented schematically for T = 85 K together with the corresponding drop profiles (Fig. 34c, d) calculated as described in

Section 3.1. The circular extension of the upper part of the first profile provides the angle $\theta_u \simeq 137°$ with the solid surface and of the second profile provides $\theta_u \simeq 60°$. Similar extensions of the lower parts of the profiles provide the angles $\theta_l \simeq 147°$, respectively. A typical dependence of θ_u on ε is presented in Fig. 35 for $T = 85$ K ($\delta = 1.409$). This dependence is almost linear in the range $0.6 < \varepsilon < 1.4$, with small deviations from linearity for $\varepsilon < 0.6$ and $\varepsilon > 1.4$. The dependence of θ_l on ε is similar. The linear parts of these dependencies can be represented by the equation

$$\theta = k\varepsilon + b, \tag{97}$$

where the subscripts u and l for θ, k, and b are omitted. For the case presented in Fig. 35, $k = k_u = -96.0°$, and $b = b_u = 194.8°$. The values of k and b for other temperatures (other δ) are

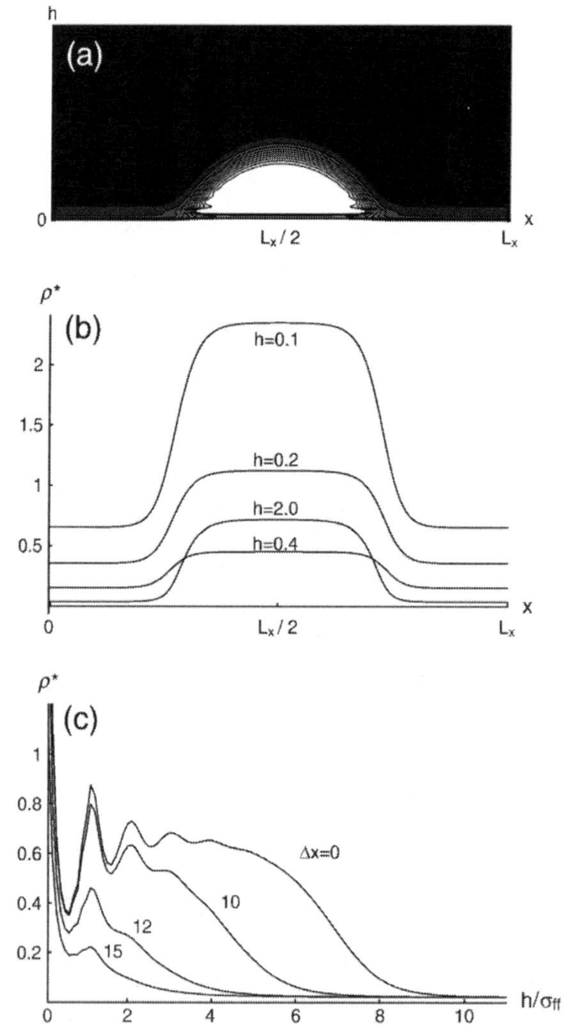

FIG. 33. (Reprinted from Ref. [17] with permission.) $\rho^*_{av} = 0.10$, $\rho^* = \rho\sigma^3_{ff}$ (a) Typical drop on a homogeneous surface. The lighter areas correspond to higher fluid densities. (b) The x dependence of the fluid density at various distances h from the surface. The values of h are indicated on the curves in units of σ_{ff} (c) The h dependence of the fluid density at various distances Δx from the plane of symmetry of the drop. The values of Δx are indicated on the curves in units of σ_{ff}.

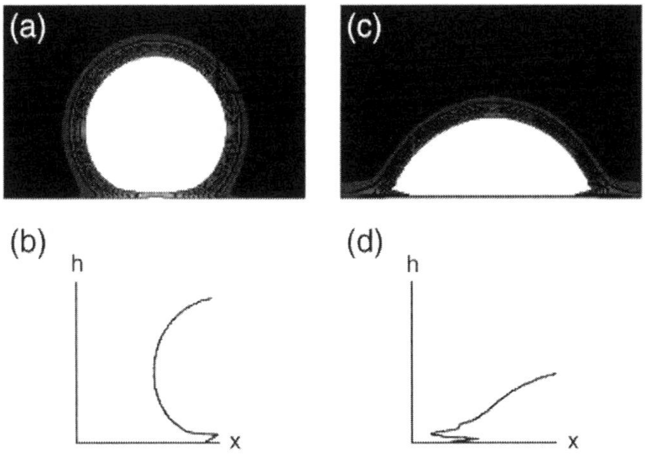

FIG. 34. (Reprinted from Ref. [19] with permission.) (a) and (c) Schematic presentations of the fluid density distribution and the left hand side of the symmetrical drop profile, respectively, for $T = 85\,K$, $\varepsilon = 0.6$, $\rho^*_{div} = 0.371$, $\rho^*_{av} = 0.1$. (b) and (d) The same as in panels (a) and (c), respectively, but for $\varepsilon = 1.4$, and $\rho^*_{div} = 0.356$. The lighter areas in panels (a) and (b) correspond to larger fluid density.

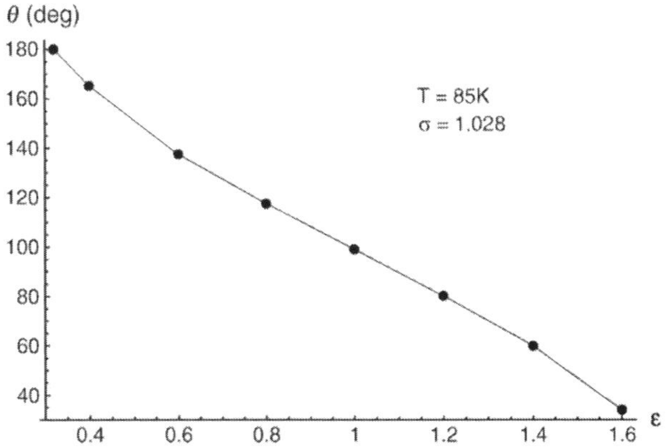

FIG. 35. (Reprinted from Ref. [19] with permission.) ε-dependence of the contact angle for T = 85 K, σ = 1.028 (σ = 3.5 Å). The points represent the results of the DFT calculations and the solid line is a guide for eyes.

listed in Table 1 for both θ_u and θ_l. The slopes k of the lines and the values of δ satisfy approximately the simple relation

$$k\delta^{\nu} = -C \tag{98}$$

where $\nu = 2.64$ and $C = 232.7°$ for θ_u and $\nu = 3.23$, $C = 317.22°$ for θ_l. In Fig. 36 the plots of θ_u vs ε are presented for the four considered values of δ. All the intersection points of the four straight lines are located inside the rectangle shown in this figure. The bold point on the plot represents the mean location of all intersection points and has the coordinates $\varepsilon = \varepsilon_{u,0} = 0.86$ and $\theta = \theta_{u,0} = 112.2°$. Since the intersection points are near to each other, one can consider that all lines intersect at the same point $(\varepsilon_0, \theta_0)$ and combine all the straight lines presented in Fig. 36 into a single line given by

TABLE 1
(Reprinted from Ref. [19] with permission.) Parameters of Eq. (97) for nanodrops at various temperatures for $\sigma = 1.028$ ($\sigma_{fs} = 3.5$ Å) and $\rho^*_{av} = 0.1$.

δ	*T* (°K)	k_u (°)	b_u (°)	k_l (°)	b_l (°)
1.409	85	–96.0	194.8	–108.1	211.7
1.331	90	–107.0	203.8	–119.0	222.8
1.261	95	–121.8	216.5	–147.5	249.4
1.198	100	–149.8	241.0	–184.8	281.2

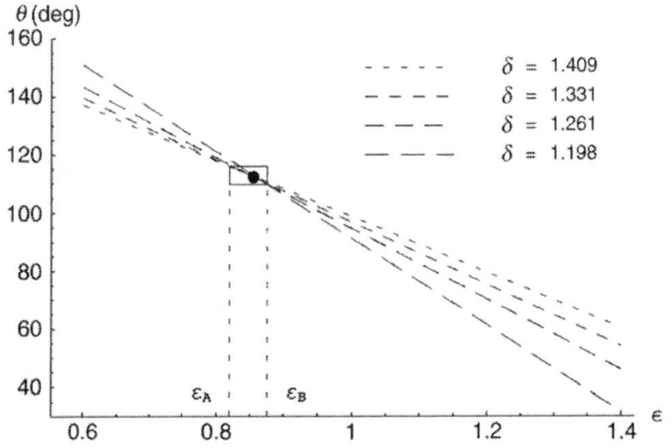

FIG. 36. (Reprinted from Ref. [19] with permission.) ε-dependence of the contact angle for $\delta = 1.409$, 1.331, 1.261, and 1.198 ($T = 85$ K, 90 K, 95 K, and 100 K, respectively) and $\sigma = 1.028$ ($\sigma_{fs} = 3.5$ Å). The bold point has the coordinates $\varepsilon_0 = 0.86$ and $\theta_0 = 112.4°$ which are equal to the mean values of the coordinates of all intersection points of the presented straight lines. All those points are inside the drawn rectangular area. $\varepsilon_A = 0.82$, $\varepsilon_C = 0.88$.

$$\theta - \theta_0 = -C\delta^{-\nu}(\varepsilon - \varepsilon_0). \qquad (99)$$

Note that in Eq. (99) and below the notation θ_0 should not be confused with the same notation used for the microscopic contact angle in Section 2. Because Eq. (99) is valid for both θ_u and θ_l, the corresponding subscripts of θ, θ_0, C, ν, and ε_0 are omitted and each of the angles θ_u and θ_l will be referred below as contact angle. Such subscripts will be used in some cases to specify if θ is θ_u or θ_l. In Eq. (99), θ_0, ε_0, and C are constants provided above and the parameter δ contains the temperature. Note, that the difference between the values of θ obtained through computations based on DFT and those obtained from Eq. (99) does not exceed 5% for the range $0.6<\varepsilon<1.4$ and 1% for the range $0.6<\varepsilon<1.1$. Under the employed conditions (constant σ_{fs} and ρ_s), Eqs. (97)–(99), provide a complete description of the fluid—solid pairs for which the parameters δ and ε are in the ranges noted above.

3.2.2. Temperature dependence of the nanodrop contact angle at constant ε for $\sigma_{fs} = 3.5$ Å

The temperature dependence of the contact angle at fixed values of ε can be easily found using Eq. (99) where $\delta = \epsilon_{ff}/k_BT$. One can see from Fig. 36, which represents this equation at various temperatures, that for $\varepsilon>\varepsilon_0$ the contact angle increases and for $\varepsilon<\varepsilon_0$ it decreases with increasing temperature (decreasing δ). The rate $\frac{d\theta}{dT}$ of change of the contact angle with temperature is provided by the equation

$$\frac{d\theta}{dT}=\frac{\nu k_B}{\epsilon_{ff}}C(\varepsilon-\varepsilon_0)\left(\frac{k_B T}{\epsilon_{ff}}\right)^{\nu-1}. \tag{100}$$

Because $\nu>1$, the absolute value of $\frac{d\theta}{dT}$ increases with increasing temperature.

An interesting implication of Eq. (99) is that for $\varepsilon=\varepsilon_0$ the contact angle does not depend on temperature and is always equal to θ_0. This observation suggests that for a given fluid one can find (or manufacture) a surface with wetting properties independent of temperature in some temperature range.

Considering $T_0 = 92.5$ K as a reference temperature, the variation $\Delta T = T - T_0$ of the temperature in the examined range $85\ K \le T \le 100\,K$ can be considered to be small $\left(\Delta T/T_0 \ll 1\right)$; therefore, Eq. (99) can be linearized with respect to ΔT

$$\theta(T)=\theta(T_0)+\alpha(T_0)\Delta T \tag{101}$$

where

$$\theta(T_0)=\theta_0-C(\varepsilon-\varepsilon_0)\left(\frac{k_B T_0}{\epsilon_{ff}}\right)^{\nu} \tag{102}$$

and

$$\alpha(T_0)=-\frac{c\nu}{T_0}(\varepsilon-\varepsilon_0)\left(\frac{k_B T_0}{\epsilon_{ff}}\right)^{\nu} \tag{103}$$

For θ_u, $\alpha(T_0) = -3.36(\varepsilon-\varepsilon_{u,0})$ and for θ_l, $\alpha(T_0) = -4.88(\varepsilon-\varepsilon_{l,0})$.

The above analysis of the temperature dependence of the contact angle involves the existence of a unique intersection point of all straight lines that represent the ε-dependence of θ at various temperatures. Even if this is not absolutely true, there are ranges of monotonous increase (decrease) of θ with temperature. They are located at $\varepsilon<\varepsilon_A$ ($\varepsilon>\varepsilon_B$), respectively (see Fig. 36).

3.2.3. Nanodrop contact angles for σ = 1.095, 0.969, and 0.910

The results for the contact angle obtained for $\sigma_{fs} = 3.727$ Å, $\sigma_{fs} = 3.3$ Å, and $\sigma_{fs} = 3.1$ Å($\sigma = 1.095$, 0.969, and 0.910, respectively) [19] are similar to those discussed in Section 3.2.1 for $\sigma_{fs} = 3.5$ Å ($\sigma = 1.028$). In the range $0.6<\varepsilon<1.4$, the plot of θ vs ε is represented by straight lines, Eq. (97). The values of the parameters k and b are listed in Table 2.

For each σ, the straight lines at various temperatures can be combined in a single one of the form of Eq. (99) with the parameters ν, C, and ε_0 dependent on σ. Such single lines are plotted in Fig. 37 for each considered value of σ, and the values of the parameters ν, C, and ε_0 are listed in Table 3 for both θ_u and θ_l. The values of the coordinates ε and θ of the bold point in Fig. 37 are ε_0 and θ_0, respectively. For other values of σ, the parameters ν, C, and ε_0 can be found by interpolating between the obtained results. Consequently, one can consider Eq. (99) as a universal equation describing the contact angle for wide ranges of fluids and solids which have intermolecular interactions of the Lennard-Jones type.

TABLE 2
(Reprinted from Ref. [19] with permission.) Parameters of Eq. (97) for nanodrops for various σ and temperatures for $\rho^*_{av} = 0.1$.

$\sigma = \sigma_{fs}/\sigma_{ff}$	σ_{fs} (Å)	T (°K)	k_u (°)	b_u (°)	k_l (°)	b_l (°)
1.095	3.727	85	−117.4	190.9	−141.3	215.1
		90	−126.8	197.5	−159.7	229.3
		95	−143.5	208.0	−184.6	250.3
		100	−171.0	226.8	−207.2	265.6
0.969	3.3	85	−77.8	195.4	−88.7	212.7
		90	−85.0	201.9	−99.8	226.6
		95	−98.0	216.8	−120.5	251.4
		100	−119.0	239.6	−140.0	272.4
0.910	3.1	85	−59.5	191.9	−73.2	215.2
		90	−65.5	199.6	−81.5	228.6
		95	−73.5	210.4	−92.6	246.2
		100	−90.2	234.2	−120.6	285.6

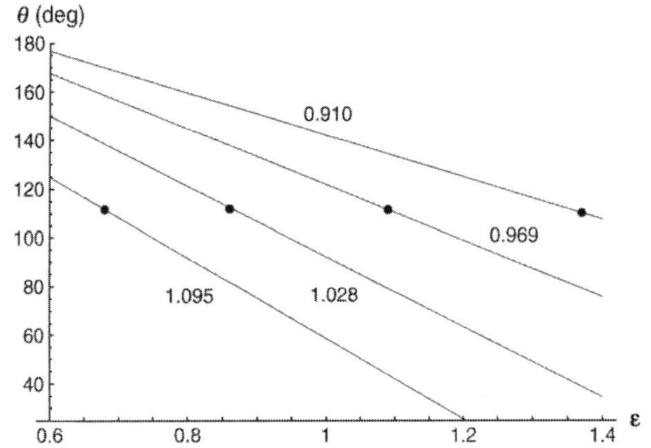

FIG. 37. (Reprinted from Ref. [19] with permission.) The ε-dependence of the contact angle $\theta = \theta_u$ provided by Eq. (99) for $\delta = \epsilon_{ff}/k_B T = 1.198$ (T = 100 K) and various values of $\sigma \equiv \sigma_{fs}/\sigma_{ff}$ indicated on each line. The parameters of Eq. (99) are listed in Table 3. The bold points on each of the lines have the coordinates $(\varepsilon_0, \theta_0)$.

3.2.4. Nanodrop contact angle at various average fluid density

In order to verify the generality of Eq. (99), in Ref. [19] calculations of the contact angle were also performed for $\rho^*_{av} = 0.2$ and all the above considered values of σ as well as for $\rho^*_{av} = 0.2$, and 0.25 and $\sigma = 1.028$. The results obtained are listed in Tables 4 and 5 which suggest that there are indeed some general features in the dependence of the contact angles of nanodrops on temperature and the fluid-solid energy parameter in the Lennard-Jones interaction potential. In particular, Table 5 shows that the calculated contact angles depend weakly on the average density of the fluid in the system. This occurs because of the weak change in the density of the vapor phase in the considered close systems when ρ^*_{av} is changed.

TABLE 3
(Reprinted from Ref. [19] with permission.) Parameters of Eqs. (98) and (99) for nanodrops for various values of $\sigma = \sigma_{fs}/\sigma_{ff}$ and $\rho^*_{av} = 0.1$.

σ	ν_u	$\theta_{u,0}$ (°)	$\varepsilon_{u,0}$	C_u (°)	ν_l	$\theta_{l,0}$ (°)	$\varepsilon_{l,0}$	C_l (°)
1.095	2.25	112.4	0.68	248.5	2.39	106.6	0.77	319.3
1.028	2.62	112.2	0.86	231.0	3.23	112.0	0.93	317.2
0.969	2.55	112.1	1.06	182.1	2.92	110.0	1.17	240.8
0.910	2.45	110.7	1.36	135.0	2.87	105.1	1.52	191.0

TABLE 4
(Reprinted from Ref. [19] with permission.) Parameters of Eqs. (98) and (99) for nanodrops for various values of $\sigma = \sigma_{fs}/\sigma_{ff}$ and $\rho^*_{av} = 0.2$.

σ	ν_u	$\theta_{u,0}$ (°)	$\varepsilon_{u,0}$	C_u (°)	ν_l	$\theta_{l,0}$ (°)	$\varepsilon_{l,0}$	C_l (°)
1.095	2.28	120.6	0.60	257.0	2.69	104.4	0.79	332.0
1.028	2.03	120.6	0.76	183.9	2.57	105.8	1.00	242.1
0.969	2.04	119.8	0.92	144.0	3.22	106.4	1.20	244.2
0.910	1.9	119.2	1.18	111.7	2.75	101.5	1.58	175.7

TABLE 5
(Reprinted from Ref. [19] with permission.) Parameters of Eqs. (98) and (99) for nanodrops for various values of the average fluid density and $\sigma = 1.028$ ($\sigma_{fs} = 3.5$ Å).

ρ^*_{av}	ν_u	$\theta_{u,0}$ (°)	$\varepsilon_{u,0}$	C_u (°)	ν_l	$\theta_{l,0}$ (°)	$\varepsilon_{l,0}$	C_l (°)
0.1	2.62	112.2	0.86	231.0	3.23	112.0	0.93	317.2
0.15	2.19	119.2	0.73	188.1	2.53	109.0	0.96	244.8
0.20	2.03	120.6	0.76	183.9	2.57	105.8	1.00	242.1
0.25	1.87	124.1	0.68	173.3	2.47	106.8	0.99	230.7

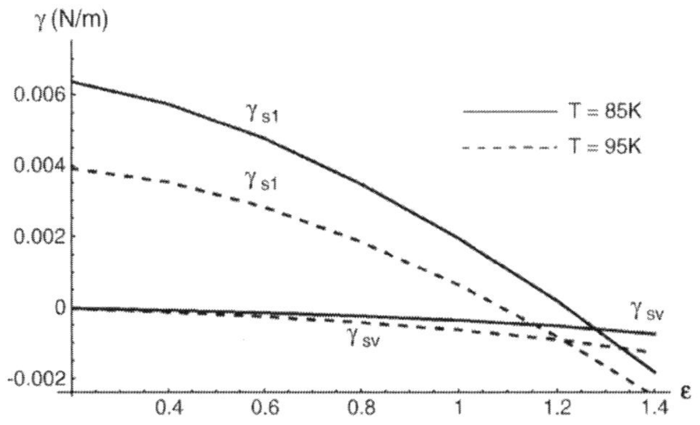

FIG. 38. (Reprinted from Ref. [21] with permission.) ε-dependence of the solid—liquid and solid—vapor surface tensions for $T = 85$ K (solid lines) and $T = 95$ K (dashed lines). The liquid—vapor surface tension is equal to 6.60×10^{-3} N/m for $T = 85$ K and 4.00×10^{-3} N/m for $T = 95$ K.

3.2.5. Dependence of the macroscopic contact angle on liquid-solid interactions and temperature

The dependence of the macroscopic contact angle on the liquid-solid interactions and temperature was examined in Ref. [21]. The DFT in this case provides the values of the surface tensions on the basis of the procedure described in Section 3.1.4. A typical dependence of γ_{sl} and γ_{sv} on ε is presented in Fig. 38 for $T = 85$ K (solid lines) and $T = 95$ K (dashed lines). In all cases, both surface tensions decrease with increasing ε, one of them, γ_{sv} remaining always negative. The solid—liquid surface tension γ_{sl} changes sign when ε increases; it is positive at smaller ε and becomes negative at larger ones. The intersection of the curves for γ_{sl} and γ_{sv} provides the value of ε at which the contact angle $\theta = 90°$. This value decreases as the temperature increases. The liquid—vapor surface tension does not depend on ε and has the values 0.006600 N/m, 0.005253 N/m, 0.003987 N/m, and 0.002893 N/m for $T = 85$ K, 90 K, 95 K, and 100 K, respectively.

The analysis of the macroscopic contact angle calculated with Eq. (1) using the obtained values of the surface tensions has shown that its behavior as function of ε and temperature is described by the same equations as those derived for nanodrops (Eqs. (97)–(100), (102), and (103)). However, the parameters of this equations for macroscopic drops are different from those for nanodrops. For comparison, the values of k and b of Eq. (97) and those of ν, θ_0, ε_0, and C of Eqs. (98) and (99) for macrodrops are presented in Tables 6 and 7, respectively.

3.2.6. General conclusions about the ε- and temperature dependence of nano and macrodrops contact angle on a smooth surface

The main conclusion which can be drawn for nano- and macrodrops is that the contact angle, characterizing the wetting properties of a fluid on a solid surface, has some general characteristics with respect to its dependence on temperature and the parameters that characterize the fluid-solid intermolecular interactions, assuming that the latter interactions are of the Lennard-Jones type. At various temperatures, the contact angle has a linear dependence on the dimensionless energy parameter $\varepsilon = \epsilon_{fs}/\epsilon_{ff}$, where ϵ_{fs} and ϵ_{ff} are the fluid—solid and fluid—fluid energy parameters, respectively. The straight lines representing these dependencies at various temperatures intersect almost in a single point (ε_0, θ_0) where ε_0 depends on σ_{fs} and θ_0 is an almost universal constant. We use the

TABLE 6

Parameters of Eq. (97) for macrodrops for various σ and temperatures.

$\sigma = \sigma_{fs}/\sigma_{ff}$	σ_{fs} (Å)	T (°K)	k (°)	b (°)
1.028	3.5	85	−70.1	180.4
		90	−75.3	184.5
		95	−83.0	191.1
		100	−91.5	197.8
1.095	3.727	85	−88.1	179.2
		90	−94.9	183.2
		95	−105.6	190.3
		100	−115.2	195.5
0.969	3.3	85	−56.3	181.1
		90	−60.2	185.0
		95	−66.4	192.0
		100	−71.6	197.1
0.910	3.1	85	−45.8	183.4
		90	−49.0	187.8
		95	−54.6	196.3
		100	−59.4	202.6

TABLE 7
(Reprinted from Ref. [21] with permission.) Parameters of Eqs. (98) and (99) for macrodrops for various values of $\sigma = \sigma_{fs}/\sigma_{ff}$.

σ	ν	θ_0 (°)	ε_0	C (°)
1.095	1.68	126.0	0.61	155.6
1.028	1.65	123.3	0.82	122.3
0.969	1.54	122.2	1.01	93.8
0.910	1.64	119.0	1.41	79.6

word "almost" because θ_0 varies between 122.5° ± 3.5°, and between 117.4°± 6.7° for macro and nanodrops, respectively. (For nanodrops we selected the angle θ_u for comparison). The contact angle θ_0 for $\varepsilon = \varepsilon_0$ is independent of temperature. For $\varepsilon<\varepsilon_0$ ($\theta>\theta_0$), the contact angle monotonously increases and for $\varepsilon>\varepsilon_0(\theta<\theta_0)$ monotonously decreases with increasing temperature. For the macroscopic drops used in experiments, this result can qualitatively explain the different temperature behaviors of the contact angle [76–80].

The temperature independent value θ_0 of the contact angle corresponding to ε_0 is independent of $\sigma = \sigma_{fs}/\sigma_{ff}$ and is therefore a general characteristic of a nanodrop on a solid surface. For $\theta<\theta_0$, the interactions of the liquid with the solid are stronger ($\varepsilon>\varepsilon_0$); whereas for $\theta>\theta_0$ the interactions are weaker ($\varepsilon<\varepsilon_0$). The existence of such a point, if confirmed, may lead to a new classification of materials in hydrophobic and hydrophilic. The conventional classification into hydrophobic and hydrophilic which selects $\theta = 90°$ as the separator between hydrophilic and hydrophobic surfaces is completely arbitrary, while the suggested classification based on θ_0 as the separator has a microscopic origin.

The general features of the contact angle identified for nanodrops can be compared with the results obtained in Refs. [58,59] via molecular dynamics simulations for truncated Lennard—Jones interactions. Although the available data [58,59] do not provide enough information for a full comparison, they are in qualitative agreement with our results. Indeed, in both treatments θ decreases with increasing σ_{fs}, ϵ_{fs}, or the size of the droplet. It should be noted that in Refs. [58,59] it was reported that cosθ has a linear dependence on ϵ_{fs} (in contrast to the linear dependence of 9 on ϵ_{fs} obtained by us). However, the replotting of the data presented in Refs. [58,59] has shown that the linear fit of θ on ϵ_{fs} is as good as that for cosθ. The quantitative differences between the values of θ can be explained by the difference in the dimensionality of the drops considered in Refs. [58,59] (three dimensional drops) and in Refs. [19,21] (two dimensional drops).

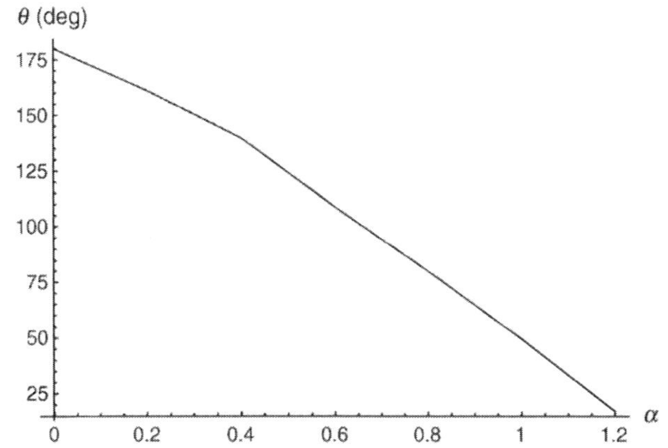

FIG. 39. Contact angle θ of a drop on a flat homogeneous surface as function of $\alpha = \epsilon_{fs}/\epsilon_{fs}^0$.

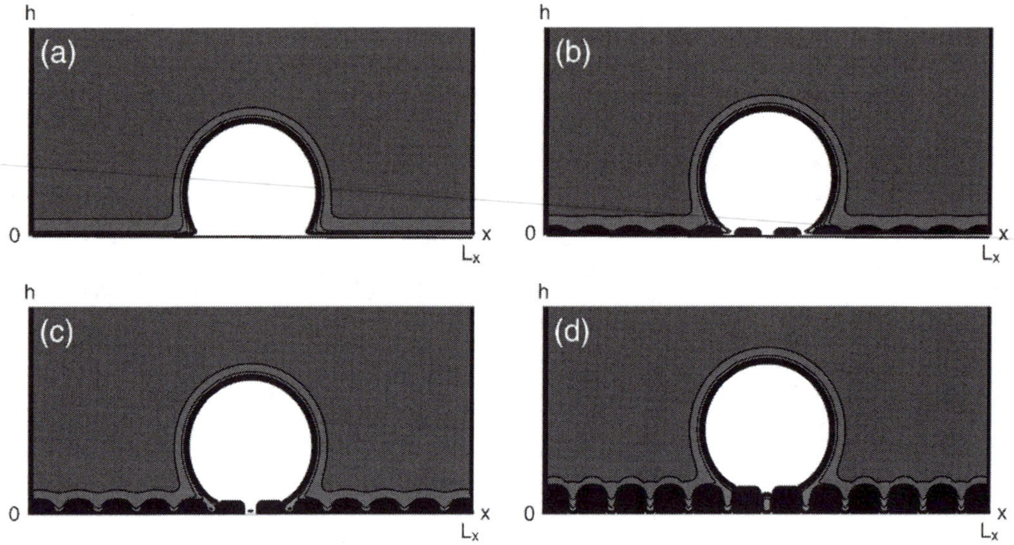

FIG. 40. (Reprinted from Ref. [17] with permission.) Drops on superhydrophobic surfaces with $\epsilon_{fs}=0.4\epsilon_{fs}^{0}$ and various pillar heights. The lighter areas correspond to higher fluid densities. $(a)\, h_p=0$, $(b)\, h_p=2\sigma_{ff}$, $(c)\, h_p=4\sigma_{ff}$, $(d)\, h_p=8\sigma_{ff}$.

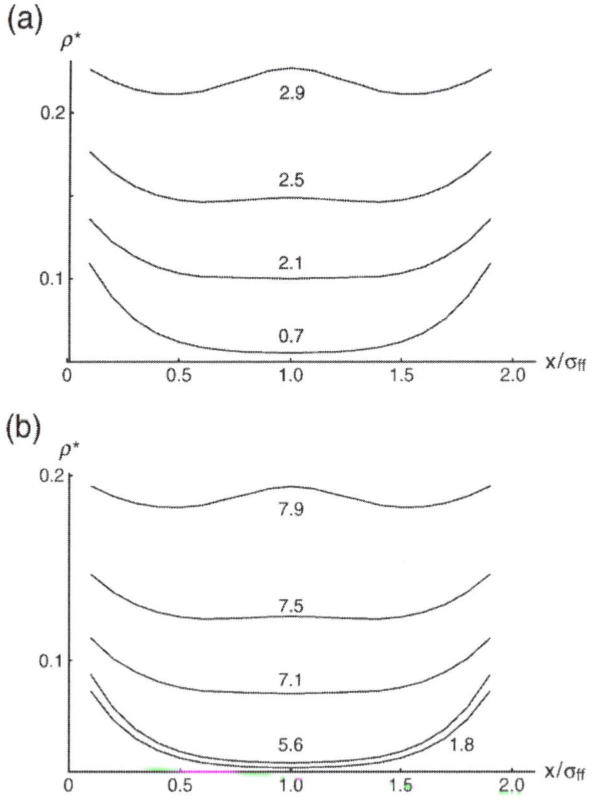

FIG. 41. (Reprinted from Ref. [17] with permission.) x-dependence of the fluid density between pillars. $\rho_{av}^{*}=0.10$. $(a)\, h_p=3\sigma_{ff}$, $(b)\, h_p=8\sigma_{ff}$. The numbers on the curves provide the distance from the solid surface in unit of σ_{ff}.

3.3. NANODROP ON A PHYSICALLY ROUGH SOLID SURFACE

In paper [17] the DFT was applied to drops of argon on a rough surface. For physical roughness, it was assumed that the pillars described in Section 3.1.3 had the width $d_p = 2\sigma_{ff}$ and the distance between them $\Delta d_p = 4\sigma_{ff}$, the height of the pillars h_p being varied between $h_p = 0$ (no pillars) to $h_p = 10\sigma_{ff}$. Hydrophobic or hydrophilic (when free of pillars) surfaces are considered, which were distinguished by the value of the energy parameter ϵ_{fs} in the Lennard—Jones potential of the fluid—solid interactions, keeping all other parameters the same. For a smooth surface, the contact angle of a drop as a function of the parameter $\alpha = \epsilon_{fs}/\epsilon_{fs}^0\left(\epsilon_{fs}^0/k_B = 153.0K\right)$ for $\sigma_{fs} = 3.727\text{Å}$ is presented in Fig. 39. From this dependence one can conclude that for $\epsilon_{fs} < 0.75\epsilon_{fs}^0$ the surface can be considered as hydrophobic (θ>90°), and for $\epsilon_{fs} > 0.75\epsilon_{fs}^0$ as hydrophilic (θ<90°). These two cases will be considered separately.

3.3.1. Droplet on a rough hydrophobic surface

For a superhydrophobic surface with $\alpha = 0.4$ which is additionally decorated with pillars, the FDDs for various heights of the pillars and $\rho_{av}^* = 0.10$ are presented in Fig. 40. For small heights of the pillars $(h_p \le 4\sigma_{ff})$ the liquid penetrates between them (Wenzel regime). With increasing heights of the pillars $(h_p \ge 6\sigma_{ff})$ most of the space between them becomes occupied by a low density (vapor-like) fluid and the drop is located on the tops of the pillars (Cassie—Baxter regime). The FDD in the x direction between pillars under the drop is presented in Fig. 41 for $h_p = 3\sigma_{ff}$ (Wenzel regime) and $h_p = 8\sigma_{ff}$ (Cassie—Baxter regime). In these two cases, the density distributions near the bottom and top of the pillars are almost the same. However, for $h_p = 8\sigma_{ff}$ a considerable part of the space ($0<h<0.7h_p = 5.6\sigma_{ff}$) is occupied by a low—density vapor—like fluid, represented in Fig. 41b by the curve for $h = 5.6\sigma_{ff}$, while for $h_p = 3\sigma_{ff}$ this part of the space is smaller ($0 < h < 0.23\sigma_{ff}$). In both cases the surface of the pillars is covered by a thin layer of fluid molecules. The dependence of the cosine of the contact angle on the pillar height h_p is presented in Fig. 42. Note, that for the selected geometry of the pillar ($d_p = 2\sigma_{ff}$, $\Delta d_p = 4\sigma_{ff}$) and their location, the roughness defined by Eq. (90) is a linear function of the pillar height ($r = 1+h_p/3$). Fig. 42 shows that for small pillar heights $(h_p \le \sigma_{ff})$, the Wenzel formula and the DFT results agree qualitatively, the contact angle provided by Wenzel formula being larger than that calculated by DFT. However, the Wenzel formula provides the maximum possible value of the contact angle, namely $\theta = 180°$ ($\cos\theta = -1$) for $h_p \sim \sigma_{ff}$, whereas the contact angle provided by the DFT at $h_p = \sigma_{ff}$ is $\theta \sim 158°(\cos\theta = -0.927)$. With increasing pillar height, the DFT provides increasing contact angles, that reached at $h_p \sim 2\sigma_{ff}$ the asymptotic value of $\theta_{max} \sim 167°$ ($\cos\theta = -0.974$). The

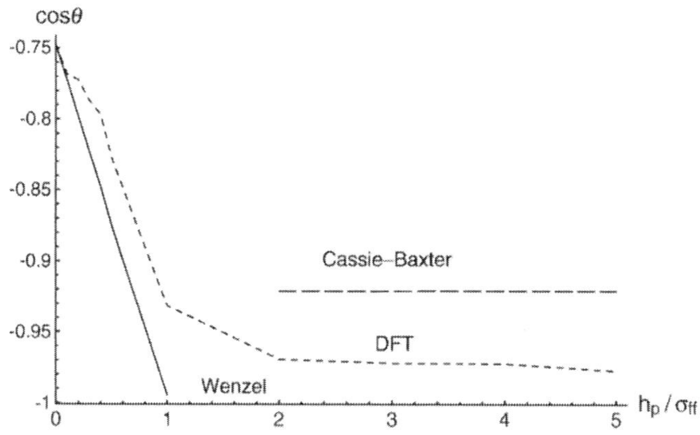

FIG. 42. (Reprinted from Ref. [17] with permission.) Dependence of $\cos\theta$ on the pillar height for $\rho_{av}^* = 0.10$. The solid line represents the Wenzel formula and the dashed line represents the DFT calculations. The long-dashed line represents the contact angle obtained for $h>2\sigma_{ff}$ using the Cassie—Baxter formula (Eq. (3))

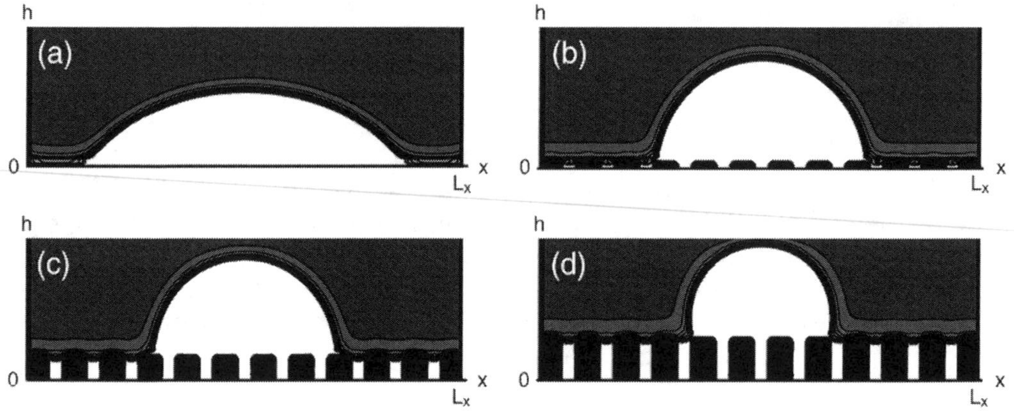

FIG. 43. (Reprinted from Ref. [17] with permission.) Drop on a hydrophilic surface with $\epsilon_{fs} = \epsilon_{fs}^{0}$, various pillars heights and $\rho_{av}^{*} = 0.20$. The lighter areas correspond to higher fluid densities. $(a)\, h_p = \sigma_{ff}$, $(b)\, h_p = 3\sigma_{ff}$, $(c)\, h_p = 6\sigma_{ff}$, $(d)\, h_p = 8\sigma_{ff}$.

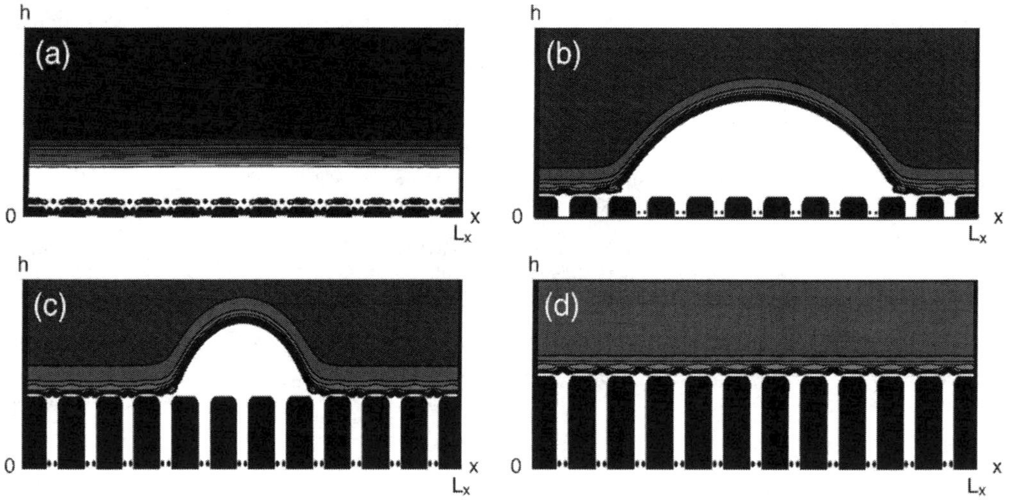

FIG. 44. (Reprinted from Ref. [17] with permission.) Drops on hydrophilic surfaces for $\epsilon_{fs} = 1.4\epsilon_{fs}^{0}$, and various pillar heights and $\rho_{av}^{*} = 0.20$. The lighter areas correspond to higher fluid densities. $(a)\, h_p = \sigma_{ff}$, $(b)\, h_p = 4\sigma_{ff}$, $(c)\, h_p = 8\sigma_{ff}$, $(d)\, h_p = 10\sigma_{ff}$.

difference in the behavior of the contact angle can be qualitatively explained by the transition from the Wenzel to the Cassie—Baxter regime. To compare the DFT results to those provided by the Cassie—Baxter formula, Eq. (3), for $h_p > 2\sigma_{ff}$ one can take $\theta_A = 139.9°$ ($\cos\theta_A = -0.76$) and $\theta_B = 180°$. These values correspond to a smooth surface and to the case of air between pillars (negligible interaction with the solid), respectively. For the selected geometry of the pillars, $f_A = 1/3$ and $f_B = 2/3$. As a result, Eq. (3) provides for the contact angle of the drop on the pillars the value $\theta = 157.2°$ ($\cos\theta = -0.92$), which is somewhat smaller than the asymptotic value $\theta = 167°$ provided by the DFT for the Cassie—Baxter regime. Note that in contrast to the nonlocal DFT, the lattice model employed in Ref. [50] provides for the Cassie—Baxter regime a contact angles smaller, and for the Wenzel regime a contact angle larger that those provided by Eqs. (3) and (4), respectively.

3.3.2. Droplet on a rough hydrophilic surface

Let us first consider a solid carbon dioxide surface with $\epsilon_{fs} = \epsilon_{fs}^{0}$. The contact angle for the drop on this surface in the absence of pillars is equal to 50°. The FDDs for $\rho_{av}^{*} = 0.20$ and various pillar heights are presented in Fig. 43. One can see that the presence of pillars provides an increase in θ as for the hydrophobic surface. This result contradicts the predictions of the Wenzel formula according to which the increase of roughness should decrease the contact angle, i.e. increase the hydrophilicity of the surface. One should also note the decrease of the drop dimensions with increasing height of the pillars. This decrease is a result of the increase of the amount of fluid accumulated between pillars. Because the total number of molecules in the system is fixed, this accumulation leads to a decrease in the number of molecules remaining to form a drop, hence, to a decrease in the size of the drop when pillars are present.

For stronger fluid—solid interactions $\left(\epsilon_{fs} = 1.4\epsilon_{fs}^{0}\right)$, no drop is formed on a smooth surface, which is covered by a liquid—like film (consequently, $\theta = 0°$, $\cos\theta = 1$). Short pillars with $h_p < 2\sigma_{ff}$ cause a periodical nonuniformity in the film in the x direction (see Fig. 44a), while large pillars generate drops on the rough surface. The size of the drop decreases with increasing pillar height because of the accumulation of liquid between the pillars. At a critical pillar height ($h_p \simeq 9\sigma_{ff}$ in the considered case), the drop disappears completely (Fig. 44d).

The DFT calculated contact angle changes nonmonotonously with increasing pillar height (see Fig. 45a) and does not follow the Wenzel formula, Eq. (4). Indeed, in this case Eq. (4) provides for

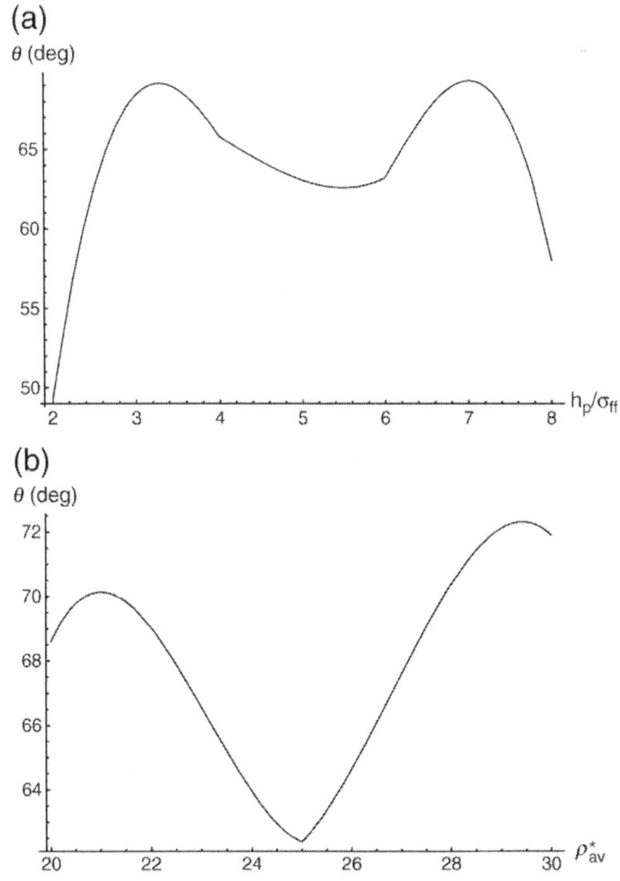

FIG. 45. (Reprinted from Ref. [17] with permission.) Dependence of the contact angle of the drop on a hydrophilic surface for $\epsilon_{fs} = 1.4\epsilon_{fs}^{0}$ (a) on the pillar height; (b) on the average density ρ_{av}^{*}.

$\cos\theta$ values larger than unity for all pillar heights because $\cos\theta_{flat} = 1$ and $r>1$, hence a drop cannot be formed on the surface. Note that at a fixed pillar height, the contact angle calculated by DFT slightly depends on the volume of the drop, which can be changed by changing the average density ρ^*_{av}. This dependence is nonmonotonous (Fig. 45b). A possible reason for the nonmonotonous dependence of the contact angle on the pillar height and on the drop volume is that this angle depends on the position of the leading edge of the nanodrop with respect to the pillar surface. With increasing (decreasing) size of the drop, the leading edge periodically changes its position which can be located either between pillars or on their tops being thus exposed to different fluid—solid interactions. This behavior is similar to the ρ^*_{av}–dependence of the contact angle on a chemically rough surface considered below in Section 3.4.

3.4. Nanodrop on a Chemically Rough Solid Surface

3.4.1. DFT calculation of the contact angle

In Ref. [17] a nanodrop on a chemically inhomogeneous solid surface, as described in Section 3.1.2 (see Fig. 28), was considered at the temperature $T = 87.3$ K. The widths of the plates A and B were selected equal to $3\sigma_{ff}$ and the energy parameter $\epsilon_{fs,A}$ was fixed at $\epsilon_{fs,A}/k_B = 153.0$ K. Six values were used for the energy parameter $\epsilon_{fs,B}$ and the ratio $\alpha_{BA} \equiv \epsilon_{fs,B}/\epsilon_{fs,A}$ was changed from 1 to 0 ($\alpha_{BA} = 1$, 0.8, 0.6, 0.4, 0.2, and 0). The case $\alpha_{BA} = 1$ corresponds to a homogeneous surface and the case $\alpha_{BA} = 0$ corresponds to a plate B with $\epsilon_{fs,B} \ll \epsilon_{fs,A}$. In all the cases considered the widths of the liquid—vapor and liquid—solid interfaces were comparable to the droplet size and the liquid density distribution had an oscillatory behavior near the latter interface.

The Euler—Lagrange equation, Eq. (81), for a drop on a chemically rough surface has generally two solutions for a given average fluid density, presented in Fig. 46. The first solution, $D1$, describes a drop symmetrical with respect to the middle of plate A, the apex of the drop being located at the local minimum of the fluid-solid interaction potential. The apex of the drop described by the second solution ($D2$) is located at the local maximum of the potential, the drop being symmetrical with respect to the middle of plate B. For the same fluid average density ρ^*_{av} in the system one of this drops is stable (i.e. corresponds to a global minimum of the free energy) and the other one is metastable (corresponds to a local minimum of the free energy). For example, in the range of fluid average densities $0.10 \le \rho^*_{av} \le 0.26$ considered in Ref. [17] and for $\alpha_{BA} = 0.2$, drop $D2$ is stable for $0.14 \le \rho^*_{av} \le 0.19$ and drop $D1$ is stable for all other values of ρ^*_{av}. Depending on specific conditions

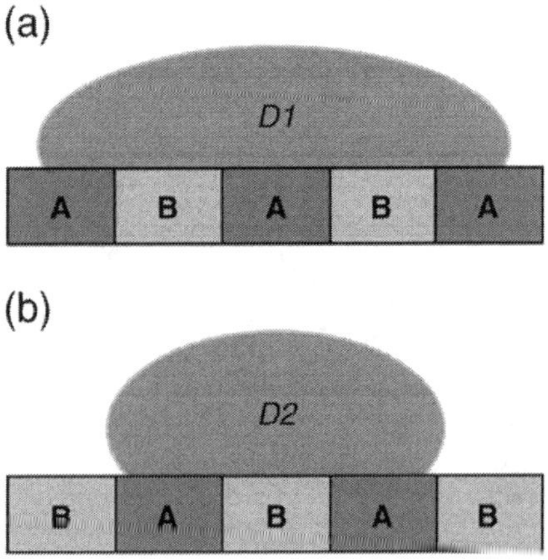

FIG. 46. (Reprinted from Ref. [20] with permission.) Two characteristic states of a nanodrop on a surface with periodic chemical roughness. The drop $D1$ ($D2$) is symmetrical with respect to the vertical plane passing through the middle of plate A (B).

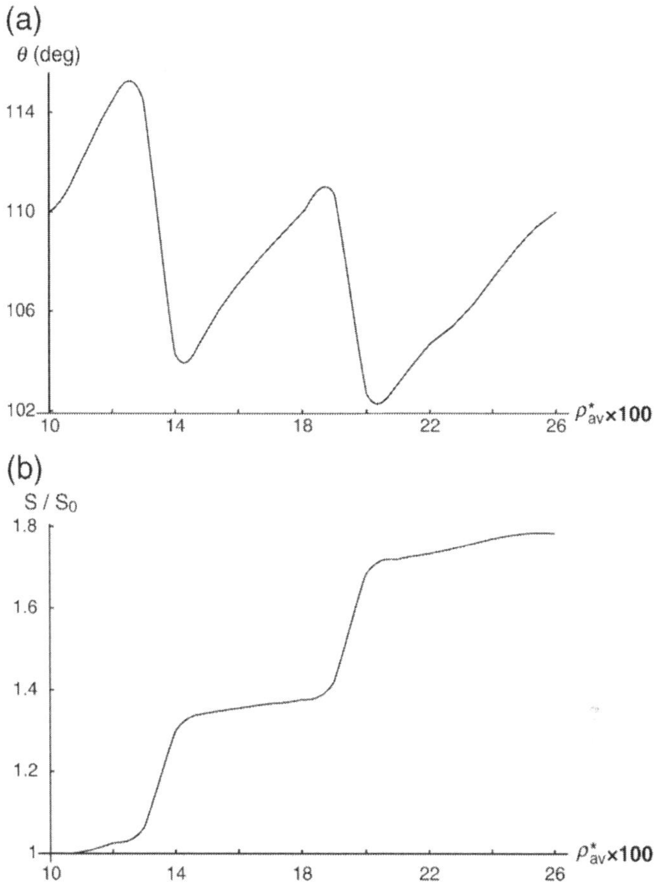

FIG. 47. (Reprinted from Ref. [17] with permission.) $\alpha_{BA} = 0.2$. (a) ρ^*_{av}–dependence of the contact angle of a drop on a chemically rough surface. (b) ρ^*_{av}–dependence of the area of the solid surface covered by the drop. S_0 is this area at $\rho^*_{av} = 0.10$.

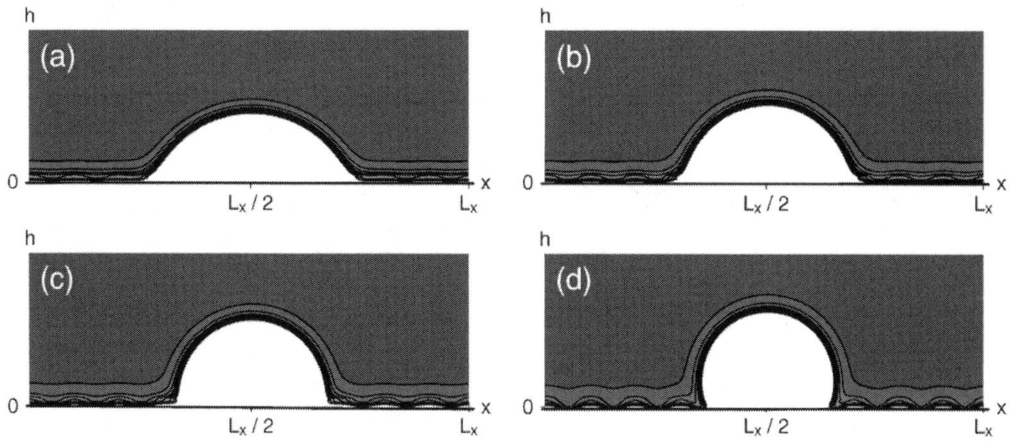

FIG. 48. Drop on chemically rough surfaces for various values of $\alpha_{BA} = \epsilon_{fs,B}/\epsilon_{fs,A}$. (a) $\alpha_{BA} = 0.8$; (b) $\alpha_{BA} = 0.4$; (c) $\alpha_{BA} = 0.2$; $\alpha_{BA} = 0$.

TABLE 8
Contact angles of a drop on various chemically rough surfaces. θ_B is the contact angle of a drop on the homogeneous surface of a solid of type B (Fig. 28) calculated by DFT, $\varepsilon_{fs,B}=\alpha_{BA}\varepsilon_{fs,A}$, θ_r is the contact angle on the chemically rough surface, and $\theta_{r,Cassie}$ is the contact angle calculated using Cassie—Baxter equation, Eq. (3).

$\alpha_{BA} = \varepsilon_{fs,B}/\varepsilon_{fs,A}$	θ_B	θ_r	$\theta_{r,Cassie}$
0.8	81.4°	68.4°	66.0°
0.6	109.7°	86.3°	80.6°
0.4	139.9°	102.7°	92.9°
0.2	161.5°	112.0°	98.2°
0.	180°	119.8°	99.7°

(chemical nature and sizes of the plates, temperature, fluid average density) the leading edges of the drop can be located either on the *A*- or *B*-plate.

As expected, the contact angle of a drop on a rough surface depends on the drop volume or, equivalently, on ρ^*_{av}. The ρ^*_{av}–dependence of θ for a stable drop is presented in Fig. 47a and the ρ^*_{av} – dependence of the area *S* of the solid surface covered by the drop in Fig. 47b. From these figures one can see that there is a correlation between the rapid decrease of the contact angle at $\rho^*_{av} \simeq 0.135$ and $\rho^*_{av} \simeq 0.195$ and the rapid increase of *S* at the same value of ρ^*_{av}. Note that when ρ^*_{av} increases, the leading edge of the stable drop is displaced along the periodic potential generated by the solid surface. For this reason one expects the contact angle to be a periodic function of the drop volume (or of ρ^*_{av}). The same values of θ for $\rho^*_{av}=0.10$ and $\rho^*_{av}=0.26$ in Fig. 47b provide results in that direction.

The effect of chemical roughness on the contact angle is illustrated in Fig. 48 where the FDDs for $\alpha_{BA} = 0.8, 0.4, 0.2$, and 0 are presented for $\rho^*_{av} = 0.10$.

The contact angles for those drops are listed in the third column of Table 8 which shows that they increase with decreasing α_{BA}, i.e. the chemical roughness increases the hydrophobicity of the solid surface.

To compare the results to those predicted by the macroscopic treatment, the contact angles for chemically rough surfaces calculated using the Cassie—Baxter equation (Eq. (3)) are listed in the last column of Table 8. (In the considered case $f_A = f_B = 1/2$.) For all considered values of α_{BA}, the Cassie—Baxter equation provides a smaller contact angle than the DFT theory, the difference being greater for smaller values of α_{BA}, i.e. for larger differences between plates *A* and *B*.

3.4.2. Qualitative comparison of DFT results with the experiments of Gao and McCarthy [81]

Recently, Gao and McCarthy [81] reported experimental results for the contact angle of a liquid drop (sodium hydroxide solution) on a chemically rough surface of PFA$(Me)_2$SiCl-modified silicon wafer, which are in obvious contradiction with the predictions of the Cassie-Baxter equation (Eq. (3)). For a cylindrical drop on a chemically heterogeneous surface the experimental setting of Ref. [81] is represented in Fig. 49. (Note that in the real experiments, axisymmetrical drops were considered). The surface can be considered as composed of two different plates *A* and B, one of them (A) being hydrophobic and the other either hydrophilic or ultrahydrophobic. At constant drop volume, the size of plate *B* was varied and the contact angle was measured and compared with the predictions of Eq. (3) in which θ_A and θ_B were considered constant and f_A and f_B were changed because the area occupied by plate (*B*) beneath the drop is changed. Experiment demonstrated that the contact angle θ remained almost constant, while Eq. (3) predicts a considerable change. [81] In the discussion that followed in the literature (see Refs. [82–86]), various arguments were brought pro and contra that conclusion. The main conclusion of Refs. [82–84] was that the area fractions f_B and $f_A = 1 - f_B$ involved in the Cassie—Baxter equation are local quantities calculated in the vicinity of the contact line. However, the considerations of Bormashenko et.al. [85] regarding the equilibrium conditions of a drop on a surface, based on the minimization of the global free energy of the drop, revealed that the entire contact area between fluid and solid contributes to the value of the contact angle. Finally, Marmur and Bittoun [86] have shown that the Cassie—Baxter as well as the Wentzel equation for a physically rough surface [10] are valid when the drop size is much larger (about three orders of magnitude larger) than the wavelength of the chemical heterogeneity or physical roughness. In Gao and McCarthy experiments the above condition is not satisfied; consequently the Cassie—Baxter equation is not applicable. Note that all above mentioned considerations were made in the framework of classical thermodynamics, which involves the surface tensions and the Young equation for the contact angle of the drop on smooth surfaces.

In paper [20], a DFT analysis of the contact angle of a nanodrop was performed for argon on a surface composed of two plates A and B (see Fig. 49) at $T = 85$ K and for $\rho^*_{av} = 0.16$. In spite of the

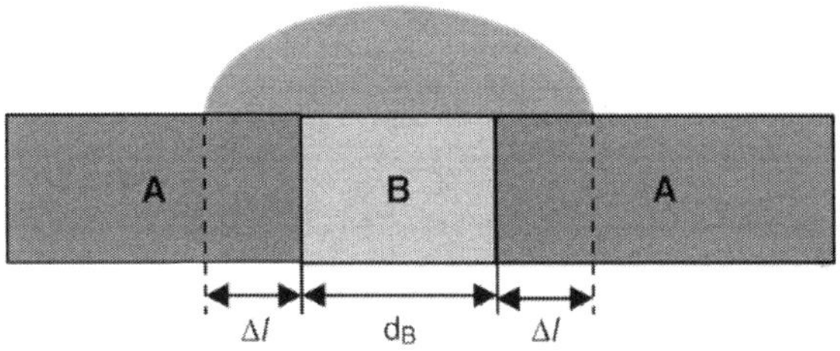

FIG. 49. Schematic representation of a hydrophilic or ultrahydrophobic area (B-plate) inside a hydrophobic surface (A-plates).

TABLE 9
Contact angles for hydrophilic area inside a hydrophobic surface.

d_B/σ_{ff}	f_B	$\Delta l/\sigma_{ff}$	$\theta(°)$	$\theta_{CB}{}^{(°)}$
6	0.239	9.5	100.6	84.3
10	0.401	7.4	100.6	73.1
14	0.560	5.4	100.6	61.4
18	0.710	3.7	99.2	49.0
22	0.821	2.4	94.2	38.0
26	0.883	1.8	85.6	30.5

TABLE 10
Contact angles for ultrahydrophobic area inside a hydrophobic surface.

d_B/σ_{ff}	f_B	$\Delta l/\sigma_{ff}$	$\theta(°)$	$\theta_{CB}{}^{(°)}$
6	0.239	9.5	100.6	84.3
10	0.401	7.4	100.6	73.1
14	0.560	5.4	100.6	61.4
18	0.710	3.7	99.2	49.0
22	0.821	2.4	94.2	38.0
26	0.883	1.8	85.6	30.5

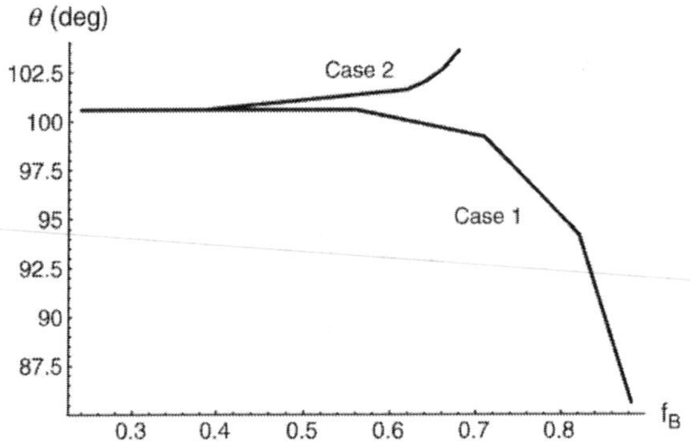

FIG. 50. (Reprinted from Ref. [20] with permission.) Dependence of the contact angles calculated via DFT on the fraction of the surface of B-type beneath the drop for a hydrophilic area inside a hydrophobic surface (Case 1) and for an ultrahydrophobic area inside a hydrophobic surface (Case 2).

different natures of the liquid and solid, one expects to have at least a qualitative similarity between the experimental and the theoretical results.

First, plate A was selected slightly hydrophobic with $\epsilon_{fs,A}/k_B = 91.8$ K. For such a surface, $\theta_A = 100.6°$. The energy parameter $\epsilon_{fs,B}$ of the B-plate was selected to be $\epsilon_{fs,B}/k_B = 183.6$ K and corresponds to a hydrophilic surface with a wetting angle $\theta_B \simeq 0°$. The thickness d_B of the B-plate was varied between $6\sigma_{ff}$ and $26\sigma_{ff}$. For $T = 85$ K and ρ^*_{av}, the width of the contact area of the drop was varied from $25\sigma_{ff}$ for the smallest thickness of the B-plate to $29\sigma_{ff}$ for the largest one. The contact angles calculated on the basis of DFT are listed in the fourth column in Table 9. The third column provides the distance Δl between the leading edge of the drop and the nearest boundary of a B-plate (see Fig. 49). The contact angle θ which the drop makes with the surface is constant ($\theta = 100.6°$) for $d_B \leq 14\sigma_{ff}$ $\left(f_B \leq 0.560,\ l > 5\sigma_{ff}\right)$ and decreases to 85.6° with increasing d_B up to $d_B = 26\sigma_{ff}\left(f_B \simeq 0.88\right)$. For comparison purposes, the contact angle θ_{CB} calculated with Eq. (3) is provided in the last column of Table 9. This angle varies from 84.3° for $d_B = 6\sigma_{ff}$ to 30.5° for $d_B = 26\sigma_{ff}$.

As a second example, the B-plate was selected as ultrahydrophobic, with $\epsilon_{fs,B}/k_B = 0$ K and $\theta_B = 180°$. (Such a situation can be achieved by replacing the B-plate by trapped air). The A-plates are as those considered in the first example. The width of the contact area of the drop with the solid surface is about $26\sigma_{ff}$. The results obtained are listed in Table 10. A drop, containing a B-plate beneath it, can exist only if the width d_B of the B-plate is smaller than $17.2\sigma_{ff}$. For $d_B > 17.2\sigma_{ff}$, the Euler—Lagrange equation, Eq. (81), does not have a stable or metastable solution with the leading edge located on the A plates.

The contact angle θ which the drop makes with the surface varies between 100.6° and 103.6° as d_B increases from 0 to $17.2\sigma_{ff}$. (The contact angle θ_{CB} calculated with Eq. (3) varies between 100.6° and 137.7°.) A visible change of the contact angle begins to take place when the distance Δl between the leading edge and the nearest boundary of a B-plate becomes smaller than $5\sigma_{ff}$.

For the above two considered cases, the dependence of the contact angle on the fraction f_B of the B-plates beneath the drop is presented in Fig. 50. The curve labeled as Case 1 corresponds to a hydrophilic area inside a hydrophobic surface. The other line labeled as Case 2 corresponds to an ultrahydrophobic area inside a hydrophobic surface.

For $f_B < 0.6$, the contact angle calculated via DFT is almost independent of f_B and remains the same regardless the nature of the B-plate (hydrophilic or hydrophobic) present beneath the drop. The distance Δl between the leading edges of the drop and the B-plates is in this case larger than $5\sigma_{ff}$. For a hydrophilic B-plate, the results are in agreement with the experimental observations of

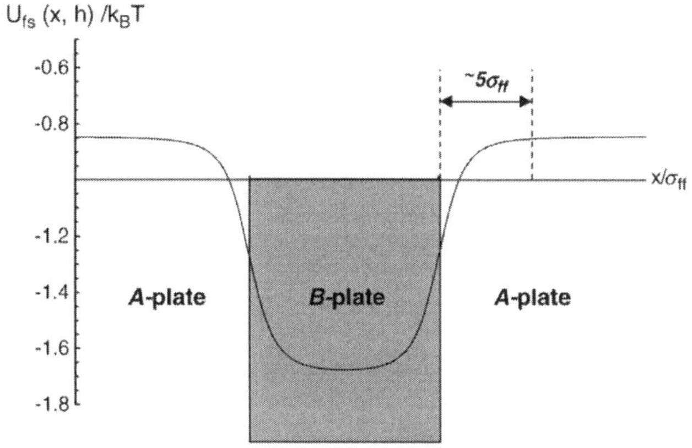

FIG. 51. (Reprinted from Ref. [20] with permission.) Potential of the fluid—solid interaction as function of x for the surface considered in Section 3.4 (Case 1) with $d_B = 10\sigma_{ff}$, calculated at a distance from the solid surface $h = 0.5\sigma_{ff}$.

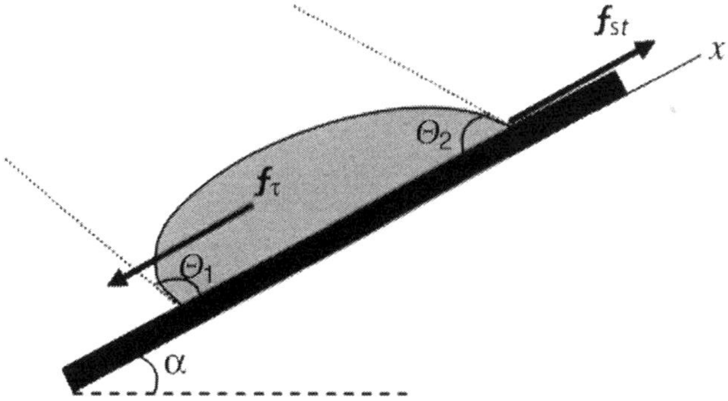

FIG. 52. (Reprinted from Ref. [18] with permission.) Liquid drop in mechanical equilibrium on an inclined surface with inclination angle α. f_τ and f_{st} $(f_\tau = f_{st})$ are the tangential components of the force of gravity and the sticking force, respectively. The front, θ_1, and rear, θ_2, contact angles for the stability limit are replaced by the advanced and receding contact angles, respectively.

Gao and McCarthy [81] which have shown that for a hydrophilic spot in a hydrophobic field (as in Case 1) the contact angle is almost independent of f_B for f_B<0.64. (Note that in their experiments surfaces with f_B>0.64 and hydrophobic surfaces have not been considered.)

For f_B> 0.6 ($\Delta l < 5\sigma_{ff}$), our DFT results for Cases 1 and 2 exhibit changes in the contact angle. One of the reasons for the changes is that for f_B> 0.6 the distance between the leading edges of the drop and the B-plates becomes very small and hence, the local conditions near the leading edges are considerably changed compared to those on a uniform surface. From a microscopic point of view those conditions are determined solely by the potential $U_{fs}(x, h)$ of the fluid-solid interactions. In Fig. 51, the change of this potential in the vicinity of the boundary between the A- and B-plates is plotted as a function of x for semi-infinite A-plates and $d_B = 10\sigma_{ff}$. The plateaus at the right and left sides of the B-plate and one in the middle of the B-plate provide potentials $U_{fs}(x, h)$ approximately equal to those for uniform infinite A- and B-plates. The transition between those values occurs within a distance of about $8\sigma_{ff}$ and begins at a distance of $5\sigma_{ff}$ when approaching the boundary between A- and B-plates from the A-plate side. Just at the latter distance begins the change of the calculated contact angle (see Tables 9 and 10).

The above results for Cases 1 and 2 support the suggestion that the contact angle is determined by the conditions near the leading edges of the drop and that the remaining area of the solid-liquid interface is irrelevant.

3.5. DFT CALCULATIONS OF THE STICKING FORCE FOR NANODROPS ON AN INCLINED SURFACE

On nonideal inclined surfaces and in the presence of gravity, two contact angles, the receding (θ_r) and the advancing (θ_a), occur for a macroscopic drop at its stability limit, which are assumed to be independent of the size of the drop. The difference between the advancing and receding contact angles (contact angle hysteresis) is caused by the interplay of the gravity and the sticking force that ensures the mechanical equilibrium of the drop and determines the largest possible inclination angle α_c of the surface. The latter angle is provided by the expression [87,88]

$$Ma \sin \alpha_c = C\gamma_{lv}\left(\cos\theta_r - \cos\theta_a\right) \tag{104}$$

where M is the mass of the drop, a is the gravitational acceleration and the constant C incorporates the geometrical characteristics of the drop. If the inclination angle α of the solid surface is smaller than α_c, the drop is in mechanical equilibrium and the tangential component, f_τ, of the force of gravity is balanced by the sticking force, f_{st}, caused by the interactions in the system and by roughness. In this case, f_τ can be calculated using the following expression similar to Eq. (104)

$$f_\tau = Ma \sin \alpha = C\gamma_{lv}\left(\cos\theta_2 - \cos\theta_1\right),\ \left(f_{st} = f_\tau\right) \tag{105}$$

where α is the inclination angle of the surface, $\theta_1 \le \theta_a$ is the contact angle at the lower edge of the drop, and $\theta_2 \le \theta_r$ is the contact angle at the upper edge (Fig. 52).

The force of gravity that acts on a molecule is extremely small. For this reason, the number of molecules in a drop which can roll down on a surface under the action of gravity should be very large. To estimate this number let us consider, for simplicity, a two-dimensional drop on a vertical surface ($\alpha = 90°$), and evaluate with Eq. (105) the order of magnitude of the sticking force. The constant C in Eq. (105) is considered equal to the unit of length, and the cosine difference is also considered unity. Using for the surface tension the value $\gamma_{lv} \simeq 1.4\times10^{-2}$ N/m for argon at $T = 85$ K [89,90], one finds that the sticking force (per unit length) is 1.4×10^{-2} N/m. Dividing this force by mg, where $m = 6.6 \times 10^{-26}$ kg is the mass of a molecule of argon, one finds that the number of molecules in

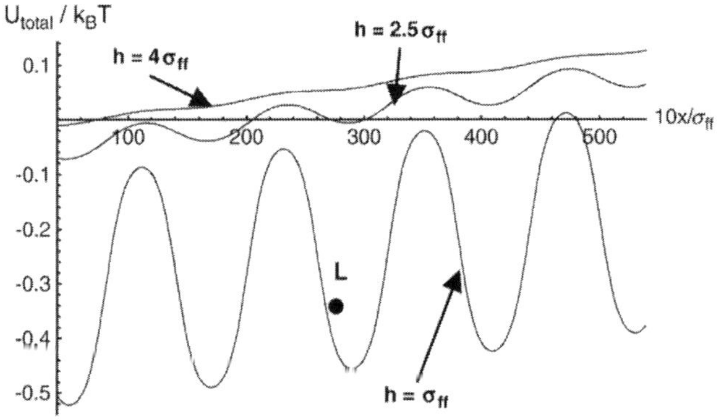

FIG. 53. (Reprinted from Ref. [18] with permission.) Dependence of the total external potential, to which a molecule of fluid is subjected, on x for $\alpha = 90°$, $a = 50a_0$ and several values of the distance h from a surface $S1$.

a drop should be greater than 2×10^{23} for the drop to be set in motion on the surface. Because the number of molecules in the nanodrops considered in the DFT calculations is usually of the order of 10^{11}–10^{12} (see Ref. [17]) it is impossible to detect the effect of gravity on such a nanodrop for the real value of a (a = g) and calculate the contact angles θ_r and θ_a. For the above values of the parameters the nanodrops will be sticked to the solid surface even for small roughnesses and the very small difference between θ_1 and θ_2 could not be calculated because of the limited precision of the numerical procedure. For this reason, in Ref. [18] the DFT calculations were performed for other values of a which are of the order of $a_0 = 10^{10}$ m/s^2. Even though these values are unrealistically large and cannot be achieved in real experiments, their usage allows to detect changes in the shape of the nanodrop profile. This extremely large value compensates for the small number of molecules which we have to consider for computational restrictions. Still we will refer to this acceleration as gravity.

To calculate the sticking force, two kinds of chemically rough surfaces consisting of plates A and B (Fig. 28) were considered in Ref. [18]. In the first (S1), the solid carbon dioxide $\left(\epsilon_{fs,A} = \epsilon_{fs}^0 = 153.0\,\mathrm{K}\right)$ was selected for the A-plates and the B-plates were made up of another (fictitious) material which differs from A only in the value of the energy parameter $\epsilon_{fs,B}$ $\left(\epsilon_{fs,B} = 0.1\epsilon_{fs}^0\right)$. The thickness of the two kinds of plates is $d = 6\sigma_{ff}$. For the second surface ($S2$), the energy parameter $\epsilon_{fs,A}$ is selected equal to $\epsilon_{fs,A} = 0.6\epsilon_{fs}^0$, all other parameters remaining those of surface $S1$.

In Fig. 53, the total potential to which a molecule of fluid is subjected is presented schematically for the surface $S1$ as a function of x for $\alpha = 90°$ (vertical surface) and several values of the distance h from the surface. The "gravitational" acceleration a was taken equal to $14.7a_0$. For $h<4\sigma_{ff}$, this potential possesses multiple minima separated by potential barriers, the height of which decreases as h increases. As a consequence of gravity, the values of the potential at points of minima gradually increase with increasing x. For $h \geq 4\sigma_{ff}$, the oscillations due to the surface inhomogeneity become smaller, the total potential increases monotonously as x increases, and no potential barriers are longer present. (The potentials for inclination angles smaller than 90° or for more realistic gravitational accelerations are expected to have similar qualitative features). One should note that, at a fixed h, the height of the potential barrier decreases with decreasing difference between the energy parameters $\epsilon_{fs,A}$ and $\epsilon_{fs,B}$; for $\epsilon_{fs,A} = \epsilon_{fs,B}$ (a smooth surface) there are no potential barriers.

The existence of potential barriers for $\epsilon_{fs,A} \neq \epsilon_{fs,B}$ (chemically rough surface) allows one to explain qualitatively the sticking force which arises due to the interaction of a fluid molecule with the rough solid. Indeed, when the leading edge of the drop on the inclined surface is located near one of the minima of U_{fs}, say at point L in Fig. 53, this potential generates a force in the positive direction of the x axis which opposes the motion of the drop (in the negative x-direction) along the inclined surface. Only a large enough gravity can overcome the potential barrier and set the drop in motion. On a smooth surface, there is no potential barrier for the drop motion and its mechanical equilibrium on an inclined surface becomes impossible

Because of the periodicity of the solid—liquid potential in the x direction, any drop in the absence of gravity has the same energy when it is displaced along the surface over an integer number of periods. However, the potential energy due to gravity decreases when the drop moves in the negative x-direction. This means that on an inclined surface any sticked drop is in a metastable state.

The above qualitative considerations remain valid for physically rough surfaces. The contribution $U_{fp,\,total}(x,h)$ of the pillars to the external potential generated by the solid is also a periodic function of x (see Section 3.1.3).

3.5.1. Calculation of the sticking force

If a drop is in mechanical equilibrium on an inclined surface with an inclination angle α (see Fig. 52), the sticking force f_{st} which prevents the drop to slide down is equal to the tangential component f_τ of the force of gravity

$$f_{st} = f_\tau = Ma\sin\alpha \qquad (106)$$

where M is the mass of the drop. By changing a, one can find the critical value α_c such that for $\alpha>\alpha_c$ the drop slides down along the surface. The corresponding critical value $f_{st}^c = Ma\sin\alpha_c$ of f_{st} depends, generally, on the natures of the liquid and solid. The critical sticking force can be also determined by changing the mass of the drop at fixed a and α, say at $\alpha = 90°$ (vertical surface), or changing the acceleration a at fixed M and α. In the first case

$$f_{st}^c = M_c a \sin\alpha \tag{107}$$

and in the second

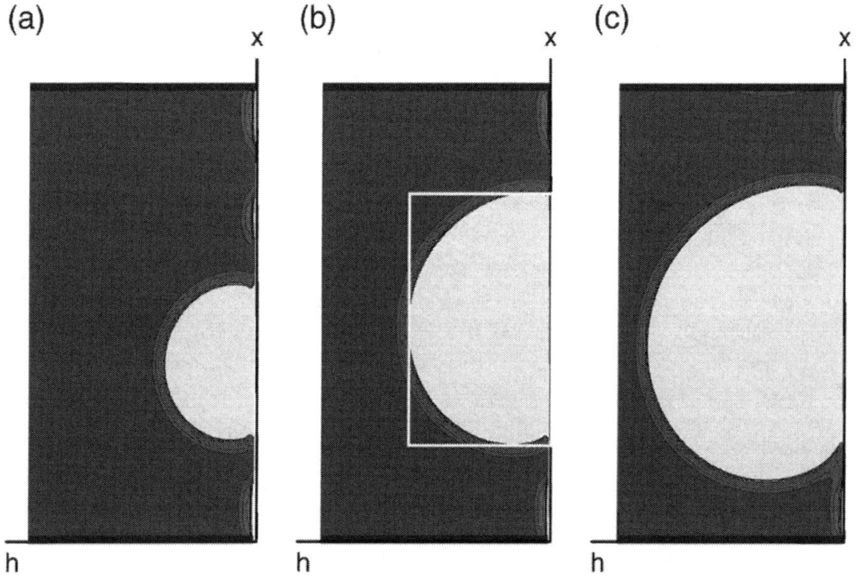

FIG. 54. (Reprinted from Ref. [18] with permission.) Drop $D2$ on a vertical surface $S1$ for $a = 3a_0$ and various average densities ρ_{av}^*. (a) $\rho_{av}^* = 0.1$, (b) $\rho_{av}^* = 0.16$, (c) $\rho_{av}^* = 0.21$. The lighter areas correspond to higher fluid densities. The white rectangle in panel (b) presents the box which is used for evaluation of number of molecules in the drop.

TABLE 11
(Reprinted from Ref. [18] with permission.) Drop D2 on a vertical chemically rough surface S1. Dependence of the contact angles at the lower (θ_1) and upper (θ_2) contact lines, sticking force f_{st}, and surface tension γ_{lv} on the average density ρ_{av}^*. The contact angles were determined by approximating the upper part of the profile by a circle.

ρ_{av}^*	θ_1 (°)	θ_2 (°)	f_{st} ×10⁴ N/m	γ_{lv} ×10³ N/m
0.10	108.3	102.6	6.7	7.0
0.13	119.4	111.7	9.5	7.8
0.16	128.0	116.8	12.2	7.4
0.18	131.5	118.9	14.0	7.8
0.19	133.5	119.4	14.9	7.6
0.20	135.0	119.7	15.9	7.5
0.21	136.2	119.8	16.7	7.3

$$f_{st}^{c} = Ma_{c}\sin\alpha \quad (108)$$

where M_c and α_c are the mass of the drop and the value of α, respectively, at the stability limit. For $M>M_c$ or for $\alpha>\alpha_c$ the drop loses its stability on an inclined surface.

It follows from the above considerations that the mass M of a drop calculated from FDD is necessary to determine the sticking force. For macroscopic drops it can be obtained by multiplying the (constant) liquid density with the drop volume. The error due to the existence of density nonhomogeneities at the vapor—liquid and liquid—solid interfaces is in this case negligible. However for nanodrops, this error can be considerable because the sizes of the interfaces and of the drop are comparable. In this case, it is convenient to characterize the size of the drop by the total number N_{total} of molecules it contains. Then the mass of the drop is equal to $M = mN_{total}$. To calculate N_{total} for a given drop, the latter was enclosed in a rectangular box of the kind shown below in Fig. 54b and N_{total} was determined by the numerical integration of FDD $\rho(x, h)$ over this box. The presence in such a box of a tiny fraction of molecules which do not belong to the drop has a small effect on N_{total} because of the small density of the fluid outside the drop.

Evidently, N_{total} should depend on the average density ρ_{av} of the fluid in the system but the form of this dependence is not clear in advance. The analysis performed in Ref. [18] has shown that in the range of $\rho_{av}^{*} \geq 0.13$ $\left(\rho^{*} = \rho\sigma_{ff}^{3}\right)$, N_{total} is a linear function of ρ_{av}^{*}. As a consequence, the sticking force is also a linear function of ρ_{av}^{*}. The coefficients of this linear function depend on the case considered and have to be specified for each of the particular cases considered.

Finally, a procedure to calculate the critical sticking force f_{st}^{c} in the framework of the DFT, which was employed in Ref. [18] consists of three steps: (i) For a selected ρ_{av}^{*}, the largest possible value a_c of the acceleration a, which provides a solution of the Euler-Lagrange equation is first obtained; (ii) The profile of the drop, the contact angles θ_1 and θ_2, and the total number of molecules are calculated from the obtained FDD using the procedures described in this section and in Section 3.1.4; (iii) The critical sticking force is calculated using Eq. (108)

3.5.2. A drop on a vertical surface S1 $\left(\epsilon_{fs,A} = \epsilon_{fs}^{0}, \epsilon_{fs,B} = 0.1\epsilon_{fs}^{0}\right)$.

As explained in Section 3.4.1, two kinds of drops, $D1$ and $D2$, presented in Fig. 46 can exist on the considered surface. The metastable FDDs corresponding to a drop of type $D2$ on a vertical surface $S1$ (inclination angle equal to 90°) are presented in Fig. 54 for various average densities and $\alpha = 3\alpha_0$. The contact angles θ_1 and θ_2 were calculated using circular approximation for the upper part of the profiles.

For $\rho_{av}^{*} = 0.1$, the drop is almost symmetrical (Fig. 54a) and the difference between the contact angles θ_1 and θ_2 is small. As the volume increases (Fig. 54b, c) this difference also increases, as one can see from Table 11.

For average densities $\rho_{av}^{*} > 0.21$ no solutions of the equation for the FDD (Eq. (81)) could be found which correspond to a drop $D2$. Hence, $\rho_{av}^{*} = 0.21$ provides a critical value for the sticking force which in this case is equal (per unit length) to 1.67×10^{-3} N/m. For $0.1\rho_{av}^{*} < 0.21$ the sticking force in the considered case can be represented by the linear expression $f_{st} = \left(91.1\rho_{av}^{*} - 2.38\right) \times 10^{-4}\ \mathrm{N/m}$.

After the determination of the sticking force and of the contact angles θ_1 and θ_2 from the calculated FDD $\rho(x, h)$, one can compare the microscopic results with the predictions of Eq. (105) by calculating the surface tension γ_{lv} from the latter equation. If Eq. (105) provides approximately the same value of γ_{lv} for all values of ρ_{av}^{*} one can consider that Eq. (105) remains valid even for nanodrops. Indeed, the analysis performed in Ref. [17] has shown that the values of the surface tension obtained using the drop profiles for various drop sizes do not differ by more than 6% from their mean value $\gamma_{lv} = 7.45 \times 10^{-3}\ \mathrm{N/m}$ and are surprisingly close to $\gamma_{lv} = 1.4 \times 10^{-2}\ \mathrm{N/m}$, value determined from experiments involving macroscopic interfaces.

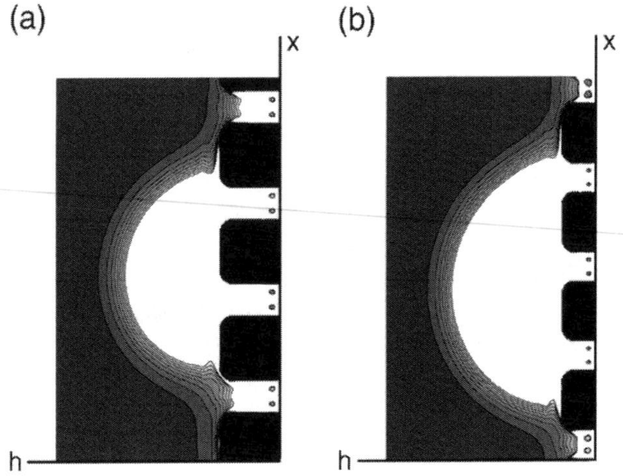

FIG. 55. (Reprinted from Ref. [18] with permission.) Drops on a physically rough vertical surfaces for $\rho^*_{av} = 0.1$. (a) $h_p = 4\sigma_{ff}$, (b) $h_p = 2.5\sigma_{ff}$. The lighter areas correspond to higher fluid densities.

A similar treatment of a $D1$ drop on a vertical surface $S1$ has shown that in the considered range of ρ^*_{av} $(0.13 \le \rho^*_{av} \le 0.34)$, the sticking force can be represented by the linear function $f_{st} = (88.7\rho^*_{av} - 1.797) \times 10^{-4}$ N/m which is slightly different from the similar expression obtained for drop $D2$.

The critical sticking force was calculated using Eq. (108) for $\alpha = 90^{\circ}$ for $\rho^*_{av} = 0.13$, $\alpha_c = 7.0a_0$ and $f_{st} = 21.3 \times 10^{-4}$ N/m. For $\rho^*_{av} = 0.24$, $a_c = 8.5a_0$ and $f_{st} = 48.2 \times 10^{-4}$ N/m. The values of the surface tension are equal to $\gamma_{lv} = 7.4 \times 10^{-3}$ N/m and $\gamma_{lv} = 6.9 \times 10^{-3}$ N/m, respectively. These values are close to those obtained for drop $D2$.

3.5.3. Nanodrops on a vertical surface S2 $\left(\epsilon_{fs,A} = 0.6\epsilon^0_{fs}, \epsilon_{fs,B} = 0.1\epsilon^0_{fs}\right)$

In the considered range of average densities $(0.1 \le \rho^*_{av} \le 0.3)$ and for $a = 3a_0$, a drop of type $D1$ could be found only for densities close to $\rho^*_{av} = 0.1$, while a drop of type $D2$ could be found in the range $0.1 \le \rho^*_{av} \le 0.22$. Due to weaker fluid—solid interactions, surface $S2$ is more hydrophobic than $S1$, hence, the contact angles θ_1 and θ_2 for $S2$ are larger than those for $S1$ (for the same average density). For example, for $\rho^*_{av} = 0.16, \theta_1 = 128.0°, \theta_2 = 116.8°$ for $S1$ and θ_1 = 140.1°, and θ_2 = 126.1° for $S2$.

For ρ^*_{av}, the critical sticking force for drop $D2$ on $S2$ has the value $f^c_{st} = 1.88 \times 10^{-3}$ N/m and is smaller than that on $S1$ $(f^c_{st} = 3.26 \times 10^{-3}$ N/m). The above results indicate that the sticking force is smaller on a hydrophobic than on a hydrophilic surface.

However, the above observations do not provide a general rule. Indeed, the sticking force provided by Eq. (105) is proportional to the difference between the cosines of the contact angles θ_1 and θ_2 at the upper and lower contact lines. For the more hydrophobic surface $S2$, both angles are larger than those on the less hydrophobic surface $S1$. Nevertheless, the difference of the cosines can be larger or smaller for a hydrophobic surface than for a hydrophilic one. For example, for ρ^*_{av}, the critical sticking force for drop $D1$ on $S1$ is equal to $f^c_{st} = 1.65 \times 10^{-3}$ but is smaller than that on S2 $(f^c_{st} = 2.03 \times 10^{-3}$ N/m) even though surface $S1$ is less hydrophobic than surface $S2$. Another example is presented in Ref. [91] on the basis of macroscopic considerations of a drop on an inclined surface. It was shown [91] that the critical value α_c of the inclination angle can be larger for a hydrophobic than for a hydrophilic surface (for the same volume of the drop). This means that in that particular case the critical sticking force on the hydrophilic was smaller than on the hydrophobic surface.

3.5.4. Sticking force on a vertical physically rough surface

The pillars used in Ref. [18] for calculating the sticking force (Fig. 29) have the width $d_p = 2\sigma_{ff}$ and the distance between them $\Delta_p = 4\sigma_{ff}$. The heights of the pillars h_p were selected $2.5\sigma_{ff}$ and $4\sigma_{ff}$. They were considered to contain the same material as the main solid (carbon dioxide). In the absence of gravity, the contact angle of a drop on a smooth (without pillars) surface was $\theta \simeq 50°$, hence such a surface is hydrophilic. The presence of pillars increases the hydrophobicity of the surface. For pillars of height $h_p = 2.5\sigma_{ff}$, $\theta = 89.3°$ and for height $h_p = 4\sigma_{ff}$, $\theta = 94.4°$. In the latter case, the rough surface becomes hydrophobic (θ>90°).

The potential of the fluid—solid interactions is a periodic function of x with a wavelength of $d_p + \Delta_p$ which resembles that presented in Fig. 53 for a chemically rough surface. For this reason, the main characteristics of a nanodrop on physically and chemically rough surfaces are similar.

Typical examples of FDDs corresponding to drops on a vertical physically rough surface are presented in Fig. 55a for $h_p = 4\sigma_{ff}$ and in Fig. 55b for $h_p = 2.5\sigma_{ff}$. In both cases $\rho^*_{av} = 0.1$ and the space between pillars is filled with a fluid of liquid like density (Wenzel regime). The critical acceleration for the first case is $a_c = 21.2a_0$ and the corresponding critical sticking force is $f^c_{st} = 1.42\times10^{-3}$ N/m. The surface tension γ_{lv} calculated with Eq. (105) is in agreement with those calculated for chemically rough surfaces.

4. CONCLUSION

Two approaches for the description of macroscopic and nanodrops on a solid surface have been presented which have obvious advantages over the phenomenological theories based on the classical thermodynamic concepts of surface tensions. (i) The observable quantities, such as the drop profile, the contact angle, the sticking force, etc. are directly connected to the microscopic parameters of the fluid-fluid and fluid-solid interaction potentials. Although only Lennard-Jones and London-van der Waals potentials were employed so far, the theory can be easily extended to other possible interaction potentials. (ii) The parameters of the macroscopic equations (for example, of the augmented Young-Laplace equation for the drop profile, the equation for the sticking force for a drop on an inclined surface) can be calculated in terms of the microscopic parameters of the interaction potentials. (iii) The microscopic theory can be applied to cases in which the traditional macroscopic concepts (such as the surface tensions) cannot be clearly defined, for example the rough surfaces.

Among the numerous results obtained on the basis of the microscopic approaches one can mention:

(i) The existence of a microscopic and a macroscopic contact angles for macrodrops.

(ii) A relationship between the microscopic and the macroscopic contact angles for a drop on a planar surface and on a fiber.

(iii) For a given liquid—solid pair, the prediction of the ranges of the interaction parameters of the interaction potentials, for which a macroscopic drop on a planar surface can have: any volume and height; any volume but a height which is smaller than some limiting value; or a drop cannot be formed on the surface (it spreads upon the surface).

(iv) Prediction of the existence of an almost universal value of a contact angle ($\theta_0 \simeq 117.4°$, for nanodrops and $\theta_0 \simeq 122.5°$ for macroscopic drops). If for a given liquid-solid pair $\theta > \theta_0$ ($\theta < \theta_0$), the contact angle increases (decreases) with increasing temperature. If confirmed experimentally, the existence of such a fixed contact angle can lead to a less arbitrary definition of hydrophobic and hydrophilic surfaces. In the conventional definition a surface is hydrophobic if $\theta > 90°$ and hydrophilic if $\theta < 90°$. The new definition should be $\theta > \theta_0$ for hydrophobic and $\theta < \theta_0$ for hydrophilic surfaces.

REFERENCES

[1] Young T. Philos Trans R Soc Lond 1805; 95:65.
[2] Everett DH, Haynes JM. J Colloid Interface Sci 1972; 38:125.
[3] Yeh EK, Newman J, Radke CJ. Colloids Surf A 1999; 156:525.
[4] Zhang XZ, Neogi P, Ybarra RM. J Colloid Interface Sci 2002; 249:134.
[5] Derjaguin BV. Acta Phys Chim USSR 1940; 12:181.
[6] Kornev KG, Shingareva IK, Neimark AV. Adv Colloid Interface Sci 2002; 96:143.
[7] Ruckenstein E, Lee PS. Surf Sci 1975; 52:298 (Section 3.1 of this volume).
[8] Berim GO, Ruckenstein E. J Phys Chem B 2004; 108:19330 (Section 3.2 of this volume).
[9] Cassie ABD, Baxter S. Trans Faraday Soc 1944; 40:546.
[10] Wenzel RN. Ind Eng Chem 1936; 28:988.
[11] IsraelachviliJ. Acc Chem Res 1987; 20:415.
[12] Tarazona P. Phys Rev A 1985;]2:]148Phys Rev A 1985; 31:2672.
[13] Berim GO, Ruckenstein E. J Phys Chem B 2004; 108:19339 (Section 3.3 of this volume).
[14] Berim GO, Ruckenstein E. J Colloid Interface Sci 2005; 286:681 ibid 2006; 295:593 (Section 3.4 of this volume).
[15] Berim GO, Ruckenstein E. J Phys Chem B 2005; 109:12515 ibid 2005; 109:22084 (Section 3.5 of this volume).
[16] Berim GO, Ruckenstein E. Langmuir 2005; 21:7743 ibid 2005; 21:12053 (Section 3.6 of this volume).
[17] Berim G, Ruckenstein E. J Chem Phys 2008; 129:014708 (Section 4.4 of this volume).
[18] Berim GO, Ruckenstein E. J Chem Phys 2008; 129:114709 (Section 4.11 of this volume).
[19] Berim GO, Ruckenstein E. J Chem Phys 2009; 130:044709 (Section 4.1 of this volume).
[20] Berim GO, Ruckenstein E. Langmuir 2009; 25:9285 (Section 4.7 of this volume).
[21] Berim GO, Ruckenstein E. J Chem Phys 2009; 130:184712 (Section 3.8 of this volume).
[22] Ebner C, Saam WF. Phys Rev Lett 1977; 38:1486.
[23] Evans R, Tarazona P. Phys Rev A 1983; 28:1864.
[24] Tarazona P, Evans R. Mol Phys 1984; 52:847.
[25] Talanquer V, Oxtoby DW. J Chem Phys 1996; 104:1483.
[26] Talanquer V, Oxtoby DW. J Chem Phys 2001; 114:2793.
[27] Bieker T, Dietrich S. Physica A 1998; 252:85.
[28] Frenkel Daan. Understanding molecular simulations. In: Smith Berend, editor. From algorithms to applications. Computational Science Series, Second editionA-cademic Press; 2001. p. 664.
[29] Lee SH, Rossky PJ. J Chem Phys 1974; 60:1545.
[30] Abramovitz M, Stegun I. Handbook of mathematic functions with formulas, graphs, mathematical tables. New York: Dover; 1972.
[31] Neimark AV. J Adhes Sci Technol 1999; 13:1137.
[32] Sharma A. Langmuir 1993; 9:3580.
[33] Elsgolc LE. Calculus of variations. New York: Pergamon Press, Addison-Wesley; 1962.
[34] Brochard F. J Chem Phys 1986; 84:4664.
[35] Carroll BJ. J Colloid Interface Sci 1976; 57:488.
[36] Adam NK. J Soc Dyers Colour 1937; 53:122.
[37] Carroll BJ. J Colloid Interface Sci 1984; 97:195.
[38] McHale G, Newton MI. Colloids Surf A Physicochem Eng Aspects 2002; 206:79.
[39] Krupenkin TN, Taylor JA, Schneider TM, Yang S. Langmuir 2004; 20:3824.
[40] Drelich J. Colloids Surf A 1996; 116:43.
[41] Swain PS, Lipowsky R. Langmuir 1998; 14:6772.
[42] vanGiessen AE, Bukman DJ, Widom B. J Colloid Interface Sci 1997; 192:257.
[43] Nilson RH, Griffiths SK. J Chem Phys 1999; 111:4281.
[44] Ustinov EA, Do DD. J Phys Chem B 2005; 109:11653.
[45] Dalfovo F, Lastri A, Pricaupenko L, Stringari S, Treiner J. Phys Rev B 1995; 52:1193.
[46] Sartarelli SA, Szybisz L. Phys Rev E 2009; 80:052602.
[47] Ancilotto F, Faccin F, Toigo F. Phys Rev B 2000; 62:17035.
[48] Berim G, Ruckenstein E. J Chem Phys 2008; 128:024704 (Section 2.4 of this volume).
[49] Berim G, Ruckenstein E. J Chem Phys 2008; 128:134713 (Section 2.5 of this volume).
[50] Porcheron F, Monson PA. Langmuir 2006; 22:1595.
[51] Porcheron F, Monson PA, Schoen M. Phys Rev E 2006; 73:041603.
[52] Ancilotto F, Barranco M, Hernandez ES, Hernando A, Pi M. Phys Rev B 2009; 79:11.

[53] Neimark AV, Ravikovitch PI, Vishnyakov A. J Phys Condens Matter 2003; 15:347.
[54] Gonzalez A, White JA, Roman FL, Velasco S, Evans R. Phys Rev Lett 1997; 79:2466.
[55] Gonzalez A, White JA, Roman FL, Evans R. J Chem Phys 1998; 109:3637.
[56] White JA, Gonzalez A, Roman FL, Velasco S. Phys Rev Lett 2000; 84:1220.
[57] White JA, Velasco S. Phys Rev E 2000; 62:4427.
[58] Maruyama S, Kurashige T, Matsumoto S, Yamaguchi Y, Kimura T. Microscale Thermophys Eng 1998; 2:49.
[59] de Ruijter MJ, Blake TD, De ConinckJ. Langmuir 1999; 15:7836.
[60] Vishnyakov A, Neimark AV. J Chem Phys 2003; 119:9755.
[61] Neimark AV, Vishnyakov A. J Chem Phys 2005; 122:234108.
[62] Giovambattista N, Debenedetti PG, Rossky PJ. J Phys Chem B 2007; 111:9581.
[63] Giovambattista N, Debenedetti PG, Rossky PJ. Proc Natl Acad Sci U S A 2009; 106:15181.
[64] Neimark AV, Ravikovitch PI, Vishnyakov A. Phys Rev E 2002; 65:031505.
[65] Mittal J, Errington JR, Truskett TM. J Chem Phys 2007; 126:244708.
[66] Lundgren M, Allan NL, Cosgrove T, George N. Langmuir 2003; 19:7127.
[67] Carnahan NF, Starling KE. J Chem Phys 1969; 51:635.
[68] Matusewicz M, Patrykiejew A, Sokolowski S, Pizio O. J Chem Phys 2007:127.
[69] Sokolowska Z, Sokolowski S. J Colloid Interface Sci 2007; 316:652.
[70] Kannan R, Sivakumar D. Colloids Surf A 2008; 317(1-3):694.
[71] Kannan R, Sivakumar D. Exp Fluids 2008; 44:927.
[72] Henderson D, Sokolowski S, Wasan D. Phys Rev E 1998; 57:5539.
[73] Tasinkevych M, Dietrich S. Eur Phys J E 2007; 23:117.
[74] Johnson Jr RE, Dettre RH. Surf Colloid Sci 1962; 2:85.
[75] Monson PA. Langmuir 2008; 24:12295.
[76] Rolley E, Guthmann C. J Low Temp Phys 1997; 108:1.
[77] Budziak CJ, Varghabutler EI, Neumann AW. J Appl Polym Sci 1991; 42:1959.
[78] Suzuki A, Kobiki Y. Jpn J Appl Phys Part 1 1999; 38:2910.
[79] She HY, Sleep BE. Ground Water Monit Rem 1999; 19:70.
[80] Bernardin JD, Mudawar I, Walsh CB, Franses EI. Int J Heat Mass Transfer 1997; 40:1017.
[81] Gao LC, McCarthy TJ. Langmuir 2007; 23:3762.
[82] McHale G. Langmuir 2007; 23:8200.
[83] Nosonovsky M. Langmuir 2007; 23:9919.
[84] Panchagnula MV, Vedantam S. Langmuir 2007; 23:13242.
[85] Whyman G, Bormashenko E, Stein T. Chem Phys Lett 2008; 450:355.
[86] Marmur A, Bittoun E. Langmuir 2009; 25:1277.
[87] Furmidge CGL.J Colloid Sci 1962; 17:309.
[88] Callies M, Quere D. Soft Matter 2005; 1:55.
[89] Upstill CE, Evans R. J Phys C Solid State Phys 1977; 10:2791.
[90] Buff FP, Lovett RA. Simple Dense Fluids. In: Frish HL, Salsburg ZW, editors. Simple dense fluids. New York: Academic; 1968. pp. 17.
[91] Krasovitski B, Marmur A. Langmuir 2005; 9:3881.

5 Theory of the Liquid and Solid Films Rupture

Eli Ruckenstein and Gersh Berim

INTRODUCTION TO CHAPTER 5

In this chapter, the rupture of liquid and solid films on a solid surface and free films are examined theoretically.

In the first part of the chapter (Secs 1–4), the focus is concentrated to thin film rupture due to small perturbations when the linear approximation can be used. In the framework of this approximation, the rupture of a thin liquid film on a solid surface and of a free liquid film have been studied using the Navier-Stokes equations (Sec. 5.1). A small amplitude perturbation analysis is used in Sec. 5.2 to determine a critical internal stress necessary for rupture and estimate the time of rupture. In Sec. 5.3, a unified theoretical framework is developed to study the stability of thin as well as thick liquid films on a solid surface immersed in a viscous fluid. On the basis of this theory, the influence of surface active agents, of surface viscosity, and of the viscosity of the semi-infinite fluid, on the growth of the perturbation and on the time of rupture, is examined. As one of the possible applications, the stability of thin (<100 nm) symmetrical and unsymmetrical membranes to short and long wavelength perturbations is investigated in Sec. 5.3. For the cells membranes, the results provide an understanding of in what manner differences in interfacial tension convey amplified or damped messages across the membrane.

In the second part of the chapter, (Secs. 5.4–5.7), the theory is extended to the case of perturbations of finite amplitude. For such a case, the nonlinear character of hydrodynamic and of the interaction force is taken into account. The essential idea of the approach is to determine the stability of a spatially nonhomogeneous stationary solution of the governing equations, thus to account for a part of the finite disturbance of the interface in the base state itself. The theory provides, in particular, conditions for the instability, the dominant wavelength of the disturbances, and the time of rupture of the thin films, all as functions of thin-film parameters (Secs. 5.4 and 5.5). The critical thickness of rupture and the lifetimes of the tangentially immobile foam and emulsion films are also calculated (Sec. 5.6). The nonlinear long-wave stability and lifetimes of thin free films subjected to the excess Lifshitz-van der Waals forces are studied in Sec. 5.7 based on numerical solutions, and a weakly nonlinear theory, which neglects mode interactions.

In the third part of the chapter, in Secs. 5.8 and 5.9, the rupture of liquid films supported on a solid surface is examined on the basis of a thermodynamic approach considering the change of the free energy of the film after formation of a hole in it. The model predictions for the film thickness at the instant of rupture are in qualitative and quantitative agreement with the available data on several solid-liquid systems. In Secs. 5.10 and 5.11, the theory is applied to practically important problem of tear film stability and rupture. The results have implications e.g to understanding the cause of adhesion of the contact lens to cornea, to estimation of the breakup times of the tear-film rupture, etc. The various possibilities for diagnosis and treatment of pathological conditions of a dry eye are also discussed.

5.1 Spontaneous Rupture of Thin Liquid Films*

Eli Ruckenstein[a]* *and Rakesh K. Jain*[b]

[a] State University of New York at Buffalo, Faculty of Engineering and Applied Sciences, Buffalo, New York 14214

[b] Department of Chemical Engineering, University of Delaware, Newark, Delaware 19711

Received 25th June, 1973

The mechanism of rupture of thin liquid films is of importance for the understanding of flotation, of foams and emulsions, of coalescence of bubbles and droplets, of vapour condensation on a solid surface, and so on. In flotation, for instance, the thinning and the rupture of the liquid film between particles and bubbles might be the rate determining step of the process.[1] Two bubbles may coalesce if the contact time between them is longer than the time needed for the thinning and rupture of the liquid film between then. In water vapour condensation on a shock-tube wall, experiment shows that on a hydrophobic surface a thin film is formed, which breaks up into droplets upon reaching a critical thickness of about 100 Å.[2]

Scheludko[3, 4] was the first to relate via a thermodynamic treatment the rupture of thin liquid films to their instability to small surface deformations. Although the surface free energy increases with the increasing surface area associated with these deformations, the total free energy of the film can decrease because of the London-van der Waals forces between molecules. A critical wavelength λ_c is predicted for which the total free energy is not changed by the corresponding perturbation. For a free film this critical wavelength is given by†

$$\lambda_c = \left(\frac{128\pi\sigma}{A_{11}}\right)^{\frac{1}{2}} h_0^2. \tag{1}$$

The film is stable for all wavelengths less than λ_c, and it is unstable for greater wavelengths.

This thermodynamic treatment of the problem gives no information about the time needed for the occurrence of rupture. Assuming laminar liquid flow parallel to the surfaces of the free film of uniform thickness, h_0, and no slip at the interfaces, Vrij[5] has established a kind of diffusion equation for the thickness of the film, which allows one to calculate the growth rate of the perturbation. His results for thin free liquid films are

$$\lambda_c = \left(\frac{4\pi^3\sigma}{A_{11}}\right)^{\frac{1}{2}} h_0^2. \tag{2}$$

and

$$\tau_m \approx 96\pi^2\sigma\mu h_0^5 A_{11}^{-2} \tag{3}$$

* *Journal of the Chemical Society. 70,* 132–147 (1974). Republished with permission.

† the nomenclature is given in Appendix 1

where τ_m is a time constant for rupture corresponding to the wavelength for which the rate of growth is maximum.

One can obtain information about the film rupture, including the time of rupture, from a more rigorous procedure than that used in the cited papers, namely hydrodynamic linear stability theory.[6] Whereas the previously cited authors have solved parts of the problem (using a specific model for each), the stability analysis permits a unified approach leading to the prediction of both the critical wavelength and time of rupture. Felderhof[7] has applied the hydrodynamic stability analysis to a thin free film accounting for the van der Waals dispersion forces and the double layer forces. His treatment is, however, restricted to the unrealistic situation of inviscid flow. In recent papers, brought to our notice by one of the referees, Lucassen *et al.*[8] and Vrij *et al.*[9] have extended the treatment to a free film of a viscous liquid.

In the present paper two situations are treated: (i) stability of a thin layer of liquid on a solid surface and (ii) stability of a thin free film. The emphasis is on the first situation because it was not examined previously. Compared to the previous ones, the present treatment has the advantage of simplicity. Because the thickness of the layer is very small, it is natural to use from the beginning the lubrication approximation of the hydrodynamic equations of motion.[10] This approximation, applied here to pure liquids and to liquids containing surface active impurities, allows one to obtain, in a simple way, information about the critical wavelength λ_c and time of rupture τ_m.

THIN PURE LIQUID LAYER ON A SOLID SURFACE

Consider a thin layer of liquid having a thickness h_0, not larger than several hundred Ångström. Since the distance over which the London dispersion forces are effective is of the same order of magnitude, the behaviour of the film is strongly influenced by such forces. Small perturbations are applied to the liquid-gas interface (fig. 1). The film is unstable and will rupture if the perturbation grows in time; the film is stable in the opposite case. If the forces of interaction between the molecules of the solid and liquid are stronger than those between the molecules of the liquid, the film will always be stable. In the opposite case it may be unstable. The perturbation generates motion in the film and the assumption will be made that the Navier-Stokes equations can be used to describe the motion. The forces which act upon an element of liquid in a thin layer differ from those in a bulk fluid because the range of intermolecular forces is larger than the thickness of the film. Compared to a bulk liquid, some liquid molecules are replaced by the atoms of the solid or of the gas within the range of intermolecular forces. The difference in forces between the thin layer and the bulk liquid is accounted for in the equations of motion by a body force.

Since the motion is very slow, the inertial terms can be neglected compared to the viscous terms. Consequently, assuming a two-dimensional motion,

$$\mu\left(\frac{\partial^2 u}{\partial x^2}+\frac{\partial^2 u}{\partial y^2}\right)=\frac{\partial p}{\partial x}+\frac{\partial \phi}{\partial x} \tag{4}$$

$$\mu\left(\frac{\partial^2 \upsilon}{\partial x^2}+\frac{\partial^2 \upsilon}{\partial y^2}\right)=\frac{\partial p}{\partial y}+\frac{\partial \phi}{\partial y} \tag{5}$$

$$\frac{\partial u}{\partial x}+\frac{\partial \upsilon}{\partial y}=0. \tag{6}$$

In eqn (4) and (5), ϕ is the potential energy function per unit volume in the liquid accounting for the difference in behaviour between a thin film and a bulk liquid. It is caused by the London-van der Waals interaction with the surrounding molecules of the liquid and with the solid and by the double layer forces. The function ϕ depends on the thickness h of the liquid film and on y

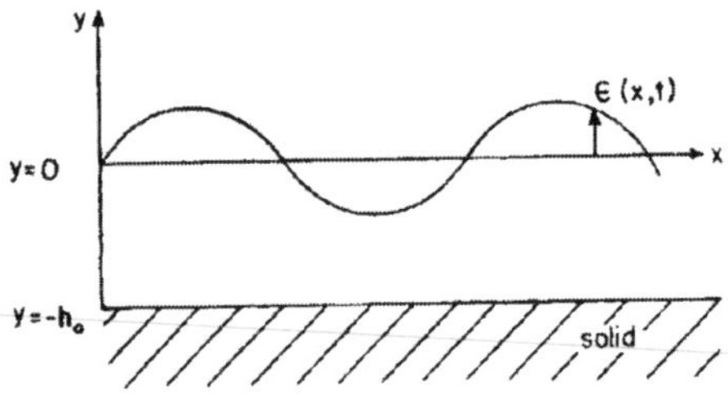

FIG. 1. Perturbation applied at the liquid-gas interface.

The boundary conditions at the solid-liquid interface are

$$u = \upsilon = 0 \text{ at } y = -h_0. \tag{7}$$

At the free surface, for deformations of small amplitude, the equality of the normal stresses leads to:

$$-p + 2\mu\frac{\partial \upsilon}{\partial y} = -p_0 + \sigma\frac{\partial^2 \varepsilon}{\partial x^2} \text{ at } y \approx 0, \tag{8}$$

and the equality of tangential stresses leads to

$$\mu\left(\frac{\partial u}{\partial y} + \frac{\partial \upsilon}{\partial x}\right) = 0 \text{ at } y \approx 0. \tag{9}$$

For ultra thin films the surface tension may depend on the thickness of the film and consequently the derivative $d\sigma/dx$ has to be introduced in the right-hand-side of the boundary condition (9). Such an effect is, however, ignored here.

The kinematic condition at the interface gives:

$$\frac{\partial \varepsilon}{\partial t} \approx \upsilon_{y=0}. \tag{10}$$

For wavelengths of the perturbation which are large compared to the thickness of the film, the lubrication approximation of the Navier-Stokes equations can be used.

Consequently, eqn (4) and (5) become

$$\mu\left(\frac{\partial^2 u}{\partial y^2} = \frac{\partial}{\partial x}(p + \phi)\right) = \frac{\partial P}{\partial x} \tag{11}$$

$$0 = \frac{\partial}{\partial y}(p + \phi) = \frac{\partial P}{\partial y}. \tag{12}$$

Using the boundary condition (8), one can write*:

$$P = p_0 + 2\mu\frac{\partial \upsilon}{\partial y} + \phi_0 - \sigma\frac{\partial^2 \varepsilon}{\partial x^2} \text{ at } y \approx 0 \tag{13}$$

and since eqn (12) shows that P is independent of y, eqn (11) and (13) lead to:

* The lubrication approximation is applied here only to the Navier-Stokes equations, but not to the boundary conditions. This leads to some smaller order terms in the final results.

$$\mu \frac{\partial^2 u}{\partial y^2} = 2\mu \frac{\partial}{\partial x}\left(\frac{\partial \upsilon}{\partial y}\right)_{y=0} - \sigma \frac{\partial^3 \varepsilon}{\partial x^3} + \frac{\partial \phi_0}{\partial x}. \tag{14}$$

Because $\phi_0 = \phi_0(h) = \phi_0(h_0 + \varepsilon)$

$$\frac{\partial \phi_0}{\partial x} = \left(\frac{\partial \phi_0}{\partial h}\right)\frac{\partial \varepsilon}{\partial x} \approx \left(\frac{\partial \phi_0}{\partial h}\right)_{h=h_0}\frac{\partial \varepsilon}{\partial x}. \tag{15}$$

and eqn (14) becomes

$$\mu \frac{\partial^2 u}{\partial y^2} = 2\mu \frac{\partial}{\partial x}\left(\frac{\partial \upsilon}{\partial y}\right)_{y=0} - \sigma \frac{\partial^3 \varepsilon}{\partial x^3} + \left(\frac{\partial \phi_0}{\partial x}\right)_{h=h_0}\frac{\partial \varepsilon}{\partial x}. \tag{16}$$

The question of interest is whether the surface perturbation grows or decays in time. The stability will be examined with respect to a small periodic perturbation because the effect of any small perturbation can be obtained by superimposing the effects of its Fourier components. Consequently,

$$\begin{bmatrix} u \\ \upsilon \\ \varepsilon \end{bmatrix} = \begin{bmatrix} \hat{u}(y) \\ \hat{\upsilon}(y) \\ \hat{\varepsilon} \end{bmatrix} e^{ikx} e^{\beta t}. \tag{17}$$

Introducing expressions (17) in eqn (6), (16), (7) and (9), one obtains

$$D\,\hat{\upsilon} + ik\hat{u} = 0 \tag{6a}$$

$$\mu D^2 \hat{u} = ik^3 \sigma \hat{\varepsilon} + \left(\frac{\partial \phi_0}{\partial h}\right)_{h=h_0} ik\,\hat{\varepsilon} + 2\mu ik\left(D\,\hat{\upsilon}\right)_{y=0} \tag{16a}$$

$$\hat{u} = \hat{\upsilon} = 0 \text{ at } y = -h_0 \tag{7a}$$

$$D\hat{u} + ik\hat{\upsilon} = 0 \text{ at } y \approx 0. \tag{9a}$$

Eliminating *u* from eqn (16*a*), (7*a*), and (9*a*) by using eqn (6*a*), one obtains:

$$D^3\,\hat{\upsilon} = \frac{1}{\mu}\left[\sigma k^4 + \left(\frac{\partial \phi_0}{\partial h}\right)_{h=h_0} k^2\right]\hat{\varepsilon} + 2k^2\left(D\,\hat{\upsilon}\right)_{y=0} \tag{18}$$

$$\hat{\upsilon} = D\,\hat{\upsilon} = 0 \text{ at } y = -h_0 \tag{19}$$

$$\left(D^2 + k^2\right)\hat{\upsilon} = 0 \text{ at } y \approx 0. \tag{20}$$

The solution of eqn (18) has the form:

$$\hat{\upsilon} = c_1 + c_2\left(y + \left(k^2 y^3\right)/3\right) + c_3 y^2 + \frac{\hat{\varepsilon}}{6\mu}\left[\sigma k^4 + \left(\frac{\partial \phi_0}{\partial h}\right)_{h=h_0} k^2\right] y^3. \tag{21}$$

The boundary conditions (19) and (20) lead to

$$c_1 = -\frac{h_0^3}{3\mu}\frac{\left[\sigma k^4 + \left(\frac{\partial \phi_0}{\partial h}\right)_{h=h_0} k^2\right]\hat{\varepsilon}}{\left[1 + \frac{3}{2}(kh_0)^2 - \frac{1}{6}(kh_0)^4\right]} \tag{22a}$$

$$c_2 = \frac{c_1}{2h_0} \frac{\left[1 + \frac{5}{6}(kh_0)^2 - \frac{1}{6}(kh_0)^4\right]}{\left[1 + (kh_0)^2\right]} \tag{22b}$$

$$c_3 = -\frac{k^2}{2} c_1. \tag{22c}$$

The kinematic condition, eqn (10), leads to the result

$$\beta = \left(\frac{\hat{\upsilon}}{\hat{\varepsilon}}\right)_{y=0} = \frac{c_1}{\hat{\varepsilon}} = \frac{(kh_0)^2}{3\mu h_0} \left[\frac{-(\partial\phi_0/\partial h)_{h=h_0} h_0^2 - \sigma(kh_0)^2}{1 + \frac{3}{2}(kh_0)^2 - \frac{1}{6}(kh_0)^4}\right]. \tag{23}$$

Rupture of the film will occur for those wave numbers for which $\beta>0$.

The lubrication approximation used here is valid only if $h_0/\lambda \ll 1$. Consequently, the denominator in eqn (23) is positive in the range of values for which the above mentioned approximation can be made.

The condition $\beta = 0$ defines a critical wavenumber k_c

$$k_c = \left[-\frac{1}{\sigma}\left(\frac{\partial\phi_0}{\partial h}\right)_{h=h_0}\right]^{\frac{1}{2}}. \tag{24}$$

The growth coefficient β has a maximum for the dominant wavenumber k_d, which is given by

$$(k_d h_0)^2 = \frac{6}{9 - k_c^2 h_0^2}\left[\left(1 + \frac{1}{6}(k_c h_0)^2 (9 - k_c^2 h_0^2)\right)^{\frac{1}{2}} - 1\right]. \tag{25}$$

Since $k_c h_0 \ll 1$, eqn (25) can be approximated by

$$k_d \sqrt{2} \approx k_c. \tag{26}$$

The coefficient for maximum rate of growth, β_m, is therefore given by

$$\beta_m = \frac{\sigma}{3\mu h_0} \frac{(k_d h_0)^4\left[(k_c/k_d)^2 - 1\right]}{\left[1 + \frac{3}{2}(k_d h_0)^2 - \frac{1}{6}(k_d h_0)^4\right]} \approx \frac{h_0^3}{12\mu\sigma}\left[\left(\frac{\partial\phi_0}{\partial h}\right)_{h=h_0}\right]^2. \tag{27}$$

Because the growth of the perturbation is dominated by the fastest growing perturbation, one may expect that the time needed for the rupture of the film will be of the order of $\beta_m^{-1} \approx \tau_m$.

To determine the values of k_c and τ_m, an explicit expression for the potential energy, $\phi_0(h)$, is needed. This potential is due to London-van der Waals dispersion forces and to the double layer forces. Neglecting the double layer forces, one obtains (see Appendix 2)

$$\phi_0(h) = \phi_B + A/6\pi h^3 \tag{28}$$

where $A = A_{11} - A_{12}$ and A_{ij} is Hamaker's constant for the interactions between molecules of type i and j (1 refers to the liquid and 2 to the solid). Interactions with the molecules of the gas are neglected. Using eqn (28), eqn (23), (24) and (27) lead to

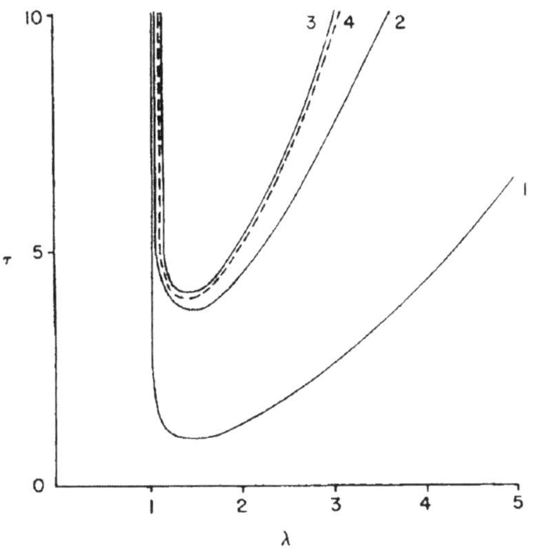

FIG. 2. Time constant ($\tau = 1/\beta$) as a function of wavelength λ for a thin film on a solid surface. The values of the parameters are given in table 1. The unit of τ is the value of τ_m calculated from eqn (31) and the unit of λ is the value of λ_c calculated from eqn (30). Curve 1, pure liquid film; curve 2, liquid film with gaseous monolayer of insoluble surfactant; curve 3, liquid film with a condensed monolayer of insoluble surfactant; curve 4, liquid film with a gaseous monolayer of soluble surfactant.

$$\beta = \frac{(kh_0)^2}{3\mu h_0} \frac{\left[\left(A/2\pi h_0^2\right) - \sigma (kh_0)^2\right]}{\left[1 + \frac{3}{2}(kh_0)^2 - \left((kh_0)^4/6\right)\right]} \quad (29)$$

$$\lambda_c = \frac{2\pi}{k_c} = \left(\frac{8\pi^3\sigma}{A}\right)^{\frac{1}{2}} h_0^2 \quad (30)$$

$$\tau_m, \approx 48\pi^2 \sigma\mu h_0^5 A^{-2}. \quad (31)$$

Eqn (29), (30) and (31) are plotted in fig. 2, 3 and 4 respectively.

EFFECT OF SURFACE ACTIVE AGENTS

Surface active agents generate surface forces which have a damping effect upon the wave motion. Levich[11] has developed a hydrodynamic theory of this wave damping for a thick film. Here the case of a thin liquid film on a solid surface is treated using the lubrication approximation. Compared to the case of a pure liquid, the boundary condition (9) at the free surface has to be replaced by

$$\mu\left(\frac{\partial u}{\partial y} + \frac{\partial \upsilon}{\partial x}\right) = \frac{\partial \sigma}{\partial x}, \quad y \approx 0. \quad (9b)$$

Using for the dynamic surface tension σ the expression

$$\sigma = \sigma_0(\Gamma) + \mu_s \frac{\partial u}{\partial x} \quad (32)$$

here μ_s is the surface viscosity, eqn (9*b*) can be rewritten as

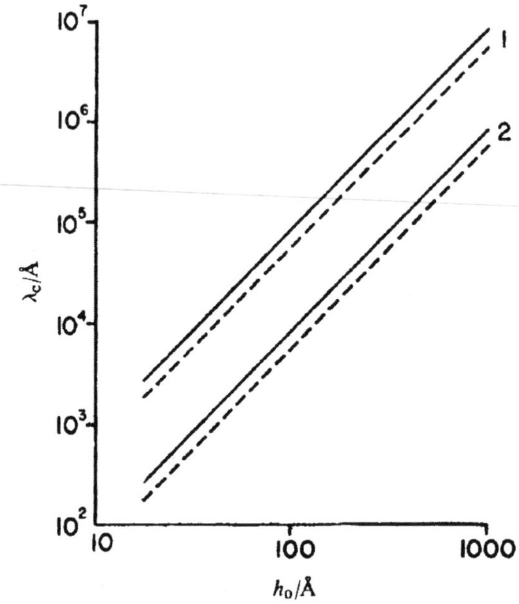

FIG. 3. Critical wavelength $\lambda_c/Å$ as a function of film thickness $h_0/Å$. Curve 1, Hamaker constant $A = 10^{-14}$ erg; $\sigma = 30$ dyn/cm; curve 2, Hamaker constant $A = 10^{-12}$ erg; $\sigma = 30$ dyn/cm; —, liquid film on a solid surface, – – – free liquid film.

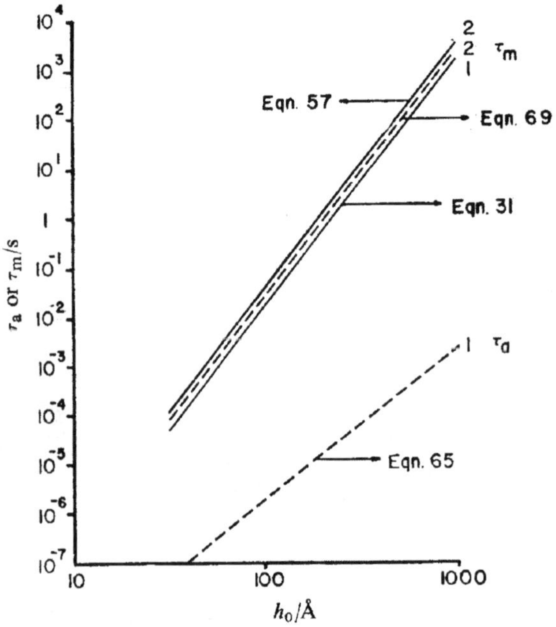

FIG. 4. Time constant τ_m/s and τ_a/s of the most rapidly growing fluctuations as a function of film thickness $h_0/Å$. The values of the parameters are given in table 1.-, Liquid layer on a solid surface;—, free liquid film. Curve 1, pure liquid film; curve 2, liquid film with a large enough concentration of surfactant. The results are practically the same for a wide range of values of the parameters (including the parameters from table 1), both for soluble and insoluble surfactants. They coincide with those given by eqn (57) for a thin film on a solid surface and by eqn (69) for a thin free film.

$$\mu\left(\frac{\partial u}{\partial y}+\frac{\partial \upsilon}{\partial x}\right)=\frac{\partial \sigma_0}{\partial \Gamma}\frac{\partial \Gamma}{\partial x}+\mu_s\frac{\partial^2 u}{\partial x^2}, \quad y\approx 0. \tag{33}$$

INSOLUBLE SURFACE ACTIVE AGENTS

In this case the surface concentration Γ of the surface active impurity satisfies the equation

$$\frac{\partial \Gamma}{\partial t}+\frac{\partial}{\partial x}(\Gamma u)=D_s\frac{\partial^2 \Gamma,}{\partial x^2} \quad y\approx 0. \tag{34}$$

Writing $\Gamma_0'+\Gamma+\Gamma'$, where Γ_0 is the surface concentration on the undeformed surface and $\Gamma' \ll \Gamma_0$, eqn (34) becomes

$$\frac{\partial \Gamma'}{\partial t}+\Gamma_0\frac{\partial u}{\partial x}=D_s\frac{\partial^2 \Gamma',}{\partial x^2} \quad y\approx 0.$$

Looking for a solution of the form

$$\Gamma' = G\ e^{ikx}e^{\beta t} \tag{35}$$

one obtains

$$G=\frac{\Gamma_0\left(D\hat{\upsilon}\right)_{y=0}}{D_s\Gamma_0 k^2+\beta}. \tag{36}$$

The boundary condition (9*b*) leads to

$$\left(D^2+k^2\right)\hat{\upsilon}=Mk^2\left(D\hat{\upsilon}\right),\ y\approx 0, \tag{20a}$$

where

$$M\equiv\frac{\Gamma_0\left(\frac{d\sigma_0}{d\Gamma}\right)_{\Gamma=\Gamma_0}}{\mu\left(D_s k^2+\beta\right)}-\frac{\mu_s}{\mu} \tag{37}$$

One obtains, using the same procedure as for pure liquid films,

$$\beta=-\frac{h_0^3}{3\mu}\frac{\left[\sigma k^4+\left(\frac{\partial \phi_0}{\partial h}\right)_{h=h_0}k^2\right]\left[1-\frac{M}{h_0}\frac{\left(kh_0\right)^2}{4}\right]}{1+\left(\frac{3}{2}-\frac{M}{h_0}\right)\left(kh_0\right)^2-\frac{\left(kh_0\right)^4}{6}} \tag{38}$$

Using eqn (28) for ϕ, eqn (38) becomes

$$\beta=-\frac{\left(kh_0\right)^2}{6\mu h_0}\frac{\frac{A}{2\pi h_0^2}-\sigma\left(kh_0\right)^2}{1+\left(\frac{3}{2}-\frac{M}{h_0}\right)\left(kh_0\right)^2-\frac{\left(kh_0\right)^4}{6}}\left[1-\frac{M}{h_0}\frac{\left(kh_0\right)^2}{4}\right]. \tag{39}$$

The effect of surface active agents is contained in the parameter *M*.

SOLUBLE SURFACE ACTIVE AGENTS

Neglecting surface diffusion, the perturbation Γ' satisfies the eqn[11]

$$-D_b\left(\frac{\partial c'}{\partial y}\right)_{y=0}=\frac{\partial \Gamma'}{\partial t}+\Gamma_0\frac{\partial u}{\partial x}. \tag{40}$$

The perturbation c' ($c=c_0+c'$) of the concentration in the film satisfies the diffusion equation

$$\frac{\partial c'}{\partial t}=D_b\left(\frac{\partial^2 c'}{\partial x^2}+\frac{\partial^2 c'}{\partial y^2}\right). \tag{41}$$

Assuming adsorption equilibrium at the free interface and using the Langmuir isotherm

$$\Gamma=\frac{k_1 c}{1+\dfrac{k_1}{\Gamma_\infty}c}, \tag{42}$$

one obtains

$$\Gamma'=\theta c',\ y\approx 0 \tag{43}$$

where

$$\theta\equiv k_1-2c_0k_1^2/\Gamma_\infty. \tag{43a}$$

Looking for a solution of the form

$$\mathbf{c}'=\xi(\mathbf{y})\mathbf{e}^{\mathbf{i}kx}\mathbf{e}^{\beta t}, \tag{44}$$

eqn (41) leads to

$$\frac{\mathrm{d}^2\xi}{\mathrm{d}y^2}=\frac{D_bk^2+\beta}{D_b}\xi. \tag{45}$$

The solution of eqn (45) is

$$\xi(y)=B_1\exp\left[\left(\frac{D_bk^2+\beta}{D_b}\right)^{\frac{1}{2}}y\right]+B_2\exp\left[-\left(\frac{D_bk^2+\beta}{D_b}\right)^{\frac{1}{2}}y\right]. \tag{46}$$

Eqn (45) has to be solved for the boundary conditions

$$-D_b\frac{\partial c'}{\partial y}=\theta\frac{\partial c'}{\partial t}+\Gamma_0\frac{\partial u}{\partial x}\quad \text{for } y\approx 0 \tag{47}$$

$$-D_b\frac{\partial c'}{\partial y}=0\qquad \text{for } y=-h_0. \tag{48}$$

One obtains

$$B_1=\frac{-\Gamma_0 ik\hat{u}(0)}{(1-B)D_b^{\frac{1}{2}}\left(D_bk^2+\beta\right)^{\frac{1}{2}}+\theta\beta(1+B)} \tag{49}$$

and

$$B_2 = B_1 B \tag{50}$$

where

$$B = \exp\left[-2\left(\frac{D_b k^2 + \beta}{D_b}\right)^{\frac{1}{2}} h_0\right]. \tag{51}$$

Hence

$$\Gamma' = \frac{-\Gamma_0 \left(\partial u/\partial x\right)_{y=0}(1+B)}{(1-B)D_b^{\frac{1}{2}}\left(D_b k^2+\beta\right)^{\frac{1}{2}}\theta^{-1}+\beta(1+B)}. \tag{52}$$

The boundary condition (9*b*) leads to

$$\left(D^2+k^2\right)\hat{\upsilon} = Nk^2\left(D\hat{\upsilon}\right),\ y \approx 0, \tag{53}$$

where

$$N = \frac{1}{\mu}\left[\frac{\Gamma_0\left(\partial\sigma_0/\partial\Gamma\right)_{\Gamma_0}(1+B)}{D_b^{\frac{1}{2}}\left(D_b k^2+\beta\right)^{\frac{1}{2}}(1-B)\theta^{-1}+\beta(1+B)} - \mu_s\right]. \tag{54}$$

The equation obtained for β is identical to eqn (38) if M is replaced by N.

The growth coefficient β has a maximum for the dominant wavelength. The time of rupture τ_m of the film can be evaluated from the maximum growth coefficient $\tau_m \approx \beta^{-1}_m$. The curves $\beta = \beta(k)$ (eqn (39)) and the minimum time of rupture as function of h_0 are presented in fig. 2 and 4. The values of various parameters are given in table 1.

If $|\Gamma_0(\partial\sigma_0/\partial\Gamma)_{\Gamma_0}|$ is large enough, the parameters —M (for insoluble) and —N (for soluble) become large and β can be approximated as for $-M$ (or $-N$)$\rightarrow \infty$: one obtains

$$\beta = \frac{(kh_0)^2}{12\mu h_0}\left(\frac{A}{2\pi h_0^2} - \sigma(kh_0)^2\right) \text{ for } -M \rightarrow \infty. \tag{55}$$

TABLE 1

	Gaseous Monolayer	Condensed Monolayer
μ	0.01 P	0.01 P
μ_s	10^{-3} g/s	10 g/s
Γ_0	8.3×10^{-12} mol/cm^2	8.3×10^{-10} mol/cm^2
$(\partial\sigma_0/\partial\Gamma)_{\Gamma_0}$	-2.4×10^{10} erg/mol	-1.34×10^{12} erg/mol
σ_0	30 dyn/cm	30 dyn/cm
D_s	10^{-5} cm^2/s	10^{-5} cm^2/s
D_b	10^{-5} cm^2/s	10^{-5} cm^2/s
θ	2×10^{-4} cm	2×10^{-4} cm
A	10^{-13} erg	10^{-13} erg
h_0	100 Å	100 Å

In this case the dominant wavenumber is given by

$$k_{\mathrm{d}} \approx 2^{-\frac{1}{2}} k_{\mathrm{c}} = (4\pi)^{-\frac{1}{2}} A^{\frac{1}{2}} \sigma^{-\frac{1}{2}} h_0^{-2} \tag{56}$$

and an explicit expression is obtained for τ_{m}

$$\tau_{\mathrm{m}} \approx 192\pi^2 \mu\sigma h_0^5 A^{-2} - N(\mathrm{or} - M) \to \infty. \tag{57}$$

For pure systems M (or N) = 0 and eqn (39) reduces to eqn (29). The time of rupture can be evaluated in this case by eqn (31). One may observe that the ratio of times of rupture in the extreme cases $-M$ (or $-N$)$\to \infty$ and M (or N) = 0 is about 4.

THIN FREE LIQUID FILMS

This situation was treated previously by Lucassen *et al.*[8] for free films of all thicknesses. The present approach is simpler, but restricted to thin films. The effect of surface active impurities is treated here in more detail.

The perturbation can be created in this case at both free surfaces. If the wave numbers of the perturbations at both the interfaces are equal, then two extreme cases of perturbations are possible, namely spatially in phase (asymmetric) and 180° out of phase (symmetric). The latter leads to the most rapid rupture. Therefore, the analysis which follows is based on the 180° out of phase perturbations.

PURE SYSTEMS

The equations of motion (4) and (5), the continuity eqn (6) and the boundary conditions (8) and (9) still hold for this situation. New boundary conditions specific to this case are,

$$\upsilon = 0 \quad \text{at} \quad y = -h_0/2 \tag{58}$$

$$\partial u/\partial y = 0 \quad \text{at} \quad y = -h_0/2. \tag{59}$$

Now $h = h_0 + 2\varepsilon$, then eqn (15) must be replaced by

$$\frac{\partial \phi_0}{\partial x} = 2\left(\frac{\partial \phi_0}{\partial h}\right)_{h=h_0} \frac{\partial \varepsilon}{\partial x}. \tag{60}$$

and eqn (16) takes the form

$$\mu \frac{\partial^2 u}{\partial y^2} = 2\mu \frac{\partial}{\partial x}\left(\frac{\partial \upsilon}{\partial y}\right)_{y=0} - \sigma \frac{\partial^3 \varepsilon}{\partial x^3} + 2\left(\frac{\partial \phi_0}{\partial h}\right)_{h=h_0} \frac{\partial \varepsilon}{\partial x} \tag{61}$$

The same procedure as that used for thin film on a solid surface leads in this situation to

$$\beta = \frac{\left[\frac{A}{\pi h_0^2} - \sigma (kh_0)^2\right]}{6\mu h_0 \left[1 - \left((kh_0)^2/18\right)\right]} \tag{62}$$

and

$$\lambda_{\mathrm{c}} = \left(\frac{4\pi^3 \sigma}{A}\right)^{\frac{1}{2}} h_0^2. \tag{63}$$

One may observe that in this case no dominant wavelength exists. For sufficiently large wavelengths, however, β becomes independent of the wavelength

$$\beta_{\mathrm{a}} = \frac{A}{6\pi\mu h_0^3}. \tag{64}$$

The time needed for the rupture of the film is in this case of the order of

$$\tau_{\mathrm{a}} \approx \frac{1}{\beta_{\mathrm{a}}} = \frac{6\pi\mu h_0^3}{A}. \tag{65}$$

The constant A is equal to the difference A_{11}–$2A_{13}$, where A_{11} is Hamaker constant for the interaction between the molecules in the film and A_{13} is the Hamaker constant for the interaction between the molecules in the film and the molecules of the other phase. Eqn (63), (64) and (65) are plotted in fig. 5, 3 and 4 respectively.

EFFECT OF SURFACE ACTIVE AGENTS ON THIN FREE LIQUID FILMS

For insoluble surface active agents one obtains

$$\beta = \frac{1}{6\mu h_0} \frac{\dfrac{A}{\pi h_0^2} - \sigma\left(kh_0\right)^2}{1 - \dfrac{2}{3}\dfrac{M}{h_0} - \dfrac{\left(kh_0\right)^2}{18}} \left[1 - \frac{M}{h_0}\frac{\left(kh_0\right)^2}{6}\right] \tag{66}$$

where M is given by eqn (37).

For soluble surface active agents the growth coefficient can be calculated from eqn (66) if M is replaced by

$$N' = \frac{1}{\mu}\left[\frac{\Gamma_0\left(\partial\sigma_0/\partial\Gamma\right)_{\Gamma_0}\left(1+B'\right)}{D_{\mathrm{b}}^{\frac{1}{2}}\left(D_{\mathrm{b}}k^2+\beta\right)^{\frac{1}{2}}\left(1-B'\right)\theta^{-1} + \beta\left(1+B'\right)} - \mu_{\mathrm{s}}\right] \tag{67}$$

where

$$B' = \exp\left[-\left(\frac{D_{\mathrm{b}}k^2+\beta}{D_{\mathrm{b}}}\right)^{\frac{1}{2}} h_0\right]. \tag{68}$$

Eqn (66) is plotted in fig. 5 for both soluble and insoluble surfactants using for the parameters the values from table 1. The time of rupture τ_{m} is evaluated as the reciprocal of β for the dominant wavenumber. Some numerical values are plotted in fig. 4.

Several effects, which are important under certain conditions,[12] have not been included in the present treatment: (i) the double layer forces (ii) the influence of surface active agents on the van der Waals interaction forces and (iii) the effect of the thickness of the layer on the surface tension.

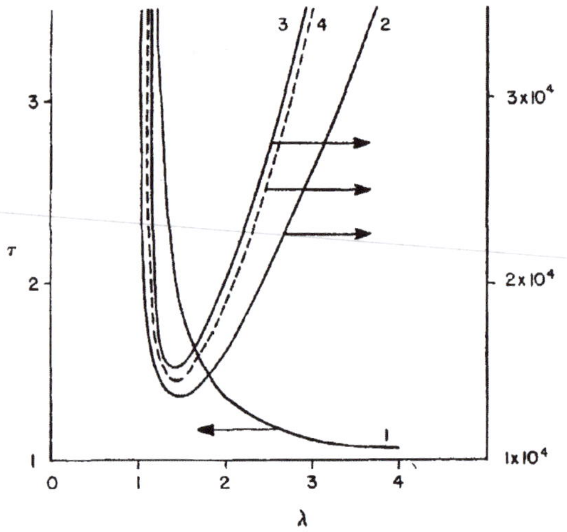

FIG. 5. Time constant ($\tau = 1/\beta$) as a function of wavelength λ for a thin free film. The values of the parameters are given in table 1. The unit of τ is the value of τ_a calculated from eqn (65) and the unit of λ is the value of λ_c calculated from eqn (63). Curve 1, pure liquid film; curve 2, liquid film with gaseous monolayer of insoluble surfactant; curve 3, liquid film with condensed monolayer of insoluble surfactant; curve 4, liquid film with a gaseous monolayer of soluble surfactant.

DISCUSSION

The hydrodynamic stability analysis allows us to interpret some of the available experimental information.

Concerning the rupture of a thin film on a solid surface, the present treatment gives some understanding of the results obtained by Goldstein [2] in a study of water vapour condensation on a shock-tube wall. For condensation on a clean hydro-phobic surface, Goldstein reports that a continuous film of condensate is formed, which, after reaching a thickness of about 100 Å in about 10 μs after the compression, begins to break up into many small droplets. In all mechanical systems there are perturbations of various wavelengths; consequently the rupture of the film can be considered in this case to be a consequence of hydrodynamic instability. It is also of interest to mention that if Hamaker's constant A is taken to be of the order of 10^{-12} erg and h_0 of the order of 50 Å, then the time of rupture computed from eqn (31) with $\sigma \approx 70$ dyn/cm and $\mu \approx 10^{-2}$ P is of the order of 10 μs, as was found experimentally.

Concerning the rupture of a free thin film, it is important to stress that in any real system there are surface active impurities. Minute quantities of them have a strong stabilizing effect, damping the waves which occur at a free surface. For this reason eqn (65), which is valid for the time of rupture in a pure liquid, gives a lower bound of the time of rupture. In the presence of surface active impurities the time of rupture can be several orders of magnitude longer than for a pure liquid (fig. 4). An upper bound for the time of rupture can be obtained from eqn (66) if $|\Gamma_0(\partial\sigma/\partial r)_{\Gamma_0}|$ is large enough so that one may assume that $-M$ (for insoluble surfactants) or $-N$ (for soluble surfactants) tends to infinity. One obtains Vrij's result[5]

$$\tau_m \approx 96\pi^2\sigma\mu h_0^5 A^{-2}, \qquad -M\left(\text{or} - N'\right) \to \infty. \tag{69}$$

The coalescence time between two bubbles in contact in a pure liquid is indeed very short.[13] In the presence of surface active impurities the coalescence time is much longer, of the order of 10^{-1} s.[13] If one assumes, as suggested by Marucci, that the drainage time is negligible and hence that the rupture of a thin film of some hundred Ångstrom thick, is the mechanism that controls the rate of coalescence, eqn (69) and (65) give an upper and lower bound of the coalescence time. One may verify that indeed the value 10^{-1} s is bounded by these equations (fig. 4).

APPENDIX 1

NOMENCLATURE

A	A_{11}–A_{12} for a film on a solid surface and A_{11}–$2A_{13}$ for a free film
A_{ij}	Hamaker constant for the interaction between molecules of type i and j. $\pi^2\alpha_{ij}n_in_j$
B, B'	quantities defined by eqn (51) and (68)
B_1, B_2	integration constants given by eqn (49) and (50)
c	bulk concentration of surfactants
c_0	concentration of surfactant in the undeformed film
c'	c-c_0
c_1, c_2, c_3	integration constants
D	$\partial/\partial y$
D_b	diffusion coefficient in the bulk
D_s	surface diffusion coefficient
G	constant defined by eqn (36)
h_0	average film thickness
h	actual film thickness ($h = h_0 + \varepsilon$ for a film on a solid surface and $h = h_0 + 2\varepsilon$ for a free film)
k	wavenumber of the perturbation
K_c	critical wavenumber
k_d	dominant wavenumber
k_1	constant in eqn (42)
M, N, N'	quantities defined by eqn (37), (54) and (67)
n_1	molecular density of the liquid
n_2	molecular density of the solid
p	hydrostatic pressure
p_o	external pressure on the film
P	$(P + \phi)$
r	distance between two molecules
t	time
u	velocity component in the $\boldsymbol{x}$ direction
u_{ij}	interaction potential between two molecules of type i and j separated by the distance r
U_{ij}	potential energy of one molecule of type i due to all molecules of type j
v	velocity in the y direction
V	volume of the solid
x, y	cartesian coordinates
Y	$h - y$
α_{ij}	London–van der Waals parameter for interaction between molecules of type i and j
β	growth parameter
β_m	maximum value of β for a thin film
β_a	maximum value of β for free films of pure liquids
ε	perturbation of the film thickness ($\varepsilon = h$–h_0 for a film on a solid surface and $2\varepsilon = h$–h_0 for a free film)
$\hat{\varepsilon}$	maximum amplitude of perturbation ε
λ	wavelength $= 2\pi/k$
λ_c	critical wavelength
λ_d	dominant wavelength
μ	viscosity of the liquid
μ_s	surface viscosity
σ	liquid-gas surface tension
σ_0	static liquid–gas surface tension
τ	$1/\beta$, characteristic time
τ_m	$1/\beta_m$
τ_a	$1/\beta_a$

ϕ	interaction potential per unit volume at a point in the liquid
ϕ_0	interaction potential per unit volume at the free surface of liquid
ϕ'_B	interaction potential per molecule of the liquid situated at the free surface in a semi-infinite liquid
ϕ_B	$n_1\phi'_B$
Γ	surface concentration of surfactants
Γ_0	value of Γ for undeformed film
Γ'	$\Gamma - \Gamma_0$
Γ_∞	constant in eqn (42)
θ	quantity defined by eqn (43*a*)

APPENDIX 2

The potential energy of a molecule in the liquid film due to the molecules of the semiinfinite solid is given by[14]

$$U_{12}(Y) = \iint_V n_2 u_{12}(r) 2\pi r^2 \mathrm{d}r \sin\theta \mathrm{d}\theta$$

where n_2 is the number of molecules per unit volume of solid and u_{12} is the potential energy due to the interaction between two molecules at distance r. V is the lower half space. For London-van der Waals type interaction

$$u_{ij} = -\alpha_{ij}/r^6$$

Consequently

$$U_{12}(Y) = -\left(\frac{\pi n_2 \alpha_{12}}{6}\right)\frac{1}{Y^3}.$$

If the molecule is located at the free surface, $Y = h$ and

$$U_{12}(h) = -\left(\frac{\pi n_2 \alpha_{12}}{6}\right)\frac{1}{h^3}.$$

Similarly, the potential due to the molecules of the liquid film on one molecule situated at the free surface of the liquid is given by

$$U_{11}(h) = \phi'_B + \left(\frac{\pi n_1 \alpha_{11}}{6}\right)\frac{1}{h^3}.$$

Therefore the total potential energy per unit volume of liquid acting at the free surface is given by:

$$\phi_0 = n_1\left[U_{11}(h) + U_{12}(h)\right]$$

$$= \phi_B + \frac{A_{11} - A_{12}}{6\pi h^3}$$

where A_{ij} is the Hamaker constant (subscript 1 refers to the liquid and 2 to the solid)

$$A_{ij} = \pi^2 \alpha_{ij} n_i n_j.$$

and

$$\phi_B = n_1 \phi'_B.$$

We are grateful to the referees for bringing ref. (8) and (9) to our attention. This work was supported in part by the National Science Foundation.

REFERENCES

1 A. Scheludko, *Kolloid Z.*, 1963, **191**, 52.
2 R. Goldstein, *J. Chem. Phys.*, 1964, **40**, 2763.
3 A. Scheludko, *Proc. K. Ned Akad., Wet.*, 1962, **B65**, 76.
4 A. Scheludko, *Adv. Colloid and Int. Sc.*, 1967, **1**, **39**
5 A. Vrij, *Disc. Faraday Soc.*, 1966, **42**, 23.
6 S. Chandrasekhar, *Hydrodynamic and Hydromagnetic Stability* (Oxford Univ. Press, Oxford, 1961).
7 B. V. Felderhof, *J. Chem. Phys.*, 1968, **49**, 44.
8 J. Lucassen, M. van den Temple, A. Vrij and F. Th. Hesselink, *Proc. K. Ned. Akad. Wet.*, 1970, **B73**, 109.
9 A. Vrij, F. Th. Hesselink, J. Lucassen and M. van den Tempel, *Proc. K. Ned. Akad. Wet.*, 1970, **B73**, 124.
10 G. K. Batchelor, *An Introduction to Fluid Dynamics* (Cambridge University Press, 1967), p. 219.
11 V. G. Levich, *Physico-Chemical Hydrodynamics* (Prentice-Hall, N.J., 1962).
12 E. Ruckenstein and R. K. Jain, to be published.
13 G. Marrucci, *Chem. Eng. Sei.*, 1969, **24**, 975.
14 J. Frenkel, *Kinetic Theory of Liquids* (Dover, N.Y., 1955).

5.2 Stability of Thin Solid Films*

Eli Ruckenstein and C. S. Dunn
Faculty of Engineering and Applied Sciences, Stale University of New York at Buffalo, Buffalo, N. Y. 14214 (U.S.A.)

(Received August 2, 1977; accepted October 10, 1977)

1. INTRODUCTION

Thin solid films on substrates are widely used as protective coatings, in microelectronics etc. Systems which utilize thin films generally fail because rupture, or hillock formation[1], ruins their useful properties. Therefore it is important to understand the mechanical stability of thin films since the mechanical stability directly affects the lifetime of these devices. Supported metal catalysts age because small clusters of metal atoms migrate on the surface of the substrate and sinter to form large crystallites[2]. Some insight into the rejuvenation of the supported metal catalyst by the fragmentation of the large crystallites back into smaller ones may also be provided if we understand the stability of very thin films. The present paper uses stability theory to predict conditions for film rupture. A small perturbation is applied to the free surface of the film, and the conditions under which the perturbation grows in time are established. Because the emphasis here is on films thinner than the range of interaction forces acting between the atoms of the film and the substrate, the forces acting on a volume element in the film differ from those on a similar element in the corresponding bulk solid. Hence, although a continuum approach is used in the analysis, a supplementary body force is included in the differential equations describing an elastic bulk solid to account for the additional forces acting in very thin films[3].

Experiment shows that thin films formed by electroplating, sputtering or evaporation techniques possess very large internal stresses which arise from two sources: (1) a thermal stress which originates when the film and substrate have different thermal expansion coefficients and the system is maintained at some temperature different from the deposition temperature and (2) an intrinsic stress caused by epitaxial growth[4], by metastable phases occurring during deposition[5] or by large defect concentrations[6] (non-stoichiometry). Since the intrinsic stress can be very large (of the order of 10^{10} dyn cm^{-2})[7], a theoretical treatment of thin film failure must account for its existence.

Because the films considered here are very thin, the boundary conditions at the film-support interface strongly affect the displacement field throughout the film and are therefore a critical factor in film stability. Exact boundary conditions at this interface should take into account the possibility of slipping or pulling away from the substrate. However, because the mere formulation of such boundary conditions is in itself a difficult problem, two limiting cases will be considered. In one the interaction forces between the film and the substrate are assumed to be negligible and therefore the film is assumed to behave like a free film with no shear at either boundary. In the second limiting case the substrate is assumed to be rigid and the displacement in the film at the solid-solid interface is taken to be zero.

2. MECHANISM OF RUPTURE

A small displacement ε is applied to the free surface of a solid film and the conditions for its growth are established (Fig. 1). The growth of the disturbance ε is affected by (1) the elastic displacements and (2) surface diffusion due to a gradient of the chemical potential along the perturbed

---

* *Thin Sofid Fifms,* 51 (1978) 43–75. Republished with permission.

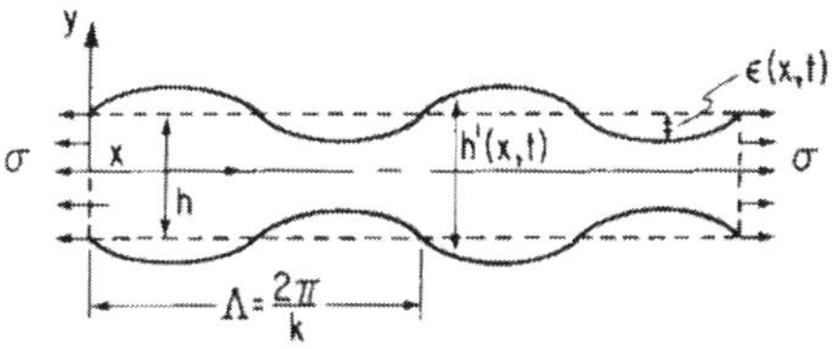

FIG. 1. Case I: asymmetric perturbations in a thin free film.

surface. (The various contributions to this gradient of the chemical potential are discussed below.) The flux J of atoms along the surface is proportional to the surface gradient of the chemical potential g per atom:

$$J = -\frac{\nu D}{k_b T}\frac{\partial g}{\partial x}$$

The growth of the disturbance is therefore given by

$$\frac{\partial \varepsilon}{\partial t} = \frac{\partial \upsilon}{\partial t} + \frac{D_v V_0}{k_b T}\frac{\partial^2 g}{\partial x^2} \tag{1}$$

where υ is the elastic component of the displacement normal to the interface. If ε grows with time the film is unstable and rupture will result, whereas if ε decays with time the film is stable. The chemical potential per atom in a volume element adjacent to a curved stressed interface of a solid is given by[8,9]

$$g = g_0 + V_0(\Delta W + p_c) \tag{2a}$$

where g_0 is a reference potential, ΔW is the strain energy per unit volume representing the stored elastic energy density near the surface of the stressed film and p_c is the energy stored per unit volume due to (and equal to) the capillary pressure. However, for very thin films a supplementary term has to be introduced into eqn. (2a) because the range of the interaction forces between the atoms of the film and the substrate is greater than the thickness of the film. For very thin films eqn. (2a) has therefore to be replaced by

$$g = g_0 + V_0(\Delta W + p_c + \phi_0) \tag{2b}$$

where ϕ_0 is the difference between the interaction potential at the free surface of the film and that at the surface of the corresponding semi-infinite solid. The excess interaction potential ϕ_0 decays to zero for sufficiently thick films but becomes increasingly important with decreasing film thicknesses. (See Section 3 for the form of the interaction potential.)

The capillary pressure is given by

$$p_c = \gamma K$$

where K, the curvature of the perturbed interface, is

$$K = -\frac{\partial^2 \varepsilon}{\partial x^2}\left\{1 + \left(\frac{\partial^2 \varepsilon}{\partial x^2}\right)^2\right\}^{-3/2}$$

and γ is the surface free energy. For small displacements $K \approx -\partial^2\varepsilon/\partial x^2$ and the expression for the growth of the perturbation of the free interface becomes

$$\frac{\partial \varepsilon}{\partial t} = \frac{\partial \upsilon}{\partial t} + \theta \frac{\partial^2}{\partial x^2}\left(\Delta W - \gamma \frac{\partial^2 \varepsilon}{\partial x^2} + \phi_0 \right) \tag{3a}$$

where

$$\theta = \frac{D\nu V_0^2}{k_b T} \tag{3b}$$

The strain energy per unit volume at the free surface of an elastic solid is given by the expression[10]

$$\Delta W = \frac{1}{2}\left(\tau_{xx}e_{xx} + \tau_{yy}e_{yy} + \tau_{zz}e_{zz} + 2\tau_{xz}e_{xz}\right)$$

where τ_{ij} and e_{ij} are the *ij*th components of the stress and strain tensors respectively.

The elastic displacements *u, υ* and *w* in the film are composed of a steady state displacement plus a small disturbance displacement which is assumed to be two dimensional, *i.e.*

$$u(x,y,z,t) = u^s(x,y,z) + u'(x,y,t) \tag{4a}$$

$$\upsilon(x,y,z,t) = \upsilon^s(x,y,z) + \upsilon'(x,y,t) \tag{4b}$$

and

$$w(x,y,z) = w^s(x,y,z) \tag{4c}$$

It is shown in Appendix B that the strain energy at the free surface can similarly be expressed as the sum of a steady state strain energy and a disturbance strain energy:

$$\Delta W = \Delta W^s + \Delta W' \tag{5}$$

where the steady state strain energy ΔW^s is given by

$$\Delta W^s = \frac{1}{2}\left\{ \tau_{xx}^s \frac{\partial u^s}{\partial x} + \tau_{yy}^s \frac{\partial \upsilon^s}{\partial y} + \tau_{zz}^s \frac{\partial w^s}{\partial z} + \tau_{xz}^s \left(\frac{\partial u^s}{\partial z} + \frac{\partial w^s}{\partial x} \right) \right\} \tag{5a}$$

Since the disturbance displacement is assumed to be small, terms of second order in perturbation quantities are neglected compared with first-order terms. It can be shown (see Appendix B) that at the free surface

$$\Delta W' \approx \left(\tau_{xx}^s + \lambda\Phi\right)\frac{\partial u'}{\partial x} - \gamma\Phi\frac{\partial^2 \varepsilon}{\partial x^2} \tag{5b}$$

where

$$\Phi = -\frac{\tau_{yy}^s}{\lambda + 2\mu} \tag{5c}$$

Here λ and μ are Lamé's coefficients, related to the Young's modulus E and the Poisson's ratio χ of the film through

$$E = \frac{\lambda(3\lambda + 2\mu)}{\lambda + \mu} \tag{5d}$$

and

$$\chi = \frac{\lambda}{2(\lambda + \mu)} \tag{5e}$$

Since υ^s is independent of time, the growth of a small perturbation of the free interface of a thin solid film is governed by

$$\frac{\partial \varepsilon}{\partial t} = \frac{\partial \upsilon'}{\partial t} + \theta \left\{ \left(\tau_{xx}{}^s + \lambda \Phi \right) \frac{\partial^3 u'}{\partial x^3} - \gamma (1 + \Phi) \frac{\partial^4 \varepsilon}{\partial x^4} + \frac{\partial^2 \varnothing_0}{\partial x^2} \right\} \tag{6}$$

where the values of u' and υ' have to be taken at the perturbed surface. It should be noted that the chemical potential as expressed by eqn. (2b) contains both steady state and perturbation terms. However, the steady state chemical potential given by $g^s = g_0 + V_0(\Delta W^s + \phi_0{}^s)$ is independent of x and therefore does not appear in eqn. (6).

Since the differential equations describing an elastic solid are linear, only the steady state stresses $\tau_{xx}{}^s$ and $\tau_{yy}{}^s$ at the free surface of the film which appear in eqn. (6) affect the perturbation displacements u' and υ'. Assuming that the pre-existing internal stress σ (pre-stress) which arises during film preparation acts uniformly in the (x, z) plane we obtain $\tau_{xx}{}^s = \tau_{zz}{}^s = \sigma$. Furthermore, a normal stress balance at the free surface yields $\tau_{yy}{}^s = -p_0$, where p_0 is the external pressure. Then eqn. (6) becomes

$$\frac{\partial \varepsilon}{\partial t} = \frac{\partial \upsilon'}{\partial t} + \theta \left\{ (\sigma + \lambda \Phi) \frac{\partial^3 u'}{\partial x^3} - \gamma (1 + \Phi) \frac{\partial^4 \varepsilon}{\partial x^4} + \frac{\partial^2 \phi_0}{\partial x^2} \right\} \tag{7}$$

where

$$\Phi = \frac{p_0}{\lambda + 2\mu}$$

Once an explicit form of the interaction potential is obtained, ϕ_0 can be determined and the equations of displacement for an elastic solid can be solved, subject to the appropriate boundary conditions, to obtain the perturbation displacements u' and υ'. Equation (7) can then be used to establish the conditions under which the perturbation grows with time.

3. THE INTERATOMIC POTENTIAL

Before an explicit expression of the interaction potential ϕ can be calculated, an appropriate form for the two-body metal-metal interatomic potential energy must be chosen. The short range repulsive forces between two metal atoms have the form of a screened Coulombic potential:

$$u_{ij}{}^{\text{rep}}(r) = \frac{z_i z_j e^2}{r} f(r) \tag{8}$$

where $z_i e$ and $z_j e$ are the nuclear charges of the interacting atoms and $f(r)$ is a screening function which accounts for the shielding effect of the atomic electrons. The screening function $f(r)$ can be computed by using various procedures such as the Thomas-Fermi approximation[11]. However, atomic scattering experiments have shown that the Molière approximation to the Thomas-Fermi screening function leads to a form of the repulsive potential between atoms which is more accurate than that obtained from the Thomas-Fermi theory alone. The Molière screening function has the form[12]

$$f(r) = 0.35\exp\left(-0.3\frac{r}{a_{ij}}\right) + 0.55\exp\left(-1.2\frac{r}{a_{ij}}\right) + 0.10\exp\left(-6.0\frac{r}{a_{ij}}\right) \tag{9}$$

where a_{ij} can be taken either as an adjustable parameter or as the Firsov screening radius

$$a_{ij} = \frac{0.885 a_0}{\left(z_i^{1/2} + z_j^{1/2}\right)^{2/3}}$$

which is comparable with the Bohr radius a_0.

It will be assumed that the attractive component of the interaction potential has the form

$$u_{ij}^{\mathrm{att}}(r) = -\frac{\beta_{ij}}{r^6} \tag{10}$$

An attractive potential of this form describes London forces which arise for neutral atoms because of the interaction of instantaneous dipoles in one individual atom with the induced dipoles in another atom. The constant β_{ij} should be chosen such that the total energy of interaction, which is found by summing all the possible pair interactions between a reference atom i in the film and all the rest of the atoms in the system, has a minimum at the nearest neighbor distance. However, phonon dispersion results together with elastic force constant data for copper and nickel have shown that the nearest neighbor interactions are dominant for these two metals[13]. Therefore we choose β_{ij} such that the total potential energy for a two-body copper–copper interaction has a minimum at the observed nearest neighbor distance. (This assumption is valid for copper and nickel but it is not true in general.) In what follows the total two-body potential energy u_{ij} of interaction between two metal atoms has the form

$$u_{ij}(r) = -\frac{\beta_{ij}}{r^6} + \frac{z_i z_j e^2}{r}\left\{0.35\exp\left(-\frac{0.3r}{a_{ij}}\right) + 0.55\exp\left(-\frac{1.2r}{a_{ij}}\right) + 0.10\exp\left(-\frac{6.0r}{a_{ij}}\right)\right\} \tag{11}$$

Figure 2 shows the interatomic potential u_{ij} as a function of separation distance r for two copper atoms. The form chosen for copper-copper interactions is adequate since the value of the two-body potential at the nearest neighbor distance ($u(2a) = -40 \times 10^{-14}$ erg where $2a = 2.56$ Å) predicts energies for vacancy formation in the correct range (1.1 – 1.2 eV)[14].

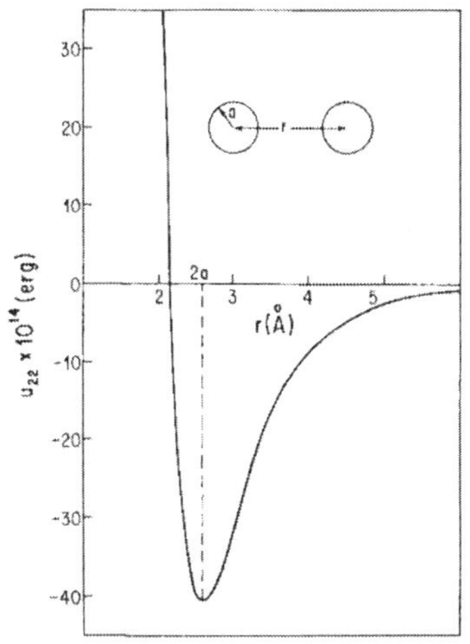

FIG. 2. The interatomic potential u_{22} for copper–copper interactions as a function of separation distance r. Here $\beta_{22} = 3.647 \times 10^{-57}$ erg cm^6 ($A_{22} = 2.6 \times 10^{-11}$ erg) and $a = 1.28$ Å.

By assuming pairwise additivity and using eqn. (11) an explicit form of the interaction potential $\phi(y, h)$ is computed (see Appendix A). We have neglected film–gas interactions because they are generally small compared with film–film and film-substrate interactions. Denoting the thickness of the film by h, we obtain

$$\phi(y,h) = \frac{A_{22}}{6\pi}\left\{\frac{1}{(h/2+y)^3} + \frac{1}{(h/2-y)^3}\right\} - \sum_{j=1}^{3}\frac{G_j}{C_j}\left[\exp\left\{-C_j\left(\frac{h}{2}+y\right)\right\} + \right.$$
$$\left. + \exp\left\{-C_j\left(\frac{h}{2}-y\right)\right\}\right] \tag{12}$$

for the free film (where y is measured from the middle of the free film) and

$$\phi(y,h) = \frac{A_{22} - A_{32}}{6\pi}\left\{\frac{1}{y^3} + \frac{1}{(h-y)^3}\right\} - \sum_{j=1}^{3}\frac{G_j}{C_j}\left[\exp(-C_j y) + \exp\{-C_j(h-y)\}\right] +$$
$$+ \frac{n_3 z_3}{n_2 z_2}\left(\frac{a_{32}}{a_{22}}\right)^2 \sum_{j=1}^{3}\frac{G_j}{C_j}\exp\left(-C_j\frac{y a_{22}}{a_{32}}\right) \tag{13}$$

for the supported film (y is now measured from the flim-substrate interface). In eqns. (12) and (13)

$$G_j = 2\pi(z_2 n_2 e)^2 \frac{c_j}{C_j} \qquad c_j = \begin{pmatrix} 0.35 \\ 0.55 \\ 0.10 \end{pmatrix} \qquad C_j = \frac{1}{a_{22}}\begin{pmatrix} 0.30 \\ 1.20 \\ 6.00 \end{pmatrix} \tag{13a}$$

$A_{22} = \pi^2 n_2^2 \beta_{22}$ and $A_{32} = \pi^2 n_2 n_3 \beta_{32}$ are constants for attractive interactions between the atoms of the film and between the atoms of the film and those of the substrate respectively. The interaction potential evaluated at the center of the first row of atoms in the perturbed surface is therefore

$$\phi_0 = \phi_0' + \frac{A_{22}}{6\pi h'^3} - \sum_{j=1}^{3} \frac{G_j}{C_j} \exp(-C_j h')$$

for the free film and

$$\phi_0 = \phi_0' + \frac{A_{22} - A_{32}}{6\pi h'^3} - \sum_{j=1}^{3} \frac{G_j}{G_j} \left[1 - \frac{n_3 z_3}{n_2 z_2} \left(\frac{a_{32}}{a_{22}} \right)^2 \exp\left\{ c_j h' \left(1 - \frac{a_{22}}{a_{32}} \right) \right\} \right] \times \exp(-C_j h')$$

for the supported film, where h' is the perturbed film thickness. Since we assume that perturbations may arise at both surfaces of a free film but only at the free surface of a supported film, $h' = h + 2\varepsilon$ for a free film while $h' = h + \varepsilon$ for the supported film. Computations show that the terms $\exp(-C_j h')$ due to repulsive forces are negligible compared with the attractive interactions $A_{22}/6\pi h'^3$, so that

$$\phi_0 \approx \phi_0' + \frac{A_{22}}{6\pi h'^3} \quad \text{with} \quad h' = h + 2\varepsilon \tag{14}$$

for the free film and

$$\phi_0 \approx \phi_0' + \frac{A_{22} - A_{32}}{6\pi h'^3} \quad \text{with} \quad h' = h + \varepsilon \tag{15}$$

for the supported film.

4. GENERAL EQUATIONS

4.1. Displacements in a Thin Elastic Film

Consider a thin supported film with a uniform pre-existing internal stress (prestress) σ which arises during film preparation. The film is assumed to be thinner than the range of interaction forces between one atom and all the other atoms of the system. Compared with a bulk solid this means that some atoms of the bulk solid are replaced by atoms of gas or substrate within the range of the interaction forces. The difference in the forces acting on a volume element of a thin film and those acting on a volume element of a bulk solid is accounted for by a body force in the equations of motion. Consequently the general differential equations describing the stress distribution in a thin solid film are

$$\rho \frac{\partial^2 u}{\partial t^2} + \frac{\partial \phi}{\partial x} = \frac{\partial \tau_{xx}}{\partial x} + \frac{\partial \tau_{yx}}{\partial y} + \frac{\partial \tau_{zx}}{\partial z} \tag{16}$$

$$\rho \frac{\partial^2 \upsilon}{\partial t^2} + \frac{\partial \phi}{\partial y} = \frac{\partial \tau_{xy}}{\partial x} + \frac{\partial \tau_{yy}}{\partial y} + \frac{\partial \tau_{zy}}{\partial z} \tag{17}$$

$$\rho \frac{\partial^2 w}{\partial t^2} + \frac{\partial \phi}{\partial z} = \frac{\partial \tau_{xz}}{\partial x} + \frac{\partial \tau_{yz}}{\partial y} + \frac{\partial \tau_{zz}}{\partial z} \tag{18}$$

where u, υ and w are the components of the displacement and ϕ is an interaction potential per unit volume equal to the difference between that in a thin film and that in a bulk solid (see Appendix A). It should be noted that eqns. (16)–(18) are also valid for thick films since ϕ always depends on the distance to an interface. In the latter case ϕ is important only near the interface, in a region with a thickness of the order of the range of the interaction forces (about several hundred ångströms).

If we assume that the film is an isotropic elastic continuum, the stress may be expressed in terms of the strain through the equations

$$\tau_{ii} = \lambda\delta + 2\mu e_{ii} \tag{19}$$

$$\tau_{ij} = 2\mu e_{ij} \tag{20}$$

where e_{ij} is the ijth component of the strain tensor $\left(e_{ij} = \frac{1}{2}(\partial u_i / \partial x_j + \partial u_j / \partial x_i)\right), \delta = e_{xx} + e_{yy} + e_{zz}$ is the change in volume of unit volume and λ and μ are Lamé's coefficients (assumed for simplicity to be those of the bulk solid).

4.2. Equations for the Perturbation Displacements

When a small perturbation is applied to the surface of a solid it generates displacements throughout the film. These displacements are composed of a steady state plus a disturbance component (eqns. (4a)–(4c)). Because the equations of an elastic solid are linear the differential equations for the perturbation displacements, assumed to be two dimensional, are

$$\rho\frac{\partial^2 u'}{\partial t^2} + \frac{\partial \phi'}{\partial x} = (\lambda + \mu)\frac{\partial \delta'}{\partial x} + \mu\nabla^2 u' \tag{21}$$

and

$$\rho\frac{\partial^2 \upsilon'}{\partial t^2} + \frac{\partial \phi'}{\partial y} = (\lambda + \mu)\frac{\partial \delta'}{\partial y} + \mu\nabla^2 \upsilon' \tag{22}$$

where $\nabla^2 = \partial^2/\partial x^2 + \partial^2/\partial y^2$ is a two-dimensional Laplacian and $\delta' = \partial u'/\partial x + \partial \upsilon'/\partial y$ is the dilatation due to the perturbation. Here ϕ' is a perturbation component of the interaction potential which arises because the interaction forces change with fluctuations in the perturbed film thickness h'. It should be noted that although we assume no perturbation in the z direction the normal stress $\tau'_{zz} = \lambda(\partial u'/\partial x + \partial \upsilon'/\partial y)$ is finite; however, this term does not contribute to the equations of motion because $\partial \tau_{zz}'/\partial z = 0$.

4.3. Equations Valid When $h/\Lambda \ll 1$

Because only very thin films are of interest in the present context, it is reasonable to consider the case when the wavelength $\Lambda = 2\pi/k$ of the perturbation is much larger than the film thickness. To evaluate the relative contributions of various terms in eqns. (21) and (22) the following scaling transformations are introduced:

$$\begin{bmatrix} u^* \\ x^* \end{bmatrix} = \frac{1}{\Lambda}\begin{bmatrix} u' \\ x \end{bmatrix} \quad \begin{bmatrix} \upsilon^* \\ y^* \end{bmatrix} = \frac{1}{h}\begin{bmatrix} \upsilon' \\ y \end{bmatrix}$$

$$t^* = \beta_m t \qquad \phi^* = \frac{\phi' h^2}{\mu \Lambda^2}$$

where β_m is the maximum value of the growth parameter β of the perturbation (see eqn. (29)). The reciprocal β_m^{-1} is a measure of the time required for film failure. The displacement in the x direction and the distance x are scaled with Λ, the wavelength of the perturbation, while the displacement in the y direction and the distance y are scaled with the film thickness h. With this scaling, eqns. (21) and (22) become

$$(\beta_m h)^2 \frac{\rho}{\mu} \frac{\partial^2 u^*}{\partial t^{*2}} + \frac{\partial \phi^*}{\partial x^*} = \frac{\lambda + \mu}{\mu} \left(\frac{h}{\Lambda} \right)^2 \frac{\partial \delta^*}{\partial x^*} + \left(\frac{h}{\Lambda} \right)^2 \frac{\partial^2 u^*}{\partial x^{*2}} + \frac{\partial^2 u^*}{\partial y^{*2}} \tag{23}$$

and

$$(\beta_m h)^2 \frac{\rho}{\mu} \frac{\partial^2 \upsilon^*}{\partial t^{*2}} + \left(\frac{\Lambda}{h} \right)^2 \frac{\partial \phi^*}{\partial y^*} = \frac{\lambda + \mu}{\mu} \frac{\partial \delta^*}{\partial y^*} + \left(\frac{h}{\Lambda} \right)^2 \frac{\partial^2 \upsilon^*}{\partial x^{*2}} + \frac{\partial^2 \upsilon^*}{\partial y^{*2}} \tag{24}$$

where $\delta^* = \partial u^*/\partial x^* + \partial \upsilon^*/\partial y^*$. When h/Λ is sufficiently small, terms of the order of $(h/\Lambda)^2$ are negligible compared with terms of order unity. Lamé's coefficient μ is generally larger than 10^{10} dyn cm^{-2} for metals, and the films under consideration are no thicker than 10^{-5} cm. For these values of μ and h the coefficient $(\beta_m h)^2(\rho/\mu)$ that multiplies the inertial term is much smaller than unity provided that $\beta_m \ll 10^{10}$ s^{-1}. Since rupture times of solid films are much greater than 10^{-10} s, the inertial term can be neglected. These assumptions greatly simplify the differential equations for the perturbations to

$$\mu \frac{\partial^2 u'}{\partial y^2} = \frac{\partial \phi'}{\partial x} \tag{25}$$

and

$$\frac{\partial}{\partial y} \left\{ (\lambda + 2\mu) \frac{\partial \upsilon'}{\partial y} + (\lambda + \mu) \frac{\partial u'}{\partial x} - \phi' \right\} = 0 \tag{26}$$

5. THIN FREE FILM

5.1. Equations

We first treat the limiting case when film–substrate interactions are sufficiently weak so that the film behaves like a free film. In this case perturbations may originate at both free surfaces of Fig. 1. For simplicity we consider only asymmetric perturbations which occur when the disturbances at the two interfaces are 180° out of phase. Equations (25) and (26) are the differential equations describing the propagation of elastic perturbations in a thin free film. The interaction potential within a perturbed film differs from that of a uniform film because of the local change of the film thickness and because of the change of the shape of the interface (curvature effect). It will be shown below that the contribution of the curvature to ϕ' is negligibly small compared with the thickness effect. Therefore the perturbation ϕ' of the interaction potential may be determined to order ε by expanding the interaction potential in a Taylor series about the average film thickness, *i.e.*

$$\phi(y, h') \approx \phi^s(y, h) + \varepsilon(x, t) \left\{ \frac{\partial \phi}{\partial (h'/2)} \right\}_{h/2} \equiv \phi^s + \phi'$$

Using eqn. (12) for $\phi(y, h')$ we obtain

$$\phi'(y,h) = -\varepsilon\left(\frac{A_{22}}{2\pi}\left\{\frac{1}{(h/2+y)^4} + \frac{1}{(h/2-y)^4}\right\} - \sum_{j=1}^{3} G_j\left[\exp\left\{-C_j\left(\frac{h}{2}+y\right)\right\} + \right.\right.$$
$$\left.\left. + \exp\left\{-C_j\left(\frac{h}{2}-y\right)\right\}\right]\right) \tag{27}$$

Since a free film is symmetric about $y = 0$, this leads to the boundary conditions

$$\upsilon' = 0 \quad \text{at} \quad y = 0 \tag{28a}$$

and

$$\frac{\partial u'}{\partial y} = 0 \quad \text{at} \quad y = 0 \tag{28b}$$

At the free surface

$$\tau_{yx'} \equiv \mu\left(\frac{\partial u'}{\partial y} + \frac{\partial \upsilon'}{\partial x}\right) = 0 \quad \text{for} \quad y \approx h/2 \tag{28c}$$

A normal stress balance at the free surface leads to

$$\tau_{yy}' \equiv (\lambda + 2\mu)\frac{\partial \upsilon'}{\partial y} + \lambda\frac{\partial u'}{\partial x} = \gamma\frac{\partial^2 \varepsilon}{\partial x^2} \quad \text{for} \quad y \approx h/2 \tag{28d}$$

The capillary pressure caused by the curvature of the perturbed interface must be included in the normal stress balance at the free surface even though the deformation of the interface is taken into account in eqns. (25) and (26). This is because the approximate method of calculating ϕ' (eqn. (27)) by accounting only for the change of the interaction potential with the local film thickness neglects changes of the interaction potential with curvature.

5.2. Stability Analysis

Equations (25), (26), (28a)–(28d) and the explicit form of the perturbed interaction potential $\phi'(x,y,t)$ given by eqn. (27) describe the perturbation displacements in a thin solid film as a result of a small asymmetric disturbance applied to the film–gas interfaces. As is customary in a linear stability analysis, the disturbance is Fourier decomposed in the x direction and a single term is retained:

$$\begin{bmatrix} u' \\ \upsilon' \\ \varepsilon \end{bmatrix} = \begin{bmatrix} \hat{u}(y) \\ \hat{\upsilon}(y) \\ \hat{\varepsilon} \end{bmatrix} \exp(\mathrm{i}kx + \beta t) \tag{29}$$

where $\mathrm{i} = \sqrt{-1}$. The insertion of eqns. (29) into eqns. (25) and (26) and in the boundary conditions (28a)–(28d) gives

$$\frac{\mathrm{d}^2\hat{u}}{\mathrm{d}y^2} = \frac{ik\hat{\varepsilon}}{\mu}\left\{\frac{\partial \phi}{\partial (h'/2)}\right\}_{h/2} \tag{30}$$

and

$$\frac{\mathrm{d}}{\mathrm{d}y}\left[(\lambda+2\mu)\frac{\mathrm{d}\hat{u}}{\mathrm{d}y}+(\lambda+\mu)\mathrm{i}k\hat{u}-\hat{\varepsilon}\left\{\frac{\partial\phi}{\partial(h'/2)}\right\}_{h/2}\right]=0 \tag{31}$$

subject to the boundary conditions

$$\hat{\upsilon}=\frac{\mathrm{d}\hat{u}}{\mathrm{d}y}=0 \quad \text{at} \quad y=0 \qquad \text{(32a), (32b)}$$

$$\hat{\upsilon}=\frac{\mathrm{i}}{k}\frac{\mathrm{d}\hat{u}}{\mathrm{d}y} \quad \text{at} \quad y\approx h/2 \tag{32}$$

and

$$(\lambda+2\mu)\frac{\mathrm{d}\hat{u}}{\mathrm{d}y}+\lambda ik\hat{u}=-\gamma k^2\hat{\varepsilon} \quad \text{at} \quad y\approx h/2 \tag{32d}$$

Equations (30) and (31) can be solved subject to the boundary conditions (32a)–(32d) to give

$$\hat{u}(y)\approx\frac{\mathrm{i}\hat{\varepsilon}}{k}\left\{\psi+\frac{k^2}{\mu}\left(\sum_{j=1}^{3}\frac{G_j}{C_j^2}\left[\exp\left\{-C_j\left(\frac{h}{2}+y\right)\right\}+\exp\left\{-C_j\left(\frac{h}{2}-y\right)\right\}\right]-\right.\right.$$
$$\left.\left.-\frac{A_{22}}{12\pi}\left\{\frac{1}{(h/2+y)^2}+\frac{1}{(h/2-y)^2}\right\}\right)\right\} \tag{33}$$

and

$$\hat{\upsilon}(y)\approx\frac{\hat{\varepsilon}}{\lambda+2\mu}\left(B_1\frac{y}{h}-\sum_{j=1}^{3}\frac{G_j}{C_j}\left[\exp\left\{-C_j\left(\frac{h}{2}+y\right)\right\}-\exp\left\{-C_j\left(\frac{h}{2}-y\right)\right\}\right]+\right.$$
$$\left.+\frac{A_{22}}{6\pi}\left\{\frac{1}{(h/2+y)^3}-\frac{1}{(h/2-y)^3}\right\}+k^2\frac{(\lambda+\mu)}{\mu}\frac{A_{22}}{12\pi}\left\{\frac{1}{(h/2+y)}-\frac{1}{(h/2-y)}\right\}\right) \tag{34}$$

$$\psi=\frac{1}{\lambda}\left\{\gamma k^2+\sum_{j=1}^{3}G_j\exp(-C_ja)-\frac{A_{22}}{2\pi a^4}\right\} \tag{34a}$$

and

$$B_1=\frac{\lambda+3\mu}{\mu}\left\{\frac{A_{22}}{6\pi a^3}-\sum_{j=1}^{3}\frac{G_j}{C_j}\exp(-C_ja)\right\} \tag{34b}$$

Care must be taken when applying the boundary conditions at the free surface because terms of the form $(h/2-y)^{-n}$ arising from the interaction potential are singular at $y = h/2$. From a physical point of view these singularities are not real. They are caused by the inaccuracies of the expression of the interaction potential for small distances. To avoid this difficulty a cut-off of the interaction potential is taken at $y = h/2 - a$ where a is the atomic radius. Consequently the boundary conditions are evaluated at the center of the first row of atoms, *i.e.* at $y = h/2 - a$.

The assumption that the radius a of an atom in the film is much less than the film thickness h is always made when the expressions for $\hat{u}(y)$ and $\hat{\upsilon}(y)$ are evaluated at the boundaries. This allows us to simplify the constants calculated on the basis of the boundary conditions since terms of order a/h can be neglected compared with unity. Furthermore, since the wavelength $2\pi/k$ of the perturbation is much greater than the average film thickness ($hk/2\pi \ll 1$), terms of the order of $(ka)^2$ are neglected compared with unity. Similarly, terms of the form $(k/C_j)^2$ and $(C_j h)^{-1}$ which are of order $(ka_{22})^2$ and a_{22}/h respectively are also neglected compared with unity.

The kinematic equation (eqn. (7)) can now be used to determine the conditions under which a small periodic disturbance of the free surface will grow with time. Noting that ϕ_0 depends upon x through the perturbed film thickness so that

$$\frac{\partial^2\phi_0}{\partial x^2} \approx \frac{\partial^2\varepsilon}{\partial x^2}\left\{\frac{\partial\phi_0}{\partial(\mathrm{h}'/2)}\right\}_{\mathrm{h}/2}$$

and inserting u', υ' and ε' as given by eqn. (29), we find

$$\beta = -\frac{\theta k^2}{\hat{\varepsilon}-\hat{\upsilon}}\left[\gamma k^2\hat{\varepsilon}(1+\Phi)+ik\hat{u}(\sigma+\lambda\Phi)+\hat{\varepsilon}\left\{\frac{\partial\phi_0}{\partial(\mathrm{h}'/2)}\right\}_{h/2}\right] \quad \text{at} \quad y \approx h/2 \tag{35}$$

The expressions of $\hat{u}(h)$ and $\hat{\upsilon}(h)$ introduced in eqn. (35) yield the following relationship between the growth parameter β and the wavenumber k:

$$\beta = -\frac{\theta k^2}{\Delta}\left[\gamma k^2(1+\Phi)+\left\{\frac{\partial\phi_0}{\partial(\mathrm{h}'/2)}\right\}_{\mathrm{h}/2}-(\sigma+\lambda\Phi)\psi\right] \tag{36}$$

where

$$\Delta = 1+\frac{1}{\mu}\left\{\sum_{j=1}^{3}\frac{G_j}{C_j}\exp(-C_j a)-\frac{A_{22}}{6\pi a^3}\right\} \tag{36a}$$

Neglecting Φ compared with unity and using eqn. (14) for $\{\partial\phi_0/\partial(h'/2)\}_{h/2}$ we obtain

$$\beta = \frac{\theta k^2}{\Delta}\left\{\frac{A_{22}}{\pi h^4}+(\sigma+\lambda\Phi)\psi-\gamma k^2\right\} \tag{36b}$$

The growth parameter has a maximum for the dominant wavenumber k_d given by

$$k_d^{\,2} \approx \frac{1}{2\gamma}\left\{\frac{A_{22}}{\pi h^4}+(\sigma+\lambda\Phi)\psi\right\} \tag{37a}$$

The growth parameter β_m for the maximum rate of growth is therefore given by

$$\beta_m \approx \frac{\theta}{4\gamma\Delta}\left\{\frac{A_{22}}{\pi h^4}+(\sigma+\lambda\Phi)\psi\right\}^2 \tag{37b}$$

We define a critical pre-stress corresponding to neutral stability ($\beta = 0$). Equation (36b) can be used to obtain the following analytical expression for the dependence of the critical stress on the wavenumber:

$$\sigma_c = -\lambda\Phi + \frac{1}{\psi}\left(\gamma k^2 - \frac{A_{22}}{\pi h^4}\right) \tag{38}$$

The expression for ψ (eqn. (34a)) can be simplified by making use of an approximate expression for the surface free energy:

$$2\gamma \approx \frac{A_{22}}{12\pi(2a)^2}$$

This is obtained by observing that the surface free energy is simply one-half the energy gained in bringing two unit surfaces together from infinity, or

$$2\gamma = \int_{2a}^{\infty} \frac{A_{22}}{6\pi y^3}\,dy$$

where y is the distance between the two surfaces. Using this expression for the surface free energy in eqn. (34a) we obtain

$$\psi \approx \frac{1}{\lambda}\left[-\frac{A_{22}}{2\pi a^4}\left\{1 - \frac{(ka)^2}{48}\right\} + \sum_{j=1}^{3} G_j \exp(-C_j a)\right]$$

which, if we neglect the terms $(ka)^2$ compared with unity, leads to

$$\psi \approx \frac{1}{\lambda}\left\{\sum_{j=1}^{3} G_j \exp(-C_j a) - \frac{A_{22}}{2\pi a^4}\right\}$$

The term γk^2 in the expression of ψ is a direct result of the capillary pressure which is included in the normal stress balance to compensate for neglecting the effect of the curvature of the interface in the evaluation of the perturbed interaction potential (eqn. (27)). As shown above γk^2 can be neglected in the expression of ψ, and consequently the effect of the curvature on ϕ', u' and v' is negligible.

5.3. Results

Equation (38) which defines neutral stability was used to obtain values of the critical pre-stress (the value of σ for which $\beta = 0$) as a function of the perturbation wavenumber for a variety of film parameters. These results are presented in Figs. 3(a), 4(a), 5(a) and 6(a). The critical internal pre-stress is tensile when σ_c is positive and compressive when σ_c is negative. In the small wavenumber limit the critical pre-stress is independent of k and is given by

$$\sigma_{c,\min} = -\left(\frac{\lambda p_0}{\lambda + 2\mu} + \frac{A_{22}}{\psi\pi h^4}\right) \tag{39}$$

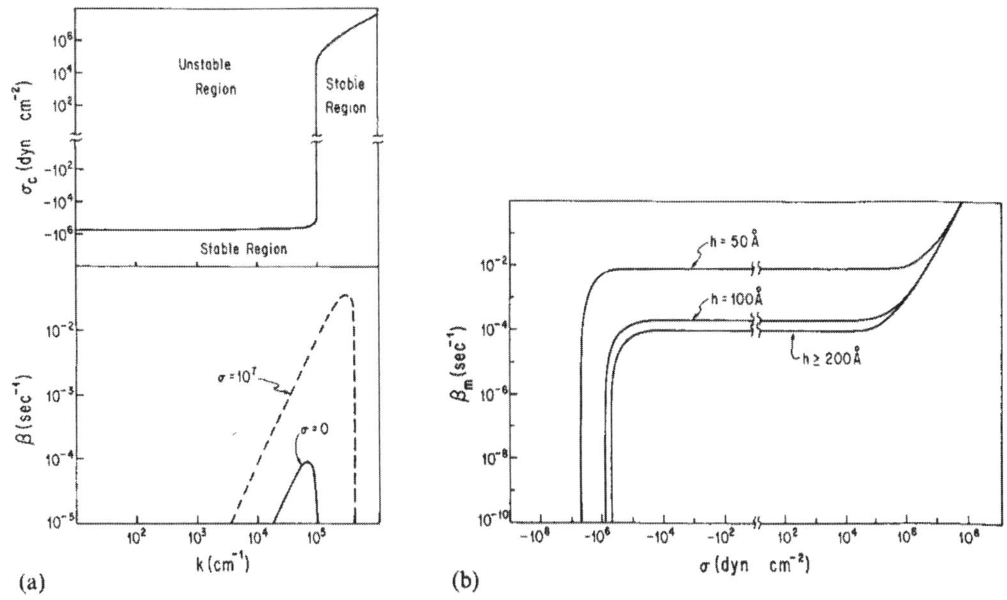

FIG. 3. (a) The critical internal pre-stress and the growth parameter as functions of the perturbation wavenumber for a free copper film 200 Å thick. Values of the parameters are given in Table 1. (b) β_m for a free copper film as a function of the internal pre-stress for various values of the film thickness. Values of the parameters are given in Table 1.

Since neutral stability ($\beta = 0$) represents the point where the perturbation neither grows nor decays with time, these curves separate the stable from the unstable regions. The growth parameter must be calculated to determine the unstable region where the perturbation grows with time (*i.e.* $\beta > 0$). For this purpose eqn. (36) is rewritten as

$$\beta = \frac{\theta k^2}{\Delta}(\sigma - \sigma_c)\psi \tag{40}$$

Since both ψ and Δ are positive quantities, the film is unstable when the internal pre-stress is greater than the critical pre-stress. The growth of the instability is dominated by the fastest growing perturbation. Therefore the inverse of the maximum value of the growth parameter (shown in Figs. 3(b), 4(b), 5(b) and 6(b) as a function of various physical parameters) gives an indication of the time needed for film rupture. Figure 3(b) shows β_m as a function of the internal pre-stress for various values of the film thickness h. As predicted by eqn. (40), an increase in the internal pre-stress above the critical pre-stress decreases the rupture time of the film. Thus Fig. 3(a) and (b) shows that a film 200 Å thick will rupture in approximately 10^4 s in the absence of an internal pre-stress, whereas a film of the same thickness with an internal pre-stress of 10^7 dyn cm^{-2} will rupture in only 40 s.

Figure 4(a) illustrates the effect of film thickness on the critical stress and on the growth parameter. A decrease in the film thickness results in a decrease in $\sigma_{c,\,min}$. For example, $\sigma_{c,\,min}$ decreases by an order of magnitude when the film thickness decreases from 200 to 50 Å. Equation (39) shows that this effect is due to the excess interaction potential ϕ_0 which becomes increasingly important with decreasing film thickness and which strongly affects the gradient of the chemical potential. Furthermore, eqn. (40) shows that a decrease in the average film thickness tends to destabilize the film since, at a fixed value of the internal pre-stress σ, a decrease in σ_c will result in a corresponding

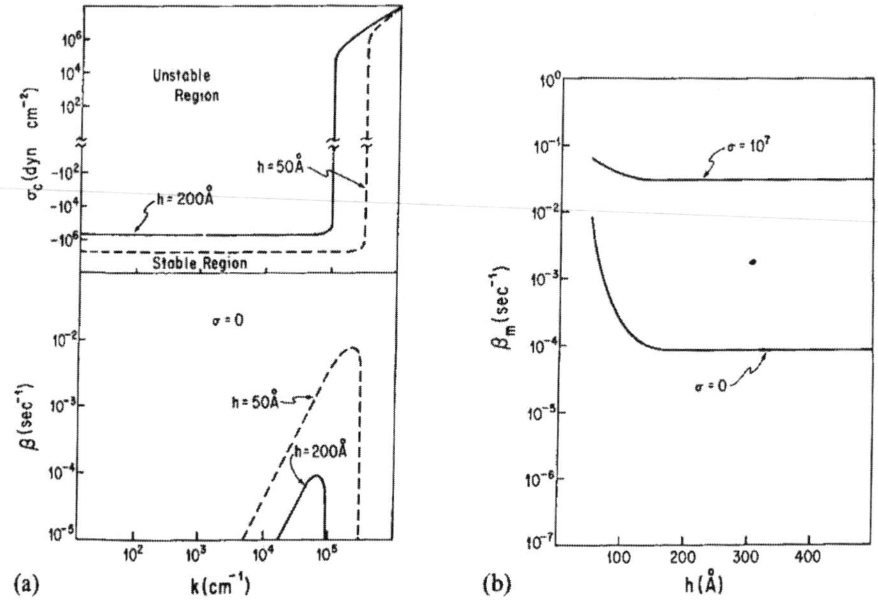

FIG. 4. (a) The critical internal pre-stress and the growth parameter as functions of wavenumber for a free copper film for various values of the film thickness. Values of the other parameters are given in Table 1. (b) β_m for a free copper film as a function of film thickness for various values of the internal pre-stress σ. Values of the parameters are given in Table 1.

TABLE 1

Typical Values of the Physical Parameters Used in Figs. 3–8 Except Where Otherwise Indicated

		Surface diffusion parameters	
λ	9.5×10^{11} dyn cm^{-2}		
μ	4.5×10^{10} dyn cm^{-2}		
a	1.28×10^{-8} cm	θ	1.87×10^{-27} s cm^{4}g^{-1}
γ	1500 dyn cm^{-1}	ν	1.93×10^{15} cm^{-2}
P_0	1.01×10^6 dyn cm^{-2} (1 atm)	V_0	1.18×10^{-23} cm^3
A_{22}	2.6×10^{-11} erg	D	1×10^{-10} cm^2 s^{-1}
Firsov a_{22} (Cu–Cu)	0.0960×10^{-8} cm	T	300 K

increase in β. Thus Fig. 4(a) shows that in the absence of an internal pre-stress ($\sigma = 0$) the rupture time decreases from 10^4 s ($\beta_m = 10^{-4}$ s^{-1}) to approximately 100 s ($\beta_m = 8 \times 10^{-3}$ s^{-1}) as the film thickness decreases from 200 to 50 Å. This result is presented graphically in Fig. 4(b) (see Fig. 3(b) also) where. β_m is plotted as a function of the film thickness for various values of the internal pre-stress. Figure 4(b) shows that the effect of film thickness on β_m becomes less pronounced at large values of the internal pre-stress. In this case the growth of the perturbation is dominated by the strain energy contribution to the chemical potential rather than by the excess interaction potential ϕ_0 (see eqn. (37b) for β_m).

Figure 5(a) which presents the critical pre-stress and the growth parameter as functions of wavenumber shows that an increase in the external pressure on a film 200 Å thick from 1 to 10 atm results in a corresponding order-of-magnitude decrease in $\sigma_{c,\,min}$ (from -5×10^5 to -5×10^6 dyn cm^{-2}). Equation (39) shows that for the relatively thick films considered in Fig. 5(a) the excess interaction

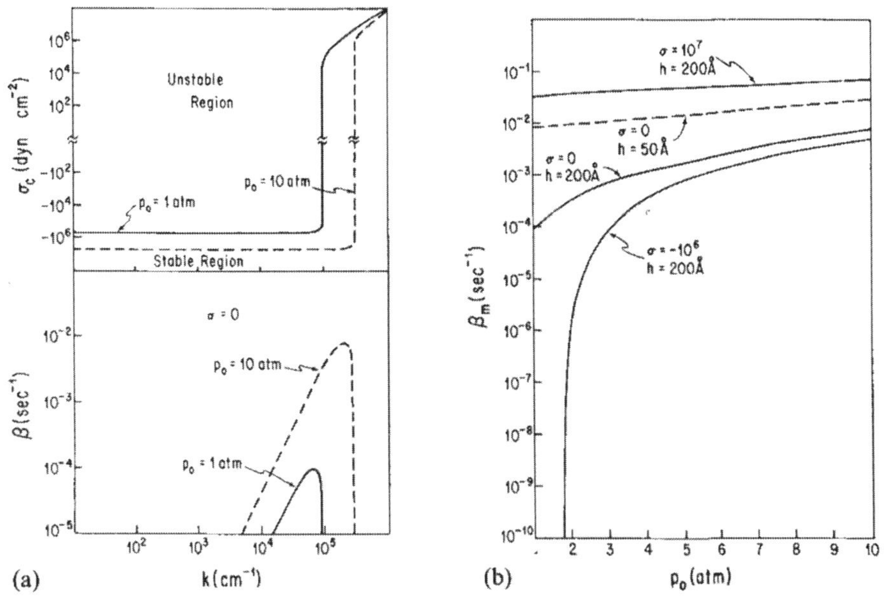

FIG. 5. (a) The critical internal pre-stress and the growth parameter as functions of the perturbation wavenumber for various values of the external pressure p_0 for a free copper film 200 Å thick. Values of the other parameters are given in Table 1. (b) β_m for a free film as a function of the external pressure p_0 for various values of the internal pre-stress and film thickness. Values of the other parameters are shown in Table 1.

potential ϕ_0 is small and $\sigma_{c,\,min}$ is proportional to $-\,p_0$. The destabilization of the film with increasing external pressure is illustrated in Fig. 5(b) where β_m is plotted as a function of the external pressure for various values of the internal pre-stress and film thickness. Equation (37b) for β_m shows that the effect of pressure is greatest for thick films and small internal pre-stresses. From Fig. 5(b) it can be seen that a film 200 Å thick with a compressive internal pre-stress of 10^6 dyn cm^{-2} will rupture in approximately 100 s at an external pressure of 10 atm but will be stable at atmospheric conditions. Equation (37b) also shows that for very thin films the effect of the external pressure is dominated by the excess interaction potential contribution to the chemical potential. Indeed, from Fig. 5(b) we can evaluate for a film 50 Å thick with $\sigma = 0$ that β_m only increases from 8×10^{-3} to 1.8×10^{-2} s^{-1} as the external pressure increases from 1 to 10 atm. In contrast β_m increases by more than one order of magnitude (from 10^{-4} to 8×10^{-3} s^{-1}) for a film 200 Å thick with $\sigma = 0$ subjected to the same pressure change.

Figure 6(a), which illustrates the effect of the surface free energy on the critical pre-stress, shows that the surface free energy has no effect on $\sigma_{c,min}$. This is because $\sigma_{c,min}$ represents the short wavenumber (long wavelength) limit of the critical pre-stress where the curvature of the interface, and therefore the capillary pressure, is negligible. However, Fig. 6(a) does show that an increase in the surface free energy stabilizes the film (decreasing β_m) in the large wavenumber region (short wavelength) where the capillary pressure is important. For example, the rupture time for a film 200 Å thick increases from approximately 10^3 to 10^4 s as the surface free energy changes from 200 to 1500 dyn cm^{-1}. This is shown more clearly in Fig. 6(b) which presents β_m as a function of the surface free energy for various values of the film thickness. These results agree with the. approximate expression for β_m (eqn. (37b)) which predicts that β_m is proportional to γ^{-1}.

The dominant wavelength Λ_d $(= 2\pi/k_d)$, which is defined as the wavelength of the perturbation which gives rise to the fastest growing perturbation, can be calculated from eqn. (37a). Since the instability is dominated by the fastest growing perturbation, the dominant wavelength Λ_d may give some approximate information about the average size of the fragments that result from film rupture.

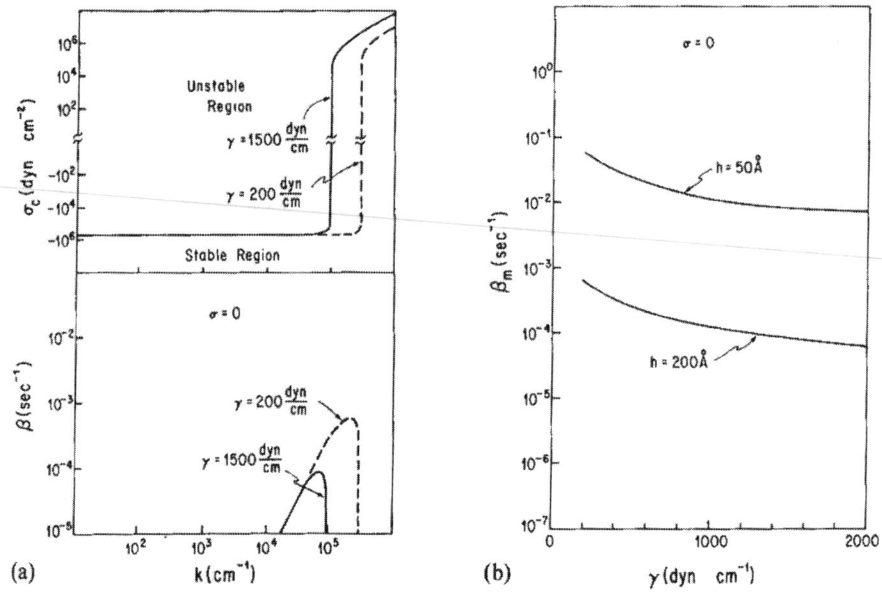

FIG. 6. (a) The critical internal pre-stress and the growth parameter as functions of the perturbation wavenumber for a copper film 200 Å thick for various values of the surface free energy. Values of the other parameters are given in Table 1. (b) β_m for a free film as a function of the surface free energy for various film thicknesses and for $\sigma = 0$. Values of the other parameters are presented in Table 1.

Equation (37a) predicts that the average size of the fragments after film rupture decreases with decreasing film thickness and increasing internal pre-stress. Thus Fig. 4(a) shows that in the absence of an internal pre-stress a copper film 50 Å thick will rupture and form fragments of the order of 2000 Å, whereas a film 200 Å thick will form fragments of approximately 6000 Å. Figure 3(a) shows that for a copper film 200 Å thick an increase of the internal pre-stress from $\sigma = 0$ to $\sigma = 10^7$ dyn cm^{-2} decreases the average fragment size from 6×10^{-5} to approximately 10^{-5} cm.

The effect of the surface diffusion parameter θ is not presented graphically because eqns. (36) and (37b) show that both β and β_m are proportional to θ.

6. THIN FILM ON A RIGID SUBSTRATE

6.1. Equations

We shall now consider the second limiting case, *i.e.* that of a thin film on a rigid substrate. In this case perturbations originate only at one interface—the free surface. Again the simplified differential equations describing the growth of perturbations in the film are

$$\mu \frac{\partial^2 u'}{\partial y^2} = \frac{\partial \phi'}{\partial x} \tag{41}$$

and

$$\frac{\partial}{\partial y}\left\{(\lambda + 2\mu)\frac{\partial v'}{\partial y} + (\lambda + \mu)\frac{\partial u'}{\partial x} - \psi' u'\right\} = 0 \tag{42}$$

where y is now the distance measured from the film–substrate interface.

Because the film is assumed to be in contact with a rigid substrate, both the x and y components of the displacement must be zero at this interface, *i.e.*

$$u' = \upsilon' = 0 \quad \text{at} \quad y = 0 \tag{43a),(43b}$$

The boundary conditions at the free surface are a tangential stress balance

$$\mu\left(\frac{\partial u'}{\partial y} + \frac{\partial \upsilon'}{\partial x}\right) = 0 \quad \text{at} \quad y \approx h \tag{43c}$$

together with the normal stress balance

$$(\lambda + 2\mu)\frac{\partial \upsilon'}{\partial y} + \lambda\frac{\partial u'}{\partial x} = \gamma\frac{\partial^2 \varepsilon}{\partial x^2} \quad \text{at} \quad y \approx h \tag{43d}$$

and the kinematic condition

$$\frac{\partial \varepsilon}{\partial t} = \frac{\partial \upsilon'}{\partial t} + \theta\left\{(\sigma + \lambda\Phi)\frac{\partial^3 u'}{\partial x^3} - \gamma(1+\Phi)\frac{\partial^4 \varepsilon}{\partial x^4} + \frac{\partial^2 \varepsilon}{\partial x^2}\left(\frac{\partial \phi_0}{\partial h'}\right)_h\right\} \quad \text{at} \quad y \approx h \tag{43e}$$

In eqn. (43e)

$$\phi_0 \approx \phi_0' + \frac{A_{22} - A_{32}}{6\pi h'^3}$$

The perturbation component of the interaction potential may again be found by expanding the interaction potential about the average film thickness and by retaining terms up to order ε to give

$$\phi' = \varepsilon(x,t)\left(\frac{\partial \phi_0}{\partial h'}\right)_h = \varepsilon\left[-\frac{A_{22}}{2\pi(h-y)^4} + \sum_{j=1}^{3} G_j \exp\{-C_j(h-y)\}\right] \tag{44}$$

where $h' = h + \varepsilon$. A periodic disturbance in the x direction of the form

$$\begin{bmatrix} u' \\ \upsilon' \\ \varepsilon \end{bmatrix} = \begin{bmatrix} \hat{u}(y) \\ \hat{\upsilon}(y) \\ \hat{\varepsilon} \end{bmatrix} \exp(\mathrm{i}kx + \beta t) \tag{45}$$

is assumed. Inserting eqn. (45) into eqns. (41) and (42) and using the boundary conditions (43a)–(43d) we obtain

$$\hat{u}(y) \approx \mathrm{i}\hat{\varepsilon}\left(\psi yhk + \frac{k}{\mu}\sum_{j=1}^{3}\frac{G_j}{C_j^2}\left[\exp\{-C_j(h-y)\} - \exp(-C_j h)\right] - \right.$$

$$\left. - \frac{kA_{22}}{12\pi\mu(h-y)^2}\left\{1 - \left(1 - \frac{y}{h}\right)^2\right\}\right) \tag{46}$$

and

$$\hat{\upsilon}(y) \approx \frac{\hat{\varepsilon}}{\lambda+2\mu}\left(-\psi(\lambda+2\mu)y\left\{1+k^2h(h-y)\frac{\lambda+\mu}{2(\lambda+2\mu)}\right\}+\sum_{j=1}^{3}\frac{G_j}{C_j}\left[\exp\{-C_j(h-y)\}\right.\right.$$
$$\left.\left.-\exp(-C_jh)\right]-\frac{A_{22}}{6\pi(h-y)^3}\left\{1-\left(1-\frac{y}{h}\right)^3+yk^2(h-y)^2\frac{\lambda+\mu}{2h\mu}\right\}\right) \tag{47}$$

where

$$\psi=\frac{1}{\lambda+2\mu+(\lambda-\mu)(hk)^2/2}\left\{\gamma k^2+\sum_{j=1}^{3}G_j\exp(-C_ja)-\frac{A_{22}}{2\pi a^4}\right\} \tag{47a}$$

Again, as in the free-film case, when applying the boundary conditions at the free surface, terms of the form $(h-y)^{-n}$ arising from the interaction potential must be evaluated at $y = h - a$. Terms of the order a/h, $(ka)^2$ and $(k/C_j)^2$ are always neglected compared with terms of order unity.

As in the free-film case, the kinematic boundary condition (43e) is used to yield an explicit expression for the growth parameter β as a function of the wavenumber k and various physical parameters. We obtain

$$\beta=\frac{\theta k^2}{\Delta}(\sigma+\lambda\Phi)(kh)^2\psi-\left(\frac{\partial\phi_0}{\partial h'}\right)_h-\gamma k^2(1+\Phi) \tag{48}$$

or, neglecting Φ with respect to unity,

$$\beta\approx\frac{\theta k^2}{\Delta}\left\{(\sigma+\lambda\Phi)(kh)^2\psi+\frac{A_{22}-A_{32}}{2\pi h^4}-\gamma k^2\right\} \tag{48a}$$

where

$$\Delta=1+\psi h \tag{48b}$$

The critical pre-stress is obtained directly from eqn. (48) as

$$\sigma_c=-\lambda\Phi+\frac{1}{\psi(kh)^2}\left\{\gamma k^2(1+\Phi)-\frac{A_{22}-A_{32}}{2\pi h^4}\right\} \tag{49}$$

6.2. Results

Using eqn. (49) we calculated the critical internal pre-stress as a function of the physical parameters. The results of these calculations are presented in Figs. 7(a), 8(a), 9(a), 10(a), 11,12 and 13. Figure 7(a) which illustrates the critical pre-stress for a film 500 Å thick shows that, depending on the sign of the quantity $A_{22} - A_{32}$, the shapes of the neutral stability curves are fundamentally different. When the constant A_{22} (due to the attractive interactions between the atoms of the film) is greater than A_{32} (due to the attractive interactions between the atoms of the film and the atoms of the substrate) the critical stress is negative (compressive) at small wavenumbers and positive (tensile) at relatively large wavenumbers. However, when A_{22} is smaller than A_{32}, the critical stress is always tensile and has a minimum value $\sigma_{c,\,min}$. The growth parameter is presented as a function of the wavenumber in Fig. 7(a) which shows that, similar to a free film, the supported film is always unstable when the internal pre-stress is greater than the critical pre-stress. This result can be derived analytically by

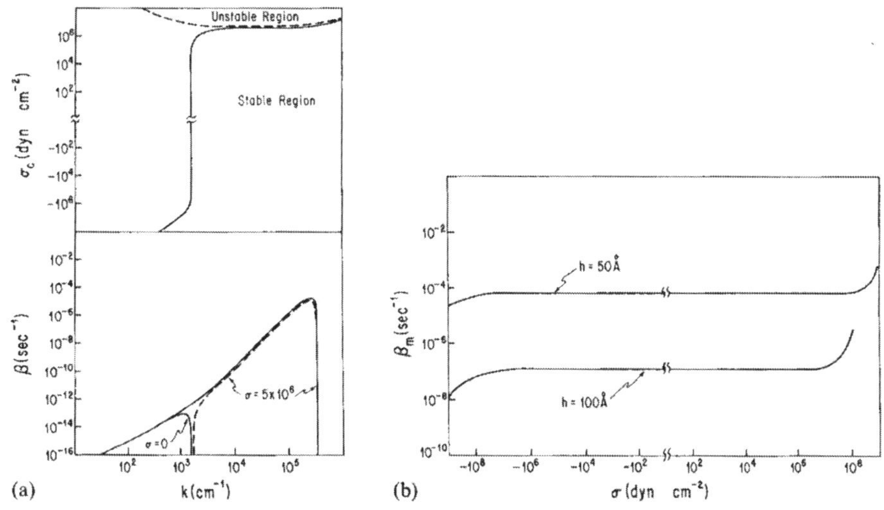

FIG. 7. (a) The critical internal pre-stress and the growth parameter for a supported copper film 500 Å thick. $A_{22} - A_{32} = 1.6 \times 10^{-11}$ erg for the full curves while $A_{22} - A_{32} = -4 \times 10^{-12}$ erg for the broken curves. Values of the other parameters are given in Table 1. (b) β_m as a function of the internal pre-stress for various film thicknesses and for $A_{22} - A_{32} = 1.6 \times 10^{-11}$ erg. Values of the other parameters are given in Table 1.

observing that the growth parameter can be written in terms of the difference between the internal pre-stress and the critical pre-stress as

$$\beta = \frac{\theta k^2}{\Delta}(\sigma - \sigma_c)\psi \tag{50}$$

Equation (50) is analogous to expression (40) for the free-film case, but ψ and Δ are now defined by eqns. (47a) and (48b). In Fig. 7(b) β_m is presented as a function of σ for various values of the film thickness. Because the critical stress is strongly dependent on the sign of $A_{22} - A_{32}$ the results of the two possible cases will be examined separately.

6.2.1. *Case 1:* $A_{22} > A_{32}$

Let us consider the case where the attractive constant A_{22} is greater than A_{32} (Figs. 8–10). In this case the kinematic boundary condition (eqn. (7)) is similar but not the same as that for the free-film case (where $A_{32} \equiv 0$). Therefore the dependence of the critical pre-stress on wavenumber is qualitatively similar to that for the free- film dependence. Figure 8(a), which presents the critical pre-stress and the growth parameter as functions of wavenumber for various values of the film thickness, shows that a decrease in film thickness shifts the critical pre-stress towards longer wavenumbers. However, the effect of the film thickness on rupture time (β_m^{-1}) is relatively complicated compared with the free-film counterpart. β_m is represented as a function of film thickness for various values of the internal pre-stress in Fig. 8(b). For very thin films the initial disturbance grows primarily because of the contribution of ϕ_0 to the chemical potential. The strain energy contribution to the chemical potential is always negligible in this region because elastic displacements are strongly damped by the rigid substrate. Therefore for very thin films eqn. (48) leads to

$$\beta \approx \frac{\theta k^2}{\Delta}\left(\frac{A_{22} - A_{32}}{2\pi h^4} - \gamma k^2\right) \tag{51}$$

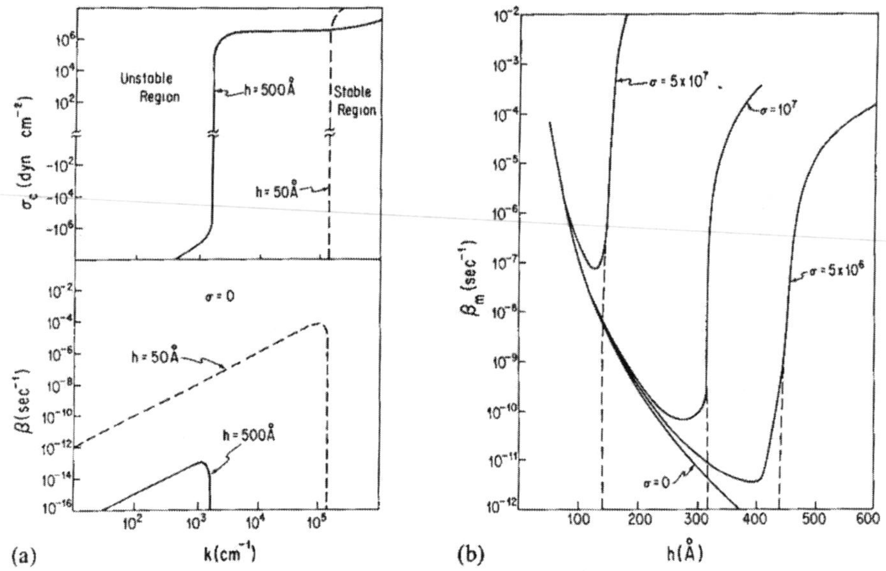

FIG. 8. (a) The critical internal pre-stress and the growth parameter as functions of the perturbation wavenumber for supported copper films with $A_{22} - A_{32} = 1.6 \times 10^{-11}$ erg. Values of the other parameters are given in Table 1. (b) β_m as a function of film thickness at various internal pre-stresses for a supported copper film. $A_{22} - A_{32} = 1.6 \times 10^{-11}$ erg for the full curves while $A_{22} - A_{32} = -4 \times 10^{-12}$ erg for the broken curves. Values of the other parameters are shown in Table 1.

and β_m decays approximately with the eighth power of the film thickness. At relatively large thicknesses the interaction potential ϕ_0 is negligible compared with the strain energy, and it is the latter effect which through surface diffusion causes growth of the disturbance. Thus for relatively thick films

$$\beta \approx \frac{\theta k^4}{\Delta}\left\{(\sigma + \lambda\Phi)h^2\psi - \gamma\right\} \tag{52}$$

and β_m increases with film thickness. Thus Fig. 8(b) shows that β_m is large for both very thin and thick films with a minimum at intermediate thicknesses. Furthermore, increasing the internal pre-stress shifts the minimum to smaller film thicknesses. A comparison of Figs. 7(b) and 8(b) with Figs. 3(b) and 4(b) shows that, in the range where both the free film and the corresponding (the same value of σ, h etc.) supported film are unstable, the supported film always has a longer rupture time than the free film. This is due to the stabilizing effect of the rigid substrate.

Figure 9(a) and (b) shows the effect of the external pressure on the critical pre-stress and on β_m. The external pressure has the greatest effect on the rupture time when the intrinsic pre-stress is small and the film thickness is relatively large (see eqn. (52)). Thus Fig. 9(b) shows that an increase in the external pressure over a film 500 Å thick (with $\sigma = 0$) from 1 to 10 atm results in a decrease of the rupture time from approximately 10^{13} to 10^5 s. However, the same pressure change produces virtually no change in the rupture time of a film 50 Å thick. Again, this dramatic change in the effect of the external pressure with film thickness is due to the stabilizing influence of the rigid substrate boundary conditions.

In Fig. 10(a) the critical pre-stress is plotted against the wavenumber for two values of the surface free energy. An effect only exists in the relatively large wavenumber region where the capillary

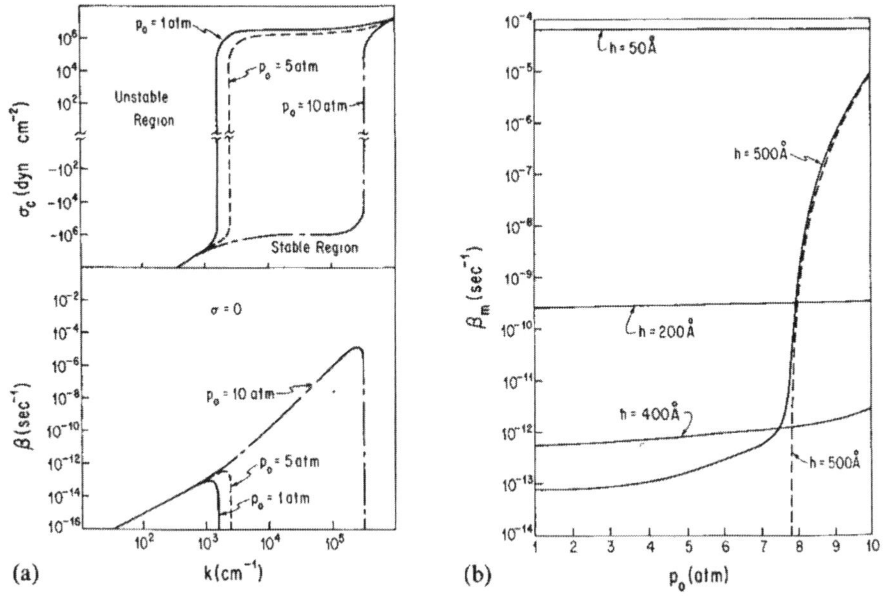

FIG. 9. (a) The critical internal pre-stress and the growth parameter at various values of the external pressure for a supported copper film 500 Å thick with $A_{22} - A_{32} = 1.6 \times 10^{-11}$ erg. Values of the other parameters are given in Table 1. (b) β_m as a function of the external pressure for various film thicknesses. $A_{22} - A_{32} = 1.6 \times 10^{-11}$ erg and $\sigma = 0$ for the full curves while $A_{22} - A_{32} = -4 \times 10^{-12}$ erg and $\sigma = 10^6$ dyn cm^{-2} for the broken curve. Values of the other parameters are presented in Table 1.

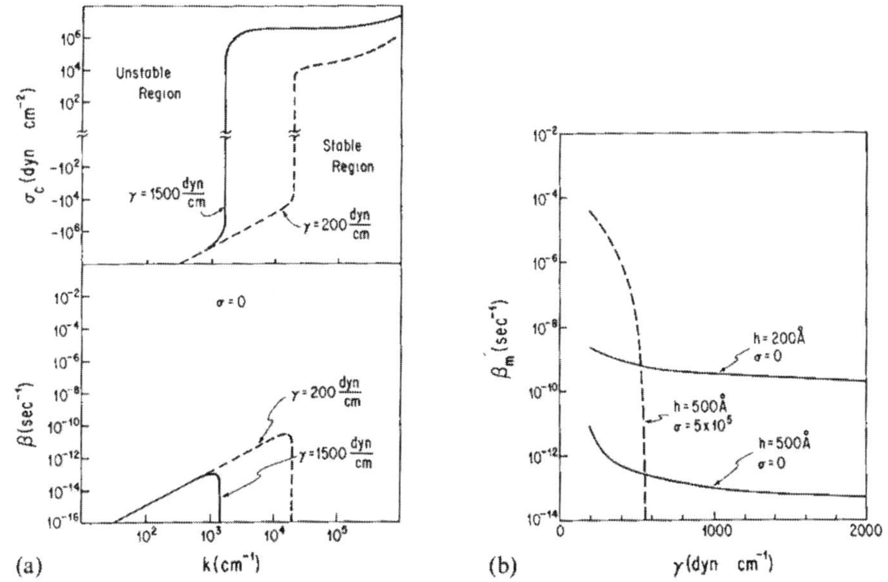

FIG. 10. (a) The critical internal pre-stress and the growth parameter at various values of the surface free energy for a supported copper film 500 Å thick with $A_{22} - A_{32} = 1.6 \times 10^{-11}$ erg. Values of the other parameters are given in Table 1. (b) β_m as a function of the surface free energy for various values of the film thickness and internal pre-stress for a supported copper film. $A_{22} - A_{32} = 1.6 \times 10^{-11}$ erg for the full curves while $A_{22} - A_{32} = -4 \times 10^{-12}$ erg for the broken curve. Values of the other parameters are presented in Table 1.

pressure is important relative to the strain energy. Figure 10(b) shows that the effect of the surface free energy on β_m is greater when the film thickness is large. For example, for a film 500 Å thick with $\sigma = 0$, β_m decreases from 8×10^{-12} to 6×10^{-14} s^{-1} as γ increases from 200 to 2000 dyn cm^{-1}; however, the same increase in γ for a film 200 Å thick only changes β_m from 2×10^{-9} to 2×10^{-10} s^{-1}.

6.2.2. *Case 2: $A_{22} < A_{32}$*

Figure 11 shows the critical pre-stress and the growth parameter as functions of film thickness for $A_{32} > A_{22}$. In this case the critical pre-stress is always tensile (positive) with a minimum value $\sigma_{c,\,min}$. Again, the film is unstable when the internal pre-stress is greater than the critical pre-stress. Figure 11 shows that a decrease in film thickness increases $\sigma_{c,\,min}$, thereby stabilizing the film (see eqn. (50)). Figure 8(b) which presents β_m as a function of film thickness for various values of the internal pre-stress shows that a decrease in film thickness decreases β_m and hence increases the rupture time. The sharp decrease in β_m shown in Fig. 8(b) occurs for two reasons: (1) when $A_{32} > A_{22}$ the gradient of the excess interaction potential, which becomes increasingly important with decreasing film thickness, stabilizes the film and (2) because of the stabilizing effect of the rigid boundary on the gradient of the strain energy. Thus Fig. 8(b) shows that, while a supported film 400 Å thick with an internal pre-stress of 10^7 dyn cm^{-2} ruptures in approximately 10^4 s, a film 300 Å thick is stable. For large film thicknesses Fig. 8(b) shows that, because the gradient of the excess interaction potential is small compared with the strain energy, β_m becomes independent of $A_{32} - A_{22}$.

Figure 12 shows the effect of the external pressure on the critical pre-stress for a film 500 Å thick. An increase in the external pressure from 1 to 10 atm decreases $\sigma_{c,\,min}$ from 3.6×10^6 to 10^6 dyn cm^{-2}. The broken curve in Fig. 9(b), which presents β_m as a function of the external pressure for a supported film 500 Å thick (with $A_{32} = 3 \times 10^{-11}$ erg and $A_{22} = 2.6 \times 10^{-11}$ erg), shows that an increase in the external pressure decreases the rupture time of the film. For example, Fig. 9(b) shows that, while a film 500 Å thick with $\sigma = 10^6$ dyn cm^{-2} under an external pressure of 10 atm will rupture in approximately 10^5 s, the film will be stable if the pressure is less than approximately 7.7 atm.

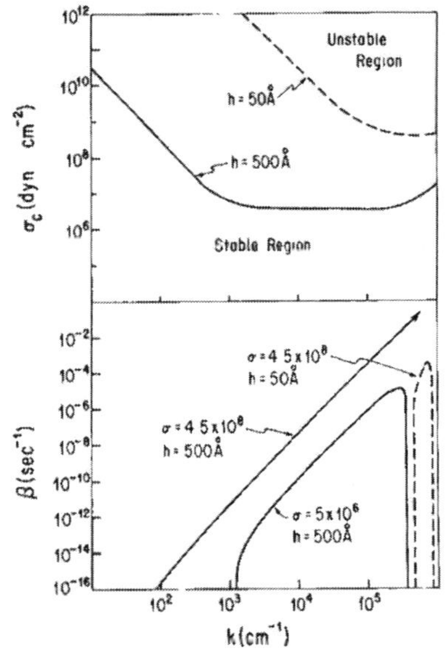

FIG. 11. The critical internal pre-stress and the growth parameter as functions of wavenumber for various film thicknesses for a supported copper film with $A_{22} - A_{32} = -4 \times 10^{-12}$ erg. Values of the other parameters are presented in Table 1.

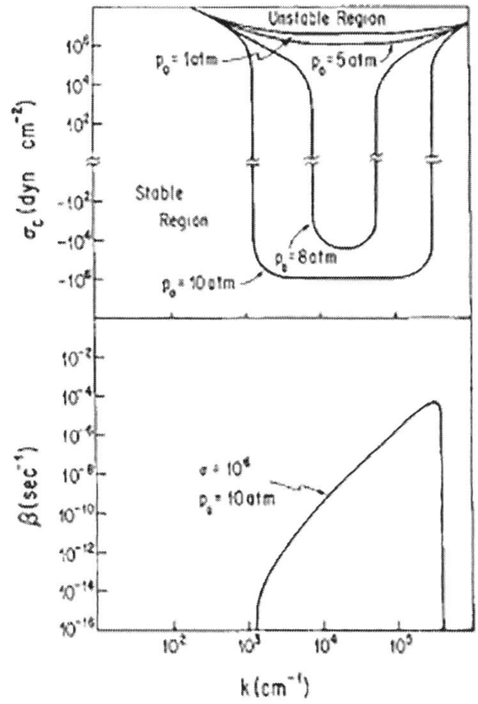

FIG. 12. The critical internal pre-stress and the growth parameter as functions of wavenumber for various values of the external pressure for a supported copper film 500 Å thick with $A_{22} - A_{22} = -4 \times 10^{-12}$ erg. Values of the other parameters are presented in Table 1.

Figure 13 shows that decreasing the surface free energy decreases both $\sigma_{c,\,min}$ and the rupture time. For the film 500 Å thick presented in Fig. 13 $\sigma_{c,\,min}$ decreases by two orders of magnitude (from 3.7×10^6 to 3.7×10^4 dyn cm^{-2}) as the surface free energy changes from 1500 to 200 dyn cm^{-1}. Representative values of the growth parameter for the above values of the surface free energy are also presented in Fig. 13. Figure 10(b) shows that β_m decreases rapidly with increasing surface free energy. For example, a film 500 Å thick with an internal pre-stress of 5×10^5 dyn cm^{-2} is always stable when the surface free energy is greater than 600 dyn cm^{-1}; the same film will rupture in approximately 10^4 s if the surface free energy is decreased to 200 dyn cm^{-1}.

A dominant wavenumber exists for the supported film. A simple analytical expression can be obtained only when the strain energy is small compared with the excess interaction potential. In this case

$$k_d \approx \left\{ -\frac{1}{2\gamma} \left(\frac{\partial \phi_0}{\partial h'} \right)_h \right\}^{1/2}$$

However, a numerical solution has to be used to calculate k_d when the strain energy is large compared with the excess interaction potential. Figure 7(a) shows that the dominant wavelength decreases with increasing internal pre-stress, both when $A_{22} > A_{32}$ and when $A_{22} < A_{32}$. For example, an increase of the internal pre-stress from zero to 5×10^6 dyn cm^{-2} in a supported film 500 Å thick with $A_{22} - A_{32} = 1.6 \times 10^{-11}$ erg decreases Λ_d from approximately 6×10^{-3} to 3×10^{-5} cm. Figure 8(a) shows that Λ_d decreases when the film thickness decreases. Thus, while $\Lambda_d \approx 6 \times 10^{-3}$ cm for a film 500 Å thick with zero pre-stress and $A_{22} - A_{32} = 1.6 \times 10^{-11}$ erg, $\Lambda_d \approx 6 \times 10^{-5}$ cm when the film is only 50 Å thick.

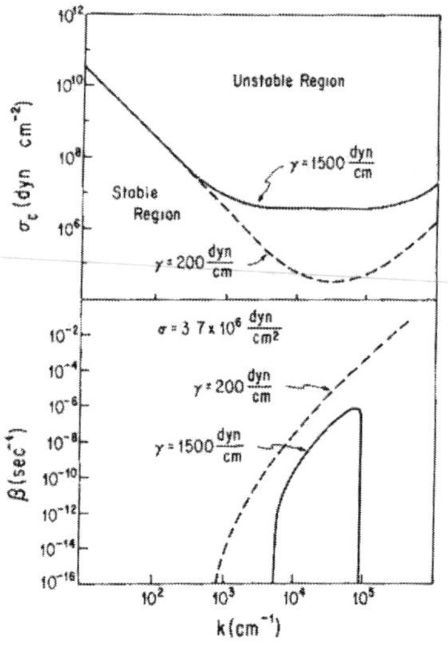

FIG. 13. The critical internal pre-stress and the growth parameter as functions of wavenumber for various values of the surface free energy for a supported copper film 500 Å thick with $A_{22} - A_{32} = -4 \times 10^{-12}$ erg. Values of the other parameters are presented in Table 1.

7. CONCLUSIONS

Equation (3a) shows that the growth of the free interface is caused essentially by surface diffusion and is proportional to the surface Laplacian of the chemical potential. Consequently, it is possible to increase the lifetime of thin film devices by affecting surface diffusion. This may be accomplished either by altering the surface diffusion parameter θ or by changing the chemical potential at the free surface of the film.

The surface diffusion coefficient, and hence the surface diffusion parameter, can be altered either by affecting the surface properties of the film through the presence of adsorbates and/or impurities or by changing the temperature. Impurities can either inhibit or enhance surface diffusion depending on the properties of the impurity. Experimental evidence suggests that impurities may be divided into two categories: (1) those which decrease the rate of surface diffusion frequently have a higher melting point (cohesive energy) than the film and (2) those which increase the surface diffusion coefficient generally have a lower melting point than the film[15]. The effect of surface impurities may be quite striking. For example, the surface diffusion coefficient of copper can be enhanced by four orders of magnitude in the presence of chlorine[16]. Similarly, the surface diffusion coefficient, and therefore β_m, can be increased by two orders of magnitude for silver by introducing oxygen into the surroundings[17].

The surface diffusion coefficient is also sensitive to temperature. An increase in temperature results in an increase of the surface diffusion coefficient. Empirical correlations for f.c.c. metals have an Arrhenius form given by[18]

$$D = 740\exp\left(-\frac{\varepsilon_{act,1}T_m}{RT}\right) \quad \text{for} \quad 0.77 < \frac{T}{T_m} < 1$$

and

$$D = 0.014\exp\left(-\frac{\varepsilon_{act,2}T_m}{RT}\right) \quad \text{for} \quad \frac{T}{T_m} < 0.77$$

where T_m is the melting point, $\varepsilon_{act,1} = 30$ cal mol^{-1} K^{-1} and $\varepsilon_{act,2} = 13$ cal mol^{-1} K^{-1}. Therefore the surface diffusion coefficient D of copper decreases by approximately four orders of magnitude from 10^{-3} to 10^{-7} cm^2 s^{-1} as T/T_m decreases from 1 to 0.5.

The growth or decay of the disturbance may also be affected by variations of the chemical potential at the free surface. The chemical potential (eqn. (2b)) contains a surface free energy (capillary pressure) term, the excess interaction potential at the free surface ϕ_0 and a strain energy contribution. The surface free energy opposes both the bending of the interface and the growth of the perturbation, and hence its increase has a stabilizing effect. The term $(\partial\phi_0/\partial h')_h$ in eqns. (36) and (48) arises from the kinematic condition which relates the growth of the perturbation at the free interface to the surface Laplacian of the chemical potential. Because the chemical potential contains the interaction potential ϕ_0, its surface Laplacian contains a term of the form

$$\frac{\partial^2\phi_0}{\partial x^2} = \frac{\partial^2\varepsilon}{\partial x^2}\left(\frac{\partial\phi_0}{\partial h'}\right)_h$$

(to order ε). When A_{22} is greater than A_{32} the variation of the interaction potential at the free surface with position causes the surface diffusion to amplify the initial disturbance; however, when A_{32} is larger than A_{22} the variation of the interaction potential opposes the growth of the initial disturbance. Thus Figs. 4(b) and 8(b) show that for very thin films, where the effect of the strain energy is small, β_m increases with the square of the gradient $(\partial\phi_0/\partial h')_h$ which is proportional to $1/h^4$. However, when A_{32} is greater than A_{22} a decrease in film thickness tends to stabilize the film (Fig. 8(b)). Strain energy contribution to the chemical potential depends on both the steady state and perturbation components of the displacement. Thus while the steady state strain energy is always positive, since it may be expressed as a sum of squares, the strain energy caused by a perturbation can be either positive or negative because it contains terms of the form $e_{ij}^s\, e_{ij}'$. Then, depending on the signs and relative magnitudes of the pre-existing internal stress and the perturbation displacements in the film, surface diffusion can either stabilize or destabilize the film. The term with the form

$$(\sigma + \lambda\Phi)\left\{\sum_{j=1}^{3}\text{constant}\times G_j\exp(-C_j a) - \frac{A_{22}}{\pi a^4}\right\} \tag{53}$$

in the expression for the growth parameter (eqns. (36) and (48)) is due to the gradient of the perturbed strain energy. Expression (53) is a product of two terms: the first, $\sigma + \lambda\Phi$, is related to the steady state stresses acting at the surface of the film; the second (the term in braces) represents the effect of a perturbation in the free surface of the film on the interaction forces acting on a volume element near the perturbed surface. The origin of this perturbation term can be understood physically by comparing the forces acting at the same point $y = h - a$ in both the perturbed and unperturbed states. The forces acting at the free surface of the unperturbed film ($y = h - a$) will differ from the forces acting at the same point in the corresponding perturbed film since a perturbation changes the distance between the chosen point ($y = h - a$) and the interface ($y = h + \varepsilon - a$). Furthermore, the difference in forces is independent of film thickness. Hence expression (53) shows that interaction forces may affect the stability of even very thick stressed films where such forces have generally been ignored.

The internal pre-stress σ affects the growth of the perturbation at the free surface via the contribution of the steady state stress to the perturbed strain energy component of the chemical potential. The influence of the internal stress on the rupture time of the film is particularly important because it may be changed by several orders of magnitude simply by controlling the deposition conditions that occur during film formation. The most straightforward method of varying the internal stress is through changes in the thermal stress contribution to the total preexisting internal stress.

Thermal stresses arise when the film and substrate have different thermal expansion coefficients and the system is maintained at a temperature that is different from that at deposition. The thermal stress contribution to the total internal stress can be calculated from the relation[4]

$$\Delta\sigma_{\text{thermal}} = (\alpha_{\text{f}} - \alpha_{\text{s}})\Delta T \frac{E}{1-\chi}$$

where α is the coefficient of thermal expansion with subscript f referring to the film and s to the substrate, E is Young's modulus, χ is Poisson's ratio and ΔT the temperature change.

It is also possible to change the internal stress through changes in the intrinsic stress. One method of achieving this is to create (or to remove) surface layers such as oxides. Large stresses will arise when the surface layer has a lattice parameter which is substantially different from that of the substrate. Tensile stresses of approximately 2×10^9 dyn cm^{-2} result when layers of aluminum oxide are prepared by anodizing aluminum[19]. Crystal defects in the film may also alter the internal stress when their presence contracts or expands adjacent areas of the film. For instance, the presence of cation interstitials will result in an expansion of the film while cation vacancies will result in a contraction of the film. The stress resulting from a defect concentration of $c(x)$ defects cm^{-3} in the film is approximately given by[20]

$$\Delta\sigma = ElRc(x)$$

where R is the volume of the perfect crystal lattice, l is the fractional change in the volume R which results from the presence of a defect in the crystal lattice and E is the Young's modulus of the film.

Rejuvenation of a supported platinum catalyst on an alumina substrate by the splitting of the sintered crystallites into smaller ones has been observed in an oxygen environment at a temperature between 450 and 600°C. This phenomenon can be explained on the basis of the theory developed here. It was suggested that a platinum-alumina complex forms in the range of temperatures where splitting occurs[21]. This complex provides an additional contribution to the internal stress normally found in sintered crystallites. This is sufficient to make the large crystallites unstable and therefore to cause them to fragment.

A dominant wavenumber, which corresponds to the wavelength ($\Lambda_{\text{d}} = 2\pi/k_{\text{d}}$) of the perturbation which gives rise to the fastest growing perturbation, is calculable for both the free-film and the supported-film cases. Since the instability is dominated by the fastest growing perturbation, the dominant wavelength Λ_{d} represents the approximate average size of the fragments resulting from film rupture. When the strain energy contribution to the chemical potential is small compared with the excess interaction potential and capillary pressure contributions (*i.e.* in the limit of small external pressures and small internal pre-stress) the dominant wavenumber is approximately given by

$$k_{\text{d}} \approx \left[-\frac{1}{2\gamma} \left\{ \frac{\partial\phi_0}{\partial(h'/2)} \right\}_{h/2} \right]^{1/2}$$

for the free film and

$$k_{\text{d}} \approx \left\{ -\frac{1}{2\gamma} \left(\frac{\partial\phi_0}{\partial h'} \right)_h \right\}^{1/2}$$

for the supported film.

Subsequent calculations have shown that retaining the inertial term in the equations of motion gives rise to an oscillatory instability, *i.e.* when inertia is taken into account the growth parameter has both real and imaginary parts ($\varepsilon = \hat{\varepsilon}\exp(ikx)\exp\{(\beta_R + i\beta_I)t\}$ where the frequency of oscillation is $\beta_I/2\pi$ s^{-1}). The calculations also show that the inclusion of inertia has a negligible effect on the magnitude of the real part β_R of the growth parameter. Since we are primarily concerned here with film rupture (or the real part of the growth parameter), inertia can be neglected even though under certain conditions (very near neutral stability) β_I, may be quite large. This has the advantage of yielding a relatively simple expression for the growth parameter while retaining the essential features of the problem.

NOMENCLATURE

a_{ij}	Firsov screening radius for atoms of types *i* and *j*, cm
A_{ij}	$\pi^2 n_i n_j \beta_{ij}$, empirical constant for attractive interactions between atoms of types *i* and *j*; 2 denotes the solid film and 3 denotes the solid support, erg
B_1,	defined by eqn. (34b), dyn cm^{-3}
c_j	constant defined by eqns. (13a)
C_J	constant defined by eqns. (13a), cm^{-1}
D	surface diffusion coefficient, $cm^2\ s^{-1}$
e	charge on an electron, $dyn^{1/2}$ cm
e_{ij}	*ij*th component of the strain tensor
E	Young's modulus, dyn cm^{-2}
g	chemical potential per atom, erg
G_j	constant defined by eqns. (13a), dyn cm^{-3}
h	average film thickness, cm
h'	perturbed film thickness, cm: for the supported film $h' = h + \varepsilon$; for the free film $h' = h + 2\varepsilon$
J	flux of atoms along the free surface, $cm^{-1}\ s^{-1}$
k	perturbation wavenumber, cm^{-1}
k_b	Boltzmann.constant, erg K^{-1}
k_d	dominant wavenumber, cm^{-1}
K	curvature of the perturbed interface, cm^{-1}
n_i	atomic density of the *i*th species, cm^{-3}
P_0	external pressure, dyn cm^{-2}
P_c	capillary pressure, dyn cm^{-2}
t	time, s
T	temperature, K
u	*x* component of the displacement, cm
U	interaction potential per atom, erg
U_B	interaction potential per atom of a bulk solid, erg
v	*y* component of the displacement, cm
V_0	atomic volume of a film atom, cm^3
w	z component of the displacement, cm
ΔW	strain energy, erg cm^{-3}
(x, y)	cartesian coordinates (for the free film y is measured from the center of the film, while for the supported film it is measured from the film-substrate interface), cm

Greek symbols

β	growth parameter, s^{-1}
β_{ij}	parameter for the attractive interactions between atoms of types i and j, erg cm^6
β_m	maximum value of β, s^{-1}
γ	$A_{22}/24\pi\,(2a)^2$, surface free energy, erg cm^{-2}
ε	perturbation of the film thickness, cm
θ	$\nu D\,V_0^2/k_b T$, surface diffusion, parameter, s $cm^4\ g^{-1}$

λ	Lamé's constant, dyn cm^{-2}
Λ	$2\pi/k$, wavelength of the perturbation, cm
Λ_d	dominant wavelength, cm
μ	Lamé's constant, dyn cm^{-2}
ν	atomic surface density, cm^{-2}
ρ	bulk density, g cm^{-3}
σ	internal pre-stress, dyn cm^{-2}
τ_{ij}	components of the stress tensor, dyn cm^{-2}
ϕ	interaction potential per unit volume of species 2 accounting for the difference in behavior of a thin film and a bulk solid, erg cm^{-3}
ϕ_0	interaction potential evaluated at the free surface, erg cm^{-3}
ϕ_0'	interaction potential at the free surface of a semi-infinite solid, erg cm^{-3}
Φ	$p_0/(\lambda + 2\mu)$, dimensionless group
χ	Poisson's ratio

Superscripts

s	steady state quantity
*	dimensionless quantity
^	function expressing the y dependence of the perturbation
'	perturbation quantity

Subscripts

c	critical quantity

REFERENCES

1. W. B. Pennebaker, *J. Appl. Phys.,40* (1969) 1.
2. E. Ruckenstein and B. Pulvermacher, *J. Catal., 29* (1973) 224.
3. E. Ruckenstein and R. K. Jain, *J. Chem. Soc., Faraday Trans. II, 70 (*1974) 132 (Section 5.1 of this volume).
4. K. L. Chopra, *Thin Film Phenomena,* McGraw-Hill, New York, 1969, p. 287.
5. K. L. Chopra, M. R. Randlett and R. H. Duff, *Philos. Mag., 16* (1967) 261.
6. J. S. Anderson, in M. W. Roberts and J. M. Thomas (eds.), *Surface and Defect Properties of Solids,* Vol. 1, Chemical Society, Burlington House, London, 1972, p. 5.
7. R. W. Holfman, *Phys. Thin Films, 3* (1966) 211.
8. C. P. Herring, in R. Gomer and C. W. Smith (eds.), *Structure and Properties of Solid Surfaces,* University of Chicago Press, Chicago, 1952.
9. J. C. M. Li, R. A. Oriani and L. S. Darken, Z. *Phys. Chem. N.F., 49* (1966) 271.
10. S. I. Sokolnikoff, *Mathematical Theory of Elasticity,* McGraw-Hill, New York, 1956.
11. L. H. Thomas, *Proc. Cambridge Philos. Soc., 23 (1927) 542.*
12. G. Moliere, Z. *Naturforsch., Teil A, 2*(1947) 133.
13. R. A. Johnson, *Radiat. Eff., 2* (1969) 1.
14. I. M. Torrens and M. T. Robinson, in P. C. Gehlen, J. R. Beeler and R. I. Jaffee (eds.), *Interatomic Potentials and Simulation of Lattice Defects,* Plenum Press, New York, 1972, pp. 423–436.
15. H. P. Bonzel, in J. M. Blakely (ed.), *Surface Physics of Materials,* Vol. 2, Academic Press, New York, 1975, pp. 280–338.
16. F. Delamare and G. E. Rhead, *Surf. Sci., 28* (1971) 267.
17. G. E. Rhead, *Acta Metall., 13* (1965) 223.
18. N. A. Gjostein, in T. T. Burke, N. L. Reed and V. Weiss (eds.), *Surfaces and Interfaces, I: Chemical and Physical Characteristics,* Syracuse University Press, Syracuse, N.Y., 1967.
19. D. H. Bradhurst and J. S. L. Leach, J. *Electrochem. Soc., 113* (1966) 1245.
20. A. T. Fromhold, *Surf. Sci., 22 (*1972) 396.
21. E. Ruckenstein and M. L. Malhotra, *J. Catal., 41* (1976) 303.

APPENDIX A

Calculation of the Interaction Potential for a Thin Film

The potential energy of an atom of species i due to a semi-infinite fluid or solid situated at a distance y from an interface (see Fig. Al) is given by

$$U_i(y) = \iint\limits_V n_j u_{ij}(r) 2\pi r^2 \sin\theta \, d\theta \tag{A1}$$

where n_i is the number of atoms per unit volume in region j and $u_{ij}(r)$ is the potential energy due to the interaction of two atoms of types i and j separated by a distance r. The integration range can be split into an integration over the film and one over the substrate:

$$\left.\begin{aligned} U_2(y) = 2\pi n_2 \Bigg(&\int_0^{\pi/2} \sin\theta \, d\theta \int_{2a}^{y/\cos\theta} u_{22}(r) r^2 dr + \\ &+ \int_{\pi/2}^{\pi} \sin\theta d\theta \int_{2a}^{-(h-y)/\cos\theta} u_{22}(r) r^2 dr \Bigg) + \end{aligned}\right\} \text{integration over flim (region II)}$$

$$\left. + 2\pi n_3 \int_0^{\pi/2} \sin\theta \, d\theta \int_{y/\cos\theta}^{\infty} u_{32}(r) r^2 dr \right\} \quad \text{integration over substrate (region III)} \tag{A2}$$

where $2a$ is the equilibrium distance of closest approach of two atoms of the film. We have neglected the integration over region I (Fig. Al) since at moderate pressures gas-solid interactions are small compared with solid-solid interactions because the atomic density of a solid is much greater than that of a gas. For the empirical form of the metal-metal interactions under consideration here we have

$$u_{22}(r) = -\frac{\beta_{22}}{r^6} + \frac{z_2 z_2 e^2}{r} \sum_{k=1}^{3} c_k \exp(C_k r) \tag{A3a}$$

and

$$u_{23}(r) = -\frac{\beta_{22}}{r^6} + \frac{z_2 z_2 e^2}{r} \sum_{k=1}^{3} c_k \exp\left(-C_k \frac{a_{22}}{a_{23}} r\right) \tag{A3b}$$

where

$$c = \begin{pmatrix} 0.35 \\ 0.55 \\ 0.10 \end{pmatrix} \qquad C = \frac{1}{a_{22}} \begin{pmatrix} 0.30 \\ 1.20 \\ 6.00 \end{pmatrix}$$

Inserting expressions (A3) into eqn. (A2) and performing the specified integrations we find that

$$U_2(y) = U_B + \frac{\pi(n_2\beta_{22} - n_3\beta_{32})}{6} \left\{ \frac{1}{y^3} + \frac{1}{(h-y)^3} \right\} - 2\pi (z_2 e)^2 n_2 \sum_{k=1}^{3} \frac{c_k}{C_k^2} \left[\exp(-C_k y) + \right.$$

$$\left. + \exp\{-C_k(h-y)\} \right] + 2\pi n_3 z_2 z_3 \left(\frac{e a_{23}}{a_{22}} \right)^2 \sum_{k=1}^{3} \frac{c_k}{C_k^2} \exp\left(-\frac{C_k y a_{22}}{a_{23}} \right) \tag{A4}$$

where U_B is the potential energy of an atom situated in the bulk solid and is given by

$$U_B = -\frac{\pi n_2 \beta_{22}}{48a^3} + 4\pi n_2 (z_2 e)^2 \sum_{k=1}^{3} \frac{c_k}{C_k^2} (C_k 2a + 1) \exp(-C_k 2a) \tag{A5}$$

Then the interaction potential per unit volume of species 2 accounting for the difference in the forces which act in a thin film and a bulk solid is given by

$$\phi(y,h) = n_2 \{U_2(y) - U_B\} = \frac{A_{22} - A_{32}}{6\pi} \left\{ \frac{1}{y^3} + \frac{1}{(h-y)^3} - 2\pi (z_2 e n_2)^2 \sum_{k=1}^{3} \frac{c_k}{C_k^2} \exp(-C_k y) \right.$$

$$\left. + \exp\{-C_k (h-y)\} \right] + 2\pi n_2 n_3 z_2 z_3 \left(\frac{e a_{23}}{a_{22}} \right)^2 \sum_{k=1}^{3} \frac{c_k}{C_k^2} \exp\left(-\frac{C_k y a_{22}}{a_{23}} \right) \tag{A6}$$

where $Aij = \pi^2 n_i n_j \beta_{ij}$.

APPENDIX B

Calculation of the Perturbed Strain Energy at the Free Surface

A general expression for the strain energy of an elastic solid is[1]

$$W = \frac{1}{2}\left(\tau_{xx} e_{xx} + \tau_{yy} e_{yy} + \tau_{zz} e_{zz} + 2\tau_{xy} e_{xy} + 2\tau_{xz} e_{xz} + 2\tau_{yz} e_{yz} \right) \tag{B1}$$

where e_{ij} is the ijth component of the strain. At the free surface the shear stresses τ_{yx} and τ_{yz} are zero. Decomposing the stress tensor τ into steady state and perturbation values, we obtain for the strain energy

$$W = \frac{1}{2}\left[\left(\tau_{xx}{}^{s} + \tau_{xx}' \right) \left(\frac{\partial u^{s}}{\partial x} + \frac{\partial u'}{\partial x} \right) + \left(\tau_{yy}{}^{s} + \tau_{yy}' \right) \left(\frac{\partial \upsilon^{s}}{\partial y} + \frac{\partial \upsilon'}{\partial y} \right) + \left(\tau_{zz}{}^{s} + \tau_{zz}' \right) \left(\frac{\partial w^{s}}{\partial z} + \frac{\partial w'}{\partial z} \right) + \right.$$

$$\left. + \left(\tau_{xz}{}^{s} + \tau_{xz}' \right) \left\{ \left(\frac{\partial u^{s}}{\partial z} + \frac{\partial w^{s}}{\partial x} \right) + \left(\frac{\partial u'}{\partial z} + \frac{\partial w'}{\partial x} \right) \right\} \right] \tag{B2}$$

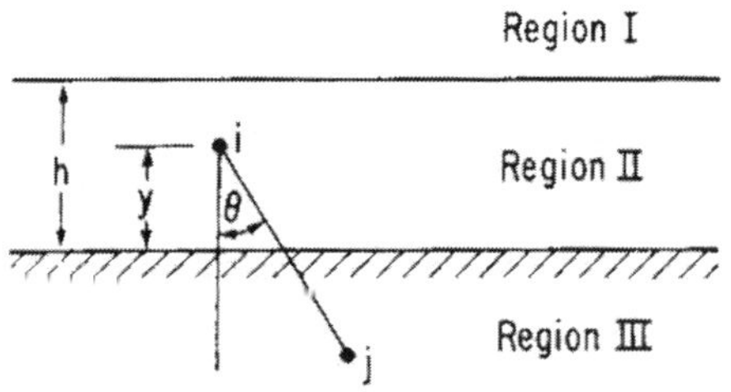

FIG. A1. The coordinate system used for the calculation of the interaction potential.

It should be noted that w' is identically zero and that u' is independent of z. Neglecting terms which are second order in perturbation quantities allows us to write eqn. (B2) as the sum of a steady state and a perturbed contribution:

$$\Delta W = \Delta W^s + \Delta W' \tag{B3}$$

where at a free surface

$$\Delta W^s = \frac{1}{2}\left\{\tau_{xx}{}^s \frac{\partial u^s}{\partial x} + \tau_{yy}{}^s \frac{\partial \upsilon^s}{\partial y} + \tau_{zz}{}^s \frac{\partial w^s}{\partial z} + \tau_{xz}{}^s \left(\frac{\partial u^s}{\partial z} + \frac{\partial w^s}{\partial x}\right)\right\} \tag{B3a}$$

and

$$\Delta W' = \frac{1}{2}\left(\tau_{xx}{}^s \frac{\partial u'}{\partial x} + \tau_{yy}{}^s \frac{\partial \upsilon'}{\partial y} + \tau_{xx}{}' \frac{\partial u^s}{\partial x} + \tau_{yy}{}' \frac{\partial \upsilon^s}{\partial y} + \tau_{zz}{}' \frac{\partial w^s}{\partial z}\right) \tag{B3b}$$

It is convenient to express eqn. (B3b) in terms of steady state stresses (which are known quantities) and perturbation strains. This can be accomplished by expanding $\tau_{xx}{}'$ and $\tau_{yy}{}'$ in terms of the displacements u' and v' to give

$$\begin{aligned} 2\Delta W' = {} & \tau_{xx}{}^s \frac{\partial u'}{\partial x} + \tau_{yy}{}^s \frac{\partial \upsilon'}{\partial y} + \left\{(\lambda + 2\mu)\frac{\partial u'}{\partial x} + \lambda \frac{\partial \upsilon'}{\partial y}\right\}\frac{\partial u^s}{\partial x} + \\ & + \left\{(\lambda + 2\mu)\frac{\partial \upsilon'}{\partial y} + \lambda \frac{\partial u'}{\partial x}\right\}\frac{\partial \upsilon^s}{\partial y} + \lambda\left(\frac{\partial u'}{\partial x} + \frac{\partial \upsilon'}{\partial y}\right)\frac{\partial w^s}{\partial z} \end{aligned} \tag{B4}$$

Rearranging slightly yields

$$\begin{aligned} 2\Delta W' = {} & \tau_{xx}{}^s \frac{\partial u'}{\partial x} + \tau_{yy}{}^s \left\{(\lambda + 2\mu)\frac{\partial u^s}{\partial x} + \lambda\left(\frac{\partial \upsilon^s}{\partial y} + \frac{\partial w^s}{\partial z}\right)\right\}\frac{\partial u'}{\partial x} + \\ & + \left\{(\lambda + 2\mu)\frac{\partial \upsilon^s}{\partial y} + \lambda\left(\frac{\partial u^s}{\partial x} + \frac{\partial w^s}{\partial z}\right)\right\}\frac{\partial \upsilon'}{\partial y} \end{aligned}$$

or

$$\Delta W' = \tau_{xx}{}^s \frac{\partial u'}{\partial x} + \tau_{yy}{}^s \frac{\partial \upsilon'}{\partial y} \tag{B5}$$

Since at the free surface $\tau_{yy}{}^s = -p_0$, the perturbation component of the strain energy at the free surface can be expressed as

$$\Delta W' = \sigma \frac{\partial u'}{\partial x} - p_0 \frac{\partial \upsilon'}{\partial y} \tag{B6}$$

where $\sigma = \tau_{xx}{}^s = \tau_{zz}{}^s$ is the internal pre-stress. Equation (B6) can be rewritten in terms of u' and ε by using the normal stress balance boundary condition

$$\tau_{yy}{}' = (\lambda + 2\mu)\frac{\partial \upsilon'}{\partial y} + \lambda \frac{\partial u'}{\partial x} = \gamma \frac{\partial^2 \varepsilon}{\partial x^2}$$

This provides an expression for $\partial v'/\partial y$ at the free surface. On inserting this result into eqn. (B6) we obtain the required result

$$\Delta W' = (\sigma + \lambda\Phi)\frac{\partial u'}{\partial x} - \gamma\Phi\frac{\partial^2 \varepsilon}{\partial x^2} \tag{B7}$$

where

$$\Phi = \frac{p_0}{\lambda + 2\mu} \tag{B7a}$$

REFERENCE FOR APPENDIX B

1. S. I. Sokolnikoff, *Mathematical Theory of Elasticity,* McGraw-Hill, New York, 1956.

5.3 Stability of Symmetric and Unsymmetric Thin Liquid Films to Short and Long Wavelength Perturbations*

Charles Maldarelli[a,1], *Rakesh K. Jain*[b,2], *Ivan B. Ivanov*[c], *and Eli Ruckenstein*[d]

[a] Department of Chemical Engineering, Columbia University, New York, New York 10027
[b] Department of Chemical Engineering, Carnegie-Mellon University, Pittsburgh, Pennsylvania 15213
[c] Department of Physical Chemistry, University of Sofia, Bulgaria
[d] Faculty of Engineering and Applied Sciences, State University of New York, Buffalo, New York 14214

Corresponding Author

[1] Present address: Department of Chemical Engineering, City College, New York, N. Y. 10031.
[2] Author to whom reprint requests should be addressed.
Received October 20, 1979; accepted March 8, 1980

INTRODUCTION

The study of surface wave-induced hydrodynamic instabilities in thin (O(10–100 nm)) liquid films has been the focus of a number of investigations by several workers (1–18). Although the motivations behind these investigations are diverse, clearly the most significant is the importance of this subject toward the understanding of particle coalescence phenomena in a continuous liquid phase and the stability of the cell membrane.

In a dispersion, the approach of two particles (e.g., gas bubbles or liquid droplets) of the dispersed phase creates a radially bounded film which subsequently drains under the combined action of capillary suction at the plateau border and the disjoining pressure. If stabilizing surfactants are not present in sufficient amounts in the film system, the draining film will generally rupture at a thickness of the order of 10–100 nm. According to deVries and later Scheludko (19), the rupture is caused by the undamped growth of mechanically or thermally induced corrugations in the film interfaces. These corrugations grow because the system becomes unstable to interfacial fluctuations at small film thicknesses.

The biological cell provides a contrasting example of a system in which a thin film plays a crucial role (13, 15, 30). The cell membrane is a highly viscous lamella bounded by two different liquid phases (i.e., the intra- and extra-cellular fluids) and having different interfacial tensions on the two faces. Certain cell processes such as mobility, ingestion (via phagocytosis), and microvilli

* *Journal of Colloid and Interface Science*. Vol. 78, No. 1. p.118 November 1980. Republished with permission.

elaboration involve a deformation of the membrane. Reasons for the onset of these membrane deformation processes have not been fully elucidated, and it is hoped that a theoretical understanding of the stability of symmetric and unsymmetric, thin liquid films to short $(\lambda \leq h)$ and long $(\lambda > h)$ wavelength perturbations may provide insight into the origins of such processes.

The goal of this investigation is to examine the stability of pure, radially unbounded thin films bounded by two different liquids and having different interfacial tensions on the two faces. A short review of previous work is in order. Felderhof (2) analyzed the dynamics of a free film composed of an inviscid fluid and considered in detail both the squeezing and stretching normal modes (Fig. 1). This study was later extended by Sche (16) and Sche and Fijnaut (17) to include the effect of liquid viscosity. Ruckenstein and Jain (8) investigated the stability of the squeezing mode of a free film, but the major part of their investigations (8, 11) was concerned with a film on a solid substrate (Fig. 1). Lucassen *et al.* (5, 6) formally treated the stability of a liquid film bounded by two different viscous phases, but only derived an explicit dispersion equation for the squeezing mode of a symmetrical film system (i.e., a system in which the film is bounded by the same phase and in which gravity is neglected). Finally, Joosten *et al.* (14) extended the analysis of Lucassen *et al.* to a free film by considering the stretching mode of the symmetrical system. However, the dispersion equations obtained by Lucassen *et al.* and Joosten *et al.* are strictly applicable only to long wavelength interfacial disturbances (i.e., disturbances with Fourier component wavelengths which are much larger than the thickness of the film) because of the expansion method they employ to account for the influence of the long-range van der Waals interaction on the film dynamics (see below). In the work presented here, we will examine the stability characteristics of thin films for short as well as long wavelengths.

An important aspect of studies concerning the dynamics of thin films is the procedure utilized to account for the influence of the long-range van der Waals force on the film stability. The consideration of this force is essential: The maximum range of the van der Waals interaction ($O(100 \text{ nm})$) is of the same order of magnitude as the characteristic film thickness, and consequently energy inhomogeneities are created by this interaction in the film and in the adjoining phases. (This same

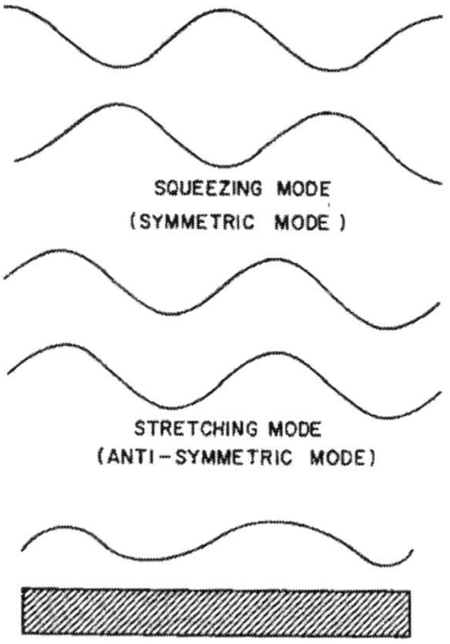

FIG. 1. Various modes of instability of a thin liquid film. In the squeezing mode, the corrugations at one interface of the film is completely out of phase with the corrugations at the other interface $(\phi = \pi)$. In the stretching mode this phase difference is zero $(\phi = 0)$. When the film is supported on a solid surface, only one interface is corrugated.

phenomenon also occurs in the transition region between two bulk phases and is usually analyzed through the concept of surface tension.) Two different procedures have been utilized to account for the influence of the van der Waals interaction on the film dynamics: a disjoining pressure approach (3, 4, 20, 21) and a body force procedure (2, 8, 11).

In the disjoining pressure approach, the film and the surrounding phases are treated as energetically homogeneous (with respect to the van der Waals interaction) up to the dividing surfaces which define the interfaces of the film. (This defines the idealized system.) The inhomogeneity caused by the van der Waals interaction is incorporated as a surface excess energy in a procedure analogous to the thermodynamic treatment of interfacial tension. Using this approach, one can show that the normal component of the stress tensor (of the idealized system) suffers a discontinuity (the disjoining pressure) at the dividing surface of the film.

The disjoining pressure approach has the advantage that all types of intermolecular forces (i.e., dispersion, Keesom, and Debye–Falkenhagen forces) which contribute to the long-range van der Waals interaction can easily be formulated in the expression for the disjoining pressure by utilizing Lifschitz's theory. However, as shown below for the stability analysis, an expression for the disjoining pressure of an interfacially deformed film is required. In all previous works (4–7, 9, 10, 12, 14), such an expression has been obtained by expanding the disjoining pressure of a plane parallel film in a power series with respect to the film thickness. This procedure is valid for long wavelength interfacial disturbances, but it is not valid for short wavelength disturbances.

In the body force approach, the influence of the van der Waals force is incorporated as a body force in the Navier–Stokes equation of motion. The force is derived from a potential which describes the energy of interaction (due to the van der Waals force) of molecules in an infinitesimal volume of continua with respect to the entire ensemble of molecules in the system. In the usual application of the body force approach, this potential is calculated via a microscopic Hamaker procedure in which the intermolecular potential is integrated over the system volume. Although an adequate approximation for describing the dispersion contribution to the van der Waals interaction, the Hamaker procedure is insufficient for incorporating the dipolar contributions (i.e., the Keesom and Debye–Falkenhagen forces) since these are incompatible with pair additivity. However, this need not be considered a serious limitation since *ad hoc* methods for incorporating the dipolar contributions into the Hamaker procedure have been developed (22).

Utilization of the Hamaker procedure to obtain the body force potential has two distinct advantages: First, in contrast to the disjoining pressure approach it can easily be formulated for any interfacially corrugated state of the film's interfaces and is therefore applicable to both short and long wavelength disturbance regimes. Second, it is computationally easier to formulate the Hamaker potential for an *unsymmetrical film* system (i.e., a film bounded by two different viscous phases) than it is to calculate the disjoining pressure of such a system.

Since this investigation is concerned with both symmetrical and unsymmetrical film systems, the body force method, in conjunction with the Hamaker procedure to calculate the intermolecular potential, is utilized. Although in all studies in which the body force method has been used (2, 8, 11, 16, 17) the Hamaker procedure has been employed to calculate the potential, no care has yet been taken regarding a systematic formulation of this potential for corrugated interfaces. Consequently this analysis begins with a systematic formulation of the Hamaker potential for an unsymmetrical, arbitrarily deformed film system. The basic equations for the first-order approximations for the field variables follow. These equations differ from those established previously through the expression of the body force which contains the effect of the corrugation of the interface and through different viscosities and interfacial tensions. A general dispersion equation is established in the next section which is valid for both short and long wavelength disturbances. The final sections of the paper examine the dispersion equation for symmetrical and unsymmetrical systems. First, the asymptotic forms of the symmetrical dispersion equations are derived for the stretching and squeezing modes in the long wavelength limit. Then, the symmetrical dispersion equations are analyzed numerically to examine the stability characteristics of the film to short wavelength perturbations. Finally, the dispersion equation for the unsymmetrical system is analyzed numerically to study the role of interfacial tension asymmetry on the film behavior. Wherever possible, results are used to explain available experimental data.

FORMULATION OF THE BODY FORCE VIA A HAMAKER PROCEDURE

To compute the long-range van der Waals interaction potential, it is useful to consider the state in which the interfaces of the film are arbitrarily deformed and located an infinite distance apart from each other (see Fig. 2). In this "infinite" system, the interfaces are isolated and can be separately treated in the manner of Gibbs as dividing surfaces of zero thickness separating phases which are homogeneous in all intensive properties (e.g., density and energy) up to the dividing surfaces. The spatial inhomogeneities in energy (which are especially pronounced in the interfacial transition regions) caused by the van der Waals force are accounted for by treating the dividing surfaces as membranes in isotropic tension; by utilizing this procedure, the idealized Gibbs system is made mechanically equivalent to the real system and no body force is considered. (In what follows, the liquid comprising the film is referred to as phase II and the fluids comprising the upper and lower semi-infinite media are denoted by phases I and III, respectively. The interfaces between phases I and II and II and III are referred to as the upper (u) and lower (l) interfaces, respectively.)

When the interfaces are brought close together by removing (to an infinite reservoir of phase II) region "A" of Fig. 2 to form the film system (Fig. 3), the molecules in the film and the adjoining phases experience a change in potential energy. The change in potential energy is derived by first computing the potential energy of interaction (due to the van der Waals force) of molecules in an

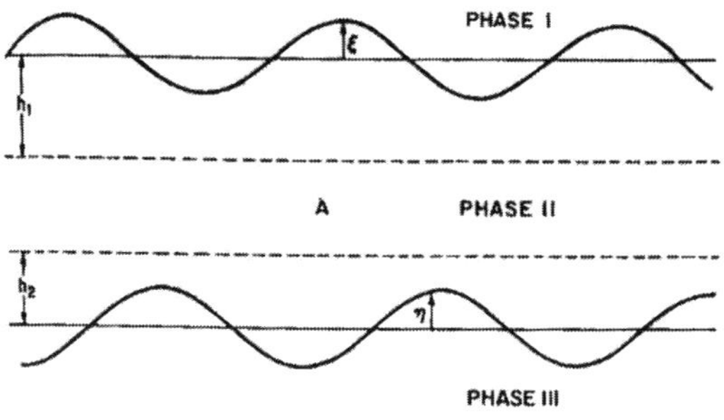

FIG. 2. The infinite system. The two interfaces, arbitrarily deformed, are separated by an infinite reservoir of phase II (region "A").

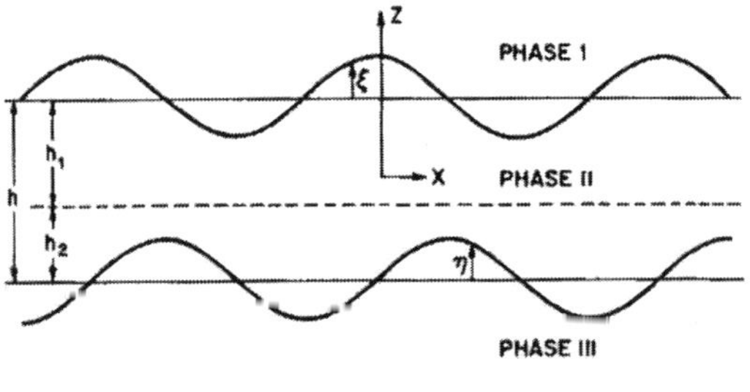

FIG. 3. The film system. Region "A" has been removed to an infinite reservoir of phase II. The origin of the coordinate system is on the midplane of the unperturbed film.

infinitesimal volume of continua with respect to (i) the rest of the molecules in the film system (Fig. 3) and (ii) the rest of the molecules in the infinite system (Fig. 2) and then subtracting the energy obtained in (ii) from that calculated in (i). This difference is termed the excess van der Waals potential; the negative gradient of this potential is the body force which is introduced into the equation of motion.

In the Hamaker procedure, the potential energy of interaction is computed by integrating the intermolecular potential $u(r)$ (the mutual potential energy of two molecules separated by a distance r) over the system volume. Thus in this procedure, the excess van der Waals potential admits the following integral representation:

$$W(\mathbf{r},t)=\int_{\substack{\text{film}\\ \text{system}}} w\left(|\mathbf{r}-\mathbf{r}'|^{1/2}\right)\rho(\mathbf{r}',t)\,d\mathbf{r}'-\int_{\substack{\text{infinite}\\ \text{system}}} w\left(|\mathbf{r}-\mathbf{r}'|^{1/2}\right)\rho(\mathbf{r}',t)\,d\mathbf{r}'. \qquad [1]$$

In Eq. [1] t represents time, $\rho(\mathrm{r}',t)$ is the mass density at the source point $\mathbf{r}'$, $W(\mathbf{r}, t)$ is the excess van der Waals potential (per unit mass) at the field point r, and $w(r)$ is the intermolecular potential $u(r)$ divided by the masses (per molecule) of the molecules located at $\mathbf{r}$ and $\mathbf{r}'$.

To formulate an explicit expression for the excess potential in terms of the Cartesian coordinate system diagrammed in Fig. 3, the following functions prove essential:

$$F_{i,j}(s)=\rho^i\rho^j\int_s^\infty 2\pi r w_{i,j}(r)\,dr \qquad [2]$$

$$H_{i,j}^1(x,y,v,t)=\rho^i\rho^j\int_{-\infty}^{\infty}\int_{-\infty}^{\infty}\int_0^{\xi(\alpha,\beta,t)} w_{i,j}\left(\left((v-\gamma)^2+(x-\alpha)^2+(y-\beta)^2\right)^{1/2}\right)d\gamma\, d\beta\, d\alpha \qquad [3]$$

$$H_{i,j}^2(x,y,v,t)=\rho^i\rho^j\int_{-\infty}^{\infty}\int_{-\infty}^{\infty}\int_0^{\eta(\alpha,\beta,t)} w_{i,j}\left(\left((v-\gamma)^2+(x-\alpha)^2+(y-\beta)^2\right)^{1/2}\right)d\gamma\, d\beta\, d\alpha. \qquad [4]$$

The variables ξ and η define the interfacial perturbations (see Fig. 3). The function $F_{i,j}(s)$ represents the potential energy (per unit volume) at a point $\mathbf{r}$ in phase i due to an infinite plane of phase j that is located a perpendicular distance s from r. The function $H_{i,j}^1(x,y,v,t)\left(H_{i,j}^2(x,y,v,t)\right)$ is the perturbation in the potential energy (per unit volume) at a point in phase i, with coordinates x, y and located a distance $|v|$ from the unperturbed upper (lower) interface, caused by the displacement of phase j with the corrugation of the interface. In Eqs. [2]–[4] density variations have been neglected since the fluids comprising the film system are assumed incompressible.

With the aid of these functions, the integral representation of the excess potential can be written with reference to the x, y, z Cartesian coordinate system of Fig. 3.

$$\begin{aligned}\rho^{\mathrm{I}}W^{\mathrm{I}}(x,y,z,t) &= \int_{z+h/2}^{\infty}\left(F_{\mathrm{I,III}}(u)-F_{\mathrm{I,II}}(u)\right)du \\ &\quad + H_{\mathrm{I,III}}^2\left(x,y,z+\frac{h}{2},t\right)-H_{\mathrm{I,II}}^2\left(x,y,z+\frac{h}{2},t\right)\end{aligned} \qquad [5]$$

$$\begin{aligned}\rho^{\mathrm{II}}W^{\mathrm{II}}(x,y,z,t) &= \int_{z+h/2}^{\infty}\left(F_{\mathrm{II,III}}(u)-F_{\mathrm{II,II}}(u)\right)du \\ &+ H^{2}_{\mathrm{II,III}}\left(x,y,z+\frac{h}{2},t\right) \\ &- H^{2}_{\mathrm{II,II}}\left(x,y,z+\frac{h}{2},t\right) \\ &\left[\xi(x,t)+\frac{h}{2} > z > \frac{h}{2}-h_1\right]\end{aligned} \qquad [6]$$

$$\begin{aligned}\rho^{\mathrm{II}}W^{\mathrm{II}}(x,y,z,t) &= \int_{h/2-z}^{\infty}\left(F_{\mathrm{I,II}}(u)-F_{\mathrm{II,II}}(u)\right)du \\ &+ H^{1}_{\mathrm{II,II}}\left(x,y,z-\frac{h}{2},t\right) \\ &- H^{1}_{\mathrm{I,II}}\left(x,y,z-\frac{h}{2},t\right) \\ &\left[\frac{h}{2}-h_1 > z > \eta(x,t)-\frac{h}{2}\right]\end{aligned} \qquad [7]$$

$$\begin{aligned}\rho^{\mathrm{III}}W^{\mathrm{III}}(x,y,z,t) &= \int_{h/2-z}^{\infty}\left(F_{\mathrm{I,III}}(u)-F_{\mathrm{II,III}}(u)\right)du \\ &+ H^{1}_{\mathrm{II,III}}\left(x,y,z-\frac{h}{2},t\right) \\ &- H^{1}_{\mathrm{I,III}}\left(x,y,z-\frac{h}{2},t\right).\end{aligned} \qquad [8]$$

In the above equations h is the film thickness; h_1 is an arbitrary parameter utilized in dividing the film volume (see Figs. 2 and 3). The roman numeraled superscripts indicate the phase in which the scripted quantity is evaluated.

It is important to note that $W^i(x,y,z,t)$ is the *excess* van der Waals potential; the total potential is the sum of the excess potential and the reference potential. Thus

$$\rho^{\mathrm{I}}\overline{W}^{\mathrm{I}} = \rho^{\mathrm{I}}W^{\mathrm{I}} + \rho^{\mathrm{I}}W^{\mathrm{I}}_{\infty(2)}$$

$$\rho^{\mathrm{II}}\overline{W}^{\mathrm{II}} = \rho^{\mathrm{II}}W^{\mathrm{II}} + \rho^{\mathrm{II}}W^{\mathrm{II}}_{\infty(2)}$$

$$\left(\xi(x,t)+\frac{h}{2} > z > \frac{h}{2} - h_1\right)$$

$$\rho^{\mathrm{II}}\bar{W}^{\mathrm{II}} = \rho^{\mathrm{II}}W^{\mathrm{II}} + \rho^{\mathrm{II}}W^{\mathrm{II}}_{\infty(1)}$$

$$\left(\frac{h}{2} - h_1 > z > \eta(x,t) - \frac{h}{2}\right)$$

$$\rho^{\mathrm{III}}\bar{W}^{\mathrm{III}} = \rho^{\mathrm{III}}W^{\mathrm{III}} + \rho^{\mathrm{III}}W^{\mathrm{III}}_{\infty(1)}.$$

Here, $\bar{W}^i$ indicates the total potential; $W^i_{\infty(1)}$ and $W^i_{\infty(2)}$ are the reference potentials (the subscript 1 indicates that the I, II interface is infinitely far away and the subscript 2 indicates that the II, III interface is at infinity). Note that since the reference potential for $z > h/2 - h_1$ is taken different than that for $z > h/2 - h_1$, the excess potential is discontinuous at $z > h/2 - h_1$ even though the total potential in the film, $\bar{W}^{\mathrm{II}}$, is continuous. However, the dispersion equation involves only the gradient of these potentials; the choice of the reference potentials does not create any problem, as will be evident in the next two sections.

STABILITY ANALYSIS

This investigation is concerned with the interfacial stability of an unbounded, non-thinning liquid film surrounded by two different viscous phases and having different interfacial tensions on the two faces. Since the film is not thinning, the base state is taken to be one in which the flow velocity is zero and the film interfaces are plane parallel. Infinitesimal perturbations are applied to the interfaces of the film, and the object of the analysis is to determine the stability of the resulting fluid motion. For simplicity, only two-dimensional motion is considered; the extension to three dimensions is discussed after Eq. [77].

The field equations which are necessary to describe the fluid motion are the continuity and the Navier–Stokes equations for an incompressible fluid (23, 24). The boundary conditions at the displaced interfaces are the continuity of velocity requirement, the kinematic relation, and the tangential and normal stress balances (24). Furthermore, the disturbance must decay to zero at an infinite distance from the perturbed interfaces.

In order to ascertain the stability of the fluid motion induced by the interfacial perturbations, the field equations and boundary conditions for the flow are solved by using perturbation methods (23). The field variables of the system are expanded in a series about their values in the base state. Since the initial disturbances are Infinitesimal it suffices to consider only linear terms. Therefore

$$p^j = p_0^j(z) + \epsilon p_1^j(x,z,t) \qquad [9]$$

$$u^j = \epsilon u_1^j(x,z,t) \qquad [10]$$

$$v^j = \epsilon v_1^j(x,z,t) \qquad [11]$$

$$\xi = \epsilon \xi_1(x,t) \qquad [12]$$

$$\eta = \epsilon \eta_1(x,t) \qquad [13]$$

$$W^j = W_0^j(z) + \epsilon W_1^j(x,z,t). \qquad [14]$$

In the above set of equations, ϵ is a small, dimensionless parameter, and the subscripts "0" and "1" indicate the base state and first-order values, respectively, of the scripted quantities. The variables p^j, u^j and v^j denote, respectively, the pressure and the x and z components of the velocity vector (all of phase j).

The base state distribution for the excess van der Waals potential is obtained from Eqs. [5]–[8] by setting the $H_{i,j}^1$ and $H_{i,j}^2$ functions equal to zero. Thus

$$\rho^{\mathrm{I}} W_0^{\mathrm{I}}(z) = \int_{z+h/2}^{\infty} \left(F_{\mathrm{I,III}}(u) - F_{\mathrm{I,II}}(u)\right) du \quad [15]$$

$$\rho^{\mathrm{II}} W_0^{\mathrm{II}}(z) = \begin{cases} \displaystyle\int_{z+h/2}^{\infty} \left(F_{\mathrm{II,III}}(u) - F_{\mathrm{II,II}}(u)\right) du & \left(\dfrac{h}{2} > z > \dfrac{h}{2} - h_1\right) \\ \displaystyle\int_{h/2-z}^{\infty} \left(F_{\mathrm{I,II}}(u) - F_{\mathrm{II,II}}(u)\right) du & \left(\dfrac{h}{2} - h_1 > z > -\dfrac{h}{2}\right) \end{cases} \quad [16]$$

$$\rho^{\mathrm{III}} W_0^{\mathrm{III}}(z) = \int_{h/2-z}^{\infty} \left(F_{\mathrm{I,III}}(u) - F_{\mathrm{II,III}}(u)\right) du. \quad [17]$$

Explicit expressions for the first-order excess van der Waals potential, W_1^j, are obtained by expanding the $H_{i,j}^1$ and $H_{i,j}^2$ functions in a power series about ξ and η equal to zero, and then substituting the first-order approximations for ξ and η (Eqs. [12] and [13]) into these expansions. The results for one-dimensional interfacial disturbance (two-dimensional flow) are:

$$\rho^{\mathrm{I}} W_1^{\mathrm{I}}(x,z,t) = Q_{\mathrm{I,III}}^2\left(x, z+\frac{h}{2}, t\right) - Q_{\mathrm{I,II}}^2\left(x, z+\frac{h}{2}, t\right) \quad [18]$$

$$\rho^{\mathrm{II}} W_1^{\mathrm{II}}(x,z,t) = \begin{cases} Q_{\mathrm{II,III}}^2\left(x, z+\dfrac{h}{2}, t\right) - Q_{\mathrm{II,II}}^2\left(x, z+\dfrac{h}{2}, t\right) & \left(\dfrac{h}{2} > z > \dfrac{h}{2} - h_1\right) \\ Q_{\mathrm{II,II}}^1\left(x, z-\dfrac{h}{2}, t\right) - Q_{\mathrm{I,II}}^1\left(x, z-\dfrac{h}{2}, t\right) & \left(\dfrac{h}{2} - h_1 > z > -\dfrac{h}{2}\right) \end{cases} \quad [19]$$

$$\rho^{\mathrm{III}} W_1^{\mathrm{III}}(x,z,t) = Q_{\mathrm{II,III}}^1\left(x, z-\frac{h}{2}, t\right) - Q_{\mathrm{I,III}}^1\left(x, z-\frac{h}{2}, t\right) \quad [20]$$

where

$$Q_{i,j}^1(x,v,t) = \rho^i \rho^j \int_{-\infty}^{\infty}\int_{-\infty}^{\infty} \xi_1(\alpha,t)\, w_{i,j}\left(\left(v^2+\xi^2+(x-\alpha)^2\right)^{1/2}\right) d\xi\, d\alpha \quad [21]$$

$$Q_{i,j}^2(x,v,t) = \rho^i \rho^j \int_{-\infty}^{\infty}\int_{-\infty}^{\infty} \eta_1(\alpha,t)\, w_{i,j}\left(\left(v^2+\xi^2+(x-\alpha)^2\right)^{1/2}\right) d\xi\, d\alpha. \quad [22]$$

The Base State

In the unperturbed (base) state the interfaces of the film are plane parallel and the flow velocity is equal to zero. Consequently the continuity equation and the entire set of boundary conditions (with the exception of the normal stress balances) reduce to identities, and the Navier–Stokes equation after integration, becomes:

$$p_0^j(z) + \rho^j g z + \rho^j W_0^j(z) = c^j \quad (j = \mathrm{I,II,III}). \quad [23]$$

In Eq. [23], g is the acceleration of gravity; it is assumed that gravity acts in the negative z direction (see Fig. 3). The only nontrivial boundary conditions, the normal stress balances, reduce to the following form:

$$p_0^{\mathrm{I}}\left(\frac{h}{2}\right)=p_0^{\mathrm{II}}\left(\frac{h}{2}\right), \qquad [24]$$

$$p_0^{\mathrm{II}}\left(-\frac{h}{2}\right)=p_0^{\mathrm{III}}\left(-\frac{h}{2}\right). \qquad [25]$$

In the base state, the total force density (excluding gravity) is equal to the sum of the pressure and the excess van der Waals potential; thus

$$P_0^j(z)=p_0^j(z)+\rho^j W_0^j(z) \qquad [26]$$

where $P_0^j(z)$ denotes the gravity-excluded total force density of phase j. Owing to the continuity in the pressure at the two interfaces (Eqs. [24] and [25]), this force density suffers a discontinuity at $z=\pm h/2$. Explicitly,

$$P_0^{\mathrm{I}}\left(\frac{h}{2}\right)-P_0^{\mathrm{II}}\left(\frac{h}{2}\right)=\rho^{\mathrm{I}}W_0^{\mathrm{I}}\left(\frac{h}{2}\right)-\rho^{\mathrm{II}}W_0^{\mathrm{II}}\left(\frac{h}{2}\right)=\int_h^{\infty}F(u)\,du \qquad [27]$$

$$P_0^{\mathrm{III}}\left(-\frac{h}{2}\right)-P_0^{\mathrm{II}}\left(-\frac{h}{2}\right)=\rho^{\mathrm{III}}W_0^{\mathrm{III}}\left(-\frac{h}{2}\right)-\rho^{\mathrm{II}}W_0^{\mathrm{II}}\left(-\frac{h}{2}\right)=\int_h^{\infty}F(u)\,du \qquad [28]$$

where

$$F(u)=F_{\mathrm{I,III}}(u)+F_{\mathrm{II,II}}(u)-F_{\mathrm{I,II}}(u)-F_{\mathrm{II,III}}(u). \qquad [29]$$

The second equality in Eqs. [27] and [28] follows from Eqs. [15]–[17]. The function *F(u)* can be expressed in the following integral form:

$$F(u)=\int_h^{\infty}2\pi\upsilon\tilde{w}(v)\,dv \qquad [30]$$

where

$$\tilde{w}(r)=\rho^{\mathrm{I}}\rho^{\mathrm{III}}w_{\mathrm{I,III}}(r)+\rho^{\mathrm{II}}\rho^{\mathrm{II}}w_{\mathrm{II,II}}(r)-\rho^{\mathrm{I}}\rho^{\mathrm{II}}w_{\mathrm{I,II}}(r)-\rho^{\mathrm{II}}\rho^{\mathrm{III}}w_{\mathrm{II,III}}(r). \qquad [31]$$

Equations [27] and [28] indicate that the discontinuities in the force density at the two interfaces of the film are equal. This conclusion is valid despite the fact that the adjoining phases have not been assumed to be identical. Hence, the equality of Eqs. [27] and [28] suggests that these localized differences in the force density can be regarded as a characteristic of the entire film system. This difference is referred to here as the disjoining pressure $(\Pi(h))$ of a plane parallel infinite film. Thus:

$$\Pi(h)=\int_h^{\infty}F(u)\,du. \qquad [32]$$

Explicit Solution for the First-Order Field Variables via Normal Mode Analysis

The first-order approximations for the field variables are substituted into the field and boundary equations, and the resulting relations are simplified to first order in ϵ. After performing these operations, the field equations and conditions at infinity become:

$$\frac{\partial u_1^j}{\partial x}+\frac{\partial v_1^j}{\partial z}=0 \quad [33]$$

$$\rho^j\frac{\partial u_1^j}{\partial t}=-\frac{\partial p_1^j}{\partial x}+\mu^j\nabla^2 u_1^j-\rho^j\frac{\partial W_1^j}{\partial x}$$

$$\rho^j\frac{\partial \upsilon_1^j}{\partial t}=-\frac{\partial p_1^j}{\partial z}+\mu^j\nabla^2 \upsilon_1^j-\rho^j\frac{\partial W_1^j}{\partial z} \quad [34]$$

$$\left(j=\mathrm{I},\mathrm{II},\mathrm{III}\right) \quad [35]$$

$$\lim_{z\to\infty} u_1^{\mathrm{I}}=\lim_{z\to\infty}\upsilon_1^{\mathrm{I}}=0 \quad [36]$$

$$\lim_{z\to-\infty} u_1^{\mathrm{III}}=\lim_{z\to-\infty}\upsilon_1^{\mathrm{III}}=0. \quad [37]$$

The boundary conditions at the displaced interfaces are composed about the mathematical surfaces $z=\xi\left(x,t\right)+h/2$ and $z=\eta\left(x,t\right)-h/2$. The value of a field variable on either one of these surfaces can be obtained by expanding this variable in a Taylor series (in powers of ξ or η) about its value on the mean plane of the interfacial displacements (i.e., z = $\pm h/2$). This series expansion can then be expressed in ascending powers of ϵ by introducing the first order approximations for ξ,η and the system variable into the series. Using this procedure, the first-order boundary conditions at the displaced surfaces become:

$$\left.\begin{aligned}
&\upsilon_1^{\mathrm{I}}=\upsilon_1^{\mathrm{II}}\\
&u_1^{\mathrm{I}}=u_1^{\mathrm{II}}\\
&\upsilon_1^{\mathrm{II}}=\frac{\partial \xi_1}{\partial t}\\
&\mu^{\mathrm{II}}\left(\frac{\partial u_1^{\mathrm{II}}}{\partial z}+\frac{\partial \upsilon_1^{\mathrm{II}}}{\partial x}\right)-\mu^{\mathrm{I}}\left(\frac{\partial u_1^{\mathrm{I}}}{\partial z}+\frac{\partial \upsilon_1^{\mathrm{I}}}{\partial x}\right)=0\\
&-p_1^{\mathrm{II}}+2\mu^{\mathrm{II}}\frac{\partial \upsilon_1^{\mathrm{II}}}{\partial z}+p_1^{\mathrm{I}}-2\mu^{\mathrm{I}}\frac{\partial \upsilon_1^{\mathrm{I}}}{\partial z}+\xi_1\left(\rho^{\mathrm{II}}DW_0^{\mathrm{II}}\right.\\
&\left.-\rho^{\mathrm{I}}DW_0^{\mathrm{I}}+g\left(\rho^{\mathrm{II}}-\rho^{\mathrm{I}}\right)\right)-\sigma_u\frac{\partial^2 \xi_1}{\partial x^2}=0
\end{aligned}\right\} z=\frac{h}{2} \quad [38\text{–}42]$$

$$\left.\begin{aligned}
&\upsilon_1^{\mathrm{II}}=\upsilon_1^{\mathrm{III}}\\
&u_1^{\mathrm{II}}=u_1^{\mathrm{III}}\\
&\upsilon_1^{\mathrm{II}}=\frac{\partial \eta_1}{\partial t}\\
&\mu^{\mathrm{III}}\left(\frac{\partial u_1^{\mathrm{III}}}{\partial z}+\frac{\partial \upsilon_1^{\mathrm{III}}}{\partial x}\right)-\mu^{\mathrm{II}}\left(\frac{\partial u_1^{\mathrm{II}}}{\partial z}+\frac{\partial \upsilon_1^{\mathrm{II}}}{\partial x}\right)=0\\
&-p_1^{\mathrm{III}}+2\mu^{\mathrm{III}}\frac{\partial \upsilon_1^{\mathrm{III}}}{\partial z}+p_1^{\mathrm{II}}-2\mu^{\mathrm{II}}\frac{\partial \upsilon_1^{\mathrm{II}}}{\partial z}+\eta_1\left(\rho^{\mathrm{III}}DW_0^{\mathrm{III}}\right.\\
&\left.-\rho^{\mathrm{II}}DW_0^{\mathrm{II}}+g\left(\rho^{\mathrm{III}}-\rho^{\mathrm{II}}\right)\right)-\sigma_1\frac{\partial^2 \eta_1}{\partial x^2}=0.
\end{aligned}\right\} z=-\frac{h}{2} \quad [43\text{–}47]$$

In the above equations, μ^j is the viscosity of phase j. The variable $\sigma_u(\sigma_l)$ denotes the tension of an interface dividing a semi-infinite phase of I (III) from a semi-infinite phase of II. [In this analysis, variation in these surface tensions due to the surface corrugations is neglected, since this effect is small even for wavelengths of the order of film thickness (31).] The symbol "D" denotes differentiation with respect to z.

The set of first-order field equations and boundary conditions (Eqs. [33]–[47]) are linear and homogeneous in the unknown functions p_1^j, u_1^j, υ_1^j, ξ_1 and η_1. (Note that the first-order potential W_1^j is an integral function of ξ_1 and η_1, cf. Eqs. [18]–[22].) This set of equations can be solved by expanding the unknown quantities in a series of normal modes (23). The boundary conditions are composed about planes of constant z, and therefore a suitable form for these normal modes is the real part of the product of the function $\exp(ikx+\omega t)$ multiplied by a scalar function of z. Owing to the linearity and homogeneity of the first-order field and boundary equations, it is sufficient to consider only one arbitrary mode; thus

$$\begin{bmatrix} p_1^j \\ u_1^j \\ \upsilon_1^j \\ \xi_1 \\ \eta_1 \end{bmatrix} = \left[\begin{bmatrix} \hat{p}^j(z,k) \\ \hat{u}^j(z,k) \\ \hat{\upsilon}^j(z,k) \\ \hat{\xi}(k) \\ \hat{\eta}(k) \end{bmatrix} \times \exp\left(ikx+\omega(k)t\right) \right]. \tag{48}$$

The parameter k is the wavenumber of the normal mode (the wavelength (λ) is equal to $2\pi/k$). The variable ω is the frequency of the motion (also referred to as the growth coefficient); the functional dependence of ω on k is the dispersion equation. If the real part of $\omega(k)$ is greater than zero for a particular mode with wavenumber k, the film system is unstable to that mode since the exponential in Eq. [48] increases without bound; similarly, if $Re(\omega(k)) < 0$, the system is stable.

The normal modes of the first order excess van der Waals potential are obtained by substituting the modal forms for ξ_1 and η_1 into Eqs. [21] and [22].

$$\rho^{\mathrm{I}}\hat{W}_1^{\mathrm{I}}(z) = \hat{\eta}\left(I_{\mathrm{I,III}}\left(z+\frac{h}{2},k\right) - I_{\mathrm{I,II}}\left(z+\frac{h}{2},k\right)\right) \tag{49}$$

$$\rho^{\mathrm{I}}\hat{W}_1^{\mathrm{II}}(z) = \begin{cases} \hat{\eta}\left(I_{\mathrm{II,III}}\left(z+\frac{h}{2},k\right) - I_{\mathrm{II,II}}\left(z+\frac{h}{2},k\right)\right)\left(\frac{h}{2} > z > \frac{h}{2} - h_1\right) \\ \hat{\xi}\left(I_{\mathrm{II,II}}\left(z-\frac{h}{2},k\right) - I_{\mathrm{I,II}}\left(z-\frac{h}{2},k\right)\right)\left(\frac{h}{2} - h_1 > z > -\frac{h}{2}\right) \end{cases} \tag{50}$$

$$\rho^{\mathrm{III}}\hat{W}_1^{\mathrm{III}}(z) = \hat{\xi}\left(I_{\mathrm{II,III}}\left(z-\frac{h}{2},k\right) - I_{\mathrm{I,III}}\left(z-\frac{h}{2},k\right)\right) \tag{51}$$

where

$$I_{i,j}(\upsilon,k) = \rho^i\rho^j \int_{-\infty}^{\infty}\int_{-\infty}^{\infty} \cos(k\beta) \times w_{i,j}\left[\left(\upsilon^2+\beta^2+\xi^2\right)^{1/2}\right] d\xi\, d\beta. \tag{52}$$

Upon substituting the normal mode expressions for the dependent variables into the first-order field equations and boundary conditions, the following relations result:

$$\left(D^4-\left(2k^2+\frac{\omega}{\nu^j}\right)D^2+k^4+\frac{k^2\omega}{\nu^j}\right)\hat{\upsilon}^j=0$$

$$(j=\mathrm{I},\mathrm{II},\mathrm{III}) \qquad [53]$$

$$\lim_{z\to\infty}\hat{\upsilon}^{\mathrm{I}}=\lim_{z\to\infty}D\hat{\upsilon}^{\mathrm{I}}=0 \qquad [54]$$

$$\lim_{z\to\infty}\hat{\upsilon}^{\mathrm{III}}=\lim_{z\to-\infty}D\hat{\upsilon}^{\mathrm{III}}=0 \qquad [55]$$

$$\left.\begin{aligned}
&\hat{\upsilon}^{\mathrm{I}}=\hat{\upsilon}^{\mathrm{II}}\\
&D\hat{\upsilon}^{\mathrm{I}}=D\hat{\upsilon}^{\mathrm{II}}\\
&\left(D^2+k^2\right)\left(\mu^{\mathrm{II}}\hat{\upsilon}^{\mathrm{II}}-\mu^{\mathrm{II}}\hat{\upsilon}^{\mathrm{I}}\right)=0\\
&\mu^{\mathrm{II}}\left(D^3-\left(3k^2+\frac{\omega}{\nu^{\mathrm{II}}}\right)D\right)\hat{\upsilon}^{\mathrm{II}}-\mu^{\mathrm{I}}\left(D^3-\left(3k^2+\frac{\omega}{\nu^{\mathrm{I}}}\right)D\right)\hat{\upsilon}^{\mathrm{I}}\\
&-\frac{k^4}{\omega}\left(\left(\sigma_u+\frac{1}{k^2}\left(-\frac{d\Pi}{dh}+g\left(\rho^{\mathrm{II}}-\rho^{\mathrm{I}}\right)\right)\right)\hat{\upsilon}^{\mathrm{II}}\left(\frac{h}{2}\right)-\frac{I(h)}{k^2}\hat{\upsilon}^{\mathrm{II}}\left(-\frac{h}{2}\right)\right)=0
\end{aligned}\right\} z=\frac{h}{2} \qquad [56–59]$$

$$\left.\begin{aligned}
&\hat{\upsilon}^{\mathrm{II}}=\hat{\upsilon}^{\mathrm{III}}\\
&D\hat{\upsilon}^{\mathrm{II}}=D\hat{\upsilon}^{\mathrm{III}}\\
&\left(D^2+k^2\right)\left(\mu^{\mathrm{II}}\hat{\upsilon}^{\mathrm{II}}-\mu^{\mathrm{III}}\hat{\upsilon}^{\mathrm{III}}\right)=0\\
&\mu^{\mathrm{III}}\left(D^3-\left(3k^2+\frac{\omega}{\nu^{\mathrm{III}}}\right)D\right)\hat{\upsilon}^{\mathrm{III}}-\mu^{\mathrm{II}}\left(D^3-\left(3k^2+\frac{\omega}{\nu^{\mathrm{II}}}\right)D\right)\hat{\upsilon}^{\mathrm{II}}\\
&-\frac{k^4}{\omega}\left(\left(\sigma_1+\frac{1}{k^2}\left(-\frac{d\Pi}{dh}+g\left(\rho^{\mathrm{III}}-\rho^{\mathrm{II}}\right)\right)\right)\hat{\upsilon}^{\mathrm{II}}\left(-\frac{h}{2}\right)-\frac{I(h)}{k^2}\hat{\upsilon}^{\mathrm{II}}\left(\frac{h}{2}\right)\right)=0.
\end{aligned}\right\} z=-\frac{h}{2} \qquad [60–63]$$

The $\hat{\rho}^j, \hat{u}^j, \hat{\xi}$ and $\hat{\eta}$ functions have been eliminated by utilizing the continuity equation, the kinematic relations and the x component of the equation of motion. The $I(h)$ function appearing in the normal stress boundary conditions is defined as:

$$I(h)=I_{\mathrm{I,III}}(h,k)+I_{\mathrm{II,II}}(h,k)-I_{\mathrm{I,II}}(h,k)-I_{\mathrm{II,III}}(h,k). \qquad [64]$$

By utilizing Eqs. [31] and [52], $I(h)$ can be expressed in the following integral form:

$$I(h)=h^2\int_{-\infty}^{\infty}\int_{-\infty}^{\infty}\cos(kh\beta')$$

$$\times\,\tilde{w}\left(h\left(1+\beta'^2+\xi'^2\right)^{1/2}\right)d\xi' d\beta' \qquad [65]$$

where $\tilde{w}(r)$ is defined by Eq. [31].

The remaining field equation (Eq. [53]) is easily integrated:

$$\hat{\upsilon}^{\mathrm{II}}(z,k) = C_1\cosh(kz) + C_2\sinh(kz) + C_3\cosh(kq^{\mathrm{II}}z) + C_4\sinh(kq^{\mathrm{II}}z) \quad [66]$$

$$\hat{\upsilon}^{\mathrm{I}}(z,k) = C_5\exp(-kz) + C_6\exp(+kq^{\mathrm{I}}z) \quad [67]$$

$$\hat{\upsilon}^{\mathrm{III}}(z,k) = C_7\exp(+kz) + C_8\exp(+kq^{\mathrm{III}}z) \quad [68]$$

where

$$q^j = \left(1 + \frac{\omega}{v^j k^2}\right)^{1/2};$$

$$Re\left[q^j\right] > 0\left(j = \mathrm{I,II,III}\right). \quad [69]$$

The variable v^j in Eq. [69] is the kinematic viscosity $\left(\mu^j/\rho^j\right)$ of phase *j*. Since the wavelength of perturbation, λ, characterizes a physical quantity, it can take only positive values. Therefore, in obtaining Eqs. [66]–[69], *k* has been restricted to positive values and the boundary conditions at infinity (Eqs. [54] and [55]) have been used.

When the integrated field equations are substituted into the boundary conditions at z = $\pm h/2$, a set of eight linear, homogeneous equations in the eight constants, C_1–C_8, are obtained. Since only nontrivial solutions to the $\hat{\upsilon}^j$ functions are desired, the determinant of the matrix of coefficients of the eight constants must be set equal to zero. The dispersion equation is derived from this latter condition since the elements of the matrix are functions of *k* and *ω*. Owing to the presence of the $\exp(-kq^{\mathrm{I}}z)$, $\exp(kq^{\mathrm{III}}z)$, $\sinh(kq^{\mathrm{II}}z)$, and $\cosh(kq^{\mathrm{II}}z)$ terms in the integrated field equations, the dispersion equation obtained will only implicitly define *ω* as a function of *k*. To obtain an explicit formulation, the "viscous liquid" inequality (25) (Eq. [70]) is introduced into the dispersion equation.

$$\frac{|\omega|}{v^j k^2} \ll 1 \;\left(j = \mathrm{I,II,III}\right). \quad [70]$$

Utilizing this inequality, the dispersion equation can be expanded in powers of $\omega/v^j k^2$; from this expansion, an explicit expression for *ω* as a function of *k* can be obtained. Once this expression for *ω*(*k*) is derived, it can then be substituted back into Eq. [70] to verify that the inequality is satisfied. In the following section, the dispersion equation for a symmetrical system is computed. The reason for initially considering this system is that it will illustrate the following important aspect of inequality [70]: When the dispersion equation is expanded to first order in $\omega/v^j k^2$, the resulting expression is the same as would have been obtained had the $\partial\upsilon_1^j/\partial t$ and $\partial u_1^j/\partial t$ terms been neglected in the first-order equations of motion (Eqs. [34] and [35]). This latter observation is then used in computing the dispersion equation (accurate to first order in $\omega/v^j k^2$) of an unsymmetrical system. The validity of neglecting all terms of second order and greater in $\left(\omega/v^j k^2\right)$ is discussed in Appendix A.

The Dispersion Equation for a Symmetrical System

As remarked in the Introduction, a symmetrical film system is one in which gravity is neglected and in which the semi-infinite phases surrounding the film are identical. Thus, in such a system $\mu^{\mathrm{I}} = \mu^{\mathrm{III}}, \rho^{\mathrm{I}} = \rho^{\mathrm{III}}, q^{\mathrm{I}} = q^{\mathrm{III}}$, and $\sigma_{\mathrm{u}} = \sigma_{\mathrm{l}} \equiv \sigma$. With these simplifications, the boundary condition at z = $\pm h/2$ admits solutions for the $\hat{\upsilon}^j(z,k)$ functions which are even (symmetric) and odd (antisymmetric) functions of z. For the even solution set $C_7 = C_5$, $C_8 = C_6$, and $C_2 = C_4 = 0$; for the odd set $C_7 = -C_5$, $C_8 = -C_6$, and $C_1 - C_3 = 0$. Substituting these solution sets into the boundary conditions at $z = h/2$ yields a set of four linear, homogeneous equations in the four remaining unknowns. (The same four equations are generated if the sets are substituted into the boundary conditions at $z = -h/2$.) Upon setting equal to zero the determinant of the matrix of coefficients of these four equations, the following dispersion equations for the even and odd solutions are obtained.

$$\frac{\psi}{k}\left(1-\frac{\mu^{\mathrm{I}}k^2}{\rho^{\mathrm{II}}\omega}\left(1+q^{\mathrm{I}}\right)\left(X-q^{\mathrm{II}}Y\right)\right)$$

$$+\frac{4\left(\mu^{\mathrm{II}}\right)^2}{\rho^{\mathrm{II}}}\left(X-q^{\mathrm{II}}Y\right)\left(1+\frac{\mu^{\mathrm{I}}}{\mu^{\mathrm{II}}}\left(q^{\mathrm{I}}-1\right)-\left(\frac{\mu^{\mathrm{I}}}{\mu^{\mathrm{II}}}\right)^2 q^{\mathrm{I}}\right)+\rho^{\mathrm{II}}\frac{\omega^2}{k^4}\left(X+\frac{\rho^{\mathrm{I}}}{\rho^{\mathrm{II}}}\right)$$

$$+\mu^{\mathrm{II}}\frac{\omega}{k^2}\left(4X-\frac{\mu^{\mathrm{I}}}{\mu^{\mathrm{II}}}\frac{\rho^{\mathrm{I}}}{\rho^{\mathrm{II}}}\left(X-q^{\mathrm{II}}Y\right)\left(1+q^{\mathrm{I}}\right)\right)$$

$$+\mu^{\mathrm{I}}\frac{\omega}{k^2}\left(\left(1+q^{\mathrm{I}}\right)\left(1+q^{\mathrm{II}}XY\right)\right.$$

$$\left.+2X\left(q^{\mathrm{I}}-1\right)\right)=0 \qquad [71]$$

where, for the even solution

$$\psi=\sigma-\frac{1}{k^2}\left(\frac{d\Pi}{dh}+I(h)\right) \qquad [72]$$

$$X=\tanh\frac{kh}{2} \qquad [73]$$

$$Y=\tanh\left(\frac{kh}{2}q^{\mathrm{II}}\right) \qquad [74]$$

and for the odd solution

$$\psi=\sigma+\frac{1}{k^2}\left(-\frac{d\Pi}{dh}+I(h)\right) \qquad [75]$$

$$X=\coth\frac{kh}{2} \qquad [76]$$

$$Y=\coth\left(\frac{kh}{2}q^{\mathrm{II}}\right). \qquad [77]$$

Following the nomenclature proposed by Felderhof, the normal modes of the even solution set are termed stretching (ST) modes and those of the odd set are classified as squeezing (SQ) modes (Fig. 1). In the limit of infinite film thickness, F(h) and $I(h)$ tend toward zero (cf. Eqs. [30] and [65]) and all four of the hyperbolic functions (see Eqs. [73], [74], [76], and [77]) tend to unity.

Consequently in this limit the dispersion equations for both the SQ and ST modes are identical and reduce to the dispersion equation for waves on the interface between two bulk fluids. Furthermore, the limiting expressions for the SQ and ST dispersion equations as μ^{I} and ρ^{I} tend towards zero (a free liquid film) are equal to those derived by Sche (16) and Sche and Fijnaut (17).

The extension to three dimensions involves only replacing k by $\left(k_x^2+k_y^2\right)^{1/2}$, where k_x and k_y are the wavenumbers in the x and y directions, respectively (23).

Expanding the q^{I}, q^{II}, and Y terms in Eq. [71] in powers of $\omega/v^j k^2$ and retaining only first-order terms yields the following expressions for the dispersion equation of the SQ (Eq. [78]) and ST (Eq. [79]) modes.

$$\omega_{\mathrm{SQ}}=-\frac{k}{2\mu^{\mathrm{II}}}\left(\sigma-\frac{1}{k^2}\left(\frac{d\Pi}{dh}-I\left(h\right)\right)\right)\Omega_{\mathrm{SQ}}\left(k,h,R\right) \qquad [78]$$

$$\omega_{\mathrm{ST}}=-\frac{k}{2\mu^{\mathrm{II}}}\left(\sigma-\frac{1}{k^2}\left(\frac{d\Pi}{dh}+I\left(h\right)\right)\right)$$
$$\times\,\Omega_{\mathrm{ST}}\left(k,h,R\right) \qquad [79]$$

where

$$\Omega_{\mathrm{SQ}}=\frac{\cosh\left(kh\right)+R\sinh\left(kh\right)-1-khR}{\left(1+R^2\right)\sinh\left(kh\right)+2R\cosh\left(kh\right)+kh\left(1-R^2\right)} \qquad [80]$$

$$\Omega_{\mathrm{ST}}=\frac{\cosh\left(kh\right)+R\sinh\left(kh\right)+1+khR}{\left(1+R^2\right)\sinh\left(kh\right)+2R\cosh\left(kh\right)-kh\left(1-R^2\right)} \qquad [81]$$

$$R=\frac{\mu^{\mathrm{I}}}{\mu^{\mathrm{II}}}. \qquad [82]$$

These results for the first-order dispersion equations are valid for both short and long wavelength disturbance; previously derived first-order equations are valid only for long wavelength perturbations (2, 4, 6, 16, 17). Equations [78] and [79] could also have been derived directly by neglecting the $\omega/v^j k^2$ terms in the normal mode field equation (Eq. [53]) and in the normal stress boundary conditions (Eqs. [59] and [63]). This latter procedure is equivalent to neglecting the $\partial u_1^j/\partial t$ and $\partial \upsilon_1^j/\partial t$ terms in the first-order equations of motion (Eqs. [34] and [35]) and is utilized in the next section to obtain the dispersion relation for an unsymmetrical system.

The Dispersion Equation for an Unsymmetrical System

After neglecting the $\omega/v^j k^2$ terms in the normal mode field equation (Eq. [53]), the following expressions for the $\hat{v}^j\left(z,k\right)$ functions are obtained upon integration:

$$\hat{\upsilon}^{\mathrm{II}}\left(z,k\right)=a_1\cosh\left(kz\right)+a_2\sinh\left(kz\right)+a_3z\cosh\left(kz\right)+a_4z\sinh\left(kz\right) \qquad [83]$$

$$\hat{\upsilon}^{\mathrm{I}}\left(z,k\right)=a_5\exp\left(-kz\right)+a_6z\exp\left(-kz\right) \qquad [84]$$

$$\hat{\upsilon}^{\mathrm{III}}\left(z,k\right)=a_7\exp\left(kz\right)+a_8z\exp\left(kz\right) \qquad [85]$$

where, as in Eqs. [66]–[69], k has been restricted to positive values and the boundary conditions at infinity (Eqs. [54] and [55]) have been used. Upon substituting these integrated solutions into the boundary conditions at $z=\pm h/2$ and then setting equal to zero the determinant of the resulting matrix of coefficients, the following two dispersion equations are obtained.

$$\omega_1=\frac{k}{2\mu^{\mathrm{II}}\delta_2}\left[\theta+\gamma\right] \qquad [86]$$

$$\omega_2 = \frac{k}{2\mu^{\mathrm{II}}\delta_2}\left[\theta - \gamma\right] \quad [87]$$

where

$$\theta = -\frac{1}{2}\left[(\chi_1 - \Phi)\Lambda_1(R_1,R_2) + (\chi_1 + \Phi)\Lambda_2(R_1,R_2) + (\chi_2 - \Phi)\Lambda_1(R_2,R_1) + (\chi_2 + \Phi)\Lambda_2(R_2,R_1)\right] \quad [88]$$

$$\gamma = \left\{\theta^2 - \delta_1\delta_2\left[(\chi_1 - \Phi)(\chi_2 - \Phi) + (\chi_1 - \Phi)(\chi_2 - \Phi)\right]\right\}^{1/2}. \quad [89]$$

The variables appearing in the above equations are defined below.

$$\begin{aligned}\Lambda_n(\alpha,\beta) = \frac{1}{2}\Big[&\big(\coth(kh) - (-1)^n \operatorname{cosech}(kh)\big)\big(1 - (-1)^n kh\operatorname{cosech}(kh)\big)\\ &+ \big(\beta + \alpha k^2h^2 \operatorname{cosech}^2(kh)\big) + \beta^2\big(\coth(kh) - kh\operatorname{cosech}^2(kh)\\ &+ \alpha\big(1 - k^2h^2\operatorname{cosech}^2(kh)\big)\big) + (R_1 + R_2)\big(\coth^2(kh) - (-1)^n\coth(kh)\big)\\ &\times \operatorname{cosech}(kh) - (-1)^n kh \operatorname{cosech}(kh)\big) + R_1R_2(2\coth(kh)\\ &- (-1)^n\operatorname{cosech}(kh) - (-1)^n kh\coth(kh)\operatorname{cosech}(kh)\Big](n = 1,2)\end{aligned} \quad [90]$$

$$\delta_1 = \frac{1}{2}\left[(R_1 + \coth(kh))(R_2 + \coth(kh)) - \operatorname{cosech}^2(kh)(1 + khR_1)(1 + khR_2)\right] \quad [91]$$

$$\begin{aligned}\delta_2 = &\left(1 - k^2h^2\operatorname{cosech}^2(kh)\right)\left(1 + R_1^2R_2^2\right) + \left(\coth^2(kh) + k^2h^2\operatorname{cosech}^2(kh)\right)\left(R_1^2 + R_2^2\right)\\ &+ 2\coth(kh)(R_1 + R_2) + 2R_1R_2\left(1 + \coth^2(kh) + (R_1 + R_2)\coth(kh)\right)\end{aligned} \quad [92]$$

$$R_1 = \frac{\mu^{\mathrm{I}}}{\mu^{\mathrm{II}}};\ R_2 = \frac{\mu^{\mathrm{III}}}{\mu^{\mathrm{II}}} \quad [93]$$

$$\chi_1 = \sigma_u + \frac{1}{k^2}\left(-\frac{d\Pi}{dh} + \left(\rho^{\mathrm{II}} - \rho^{\mathrm{I}}\right)g\right) \quad [94]$$

$$\chi_2 = \sigma_1 + \frac{1}{k^2}\left(-\frac{d\Pi}{dh} + \left(\rho^{\mathrm{III}} - \rho^{\mathrm{II}}\right)g\right) \quad [95]$$

$$\Phi = I(h)/k^2. \quad [96]$$

These results are new in that explicit first-order dispersion equations for an unsymmetrical film system have yet to be derived; they lead to dispersion equations for a symmetrical system and for a film on a solid substrate when appropriate limits are taken. For the parameters of a symmetrical system (i.e., $\sigma_1 = \sigma_u$, $\mu^{\mathrm{I}} = \mu^{\mathrm{III}}, g = 0$) Eqs. [86] and [87] reduce to Eqs. [78] and [79]. Furthermore, from the equations for an unsymmetrical system, the dispersion relation for a wetting film (i.e., a film on a solid substrate) can be derived by computing the limiting expressions of Eqs. [86] and [87] as R_2 tends to infinity (Fig. 1). Depending on the sign of the $\lim_{R\to\infty}\theta$, one of the above two equations (either [86] or) simplifies to $\omega^2 = 0$ as R_2 tends to infinity and the other reduces to:

$$\omega = -\frac{k}{2\mu^{\mathrm{II}}}\left(\sigma_u + \frac{1}{k^2}\left(-\frac{d\Pi}{dh} + \left(\rho^{\mathrm{II}} - \rho^{\mathrm{I}}\right)g\right)\right)\Xi(k,h,R_1) \quad [97]$$

where

$$\Xi(k,h,R_1)=\frac{\sinh(kh)\cosh(kh)-kh+R_1\left(\sinh^2(kh)-k^2h^2\right)}{\left(R_1\sinh(kh)+\cosh(kh)\right)^2+k^2h^2\left(1-R_1^2\right)}. \qquad [98]$$

Equation [97] is analogous to the result obtained by Jain and Ruckenstein (11), except for the gravitational term which was neglected in their analysis.

RESULTS AND DISCUSSION

In this section, the stability characteristics of symmetrical and unsymmetrical systems are examined, both in the long and short wavelength limits. Wherever possible, explicit asymptotic results are obtained from the general dispersion relations (Eqs. [78], [79], [86], and [87]) derived in the previous section. The results are, then, used to explain available experimental observations.

Symmetrical Systems

In this subsection, first the asymptotic forms of the dispersion equations for the squeezing and stretching modes are derived for long wavelengths (i.e., for wavenumbers satisfying the inequality $kh \ll 1$). Then, the complete dispersion equations (Eqs. [78] and [79]) are analyzed numerically.

(a) Asymptotic relations.

A necessary prerequisite for this asymptotic development is to obtain an expression for $I(h)$ as $k \to 0$. Such an expression is obtained in Appendix B and is given here in terms of Π (Eq. [32]):

$$I(h)=-\frac{d\Pi}{dh}-\frac{k^2h^2}{2}\left(\frac{\Pi(h)}{h}+\frac{1}{h^2}\int_h^\infty \Pi(v)\,dv\right)+O\left(k^4\right)\ (k\to 0). \qquad [99]$$

For a symmetrical film system Eq. [99] can be written in terms of the film tension Δ, where

$$\Delta = 2\sigma + h\Pi(h)+\int_h^\infty \Pi(v)\,dv. \qquad [100]$$

Combining Eqs. [99] and [100] yields:

$$I(h)=-\frac{d\Pi}{dh}+k^2\left(\sigma-\frac{\Delta}{2}\right)+O\left((k)^4\right)\ (k\to 0). \qquad [101]$$

With the aid of the asymptotic development for $I(h)$, the dispersion equations can easily be simplified for the long wavelength limit.

Expanding the hyperbolic functions in the defining relations for Ω_{SQ} and Ω_{ST} in a Taylor series in powers of kh yields the following approximate equations for ω_{SQ} and ω_{ST}:

$$\omega_{SQ}\approx-\frac{k^3h^2}{8\mu^{II}}\left(2\sigma-\frac{\Delta}{2}-\frac{2}{k^2}\frac{d\Pi}{dh}\right)\left[\frac{1+\frac{kh}{12}R}{kh+R+\frac{k^3h^3}{12}R^2}\right] \qquad [102]$$

$$\approx -\frac{k^3h^2}{8\mu^{\text{I}}}\left(2\sigma - \frac{\Delta}{2} - \frac{2}{k^2}\frac{d\Pi}{dh}\right)$$

$$\text{for}\ \frac{1}{kh} \gg R \gg kh \qquad [103]$$

$$\omega_{\text{ST}} \approx -\frac{k\Delta}{4\mu^{\text{II}}}\left[\frac{1+khR}{\frac{k^3h^3}{12}+R+khR^2}\right] \qquad [104]$$

$$\approx -\frac{k}{4\mu^{\text{I}}}\Delta\ \left(R \gg kh\right). \qquad [105]$$

In the above equations, only the first two terms in the asymptotic development of $I(h)$ are used. From Eqs. [102] and [104], the criteria for stability of the SQ and ST modes in the long wavelength limit can be formulated.

For the squeezing mode, the condition for marginal or neutral stability $\omega = 0$ is defined by:

$$\sigma - \frac{1}{2}h\Pi(h) - \frac{1}{2}\int_h^{\infty}\Pi(v)dv - \frac{2}{k^2}\frac{d\Pi}{dh} = 0. \qquad [106]$$

The value of k which satisfies Eq. [106] is termed the critical wavenumber (k_c). If London's law $\left(w_{i,j}(r) = -C_{i,j}r^{-6}\right)$. is utilized for the intermolecular potential, $\Pi(h)$, calculated from Eqs. [30] and [32], has the form

$$\Pi(h) = -\frac{A}{6\pi h^3}. \qquad [107]$$

The constant A is the Hamaker constant of the film system and is defined as:

$$A = A_{\text{I,III}} + A_{\text{II,II}} - A_{\text{I,II}} - A_{\text{II,III}} \qquad [108]$$

where

$$A_{i,j} = \pi^2\rho^i\rho^jC_{i,j}. \qquad [109]$$

With $\Pi(h)$ defined by Eq. [107], the following equation for the critical wavelength $\lambda_c\left(\lambda_c = 2\pi/k_c\right)$. is obtained:

$$\lambda_c = 2\pi\left[\frac{\pi h^4\sigma}{A}\right]^{1/2}\left[1+\frac{A}{8\pi\sigma h^2}\right]^{1/2}. \qquad [110]$$

Using typical values for σ (10^{-3}–10^{-2} N/m), A (10^{-20} J), and h (10 nm) yield values for λ_c which satisfy the long wavelength restriction ($k_ch \ll 1$). For $\lambda > \lambda_c, \omega_{\text{SQ}} > 0$ and the SQ mode is unstable; alternatively, for $\lambda < \lambda_c$, $\omega_{\text{SQ}} < 0$ and the SQ mode is stable.

Computer simulations of $\omega_{\text{SQ}}^{-1}\left(=\tau_{\text{SQ}}\right)$ as a function of λ for $\lambda > \lambda_c$ and for values of R = 1, 2, and 10 are given in Fig. 4. These simulations indicate the existence of an inflection point minimum of λ_{SQ}. An analytical expression for the wavelength at which this minimum occurs is obtained by differentiating the expression for ω_{SQ} (Eq. [78]) with respect to k and then obtaining the roots of the equation $d\omega_{\text{SQ}}/dk = 0$ which satisfy the long wavelength limit. For values of k_c such that $k_ch \ll 1$ and

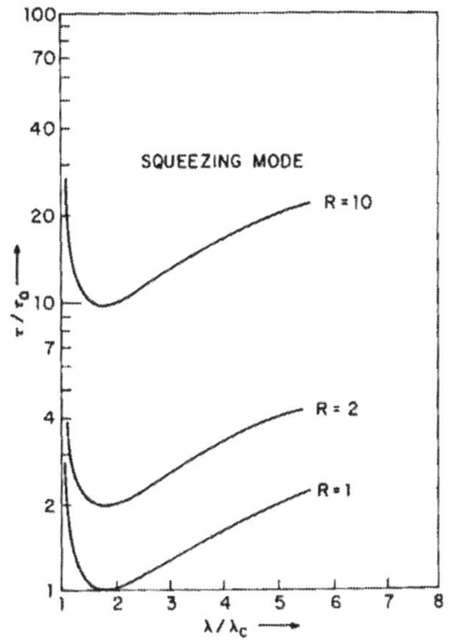

FIG. 4. Dependence of $1/\omega$ $(=\tau)$ on λ for the SQ mode for various values of R as calculated from Eq. [78] with Eq. [107] used for $\Pi(h)$. Note that τ is normalized with respect to τ_0. τ_0 is equal to τ evaluated at $\lambda=\lambda_d$ and $R = 1$ for the SQ mode. [$h = 50$ mn. $\sigma = 0.05\,\text{N/m}$, $A = 10^{-20}\,\text{J}$, $\mu^{\text{II}} = 10^{-3}\,\text{kg/m-sec.}$]

$1/k_c h > R > k_c h$, the wavelength (termed the dominant wavelength and denoted by λ_d) at which the inflection point occurs is given by:

$$\lambda_d \approx 3^{1/2}\lambda_c. \qquad [111]$$

In the limit, $R \to 0$, the dominant wavelength tends to infinity. The minimum value of τ_{SQ}, since it represents the largest value of ω_{SQ}, will dominate the growth of instabilities in the linear regime of films vibrating in the squeezing mode.

The dispersion equation given by Eq. [105] is a new result. Equations [102], [103], [110], and [111] are, however, identical with the results of Vrij *et al.* (when expanded to first order in ω) except for the term $1/2\Pi(h)h + 1/2\int_h^\infty \Pi(v)dv$ which, for $\Pi(h) = -A/6\pi h^3$, is $-A/8\pi h^2$. This term represents the second-order term in the asymptotic development of $I(h)$ as $k \to 0$ (see Eq. [B–12]); the term is not present in the paper of Vrij *et al.* because their work uses the approximate disjoining pressure approach to formulate the influence of the van der Waals interaction on the film stability (see the Introduction).

From Eq. [104], the condition for marginal stability of the stretching mode is $\Delta = 0$. The system becomes unstable to the ST mode when $\Delta < 0$. With $\Pi(h)$ defined by Eq. [107], the expression for the film tension is:

$$\frac{\Delta}{2} = \sigma - \frac{A}{8\pi h^2}. \qquad [112]$$

Thus for systems in which $\sigma < A/8\pi h^2$, the ST mode is unstable, and conversely, if $\sigma < A/8\pi h^2$, the ST mode is stable. As remarked earlier, the Hamaker constant A is on the order of 10^{-20} J; using a typical value for h of 10 nm yields a value for $A/8\pi h^2$ which is of the order of 10^{-6} N/m. Thus the stretching mode will become unstable only for systems with extremely low interfacial tensions, such as the cell membrane.

Note that the condition for marginal stability, Eq. [112], is independent of the wavelength. It should be interpreted with caution. This condition was derived using the long wavelength approximation for *I(h)*, and, therefore, cannot be used to rule out the possibility of a critical wavelength in the short wavelength limit. As a matter of fact, we will show numerically in the next subsection that a critical wavelength exists for the stretching mode in the short wavelength limit, and convince the reader that restricting the analysis to long wavelengths can, sometime, lead to misleading or incomplete results.

For the stretching mode, the dominant wavelength, in the long wavelength limit, is given by

$$\lambda_d \approx \frac{2\pi h}{(6R)^{1/3}} \quad [113]$$

for values of R much less than one $(R \ll 1)$. Thus for the ST mode, a dominant wavelength will exist for long waves only for values of μ^{I}/μ^{II} which are much less than one. A significant, new conclusion evident from Eq. [113] is that the dominant wavelength of the ST mode is a function of the hydrodynamic parameter R (the ratio of viscosities) and the film thickness h. For the SQ mode (cf. Eq. [111]), however, λ_d is independent of R and is a function of h, σ, and the intermolecular interaction parameter A.

Using parameter characteristics of a biological membrane ($R = 10^{-6}$ and $h = 10$ nm), one finds that λ_d is in the order of $1\,\mu$m, which is the characteristic diameter of microvilli (13). The present analysis also suggests that the rate of growth of perturbations, ω_{ST}, is inversely proportional to membrane viscosity (Eq. [104]). Consequently, cells with a low value of membrane viscosity will form villi at a rate faster than cells with large viscosity. As a matter of fact, several investigators have suggested that the membrane of a neoplastic cell is more "fluid" (less viscous) than that of a normal cell. This increased growth rate, due to a low viscosity, could account, at least in part, for the greatly increased formation of villi in malignant cells compared to those in normal cells (13).

The stretching mode instability of a membrane can also explain ingestion and phagocytosis phenomena, in which a biological membrane bends to engulf the extracellular matter. Deformation of a biological membrane is also a key step in cellular movement. Since time constants of these processes are not available, it is not possible to compare quantitatively the results of our analysis with experimental observations, however, qualitatively our analysis offers plausible explanations of these complex biological phenomena.

(b) Numerical simulations.

The cell membrane provides an example of a system in which both the squeezing and stretching modes can become unstable. Therefore, we will use parameters characteristic of the cell membrane in these numerical simulations using Eqs. [78] and [79]. The interfacial tension of the cell membrane is characteristically small (28), with typical values ranging between 10^{-6} and 10^{-3} N/m. Furthermore, extending into the intra- and extracellular fluids of the membrane are electrical double layers: These double layers have a destabilizing effect and can be at least qualitatively accounted for in this analysis by adding to the interfacial tension (σ) the double layer tension σ_{dl} (29). The double layer tension is of the order of -10^{-3} N/m for biological systems (13), and it can reduce the value of both σ_u and σ_l considerably. Therefore, first, we will examine the role of interfacial tension $\sigma = \sigma_u = \sigma_l$ on the stability characteristics of a membrane. Specifically, we will examine the effect of σ on λ_c for both squeezing and stretching modes.

Owing to the positive definite nature of Ω_{SQ} and Ω_{ST} (Eqs. [78] and [79]), critical wavenumbers $k_c \left(= 2\pi/\lambda_c\right)$ must satisfy the following equations:

$$\sigma - \frac{A}{2\pi k_{c,SQ}^2 h^4} - \frac{A}{4\pi h} K_2\left(k_{c,SQ} h\right) = 0 \quad [114]$$

And

$$\sigma - \frac{A}{2\pi k_{c,ST}^2 h^4} + \frac{A}{4\pi h^2} K_2\left(k_{c,ST}h\right) = 0 \qquad [115]$$

where K_2 is the modified Bessel function of second kind. In deriving Eqs. [114] and [115] from Eqs. [78] and [79], we have used London's law $\left(w_{i,j} = -C_{i,j}/r^6\right)$ to obtain expressions for $\Pi(h)\left(= -A/6\pi h^3\right)$ and $I(h)\left(= -\left(A/4\pi h^2\right)k^2 K_2(kh)\right)$. Since $K_2(kh)$ is a monotonically decreasing, positive definite function of kh, a value of k_c will always exist for the squeezing mode for all values of k_c. However, k_c will exist for the stretching mode only when $\sigma < A/8\pi h^2$ (or, $\Delta < 0$). The film will become stable to the stretching mode for $\Delta > 0$ (Eq. [112]). In addition, if there is a critical wavenumber for the stretching mode, then one can show that for a given value of σ

$$k_{c,SQ} > k_{c,ST} \text{ or } \lambda_{c,SQ} < \lambda_{c,ST}. \qquad [116]$$

This condition can be obtained by subtracting Eq. [114] from Eq. [115], and realizing that $K_2(kh)$ is positive-definite.

The dependence of λ_c/h on σ for squeezing and stretching modes is plotted in Fig. 5. Note that $\lambda_{c,ST}$ ceases to exist after $\sigma = 4\times10^{-6}$ N/m when $\Delta > 0$. When both the squeezing and stretching modes are unstable, $\lambda_{c,ST} > \lambda_{c,SQ}$, though the difference in $\lambda_{c,ST}$ and $\lambda_{c,SQ}$ is small. It must be pointed out here that $\lambda_{c,SQ}/h$ agrees exactly with the value calculated using Eq. [110] for $\lambda_c/h > 10$. As mentioned in the previous section, it is not possible to calculate $\lambda_{c,ST}$ using the asymptotic expression (Eq. [112]).

Now, it is worthwhile to examine the stability characteristics of a film to perturbations of wide range of wavelengths.

In Fig. 6 is plotted computer simulations of $\omega_{SQ}^{-1}\left(= \tau_{SQ}\right)$ and $\omega_{ST}^{-1}\left(= \tau_{ST}\right)$ as a function of λ/h for λ/h for $1 \le \lambda/h \le 3000$. The maximum wavelength considered (30 μm) is a typical value for the circumference of a cell membrane. A value of 10^{-8} N/m was used for the total tension $\left(\sigma + \sigma_{dl}\right)$. As expected, the numerical simulations shown in Fig. 6 are in exact agreement with the asymptotic results obtained for $\lambda/h \gg 1$ (Eqs. [102] and [104]). They, however, also lead to some interesting and new results in the limit $\lambda/h \sim 1$. Most importantly, we observe a critical wavelength for the stretching mode, which was not obvious from the long wavelength analysis (Eq. [112]). Specifically for these parameter values, $\lambda_c/h \cong 1.5$ for the squeezing mode, and $\lambda_c/h \cong 1.75$ for the stretching mode. The minimum value of τ_{ST} occurs at $\lambda_d/h \cong 345$, exactly as predicted from the long wavelength analysis (Eq. [113]). While it

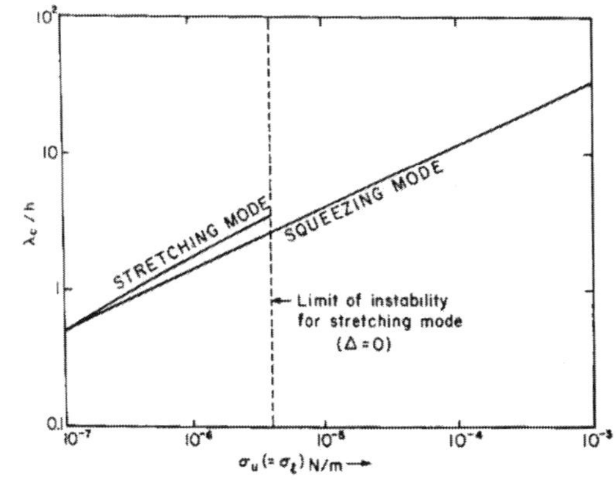

FIG. 5. Dependence of λ_c/h on the surface tension for squeezing and stretching modes of vibration. Note that the stretching mode is stable for $\sigma > A/8\pi h^2 = 4\times10^{-6}$ N/m, and that $\lambda_{c,SQ} < \lambda_{c,ST}$. $\left[h = 10 \text{ nm}, A = 10^{-20} \text{ J}, \mu^{II} = 10^3 \text{kg/m-sec}, \mu^{I} = \mu^{III} = 10^{-3} \text{kg/m-sec.}\right]$

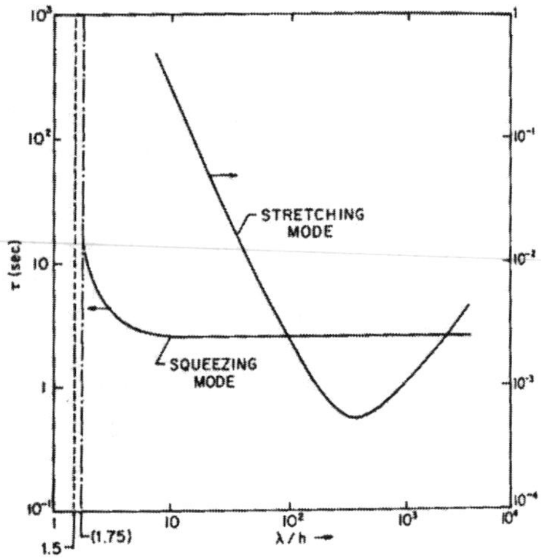

FIG. 6. Dependence of $1/\omega\ (=\tau)$ on λ for SQ and ST modes using biological parameters as calculated from Eqs. [78] and [79] with Eq. [107] used for $\Pi(h)$. [h = 10 nm, σ = 10^{-6} N/m, A = 10^{-20} J, $\mu^{\mathrm{II}} = 10^3\,\mathrm{kg/m-sec}$, $\mu^{\mathrm{I}} = 10^{-3}\,\mathrm{kg/m-sec}$.] Note that the minimum value of τ_{ST} is four orders of magnitude less than the minimum value of τ_{SQ}.

is not obvious from Fig. 6 that τ_{SQ} exhibits a minimum similar to those shown in Fig. 4 for $R = 1$, 2, and 10, its existence can be proved numerically by plotting τ versus λ/h for values of $\lambda/h > 10^4$. As a matter of fact, $d^2\tau_{SQ}/d\lambda^2$ near λ_d increases as R increases. This shallow minimum in the τ_{SQ} versus λ/h curve is expected since for a biological membrane, R ($=10^{-6}$) is much less than one.

Finally, the results indicate that both the SQ and ST modes are unstable for these parameter values and that the growth of the instabilities is dominated by the stretching mode since the minimum value of τ_{ST} is four orders of magnitude less than the smallest value of τ_{SQ}. Physically, these simulations imply that a biological membrane will bend more readily to form microvilli or to ingest external material via phagocytosis than have "sausage type" local variation in its thickness as a result of instability (Fig. 1).

Unsymmetrical Systems

The study of the stability characteristics of unsymmetrical films is difficult because of the complexity of the general dispersion equations. A convenient starting point for such a study is the examination of the modal forms of the first-order interfacial displacement functions ξ_1 and η_1. The modal forms of ξ_1, and η_1, are characterized by the amplitude ratio k and the phase difference ϕ between the sinusoidal vibrations of the interfaces of the film. From the modal expressions for ξ_1 and η_1, (Eq. [48]), it is evident that

$$\kappa = \frac{|\hat{\xi}|}{|\hat{\eta}|};\ \phi = \arctan\left(\frac{Im[\ \hat{\xi}/\hat{\eta}\ ]}{Re[\ \hat{\xi}/\hat{\eta}\ ]}\right). \qquad [117]$$

The ratio $\xi/\hat{\eta}$ is equal to $\hat{\upsilon}^{\mathrm{I}}\left(h/2\right)/\hat{\upsilon}^{\mathrm{III}}\left(-h/2\right)$ (cf. Eqs. [40] and [45]). An analytical expression for $\hat{\upsilon}^{\mathrm{I}}\left(h/2\right)/\hat{\upsilon}^{\mathrm{III}}\left(-h/2\right)$ can be obtained from the integrated field equations. To simplify the calculations, all terms containing ω/v^jk^2 are neglected in the normal mode field equations and boundary conditions (Eqs. [53]–[63]). With this approximation, the ratio $\hat{\upsilon}^{\mathrm{I}}\left(h/2\right)/\hat{\upsilon}^{\mathrm{III}}\left(-h/2\right)$ is, from Eqs. [84] and [85], equal to:

$$\frac{\hat{\upsilon}^{\mathrm{I}}\left(\dfrac{h}{2}\right)}{\hat{\upsilon}^{\mathrm{III}}\left(-\dfrac{h}{2}\right)} = \frac{1}{\dfrac{a_7}{a_5} - \dfrac{h}{2}\dfrac{a_8}{a_5}} + \frac{\dfrac{h}{2}}{\dfrac{a_7}{a_6} - \dfrac{h}{2}\dfrac{a_8}{a_6}}. \qquad [118]$$

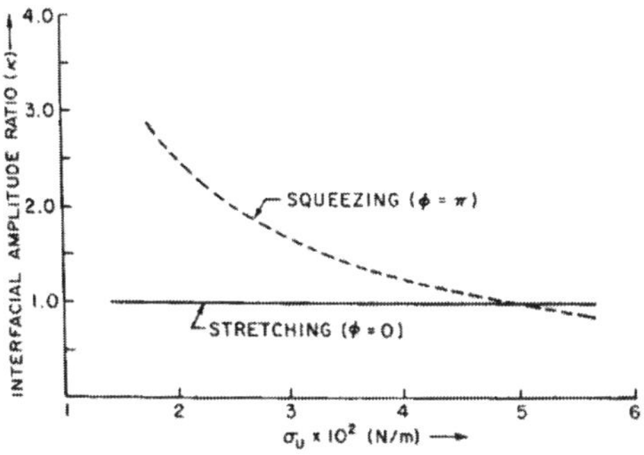

FIG. 7. Dependence of k on σ_u in a large surface tension system for the two modes of vibration described by Eqs. [86] and [87] with Eq. [107] used for $\Pi(h)$. [h = 10 nm, $\sigma_l = 0.05\,N/m, A = 10^{-20}\,J$, $\mu^I = \mu^{II} = \mu^{III} = 10^{-1}$ kg/m-sec; $\lambda = 10^2\,\mu m$.]

From Eq. [118], it is clear that to determine the variables k and ϕ, expressions for certain quotients of the integration constants are required. These expressions are obtained through the matrix of coefficients of the boundary conditions; details of the calculations are provided elsewhere (15). Numerical computations indicate that the variables κ and ϕ depend on the interfacial tensions σ_l and σ_u and the inter-molecular forces. When the interface tension $(\sigma_l \text{ or } \sigma_u)$ is large, the result is a decrease in the amplitude of the disturbance. When intermolecular forces dominate the film dynamics, the result is an increase in the amplitude of perturbation. In Figs. 7 and 8 are illustrated the variation in κ with changes in σ_u (σ_l was held constant in the simulation) for the two modes of vibration described by Eqs. [86] and [87]. For $\sigma_l = 0.05\,N/m$, $\lambda = 100\,\mu m$, and h = 10 nm, ϕ for one mode is always equal to π (a squeezing vibration) and ϕ for the other mode is always equal to 0 (a stretching vibration) (Fig. 7). For the mode corresponding to a squeezing vibration ($\phi = \pi$), κ decreases as (σ_u increases. This is an expected result since the interfacial tension is inversely proportional to the deformability of an interface. For the mode corresponding to a stretching vibration ($\phi = 0$), κ is independent of σ_u and remains equal to one. For $\sigma_l = 10^{-6}\,N/m$, $\lambda = 50$ nm, and h = 10 nm, ϕ for one mode is always equal to 0 (a stretching vibration) and ϕ for the other mode is equal to 0 (a squeezing vibration) up to a certain value of σ_u ($= 2.5 \times 10^{-6}\,N/m$) and beyond that value of surface tension, ϕ becomes zero again (Fig. 8). Thus, an unsymmetric membrane can have $\phi = 0$ for both modes of vibration depending upon the values of surface tensions and wavelengths (Fig. 8). In this case, κ decreases as increases at large values of σ_u as expected. At low values of σ_u, intermolecular forces lead to enhanced attraction at the crests of perturbations causing the amplitude ratio to increase (Fig. 8). Numerical simulations carried out for disturbances of various wavelengths showed similar trends. It must be pointed out

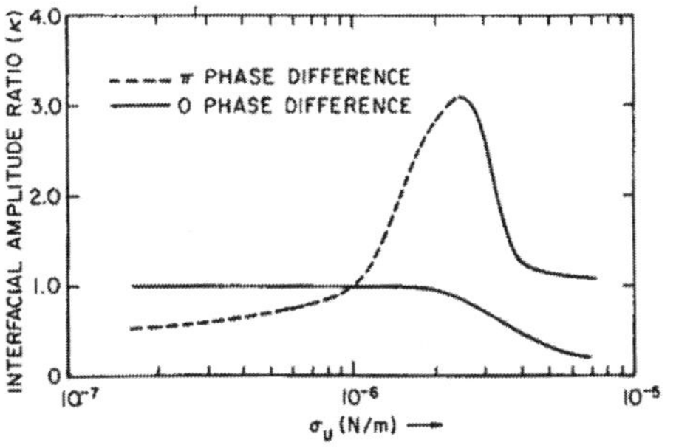

FIG. 8. Dependence of κ on σ_u in a low surface tension system for the two modes of vibration described by Eqs. [86] and [87] with Eq. [107] used for $\Pi(h)$. [h = 10 nm, $\sigma_l = 10^{-8}\,N/m$, $A = 10^{-20}\,J$, $\mu^I = \mu^{III} = 10^{-3}$ kg/m-sec, $\mu^{II} = 10^3$ kg/m-sec, $\lambda = 50$ nm.] Note that for $\sigma_u > 2.5 \times 10^{-6}\,N/m$, both modes of instability are stretching vibrations ($\phi = 0$).

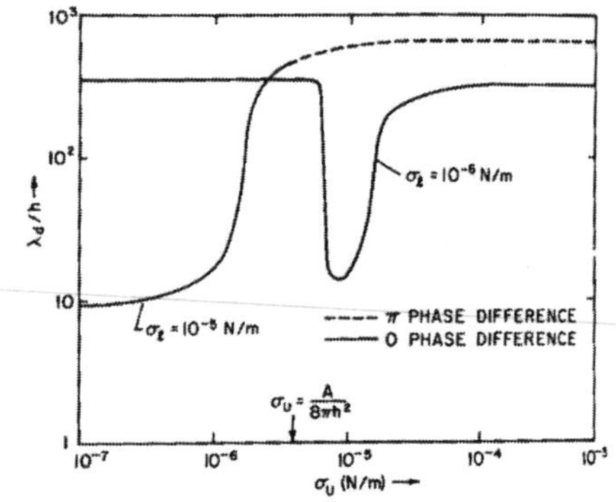

FIG. 9. Dependence of λ_d/h on σ_u for fixed values of σ_l for the two modes of vibration described by Eqs. [86] and [87] with Eq. [107] used for $\Pi(h)$. [$h = 10$ nm, $A = 10^{-20}$ J, $\mu^{I} = \mu^{III} = 10^{-3}$ kg/m−sec, $\mu^{II} = 10^{3}$ kg/m−sec] Note that for $\sigma_l = 10^{-5}$ N/m, the dominant mode is the stretching vibration for $\sigma_u < 4 \times 10^{-6}$ N/m, and the squeezing vibration for $\sigma_u > 4 \times 10^{-6}$ N/m.

here that in all these simulations the phase angle, ϕ, was either 0 or π; no other values of ϕ were obtained since the growth coefficient, ω, was always real.

Since the growth of disturbances, following the onset of instability, is governed by the dominant wavelength, λ_d, the latter was evaluated numerically as a function of σ_u for a fixed value of σ_l (Fig. 9). In these simulations parameters characteristic of a biological membrane were used to study the role of interfacial tension asymmetry on the membrane deformation. Since the dispersion equations for the unsymmetric system are complex, specific analytical generalizations are not possible. However, simulations shown in Fig. 9 can easily be interpreted if one recalls that the value of $A/8\pi h^2$ is 4×10^{-6} N/m. When *either* σ_l or σ_u is less than $A/8\pi h^2$, one would expect the stretching mode to be the dominant one on the basis of our analysis of a symmetric system (Fig. 6). This is precisely what we observe in Fig. 9. When $\sigma_l = 10^{-6}$ N/m, $A/8\pi h^2$ is always greater than σ_l, and the dominant mode is always the stretching vibration. When $\sigma_l = 10^{-5}$ N/m, the dominant mode is the stretching vibration until $\sigma_u = 4 \times 10^{-6}$ N/m, and thereafter the dominant mode is the squeezing vibration. In addition, when $\sigma_l = \sigma_u = 10^{-6}$ N/m, λ_d/h is equal to 350, as shown in Fig. 6. When $\sigma_u = \sigma_l = 10^{-5}$ N/m, σ is greater than $A/8\neq h^2$, and the stretching mode is stable. In this case, the dominant mode is squeezing mode, and λ_d/h is equal to 650.

Figure 10 shows the amplitude ratio, κ_d, at the dominant wavelength as a function of σ_u for the same fixed values of σ_l (10^{-6} and 10^{-5} N/m) as in Fig. 9. Consider first the case when $\sigma_l = 10^{-6}$ N/m

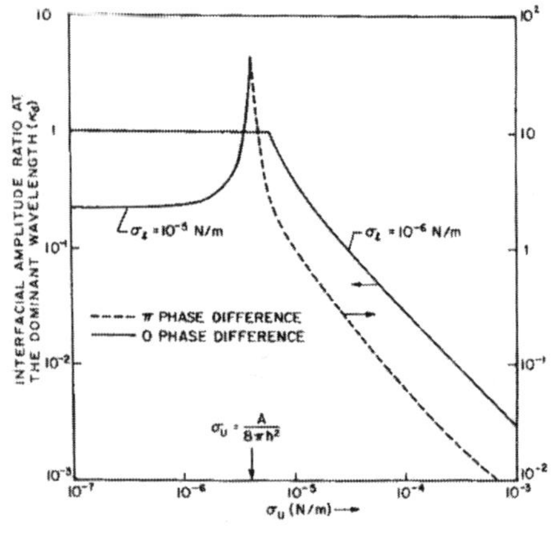

FIG. 10. Dependence of κ_d on σ_u for fixed values of σ_l. [Parameter values are the same as in Fig. 9.]

and the stretching mode is dominant for all values of σ_u. The graph shows a critical behavior at $\sigma_u = 7 \times 10^{-6}$ N/m: for $\sigma_u < 7 \times 10^{-6}$ N/m $k_d = 1$, while for $\sigma_u > 7 \times 10^{-6}$ N/m k_d decreases monotonically with increasing σ_u, a logical result. Note also that this same critical value of σ_u signals the rapid decline in λ_d (see Fig. 9). A possible explanation for this change in the system characteristics is the fact that the critical value of σ_u (7×10^{-6} N/m) is just greater than $A/8\pi h^2$ (4×10^{-6} N/m). This same phenomenon also occurs in the $\sigma_l = 10^{-5}$ N/m simulation, but right at $\sigma_l = A/8\pi h^2 = 4\times10^{-6}$ N/m. After this value of σ_u, both σ_u and σ_l are greater than $A/8\pi h^2$, and consequently, as explained previously, the squeezing mode begins to dominate.

While experimental data for unsymmetrical systems are not available, the results we have obtained can, perhaps, suggest a mechanism which a cell uses to transmit information across its membrane. For instance, let us assume that a chemical reaction in the extra- or intracellular fluid causes the surface tension of one of the interfaces of the membrane to decrease to a value that the membrane becomes unstable. Following our analysis, we would expect that due to asymmetry in the surface tension, the amplitude ratio of disturbances (k_d) may not be equal to one. The extent of deviation from one will depend upon the asymmetry in the system, and will characterize the event which has occurred in the intra- or extracellular fluid. Thus, unequal deformation of the membrane may be one possible mechanism by which a cell transmits or receives information from its surroundings.

CONCLUSIONS

The stability of thin liquid films has been studied to small perturbations using the body force approach. The analysis is more general than previous ones because it accounts for the difference in the interfacial tensions of two faces of the film and difference in the viscosities of the bounding media; it is complete because it is valid for both short and long wavelengths; and, it is systematic because it accounts for the effect of corrugations of the interface on the body force.

When the bounding phases are identical and gravity is neglected (symmetrical case), the dispersion equations suggest two modes of instability: squeezing and stretching vibrations (Fig. 1). In this case, analytical expressions are derived for the critical and dominant wavelengths, and numerical simulations are obtained for a wide range of wavelengths. The results indicate that for parameters characteristic of a biological membrane, the growth of the perturbations is dominated by the stretching mode. Physically, these simulations suggest that a biological membrane will bend more readily to form microvilli or to engulf external material, than have local variation in its thickness as seen in a squeezing vibration.

When the interfacial tensions of the two faces of the membrane are not identical, the amplitude ratio of disturbances may not be equal to one. We suggest that this deviation of the amplitude ratio from one is a possible mechanism which a cell employs to amplify or dampen a message across its membrane.

Finally, contrary to previous assertions (16, 17), for most cases of practical interest, the acceleration terms in the equations of motion can be neglected in a linear stability analysis of thin liquid films (Appendix A).

APPENDIX A

The objective of this appendix is to discuss the admissibility of neglecting all terms of second order and greater in $\omega/v^j k^2$ to obtain the first-order dispersion relations. The validity of this procedure with respect to the SQ and ST dispersion equations will be discussed first. For the wavelength ranges used in the simulations exhibited in Figs. 4 through 10, the parameter $\omega/v^j k^2$ (j = I, II, III) was computed and found to be much less than one. However, as is evident from the approximate dispersion equations of the SQ and ST modes (Eqs. [102]–[105]), the $\lim_{k\to0} \omega(k)/k^2$ does not converge. Consequently for extremely long waves, the approximation $\omega / v^j k^2$ would not be valid. Nevertheless, the values of k at which $\omega / v^j k^2$ begins to diverge usually represent wavelengths much larger than the characteristic radii of a particle in a dispersion (100 μm) or of the circumference of a biological cell (10 μm) and are therefore not of interest.

Numerical simulations also indicate that the parameter $\omega / v^j k^2$ is not less than one for values of $R \ll 1$ and for values of the film viscosity less than or equal to the viscosity of water (10^{-3} kg/m-sec). For these limiting cases (of which a free film is a most important example), at least the second-order term must be retained in the expansion.

The verification that the parameter $\omega / v^j k^2$ is much less than one is not a sufficient condition for neglecting the second-order and greater terms of this parameter because the coefficients of these higher powers of $\omega / v^j k^2$ may be large (especially in the long wavelength limit). Therefore the exact dispersion relations were expanded to third order in $\omega / v^j k^2$ and kh. The results indicate that the neglect of the second and third powers of $\omega / v^j k^2$ (when $\omega / v^j k^2 \ll 1$) is unconditionally valid for the SQ mode; for the ST mode this procedure is valid without qualification only for $R \gg kh$, but it is only admissible otherwise when

$$\frac{kh}{4R}\left(\frac{\omega}{v^{\mathrm{II}} k^2}\right) \ll 1 \left(\frac{k^3 h^3}{12} \ll R \ll kh\right)$$

and

$$\frac{3}{k^2 h^2}\left(\frac{\omega}{v^{\mathrm{II}} k^2}\right) \ll 1 \left(R \ll \frac{k^3 h^3}{12}\right). \qquad \text{[A-1]}$$

Inequality [A-1] is satisfied when the kinematic viscosity of the fluid comprising the film is very large. In particular, for the stability of the stretching mode of the cell membrane (Fig. 6), inequality [A-1] is satisfied because of the large value (10^3 kg/m-sec) of the viscosity of the membrane.

When R is not much greater than kh and inequality [A-1] is unsatisfied, the second-order term $\omega / v^{\mathrm{II}} k^2$ must be retained in the long wavelength limit dispersion equation for the ST mode; with this correction Eq. [104] becomes:

$$\omega_{\mathrm{ST}} = -\frac{k}{4\mu^{\mathrm{II}}} \Delta \frac{1 + khR}{\frac{k^3 h^3}{12}\left(1 + 3\frac{\omega_{\mathrm{ST}}}{v^{\mathrm{II}} k^4 h^2}\right) + R + khR^2}. \qquad \text{[A-2]}$$

In general, these conclusions can be extended to the unsymmetrical system, and therefore the first-order dispersion relations for this system (Eqs. [86] and [87]) are valid (in the long wavelength limit) when $\omega / v^j k^2 \ll 1$ and $R_1, R_2 \gg kh$ and when inequality [A-1] is satisfied for R_1 and R_2 not much greater than kh.

APPENDIX B

The objective of this appendix is to obtain, an asymptotic relation for $I(h)$ as $k \to 0$. Such a relation can be obtained by substituting for the cos $(kh\beta')$ in the defining relation for $I(h)$ (Eq. [65]) the following identity:

$$\cos(kh\beta') = 1 - \frac{(kh\beta')^2}{2} + \left(\cos(kh\beta') - 1 + \frac{(kh\beta')^2}{2}\right), \qquad \text{[B-1]}$$

With the above substitution, Eq. [65] becomes:

$$I(h) = h^2 \int_{-\infty}^{\infty}\int_{-\infty}^{\infty} \tilde{w}\left(h\left(1+\beta'^2+\xi'^2\right)^{1/2}\right) d\xi' d\beta' - \frac{1}{2}k^2h^4 \int_{-\infty}^{\infty}\int_{-\infty}^{\infty} \beta'^2$$
$$\times \tilde{w}\left(h\left(1+\beta'^2+\xi'^2\right)^{1/2}\right) d\xi' d\beta' + h^2 \int_{-\infty}^{\infty}\int_{-\infty}^{\infty} \left(\cos(kh\beta') - 1 + \frac{(kh\beta')^2}{2}\right) \quad \text{[B-2]}$$
$$\times \tilde{w}\left(h\left(1+\beta'^2+\xi'^2\right)^{1/2}\right) d\xi' d\beta'.$$

The first integral on the right-hand side (rhs) of Eq. [B-2] can be simplified by transforming to polar coordinates:

$$h^2 \int_{-\infty}^{\infty}\int_{-\infty}^{\infty} \tilde{w}\left(h\left(1+\beta'^2+\xi'^2\right)^{1/2}\right) d\xi' d\beta' = 2\pi h^2 \int_0^{\infty} \tilde{w}\left(h\left(1+r^2\right)^{1/2}\right) r dr. \quad \text{[B-3]}$$

Defining a new variable v by the relation $v = h(1 + r^2)^{1/2}$ and changing the variable of integration in the integral on the rhs of Eq. [B-3] from r to v yields:

$$h^2 \int_{-\infty}^{\infty}\int_{-\infty}^{\infty} \tilde{w}\left(h\left(1+\beta'^2+\xi'^2\right)^{1/2}\right) d\xi' d\beta' = \int_h^{\infty} 2\pi v \tilde{w}(v) dv = -\frac{d\Pi}{dh} \quad \text{[B-4]}$$

where the second equality in Eq. [B-4] follows from Eq. [32].

After utilizing the same two transformations to rewrite the second integral on the rhs of Eqs. [B-2], the following expression is obtained:

$$-\frac{1}{2}k^2h^4 \int_{-\infty}^{\infty}\int_{-\infty}^{\infty} \beta'^2$$
$$\times \tilde{w}\left(h\left(1+\beta'^2+\xi'^2\right)^{1/2}\right) d\xi' d\beta' = -\frac{k^2h^2}{4}$$
$$\times \left[\frac{1}{h^2}\int_h^{\infty} 2\pi v^3 \tilde{w}(v) dv + \frac{d\Pi}{dh}\right]. \quad \text{[B-5]}$$

Integrating by parts twice the integral on the rhs of Eq. [B-5] yields:

$$\int_h^{\infty} 2\pi v^3 \tilde{w}(v) dv = -h^2 \frac{d\Pi}{dh} + 2h\Pi(h) + 2\int_h^{\infty} \Pi(v) dv \quad \text{[B-6]}$$

subject to the condition that

$$\lim_{\alpha \to \infty} \alpha^4 \tilde{w}(\alpha) = 0. \quad \text{[B-7]}$$

Combining Eqs. [B-5] and [B-6] results in the following:

$$-\frac{1}{2}k^2h^4 \int_{-\infty}^{\infty}\int_{-\infty}^{\infty} \beta'^2$$
$$\times \tilde{w}\left(h\left(1+\beta'^2+\xi'^2\right)^{1/2}\right) d\xi' d\beta' = -\frac{k^2h^2}{2}$$
$$\times \left(\frac{\Pi(h)}{h} + \frac{1}{h^2}\int_h^{\infty} \Pi(v) dv\right). \quad \text{[B-8]}$$

For the last integral on the rhs of Eq. [B-2] it can easily be shown by repeated use of L'Hôpital's rule that

$$h^2 \int_{-\infty}^{\infty} \int_{-\infty}^{\infty} \left(\cos\left(kh\beta'\right) - 1 + \frac{\left(kh\beta'\right)^2}{2} \right)$$

$$\times \tilde{w}\left(h\left(1+\beta'^2+\xi'^2\right)^{1/2} \right) d\xi' d\beta'$$

$$= O\left(k^4\right) \quad \text{[B-9]}$$

when the integral A defined by

$$A = \int_{-\infty}^{\infty} \int_{-\infty}^{\infty} \beta'^4$$

$$\times \tilde{w}\left(h\left(1+\beta'^2+\xi'^2\right)^{1/2} \right) d\xi' d\beta' \quad \text{[B-10]}$$

is bounded. In Eq. [B-9], the upper case O is the "large O" order symbol (26). If

$$\lim_{\alpha \to \infty} \alpha^{5+\delta} \tilde{w}\left(\alpha\right) = 0 \;\; \delta > 0, \quad \text{[B-11]}$$

then the boundedness of A is guaranteed from the "μ test" for convergence (27). Equation [B-11] is a stronger condition on the composite intermolecular potential $\tilde{w}(r)$ than Eq. [B-8] and therefore this asymptotic development is restricted to intermolecular potentials which fall off faster than $r^{5+\delta}, \delta > 0$. In particular, London's law ($w_{i,j}(r) = -C_{i,j}/r^6$) satisfies Eq. [B-11].

Combining Eqs. [B-3], [B-8], and [B-9] yields:

$$I\left(h\right) = -\frac{d\Pi}{dh} - \frac{k^2h^2}{2}$$

$$+ \left(\frac{\Pi\left(h\right)}{h} + \frac{1}{h^2} \int_h^{\infty} \Pi\left(v\right) dv \right)$$

$$+ O\left(\left(k\right)^4\right)\left(k \to 0\right) \quad \text{[B-12]}$$

subject to Eq. [B-11].

APPENDIX C: NOMENCLATURE

A	Hamaker constant of the film system, defined by Eq. [108]
$A_{i,j}$	Hamaker constant defined by Eq. [109]
$C_{i,j}$	Interaction constant (between molecules i and j) in London's law
D	Differentiation with respect to z
F	Function defined by Eq. [29]
$F_{i,j}$	Function defined by Eq. [2]
g	Acceleration of gravity
h	Film thickness

h_1, h_2	Lengths defined in Fig. 2
$H^1_{i,j}$, $H^2_{i,j}$	Functions defined by Eqs. [3] and [4], respectively
$I_{i,j}$	Function defined by Eq. [52]
$I(h)$	Function defined by Eq. [64]
Im	Imaginary part of a function
k	Wavenumber of disturbance
p	Pressure
P	Gravity-excluded total force density (defined by Eq. [26] for the base state)
q^{I}, q^{II}, q^{III}	Variables defined by Eq. [69]
$Q^1_{i,j}$, $Q^2_{i,j}$	Functions defined by Eqs. [21] and [22], respectively
R_1, R_2	Viscosity ratios defined by Eq. [93]
R	Viscosity ratio for a symmetrical system, defined by Eq. [82]
Re	Real part of a function
SQ	Squeezing mode
ST	Stretching mode
t	Time
u	x component of the velocity vector
$u_{i,j}(r)$	Intermolecular potential of two molecules i and j separated by a distance r
υ	z component of the velocity vector
$w_{i,j}(r)$	$u_{i,j}(r)$ divided by the masses of molecules i and j
$\bar{w}(r)$	Function defined by Eq. [31]
W	Excess van der Waals potential
x	Spatial coordinate (see Fig. 3)
X	Variable defined by Eq. [73] or [76]
Y	Variable defined by Eq. [74] or [77]
z	Spatial coordinate (see Fig. 3)
Δ	Film tension defined by Eq. [100]
ϵ	Dimensionless arbitrary parameter (much less than one) appearing in Eqs. [9]–[14]
η	Perturbation in lower interface
κ	Interfacial amplitude ratio defined by Eq. [117]
λ	Wavelength of disturbance ($\lambda = 2\pi/k$)
μ	Viscosity
ν	Kinematic viscosity ($\nu = \mu/\rho$)
ξ	Perturbation in upper interface
Π	Disjoining pressure of a plane parallel, unbounded film (defined by Eq. [32])
ρ	Mass density
σ	Interfacial tensions
σ_{dl}	Double layer tension
τ	Reciprocal of frequency of motion $(\tau = 1/\omega)$
ϕ	Phase difference defined by Eq. [117]
ω	Frequency of motion

Subscripts

0	Indicates base state value
1	Indicates first-order approximation
c	Indicates critical value
d	Indicates dominant value
l	Indicates a property of the lower (II–III) interface
u	Indicates a property of the upper (I–II) interface
i, j	Scripted quantity is a char acteristic of the interaction between molecules i and j

Superscripts

I	Indicates upper semi-infinite phase
II	Indicates film volume
III	Indicates lower semi-infinite phase
	Preexponential function (see Eq. [48])

ACKNOWLEDGMENTS

This work was partially supported by the National Science Foundation, and is based on the M.S. Thesis of Charles Maldarelli.

REFERENCES

1. Vrij, A., *Disc. Faraday Soc.* **42**, 23 (1966).
2. Felderhof, B. *V.,J. Chem. Phys.* **49**, 44 (1968).
3. Vrij, A., and Overbeek, J. Th. G., *J. Am. Chem. Soc.* **90**, 3074 (1968).
4. Ivanov, I., Radoev, B., Manev, E., and Scheludko, A., *Trans. Faraday Soc.* **66**, 1262 (1970).
5. Lucassen, J., van den Tempel, M., Vrij, A., and Hesselink, F., *Proc. Kon. Ned. Akad. Wet.* **B73**, 108 (1970).
6. Vrij, A., Hesselink, F., Lucassen, J., and van den Tempel, M., *Proc. Kon. Ned. Akad. Wet.* **B73**, 124 (1970).
7. Ivanov, I., and Dimitrov, D., *Colloid Polym. Sci.* **252**, 982 (1974).
8. Ruckenstein, E., and Jain, R. *K., Faraday Trans. II* 70, 132 (1974) (Section 5.1 of this volume).
9. Gumerman, R., and Homsy, G., *Chem. Eng. Commun.* 2, 27 (1975).
10. Patzer, J., and Homsy, G., *J. Colloid Interface Sci.* **51**, 499(1975).
11. Jain, R. K., and Ruckenstein, E., *J. Colloid Interface Sci.* **54**, 108 (1976).
12. Ivanov, I., Doctoral Dissertation, Univ. of Sofia, Bulgaria, 1977.
13. Jain, R. K., Maldarelli, C., and Ruckenstein, E., *A.I.Ch. E. Symp. Ser. Biorheol.* **74**, 120 (1978).
14. Joosten, J. G. H., Vrij, A., and Fijnaut, H. M., *Proc. Int. Conf. on Phys. Chem. and Hydrodynamics, 1977, Hemisphere, Washington, D.C.* (in press).
15. Maldarelli, C., M.S. Thesis, Columbia Univ., New York, 1978.
16. Sche, S., *J. Electrostal.* 5, 71 (1978).
17. Sche, S., and Fijnaut, H. M., *Surface Sci.* 76, 186 (1978).
18. (a) Jain, R. K., and Ivanov, I., *Faraday Trans. II* **76**, 250 (1980). (b) Jain, R. K., Ivanov, I. B., Maldarelli, C., and Ruckenstein, E., *in* "Dynamics and Instability of Fluid Interfaces" (T. S. Sorensen, Ed.), pp. 140–167. Springer-Verlag, Berlin, 1979.
19. Scheludko, A., *Proc. Kon. Ned. Akad. Wet.* **B65**, 87(1962).
20. Toshev, B. V., and Ivanov, I. B., *Colloid Polym. Sci.* **253**, 558 (1975).
21. Ivanov, I. B., and Toshev, B. V., *Colloid Polym. Sci.* **253**, 593 (1975).
22. Nir, *S., Prog. Surface Sci.***8**, 1 (1976).
23. Chandrasekhar, S., "Hydrodynamic and Hydro-magnetic Stability." Clarendon Press, Oxford, 1961.
24. Aris, R., "Vectors, Tensors and the Basic Equations of Fluid Mechanics." Prentice Hall, Englewood Cliffs, New Jersey, 1962.
25. Levich, V., "Physico-chemical Hydrodynamics." Prentice-Hall, Englewood Cliffs, New Jersey, 1967.
26. Nayfeh, A., "Perturbation Methods." Wiley-Interscience, New York, 1973.
27. Carslaw, H. S., "An Introduction to the Theory of Fourier's Sines and Integrals." Dover Publications, New York, 1950.
28. Dolowy, K., *J. Theor. Biol.* **52**, 83 (1975).
29. Miller, C., Ph. D. Thesis, Univ. of Minnesota, Minneapolis, Minn., 1968.
30. Sanfeld, A., Steinchen, A., Dame Vedove, W., and Bisch, P. M., "Reactions in Models of Biomembranes in Connection with Cell Motion," paper presented at the American Chemical Society Meeting, Honolulu, April 1979.
31. Ono, S., and Kondo, S., "Molecular Theory of Surface Tension in Liquids," Handbuch Der Physik, Vol. 10 (S. Flugge-Marburg, Ed.), Springer-Verlag, Berlin, 1960.

5.4 An Analytical Nonlinear Theory of Thin Film Rupture and Its Application to Wetting Films*

Ashutosh Sharma and Eli Ruckenstein[†]

Department Chemical Engineering, State University of New York at Buffalo, Buffalo, New York 14260

Corresponding Author

[†]To whom correspondence should be addressed.
Received July 31, 1985; accepted January 6, 1986

I. INTRODUCTION

The study of hydrodynamic instabilities induced by intermolecular forces in thin film has been motivated by their industrial application in disperse and colloid systems on one hand and the understanding of diverse biological phenomena on the other. Some typical industrial applications of thin film rupture and particle coalescence are in foams and emulsions, flotation, thin coatings, oil-recovery, vapor condensation on a tube wall, detergent action in cleaning of oil films, and in the determination of the rate of attachment of bubbles and droplets to collectors. The various models of thin films analyzed for their stability are motivated by these applications (1–6). On the biological front, the phenomena involving the small scale deformations of membranes, such as the onset of microvilli in normal and neoplastic cells, the deformation and rhythmic movement (flickering) of red blood cells, adhesion, fusion, and attachment of cells and microorganisms and phagocytosis, require an understanding of the thin membrane instabilities (7–10). Other areas of interest in this connection are morphogenesis, interfacial interactions in immunology, and invasion and metastasis in cancer. A novel application of thin film stability was recently suggested for the explanation of tear film breakup and dependence of the time of rupture on various physicochemical characteristics of the external eye (11, 12). The solution methodology underlying all of the studies of the thin film stability, with an exception of the numerical computations of Williams and Davis (13), is to employ the *linear* stability analysis for obtaining estimates of the conditions for the rupture of the film and its time of rupture. Although the linear stability analysis is relatively straightforward, at least conceptually, it is valid only for disturbances that have vanishingly small amplitudes. In practice however, the mechanically generated initial interfacial disturbances may be large in view of the large energy dissipation and turbulent, ill-defined flow patterns. Even the thermally induced corrugations of the interface may have nonnegligible relative amplitudes for very thin films (<100 Å) and low interfacial tensions (<1 dyn/cm), such as those encountered for the thin mucus layer coating of the corneal epithelium and proteinous coat of the cell walls.

* *Journal of Colloid and Interface Science*. Vol. 113, No.1, p.456. October 1986. Republished with permission.

The numerical results of Williams and Davis (13) for a thin wetting film devoid of surfactants show that the nonlinearities of the hydrodynamic equations accelerate the film thinning and thus reduce their time of rupture up to an order of magnitude compared to the time of rupture derived from the linear theory. However, the use of computer simulations for investigating the nonlinear dynamics of a diversity of thin film models with varying amount of complexities, is not always an easy task because of the following reasons. First, it is not always possible to reduce the set of governing transport equations to a single evolution equation for the location of the interface, as was done by Williams and Davis (13), even for a single film subjected to long wavelength perturbations. In general, one needs to solve then a set of coupled, strongly nonlinear partial differential equations with moving boundaries. Typical examples are wetting films (films in contact with a solid) with surfactants and free film with or without surfactants. Second, the numerical search for the wave number of the fastest growing and the neutrally stable perturbations is a tedious process, since this has to be found by trial and error for a given set of thin film parameters. Lastly, the effect of various parameters on the instability domain and on the time of rupture would often require extensive computations for a given model before it becomes physically transparent. It is in view of the above considerations and a diversity of existing thin film models that it is desirable to develop an analytical, albeit simple, method of analysis that yields information about the set of parameter values for which a film is unstable and the growth rate of initial disturbances when they have a finite amplitude and thus, the nonlinearities are important.

The nonlinear problem of the determination of the dominant mode (dominant wave number and the corresponding growth rate) and of the interactions between different modes have received considerable attention for wave propagation (14–17). These cases are characterized by the "saturation" of the growing waves which is caused by nonlinearities (e.g., formation of ripples on the film surface). The problems of thin film stability, however, are typically not characterized by the saturation of growing waves, but by the unrestrained growth of the initial perturbations leading to rupture. The nonlinearities alter, of course, the growth rate of the initial perturbations. By comparing the numerical, nonlinear treatment of Williams and Davis (13) with the linear theory for a thin film, one can readily conclude that the latter overestimates the stabilizing influence of the interfacial tension resistance and underestimates the destabilizing contribution of intermolecular dispersion forces.

The purpose of this paper is to develop a perturbative analysis for a finite amplitude instability of the hydrodynamically thin films (typically less than or of the order of 1000 Å), for which the intermolecular forces, such as London van der Waals dispersion forces, are important. The information sought is the wave numbers of the neutrally stable and the fastest growing perturbations and the growth rate (eigen mode) of the latter, as functions of various thin film parameters and the amplitude of the initial disturbance.

As is shown later, the set of hydrodynamic and transport equations describing a thin film, in general have two types of time stationary solutions. The first of these is a spatially homogeneous solution that corresponds to a planar interface and may thus be called "trivial solution." The linear stability analysis is concerned with the stability of this trivial state when the effects of all but the first order harmonic are ignored and all of the nonlinearities are linearized around the planar interface. The other stationary solution is spatially nonhomogeneous and exists for values of parameters (wavenumbers etc.) at which the trivial solution loses its stability. The *leading* order term of this solution therefore corresponds to the neutrally stable wave of the linear stability analysis. The onset of time dependent flow beyond the conditions of neutral stability is thus investigated by further perturbing the nonhomogeneous stationary state. The spatially nonhomogeneous state is determined up to a constant however, as a solution of the steady state equations. For any given disturbance, the constant is specified by matching the overall response of the interface at initial time with the imposed perturbation. This then establishes a correspondence between the amplitudes of the initial disturbance and the stationary state. It is in view of this procedure that the general spatially nonhomogeneous stationary solution reduces to the trivial state when the amplitude of disturbance is vanishingly small. As is expected, the results of the linear stability analysis are readily recovered in such an instance. For any finite initial amplitude however, it becomes natural to address the issue of the stability of the *corresponding* stationary, albeit spatially nonhomogeneous, solution of the governing equations. This, together with the inclusion of higher harmonics (if significant) lays the groundwork for the present analysis. In essence then, the planar interfacial configuration is not considered as a base state for a finite amplitude perturbation. The base state for a finite initial disturbance is a finite amplitude corrugation of the interface, with the amplitude of the disturbance determining the amplitude of the base state. In this way, the initial disturbance is envisaged as a perturbation of the corresponding, finite amplitude stationary state. One of the advantages here of course, is that the base state that accounts for the interfacial corrugations is closer to the finite amplitude initial disturbance. This is in contrast to the choice of planar, undisturbed state as the base state, for which the deviation is large for a finite initial amplitude. The methodology developed is applicable to films that either thin until they rupture or attain a new steady state due, for instance, to double layer repulsion. The analysis in its present form is not expected to work when the neutrally stable states of the system are oscillatory in time, i.e., the system is dispersive.

The analytical results are first obtained for a model system of a wetting (supported on a solid) thin film devoid of surfactants. These are compared with the numerical simulations of Williams and Davis (13). The analysis is then extended to a wetting film bounded by another fluid with soluble surfactants in both the bounding fluid and the thin film. In addition to the nonlinear effects of the interfacial tension and dispersion forces, the nonlinear effects of surfactant concentration driven Marangoni-motion, the interfacial shear viscosity and the surface diffusion are also investigated. Finally, analytical expressions are derived for a finite amplitude instability of the mucous layer covering the corneal epithelium. The dependence of the tear breakup time on various parameters of the external eye is then discussed within this framework.

II. A WETTING THIN FILM WITH SURFACTANTS

Figure 1 depicts the thin film model being considered. Both the thin film and the bulk fluid surrounding it are considered Newtonian and contain soluble surfactants. This may be considered as a convenient model for studying the rate of attachment of a bubble and a solid particle (flotation), breakup of condensed vapor films on walls (4), the rate of adhesion of microorganisms and cells to surfaces and polymolecular adsorption (17). The stability of thin alveolar lining layer and the mucus coating of corneal epithelium provide some of the physiological examples.

II.1. The Hydrodynamic Equations

The hydrodynamics of the thin film is described by the usual Navier-Stokes equations with the inclusion of an extra body force term that arises due to the intermolecular interactions, if the film thickness is smaller than the spatial range of these forces. Denoting the kinematic viscosity and density of the thin film fluid by ν and ρ, respectively, and its initial mean thickness by h_0, the governing equations and the boundary conditions are nondimensionalized on the following scales: lengths $\sim h_0$, time $\sim h^2{}_0/\nu$, velocities $\sim \nu/h_0$ and pressures $\sim \rho\nu^2/h_0{}^2$. Denoting by unprimed and primed variables the nondimensional velocity components (u, w, u', w'), pressures (p and p') and the interaction potentials (ϕ and ϕ') in the thin film and the bounding fluid, respectively, we have the following nondimensional Navier-Stokes equations for the thin film and the bounding fluid:

$$u_t + uu_x + wu_z = \nabla^2 u - p_x - \phi_x \qquad [1a]$$

and

$$w_t + uw_x + ww_z = \nabla^2 w - p_z - \phi_z \qquad [1b]$$

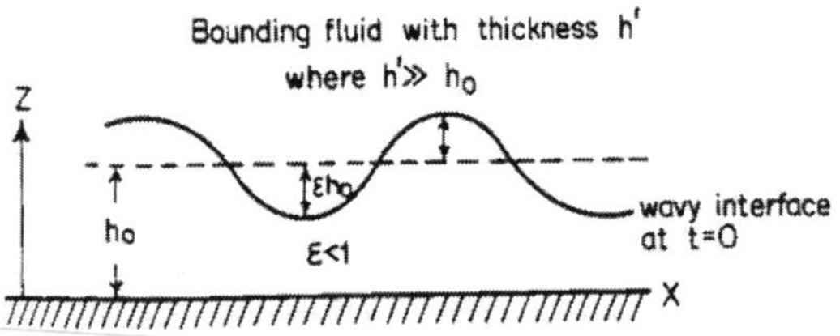

FIG. 1. A thin wetting film bounded by a thick fluid layer. The initial interfacial disturbance has a finite amplitude, ϵh_0, where $\epsilon < 1$. Both the thin film and the bounding media contain soluble surfactants.

for $0 \leqslant z \leqslant h(x,t)$ and $< -\infty < x < \infty$, where subscripts denote partial differentiations and $h(x, t)$ is the dimensionless location of the film-bounding fluid interface;

$$u'_t + u'u'_x + w'u'_z = (v'/v)\nabla^2 u' - (\rho/\rho')\, p'_x - \phi'_x \quad [2a]$$

and

$$w'_t + u'w'_x + w'w'_z$$

$$= (v'/v)\nabla^2 w - (\rho/\rho')\, p'_z - \phi'_z, \quad [2b]$$

for $h(x,t) \leqslant z \leqslant h'$ and $-\infty < x < \infty$, where h' is the dimensionless location of the free surface of the bounding fluid. We will consider only the cases where $h' \gg h$ and thus ignore the interaction potential at h'. The continuity equations for the two phases are

$$u_x + w_z = 0 \quad [3a]$$

and

$$u'_x + w'_z = 0. \quad [3b]$$

If the interaction potential arises due to the van der Waals dispersion forces, the functional dependence of the nondimensional interaction potentials (acting on a unit volume located at the interface $z = h$) on the film thickness is given by

$$\phi(h) = (A_{22} - A_{23}) / 6\pi h_0 \rho v^2 h^3 \quad [4a]$$

and

$$\phi'(h) = (A_{12} - A_{13}) / 6\pi h_0 \rho' v^2 h^3, \quad [4b]$$

where A_{ij} are the Hamaker constants for the interactions between molecules of type i and j and subscripts 1, 2, and 3 refer to the molecules of the bulk bounding fluid, thin film and solid, respectively. The following nondimensionalized boundary conditions are specified at various interfaces:

$$\text{at } z = 0;\ u = w = 0. \quad [5a]$$

The balance of tangential forces at the film-bounding fluid interface:

$$\text{at } z = h(x,t);$$

$$\left(1-h_x^2\right)\left[r\left(u_z+w_x\right)-\left(u_z'+w_x'\right)\right]$$

$$+2h_x\left[r\left(w_z-u_x\right)-\left(w_z'-u_x'\right)\right]$$

$$-\left(h_0/\mu'v\right)\sigma_x\left(1+h_x^2\right)^{-1/2}=0 \qquad [5b]$$

where $r=\left(\mu/\mu'\right)$, σ is the interfacial tension and the last term represents the stress generated due to the interfacial tension gradients.

The relation of the pressure jump to the interfacial tension:

$$\text{at } z=h\left(x,t\right);$$

$$p'-p+2\left[\left(1-h_x^2\right)\left(w_z-w_z'/r\right)\right]\left(1+h_x^2\right)^{-1}$$

$$=3Sh_{xx}\left(1+h_x^2\right)^{-3/2}, \qquad [5c]$$

where $S=\left(h_0\sigma/3\rho v^2\right)$ is the inverse of a capillary number and may be interpreted as the ratio of surface to viscous forces. In addition,

$$\text{at } z=h\left(x,t\right);\ u=u' \text{ and } w=w' \qquad [5d]$$

$$\text{at } z=h';\ u'\to 0 \text{ as } h'\gg h \qquad [5e]$$

$$\text{at } z=h';\ u_z'+w_x'=0. \qquad [5f]$$

The following kinematic condition now completes the hydrodynamic description of the model:

$$\text{at } z=h\left(x,t\right);\ h_t+uh_x=w. \qquad [6]$$

The mass transport equations for the surfactants should be added to this set of hydrodynamic equations. Although, the nonlinear stability analysis developed later does not warrant the reduction of hydrodynamic equations to any simpler set, we undertake the stability analysis for the long wavelength perturbations only. It is known from the linear theory that the fastest growing perturbations for this model of thin film are those that have wavelengths far exceeding the thickness of the film (4, 13). As is shown later, this also holds for the dominant wavelength derived from the nonlinear theory. Williams and Davis (13) employed the long wavelength analysis of Benney (18) and Atherton and Homsy (19), to derive a nonlinear equation of evolution for a single film devoid of surfactants. The long wavelength reduction procedure followed here is in the spirit of Williams and Davis (13).

II.2. The Nonlinear Evolution Equation in the Long Wave Limit

The nondimensional wave number, *k*, is a small parameter for the interfacial waves with wavelengths large compared to the mean initial thickness of the film. The hydrodynamic equations are then rescaled by stretching the lateral direction, *x* and time, *t*, by the transformations

$$\xi=kx \text{ and } T=kt. \qquad [7a]$$

Assuming that all derivatives and the velocity components in the x–direction (u, u') in the transformed coordinate system (ξ, z, T) are of the order of unity, the continuity equations [3a] and [3b] give

$$w = kw_0 \text{ and } w' = kw_0',$$

$$\text{where } w_0, w_0' = O(1). \qquad [7b]$$

Further, in order to retain the pressures and the interaction forces in the lowest order x components of the equations of motion, we introduce

$$p_0 = kp,\ p_0' = kp',\ \phi_0 = k\phi,$$

$$\text{and } \phi_0' = k\phi' \qquad [7c]$$

Obviously, it is essential that the pressure and the interaction forces be retained for investigating the instability of the thin film, for in their absence, the film is always stable.

It may be noted that the long wave length scalings given above are not universal for all thin film models and have to be chosen so as to be consistent with the results of linear stability analysis. Substituting the rescaled variables [7a] to [7c] in Eqs. [la] to [2b] and equating the constant (k–free) terms, one obtains

$$u_{zz} - p_{0\xi} - \phi_{0\xi} = 0, \qquad [8a]$$

$$p_{0z} + \phi_{0z} = 0, \qquad [8b]$$

$$u_{zz}' - rp_{0\xi}' - r\alpha\phi_{0\xi}' = 0, \qquad [8c]$$

and

$$p_{0z}' + \alpha\phi_{0z}' = 0, \qquad [8d]$$

where $\alpha = (\rho'/\rho)$. The boundary conditions [5a] to [5f] are simplified, respectively, to

$$\text{at } z = 0;\ u = w_0 = 0, \qquad [9a]$$

$$\text{at } z = h;\ \mu u_z = \mu' u_z' + \sigma_\xi^*, \qquad [9b]$$

where

$$\sigma^* - \left(\frac{h_0}{v}\right)k\sigma = O(1),$$

$$\text{at } z = h;\ p' - p = 3\overline{S}h_{\xi\xi}, \qquad [9c]$$

where $\overline{S} = k^3 S = O(1)$,

$$\text{at } z = h;\ u = u' \text{ and } w_0 = w_0', \qquad [9d]$$

$$\text{at z} = h';\ u' = 0 \qquad [9e]$$

and

$$\text{at } z = h';\ u_z' = 0. \qquad [9f]$$

The solution of Eq. [8d] is

$$\psi_1' \equiv p_0' + \alpha\phi_0' = p_0'(h) + \alpha\phi_0'(h), \qquad [10a]$$

and the solution of Eq. [8b] which satisfies condition [9c] and expression [10a] is

$$\psi_2 \equiv p_0 + \phi_0 = p_0(h) + \phi_0(h)$$

$$= \phi_0(h) - \alpha\phi_0'(h) - 3\bar{S}h_{\xi\xi} + \psi_1'. \qquad [10b]$$

As shown by Eqs. [10a] and [10b], the modified pressure consisting of the sum of the hydrodynamic pressure and the dispersion potential is constant across the entire thickness of the film. It is thus sufficient to evaluate it at the interface. The solution of Eqs. [8a] and [8c] are now constructed by using expressions [10a] and [10b], to give

$$u' = \psi_1 z^2/2 + a_1 z + a_2 \qquad [11a]$$

and

$$u = \frac{\psi_2 z^2}{2} + a_3 z + a_4, \qquad [11b]$$

where $\psi_1 = r\psi_1'$. The boundary conditions [9a], [9b], [9d], [9e], and [9f], when used in conjunction with expressions [11a] and [11b], provide

$$\psi_1 = \left(2h\sigma_\xi^* - \mu h^2 P_0\right)/\,\mu h'^2 \qquad [12a]$$

and

$$a_3 = \left(2h\sigma_\xi^* - \mu h^2 P_0\right)/\,2\mu h - \left(P_0 h/2\right), \qquad [12b]$$

where

$$P_0 \equiv \phi_{0\xi} - \alpha\phi_{0\xi}' - 3\bar{S}h_{\xi\xi\xi}. \qquad [13]$$

Equating the coefficients of order k in the continuity equation [3a], gives

$$w_{0z} = -u_\xi, \qquad [14]$$

which, when solved with the help of expressions [11b], [10b], [12a], [12b], and the boundary condition [9a], determines w_0 as

$$w_0 = -\psi_{2\xi} z^3/6 - a_{3\xi} z^2/2. \qquad [15]$$

The first nontrivial equation of evolution is obtained by rescaling the kinematic condition [6] and substituting in it the expressions for the velocities u and w_0. In this manner, one obtains

$$h_T + \left(\psi_2 h^3/6\right)_\xi + \left(a_3 h^2/2\right)_\xi = 0$$

$$\text{at } z = h(\xi, T), \qquad [16a]$$

which after substituting expressions for ψ_2 and a_3 and noting that $h' \gg h$, becomes

$$h_T - \frac{1}{3}\left(P_0 h^3\right)_\xi + \left(\sigma_\xi^* h^2/2\mu\right)_\xi = 0. \qquad [16b]$$

Having obtained this nonlinear equation of evolution for the long wave length perturbations, we may now make use of the interaction potentials [4] and revert to the original nondimensional variables:

$$h_t + A^*\left(h^{-1}h_x\right)_x + \left(Sh^3 h_{xxx}\right)_x$$

$$+\left(h_0 / 2\mu v\right)\left(h^2\sigma_x\right)_x = 0, \qquad [16c]$$

where $A^* = A / 6\pi h_0 \rho v^2$ and A is an effective Hamaker constant that is given by

$$A = A_{22} - A_{23} - A_{12} + A_{13}. \qquad [17]$$

The equation provides the evolution of the initial interfacial disturbances, $h(x, 0)$, with time and the lateral spatial coordinate, x. The equation reduces to that obtained earlier (13), if the effect of surfactants (as reflected in σ_x and a nonconstant S due to the surface viscosity) and the presence of the bounding fluid are neglected.

Finally, Eq. [16c] in the dimensional form is

$$H_t + \left(A / 6\pi\mu\right)\left(H^{-1}H_x\right)_x + \left(1/3\mu\right)\left(\sigma H^3 H_{xxx}\right)_x$$

$$+\left(1/2\mu\right)\left(H^2\sigma_x\right)_x = 0, \qquad [18]$$

where $H(x, t)$ is the dimensional thickness of the film and x and t are now the *dimensional* lateral coordinate and time, respectively. The above hydrodynamic equation has to be supplemented by a mass transport equation that determines the distribution of surfactants and thus the interfacial tension gradient, σ_x.

II.3. The Surfactant Transport Equations

The dimensional convective-diffusion equations for the concentration in the thin film, C, and that in the bulk fluid, C', are

$$C_t + u^* C_x + w^* C_z = D\nabla^2 C \qquad [19a]$$

and

$$C'_t + u'^* C'_x + w'^* C'_z = D'\nabla^2 C', \qquad [19b]$$

where the dimensional velocity components, u^*, w^*, u'^*, and w'^* correspond to the nondimensional components u, w, u', and w', respectively. The surface excess concentration of surfactant at interface, Γ, is represented by the Langmuir adsorption isotherm, viz.,

$$\Gamma = k_1 C \Big/ \left(1 + \frac{k_1}{\Gamma_\infty} C\right)$$

$$= k'_1 C' \Big/ \left(1 + \frac{k'_1}{\Gamma_\infty} C'\right). \qquad [20a]$$

The equality of chemical potentials for the equilibrium state, $C = C_0$ and $C' = C'_0$, gives

$$C'_0 = k_e C_0, \qquad [20b]$$

which, when combined with Eq. [20a] shows that $k_1 = k_e k_1'$. If the surfactant is insoluble in the thin film, $k_e \rightarrow \infty$ and if it is insoluble in the bounding fluid, $k_e \rightarrow 0$. Γ_∞ is the asymptotic surface excess concentration corresponding to the saturation of the interface. The entire analysis, however, remains valid for any isotherm of the form

$$\Gamma = \Gamma(C).$$

The boundary conditions for the long wavelength perturbations are

at $z = H$;

$$-DC_z + D'C_z' = \Gamma_t + \left(\Gamma u_s^*\right)_x - D_s \Gamma_{xx}, \qquad [21a]$$

where D_s is the surface diffusion coefficient and u_s^* is the interfacial velocity component.

$$\text{at } z = H';\ D'C_z' = 0, \qquad [21b]$$

$$\text{at } z = 0;\ DC_z = 0. \qquad [21c]$$

The coupling between the hydrodynamic equation and the mass transport equations arises because the concentration distribution depends on the velocity field, which in turn depends on the interfacial tension gradient. The interfacial tension gradient, however, depends on the concentration field and the velocity distribution, since the dynamic interfacial tension, σ, is given by

$$\sigma = \sigma_0(\Gamma) + \mu_s\left(u_s^*\right)_x, \qquad [22]$$

where $\sigma_0(\Gamma)$ is the concentration dependent part of the interfacial tension, μ_s is the surface shear viscosity, and $(u_s^*)_x$ is the gradient of the surface velocity. With this formulation of the model, we now proceed to the nonlinear stability of these equations.

III. THE NONLINEAR STABILITY

III.1. Wetting Film Devoid of Surfactants

Before entering into the analysis of the full model outlined, we first consider the prototype problem of a wetting film devoid of surfactants, which was numerically solved by Williams and Davis (13). In this case, the nondimensional hydrodynamic equation [16c] becomes

$$h_t + A^*\left(h^{-1}h_x\right) + S\left(h^3 h_{xxx}\right)_x = 0. \qquad [23a]$$

The parameters can be removed by a simple rescaling of the independent variables, viz.,

$$X = \left(A^*/S\right)^{1/2} x \text{ and } \tau = \left(A^{*2}/S\right)t$$

which reduce Eq. [23a] to

$$h_\tau + \left(h^{-1}h_X\right)_X + \left(h^3 h_{XXX}\right)_X = 0. \qquad [23b]$$

The linear stability of the trivial state of this equation is readily ascertained by linearizing it around the initial mean value of the thickness (the trivial state) (13). Thus, we substitute $h = 1 + h_1(X, T)$ in Eq. [23b] and discard all but linear terms to arrive at

$$h_{1\tau} + h_{1XX} + h_{1XXXX} = 0. \quad [24]$$

Substituting the normal mode decomposition, $h_1 = \epsilon \exp(\omega\tau + iqX)$ in the above equation gives the following dispersion or characteristic relation, relating the disturbance growth rate (eigenvalue), ω, to the wave number, q:

$$\omega = q^2\left(1 - q^2\right). \quad [25]$$

Since the growth coefficient is always real in this case, the system is purely dissipative. The dissipation occurs for all wave numbers with $q > 1$, for which ω is negative, and the disturbances are attenuated. As the dissipation is further increased, viz., $q \to 1$ from above, the decay rate of the perturbations decreases and eventually, for all $0 < q < 1$, the trivial state of system becomes linearly unstable, viz., $\omega > 0$. The trivial state, $h = 1$, is neutrally stable when the imposed disturbances are stationary in time, viz., $\omega = 0$ and therefore $q^2 = 1$. It is at a certain wave number ($q = 1$) that the stability of solution changes. It is, therefore, obvious that $h = 1 + \epsilon \sin X$ is a solution of the steady-state part of Eq. [24] and hence, a leading order term of the solution of Eq. [23b] with $h_\tau = 0$. As will be shown, it is also in the vicinity of this wave number ($q = 1$) that the bifurcation to a time dependent solution is studied. Within the linear theory framework then, the maximum growth rate, ω_m occurs for the disturbance for which $(\partial\omega/\partial q)_m = 0$. This gives

$$q_m^2 = (1/2) \quad [26]$$

$$\omega_m = (1/4) \quad [27]$$

and the time of rupture as determined from the linear theory, τ_L is

$$\tau_L = \frac{1}{\omega_m} \ln(1/\epsilon) = 4\ln(1/\epsilon). \quad [28]$$

The time of rupture may be affected by the contribution of higher harmonics. To see if that is indeed the case, the film thickness may be expanded as the following naive perturbation series:

$$h = 1 + \mu^* h_1(X,\tau) + \sum_{i=2} (\mu^*)^i h_i(X,\tau), \quad [29]$$

where $\mu^* < 1$ is a small parameter which is related to the initial nondimensional amplitude of the perturbation, ϵ, and the coefficients h_1, h_i are to be determined through successive approximations by the substitution of expansion [29] into Eq. [23b]. The first harmonic, h_1, as shown before, corresponds to the most unstable linear mode, i.e.,

$$h_1 = \sin q_m X e^{\omega m \tau}. \quad [30a]$$

The second harmonic is determined by the following linear, nonhomogeneous equation

$$h_{2\tau} + \left(h_{2X} - h_1 h_{1X}\right)_X$$

$$+ \left(h_{2XXX} + 3h_1 h_{1XXX}\right)_X = 0. \quad [30b]$$

This is solved with the help of [30a] to give

$$h_2 = \frac{q_m^2\left(1+q_m^2\right)}{2\omega_m - 4q_m^2\left(1-4q_m^2\right)}\cos 2q_m X e^{2\omega m\tau}. \qquad [30c]$$

Similarly, the third order problem,

$$h_{3\tau} + \left\{h_{3X} - h_1 h_{2X} + h_{1X}\left(h_1^2 - h_2\right)\right\}_X$$

$$+ \left\{h_{3XXX} + 3h_1 h_{2XXX} + 3h_{1XXX}\right.$$

$$\left.\times\left(h_1^2 + h_2\right)\right\}_X = 0, \qquad [30d]$$

has the solution

$$h_3 = \frac{-3q_m^2\left(1+6q_m^2\right)}{3\omega_m - 9q_m^2\left(1-9q_m^2\right)}\sin 3q_m X e^{2\omega m\tau}. \qquad [30e]$$

Thus the time of rupture, τ, up to this approximation is determined by setting $h = 0$ in Eq. [29]:

$$1 - \mu^* e^{\tau/4} - \frac{\mu^{*2}}{2} e^{\tau/2} - \frac{12}{33}\mu^{*3} e^{3\tau/4} = 0, \qquad [31a]$$

where the parameter μ^* is determined by matching the maximum amplitude of the initial perturbation with the amplitude of the perturbed series [29] at time $\tau = 0$, viz.,

$$\mu^*\left(1 + \frac{\mu^*}{2} + \frac{12}{33}\mu^{*2}\right) = \epsilon. \qquad [31b]$$

The ratio of the time of rupture as computed from Eq. [31a] and that given by the linear Eq. [27] is shown as the dashed line in Fig. 2. The curve labeled I represents the computer simulations of Williams and Davis (13). The discrepancy between the numerical simulations and this procedure indicates that the contribution of the higher harmonics, which is relatively unimportant, cannot explain the large reduction in the time of rupture due to nonlinearities. This observation is, in fact, also in line with the spatial step size that was deemed adequate for the computer simulations—a step size that can barely keep track of the second and higher order harmonics (13). The problem is thus essentially that of mode modification, viz., the dominant wave number and the growth coefficient of the nonlinear system differ from those inferred from the linear theory. The key idea in the subsequent analysis is the examination of the stability of the stationary, albeit, *spatially nonuniform* solution of the evolution equation, which is in contrast to the investigation of the stability of the trivial solution, $h = 1$. As is shown later, this spatially nonhomogeneous, finite amplitude, stationary solution is a more general solution of the steady-state equations and it reduces to the trivial solution only in the limit of an infinitesimal perturbation. Thus we now address the issue of the bifurcation to a time dependent solution from the spatially nonuniform stationary solutions of Eq. [23b], with the wavenumber as a bifurcation parameter. To this end, we examine the perturbation expansion

$$h = h_0(X) + \sum_{i=1}\left(\mu^*\right)^i h_i(X,\tau), \qquad [32]$$

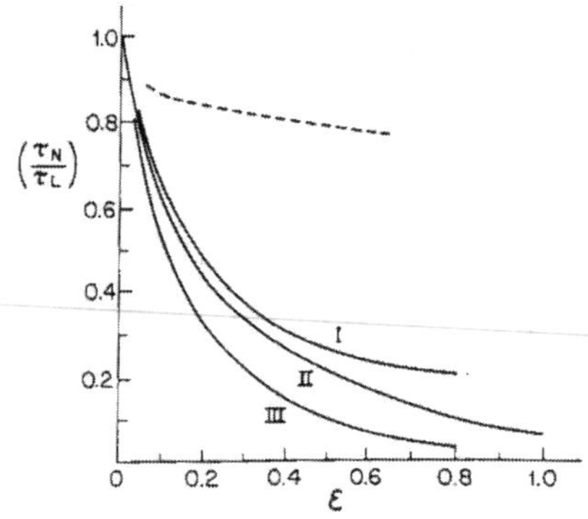

FIG. 2. The ratio between the time of rupture as computed by nonlinear and linear theories, as a function of the nondimensional initial amplitude, ϵ. The dashed curve is for a theory that accounts for the effect of higher harmonics but no mode modification, the wave labeled I is the numerical simulations and curve labeled II is derived from the present analysis. Curve III corresponds to the retarded dispersion interactions.

where $h_0(X)$ is now a stationary, spatially quasiharmonic solution of the stationary evolution equation

$$\left(h_0^{-1}h_{0X}\right)_X+\left(h_0^3h_{0XXX}\right)_X=0. \quad [33]$$

The base solution may again be constructed as a perturbed Fourier series

$$h_0(X)=1+\mu^*\sin q_1X+\sum_{i=2}\left(\mu^*\right)^i f_i(X). \quad [34]$$

The substitution of expansion [34] in Eq. [33] gives the compatibility condition

$$q_1^2\left(1-q_1^2\right)=0 \quad [35a]$$

and an approximate stationary, quasiharmonic solution

$$h_0(X)=1+\mu^*\sin q_1X$$

$$-\frac{1}{3}\mu^{*2}\cos 2q_1X+O\left(\mu^{*3}\right). \quad [35b]$$

It may be noted that the small parameter μ^* is as yet unknown and is later uniquely determined by matching the overall solution at initial time with the imposed perturbation. The spatially nonhomogeneous base state [35b] already accounts for a finite corrugation of the interface and the distance between this base state and an imposed perturbation with amplitude ϵ, is of the order of $(\epsilon-\mu^*)$. For the trivial state, $h=1$, this distance is of the order of ϵ. The base state [35b] is therefore, closer to a finite amplitude disturbance and thus the latter may conveniently be envisaged as a smaller perturbation of the former. For the trivial root of Eq. [35a], $q_1=0$, the linear theory results are recovered inasmuch as one now investigates the stability of the trivial state, $h_0=1+O(\mu^{*2})$. However, as was shown earlier, waves with all wave numbers above and below $q_1=0$ are unstable and hence the stability property of the solution remains unchanged in the vicinity of $q_1=0$. Such a change however occurs for the nontrivial solution $q=1$, as the trivial state is stable for all wave numbers satisfying $|q_1|>1$ and is unstable for all wave numbers in the range $0<|q_1|<1$. Thus, our interest lies in studying the evolution of time dependent solutions as the wave number crosses its threshold value of 1 and becomes supercritical, i.e., $|q_1|$ becomes less than one. Taking therefore the base state to be the one corresponding to $q_1^2 = 1$, the solution [35b] is a nontrivial, spatially nonhomogeneous stationary state. Moreover, it is obvious that it is only in the limit of vanishingly small amplitudes, $\mu^* \to 0$, that the nontrivial solution coincides with the trivial solution, $h = 1$. For any finite amplitude perturbation however, there is a corresponding stationary, spatially nonhomogeneous solution [35b], which may further lose its stability to disturbances with certain wave numbers. These wave numbers and the dominant growth coefficient is what we wish to determine in the subsequent analysis. Thus, the first step in the proposed analysis is to construct a nontrivial stationary solution of the evolution equation and to find a wave number for which this solution exists. As is apparent, this wave number is the wave number corresponding to the neutral stability of the trivial solution, viz., setting $\omega = 0$ in the linear theory dispersion relation provides this wave number. Now, substituting expansion [32] in Eq. [23b] and equating the terms of order μ^*, gives the following homogeneous equation

$$h_{1\tau} + \left(h_0^{-1}h_{1X} - h_0^{-2}h_{0X}h_1\right)_X$$

$$+\left(h_0^3 h_{1XXX} + 3h_0^2 h_{0XXX}h_1\right)_X = 0. \qquad [36]$$

The solution of this equation is given by using expression [35b] for $h_0(x)$. This equation may be solved either by a finite mode Galerkin approximation or a further reordering of the problem with the construction of normal mode solutions for successive problems. The former is numerical and the latter is not expected to be valid for moderate to large amplitudes.

Equation [36] is still complex enough to defy analytical solutions. This difficulty can be circumvented by noting that in order to investigate the interaction force mediated *rupture* of the film, it is sufficient to study the instability of the basic solution where the film is locally thinnest, i.e., where the growth of the first harmonic is maximum. This procedure, which reduces Eq. [36] to an equation with constant coefficients, is physically motivated and as is seen later, retains the full description of the nonlinearities. Introducing thus the estimates, $\sin q_1 X = -1$, viz., $h_0 \approx 1 - \mu^*$ and hence $h_{0X} = \mu^* q_1 \cos q_1 X = 0$, $h_{0XX} = \mu^* q_1^2$, $h_{0XXX} = 0$, and $h_{0XXXX} = -\mu^* q_1^4$ in Eq. [36], the following normal mode solution is constructed.

$$h_1 = \sin qX e^{\omega_N \tau}, \qquad [37a]$$

where ω_N is a modified growth rate given by

$$\omega_N = \left\{\left(1-\mu^*\right)^{-1} q^2 - \left(1-\mu^*\right)^3 q^4\right\}$$

$$+\left\{\mu^*\left(1-\mu^*\right)^{-2} q_1^2 + 3\mu^*\left(1-\mu^*\right)^2 q_1^4\right\},$$

where

$$q_1^2 = 1. \qquad [37b]$$

The critical wave number, q_c, is determined by setting the growth coefficient to zero, viz.,

$$2q_c^2 = \left(1-\mu^*\right)^{-4} + \left\{\left(1-\mu^*\right)^{-8}\right.$$

$$\left. + 4y\left(1-\mu^*\right)^{-3}\right\}^{1/2}, \qquad [37c]$$

where

$$y = \mu^* q_1^{*2}\left\{\left(1-\mu^*\right)^{-2} + 3\left(1-\mu^*\right)^2 q_1^{*2}\right\}. \quad [37d]$$

The wave number of the fastest growing wave satisfies the equation $(\partial\omega/\partial q) = 0$. The solution is given by

$$q_{\mathrm{m}}^2 = \left\{2\left(1-\mu^*\right)^4\right\}^{-1}, \quad [37e]$$

and the dominant eigenvalue corresponding to the dominant wave number q_m is

$$\omega_{\mathrm{N}}^* = \left\{4\left(1-\mu^*\right)^5\right\}^{-1} + \mu^*\left(1-\mu^*\right)^{-2}$$

$$+ 3\mu^*\left(1-\mu^*\right)^2. \quad [37f]$$

Finally, μ^* is determined by matching the amplitude of the initial perturbation with the amplitude of expansion [32] at time $\tau = 0$, i.e., up to terms of order μ^*,

$$\mu^* = \epsilon/2. \quad [37g]$$

In the analysis pursued so far, we have accounted for zeroth and first order harmonics. The inclusion of higher order harmonics is also rather straightforward by going to the terms of order μ^{*2} and making use of the lower order solutions. This, however, was not deemed necessary here due to a very rapid convergence. The rapid convergence of this solution appears to be a generic property of thin films that rupture, but more examples are required to establish it.

As is apparent from Eqs. [37b] to [37f], the critical wave number, the wave number of the dominant mode, and the dominant growth rate, all are functions of the initial amplitude of the perturbation, ϵ, as opposed to the case of linear stability analysis. They all reduce to the proper linear limits as $\epsilon \to 0$. For any finite amplitude however, q_c, q_{m} and ω_{N}^* are always greater than their respective linear theory counterparts and all of them increase as ϵ is increased. The nonlinear effects thus select a shorter wavelength as compared to the linear theory and produce a higher rate of growth of perturbations.

Another issue that deserves attention here is that owing to the linearity and homogeneity of the leading order Eq. [36], the principle of superposition is valid and thus a growth coefficient may be determined corresponding to each Fourier mode of the initial disturbance. The most prominent growth coefficient now corresponds to the Fourier mode for which the combination of the wave number and amplitude is such so as to make its growth coefficient larger than that of any other mode. This is in contrast to the linear theory for which the growth coefficient is only a function of wave number and thus the dominant mode has a wave number corresponding to one that maximizes the growth coefficient. Finally, the linear theory predicts that all infinitesimal initial disturbances with wave numbers greater than one decay, whereas Eq. [37b] and [37c] show that some of these finite amplitude disturbances that have wave numbers falling in the domain $1 < q < q_c$, grow.

The ratio between the dominant wave number selected by the nonlinear theory (Eq. [37e]) and that selected by the linear theory is given by $(1 - \mu^*)^{-2}$. This ratio is 1.11 for $\epsilon = 0.1$, which is in entire agreement with the numerical simulations (13), that report it to be approximately 1.1. For the extreme case of $\epsilon \to 1.0$, the dominant wavelength is reduced by about a factor of 4 (it is still of the same order of magnitude as in the case of $\epsilon \to 0$) and hence the long wavelength theory holds for large initial amplitudes as well.

The time of rupture τ_N, is now computed as a solution of $h_0 + \mu^* h_1 = 0$, which gives τ_N as

$$\tau_N = \frac{1}{\omega_n^*} \ln\left(\frac{1-\mu^*}{\mu^*}\right). \quad [37h]$$

The ratio of the time of rupture as computed from Eq. [37h] and that given by the linear theory, Eq. [28], is plotted as a function of amplitude in Fig. 2 (curve labeled II). The discrepancy between the numerical solutions and the present solution is less than 15% for initial amplitudes as large as 0.6 of the initial film thickness. Moreover, the numerical solutions have been reported to overestimate the actual time of rupture even for the linear equation with relatively small amplitude ($\epsilon = 0.1$) (13). The basic advantage of the method resides in the fact that its computational aspects are only slightly more involved than the linear stability analysis itself. It thus becomes attractive to investigate the nonlinear stability of numerous thin film problems with an ease that has been offered hitherto only by the linear stability analysis. As is clear from the preceding analysis, the theory may be generalized for any potential of the form, $\phi = \phi(h)$. For example, if one uses the dimensionless potential pair ($c \neq 0$),

$$\phi(h) = B_1 / h_0^{c-2} \rho v^2 h^c$$

and

$$\phi'(h) = B_2 / h_0^{c-2} \rho' v^2 h^c, \quad [37i]$$

instead of potentials [4a] and [4b], one obtains the following equation of evolution, in lieu of Eq. [16c],

$$h_t + \left(cB^*/3\right)\left(h^{2-c} h_x\right)_x + S\left(h^3 h_{xxx}\right)_x$$

$$+\left(h_0 / 2\mu v\right)\left(\sigma_x h^2\right)_x = 0, \quad [37j]$$

where $B^* = (B_1 - B_2)$.

For a film devoid of surfactants, it may again be brought to a parameterless form by defining $X = \left(cB^*/3S\right)^{1/2} x$ and $\tau = \left(cB^*/3\right)^2 S^{-1} t$, which reduces Eq. [37j] to

$$h_\tau + \left(h^{2-c} h_X\right)_X + \left(h^3 h_{XXX}\right)_X = 0. \quad [37k]$$

The first order equation now is

$$h_{1\tau} + \left[h_0^{2-c} h_{1X} - (c-2) h_1 h_{0X} h_0^{1-c}\right]_X$$

$$+\left[h_0^3 h_{1XXX} + 3h_0^2 h_{0XXX} h_1\right]_X = 0. \quad [37l]$$

Following the same analysis as that pursued for Eq. [36], the following expression for the growth coefficient may now be derived.

$$\omega_N = q^2\left[\kappa^{2-c} - \kappa^3 q^2\right]$$

$$+ \mu^*\left[(c-2)\kappa^{1-c} + 3\kappa^2\right], \quad [37m]$$

where $\kappa = (1-\mu^*)$.

The wave number of the neutrally stable wave is obtained by setting $\omega_N = 0$ in the above equation and further, the dominant growth coefficient is ($c \neq 0, 2$).

$$\omega_N^* = \frac{1}{4}\left[\left(1-\mu^*\right)^{1-2c} + 4\mu^*\left\{\left(c-2\right) \times \left(1-\mu^*\right)^{1-c} + 3\left(1-\mu^*\right)^2\right\}\right], \quad [37n]$$

which may be contrasted with the dominant growth coefficient of the linear theory, viz., $\omega_L^* = 1/4$. For any $c > 1$, as the singularity of the interaction potential, c, increases, the dominant growth coefficient derived from the present theory shows greater deviation from its linear theory counterpart. For example, if the interaction potential is assumed to be due to the retarded dispersion interactions, viz., $c = 4$, the dominant growth coefficient is

$$\omega_n^* = \frac{1}{4}\left[\left(1-\mu^*\right)^{-7} + 4\mu^*\left\{2\left(1-\mu^*\right)^{-3} + 3\left(1-\mu^*\right)^2\right\}\right]. \quad [37o]$$

Thus the effect of nonlinearities here is even more pronounced than in the case of nonretarded dispersion interactions. This may be seen in Fig. 2, where curve III shows the ratio between the nonlinear theory time of rupture and the linear theory rupture time, as a function of the amplitude of disturbances.

With this background, the nonlinear effects of Marangoni motion on the stability of thin films may now be considered.

III.2. The Effect of Soluble Surfactants on the Rupture of a Wetting Film

III.2.1. The Growth Coefficient.

Henceforth, we will consider the set of dimensional equations [18] to [22]. The analysis pursued here also puts in perspective the generalization of the methodology developed in the last section to a vector (a set of equations). The first step is to expand all system variables around their values corresponding to the neutrally stable state of the linear theory, viz.,

$$H = H_0 + \epsilon^* H_1 + O\left(\epsilon^{*2}\right), \quad [38a]$$

$$C = C_0 + \epsilon^* C_1 + O\left(\epsilon^{*2}\right), \quad [38b]$$

$$C' = C_0' + \epsilon^* C_1' + O\left(\epsilon^{*2}\right), \quad [38c]$$

and

$$\Gamma = \Gamma_0 + \epsilon^* \Gamma_1 + O\left(\epsilon^{*2}\right), \quad [38d]$$

where H_0, C_0, and C'_0 are, respectively, the base case (corresponding to the linear neutral stability), dimensional thickness, and concentrations in the two phases and ϵ^* is a dimensional small parameter related to the initial amplitude of the perturbation ($\epsilon^* = x\mu^* h_0$), where h_0 is the mean initial thickness of the thin film. To determine the wavenumber of the linearly neutral stable wave in this case, it is sufficient to note that for a stationary wave, all of the velocity components and their derivatives vanish. In such an event, there is no distribution of surfactants and consequently the interfacial tension gradients also vanish, viz., $\sigma_x = 0$. The neutrally stable waveform in this case is thus the same as that in the absence of surfactants and is computed directly from Eq. [18] by substituting the stationary waveform

$$H_0 = h_0 + \epsilon^* \sin K_1 x + O\left(\epsilon^{*2}\right). \quad [39a]$$

This gives the following dimensional compatibility condition, corresponding to its nondimensional counterpart, $q^2 = 1$.

$$K_1^2 = A / 2\pi h_0^4 \sigma_0\left(C_0\right) \quad [39b]$$

where $\sigma_0(C_0)$ is the interfacial tension evaluated at the equilibrium value of the surfactant concentrations, C_0 and C'_0. In what follows, we will use the symbol σ_0 to denote the equilibrium value of the interfacial tension. The first order problem corresponding to the hydrodynamic equation [18] is

$$H_{1t} + \left(A/6\pi\mu\right)\left(H_0^{-1} H_{1xx} - H_0^{-2} H_1 H_{0xx}\right)$$

$$+\left(\sigma_0/3\mu\right)\left(H_0^3 H_{1xxxx} + 3H_0^2 H_1 H_{0xxxx}\right)$$

$$+\left(\sigma_1/3\mu\right)\left(H_0^3 H_{0xxxx}\right) + \left(1/2\mu\right)\left(H_0^2 \sigma_{1x}\right)_x = 0. \quad [40]$$

The first order surfactant transport equations are

$$C_{1t} = D\nabla^2 C_1, \quad [41a]$$

and

$$C'_{1t} = D'\nabla^2 C'_t, \quad [41b]$$

and the boundary conditions reduce to the following equilibrium condition;

$$\theta C_1 = \theta' C'_1, \quad [42a]$$

where θ and θ' are the derivatives of isotherms [20], as evaluated at the zeroth order (equilibrium) concentrations. For the Langmuir type isotherms [20a], they are given by

$$\theta = k_1 / \left(1 + \frac{k_1}{\Gamma_\infty} C_0\right)^2 \text{ and } \theta' = \left(\theta / k_e\right).$$

In addition,

at $Z = H_0$;

$$-DC_{1z} + D'C'_{1z} = \Gamma_{1t} + \Gamma_0 u^*_{1x} - D_s \Gamma_{1xx}, \tag{42b}$$

$$Z \to \infty;\ D'C'_{1z} = 0, \tag{42c}$$

and

$$Z = 0;\ \ DC_z = 0. \tag{42d}$$

The first order gradient of the surface velocity, u^*_{1x}, is readily obtained from [11b] as

at $Z = H_0$;

$$u^*_{1x} = (A/4\pi\mu)\left(H_0^{-2} H_{1xx} - 2H_0^{-3} H_1 H_{0xx}\right)$$

$$+ (\sigma_0/2\mu)\left(H_0^2 H_{1xxxx} + 2H_0 H_1 H_{0xxxx}\right)$$

$$+ (\sigma_1/2\mu) H_0^2 H_{0xxxx} + (1/\mu) H_0 \sigma_{1xx}. \tag{43a}$$

Additionally, from Eq. [22] one has

$$\text{at } z = H_0;\ u^*_{1x} = (\sigma_1 - M\Gamma)/\mu_s, \tag{43b}$$

where $M = (\partial\sigma/\partial\Gamma)_0$. Again postulating for the first order variables, solutions of the form

$$H_1 = \sin Kx e^{\omega t}, \tag{44a}$$

$$C_1 = \eta \sin Kx e^{\omega t}, \tag{44b}$$

$$C'_1 = \eta' \sin Kx e^{\omega t}, \tag{44c}$$

and

$$\sigma_1 = \hat{\sigma}_1 \sin Kx e^{\omega t}, \tag{44d}$$

and making use of them in conjunction with the definitions [43a] and [43b], one obtains

$$\hat{\sigma}_1 = (N\mu_s + M\theta\eta)/(1 + P) \tag{45a}$$

where

$$N \equiv \frac{\sigma_0}{2\mu}\left(H_0^2 K^4 - 2H_0 h_0 \mu^* K_1^4\right)$$

$$- \frac{A}{4\pi\mu}\left(H_0^{-2} K^2 + 2H_0^{-3} h_0 \mu^* K_1^2\right) \tag{45b}$$

and

$$P \equiv \frac{H_0 K^2 \mu_s}{\mu} + \frac{H_0^2 \mu^* h_0 K_1^4 \mu_s}{2\mu}. \tag{45c}$$

Equation [43b] now gives

$$u_{1x}^{*} = \left[N/(1+P) - M\theta\eta Q/(1+P) \right] \sin Kx e^{\omega t}, \qquad [45d]$$

where

$$Q = P / \mu_s. \qquad [45e]$$

Finally, the boundary condition [42b], in conjunction with Eqs. [41], perturbed forms [44] and expressions [45], determines η at the interface as

$$\eta = -\Gamma_0 N / R(1+P), \qquad [46a]$$

where

$$R \equiv Da\tanh(aH_0) + D'a'\theta/\theta' + \theta\omega$$

$$+\Gamma_0 |M| \theta Q/(1+P) + D_s\theta K^2, \qquad [46b]$$

$$(\theta/\theta') = k_e, \qquad [46c]$$

$$a = \left(\frac{\omega}{D} + K^2 \right)^{1/2}, \qquad [46d]$$

and

$$a' = \left(\frac{\omega}{D'} + K^2 \right)^{1/2}. \qquad [46e]$$

This completes the determination of the interfacial tension (and its gradient) from Eq. [45a], giving

$$\hat{\sigma}_1 = \frac{N}{(1+P)} \left\{ \mu_s + \frac{|M| \theta \Gamma_0}{R(1+P)} \right\}. \qquad [47]$$

The modified dispersion equation is inferred from Eq. [40] by making use of the perturbed forms [44] and the expression [47], which, after considerable rearrangement gives

$$\omega = K^2 \left(\frac{A}{6\pi\mu H_0} - \frac{\sigma_0 H_0^3 K^2}{3\mu} \right) \left(1 - \frac{3}{2} Z^* \right)$$

$$+ \frac{\mu^* h_0 K_1^2 A}{6\pi\mu H_0^2} \left(1 - 3Z^* \right)$$

$$+ \frac{\sigma_0 H_0^2 h_0 \mu^* K_1^4}{\mu} \left(1 - Z^* \right), \qquad [48a]$$

where

$$Z^* = \frac{1}{\mu H_0 (1+P)} \left\{ \mu_s + \frac{|M| \theta \Gamma_0}{R(1+P)} \right\}$$

$$\times \left(\frac{H_0^2 K^2}{2} + \frac{H_0^3 \mu^* h_0 K_1^4}{3} \right). \qquad [48b]$$

The dispersion relation [48a] can be rewritten in a more compact form by substituting for K_1^2 from Eq. [39b] and noting that for long wavelength perturbations, (KH_0) and $(K_1H_0) \ll 1$ and thus $H_0^2K^2 \gg H_0^3\,\mu^* h_0 K_1^4$. This yields

$$\omega = \omega_0\left(1-\frac{3}{4}\bar{Z}\right)+\frac{\mu^* A^2}{4\pi^2\mu\sigma_0 h_0^5}\left\{\frac{1}{3\left(1-\mu^*\right)^2}\right.$$

$$\left.\times\left(1-\frac{3}{2}\bar{Z}\right)+\left(1-\mu^*\right)^2\left(1-\frac{1}{2}\bar{Z}\right)\right\}, \qquad [49]$$

where

$$\bar{Z}=\frac{H_0K^2}{\mu(1+P)}\left\{\frac{|M|\theta\Gamma_0}{R(1+P)}+\mu_s\right\}, \qquad [50a]$$

$$\omega_0=\frac{K^2}{3\mu}\left\{\frac{A}{2\pi h_0\left(1-\mu^*\right)}-\sigma_0 h_0^3\left(1-\mu^*\right)^3K^2\right\}, \qquad [50b]$$

P and Q as given by Eqs. [45c] and [45e], respectively, are also simplified to

$$P=K^2h_0\left(1-\mu^*\right)\mu_s/\mu, \qquad [50c]$$

$$Q=h_0\left(1-\mu^*\right)K^2/\mu. \qquad [50d]$$

It is easy to verify that the dimensionless parameter $\bar{Z}$ goes to zero as both $|M|\Gamma_0$ and μ_s tend to zero and it is close to unity if either $|M|\Gamma_0$ or μ_s is large. Thus, $\bar{Z}\in(0.1)$ for all parameter values.

In the event the amplitude of perturbations is vanishingly small, viz., $\epsilon \to 0$, the last term in Eq. [49] becomes negligible and one recovers the linear stability results of Jain and Ruckenstein (20) in the long wavelength limit. In general, it is not possible to establish analytically the wavenumber for which the growth coefficient as given by Eq. [49] is maximum. This may however be done numerically and then the time of rupture is given by

$$\tau_N=\frac{1}{\omega^*}\ln\left(\frac{1-\mu^*}{\mu^*}\right), \qquad [51]$$

where ω^* is the maximum growth rate for all $K\in(0,\infty)$ and a given set of parameters.

III.2.2. Computations and Discussions.

Figure 3 shows the effects of the initial amplitude of the disturbance on the time of rupture for different values of the Marangoni-parameters, $|M|\Gamma_0$. All other parameters are the same as those of Jain and Ruckenstein (20) and are reported in the figure caption. Just as in the linear theory, increasing the Marangoni-parameter increases the time of rupture. This happens because for a thinning film, the redistribution of surfactant molecules makes their surface excess concentration higher at elevated regions as compared to the depressed regions. This in turn results in an interfacial

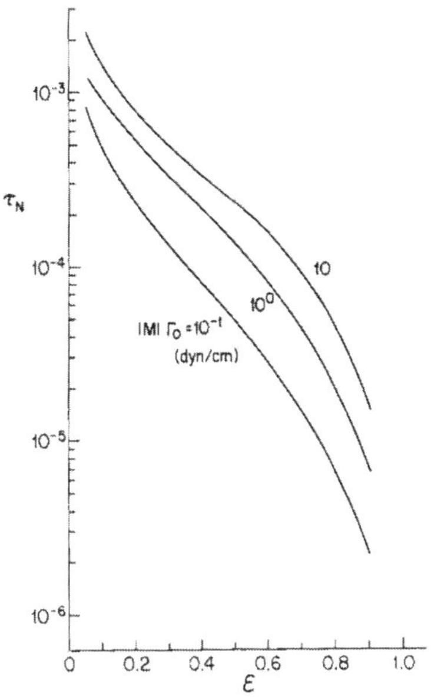

FIG. 3. The nonlinear theory rupture time (in seconds) as a function of the nondimensional initial amplitude for three different values of the parameter, $|M|\Gamma_0$. The other parameters are fixed as $h_0 = 100\,Å$, $A = 10^{-12}$ erg, $\sigma_0 = 50$ dyn/cm, $\mu_s = 0$, $D_s = D = D' = 10^{-5}$ cm²/s, $\theta = \theta' = 2 \times 10^{-4}$ cm, and $\mu = 0.01$ g/cm s.

tension gradient, with the interfacial tension being lower at the elevated regions. The flow of fluid from lower to the higher interfacial tension regions is the Marangoni-flow. Since this flow is in a direction opposite to the flow generated due to the dispersion force mediated thinning of the film, it retards the process of thinning and thus prolongs the time of rupture.

As is shown in the figure, the initial amplitude of the interfacial perturbations has a profound effect on the time of rupture. The time of rupture decreases by about two orders of magnitude as the dimensionless amplitude of the perturbation increases from 0.05 to 0.75. Thus even an order of magnitude prediction of the time of rupture of a thin film requires some a priori knowledge concerning the amplitude of the initial interfacial perturbations, especially in the event when they have a mechanical origin.

The extensive computations of the maximum growth coefficient from Eq. [49] for a large number of parameter values showed that the maximum in the growth coefficient always occurs at or in the immediate vicinity of the wavenumber,

$$K^2 = A / 4\pi\sigma_0 h_0^4 \left(1 - \mu^*\right)^4. \qquad [52]$$

This wavenumber is the same as that which maximizes the growth coefficient in the absence of the Marangoni-motion, viz., it corresponds to the solution of $(\partial\omega_0/\partial K) = 0$. It is verified through direct numerical computations that evaluating ω from Eq. [49] based on this wavenumber typically results in less than 5% error in the prediction of the dominant growth rate. Theoretically, K^2 as given by Eq. [52] is indeed a rigorous solution of $(\partial\omega/\partial K) = 0$. for two asymptotic cases, namely, for a film devoid of surfactants, $\bar{Z} \to 0$, and also when surfactants are in excess, viz., $|M|\Gamma_0 \to \infty$ and consequently, $\bar{Z} \to 1$. It may thus be expected that Eq. [52] is indeed a good representation for the dominant wavenumber for all values of parameter $|M|\Gamma_0$. The substitution of Eq. [52] in Eqs. [49], [50a], and [50b] leads to the following compact expression for the dominant eigenmode, ω^*:

$$\omega^* = \left(\frac{A^2}{48\pi^2\mu\sigma_0 h_0^5}\right)\left[\left(1-\frac{3}{4}\bar{Z}\right)\left(1-\mu^*\right)^{-5} + 4\mu^*\left(1-\frac{3}{2}\bar{Z}\right)\left(1-\mu^*\right)^{-2} + 12\mu^*\left(1-\mu^*\right)^2\left(1-\frac{1}{2}\bar{Z}\right)\right]. \quad [53]$$

The expression for $\bar{Z}$ is simplified by substituting the approximation [52] for the wave number in Eq. [50a] and in Eqs. [46b], [46c], [46d], [50c], and [50d], that represent, respectively, the definitions of $\bar{Z}$, R, a, a', P, and Q. The advantage of such a simplification is that now the time of rupture can be directly computed from Eq. [51], and the ratio τ_N/τ_0, where $\tau_0 = 48\pi^2\mu\sigma_0 h_0^5 A^{-2}$, is represented in a universal form as a function of the parameters μ^* and $\bar{Z}$ alone. This is shown in Fig. 4. The time of rupture as given by the nonlinear theory may now be directly computed from Fig. 4 for any given set of parameters and initial amplitude of the perturbation.

It is of interest to determine the maximum stabilizing effect of surfactants on the time of rupture, compared to the case when surfactants are absent. The dominant growth coefficient for the case when surfactants are in excess is determined from Eq. [53] by letting $\bar{Z} \to 1$. In the event of the absence of surfactants it is determined by letting $\bar{Z} \to 0$. The ratio of the times of rupture when $\bar{Z} \to 1$ and when $\bar{Z} \to 0$ is depicted in Fig. 5. As is well known, the linear stability analysis predicts this ratio to be four (4, 20). As is indeed seen in Fig. 5, this corresponds to a vanishingly small amplitude of the initial perturbation. As the amplitude of the perturbation is increased, the Marangoni-motion prolongs the time of rupture by a factor greater than 4. For the nondimensional amplitude of 0.7, the presence of a large amount of surfactants prolongs the time of rupture by a factor of 10 compared to the case of a film devoid of surfactants. The same curve is also obtained

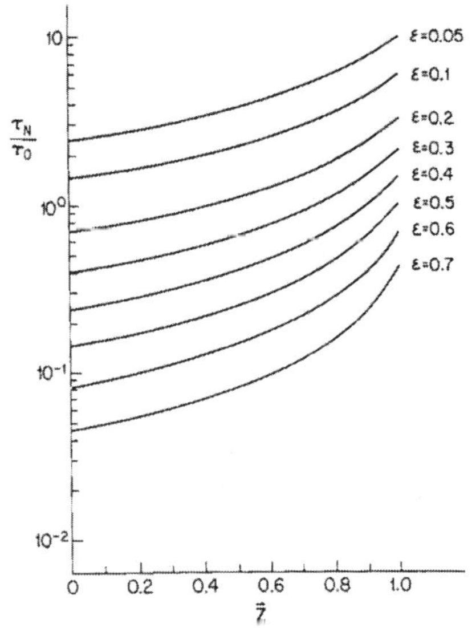

FIG. 4. A generalized plot to determine the nonlinear rapture time τ_N, when both the thin film and the bounding media have soluble surfactants. τ_0 is defined as $48\pi^2\mu\sigma_0 A^{-2} h_0^5$ and the dimensionless parameter $\bar{Z}$ is defined by Eq. [50a]. The ratio, τ_N/τ_0, is a nondimensional quantity.

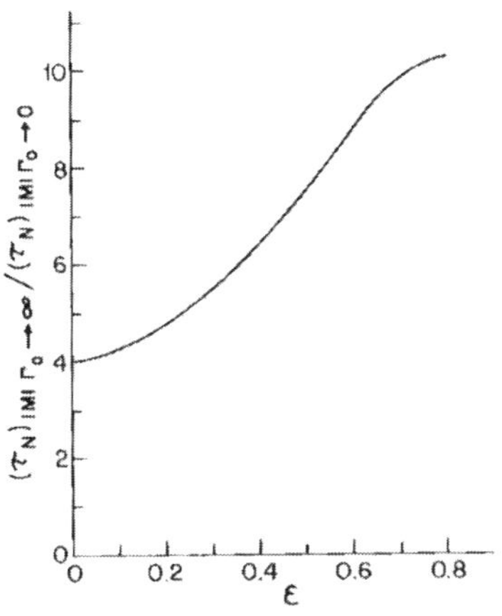

FIG. 5. The effect of an excess amount of surfactants on the time of rupture as predicted by the nonlinear theory. $(\tau_N)|M|\Gamma_0 \to \infty$ is the time of rupture in the presence of an excess amount of surfactant and $(\tau_N)|M|\Gamma_0 \to 0$ is the time of rupture in their absence. The asymptotic value of 4 that is predicted by the linear theory is attained in the limit of $\epsilon \to 0$.

when the time of rupture for a film with large surface shear viscosity ($\mu_s \to \infty$) is compared with the time of rupture when $|M|\Gamma_0$ and $\mu_s \to 0$.

Finally, the ratio of the time of rupture as computed by the present analysis and that given by the linear theory, is shown in Fig. 6 as a function of the initial amplitude of the perturbation. The upper curve corresponds to an excess of surfactants and the lower one to the absence of surfactants. Thus, the ratio between the times of rupture as computed by nonlinear and linear theories always lies between these two curves for all values of parameters characterizing the surfactant and its effect on the interface.

In conclusion, the stabilizing influences of the Marangoni-motion (or surface elasticity) and the surface shear viscosity in prolonging the time of rupture are underestimated by the linear stability analysis. Also, as is shown earlier, the destabilizing influence of dispersion forces and the stabilizing effect of interfacial tension are, respectively, under-and overestimated by the linear theory.

IV. APPLICATION OF THE NONLINEAR ANALYSIS TO BREAKUP OF THE TEAR FILM

The tear film, that covers the conjunctiva and cornea, is made up of three distinct films (21). A thin mucus layer coats the corneal epithelium and supports a much thicker (1–10 μm) aqueous layer, which in turn is bounded by a lipid layer about 1000 Å thick. It is well known that the aqueous tear film breaks in about 20 to 200 s if the blinking is prevented or in shorter time intervals if the eye is pathological (21, 22). We had earlier proposed a "two-step, double film" mechanism of this rupture that traces the instability of the aqueous layer in the van der Waals force mediated rupture of the thin mucus layer (11, 12). Once the mucus layer is ruptured, the aqueous layer comes into direct contact with the patches of the underlying non-water wet-table corneal epithelium (21). This second step results in a rather sudden dewetting of epithelial surfaces. For a more elaborate discussion of the physiological aspects of the proposed mechanism and its implications, the reader is referred to our earlier papers (11,23). For now, it suffices to note that the study of the rupture of the thin corneal mucus layer may be pursued along the lines contained in earlier sections.

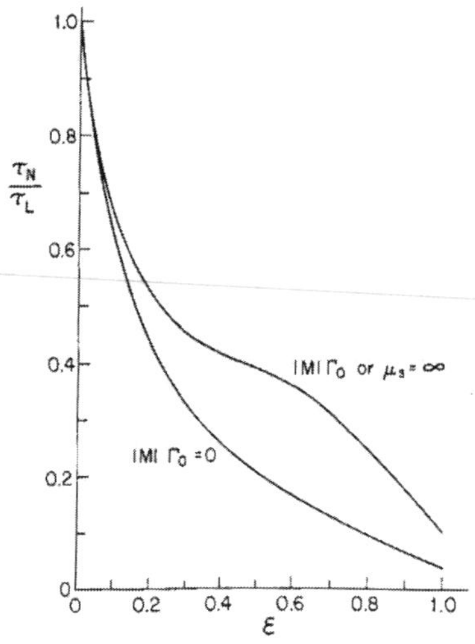

FIG. 6. The ratio of nonlinear theory time versus linear theory rupture time as a function of the nondimensional initial amplitude, ϵ. The upper curve is when an excess amount of surfactants is present and the lower one is in their absence.

IV.1. A Model of Tear Film and Its Nonlinear Stability Analysis

The aqueous layer bounding the mucus film bears a high surface tension and is thick enough (micrometer sized) for the van der Waals interactions induced thinning to be insignificant for this layer. Thus the model of a thin film sketched in Fig. 1 and formulated in Section II, is also applicable to the thin mucus coating of the epithelium. There are two important differences however. First, we will consider here the effects of antisurfactants (or inverse surfactants) rather than surfactants and second, the no flux boundary condition at $Z = H'$ (Eq. [21b]) is replaced by a constant concentration boundary condition for the concentration of lipids. The tear-film antisurfactants are those solutes that *augment* the mucous-aqueous interfacial tension from its rather low value of 1 dyn/cm or less that prevails in the absence of these solutes. Although this interfacial tension has not been measured, it may be taken to be in the same range as that observed for a cell membrane with an external glycoprotein coat. This has been reported to be in the range of 10^{-3} to 1 dyn/cm (24). A typical example of antisurfactant for the mucous-aqueous interface is the nonpolar eye lipids (21). With these differences, the linear stability of this model of tear film has been carried out elsewhere and the effect of various alterations in the physiology of the tear film on the breakup times (BUT) has been studied (23). Now, we turn towards the nonlinear stability of the mucus layer.

In view of the fact that the thickness of the aqueous layer, H' is much larger than the thickness of the mucus layer, h_0, the first order hydrodynamic equation is the same as Eq. [40]. Neglecting the presence of lipids in the mucus layer, the first order transport equation for lipids in aqueous tears is

$$C'_{1t} = D'\nabla^2 C'_1, \qquad [54a]$$

and the boundary conditions are

$$\text{at } Z = H_0;\ D'C'_{1z} = \Gamma_{1t} + \Gamma_0 u^*_{1x}, \qquad [54b]$$

$$\text{at } Z = H';\ C'_1 = 0. \qquad [54c]$$

The boundary condition [54c] arises because the concentration of lipids at the aqueous layer-lipid film boundary is assumed to be constant. One may now pursue the same analysis as that presented in Section III.2.1 with the modified transport equations [54]. It is to be noted that in these derivations, the surface excess concentration, Γ_0, is now a negative quantity since antisurfactants augment the interfacial tension, as opposed to surfactants that reduce the interfacial tension. Consequently, the isotherm constant, θ' is also negative. The parameter, $M = (\partial\sigma/\partial\Gamma)_0$ is, however, still negative, just as in the case of surfactants. This fact may be easily verified from the Gibbs equation

$$d\sigma = -RT\Gamma d\ln c \quad [55a]$$

which, with the help of the isotherm $\Gamma = \theta'C'$; $\theta' < 0$, gives (for a linear isotherm)

$$(\partial\sigma/\partial\Gamma) = -RT. \quad [55b]$$

Keeping note of these sign changes, the final dispersion relation of the nonlinear theory is given by

$$\omega = \omega_0\left(1-\frac{3}{4}Z_1\right)+\frac{\mu^* A^2}{4\pi^2\mu\sigma_0 h_0^5}\left\{\frac{1}{3\left(1-\mu^*\right)^2}\right.$$

$$\left.\times\left(1-\frac{3}{2}Z_1\right)+\left(1-\mu^*\right)^2\left(1-\frac{1}{2}Z_1\right)\right\}, \quad [56]$$

where

$$Z_1 = \frac{H_0 K^2}{\mu(1+P_1)}\left\{\frac{|M\theta'\Gamma_0|}{R_1}+\mu_s\right\}, \quad [57a]$$

$$P_1 = K^2 h_0\left(1-\mu^*\right)\mu_s/\mu, \quad [57b]$$

$$R_1 = D'a'\coth\{a'(H'-h_0)\} - |\theta'|\omega$$

$$+|M\theta'\Gamma_0|Q_1/(1+P_1) \quad [57c]$$

and

$$Q_1 = h_0\left(1-\mu^*\right)K^2/\mu. \quad [57d]$$

The difference in the definition of R_1 as compared to R (given by Eq. [46b]) is due to different mass transport boundary conditions and because of the replacement of the surfactant by the antisurfactant. The time of rupture of the mucus layer is again computed by maximizing the growth coefficient from Eq. [56] and then using Eq. [51].

IV.2. Results and Discussions

The solutes being considered here are antisurfactants and thus their influence on the mucous-aqueous interface that is devoid of solutes may be represented as

$$\sigma_0(\Gamma_0) \approx \sigma_0(0) + (\partial\sigma_0/\partial\Gamma_0)(\Gamma_0). \quad [58a]$$

Since both $\partial\sigma_0/\partial\Gamma_0$ and Γ_0 are negative, Eq. [58a] becomes

$$\sigma_0(\Gamma_0) \approx \sigma_0(0) + |M\Gamma_0|. \qquad [58b]$$

In all of the subsequent calculations, the mucous-aqueous interfacial tension in the absence of solutes is taken to be 1 dyn/cm, viz., $\sigma_0(0) = 1$ dyn/cm. Although, the Hamaker constant for the epithelium-mucus-aqueous system has never been measured and there is some uncertainty regarding the thickness of the mucus layer (23) prevailing under dynamic conditions of tear film and blinking, we will take some reasonable estimates of both, principally in order to demonstrate the importance of the nonlinear effects. The times of rupture shown in Fig. 7 are drawn for $|\theta'| = 2\times10^{-4}$ cm, thickness of mucus layer (h_0) = 400 Å, thickness of aqueous layer (H') = 10 μm, $A = 7\times10^{-14}$ erg, $|M_0| = 3$dyn / cm, $D' = 10^{-5}$ cm^2/s, $\mu_s = 0$, and $\mu = 0.1$ g/cm s. The curve labeled I is the time of rupture of the mucus layer as calculated from the nonlinear theory without accounting for the Marangoni-convection. The curve II accounts also for the stabilizing effect of the Marangoni-flow. Curve III is the time of rupture as computed from the linear theory and accounts for the Marangoni-flow. As is seen from the figure, neglecting either the nonlinearities or the Marangoni-flow may result in a misleading estimate of the time of rupture. The former is especially important for relatively large initial interfacial perturbations. Although, the amplitude of the initial interfacial nonhomogeneities for the mucus layer has not been determined, it is likely that they are large because mucus is a "sloppy" gel, bears a low interfacial tension and the dynamics of tear film and blinking encourage the shearing at the interface.

Figure 8 depicts the same curves for $h_0 = 1200$ Å and $A = 7\times10^{-13}$ erg and illustrates the effect of the amplitude of the initial nonhomogeneities on the time of rupture. Figure 9 shows the region of compatibility in the Hamaker constant and the Marangoni-parameter, $|M\Gamma_0|$, representation, viz. for all values of these parameters corresponding to the shaded region, the time of rupture as predicted by this model is compatible with the observed times of rupture. The range of the Hamaker constant falling within the shaded region is in agreement with the values reported for several materials. The uppermost and the lowermost curves correspond to the compatibility range as calculated by the linear theory. Any number of such compatibility domains may be constructed for a set of feasible parameters values by using the proposed nonlinear theory. An a priori prediction of the tear breakup times has to however, wait until the experimental values of various parameters become known.

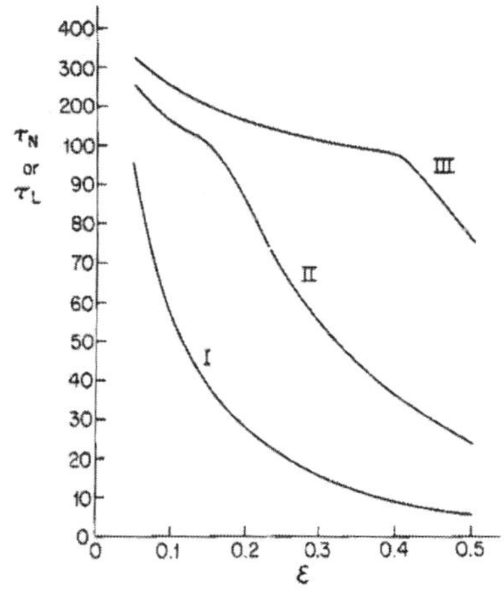

FIG. 7. The effect of the initial amplitude of the mucous-aqueous interfacial nonhomogeneities on the time (in seconds) of tear film rupture. The parameters are: $\sigma(0) = 1$ dyn/cm, $\theta = 2\times10^{-4}$, $h_0 = 400$ Å, $H' = 10$ μm, $|M\Gamma_0| = 3$dyn / cm, $A = 7\times10^{-14}$, $\mu = 0.1$ g/cm s, and $D' = 10^{-5}$ cm^2/s. Curves II and I are the nonlinear theory rupture times with and without the Marangoni-convection, respectively, and Curve III is the linear theory rupture time with the Marangoni-convection.

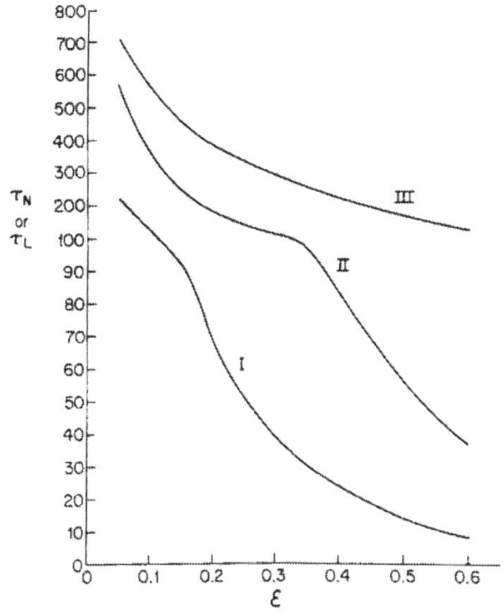

FIG. 8. The same as Fig. 7, except $h_0 = 0.12\ \mu$m and $A = 7 \times 10^{-13}$ erg.

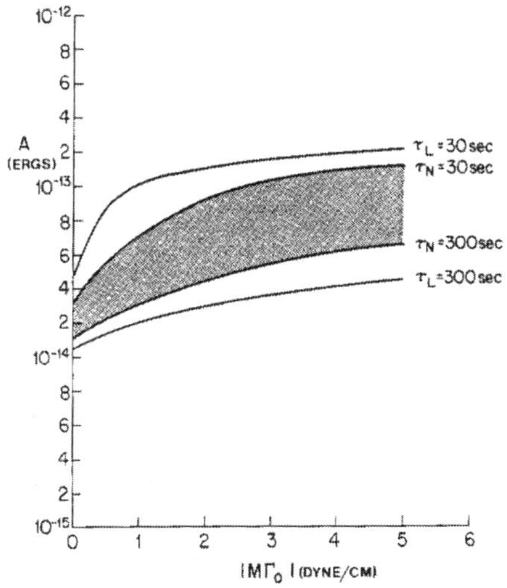

FIG. 9. The compatibility region for the tear film rupture for $\epsilon = 0.2$ and other parameter values as reported in Fig. 7. The shaded region corresponds to the parameter values for which nonlinear theory predicts a time of tear film rupture that is in agreement with the clinical results. The uppermost and the lowermost curves correspond to the predictions of the linear theory. It is to be noted that this compatibility domain is not unique and several others may be constructed for different feasible sets of physicochemical parameters (23).

Finally, we consider the effect of the thickness of the aqueous layer H', on the tear film breakup times. As the thickness of the aqueous layer decreases from its average value within the interblink period (about 4 – 8 μm), $\coth\{a'(H' - h_0)\}$ starts to increase. In the extreme case of $H' \to h_0$, $\coth\{a'(H' - h_0)\} \to \infty$ and thus as seen from Eqs. [57a] and [57c], the parameter, Z_1, tends to zero. The stabilizing influence of the Marangoni-convection is thus completely wiped out in such an event. Physically, it happens because a decrease in the aqueous layer thickness leads to a more rapid diffusional redistribution of solutes and thus undermines the concentration gradient

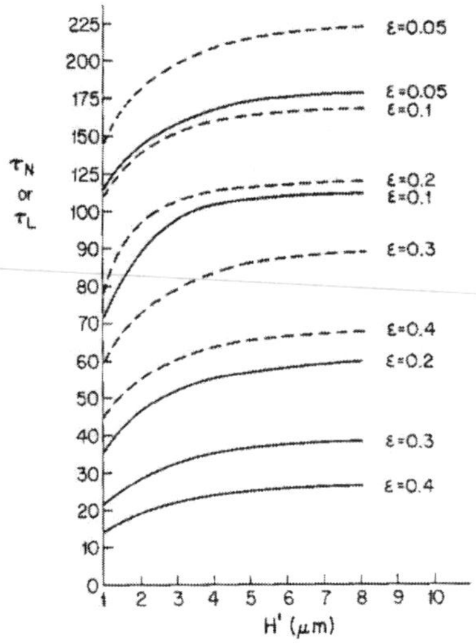

FIG. 10. The effect of the aqueous layer thickness on the tear film breakup times (in seconds). The parameters are $A = 1.1 \times 10^{-13}$ erg, $h_0 = 400$ Å, $|M\Gamma_0| = 5$dyn / cm, and other parameters are as reported in Fig. 7. The solid lines and the dashed lines are the predictions of the non-linear and the linear theories, respectively.

and consequently, the interfacial tension gradient. Figure 10 shows the effect of the aqueous layer thickness on the time of rupture for $A = 1.1 \times 10^{-13}$ erg, $h_0 = 400$ Å, and $|M\Gamma_0| = 5$ dyn / cm and for various values of the amplitude of the initial interfacial corrugations. The solid curves correspond to the time of rupture as given by the nonlinear theory and dashed curves represent the linear theory results. About 40% decrease is predicted in the time of the tear film rupture as the average aqueous layer thickness decreases from 4 to 1 μm. Clinically, such a reduction in the time of tear film breakup is well known in the event of a deficiency of the aqueous tears and is encountered in many pathological conditions of a dry eye (21, 23). Again, large discrepancies in the times of rupture predicted by the nonlinear and linear theories are apparent from this figure.

It is interesting to note that the successively performed clinical tests report significant statistical variations (21) in the times of tear film rupture for the same patient. According to this model of the tear film rupture, such statistical variations seem imminent because, after the restoration of the mucus layer after each blink, the amplitude of the perturbations would be different depending on the quality and frequency of blinking. This then results in different times of rupture even if the mean thickness of the mucus layer is approximately constant after each blink.

In conclusion, the earlier proposed mechanism of the tear film rupture (11) and the qualitative conclusions derived from the linear stability analysis (11, 23) are not altered if the nonlinear theory is used instead. The nonlinear theory is, however, important for quantitative predictions and may be useful when the various parameters characterizing the tear film become available. It may be noted that qualitative conclusions drawn herein remain the same if instead a different feasible set of parameter values (23) are used for the illustrative calculations.

V. DISCUSSION

The present work deals with the effect of the finite sized initial perturbations on the stability and rupture of the thin films, which is to be contrasted with the linear stability analysis that is valid for infinitesimal disturbance only. The nonlinearities of the governing transport equations become significant when the amplitude of the perturbations is large.

The key idea in the approach developed is to investigate the stability of the time stationary, spatially nonuniform solution of the governing equations, as opposed to the linear stability analysis that concerns itself with the stability of the trivial solution (spatially uniform stationary state). As opposed to the trivial base state then, the spatially nonhomogeneous base state accounts for a finite corrugation of the interface and is therefore, closer to an imposed, finite amplitude disturbance. Moreover, the amplitude of base state is in direct proportion to the amplitude of disturbance, which is in contrast to a fixed value of base state, $h = 1$, chosen in linear stability analysis.

The only stationary spatially quasiharmonic solution is that corresponding to the neutral stability of the trivial state. Thus for the proposed analysis to be successful in its present form, it is necessary that the neutrally stable state of the trivial solution be nonoscillatory in time, or in other words, a time stationary, space periodic solution of the governing equations be available for some nontrivial wave number. It is always possible to find such a nontrivial stationary solution if the governing equations are invariant under the inversion of the sign of the lateral spatial coordinate, x (see, for instance, Eq. [18]). A rigorous proof of the existence of such solutions when the governing equations satisfy the above condition has been provided by V. I. Arnold in a paper of Malomed and Tribelsky (16). Such a condition is, however, automatically satisfied for all dissipative systems (systems with a real growth coefficient) and it is sometimes possible to reduce a dissipative-dispersive system to a purely dissipative one by a suitable change of moving coordinates (see, for instance, Shlang and Sivashinsky (25)). More importantly, the proposed method in its present form is expected to be applicable only to the problems where the growth of the initial perturbations is not saturated by the nonlinear terms. Such a saturation occurs for instance in the case of formation of ripples on falling films (18, 24). This, however, does not occur for thin films that either eventually rupture or attain a new equilibrium state due to double layer repulsion (black films).

Based on this analysis, we investigate the nonlinear stability of a wetting thin film with surfactants in both the film and the bounding media. The nonlinear effects of van der Waals dispersion forces, interfacial tension, surfactant concentration gradient driven Marangoni-motion and interfacial shear viscosity are derived analytically as a function of the disturbance amplitude. The analytical results obtained for the dominant wavelength and the time of rupture agree well with the numerical simulations of Williams and Davis (13) for a simple model system—a thin film devoid of surfactants. It is shown that the linear stability results underestimate the role of Marangoni motion (or surface elasticity) and surface viscosity in prolonging the time of rupture. Both of these factors can enhance the stability of a thin film by a factor of about ten, which is in contrast to the linear theory that predicts this factor to be 4. In general, the time of rupture derived by the present theory is always less than that derived from the linear analysis, because the latter underestimates the destabilizing effects of the dispersion forces and overestimates the stabilizing influence of the interfacial tension (this, in fact is clear from the nonlinear Eq. [18] which predicts that as the film gets thinner, viz., $H \rightarrow 0$, the term corresponding to the dispersion attraction becomes unbounded whereas the one reflecting the interfacial tension effects goes to zero).

Finally, the methodology is applied to a previously proposed model of the tear film rupture. It is concluded that all of the qualitative features of the model remain the same as that given by the linear stability analysis. Important quantitative differences in the time of rupture are, however, pointed out and will be of definite interest when a better characterization of the tear film parameters becomes available. The present approach offers an explanation of the enigma surrounding the large variability in the repeated measurements of the tear film breakup time.

In conclusion, the paramount strength of the nonlinear analysis developed here resides in the fact that the derivations and computations using this theory are only slightly more involved than the linear stability analysis, and yet it gives analytical expressions for the growth rate of perturbations as a function of various system parameters and amplitude of perturbations. In view of this, it becomes rather straightforward and attractive to exploit this method for other problems of thin film stability.

APPENDIX: NOMENCLATURE

A	Overall Hamaker constant of the film system, defined by Eq. [17]
A^*	Dimensionless Hamaker constant, Eq. [16c]
A_{ij}	Hamaker constant for interaction between molecules of type i and j
a, a'	Functions defined by Eqs. [46d] and [46e], respectively
a_1, a_2, a_3, a_4	Integration constants, Eqs. [11a] and [11b]
B^*, B_k	Hamaker constants for a general potential pair, defined by Eqs. [37i] and [37j]
C, C'	Concentration of solutes in film and the bounding medium, respectively
c	Exponent of film thickness in a general interaction potential, defined by Eq. [37i]
D, D', D_s	Diffusion coefficients for the film, bounding media and the interface, respectively
H, H'	Thicknesses of the film and the bounding media, respectively
h, h'	Nondimensional thicknesses of the film and the bounding media, respectively
h_0	Initial mean thickness of the film
K, K_1	Wave numbers corresponding to the first order and the zeroth order solutions, respectively
k	Nondimensional wave number (a small parameter for long waves)
k_1, k_1', k_c	Constants of the Langmuir adsorption isotherm, Eqs. [20a] and [20b]
M	Differential of interfacial tension w.r.t. the surface excess concentration, Eq. [43b]
N	Function defined by Eq. [45b]
P, P_0, P_1	Functions defined by Eqs. [45c], [13], and [57b], respectively
p	Nondimensional pressure
Q, Q_1	Functions defined by Eqs. [45e] and [57d], respectively
$q, q_1,$	Nondimensional wave numbers
q_m, q_c	corresponding, respectively, to the first order solution, the zeroth order solution, dominant wave, and the neutrally stable wave
R, R_1	Functions defined by Eqs. [46b] and [57c], respectively
r	Ratio of film to bounding media viscosity
S	Inverse of capillary number, Eq. [5c]
$\bar{S}$	Defined as k^3S
T	Rescaled time (kt)
t	Time (both dimensional and nondimensional)
u, u^*	x components of the nondimensional and dimensional velocity vectors, respectively
w, w^*	z components of the nondimensional and dimensional velocity vectors, respectively.
X	Rescaled x coordinate, Eq. [23a] or Eq. [37k]
x	Spatial coordinate (see Fig. 1)
y	Function defined by Eq. [37d]
$Z^*, \bar{Z}, Z_1$	Functions defined by Eqs. [48b], [50a], and [57a], respectively
z	Spatial coordinate (see Fig. 1)

Greek Symbols

α	Ratio of the bounding media density to film density
Γ, Γ_∞	Surface excess concentration and the maximum surface excess concentration, respectively
ϵ	Nondimensional amplitude of interfacial perturbation
ϵ^*	Dimensional small parameter
η, η'	Parameters defined by perturbed forms, Eqs. [44b] and [44c]
κ	$\left(1-\mu^*\right)$
μ, μ_s	Bulk and surface viscosities, respectively
μ^*	Nondimensional small parameter
v	Kinematic viscosity
ξ	Rescaled x coordinate, Eq. [7a]
ρ	Mass density
σ	Interfacial tension

σ^*	Defined by Eq. [9b]
$\hat{\sigma}_1$	Parameter defined in perturbed form [44d]
τ	Rescaled time, Eqs. [23a] or [37k]
τ_0	Defined as $48\pi^2\mu\sigma_0 h_0^5 A^{-2}$
ϕ	Interaction potential per unit volume at a point on the interface
θ, θ'	Isotherm constants, Eq. [42a]
ω	Growth coefficient
ω^*	Dominant growth coefficient
ω_0	Growth coefficient defined by Eq. [50b]
ψ_1, ψ_1', ψ_2	Functions defined by Eqs. [12a], [10a], and [10b], respectively

Subscripts

0	Indicates base state or the zeroth order solution
1	Indicates first order (perturbed) solution
L	Indicates a quantity as given by the linear theory
m	Refers to dominant value of a quantity
N	Indicates a quantity evaluated by the present (nonlinear) theory

Superscripts

Refers to the property related to the bounding media

REFERENCES

1. Vrij, A., and Overbeek, J. Th. G., *J. Amer. Chem. Soc.* **90**, 3074 (1968).
2. Felderhof, B. U., *J. Chem. Phys.* **49**, 44 (1968).
3. Vrij, A., Hesselink, F. Th., Lucassen, J., and Van Den *Tempel, M.*, Proc. K. Ned. Akad. Wet. Ser. B: Paleontol., Geol, Phys., Chem. **73**, 124 (1970).
4. Ruckenstein, E., and Jain, R. K., *Faraday Trans. 2* **70**, 132 (1974) (Section 5.1 of this volume).
5. Gumerman, R., and Homsy, G., *Chem. Eng. Commun.* **2**, 27 (1975).
6. Jain, R. K., Ivanov, I., Maldarelli, C., and Ruckenstein, E., *in* "Dynamics and Instability of Fluid Interfaces" (T. S. Sorensen, Ed.), pp. 140–167. Springer-Verlag, Berlin, 1979.
7. Jain, R. K., Maldarelli, C., and Ruckenstein, E., AIChE Symp. Ser. Biorheol. **74**, 120 (1978).
8. Bisch, P. M., and Sanfeld, A., *Bioelectrochem. Bio-energ.* **5**, 401 (1978).
9. Maldarelli, C., and Jain, R. K., *J. Colloid Interface Sci.* **90**, 263 (1982).
10. Poste, G., and Nicolson, G. L., "Membrane Fusion." North-Holland, Amsterdam, 1978.
11. Sharma, A., and Ruckenstein, E., *J Colloid Interface Sci.* **106**, 12 (1985) (Section 5.10 of this volume).
12. Ruckenstein, E., and Sharma, A., Paper presented at the International Tear Film Symposium, Lubbock, Texas (published in the Symposium proceedings).
13. Williams, M. B., and Davis, S. H., *J. Colloid Interface Sci.* **90**, 220 (1982).
14. Jeffrey, A., and Kawahara, T., "Asymptotic Methods in Nonlinear Wave Theory," Pitman Advanced Publishing Program, 1982.
15. Sivashinsky, G. I., *Physica 4D*, 227 (1982).
16. Malomed, B. A., and Tribelsky, M. I., *Physica 4D*, 67 (1984).
17. Scheludko, A., *Adv. Colloid Interface Sci.* **1**, 391 (1967).
18. Benney, D. J., *J. Math. Phys.* **45**, 150 (1966).
19. Atherton, R. W., and Homsy, G. M., *Chem. Eng.Commun.* **2**, 57 (1976).
20. Jain, R. K., and Ruckenstein, E., *J. Colloid Interface Sci.* **54**, 108 (1976).
21. Records, R. E., *in* "Physiology of Human Eye and Visual System" (R. E. Records, Ed.), Harper & Row, New York, 1979.
22. Mengher, L. S., Tonge, S., Gilbert, D., and Bron, A. J., Paper presented at the 24th meeting of the Association for Eye Research, Vienna, Austria, 1983.
23. Sharma, A., and Ruckenstein, E., *J. Colloid Interface Sci.* **111**, 8 (1986) (Section 5.11 of this volume).
24. Dolowy, K., *J. Theor. Biol.* **52**, 83 (1975).
25. Shlang, T., and Sivashinsky, G. I., *J. Phys. (Paris)* **43**, 459 (1982).

5.5 Finite-Amplitude Instability of Thin Free and Wetting Films

Prediction of Lifetimes[†]

*Ashutosh Sharma and Eli Ruckenstein**
Department of Chemical Engineering, State University of New York at Buffalo, Buffalo, New York 14260

Received January 7, 1986

1. INTRODUCTION

The numerous and varied investigations of thin-film stability over the last 2 decades[1–15] attest to their importance in dispersed systems on one hand and in the explanation of a myriad of biological phenomena on the other. The determination of the kinetics of coalescence in foam, emulsion, and flotation systems demands a quantitative understanding of the thin-film breakup. The intermolecular force induced instabilities in thin membranes/films have been shown to be related to the onset of microvilli in normal and neoplastic cells,[11,12] the flickering of red blood cells, the adhesion and fusion of membranes,[12,13] and the breakup of the tear film over the corneal epithelium.[14,15]

A principal difficulty in quantifying the rate of film thinning stems from the highly nonlinear nature of the governing hydrodynamic equations, the boundary conditions, and the intermolecular interaction potentials, as well as from the movement of the boundaries. A widespread strategy has been to linearize these around the *unperturbed state* of the system and then seek the normal-mode solutions to these linearized set of equations. If at least one eigenvalue (also referred to as the growth rate) so determined is positive for a set of wavenumbers, the *planar configuration* of the thin film is unstable to disturbances with those wavenumbers. Usually, the time of rupture is then interpreted as the inverse of the dominant (maximum) eigenvalue. Although the linear stability analysis is straightforward, at least conceptually, it is expected to be quantitative only for the disturbances that are vanishingly small. Williams and Davis[16] performed numerical simulations for the long-wavelength instability of a wetting (i.e., in contact with a solid) thin film devoid of surfactants and bounded by a gas phase. Large discrepancies were observed between the times of rupture as computed from simulations and the linear theory. The results of the linear analysis were found to be quantitatively correct only for infinitesimal disturbances for which the initial state of the film is indeed closely approximated by an undisturbed planar interface. Thus even though the linear stability analysis is satisfactory for discerning the qualitative dependence of the time of rupture on various thin-film parameters, the design of flotation, foam, and emulsion systems requires a quantitive analysis as, in practice, the mechanically generated disturbances are quite large. A recent experimental study[17] of foam collapse concluded that the observed critical holdup was 1–2 orders of magnitude higher than that arrived at by the linear analysis;[18] i.e., the breakup time was substantially smaller. The authors also attributed this discrepancy to the finite-amplitude mechanical perturbations. The amplitude of both the mechanical and the thermal perturbations is even larger for extremely low interfacial tension (less than 1 dyn) systems

* *Langmuir, Vol. 2, No. 4, 1986* 481. Republished with permission.

such as proteinaneous coatings (e.g., mucous coating of the corneal epithelium) and certain oil-water interfaces which are encountered, for instance, in oil recovery. It is in view of these considerations that it is desirable to develop an analytical, albeit simple, method of analysis that gives the growth rate of finite-amplitude disturbances and, therefore, accounts for the nonlinearities of the governing equations. We have earlier proposed such a formalism[19] that predicts the response of a thin film to large external disturbances. This formalism is outlined in the next section. It is then applied to a thin film in contact with a solid and a thin free film. The information sought is the wavenumbers of the neutrally stable and the fastest growing perturbations and the growth rate of the latter, all as functions of the thin-film parameters and the *amplitude* of the initial disturbances. We are thus able to delineate the effects of the nonlinearities associated with the van der Waals interactions, the surface tension restoring force, the Marangoni effect, and the surface viscosity, on the lifetimes of thin films.

2. METHODOLOGY

The objective is to develop a perturbative analysis for the finite-amplitude instability of thin films (typically less than or of the order of 1000 Å), for which the long-range van der Waals dispersion forces are important. The essential idea of any stability analysis is to envisage the external disturbances as perturbations of a time-stationary solution of the governing equations. Hitherto, this timestationary solution was taken to be the same as the "trivial solution" or the undisturbed interface and all of the nonlinearities were linearized around a thin "planar" interface. All of the studies of thin-film stability have been concerned with the stability of this trivial, spatially homogeneous steady state when, in addition, the effects of all but the first-order, dominant harmonic are ignored. As is shown by the numerical simulations of Williams and Davis,[16] such a procedure (the linear stability analysis) indeed predicts the correct results for all times *up to time of rupture*, whenever the amplitude of the initial disturbance is small compared to the mean thickness of the film. From this simple observation, it may be inferred that the growing perturbations experience very little distortion and alteration in their growth coefficients from their initial values, even when their amplitudes become comparable to the thickness of the film. That the inclusion of higher order harmonics is insignificant may be easily shown by going to the terms of order of amplitude squared in the linearized thin-film equations. This was indeed shown to be the case for a wetting film.[19] Also, as is shown later, this conclusion automatically holds for a free film with surfactants, because the functional form of the governing equation for this case is the same as that for a pure wetting film.

It may, therefore, be argued that in the case of finite-amplitude disturbances, the dominant growth coefficient associated with the first-order harmonic itself is different from that inferred by linearizing the equations around the trivial state. Let us denote by $\omega_n(\epsilon)$ the growth coefficient associated with a finite-amplitude disturbance with an initial nondimensional amplitude ϵ. Here, ϵ represents the amplitude of the initial disturbance which is nondimensionalized with the initial mean thickness of the film, h_0. It is then expected that for any finite ϵ, $\omega_n(\epsilon) \neq \omega_L$, where ω_L is independent of ϵ and is obtained by linearizing the transport equations around the trivial state. Further, it is expected that in the limit of infinitesimal perturbations, i.e., $\epsilon \to 0$, $\omega_n(\epsilon)$ approaches ω_L and the results of the linear stability analysis are recovered. It is thus obvious that the time-stationary solution (the base state) corresponding to a finite-amplitude perturbation is different from the trivial state, and it reduces to the latter only in the event of vanishingly small amplitudes. As is shown later, a nontrivial stationary solution of the thin-film equations indeed exists for any finite-amplitude external disturbance. The first step of the proposed formalism is, therefore, to construct this spatially nonuniform time-stationary solution of the governing equations. This is easily achieved by setting all time derivatives, velocities, and the concentration gradients equal to zero in the transport equations and solving them for the film thickness. The *leading order* term of this stationary solution corresponds to the neutrally stable wave of the linear stability analysis. This is so because a neutrally stable solution satisfies the requirements of being both time stationary and spatially nonuniform. The onset of time-dependent flow beyond the condition of neutral stability may be investigated by further perturbing the spatially nonuniform steady state. The neutrally stable wave of the linear stability is determined up to a multiplicative constant, however, and thus its amplitude remains to be determined. Finally, the initial disturbance is envisaged as the total solution (base state plus the perturbation) at the initial time. This "matching" of the overall response of the interface at the initial time with the imposed perturbation then establishes a relation between the amplitudes of the base state and the external perturbation. In summary, there exists a unique spatially nonhomogeneous steady state for a given amplitude of the external perturbation. For a plane parallel film, this is of the form $h_0(1 + f(\epsilon)\sin k_1 x)$, where h_0 is the mean thickness of the film and $f(\epsilon)$ is a function of the nondimensional amplitude of the disturbance. For other geometries, other appropriate orthogonal functions replace the term $\sin k_1 x$. Taking this nonuniform steady state as the base state then provides a growth coefficient that is a function of ϵ, and as $\epsilon \to 0, f(\epsilon) \to 0$. In this event, the base state reduces to the trivial state, h_0. The results of linear stability are of course recovered in such an event. From a purely computational point of view, a spatially nonuniform steady state is closer to a finite-amplitude initial disturbance as compared to the trivial state. This consideration intuitively favors the selection of the nonuniform state as the base state. On a more formal level, the onset of time-dependent flow may be studied only by perturbing the neutrally stable state and not the trivial state, the latter being no longer stable beyond the conditions of neutral stability. In other words, the spatially nonuniform base state exists for that value of the wavenumber at which the trivial solution loses its stability. The bifurcation of the time-dependent solution, therefore, occurs from this nonhomogeneous steady state.

In essence, the response of the film is analyzed by the following type of perturbation series:

$$h = H_0(x) + f_1(\epsilon)H_1(x,t) + \sum f_i(\epsilon)H_i(x,t)$$

where h is the location of the film interface, x and t are the lateral space coordinate and time, respectively, and $H_0(x)$ is a spatially nonuniform stationary solution of the governing equations. The steady-state version of the governing equations yields a nonlinear equation for the base case, $H_0(x)$. It is, therefore, possible to construct the base case solution only by a perturbative analysis. The leading order term of this solution is the same as that of the neutrally stable solution that is obtained by the linear stability analysis. The base case solution, $H_0(x)$, reduces to the trivial solution only in the limit of $\epsilon \to 0$. H_1 is the leading order coefficient that determines the growth/decay of disturbances that are envisaged to be imposed upon the stationary solution, $H_0(x)$. The function $f_1(\epsilon)$ is determined by matching the initial perturbation with h at time $t = 0$.

Expressions are derived for the wavenumber of the fastest growing finite-amplitude disturbance as well as for the associated growth rate. In contrast to the linear theory, the neutrally stable wavenumber, the dominant wavenumber, and the growth rate of a disturbance all are functions of the amplitude of the disturbance now. This approach, when applied to a wetting film devoid of solutes, predicted the time of rupture that showed less than 15% deviation from the numerical simulations even for initial amplitudes as large as 0.6 of the film thickness.[19] Here we apply this formalism to infer the stability conditions and the times of rupture of both the wetting and the free films that are subjected to long-wavelength perturbations. In addition to the effects of hydrodynamic and the van der Waals interaction force nonlinearities, the nonlinear influences of the surfactant concentration driven Marangoni motion and the surface viscosity are also investigated. The effects of a difference in the densities of the thin film and the bounding bulk fluid (Rayleigh-Taylor effect) (see Figure 1) and its synergism with the intermolecular interactions and the Marangoni motion are studied for the case of wetting films. The consideration of Rayleigh- Taylor effect[20,21] is of interest in the study of the attachment of a bubble to a solid surface (flotation) and of hydrodynamic aspects of boiling heat transfer.

Next, we derive equations of evolution appropriate for a wetting film subjected to long-wavelength perturbations and then carry out the stability analysis. In what follows, the details of the formalism are further clarified.

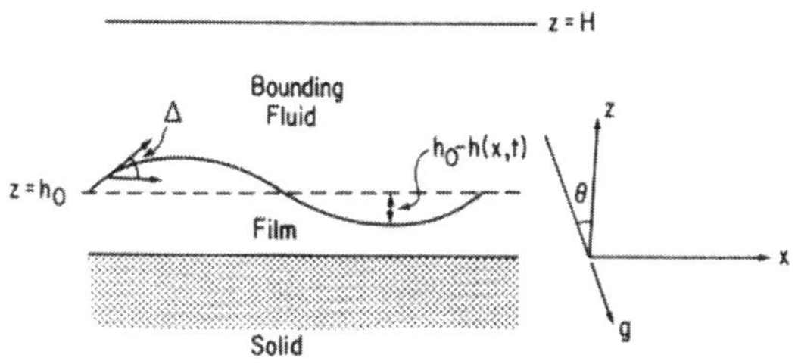

FIGURE 1. Thin wetting film sandwiched between a thick fluid layer and solid. The initial disturbance has an amplitude ϵh_0 where $\epsilon < 1$ and both the thin film and the bounding film contain surfactants.

3. WETTING THIN FILM WITH SURFACTANTS

Figure 1 depicts a wetting thin film sandwiched between a thick solid and a bulk fluid phase. The bounding fluid is assumed to be thick enough for the excess interaction forces to be negligible at its free boundary, $z = H$ ($H > 1$ μm), but not necessarily thick enough for the surfactant diffusion to be ignored. Both the thin film and the bounding fluid contain soluble surfactants. This may be regarded as a convenient prototype for the later stages of film thinning in flotation, attachment of a drop to a solid surface, breakup of a condensed thin film on a shock tube[5] or condenser wall, and hydrodynamics of thin-film boiling.

The governing hydrodynamic equations of motion are the usual Navier-Stokes equations with the inclusion of an extra body force term which arises because the film is thinner than the spatial range of the van der Waals interactions. The continuity and the Navier-Stokes equations for the thin film are (for $0 \le z \le h(x, t)$)

$$\nabla \cdot \mathrm{V} = 0 \tag{1}$$

$$(\partial \mathrm{V}/\partial t) + (\mathrm{V} \cdot \nabla)\mathrm{V} = -(1/\rho)\nabla p + \nu\nabla^2 \mathrm{V} - (1/\rho)\nabla\phi \tag{2}$$

where V is the velocity vector with x and z components denoted by u and υ respectively; ϕ, p, ν, ρ, and t are the intermolecular potential per unit volume, pressure, kinematic viscosity, film fluid density, and time, respectively. Similar equations also hold for the bounding fluid velocity vector, V′. The interfacial conservation equation for the soluble surfactants is

$$\Gamma_t + (\dot{a}/2a)\Gamma + \nabla_s \cdot (\Gamma \mathrm{V}_s) = D_s \nabla^2 \Gamma_s + D_1(\nabla C)\cdot \hat{n} + D_2(\nabla C')\cdot \hat{n} \tag{3}$$

at $z = h(x, t)$. Here Γ is the surface excess concentration, D_s is the surface diffusion coefficient, V_s is the surface velocity vector, C and C' are the bulk concentrations of surfactant in the film and the bounding fluid, respectively, $\dot{a}$ is the time derivative of the determinant of the metric tensor that relates the surface coordinates to the fixed Cartesian coordinates, and ∇_s denotes the differentiation along the surface coordinates. The last two terms represent the exchange of surfactant between the interface and the "bulks" of the film and the bounding fluid, respectively, and $\hat{n}$ is the unit normal vector to the interface, $z = h(x, t)$. The interfacial equation 3 may be regarded as a boundary condition to the following transport equations that describe the concentration distribution of surfactant in the "bulks" of the film and bounding media.

$$C_t + \mathrm{V} \cdot \nabla C = D_1 \nabla^2 C \quad \text{for } 0 \le z \le h(x,t) \tag{4}$$

$$C'_t + \mathrm{V}' \cdot \nabla C' = D_2 \nabla^2 C' \quad \text{for } h(x,t) \le z \le H \tag{5}$$

where C and C' are the concentrations in the bulks of the film and the bounding fluid, respectively, and subscripts denote differentiation. The stress boundary conditions at the interface $z = h(x, t)$ are highly nonlinear due to the presence of curvature terms and thus only their appropriate forms for the long-wavelength perturbations are reported later. A long-wavelength perturbation is the one for which the wavelength is much larger than the mean film thickness. As has been shown by the linear theories,[11,20–23] the dominant growth rate occurs for the long-wavelength disturbances and thus we confine our attention to the long-wavelength perturbations from the onset. This has the advantage of simplifying the hydrodynamic equations without placing any restrictions on the amplitude of the perturbation. The long-wavelength reduction procedure was developed and applied by Benney[24] and by Atherton and Homsy[20] for the falling films. Williams and Davis[16] extended the analysis to a thin wetting film devoid of solutes and derived a nonlinear equation of evolution for the location of the free interface. While the essentials of a formal perturbative analysis may be found in these references and were pursued by us[19] for a wetting film (without the Rayleigh-Taylor effect) elsewhere, it suffices to note that the angle Δ in Figure 1 is small for any finite-amplitude, albeit long-wavelength, disturbance. It is in view of this that the curvature effects are negligible in the leading order equations,[19,26] as are the transient and the inertia terms in the leading order Navier-Stokes equations. The following simplified Navier-Stokes equations thus describe the hydrodynamics of the thin film subjected to long-wavelength disturbances[19] (these are, in effect, the equations in the "lubrication approximation").

$$\mu u_{zz} = p_x + \phi_x \tag{5a}$$

$$p_z + \phi_z = 0 \tag{5b}$$

The bounding fluid is not necessarily thin enough for the lubrication approximation to hold but is assumed to be nonviscous. This, however, is not very restrictive, as the effects of bounding medium viscosity are not significant for a wetting film.[14,22] If the inertia and the transient terms for the bonding fluid are neglected (as is shown later, inertia is indeed negligible in the first-order equations, because the base state is a time-stationary solution of the governing equations and hence the base case velocities vanish identically), the z component of motion reduces to

$$p'_z + \phi'_z = 0 \tag{6}$$

p' and ϕ' are the pressure and the interaction potential in the bounding fluid. The no slip and equality of stress conditions simplify respectively to (in the leading order equations)[19]

$$u = \upsilon = 0 \quad \text{at } z = 0 \tag{7a}$$

$$-p + p' = \sigma h_{xx} + \delta\left(h - h_0\right) \quad \text{at } z = h\left(x,t\right) \tag{7b}$$

and

$$\mu u_z = \sigma_x \quad \text{at } z = h\left(x,t\right) \tag{7c}$$

The parameter δ is defined as $g(\rho' - \rho)\cos\theta$, where ρ' is the density of the bounding fluid, g is the acceleration due to gravity, θ is the angle depicted in Figure 1, and σ is the film-bounding fluid interfacial tension. Since the interfacial tension is not a constant due to the presence of surfactants, the excess stress, σ_x, that arises due to the uneven interfacial distribution of surfactants is incorporated in boundary condition 7c. The parameter δ may be positive or negative. It is positive when either the bounding fluid is heavier than the film fluid and the angle $\theta < \pi/2$ (bounding fluid is on top) or when the bounding fluid is lighter, but $\pi > \theta > \pi/2$, viz., the geometry of Figure 1 is inverted and the thin film is on top of the bounding fluid. The parameter δ is negative in other two circumstances when the lighter fluid is on top of the heavier fluid. For example, δ is positive and negative respectively for the attachment of a bubble to a solid particle from below and from above. Finally, the location of the film interface is given by the following kinematic condition

$$h_t + u h_x = \upsilon \quad \text{at } z = h(x,t) \tag{8}$$

The set of equations 5a–7c may be solved and then the interfacial velocities may be evaluated in terms of the unknown functions $h(x, t)$ and σ_x. The substitution of velocity components in kinematic condition 8 provides a nonlinear equation of evolution for the location of the interface, $h(x, t)$.

The integration of eq 5b and 6 shows that the sum of pressure and the interaction potential is independent of the z coordinate and may thus be evaluated at any z = constant plane, viz.,

$$p + \phi = p(h) + \phi(h) \tag{9a}$$

and

$$p' + \phi' = p'(h) + \phi'(h) = p'(H) + \phi'(H) = p_0 + \phi'_0 \tag{9b}$$

The pressure outside the thick bounding fluid is denoted by p_0 and ϕ'_0 is the constant part of the interaction potential in the bounding fluid. Equation 5a may be rewritten with the help of conditions 7b, 9a, and 9b as

$$\mu u_{zz} = \psi \tag{10}$$

where

$$\psi \equiv p_x(h) + \phi_x(h) = \phi_x(h) - \phi'_x(h) - \delta h_x - (\sigma h_{xx})_x \tag{11a}$$

The solution of eq 10 is constructed with the help of conditions 7a and 7c and is given by

$$\mu u = \psi z\left((z/2) - h\right) + \sigma_x z \tag{11b}$$

The continuity equation (1) and the condition 7a now determine the normal velocity component as

$$6\mu\upsilon = -\psi_x z^3 + 3z^2 (\psi h)_x - 3\sigma_{xx} z^2 \tag{11c}$$

Finally, the evaluation of the velocity components u and at $z = h(x, t)$ and their substitution in the kinematic condition 8 yields the sought after equation of evolution.

$$h_t - \frac{1}{3\mu}\left(\psi h^3\right)_x + \frac{1}{2\mu}\left(\sigma_x h^2\right)_x = 0 \tag{12a}$$

The definition of ψ eq 11a, may be made explicit by substituing the following dependence of the excess van der Waals potentials on the film thickness.

$$\phi(h) - \phi'(h) = A / 6\pi h^3 \tag{12b}$$

The effective Hamaker constant, A, is defined as

$$A = A_{22} - A_{23} - A_{12} + A_{13} \tag{12c}$$

The Hamaker constant A_{ij} is for interactions between the molecules of type i and j and subscripts 1, 2, and 3 refer to the molecules of the bounding fluid, thin film, and the solid support, respectively. Equation 12a reduces to that of Williams and Davis[16] if the film is devoid of surfactants; viz., $\sigma_x = 0$ and the gravity is neglected, i.e., $\delta = 0$. Equations 12 and 11a describe the dynamics of a thin wetting film when it is subjected to an initial interfacial deformation, $h(x,0) = f(x)$. We assume it to be of a form $f(x) = \epsilon h_0 \sin kx$, where ϵ is the nondimensional amplitude of the perturbation ($\epsilon < 1$) and k is the wavenumber. The choice of a single Fourier component to represent the initial disturbance is, however, not restrictive as is shown later.

The solution methodology consists of finding first the stationary solutions of the equation of evolution (eq 12a). The investigation of the stability of the nonhomogeneous stationary solution then provides the conditions for the neutral stability and the growth rate of the finite-amplitude perturbations. Although in certain cases a rigorous stationary solution may be found, we pursue here a perturbative analysis to construct this solution. This approach is more general and has the advantage that the leading order term of the solution may be shown to coincide with the neutrally stable wave of the linear stability analysis. In most instances, it is either known or is easily determined. For example, the stationary equation corresponding to the dynamic equation 12a is readily obtained by noting that both h_t and σ_x vanish for a time-stationary solution. Representing the steady solution as H_0 and the constant part of the interfacial tension as $\sigma_0(\Gamma = \Gamma_0)$ and making use of the definition of ψ, one obtains

$$\sigma_0\left(H_0^3 H_{0xxx}\right)_x + \delta\left(H_0^3 H_{0x}\right)_x + (A/2\pi)\left(H_0^{-1} H_{0x}\right)_x = 0 \tag{13a}$$

The solution of eq 13a may be constructed as the following perturbation series involving a small parameter, μ^*, and as yet undetermined coefficients f_i. The ordering parameter, μ^*, is related to the nondimensional amplitude of the perturbation, ϵ, later.

$$H_0(x)/h_0 = 1 + \mu^* \sin k_1 x + \sum_{i=2} \left(u^*\right)^i f_i \tag{13b}$$

Substituting the above in eq 13a and equating the terms of order μ^* gives the following compatibility condition

$$k_1^2\left[k_1^2 - \frac{1}{\sigma_0}\left(\frac{A}{2\pi h_0^4} + \delta\right)\right] = 0 \tag{13c}$$

and hence an approximate stationary solution

$$H_0(x)/h_0 \approx 1 + \mu^* \sin k_1 x + 0\left(\mu^{*2}\right) \tag{13d}$$

For the trivial root of eq 13c, viz., for $k_1 = 0$, the results of the linear theory are recovered inasmuch as the stationary solution now is but the trivial state, $H_0 = h_0$. For $k_1^2 = (A/2\pi h_0^4 + \delta)\sigma_0^{-1}$, however, the stationary solution (13d) is a nontrivial, spatially nonhomogeneous steady state. It is easily recognized that k_1^2 is the wavenumber of the neutrally stable wave *as determined from the linear stability analysis.* It may be noted that the finite-amplitude solution 13d reduces to the trivial steady state only in the limit of $\mu^* \to 0$, or as is shown later, in the limit of an infinitesimal disturbance. As is obvious, the nontrivial solution exists for the wavenumber at which the trivial state of the film becomes unstable. The onset of instability beyond the conditions of neutral stability is therefore studied by further perturbing the nontrivial solution 13d. In other words, we study the instability as the wavenumber crosses its threshold value of $[(A/2\pi h_0^4) + \delta]^{1/2} \sigma_0^{-1/2}$ and hence the trivial state becomes unstable. The determination of the stability of the nontrivial solution 13d is achieved by expanding the film thickness and the interfacial tension as the following perturbation series around the basic solutions, H_0 and σ_0.

$$h(x,t)/h_0 = H_0(x) + \sum_{i=1} h_0^{(i-1)} H_i(x,t)\left(\mu^*\right)^i \tag{14a}$$

and

$$\sigma = \sigma_0(\Gamma_0) + \sum_{i=1} \sigma_i \left(\mu^*\right)^i \tag{14b}$$

The successive coefficients, H_i, are found by substituting expansions 14 into eq 12a and by equating the coefficients of the same order. The first-order equation is the following linear, homogeneous equation.

$$H_{1t} + (A/6\pi\mu)\left(H_0^{-1}H_{1x} - H_0^{-2}H_{0x}H_1\right)_x + (\delta/3\mu)\times$$

$$\left(H_0^3 H_{1x} + 3H_0^2 H_1 H_{0x}\right)_x + (1/3\mu)[H_0^3\left(\sigma_1 H_{0xx} + \sigma_0 H_{1xx}\right)_x$$

$$+ 3H_0^2 H_1 \sigma_0 H_{0xxx}]_x + (1/2\mu)\left(H_0^2 \sigma_{1x} + \sigma_{0x} H_0 H_1\right)_x = 0 \tag{15}$$

Recalling that the basic solution, H_0, is a function of x, one finds that the solution of eq 15 is perhaps only slightly simpler than the original equation (eq 12a) itself. Of course, the nonlinear functions of the basic solution H_0 that appear in eq 15 may further be linearized at this point and a truncated normal mode analysis may be pursued. This, however, would destroy the essential effects of the nonlinearities we wish to investigate. This problem may be circumvented by noting that since our primary interest is in the study of the rupture of the film, it is sufficient to investigate the instability of the basic solution where the film is locally thinnest, viz., where the growth rate is maximum. We may thus introduce the estimates $\sin k_1 x = -1$ and $H_0 = h_0(1-\mu^*)$ and hence $H_{0x} = h_0\mu^* k_1 \cos k_1 x = 0$, $H_{0xx} = -h_0\mu^* k_1^2 \sin k_1 x = h_0\mu^* k_1^2$, $H_{0xxx} = 0$, and $H_{0xxxx} = -h_0\mu^* k_1^4$ in eq 15 and hence simplify it to the following equation with constant coefficients.

$$H_{1t} + (A/6\pi\mu)\left(H_0^{-1}H_{1xx} - H_0^{-2}H_1 H_{0xx}\right) + (\delta/3\mu)\times$$

$$\left(3H_1 H_0^2 H_{0xx} + H_0^3 H_{1xx}\right) + (1/3\mu)[H_0^3(H_{0xx}\sigma_{1xx} + \sigma_1 H_{0xxxx})$$

$$+ \sigma_0 H_0^2\left(H_0 H_{1xxxx} + 3H_1 H_{0xxxx}\right)] +$$

$$(1/2\mu) H_0^2 \sigma_{1xx} = 0 \tag{16}$$

The gradients of the interfacial tension are evaluated with the help of the mass transport equations 3–5. In view of the fact that we are dealing with the long-wavelength perturbations and also that the base case velocities and concentration gradients (corresponding to the stationary solution) vanish identically, the first-order surfactant transport equations are

$$C_{1t} = D_1\left(C_{1xx} + C_{1zz}\right); \quad 0 \le z \le h(x,t) \tag{17a}$$

$$C'_{1t} = D_2\left(C'_{1xx} + C'_{1zz}\right); \quad h(x,t) \le z \le H \tag{17b}$$

The boundary conditions simplify to

$$\Gamma_{1t} + \Gamma_0 u_{1x}^{\mathrm{s}} - D_S \Gamma_{1xx} = -D_1 C_{1z} + D_2 C'_{1z} \quad \text{at } z = h(x,t) \tag{17c}$$

where u_{1x}^{s} is the first-order tangential interfacial velocity, Γ_0 is the surface excess concentration corresponding to the base state of a nonthinning film, and subscript 1 refers to the first-order (perturbed) variables.

$$-D_1 C_{1z} = 0 \quad \text{at } z = 0 \tag{17d}$$

$$-D_2 C'_{1z} = 0 \quad \text{at } z = H \tag{17e}$$

It is also possible to have a constant-concentration boundary condition at $z = H$, instead of the no flux condition 17e. This, for instance, is the case with the thin mucous layer sandwiched between the epithelium and the aqueous tear film.[27] The thick aqueous layer has a constant concentration of eye lipids at the tear-air interface. Thus, we will consider also the boundary condition

$$C'_1 = 0 \quad \text{at } z = H \tag{17f}$$

The hydrodynamic and the surfactant transport equations are coupled via the following constitutive equation for the dynamic interfacial tension (valid for a purely viscous adsorbed layer and the long-wavelength perturbation only).

$$\sigma_1 = M\Gamma_1 + \mu_s u_{1x}^{\mathrm{s}} \tag{18}$$

where $M\Gamma_1$ is the surface excess concentration dependent part of the interfacial tension, $M = (\partial\sigma/\partial\Gamma)_0$, and μ_s is the surface viscosity. The parameter M characterizes the influence of surface excess concentration on the interfacial tension and is negative for a surfactant. Finally, the surface excess concentration is related to the bulk concentrations by some equilibrium isotherms. A realistic choice in many instances is the Langmuir adsorption isotherm, for which the first-order variables satisfy the following relations:

$$\Gamma_1 = \theta C_1 = \theta' C'_1 \tag{19a}$$

where

$$\theta = \theta_1/(1 + (\theta/\Gamma_\infty)C_0)^2, \text{ and } \theta' = (\theta/k_e) \tag{19b}$$

The constants θ_1, Γ_∞, C_0, and k_e are the adsorption isotherm constant, the asymptotic surface excess concentration corresponding to saturation, the unperturbed "bulk" concentration in the film, and the equilibrium partition coefficient, respectively. If the surfactant is insoluble in the film, $k_e \to \infty$, and if it is insoluble in the bounding media, $k_e \to 0$.

The first-order interfacial velocity gradient appearing in condition 17c is evaluated from expression 11b as

$$u_{1x}^{\mathrm{s}} = (A/4\pi\mu)\left(H_0^{-2} H_{1xx} - 2H_0^{-3} H_1 H_{0xx}\right) +$$

$$(\delta/2\mu)\left(H_0^2 H_{1xx} + 2H_1 H_0 H_{0xx}\right) + (1/2\mu)[\sigma_0(H_0^2 H_{1xxxx} +$$

$$2H_0 H_1 H_{0xxxx}) + \sigma_1 H_0^2 H_{0xxxx}] + (1/2\mu)\left(H_0^2 H_{0xx}\sigma_{1xx} + 2H_0\sigma_{1xx}\right) \tag{20}$$

Postulating the following normal modes for the first-order variables

$$\begin{bmatrix} H_1 \\ C_1 \\ C_1' \\ \sigma_1 \end{bmatrix} = \begin{bmatrix} 1 \\ \eta^* \\ \eta' \\ \hat{\sigma}_1 \end{bmatrix} (\sin kx) e^{\omega t} \tag{21}$$

and making use of them in conjunction with expressions 18 and 20 give

$$\hat{\sigma}_1 = \left(N \mu_s + M\theta\eta^*\right)/\left(1+P\right) \tag{22a}$$

where

$$N = (\sigma_0/2\mu)\left(H_0^2 k^4 - 2H_0 h_0 \mu^* k_1^4\right) - \left(A/4\pi\mu\right)\times$$

$$\left(H_0^{-2}k^2 + 2H_0^{-3}h_0\mu^* k_1^2\right) + \left(\delta/2\mu\right)\left(2H_0\mu^* h_0 k_1^2 - k^2 H_0^2\right) \tag{22b}$$

$$2\mu P = 2\mu_s H_0 k^2 + \mu_s H_0^2 \mu^* k_1^2 h_0 \left(k_1^2 + k^2\right) \tag{22c}$$

Equation 18 may now be solved as

$$\left(1+P\right) u_{1x}^{s} = N - PM\theta\eta^*/\mu_S \tag{23}$$

Finally, boundary condition 17c in conjunction with the solutions of transport equations 17a and 17b and the expression 23 gives

$$\eta^* = -\Gamma_0 N/R\left(1+P\right) \tag{24}$$

When the zero-flux boundary condition 17e is employed, the parameter R is defined as

$$R = D_1 a \tanh\left(aH_0\right) + D_2 a' k_e \tanh\left[a'\left(H - H_0\right)\right] + \omega + \Gamma_0 P|M|\theta/\mu_s\left(1+P\right) + D_s\theta k^2 \tag{25}$$

For the constant concentration boundary condition 17f, the parameter R is also given by eq 25 with tanh $[a'(H - H_0)]$ being replaced by coth $[a'(H - H_0)]$. The parameters a and a' are defined as

$$a = \left(k^2 + \omega/D_1\right)^{1/2} \quad a' = \left(k^2 + \omega/D_2\right)^{1/2} \tag{26}$$

This completes the determination of the interfacial tension (and hence its gradients) from eq 22a as

$$\hat{\sigma}_1 = \frac{N}{\left(1+P\right)}\left[\mu_s + \frac{|M|\theta\Gamma_0}{R\left(1+P\right)}\right] \tag{27}$$

The dispersion equation is obtained from the equation of evolution 16 with the help of perturbed forms 21 and expression 27. After considerable rearrangement, it may be written in the following compact form

$$\omega = k^2\left(\frac{A}{6\pi\mu H_0} - \frac{\sigma_0 H_0^3 k^2}{3\mu} + \frac{\delta H_0^3}{3\mu}\right)\left(1 - \frac{3}{4}z^*\right) +$$

$$\frac{\mu^* h_0 k_1^2 A}{6\pi\mu H_0^2}\left(1 - \frac{3}{2}z^*\right) + \left(\frac{\sigma_0 H_0^2 \mu^* k_1^4 h_0}{\mu} -\right.$$

$$\left.\frac{\delta H_0^2 \mu^* k_1^2 h_0}{\mu}\right)\left(1 - \frac{1}{2}z^*\right) \tag{28a}$$

Where

$$z^* = \frac{H_0 k^2}{\mu(1+P)}\left[\mu_s + \frac{|M|\theta\Gamma_0}{R(1+P)}\right] \tag{28b}$$

$$k_1^2 = \left(\frac{A}{2\pi h_0^4 \sigma_0}\right)(1+\eta) \tag{28c}$$

and

$$\eta = \delta/(A/2\pi h_0^4) \tag{28d}$$

It is easily verified that the parameter z^* is zero when the film is devoid of surfactants, i.e., *both* μ_s and $|\boldsymbol{M}|\Gamma_0$ vanish, and z^* approaches unity when *either* μ_s or $|M|\Gamma_0$ is large. Thus, $z^* \in (0,1)$ for all parameter values and for the wavenumbers, $k \in (0, \infty)$. The parameter μ^* is determined by matching the maximum amplitude of the perturbed form 14a at time $t = 0$ with the maximum amplitude of the imposed disturbance. This procedure gives

$$\mu^* = \epsilon/2 \tag{29}$$

In the event of a vanishingly small initial amplitude, viz., $\mu^* \to 0$, the last two terms of the modified dispersion relation 28a are negligible and the linear theory dispersion relation is recovered. The conditions for the neutral stability of a finite-amplitude disturbance are directly inferred from the dispersion relation by setting the growth coefficient ω to zero. For illustrative purposes, the following two asymptotic cases are explicitly considered: (a) the absence of surfactants, viz., $z^* \to 0$, and (b) either large surface viscosity, μ_s, or large Marangoni parameter, $|M|\Gamma_0$, viz., $z^* \to 1$. The following quadratic equation is obtained for the critical (or neutrally stable) wavenumber if the film is devoid of solutes:

$$\left(\frac{y}{y_L}\right)^2 - \frac{\left[1+\eta\left(1-\mu^*\right)^4\right]}{(1+\eta)(1-\mu^*)^4}\left(\frac{y}{y_L}\right) -$$

$$\frac{3\mu^*}{(1+\eta)}\left[(1-\mu^*)^{-1} + \frac{(1-\mu^*)^{-5}}{3}\right] = 0 \tag{30a}$$

where $y = k_{cn}^2$, $y_L = k_{cL}^2 = k_1^2$ and k_{cn} and k_{cL} are the wavenumbers of a neutrally stable wave as derived from the nonlinear and the linear theories, respectively. The ratio y/y_L is plotted in Figure 2 (solid lines) as a function of the perturbation amplitude and the parameter η. It shows that the wavelength (reciprocal of the wavenumber) of a finite amplitude, neutrally stable wave is always smaller

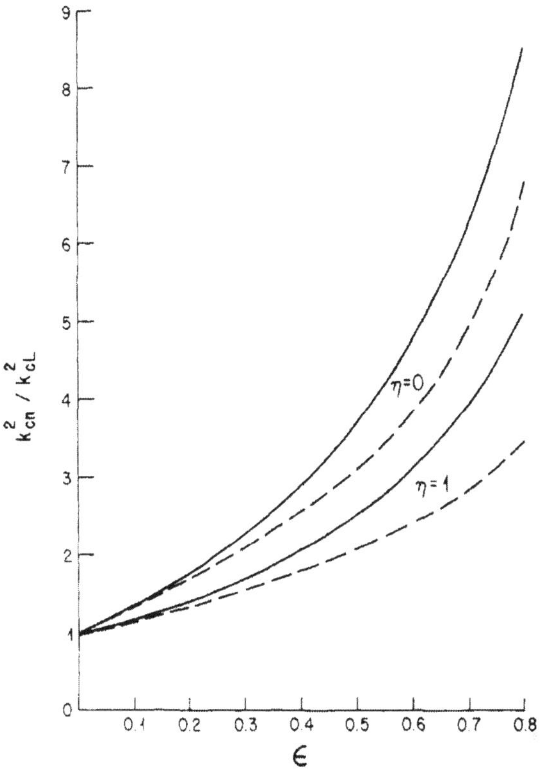

FIGURE 2. Ratio $(k_{cn}/k_{cL})^2$ as a function of initial amplitude. k_{cn} and k_{cL} are the wavenumbers of a neutrally stable wave as determined from the nonlinear and the linear theories, respectively. Solid curves are for a wetting film devoid of solutes and the dashed curves are for a film with excess surfactants.

than the one derived from the linear stability analysis. The parameter η denotes the importance of Rayleigh-Taylor parameter δ as compared to the group $(A/2\pi h_0^4)$. It may be verified that for Hamaker constants in the range 10^{-12} to 10^{-16} ergs and for $\delta \sim 10^3$ dyn/cm^3, the range of h_0 for which $\eta \approx 1$ is between 10^3 and 10^4 Å. For any film thickness smaller than this range, η is less than one and hence $\eta = 1$ may be regarded as an upper estimate.

For the second asymptotic case of the excess surfactants ($z^* \rightarrow 1$), the following equation holds for the wavenumber of the neutrally stable wave.

$$\left(\frac{y}{y_L}\right)^2 - \frac{\left[1+\eta(1-\mu^*)^4\right]}{(1+\eta)(1-\mu^*)^4}\left(\frac{y}{y_L}\right) - \frac{6\mu^*}{(1+\eta)(1-\mu^*)^3}\left[(1-\mu^*)^2 - \frac{(1-\mu^*)^{-2}}{3}\right] = 0 \tag{30b}$$

Again, the ratio y/y_L is plotted in Figure 2 (broken curves). The presence of surfactants increases the wavelength of a neutrally stable wave as compared to the case of a film devoid of solutes. Also an increase in the parameter η has the effect of increasing the wavelength of a neutrally stable wave. Of course, as $\epsilon \rightarrow 0$, all of these nonlinear effects vanish and the ratio y/y_L approaches its lower limit of unity.

In conclusion, the linear theory predicts that all initial disturbances with wavenumbers greater than k_{cL} decay, whereas eq 30a, b show that finite amplitude disturbances with wavenumbers greater than k_{cL} but smaller than k_{cn} are unstable.

The dominant growth coefficient is computed from expression 28a by maximizing the growth coefficient with respect to the wavenumber k. It cannot, however, be done analytically due to a highly nonlinear dependence of z^* on the wavenumber. As it turns out though, the dependence of ω on k through z^* is quite weak compared to its dependence on the term $k^2[A(\pi H_0)^{-1} - 2\sigma_0 H_0{}^3 k^2 + 2\delta H_0{}^3]$. It is confirmed by the direct computer simulations that the parameter z^* may be regarded as a constant while solving the equation $(\partial\omega/\partial k) = 0$ for the dominant wavenumber. This approximation yields the dominant wavenumber for a finite-amplitude perturbation to be the following

$$k_{mn}{}^2 = \frac{k_{mL}{}^2}{(1+\eta)}[\eta + (1-\mu^*)^{-4}] \tag{31a}$$

where k_{mn} and k_{mL} are the dominant wavenumbers as derived from the nonlinear and the linear theories, respectively. The latter is defined as

$$k_{mL}{}^2 = \left(A/2\pi\sigma_0 h_0{}^4\right)(1+\eta) \tag{31b}$$

As is obvious from eq 31a, the nonlinearities of the governing equations always select a shorter dominant wave ($k_{mn} > k_{mL}$) compared to the linear theory. For $\eta = 0$ and $\epsilon = 0.1$, eq 31a predicts the ratio k_{mn}/k_{mL} to be 1.1 which is in complete agreement with the numerical simulations of Williams and Davis[16] for a thin wetting film devoid of surfactants. The advantage here is that the nonlinear partial differential equations (12) need not be solved for each wave with a certain wavenumber and thus the determinations of the neutrally stable and the dominant wavenumbers do not require a large number of computations involving trial and error. Moreover, the generalization to include other nonlinear effects, such as the Marangoni motion, surface viscosity, and the Rayleigh-Taylor instability is rather straightforward. Finally, due to the availability of the analytical results, the dependence of various quantities of interest on various film parameters and the amplitude becomes physically transparent.

The dominant growth rate is determined by substituting the expression for the dominant wavenumber in the expression for the growth coefficient. In this way, the maximum growth rate, ω_n, is obtained as

$$\omega_n = \omega_0\left(1 - \tfrac{3}{4}z^*\right) + 12\mu^*\psi_0(1+\eta)\times$$

$$\left[\tfrac{1}{3}(1-\mu^*)^{-2}\left(1-\tfrac{3}{2}z^*\right) + (1-\mu^*)^2\left(1-(z^*/2)\right)\right] \tag{32a}$$

The parameter z^* is now evaluated for $k = k_{mn}$ and ψ_0 is the dominant growth coefficient corresponding to the linear theory in the absence of gravity effects, viz.,

$$\psi_0 = A^2/(48\pi^2\mu\sigma_0 h_0{}^5) \tag{32b}$$

and ω_0 is defined as

$$\omega_0 = \psi_0(1-\mu^*)^{-5}[1+\eta(1-\mu^*)^4]^2 \tag{32c}$$

In contrast to the dominant growth rate derived from the linear theory ($\mu^* \to 0$), the one derived from the present theory is a function of the amplitude of the initial perturbation. The ratio $[\omega_n/\psi_0(1+\eta)^2]$

may be compactly represented as a function of three parameters—the amplitude of the disturbance ϵ, z^*, and η. This is shown in Figure 3. The solid curves correspond to the case when the film and the bounding media have the same density or the attachment is horizontal, viz., $\eta = 0$. The uniformly broken lines are for $\eta = 1.0$ (the positivity of η implies that the heavier fluid is on top). The nonuniformly broken lines represent the case of $\eta = -0.5$ (the lighter fluid is on top). The ratio $\left[\omega_n/\psi_0(1+\eta)^2\right]$ is one in the limit of z^* and μ^* being zero, because $\psi_0(1+\eta)^2$ is the dominant growth rate inferred from the linear theory for a pure film. As is apparent, the deviation of this ratio from unity is considerable at $z^* = 0$ for finite-amplitude disturbances. The deviation from the linear theory is maximum when the lighter fluid is on top. This is so because the linear theory underestimates both the effects of the dispersion forces and the Rayleigh-Taylor parameter δ on the growth coefficient when δ is negative. For the case of δ positive, the linear theory overestimates the contribution of Rayleigh-Taylor effect on the growth rate. Finally, the discrepancy between the linear and nonlinear theories is maximum when the film is pure.

The time of rupture of the thin film, which is subjected to a finite-amplitude disturbance, is deduced from expansion 14a by setting $h = 0$. This gives the time of rupture t_n as

$$t_n = \frac{1}{\omega_n} \ln\left(\frac{2-\epsilon}{\epsilon}\right) \tag{33a}$$

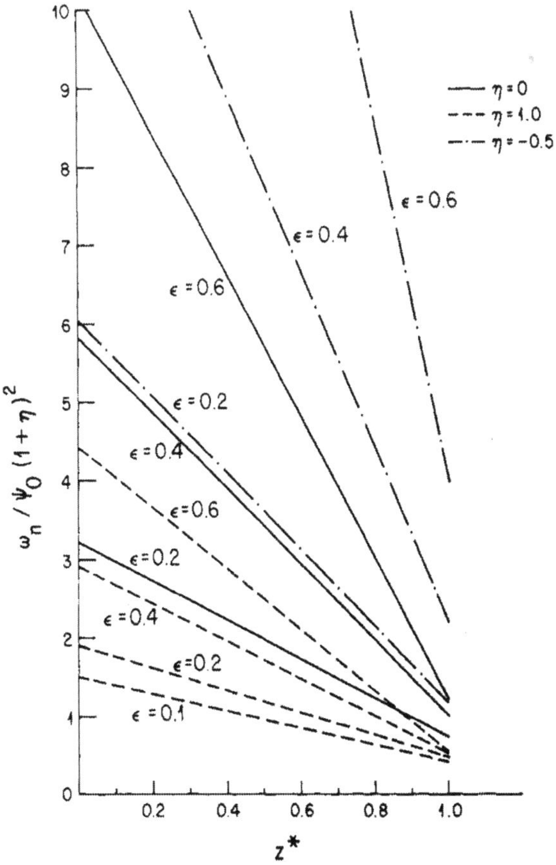

FIGURE 3. Generalized plot for the dominant growth rate $[\omega_n/\psi_0(1+\eta)^2]$ as a function of z^*, ϵ, and the parameter η.

This may be contrasted to the time of rupture given by the linear theory

$$t_L = \frac{1}{\omega_L} \ln\left(\frac{1}{\epsilon}\right) \tag{33b}$$

where

$$\omega_L = \lim\left(\mu^* \to 0\right)\omega_n \tag{33c}$$

It may be noted that usually t_L is interpreted as the inverse of the dominant growth coefficient, thus leaving out its dependence on the amplitude which is reflected in the term $\ln(1/\epsilon)$.

The ratio t_n/t_L is plotted in Figure 4 for a film devoid of surfactants ($z^* = 0$). The deviation of this ratio from unity increases as the perturbation amplitude is increased. The discrepancy also increases as the density of the fluid on top decreases, i.e., the discrepancy is less for a positive δ and more for a negative δ. The overall trend is that the dispersion force and the interfacial tension related nonlinearities accelerate the film thinning, i.e., reduce the time of rupture (curve labeled $\eta = 0$). The nonlinearities of the Rayleigh-Taylor effect (difference in densities) accelerate the film thinning when the lighter fluid is on top and retard it when the heavier fluid is on top. The same ratio is plotted in Figure 5 for the case when either the surface viscosity or the Marangoni parameter $|M|\Gamma_0$ is large. Here, a positive value of the parameter η has a more profound influence on the ratio t_n/t_L. This is so, because the nonlinear effects of the dispersion forces and the interfacial tension are less profound for a film with excess surfactants, as compared to the case of a pure film (Figure 4). It is also interesting to note that the ratio t_n/t_L is identically zero for $\eta = -1$. This happens because according to the linear stability theory, such a film never breaks ($t_L \to \infty$) due to the balancing of the destabilizing influence of molecular interactions with the stabilizing contribution of the difference in the densities. The nonlinearities of the system, however, make such a film unstable to any

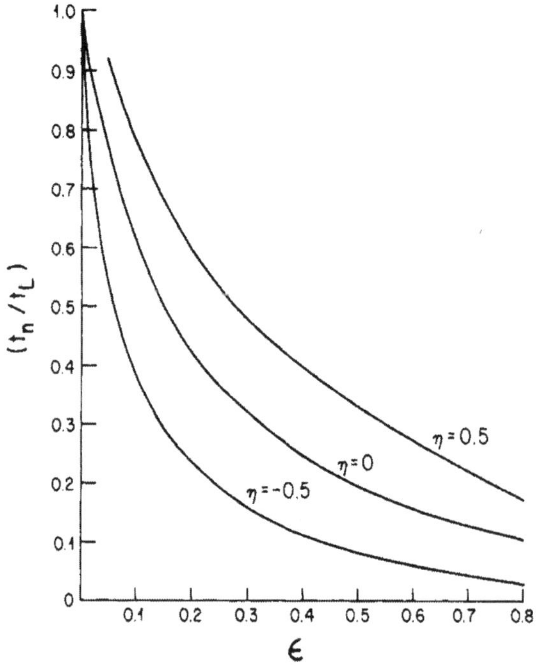

FIGURE 4. Ratio between the times of rupture of a wetting film as derived from the nonlinear and the linear theories. The wetting film is devoid of surfactants.

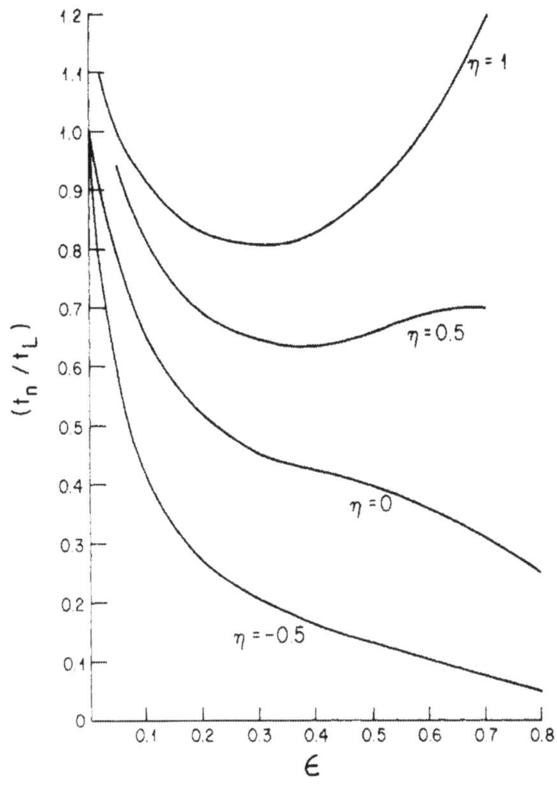

FIGURE 5. Ratio between the times of rupture as derived from the nonlinear and the linear theories, respectively. The film contains an excess of surfactants.

finite-amplitude perturbation. The curve labeled $\eta = 0$ in Figure 4 corresponds to the case treated by Williams and Davis[16] using numerical simulations. We have earlier shown that the discrepancy between their numerical simulations and this curve is less than 15% for initial amplitudes as large as 0.6 of the film thickness.[19]

Finally, we consider the maximum stabilizing influence of surfactants by comparing the times of rupture of a film with excess surfactants and a film devoid of surfactants. The ratio of the times of rupture in the presence of excess surfactants and in their absence is plotted in Figure 6 as a function of the disturbance amplitude. As is expected, this ratio is 4 when the amplitude is vanishingly small and hence the linear stability theory holds. For large initial amplitudes, however, the presence of surfactants may prolong the time of rupture by as much as a factor of 10–20, depending on the magnitude of parameter η. Thus, the stabilizing influence of the Marangoni motion and the surface viscosity is underestimated by the linear theory and the discrepancy increases with an increase in parameter η.

Next, the finite amplitude instability of a thin free film is investigated.

4. THIN FREE FILMS

The theory of the stability of a free film and the determination of its lifetime finds applications[28] in the foam fractionation as a separation technique, in fire fighting applications, and in applications requiring a large gasliquid interfacial area. The selective adsorption and the subsequent removal of impurities from the waste water, elimination of radioactive contaminants from dilute effluents, and

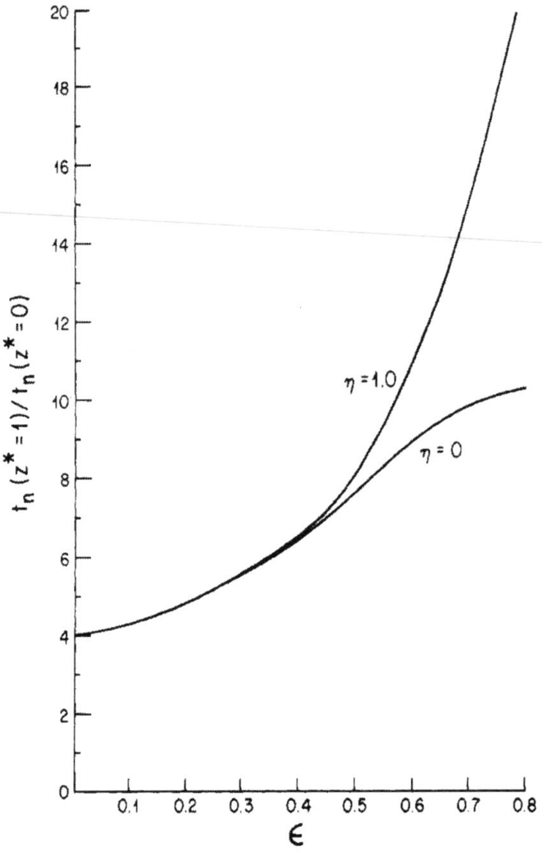

FIGURE 6. Effect of surfactants in prolonging the time of rupture of a wetting film.

selective separation of certain complex organic molecules and proteins typify the role of foams in separations. An understanding and the design of all these processes require the calculation of the time for the collapse of foam. Another issue of interest here is the effect of foam stabilizers (usually surfactants) in prolonging the time of collapse. The previous investigations of the instability of thin free films have used exclusively the linear stability analysis, which may, for instance, be applicable when the interfacial perturbations arise solely due to the thermal fluctuations and thus have vanishingly small amplitudes. The present analysis is now used to delineate the role of the finite amplitudes of the perturbations and the nonlinearities of the governing equations. The film is assumed to be thin enough for the van der Waals forces to be important and the drainage due to gravity or the capillary pressure is neglected. The change in the mean thickness of the film due to the drainage may, however, be combined with the present analysis by making a quasi-steady assumption for the drainage, as has been done also for the linear stability analyses.[18,29,30] The initial perturbations considered have long wavelength and are symmetric, i.e., out of phase by 180° on the two free surfaces of the film. As is well-known, this form of the perturbation leads to the most rapid rupture of a thin free film. The Navier-Stokes equations for the thin free film in the long-wavelength limit are

$$\mu u_{zz} = p_x + \phi_x \tag{34a}$$

$$p_z + \phi_z = 0 \tag{34b}$$

Taking the midplane of the film to coincide with plane $z = 0$, the following boundary conditions hold for the long-wavelength perturbations:

$$\upsilon = \partial u/\partial z = 0 \text{ at } z = 0 \tag{35a}$$

$$\mu u_z = \sigma_x \text{ at } z = h(x,t)/2 \tag{35b}$$

and

$$-p + 2\mu\upsilon_z = \sigma h_{xx}/2 \text{ at } z = h(x,t)/2 \tag{35c}$$

The solution strategy here, as before, would be to first reduce the set of governing equations to a minimum number of independent nonlinear equations and then seek their solutions.

The integration of eq 34b shows that the quantity $(p + \phi)$ is independent of z and hence may be evaluated at a free interface, i.e.,

$$p + \phi = p(h/2) + \phi(h/2) \tag{36}$$

Equation 34a is now integrated once with the help of eq 36 and 35a to yield

$$\mu u_z = \left[p_x(h/2) + \phi_x(h/2) \right] z \tag{37}$$

The use of condition 35b gives

$$p_x + \phi_x = 2\sigma_x/h \tag{38}$$

The tangential velocity component is now determined from eq 37 as

$$u = \frac{z^2}{\mu}\frac{\sigma_x}{h} + a(x,t) \tag{39a}$$

where a is an integration constant. The normal velocity component is obtained from the continuity equation $u_x + \upsilon_z = 0$, in conjunction with expression 39a and condition 35a:

$$\upsilon = -\frac{z^3}{3\mu}\left(\frac{\sigma_x}{h}\right)_x - a_x z \tag{39b}$$

The substitution of the interfacial velocities $u(h/2)$ and $\upsilon(h/2)$ in the kinematic condition

$$h_t + h_x u - 2\upsilon = 0 \text{ at } z = h/2 \tag{40}$$

formally establishes an equation of evolution for the location of the free interface, $h(x, t)$. This may be brought to the following compact form

$$h_t + (ah)_x + \frac{1}{12\mu}(h^2\sigma_x)_x = 0 \tag{41}$$

Determination of unknowns $a(x, t)$ and σ_x as functions of $h(x, t)$ requires the solution of the following mass transport equations for the surfactant:

$$C_t = D\left(C_{xx} + C_{zz}\right) \text{ for } -h/2 < z < h/2 \tag{42a}$$

$$C_z = 0 \text{ at } z = 0 \tag{42b}$$

and

$$\Gamma_t + (\Gamma u)_x - D_s\Gamma_{xx} = -DC_z \text{ at } z = h/2 \tag{43c}$$

Further, the dynamic interfacial tension is defined as

$$\sigma = \sigma_0(\Gamma) + \mu_s u_x \text{ at } z = h/2 \tag{44}$$

and the adsorption isotherm is again the same as that given by eq 19a, b.

We now seek a solution of the form

$$C = C_0 + \eta(z) f(x) e^{\omega t} \tag{45}$$

where $f(x)$ is a harmonic function that is compatible with the functional form of the initial perturbation. We study here a Fourier component of the disturbance, sin kx and thus take $f(x)$ to be sin kx. Diffusion equation 42a is solved with the help of expression 45 and boundary condition 42b and the flux at the free surface is then given by

$$-DC_z(h/2) = -\frac{D\alpha}{\theta}(\Gamma - \Gamma_0)\tanh(\alpha h/2) \tag{46a}$$

where

$$\alpha^2 = \left[k^2 + (\omega/D)\right] \tag{46b}$$

The velocity gradient at the surface, $u_x(h/2)$ is obtained from eq 44 as

$$u_x(h/2) = \frac{\sigma - \sigma_0}{\mu_s} \tag{47}$$

Recognizing that $u_x(h/2) = -v_z(h/2)$ from the continuity equation, eq 38 may be arranged with the help of boundary condition 35c as

$$\frac{2\sigma_x}{h} - \phi_x + \frac{1}{2}(\sigma h_{xx})_x + \frac{2\mu}{\mu_s}(\sigma_x - M\Gamma_x) = 0 \tag{48a}$$

This is solved for σ_x by noting that the inequality $hh_{xx} << 1$ holds for the long-wavelength perturbations, viz.,

$$\sigma_x = \frac{h}{2}\left(\phi_x - \frac{1}{2}\sigma h_{xxx}\right)\left(1 + \frac{\mu h}{\mu_s}\right)^{-1} + M\Gamma_x\left(1 + \frac{\mu_s}{\mu h}\right)^{-1} \tag{48b}$$

An equation for the surface excess concentration is derived by combining eq 43c and 46a, by differentiating the resultant equation with respect to x and then substituting for u_x from eq 47:

$$\Gamma_{tx} + \frac{\Gamma}{\mu_s}(\sigma_x - M\Gamma_x) + (\mu\Gamma_x)_x - D_s\Gamma_{xxx} +$$

$$\frac{D\alpha}{\theta}\left[(\Gamma - \Gamma_0)\tanh\left(\frac{\alpha h}{2}\right)\right]_x = 0 \tag{49a}$$

σ_x may be eliminated from the above equations by combining it with eq 48b.

$$\Gamma_{tx}+\left(1+\frac{\mu h}{\mu_s}\right)^{-1}\frac{h}{2\mu_s}\left(\phi_x-\frac{1}{2}\sigma h_{xxx}\right)-$$

$$\left(1+\frac{\mu h}{\mu_s}\right)^{-1}\frac{M\Gamma\Gamma_x}{\mu_s}+(u\Gamma_x)_x+$$

$$\frac{D\alpha}{\theta}\left[(\Gamma-\Gamma_0)\tanh\left(\frac{\alpha h}{2}\right)\right]_x=0 \tag{49b}$$

The equation of evolution, eq 41, is recast into a convenient form by observing that substituting the interfacial velocity $u(h/2)$ from expression 39a into eq 43c yields

$$a_x=-\frac{1}{4\mu}(h\sigma_x)_x+$$

$$\frac{1}{\Gamma}\left[D_s\Gamma_{xx}-u\Gamma_x-\Gamma_t-\frac{D\alpha}{\theta}(\Gamma-\Gamma_0)\tanh\left(\frac{\alpha h}{2}\right)\right] \tag{50}$$

Equation 41 is thus transformed to

$$h_t+\frac{1}{12\mu}(\sigma_x h^2)_x-\frac{1}{4\mu}h(\sigma_x h)_x+ah_x+$$

$$\frac{h}{\Gamma}\left[D_s\Gamma_{xx}-\Gamma_t-\Gamma_x u\right]-h(\Gamma-\Gamma_0)\frac{D\alpha}{\theta\Gamma}\tanh\left(\frac{\alpha h}{2}\right)=0 \tag{51}$$

Finally, substitution of σ_x from eq 48b in eq 51 gives a nonlinear equation involving only h and Γ:

$$h_t-\frac{1}{12\mu}\left[h^3\left(\phi_x-\frac{1}{2}\sigma h_{xxx}\right)\left(1+\frac{\mu h}{\mu_s}\right)^{-1}\right]_x-$$

$$\frac{1}{12\mu}\left[2h^2\left(1+\frac{\mu_s}{\mu h}\right)^{-1}M\Gamma_x\right]_x+\frac{1}{4\mu}(hh_x\sigma_x)+ah_x+$$

$$\frac{h}{\Gamma}\left(D_s\Gamma_{xx}-\Gamma_t-\Gamma_x u-\frac{(\Gamma-\Gamma_0)D\alpha}{\theta}\tanh\left(\frac{\alpha h}{2}\right)\right]=0 \tag{52}$$

The present formalism requires only the simultaneous solution of eq 49b and 52 for obtaining the nonlinear dynamics of the thin free film. This is so because, as is shown in some detail later, the terms $u\Gamma_x$, $h_x\sigma_x$, and ah_x vanish identically in the first-order equations corresponding to eq 49b and 52. Equations 49b and 52 may thus be regarded as a set of coupled equations determining $h(x, t)$ and $\Gamma(x, t)$ only. However, before pursuing the solution of the problem in its entirety, it is instructive to examine the form of these equations for some limiting cases.

Case I. Film Devoid of Solutes.

In this case, as it is in the case of surface-inactive solutes with vanishing surface viscosity, both μ_s and $M\Gamma_x$ are zero. Consequently, σ_x=0 and the evolution equation (41) reduces to

$$h_t + (ah)_x = 0 \tag{53a}$$

The coefficient $a(x, t)$ is determined with the help of eq 36a, 39b, and 38 by substuting $\sigma_x = 0$. This gives

$$4\mu a_{xx} = 2\phi_x - \sigma h_{xxx} \tag{53b}$$

Case II. Tangentially Immobile Film.

If either the surface viscosity or the Gibbs elasticity is large, the surface of a thin free film is rendered tangentially immobile and the stabilizing influence of surfactant is maximum. In this event, eq 39a may be solved for the coefficient $a(x, t)$ by setting the tangential velocity at the interface to be zero, i.e.,

$$a = -\frac{1}{4\mu}(\sigma_x h) \tag{54a}$$

This, in conjunction with the expression for σ_x from eq 48b, allows us to write the kinematic condition, eq 41, as

$$h_t + \left(\frac{A}{24\pi\mu}\right)(h^{-1}h_x)_x + \left(\frac{\sigma_0}{24\mu}\right)(h^3 h_{xxx})_x = 0 \tag{54b}$$

To arrive at the above equation, the functional form of the London-van der Waals interaction potential

$$\phi = A/6\pi h^3 \tag{54c}$$

has also been substituted. It is interesting to note that eq 54b may be recast into a parameterless form by the following transformations: $H = (h/h_0)$, $\xi = h_0^{-2}(A/\pi\sigma_0)^{1/2}x$, and $\tau = (A^2/24\pi\mu\sigma_0^2 h_0^5)t$, which transform eq 54b to

$$H_\tau + (H^{-1}H_\xi)_\xi + (H^3 H_{\xi\xi\xi})_\xi = 0 \tag{54d}$$

This parameterless equation is the same as that obtained by Williams and Davis[16] for a thin film in contact with a solid and devoid of solutes. A comparison of eq 12a with $\sigma_x = 0$ and eq 54b reveals that the two differ only in the multiplicative constants. This consideration establishes the equivalence between the nonlinear dynamics of a wetting film that is devoid of solutes and a free film with excess of surface-active solutes. That one obtains the same functional form of the nonlinear equations for two different thin film models is perhaps not surprising in view of the fact that the linear dispersion relations for these two models are also the same except the multiplicative constants.[5]

Since the asymptotic cases have been obtained, the stability analysis of the general equations 49b and 52 is now undertaken.

As was discussed earlier, the base state for a finite-amplitude perturbation is the corresponding stationary, albeit spatially nonhomogeneous, solution of the governing equations. The film thickness and the concentration may thus be expanded around their respective base solutions as

$$h = H_0(x) + 2\mu^* H_1(x,t) + O(\epsilon^2) \tag{55a}$$

and

$$\Gamma = \Gamma_0 + \Gamma_1 \sin kx\, e^{\omega t} + O\left(\epsilon^2\right) \tag{55b}$$

where μ^* is as yet an undetermined coefficient which is later related to the amplitude of the perturbation. The time independence of the zeroth-order solution $H_0(x)$ implies that all of the velocities and the interfacial tension gradients associated with the base case vanish identically. It is in view of this that a combination of zeroth-order equations corresponding to eq 48b and 52 determine the base state $H_0(x)$ as

$$\left[\left(1+\frac{\mu H_0}{\mu_s}\right)^{-1}\left[H_0^{-1}H_{0x}+\left(\frac{\pi\sigma_0}{A}\right)H_0^{3}H_{0xxx}\right]\right]_x = 0 \tag{56}$$

As was done for the wetting film, an approximate stationary solution may be found by representing $H_0(x)$ as

$$H_0 = h_0\left(1+2\mu^* \sin k_1 x\right) \tag{57a}$$

and equating the terms of order μ^* in eq 56. This provides the compatibility condition

$$k_1^2 = A/\pi h_0^4 \sigma_0 \tag{57b}$$

As is apparent, this is the same as the condition for neutral stability of the trivial state, $h = h_0$. Therefore, expression 57a for H_0 is an approximate time-stationary, albeit nontrivial, solution of the governing equations. The only condition under which it reduces to the trivial state is when μ^* is zero or, as is shown later, when the perturbation amplitude is vanishingly small. The bifurcation of a time-dependent solution is now investigated by perturbing $H_0(x)$ according to expansions 55.

Substituting eq 55b in eq 49b and recalling that, for the base case, $u_0 = \Gamma_{0x} = 0$ give the first-order equation

$$\Gamma_{1x} = -\frac{\Gamma_0}{2R}\left[\frac{h}{\mu_s+\mu h}\left(\phi_x - \frac{1}{2}\sigma h_{xxx}\right)\right]_1 \tag{58a}$$

Where

$$R = \omega + \frac{|M|\Gamma_0}{\mu_s+\mu H_0} + D_s k^2 + \frac{D\alpha}{\theta}\tanh\left(\frac{\alpha H_0}{2}\right) \tag{58b}$$

and the subscript 1 on the square braces implies that only the first-order terms are retained.

The first-order equation corresponding to eq 52 may now be written with the help of expression 58a and eq 56 as

$$H_{1t} + \frac{1}{24\mu}\left[\left[\frac{\mu H_0 \Gamma_0 |M|}{R(\mu_s+\mu H_0)^2} + \frac{\mu_s}{\mu_s+\mu H_0}\right]\times\right.$$

$$\left[\frac{A}{\pi}\left(H_0^{-1}H_{1x} - H_{0x}H_0^{-2}H_1\right) + \sigma_0(H_0^{3}H_{1xxx} + \right.$$

$$\left.3H_0^2 H_1 H_{0xxx})\Big]\Big]_x + \frac{1}{4\mu}H_0 H_{0x}\sigma_{1x} + a_1 H_{0x} -\right.$$

$$\frac{H_0\Gamma_1}{2\mu^*\Gamma_0}\left[D_s k^2 + \omega + \frac{D\alpha}{\theta}\tanh\left(\frac{\alpha H_0}{2}\right)\right] = 0 \tag{59}$$

Further analytical progress is made by investigating the growth of perturbations at a point where the film is locally thinnest and hence the growth rate due to the van der Waals interactions is maximum. For the out of phase or the symmetric perturbations that lead to the film rupture, the initial configuration of the film is thinnest at a point where $\sin k_1 x = -1$. Therefore, just as in the case of wetting film, the following estimates are derived from eq 57a:

$$H_0 \approx h_0\left(1-2\mu^*\right) \tag{60a}$$

$$H_{0x} = 2\mu^* h_0 k_1 \cos k_1 x \approx 0 \tag{60b}$$

$$H_{0xx} = -2\mu^* h_0 k_1^2 \sin k_1 x \approx 2\mu^* h_0 k_1^2 \tag{60c}$$

$$H_{0xxx} = -2\mu^* h_0 k_1^3 \cos k_1 x \approx 0 \tag{60d}$$

$$H_{0xxxx} = 2\mu^* h_0 k_1^4 \sin k_1 x \approx -2\mu^* h_0 k_1^4 \tag{60e}$$

These estimates, together with the determination of Γ_1 from eq 58a, simplify eq 59 to the following linear, homogeneous equation with constant coefficients:

$$H_{1t} + \frac{1}{24\mu}\left[\frac{\mu H_0 \Gamma_0 |M|}{R(\mu_s + \mu H_0)^2} + \frac{\mu_s}{\mu_s + \mu H_0}\right] \times$$

$$\left[\frac{A}{\pi}\left(H_0^{-1} H_{1xx} - H_0^{-2} H_1 H_{0xx}\right) + \sigma_0 (H_0^3 H_{1xxxx} + \right.$$

$$\left. 3H_0^2 H_1 H_{0xxxx})\right] - \frac{H_0^2 R_0}{4\left(\mu_s + \mu H_0\right) R}\left(\frac{A}{\pi H_0^4} H_1 - \sigma_0 H_{1xx}\right) = 0 \tag{61}$$

where

$$R_0 - D_s k^2 + \omega + \frac{D\alpha}{\theta}\tanh\left(\frac{\alpha H_0}{2}\right) \tag{62}$$

The unknown small parameter μ^* may be related to the nondimensional amplitude of the initial disturbance ϵ, by matching the location of the interface $h(x, t)$ at time $t = 0$ with the amplitude of the imposed perturbation. Therefore, using expansions 55a and 57a gives the relation

$$2\mu^* = \epsilon < \tfrac{1}{2} \tag{63}$$

where ϵ is the nondimensional amplitude (nondimensionalized with respect to h_0) of the disturbance at *each face* of the free film.

Seeking now a solution of the form

$$H_1 = \sin kx\, e^{\omega t} \tag{64}$$

and using it in eq 61 give the following expression for the growth rate associated with a finite-amplitude perturbation:

$$\omega = \left[\frac{A}{\pi h_0^4}(1-\epsilon)^{-4} - \sigma_0 k^2\right]\left[\frac{Nk^2 h_0^3 (1-\epsilon)^3}{24\mu} + \right.$$

$$\left.\frac{h_0^2(1-\epsilon)^2 R_0}{4\left[\mu_s + \mu h_0 (1-\epsilon)\right] R}\right] +$$

$$\frac{NA^2\epsilon}{8\pi^2 \mu\sigma_0 h_0^5}\left[(1-\epsilon)^2 + \frac{(1-\epsilon)^{-2}}{3}\right] \tag{65a}$$

where

$$N = \frac{\mu H_0 \Gamma_0 |M|}{R(\mu_s + \mu H_0)^2} + \frac{\mu_s}{\mu_s + \mu H_0} \tag{65b}$$

and parameters R and R_0 are defined by expressions 58b and 62, respectively. In arriving at relation 65a, k_1^2 from eq 57b has been used for evaluating expressions 60c and 60d.

It is easy to verify that the growth coefficient for a finite-amplitude perturbation reduces to that derived from the linear analysis when the initial disturbances have a vanishingly small amplitude, viz., $\epsilon \to 0$. This is so because the spatially nonhomogeneous, stationary solution (57a) reduces to the trivial solution $H_0 = h_0$ in such an instance.

The time of rupture for a finite-amplitude perturbation is found by maximizing the growth rate with respect to the wavenumber and then setting $h = 0$ in expansion 55a. This procedure gives

$$t_N = (\omega_m)^{-1} \ln\left[(1-\epsilon)/\epsilon\right];\ \epsilon < 1/2 \tag{66}$$

where t_N is the time of rupture derived from the present analysis and ω_m is the dominant growth rate corresponding to the dominant wavenumber k_m. The maximization of ω is, in general, not possible analytically because of the highly nonlinear dependence of ω on k. This may, however, be achieved with some accuracy by assuming R_0 and R to be weak functions of k and then solving $\partial\omega/\partial k = 0$. This procedure gives the following expression for the dominant wavenumber

$$k_m^2 \simeq \frac{1}{2}\left[\frac{A}{\pi H_0^4 \sigma_0} - \frac{6\mu}{NH_0\left(|M|\Gamma_0 R_0^{-1} + \mu_s + \mu H_0\right)}\right] \tag{67}$$

where N and R_0 are now evaluated for the dominant wavenumber. This solution holds whenever k_m^2 as computed from expression 67 is positive or else the dominant wavenumber is zero. For a free thin film that is devoid of surfactants, k_m^2 as determined from expression 67 is a negative quantity and hence the only feasible dominant wavelength is a very large one, viz., $k_m = 0$. It may, however, be readily verified that for the majority of cases involving a thin free film with surfactants, the first term of expression 67 dominates and k_m^2 is positive. The approximation

$$k_m^2 \approx \left(A/2\pi h_0^4\right)(1-\epsilon)^{-4} \tag{68}$$

holds even for a "gaseous" monolayer[5] ($\mu_s = 10^{-3}$ g/s, $\Gamma_0|M| = 10^{-2}$ erg/cm^2, $\theta = 2 \times 10^{-4}$ cm) and for a film characterized by $h_0 = 100$ Å, $A = 10^{-13}$ erg, and $\sigma_0 = 30$ dyn/cm. It is interesting to note that for a given set of physicochemical parameters, increasing the film thickness reduces the term $(A/\pi H_0^4\sigma_0)$ greatly and hence k_m^2 decreases. Thus, there is a critical thickness such that for all films thicker than the critical thickness, the dominant wavenumber is close to zero and hence the dynamics is closely approximated by that of a film devoid of surfactants. The stabilizing influences of the Marangoni motion and the surface viscosity are thus minimal for a film having an initial thickness greater than the critical thickness. As is also apparent from expression 67, the critical thickness (for which $k_m^2 = 0$) increases with an increase in surface viscosity or surface elasticity, both of which usually increase with increasing surfactant concentration.

While an approximate dominant growth rate for a general case may be obtained by the substitution of k_m^2 from expression 67 into expression 65, the dominant growth coefficients are exactly determined for the following two important asymptotic cases.

Case I. Film Devoid of Surfactants.

In this case, both μ_s and $|M|\Gamma_0$ vanish and k_m^2 tends to zero. The dominant growth coefficient is

$$\omega_m(N \to 0) = \left(\frac{A}{4\pi\mu h_0^4}\right)(1-\epsilon)^{-3} \tag{69a}$$

This is a factor of $(1-\epsilon)^{-3}$ larger than the corresponding growth rate derived from the linear theory. The wavenumber of a neutrally stable finite-amplitude perturbation is obtained by solving eq 65a with $\omega = 0$, $R_0 = R$, and $\mu_s = N = 0$. This gives

$$k_c^2 = \left(\frac{A}{\pi h_0^4\sigma_0}\right)(1-\epsilon)^{-4} \tag{69b}$$

Therefore, if the wavenumber of a finite-amplitude perturbation is less than k_c, it grows with time.

These results may also be obtained directly from 53a and 53b.

Case II. Film with Excess of Surfactant.

If either μ_s or $|M|\Gamma_0$ is large, the parameter N tends to one and the dominant wavenumber is

$$k_m^2 = \left(\frac{A}{2\pi\sigma_0 h_0^4}\right)(1-\epsilon)^{-4} \tag{70a}$$

The dominant wavelength of the finite-amplitude disturbance is therefore a factor of $(1-\epsilon)^2$ smaller than the dominant wavelength derived from the linear theory. The dominent growth coefficient now is

$$\omega_m(N \to 1) =$$

$$\left(\frac{A^2}{96\pi^2\mu\sigma_0 h_0^5}\right)\left[(1-\epsilon)^{-5} + 4\epsilon(1-\epsilon)^{-2} + 12\epsilon(1-\epsilon)^2\right] \tag{70b}$$

The linear growth coefficient is recovered by taking the limit $\epsilon \to 0$ in the above expression.

These results also may directly be obtained from the nonlinear equation 54b. In view of the equivalence between the models of a wetting film devoid of solutes and a free film with excess of solutes (see eq 54d), the dependence of ω_m on ϵ in eq 70b is the same as the dependence of ω_m

on μ^* in eq 32a with $z^* = 0$. The equation for the neutrally stable wavenumber is therefore also the same as eq 30a, viz.,

$$(y/y_L)^2 - (1-\epsilon)^{-4}(y/y_L) - 3\epsilon\left[(1-\epsilon)^{-1} + \frac{(1-\epsilon)^{-5}}{3}\right] = 0 \tag{70c}$$

where $y = k_c^2$ and $y_L = A/\pi h_0^4\sigma_0$. The uppermost solid curve in Figure 2 therefore depicts also the ratio (y/y_L) for a tangentially immobile free film if ϵ is interpreted there as the total perturbation amplitude.

The present analysis is readily extended to a general potential of the form

$$\phi = B/h^c$$

In this case, the dominant growth coefficient has the following form (c $\neq$ 0, 2)

$$\omega_m = \omega_L\left[(1-\epsilon)^{1-2c} + 4\epsilon\left[(c-2)(1-\epsilon)^{1-c} + 3(1-\epsilon)^2\right]\right] \tag{71}$$

where ω_L is the corresponding growth coefficient derived from the linear theory. For a tangentially immobile film, ω_L is given by

$$\omega_L = c^2B^2h_0^{1-2c}/24\mu\sigma_0 \tag{72}$$

Expression 71 also holds for a wetting film devoid of solutes if ω_L is the appropriate linear theory growth rate and ϵ is replaced by $\epsilon/2$.

Thus for any $c > 1$, the dominant growth coefficient for a finite-amplitude perturbation shows a greater deviation from its linear theory counterpart as c is increased. Thus the effects of nonlinearities is even more pronounced for retarded van der Waals interactions for which $c = 4$.

Having obtained the dominant growth coefficients, the times of rupture as derived from the linear and nonlinear theories may be compared. The time of rupture as given by the linear theory is

$$t_L = [\omega_m(\epsilon \to 0)]^{-1}\ln(1/2\epsilon) \tag{73}$$

The ratio of the time of rupture as determined from the present analysis (eq 66) and the linear theory is shown in Figure 7 as a function of the total disturbance amplitude. The curve labeled $N = 0$ is for a free film that is devoid of surfactants and that labeled $N = 1$ is for a film with excess of surfactants. Just as in the case of a wetting film, the nonlinearities of the governing equations accelerate the process of rupture for a free film. The effects become quite pronounced as the perturbation amplitude increases.

It is well-known that the presence of surfactants retards the growth of perturbations and thus prolongs the time of rupture of a free film significantly. The ratio of the times of rupture when the film has an excess of surfactants and when it is devoid of surfactants may be calculated by dividing eq 70b by eq 69a. Therefore, the factor by which the time of rupture is prolonged by surfactants is given by

$$F_n = F_L F(\epsilon) \tag{74a}$$

where F_L is the stabilization factor as predicted by the linear theory, viz.,

$$F_L = 24\pi\sigma_0 h_0^2/A) \tag{74b}$$

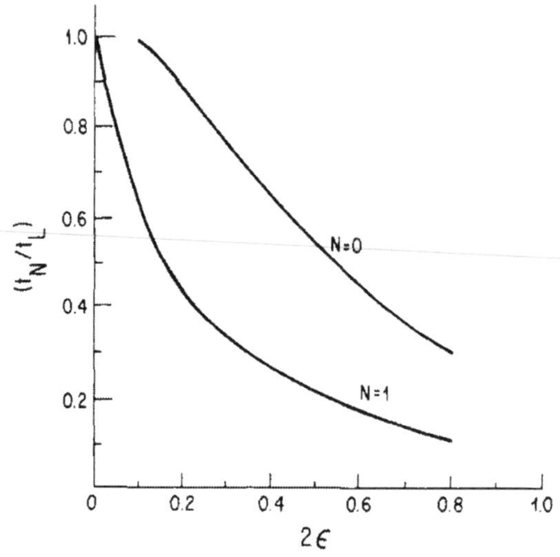

FIGURE 7. Ratio between the times of rupture of a free film as derived from the nonlinear and the linear theories. The curve labeled $N = 0$ is for a pure film and that labeled $N = 1$ is for a tangentially immobile film.

and $F(\epsilon)$ is a function only of the initial amplitude of the perturbation, which is given by

$$F(\epsilon)=\left[(1-\epsilon)^{-2}+4\epsilon(1-\epsilon)+12\epsilon(1-\epsilon)^{5}\right]^{-1} \tag{74c}$$

It may be noted that F_L is always much larger than unity for the long-wavelength perturbations. This is so because $k^2h_0^2 << 1$ for long-wavelength disturbances and hence it may be concluded from expression 69b or 70a that $\pi\sigma_0 h_0^2/A >> 1$.

The finite size of perturbations, however, decrease F_L by a factor of $F(\epsilon)$. The amplitude dependence of the stabilization factor $F(\epsilon)$ is plotted in Figure 8 as a function of the total initial amplitude, 2ϵ. It shows that $F(\epsilon)$ is always less than unity for very small disturbances. It may thus be concluded that the stabilizing influences of the Marangoni motion and the surface viscosity are *overestimated* by the linear theory for a free film.

In conclusion, the dynamics of free films with surfactants is analyzed when they are subjected to large external disturbances. The substitution of expression 67 in expressions 65 provides the dominant growth coefficient and the time of rupture is then evaluated from expression 66. Explicit expressions are obtained for a film devoid of solutes (eq 69a) and a tangentially immobile film (expression 70b).

5. DISCUSSION AND CONCLUSIONS

A first-order theory is developed to account for the effects of the amplitude of disturbances on the conditions for the instability of a thin film and its time of rupture. This is to be contrasted with the linear stability analysis which is valid when the initial deviation of the interface from the plane-parallel configuration is minimal, i.e., disturbances are infinitesimal. The key idea of the present approach is to investigate the stability of the time-stationary, albeit spatially nonhomogeneous, solution of the governing equations. The general finite-amplitude stationary solution reduces to the trivial solution for infinitesimal disturbances (planar interface), and thus the results of linear stability analysis are recovered. The nonlinearities of the evolution equations are retained in the first-order equations by evaluating the basic solution at those points where the film thickness is the least and, consequently, the rate of thinning is maximum. In this way, the conditions are derived

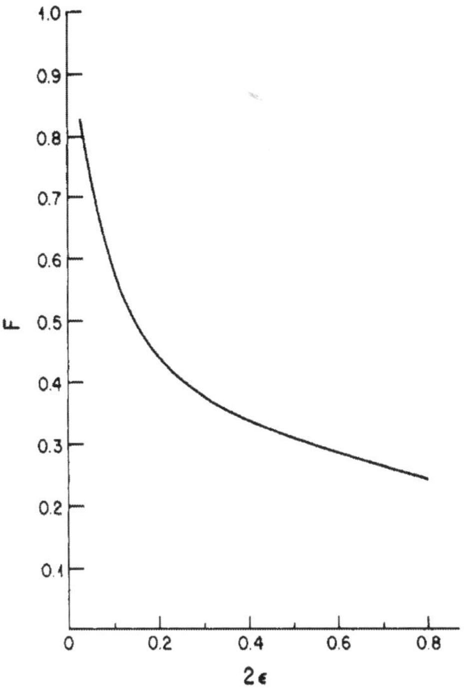

FIGURE 8. Maximum influence of surfactants in prolonging the time of rupture of a free film. F is the ratio of the stabilization factors as predicted by the nonlinear and the linear theories, respectively.

for the neutral stability of finite-amplitude perturbations imposed on a thin film. The expressions for the dominant growth rate and the time of rupture are obtained as functions of various thin film parameters and the amplitude of disturbances.

Three issues related to this methodology that deserve attention are the generalization of analysis to an arbitrary initial perturbation, $f(x)$, the generation of higher harmonics and their influence on the stability, and, last, the restrictions imposed on the analysis. These are now discussed.

The rationale behind studying the response of the thin-film equations to a single Fourier mode of the perturbation is established by noting that the first-order equations that are derived by perturbing the basic solution constitute a set of linear, homogeneous partial differential equations. It is thus possible to Fourier decompose any general interfacial perturbation by using an appropriate solution set of orthonormal functions. What is essentially done here is to investigate the response of the film to each of these Fourier modes with a certain wavenumber and amplitude. In view of the linearity and the homogeneity of the leading order equations, the total response may then be constructed as the sum of the individual component responses. A Fourier component of the initial disturbance may be regarded as the dominant one only if its response, $f(\epsilon)\exp[\omega\ (k,\ \epsilon)t]$ exceeds greatly the response of any other Fourier mode. Here, $f(\epsilon)$ denotes the amplitude of the response corresponding to a particular Fourier mode of the initial disturbance with wavenumber k and amplitude ϵ. Since the growth rate, ω, now is a function both of k and ϵ, it enhances the possibility that many Fourier modes of the initial disturbance have comparable reponses. In such an event, the total solution is a sum of all significant individual responses. Of course, if it is deemed sufficient to represent the initial disturbance by a single Fourier mode, then there is a unique growth coefficient determined by the wavelength and the amplitude of the initial disturbance.

Within the framework of the present perturbation scheme, the higher order harmonics ($\cos 2kx$, $\sin 3kx$, etc.) are associated with the solution of higher order equations and thus have amplitudes that are at least of the order of μ^{*2}. Since the ordering parameter μ^* (or $2\mu^*$ for a free film) is always less than 0.5, the convergence of the solution is expected to be rather rapid, even for moderate to large amplitudes. This conjecture was proved in an earlier paper[19] for the wetting films without the Rayleigh-Taylor effect. This observation is also fully compatible with the numerical simulations,[16] where the step size deemed satisfactory for the convergence was not small enough to keep track of higher harmonics adequately. The equivalence between the nonlinear equations of a pure wetting film and a tangentially immobile free film leads to the conclusion that higher harmonics are unimportant for the latter case also. In conclusion, although the exact magnitude of the error incurred by neglecting the higher order terms cannot be asserted a priori for all thin-film models, it is expected to be of the order of μ^{*2}, and also since even the solution of the first-order equation given here is approximate, the inclusion of higher order terms is not warranted.

Finally, while summarizing the present methodology, it is probably easy to give the impression that it is valid for any thin-film instability. This, however, is not the case, because, for the proposed analysis to be successful in its present form, it is necessary that the instability be of a purely dissipative type, i.e., the growth coefficient be real. This condition automatically guarantees that a time-stationary, quasi-harmonic solution of the governing equations exists for a nontrivial wavenumber and rules out the possibility of temporal oscillations. Also, it is of interest to note[31] that a nontrivial stationary solution always exists when the governing equations of evolution are invariant under the inversion of the sign of the lateral spatial coordinate x (eq 16 and 52, for instance). The methodology in its present form is not suitable for studying the saturation of the initially growing waves due to the nonlinearities. A classical example is of course that of formation of ripples on a falling film. These considerations, however, do not seem restrictive for studying the thin films that either rupture or form the so-called "black" films.

The application of this nonlinear theory to the *wetting films* shows that the linear theory (a) underestimates the destabilizing effects of van der Waals interactions on the stability and the time of rupture, (b) overestimates the stabilizing influence of the interfacial tension, (c) under estimates the role of surfactant concentration driven Marangoni motion and the surface viscosity in prolonging the time of rupture, (d) overestimates the stabilization due to the presence of a lighter fluid on top of a denser fluid, (e) overestimates the destabilization due to the Rayleigh- Taylor instability (heavier fluid on top), and (f) overestimates the critical and dominant wavelengths of the disturbances. It is found that for almost all situations of practical interest, viz., $|\eta| < 1$, factors a and b are dominant over factors c and e and, consequently, the time of rupture is always found to be less than that given by the linear theory. For example, a wetting film with an initial amplitude of disturbance equaling 0.4 of the thickness and $\eta = -0.5$ ruptures in about one-tenth of the time predicted by the linear theory (Figure 4). The result that the surfactants are able to prolong the time of rupture of a wetting film by an order of magnitude appears to be of particular interest in controlling the rate of flotation and the rate of cooling of a hot metal that is cooled by quenching it in a liquid that boils due to the release of heat. Conclusions a and b were also arrived at by Williams and Davis[16] through direct numerical simulations for a single film devoid of solutes. Quantitatively, the analytical expression from the present theory yields results for a pure wetting film that are within 15% of the numerical simulations even for amplitudes as large as 0.6 of the film thickness.

The methodology is applied also to thin free films and thus finds application in predicting the lifetime of foams. Again, it is deduced that the nonlinearities here prefer shorter wavelengths (as compared to the linear theory) both for the finite-amplitude neutrally stable and dominant waves. Just as in the case of a wetting film, the nonlinear theory predicts a greater destabilization due to the van der Waals interactions as compared to the linear theory. In contrast to a wetting film, however, the nonlinearities act to diminish the stabilizing influence of the Marangoni motion and the surface viscosity. The linear theory thus overestimates the stabilizing effects of surfactants on the time of rupture by as much as a factor of about 4. The overall effect of nonlinearities is thus always to reduce the time of rupture as compared to the one given by the linear theory. For a free film with surfactants, the time of rupture may be as much as about 1 order of magnitude smaller than that given by the linear theory, depending on the amplitude of the perturbation.

Finally, the paramount strength of the method is in the ease of derivations and computations that it offers. These are only slightly more involved than the linear stability analysis and yet give explicit analytical expressions for all of the quantities of interest. It is in view of this that it appears attractive to exploit the present methodology to other thin-film situations.

REFERENCES

(1) Scheludko, A. *Adu. Colloid Interface Sci.* **1967**, *1*, 391.
(2) Vrij, A.; Overbeek, J. Th. G. *J. Am. Chem. Soc.* **1968**, *90*, 3074.
(3) Felderhof, B. U. *J. Chem. Phys.* **1968**, *49*, 44.
(4) Vrij, A.; Hesselink, F. Th.; Lucassen, J.; Van Den Tempel, M. *Proc. K. Ned. Akad. Wet., Ser. B:Phys. Sci.* **1970**, *73*, 124.
(5) Ruckenstein, E.; Jain, R. K. *Faraday Trans. 2* **1974**, *70*, 132 (Section 5.1 of this volume).
(6) Gumerman, R.; Homsy, G. *Chem. Eng. Commun.* **1975**, 2, 27.
(7) Sanfeld, A.; Steinchen, A. *Biophys. Chem.* **1975**, *3*, 99.
(8) Bisch, P. M.; Sanfeld, A. *Bioelectrochem. Bioenerg.* **1978**, *5*, 401.
(9) Jain, R. K.; Ivanov, I.; Maldarelli, C.; Ruckenstein, E. In *Dynamics and Instability of Fluid Interfaces;* Sorensen, T. S., Ed.; Springer-Verlag: Berlin, **1979**; p 140.
(10) Maldarelli, C.; Jain, R. K. *J. Colloid Interface Sci.* **1982**, *90*, 233.
(11) Jain, R. K.; Maldarelli, C.; Ruckenstein, E. *AIChE Symp. Ser.* 1978, *74*, 120; Biorheol.
(12) Maldarelli, C.; Jain, R. K.; Ivanov, I.; Ruckenstein, E. *J. Colloid Interface Sci.* **1980**, *78*, 118 (Section 5.3 of this volume).
(13) Dimitrov, D. S.; Zhelev, D. V. *J. Colloid Interface Sci.* **1984**, *99*, 324.
(14) Sharma, A.; Ruckenstein, E. *J. Colloid Interface Sci.* **1985**, *106*, 12 (Section 5.10 of this volume).
(15) Sharma, A.; Ruckenstein, E. *Am. J. Optom. Physiol. Opt.* **1985**, *62*, 246.
(16) Williams, M. B.; Davis, S. H. *J. Colloid Interface Sci.* **1982**, *90*, 220.
(17) Desai, D.; Kumar, R. *Chem. Eng. Sci.* **1985**, *40*, 1305.
(18) Ivanov, I. B.; Dimitrov, D. S. *Colloid Polym. Sci.* **1974**, *252*, 982.
(19) Sharma, A.; Ruckenstein, E. *J. Colloid Interface Sci.,* **1986**, *113*, 456. (Section 5.4 of this volume).
(20) Taylor, G. I. *Proc. R. Soc. London Ser. A* **1950**, *201*, 192.
(21) Chandrasekhar, S. *Hydrodynamic and Hydromagnetic Stability*; Oxford University Press: London, **1968**.
(22) Jain, R. K.; Ruckenstein, E. *J. Colloid Interface Sci.* **1976**, *54*, 108.
(23) Babchin, A. J.; Frenkel, A. L.; Levich, B. G.; Sivashinsky, G. I. *Phys. Fluids* **1983**, *26*, 3159.
(24) Benney, D. J. *J. Math. Phys.* **1966**, *45*, 150.
(25) Atherton, R. W.; Homsy, G. M. *Chem. Eng. Commun.* **1976**, *2*, 57.
(26) Dagan, Z.; Pismen, L. M. *J. Colloid Interface* Sci. **1984**, *99*, 215.
(27) Sharma, A.; Ruckenstein, E. *J. Colloid Interface Sci.* **1986**, *111*, 834 (Section 5.11 of this volume).
(28) Bikerman, J. J. *Foams: Theory and Industrial Applications;* Reinhold: New York, 1953. (Also, Springer-Verlag: New York, 1973.)
(29) Radoev, B. P.; Dimitrov, D. S.; Ivanov, I. B. *Colloid Polym. Sci.* **1974**, *252*, 50.
(30) Radeov, B. P.; Scheludko, A. D.; Manev, E. D. *J. Colloid Interfaces sci.* **1983**, *95, 254.*
(31) Malomed, B. A.; Tribelsky, M. I. *Physica* D **1984**, *4D,* 67.

5.6 Stability, Critical Thickness, and the Time of Rupture of Thinning Foam and Emulsion Films*

Ashutosh Sharma and Eli Ruckenstein†

Department of Chemical Engineering, State University of New York at Buffalo, Buffalo, New York 14260

Received November 3, 1986. In Final Form: March 11, 1987

I. INTRODUCTION

The intermolecular forces induce the instability of thin films in such diverse settings as foams and emulsions,[1–5] deformation of biological membranes,[6,7] electrostatic powder coating,[8] paper coating,[9] thin solid films used in microelectronics,[10] and the corneal tear film.[11,12] In essence, a thin film becomes unstable when the destabilizing influence of the van der Waals forces overcomes the stabilizing effects of the surface tension, Marangoni convection, surface viscosity, double-layer repulsion, and other stabilizing factors, if present (as is shown later, drainage of the fluid from the film also exerts a stabilizing influence). The velocity of film thinning due to drainage is inversely proportional to a certain power (<2) of the film radius.[13,14] (The celebrated Reynolds' law of film thinning predicts the drainage velocity to be inversely proportional to the square of the film radius.) Thus, for a film of large lateral dimensions, the mean thickness of the film remains constant and equal to its initial thickness. Various studies have explored the conditions for instability and the time of rupture of such nondraining thin films by linearizing the governing equations.[1–3,5,7,15,16] The nonlinear effects have also been studied for nonthinning thin films, both numerically[17] and analytically.[18,19] However, the finite sized interstitial thin films occurring in foams and emulsions drain rather quickly because of the capillary pressure, Plateau border suction, and disjoining pressure. For dispersions such as foams and emulsions, the film radius is a complicated function of the radii of the adjacent bubbles and drops. In an experimental setting[2–5] however, the film radius is directly determined as the radius beyond which the meniscus curvature becomes significant. The thermally and mechanically generated corrugations of the interface first begin to amplify when the film is thin enough for the van der Waals forces to exactly counter the stabilizing effects. This is the so-called neutrally stable or transition thickness, h_t. The van der Waals forces become stronger on further thinning of the film, and the growth of the interfacial corrugations eventually either ruptures the film or leads to a black film. The mean thickness of the film (averaged over the thermal disturbances) at this stage is referred to as the critical thickness, h_c. The purpose of this paper is to explore the interactions between the drainage and the growth of perturbations and to predict both the conditions for instability and the critical thickness of foam and emulsion films as functions of the drainage velocity. Alternatively, the theory predicts the lifetimes of thin films when only the

* *Langmuir, Vol.* 3, *No.* 5, 1987 761. Republished with permission.

critical thickness (but not the thinning velocity) is known. These are presented after a brief review of the relevant information on the topography and rupture characteristics of a thinning film.

II. PRELUDE

While the critical thickness of foam films was measured as early as in 1960,[20] it was the study of Manev et al.[4] that photographically demonstrated that the thin films are not plane-parallel, but are nonhomogeneous in thickness. Radoev et al.[5] quantified the extent and the amplitude of these "hydrodynamic" nonhomogeneities by using an oscillating fiber optic probe. The wavelength of these nonhomogeneities was indirectly inferred to be about 5×10^{-3} cm (which probably increases with an increase in the film radius), and the amplitude ranged from about 40 to 250 Å as the film radius increased from 10^{-2} to 10^{-1} cm. That the amplitude increases with an increase in the film radius is also evident from an increased scatter in the measured mean critical thicknesses as the film radius increases.[4] Experiments also demonstrated that the amplitude of the hydrodynamic nonhomogeneities (2ϵ) is preserved as the film thins, until the growth of thermal perturbations destroys the primary film.[5] Due to the presence of finite- amplitude hydrodynamic nonhomogeneities, the film is not plane-parallel but contains several thinner and thicker regions. Figure 1 a, b depicts the topography of a part of a thinning foam/emulsion film for one wavelength of the hydrodynamic nonhomogeneities and the superimposed thermal corrugations of the interfaces. Clearly, the growth of thermal disturbances ruptures the film first where the film is thinnest, and, therefore, the theoretical calculations predict the *minimum* critical thickness (h_c in Figure 1b). The *mean* critical thickness, h_{cm}, is related to h_c by

$$h_{cm} = h_c(R) + 2\epsilon(R) \tag{1}$$

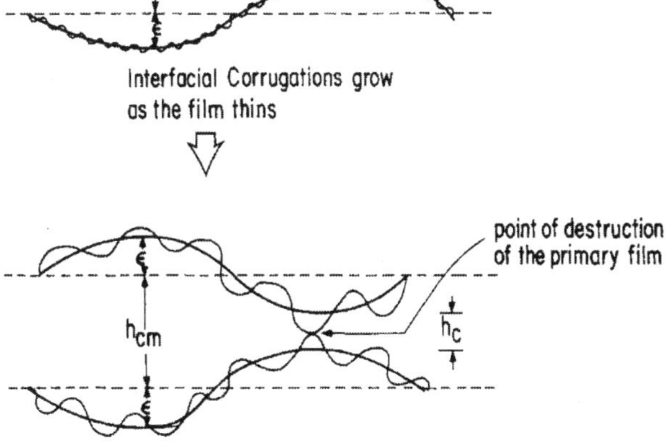

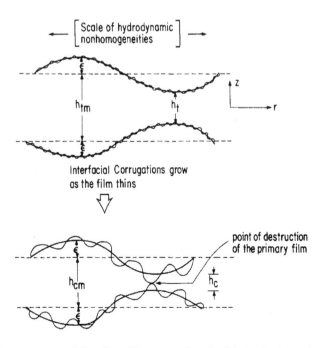

FIGURE 1 Thinning foam or emulsion film with one wavelength of the hydrodynamic nonhomogeneity and the superimposed thermal perturbations. The thermal disturbances start to grow when the minimum thickness of the film becomes less than h_t. The growing corrugations eventually rupture the film at the minimum critical thickness, h_c. h_{cm} and ϵ denote, respectively, the mean critical thickness and the amplitude of the hydrodynamic nonhomogeneities.

Experiments show that for all radii greater than 2×10^{-2} cm, $2\epsilon(R)$ is of the same order (100 Å) as the minimum critical thickness, h_c, itself.[5] Thus, the mean critical thickness may be substantially higher than the minimum critical thickness, and, in fact, it is the former that was measured by all of the experimental studies except that of Radoev et al.[5] That the theoretical predictions,[1,3] which are supposed to predict h_c, partially matched the observed *mean* critical thickness h_{cm} (instead of h_c) is due in part to the following reasons. These models employ the Reynolds' law of film thinning, which always underestimates (even for the tangentially immobile films) the velocity of thinning as compared to that which is obtained experimentally. The discrepancy becomes substantial for films as small as 0.5×10^{-2} cm in radius and increases further with an increase in the film radius.[13,14] This happens because the experimentally obtained velocities of the film thinning vary inversely with about the first power of the film radius and *not* with the square of the film radius, as predicted by the Reynolds' equation.[13,14] A lower velocity of thinning implies a higher critical thickness (ref 1, 3, and 5 and as is also inferred later); therefore, the model predictions tend to agree with the larger mean critical thickness h_{cm} rather than with h_c. This coincidence may be understood quantitatively by noting that the models incorporating the Reynolds' law of thinning predict the slope of log h_c vs. log R to be between 0.25 and 0.5. In contrast, the recent experiments of Radoev et al.[5] show this slope to be only about 0.13. However, it so happens that the slope of log h_{cm} vs. log R (from eq 1) is also about 0.25 for all $R > 10^{-2}$ cm, because of the inclusion of the amplitude, $\epsilon(R)$.[5] Thus, the models employing the Reynolds' law predict a critical thickness that coincidently is in partial agreement with h_{cm} and not with h_c. The conclusion, as was also pointed out by Radoev et al.,[5] is to employ the experimentally observed velocities of film thinning for predicting the critical thickness and for the purposes of model discrimination until a suitable model for the film thinning becomes available. This requirement is not very stringent, because, as is shown later, only one value of the velocity is all that is needed to predict the critical thickness for a given film radius.

Having outlined the central experimental observations, the various theoretical approaches are now summarized. The earliest model of Vrij,[1] while conceptually sound, did not account for the influence of drainage on the growth of surface corrugations, nor did it combine the wave motion with drainage. Ivanov et al.[21] and Ivanov and Dimitrov[3] combined the wave motion with the drainage in the so-called quasi-static approximation. In this approximation, the details of the wave motion are inferred at each instant of time by freezing the relatively slow motion of the interface; hence, in essence, the time increments are replaced by dh/V. Here, dh is the change in the mean thickness of the film and V is the velocity of thinning. This model neglects the influence of drainage on the growth of waves and on the condition of neutral stability. That the drainage may affect profoundly the conditions for neutral stability has been shown by Gumerman and Homsy[22] for a particular wave—a wave with a wavelength twice the diameter of the thin film. The procedure of Gumerman and Homsy[22] was also exploited by Malhotra and Wasan[23] for a more realistic velocity distribution. These results are, however, only indicative of the possible influence of the drainage since the wave responsible for the destruction of the film is at least 1 order of magnitude smaller than the film radius and, in addition, the critical thickness is always smaller than the neutrally stable thickness. It is only the latter that is obtained by this approach[22] and also by the procedure used by Manev et al.[4] The model of Manev et al.[4] appeared to fit their data because the model predictions were compared against the mean value of the critical thickness, and both the model predictions and the mean critical thickness overestimate the minimum critical thickness. The approach of Radoev et al.[5] used the experimental velocities of film thinning but employed the same growth coefficient and the neutrally stable thickness as for a nonthinning film. Further, the wave motion and the drainage were not combined, and (as is shown later) their results are rather sensitive to the amplitude of the perturbations. As is discussed later, it appears that an unrealistically high value of thermal amplitude is needed to fit the experimental data when their theory is used. A consistent picture that emerges from the above considerations is that a model predicting the critical thickness should (a) incorporate the effects of drainage on the growth of surface corrugations and on the conditions for the neutral stability (this should be done for a wave (the dominant wave) that causes the fastest destruction of the film, i.e., a wave for which the corresponding critical thickness is maximum), (b) combine the drainage and the wave motion, and (c) employ the experimentally obtained velocity of film thinning. The effects of various phenomena may then be checked a posteriori. In what follows a theory is formulated, and efforts are made to keep it as simple as possible without sacrificing either the details of the various phenomena or the accuracy. The predictions of this theory are then compared with the recent experimental results of Radoev et al.[5] Only theoretical estimates of the Hamaker constant and of the amplitude of the thermal perturbations are employed, and, therefore, the predictions are made without any adjustable parameter.

III. THEORY

III.1. Influence of Drainage on the Growth of Waves

The concurrent processes of film thinning and the growth of thermal surface waves are depicted in Figures 1 parts a and b. The dominant thermal wave first becomes unstable at thickness h_t and grows until it destroys the primary film at a certain minimum thickness, h_c. As is well-known, the wavelength of the dominant wave has a wavelength far exceeding the thickness of the neutrally stable film, h_t, and, therefore, the following Navier—Stokes equations in the "thin film" approximation describe the hydrodynamics of the thin film:[16,19]

$$\frac{\partial \phi}{\partial z} + \frac{\partial p}{\partial z} = 0 \tag{2}$$

$$\frac{\partial \phi}{\partial r} + \frac{\partial p}{\partial r} = \mu \frac{\partial^2 \upsilon_r}{\partial z^2} \tag{3}$$

where p is the hydrodynamic pressure, υ_r is the radial component of the velocity, and ϕ is the van der Waals interaction potential per unit volume of the liquid. The continuity equation has the form

$$\frac{\partial \upsilon_z}{\partial z} + \frac{1}{r}\frac{\partial}{\partial r}(r\upsilon_r) = 0 \tag{4}$$

where υ_r is the normal component of the velocity. The boundary conditions at the interface, $z = H(r, t)/2$, are

$$\frac{\partial H}{\partial t} + \upsilon_r \frac{\partial H}{\partial r} = 2\upsilon_z \tag{5}$$

$$\upsilon_r = 0 \tag{6}$$

$$-p_g + p = -(\sigma/2)\Delta_r H + 2\mu \frac{\partial \upsilon_z}{\partial z} \tag{7}$$

where p_g is the pressure outside the film, σ is the interfacial tension, and Δ_r is the surface Laplacian. Equations 5, 6, and 7 are the kinematic condition, the condition for the tangential immobility of the interface, and the condition for the pressure jump, respectively. In addition, the symmetry of the film at the $z = 0$ plane allows us to write

$$v_z = 0 \text{ at } z = 0 \tag{8}$$

and

$$\frac{\partial v_r}{\partial z} = 0 \text{ at } z = 0 \tag{9}$$

The solution of eq 2 is

$$p + \phi = p(H/2) + \phi(H/2) \tag{10}$$

where $p(H/2)$ is determined from the boundary condition 7 and $\phi(H/2)$ is given by eq 11,

$$\phi(H/2) = A/6\pi H^3 \tag{11}$$

where A is the Hamaker constant for the interactions between the molecules of the thin film. Equation 3 may now be solved with the help of boundary conditions (9) and (6) to give

$$v_r = \frac{1}{2\mu}\left(z^2 - \frac{H^2}{4}\right)\left(\frac{\partial p}{\partial r} + \frac{\partial \phi}{\partial r}\right) \tag{12}$$

The continuity equation (eq 4) in conjunction with the above expression and the boundary condition (8) yields the normal component of the velocity as

$$v_z = -\frac{1}{2\mu}\frac{1}{r}\frac{\partial}{\partial r}\left[r\left(\frac{z^3}{3} - \frac{H^2 z}{4}\right)\left(\frac{\partial p}{\partial r} + \frac{\partial \phi}{\partial r}\right)\right] \tag{13}$$

Finally, the kinematic condition (5) gives the sought after equation of evolution, viz.,

$$\frac{\partial H}{\partial t} = \frac{1}{12\mu}\frac{1}{r}\frac{\partial}{\partial r}\left[rH^3\left(\frac{\partial p}{\partial r} + \frac{\partial \phi}{\partial r}\right)\right] \tag{14}$$

The above equation describes both the drainage of the film and the growth/decay of thermal corrugations of the interface. However, these two motions may be decoupled because of the widely separated length and time scales governing the two phenomena, and the location of the interface may be decomposed into two parts, viz.,

$$H(r,t) = h(t) + 2\xi_0\xi(r,t) \tag{15}$$

where $h(t)$ is the mean (averaged over the surface corrugations) thickness of the film that decreases due to the drainage, ξ_0 is the amplitude of the thermal perturbations, and $\xi(x, t)$ describes the spatial and temporal dependence of the thermal surface corrugations. It may be noted that until the very last stages of the process of rupture, viz., $h \approx h_c$, the inequality $2\xi_0\xi(r, t) \ll h(t)$ holds. Similarly, the total pressure may be represented as

$$p(r,t) = p_d + 2p_\xi\xi_0 \tag{16}$$

where p_d corresponds to the pressure for the film drainage and p_ξ is related to the pressure corresponding to the growth of surface waves. It is now well established that the linear decomposition (15) leads to the same time of rupture as the nonlinear theories, as long as the amplitude of the initial corrugations, ξ_o, is small (less than 5 Å) compared to the mean thickness of the film.[17–19] Physically, this happens because the slow initial growth of waves is the rate-determining step, and this stage is well represented by the decomposition (15) with the assumption that $\xi_o\xi(x, t) < h(t)$. Therefore, it may be noted that the one term perturbation expansions (15) and (16) *are accurate* for perturbations of *thermal origin,* and, in view of recent nonlinear studies,[17–19] the nonlinearities are not expected to play any role for very small amplitude thermal perturbations.

Introducing the expansions (15) and (16) in eq 14 and comparing the zero-order and the first-order terms with respect to the amplitude ξ_o give

$$\frac{\partial h}{\partial t}=\frac{h^3}{12\mu}\frac{1}{r}\frac{\partial}{\partial r}\left(r\frac{\partial p_d}{\partial r}\right)=-V \tag{17}$$

and

$$\frac{\partial \xi}{\partial t}=\frac{1}{12\mu}\frac{1}{r}\frac{\partial}{\partial r}\left[h^3 r\left(\frac{\partial p_\xi}{\partial r}-\frac{A}{2\pi h^4}\frac{\partial \xi}{\partial r}\right)+3rh^2\xi\frac{\partial p_d}{\partial r}\right] \tag{18}$$

Equation 17 describes the process of film thinning, and an explicit form of the velocity, V, of film thinning may be obtained for a given pressure distribution, $p_d = p_d(r)$. The choice

$$\frac{1}{r}\frac{\partial}{\partial r}\left(r\frac{\partial p_d}{\partial r}\right)=\frac{8}{R^2}\left(P_\lambda+\frac{A}{6\pi h^3}\right)=\frac{8}{R^2}\Delta P \tag{19}$$

corresponds to the Reynolds' law of film thinning. Here P_λ is the capillary pressure and the term $P_\lambda + A/6\pi h^3$ is the pressure causing the drainage. However, as discussed earlier, the dependence of V on R as given by Reynolds' law is not correct, and thus, we will not employ eq 19 but will leave the functional form of p_d open. The pressure distribution for the wave motion is evaluated from the boundary condition (7) by making a Taylor series expansion with the help of eq 15. This gives (ref 22 and as is shown in the Appendix) eq 20. The last two terms of eq 20

$$2p_\xi=-\sigma\Delta_r\xi-\frac{12\mu}{h^2}\frac{\mathrm{d}h}{\mathrm{d}t}\xi-\frac{6\mu}{h^2}\frac{\mathrm{d}h}{\mathrm{d}t}\frac{1}{r}\frac{\partial}{\partial r}\left(r^2\xi\right) \tag{20}$$

and the last term in eq 18 reflect the coupling between the drainage and the wave motion. Only the former was accounted for by Gumerman and Homsy,[22] not the latter. It may be noted that the first term in eq 20 is of the order of $\sigma k^2\xi$ where k is the wavenumber of the perturbation, whereas the second and the third terms are of the order of $(10h/R^2)\Delta P\xi$ (here, we have used for $\mathrm{d}h/\mathrm{d}t$ the Reynolds' law for illustrative purposes). The minimum value of $\sigma k^2\xi$ is obtained for the largest possible wavelength, viz., for $\lambda = 4R$, and this value is about 3 orders of magnitude greater than the maximum value of the last two terms. The discrepancy is even greater for the dominant wave for which the wavelength is usually less than one-tenth of the film radius. The last two terms of eq 20 may, thus, be neglected, and the main effect of drainage on the wave motion is confined to the last term of eq 18.

Seeking now a solution of the form

$$\xi=J_o(kr)e^{\omega t} \tag{21}$$

by eliminating $\partial p_d/\partial r$ from eq 17 and 18 and comparing the terms multiplied by $J_o(kr)$, one obtains

$$\omega = \frac{k^2 h^3}{24\mu}\left(\frac{A}{\pi h^4} - \sigma k^2\right) - \frac{3V}{h} = \omega_o - \omega_d \tag{22}$$

In arriving at eq 22 a single term $\left((\partial(r\xi)/r\partial r)(\partial p_d/\partial r)\right)$ from eq 18 that is multiplied by $J_1(kr)$ is neglected because the film first ruptures where $J_o(kr) \rightarrow 1$, and at this point, $J_1(kr) \rightarrow 0$. Moreover, if the general solution (eq 21) is sought in terms of Hankel functions ($H_o^{(1)}$ and $H_o^{(2)}$), the neglected term may be shown to contribute only an imaginary part to the growth coefficient.

Equation 22 gives the modified growth coefficient for the thermal corrugations, the last term of which reflects the stabilizing influence of the drainage on the growth of surface waves. This stabilizing effect has a very simple physical origin; namely, interfacial perturbations cause an unevenness in the thickness of the film, and, the velocity of thinning being proportional to the third power of the film thickness, the velocity of film thinning at the depressed regions is smaller than that at the elevated regions. This results in the dampening of the imposed perturbations. Quantitatively, if one envisages the relation (see eq 17)

$$-V = h^3 f\left(\Delta P, \mu, R\right) \tag{23}$$

to hold at each point of the film, than the change in the local velocity due to the perturbation dh is

$$-dV = 3h^2 f\left(\Delta P, \mu, R\right)\mathrm{d}h = \frac{3V}{h}\mathrm{d}h \tag{24}$$

This represents the rate of decay of the growing perturbations due to the drainage and is exactly the phenomenon captured by the last term of eq 18.

III.2. Conditions for the Neutral Stability

A surface wave with a certain wavenumber k_m becomes neutrally stable at a thickness h_t when the corresponding growth coefficient becomes zero, viz.,

$$\left[\frac{A}{\pi h_t^{\,4}} - \sigma k_m^{\,2}\right] = \frac{72\mu V_t}{k_m^{\,2} h_t^{\,4}} \tag{25}$$

where V_t is the velocity of film thinning at the neutrally stable or transition thickness h_t. This may be contrasted with the condition for neutral stability in the absence of drainage where the right-hand side of eq 25 vanishes. Equation 25 shows that the neutrally stable thickness for a thinning film is always smaller than that for a nonthinning one, and the discrepancy increases as the wavelength of the perturbation increases or the wavenumber k_m decreases. By use of, for illustrative purposes, the Reynolds' law

$$V_t = \frac{2h_t^{\,3}}{3\mu R^2}\left(P_\lambda + \frac{A}{6\pi h_t^{\,3}}\right) \tag{26}$$

and its combination with eq 25, it is readily inferred that all disturbances with wavelengths exceeding the film diameter are *unconditionally stable* (regardless of the magnitude of the Hamaker constant) down to zero thickness. The experimentally observed departures from the Reynolds' law increase even further the effect of drainage in determining the transition thickness. This observation also partially accounts for the fact[5] that the large-scale hydrodynamic nonhomogeneities do not grow in amplitude even down to the thickness where the film ruptures. In extreme contrast to this is a nonthinning film, for which the disturbances with wavelengths $\lambda = 2R$ and $4R$ become unstable at thicknesses of $(R^2A/\pi^3\sigma)^{1/4}$ and $(4R^2A/\pi^3\sigma)^{1/4}$, respectively. For realistic values of parameters ($A = 2 \times 10^{-13}$ erg, $\sigma = 35$ dyn/cm) and even for very small films ($R \sim 5 \times 10^{-3}$ cm), these perturbations make a nonthinning film unstable at a thickness of about 1000 Å. The effect of drainage is, however, not as drastic for small wavelength perturbations. The wavelength of the "dominant" wave and the corresponding critical thickness are now derived.

III 3 Critical Thickness of a Thinning Film

Ivanov et al.[21] and Ivanov and Dimitrov[3] combined the wave motion with the drainage in the quasi-static approximation. The essential idea of their approach is to replace the time increments dt by dh/V. The wave velocity may then be written as

$$\upsilon_\xi = -(\partial\xi/\partial h)(\mathrm{d}h/\mathrm{d}t)\xi_o = V(\partial\xi/\partial h)\xi_o \tag{27a}$$

$$\upsilon_\xi = \omega\xi\xi_o \ (\text{from eq 21}) \tag{27b}$$

From eq 27a, 27b, and 21, one obtains

$$\xi(h) = J_o(kr)\exp\left[\int_{h_t}^{h}\frac{\omega}{V}\mathrm{d}h\right] \tag{28}$$

The condition for the rupture of the film is obtained by setting $H(r, t)$ equal to zero in the expansion (15), viz.,

$$h_c = |2\xi_o\xi(h_c)| \tag{29}$$

The dominant wave that is responsible for the film destruction is the one that causes the fastest rupture of the film or, alternatively, the one for which the critical thickness is maximum. The wave-number, k_m, of such a wave is determined by the solution of

$$\frac{\partial h_c}{\partial k} = \frac{\partial\xi}{\partial k} \cong \frac{\partial}{\partial k}\exp\left[\int_{h_t}^{h_c}\frac{\omega}{V}\mathrm{d}h\right] = 0 \tag{30}$$

The simultaneous solution of eq 25, 29, and 30 in conjunction with the expression for ω (eq 22) provides the three unknowns (k_m, h_t, and h_c) for a given velocity distribution $V = V(h)$. However, we wish to generalize the theory when only one value of the velocity at any thickness is known. This is readily obtained from the experiments, and, in fact, Radoev et al.[5] measured the velocity of film thinning at the critical thickness. It may be noted that the largest contribution to the integral $\int_{h_t}^{h_c}\omega/V\,\mathrm{d}h$ comes from the neighborhood of the critical thickness h_c, because the growth coefficient there is the maximum and the velocity of thinning is minimum. Thus, in the limited domain $h \in (h_t, h_c)$, this integral may be approximated by (see eq 22)

$$\int_{h_t}^{h_c}\frac{\omega}{V}\mathrm{d}h \cong \frac{1}{V_c}\int_{h_t}^{h_c}\omega_o\mathrm{d}h - \int_{h_t}^{h_c}\frac{3}{h}\mathrm{d}h \tag{31}$$

where V_c is the velocity of film thinning at the critical thickness. That the aforementioned procedure causes less than 10% error in the estimation of V_c for a given h_c and less than 2% error in the prediction of the critical thickness for a given V_c was confirmed by direct numerical simulations. This is due, in part, to the fact that a slight underestimation of the velocity causes also a somewhat underestimation of the growth coefficient, and these two have a compensating effect on the critical thickness. In addition, as is shown later, the critical thickness is a rather weak function of the velocity of thinning. It is now possible to determine k_m explicitly with the help of eq 22, 30, and 31 in the form

$$k_m^{\,2} = \left(\frac{A}{\pi\sigma h_t^{\,4}}\right)\left[\frac{2\ln(1/x)}{1-x^4}\right] \tag{32}$$

where

$$x = (h_c/h_t) < 1 \tag{33}$$

The condition for neutral stability of the dominant wave is obtained from eq 25 as

$$\frac{V_t h_t^{\,4}}{\alpha} = \frac{y(1-y)}{3} = \beta \tag{34}$$

where

$$\alpha = \left(A^2/24\pi^2\mu\sigma\right) \tag{35}$$

$$y = 2\ln(1/x)/(1-x^4) \tag{36}$$

Finally, the condition for film rupture (eq 29) is transformed to the following simple form with the help of eq 22, 28, 31, and 32:

$$\frac{h_c}{2\xi_o} = x^3\left(\frac{1}{x}\right)^{y/2\beta\psi} \tag{37}$$

In arriving at eq 37, we have also made use of the fact that

$$\psi = \frac{V_c}{V_t} = \frac{h_c^{\,3} + (A/6\pi P_\lambda)}{h_t^{\,3} + (A/6\pi P_\lambda)} < 1 \tag{38}$$

It is to be noted that the above relation does not suppose the Reynolds' law but only the general relation $V \propto h^3\Delta P$. In any event, it will be shortly shown that the results are quite independent of the actual value of ψ even in the extreme range of $\psi \in \left((h_c/h_t)^3, 1\right)$

Equations 34 and 37 are our final coupled nonlinear equations that predict h_t and V_c for a given critical thickness. Fortunately, it was possible to simplify these results even further by the following procedure. First, ψ from eq 38 may be written as

$$\psi = \frac{(h_c/2\xi_o)^3 + P}{(h_c/2\xi_o)^3(1/x^3) + P} \tag{39}$$

TABLE 1
Numerical Representation of the General Dimensionless Correlation $h_c/2\xi_o = f(V_e, P)$

	$V_c(2\xi_o)^4/\alpha = V_e$		
$h_c/2\xi_o$	$P = 10^5$	$P = 10^7$	$P = 10^9$
20	0.623×10^{-7}	0.623×10^{-7}	0.623×10^{-7}
40	0.330×10^{-8}	0.331×10^{-8}	0.331×10^{-8}
60	0.595×10^{-9}	0.602×10^{-9}	0.602×10^{-9}
80	0.177×10^{-9}	0.180×10^{-9}	0.180×10^{-9}
100	0.697×10^{-10}	0.709×10^{-10}	0.710×10^{-10}
120	0.326×10^{-10}	0.331×10^{-10}	0.331×10^{-10}
140	0.171×10^{-10}	0.174×10^{-10}	0.174×10^{-10}
160	0.981×10^{-11}	0.997×10^{-11}	0.999×10^{-11}
180	0.601×10^{-11}	0.610×10^{-11}	0.612×10^{-11}
200	0.388×10^{-11}	0.393×10^{-11}	0.395×10^{-11}
220	0.261×10^{-11}	0.264×10^{-11}	0.266×10^{-11}
240	0.182×10^{-11}	0.184×10^{-11}	0.185×10^{-11}
260	0.130×10^{-11}	0.132×10^{-11}	0.133×10^{-11}
280	0.960×10^{-12}	0.966×10^{-12}	0.976×10^{-12}
300	0.721×10^{-12}	0.725×10^{-12}	0.733×10^{-12}
320	0.552×10^{-12}	0.555×10^{-12}	0.561×10^{-12}
340	0.429×10^{-12}	0.431×10^{-12}	0.436×10^{-12}
360	0.340×10^{-12}	0.340×10^{-12}	0.344×10^{-12}
380	0.271×10^{-12}	0.272×10^{-12}	0.275×10^{-12}
400	0.219×10^{-12}	0.220×10^{-12}	0.222×10^{-12}
420	0.179×10^{-12}	0.179×10^{-12}	0.182×10^{-12}
440	0.148×10^{-12}	0.148×10^{-12}	0.150×10^{-12}
460	0.123×10^{-12}	0.123×10^{-12}	0.125×10^{-12}
480	0.103×10^{-12}	0.103×10^{-12}	0.105×10^{-12}
500	0.869×10^{-13}	0.871×10^{-13}	0.883×10^{-13}
520	0.739×10^{-13}	0.740×10^{-13}	0.750×10^{-13}
540	0.632×10^{-13}	0.633×10^{-13}	0.642×10^{-13}
560	0.544×10^{-13}	0.545×10^{-13}	0.552×10^{-13}
580	0.471×10^{-13}	0.471×10^{-13}	0.478×10^{-13}
600	0.409×10^{-13}	0.410×10^{-13}	0.415×10^{-13}

where P is the dimensionless group $(A/6\pi P_\lambda)(1/2\xi_o)^3$. Second, for a given $h_c/2\xi_o$ and $(A/6\pi P_\lambda)(1/2\xi_o)^3$, eq 37 may be solved for the single unknown x, and then eq 34 yields

$$\frac{V_c(2\xi_o)^4}{\alpha} = \frac{x^4\beta\psi}{(h_c/2\xi_o)^4} = V_e \tag{40}$$

A typical range for the dimensionless parameter P is $10^5 - 10^9$ and that for $h_c/2\xi_o$ is 20–600. Table 1 summarizes the general correlation $h_c/2\xi_o = f(V_c(2\xi_o)^4/\alpha, P)$ for the feasible range of the parameter values. The influence of the dimensionless group P is insignificant (it changes only the value of x somewhat but not that of $h_c/2\xi_o$ or V_e), and, in fact, extensive computer simulations revealed this to be the case for all $P \in (10^2, 10^{14})$! Further, fitting the data of Table 1 on a log—log scale gave the correlation

$$\left(\frac{h_c}{2\xi_o}\right)^n\left[\frac{V_c(2\xi_o)^4}{\alpha}\right]=C \tag{41a}$$

where the parameters n and C are given, respectively, by

$$n=\frac{4.37286}{\left(h_c/2\xi_o\right)^{0.0088787}} \tag{41b}$$

$$C=\frac{0.03907}{\left(h_c/2\xi_o\right)^{0.18067}} \tag{41c}$$

The above correlation fits the numerical simulations (Table 1) with an accuracy of more than 99%. The exponent n ranges from 4.258 to 4.13 as $h_c/2\xi_o$ increases from 20 to 600, and the value of C is between 0.0227 and 0.0123 for the same interval of $h_c/2\xi_o$. Averaging the exponent n and the quantity C with respect to $h_c/2\xi_o$ $(20<h_c/2\xi_o<600)$, eq 41 becomes

$$h_c^{4.164}V_c=0.015\alpha\left(2\xi_o\right)^{0.164} \tag{42}$$

which fits the numerical simulations with an average accuracy of more than 97%. This excellent fit results because of a compensating effect of changes in both n and C as $h_c/2\xi_o$ is varied. Equation 41 or 42 predicts the critical thickness for a given velocity of thinning at the critical thickness V_c, the amplitude of the disturbances ξ_o, and the dimensionless group $\alpha=(A^2/24\pi^2\mu\sigma)$. It is to be emphasized that eq 42 is a theoretical correlation between h_c and V_c and that one of these quantities is to be determined experimentally in order to predict the other. It may be noted that the prediction of the critical thickness requires only the knowledge of the velocity of thinning at any given thickness, h_o, because V_c in eq 41 may perhaps be replaced by

$$V_c=V_o\frac{\left(h_c^3+A/6\pi P_\lambda\right)}{\left(h_o^3+A/6\pi P_\lambda\right)} \tag{43}$$

where V_o is the velocity of thinning at a given thickness, h_o. Equation 43 in conjunction with eq 41 or 42 may then again be used to evaluate the critical thickness. We now compare our results with the experimental data of Radoev et al.[5]

IV. COMPARISON AND DISCUSSION

As pointed out earlier, Radoev et al.[5] employed an oscillating fiber optic probe to measure the smallest critical thickness of rupture and the velocity of thinning at the critical thickness. Their experimental data still overestimate, however, the smallest critical thickness somewhat, because the probe may hit the smallest thickness only by chance.[5] This error in overestimating the critical thickness should increase as the radius of the film increases, and, therefore, the amplitude of the hydrodynamic nonhomogeneities also increases. We found that the correlation (where ϵ has the units of angstroms and R is measured in centimeters)

$$2\epsilon=\left(797R^{0.25}-209\right) \tag{44}$$

fits very well the observed variation of amplitude with the film radius (Figure 3 of Radoev et al.[5]). Thus, the films with radii less than 5×10^{-3} cm may be considered plane-parallel, whereas those with radii greater than 10^{-2} cm are considerably uneven in thickness with $2\epsilon > 40$ Å. In view of this, the data for the films with radii less than 10^{-2} cm appear to be the most reliable, and the best fit of this data (from Figures 4 and 5 of Radoev et al.[5]) is shown in Figure 2 as circles. The data for films with radii in the range 10^{-2}–10^{-1} cm are depicted as triangles in the same figure. The amplitude of the hydrodynamic nonhomogeneity for the film of radius 10^{-1} cm is as large as 240 Å. The theoretical dependence was calculated from eq 34 and 37, or it may be calculated also on the basis of the correlations (41) or (42). The Hamaker constant for this system was reported by Radoev et al.[5] and was calculated on the basis of Lifshitz theory.[24] While the theory provides upper and lower bounds for the Hamaker constant at any given thickness, we used the *average* Hamaker constant that is given by

$$\left(A/6\pi\right)=\left(1.475-1.8\times10^{5}h\right)\times10^{-14}\,\text{erg} \tag{45}$$

where h is the thickness of the film in centimeters. The viscosity and the surface tension for this system are reported as 0.89 cP and 34.5 dyn/cm, respectively.[5] Radoev et al.[5] take the amplitude of the thermal perturbations to be

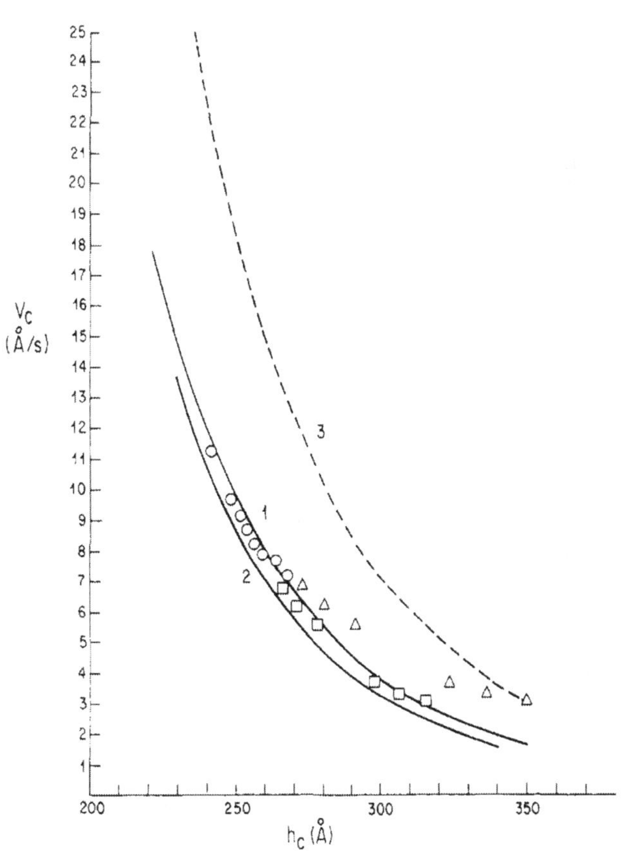

FIGURE 2 Relation between the critical thickness and the critical velocity of film thinning. The experimental data[5] for all film radii less than 10^{-2} cm are depicted as circles, whereas the less reliable data for film radii greater than 10^{-2} cm are shown by triangles. The squares correspond to the data depicted by triangles after the experimental overestimation of the smallest critical thickness has been accounted for. Curves 1 and 2 are the predictions of the present theory for $\xi_o = 1.7$ Å and ξ_o as given by eq 47, respectively. Curve 3 is obtained if the stabilizing effects of drainage on the growth of perturbations are neglected.

$$\langle \xi_o^{\ 2} \rangle = (k_B T / \sigma) \tag{46}$$

whereas Ivanov and Dimitrov[3] derive for it the expression

$$\langle \xi_o^{\ 2} \rangle = k_B T / \pi \sigma k_m^{\ 2} R^2 J_1^{\ 2}(k_m R) \sim \frac{k_B T}{\sigma} \frac{1}{k_m R} \text{ for large } k_m R \tag{47}$$

where k_B is the Boltzmann constant, T is the absolute temperature, $k_m^{\ 2}$ is given by eq 32, and J_1 is the Bessel function of order 1. Equation 47 appears to be more accurate inasmuch as it accounts for the dependence of amplitude on the wavelength according to Einstein's theorem[25] and is also supported by other investigations.[1,26] In any event, the results of the present theory are not very sensitive to the exact value of the thermal amplitude (eq 42), and curve 1 in Figure 2 depicts the theoretical calculations for $\xi_o = 1.7$ Å, which is about the mean of what is predicted by eq 46 and 47. Curve 2 of Figure 2 shows the results of the present theory for the amplitudes calculated from eq 47. The data for films smaller than 10^{-2} cm show excellent agreement with the theory. As pointed out earlier, the data for film radii greater than 10^{-2} cm begin to systematically overestimate the actual smallest critical thickness. However, it is reasonable to assume that the degree of overestimation is directly proportional to the amplitude of the hydrodynamic nonhomogeneity, 2ϵ. Further, if it is assumed that the data for the film of radius 10^{-1} cm are overestimated by 35 Å, then by use of eq 44 it may be shown that the experimental data overestimate the actual smallest critical thickness by $0.146(797R^{0.25}-209)$ Å for a film of radius R. This correction becomes significant only for films of radii greater than 10^{-2} cm, and the points denoted by squares in Figure 2 correspond to the experimental data depicted by triangles after this correction has been applied. While this estimation is somewhat crude, it does show that a proper interpretation of the experimental data for $R > 10^{-2}$ cm leads to an even better agreement with the theory. Of course, for $R < 10^{-2}$ cm, the experimental data never overestimate the critical thickness by more than 5 Å; hence, the points depicted by the circles are in entire agreement with the theory.

The broken curve (curve 3) in Figure 2 corresponds to the case when the effects of film thinning on the growth of perturbations are neglected, but all other parameters are the same as for curve 1. In this case, the growth coefficient is given by the first term of eq 22 only. The latter approach overestimates the critical velocity of film thinning by about a factor of 2 and also overestimates the critical thickness by about 30 Å for this system. As is shown shortly, the determination of the velocity of thinning is important for calculating the lifetime of a free film. The dominant wavelength, $\lambda_m = (2\pi/k_m)$, is about one- tenth of the film radius for the system considered here. The influence of film thinning on the wave motion is, however, more important for large wavelength perturbations (see section III.2).

The estimates of the time of rupture of a foam film are needed to predict the time of collapse of foams and for the design of foam fractionators.[27] The lifetime of a foam film with initial thickness h_o is readily determined as

$$\Delta t = \int_{h_c}^{h_o} \frac{dh}{V} \tag{48}$$

This may be evaluated to a first approximation by noting that

This may be evaluated to a first approximation by noting that

$$V(h) = \frac{V_c(h^3 + a^3)}{(h_c^{\ 3} + a^3)} \tag{49}$$

where $a^3 = (A/6\pi P_\lambda)$.

Substituting the above expression in eq 48 gives

$$\Delta t = \left(\frac{h_c^3 + a^3}{V_c a^2}\right)\left(\frac{1}{3^{1/2}}\left\{\tan^{-1}\frac{2h_o - a}{a(3^{1/2})} - \tan^{-1}\frac{2h_c - a}{a(3^{1/2})}\right\} + \frac{1}{6}\ln\left[\frac{(h_c^2 - ah_c + a^2)(h_o + a)^2}{(h_o^2 - ah_o + a^2)(h_c + a)^2}\right]\right) \quad (50)$$

The time of rupture can be computed from eq 50 in conjunction with eq 42 if the critical thickness h_c is experimentally determined. Note that in the absence of the theoretical result, eq 42, both the velocity and the critical thickness are to be measured experimentally for obtaining the time of rupture. The entries in the fourth column of Table 2 are the times of rupture as computed from eq 50 and 41 with the data of Radoev et al.[5] for the critical thickness and for an initial thickness of 2000 Å. The critical velocity is then obtained from eq 41 or 42 and the time of rupture from eq 50. Manev et al.[14] have determined the times of rupture of a foam film for a similar system, and their experimental results are also reported in Table 2 (these are also for an initial thickness h_o = 2000 Å). The theoretical predictions are in good agreement with the experimental values, especially in view of the fact that none of the parameters is adjusted, and the conditions for the systems of Manev et al.[14] and Radoev et al.[5], although similar, are not entirely identical. The critical thicknesses used are taken from the latter source because Manev et al.[14] report only the mean critical thickness. In contrast, neglecting the stabilization effect due to the drainage decreases the time of rupture by about a factor of 2, and the comparison with the experimental values becomes poor (Table 2). Finally, as is apparent from entries in the last column of Table 2, if the Reynolds' law of film thinning is used for calculating Δt, the results are highly unrealistic.

In conclusion, the process of film thinning has an important stabilizing influence on the process of rupture, and its neglect leads to a decrease in the time of rupture by about a factor of 2. The use of Reynolds' law of film thinning certainly cannot be justified, as eq 42 gives the following dependence of the critical thickness on the film radius:

TABLE 2

Comparison of Theoretical and Experimental Values of the Time of Rupture (Drainage Time Δt (s)) Using Various Approaches

film radius $10^2 R$, cm	exptl values from ref 14 *a*	exptl values from ref 14 *b*	predictions of present theory (eq 41 and 50)[c]	calculated by neglecting stabilization due to drainage[d]	calculated by Reynold's law of film thinning[e]
0.5	19 ± 1	31 ± 1	21	11	>40
1.0	29 ± 1	45 • 1	35	18	>120
2.0	43 ± 1	63 ± 2	46	24	>380
5.0	67 ± 3	105 ± 3	73	38	>1500

[a] 4.3×10^{-4} mol/L SDS, +0.1 mol/L NaCl, $P_\lambda = 490$ dyn/cm.
[b] 3.5×10^{-3} mol/L SDS, +0.25 mol/L NaCl, $P_\lambda = 380$ dyn/cm.
[c] 10^{-3} mol/L SDS, +0.1 mol/L NaCl, $P_\lambda \sim 400$ dyn/cm.
[d] 10^{-3} mol/L SDS, +0.1 mol/L NaCl, $P_\lambda \sim 400$ dyn/cm.
[e] 4.3×10^{-4} mol/L SDS, +0.1 mol/L NaCl, $P_\lambda = 490$ dyn/cm.

$$h_c^{4.164} \propto \frac{1}{V_c} \propto \frac{R^m}{P_\lambda\left(h_c^3 + a^3\right)} \tag{51}$$

where it is assumed that the velocity of thinning is inversely proportional to the *m*th power of the film radius. According to the Reynolds' law of thinning, $m = 2$ and, therefore, the slope of h_c vs. R on a log—log plot varies between 0.28 and 0.48. The experimentally observed slope is between 0.11 and 0.14 (Figure 5 of Radoev et al.[5]). However, the experimentally observed value of the exponent m is much smaller than 2 and is about 0.8.[13,14] It is in view of this that the present theory predicts the slope of h_c vs. R on a log—log plot to be 0.11 when $P_\lambda \gg A/6\pi h_c^3$ and about 0.19 when $P_\lambda \ll A/6\pi h_c^3$.

The theory proposed by Radoev et al.[5] leads to the following counterpart of eq 41 or 42 (eq 18 of Radoev et al. in the present notations):

$$\left(h_c^5 V_c\right)_1 = \frac{3\left(3^{1/2}\right)}{8}\xi_o\alpha \tag{52}$$

where the subscript 1 denotes the quantity as evaluated by the theory of Radoev et al.[5] A comparison of eq 42 and 52 suggests that the latter shows a rather strong dependence of the critical thickness (or the critical velocity) on the amplitude ξ_o of the thermal perturbations. For a given critical velocity, the ratio of the critical thicknesses as predicted from eq 52 and 42 is

$$h_{c1}/h_c = 2.08\left(\xi_o/h_c\right)^{0.1672} \tag{53}$$

The reason that eq 52 appeared to fit the experimental data is that a rather large value of ξ_o (about 3.4 Å from eq 46) was employed in the calculations.[5] It is readily verified that for $\xi_o = 3.4$ Å and h_c in the range 242–350 Å, the ratio h_{c1}/h_c is close to 1, or in other words, both eq 42 and 52 coincidently predict the same critical thickness. However, eq 47 shows that the amplitude of the dominant thermal fluctuations varies from about 0.9 to 0.3 Å as the critical thickness varies from 242 to 340 Å. For these values of the thermal amplitudes, the ratio h_{c1}/h_c is between 0.6 and 0.8, and, hence, eq 52 predicts critical thicknesses that are not borne out by the experiments. Further, for a given critical thickness, eq 52 predicts critical velocities that ate up to 1 order of magnitude smaller than those given by the present theory. In general, the discrepancy between the present theory and eq 52 increases as the ratio ξ_o/h_c deviates from a value of 0.0125, which usually happens as the critical thickness increases and ξ_o decreases.

In contrast, the present theory shows a much weaker dependence on the amplitude of the thermal perturbations, which is apparent from curves 1 and 2 of Figure 2.

Finally, the present theory also confirms indirectly the dependence

$$V = \frac{V_o}{\mu R^{0.8}} h^3\left(P_\lambda + \frac{A}{6\pi h^3}\right) \tag{54}$$

This dependence on R was arrived at experimentally for the time taken for the thinning of a free film.[13,14] Here, the best fit for the parameter V_o may be evaluated by knowing only a single value of the velocity for any thickness and film radius. That the functional form eq 54 is consistent with the predictions of the present theory may be seen by evaluating V_o from the data of Radoev et al.[5] at $P_\lambda = 400$ dyn/cm^2, $h_c = 242$ Å, and $V = 11.2$ Å/s. This occurs for a film of radius $R = 5 \times 10^{-3}$ cm, and this point of the data is the most reliable due to the fact that $2\epsilon \rightarrow 0$ here. Thus, the velocity V is given by

$$V = \frac{1.032\times10^5}{R^{0.8}} h^3 \left(P_\lambda + \frac{A}{6\pi h^3} \right) \tag{55}$$

where all of the variables are given in the cgs system of units. Combining eq 42 and 55 gives the following equation for the critical thickness

$$h_c^{7.164}\left(P_\lambda + \frac{A}{6\pi h_c^3} \right) = \left(7.725\times10^{-8}\right)\left(\xi_o\right)^{0.164} R^{0.8}\left(A^2/\sigma\right) \tag{56}$$

where the Hamaker constant $A = A(h_c)$ is given by eq 45 and ξ_o is determined from eq 47. Equation 56 predicts the critical thickness for radii smaller than 10^{-2} cm with an accuracy of 98%. As explained earlier, the data for $R > 10^{-2}$ cm are always overestimated by the experiments. Thus, the theory presented here for the critical thickness lends additional support to the veracity of the functional form eq 54 as a combination of eq 54 and 42 leads to critical thicknesses that are in excellent agreement with the experimental data.

V. CONCLUSIONS

A theory for predicting the critical thickness of tangentially immobile foam and emulsion films is presented that takes into account the effects of drainage on the growth of interfacial corrugations. The stabilization of the film by the process of film thinning increases as the wavelength of the perturbation increases. If the disturbance wavelength is greater than the film diameter, the perturbation remains stable down to zero thickness, regardless of the magnitude of the Hamaker constant. In contrast to many earlier investigations, the proposed theory does not incorporate Reynolds' law of film thinning, as the Reynolds' law grossly underestimates the experimental velocities of film thinning. The present theory, in con junction with the experimentally determined velocity of thinning at the critical thickness, predicts the critical thickness of foam and emulsion films. An application of the semiempirical eq 54 in conjunction with eq 42 predicts the experimentally observed critical thicknesses with a good accuracy, and, therefore, the dependence on film radius as given by eq 54 appears to be sound. The theoretical predictions fit the experimental data of Radoev et al.[5] with an average accuracy of 99%. This is remarkable in view of the simplicity of the proposed theory (eq 41 and 45) and the fact that only the theoretical estimates of the Hamaker constant and the amplitude of the thermal perturbations are used. The present theory in conjunction with the data for critical thickness[5] predicts lifetimes of foam films that are in good agreement with the experiments of Manevet al.[14]

The critical thickness is shown to be a weak function of the amplitude of disturbances, viz., $h_c \propto \xi_o^{\,p}$, where p is in the range 0.023–0.04. Varying ξ_o by a factor of 2 changes the critical thickness by less than a factor of 1.03. Finally, an excellent agreement of the present theory with the experimental data supports the most probable value of the Hamaker constant as obtained from the Lifshitz theory in conjunction with optical data.[24]

Note Added in Proof. Our more recent studies[28,29] have shown that the velocity of thinning (as given by Reynolds' model) is greatly enhanced by the "pumping" action of the observed "hydrodynamic" nonhomogeneities. Instead of employing the experimental velocities of thinning, this new theory of film thinning[28] may be used in conjunction with the theory developed here for predicting the critical thickness[29] and lifetime[29] of foam and emulsion films. This recent theory of film thinning[28] also supports a conclusion given here, namely, that the velocity of film thinning is approximately inversely proportional to the film radius (eq 51, 55) and not to the square of the film radius. While the expression for ψ (defined by eq 38) has a different functional form if this new theory of the film thinning is used, the resulting correlations (41) remain valid due to an extremely weak dependence of the results on $\psi \lesssim O(1)$.

APPENDIX

The total velocities may be represented as

$$\upsilon_z = \upsilon + W \tag{57}$$

$$\upsilon_r = u + U \tag{58}$$

where U and W are the base-case tangential and the normal velocity component for the film thinning and u and υ are the corresponding components for the growth of perturbations. The solution of the base case problem

$$\frac{\partial W}{\partial z} + \frac{1}{r}\frac{\partial}{\partial r}(rU) = 0 \tag{59}$$

$$\frac{\partial p_\mathrm{d}}{\partial r} = \mu \frac{\partial^2 U}{\partial z^2} \tag{60}$$

$$\frac{\partial p_\mathrm{d}}{\partial z} = 0 \tag{61}$$

$$W = \frac{1}{2}\frac{\mathrm{d}h}{\mathrm{d}t}; U = 0 \text{ at } z = \frac{h}{2} \tag{62}$$

is

$$W = -z\left(\frac{z^2}{3} - \frac{h^2}{4}\right)\frac{6}{h^3}\frac{\mathrm{d}h}{\mathrm{d}t} \tag{63}$$

$$U = \frac{3r}{h^3}\left(z^2 - \frac{h^2}{4}\right)\frac{\mathrm{d}h}{\mathrm{d}t} \tag{64}$$

The boundary condition (7), with the help of eq 57, 15, and 16, becomes

$$2\xi_\mathrm{o} p_\xi = -\xi_\mathrm{o}\Delta_\mathrm{r}\xi + 2\mu\frac{\partial \upsilon}{\partial z} + 2\mu(\xi_\mathrm{o}\xi)\frac{\partial^2 W}{\partial z^2} \tag{65}$$

where we have used the expression

$$\frac{\partial W}{\partial z}(H/2) = \frac{\partial W}{\partial z}(h/2) + \frac{\partial^2 W}{\partial z^2}(h/2)\frac{1}{2}(H-h) \tag{66}$$

and the fact that $(\partial W/\partial z)(h/2)$ is zero from eq 63. The last term of eq 65 is the second term of eq 20. Further, $\mathrm{d}\upsilon/\mathrm{d}z$ is determined from eq 58 and 64, the condition for the tangential immobility, and the first-order continuity equation

$$\frac{\partial \upsilon}{\partial z} = -\frac{1}{r}\frac{\partial}{\partial r}(ru) \tag{67}$$

which give

$$\frac{\partial \upsilon}{\partial z}\left(h/2\right)=\frac{1}{r}\frac{\partial}{\partial r}\left(r^2\xi\right)\frac{3}{h^2}\frac{\mathrm{d}h}{\mathrm{d}t} \tag{68}$$

Substitution of the above in eq 65 finally gives the last term of eq 20.

REFERENCES

(1) Vrij, A. Discuss. Faraday Soc. **1966**, 42, 23.
(2) Scheludko, A. Adv. Colloid Interface Sci. **1967**, 1, 391.
(3) Ivanov, I. B.; Dimitrov, D. S. *Colloid Polym. Sci.* **1974**, *252,* 982.
(4) Manev, E.; Scheludko, A.; Exerowa, D. *Colloid Polym. Sci.* **1974**, *252,* 586.
(5) Radoev, B.; Scheludko, A.; Manev, E. *J. Colloid Interface Sci.* **1983**, *95,* 254.
(6) Jain, R. K.; Maldarelli, C.; Ruckenstein, E. *AIChE Symp. Ser.* **1978**, *74,* 120.
(7) Maldarelli, C.; Jain, R. K. *J. Colloid Interface Sci.* **1982**, *90,* 263.
(8) Cross, J. A. In *Surface Contamination;* Mittal, K. L., Ed.; Plenum: New York, **1979**; Vol. 1, p 89.
(9) Babchin, A. J.; Clish, R. L.; Warren, D. *Adv. Colloid Interface Sci.* **1981**, *14,* 251.
(10) Ruckenstein, E.; Dunn, C. S. *Thin Solid Films* **1978**, *51,* 43 (Section 5.2 of this volume).
(11) Sharma, A.; Ruckenstein, E. *Am. J. Optom. Physiol. Opt.* **1985**, *62,* 246.
(12) Sharma, A.; Ruckenstein, E. *J. Colloid Interface Sci.* **1986**, *111,* 8 (Section 5.11 of this volume).
(13) Rao, A. A.; Wasan, D. T.; Manev, E. *Chem. Eng. Commun.* **1982**, *15,* 63.
(14) Manev, E.; Sazdanova, S. V.; Wasan, D. T. *J. Colloid Interface Sci.* **1984**, *97,* 591.
(15) Vrij, A.; Hesselink, F. Th.; Lucassen, J.; van den Tempel, M. *Proc. K. Ned. Akad. Wet., Ser. B: Phys. Sci.* **1970**, *B73,* 124.
(16) Ruckenstein, E.; Jain, R. K. *J. Chem. Soc., Faraday Trans. 2* **1974**, *70,* 132 (Section 5.1 of this volume).
(17) Williams, M. B.; Davis, S. H. *J. Colloid Interface Sci.* **1982**, *90,* 220.
(18) Sharma, A.; Ruckenstein, E. *J. Colloid Interface Sei.* **1986**, *113,* 456 (Section 5.4 of this volume).
(19) Sharma, A.; Ruckenstein, E. *Langmuir* **1986**, *2,* 480 (Section 5.5 of this volume).
(20) Scheludko, A.; Exerowa, D. *Violloidn. Zh.* **1960**, *168,* 24.
(21) Ivanov, I.; Radoev, B.; Manev, E.; Scheludko, A. *Trans. Faraday Soc.* **1970**, *66,* 1262.
(22) Gumerman, R. J.; Homsy, G. M. *Chem. Eng. Commun.* **1975**, *2,* 27.
(23) Malhotra, A. K.; Wasan, D. T. *Chem. Eng. Commun.,* in press.
(24) Vassilieff, C. S.; Ivanov, I. B. *Z. Naturforsch. A: Phys., Phys. Chem., Viosmorphys.* **1976**, *31,* 1584.
(25) Einstein, A. *Ann. Phys.* **1910**, *33,* 1275.
(26) Hajiloo, A.; Slattery, J. C. *J. Colloid Interface Sci.* **1986**. *112,* 325.
(27) Narsimhan, G.; Ruckenstein, E. *Langmuir* **1986**, *2,* 230.
(28) Ruckenstein, E.; Sharma, A. *J. Colloid Interface Sci.,* **1987**, *119,* 1.
(29) Sharma, A.; Ruckenstein, E. *J. Colloid Interface Sci.,* **1987**, *119,* 14.

5.7 Nonlinear Stability and Rupture of Ultrathin Free Films*

Ashutosh Sharma[a], *C. S. Kishore*[a], *S. Salaniwal*[a], *and Eli Ruckenstein*[b]

[a] Department of Chemical Engineering, Indian Institute of Technology at Kanpur, Kanpur 208016, India

[b] Department of Chemical Engineering, State University of New York at Buffalo, Buffalo, New York 14260

(Received 24 August 1994; accepted 24 March 1995)

I. INTRODUCTION

Symmetric thin (<100 nm) films (e.g., foam and emulsion films) are inherently unstable,[1–9] and in the absence of significant double layer repulsion,[10,11] rupture occurs due to the excess negative disjoining (conjoining) pressure engendered by the Lifshitz-van der Waals (LW) interactions.[12–14] Negative disjoining pressure[15] signifies an increased pressure in the film compared to the bulk, and the LW component of pressure increases with a decrease in the film thickness. The flow in the film thus occurs from thinner to the thicker regions, leading to a spontaneous growth of interfacial perturbations. In contrast, films of asymmetric systems (e.g., films on solids)[3–5,16–20] may be perfectly wetting (stable) whenever the effective Hamaker constant is negative, which implies a positive LW component of the disjoining pressure and a positive LW component of the spreading coefficient on the substrate.[19,20] Another important difference between the supported and free films is the presence of two stress free interfaces for the latter, which greatly diminish the viscous resistance to the flow in free films without surfactants.[5] However, surfactants exert stabilizing influences[3,5–7,17,21–23] by engendering the Marangoni flow (surface elasticity) and surface viscosity. Small amounts of surfactants can stabilize the film to the extent of rendering the free interfaces tangentially immobile, in which case the free film behaves much like the film on a solid.[6,7,21,22] Analysis of equations of motion in the thin film (lubrication) approximation shows that both for supported films and for free films with tangentially immobile surfaces, viscous forces dominate and the effects of acceleration (i.e., unsteady and inertial forces) are relatively unimportant.[3–7,16–22] However, for pure free films, inertia and unsteady effects may exert profound influences, as is also demonstrated by studies of thin film drainage based on generalized thin film equations.[5] The purpose of this paper is to assess the role of various nonlinear forces (viscous, inertial, LW, surface tension, and unsteady) in the growth of interfacial instability and rupture of free films devoid of surfactants. While the linear stability analyses of supported films, as well as free films, are quiet extensive, the linear theory is clearly not suitable for our purposes, because inertial effects are inherently nonlinear, and the other significant nonlinearities associated with the Lifshitz-van der Waals, viscous, unsteady, and surface tension forces, are also not retained in the linear theory. In essence, the linear theory assumes the force field to be a constant during all stages of the

* *Phys. Fluids* 7 (8), p, 1832 August 1995. Republished with permission.

film deformation and rupture, whereas in reality, forces experienced by different portions of an uneven film are nonlinear functions of the local film thickness.

For thin films, the growth of instability occurs on a length scale (wavelength) that is very large compared to the film thickness.[3,5] This feature allows considerable simplifications of the governing Navier-Stokes equations and the boundary conditions at the deformable interface. The long wave reduction procedure was pioneered by Benny[24] and by Atherton and Homsy[25] to obtain equations for evolution of interfacial instability in falling films. Beginning with the work of Williams and Davis,[16] several investigations addressed the nonlinear aspects of instability in ultrathin supported films[17–21,26–30] and free films with surfactants, when viscous effects dominate.[6,7,21,22] Recently, Emeux and Davis[9] obtained a pair of coupled equations to describe the nonlinear evolution of instability in free films devoid of surfactants. The inertial and unsteady terms were retained by a higherorder analysis of equations of motion. The resulting set of equations were also subjected to a weakly nonlinear analytical analysis, which correctly identified the qualitative role of LW nonlinearities in accelerating the film rupture. However, it may be anticipated that weakly nonlinear analytical schemes[9,17,21] may not be quantitatively accurate for pure free films because: (a) only the first Fourier mode is retained even though higher order modes also receive energy, especially during the later stages of the nonlinear growth, and (b) the higher-order corrections in amplitude are not easily retained for analytical purposes. This limitation is especially serious for the strong nonlinearity associated with the LW forces, which scales as h^{-4}, where h is the local film thickness. In contrast to rather complete nonlinear results available for supported films,[16,17,20,21,26–30] no extensive computational study of the pure free film is available. The recent study of De Wit *et al.*[23] is concerned largely with the role of the Marangoni flow in prolonging the time of rupture of free films with insoluble surfactants. In contrast, we address the following problems of the nonlinear stability of pure films: the effects of wave number and initial amplitude on film breakup; problem of the nonlinear mode selection; possibility of subcritical instabilities; and possible simplifications of equations of Erneux and Davies.[9] In what follows, we report the numerical solutions of the complete set of equations of evolution for a free film, as well as solutions to simplified equations neglecting the inertial and unsteady effects. A generalization of the simplified weakly nonlinear results of Erneux and Davis[9] is also attempted, which in conjunction with numerical solutions, should help us delineate the influence of different nonlinearities on the film breakup.

II. LONG WAVE EQUATIONS

As shown in Fig. 1, we investigate the evolution of interfacial instability in a free symmetric liquid film bounded by nonviscous media. For symmetric squeezing mode leading to the film rupture, the Lifshitz-van der Waals potential energy (per unit volume) of material at the interface $(z = \pm h)$ is given by

$$\phi' = A'/6\pi\left(2h\right)^3 h_0^3, \tag{1}$$

where A' is the Hamaker constant and h is nondimensional half-film thickness, which is scaled by the total mean film thickness, h_0. The continuum description of the film flow is obtained by introduction of the body forces due to potential [Eq. (1)] in the Navier-Stokes equations,[3,4] which are to be solved in conjunction with the equation of continuity, kinematic condition, normal (pressure jump), and shear stress conditions at the free interface, and conditions of the symmetry at the central plane $(z = 0)$.

The complete set of nonlinear equations may be simplified considerably by noting that whenever the squeezing mode is unstable, the instability evolves on a length scale, λ' that is large compared to the mean film thickness, h_0. The leading order nonlinear equations are then obtained by introducing the nondimensional wave number, $k = \left(2\pi h_0/\lambda'\right) \ll 1$, as a small ordering parameter and by rescaling all variables and parameters of the problem to make them of $\mathbf{0}(1)$. Velocities and pressure are then expanded in power series of k. Detailed derivations for pure free film are given by Erneux and

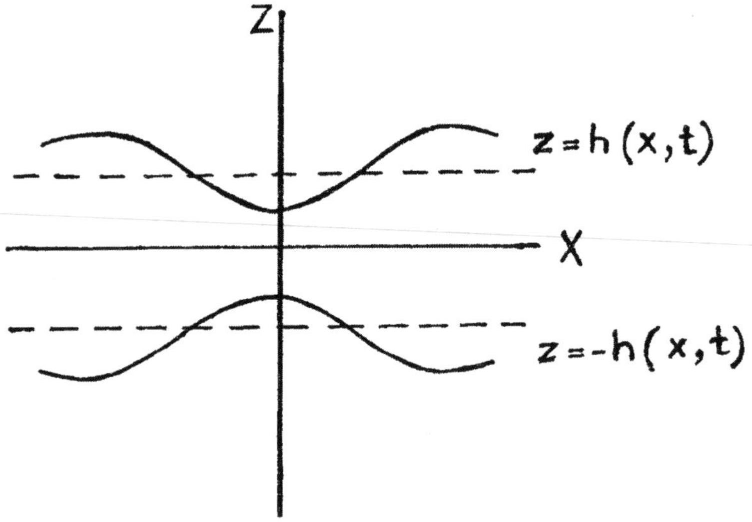

FIG. 1 Schematic of the deforming free film. Dashed lines indicate the mean position of interfaces.

Davis.[9] The long wave analysis for supported film reduces the entire set of equations with moving boundaries to a single, nonlinear equation of evolution for the film thickness,[16,17] since inertia and unsteady effects are not important. However, in the case of the free film, retention of these effects requires different scalings of variables in k,[9] which lead to the following two coupled nonlinear evolution equations,[9] in which length, x and time, t are scaled by h_0, and $\left(h_0^2/v\right)$, respectively:

$$h_t+\left(uh\right)_x=0, \tag{2}$$

$$(u_t+uu_x+\phi_x-3Sh_{xxx})=4\left(u_{xx}+h^{-1}h_xu_x\right), \tag{3}$$

where subscripts denote differentiation, and h and u are nondimensional half-film thickness (scaled by h_0) and longitudinal (x component) velocity (scaled by v/h_0; v is kinematic viscosity). As is clear from Eqs. (2) and (3), the film thickness as well as the leading-order longitudinal velocity are functions of x and t alone; S is a nondimensional surface tension defined as $S=\mathrm{h}_0\sigma/3\rho v^2$, where σ is surface tension and ρ is density. The nondimensional potential, ϕ in Eq. (3) is defined as

$$\phi'=A\left(2h\right)^{-3},\ \ \phi_x=-\left(3A/8\right)h^{-4}h_x; \tag{4}$$

$$A=A'/\pi h_0\rho v^2$$

Equation (2) is a rearranged form of kinematic condition, viscous forces appear on the right-hand side (RHS) of Eq. (3), and the left-hand side (LHS) represents (from left to right) terms due to unsteady, inertia, van der Waals, and surface tension forces, respectively. The last term of Eq. (3) will be referred to as the nonlinear viscous correction. As the film thins locally, the destabilizing LW forces, as well as the stabilizing viscous forces grow stronger in Eq. (3).

The evolution of the film thickness is described by Eqn. (2)–(4) together with the periodic boundary conditions over the periodicity interval $x\in\left[0,\lambda\right]$, and space periodic initial conditions of the form

$$h\left(0,x\right)=\left(1/2\right)+\epsilon\,h_i\left(kx\right), \tag{5a}$$

$$u\left(0,x\right)=u_i\left(kx\right), \tag{5b}$$

where ϵ is nondimensional amplitude of initial disturbance scaled with $h_0, k(=2\pi/\lambda)$ is the wave number, and h_i, u_i, are periodic functions with wavelength λ. Before reporting the full numerical solutions, we examine some of the asymptotic simplifications of Eqs. (2)–(4) which are offered by the small-ordering parameter $\epsilon < 0.5$.

III. SIMPLIFIED MODELS

A. Weakly Nonlinear Theory

In order to examine the physics and semiquantitative aspects of film rupture, we first analyze the weakly nonlinear evolution of small amplitude $(\epsilon \ll 1)$ disturbances superimposed on the basic uniform solution [Eqs. (5a) and (5b)]. In the weakly nonlinear regime, we retain only the first Fourier mode consistent with the initial conditions Eq. (5), so that analytical manipulations are possible.

The half-film thickness is thus represented by

$$h(x,t)=(1/2)+\epsilon\left[Z(t)e^{ikx}+Z_c(t)e^{-ikx}\right]+\mathbf{0}\left(\epsilon^2 e^{\pm 2ikx}\right)$$

$$=(1/2)+\epsilon\left(y_1 \cos kx + y_2 \sin kx\right), \tag{6}$$

where Z and Z_c are the complex conjugate amplitudes with real and imaginary parts Z_r and Z_i, respectively, which are related to the real amplitude functions (y_1 and y_2) by $y_1 = 2Z_r$, $y_2 = -2Z_i$. The higher-order modes are assumed to be unexcited. In the instability of supported thin films engendered by LW forces, the primary mode remains dominant and higher modes usually contribute only during the late explosive phase of growth, which is not rate limiting.[6,7,9,16,20,22,26–29] Introduction of Eq. (6) in Eq. (2) yields the leading-order velocity up to $\mathbf{0}(\epsilon^3)$ terms as (primes denote differentiation with respect to t):

$$u(x,t)=(2\epsilon i/k)\Big[\left(Z'+8\epsilon^2 ZZ_cZ'-4\epsilon^2 Z^2Z_c'\right)e^{ikx}$$

$$+\left(-Z_c'-8\epsilon^2 ZZ_cZ_c'-4\epsilon^2 Z_c^2Z'\right)e^{-ikx}$$

$$-2\epsilon ZZ'e^{2ikx}+2\epsilon Z_cZ_c'e^{-2ikx}+2\epsilon\left(ZZ_c'-Z_cZ'\right)\Big]+\mathbf{0}\left(e^{\pm 3ikx}\right). \tag{7}$$

From Eqs. (6) and (7), it is clear that since $h = \mathbf{0}(1)$, spatial derivatives of h are $\mathbf{0}(\epsilon)$, and $u = \mathbf{0}(\epsilon)$, the inertial term (uu_x) and the nonlinear correction to the viscous term $(h^{-1}h_xu_x)$ in Eq. (3) are smaller $\left[\mathbf{0}(\epsilon^2)\right]$ compared to other terms of Eq. (3), all of which are of $\mathbf{0}(\epsilon)$. A coupled set of weakly nonlinear equations for amplitudes [(y_1,y_2) or (Z, Z_c)] are obtained up to terms of ϵ^2 by substitution of Eqs. (6) and (7) in Eq. (3). It turns out that if either the cosine or the sine mode is initially absent [$y_1(0)$ or $y_2(0) = 0$], it remains unexcited for all times during evolution ($y_t' = 0$). Thus without loss of generality, we assume $y_2(0)$ to be zero, and obtain an equation for the amplitude, $y_1 = y$

$$y''+4k^2y'-3Ak^2y\left(1+10\epsilon^2y^2\right)+(3/2)Sk^4y$$

$$+\epsilon^2\left[y^2y''+y\left(y'\right)^2+8k^2y^2y'\right]=0. \tag{8}$$

The above equation is consistent with the initial conditions

$$h(0)=(1/2)+\epsilon\cos kx,\ y(0)=1, \tag{9a}$$

$$u(0) = -(2\epsilon/k)\,y'(0)\sin kx. \tag{9b}$$

The first two terms multiplied by ϵ^2 in Eq. (8) in square braces, represent the nonlinear corrections due to acceleration (unsteady and inertial terms), and the last term is the nonlinear correction from the viscous terms. As is clear from Eq. (8) and from numerical results reported later, all of these nonlinear corrections exert stabilizing influences since y', $y'' > 0$ in Éq. (8). Intermediate steps in the derivation of Eq. (8) showed that the inertial nonlinearities are destabilizing, but the net contribution of acceleration, $u_t + uu_x$, has a stabilizing influence. The nonlinearity associated with the van der Waals force, however, has a more profound destabilizing influence [third term in Eq. (8)].

While the initial value problem posed by Eq. (9) is considerably simpler than the original evolution equations, it is still not amenable to an analytical solution. Equation (8) will be referred to as the weakly nonlinear theory (WNT).

B. Long-time evolution in weakly nonlinear regime

The linearized version of Eq. (8) admits solution of the form $y = e^{\omega t}$, where the growth coefficient ω is given by the characteristic equation[9]

$$\omega^2 + 4k^2\omega + (3/2)k^2\left(Sk^2 - 2A\right) = 0. \tag{10}$$

The film is therefore stable to small amplitude disturbances with wave numbers less than a critical "cutoff" wave number $k_c = (2A/S)^{1/2}$. The fastest growing (dominant) wave of the linear theory has a wave number, k_m obtained from $(\partial\omega/\partial k) = 0$. The variation of the growth coefficient with the wave number is shown in Fig. 2, both with and without the inclusion of the unsteady term, ω^2 of Eq. (10). While the inclusion of the unsteady term is necessary for existence of the dominant wave, its influence on ω is rather unimportant for waves with $k > k_m$. If we neglect the unsteady term, ω^2 in Eq. (10) for $k_c \geqslant k > k_m$, the simplified growth coefficient, ω_0 is same as that obtained by Ruckenstein and Jain[3]

$$\omega_0 = (3/8)\left(2A - Sk^2\right). \tag{11}$$

Further, in the linear approximation, the time of rupture is given by (from $h = 0$ or $y = 1/2\epsilon$)

$$t_L = \left[1/\omega(k)\right]\ln(1/2\epsilon). \tag{12}$$

Clearly, rupture occurs on a long-time scale such that $t_L\omega \sim \mathbf{0}(1)$, where ω is a small parameter (Fig. 2). Thus the long term evolution may be described by introducing a small parameter $\mu > \omega$ and a rescaled time, $\eta = \mu t \sim \mathbf{0}(1)$. The evolution after a time $t \sim \mathbf{0}(1/\mu)$ is therefore described by a simpler equation, since terms containing y'' and $(y')^2$ become small $\left[\mathbf{0}\left(\mu^2\right)\right]$;

$$y'\left(1 + 2\epsilon^2 y^2\right) = \omega_0 y + (30/4)A\epsilon^2 y^3. \tag{13}$$

The above equation can be integrated analytically with $y(0) = 1$, to give

$$t = \frac{1}{2\omega_0}\ln\left[\frac{y^2\left(1 + a^2\right)}{y^2 + a^2}\right] + \left(\frac{4}{30A}\right)\ln\left[\frac{y^2 + a^2}{a^2}\right], \tag{14a}$$

where

$$a^2 = 4\omega_0/30\epsilon^2 A. \tag{14b}$$

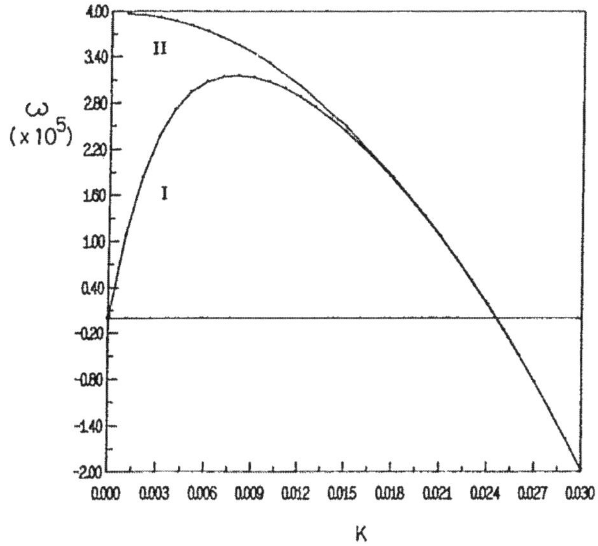

FIG. 2 Variation of the growth coefficient from linear theories $\left(A=5.3\times10^{-5}, S=0.176, k_c=0.0245, k_m=0.008\right)$. Curves labeled I and II are from Eqs. (10) and (11), respectively. Neglect of unsteady effects (curve II) introduces an error of <15% for $k>k_m$.

The above result corresponds to the weakly nonlinear regime of Eq. (3) when the acceleration is neglected and the dominant wave number does not exist. Equation (14a) is, therefore, not expected to be good for $k<k_m$. Interestingly, the WNT [Eq. (8)], as well as the simplified WNT [Eq. (14)] show the possibility of subcritical instabilities with $k>k_c$, which are engendered by increased LW forces for finite amplitude disturbances. The cutoff wave number, k_{cw} of the WNT (or simplified WNT) is obtained by setting $y' = y'' = 0$ and $y(0) = 1$ in Eq. (8) or Eq. (13);

$$k_{cw}^2 = 2A\left(1+10\epsilon^2\right)/S > k_c^2. \tag{14c}$$

Thus linearly stable (subcritical) finite amplitude disturbances with $k_c<k<k_{cw}$ should also lead to the instability and film rupture. This prediction of the WNT is tested later with direct numerical solutions of the full model.

The result [Eq. (14a)] without the last nonlinear viscous term, and a slightly different numerical coefficient in definition of a^2 was also obtained by Erneux and Davis.[9] These authors, however, used a different asymptotic technique which is valid for very small deviations from the critical condition, for $k^2/k_c^2 = 1-\mathbf{0}\left(\epsilon^2\right)$. As is shown later in the Results and Discussion section, however, $k \geq k_c$ is a truly nonlinear regime, where mode interactions cannot be neglected, and results based on a single mode [Eqs. (8) and (14)] are largely inapplicable. There is, however, a possibility that the simplified weakly nonlinear theory, Eq. (14a) may work for regions close to the most important case of the dominant wave number. This possibility exists because Eqs. (13) and (14a) underestimate destabilization due to the LW term $(h^{-4}h_x)$. At the same time, stabilization due to the acceleration is neglected altogether, which becomes important in the vicinity of dominant wave numbers (Fig. 2). Whether a fortuitous compensation of errors does indeed occur is tested based on calculations reported later.

An analytical estimate for the time of rupture from Eqs. (14a) and (14b) is obtained at $y_m = 1/2\epsilon$, i.e.,

$$t_R = t; \quad \text{at } y_{\max} = \left(1/2\epsilon\right). \tag{15}$$

The discrepancy between the linear theory [Eq. (12) with $\omega = \omega_0$] and predictions of Eq. (14) decreases as the parameter a^2 increases, and formally, the linear theory results are recovered for $a \to \infty$, but $y_m^2 = 1/4\epsilon^2$ remaining bounded. Clearly, this condition can never be exactly satisfied even for $\epsilon \to 0$, and the rupture is always accelerated during the later stages of growth due to greatly enhanced LW forces, even if initially ϵ is small. This expectation is later confirmed by numerical simulations even for small initial dimensional amplitudes characteristic of thermal perturbations $(\sim 1\ \AA)$. In general, Eq. (14a) predicts that y grows faster than the exponential growth of the linear theory, and in particular, an èxplosive phase of growth commences before $t \sim (1/2\omega_0)(1+a^2)$. This is easily concluded by neglecting the second term of Eq. (14a) and solving it for y, which gives

$$y = a^2 \exp(\omega_0 t)/\left[1 + a^2 - \exp(2\omega_0 t)\right]^{1/2}.$$

As was identified earlier,[9] the "blowoff" occurs when the denominator becomes small. The time of rupture from Eqs. (14) and (15) nearly equals, the blowoff time, $t = (1/2\omega_0)\ln(1+a^2)$, whenever $y^2_{\max} \gg a^2$ This approximation is, however, not good for large ω_0 $(k \sim k_m)$, and the time of rupture should be obtained directly from Eqs. (14a) and (15), which will be referred to as the simplified weakly nonlinear theory (SWNT).

In summary, while the weakly nonlinear results [Eqs. (8) and (14)] do capture the essential physics of the problem, they may not be quantitatively accurate, in that even the more comprehensive equation (8) neglects mode interactions, and nonlinearities are retained up to order of ϵ^2. It does not do full justice to the very strong nonlinearity of LW forces which grows stronger as $\sim h^{-4}h_x$ [Eq. (4)], at locally thin spots. In addition, further simplifications leading to Eq. (14) neglect all contributions of the acceleration as well. In order to systematically quantify the role of various factors in Eqs. (2) and (3), we now report numerical solutions of the following three models, and compare them with much simpler weakly nonlinear models:

1. The general model (GM) consisting of the full set, Eqs. (2) to (4).
2. A simplified model (SM I) in which $\mathbf{0}(\epsilon^2)$. terms of Eq. (3) $-uu_x$ (inertia) and $h^{-1}h_x u_x$ (nonlinear viscous correction) are neglected. The linear theory results of SM I are identical to that for GM [Eq. (10)].
3. A model (SM II) further simplified from SM I by omitting the unsteady term $-u_t$, as is done in the weakly nonlinear analysis which is valid for the long time evolution [Eq. (14)]. In addition, a $\mathbf{0}(\epsilon^2)$ order correction $-uh_x$ in the kinematic condition [Eq. (2)] was also found to be small and neglected. The set of equations are the same as those derived by Sharma and Ruckenstein[6,21] as an asymptotic case of the more general problem of free film with surfactants. The linear results of SM II are given by Eq. (11) and the weakly nonlinear results by Eq. (14).

IV. NUMERICAL METHOD

Equations (2) and (3) are to be considered with periodic boundary conditions in the interval $x \in [\mathbf{0}, \lambda]$ and the set of initial conditions given by Eqs. (9a) and (9b). The effect of initial condition for u [Eq. (9b)] was explored by employing two different initial conditions: $(a)\, y'(0) = \omega$, which is consistent with the linear theory, and (b) $y'(0)=0$.

A pseudospectral method—the Fourier collocation,[31] which has been used previously for a variety of thin films,[20,29] was employed for the numerical solutions. Briefly, the periodicity interval is normalized to $(0, 2\pi)$ by the scaling $\xi = kx$, and Eqs. (2) and (3) are then exactly satisfied at N (even integer) number of Fourier collocation grid points, $\xi_j = 2\pi(j-1)/N, j = 1, \ldots N$. Fourier collocation differentiation is represented by a $N \times N$ skew-symmetric matrix, $\mathbf{D}$ with components[31]

$$D_{ij} = (1/2)(-1)^{i+j}\cot\left[(i-j)\pi/N\right],\ i \neq j.$$

$$= 0,\ i = j.$$

The higher-order spatial derivative matrices $\mathbf{D}^2$ and $\mathbf{D}^3$ are simply obtained by appropriate number of multiplications of the matrix **D**. The requirement that Eqs. (2) and (3) be satisfied at each of the collocation points, gives a coupled set of $2N$ ordinary differential equations written in the scaled time, $\theta = kt$:

$$\frac{d\mathbf{H}}{d\theta} + H \boxtimes \mathbf{DU} + \mathbf{U} \boxtimes \mathbf{DH} = 0, \tag{16a}$$

$$\frac{d\mathbf{U}}{d\theta} + \mathbf{U} \boxtimes \mathbf{DU} - (3A/8)\mathbf{H}_1 \boxtimes \mathbf{DH} - 3Sk^2\mathbf{D}^3\mathbf{H}$$

$$= 4k\left[\mathbf{H}_2 \boxtimes \mathbf{DH} \boxtimes \mathbf{DU} + \mathbf{D}^2\mathbf{U}\right], \tag{16b}$$

where $\boxtimes$ denotes the pointwise product of two vectors, **H** and **U** are the column vectors with jth components equal to hj and u_j, respectively, and the jth components of $\mathbf{H}_1$ and $\mathbf{H}_2$ are h_j^{-4} and h_j^{-1}, respectively. Simplified models (SM I and SM II) were similarly solved by Fourier collocation. In particular, SM II reduces to a single nonlinear equation,

$$h_t + (h/4)\left[(A/8)\left(h^{-3} - 8\right) - 3Sh_{xx}\right] = 0,$$

which results in a considerable saving of computational effort.

Equations were integrated in the nonconservative form Eq. (16), as well as in the conservative form by the use of GEAR algorithm for stiff equations. The time integration was continued until the film ruptured, or the thickness declined below the molecular cutoff ~1 Å. Results from the conservative and nonconservative forms matched to within 2% and became insensitive to the number of grid points for $N \geqslant 64$. The grid convergence could be obtained with smaller number of grids ($N \geqslant 32$) when inertial effects were not dominant ($k \lll k_m$), but larger number of grids were required near k_c (including the case of subcritical instabilities), where large number of modes participated. All of the results reported here are for $N = 64$. The computation time required (on HP-9000 supermini) was comparable for GM and SMI, but at least a factor of 10 smaller for SM II. While simulations were performed[32] for a large number of different feasible combinations of parameters ϵ, k, A, and S, for three models (GM, SM I, and SM II) and for two different initial conditions for u [$y'(0) = 0$ and ω], only a select number of cases are reported here to illustrate prominent features and then underlying physics.

V. RESULTS AND DISCUSSION

For nonretarded LW interactions to be significant, we consider the mean film thickness, $h_0 \sim 10$ nm, the surface tension in the range of 20–70 mN/m², and the Hamaker constant, A' for the free film to be of the order of 10^{-20} J. The most realistic values of nondimensional parameters therefore fall in the range of $A \sim 10^{-5}$ to 10^{-3} and $S \geqslant 10^{-1}$. For $h_0 \sim 10$ nm, a nondimensional amplitude of 0.01 corresponds to the scale of thermal perturbations (~1Å). Some results for relatively thick films ($h_0 \sim 100$ nm, $A \sim 10^{-6}$, $S \sim 1$) are also included for comparison.

A. Nonlinear models: GM, SM I, and SM II

Figures 3(a) and 3(b) depict the evolution of instability for the GM, which shows a superexponential (an almost explosive) nonlinear growth during the last stages of rupture. The explosive phase of the growth, fueled by the nonlinear intermolecular interactions, is a characteristic shared by all thin films that are unstable to the extent of rupture.[16,20,22,27–29]

For the GM, the variation of time of rupture with the wave number and initial amplitude are shown in Fig. 4 ($A = 5.3 \times 10^{-5}$, $S = 0.176$, $k_m = 0.008$, $k_c = 0.0245$). Triangles correspond to an initial condition consistent with the linear theory [$y'(o) = \omega$], whereas circles are for vanishing initial longitudinal velocity [$y'(0) = 0$]. Solid lines are drawn to guide the eye.

There is no significant difference between times of rupture for two different initial conditions. This is because the time during which u evolves to become $0(\epsilon)$ is very small compared to the time of rupture. As expected, the time of rupture for small amplitudes, $\epsilon \leqslant 0.02$ displays a minimum near k_m, due to the existence of a dominant mode with wave number k_m for linear waves. The wave number, k_{mn} of the fastest growing wave shifts towards k_c and beyond (for $\epsilon = 0.2$) as the initial amplitude increases.

For supported films also, it is known that the wave number of the dominant nonlinear wave is usually larger than the wave number of the fastest growing wave of the linear theory.[16,17,20,27–29] Table 1 summarizes our findings for the nonlinear mode selection in free films. The nonlinear waves responsible for the fastest rupture become increasingly shorter ($k_{mn} > k_m$) as the initial amplitude is increased. The more efficient penetration of the narrower fronts due to LW forces is a feature common to all sorts of thin films,[16,20,29] which offer increased viscous resistance (and also other repulsive forces, if present) as the local thinning occurs. An inspection of Eq. (8) shows that destabilizing LW forces, $\phi_x \sim h^{-4}h_x$, grow stronger at locally thin regions of large slopes (shorter wavelengths). Results of the WNT also show (qualitatively) that the LW nonlinearities prefer increasingly shorter wavelengths as the amplitude increases. For large initial amplitudes, the increased nonlinear LW forces are experienced during the early, slow rate determining step itself. The dominant nonlinear mode therefore becomes substantially shorter than $\lambda_m = 2\pi/k_m$.

In Fig. 4, nonlinear LW forces, inertia and mode interactions become very important for large amplitudes ($\epsilon > 0.05$) and for $k > k_m$. Subcritical instabilities witnessed for $k > k_c (= 0.0245)$ and $\epsilon > 0.05$ are largely engendered by these factors. While grid convergence was found to be adequate for N=32 when $k \lesssim k_m$, larger number of grids (N=64) were required to properly account for the higher modes when $k \gtrsim k_c$. While the linear theory (as well as WNT) predicts a sharp variation of the time of rupture near k_m (Fig. 2), the variation of t_N with k in Fig. 4 is rather weak, and the minima of t_N are broad. This happens because the nonlinear destabilization due to LW forces and inertia becomes stronger for increasing e and for $k > k_m$. These nonlinearities are neglected in the linear theory. Thus, with increased ϵ and k, increasingly larger number of modes conspire to produce the film breakup on comparable time scales.

The role of inertia is best examined by comparing the results of the SM I (which neglects inertia) to the results of GM. In order to assess the role of nonlinearities, the ratio (t_n/t_L), of the time of rupture from the nonlinear and linear theories is shown in Fig. 5. Solid lines are for the GM, whereas broken lines are for the SM I. The time of rupture computed from SM I is always larger than predictions of the GM because destabilization due to the inertia is neglected in SM I. While the stabilization due to nonlinear viscous correction is also neglected, its influence is evidently less than that of inertia. Comparison of the GM and the SM I in Fig. 5 reveals that the destabilization due to inertia is most significant for $k \gtrsim 0.012$, where higher modes also become important.

However, it appears that the inertia and the nonlinear viscous corrections may be neglected, at least for the most interesting case of small amplitude dominant waves with $k \sim k_{mn}$, and also for all amplitudes with $ks \lesssim k_m$.

Due to a strong nonlinearity of the LW forces ($h^{-4}h_x$ term) and the inertia, the linear theory always overestimates the actual time of rupture, and the discrepancy increases with increased initial amplitude and wave number. The ratio approaches zero as $k \rightarrow k_c$, because $t_L \rightarrow \infty$, but t_N remains bounded for $k \rightarrow k_c$. As is clear from Eq. (3) or (8) the nonlinear destabilization due to LW nonlinearities does not vanish as $k \rightarrow k_c$. As shown earlier in Fig. 4, subcritical instabilities with $k > k_c$ also occur.

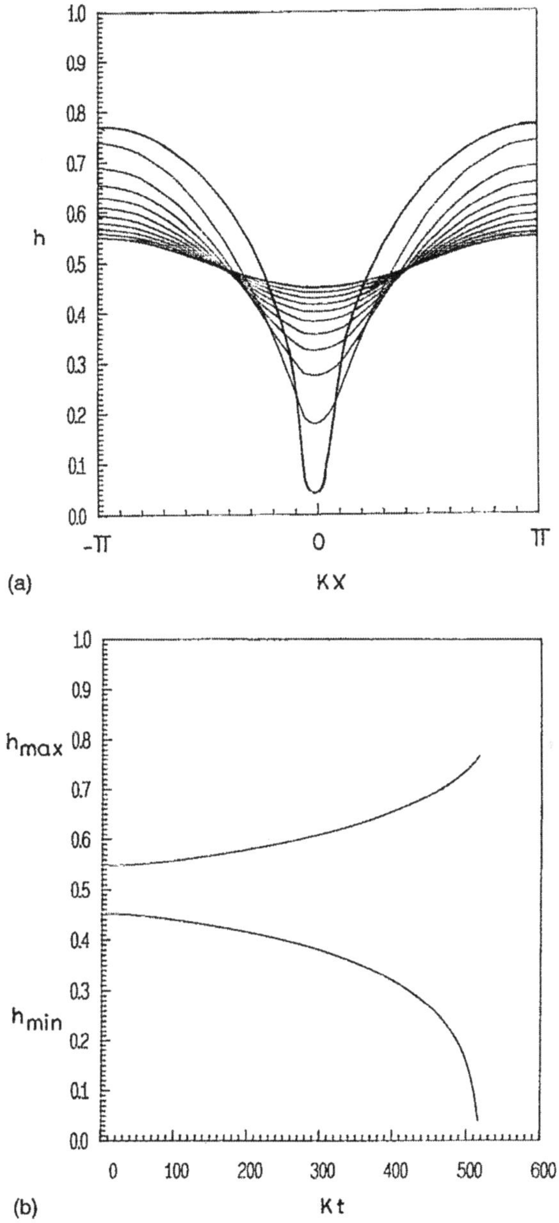

FIG. 3 Evolution of the interfacial instability for $k = 0.01$, $\epsilon = 0.05$, $N = 32$, $A = 5.3\times10-5$, and $S = 0.176$. (a) Film profiles at different times (kt = 0, 100, 150, 200, 250, 300, 350, 400, 450, 500, 515) until rupture occurs, (b) evolution of the minimum and maximum wave amplitudes, showing explosive growth for $kt > 300$.

Results for other feasible values of A and S (not shown) are qualitatively similar to Figs. 4 and 5. It was found[32] that the ratio t_n/t_L declines with increase in A and S (results not shown). This is because the linear theory underestimates the destabilization due to the LW nonlinearities, and overestimates the stabilizing influence of the surface tension. As an example, t_n/t_t for $A = 4.24 \times 10^{-4}$ was found to be 10%-30% less compared to the case of $A = 5.3 \times 10^{-5}$ (Fig. 5). The decline in t_n/t_L was less (about 10%) for small initial amplitude $(\epsilon \sim 0.02)$ and for $k{\sim}k_m$.

A comparison of results of the GM (solid lines) and SM II (broken lines) is shown in Fig. 6 for $A = 4.24 \times 10^{-4}$, $S = 0.121$, $k_m = 0.025$, and $k_c = 0.084$. Since the SM II neglects the strong retardation

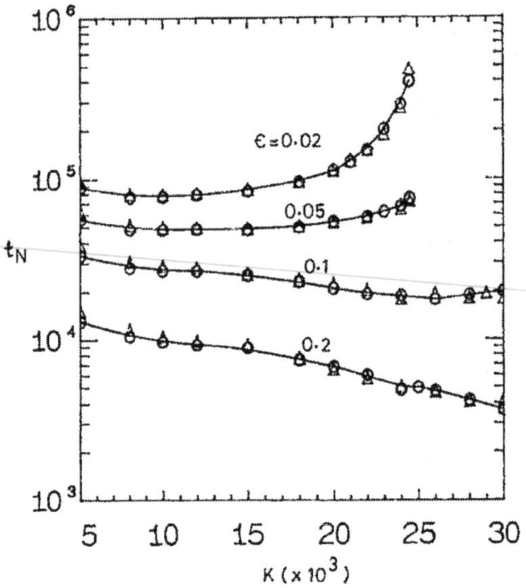

FIG. 4 Variation of the nonlinear time of rupture [Eqs. (2)–(4); GM; A-5.3 × 10^{-5}, S = 0.176]. Circles are for $y'(0=0)$, and triangles are for $y'(0)=\omega$. Data represented by circles are joined using cubic splines to guide the eye. The minimum time of rupture occurs for the dominant waves with $k > k_m$,(=0.008). For large amplitudes $(\epsilon = 0.1, 0.2)$, the subcritical instabilities ($k > k_c$,=0.0245) become very pronounced.

TABLE 1
Wave numbers of the dominant nonlinear waves.

Parameters	ϵ	k_m (nonlinear)
$A = 5.3 \times 10^{-5}$	0.02	0.010
$S = 0.176$	0.05	0.015
k_m (linear) = 0.008	0.10	0.026
$A = 4.24 \times 10^{-4}$	0.02	0.031
$S = 0.121$	0.05	0.055
k_m (linear) = 0.025	0.10	>0.08

due to the unsteady term, it underestimates the time of rupture for $k \lesssim 0.05$, where unsteady effects are important. For $k \gtrsim 0.05$ and $\epsilon \geqslant 0.05$, inertia becomes very important and its neglect in the SM II leads to the overprediction of the time of rupture. In any case, SM II is not likely to be a good approximation for $k<k_m$, since even the linear results of the GM and SM II diverge in this region (Fig. 2). An apparently surprising finding from Fig. 6 is that for $\epsilon \leqslant 0.05$ the error incurred by SM II is minimum for wave numbers in the vicinity of the dominant wave numbers of the nonlinear theory (Table 1). This is because the SM II neglects the stabilization due to unsteady term, and also the destabilization due to inertia. The unsteady effects are most important for $k<k_m$, but inertia becomes pronounced for $k>k_m$. Thus, in intermediate region close to k_{mn} (Table 1), both of these effects are relatively small and oppositely acting, leading to cancellation of errors.

In summary, SM I always overestimates, but SM II can overestimate or underestimate the time of rupture. However, errors are the least for the most interesting cases of the small amplitude $(\epsilon \leqslant 0.05)$ dominant waves, and may be considered tolerable in view of uncertainties in determination of ϵ and A.

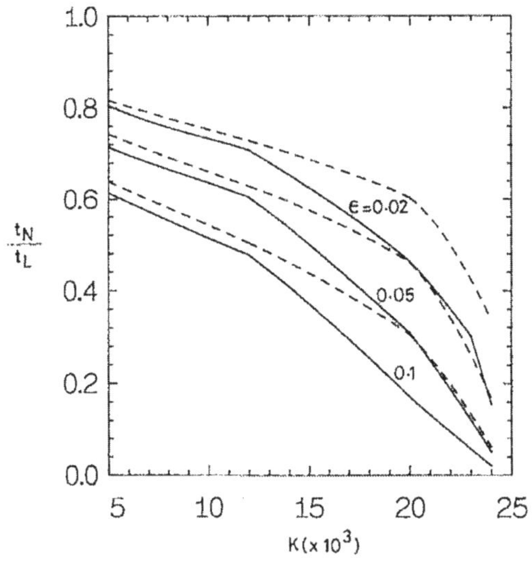

FIG. 5 Ratio of film lifetimes from the nonlinear and linear theories, respectively [$A = 5.3 \times 10^{-5}$, $S = 0.176$, $y'(0) = 0$] Solid lines and broken lines represent nonlinear results of the GM and SM I, respectively.

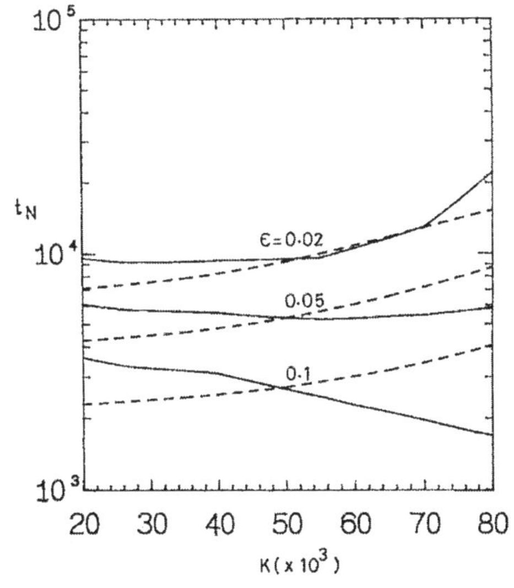

FIG. 6 Comparison of the film lifetimes obtained from the GM with initial condition $y'(0) = 0$ (solid lines) and SM II (broken lines). $A = 4.24 \times 10^{-4}$, $S = 0.121$, $k_c = 0.084$, $k_c = 0.025$.

B. Weakly Nonlinear Theories

We may now address the question of how weakly nonlinear theories (WNT), Eq. (8) and SWNT, Eqs. (14)–(15) fare quantitatively.

Figures 7 and 8 show the comparison of the linear (t_L) and fully nonlinear (t_n) models with the weakly nonlinear models [t(WNT) and t(SWNT)]. The weakly nonlinear model (WNT) generally overpredicts, which is largely due to its underestimation of the strongly destabilizing LW nonlinearity which scales as h^{-4}. As discussed earlier, for large amplitudes ($\epsilon = 0.1$ in Figs. 7 and 8) and for $k \sim k_c$, inertia and mode interactions also become very significant and the WNT becomes completely inadequate. As far as the problem of the nonlinear mode selection is concerned, WNT does correctly predict an increase in the wave number of the dominant mode comparable to results of GM for small

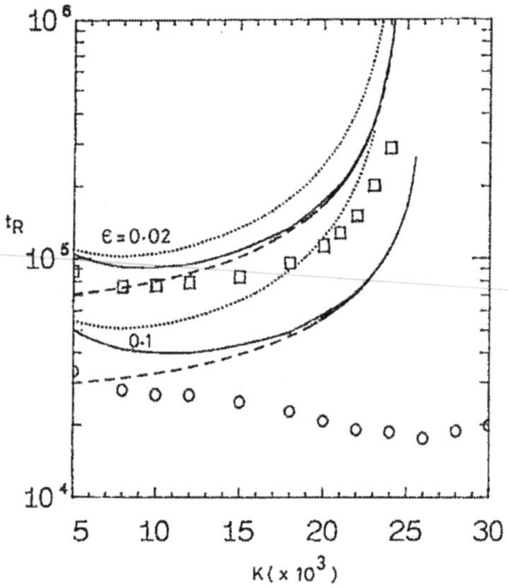

FIG. 7 Comparison of the linear, nonlinear (GM) and weakly nonlinear theories in prediction of lifetimes. (···) linear theory (—) weakly nonlinear theory, WNT, with $y'(0) = 0$; (---) simplified weakly nonlinear theory, SWNT; (□) nonlinear theory, GM with e = 0.02 and $y'(0) = 0$; and (O) GM with $\epsilon = 0.1$ and $y'(0) = 0$. $A = 5.3 \times 10^{-5}$, $S = 0.176$.

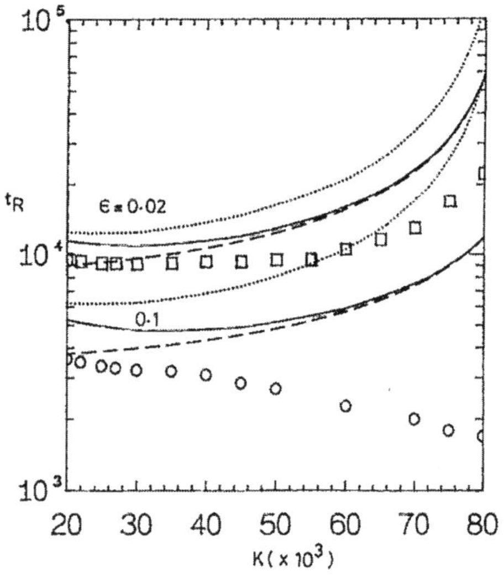

FIG. 8 Same as Fig. 7, but for different set of parameters, $A = 4.24 \times 10^{-4}$, $S = 0.121$. Square and circles represent results of GM with $y'(0) = 0$ and $\epsilon = 0.02$ and 0.1, respectively.

amplitudes (Table 1). Surprisingly, the simplified weakly nonlinear theory [Eqs. (14)–(15)] does an even better job than WNT for dominant waves, $k \sim k_m$. This is largely due to a compensation of errors in SWNT. As in the case of WNT, SWNT also underestimates destabilization due to LW forces and inertia, but at the same time, neglects the stabilizing influence of acceleration altogether. However, SWNT [as well as its linear counterpart, Eq. (11)] cannot address the problem of mode selection, because the dominant wave number does not exist if acceleration is neglected. Thus the theory may be used for quantitative predictions only for $k \geqslant k_m$ and small amplitudes.

Finally, Fig. 9 shows the comparison of the GM with WNT for relatively thick films with weaker LW and stronger surface tension forces ($A = 5.3 \times 10^{-6}$, $S = 1.76$, $k_c = 2454 \times 10^{-3}$, $k_m = 1.16 \times 10^{-3}$).

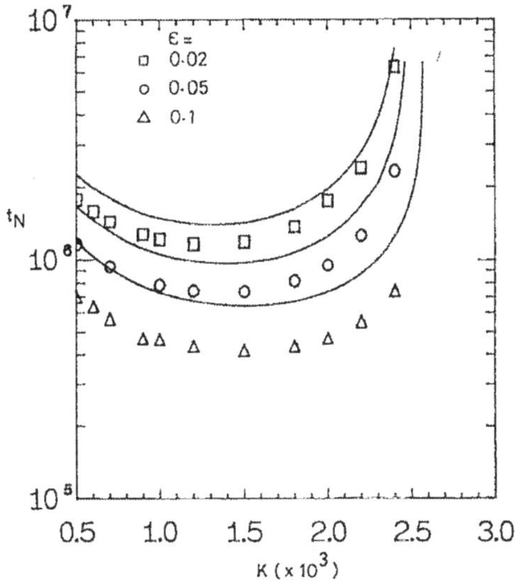

FIG. 9 Comparison of the GM and weakly nonlinear theories for relatively thick films, $A = 5.3 \times 10^{-6}$, $S = 1.76$, $k_c = 2.454 \times 10^{-3}$, and $k_m = 1.16 \times 10^{-3}$. □, O, and Δ, represent the results of the GM with $\epsilon = 0.02$, 0.05, and 0.1, respectively, and solid lines (from top to bottom) are the results of the WNT for these amplitudes.

As in the case of thinner films, the WNT overestimates the time of rupture, but the results of the GM and WNT are qualitatively similar. In particular, the nonlinear dominant wave number and cutoff wavenumber are adequately represented by the WNT. It may, therefore, be concluded that the mode interactions are not very severe for the weaker instability in relatively thick films.

In summary, we have addressed the role of various nonlinearities in the stability and rupture of thin free films devoid of surfactants, based on the long-wave evolution equations of Erneux and Davis. The influence of $\mathbf{0}(\epsilon^2)$ terms—inertia and nonlinear viscous correction, are found not to be very significant for the small amplitude $(\epsilon \leqslant 0.05)$, fastest growing modes of the nonlinear theory (Fig. 5). Destabilization due to inertia is, however, usually important for $k>k_m$; especially near k_c (Fig. 5) and for relatively large amplitude disturbances. The nonlinearity of LW forces and inertia conspire to produce strong subcritical instabilities with $k>k_c$. Neglect of unsteady effects is not uniformly satisfactory (Fig. 6). For small amplitudes, the dominant nonlinear modes and the time of rupture are adequately predicted by the full weakly nonlinear theory [Eq. (8)], use of which reduces the computational time by orders of magnitude. Finally, an analytical result similar (but not identical) to that of Erneux and Davis[9] may also be used for a quick estimate of lifetimes, due to a favorable cancellation of errors. However, the analytical result cannot be used for prediction of dominant wave number, and for studying the influence of the initial velocity, $u(0)$.

REFERENCES

1 A. Vrij, "Possible mechanism for the spontaneous rupture of thin free liquid films," Discuss. Faraday Soc. **42**, 23 (1966).
2 A. Sheludko, "Thin liquid films," Adv. Colloid Interface Sei. **1**, 391 (1967).
3 E. Ruckenstein and R. K. Jain, "Spontaneous rupture of thin liquid films," J. Chem. Soc. Faraday Trans. II **70**, 132 (1974) (Section 5.1 of this volume).
4 C. Maldarelli, R. K. Jain, I. Ivanov, and E. Ruckenstein, "Stability of symmetric and unsymmetric thin liquid films to short and long wavelength perturbations," J. Colloid Interface Sei. **78**, 118 (1980) (Section 5.3 of this volume).
5 D. S. Dimitrov, "Dynamic interactions between approaching surfaces of biological interest," Progr. Surface Sei. **14**, 295 (1983).

6 A. Sharma and E. Ruckenstein, "Rupture of thin free liquid films with insoluble surfactants: Nonlinear aspects," AIChE Symp. Ser. **252**, 130 (1986).
7 M. Prevost and D. Gallez, "Nonlinear stability of thin free liquid films: Rupture and Marangoni effects," AIChE Symp. Ser. **252**, 123 (1986).
8 A. K. Malhotra and D. T. Wasan, "Effect of film size on drainage of foams and emulsion films," AIChE J. **39**, 1533 (1987).
9 T. Emeux and S. H. Davis, "Nonlinear rupture of free films," Phys. Fluids A**5**, 1117 (1993).
10 J. T. G. Overbeek, "Black soap films," **64**, 1178 (1960).
11 B. U. Felderhof, "Dynamic of free liquid films," J. Chem. Phys. **49**, 44 (1968).
12 H. C. Hamaker, "The London van der Waals attraction between spherical particles," Physica **4**, 1058 (1937).
13 I. E. Dzyaloshinskii, E. M. Lifshitz, and L. P. Pitaevskii, "The general theory of van der Waals forces," Adv. Phys. **10**, 165 (1965).
14 J. H. Israelachvili, *Intermolecular and Surface Forces* (Academic, New York, 1985).
15 B.V. Deijaguin, "The definition and magnitude of disjoining pressure and its role in the statics and dynamics of thin liquid films," Kolloid Zh. **17**, 207 (1955).
16 M. B. Williams and S. H. Davis, "Nonlinear theory of film rupture," J. Colloid Interface Sei. **90**, 220 (1982).
17 A. Sharma and E. Ruckenstein, "An analytical nonlinear theory of thin film rupture and its application to wetting films," J. Colloid Interface Sei. **113**, 456 (1986) (Section 5.4 of this volume).
18 G. F. Teletzke, H. T. Davis, and L. E. Scriven, "Wetting hydrodynamics," Rev. Phys. Appl. **23**, 989 (1988).
19 A. Sharma, "Relationship of thin film stability and morphology to macroscopic parameters of wetting in the apolar and polar systems," Langmuir **9**, 861 (1993).
20 A. Sharma and A. T. Jameel, "Nonlinear stability, rupture and morphological phase separation of thin liquid films on the apolar and polar substrates," J. Colloid Interface Sci. **161**, 190 (1993).
21 A. Sharma and E. Ruckenstein, "Finite amplitude instability of thin free and wetting films: prediction of lifetimes," Langmuir **2**, 480 (1986) (Section 5.5 of this volume).
22 M. Prevost and D. Gallez, "Nonlinear rupture of thin free liquid films," J. Chem. Phys. **84**, 4043 (1986).
23 While the present paper was under review, numerical solutions for the free film with surfactants have appeared in A. De Wit, D. Gallez, and C. I. Christov, "Nonlinear evolution equation for thin liquid films with insoluble surfactants," Phys. Fluids A **6**, 3256 (1994).
24 D. J. Benney, "Long waves on liquid films," J. Math. Phys. **45**, 150 (1966).
25 R. W. Atherton and G. M. Homsy, "On the derivation of evolution equations for interfacial waves," Chem. Eng. Commun. **2**, 57 (1976).
26 H. S. Kheshgi and L. E. Scriven, "Dewetting: Nucléation and growth of dry regions," Chem. Eng. Sci. **46**, 519 (1991).
27 J. P. Burelbach, S. G. Bankoff, and S. H. Davis, "Nonlinear stability of evaporating/condensing liquid films," J. Fluid Mech. **195**, 463 (1988).
28 S. G. Yiantsios and B. G. Higgins, "Rupture of thin films: Nonlinear stability analysis," J. Colloid Interface Sci. **147**, 341 (1991).
29 A. Sharma and A. T. Jameel, "Stability of thin polar films on nonwettable substrates," J. Chem. Soc. Faraday Trans. **90**, 625 (1994).
30 V. S. Mitlin, "Dewetting of solid surface: Analogy with spinodal decomposition," J. Colloid Interface Sci. **156**, 491 (1993).
31 C. Canuto, M. Y. Hussaini, A. Quarteroni, and T. A. Zang, *Spectral Methods in Fluid Dynamics* (Springer-Verlag, New York, 1988).
32 C.S. Kishore, "Nonlinear stability of thin free films," M. S. thesis, I.I.T. Kanpur, 1994.

5.8 Dewetting of Solids by the Formation of Holes in Macroscopic Liquid Films*

Ashutosh Sharma and Eli Ruckenstein[†, ‡]

Department of Ophthalmology, School of Medicine and Biomedical Sciences, State University of New York at Buffalo, Buffalo, New York 14214, and [†] Department of Chemical Engineering, State University of New York at Buffalo, Buffalo, New York 14260

Corresponding Author

[‡]To whom correspondence should be addressed.

Received November 14, 1988; accepted March 2, 1989

INTRODUCTION

If the thickness of a liquid film that is supported on a horizontal solid surface is gradually decreased, one of several interesting phenomena can occur depending on the physicochemical properties of the solid surface and the liquid. The liquid films that display a finite angle of contact with the solid surface rupture suddenly due to the formation of holes at some "critical" thickness (1, 2). The formation of holes cannot occur spontaneously by van der Waals interactions in liquid films that are several hundred micrometers thick because the disjoining pressure is inoperative at such large distances. However, once a hole forms due to external vibrations or particles hitting the surface, it may be stable or unstable depending upon the various interfacial energies involved and the gravity. If the liquid film is relatively thick (thicker than about 0.5 cm for water) the film remains unconditionally stable; i.e., it cannot be disrupted and made to recede from a part of the solid surface regardless of the diameter of the hole produced in the liquid film. This is so because, for thick films, the stabilization due to gravity overwhelms the maximum possible destabilization due to interfacial tension forces and the fluid promptly rewets the surface following the formation of holes. Lamb (3) and, later, Taylor and Michael (4) derived this limiting film thickness from the Young-Laplace equation of capillarity. This limiting thickness, h_1, beyond which films are unconditionally stable, is given by

$$h_1 = 2\left(\frac{\gamma_1}{\rho g}\right)^{1/2} \sin\frac{\theta}{2}, \qquad [1]$$

where γ_1, ρ, g, and θ are the liquid surface tension, fluid density, gravitational constant, and equilibrium contact angle, respectively. For fluid films somewhat thinner than the limiting thickness, a competition between gravitational and interfacial tension forces determines a critical hole radius such that all holes with radii larger than the critical value open out and an irreversible dewetting

* *Journal of Colloid and Interface Science*. Vol. 133, No.2, p.358 December 1989. Republished with permission.

of solid occurs (4). Experiments of Padday (1) and later Doughman *et al.* (2) have shown that both aqueous and hydrocarbon fluid films rupture rather spontaneously on a variety of surfaces when their thicknesses are reduced to several hundred micrometers. A high-speed cine camera recording indicated that the rupture of films occurs by formation of holes (1). Further, experimental observations suggest that the film thickness at the instant of rupture—we will refer to it as the critical thickness—depends on both the liquid and the solid surface being studied. For example, water films 510, 420, and 310 μm thick, or thinner, rupture on Teflon, polyethylene, and paraffin wax surfaces, respectively, whereas films of ethylene glycol and 1-decene rupture at and below 560 and 260 μm, respectively, on a Teflon surface (1, 2).

The general equation of capillarity and also the energetic considerations given later show that the effect of gravity is negligible for films that are only several hundred micrometers thick and, hence, the instability is governed only by a competition between the various interfacial tensions involved. The relationship between the critical thickness and the interfacial properties of the fluid and solid surface have, however, remained obscure. This problem is of interest in wetting/dewetting of solids in general and in breakup of tear film on corneal epithelium and contact lenses in particular (5–9). The purpose of this paper is to derive a simple criterion for the breakup (dewetting) of a liquid film supported on a solid surface. The essential idea of the present approach is to compare the free energy of the unbroken liquid film with the free energy of the film with a hole. Dewetting of solid is favored whenever the free energy decreases upon formation of a hole and vice versa. In this way, we obtain a relationship between the critical film thickness and the interfacial tensions of the solid and liquid. Finally, the theoretical results are compared with the available data on the breakup of liquid films and some insights offered by this theory for the process of dewetting are discussed.

PRELUDE

Figure 1 depicts a liquid film of thickness h, which is of the order of several hundred micrometers, but much thinner than the limiting thickness as given by Eq. [1]. The lateral extents of the solid surface and liquid film are assumed to be much larger than both the fluid thickness and the radius of the hole produced in the film. It is clear that for the relatively thick films considered here, the rupture or hole formation cannot occur due to the action of long-range intermolecular van der Waals forces, which can destabilize only the liquid films that are thinner than about 0.1 μm (10). Prior to the formation of a hole that exposes part of the solid surface to air, a deformation of the liquid surface must occur for which the energy is to be supplied by an external source. The entire process is analogous to a particle that may escape to a lower energy state only if it has sufficient energy to overcome a potential barrier. The initial process of formation of a hole is, therefore, mediated by

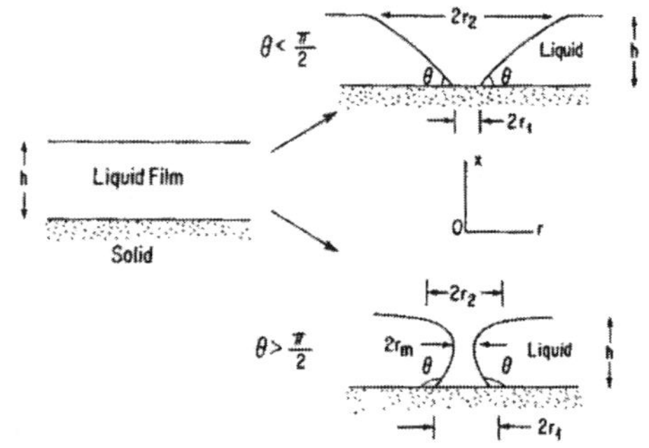

FIG. 1. Schematic representation of hole formation in a liquid film supported on a horizontal solid surface. θ, r_1, and r_2 are the equilibrium solid-liquid contact angle, the radius of the hole at its base, and the radius at the liquid-vapor interface, respectively. r_m is the minimum thickness of the hole for $\theta > \pi/2$.

such "external" stimuli as vibrations, trapped gaseous bubbles, impurities, and particles hitting the film surface. For relatively thick films of high surface tensions that are formed in a relatively isolated and quiescent environment, such as those considered by Taylor and Michael (4), holes were formed by an external probe or a jet of air. Regardless of the mechanism of initial hole formation, however, a hole that is once initiated will rapidly attain one of the two possible configurations shown in Fig. 1 (depending on the magnitude of the equilibrium contact angle, θ), if the free energy of the hole configuration is less than the free energy of the film system. Conversely, if the uninterrupted film has the lower free energy, the film will rewet the surface, the hole will close, and the initial film configuration will be restored. The process may again be viewed as a particle seeking the lower free energy state under the influence of external noise.

Now, if the area of contact between the liquid film and the solid is denoted by $A \gg \pi r^2,\ 2\pi rh$ and the area of the liquid surface of the hole by S, the total change in free energy produced by the creation of a hole (as shown in Fig. 1) is given by

$$\Delta F = F_{\text{hole}} - F_{\text{film}} = \left(A - \pi r_2^2\right)\gamma_1 + \left(A - \pi r_1^2\right)\gamma_{\text{sl}}$$

$$+ S\gamma_1 + \pi r_1^2 \gamma_{\text{s}} - A\gamma_1 - A\gamma_{\text{sl}}, \tag{2}$$

where r_1, r_2 are the radii of the hole at the base and film surface, respectively, and γ_s, γ_{sl} are interfacial tensions of the solid against air and liquid, respectively. An increase in the free energy due to the work done against the gravity also occurs, but it is negligible for the film thicknesses considered here. This is shown in the appendix with a simplified hole profile model based on Eqs. [18]. In addition, from Young's equation

$$\gamma_1 \cos\theta + \gamma_{\text{sl}} = \gamma_{\text{s}}. \tag{3}$$

Elimination of $(\gamma_s - \gamma_{sl})$ between Eqs. [2] and [3] simplifies the expression for the free energy change to

$$\Delta F = \gamma_1\left[S + \pi r_1^2 \cos\theta - \pi r_2^2\right]. \tag{4}$$

Expression [4] gives the difference between the free energy of two equilibrium states, namely, the hole and the film configurations. The condition $\Delta F = 0$ represents the threshold of stability; i.e., this condition gives the maximum film thickness that may rupture for a given hole radius. The hole will open out only if the film thickness is smaller than this critical thickness. As is shown in the next section, r_2 and S depend on r_1, θ, and h. Thus, the critical thickness, h_c may be evaluated from Eq. [4] with $\Delta F = 0$ as a function of any characteristic radius (r_1 or r_2) of the hole and the equilibrium contact angle, θ.

MODEL AND RESULTS

The surface area of the hole may be evaluated if its shape is known. The equilibrium shape of the hole is governed by the equation of capillarity (4, 11),

$$\frac{1}{p^3}\frac{d^2r}{dx^2} - \frac{1}{rp} - \left(\frac{g\rho}{\gamma_1}\right)\left(h - x\right) = 0, \tag{5a}$$

where

$$p = \left[1 + \left(dr/dx\right)^2\right]^{1/2}. \tag{5b}$$

Here r and x are the radial and normal space coordinates (Fig. 1) and the origin of the coordinate system coincides with the center of the circular hole at the level of the solid surface. Equation [5] is to be solved in conjunction with the boundary conditions

$$\text{at } x = 0; \quad r = r_1,$$

$$\text{and } (dx/dr) = \tan\theta, \tag{6}$$

where r_1 is the hole radius at its base and θ is the three-phase contact angle as determined from Young's equation.

If Eq. [5] is nondimensionalized by scaling r and x by the respective characteristic dimensions of the hole, r_1 and h, the second and the third terms of Eq. [5a] are multiplied by the nondimensional quantities (h^2/r_1^2) and $(\rho gh^3/\gamma_1 r_1)$, respectively. For the film thicknesses and hole radii of interest here, inequalities $(\rho gh^3/\gamma_1 r_1) \ll 1$ and $(g\rho hr_1/\gamma_1) \ll 1$ are satisfied and, hence, the gravitational contribution (third term of Eq. [5a]) is neglected. The solution of the simplified equation

$$r = \frac{d^2r}{dx^2} = \left[1 + \left(\frac{dr}{dx}\right)^2\right] \tag{7}$$

is easily obtained by making the substitutions

$$y = (dx/dx), \qquad [8a]$$

$$\frac{d^2r}{dx^2} = \frac{dy}{dx} = \frac{dy}{dr}y, \qquad [8b]$$

and by integrating the resulting equation twice with the help of boundary conditions (Eq. [6]). The solution is

$$r(x) = r_1[\cosh(ax) + \cos\theta \sinh(ax)], \qquad [9a]$$

where

$$a = (1/r_1 \sin\theta) \qquad [9b]$$

The radius of the hole at the film surface, r_2, is determined from the solution [9] as

$$r_2 = h(h) = r_1\phi$$
$$= r_1[\cosh(ah) + \cos\theta \sinh(ah)]. \qquad [10a]$$

It may be verified from Eq. [9a] that for $\theta < \pi/2$, the minimum hole radius occurs at its base and it equals r_1, whereas for $\theta \geqslant \pi/2$, the minimum hole radius occurs at some $x > 0$ where the tangents to the hole surface are vertical (Fig. 1). The minimum hole radius, r_m for $\theta > \pi/2$ is given by the solution of $(dr/dx) = 0$ and this solution is

$$r_m = r_1 \sin\theta. \qquad [10b]$$

Finally, the surface area of the hole ($0 < x < h$; $r_1 < r < r_2$) is determined from

$$S = \int_0^h 2\pi r \left[1 + \left(\frac{dr}{dx}\right)^2\right]^{1/2} dx, \tag{11a}$$

which may be rearranged to the following simple form with the help of Eqs. [7] and [9a]:

$$S = 2\pi a \int_0^h r^2 dx \tag{11b}$$

$$= (1/2)\pi r_1^2 \Big[\left(1+\cos^2\theta\right)\sinh(2ah)$$

$$+ 2\cos\theta\ (\cosh(2ah) - 1)$$

$$+ 2ah\ \sinh^2\theta\Big] \tag{12}$$

It may now be noted that the influence of gravity in Eq. [5a] is confined to a narrow boundary layer near the film interface ($x \rightarrow h$) for thin films (4). This contribution increases the slope (dr/dx) near $x = h$ and allows for a more gradual (smooth) transition of the hole profile into the free film surface. Thus, for $r > r_2$, the hole profile closely follows the asymptotic flat film surface and, hence, the profile here is truncated to $x = h$ for $r > r_2$. Since we are interested only in the total change in the free energy (Eq. [2]), the contribution of gravity near $x \rightarrow h$, $r > r_2$ does not seem important for free energy considerations. Substituting for r_2 (Eq. [10a]) and S (Eq. [12]) in Eq. [4] with $\Delta F = 0$ gives the following compatibility condition which provides the maximum unstable film thickness for a given hole radius, r_1, and contact angle, θ:

$$2ah_c = 1 + \frac{\left(1-\cos\theta\right)^2}{\sin^2\theta}\left[\cosh\left(2ah_c\right) - \sinh\left(2ah_c\right)\right]. \tag{13}$$

Recalling that $a = (1/r_1 \sin\theta)$, the ratio $(h_c/r_1) = f(\theta)$ may be computed from Eq. [13] for any equilibrium contact angle, θ. Similarly, the ratios (h_c/r_2) and (h_c/r_m) are evaluated by noting that

$$\left(h_c/r_2\right) = \left(h_c/r_1\phi\right) \quad \text{and}$$

$$\left(h_c/r_m\right) = \left(h_c/r_1 \sin\theta\right). \tag{14}$$

The solution of Eq. [13] for $0 < \theta \leqslant 110°$ is obtained explicitly with an error of 2% or less by noting that $2ah_c = 1 + \epsilon$, where $\epsilon < 1$, and by expanding in series the hyperbolic functions up to the linear terms. This gives

$$2ah_c = 1 + \frac{\left(1-\cos\theta\right)^2}{e\sin^2\theta + \left(1-\cos\theta\right)^2} \tag{15a}$$

and

$$f(\theta) \equiv \left(\frac{h_c}{r_1}\right) = \frac{1}{2}\sin\theta$$

$$\times\left[1 + \frac{(1-\cos\theta)^2}{e\sin^2\theta + (1-\cos\theta)^2}\right]. \quad [15b]$$

For $\theta < \pi/4$, the leading order term of the solution is found to be entirely adequate, viz.,

$$(h_c/r_1) = 0.5 \sin \theta, \ \text{ for } \theta < (\pi/4). \quad [15c]$$

It is clear from Eqs. [4] and [13] that for film thicknesses in excess of the critical thickness (Eq. [13]), ΔF is positive and hence the film configuration is stable; i.e., dewetting cannot occur. From Eqs. [13] to [15] it may be noted that the critical thickness, h_c, depends only on the equilibrium contact angle, θ, and the hole radius but does not depend explicitly on the solid and liquid interfacial tensions. This means that any combination of different interfacial tensions γ_l, γ_s and γ_{sl} that result in the same θ (from Eq. [3]) corresponds to the same critical thickness for a given hole. As may be physically anticipated, the critical thickness increases with increased nonwettability (increased θ) of the solid surface. Clearly, for a completely wettable surface ($\theta \rightarrow 0°$), the critical thickness also becomes vanishingly small and the film configuration should be stable down to molecular dimensions on an intensely hydrophilic surface.

The dependence of the critical thickness on the contact angle implies that the critical thickness may also be alternatively viewed as a function of the spreading coefficient, S, that is defined as ($S < 0$)

$$S = \gamma_s - \gamma_{sl} - \gamma_l = -\gamma_l(1-\cos\theta),$$

from the Young equation.

In addition to its dependence on the equilibrium contact angle, the critical thickness depends also on the radius of the hole. The radius of the hole may be relatively constant in a given environment if the hole is created by an external probe, the impact of particles, surface defects, and/or coalescence of trapped air bubbles. The other limiting case may be the one in which the hole formation occurs due to the energy of external vibrations or disturbances. Clearly, an energy barrier of the order of $2\pi r_2 h\gamma_l$ has to be overcome for each hole created in the film. If the maximum energy available to create holes is of the order of ΔE^* and n is the number of holes that are formed, the following equality should hold:

$$\Delta E^* = 2\pi n r_2 h\gamma_l = 2\pi n r_1 \phi h \gamma_l$$

$$(\text{from Eq. [10a]}). \quad [16]$$

In such a case, eliminating $\boldsymbol{r_1}$ from Eqs. [13] and [16] gives the equation for the critical thickness

$$\left(\frac{n\gamma_l}{\alpha}\right)h_c^2 = f(\theta)/\phi(\theta), \quad [17a]$$

where

$$\alpha = \left(\Delta E^{*}/2\pi\right) \quad [17b]$$

and $f(\theta) = (h_c/r_1)$ is the solution of Eq. [13] or of Eq. [15b]. The corresponding $\phi(\theta)$ is evaluated from Eq. [10a] as

$$\phi(\theta) = \cosh\left(f(\theta)/\sin\theta\right) + \cos\theta \sinh\left(f(\theta)/\sin\theta\right). \quad [17c]$$

For a given environment, then, if the maximum work done by the perturbation in creating holes is relatively constant, the critical thickness also depends explicitly on the liquid surface tension, γ_l. This may be physically anticipated because, for a given ΔE^*, the larger the liquid surface tension, the smaller the hole radius and, consequently, the critical thickness would also decrease. In either case (constant r_1 or constant a), the general conclusion is that the critical thickness increases with an increase in the equilibrium contact angle. In the case of a constant hole radius, the functional dependence on θ is of the form, $h_c \propto f(\theta)$, whereas for the constant energy case it is of a different form, namely, $h_c \propto (f(\theta)/\phi)^{1/2}$.

Thus far, the free energy calculations have been carried out by assuming the hole profile (shape) to be that given by the equilibrium solution of the equation of capillarity. In order to determine the sensitivity of the results to presumed hole shapes, the following simple model for the initial shape of the hole is also considered.

For thin films and $\theta < \pi/2$, if the surface profile of the hole is considered to be linear, the hole is represented by a conical section, for which

$$r_2 = r_1 + \left(h/\tan\theta\right) \quad [18a]$$

And

$$S = \pi\left(r_1 + r_2\right)h/\sin\theta \quad [18b]$$

From Eqs. [15b], [10a], and [17c], it may be noted that for $\pi/2 < \theta < 110°$, all of the characteristic hole radii, r_1, r_2, and r_m, are approximately the same and, thus, the surface area may be approximated by that of a perfect cylinder with radius r_1. Now, Eq. [4] with $\Delta F = 0$ determines the critical film thickness for the above-simplified model, which is given by

$$\left(h_c/r_1\right) = \tan\theta\left[\sqrt{1+\cos} - 1\right] \equiv F_1\left(\theta\right), \text{ for } \theta \leqslant \pi/2, \quad [19a]$$

$$r_2 = \sqrt{1+\cos\theta}\, r_1 = r_1\phi\left(\theta\right), \text{ for } \theta \leqslant \pi/2, \quad [19b]$$

$$\left(h_c/r_2\right) = F_1\left(\theta\right)/\sqrt{1+\cos\theta} \equiv F_2\left(\theta\right), \text{ for } \theta \leqslant \pi/2, \quad [19c]$$

$$(h_c/r_2) = (1-\cos\theta)/2 = (h_c/r_2) \equiv F_3(\theta),$$

$$\text{for}\ \pi/2 \leqslant \theta < 110°. \quad [19d]$$

The ratios evaluated from Eqs. [19] are in remarkable qualitative and quantitative agreement with the same ratios calculated from the solution of Eq. [13] or Eq. [15b]. For example, (h_c/r_1) from Eqs. [19] equals 0.21, 0.34, 0.44, 0.5, and 0.67 for $\theta = 30°, 50°, 70°, 90°$ and 110°, respectively. The ratio (h_c/r_1) evaluated from Eq. [13] for the same θ values equals 0.25, 0.41, 0.54, 0.64, and 0.69, respectively. Further, for small contact angles ($\theta < 45°$), the asymptotic expression for (h_c/r_1) from Eq. [19a] is

$$(h_c/r_1) = 0.414\ \sin\theta, \quad [19e]$$

which displays the same dependence on θ as the corresponding Eq. [15c], with a slight numerical difference in the multiplicative constant. It is thus concluded that the critical thickness derived from the free energy considerations is not extremely sensitive to the exact shape of the hole. In what follows, the theoretical results are compared with the available data on the breakup of thin liquid films on nonwettable solids.

COMPARISON WITH EXPERIMENT AND DISCUSSION

The experimental values of the critical thicknesses for several solid-liquid systems are summarized in Table 1 (1, 2). For some of these systems, the contact angle values have not been reported and these are evaluated based on the Girifalco-Good-Fowkes-Young equation (12)

$$(1+\cos\theta)\gamma_l = 2\left(\gamma_s^d \gamma_l^d\right)^{1/2}, \quad [20]$$

where γ_l is the total (dispersion + polar) surface tension of the liquid and γ_s^d and γ_l^d are effective surface tensions of solid and liquid due to the dispersion component of the interfacial tension. For relatively nonpolar liquids such as benzene and decenes, the total surface tension, γ_l is the same as γ_l^d. For water, $\gamma_l = 72.8$ and $\gamma_l^d = 21.8$ dyn/cm and for ethylene glycol, the values of γ_l and γ_l^d are 49 and 28.6 dyn/cm, respectively (12). γ_s^d corresponds to Zisman's critical surface tension and this is about 19.5 dyn/cm for Teflon (13). Values of the contact angle for the experimental systems are also summarized in Table 1.

It is seen from Table 1 that the critical thickness for the Teflon-water system studied by Doughman *et al.* (2) is substantially larger than that observed by Padday (1) for the same system. Also, while the Teflon-benzene system is essentially similar in physical properties (e.g., contact angle, surface tension, density) to Teflon-1-decene, tetradecene, and octadecene systems, the critical thickness for benzene is substantially smaller than for the other three systems. These deviations from the general trend may be due to some inadvertent alterations in the "environment" or contamination. Thus, for reasons of internal consistency, these two data points seem to represent aberrations from the assumption of a fixed environment.

As was pointed out in the previous section, the theoretical results may be compared with experiment, either by assuming a constant hole radius r_1 or by assuming a fixed work done by the surroundings in the process of hole formation.

Figure 2 depicts the results for the case when the hole radius at the base, r_1, is assumed to be independent of the solid-liquid system under investigation. As is clear from Eqs. [13] to [15], the ratios (h_c/r_1) and (h_c/r_2) are functions only of the contact angle. The broken and the solid curves in this figure represent the theoretical prediction of (h_c/r_2) ($f(\theta)$ from Eq. [13]) and (h_c/r_2) ($f(\theta)/\phi$ from

TABLE 1
Critical Thickness and Contact Angles of Polar and Nonpolar Liquids on Various Solid Surfaces

System (solid and liquid)	Liquid surface tension γ_1 (dyn/cm)	Equilibrium contact angle θ (degrees)	Critical Thickness h_c (μm)	Symbolic representation of the system in the figures
Teflon-water (1)	72.8	110 (12, 13)	510	○
Teflon-ethylene glycol (1)	49	92[a]	560	†
Teflon-1-decene (1)	23	33[a]	260	▽
Teflon-1-tetradecene (1)	27	45[a]	270	*
Teflon-1-octadecene (1)	28.4	50[a]	270	□
Teflon-benzene (1)	29	45 (12, 13)	160	△
Paraffin-water (1)	72.8	65[b], 110(4, 14)	310	◇
Teflon-water (2)	72.8	110 (12, 13)	787	○
Polyethylene-water (2)	72.8	94 (15)	420	•
Polymethyl metacrylate-water (2)	72.8	70–80[c](2)	393	▲

[a] Calculated values (see Eq. [20]).
[b] Receding contact angle (4).
[c] Range of contact angle values, depending upon PMMA gel characteristic (2).

Eq. [14]), respectively. In order to evaluate the same ratio for the data reported in Table 1, one value of the hole radius, r_1, is needed. Padday (1) has provided a photograph for the rupture of water on paraffin wax surface. In this photo, the hole radius appears to be between 500 and about 1000 μm. Assuming r_1 to be 750 μm for all systems, the experimental value of the ratio (h_c/r_1) is calculated by using the data for the critical thickness given in Table 1. The upper ends of the vertical bars drawn in Fig. 2 correspond to this ratio, (h_c/r_1). A comparison of these data with the theoretical broken curve for (h_c/r_1) shows that the mean error in the prediction of the data is about ±12% and the maximum

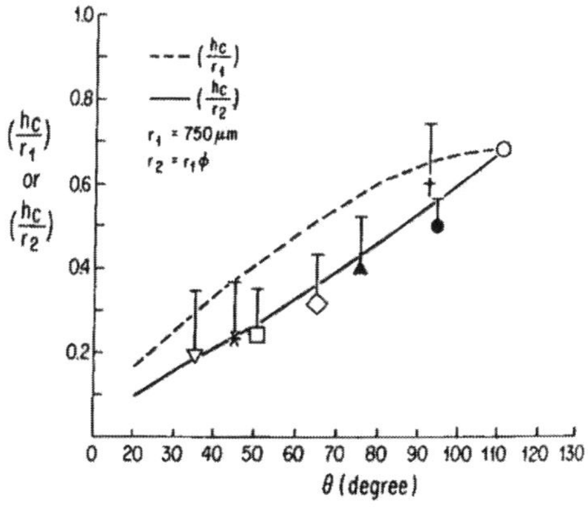

FIG. 2 Variation of the ratio (h_c/r_1) and (h_c/r_2) with the equilibrium contact angle. The dashed curve represents (h_c/r_1) evaluated from Eq. [13] and the solid curve is for (h_c/r_2) obtained by Eq. [14]. The data points correspond to the ratio (h_c/r_2) as evaluated with the help of critical thicknesses reported in Table 1 and the definition of r_2, Eq. [10a]. The hole radius at the base, r_1, is assumed to be equal to 750 μm for all solid-liquid systems. The upper ends of the vertical bars correspond to the experimental ratio (h_c/r_1). The particular solid-liquid system corresponding to each symbol is reported in Table 1.

deviation is 20% (for the wax-water system). Further, the anticipated experimental values of the ratio (h_c/r_2) are evaluated by calculating r_2 from Eq. [10a], with the help of the data for the critical thickness (Table I) and assuming $r_1 = 750\ \mu$m for all of the systems. The results for the experimental value of (h_c/r_2) are also shown in Fig. 2 with the help of various symbols. The solid-liquid systems corresponding to these symbols are listed in Table 1. Thus, the interval between each symbol and the upper end of its vertical bar represents the interval within which the experimental value of the ratio (h_c/r) is confined. Again, a comparison of experimental (h_c/r_2) with theoretical values of (h_c/r_2) (solid curve) reveals a mean error of about 8% in the prediction of the data.

It is thus apparent from Fig. 2 that the qualitative and quantitative dependence of the critical thickness on the contact angle is very well predicted even if the base radius is assumed to be a constant. The overall agreement between theory and experiment is quite striking if one also notes that the exact values of the contact angle are very sensitive to impurities and the solid sample used and that some statistical variations in the hole radius are anticipated from its mean value. In addition, effects of film pressure, π° in the Young equation, as well as in the evaluation of the free energy charge, have been neglected. The data for this term are presently neither conclusive nor available for most systems (12).

Next, the experimental values of (h_c/r_1) and (h_c/r_2) are compared with the predictions of the simple, linear hole model (Eqs. [18]). The solid and the broken curves of Fig. 3 represent the theoretical values of the ratio (h_c/r_2) (Eqs. [19c], [19d]) and (h_c/r_1) (Eqs. [19a], [19d]), respectively, for this model of the hole. A comparison with the experimental data for $r_1 = 750\ \mu$m reveals a satisfactory fit. The mean and the maximum errors in prediction of data are about 17 and 30%, respectively, both of which are substantially larger than the corresponding values for the model in which the hole profile is derived from the equation of capillarity. It is thus concluded that the equation of capillarity is perhaps more accurate for describing the shape of the hole and for predicting the critical thickness with the help of energetic arguments.

Next, we assess whether the assumption of a constant perturbation energy (Eqs. [17]) describes better the experimental situation encountered in the experiments of Padday (1) and Doughman *et al.* (2). For this case, the prediction of the critical thickness requires knowledge regarding two parameters, namely, the maximum external energy α and the number of holes n. Again, in the experiments of Padday (1) with water films, initially only one dimple evolves to create a single hole, i.e., $n = 1$ for water films. This may have been due to the fact that water has high surface

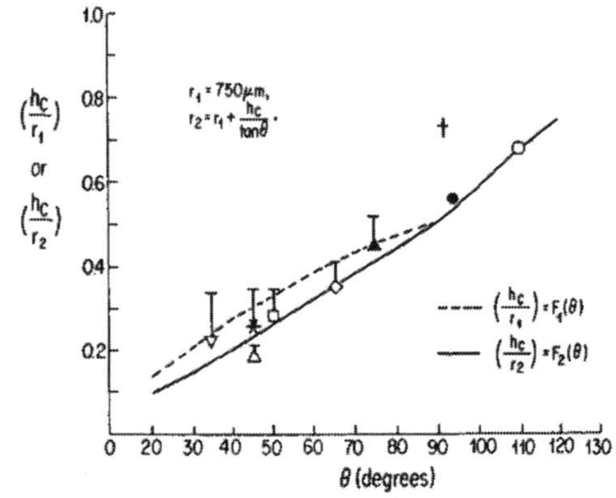

FIG. 3 Variation of the ratio of the critical thickness and hole radius $[(h_c/r_1)$ and $(h_2/r_2)]$ with the contact angle θ. Calculations displayed here are for the simplified hole shape as represented by Eqs. [18]. The dashed curve corresponds to the ratio $(h_c/r_2) = F_1(\theta)$ from Eq. [19a] and the solid curve represents the ratio $(h_c/r_2) = F_2(\theta)$ from Eq. [19c], Various data points correspond to the ratio (h_c/r_2) as evaluated from the experimental critical thicknesses given in Table 1 and the definition of r_2, Eq. [19b]. The hole radius at the base, r_1 is assumed to be constant and equal to 750 μm for all systems. The upper end points of the vertical bars correspond to the experimental ratio (h_c/r_1).

tension and all of the vibrational energy available was used up for creating a single hole. While the number of holes for the low surface tension decene films was not reported, the possibility of multiple holes cannot be ruled out for these cases. Clearly, if ΔE^* is a constant in a given environment for two different liquids that are represented by subscripts 1 and 2, then, from Eq. [16],

$$\alpha = n_1 r_{21} h_1 \gamma_{11} = n_2 r_{22} h_2 \gamma_{12}. \qquad [21]$$

Thus, for decene films (γ_{12} = 23–28.4 dyn/cm), multiple holes may coexist.

The solid curve of Fig. 4 depicts the theoretical variation of parameter $h_c\sqrt{n\gamma_1/\alpha} = (f/\phi)^{1/2}$ (Eq. [17]) with the contact angle. If the theoretical dependence given by Eq. [17] holds, the dimensionless groups $h_c\sqrt{n\gamma_1/\alpha}$ as determined from experiments should fall on the theoretical curve. An excellent agreement between theory and experiment indeed results for water and ethylene glycol films, for which n = 1 and the best fit for the parameter α is 0.2704 ergs. Obviously, if the vibrational energy is a constant, the parameter α may be evaluated based on a single system (Eq. [16]), say Teflon and water, and the same value should hold for other systems as well. Indeed, predictions of the critical thickness from [17] for α = 0.2704 ergs show a mean error of about 6% from the experimental values for systems as diverse as Teflon-water, Teflon- ethylene glycol, polyethylene-water, PMMA- water, and paraffin-water. This is commensurate with the reported accuracy of the experimental measurements (1). However, for low surface tension decene films, predictions of the critical thickness are in agreement with the data for the same α only if n = 3.

Presently, we admittedly do not know the factor that may determine the number of holes in low energy liquids or how this issue relates to the most efficient mechanism of dissipation of available energy. A possible scenario is that if a is a constant, then the product $n_2 r_{22}$ must be large for small values of the surface tension, γ_{12} (from Eq. [21]). In fact, if it is assumed that n_2 is always 1, then the hole radius be comes as large as 4500 μm for decene systems! It may not always be possible to create such large holes due to the large amount of liquid that would initially collect around the rim of the hole. Thus, the second possibility is that the hole radius remains relatively constant, but the number of holes increases with a decrease in γ_{12}. In fact, if it is assumed that the hole radius remains comparable to its value corresponding to a fixed base radius of 750 μm, viz., r_{22} is comparable to [750 + (h_c/tan θ)] μm, then Eq. [21] with α = 0.2704 shows that n_2 must be between 3 and 4 for various decenes. Assuming the integer value of 3, the adjusted hole radius may be evaluated from Eq. [21]. For various decenes, this procedure also leads to the data shown in Fig. 3. This idea

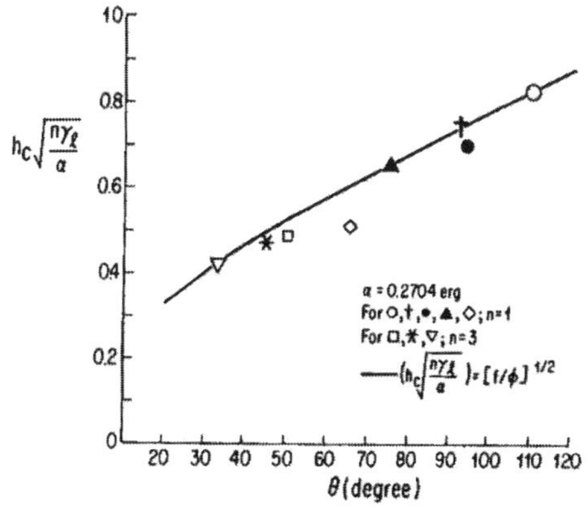

FIG. 4 Comparison of theoretical and experimental critical thicknesses for a constant external energy, α = 0.2704 ergs. The solid curve represents the theoretical predictions for the quantity $h_c\sqrt{n\gamma_1/\alpha} = (f/\phi)^{1/2}$ as a function of the contact angle. The number of holes for water and ethylene glycol films is one (n = 1) and for low surface tension decene systems, n = 3. Various data points shown are for the experimental values of the quantity $h_c\sqrt{n\gamma_1/\alpha}$, as evaluated from the critical thicknesses reported in Table 1.

seems physically reasonable in that the hole radii for all of the systems studied here turn out to be comparable in magnitude and, at the same time, α also remains a constant.

To summarize, the critical thickness depends both on the interfacial properties of the solid-liquid system (contact angle value) and on environmental factors that determine the initial size of the hole. A conclusion from the above discussions and Figs. 2 and 4 is that the base radii of holes created in different liquid-solid systems are essentially the same and are probably determined solely by the environment and the mechanism of hole formation. It is because of this reason that the theory provides the dependence of the critical thickness on the contact angle. An excellent degree of correlation between theory and experiments shows that the predicted dependence of the critical thickness on the contact angle is essentially correct. Some scatter in the data may be ascribed to the uncontrolled nature of the experiments and a lack of information regarding the hole size and contact angle values. These considerations are important for a more careful design of experiments in the future.

It is of interest to note that the theory leads immediately to a new operational definition of hydrophilicity/hydrophobicity of a solid surface. Clearly, the phenomenon of dewetting is not simply a reversal of wetting, but it does depend on the contact angle in a nonlinear fashion. Depending on the context, the surface hydrophilicity has been considered a spreading phenomenon, a capillary phenomenon, or a two-phase phenomenon. In terms of spreading, a solid is considered hydrophilic if water spontaneously spreads on its surface; i.e., the advancing contact angle of water is zero. Hydrophilicity as a capillary phenomenon implies a positive capillary rise of water; i.e., the advancing contact angle of water on the inner wall of the capillary should be less than 90°. The relative hydrophilicity as a two-phase phenomenon means that the water-phase contact angle is less than 90° in the presence of a distinct nonpolar liquid phase; i.e., the surface exhibits a preference toward water. Now, in view of the theory presented here, hydrophilicity/hydrophobicity may also be defined as a *dewetting* phenomenon. Clearly, such an operational definition would be useful for ranking the stability of liquid films (water) on different solid surfaces. In a given environment, then, the affinity of a liquid for a solid as evidenced by the *dewetting phenomenon* is directly related to its affinity in terms of *spreading* (equilibrium contact angle). The lower the contact angle of water, the smaller the critical thickness, and the solid surface is also more hydrophilic by this definition.

CONCLUSION

Based on energetic considerations, a theory is developed for predicting the thickness at which liquid films supported on solid surfaces rupture and dewetting occurs. The rupture of the film occurs by hole formation when the free energy of the film-solid system becomes equal to the free energy of the hole-liquid-solid system. The critical thickness depends nonlinearly on the equilibrium solid-liquid contact angle and increases with an increase in the contact angle. The critical thickness is also found to be proportional to the hole radius for liquid films that are up to several hundred micrometers thick. The radius of the hole is governed by the external mechanism of hole formation, which may be due to the energy of vibrations, trapped gaseous bubbles, or colloidal particles hitting the free surface of the liquid. In a given environment, then, the radius of the hole at the solid may be either a constant or may have a value determined by the maximum external energy available for the formation of holes.

It is shown that even if a surface is hydrophobic (i.e., the contact angle of water is greater than 90°), a water film supported on this surface will not always disjoin to create dry surface patches. The film rupture will only occur if the film thickness is smaller than the corresponding critical thickness. This suggests an alternative operational definition of surface hydrophilicity/hydrophobicity as a *dewetting phenomenon,* in contrast to the definition in terms of a *spreading phenomenon.*

The experimentally observed variation of the critical thickness with the equilibrium contact angle is well predicted by the present theory for a variety of solid-liquid systems which display a wide range of contact angle values (30° to 110°).

Finally, in many biomedical and industrial applications, it is desired that the solid surface be wettable by a stable liquid film of given thickness. An example is the wetting of corneal epithelium and contact lenses by the tear film, which is only about 4 to 10 μm thick in a normal eye and is even thinner on the surface of contact lenses. The theory presented here provides a criterion for the design of solid surfaces which remain wettable by a liquid film of desired thickness.

APPENDIX

If the work done against the gravity in creating a hole is also accounted for, the overall change in the free energy of the system in the process of hole formation is given by the following expression, instead of by Eq. [2] of the text:

$$\begin{aligned}\frac{\Delta F}{\pi} &= r_l^2\left(\gamma_s - \gamma_{sl} - \gamma_l\right)\\ &+ h\left(\frac{1-\cos\theta}{\sin\theta}\right)\gamma_l\left(2r_l + \frac{h}{\tan\theta}\right)\\ &+ \rho g h^2\left(\frac{r_l^2}{2} + \frac{r_l h}{3\tan\theta} + \frac{h^2}{12\tan^2\theta}\right).\end{aligned} \qquad [A.1]$$

The above equation holds for a conical hole. The critical thickness is again evaluated by setting $\Delta F = 0$ and eliminating $\gamma_s - \gamma_{sl}$ with the help of the Young equation, Eq. [3]. This leads to the following relation between the critical thickness, h_c, and the hole radius, r_l:

$$\begin{aligned}&r_l^2\left[\frac{\rho g h_c^2}{2} - (1-\cos\theta)\gamma_l\right]\\ &+ r_l\left[\frac{\rho g h_c^3}{3\tan\theta} + 2h_c\left(\frac{1-\cos\theta}{\sin\theta}\right)\gamma_l\right]\\ &+ \frac{h_c^2}{\tan\theta}\left[\frac{\rho g h_c^2}{12\tan\theta} + \gamma_l\left(\frac{1-\cos\theta}{\sin\theta}\right)\right] = 0\end{aligned} \qquad [A.2]$$

It is now easily verified that for critical thicknesses of the order of several hundred micrometers and $\gamma_l > 20$ dyn/cm, the gravitational terms in each of the square braces of Eq. [A.2] are negligible compared to the interfacial energy contributions. Thus for all the cases considered in this paper, Eq. [A.2] reduces to Eq. [19] of the text.

It is interesting to note that Eq. [A.2] also predicts the limiting thickness h_l correctly (see Eq. [19]). If the quantity in the first square braces becomes zero, viz.,

$$\rho g h_c^2 = 2\gamma_l(1-\cos\theta) \qquad [A.3]$$

the hole radius r_l becomes infinite. Physically, it means that for all thicknesses in excess of a limiting thickness, given by Eq. [A.3], the film configuration is always stable. Equation [A.3] is easily rearranged in the form of Eq. [1]. The thickness given by Eq. [A.3] are, however, at least an order of magnitude larger than the critical thicknesses considered here.

Note added in proof. In this paper, we have employed a conservative thermodynamic criterion, $\Delta F = 0$, for film rupture. We have recently shown (paper submitted to *J. Colloid Interface Sci.*) that ΔF against r_1 passes through a maximum before becoming zero. The criterion for film rupture based on maximization is consistent with the treatment of Taylor and Michael (4), and leads to a qualitative dependence of h_c on θ which is similar to that derived here on the basis of $\Delta F = 0$. However, the maximum of the free energy change is found to be extremely large, which is due to a large surface free energy of the hole con\uration near the maximum $\partial \Delta F / \partial r_1 = 0$. This large amount of energy cannot be supplied by the low energy vibrations in a quiescent environment. Thus, our recent results show that the observed spontaneous breakup of thin films occurs with greater probability in the vicinity of $\Delta F = 0$ than in the vicinity of the maximum of ΔF.

REFERENCES

1. Padday, J. F., *Spec. Discuss. Faraday Soc.* **1**, 64 (1970).
2. Doughman, D. J., Holly, F. J., and Dohlman, C. H., paper presented at the Association for Research in Vision and Ophthalmology meeting, Sarasota, Florida, 1971.
3. Lamb, H., "Statics," 2nd ed. Cambridge University Press, London, 1916.
4. Taylor, G. I., and Michael, D. H., *J. Fluid Mech.* **58**, 625 (1973).
5. Holly, F. J., and Lemp, M. A., *Exp. Eye Res.* **11**, 239 (1971).
6. Lemp, M. A., Holly, F. J., Iwata, S., *et al., Arch.Ophthalmol.* **83**, 89 (1970).
7. Sharma, A., and Ruckenstein, E., *J. Colloid Interface Sci.* **106**, 12 (1985) (Section 5.10 of this volume).
8. Sharma, A., and Ruckenstein, E., *J. Colloid Interface Sci.* **111**, 8 (1986) (Section 5.11 of this volume).
9. Cope, C., Dilly, P. N., Kaura, R., *et al., Curr. EyeRes.* **5**, 777 (1986).
10. Scheludko, A., *Adv. Colloid Interface Sci.* **1**, 391 (1967).
11. Padday, J. F., and Pitt, A., *J. Colloid Interface Sci.* **38**, 323 (1972).
12. Adamson, A. W., "Physical Chemistry of Surfaces." Wiley, New York, 1982.
13. Zisman, W. A., *Adv. Chem. Ser.* **43** (1964).
14. Dann, J. R., *J. Colloid Interface Sci.* **32**, 302 (1970).
15. Fox, H. W., and Zisman, N. A., *J. Colloid Sci.* **7**, 428 (1952).

5.9 Energetic Criteria for the Breakup of Liquid Films on Nonwetting Solid Surfaces*

Ashutosh Sharma[a,1] *and Eli Ruckenstein*[b,2]

[a]Department of Ophthalmology, School of Medicine and Biomedical Sciences, State University of New York at Buffalo, Buffalo, New York 14214

[b]Department of Chemical Engineering, Carnegie Mellon University, Pittsburgh, Pennsylvania 15213

Corresponding Author

[1]Present address: Department ofChemica1 Engineering, Indian Institute of Technology, Kanpur 208016, India.

[2]On leave from the Department of Chemical Engineering, State University of New York, Buffalo, NY 14260.To whom correspondence should be addressed at this address.

Received August 23, 1989; accepted November 2, 1989

Thin liquid films supported on a variety of solid surfaces (e.g. Teflon, polyethylene, PMMA, wax) rupture spontaneously and recede when their thicknesses are gradually reduced to several hundred micrometers (1,2). The film breakup is certainly not engendered by intermolecular van der Waals forces, the destabilizing influence of which becomes negligible for films thicker than 0.1 μm (3–5). Microscopic observations showed that destruction of the film occurs due to formation of a hole at a certain well-defined and reproducible thickness (1). The maximum thickness at which the film first becomes unstable depends on the properties of the liquid, as well as on the surface properties of the underlying surface (1, 2). However, neither the mechanism of dewetting nor the relationship between the film thickness and the physico—chemical parameters is completely understood. Our approach to the problem is based on the idea that the film breakup is energetically preferred beyond a point where the free energy of the broken film—solid system becomes maximum and starts to decline with a further reduction in the film thickness. This is a necessary condition of film instability. Eventually, a state is reached where the free energy of the system (i.e., hole plus film plus solid surface) becomes equal to the free energy of the initial state of a continuous, unruptured film. This is a sufficient condition of film instability as the film breakup is, in all probability, guaranteed upon reaching this state. On the basis of these energetic arguments, we establish conditions for the instability of liquid films and correlate the film thickness-at-rupture with such solid—liquid properties as the surface and interfacial tensions and the effects of gravity (liquid density). Theoretical predictions are then compared with the available data for the breakup of liquid films on solid supports (1,2).

* *Journal of Colloid and Interface Science*, vol. 137, No.2, p. 433 July 1990. Republished with permission.

Determination of the critical thickness and its relationship to surface properties is important for processes involving wetting and dewetting of surfaces. Examples include heat and mass transfer across thin liquid films supported on surfaces, foam generation in porous media during oil recovery, and coating of surfaces and attachment of bubbles and drops to solid surfaces following breakup of intervening liquid films. Frequently, the goal is to maintain the structural integrity of the thin film, but in some cases (e.g., cleaning of surfaces, froth flotation) its elimination is desirable.

An important application of the theory is in understanding the mechanism of tear-film breakup on the corneal epithelium and contact lens surfaces. A normal tear film is about a 4-to 10-μm-thick aqueous film that provides uninterrupted protection to the external corneal and conjunctival surfaces. A rapid breakup of the tear film followed by the corneal drying and damage occurs in some dry eye conditions (6). Given the present theory of rupture of thin films, a possible mechanism of the tear film breakup is formulated in a companion paper (7).

I. PRELUDE

It is known that for film thicknesses in excess of a limiting thickness, the film cannot be made to rupture and recede regardless of the size of the hole produced by external means (8, 9). This limiting thickness, h_l, beyond which films are unconditionally stable, is obtained from the expression (8, 9)

$$h_l = 2\left(\gamma_L/\rho g\right)^{1/2} \sin\left(\theta/2\right), \quad [1]$$

where γ_L, ρ, g, and θ are the liquid surface tension, the fluid density, the gravitational constant, and the angle of contact between the fluid drop and the surface, respectively. For liquid films thinner than the limiting thicknesses, Taylor and Michael (9) obtained an energetically metastable solution of the Young—Laplace equation of capillarity, which determines a transitional hole size for a film of known thickness. Holes larger than the transitional size expand, whereas smaller holes close promptly and small discontinuities of the film are thus healed. Predictions of this theory were tested by creating holes in water films on a paraffin wax surface (9). The water film tested were thicker than 0.25 cm (2500 μm) and relatively large holes had to be created by blowing an air jet in order to permanently rupture the film. A spontaneous rupture of much thinner (160–510 μm) liquid films was, however, observed by Padday (1) and Doughman *et al.* (2), even when no effort was made to create holes by external means. Sharma and Ruckenstein (10) proposed an energetic criterion of the film breakup to explain the observed (1, 2) dependence of the critical thickness on the surface properties. The essential idea of that approach was to compare the free energy of the unbroken liquid film system with the free energy after creation of a hole. In this approach, a critical state is reached when the change in free energy becomes zero during the process of hole formation. Further, it was assumed that transient holes produced in thin liquid films by vibrations, trapped gaseous bubbles, impurities, or particle impact had a relatively constant size in a given environment. Values of the critical thickness for different solid—liquid systems were computed by assuming a constant hole size (radius, 750 μm) and these predictions were found to be in good qualitative and quantitative agreement with the data (1,2). The theory presented in Ref. (10) needs to be developed in two important regards. First, the increase in the gravitational energy during the hole formation was neglected and second, the effect of gravity on the hole profile (shape) was also neglected in the Young—Laplace equation of capillarity. While these assumptions appear justifiable for the thin film systems considered earlier, their range of applicability to other thin film situations cannot be ascertained *a priori*. The asymptotic solution obtained by neglecting gravity predicts a monotonie increase in the critical thickness with increasing hole size (10) which is clearly invalid for large hole sizes (or thicker films) because a limiting critical thickness h_l exists for infinitely large holes (Eq. [1]). In what follows, we first numerically obtain energetically metastable solutions of the equation of capillarity for a wide range of parameters and extend the range of solutions reported by Taylor and Michael (9) to the parameter values that characterize the thin films studied by Padday (1) and Doughman *et al.* (2). Predictions of this approach are then compared with experiments (1, 2). We then generalize the previous results (10), by including the effect of gravity in the equation of capillarity, and by solving it in conjunction with the criterion that the change in free energy is zero during the process of film breakup. Results are again compared with the data (1,2). In this way, new insights are gained regarding the mechanism of spontaneous breakup of thin films, the stabilizing effects of gravity, and the relationship between the two different approaches (9, 10) that have been used to quantify breakup of thin films.

II. MODELS AND SOLUTIONS

1. Derivation and Qualitative Aspects of Film Breakup

Creation of small size, transient holes in thin films becomes possible due to external vibrations and disturbances of the film surface. Not all transient holes can, however, lead to surface dewetting, which can occur only if the free energy change is such that closure of the hole and the subsequent restoration of the unruptured film become energetically unfavorable. The free energy arguments have also been used by Ruckenstein (11) in the analysis of the film vs crystallite stability in supported metal catalysts. Depending on the magnitude of the contact angle, θ, two possible configurations of the hole are shown in Fig. 1. The change in the free energy (interfacial, surface, and gravitational energies) produced by the creation of a hole is given by

$$\begin{aligned}\Delta F &= F_{\mathrm{hole}} - F_{\mathrm{film}} \\ &= \left(A-\pi r_2^2\right)\gamma_{\mathrm{L}} + \left(A-\pi r_1^2\right)\gamma_{\mathrm{SL}} + S\gamma_{\mathrm{L}} \\ &\quad + \pi r_1^2\gamma_{\mathrm{S}} - A\gamma_{\mathrm{L}} - A\gamma_{\mathrm{SL}} \\ &\quad + \pi\rho g \int_0^{h_r 2} (h-x)\,dx, \end{aligned} \qquad [2]$$

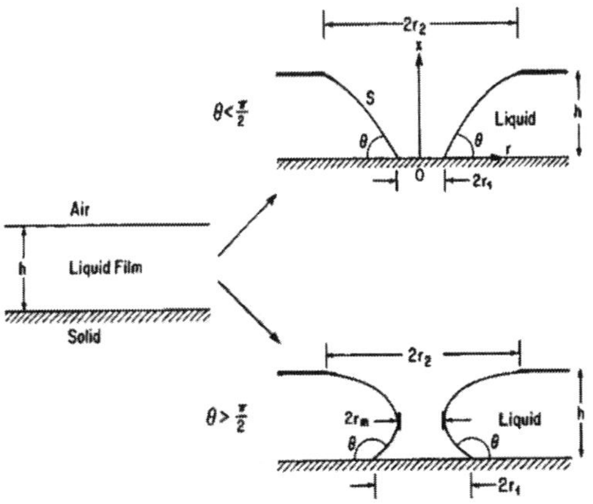

FIG. 1. Formation of a hole in a liquid film supported on a horizontal surface, r_1, and r_2 are the hole radii at the solid surface (x = 0) and at the hole mouth ($x = h$), respectively, and S is the surface area of the hole bounded between planes $x = 0$ and $x = h$. r_m is the minimum radius of the hole for $\theta > \pi/2$. The hole configurations shown are for the case when either $r_1 > r_{1t}$ or $h < h_t$.

where A denotes the initial area of contact between the liquid film and solid surface, and S denotes the surface area of the hole bounded between the horizontal planes $x = 0$ and $x = h$ (see Fig. 1). r_1 and r_2 are the radii of the hole at its base (x = 0) and at the top (x = h), respectively, and γ_L, γ_S, and γ_{SL}, are the liquid surface tension, the solid surface tension, and the solid—liquid interfacial tension, respectively. $r = r(x)$ denotes the hole profile as a function of the vertical coordinate, x. The last term in expression [2] reflects the increase of the gravitational energy due to removal of fluid in creation of the hole. Further, the surface area of the hole is determined from the formula of analytical geometry,

$$S = \int_0^h 2\pi r \left[1 + \left(\frac{dr}{dx}\right)^2\right]^{1/2} dx. \qquad [3]$$

The equilibrium profile of the hole, r = r(x), is determined by the minimization of the free energy from Eq. [2]. In this way, one obtains the well-known equation of capillarity (9, 12), by making use of the calculus of variations,

$$\frac{d^2r}{dx^2} - \frac{1}{r}\left[1 + \left(\frac{dr}{dx}\right)^2\right] = \left(\frac{\rho g}{\gamma_L}\right)(h - x)\left[1 + \left(\frac{dr}{dx}\right)^2\right]^{3/2}. \qquad [4]$$

The above equation is to be solved together with the boundary conditions

$$\text{at } x = 0, \; r = r_1, \qquad [5a]$$

$$\text{at } x = 0, \; (dx/dr) = \tan\theta. \qquad [5b]$$

The angle of contact, θ, depends only on surface and interfacial properties. θ may be either measured or evaluated from the Young equation (13),

$$\gamma_L \cos\theta + \gamma_{SL} = \gamma_S, \qquad [6]$$

when the appropriate solid and liquid properties are known.

The solution of the nonlinear equation [4] cannot be obtained analytically, but numerical simulations are performed by introducing a new variable tan ψ = (dr/dx) and thus decomposing the equation into a set of two nonlinear first-order differential equations (9) with appropriate initial values from conditions [5]. The transformed equations are

$$\frac{dr}{dx} = \tan\psi \qquad [7a]$$

and

$$\frac{d\psi}{dx} = \frac{1}{r} + \frac{(h - x)}{\cos\psi}\frac{1}{\kappa^2}, \qquad [7b]$$

Where k^2 = $(\gamma_L/\rho g)$ Numerical solutions of Eqs, [17] were obtained by a fourth-order Runge—Kutta method after elimination of parameter k^2 from Eqs. [7] by dividing x, r, and h by k, and thus rendering them nondimensional (9).

For each of the hole configurations, the change in the free energy was calculated from a rearranged version of expression [2], which also incorporates condition [6],

$$\frac{\Delta F}{\rho g} = k^2\left(S + \pi r_1^2 \cos\theta - \pi r_2^2\right) + \pi \int_0^h r^2 (h - x)\, dx, \qquad [8]$$

where S is given by Eq. [3].

For a given hole radius r_1 and contact angle θ, the following key features emerged from the numerical solution of Eq. [4]:

(i) No feasible solution for the hole profile exists in the entire range of $0 < r_1 < \infty$ whenever the film thickness is greater than a limiting thickness given by Eq. [1]. This is in accord with earlier observations (8, 9) that a film thicker than h_l cannot be made to rupture even by creation of a large hole.

(ii) For a given hole radius r_1 and a given θ, there exists a "transitional" film thickness h_t for which the hole profile asymptotically approaches its asymptote $x = h_t$ as $r \to \infty$ (see Fig. 2). No characteristic hole radius r_2 (as depicted in Fig. 1) exists for this configuration. Whenever the film thickness exceeds h_t a feasible solution cannot be found, whereas if the film thickness is smaller than h_t, the hole profile corresponds to that shown in Fig. 1. The configuration shown in Fig. 1 has a finite hole radius at its mouth, i.e., at $x = h$. The transitional film thickness is the solution of the capillarity equation with the condition $(dr/dx) \to \infty$ at $x = h$, and this solution is reported by Taylor and Michael (9). Clearly, if the hole radius is very large ($r \to \infty$), the transitional film thickness approaches the limiting film thickness as defined in the preceding conclusion (i).

(iii) The change in free energy (computed numerically using Eq. [8]), corresponding to the hole configuration at $h = h_t$ (Fig. 2) is an extremely large, positive quantity; viz, a large amount of external energy is required to create a transitional hole. This is due to an extremely large radius of the hole mouth near the asymptotic free surface, $x \cong h_t$. Indeed, qualitatively, for a thin film ($h \ll r_2$), the surface of the hole may be regarded as being close to a flat liquid sheet of radius r_2. The surface area, S, as computed from Eq. [3] can be obviously approximated by. πr_2^2. Thus, at the transitional state, the first (positive) term of Eq. [8], and the third (negative) term of Eq. [8] are of comparable magnitudes, whereas the last (positive) term due to gravity is large. The free energy change, however, decreases if the hole radius r_1 increases (for the same film thickness) or if the film thickness decreases below h_t (for the same r_1). On the basis of a similar energetic argument, Taylor and Michael (9) identified h_t to be the maximum unstable thickness for a given hole size r_1 or, alternatively, identified the minimum unstable hole size for a given film thickness. This criterion of instability seems to be most meaningful for relatively thick films in a quiescent environment, where, initially, holes have to be produced by external

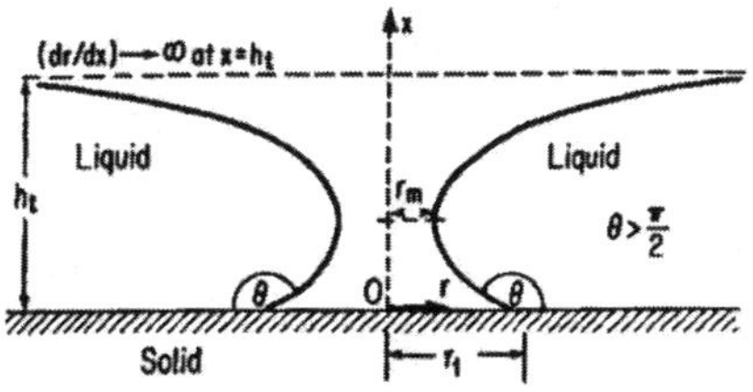

FIG. 2. Profile of a hole ($\theta > \pi/2$) at the transitional state. The hole profile approaches its asymptote, $r \to \infty$ at $x = h_t$ or in other words, $r_2 \to \infty$ and $(dx/dr) \to 0$ at $x = h_t$.

probes, at a considerable expense of energy. Once formed by external means, a hole larger than the transitional hole is likely to expand in order to reduce the free energy of the system. This process is depicted in Fig. 3, where the variations of the change in the free energy (ΔF) with the hole radius (r_1) and the film thickness (h) are shown. Note that for $r_1 < r_{1t}$, or for $h > h_t$, the film configuration is the only possibility and hence, the equilibrium value of ΔF may be considered to be zero under these conditions. A discontinuity occurs in ΔF at the transitional state, such that $\Delta F = 0$ for $r_1 = r_{1t} - \epsilon$, and $\Delta F = \Delta F_{max}$ for $r_1 = r_{1t} + \epsilon$. The change in free energy attains the maximum at the transitional thickness, then declines rapidly and eventually reaches a "critical" state ($r_{1c} > r_{1t}$; $h_c < h_t$) where $\Delta F = 0$. Clearly, producing a hole in the vicinity of the transitional point requires a large amount of energy and such holes have to be initiated artificially by external probes or jets of air (9). On the other hand, initial formation of holes in films sufficiently thin compared to h_t is very likely to occur spontaneously, because of the low energy requirements. Indeed, small external perturbations which are always present in any fluid mechanical system can, in such cases, provide the required relatively low energy input for hole formation. Thus, the spontaneous creation of a critical hole, as compared to the energetically expensive transitional hole, is a more probable event. Moreover, it is clear that in the absence of continuing disturbances, an externally produced hole with radius $r_1^* > r_{1t}$ will spontaneously open up to decrease its free energy. However, in the presence of disturbances, there is no *a priori* reason to believe that a hole with radius $r_1^* > r_{1t}$, but sufficiently near to r_{1t}, will necessarily grow, because the subsequent disturbances of the liquid surface can as easily roll back the boundaries of the hole to a point ($r_1 < r_{1t}$) where the film configuration again becomes energetically favorable. It is clear from Fig. 3 that the free energy of the system is also decreased when a hole of radius $r_{1t} < r_1^* < r_{1c}$ collapses and the film configuration is restored. For thin films that display "spontaneous" rupture, which are up to only several hundred micrometers thick, small holes are created by the low energy dynamic disturbances of the film surface. As already noted, for a given h and θ, there is a hole radius (denoted by r_{1c}) at which the film and the hole configurations have the same free energy, i.e., $\Delta F = 0$. The probability of

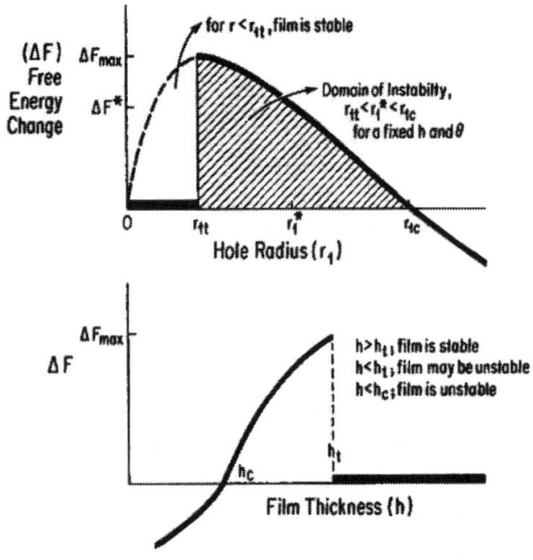

FIG. 3. The change in the total free energy ($\Delta F = F_{hole} - F_{film}$) is positive and maximum at the transitional state and decreases away from it ($h < h_t$ or $r_1 > r_{1t}$). ΔF eventually becomes zero at the critical state. ΔF is considered zero for $r < r_t$, because all transient holes in this region are spontaneously healed, and because an equilibrium solution for the hole profile does not exist. The only equilibrium solution here is the film configuration ($\Delta F = 0$). The dashed curve for $r < r_t$ acknowledges the fact that positive values of ΔF do occur for dynamic profiles of holes that are spontaneously healing.

a critical hole reverting back to the film configuration becomes remote, even in the presence of disturbances, because of the large expense of energy needed. In contrast, the probability of a transitional metastable hole being destroyed is higher.

In conclusion, spontaneous formation of a hole, due to low energy vibrations, becomes increasingly probable, and subsequent reversal to the film configuration becomes increasingly improbable, as the film thickness decreases below the transitional thickness and approaches the critical thickness. In view of these arguments (see also Fig. 3), the condition $h \leqq h_c \left(\text{or } r_l \geqq r_{lc}\right)$ may be identified to be a *sufficient* condition for film instability, whereas the condition $h < h_t (\text{or } r_l > r_{lt})$ constitutes a *necessary* condition of film instability. In general, then, the domain of *initial* instability for thin films is, most likely, confined to the parameter values,

$$r_{lc} \geqq r_l > r_{lt}$$

$$\left(\text{for a given film thickness}\right) \qquad [9a]$$

or

$$h_c \leqq h < h_t \quad \left(\text{for a given hole size}\right). \qquad [9b]$$

We now briefly recall that the critical state (h_c or r_{lc}) is determined by a solution of the capillarity equation for which $\Delta F = 0$, whereas the transitional state (h_t or r_{lt}) is a particular solution of the capillarity equation [4] for which the hole profile approaches the plane $x = h_t$ asymptotically (Fig. 2). Alternatively, the transitional state may also be defined as the solution for which the one sided derivative ($\partial \Delta F/\partial r_l$) vanishes. The transitional state is a metastable equilibrium, departures from which decrease the free energy in two distinct ways, namely, either by reverting back to the film configuration or by further expansion of the hole. The path taken depends upon details of the external disturbances of the liquid surface. However, a further reduction in the total free energy from the critical state is possible only by the continued expansion of the hole.

While we have obtained equilibrium shapes of holes numerically from the equation of capillarity, all of the qualitative aspects of the discussion presented above can be analytically illustrated by approximating the hole with a cylinder of radius r. The change in the free energy in this simplified case may be easily shown to have the analytical form

$$\Delta F = \pi \gamma_L \left[2rh - r^2\left(1 - \cos\theta\right)\right].$$

Clearly, ΔF becomes zero for a vanishingly small hole; ΔF increases with an increase in the hole radius until it attains its maximum value (the transitional state), and then declines until it becomes zero once again (the critical state). While ΔF is positive in the entire domain $0 < r < r_c$ (i.e., free energy is required for hole formation), the condition $(\partial \Delta F/\partial r) > 0$ implies stability, whereas the condition $(\partial \Delta F/\partial r) < 0$ implies instability.

2. Characterization of the Transitional State

The metastable equilibrium or the transitional state is a particular solution of Eqs. [7a] and [7b] with boundary conditions [5a] and [5b], for which the additional asymptotic condition,

$$\frac{dr}{dx} = \tan\psi \to \infty \text{ or } \psi \to \pi/2,$$

$$\text{at } x = h, \qquad [10]$$

is also satisfied (see Fig. 2).

For a given radius of the hole at its base (r_1) and angle of contact (θ), Eqs. [7] were integrated by assuming a value of the film thickness (h) smaller than the limiting thickness (h_1). In general, this integration leads to a value of ψ at the boundary $x = h$ which is different from $\pi/2$. The thickness at which ψ equals $\pi/2$ at $x = h$ is the transitional film thickness for the given r_1 and θ. Using a slightly different computational approach, Taylor and Michael (9) have given plots of (h_t/k) versus (r_m/k), the minimum hole radius, for holes larger than $0.1k$. This range of hole radii corresponds to transitional film thicknesses larger than those studied by Padday (1) and Doughman *et al.* (2). We, therefore, report results of our computations for smaller hole sizes which are germane to the experimental results (1,2), to the investigation of tear films (7), and to potential future studies of thin liquid films.

Figures 4 and 5 depict the variation of the nondimensional transitional film thickness (h_t/k) with the nondimensional minimum hole radius (r_m/k) for a wide range of contact angle values. Results for the hole radii in the range of $10^{-3}k$ to $10^{-1}k$ that correspond to initial hole diameters of about 5 to 540 μm for pure water are shown. Recall that r_m is the minimum hole radius. For $\theta > \pi/2$, this quantity is physically meaningful (Figs. 1 and 2). For $\theta < \pi/2$, the minimum hole radius is a fictitious quantity which is reached below the solid surface ($x < 0$) if the solution of the capillarity equation is continued for $x < 0$. Experimentally, the radius of the hole at its base (r_1) is a more meaningful parameter, especially for $\theta < \pi/2$, and r_1 is related to r_m as shown in Fig. 6. Note that due to a pronounced symmetry of the hole profile around the plane where $r = r_m$ (Fig. 2), r_1, is about

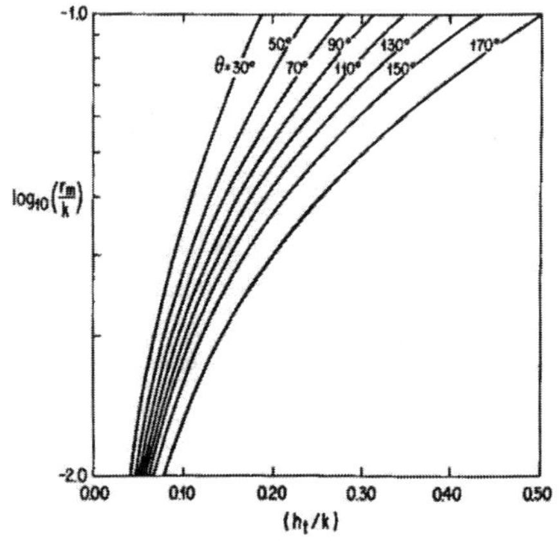

FIG. 4. Variation of nondimensional transitional thickness with the minimum hole radius for several values of the equilibrium contact angle. Parameter k is defined as $(\gamma_L/\rho g)^{1/2}$.

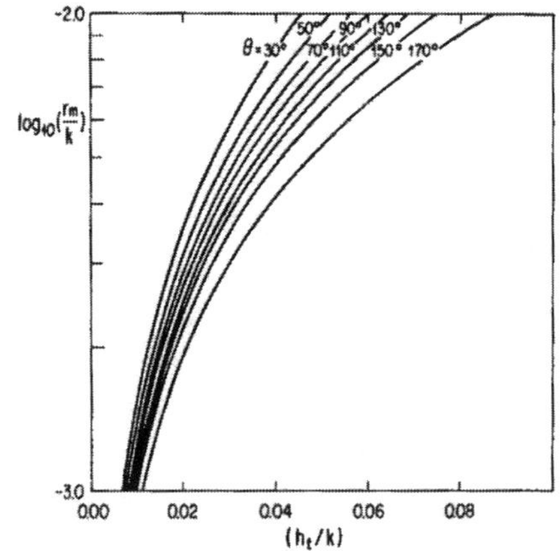

FIG. 5. Variation of nondimensional transitional thickness with the minimum hole radius.

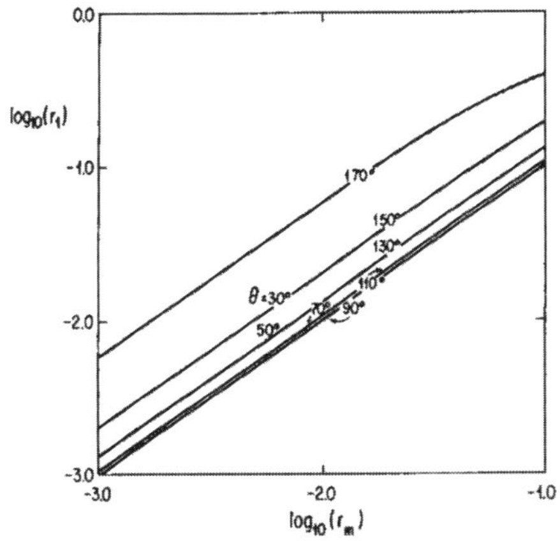

FIG. 6. Relationship between the minimum hole radius (r_m) and the hole radius at its base (r_1). Note that the same relationship holds for any pair of contact angles that are given by ($\pi/2 + \beta$) and ($\pi/2 - \beta$) (e.g., 70°, 110°; 50°, 130°). The minimum hole radius is a physical quantity for $\theta > \pi/2$ (Figs. 1, 2), whereas for $\theta < \pi/2$, r_m is a hypothetical quantity obtained by continuing the solution below the solid surface ($x < 0$).

the same for $\theta = \pi/2 + \beta$ and for $\theta = \pi/2 - \beta$, for any $\beta < \pi/2$. It is physically obvious that $r_1 \cong r_m$. for small deviations from $\theta \cong \pi/2$. However, for very large or very small angles (e.g. 150° and 30°), the base radius (r_1) is substantially larger than the minimum hole radius.

In essence, a transitional film thickness may be found from Figs. 4–6 when θ and either r_1 or r_m are known. Alternatively, a transitional hole size (r_1 and r_m) may be determined for a given film thickness and θ. This completes the description of one of the boundaries of the domain of instability (see Eqs. [9]). Recall that film instability ensues only for thicknesses smaller than the transitional thickness. While the computations reported here are for small hole sizes and thin films, the computations of Taylor and Michael (9) for large hole sizes ($r_1 \to \infty$) show that the transitional film thickness approaches its limiting value h_1 as given by Eq. [1]. This asymptotic result is easily obtained from Eq. [4] by neglecting the second term when $r \to \infty$, by introducing a new variable $\eta = (dr/dx)$, and by integrating the resulting equation once with the help of condition [10]. This gives

$$\frac{\eta}{\sqrt{1+\eta^2}} = 1 - \frac{1}{2k^2}(h-x)^2. \qquad [11]$$

Now making use of the boundary condition [5b], i.e., $\eta = \cot\theta$ at $x = 0$, one directly obtains the formula for the limiting film thickness, Eq. [1].

3. Characterization of the Critical State

As discussed earlier, we define a critical state such that it has the same free energy as an uninterrupted film, i.e., $\Delta F = 0$ in Eq. [8]. Computations of the critical thickness proceed by solving the capillarity equation [7] in conjunction with Eq. [8] with the condition that ΔF is zero. For given r and θ, this is achieved by obtaining the hole profiles from Eq. [7] for $h < h_t$. ΔF is then calculated from Eq. [8] for each of these hole profiles and the film thickness at which ΔF changes its sign from positive to negative is identified as the critical thickness (see also Fig. 3). The change in free energy (ΔF) is a large positive quantity at $h = h_t$. and decreases with decreasing h, eventually becoming zero at $h = h_c$.

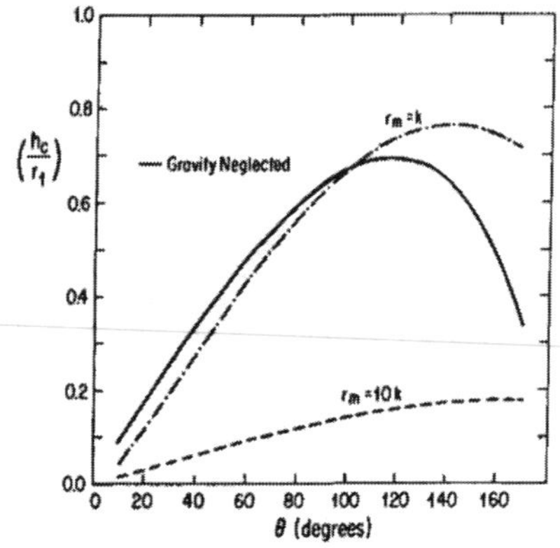

FIG. 7. Variation of the critical thickness with the solid-liquid contact angle. The solid curve represents an analytical solution [14] obtained by neglect of gravity. The neglect of gravity is justified for small hole radii not exceeding about 0.3k and for contact angles in exces of 20°. For large hole sizes ($r_m \to \infty$), the critical thickness approaches its upper limit of the limiting thickness and the analytical solution is no longer valid.

Numerical solutions for the ratio of the critical thickness to the base hole radius are shown in Fig. 7 as a function of the contact angle for different values of the minimum radius, r_m. The solid curve corresponds to an analytical solution obtained earlier by Sharma and Ruckenstein (10) by neglecting gravity in Eqs. [4] and [8]. The reason that progress of the curve slows down after $\theta > \pi/2$ and reverses itself after $\theta > 120°$, is that for $\theta > \pi/2$, r_1 becomes larger than r_m and the discrepancy between the two increases with an increase in θ (Fig. 2). Thus, even though (h_c/r_m) increases monotonically for $\theta > \pi/2$, (h_c/r_1) begins to decrease. The critical thickness of course increases with an increase in surface nonwettability (increase in θ) for a fixed minimum hole radius. This is easily verified by converting the ratio (h_c/r_1) into the ratio (h_c/r_m) for $\theta > \pi/2$ with the help of the relation between r_1 and r_m as reported in Fig. 6. In conclusion, the base radius (r_1) is the most meaningful characteristic hole dimension for $\theta < \pi/2$, whereas for $\theta > \pi/2$, the minimum radius (r_m) may be employed for comparisons among systems with different contact angles (θ).

Further, computer simulations showed that for the hole radii (r_m) less than 0.2k and for $20° < \theta < 170°$, the computed critical thicknesses are virtually the same as those obtained by neglecting gravity. Thus, the analytical solutions reported earlier (10) are satisfactory for moderate to large contact angles whenever the critical hole size or the critical film thickness is small. As an example, gravity may be neglected for holes of radius less than 540 μm in water films, which corresponds to a critical thickness of less than 350 μm for $\theta = 90°$. An important conclusion is that the ratios (h_c/r_1) and (h_c/r_m) are independent of r_1 or r_m for small radii and depend only on the contact angle, i.e., the critical thickness increases monotonically with the hole radius. However, as is shown in Fig. 7, the ratio $(h_c/r_1,)$ decreases for large holes and, in fact, h_c approaches a maximum value for r_m or $r_1 \to \infty$. The maximum value of h_c is easily deduced from Eq. [8] with $\Delta F = 0$ in the limit of $r_1 = r_2 = r \to \infty$ Interestingly, this also leads to the earlier limit

$$h_c = h_l = 2k \sin(\theta/2) \text{ for } r \to \infty. \qquad [12]$$

As can be physically anticipated, for large holes, the critical thickness approaches the transitional thickness and both of these equal the limiting film thickness. The critical state and the transitional state are of course substantially different for thin films with small diameter holes. As discussed earlier, the radius of a hole at its mouth (r_2 in Fig. 1) is finite when the thickness is less than the transitional thickness. Thus, for a given r_1, or r_m, a unique r_2 exists for a critical hole size. The variation of the ratio (h_c/r_2) is depicted in Fig. 8. The analytical results (10) were again found to be satisfactory for $r_m < 0.2k$, whereas the asymptote of h_c for large hole sizes is given by Eq. [12].

In summary, the critical thickness for a given solid—liquid system may be evaluated from Figs. 6–8 if any of the characteristic hole radii, r_1, r_m, or r_2, are known.

III. COMPARISON OF THE MODEL WITH EXPERIMENTS

Experimental values of the film thickness at the instant of rupture are summarized in Table 1 for several solid-liquid systems (1,2).

Values of the contact angle have not been reported for some of these systems and these are estimated from the equation for nonpolar solids (13),

$$\left(1+\cos\theta\right)\gamma_{\mathrm{L}} = 2\left(\gamma_{\mathrm{s}}^{\mathrm{d}}\gamma_{\mathrm{L}}^{\mathrm{d}}\right)^{1/2}, \qquad [13]$$

where γ_L is the total (dispersion plus polar) surface tension of the liquid and γ_s^d and γ_L^d are the surface tension components of the solid and liquid due to the dispersion (apolar) interactions. In the absence of any information regarding the type of Teflon used by Padday (1), we assume a "classical" value of 19.5 dyn/ cm for γ_s^d of Teflon (12). For ethylene glycol (12), $\gamma_L = 49$ dyn/cm and $\gamma_L^d = 28.6$ dyn/cm, and for tetradecene and octadecene, $\gamma_L \cong \gamma_L^d = 27$ and 28.4 dyn/cm, respectively (1). As is explained elsewhere (10), the data of Padday (1) for the Teflon—benzene system and the data of Doughman *et al.* (2) for the Teflon-water system are excluded from the analysis. The former because it is not consistent with the other data of Padday, and the latter because it is not consistent with the data of Padday for the same system.

In reality, the film thickness at the instant of rupture, h_R, is expected to lie between the transitional thickness and the critical thickness (see Eq. [9]). Two approaches for comparing the theory with the experiment are possible. If the film thickness at the instant of rupture is assumed to be the same as the transitional film thickness, the corresponding hole radius, r_{1t} may be evaluated from the theory of Section II.2 (see Figs. 4–6). If the film thickness at breakup is identified as the critical thickness, the corresponding critical hole size, r_{1c}, may be computed from the theory of Section II.3 (see Fig. 7–8). The computed transitional and critical hole radii corresponding to the film thickness data of Padday (1) and Doughman *et al.* (2) are listed in the last two columns of Table 1. The hole size is expected to be relatively invariant in a given environment and indeed, this appears to be the case with the calculated hole radii as given in Table 1. For the data of Padday (1), the mean hole radius ± standard deviation for the transitional hole is 127.5 ± 27 μm, and for the critical hole this quantity is 719.5 ± 84.5 μm. However, the relative spread (= SD/mean) of the hole radii around their mean value is larger (21.2%) for transitional holes as compared to critical holes (11.7%).

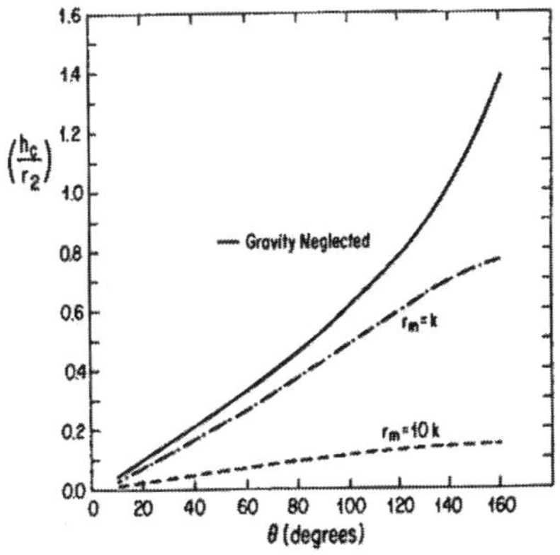

FIG. 8. Variation of the ratio (h_c/r_2) with the contact angle. The information contained in this figure is essentially the same as that in Fig. 7, but this representation is better suited for evaluation of the critical thickness when the hole radius at the mouth (r_2), rather than the radius at the base (r_1), is known.

TABLE 1
Predictions of the Hole Size for Various Solid—Liquid Systems

Solid—liquid system	Contact angle, θ (degrees)	Parameter. $k = (\gamma_L/\rho g)^{1/2}$ (cm)	Observed film thickness at rupture, h_R (μm)	Theoretical predictions for hole radius at its base, r_1 (μm)	
				Transitional size	Critical size
Data of Padday (1)	110(13)	0.27	510	126	713
Teflon—water	92[a]	0.21	560	168	840
Teflon—ethylene glycol	40(13)	0.176	260	146	783
Teflon-1-decene	45[a]	0.187	270	127	720
Teflon-1-tetradecene	50[a]	0.19	270	111	656
Teflon-1-octadecene	65[b]	0.27	310	87	605
Paraffin wax-water					
Data of Doughman *et al.* (2)					
Polyethylene—water	94 (2, 13)	0.27	420	101	632
Polymethyl methacrylate-water	50-50–75[c]	0.27	393	107–65	683–956

[a] Estimated values from Eq. [13].
[b] While the paraffin wax used was not characterized for contact angles, a possible value of receding angle is reported (9).
[c] Probable range of receding-advancing contact angles for water on PMMA (14).

This would argue that the actual hole size is closer to the critical value, but whether the difference is statistically significant is uncertain. Photographs of a freshly created hole (1) certainly seem to indicate a hole radius between the predicted upper and lower bounds (719 and 127 μm). Even if the data of Doughman *et al.* (2) are included in the analysis, the mean hole radius and standard deviation remain about the same as those reported for Padday's data. These calculations support the hypothesis that the initial size of the holes created in different solid—liquid systems is relatively invariant, at least in a given environment. This means the film thicknesses at the instant of rupture are different for different liquid—solid combinations due largely to differences in the contact angle and in the parameter $k^2 = (\gamma_L/\rho g)$.

It is clear from Table 1 that the film thickness at instability may be predicted for any solid—liquid system whenever it is known for a single well-defined solid—liquid pair. This is accomplished by first evaluating the transitional hole size or the critical hole size and then assuming the same hole radius for other systems. This test of the theory is confirmed in Table 2. Predictions for the transitional film thicknesses are reported in column 4 for a fixed base hole radius of 127.5 μm, which is the average hole radius evaluated from column 5 of Table 1. Predictions of the critical thicknesses are reported in column 5 of Table 2 for a fixed hole radius of 719.5 μm. The predicted values agree well with the experimental values; the mean errors in prediction being 12.4% for the theory based on the transitional hole size and 9.1% for the theory based on the critical hole size. Further, the degree of linear correlation between the predicted and experimental values was assessed by the correlation coefficient. A high degree of correlation was found between the predicted and experimental thicknesses, the correlation coefficients being 0.86 for the predicted transitional thicknesses and 0.93 for the predicted critical thicknesses. The formalism based on the critical state offers, however, a distinct computational advantage since an analytical solution is possible for the critical thickness of thin films. The critical thickness is obtained from the following equation (10) whenever the influence of gravity is negligible in the equation of capillarity, Eq. [4]:

TABLE 2
Prediction of the Film Thickness at Rupture

System	θ (degree)	h_R μm	$h_t(\mu m)$ $r_t = 127.5\ \mu$m	$h_c(\mu m)$ $r_t = 719.5\ \mu$m
Teflon-l-decene	40	260	238	240
Teflon-1-tetradecene	45	270	269	270
Teflon-1-octadecene	50	270	298	295
Paraffin wax-water	65	310	407	366
PMMA-water	50–75	393	33–448[a]	296–414[a]
Teflon-ethylene glycol	92	560	466	480
Polyethylene-water	94	420	500	478
Teflon-water	110	510	513	512

Note: A constant value of the hole radius at its base is assumed for all solid—liquid systems. The correlation coefficient of the predicted transitional thicknesses (h_t) with the data (h_R) is 0.86. The percentage mean error in prediction is 12.4%. The correlation coefficient of the predicted critical thicknesses (h_c) with the data (h_R) is 0.93. The percentage mean error in prediction is 9.1%. Data for PMMA—water system are excluded from the statistical analysis.

[a] Based on a range of receding—advancing contact angle values. The experimental value lies in the predicted range.

$$\frac{2h_c}{r_1 \sin\theta} = 1 + \frac{(1-\cos\theta)^2}{\sin^2\theta} \exp\left(\frac{-2h_c}{r_1 \sin\theta}\right). \quad [14]$$

Note that in this approximation, the ratio (h_c/r_1) depends only on the contact angle θ and not on the parameter k. The critical thicknesses as computed from Eq. [14] by assuming $r_1 = 719.5\ \mu$m show a mean deviation of about 1% from the values reported in Table 2, which are based on numerical solutions of Eq. [4] with the inclusion of gravity. The approximation [14] is, however, not accurate for $\theta < 20°$ and for large holes/thick films that are comparable to the limiting thickness of the film (see Figs. 7 and 8).

In conclusion, the initial radius of the hole responsible for film destruction lies between the radii of the transitional hole and the critical hole. The actual hole size appears to be relatively invariant for different combinations of solid surfaces and liquids forming the film. The film thickness at the instant of rupture and its variation with surface properties are well predicted by assuming the hole to be either a metastable equilibrium (transitional state as determined from the equation of capillarity) or a critical point where the change in the free energy is zero. The latter approach is simpler and appears to offer a slightly better fit to the available data (1, 2) on the breakup of the thin liquid films.

Both approaches provide a satisfactory explanation of the available data on the rupture of thin liquid films up to 560 μm thick. This is because the films investigated are thin and show little variation in the parameter k. In these circumstances, the variations of the transitional thickness and the critical thickness with the contact angle are essentially similar. The overall agreement between the theory and the experiments is quite striking if one also notes that the exact value of the contact angle is very sensitive to the solid sample used, and that some variations of the hole radius from its mean value are also quite likely.

IV. CONCLUSIONS

We evaluate changes in the free energy during the process of hole formation and provide energetic criteria for the instability of liquid films supported on horizontal solid surfaces. The equilibrium profiles of holes are determined from the Young—Laplace equation of capillarity. For film thicknesses in excess of a limiting thickness (see Eq. [1]), the formation of a hole *always* increases the free energy. This is due to an overhwelming stabilizing influence of the gravity which dwarfs the destabilizing changes in the surface and interfacial free energies. A thick film is therefore unconditionally stable even on the most lyophobic surface ($\theta \rightarrow \pi$) and even the largest hole created in such a film is spontaneously healed.

For a finite-sized hole, there is a corresponding transitional film thickness (h_t) at which the change in the free energy is positive and maximum. A further reduction in the film thickness decreases the free energy and a critical film thickness (h_c) is eventually reached when the change in the free energy becomes zero, or, in other words, the free energy after creation of the hole equals the free energy of the unbroken film. For a given hole size, then, the initial instability and rupture of the film is likely to occur at an intermediate thickness, h_R, such that

$$h_c < h_R < h_t.$$

Both the transitional thickness and the critical thickness increase with increased hole radius and increased surface nonwettability as reflected in the contact angle, θ. For extremely large holes ($r \rightarrow \infty$), both h_c and h_t, approach the maximum asymptotic value (the limiting thickness), which depends only on the contact angle (Eq. [1]).

The exact film thickness at rupture is intimately related to the initial hole radius as well as to the mechanism of hole formation. The maximum unstable film thickness is expected to be only slightly smaller than the transitional thickness $\left(h_R \cong h_t\right)$ for thick films in a quiescent environment, where holes have to be formed by an external probe or wisp of air (2). However, for thin films (1, 2), near the critical state, the energy barrier for hole formation is reduced and transient holes may form spontaneously due to external disturbances, small particle impact, trapped bubbles, and impurities. In this event, the unstable film thickness may be closer to the critical thickness; both because it is easier to form holes and because it is difficult to close them after their formation. In the absence of a well-defined external constraint causing hole formation, the hole radius cannot be ascertained a priori. It may, however, be anticipated that in a given environment, the hole radius may be relatively invariant for different solid—liquid systems. On the basis of the theory presented here, this indeed turns out to be the case for the available data (1, 2) on the film thickness at instability (see Table 1). This is fortunate, because it paves the way for predicting the film thickness for any solid—liquid system whenever data are available for one well-defined system.

Further, we have shown that for thin films (up to several hundred micrometers thick) with comparable values of the parameter k the variations of h_c and h_t with the contact angle θ are very similar. It is because of this that the experimental values of the film thickness are about equally well predicted by computations of the critical or the transitional thickness (see Table 2). For thin films, computation of the critical thickness is more straightforward since an analytical solution Eq. [14] is available.

In conclusion, the energetic criteria are able to predict the variation of the film thickness at instability with changes in the surface and in the interfacial properties of the liquid—solid system. The exact magnitudes of the unstable film thicknesses for different systems may, however, be predicted only if data are available for one solid—liquid system, from which the hole radius may be evaluated. The theory suggests an alternative operational definition of surface hydrophilicity/hydrophobicity as a dewetting phenomenon (10) and provides a criterion for design of surfaces which remain wettable by a liquid film of desired thickness.

ACKNOWLEDGMENT

We thank Mr. Sanjeev Narayan for obtaining numerical solutions of Eqs. [7] that are reported in Figs. 4–8.

REFERENCES

1. Padday, J. F., *Spec. Discuss. Faraday Soc.* **1**, 64 (1970).
2. Doughman, D. J., Holly F. J., and Dohlman, C. H., presented at the Association for Research in Vision and Ophthalmology Meeting, Sarasota, FL (1971); Holly, F. J., *Exp. Eye Res.* **15**, 515 (1973).
3. Scheludko, A., *Adv. Colloid Interface Sci.* **1**, 391 (1967).
4. Ruckenstein, E., and Jain, R. K., *J. Chem. Soc. Far*aday Trans. *II* **70**, 132 (1974) (Section 5.1 of this volume).
5. Sharma, A., and Ruckenstein, E., *J. Colloid Interface Sci.* **111**, 8 (1986) (Section 5.11 of this volume).
6. Ruckenstein, E., and Sharma, A., "Preocular Tear Film" (Holly, F. J., Ed.), p. 697. Dry Eye Institute, Inc., Lubbock, 1986.
7. Sharma, A., and Coles, W. H., *Invest. Ophthalmol. Visual Sci.* (Suppl), in press.
8. Lamb, H., "Statics," 2nd ed. Cambridge Univ. Press, London, 1916.
9. Taylor, G. I., and Michael, D. H., *J. Fluid Mech.* **58**, 625 (1973).
10. Sharma, A., and Ruckenstein, E., *J. Colloid Interface Sci.*, **133**, 358 (1989) (Section 5.8 of this volume).
11. Ruckenstein, E., *in* "Metal-Support Interactions in Catalysis, Sintering and Redispersion" (S. A. Stevenson, J. A. Dumesic, R. T. K. Baker, and E. Ruckenstein, Eds.), p. 236. Van Nostrand, New York, 1987.
12. Padday, J. F., and Pitts, A., *J. Colloid Interface Sci.* **38**, 323 (1972).
13. Adamson, A. W., "Physical Chemistry of Surfaces," 4th ed., pp. 332–368. Wiley, New York, 1982.
14. Holly, F. J., *in* "Physico-Chemical Aspects of Polymer Surfaces" (K. L. Mittal, Ed.), Vol. 1, pp. 141–154. Plenum, New York, 1983.

5.10 Mechanism of Tear Film Rupture and Formation of Dry Spots on Cornea*

Ashutosh Sharma and Eli Ruckenstein[1]

Department of Chemical Engineering, State University of New York, Buffalo, New York 14260 Received June 8, 1984; accepted October 4, 1984

INTRODUCTION

The stability of the lacrimal liquid film covering the cornea (or the contact lens) has attracted much attention (1–11) because of the important role it plays in the optimal conditions of vision; both from the optical and physiological points of view. The normal tear film forms a continuous smooth cover over the corneal surface. The structural integrity of this continuous film is maintained by involuntary periodic blinking, with a normal interblink interval of about 5 to 10 s. If, however, the eyelids are held open for a longer time period, the stability of the tear film is threatened, resulting in the eventual formation of randomly distributed holes in the lacrimal liquid.

The average time elapsed between a blink and the first hole to appear is defined as the tear breakup time (BUT). This is found to be in the range of 20 to 50 s for a normal adult eye (2). For a pathological eye, however, BUT is found to be considerably less than the normal BUT. A breakup time of less than 10 s is usually considered abnormal and may have mild to serious consequences, such as: an instantaneous sensation of local irritation, a persistent feeling of ocular irritation, or a foreign body sensation. In severe cases, a breakup time of less than the interblink period may result in an irreversible dewetting of cornea, epithelial damage, and corneal ulceration. An abnormally rapid tear breakup also plays an important role in determining the contact lens tolerance. It has been shown to be responsible for the adhesion of the lens to cornea (and consequent epithelial damage) and excessive, rapid deposit formation on the anterior lens surface (12, 13).

The measurement of BUT thus plays an important role in the characterization and differentiation of a normal cornea from a pathological one. In view of this, it is not surprising that the clinical measurements of BUT are widespread in assessing the severity of conditions associated with a dry eye. It is, however, surprising that in spite of its significance, none of the models proposed for the instability of the tear film address the issue of the time of rupture. This information may be used as an important criterion for model discrimination. The following two mechanisms have been suggested as possible causes for the rupture of the tear film.

MARANGONI INSTABILITY

It has been suggested that the lipids secreted from the tarsal glands decrease the overall wetting ability of the aqueous layer which rests on a mucus-coated corneal epithelium (12–15). The tear film consists of a mucous layer covering the epithelium, an aqueous layer and a lipid—mucin bilayer located at the air—tear interface (a brief exposure to the structure of the tear film is provided later). In a series of papers,

* *Journal Colloid and Interface Science*. Vol. 106, No. 1, p. 12 July 1985. Republished with permission.
[1] To whom correspondence should be addressed.

Holly and his associates (1–3, 12–17) have advanced a mechanism of film rupture based on the assumption that the lipids present at the tear—air interface migrate rather rapidly to the mucous—aqueous interface and eventually overwhelm the protective capacity of the mucous layer, thus creating areas of high hydrophobicity. The surface-tension driven motion (Marangoni effect) was thought to be responsible for such a process (14). This may happen if the convective motion along the lipid—air interface (from the regions of low surface tension to those with high surface tension) is in the same direction as the convective flow induced by the interfacial perturbations. However, any uneven distribution of interfacial lipids, due to interfacial perturbations, has a tendency to equilibriate. This is so, because the surface-tension gradient driven Marangoni motion is in the direction opposite to the convective flow caused by interfacial perturbations. This stabilizing effect of surface active agents on the stability of a falling film (18) and a thin film subjected to dispersion forces (19) has been reported. It should however be noted that even in the absence of Marangoni convection, the diffusion of lipids in the aqueous layer is present. Based on the magnitudes of the thickness of aqueous layer ($h_0 \sim 6$–$9\ \mu$m) and the diffusion coefficient, $D \sim 10^{-5}$ cm²/s., the penetration time (h_0^2/D), for the lipids is easily shown to be of the order of 10^{-2} s. Thus the aqueous layer becomes saturated with lipids in less than a second. The solubility of lipids, which are made of waxy and cholesteryl esters (2), in the aqueous medium is rather low. In addition, since the lipids are much less surface active than the mucous material covering the epithelium, their adsorption on the mucous- aqueous interface is thermodynamically unfavorable, viz., it increases the mucous—aqueous interfacial tension. As shown later, an increased mucous—aqueous interfacial tension makes the mucous layer more resistant to the interfacial deformations and thus contributes to its stability. In addition, it may be noted that the presence of lipids is not necessary for the tear film breakup which is observed even in the event of a complete destruction of meibomian gland openings (2).

Based on an upper estimate for the mucin-concentration gradient between the epithelium and the air—tear interface, Lin and Brenner (5) show that a mucin-concentration driven Marangoni motion may exist in the aqueous tear film. The presence of surface-active agents (in this case, interfacial lipids), however, makes it less likely.

RUPTURE DUE TO DISPERSION FORCES

Lin and Brenner (6) carried out a linear stability analysis of a nondraining, micrometer-sized tear film, under the influence of retarded van der Waals dispersion forces. The thickness of neutrally stable tear film was computed for various values of the retarded Hamaker constant, by taking the wavelength of the perturbation to be the same as the linear dimension of the eye (~ 1 cm). A tear film with a thickness less than this critical thickness eventually ruptures because of the dominance of the dispersion forces over the viscous and surface tension resistances of the fluid to any interfacial deformation. Additional information about the time of rupture is, however, needed for the purpose of model differentiation.

The main purpose of this work is thus to formulate a mechanism of the tear-film rupture which is consistent with the observed breakup times and other characteristics of the rupture process. For this purpose, we establish a unified linear instability criterion which incorporates the van der Waals interactions and the hydrodynamics of the tear flow. An estimate for the time of rupture is obtained from this linear stability analysis. An application of this formalism to the entire micrometer-sized tear film (considered as a homogeneous film) shows that such a rupture cannot occur within the observed breakup times. Therefore a different mechanism of the tear-film rupture is required to explain the observed BUT. The fact that the tear film is, in reality, composed of a hydrophobic epithelium, a mucous layer, and an aqueous layer, naturally suggests a mechanism which should incorporate this heterogeneity. The cause for the instability of the entire tear film is sought in the instability and the rupture of the thin mucous layer, which covers the epithelium, under the influence of van der Waals dispersion forces. This mechanism is shown to be consistent with the observed breakup times and other characteristics of the rupture process. In addition, it aids in a rational understanding of a diversity of clinical observations about the pathological conditions of a mucus-deficient eye and the tolerance of contact lens. A clear understanding of the proposed mechanism and model differentiation thus necessitates a brief exposure to the structure and functioning of the tear film, as well as to the observed characteristics of the rupture process. This is followed by linear stability analysis and derivation of expressions for BUT. The implications of the suggested mechanism are then discussed.

THE STRUCTURE AND THE RUPTURE CHARACTERISTICS OF A TEAR FILM

The tear film which covers conjunctiva and cornea consists of at least three distinct fluid layers. The outermost layer making the air—tear interface is a lipid—mucin layer, which is about 1000 to 2000 Å thick (2). Lipids present in this layer retard the evaporation of the aqueous phase so that only about 7% of the aqueous layer evaporates in 1 min under the normal circumstances (13). An increased rate of evaporation has an adverse effect both on the thinning of the tear film and the dehydration of cornea. In the presence of a normal lipid layer, the drying of the epithelial surface cannot be caused by the evaporation, since it would take about 10 to 20 min. Underneath this lipid layer is an aqueous electrolyte middle phase which is about 6–9 μm thick immediately after blinking. The thickness then decreases in an almost linear manner because of evaporation and osmotic transfer of tears across the cornea, and in approximately 20 to 50 s (BUT), its thickness is reduced to about 4 μm. At this point, the film has been observed to rupture almost instantaneously (4). Sandwiched between the aqueous phase and the corneal epithelium is a mucoid layer, which, in a normal eye, is about 200 to 500 Å thick (2). Most of the mucous material covering the superficial epithelium originates in the conjunctival goblet cells and is distributed over the preocular surface by the shear created by the lid motion during blinking. The renewal rate of this layer is very small and only a small fraction of it is removed during each blink. In addition to serving such vital functions as the maintenance of corneal and conjunctival surfaces in the proper state of hydration and lubrication, the mucous layer provides a hydrophilic base for an even spreading of the aqueous tear film (15).

The corneal and conjunctival surfaces are highly hydrophobic and indeed, they are incapable of sustaining a continuous, aqueous tear film without the presence of the mucous layer coating the epithelium (2, 3, 15). There is even some experimental evidence for a highly hydrophobic lipid monolayer sandwiched between the epithelium and the mucous layer (14). The presence of the mucous layer is thus necessary to effectively mask the hydrophobic character of the epithelial surface and to impart a stable, hydrophilic base to the tear film. Experiments of Dolhman *et al.* (8) also demonstrate that a reduction in the glycoproteins contents (a mucous-layer constituent) is associated with a decrease in the tear-film breakup time and was detected in some pathological states of the dry eye. The various components of a typical, normal tear film are depicted in Fig.1.

The production rate of tear is approximately 1.2 μl/min, which is drained continuously between each blink. Calculations based on the cross-sectional area of the tear film (~2×10^{-3} cm^2) and the production rate give an average tear velocity of about 10^{-2} cm/s for such a flow (6).

Perhaps, it is also important to note that a meniscus extends along the entire margin of both the upper and lower eyelids. Similar menisci are present surrounding the bubbles and debris found in the tear film (11) and contact lenses. The locally thin areas, the so-called "black lines," appear adjacent to these thick menisci. Such locally thin (couple of micrometers thick) areas have been thought to be instrumental in accelerating the process of film rupture (6).

With this brief exposure to the tear film, we now proceed to derive a stability criterion for a tear film.

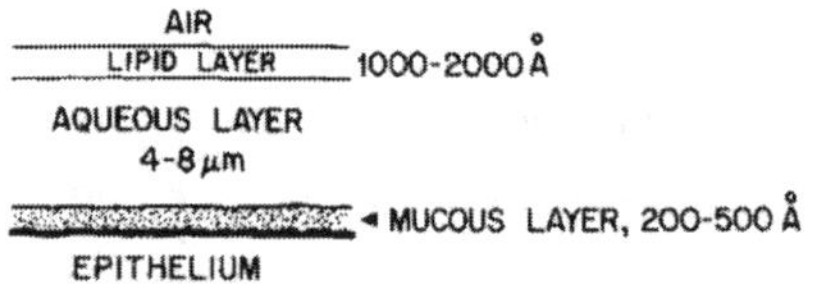

FIG. 1 Structure of the precorneal tear film.

THE INSTABILITY CRITERION AND THE TIME OF RUPTURE OF A THIN FILM

From the geometry and the dimensions of eye as shown in Fig. 2, it is readily shown that the angle β_0 is less than 110°. Thus the deviation of the tear film from a vertical film ($\beta_0 = \pi/2$) is only slight and the small curvature of the vertical plane may be neglected. Thus we essentially consider the stability of the tear film on a two-dimensional cylindrical surface as shown in Fig. 3. The relevant equations describing the hydrodynamics are the Navier—Stokes equations which include the van der Waals dispersion forces acting on the tear film as a body force. This body force becomes important when the film is thinner than the range of interaction of the dispersion forces between the epithelium and the film resting on it. The effect of van der Waals dispersion forces is incorporated by using a Hamaker-type approximation, which has been used by Scheludko (20), Ruckenstein and Jain (19), and Gumerman and Homsey (21), with many others for various problems of thin-film stability. For thin films ($h_0 < 500$ Å), the van der Waals dispersion forces acting on a unit volume located at the interface are derived from the potential

$$\phi' = \bar{A}/6\pi h'^3, \qquad [1]$$

where $\bar{A}$ is an effective Hamaker constant and h' is the thickness of the film. For films of higher thicknesses, the retardation effect has to be included. Consequently, while investigating the stability of the micrometer sized homogeneous tear film, the following retarded potential may be used:

$$\phi' = B/h'^4, \qquad [2]$$

where $B = B_{11} - B_{12}$ and B_{ij} is the retarded Hamaker constant for the interactions between molecules of type i and j (1 refers to the tear film and 2 to the epithelium).

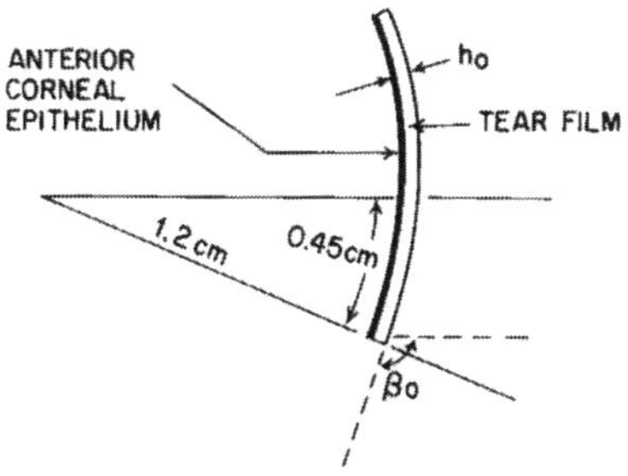

FIG. 2 Side view of a human eye.

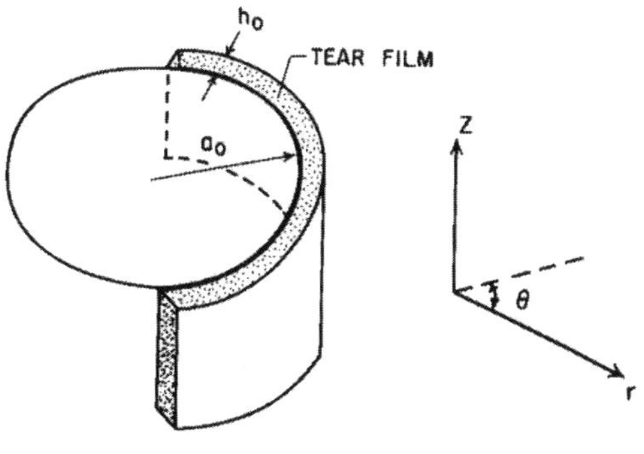

FIG. 3 Definition sketch.

Selecting the initial film thickness, h_0, and the free interface velocity of a film flowing down a flat vertical wall, $\upsilon_{z0}=\rho g h_0^2/2\mu$, as the basic units of length and velocity, the time and pressure may be scaled by h_0/υ_{z0} and $\rho\upsilon_{z0}^2$, respectively. In this way, one obtains the following nondimensional form of the Navier—Stokes equations in the cylindrical coordinates

$$\frac{\partial\upsilon_r}{\partial t}+\upsilon_r\frac{\partial\upsilon_r}{\partial r}+\frac{\upsilon_\theta}{r}\frac{\partial\upsilon_r}{\partial\theta}+\upsilon_z\frac{\partial\upsilon_r}{\partial z}-\frac{\upsilon_\theta^2}{r}$$
$$=-\frac{\partial p}{\partial r}-\eta\frac{\partial\phi}{\partial r}+\frac{1}{Re}\left(\frac{\partial^2\upsilon_r}{\partial r^2}+\frac{1}{r^2}\frac{\partial^2\upsilon_r}{\partial\theta^2}+\frac{\partial^2\upsilon_r}{\partial z^2}+\frac{1}{r}\frac{\partial\upsilon_r}{\partial r}-\frac{2}{r^2}\frac{\partial\upsilon_\theta}{\partial\theta}-\frac{\upsilon_r}{r^2}\right), \quad [3]$$

$$\frac{\partial\upsilon_\theta}{\partial t}+\upsilon_r\frac{\partial\upsilon_\theta}{\partial r}+\frac{\upsilon_\theta}{r}\frac{\partial\upsilon_\theta}{\partial\theta}+\upsilon_z\frac{\partial\upsilon_\theta}{\partial z}+\frac{\upsilon_r\upsilon_\theta}{r}$$
$$=-\frac{1}{r}\frac{\partial p}{\partial\theta}-\frac{\eta}{r}\frac{\partial\phi}{\partial\theta}+\frac{1}{Re}\left(\frac{\partial^2\upsilon_\theta}{\partial r^2}+\frac{1}{r^2}\frac{\partial^2\upsilon_\theta}{\partial\theta^2}+\frac{\partial^2\upsilon_\theta}{\partial z^2}+\frac{1}{r}\frac{\partial\upsilon_\theta}{\partial r}+\frac{2}{r^2}\frac{\partial\upsilon_r}{\partial\theta}-\frac{\upsilon_\theta}{r^2}\right) \quad [4]$$

and

$$\frac{\partial\upsilon_z}{\partial t}+\upsilon_r\frac{\partial\upsilon_z}{\partial r}+\frac{\upsilon_\theta}{r}\frac{\partial\upsilon_z}{\partial\theta}+\upsilon_z\frac{\partial\upsilon_z}{\partial z}$$
$$=-\frac{2}{Re}-\frac{\partial p}{\partial z}-\eta\frac{\partial\phi}{\partial z}+\frac{1}{Re}\left(\frac{\partial^2\upsilon_z}{\partial r^2}+\frac{1}{r^2}\frac{\partial^2\upsilon_z}{\partial\theta^2}+\frac{\partial^2\upsilon_z}{\partial z^2}+\frac{1}{r}\frac{\partial\upsilon_z}{\partial r}\right). \quad [5]$$

The continuity equation is

$$\frac{\partial\upsilon_r}{\partial r}+\frac{\upsilon_r}{r}+\frac{1}{r}\frac{\partial\upsilon_\theta}{\partial\theta}+\frac{\partial\upsilon_z}{\partial z}=0. \quad [6]$$

In Eqs. [3] to [6], υ_r, υ_θ, υ_z, and p are the three nondimensional velocity components and the nondimensional pressure, respectively. The coordinate system is shown in Fig. 3. The Reynolds number, Re, is defined as $\rho\upsilon_{z0}h_0/\mu$. Although the interaction potential ϕ is in general a function of the spatial coordinates, it is only the potential at the film interface that is required within the framework of the approximation employed here (Appendix A). At the interface, the parameter η is given by $B'/h_0^c\rho\upsilon_{z0}^2$ and $\phi=1/h^c$. The quantity ϕ is related to the potential $\phi'=B'/h'^c$ by the relation $\phi=h_0^c\phi'/B'$. Clearly, $B'=B$ and $\boldsymbol{c}=4$ apply to the retarded case, and $B'=\bar{A}/6\pi$ and $\boldsymbol{c}=3$ to the nonretarded case. Here h is the nondimensional thickness of the film.

On the anterior epithelium, the following adhesion conditions are satisfied:

$$\text{at } r=\left(\frac{a_0}{h_0}\right)=a,\ \upsilon_r=\upsilon_\theta=\upsilon_z=0. \quad [7]$$

The following three boundary conditions are to be satisfied at the tear—air interface, viz., at $r=a+h(z,\theta,t)$: two conditions of zero shearing stresses,

$$\frac{p_{rr}}{r}\frac{\partial h}{\partial\theta}+p_{r\theta}\left[1-\frac{1}{r^2}\left(\frac{\partial h}{\partial\theta}\right)^2\right]$$
$$-\frac{p_{rz}}{r}\frac{\partial h}{\partial\theta}\frac{\partial h}{\partial z}-\frac{p_{\theta z}}{r}\frac{\partial h}{\partial\theta}-p_{\theta z}\frac{\partial h}{\partial z}=0 \quad [8]$$

and

$$p_{rr}\frac{\partial h}{\partial z}-\frac{p_{r0}}{r}\frac{\partial h}{\partial z}\frac{\partial h}{\partial \theta}$$

$$+p_{rz}\left[1-\left(\frac{\partial h}{\partial z}\right)^2\right]-\frac{p_{r0}}{r}\frac{\partial h}{\partial \theta}$$

$$-p_{zz}\frac{\partial h}{\partial \theta}=0, \qquad [9]$$

and the relation between the pressure jump and the surface tension

$$\left[p_{rr}+\frac{p_{00}}{r^2}\left(\frac{\partial h}{\partial \theta}\right)^2+p_{zz}\left(\frac{\partial h}{\partial z}\right)^2-\frac{2p_{r0}}{r}\frac{\partial h}{\partial \theta}+\frac{2p_{0z}}{r}\frac{\partial h}{\partial \theta}\frac{\partial h}{\partial z}-2p_{rz}\frac{\partial h}{\partial z}\right]\left[1+\frac{1}{r^2}\left(\frac{\partial h}{\partial \theta}\right)^2+\left(\frac{\partial h}{\partial z}\right)^2\right]^{-1}$$

$$=-p_0-S\left\{\frac{2}{r^3}\left(\frac{\partial h}{\partial \theta}\right)^2+\frac{1}{r}\left[1+\left(\frac{\partial h}{\partial z}\right)^2\right]-\frac{1}{r^2}\frac{\partial^2 h}{\partial \theta^2}\left[1+\left(\frac{\partial h}{\partial z}\right)^2\right]+\frac{2}{r^2}\frac{\partial h}{\partial \theta}\frac{\partial h}{\partial z}\frac{\partial^2 h}{\partial \theta \partial z}\right.$$

$$\left.-\frac{\partial^2 h}{\partial z^2}\left[1+\frac{1}{r^2}\left(\frac{\partial h}{\partial \theta}\right)^2\right]\right\}\left[1+\frac{1}{r^2}\left(\frac{\partial h}{\partial \theta}\right)^2+\left(\frac{\partial h}{\partial z}\right)^2\right]^{-\frac{3}{2}}. \qquad [10]$$

Here p_0 is the static nondimensional pressure of the nonviscous, semi-infinite medium (air), S is a nondimensional parameter defined as $S=\sigma/\rho h_0 \upsilon_{z0}^2$, and σ is the interfacial tension. Finally, the description of the interface is made complete by adding the kinematic condition

$$\frac{\partial h}{\partial t}-\upsilon_r+\frac{\upsilon_\theta}{r}\frac{\partial h}{\partial \theta}+\upsilon_z\frac{\partial h}{\partial z}=0,$$

$$\text{at } r=a+h. \qquad [11]$$

For a Newtonian fluid, the various components of the nondimensional pressure tensor are related to the velocity gradients by the following constitutive relations:

$$p_{rr}=-p+\frac{2}{Re}\frac{\partial \upsilon_r}{\partial r},$$

$$p_{00}=-p+\frac{2}{Re}\left(\frac{1}{r}\frac{\partial \upsilon_\theta}{\partial r}+\frac{\upsilon_r}{r}\right),$$

$$p_{zz}=-p+\frac{2}{Re}\frac{\partial \upsilon_z}{\partial z},$$

$$p_{r0}=\frac{1}{Re}\left(\frac{1}{r}\frac{\partial \upsilon_r}{\partial \theta}+\frac{\partial \upsilon_\theta}{\partial r}-\frac{\upsilon_\theta}{r}\right),$$

$$p_{0z}=\frac{1}{Re}\left(\frac{\partial \upsilon_\theta}{\partial z}+\frac{1}{r}\frac{\partial \upsilon_z}{\partial \theta}\right),$$

$$p_{rz} = \frac{1}{Re}\left(\frac{\partial \upsilon_r}{\partial z} + \frac{\partial \upsilon_z}{\partial r}\right). \qquad [12]$$

The domain of the solution is bounded by the layer

$$a \leqslant r \leqslant a + h(z,\theta,t).$$

The basic, steady, laminar flow solution to the set of Eqs. [3] to [12] is given by

$$\upsilon_{rb} = \upsilon_{\theta b} = 0,\ \upsilon_{zb} = \frac{1}{2}\left(r^2 - a^2\right)$$

$$-(1+a)^2 \ \ln(r/a),\ h = 1 \qquad [13a]$$

and

$$p_b = p_0 + S/(1+a), \qquad [13b]$$

where the subscript b denotes the base case velocity and pressure distributions. Now in order to investigate the instability of the basic solution [13], which is caused by the dispersion forces and flow, we make use of a perturbation approach similar to that of Shlang and Sivashinsky (22). This approach has the advantage of yielding a nonlinear equation of evolution for the interface, from which the information about the linear stability of the interface may also be extracted. In what follows, we make use of the fact that $S = \sigma/\rho h_0 \upsilon_{z0}^2$ is a large parameter for thin liquid films ($S \sim 10^8$ for the tear film) and thus defining $S = 1/\epsilon^2$, the estimate $\epsilon \ll 1$ holds. This may be referred to as the "large-surface tension approximation" combined with the "thin-film approximation." At this stage, the problem involves two time scales which are related to the deformation of the free surface at a given cross section and to the vertical wave motion, respectively. This complication may be circumvented by introducing a moving coordinate system and thus eliminating the time scale associated with the vertical wave motion

$$z' = z + \gamma t, \qquad [14]$$

where γ is a nondimensional velocity, which is determined in what follows. The following space-time stretching transformations are now introduced

$$x = r - a,\ z_1 = \epsilon z',\ \alpha = a\epsilon,$$

$$y = \alpha\theta,\ T = \epsilon^2 t. \qquad [15]$$

The film thickness, velocities, and the pressure may thus be expanded around their base values as the following power series perturbation expansions

$$h = 1 + \sum_{n=1} \epsilon^n h_n(y, z_1, T), \qquad [16a]$$

$$\upsilon_r = \sum_{n=1} \epsilon^n \upsilon_{rn}(x, y, z_1, T), \qquad [16b]$$

$$\upsilon_\theta = \sum_{n=1} \epsilon^n \upsilon_{\theta n}(x, y, z_1, T), \qquad [16c]$$

$$\upsilon_z = \upsilon_{zb}\left(x,\epsilon\right) + \sum_{n=1} \epsilon^n \upsilon_{zn}\left(x, y, z_1, T\right), \qquad [16d]$$

and

$$p = p_b\left(\epsilon\right) + \sum_{n=1} \epsilon^n p_n. \qquad [16e]$$

The dispersion potential, ϕ, is, however, not expanded in power series, since we wish to retain the dispersion nonlinearities in the subsequent derivation. The derivation pursued here retains both the convective and the dispersion force contributions to the instability of the film and allows one to study the synergism between the two. Transforming to the coordinate system [14]–[15] and equating the like powers of ϵ, one obtains a hierarchy of perturbed equations, which are solved successively for the deviation in the film thickness, h_1. Details of these calculations are given in Appendix A. The nonlinear equation of evolution for the deviation in the film thickness, h_1, is obtained,

$$\frac{\partial h_1}{\partial t} - \frac{2}{3a}\frac{\partial h_1}{\partial z'} - \frac{4h_1}{S^{1/2}}\frac{\partial h_1}{\partial z'} + \frac{8Re}{15}\frac{\partial^2 h_1}{\partial z'^2} + \frac{Re}{3a^2}\nabla^2 h_1$$
$$+ \frac{Re}{3S}\nabla^4 h_1 - \frac{Re\eta}{S^{1/2}}\nabla^2\phi = 0, \qquad [17]$$

where

$$\nabla^2 = \left(\frac{\partial^2}{\partial z'^2} + \frac{1}{a^2}\frac{\partial^2}{\partial \theta^2}\right).$$

This may be solved numerically to obtain an accurate information about the kinetics of rupture, for a given initial interfacial disturbance. We, however, do not undertake this calculation here, since such an endeavor demands that an equally accurate information be available for various physical parameters (most crucially, the Hamaker constant) and the amplitude of interfacial perturbations. As only an order of magnitude information is available presently for various parameters and the breakup times to be predicted, an estimate of the breakup times obtained by linearizing Eq. [17] is deemed satisfactory for model discrimination.

The linear stability of the interface is determined by examining the response of the interface to a Fourier component of an arbitrary traveling wave perturbation, viz.,

$$h_1 \sim \exp\left(\beta t + ik'z' + il'\theta\right), \qquad [18]$$

where k' and l' are the wavenumbers of the disturbances in the z' and θ coordinates. Note that k' is dimensionless. Substituting [18] in Eq. [17] and discarding the nonlinear terms in Eq. [17], one arrives at the dispersion relation in terms of the original variables,

$$\beta = \frac{2}{3}\left(\frac{\upsilon_{z0}h_0}{a_0}\right)ik + \frac{1}{3\mu h_0}\left[\frac{8}{5}h_0\upsilon_{z0}^2\rho\right.$$
$$\left. + \rho h_0\left(\frac{\upsilon_{z0}h_0}{a_0}\right)^2 + \frac{cB'}{h_0^{c-1}} - \sigma\left\{\left(kh_0\right)^2 + \left(lh_0\right)^2\right\}\right] \qquad [19]$$
$$\times\left\{\left(kh_0\right)^2 + \left(lh_0\right)^2\right\},$$

where $l = (l'/a_0)$, $k = (k'/h_0)$, β is the rate of instability parameter, and k and l are the wavenumbers of perturbations in the dimensional coordinates corresponding to z' and θ coordinates, respectively. Any arbitrary perturbation imposed on the surface may be Fourier-decomposed into various wave-like components (Eq. [18]) with different wavenumbers. We seek to determine the wavenumber for which the growth rate, β, is a maximum. The waves with these wavenumbers are selectively amplified at a faster rate, thus causing the eventual rupture of the film. The wavenumber of this fastest growing perturbation is thus obtained by maximizing the real part of β with respect to $(k^2 + l^2)$, viz., from

$$d\beta/d\left(k^2+l^2\right)=0. \qquad [20]$$

Denoting the wavenumbers of the most dangerous perturbations by k_m and l_m and noting that for the parameters of tear film (Table 1),

$$\rho h_0\left(\frac{\upsilon_{z0}h_0}{a_0}\right) \ll \frac{8}{5}h_0\upsilon_{z0}^2\rho,$$

one obtains

$$k_m^2+l_m^2=\left(\frac{8}{5}\rho h_0\upsilon_{z0}^2+\frac{cB'}{h_0^{c-1}}\right)/\left(2\sigma h_0^2\right). \qquad [21]$$

All of the two-dimensional wave disturbances which satisfy Eq. [21] are now selectively amplified at the fastest rate. A characteristic wavelength for these perturbations may be defined by letting $l_m \to 0$ (one-dimensional disturbance) in which case

$$\lambda_m=2\pi/k_m. \qquad [22]$$

The above definition is chosen only for convenience, since only an order of magnitude for the wavelength is desired. An alternate choice such as $k_m \sim l_m$ may also be used to define the characteristic wavelength. We compute this characteristic wavelength to show that the wavelength of the fastest growing perturbation is smaller than the linear dimension of the eye.

Substituting the wavenumbers of the fastest growing perturbations (Eq. [21]), in the real part of β, the maximum growth coefficient, β_m, may be obtained. Finally, an estimate for the time of rupture is obtained from $\tau_m \approx \beta_m^{-1}$, which leads to the expression for the time of rupture:

TABLE 1
Physical Properties of the Tear Film

Property	Average value	Ref.
Maximum thickness of the tear film	7.8 μm	(2, 4, 5)
Minimum thickness of the tear film	4.0 μm	(4)
Mucous layer thickness	200–500 Å	(2)
Surface tension for the air—tear interface (σ)	40 dyn/cm	(5)
Density (ρ) of the tear film	1.0 g/cm^3	(5)
Viscosity (μ) of the tear film	0.035 g/cm-s	(5)
($\mu\sigma$) for the mucous layer	1.0 dyn g/cm^2-s	
S (dimensionless)	5.1×10^8	
Re (dimensionless)	3.1×10^{-4}	

$$\tau_m = 12\mu\sigma h_0 \Big/ \left[\frac{2}{5}\frac{g^2\rho^3 h_0^5}{\mu^2} + \frac{cB'}{h_0^{c-1}} \right]^2. \qquad [23]$$

The expression, $\upsilon_{z0} = \rho g h_0^2/2\mu$, has been used in the derivation of Eq. [23].

As is expected, our results without the van der Waals interactions ($B' = 0$) recover the results of Benjamin (23), who considered the stability of the flow of a viscous liquid film down a vertical plate. For a nonretarded potential, viz., $C = 3$ and $B' = \bar{A}/6\pi$ and in the absence of flow, it reduces to the result derived by Ruckenstein and Jain (19) for the rupture of thin films. It may be pointed out that the interfacial instability of the film due to the flow (which is reflected in the first term of the denominator of Eq. [23]) does not lead to the film rupture, but results in a stable self-fluctuating wave or the formation of ripples on the film surface (22). The van der Waals dispersion forces on the other hand are capable of rupturing a thin film. This happens because, although the surface free energy increases with the increasing surface area associated with the deformation of the interface, the total free energy of the system decreases due to van der Waals dispersion interactions.

Williams and Davis (24) obtained the time of rupture of a nonflowing, thin film with a nonretarded potential ($C = 3$, $B' = \bar{A}/6\pi$), by solving a nonlinear equation which they derived by a long-wavelength approximation. Their results indicate that the rupture times computed after retaining the nonlinearities are no more than a factor of two lower than that derived from a linear theory, even for amplitudes of perturbations as large as one-third of the film thickness. It is in view of this, and of the fact that only an order of magnitude is available for the Hamaker constant, that Eq. [23] can be used to evaluate the breakup time, τ_m.

RUPTURE OF A HOMOGENEOUS TEAR FILM

Equation [23] is now applied to the entire tear film considered as a homogeneous film of several micrometers in thickness. The shaded region in Fig. 4 depicts the feasible ranges for the retarded Hamaker constant (B) and the tear-film thickness which are consistent with the observed breakup

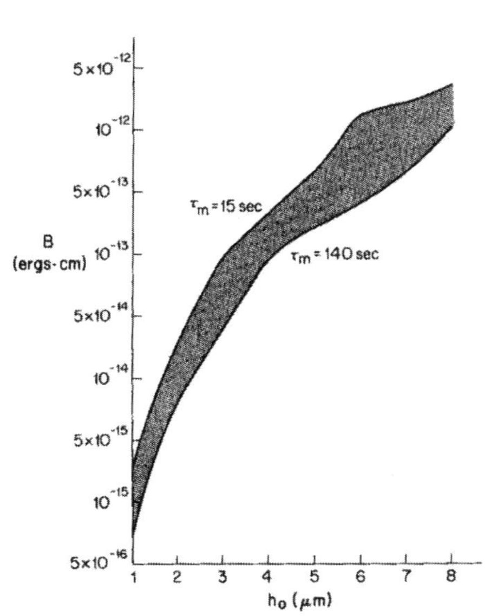

FIG. 4 Feasible region for the tear-film rupture in the Hamaker constant—tear-film thickness parameter space.

times. Although the retarded Hamaker constant has not been measured for the epithelium—tear fluid or the contact lens—tear fluid systems, various experiments for other substances indicate an upper bound of 10^{-19} erg-cm for B (25–28). Thus even assuming local micrometer-sized inhomogenities in the tear film and a rather large time for the tear breakup (~140 s), the estimated Hamaker constant by this mechanism is about three orders of magnitude higher than that expected. For a more realistic tear-film thickness of 4 μm and a breakup time of 50 s, the rupture of the tear film requires a Hamaker constant of about 3×10^{-13} erg-cm, which is about six orders of magnitude higher than that reported. In other words, even assuming a retarded Hamaker constant of 10^{-18} erg-cm and a micrometer-sized local inhomogenity in the tear film, Eq. [23] predicts the time of rupture of such a film to be about 3 years!

In view of these considerations, it may be concluded that the retarded van der Waals interactions are not strong enough to break the micrometer-sized tear films within the observed times of rupture. In addition, such a model of rupture predicts a continuous process of thinning prior to the rupture and thus cannot explain the observed instantaneous rupture or "beading-up" of tear film at a thickness of about 4 μm (4). Finally, it fails to recognize the key role played by the mucous layer in determining the stability characteristics of the tear film. This last point is amply supported by various experiments (8, 29) and clinical tests (12–17). Thus, although the instability of the air—tear interface due to the dispersion forces and flow exists, it is not sufficient to cause the observed rapid rupture.

THE RUPTURE OF UNDERLYING MUCOUS LAYER: A NEW MECHANISM OF TEAR-FILM INSTABILITY

It is well known that the presence of the mucous layer on the epithelium is absolutely necessary for the wetting of cornea (2, 14–16, 30). These studies indicate that the epithelial mucus is capable of adsorbing onto the low-energy surfaces, converting them into hydrophilic surfaces. Even Teflon, which cannot be rendered hydrophilic by any commercially available surfactant, becomes water wettable when exposed to a layer of mucin. The experiments of Doughman *et al.* (17) and Padday (31) demonstrate the rupture of aqueous films of several hundred micrometers thick on hydrophobic surfaces such as Teflon, paraffin, and polyethylene. The hydrophobicity and water-wetting characteristics of a *clean* (without mucus) anterior epithelium are found to be similar (32).

The tear film is seen to rupture spontaneously only after it thins from its initial thickness of 7 to about 4 μm. This is much smaller than the critical thickness of water over hydrophobic surfaces at which rupture occurs (17, 31) and much too large for the retarded dispersion forces to cause a rupture within observed BUT. A plausible mechanism which emerges from this discussion is that the initially mucus coated cornea converts its wettability characteristics within the observed breakup times, thus effectuating a spontaneous rupture of the aqueous layer.

The rupture of the thin mucous layer, which coats the hydrophobic epithelium and provides a hydrophilic base for the aqueous tears, due to dispersion interactions indeed leads to such a hydrophilic—hydrophobic transition. The issue whether the time of rupture of the mucous layer is consistent with the observed range of BUT is addressed next.

The thickness of the aqueous film is at least two orders of magnitude larger than the mucous layer, and hence a slow deformation of the air—aqueous interface may be neglected while investigating the stability of the mucous—aqueous interface. The superficial epithelial cell walls, which are coated by the mucous layer, have a complex structure of protrusions known as microvilli (2). The characteristic dimensions of these protrusions are, however, about an order of magnitude larger than the thickness of the mucous layer (2) and thus the mucous—epithelial interface may be considered to be planar while investigating the rupture of the mucous layer. In other words, we consider the instability of a thin (200–500 Å) mucous layer sandwiched between epithelium and a semi—infinite viscous, aqueous phase (Fig. 1). Since the film is thin, the following potential is used for the dispersion forces:

$$\phi' = \bar{A} / 6\pi h'^3. \quad [24]$$

The effective Hamaker constant for this geometry is given by

$$\bar{A} = A_{12} + A_{22} - A_{13} - A_{23}, \quad [25]$$

where A_{ij} is the Hamaker constant for the interaction between the molecules of type i and j (1 refers to the aqueous phase, 2 to the mucous layer, and 3 to the anterior epithelium). The aqueous layer is viscous and thus the boundary conditions [8]–[10], which are established for a nonviscous (air) bounding medium, do not hold rigorously. However, the ratio of the viscosities of the aqueous layer and the mucous layer is much smaller than unity. In this case, the effects of the viscosity of the bounding medium on the stability of the mucous layer may be shown to be insignificant (less than 1%) (Appendix B). Thus, the entire formalism developed earlier holds and we now investigate the amplification of fluctuations or the initial nonhomogeneities arising at the mucous—aqueous interface. In addition, $v_{z0} \approx 0$ since the mucous layer is very thin and viscous. The time of rupture for such a layer is derived from Eq. [23] by making use of the potential [24], Thus the substitutions $C = 3$, $B' = \bar{A}/6\pi$, and $g = 0$ (no-flow) in Eq. [23] give

$$\tau_m = 48\pi^2 \sigma \mu h_0^5 / \bar{A}^2. \quad [26]$$

This result has been derived earlier by Ruckenstein and Jain (19) by making a lubrication approximation for a thin film.

The discrepancy between the time of rupture as determined from the linear and nonlinear equations depends on the magnitude of the perturbations (24). Since no a priori information is available about the amplitude of perturbations at the mucous—aqueous interface, a factor of 2 may again be used to convert the times of rupture as computed by Eq. [26] to more realistic times of rupture computed from the solution of a nonlinear equation (24). The feasible range of Hamaker's constant for the rupture of the mucous film to occur may thus be obtained by rewriting Eq. [26] as

$$\bar{A}_{max,min} = \left(24\sigma\mu\pi^2 h_0^5 / \tau_{min,max}\right)^{1/2}. \quad [27]$$

The breakup time estimates $\tau_{min} = 20$ s and $\tau_{max} = 50$ s are used to find the range of the Hamaker constant which is consistent with the observed BUT. These are shown as the shaded region in Fig. 5. The wavelength of the most dangerous perturbation is

$$\lambda_m = \left(8\pi^3 \sigma / \bar{A}\right)^{1/2} h_0^2, \quad [28]$$

which, for the parameters values falling in the shaded region of Fig. 5, is easily verified to be several orders of magnitude smaller than the dimensions of the eye.

The Hamaker constant thus obtained for a mucous-film thickness of 200 to 500 Å is in the range of 10^{-14} to 4×10^{-13} erg. This magnitude of the Hamaker constant is realistic (19, 20, 27) and thus the dispersion forces acting on the mucous layer are indeed capable of rupturing it within a time period consistent with the observed breakup times. With this understanding, the basic steps of the proposed mechanism may now be summarized:

i. The blinking movement exerts shear across the thin, aqueous film located between the eyelid and the ocular globe and thus redistributes and smoothens the mucous layer on the corneal epithelium. Immediately after its structural integrity is restored, this mucous layer

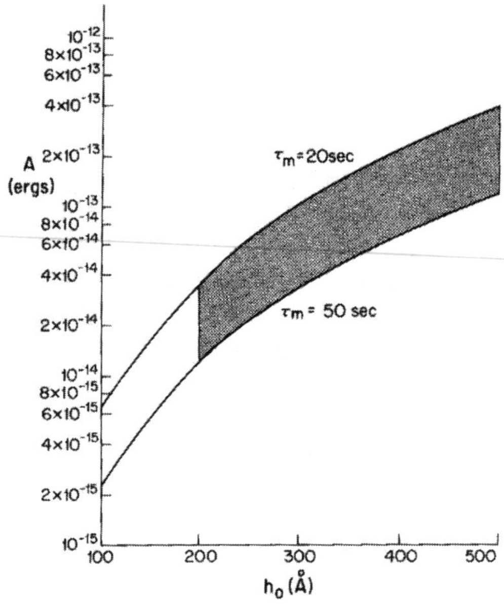

FIG. 5 Feasible region for the mucous-film rupture in the Hamaker constant—tear-film thickness parameter space.

is responsible for effectively masking any lipid contamination as well as the basic hydrophobic nature of the epithelium. At this stage, the overlying aqueous film is complete and is slowly thinning due to the evaporation and the osmotic transfer across the cornea.

ii. Upon its restoration, the small inhomogeneities of the mucous layer begin to amplify at several places under the influence of the dispersion forces. If this process is not reversed by an intermediate blink, the growing interfacial perturbations cause the rupture of the mucous layer in about 15 to 50 s in a normal eye. At this point, the inherent hydrophobicity of the exposed epithelium is responsible for an instantaneous (fast compared to the previous step) rupture of the aqueous tear film.

Due to its sequential nature and the fact that the model incorporates the structure of the tear film as made up of two distinct films, the proposed mechanism may be referred to as a "two-step, double-film" mechanism of tear-film rupture.

DISCUSSION AND IMPLICATIONS

In a normal eye, tear-film rupture does not occur because the interblink time is small compared to the breakup time and the blinking is instrumental in restoring the structural integrity of the mucous layer. This rupture and the consequent formation of dry spots on cornea, however, are crucial for a mucus-deficient pathological eye, for which the rupture would occur rather frequently within the interblink period. For instance, assuming an effective Hamaker constant of 10^{-14} erg, an abnormally thin (~100 Å) mucous layer would break within 4–5 s, thus exposing the hydrophobic epithelium to the aqueous layer. This may result in the extensive dewetting and desiccation of cornea. Indeed, in a clinical study (3), patients with chronic conjunctival inflammation were observed to have an abnormally short breakup time (~ 3 to 5 s) of the tear film, despite the normal amounts of aqueous tears. Conjunctival biopsy revealed a marked decrease in the population of conjunctival goblet cells which are responsible for the production of conjunctival mucous. The crucial role of the mucus in the tear-film stability and thus in determination of BUT, has also correlated well in other clinical findings.

The deficiency of vitamin A has been observed to catalyze the disappearance of conjunctival goblet cells and the consequent appearance of the dry-eye syndromes (2.) Some other conditions such as ocular pemphigoid, Stevens—Johnson syndrome, trachoma, chemical burns, and certain forms of drug-induced diseases also decrease the goblet cell population, resulting in decreased BUT, even in the presence of normal tear volume (2, 33).

An important consequence of the stability of the tear film is in the normal wearing and functionality of the contact lenses (12, 13, 34). A well-fitted contact lens rests on a continuous tear film sandwiched between the epithelium and the lens and is also coated with a continuous tear film on the outside. The stability of both the prelens (in-front-of- the-lens) and the postlens (behind-the-lens) tear film is important for good contact lens wearing performance. Most of the silicone and hydrogel lenses being used presently are not completely water wettable. Thus, it is most likely that the conjunctival mucus soon coats the anterior surface of a lens placed in the eye and contributes to the wettability of the prelens film. The premature rupture of this mucous layer would result in the prelens film breakup and the accumulation of denaturated proteins contaminated with the lipids which are observed on the lens surface (13). This arises due to the high interfacial tension between the aqueous layer and the exposed (without the mucous coating) prelens surface and results in a deposition of proteins and lipids because they tend to lower the interfacial tension at the aqueous—lens interface. The strong interactions between the lens surface and the proteins give rise to multiple site adsorption, resulting in less exposure of hydrophilic groups of the proteins to the aqueous tears and consequently, their denaturation. While the interfacial tension is somewhat lowered by this process, it may not be sufficient to stabilize the aqueous tear film due to the unavailability of enough hydrophilic groups which face the tear film. Indeed, some successful attempts have been made to correlate the length of deposit-free wear time with the tear-film breakup time (35), which show that a rapid deposit formation is associated with a short BUT. This is expected from the proposed mechanism, because both the BUT and the prelens film breakup time depend on the thickness of the mucous layer coating the epithelium and the lens surface, respectively. The mucus deficiency would thus decrease both the BUT and the prelens breakup time and the latter would encourage deposit formation. The cause for a condition called "giant papillary conjunctiviries" has been linked with an immune-type reaction to antigenic, denaturated proteins accumulated on the lens surface (36). Thus the stabilization of the prelens tear film by coating the lens surface with mucus or mucus-like substances seems a logical step for the prevention of lens surface contamination.

According to our mechanism, the stability of *postlens* tear film is determined by the structural integrity of the epithelial mucous layer. As the epithelial mucous layer ruptures, the postlens film would also break if the lens surface is not wettable and the adhesion of the contact lens to cornea can take place. This may result in excessive ocular discomfort, irritation, or even epithelial damage. In fact, a clinical study (12) shows that five of the six cases of corneal erosion observed occurred under the stationary lens, which results due to the premature rupture of postlens film. In view of this, wearing of contact lens in a normally mucus-deficient eye may result in these complications. In addition, in the presence of contact lens, even a marginally mucus-deficient eye may be affected by these conditions because of two reasons: (i) contact lenses adversely affect the lid—globe congruity (13) which is important for the formation and renewal of an evenly distributed mucous layer; (ii) contact lens wearers often become lazy blinkers and in addition the blinking becomes incomplete (13), resulting in incomplete restoration of mucous layer. Both of these factors may eventually contribute toward a poor lens tolerance and various conditions of a pathological eye, because they encourage a thinner than the normal mucous layer and hence reduce its time of rupture.

The only objective clinical measurement which reflects the relative stability of the tear film is the measurement of BUT under properly controlled environmental conditions. The proposed mechanism actually correlates BUT with the thickness of mucous layer and the mucous-aqueous interfacial tension. Tests such as the lack of mucus in inferior fornix and the absence of goblet cells as determined by conjunctival biopsy (3), may be readily used to diagnose the mucus deficiencies. Thus, in the absence of other factors such as lipid abnormalities, large aqueous-tear deficiency, impaired lid functions, and gross irregularities in the corneal epithelium, the replacement of mucus-like substances in the eye seems to be the logical starting point for the therapy of short BUT and its consequences.

Finally, the model proposed does not account for the non-Newtonian rheology of the mucous layer, since its rheological properties have never been measured. It seems likely that a better characterization of both the Hamaker constant (for epithelium—tear and contact lens—tear systems) and rheological behavior of mucous material would result in improved knowledge of the kinetics of rupture. Thus there is a need for both *in vivo* and *in vitro* experiments on the rupture of "thin" mucous layers, for this indeed appears to play a central role in the pathological states associated with a dry eye.

CONCLUSIONS

We have derived a nonlinear evolution equation for the interface, a dispersion relation, and an estimate for the time of rupture of a thin film. The formalism developed incorporates the effects of van der Waals dispersion forces, the convective motion of the film and the two-dimensional perturbations arising at the interface. Application of this model to the tear film, considered as a several-micrometer-thick homogeneous film, fails to predict the observed kinetics of rupture. Based on many clinical and experimental observations about the role of mucus in determining the stability characteristics of the entire tear film, we propose a new mechanisms for the formation of dry spots on the cornea. Immediately after blinking, a thin (200-500 Å) mucous layer is restored on the corneal epithelium. The van der Waals dispersion forces acting on the mucous layer act to destabilize this layer, resulting in its eventual rupture. This, in turn, exposes the hydrophobic epithelium to the aqueous film and to the omnipresent lipids. It is at these sites that the aqueous tear film breaks rather rapidly causing the dewetting of the cornea. The time of rupture computed by this mechanism is in agreement with the clinically observed tear-film breakup times of about 15 to 50 s for a healthy eye. This rupture does not occur in a normal eye because the interblink time is small compared to the time of rupture of the tear film.

The mechanism proposed is also consistent with other observed characteristics of the tear-film rupture, such as the almost spontaneous rupture of the tear film after it thins from an initial thickness of about 7 μm to a thickness of 4 μm, an abnormal decrease of BUT in mucus-deficient eye and the adhesion of the contact lens to the cornea in mucus-deficient patients or lazy blinkers. The mechanism proposed thus have implications for the diagnosis and treatment of pathological eye, as well as in determining the contact lens tolerance. It also lends support to a number of clinical and experimental observations about the crucial role of mucus in determining the stability characteristics of the tear film.

APPENDIX A

Making use of the set of transformations [14], [15] and the perturbation expansions [16a]–[16e], the set of governing Eqs. [3] to [12] may be manipulated to yield a hierarchy of equations.

Equating the terms of order ϵ gives the first nontrivial set of equations and the corresponding boundary conditions as

$$\eta \frac{\partial \phi}{\partial x} + \frac{\partial p_1}{\partial x} = 0, \quad \frac{\partial^2 \upsilon_{01}}{\partial x^2} = 0, \quad \frac{\partial^2 \upsilon_{z1}}{\partial x^2} = 0 \qquad \text{[1a]}$$

and

$$\frac{\partial \upsilon_{r1}}{\partial x} = 0, \qquad \text{[1b]}$$

together with

$$\upsilon_{r1} = \upsilon_{01} = \upsilon_{z1} = 0 \text{ at } \boldsymbol{x} = 0 \tag{1c}$$

and

$$\upsilon_{r1} = 0, \quad 2h_1 + \frac{\partial \upsilon_{z1}}{\partial x} = 0, \ \frac{\partial \upsilon_{01}}{\partial x} = 0 \tag{1d}$$

and

$$p_1 = -\frac{1}{\alpha^2} h_1 - \nabla^2 h_1 \text{ at } x = 1. \tag{1e}$$

The solution to this first approximation is immediately obtained as

$$\upsilon_{r1} = \upsilon_{01} = 0 \text{ and } \upsilon_{z1} = -2h_1 x \tag{2a}$$

and

$$\eta\phi + p_1 = -\frac{1}{\alpha^2} h_1 - \nabla^2 h_1 + \eta\phi\,|_{x=1} = \eta\phi(y, z_1) + p_1(y, z_1), \tag{2b}$$

where

$$\nabla^2 = \left(\frac{\partial^2}{\partial y^2} + \frac{\partial^2}{\partial z_1^2} \right).$$

As shown by Eq. [2b], the further hydrodynamic calculations require only the interaction potential at the interface, viz. at $\boldsymbol{x} = 1$. The second approximation is obtained by equating the terms of order ϵ^2, giving

$$\frac{\partial \upsilon_{r2}}{\partial x} + \frac{\partial \upsilon_{z1}}{\partial z_1} = 0, \tag{3a}$$

$$-\frac{\partial p_1}{\partial y} - \eta\epsilon^{-1} \frac{\partial \phi}{\partial y} + \frac{1}{Re} \frac{\partial^2 \upsilon_{02}}{\partial x^2} = 0 \tag{3b}$$

and

$$\gamma \frac{\partial \upsilon_{z1}}{\partial z} + 2(x-1)\upsilon_{r2} + x(x-2)\frac{\partial \upsilon_{z1}}{\partial z_1} = -\frac{\partial p_1}{\partial z_1} - \eta\epsilon^{-1} \frac{\partial \phi}{\partial z_1} + \frac{1}{Re} \frac{\partial^2 \upsilon_{z2}}{\partial x^2} + \frac{1}{\alpha Re} \frac{\partial \upsilon_{z1}}{\partial x}. \tag{3c}$$

In the derivation of [3b] and [3c], we take $\eta\epsilon^{-1}(\partial\phi / \partial y)$ and $\eta\epsilon^{-1}(\partial\phi / \partial z_1) \sim \mathrm{O}(1)$, in order to retain the dispersion force nonlinearities in the lower-order equations. This point is better understood by noting that if the potential, ϕ, is expanded in a power series as

$$\phi = 1 + \epsilon\phi_1 + \cdots,$$

then we would have terms $\eta(\partial\phi_1 / \partial y)$ and $\eta(\partial\phi_1 / \partial z_1)$, instead of $\eta\epsilon^{-1}(\partial\phi / \partial y)$ and $\eta\epsilon^{-1}(\partial\phi / \partial z_1)$, respectively, in the second-order equations. Such an expansion would, however, destroy the dispersion-force nonlinearities which we wish to retain in the first-order equation of evolution for the interface. The boundary conditions for the second-order problem are

$$\text{at } x = 0, \ \upsilon_{r2} = \upsilon_{02} = \upsilon_{z2} = 0 \tag{3d}$$

$$\text{at } x = 1,\ \frac{\partial \upsilon_{02}}{\partial x} = 0,\ 2h_2 + \frac{\partial \upsilon_{z2}}{\partial x} = 0,$$

$$(\gamma - 1)\frac{\partial h_1}{\partial z_1} - \upsilon_{r2} = 0. \qquad [3e]$$

The solutions to the set of Eqs. [3a]–[3e] are

$$\upsilon_{r2} = x^2 \frac{\partial h_1}{\partial z_1}, \qquad [4a]$$

$$\upsilon_{02} = -\frac{1}{2} Re\, x(x-2)$$

$$\times \left[\frac{1}{\alpha^2}\frac{\partial h_1}{\partial y} + \nabla^2 \frac{\partial h_1}{\partial y} - \eta\epsilon^{-1}\frac{\partial \phi}{\partial y} \right], \qquad [4b]$$

$$\upsilon_{z2} = Re \frac{\partial h_1}{\partial z_1}\left(\frac{1}{6}x^4 - \frac{2}{3}x^3 + \frac{4}{3}x \right) - \frac{1}{2} Re\, x$$

$$\times (x-2)\left[\frac{1}{\alpha^2}\frac{\partial h_1}{\partial z_1} + \nabla^2 \frac{\partial h_1}{\partial z_1} - \eta\epsilon^{-1}\frac{\partial \phi}{\partial z_1} \right]$$

$$+\frac{1}{\alpha} x(x-2)h_1 - 2xh_2, \qquad [4c]$$

and finally,

$$\gamma = 2. \qquad [4d]$$

Equation [4d] reflects the well-known fact that the wave velocity for a falling film on a flat vertical plane is twice the velocity of the free interface.

Going to the terms of order ϵ^3, the kinematic condition gives the following equation of evolution for the interface:

$$\frac{\partial h_1}{\partial T} + 2\frac{\partial h_2}{\partial z_1} - 2h_1\frac{\partial h_1}{\partial z_1} - \upsilon_{r3}\big|_{x=1} - \frac{\partial h_2}{\partial z_1} + \left(\upsilon_{z1}\big|_{x=1} - \frac{1}{3\alpha} \right)\frac{\partial h_1}{\partial z_1} = 0. \qquad [5]$$

The velocity component, v_{r3}, appearing in the above equation is determined by equating the terms of order ϵ^3 in the continuity equation, to obtain

$$\frac{\partial \upsilon_{r3}}{\partial x} + \frac{1}{\alpha}\upsilon_{r2} + \frac{\partial \upsilon_{02}}{\partial y} + \frac{\partial \upsilon_{z2}}{\partial z_1} = 0. \qquad [6a]$$

With the help of boundary condition

$$\upsilon_{r3} = 0 \ \text{at} \quad x = 0, \qquad [6b]$$

the solution to Eq. [6a] is given by

$$\upsilon_{r3} = -\frac{2}{\alpha}\left(\frac{1}{3}x^3 - \frac{1}{2}x^2 \right)\frac{\partial h_1}{\partial z_1} + \frac{1}{2} Re\left(\frac{1}{3}x^3 - x^2 \right)\left(\frac{1}{\alpha^2}\nabla^2 h_1 + \nabla^4 h_1 + \nabla^2 \phi \right)$$
$$- Re\left(\frac{1}{30}x^5 - \frac{1}{6}x^4 + \frac{2}{3}x^3 \right)\frac{\partial^2 h_1}{\partial z_1^2} + \frac{\partial h_2}{\partial z_1}x^2. \qquad [7]$$

Making use of the expressions [2a] and [7] for υ_{z1} and υ_{r3}, respectively, the kinematic condition [5] is manipulated to yield the following closed-form equation of evolution for the deviation in the interface from its equilibrium value:

$$\frac{\partial h_1}{\partial T}-\frac{2}{3\alpha}\frac{\partial h_1}{\partial z_1}-4h_1\frac{\partial h_1}{\partial z_1}+\frac{8Re}{15}\frac{\partial^2 h_1}{\partial z_1^2}+\frac{Re}{3\alpha^2}\nabla^2 h_1+\frac{Re}{3}\nabla^4 h_1-Re\eta\epsilon^{-1}\nabla^2\phi=0. \quad [8]$$

For a nondimensional potential of the form

$$\phi = 1/h^c,$$

$\nabla^2\phi$ in Eq. [8] is determined as

$$\nabla^2\phi=\left(\partial^2/\partial y^2+\partial^2/\partial z_1^2\right)\phi = ch^{-(1+c)}$$

$$\times\left[\epsilon\nabla^2 h_1-\epsilon^2 h^{-1}(1+c)\left\{\frac{\partial h_1}{\partial y}+\frac{\partial h_1}{\partial z_1}\right\}^2\right]. \quad [9]$$

Equation [8] is the desired nonlinear equation of evolution which retains both the convective and the dispersion-forces nonlinearities and may be solved numerically for a given initial disturbance.

APPENDIX B

In order to evaluate the effects of viscous aqueous layer on the stability of thin mucous film, we employ the results of Jain and Ruckenstein (37). Their final dispersion relation for the linear stability of a thin film on a solid support and bounded by a semiinfinite viscous fluid is

$$\beta=-\frac{\sigma_{\text{eff}}kh_0}{2\mu h_0}\left[\frac{\left(\sinh(kh_0)\cosh(kh_0)-kh_0\right)+R\left(\sinh^2(kh_0)-(kh_0)^2\right)}{\left(1-R^2\right)(kh_0)^2+\left(\cosh(kh_0)+R\sinh(kh_0)\right)^2}\right], \quad [1]$$

where $\sigma_{\text{eff}} = \sigma + (\partial\phi'/\partial h')_{h=h0}\,(1/k^2)$ and R is the ratio of the viscosities of the aqueous and mucous layers.

Since the mucous film is thin and the critical wavelength is orders of magnitude higher than the film thickness, Eq. [1] may be simplified by noting that

$$\sinh(kh_0)\approx(kh_0)+\frac{(kh_0)^3}{6}$$

and

$$\cosh(kh_0)\approx 1+\frac{(kh_0)^2}{2},$$

$$\text{for } (kh_0)<0.3. \quad [2]$$

These estimates simplify Eq. [1] to

$$\beta=\frac{(kh_0)^2}{3\mu h_0}\left[\frac{\bar{A}}{2\pi h_0^2}-\sigma(kh_0)^2\right]\frac{1}{\left(1+2R(kh_0)\right)}. \quad [3]$$

The wavenumber of the fastest growing perturbation is the solution of $d\beta/dk = 0$, viz.,

$$\left(3\sigma R h_0^3\right) k_{\mathrm{m}}^3 + \left(2\sigma h_0^2\right) k_{\mathrm{m}}^2 - \frac{R_0 \bar{A}}{2\pi h_0} k_{\mathrm{m}} - \frac{\bar{A}}{2\pi h_0^2} = 0, \quad [4]$$

which in the event of an inviscid semiinfinite fluid gives the well-known result

$$\lim_{R\to 0} k_{\mathrm{m}}^2 = \frac{\bar{A}}{4\pi\sigma h_0^4}. \quad [5]$$

The other limit of infinitely viscous, semiinfinite fluid is easily shown to be

$$\lim_{R\to \infty} k_{\mathrm{m}}^2 = \frac{\bar{A}}{6\pi\sigma h_0^4}. \quad [6]$$

Thus the fastest growing wavenumber is bounded by

$$\frac{0.41}{h_0^2}\left(\frac{\bar{A}}{\pi\sigma}\right)^{1/2} \leqslant k_{\mathrm{m}} \leqslant \frac{0.5}{h_0^2}\left(\frac{\bar{A}}{\pi\sigma}\right)^{1/2}$$

for all $R \in (0, \infty)$.

For the tear film, $R < 1$, and its inclusion in Eq. [3] makes negligible contribution to the final result.

REFERENCES

1. Holly, F. J., "Wetting, Spreading and Adhesion" (J. F. Paddy, Ed.), p. 439. American Press, New York, 1978.
2. Holly, F. J., and Lemp, M. A., *Surv. Ophthalmol.* **22**, 69 (1977).
3. Lemp, M. A., Dohlman, C. H., and Holly, F. J., Ann. Ophthalmol. **2**, 259 (1970).
4. Nom, M. S., *Acta Ophthalmol.* **47**, 865 (1969).
5. Lin, S. P., and Brenner, H., *J. Colloid Interface Sci.* **85**, 59 (1982).
6. Lin, S. P., and Brenner, H., *J. Colloid Interface Sci.* **89**, 226 (1982).
7. Nom, M. S., *Acta Ophthalmol.* **57**, 766 (1979).
8. Dohlman, C. H., Friend, J., Kalevar, V., Yadoga, D., and Balazs, E., *Exp. Eye Res.* **22**, 359 (1976).
9. Nom, M. S., *Acta Ophthalmol.* **41**, 531 (1963).
10. Maurice, D. N., *Invest. Ophthalmol.* **4**, 464, (1967).
11. Ehlers, N., *Acta Ophthalmol. Suppl.* **81**, 1 (1965).
12. Fanti, P., and Holly, F. J., *Contact Intraocular Lens Med. J.* **6**(2), 111 (1980).
13. Holly, F. J., *Amer. J. Optom. Physiol. Opt.* **58(4)**, 331 (1981).
14. Holly, F. J., *Exp. Eye Res.* **15**, 515 (1973).
15. Holly, F. J., and Lemp, M. A., *Exp. Eye Res.* **11**, 239 (1971).
16. Holly, F. J., *Int. Ophthalmol. Clinics* **3**, 171 (1980).
17. Doughman, D. J., Holly, F. J., and Dohlman, C. H., Presented at the ARVO Spring Meeting. Sarasota, Fla., 1971.
18. Whitaker, S., *I & EC Fundam.* **2**, 132 (1964).
19. Ruckenstein, E., and Jain, R. K., *Faraday Trans. 2* **70**, 132 (1974) (Section 5.1 of this volume).
20. Sheludko, A., *Adv. Colloid Interface Sei.* **1**, 392 (1967).
21. Gumerman, R., and Homsey, G. (*Chem. Eng. Commun.* **2**, 27 (1975).
22. Shlang, T., and Sivashinsky, G. I., *J. Phys. Colloq. (Orsay, Fr.)* **43**, 459 (1982).
23. Benjamin, T. B., *J. Fluid Mech.* **2**, 554 (1957).
24. Williams, M. B., and Davis, S. H., *J. Colloid Interface Sci.* **90**, 220 (1982).

25. Spamaay, M. J., and Jochems, P. W., Int. Congr. Surface Activity, 3rd 2, B/III/1, 375 (1960).
26. Van Silfout, A., *Proc. K. Ned. Akad. Wet. Ser. B: Palaentol. Geol. Phys. Chem.* **69**, 501 *(1966).*
27. Rouweler, G. C. J., and Overbeek, J., Th. G., *Trans.Faraday Soc.* **67**, 2117 (1971).
28. Hunklinger, S., Geisselmann, H., and Arnold, W., *Rev. Sci. Instrum.* **43**, 584 (1972).
29. Proust, J. E. et al., *J. Colloid Interface Sei.* **98**, 319 (1984).
30. Lemp, M. A., Holly, F. J., Iwata, S., and Dohlman, C. H., *Arch. Ophthalmol.* **83**, 89 (1970).
31. Padday, J. F., *Spec. Discuss. Faraday Soc.* **1**, 64 (1970).
32. Mishima, S., *Arch. Ophthalmol.* **73**, 233 (1965).
33. Lemp, M. A., Dohlman, C. H., Kuwabara, T., et al., Trans. Amer. Acad. Ophthalmol. Otolaryngol. **75**, 1223 (1971).
34. Holly, F. J., *Amer. J. Optom. Physiol. Opt.* **58**(*4*), 324 (1981).
35. Lowther, G. E., Depots Surles Lentilles Hydrophiles. Conference faite au 5[e] Congres Français d'Optique de Contact, Bordeaux, France, 1976.
36. Alansmith, M. R., Korb, D. R., Greiner, J. V., et al. Amer. J. Ophthalmol. **83**, 697 (1977).
37. Jain, R. K., and Ruckenstein, E., *J. Colloid Interface Sci.* **54**, 108 (1976).

5.11 The Role of Lipid Abnormalities, Aqueous and Mucus Deficiencies in the Tear Film Breakup, and Implications for Tear Substitutes and Contact Lens Tolerance*

Ashutosh Sharma and Eli Ruckenstein†
Department of Chemical Engineering, State University of New York at Buffalo, Buffalo, New York 14260
Corresponding Author
†To whom correspondence should be addressed.

Received January 8, 1985; accepted July 22, 1985

INTRODUCTION

A rational understanding of the factors affecting the rupture of the tear film, which covers the conjunctival and corneal surfaces, is important in various pathological conditions associated with a dry eye, their diagnosis and treatment. A widespread use of contact lenses has catalyzed research on the contact lens—tear film interactions, inasmuch as the instability and premature rupture of the tear film manifests itself in poor contact lens tolerance. The formulation of wetting and cushioning solutions, which are used to coat the surface of the contact lens before its insertion into eye, is another area which may benefit from a fundamental understanding of the rupture of the tear film.

To this day, one of the single most important treatment of a dry eye is the use of tear substitutes. The tear substitutes are aqueous based formulations which, when periodically instilled in a pathological eye, increase the time of the tear film rupture and consequently, alleviate discomfort and epithelial damage. It is hoped that a basic understanding of the tear film breakup and of the various factors affecting it, would also result in delineating the role of the tear substitutes and guidelines for the selection of their ingredients.

A normal tear film completely covers the corneal and conjunctival surfaces and keeps them in the proper conditions of lubrication and hydration. The structural continuity of this tear film is maintained by involuntary periodic blinking, with a typical interblink period ranging from 5 to 10 s (1). It is, however, observed that if either the eyelids are held open for a longer time period or if the eye is pathological, the tear film breaks (usually first over the corneal surface), resulting in

* *Journal of Colloidal and Interface Science*. Vol. III, No. I, p.8, May 1986. Republished with permission.

the appearance of increasingly dewetting areas on the epithelial surface. The time elapsed between a blink and the first hole to appear is defined as the breakup time (BUT) and is found to be in the range of 15–45 s for a normal adult eye (2a). This estimate of BUT is, however, based on a conventional technique that requires the instillation of fluorescein in the tear film. Some recent *in vivo* tests of tear film stability have employed a noninvasive technique, the Toposcope and have concluded that the instillation of fluorescein reduces the actual BUT by as much as a factor of 30 (2b). A tear film breakup time of less than 10 s (more precisely, a BUT less than the interblink period) is considered abnormal and may result in epithelial tissue damage and corneal ulceration (1). The adhesion of contact lens to cornea (3, 4) and the rapid buildup of deposits on the lens surface (5) have also been observed to be related to a premature rupture of the tear film. It is in view of these considerations that the clinical measurements of BUT, although statistical (6), serve as a useful yardstick for differentiating a normal eye from a pathological one. The tear film is, in reality, composed of at least three distinct layers, which are: (a) a thin (200–400 Å) (7a)* mucus coating of the epithelial surface, (b) an aqueous layer which rests on the mucus-coated epithelium and which is about 4–10 μm thick, and (c) a 1000-Å-thick lipid layer that makes the tear—air interface. For a more elaborate discussion of the physiology of the eye, the structure of the precorneal tear film and other related issues, the reader is referred to reviews (1, 7a).

The structure of the tear film and various plausible physicochemical interactions occurring therein form a complex and delicately balanced system. It is therefore not surprising that a systematic investigation of certain aspects of the hydrodynamics of the tear film is of very recent origin. Based on a linear stability analysis, Lin and Brenner (8) showed that the temperature or mucin concentration gradient driven Marangoni convection in the aqueous tears appears unlikely under the normal circumstances. They also proposed the possibility of the rupture of micron-sized *aqueous tear film* due to retarded van der Waals dispersion interactions (9). An evaluation of the time of rupture based on their mechanism, however, showed that this rupture would take at least tens of days (10).

In our earlier papers (10, 11), we proposed a "two-step, double film" mechanism of the tear film rupture. Briefly, the mechanism consisted of the following two events in succession:

Step 1: the rupture of the mucus layer. Immediately after blinking, the shear created by rapid lid motion restores the structural integrity of a thin (~200-to 400-Å-thick) mucus layer on the corneal epithelium. The overlying aqueous tear film completely wets this mucus—coated epithelium because of a low interfacial tension at the mucous—aqueous interface and because of the surface tension depression at the tear—air interface due to surface active mucin and lipids (12). The underlying thin mucus layer, however, begins to thin because of the van der Waals interactions and eventually, ruptures in a time period consistent with the observed range of breakup time (~ 15–100 s). The linear stability analysis showed that the time of rupture is directly proportional to the mucous—aqueous interfacial tension and is proportional to the fifth power of the mucus layer thickness.

* Contrary to the consensus so far regarding to the thickness of the mucus layer (7a), a recent study conducted on the exercised guinea pig cornea has concluded that the *maximum* thickness of the mucus layer covering the corneal epithelium is about 0.4–0.8 μm (7b). The corneal mucus layer thickness, however, was not observed to be uniform in this study because of the presence of epithelial protrusions, the so-called microvilli. These microvilli have a characteristic dimension of about 0.2–0.5 μm over the corneal epithelium. The thickness of the mucus layer thus typically varies from about 0.05 to 0.2 μm over the tip of the microvilli to the maximum thickness cited above. Although, much of the computations illustrated in the paper are carried out for a mucus layer of thickness 400 Å, we show in Appendix B that the proposed mechanism of the tear film rupture is consistent with the thickness reported in Ref. (7b) for the feasible range of physicochemical parameters. Also, it is to be noted that various conclusions drawn from calculations based on the mucus layer thickness of 400 Å remain valid when instead a different set of feasible parameters is used for computations.

Step 2: wetting instability. The rupture of the mucus layer exposes the corneal epithelium to the aqueous tears at various places. There is ample experimental and clinical evidence which suggests that the mucus free corneal epithelium is a low energy surface and is nonwettable by aqueous tears (12–14). This wetting to nonwetting transition of the support of the aqueous layer is thus responsible for effectuating the rupture of the overlying aqueous film. Although the equilibrium considerations related to the stability or instability of a liquid layer on a solid support are known for a long time (15), the dynamics of this rupture is not well understood. However, clinical observations suggest that in the absence of mucus—secreting goblet cells, the tear film ruptures in about 3–5 s following a blink (16). This time, which may be referred to as the "dewetting time," is short compared to the time taken for the rupture of the mucus layer in a normal eye and it may be assumed to be constant, pending the development of an appropriate theory. In certain pathological conditions, most notably, in the event of an altered morphology of the epithelium (epitheliopathies such as dellen, erosion, and bullae), the tear film may break immediately following a blink (17). This is referred to as a permanent discontinuity, wetting time zero, or alternately, a permanent dry spot. In such cases and perhaps also in the case of an absolute deficiency of the aqueous tears, when a continuous aqueous film does not form following a blink, the dewetting time is close to zero.

This mechanism is shown to be in accord with several clinical and experimental observations related to the phenomenon of tear film rupture (10, 11). It clarified as to why a decreased amount of mucus production leads to a pathologically short BUT.

It is well known that apart from the mucus deficiency, the lipid abnormalities and the aqueous tear deficiency also cause a rapid rupture of the tear film and the dry eye syndromes (see, for instance, Ref. (7a)). One of the purposes of this work is to elucidate the role of normal lipids in the stability of the thin mucus layer and consequently, the effect of various types of lipid abnormalities on BUT. The effect of the aqueous tear deficiency and the action of the tear substitutes is also considered within this unified framework. For this purpose, we first discuss (albeit briefly) some of the relevant clinical observations which we wish to explain and then examine the instability of the mucus layer in the presence of lipids and aqueous tears. An estimate for the time of rupture is obtained and the parametric dependence of BUT on various factors involved is discussed. We also explain numerous clinical observations within this framework and discuss various implications of this analysis for the diagnosis and treatment of dry eyes and for the tolerance of contact lenses.

DEFICIENCIES OF A PATHOLOGICAL EYE

Numerous clinical observations accumulated so far indicate the following eye deficiencies to be primarily responsible for a rapid tear film breakup and various dry eye syndromes. The classification of dysfunctions of a dry eye as mucus deficiency, aqueous tear deficiency and lipid abnormalities follows that of Holly and Lemp (7a).

A. Mucus Deficiency

The corneal and conjunctival surfaces are coated by a thin film of mucus, which is about 200 to 400 Å thick in a normal eye (7a). The maximum thickness is reported to be an order of magnitude higher recently (7b). About 2.2 μl mucus per day is secreted by the goblet cells (18), of which about 1.5 million are distributed over the conjunctival epithelial surface (1). An even distribution of the mucus on the corneal surface is most probably facilitated by the shear created during the lid motion. A small amount of mucus is removed from the eye during each blink. Several conditions such as hypovitaminosis A (vitamin A deficiency), ocular pemphigoid, Stevens—Johnson syndrome, trachoma, and chemical burns affect the goblet cell density adversely and consequently reduce the epithelial mucus production (19–21). A reduced goblet cell count has been correlated with a short BUT (21) and corneal desiccation is sometimes observed in such instances, despite the presence of the normal amount of aqueous tears (16).

B. Aqueous Tear Deficiency

The aqueous tear film sandwiched between the mucus-coated epithelium and the superficial lipid layer is about 7–10 μm thick immediately after blinking, in a normal eye (1, 7a). The thickness then decreases in an almost linear manner with time due to the evaporation, drainage, and sometimes, the osmotic transfer across the cornea. Normal tear film is observed to break rather suddenly when its thickness is reduced to about 4 μm within the breakup time (6). The aqueous tears contain dissolved proteins, glycoproteins, carbohydrates, inorganic salts, lipids, and bactericidal agents (lysozyme, etc.) (1, 7a, 22). Some of the glycoprotein fractions (mucin) are highly surface active for the aqueous—air interface. A common cause of decreased BUT is identified to be an absolute or partial deficiency of aqueous tears (1, 7a) as determined from Schirmer (1, 7a) or fluorescein dilution (23) tests.

C. Lipid Abnormalities

The outermost layer of the tear film (located at the tear—air interface) is made up of low polarity waxy and cholesteryl esters (24). This superficial lipid layer is about 1000 Å thick in a normal eye (7a). The amount of high polarity lipids like triglycerides, free fatty acids and phospholipids is negligible in a normal eye (25). The following two distinct types of lipid abnormalities are identified to affect the BUT adversely:

1. A complete absence of the lipid layer may occur if either the lipid-secreting meibomian gland openings are absent or these glands are destroyed. The occurrence of this condition is, however, very rare (7a).
2. An alteration in the chemical composition and the consequent high surface activity of the eye lipids, such as that seen in the case of chronic blepharitis and facial skin infections like *acne rosacea*, leads to a marked decrease in BUT (26,27). A significant amount of polar lipids, viz., free fatty acids and triglycerides, are present in such instances.

In addition to the above-mentioned pathological conditions of the eye, the instillation of many low molecular weight surface active agents (artificial tear preservatives and anesthetic agents) also reduces BUT (28,29). Even the past history of blinking and the quality of blink immediately preceding the BUT measurement also results in some variability in the measurement of BUT (28), as do the extreme environmental conditions. Thus it is not an isolated measurement of BUT, but the robust mean of a statistical sample that appears to be more relevant for the diagnosis of a pathological eye (6).

We now turn towards a suitable model of the tear film and the analysis of its stability characteristics.

THEORY AND ANALYSIS

I. A Tear Film Model

A normal tear film consists of a mucus-coated microvillus structure of the epithelium, an overlying aqueous film, and a superficial lipid layer making the tear—air interface. The characteristic dimensions of epithelial protrusions, the so called microvilli and microplicae, are about 0.5 μm over the corneal epithelium. If the thickness of the mucus layer is an order of magnitude smaller than these characteristic dimensions (as is believed to be the case, except for one study (7b)), the support of the mucus layer may essentially be viewed as flat. This model of the precorneal tear film is depicted in

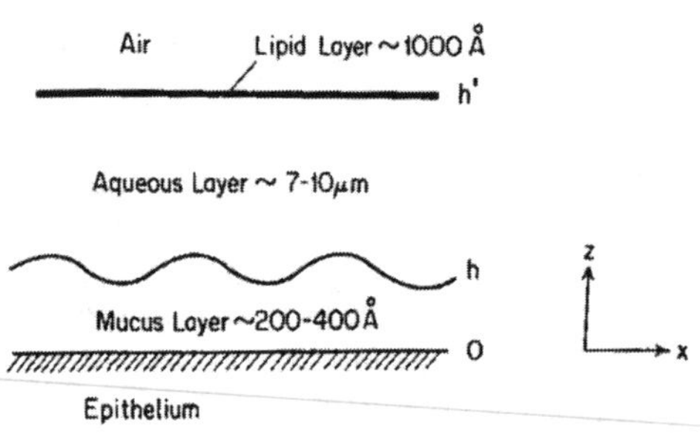

FIG. 1. A model of the tear film. The thickness of the mucus layer cited here should be interpreted as local thickness, as the thickness may vary owing to the presence of microvilli.

Fig. 1. If on the other hand, the maximum thickness of the mucus layer is of the same order as the dimension of microvilli (7b) the thickness of the mucus layer, as measured from the surface supporting it, is not uniform. It is minimum over the tip of a microvilli and is maximum over the valleys existing between two adjacent microvilli. Thus, although the incorporation of the morphology of the epithelium in this event is desirable, the essential physics is retained in a manageable description of a flat epithelium. Indeed, the mucus layer thickness may now be interpreted as a local quantity, which is minimum over the microvilli and thus the rupture first occurs at these spots.

The only significant aqueous tear flow occurring during the interblink period is confined to the tear meniscus that extends along the entire margin of the upper and lower eyelids, the so called "lacrimal river" (7a, 28). This flow, however, does not affect the stability of the mucus layer as showed earlier (10) and is thus neglected. The normal lipid layer located at the tear—air interface acts as a lipid reservoir and thus maintains a constant concentration of lipids at the lipid—aqueous interface. This interfacial concentration equals the solubility of normal lipids in the aqueous tears. As discussed earlier, the shear created during blinking restores the structural integrity of the mucus layer on the corneal epithelium and as the eyelids open, the mucous—aqueous interface is *relatively* homogeneous in appearance. This sheetlike mucus smeared over the epithelial surface (immediately after a blink) does not have a completely planar mucous—aqueous interface, because of the uneven shear and surface characteristics and also because the mucus layer is expected to be thicker at the mouths of the goblet cells. The initial nonhomogeneities begin to amplify because of the van der Waals forces acting on the molecules of the mucus layer. At the same time, the resulting flow (or velocities) redistributes the dissolved solutes such as lipids, ions, etc., which results in the generation of concentration gradients along the interface and consequently in an interfacial tension gradient along the mucus—aqueous interface. The interfacial tension gradient so generated induces a flow from the lower interfacial tension regions to regions with higher interfacial tension. This is the so called Marangoni flow which retards the thinning of the film when it opposes the flow produced by the perturbations. Finally an interplay of the factors delineated above determines the overall stability or instability of the mucous—aqueous interface and the possible rupture of the mucous layer. Thus we investigate the stability of the mucous—aqueous interface with the following initial state:

$$C = C_0 \text{ and various velocities} = 0,$$

where C_0 is the solubility of lipids in the aqueous tears.

II. The Hydrodynamics of the Tear Film

The attractive, intermolecular London—van der Waals forces between a molecule of the film and the other molecules of the film and the molecules of the surrounding phases are important when the range of these forces, which is usually a few thousand Ångstroms, is larger than the thickness of the film. The London—van der Waals forces decay rather rapidly as the thickness of the film increases. Their effect is thus quite pronounced for thin films, such as the mucus coating of the conjunctival and corneal epithelium (10).

The relevant equations describing the hydrodynamics are the Navier–Stokes equations, which incorporate the van der Waals forces acting on the unit volume of materials as body forces. This body force approach is rigorous and has been used extensively in investigating the stability of thin films (31–35). The dispersion forces are assumed to be nonretarded for the mucus layer because of its small thickness (<400 Å) and retarded for the aqueous layer. However, the Keesom interactions arising due to the interactions between permanent dipoles also appear important here, as is the case for water molecules. The retardation effect is accounted for while investigating the stability of a mucus layer thicker than 0.1 μm (Appendix B). The mucus layer is thin and thus the most effective or selectively amplified interfacial perturbations are those that have wavelength far exceeding the thickness of the film. It is because of this that one needs to consider Navier—Stokes equation in the "thin film" or the "lubrication-approximation" only. This approach has been used for a single film by Ruckenstein and Jain (33) and more formally by Williams and Davis (34).

Denoting by h_0 and h_0' as the initial locations of mucous—aqueous and aqueous—lipid interfaces and selecting h_0 and (v/h_0) as the basic units of lengths and velocities, the time and pressures may be scaled by $(h_0^2/v)$and $(\rho v^2/h_0^2)$, respectively, v and ρ are the kinematic viscosity and density, respectively. In view of the above scalings and denoting the mucus and aqueous properties by unprimed and primed variables, one may write the following simplified nondimensional Navier—Stokes equations in the lubrication-approximation:

$$u_{zz} - p_x - \phi_x = 0, \quad [1]$$

$$-p_z - \phi_z = 0, \quad [2]$$

$$\text{for } 0 \leqslant z \leqslant h;\ -\infty < x < \infty$$

and

$$(v'/v)u'_{zz} - (\rho/\rho')p'_x - \phi'_x = 0, \quad [3]$$

$$-(\rho/\rho')p'_z - \phi'_z = 0, \quad [4]$$

$$\text{for } h \leqslant z \leqslant h', -\infty < x < \infty.$$

Here the subscripts refer to partial differentiations, u and w are the dimensionless mucus layer velocities in the x and z directions and u' and w' are the corresponding dimensionless velocities in the aqueous layer, ρ' and v' are the aqueous layer density and kinematic viscosity, respectively; ϕ and ϕ' are the dimensionless van der Waals interaction potentials in the mucous and aqueous layers and are specified later. The derivatives of the potentials thus represent the component of the forces per unit volume acting on the corresponding layers.

The following two continuity equations hold for the mucous and aqueous layers, respectively:

$$u_x + w_z = 0 \quad [5]$$

$$u'_x + w'_z = 0. \quad [6]$$

In addition, the following linearized boundary conditions are specified at various interfaces.

The no slip boundary condition at the epithelium—mucous interface:

$$\text{at } z = 0,\ u = w = 0. \tag{7a}$$

The balance of tangential forces at the mucous—aqueous interface:

$$\text{At } z = h(x,t),\ \mu u_z = \mu' u'_z + (h_0/v)\sigma_x, \tag{7b}$$

where $h(x, t)$ is the dimensionless location of the mucus—aqueous interface and μ. and μ' are the dynamic viscosities of the mucous and aqueous layers, respectively.

The last term in the right hand side of the above equation represents the stress generated due to the mucous—aqueous interfacial tension gradient.

The relation of the pressure jump to the interfacial tension:

$$\text{At } z = h(x,t),\ p' - p = 3Sh_{xx}, \tag{7c}$$

where $S = (h_0\sigma/3\rho v^2)$ is a dimensionless number and σ is the mucous—aqueous interfacial tension.

The continuity of velocities across the mucous—aqueous interface:

$$\text{At } z = h(x,t),\ u = u' \text{ and } w = w'. \tag{7d}$$

Similarly, the following boundary conditions hold at the tear—air interface:

$$\text{At } z = h'(x,t),\ u'_z = 0 \tag{7e}$$

and

$$-p' = 3S'h'_{xx}, \tag{7f}$$

where h' is the dimensionless location of tear—air interface, $S' = (h_0\sigma'/3\rho v^2)$, and σ' is the air—tear surface tension.

Integrating Eqs. [2] and [4] gives

$$p + \phi = p(x,t) + \phi(x,t) = p(h) + \phi(h) \tag{8}$$

and

$$p' + \alpha\phi' = p'(h') + \alpha\phi'(h') = p'(h) + \alpha\phi'(h). \tag{9}$$

Using boundary conditions [7f] and [7c] in conjunction with Eqs. [8] and [9] shows that

$$p' + \alpha\phi' = -3S'h'_{xx} + \alpha\phi'(h') \tag{10}$$

and

$$p + \phi = \phi(h) - \alpha\phi'(h) + \alpha\phi'(h') - 3S'h'_{xx} - 3Sh_{xx}, \tag{11}$$

where $\alpha = (\rho'/\rho)$ and various potentials are evaluated at the interfaces indicated within the parenthesis. Equations [1] and [3] are now easily solved with the help of expressions [10] and [11] and the remaining boundary conditions, giving

$$u' = \psi_1 z^2/2 + a_1 z + a_2,\ h \leqslant z \leqslant h' \quad [12]$$

and

$$u = \psi_2 z^2/2 + a_3 z,\ 0 \leqslant z \leqslant h, \quad [13]$$

where

$$r\psi_1 = \left(-3S'h'_{xxx} + \alpha\phi'_x\right),\ r = \left(\mu'/\mu\right), \quad [14]$$

$$\psi_2 = \left(\phi_x(h) - \alpha\phi'_x(h)\right. \quad [15]$$

$$\left. + \alpha\phi'_x(h') - 3Sh_{xxx} - 3S'h'_{xxx}\right)$$

$$a_1 = -\psi_1 h', \quad [16]$$

$$a_2 = -\psi_2 h^2/2 + \psi_1 h\left[r\left(h - h'\right)\right.$$

$$\left. -(h/2) + h'\right] + h\left(h_0/\mu v\right)\sigma_x \quad [17]$$

and

$$a_3 = r\left(h - h'\right)\psi_1 - \psi_2 h + \left(h_0/\mu v\right)\sigma_x. \quad [18]$$

Now the solutions of the continuity Eqs. [5] and [6] are straightforward with the help of boundary conditions [7a] and [7d] and the expressions [12] and [13]. This gives

$$w = -\psi_{2x} z^3/6 - a_{3x} z^2/2 \quad [19]$$

and

$$w' = -\psi_{1x}\left(z^3/6 - h'z^2/2\right)$$

$$+ \psi_1 h'_x z^2/2 - a_{2x} z + a_4, \quad [20]$$

where

$$a_4 = \left(\psi_{1x} - \psi_{2x}\right)h^3/6 \quad [21]$$

$$-h^2\left(a_{3x} + \psi_1 h'_x + h'\psi_{1x}\right)/2 + ha_{2x}.$$

The evolution equations for the mucus—aqueous interface and the aqueous surface are now derived by using the following kinematic conditions:

$$h_t + h_x u - w = 0 \text{ at } z = h \quad [22]$$

and

$$h'_t + h'_x u' - w' = 0 \text{ at } z = h'. \quad [23]$$

Substitution of various velocities, viz. expressions [12], [13], [19], and [20], in the above equations give the lowest order, rigorous equations of evolution for the interfaces. These are, however, highly nonlinear and henceforth we focus our attention to the analysis of the linear equations only, awaiting quantitative information regarding various physicochemical and especially morphological parameters characterizing the mucus layer and the epithelium. Linearizing Eqs. [19], [20], and [22], and [23] around the planar locations of the interfaces, viz. $h = 1$ and $h' = (h'_0/h_0) = \beta$, and combining them, gives the following linear equations in the dimensionless variables:

$$h_t + (1-\beta) r\psi_{1x}/2 - \psi_{2x}/3 + h_0\sigma_{xx}(h)/2\mu v = 0 \tag{24}$$

and

$$h'_t + \psi_{1x}\theta/6 + (1-3\beta)\psi_{2x}/6 + h_0\sigma'_{xx}(h')\left(\beta - \frac{1}{2}\right)\Big/\mu v = 0, \tag{25}$$

where

$$\theta = (\beta-1)\left[1 - 2\beta(\beta+1) + 6\left(\beta - \frac{1}{2}\right)(1-r)\right]. \tag{26}$$

It may be noted that for the tear film, $\sigma'_{xx}(h') = 0$, because the aqueous—lipid surface has a constant concentration of lipids and thus an interfacial tension gradient does not exist at this interface.

The coefficients, ψ_{1x} and ψ_{2x}, may be now written explicitly by making use of expressions [14] and [15] and by prescribing the Hamaker type approximations for the dispersion potentials arising at the mucous—aqueous and the aqueous—lipid interfaces. The retarded dispersion potentials are prescribed for the aqueous layer, since it is quite thick (~ 10 μm) (9). The derivation of the dispersion potential arising at the two interfaces is straightforward and proceeds on the same lines as those pursued in Ref. (33) for a single film. In view of the scalings used to nondimensionalize the Navier—Stokes equations, the second derivatives of the linearized dimensionless potentials may be written as

$$\phi_{xx}(h) = -3e_2(A_{22} - A_{23})h_{xx} + 4B_{21}e_1(h'_{xx} - h_{xx}), \tag{27a}$$

$$\phi'_{xx}(h') = 4e_1(B_{12} - B_{11})(h'_{xx} - h_{xx}) + 4(B_{13} - B_{12})h'_{xx}e_3 \tag{27b}$$

and

$$\phi'_{xx}(h) = -3e_2(A_{12} - A_{13})h_{xx}/\alpha + 4B_{11}(h'_{xx} - h_{xx})e_1, \tag{27c}$$

where

$$e_1 = (\beta - 1)^{-5}\left(\rho' h_0^2 v^2\right)^{-1}, \quad e_2 = \left(6\pi\rho h_0 v^2\right)^{-1}$$

$$\text{and} \quad e_3 = \beta^{-5}\left(\rho' h_0^2 v^2\right)^{-1}.$$

A_{ij} and B_{ij} are the nonretarded and the retarded Hamaker constants for the interactions between the molecules of type *i* and *j*. The subscripts 1, 2, and 3 refer to the aqueous layer, mucus layer, and the epithelium molecules, respectively.

The evaluation of the mucous—aqueous interfacial tension gradient, $\sigma_{xx}(h)$, requires the solution of a mass transfer problem in conjunction with the hydrodynamic equations [24] and [25].

III. Interfacial Tension Gradients

The mucous—aqueous interfacial tension is expected to be very low, in view of the extreme hydrophilicity of mucus molecules and the highly hydrated state of the mucus layer. It is reasonable to assume it to be of the same order of magnitude as that encountered for cell membranes with a glycoprotein coat. This has been reported to be in the range of 10^{-3} to 1 dyn/cm (36a). The presence of dissolved lipids tends to augment this interfacial tension (28, 36b). In view of the fact that the presence of lipids in the vicinity of mucus—aqueous interface is thermodynamically unfavorable, their concentration in the aqueous layer increases away from the mucous—aqueous interface. This results in a negative surface excess concentration, as opposed to the positive surface excess concentration in the case of energetically favorable adsorption of surfactants (15). The lipids (and some other constituents of the aqueous phase, a possible exception being the dissolved mucin) may thus be characterized as antisurfactants (sometimes referred to as inverse surfactants), *with reference* to the mucous—aqueous interface, in recognition of their interfacial tension augmenting capacity. Some aspects of the role of surfactant adsorption on the van der Waals force mediated rupture of thin films have been investigated (33), but no such study has been undertaken for the antisurfactants to the best of our knowledge.

The nonuniform solute concentration distribution normal to the interface may be ascribed to the presence of an interaction potential, which acts within a short distance, δ, from the interface and it is because of this that the concentration distribution in the boundary layer, $h \leqslant z \leqslant h + \delta$, is expected to differ from the concentration in the bulk, $h + \delta < z < h'$.

The excess concentration of a component is defined here as the difference between the actual moles contained in the boundary layer and the moles contained assuming that the concentration is that governed by the convective diffusion equation without the presence of the interaction potential

$$\Gamma = \int_{h}^{h+\delta} \left(C^* - C\right) dz, \qquad [28]$$

where Γ is the surface excess concentration, C^* is the solute concentration in the boundary layer, $h \leqslant z \leqslant h + \delta$, and C is the concentration distribution in the absence of the interaction potential. δ is the characteristic thickness of the boundary layer in the vicinity of the interface over which C^* differs from C due to the interaction potential, $\psi(z)$. This potential is operative within a few tens of Ångstroms from the interface and is attractive for surfactants and repelling for antisurfactants. For surfactants, the concentration in the vicinity of the interface is usually very large compared to their concentration in the bulk, viz. $C^* \gg C$. Therefore, this thin boundary layer, $h \leqslant z \leqslant h + \delta$, is often envisaged as a two-dimensional surface and the excess concentration is viewed as the actual interfacial concentration of surfactants. In the case of antisurfactants, however, this picture is inaccurate inasmuch as $C^* < C$ here and thus $\Gamma < 0$, viz. their concentration in the bulk is the highest.

The determination of the surface excess concentration is important because the change in the interfacial tension does not depend on the solute concentration present in the vicinity of the interface per se, but on the excess surface concentration, Γ (15). In what follows, we first formulate an ideal hydrodynamic description of the interface, applicable both to surfactants and antisurfactants.

Consider an interface whose location is given by $h(x,t)$ and a thin boundary layer of thickness $\delta(x,t)$ adjacent to it over which the interaction potential, $\psi(z)$, is operative. The following convective diffusion equation may be written for the solute concentration within the boundary layer, C^*:

$$C_t^* + u'C_x^* + w'C_z^*$$

$$= (D/v)\nabla^2 C^* + (D/k_b T_v)\left(C^* \psi_z\right)_z, \qquad [29]$$

where k_b is Boltzmann's constant and T is the temperature.

The contribution of the potential $\psi(z)$ is negligible outside the boundary layer and thus the convective diffusion equation for the bulk is

$$C_t + u'C_x + w'C_z = (D/v)\nabla^2 C. \qquad [30]$$

Subtracting Eq. [30] from Eq. [29], integrating the resulting equation over the thickness of the boundary layer, viz. the domain, $h \leqslant z \leqslant h+\delta$, and simplifying, gives the following equation relating the variations in the surface excess concentration to the solute flux from the bulk phase (the derivation is outlined in Appendix A):

$$\theta_1 C_z = \Gamma_t + (u'\Gamma)_x \quad \text{at} \quad z = h(x,t), \qquad [31]$$

where $\theta_1 = (Dh_0/v)$.

The equilibrium may be assumed to prevail in the boundary layer because it is extremely thin and thus the equilibrium isotherm is derived by noting that the concentration in the boundary layer (Eq. [29]) satisfies

$$k_b T C_{zz}^* + \left(C^* \psi_z\right)_z = 0, \qquad [32]$$

which, when integrated twice and combined with the zero flux condition and the boundary condition, $C^* = C$ at $z = h+\delta$, gives

$$C^* = C \exp\{-\psi(z)/k_b T\}. \qquad [33]$$

It is instructive to note that the same result may also be obtained by thermodynamic considerations, viz. by making use of the equality of the chemical potentials at different locations.

The definition of Γ, (Eq. [28]), in conjunction with expression [33] yields the desired isotherm, viz.

$$\Gamma = C\int_0^{\delta}\left(e^{-\psi/k_b T} - 1\right)dz \equiv \theta_0 C, \qquad [34a]$$

where

$$\theta_0 = \int_0^{\delta}\left(e^{-\psi/k_b T} - 1\right)dz. \qquad [34b]$$

For surfactants, $\psi < 0$ and consequently, $\theta_0 > 0$, whereas for antisurfactants, $\psi > 0$ and thus $\theta_0 < 0$. The linear isotherm, Eq. [34], may be recognized to be the isotherm for a "gaseous monolayer" type adsorption. This behavior is somewhat restrictive for surfactants, as it cannot predict the surface (or boundary layer) saturation obtained at higher bulk concentrations. This may be remedied by assuming a more realistic isotherm, such as the Langmuir isotherm, that accounts for the "surface exclusion" and thus predicts the saturation of the interface at higher bulk concentration of the surfactants. This, however, is of less concern for antisurfactants for which the concentration in the boundary layer is lower than in the bulk and thus the saturation is more likely to occur first in the bulk (when the bulk concentration of antisurfactants attains its solubility limit).

The linearized mass transport equation for the diffusion of lipids in the bulk is now written as

$$C_t = \left(D/v\right)\nabla^2 C, \qquad [35]$$

which is to be solved together with the boundary conditions

$$C = C_0 \quad \text{at} \quad z = \beta \qquad [36a]$$

and

$$\theta_1 C_z = \Gamma_t + \Gamma_0 u_x' \quad \text{at} \quad z = 1, \qquad [36b]$$

where $\Gamma_0 = \Gamma(C_0)$.

The dynamic interfacial tension, which accounts for the surface viscosity, μ_s, is defined as

$$\sigma(\Gamma) = \sigma_0(\Gamma) + \left(\mu_s v/h_0^2\right)u_x \quad \text{at} \quad z = 1. \qquad [37]$$

Substitution of $(du/dx)_{z=1}$ from Eq. [13] in Eq. [37] yields the following differential equation for the interfacial tension σ:

$$\sigma_{xx} - \left(h_0\mu/\mu_s\right)\left(\sigma - \sigma_0\right)$$

$$+\left(v\mu/h_0\right)\left[r\psi_{1x}\left(1-\beta\right) - \psi_{2x}/2\right] = 0. \qquad [38]$$

Having completed the description of the hydrodynamic and the mass transport problems and the determination of interfacial tension gradient, we now turn towards the linear stability analysis of these equations.

IV. Linear Stability

In reality, the location of the mucous—aqueous interface is not absolutely flat, but is nonhomogeneous because of the uneven shear created during the lid motion, the thermal perturbations, nonhomogeneous epithelium, and the presence of the goblet cells. In what follows, we seek to determine the growth of local inhomogeneities due to dispersion interactions and the surfactant or antisurfactant concentration driven Marangoni motion. Any arbitrary disturbance originating at the interfaces may be represented by a sum of Fourier components of the form

$$h \approx 1 + \epsilon_1 e^{ikx+\omega t}$$

and

$$h' \approx \beta + \epsilon_2 e^{ikx+\omega t}, \qquad [39a]$$

where ϵ_1 and ϵ_2 are the nondimensional initial amplitudes of the perturbations, k is a dimensionless wavenumber, and ω is a dimensionless growth coefficient which shows the rate of amplification (for ω positive) or decay (if ω is negative) of the initial interfacial perturbations. The perturbations in the location of the interface induce convective velocities and, consequently, a deviation in the concentration from its equilibrium value. This may be represented as

$$C = C_0 + \eta(z)e^{ikx+\omega t} \quad [39b]$$

and consequently from Eq. [34a],

$$\Gamma - \Gamma_0 = \theta_0 \eta(z)e^{ikx+\omega t}. \quad [39c]$$

Perturbation in interfacial tension is now obtained as

$$\sigma_0 - \sigma_0(\Gamma_0) = (\partial\sigma_0/\partial\Gamma)_{\Gamma=\Gamma_0}(\Gamma - \Gamma_0) \quad [39d]$$

$$= M\eta(z)\theta_0 e^{ikx+\omega t}, \quad [39e]$$

where $M = (\partial\sigma_0/\partial\Gamma)_{\Gamma=\Gamma_0}$, is a parameter that shows the variation of the interfacial tension with the surface excess concentration. An estimate for this may be arrived at by using the Gibbs adsorption equation

$$d\sigma_0 = -\Gamma d\mu, \quad [40a]$$

where the concentration dependent part of the chemical potential, μ, is given by RT ln C, and consequently,

$$d\sigma_0 = -RT\Gamma d(\ln C). \quad [40b]$$

Making use of the isotherm [34a], one obtains

$$M = (d\sigma_0/d\Gamma) = -RT. \quad [40c]$$

Thus M is about -2.4×10^{10} erg/mole which is consistent with the values reported for a gaseous monolayer adsorption.

Substituting expressions [39] in Eq. [38] and solving gives

$$\alpha_4(h_0/\mu v)\sigma_{xx} = e^{ikx+\omega t}\left[-M\eta\theta_0 h_0 - \epsilon_1\mu_s \right.$$

$$\times\left(3Sk^4/2 + b_1k^2/2 + (1-\beta)\alpha a_{31}k^2\right)$$

$$-\epsilon_2\mu_s\left\{3S'k^4/2 + b_2k^2/2 + (\beta - 1)\right.$$

$$\left.\left.\times\left(3S'k^4 + \alpha a_{32}k^2\right)\right\}\right], \quad [41]$$

where the various coefficients are defined by

$$b_1 = 4\alpha B_{11}e_1 + 3\alpha e_2\left(A_{12} - A_{13}\right)$$

$$-4\left(B_{12} - B_{11}\right)\alpha e_1$$

$$-3e_2\left(A_{22} - A_{23}\right) - 4B_{21}e_1, \qquad [42a]$$

$$b_2 = 4B_{21}e_1 - 4\alpha B_{11}e_1 + 4\alpha\left(B_{12} - B_{11}\right)e_1$$

$$+ 4\alpha\left(B_{13} - B_{12}\right)/\beta^5\rho' h_0^2 v^2, \qquad [42b]$$

$$a_{31} = 4\left(B_{12} - B_{11}\right)e_1, \qquad [42c]$$

$$a_{32} = 4\left(B_{12} - B_{11}\right)e_1 + 4\left(B_{13} - B_{12}\right)/\beta^5\rho' h_0^2 v^2 \qquad [42d]$$

and

$$\alpha_4 = \left(\mu_s + h_0\mu/k^2\right). \qquad [42e]$$

With the help of Eqs. [13], [39b], and [39c], the diffusion Eq. [35] and boundary conditions [36a] and [36b] become

$$\eta_{zz} = a^2\eta, \text{ where } a = \left(k^2 + \omega v/D\right)^{1/2}, \qquad [43a]$$

$$\text{at } z = \beta, \eta = 0, \qquad [43b]$$

$$\text{at } z = 1, \theta_1\eta_z = \alpha_1\eta + \epsilon_1\alpha_2 + \epsilon_2\alpha_3, \qquad [43c]$$

where

$$\alpha_1 = \theta_0\omega - \left(\Gamma_0 M\theta_0 h_0/\alpha_4\right), \qquad [44a]$$

$$2\alpha_2 = \Gamma_0\left\{3Sk^4 + b_1k^2\right.$$

$$\left. + 2(1-\beta)\alpha a_{31}k^2\right\}\left(1 - \mu_s/\alpha_4\right) \qquad [44b]$$

and

$$2\alpha_3 = \Gamma_0\left\{3S'k^4 + b_2k^2 + 2\left(\beta - 1\right)\right.$$

$$\left.\times\left(3S'k^4 + \alpha a_{32}k^2\right)\right\}\left(1 - \mu_s/\alpha_4\right). \qquad [44c]$$

The solution of Eqs. [42a]–[42c] is given by

$$\eta = 2\left(\epsilon_1\alpha_2' + \epsilon_2\alpha_3'\right)e^{a\beta}\sinh\left\{a\left(1-\beta\right)\right\}, \qquad [45]$$

where

$$2\alpha_2' = \alpha_2 e^{-a\beta}\Big[a\theta_1 \cosh\{a(\beta-1)\}$$

$$+\alpha_1 \sinh\{a(\beta-1)\}\Big]^{-1} \quad [45a]$$

and

$$\alpha_3' = \alpha_2'\alpha_3/\alpha_2. \quad [45b]$$

The determination of η completes the mass transfer problem since the interfacial tension gradient, σ_{xx}, is now immediately determined from Eq. [41]. The dispersion relation is obtained by combining Eqs. [24], [25], [39a], [41], and [45], giving

$$\epsilon_1\Big[\omega + \alpha a_{31}(1-\beta)k^2/2$$

$$+ Sk^4 + b_1k^2/3 + N_1\Big]$$

$$+\epsilon_2\Big[(1-\beta)(-3S'k^4 - \alpha a_{32}k^2)/2$$

$$+ S'k^4 + b_2k^2/3 + N_2$$

$$\equiv a_{11}\epsilon_1 + a_{12}\epsilon_2 = 0 \quad [47a]$$

and

$$\epsilon_1\Big[\alpha a_{31}k^2\theta/6r + (1-3\beta)(-3Sk^4 - b_1k^2)/6\Big]$$

$$+\epsilon_2\Big[\omega - \theta(-3S'k^4 + \alpha a_{32}k^2)/6r$$

$$+(1-3\beta)(-3S'k^4 + b_2k^2/6)\Big]$$

$$\equiv a_{21}\epsilon_1 + a_{22}\epsilon_2 = 0 \quad [47b]$$

where

$$2\alpha_4 N_1 = \mu_s\{(\beta-1)\alpha a_{31}k^2 - 3Sk^4/2 - b_1k^2/2\}$$

$$-2M\theta_0 h_0 \alpha_2' e^{a\beta}\sinh\{a(1-\beta)\}, \quad [48a]$$

and

$$2\alpha_4 N_2 = \{(1-\beta)(3S'k^4 + \alpha a_{32}k^2)$$

$$-3S'k^4/2 - b_2k^2/2\}\mu_s$$

$$-2M\theta_0 h_0 \alpha_3' e^{a\beta}\sinh\{a(1-\beta)\}. \quad [48b]$$

The coefficients N_1 and N_2 contain the effects of the interfacial tension gradient driven convective motion on the stability of the mucus layer.

The set of homogeneous, algebraic equations [47a] and [47b] have a nontrivial solution if the following solvability condition is satisfied:

$$\begin{vmatrix} a_{11} & a_{12} \\ a_{21} & a_{22} \end{vmatrix} = 0. \quad [49]$$

This equation, when written out in full, is highly nonlinear and somewhat unwieldly. However, a simplification may be noted when the thickness of the aqueous layer is large compared to the thickness of the underlying mucus layer, which is almost always the case for a tear film. Due to a rapid decay of the dispersion forces with an increased thickness of the aqueous layer and a high surface tension of the aqueous layer, we found that the thickness dependence of the dispersion interactions at the tear—air interface may be neglected for the aqueous layer thicknesses in excess of 5000 Å. In this case, $a_{21}/a_{22} \rightarrow 0$ and the dispersion relation, Eq. [49], reduces to

$$a_{11} = 0. \quad [50]$$

This is tantamount to neglecting the thinning of the aqueous film due to dispersion forces during the time in which the mucus layer ruptures.

Noting this simplification and reverting back to the original dimensional variables, the dispersion relation [50] becomes

$$\lambda = \left[\left(A/2\pi h_0^4 - \sigma_0(\Gamma_0)K^2\right)h_0^3 K^2/3\mu\right] \times \left\{1 - 3\left(N + \mu_s/2\right)/2\alpha_4'\right\}, \quad [51]$$

where λ is the dimensional growth coefficient, K is the dimensional wavenumber, and A is an effective Hamaker constant defined as

$$A = A_{22} + A_{13} - A_{23} - A_{12}. \quad [52]$$

Further, parameters N, α_4', and a' are defined as

$$2N = -M\Gamma_0\left(1 - \mu_s/\alpha_4\right)\left[a'D\coth\left\{a'\left(h_0' - h_0\right)\right\} \times \theta_0 + \lambda - \Gamma_0 M/\alpha_4\right], \quad [53]$$

$$\alpha_4' = \mu_s + \mu/K^2 h_0 \quad [54]$$

and

$$a' = \left(K^2 + \lambda/D\right)^{1/2}. \quad [55]$$

The static part of the interfacial tension in the presence of a surfactant or antisurfactant may be expressed by a MacLaurin series approximation, viz.

$$\sigma_0(\Gamma_0) \approx \sigma_0(0) + M\left(\Gamma_0 - \Gamma_0(0)\right) \quad [56]$$

Assuming that $\sigma_0(0)$ represents the mucus—aqueous interfacial tension when the aqueous phase is pure, viz., $\Gamma_0(0) = 0$, the dispersion relation [50] is rewritten as

$$\lambda = \left[\lambda_0 - M\Gamma_0 h_0^3 K^4/3\mu\right]$$

$$\times\left[1-3\left(N+\mu_s/2\right)/2\alpha_4'\right], \qquad [57]$$

where λ_0 is the growth coefficient of a clean mucus interface when the aqueous phase is devoid of any surfactants or antisurfactants, viz.

$$\lambda_0 = \frac{h_0^3 K^2}{3\mu}\left(\frac{A}{2\pi h_0^4} - \sigma_0(0)K^2\right). \qquad [58a]$$

If the London dispersion interactions are assumed to be retarded for the mucus layer thicknesses in excess of 0.1 μm, viz. one uses the potential of the form B/h^4 instead of $A/6\pi h^3$, the growth coefficient is given by

$$\lambda = \frac{h_0^3 K^2}{3\mu}\left[\frac{4B}{h_0^5} - \sigma_0 K^2\right]$$

$$\times\left[1-3\left(N+\mu_s/2\right)/2\alpha_4'\right]. \qquad [58b]$$

At this point, it is instructive to recall that for antisurfactants, $\theta_0<0$ and Γ_0 is a negative quantity, whereas for surfactants, $\theta_0 > 0$ and Γ_0 is positive. Thus, the parameter N may in general be written as

$$2N = \left|\Gamma_0 M\right|\left(1-\mu_s/\alpha_4'\right)\left[\left|\theta_0^{-t}\right|Da'\right.$$

$$\left.\times\coth\left\{a'\left(h_0'-h_0\right)\right\} \pm \lambda + \left|M\Gamma_0\right|/\alpha_4'\right]^{-1}. \qquad [59]$$

The sign in $\pm\lambda$ is determined by whether the material is a surfactant (positive sign) or an antisurfactant (negative sign).

An estimate for the time of rupture of the mucus layer is given by

$$\tau \approx \lambda_{\mathrm{m}}^{-1}, \qquad [60]$$

where λ_m is the maximum growth rate for a given set of parameters that is maximized with respect to the wave number, $K\epsilon(0, \infty)$. The Fourier component of the initial disturbance with the wavenumber corresponding to the solution of $(d\lambda/dK) = 0$ is selectively amplified until the layer ruptures.

The analytical determination of the wavenumber for which λ is a maximum is not possible due to the transcendental nature of the dispersion equation [57] and is thus done numerically. Before undertaking this however, it is instructive to synopsize certain asymptotic cases and the physics of the phenomena involved, lest they are obscured by the mathematical derivations.

ASYMPTOTIC CASES AND PHYSICAL EXPLANATIONS

Marangoni Flow

For large values of the parameter $|\Gamma_0 M|$, the coefficient N, given by Eq. [59], reduces to

$$N = (\alpha_4' - \mu_s)/2 \tag{61a}$$

which, when substituted into Eq. [57], yields

$$\lambda = \left[\lambda_0 - M\Gamma_0 h_0^3 K^4/3\mu\right]\left(\frac{1}{4}\right). \tag{61b}$$

Thus, the maximum stabilizing effect of surfactant or antisurfactant concentration gradient induced Marangoni flow is to increase the time of rupture by a factor of 4, compared to the case of a thin film with an apparent interfacial tension of $\{\sigma_0(0) + M\Gamma_0\} = \sigma_{app}$. By maximizing λ as given by Eq. [61b] with respect to K, the time of rupture from Eq. [60] is given as

$$\tau = 192\pi^2 \mu h_0^5 \{\sigma_0(0) + M\Gamma_0\} A^{-2}, \tag{62}$$

in contrast to a film with $|M\Gamma_0| \to 0$ and consequently, $N \to 0$, where it is given by

$$\tau = 48\pi^2 \mu h_0^5 \sigma_0(0) A^{-2}. \tag{63}$$

Thus the presence of dissolved material in the aqueous phase stabilizes the film due to Marangoni convection and at the same time, alters the interfacial tension, $\sigma_0(0)$, to σ_{app}. This later effect, as shown by Eq. [62], is stabilizing for antisurfactants as $M\Gamma_0$ is positive (the interfacial tension is augmented) and destabilizing for surfactants.

It is important to note that $|M\Gamma_0|$ cannot be increased indefinitely, because of the saturation of the interface for surfactants and the saturation of the bulk for antisurfactants. In conclusion, the overall effect of antisurfactants is always stabilizing and that of surfactants may be stabilizing or destabilizing, depending on whether the stabilizing effect of Marangoni flow is strong enough to compensate for the destabilizing effect of reduced interfacial tension.

It is important to note that the stabilizing effects of the Marangoni flow as delineated above, hold only for a strictly linear isotherm, Eq. [34], where the isotherm constant, θ_0, may be assumed to be independent of the bulk concentration. In the event of the saturation of the interface with solute molecules, the surface excess concentration becomes constant (independent of the bulk concentration) and thus cannot be perturbed, viz. $\theta \to 0$ in Eq. [37c]. As is apparent from Eq. [59], N tends to zero in such an event and the stabilizing effect of the Marangoni flow is wiped out. Thus, in reality, there exists an intermediate solute concentration for which the stabilizing influence of the Marangoni flow is the maximum.

The physical origin of concentration differential driven Marangoni motion is explained in Figs. 2a and b. Figure 2a shows that any interfacial perturbation of the mucus layer induces a flow from the depressed regions to the elevated regions. Due to this convective transfer, the concentration, and consequently, the excess concentration of surfactant or antisurfactant becomes larger at the elevated locations compared to the depressed ones. In view of the Gibbs equation, Eq. [40a], the interfacial tension at the elevated regions is thus lower than that prevailing at the depressed regions. The interfacial tension gradient so generated induces a convective Marangoni flow from the low interfacial tension regions to the regions with higher interfacial tension (Fig. 2b). As shown in Fig. 2b, this flow tends to counter the flow generated due to the film thinning and thus exerts a stabilizing effect.

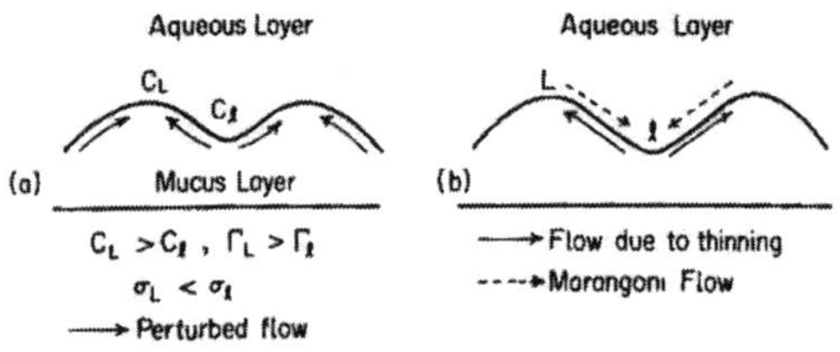

FIG. 2 The physical origin of antisurfactant or surfactant concentration differential driven Marangoni flow.

Surface Viscosity

The effect of surface viscosity is also delineated within the above framework by noting that the dynamic part of the surface tension, σ_d, is related to the velocity differential by

$$\sigma_d = \mu_s u_x \text{ at } z = 1, \tag{64a}$$

where

$$\sigma = \sigma_0 + \sigma_d. \tag{64b}$$

The dynamic part of the interfacial tension, σ_d, vanishes when either the surface viscosity, μ_s is zero or the surface is unperturbed, viz. $u_x = 0$. For a finite μ_s and a perturbed interface, however, the variation of the lateral velocity in x direction, viz. u_x, is negative at the elevated points as opposed to a positive u_x at the depressed points. This process is depicted in Fig. 3. Again as is clear from Eq. [64b], the interfacial tension at elevated regions is lower compared to depressed regions and a stabilizing flow ensues. It is easy to verify that when $\mu_s \rightarrow \infty$, there is a fourfold increase in the time of rupture as compared to the case of a film with an apparent interfacial tension, σ_{app}, viz. Eq. [62] applies to this case as well.

Thickness of Aqueous Layer

The effect of the thickness of the aqueous layer is contained in the parameter, $L = \coth\{a'(h_0' - h_0)\}$, which appears in the definition of N, Eq. [59], As the thickness, $(h_0' - h_0)$, decreases, L increases and consequently, N decreases. In the limiting case of $h_0' \rightarrow h_0$, $L \rightarrow \infty$ and thus $N \rightarrow 0$. The stabilizing effect of the Marangoni convection is completely wiped out in this case. In general, a decreased aqueous layer thickness makes the stabilizing effect of Marangoni flow less pronounced. This is expected physically because since the concentration of lipid is constant at the aqueous—lipid interface, a decreased aqueous layer thickness facilitates a faster redistribution of lipids in the layer because of the diffusion. This results in a tendency to even out the concentration gradients generated due to the convective flow and consequently, the interfacial tension gradients are also reduced. The stabilizing Marangoni flow resulting from the interfacial tension gradients is thus minimized.

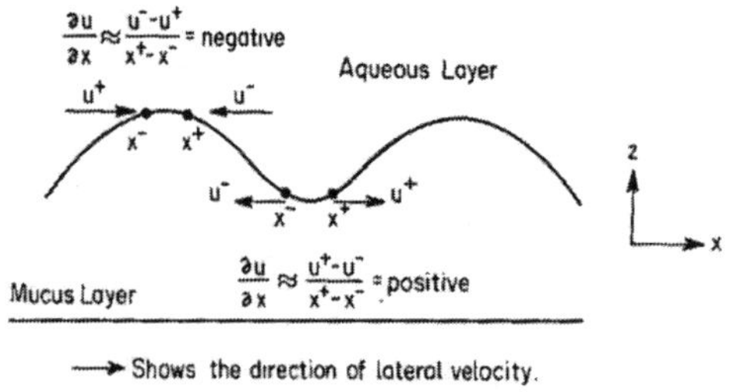

FIG. 3 The physical origin of the stabilizing effect of the mucus surface viscosity.

Having expounded on the qualitative aspects of the rupture of the thin mucus layer, we now see how it enhances our understanding of the issues (which have been discussed earlier) related to the breakup of tear film.

RESULTS AND DISCUSSION

An exact *a priori* prediction of BUT is not possible, not only because of the complexity of the eye and the present limited knowledge concerning the value of the various physicochemical and morphological parameters, but also because the breakup of the mucus layer is a nonlinear phenomenon and is affected, among other factors, by the extent and the nature of the initial interfacial perturbations or inhomogeneities. In view of this, and some other factors that are explained later, it is not surprising that the repeated clinical measurements of BUT tend to show a deviation (6, 37) even for the same patient. Fortunately, what *is* of paramount importance here is a rational understanding of the causes affecting the tear—film breakup (and BUT), which albeit semiquantitative, paves a way to a causal approach to the explanation of the numerous clinical and experimental observations, selective experimentation, and the diagnosis and treatment of dry eyes. The discussion that follows should be viewed in this spirit.

The interfacial tension of a *clean* mucous—aqueous interface is low and is assumed to be in the range of 10^{-3} dyn/cm to 1 dyn/cm, viz. $\sigma_0(0) = 10^{-3}$ to 1 dyn/cm, which is the same range as that obtained for the interfacial tension of the cell membranes with a glycoprotein coat (36a). The normal lipids are probably the most antisurface active material present in the eye, i.e., they augment the interfacial tension the most. Based on these estimates, and the physical parameter values reported earlier (10) (also provided in the figure captions), Eq. [57] may be maximized numerically with respect to the wavenumber K, and an estimate for the time of rupture may be computed from Eq.[60]. The region of compatibility for the mucus layers of thickness 400 Å and a rather high interfacial tension of 1 dyn/cm is depicted as the shaded region in Fig. 4. The values of the effective Hamaker constant, A, and the parameter $M\Gamma_0$ corresponding to the shaded regions are the ones that are compatible

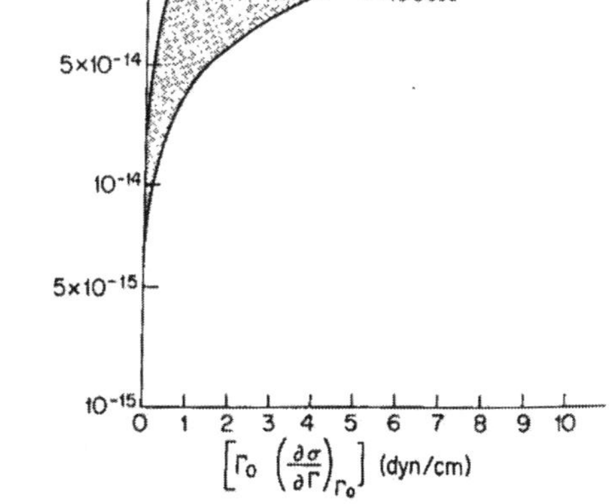

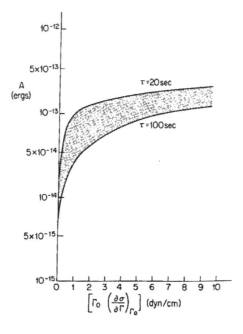

FIG. 4 The region of compatibility for the rupture of the mucus layer. For the parameter values falling in the shaded region, the mucus layer ruptures within 20 to 100 s. Other parameters are fixed at $\sigma_0(0) = 1$ dyn/cm, $\mu = 0.1$/cm · s, $\mu_s = 10^{-3}$ g/s, $h_0 = 400$ Å, $h'_0 = 10$ μm, $D = 10^{-5}$cm²/s, and $\theta_0 = 2\times10^{-4}$cm.

with the observed kinetics of tear film rupture (BUT) for a normal eye. The magnitudes of the effective Hamaker constant belonging to the shaded regions are entirely within a realistic range (31, 33, 38–40). It is thus concluded that even relatively weak van der Waals interactions break a mucus layer of thickness 400 Å and a relatively high effective interfacial tension (about 5 dyn/cm) within a short period of about 20–100 s. The compatibility of the proposed mechanism with the observed range of BUT for thicknesses higher than 400 Å is demonstrated in Appendix B for the feasible values of the Hamaker constant and the mucous—aqueous interfacial tension.

Henceforth, we will illustrate various ramifications of the proposed mechanism with the help of a particular set of parameters (those corresponding to Fig. 4). However, it is to be understood that these conclusions remain the same if instead a different feasible set of parameters are chosen.

A more precise prediction of BUT, however, awaits the experimental availability of physicochemical parameters such as the effective Hamaker constant for the epithelium—mucous—aqueous layer system, viscosity and rheology of epithelial mucus layer, the mucous—aqueous interfacial tension, $M\Gamma_0$, and the thickness of the corneal mucus layer that prevails under the normal dynamic conditions of the tear film.

With this general background, we now go on to relate the specific deficiencies of the eye, the role of the tear substitutes and the contact lens tolerance to the proposed model of tear film breakup.

Mucus Deficiency

A decreased population of goblet cell density or a deficiency of mucus production in eye would decrease the thickness of the mucus coating of the corneal epithelium. A thinner mucus layer ruptures faster. This has been discussed in some detail in our previous papers (10, 11). This deficiency of the eye appears to be the most difficult to treat as there does not exist any direct method of replenishing the required mucus on the epithelium on a regular basis. The polymeric ingredients of the ophthalmic solutions may increase it by adsorption and absorption and may thus prolong the time of rupture. However, in the case of a moderate mucus deficiency, it appears possible to offset its effect on BUT by manipulating certain other factors which are discussed next.

The Stabilizing Role of Normal Lipids

The normal lipids, in conjunction with other dissolved materials of the tear film, augment the low mucous—aqueous interfacial tension and thus make the mucus layer more resistant to the interfacial deformations caused by the dispersion forces. In addition, they contribute toward the stability by enhancing the interfacial tension driven Marangoni motion. To see this dual stabilizing role of normal lipids in the most transparent way, their effect is displayed for a typical set of parameter values, $A = 10^{-3}$ erg and 1.5×10^{-3} erg, $h_0 = 400$ Å, $\sigma_0(0)$dyn/cm, $\mu = 0.1$ g/cm·s and $\mu_s = 10^{-3}$ g/s. The time of rupture of the mucus layer is shown as a function of antisurfactant parameter, $M\Gamma_0$, in Fig. 5. The time of rupture increases with an increase in the parameter, $M\Gamma_0$. For illustrative purposes, we assume that $M\Gamma_0 = 3$ for a healthy tear film complete with superficial lipid layer and that in the absence of lipids, $M\Gamma_0 = 1$. This reduces BUT by about a factor of 2 (Fig. 5) and thus the presence of normal lipids seems imperative for maintaining a healthy tear film breakup time. Also in view of this, a complete destruction of meibomian glands or the absence of normal lipids may result in an unstable tear film.

Lipid Abnormality (Increased Surface Activity of Normal Lipids)

As discussed earlier, pathological conditions such as chronic blepharitis, facial skin infections, and increased activity of the lipase-secreting bacteria because of any number of reasons, alter the normal interfacial activity of lipids, making them much more surface active. The presence of these surface active lipids (free fatty acids, etc.) in the vicinity of the mucus—aqueous interface is energetically

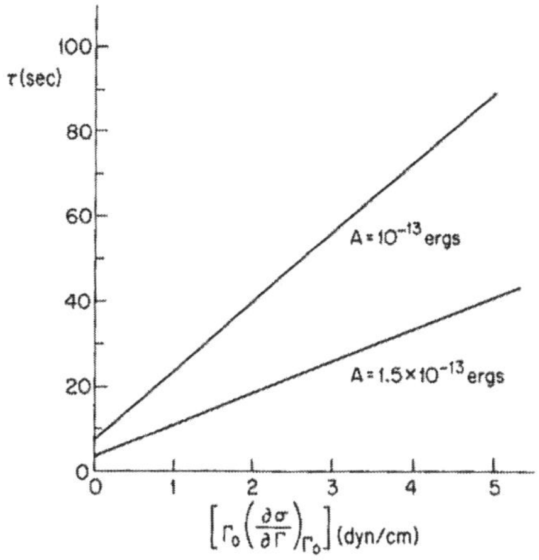

FIG. 5 The effect of antisurfactant parameter, $M\Gamma_0$, on the time of rupture of the mucus layer for parameter values as given in Fig. 4.

favored *compared* to the normal, less surface active lipids of the tear film. This is so because the polar lipids like fatty acids, triglycerides, and phospholipids have a nonpolar hydrocarbon tail and a polar carboxyl head. Their minimum free energy configuration at the mucus—aqueous interface thus corresponds to the polar, hydrophilic group being immersed in the aqueous tears and the hydrocarbon tail being imbedded in the mucus layer. The possible effect it has on the stability of the tear film is best understood by noting that:

(a) Such a change in the chemical composition of the normal lipids decreases the mucus—aqueous interfacial tension from the value that prevails in the presence of normal lipids. In other words, the interfacial tension decreases to a value of $\{\sigma_{\text{app}} - |M\Gamma_0| s\}$ from its normal value of $\sigma_{\text{app}} = \sigma_0(0) + M\Gamma_0$, where $|M\Gamma_0|_s$ is the decrease in the magnitude of interfacial tension because of the more surface active lipid components (free fatty acids, etc.).
(b) It contributes to the stabilizing effect of interfacial tension driven Marangoni motion as the Marangoni motion is enhanced both by the surface active and antisurfactant solutes alike. Thus assuming no synergistic or interactive effects between the normal and surface active lipids, the Marangoni parameter, *N*, in Eq. [59] is modified by replacing $|M\Gamma_0|$ with $|M\Gamma_0| + |M\Gamma_0|_s$.

The overall effect of the surface active lipids on the time of rupture is depicted in Fig. 6 for a typical set of parameter values and for varying parameter, $|M\Gamma_0|_s$. Curves 1 and 2 show the decrease in the time of rupture as the concentration of surface active lipids is increased. As is apparent from this figure, the marginal increase in the stabilizing effect of the Marangoni motion is completely dominated by the destabilizing effect of the reduction in the interfacial tension. This is expected because as is shown by Eqs. [62] and [63], the Marangoni motion can stabilize the mucus layer only to a certain extent and increasing the parameter, *N*, by further adding $|M\Gamma_0|_s$ to the already existing value of $|M\Gamma_0|$, does not affect the Marangoni motion significantly. This point is also clear by comparing curves 1 and 3 of Fig. 6 for $|\Gamma_0 M|_s = 0$. Curve 3 is drawn for parameter values which are the same as that of curve 1 but by ignoring the stabilizing effect of the Marangoni convection. It is apparent that even in the absence of any surface active lipids, the stabilizing effect of the Marangoni convection has reached its asymptotic value due to the presence of normal lipids and other solutes of tear film. The further presence of chemically altered or contaminating lipids thus does not enhance it any further.

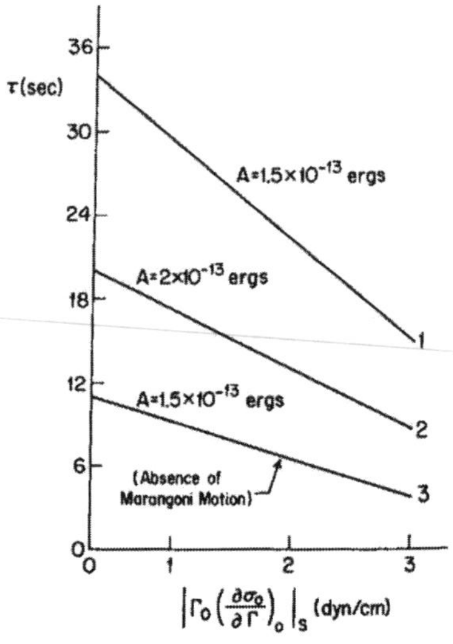

FIG. 6 The effect of added surfactant characterized by parameter, $|M\Gamma_0|_s$, on the time of rupture of the mucus layer for $\sigma_{app} = 5\,\text{dyn/cm}$, where σ_{app} is the mucous—aqueous interfacial tension in the absence of surfactants. The other parameter values are those indicated in Fig. 4.

It appears that the surface activity of the normal tear components is greatly augmented in a variety of other conditions of immunologic origins as well. Holly *et al.* (41) detected a pathological increase in the surface activity of the tear components in tears from Stevens-Johnson syndrome patients. The surface activity of these components, which was probably due to highly surface active inflammation products, surpassed even the surface activity of a pure mucin solution (mucin is known to be the most surface active component of the normal tears). A pathologically short BUT of less than 5 s was observed for these patients (41). It appears likely that the rupture of the mucus layer was rather instantaneous in these subjects because of a synergism between all three major deficiencies of a dry eye. The observed breakup times in these patients thus reflect essentially the dewetting times for the tear film supported on a hydrophobic epithelium.

In conclusion, the increased surface activity of normal eye lipids as in chronic blepharitis or the contamination of eye with surface active lipids of other origins (such as the lipids secreted from the sebaceous glands of skin and lipids of cosmetic aids) affect the stability of the tear film and BUT adversely by decreasing the mucous—aqueous interfacial tension that exists in the presence of nonpolar lipids. The same holds true for other conditions in which highly surface active products are formed due to inflammation or immune reaction.

The Role of Tear Substitutes and Surfactants

Closely related to the above discussion are the effects of highly surface active polymers, polymeric surfactants, and cationic surface active preservative such as benzalkonium chloride (BAC), instillation of which in the eye has been shown to affect BUT adversely at concentration levels as low as 0.01% (29, 42, 43). The polymeric substances dissolved in tear formulations presently available are usually substituted cellulose ether group, polyoxyethylenes, and polysaccharides, all of which are hydrophilic and exhibit only a slight surface activity (44). The most surface active component of a group of tear substitutes with trade names Ultratears, Tears naturale, Lyteers, Isopototears, and Tearisol, is BAC which is used as preservative at 0.01% or lower concentration levels (44). As shown in Fig. 6, the incorporation of a highly surface active material such as BAC, in the eye, would actually tend to decrease the BUT, thus rendering the tear substitute less effective. In addition, small molecule solutes like BAC may also bring about an increased solubility of the mucus glycoproteins

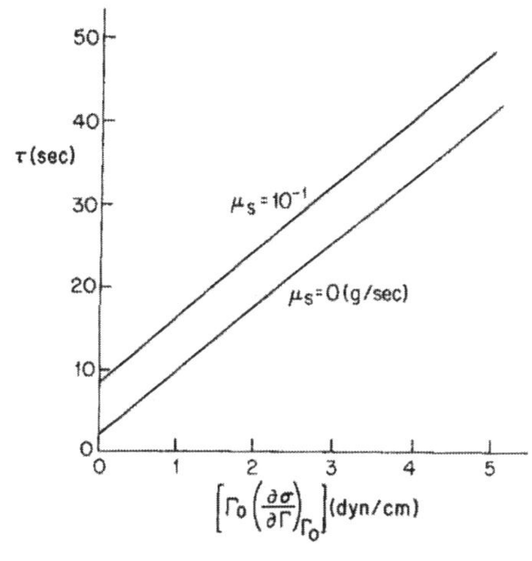

FIG. 7 The effect of the surface viscosity on the time of rupture of the mucus layer; the other parameter values are given in Fig. 4.

into the aqueous tears by the mechanism of polymer binding (45). This in turn will accelerate the rupture of the mucus layer. The principal role of tear substitutes, in addition to replacing the aqueous tears, appears to be in augmenting the thickness of the mucus layer by absorption and adsorption, increasing the mucous—aqueous interfacial tension, and aiding in Marangoni motion. The latter are important when there is a deficiency of naturally occurring antisurfactants in the tear film.

Based on these criteria, the tear substitutes with trade names Liquifilm and adsorbotears appear to be superior formulations, inasmuch as they are preserved with chlorobutanol (a bacteriocidal agent) and thimerosal, respectively (44), both of which display very little surface activity.

In addition to the effects of the added tear film solutes on the mucous—aqueous interfacial tension, the Marangoni-motion and the thickness of the mucus layer, their effects on the structure and the volume occupied by the mucus molecules may also be significant. It is well known that the hydrated mucus is a rheo-especially for the extreme conditions of aqueous deficiency.

Epitheliopathy and Impaired Lid Functions

The mechanism of tear film rupture suggests that if any of several factors such as the uneven or insufficient shear distribution during blinking, incomplete blinking, and the gross epithelial irregularities, result in thinner than normal and uneven mucus coating at some localized sites, the support of tear film will become nonwettable at these spots faster than the normal BUT. In addition, due to a decreased thickness of the aqueous layer over gross epithelial protrusions, the stabilizing effect of the Marangoni motion and consequently, BUT, would decrease. That the rupture of the tear film in these cases is essentially due to the formation of locally dewetting regions devoid of mucus coating is thus rightly identified in a study (16).

The concepts delineated above also help understand as to why the quality and frequency of blinks prior to the clinical test of BUT may also affect the tear film breakup times, as these are some of the factors that determine the thickness and the morphology of the thin mucus film coating the epithelium.

Finally, the stabilizing role of mucus layer surface viscosity is shown in Fig. 8. The effect is most pronounced at low values of $M\Gamma_0$. However, it appears that it is probably difficult to manipulate surface viscosity by any direct means when the normal antisurfactants are present in the tear film. In the absence or deficiency of dissolved components, however, the tear substitutes would also contribute towards a higher surface viscosity. This is another therapeutic value of tear substitutes in prolonging the time of tear film rupture.

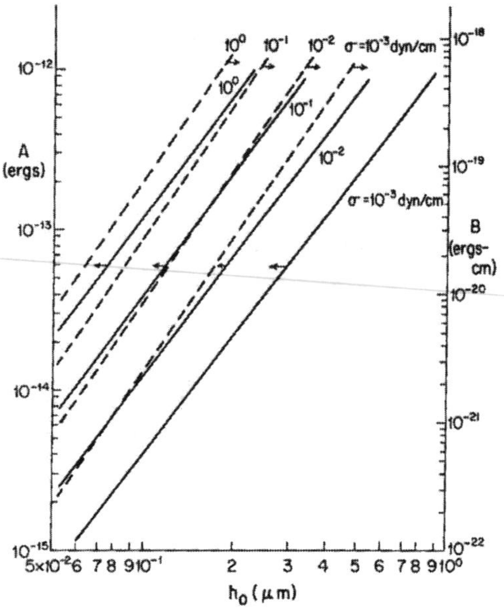

FIG. 8 The solid curves represent the nonretarded Hamaker constants and the mucus layer thicknesses that are compatible with a BUT of 300 s. The dashed curves are drawn for the retarded van der Waals interactions and a BUT of 100 s.

The Contact Lens Tolerance

An important application of the understanding of tear film rupture is in the normal wearing and functionality of contact lenses. A well-fitted contact lens floats on a continuous tear film sandwiched between the corneal epithelium and the lens (postlens tear film) and is coated with a tear film on the outside (prelens tear film), which is complete with a superficial lipid layer (4, 28). It has been suggested that a premature rupture of the postlens tear film would result in the adhesion of the contact lens to cornea, and epithelial damage is sometimes seen to occur under such circumstances (3). It thus seems reasonable that the tolerance of a contact lens would depend on the postlens and prelens tear film BUTs, but, and it is important, this BUT is not necessarily the same as the tear film breakup time in the absence of a contact lens. The following are the known physiological changes that occur in the tear film due to the insertion of a contact lens:

(a) The adverse effect of contact lens on the lid—globe congruity makes the formation and rejuvenation of an evenly distributed mucus layer on the epithelium more difficult (4).
(b) The contact lens wearers sometimes become lazy blinkers and incomplete blinks are observed. This again affects the even distribution of mucus layer on the epithelium (4).
(c) The wetting characteristics of a lens surface are usually not the same as those of the clean corneal epithelium. Thus in the case of a relatively wettable lens surface, the postlens tear film may not break even if the corneal epithelium becomes nonwettable. For a hydrophobic lens surface, however, the postlens film would break when the cornea becomes nonwettable (4).

In the case of a nonwettable lens surface (silicone lens, etc.), the factors (a) and (b) would result in a faster than normal rupture of the mucus layer and thus of the postlens film. In view of this, a marginally mucus deficient eye, as characterized by a short BUT, stands greater chances for the adherence of the lens to the cornea.

In addition to the changes outlined above, the following changes in the physiology of the tear film seem imminent after the insertion of a contact lens which is tightly or steeply fitted with no congruity between the lens and the cornea. In this case, the transport of normal lipids and other antisurfactants of the tear film (some of which are metabolites) to the mucous—postlens tear film interface is hindered and thus their capacity to impart stability to the mucous—aqueous interface is also undermined. An increased pumping (the extent of mixing of prelens and postlens tear films during blinking), thus seems necessary to replenish the postlens tear film with antisurfactants. It is indeed a clinical experience that steeply fitted silicone lenses settle tightly sooner (3). This is one of the factors that makes the proper fitting of the contact lens imperative. It has been shown that the pumping rate for PMMA hard lenses is 10–15 times higher than that for the hydrogel lenses (55). It is, however, not always possible to strike a balance between various conflicting demands by using presently available lens materials. For instance, although the PMMA hard lenses have a higher pumping rate, they are less wettable than the hydrogels and due to their initially higher interfacial tension against aqueous tears (4), they would be less biocompatible (56, 57).

A second important lens mediated change that occurs in the physiology of the tear film is that the contact lens divides the tear film into two parts. It is likely that the stability of the mucus layer is now determined by the thickness of the postlens tear film and not by the thickness of the entire tear film. As is shown in the discussion related to the destabilizing effects of the aqueous deficiency (Fig. 7), a division of the tear film would reduce the postlens film BUT compared to the BUT in the absence of the contact lens. The closer the lens to the cornea, the greater would be this effect and it is likely that in the event of aqueous tear deficiency and the lack of reflex tearing, the minimum thickness of the postlens film formed is only about 1 μm or less. The resulting 30% decrease in the time of rupture (Fig. 7) may be devastating for an already marginally deficient eye. Indeed, it is interesting to note that in a rather novel clinical study (3), it was found that the tolerance and adhesion of the contact lens could be correlated well with a composite linear function that includes both the BUT in the absence of contact lens and the Schirmer value. Adhesion of the contact lens to cornea occurred in 50% of the cases when both the BUT and the Schirmer value were low, which might have together compounded the problem. In other groups, either the BUT was somewhat low but Schirmer value was high and thus this BUT would not change significantly due to the insertion of contact lens in the eye, or the BUT was well above the pathological value and thus would not alter to a great extent because of the aqueous tear deficiency. Only 1 lens out of a total of 11 lenses showed the adhesion of lens to cornea in this group. Lens wearers with both high BUT and high Schirmer value showed excellent lens tolerance.

The qualitative discussion presented above is a plausible explanation that results from the proposed model and conclusions drawn therein. It would, however, be of great interest to formulate a more rigorous hydrodynamic model of tear film in the presence of a contact lens and thus to delineate the tear film—contact lens interactions more fully.

Morphology of Ruptured Mucus Layer

Adams (58) investigated the morphology of the conjunctival mucus by "filter surface biopsy" and found essentially three types of structures displayed by the conjunctival mucus: (a) the sheet structure, (b) the clusters, and (c) the strands. It appears that these are generated due to the breakup of a relatively homogeneous mucus film caused by the van der Waals forces. The collapse of a relatively homogeneous mucus layer would cause a network like structure, the retracting mucus being accumulated in the places where the mucus layer was initially thicker. The mucus layer is expected to be locally thicker at the mouths of the mucus secreting goblet cells and in the space between two adjacent goblet cells. The mucus islands and strands were indeed seen to appear on and in the immediate vicinity of the goblet cells, as well as filling the spaces between them and thus forming a network structure (58). A systematic *in vivo* examination of the alterations in the mucus layer, starting with a blink, would be very helpful in further clarifying these and the related issues. Care must, however, be taken so as not to alter the normal rate of thinning and breakup of mucus layer by instillation of surface active stains or fixation techniques. Any *in vitro* study of the rupture of the thin mucus layer should reproduce the physicochemical eye environment faithfully. The duplication of the effective Hamaker constant and the chemical composition of the aqueous tears appears to be necessary while studying the rupture of thin mucus layer *in vitro.*

CONCLUSIONS

Based on our proposed "two-step, double film" mechanism of tear film rupture, a self– consistent hydrodynamic formalism is developed that, aside from the role of mucus layer in the stability of tear film, also accounts for the effects of antisurfactants (or surfactants) and aqueous tear deficiency on the time of tear film rupture.

A reduced mucus production by goblet cells is identified to be one of the principal reasons for the rapid rupture of tear film.

It is shown that the central role of antisurfactants (with respect to the mucous—aqueous interface) of tear film, such as normal lipids, is to augment the interfacial tension of mucous—aqueous interface and to generate an interfacial tension driven Marangoni flow, both of which tend to oppose the thinning of mucus layer caused by the van der Waals forces.

An increased surface activity of eye lipids, that may be caused by various pathological conditions, is shown to have an adverse effect on BUT. Some highly surface active ingredients of artificial tear substitutes, various staining agents and dyes used for clinical testings, and topical anesthetics also undermine the stability of the tear film. This is so because the increased surface activity of dissolved material reduces the mucous—aqueous interfacial tension from that existing in the presence of normal lipids and other naturally occurring antisurfactants of a tear film. In addition, they also appear to encourage an increased dissolution of the mucus layer and make it thinner by inducing conformational changes in mu-coprotein molecules (47). These, in turn, lead to a faster rupture of the mucus layer and consequently, a shorter BUT. An abnormally high surface activity of tear components (possibly due to inflammation products) such as that seen in Stevens–Johnson syndrome patients, thus also reduces the BUT.

The deficiency of aqueous tears, among other things, leads to a rapid diffusion or redistribution of normal lipids present in the tear film and thus tends to even out the concentration and interfacial tension gradients caused by the thinning mucous layer. This again undercuts the stabilizing effect of the Marangoni motion and leads to a shorter BUT. The general trend is that the BUT is nearly independent of the aqueous film thickness for all thicknesses in excess of 4 μm. About 30% reduction in BUT is predicted if the average thickness of the tear film within the interblink period is about 1 μm. In addition to this direct effect of the aqueous tear deficiency, we have also accentuated certain synergisms that may cause a lipid abnormality and a reduced mucus production in the event of decreased aqueous tear secretion.

The various factors identified to cause the dry eye syndromes, and the semiquantitative predictions made are in line with numerous experimental and clinical findings.

Based on this mechanism of tear film rupture, it is suggested that the most active ingredient of a tear substitute should be such so as not to decrease the mucous—aqueous interfacial tension from its normal value prevailing in the presence of naturally occurring antisurfactants of tear film.

Implications of this mechanism in the normal wearing and functionality of contact lenses are examined with an emphasis on the factors affecting the adhesion of contact lens to cornea.

Finally, it is to be noted that even a carefully conducted clinical test of BUT is only an overall manifestation of several factors discussed and their interactions thereof. Some variability in these measurements is expected to occur because, depending on the quality of the blink, the extent to which the mucus layer is restored or smeared over the epithelium may be different. A robust mean of a statistical sample should, however, be a good diagnostic indicator for a dry eye and its consequences. A widespread use, and refinement of existing techniques to isolate the factors causing the dry eye in a particular instance would lead to more rational and specific treatment policies.

APPENDIX A

Subtracting Eq. [30] from Eq. [29], neglecting the lateral diffusion and integrating over the depth of the boundary layer gives

$$\int_h^{h+\delta} \frac{\partial}{\partial t}\left(C^* - C\right)dz$$

$$+\int_h^{h+\delta} u' \frac{\partial}{\partial x}\left(C^* - C\right)dz + \int_h^{h+\delta} w' \frac{\partial}{\partial z}\left(C^* - C\right)dz$$

$$=\left(D/v\right)\int_h^{h+\delta} \frac{\partial^2}{\partial z^2}\left(C^* - C\right)dz$$

$$+\left(D/k_b Tv\right)\int_h^{h+\delta} \frac{\partial}{\partial z}\left(C^* \frac{\partial \psi}{\partial z}\right)dz. \qquad [A1]$$

Now making use of the definition of surface excess concentration, viz.

$$\Gamma = \int_h^{h+\delta} \left(C^* - C\right)dz \qquad [A2]$$

and the Leibnitz's rule for the differentiation of an integral of an arbitrary function $F(\alpha, x)$, viz.

$$\frac{\partial}{\partial \alpha}\int_{\phi_1(\alpha)}^{\phi_2(\alpha)} F\left(x,\alpha\right)dx = \int_{\phi_1}^{\phi_2} \frac{\partial F}{\partial \alpha}dx$$

$$+ F\left(\phi_2,\alpha\right)\frac{\partial \phi_2}{\partial \alpha} - F_1\left(\phi_1,\alpha\right)\frac{\partial \phi_1}{\partial \alpha}, \qquad [A3]$$

one may transform Eq. [Al] to

$$\frac{\partial \Gamma}{\partial t} - \left(C^* - C\right)_{z=h+\delta}\frac{\partial\left(h+\delta\right)}{\partial t}$$

$$+\left(C^* - C\right)_{z=h}\frac{\partial h}{\partial t} + \int_h^{h+\delta}\left[\frac{\partial}{\partial x}u'\left(C^* - C\right)\right.$$

$$\left. + \frac{\partial}{\partial z}w'\left(C^* - C\right)dz\right.$$

$$-\int_{h}^{h+\delta}\left(C^{*}-C\right)\left(\frac{\partial u'}{\partial x}+\frac{\partial w'}{\partial z}\right)dz$$

$$=\left(D/v\right)\left[\left(\frac{\partial C^{*}}{\partial z}\right)_{z=h+\delta}-\left(\frac{\partial C^{*}}{\partial z}\right)_{z=h}\right.$$

$$\left.-\left(\frac{\partial C}{\partial z}\right)_{z=h+\delta}+\left(\frac{\partial C}{\partial z}\right)_{z=h}\right]+\left(D/vk_{b}T\right)$$

$$\times\left[\left(C^{*}\frac{\partial\psi}{\partial z}\right)_{z=h+\delta}-\left(C^{*}\frac{\partial\psi}{\partial z}\right)_{z=h}\right]. \quad [A4]$$

Further, the following simplifications may be noted

$$\frac{\partial u'}{\partial x}+\frac{\partial w'}{\partial z}=0, \quad [A5]$$

$$\left(\frac{\partial\psi}{\partial z}\right)_{z=h+\delta}=0, \quad [A6]$$

$$\left(\frac{\partial C^{*}}{\partial z}\right)_{z=h}+\left(1/k_{b}T\right)\left(C^{*}\frac{\partial\psi}{\partial z}\right)_{z=h}=0. \quad [A7]$$

The matching of concentrations and the fluxes at the end of the boundary layer provides

$$\left(C^{*}\right)_{z=h+\delta}=\left(C\right)_{z=h+\delta}, \quad [A8]$$

$$\left(\frac{\partial C^{*}}{\partial z}\right)_{z=h+\delta}=\left(\frac{\partial C}{\partial z}\right)_{z=h+\delta}. \quad [A9]$$

These observations transcribe Eq. [A4] to the following simple form

$$\frac{\partial\Gamma}{\partial t}+\left(C^{*}-C\right)_{z=h}\frac{\partial h}{\partial t}+\int_{h}^{h+\delta}\frac{\partial}{\partial x}$$

$$\times u'\left(C^{*}-C\right)dz+\int_{h}^{h+\delta}\frac{\partial}{\partial z}w'\left(C^{*}-C\right)dz$$

$$=D\left(\frac{\partial C}{\partial Z}\right)_{z=h}. \quad [A10]$$

Now assuming that the variation of u over the length scale δ is negligible, the first integral becomes

$$\int_{h}^{h+\delta}\frac{\partial}{\partial x}u'\left(C^{*}-C\right)dz$$

$$\frac{\partial}{\partial x}\left(u'\Gamma\right)+u'\left(C^{*}-C\right)_{z=h}\frac{\partial h}{\partial x} \quad \text{[A11]}$$

and

$$\int_{h}^{h+\delta}\frac{\partial}{\partial z}w'\left(C^{*}-C\right)dz$$

$$=-w'\left(C^{*}-C\right)_{z=h}. \quad \text{[A12]}$$

The kinematic condition at the mucus—aqueous interface is

$$\frac{\partial h}{\partial t}=-w-u\frac{\partial h}{\partial x}. \quad \text{[A13]}$$

Combining Eqs. [A10]–[A13] finally yields the following equation for the surface excess concentration (valid both for surfactants and antisurfactants):

$$\frac{\partial \Gamma}{\partial t}+\frac{\partial}{\partial x}\left(u'\Gamma\right)=D\left(\frac{\partial C}{\partial z}\right)_{z=h} \quad \text{[A14]}$$

where $u'=u'\left(x,t\right)=u\left(h\right)$.

APPENDIX B

As discussed earlier, some recent experiments indicate different estimates of both the thickness of the mucus layer (7b) and the tear film breakup times (2b). A study of excised guinea pig cornea employed chemical fixation and freeze substitution techniques in conjunction with electron microscopy to show that the maximum thickness of the mucus layer over the "valleys" of the corneal epithelium is about 0.4–0.8 μm (7b). The minimum thickness is seen to be in the range of 0.05–0.2 μm in the same study (the thickness over the microvilli as shown in Fig. 4 of Nicholis *et al*). This minimum thickness is a few times larger than that determined by earlier experiments. According to the proposed mechanism, the van der Waals force mediated rupture would first occur where the mucus layer is locally thin, i.e., on and in the vicinity of the tips of the microvilli. The solid curves in Fig. 9 depict the values of the Hamaker constants and the mucus layer thicknesses that are compatible with the breakup of tear film in 300 s (an estimate for the time of rupture as provided in Ref. (2b)). Separate curves are drawn for different feasible magnitudes of the mucus—aqueous interfacial tension. The dashed curves in the same figure represent the compatible values of the *retarded* Hamaker constants that can rupture the mucus layer in 10^2 s (a lower estimate of BUT). The dashed curves are drawn by assuming that the interactions remain retarded for all thicknesses up to the rupture of the film. It is in view of this and a low value of BUT corresponding to the dashed curves that the mucus layer thicknesses as computed by the dashed curves provide a lower bound. By the inverse reasoning, the solid curves provide an upper bound for the mucus layer thickness that is consistent with the proposed mechanism of tear film rupture. It may be inferred from the figure that the mucus layer would readily rupture and retract from the tips of microvilli (where it is 0.05–0.2 μm thick) within a time period consistent with the observed range of BUTs and thus give rise to nonwettable patches on the denuded corneal epithelium. Even the deeper layers of mucus over the valleys (0.2–0.8 μm) may be unstable depending on the magnitude of the Hamaker constant and the mucus—aqueous interfacial tension. Finally, it may be noted that although the orders of magnitudes reported for the mucus layer thickness in the recent study of pig cornea (7b) are perhaps correct, their relation with the thickness that exists under the dynamic conditions of the tear film in human subjects has not yet been established. Further, the possible effects of the chemical fixatives, staining agents, and the precipitation procedure on the thickness of the mucus layer remain to be clarified. The *in vivo* tests of the mucus layer thickness would be very useful in this context.

Note added in proof. While the analysis of the tear film breakup is based in the present paper on the linear stability theory, we have shown recently (59) that the inclusion of nonlinearities does not alter any of the qualitative conclusions. Nonlinearities do, however, accelerate the growth of perturbations and lead therefore to a shorter BUT. The discrepancy between the times of rupture as computed from the present analysis and the nonlinear analysis increases as the amplitude of the initial perturbations (i.e., the corrugation of the mucous—aqueous interface after a blink) increases. Since the extent of interfacial corrugations is expected to depend on the quality and frequency of blinking, this constitutes yet another reason for variability in the clinical measurement of BUT for the same patient.

REFERENCES

1. Records, R. E., *in* "Physiology of Human Eye and Visual System" (R. E. Records, Ed.). Harper & Row, New York, 1979.
2. (a) Lemp, M. A., and Hamill, J. R., *Arch. Ophthalmol.* **89**, 103 (1973); (b) Mengher, L. S., Tonge, S., Gilbert, D., and Bron, A. J., Paper presented at 24th Meeting of the association for Eye Research, Vienna, Austria, Sept. 5–9, 1983.
3. Fanti, P., and Holly, F. J., *Contact Intraocular Lens Med. J.* **6**(2), 111 (1980).
4. Holly, F. J., *Amer. J. Optom. Physiol. Opt.* **58**(4), 331 (1981).
5. Lowther, G. E., Depots Surles Lentilles Hydrophiles. Conference faite au 5' Congrès Fançais d'Optique de Contact, Boredeaux, France, 1976.
6. Norn, M. S.,*Acta. Ophthalmol.* (*Kbh*) **47**, 865 (1969).
7. (a) Holly, F. J., and Lemp, M. A., *Survey Ophthalmol. 22,* 69 (1977); (b) Nichols, B. A., Chiappino, M. L., and Dawson, C. R., *Invest. Ophthalmol. Vis. Sci. 26*, 464 (1985).
8. Lin, S. P., and Brenner, H., *J. Colloid Interface Sci.* **85**, 59 (1982).
9. Lin, S. P., and Brenner, H., *J. Colloid Interface Sci.* **89**, 226 (1982).
10. Sharma, A., and Ruckenstein, E., *J. Colloid Interface Sci.* **106**, 12 (1985) (Section 5.10 of this volume).
11. Sharma, A., and Ruckenstein, E., *Amer. J. Optom.Physiol. Opt. 62*, 246 (1985).
12. Holly, F. J., and Lemp, M. A., *Exp. Eye Res.* **11**, 239 (1971).
13. Mishima, S., *Arch. Ophthalmol.* **73**, 233 (1965).
14. Lemp, M. A., Holly, F. J., Iwata, S., and Dohlman, C. H., *Arch. Ophthalmol.* **83**, 89 (1970).
15. Adamson, A. W., "Physical Chemistry of Surfaces." Wiley–Interscience, New York, 1982.
16. Lemp, M. A., Dohlman, C. H,, and Holly, F. *J.,Ann. Ophthalmol.* **2**, 258 (1970).
17. Norn, M. S., External Eye: Methods of Examination." Scriptor, Copenhagen, 1974.
18. Ehlers, N., Kessing, S. V., and Nom, M. S., *Acta Ophthalmol.* **50**, 210 (1972).
19. Oomen, H. A., *Int. Rev. Trop. Med.* **1**, 131 (1961).
20. Kreiker, A., *Albrecht von Graefes, Arch. Ophthalmol.* **124**, 191 (1930).
21. Lemp, M. A., Dohlman, C. H., and Kuwabara, T., *et al., Trans. Amer. Acad. Ophthalmol. Otolaryngol.* **75**, 1223 (1971).
22. Iwata, S., *in* "The Preocular Tear Film and Dry Eye Syndromes" (F. J. Holly and M. A. Lemp, Eds.). *Int. Ophthalmol. Clin.* **13**(1), 97 (1973).
23. Norn, M. S., *Acta Ophthalmol.* (*Kbh*) **43**, 557 (1965).
24. Andrews, J. S., *Exp. Eye Res.* **10**, 223 (1970).
25. Andrews, J. S., *in* "The Preocular Tear Film and Dry Eye Syndromes" (F. J. Holly and M. A. Lemp, Eds.). *Int. Ophthalmol. Clin.* **13** (1), 23 (1973).
26. McCulley, J. P., and Sciallis, G. F., *Amer. J. Ophthalmol.* **86**, 788 (1977).
27. McDonald, J. E., *Trans. Amer. Ophthalmol. Soc.* **66**, 905 (1968).
28. Holly, F. J., *Amer. J. Optom. Physiol. Opt.* **58** (4), 324 (1981).
29. Wilson, W. S., Duncan, A. J., and Jay, J. L., *Brit. J.Ophthalmol.* **59**, 667 (1975).
30. Records, R. E., *in* "Physiology of the Human Eye and Visual System" (R. E. Records, Ed.). Harper & Row, New York, 1979.

31. Sheludko, A., *Adv. Colloid Interface Sei.* **1**, 392 (1967).
32. Vrij, A., Hesselink, F. Th., Lucassen, J., and van den Tempel, M., *Proc. K. Ned. Akad. Wet. B* **73**, 124 (1970).
33. Ruckenstein, E., and Jain, R. K., *Faraday Trans. II* **70**, 132 (1974) (Section 5.1 of this volume).
34. Williams, M. B., and Davis, S. H., *J. Colloid Interface Sci.* **90**, 220 (1982).
35. Maldarelli, C., Jain, R. K., Ivanov, I. B., and Ruckenstein, E., *J. Colloid Interface Sei.* **78**, 118 (1980) (Section 5.3 of this volume).
36. (a) Dolowy, K., *J. Theor. Biol.* **52**, 83 (1975); (b) Holly, F. J., *Exp. Eye Res.* **15**, 515 (1973).
37. Vanley, G. T., Leopold, I. H., and Gregg, T. H., *Arch.Ophthalmol.* **95**, 445 (1977).
38. Parfitt, G. D., and Peacock, J., *in* "Surface and Colloid Science" (E. Matijevic, Ed.), Vol. 10, p. 171. Plenum, New York, 1978.
39. Nir, S., *Prog. Surf. Sci.* **8**, 1, 1976.
40. Weiss, L,, *Exp. Cell Res.* **53**, 603, 1968.
41. Holly, F. J., Patten, J. T., and Dohlman, C. H., *Exp.Eye. Res.* **24**, 479 (1977).
42. Holly, F. J., and Lemp, M. A., *Contact Lens Soc.Amer. J.* **5**, 12 (1971).
43. Lemp, M. A., *in* "The Preocular Tear Film and Dry Eye Syndromes" (F. J. Holly and M. A. Lemp, Eds.). *Int. Ophthalmol. Clin.* **13** (1), 221 (1973).
44. Holly, F. J., *Contact Intraocular Lens Med. J.* **4**,14 (1978).
45. Molyneux, P., *in* "Water: A Comprehensive Treatise" (F. Frank, Ed.), Vol. 4. Plenum, New York, 1975.
46. Allen, A., *Trends Biochem. Sci.* **8**, 169 (1983).
47. Allen, A., Pain, R. H., and Snary, D., *Faraday Discuss.Chem. Soc.***57**, 210 (1974).
48. Holly, F. J., *Contacto* **26**, 9 (1982).
49. Rengstorlf, R. H., *Amer. J. Optom. Physiol. Opt.* **51**, 765 (1974).
50. de Roetth, A., *Arch. Ophthalmol.* **49**, 185 (1953).
51. Maurice, D. N., *Invest. Ophthalmol.* **6**, 464 (1967).
52. Henderson, J. W., and Progugh, W. A., *Arch.Ophthalmol.* **43**, 224 (1950).
53. Ralph, R. A., *Invest. Ophthalmol.* **14**, 299 (1975).
54. Meyer, K., *in* "Modem Trends in Ophthalmology" (A. Sorsby, Ed.), p. 71. Butterworth, London, 1948.
55. Poise, K. A., *Invest. Ophthalmol.* **18**, 409 (1979).
56. Andrade, J. D., *Med. Instrum.* **7**, 110 (1973).
57. Ruckenstein, E., and Gourisankar, S. V., *J. Colloid Interface Sci.* **101**, 436 (1984) (Section 1.2 of the second volume).
58. Adams, A. D., *Arch. Ophthalmol.* **97**, 730 (1979).
59. Sharma, A., and Ruckenstein, E., *J. Colloid Interface Sci.,* **113**, 456 (1984) (Section 5.4 of this volume).

Index

Note: Page numbers followed by f and t refer to figures and tables respectively.

D

P

Q

R

S

U

V

W

Y